P9-APS-504

THE LINEMAN'S AND CABLEMAN'S HANDBOOK

THE LINEMAN'S AND CABLEMAN'S HANDBOOK

Thomas M. Shoemaker, P.E., B.S.E.E.

Life Senior Member, IEEE; Consulting Engineer

Formerly: *Manager, Distribution Department, Iowa-Illinois Gas & Electric Company; Member, Transmission and Distribution Committee, Edison Electric Institute; Captain, Signal Corps, U.S. Army*

James E. Mack, P.E., B.S.E.E., M.B.A.

Manager, Electric Reliability, MidAmerican Energy Company; Member, IEEE; Member, NSPE

Eleventh Edition

McGRAW-HILL

New York Chicago San Francisco Lisbon London Madrid Mexico City Milan New Delhi San Juan Seoul Singapore Sydney Toronto

The McGraw-Hill Companies

Cataloging-in-Publication Data is on file with the Library of Congress

2 3 4 5 6 7 8 9 0 DOC/DOC 0 1 2 1 0 9 8 7

ISBN-13: 978-0-07-146789-6
ISBN-10: 0-07-146789-0

The sponsoring editor for this book was Stephen S. Chapman and the production supervisor was Richard C. Ruzycka. It was set in Times by International Typesetting and Composition. The art director for the cover was Anthony Landi.

Printed and bound by RR Donnelley.

This book is printed on acid-free paper.

ABOUT THE AUTHORS

THOMAS M. SHOEMAKER is a Life Senior Member of the IEEE.

JAMES E. MACK is the Manager of Electric Reliability at the MidAmerican Energy Company. He is a member of the IEEE and NSPE.

CONTENTS

PREFACE

This Handbook is written for the apprentice, the lineman, the cableman, the foreman, the supervisor, and other employees of electric line construction contractors and transmission and distribution departments of electric utility companies. It is primarily intended to be used as an apprenticeship textbook and a home-study book to supplement daily work experiences.

All chapters in this Eleventh Edition have been revised where necessary to be consistent with the newest equipment, techniques, and procedures. A special effort was made to present all discussions clearly and in simple language. As in previous editions, a large number of illustrations showing the construction and maintenance processes are provided to assist the reader in a better understanding of the text. The illustrations clarify many details that would require additional words to express. Many of the photographs were taken specifically for use in this edition. They portray the practices in use by some of the foremost electric utility and contracting companies in the United States.

Methods of transmission-, distribution-, and rural-line construction have become quite standardized since the First Edition of the Handbook was published in 1928. The construction procedures described and illustrated are in most instances representative of general practice. While each operating company has its own standards of construction to which its linemen and cablemen must adhere, the procedures described explain why things are done in a given way. Such basic knowledge will be helpful to the lineman or cableman who is interested in learning the whys and wherefores of doing things one way or another.

Safety is emphasized throughout this book. Of course, understanding the principles involved in any operation and knowing the reasons for doing things a given way are the best aids to safety. The opinion has become quite firmly established that a person is not a good lineman unless he does his work in accordance with established safety procedures and without injury to himself or others. It is necessary for those engaged in electrical work to know the safety rules and the precautions applicable to their trades, as specified in the *National Electrical Safety Code,* Occupational Safety and Health Act (OSHA) standards, and their employer's safety manuals and standards, and to make the observance of safety rules and procedures an inseparable part of their working habits.

This Handbook places emphasis on the *National Electrical Safety Code*, OSHA standards, ANSI standards, and ASTM standards. Important requirements of all of these are discussed, but the *National Electrical Safety Code*, OSHA standards, and ANSI standards should be studied for detailed work procedures. Many applicable codes and standards are specified throughout the text to assist the reader.

The lineman and the cableman must become acquainted with the minimum construction requirements and maintenance and operating procedures in the various codes and standards to ensure the safety of the public and all workers. A copy of the *National Electrical Safety Code* (ANSI C2) can be secured for a fee from the Institute of Electrical and Electronics Engineers, Inc., 345 East 47th Street, New York, NY 10017. It is necessary that all linemen and cablemen know the information in the *National Electrical Safety Code* and that they adhere to the rules and procedures while performing their work assignments.

The *National Electrical Code* details the rules and regulations for electrical installations, except those under the control of an electric utility. It excludes any indoor facility used and controlled exclusively by a utility for all phases from the generation through distribution of electricity and for communication and metering, as well as outdoor facilities on a utility's own or a leased site or on public or private (by established rights) property.

Reference material for the interested reader includes *Standard Handbook for Electrical Engineers*, edited by H. Wayne Beaty and Donald G. Fink and published by McGraw-Hill; Edison Electric Institute publications; *Underground Power Transmission* by Arthur D. Little, Inc., for the Electric Power Research Council; *Electric Power Transmission and Environment*, published by the Federal Power Commission; *IEEE Standard Dictionary of Electrical and Electronic Terms; User's Manual for the Installation of Underground Plastic Duct*, published by the National Electrical Manufacturers Association; and the Electric Power Research Institute's (EPRI) publications.

The editors are well aware that one cannot become a competent lineman or cableman from a study of the pages of this book alone. However, diligent study along with daily practical experience and observation should give the apprentice an understanding of construction and maintenance procedures—and a regard for safety—that should make his progress and promotion rapid.

Thomas M. Shoemaker
James E. Mack

ACKNOWLEDGMENTS

The editors are deeply indebted to two individuals deserving of special recognition: Edwin B. Kurtz, E.E., P.E., Ph.D. (deceased), and Ruth Shoemaker (deceased). Dr. Kurtz was the original author and editor of *The Lineman's and Cableman's Handbook*. He was a former Emeritus Professor of Electrical Engineering and Head of Department at the University of Iowa. Dr Kurtz's work established the solid foundation that this textbook provides to the lineman's growth and development in the profession. The Handbook has stood the test of time, which is a testament to the foresight that Professor Kurtz displayed. Ruth Shoemaker devoted many, many hours of typing and proofreading to enhance the quality of the Handbook.

The editors also wish to thank several companies and their representatives for contributions to this or previous editions of the Handbook. We are especially grateful to Anita Fisher of Altec; Aaron Howell and Alan Drew of Northwest Lineman College; Phil Bergman of Raytek; Richard Erdel of A. B. Chance Division of Ohio Brass; Kristin Wild of Asplundh Tree Expert Company; Peter Simpson of Eastern Utilities; Paul Moehn of Moehn Electrical Sales; Kurt Batten, Karen Vaglia, D. Reinke, and R. Reardon of ABB Power T&D Company Inc.; Jarrett Cowden and Deb McClurg of Vermeer Manufacturing; W. J. Brian Ackley of A.D. Ventures; Kay Laporte of PEPCO Technologies; Kerry Diehl of Power Quality Systems; Jeremy Adcock of Hendrix Wire & Cable Inc.; Doug Parentice of Maysteel Electric Utility Products; Brian Harris of Federal Pacific; Irene Santoyo of Composite Technology Corporation; Joni Mack; and others.

Finally, the editors are also indebted to the Institute of Electrical and Electronics Engineers, Inc., publisher of the *National Electrical Safety Code; Electric Light and Power;* the Rural Electrification Administration; McGraw-Hill, publisher of the *Standard Handbook for Electrical Engineers*, edited by Beaty and Fink, and of *Electrical World; Transmission & Distribution*; and the Edison Electrical Institute for permission to reprint various items from its literature.

Thomas M. Shoemaker
James E. Mack

Lineman working from insulated bucket. The energized primary is covered with line hose. The lineman has positioned himself for convenient access to the work being completed. (*Courtesy Pepco Technologies.*)

The terms *lineman* and *cableman*, long-established and still current in the industry, are beginning to be replaced by nonsexist titles in official documents and government publications. Both men and women are employed in these capacities in the military and in the industry. To avoid awkwardness, this Handbook uses the masculine pronoun, but it in no way implies that the jobs involved are held only by men.

INTRODUCTION

Linemen and cablemen construct and maintain the electric transmission and distribution facilities that deliver electrical energy to our homes, factories, and commercial establishments. They provide important skilled services to the electrical industry—important because the health and welfare of the public are dependent on reliable electric service.

When emergencies develop as a result of lightning, wind, or ice storms, linemen and cablemen respond to restore electric service at any time of the day or night.

An understanding of electrical principles and their application in electrical construction and maintenance work is essential to completing the work safely, efficiently, and reliably. New equipment and the public's increased dependence on continuous electric service require that all linemen and cablemen be highly skilled.

The time and effort spent studying these pages will increase the reader's knowledge of electric distribution and transmission facilities and improve his skills. Every lineman, cableman, and groundman should develop the highest possible level of skills so that he will be able to meet the challenges of the work and be qualified for promotion when the opportunity becomes available.

As a testament to the significance of the role that the lineman has had in the United States, the following summary was submitted by Alan Drew, Northwest Lineman College.

"A hot crew with their hot wagon." This photograph is from the 1930s when live line work was in its infancy and was specialized at many companies. Lineman working on these crews developed many effective methods and tools that are still used today.

EVOLUTION OF THE LINEMAN

In the 1840s, the use of the telegraph as a means of communication began in the United States. To obtain the benefits of this enhanced method of communication, it meant that lines would have to be constructed and maintained. It was found that telegraph lines could be strung on trees if they were available, but that wood poles provided the best method of supporting the lines. The expansion of the telegraph system required men who could set poles and then climb them to string the wire. The term "lineman" quickly evolved as a title for those who worked on telegraph lines.

In the late 1870s, the telephone was invented and it began to replace the telegraph as a means of communication. The telephone also needed lines to be constructed, which were similar to the telegraph lines except that they utilized more wires for the needed circuits. The term "lineman" was well established and carried on into the telephone era.

In the late 1890s, electric power started to become a useful form of energy and power plants and lines were built. This new form of energy immediately proved to be considerably more hazardous than the telegraph or telephone systems. A new breed of the "lineman" was now needed to work on these lines. These linemen took, and were expected to take, many risks in working on power lines and equipment. A high level of injuries occurred because of the limited training, and lack of proper equipment, construction standards, and safety rules. Although records are sketchy, in some areas it has been said that one of every three lineman was killed on the job, and mostly from electrocution.

As the power system evolved across the country, there were many large line building projects which resulted in linemen "booming around" from project to project. These early linemen quickly established a reputation as individuals who worked hard, took many risks, played hard, and took pride in their work.

In the late 1930s, the complexity of the lineman's job was recognized as a good fit for apprenticeship training. This resulted in the establishment of the apprentice lineman, and soon programs started to evolve across the country. This was a significant step in the establishment of more formalized training to develop competent linemen.

The use of electricity had brought with it a higher quality of life at home and the ability to be more productive for businesses. When power outages occurred, the impact on customers started to become significant. Linemen would quickly respond to these outages and they would make Herculean efforts in all types of weather to restore power. They soon established a reputation as "heroes" in the eyes of many customers.

The early linemen established their reputation while working mainly on wood pole lines. In the late 1950s, the underground started to evolve, as a popular and reliable way to deliver power and it soon become part of the lineman's work. This added a new dimension and more complexity to the lineman's job.

Today's linemen face similar challenges of the past; however, today, customers rely heavily on the continuous delivery of power and are less tolerant of outages. This results in considerably more work on energized lines than in the past. The lineman of today has a wide array of enhanced vehicles, tools, equipment, and training to meet these challenges. In addition, OSHA rules, standards, and procedures have greatly improved.

Today's linemen take pride in their work and they remain heroes to customers when their power is restored after a long outage. As we look back at the vast amount of experience and knowledge that has been gained from the efforts of these pioneering linemen, it is appropriate to maintain a commitment to their legacy.

THE LINEMAN'S AND CABLEMAN'S HANDBOOK

CHAPTER 1
ELEMENTARY ELECTRICAL PRINCIPLES

Electron Theory. The basis of our understanding of electricity is the electron theory. This theory states that all matter, that is, everything that occupies space and has weight, is composed of tiny invisible units called atoms. Atoms, in turn, are subdivided into still smaller particles called protons, neutrons, and electrons. The protons and neutrons make up the central core, or nucleus, of the atom, while the electrons spin around this central core in orbits as illustrated in Fig. 1.1.

The protons and electrons are charged with small amounts of electricity. The proton always has a positive charge of electricity on it, while the electron has a small negative charge. The magnitude of the total positive charge is equal in amount to the sum of all the negative charges on all the electrons. The neutron has no charge on it, either positive or negative, and is therefore neutral and hence called neutron.

Atoms differ from one another in the number of electrons encircling the nucleus. Some atoms have as many as 100 electrons spinning around the nucleus in different orbits. The atom of hydrogen gas has only 1 electron. The atom of lead has 82 electrons.

Positive and negative charges of electricity attract each other, that is, protons attract electrons. But the atom does not collapse because of this attraction. The spinning of the electron around the nucleus causes a centrifugal force that just balances the force of attraction and thus keeps them apart.

Electric Current. The electrons in the outermost orbit of an atom are usually not securely bound to the nucleus and therefore may fly off the atom (see Fig. 1.2) and move into an outer orbit of another atom. These relatively free electrons normally move at random in all

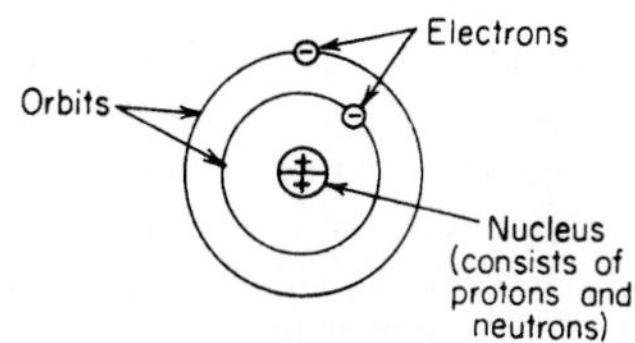

FIGURE 1.1 Typical atom consisting of nucleus and revolving electrons. The nucleus is composed of protons and neutrons. The protons carry a positive electrical charge, the electrons carry a negative electrical charge, and the neutrons are neutral; that is, they carry neither a positive nor a negative charge.

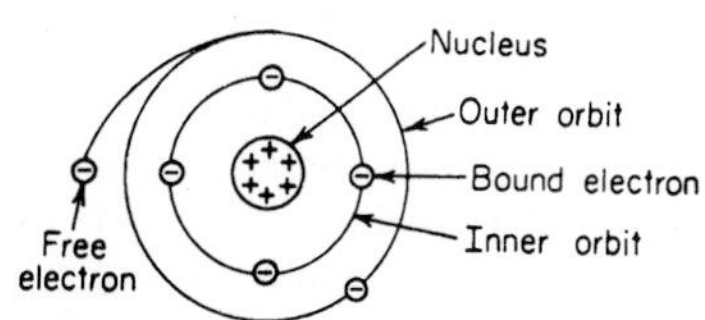

FIGURE 1.2 Atom showing electron in outer orbit leaving atom. The atom then has more positive charge than negative charge. The nucleus will therefore attract some other free electron that moves into its vicinity.

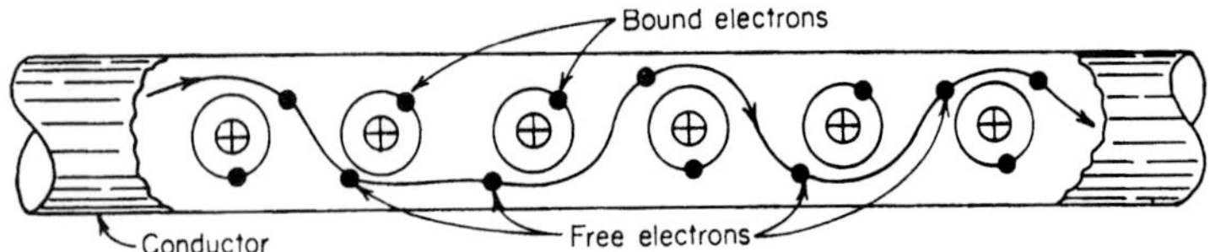

FIGURE 1.3 Flow of free electrons in a conductor. Only electrons in outer orbit are free to move from one unbalanced atom to another unbalanced atom. This flow or drift of free electrons is called an electric current.

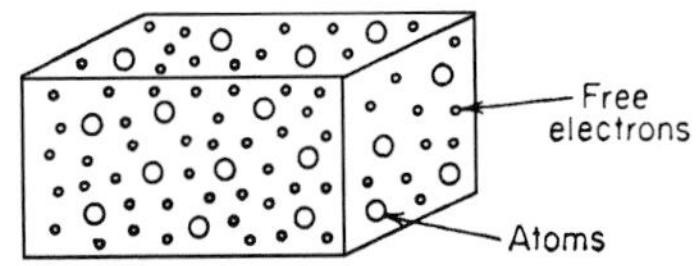

FIGURE 1.4 Material having many free electrons makes a good conductor. Copper and aluminum have many free electrons and therefore are widely used as conductors of electricity.

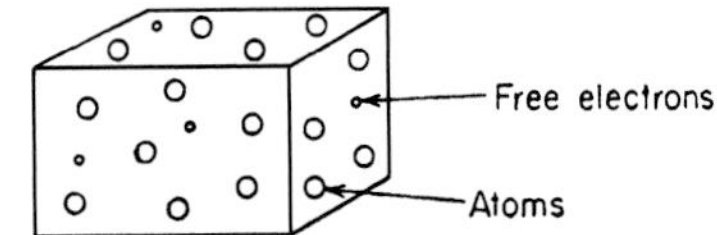

FIGURE 1.5 Material having few free electrons makes a good electrical insulator. Porcelain, glass, rubber, polyethylene, fiber-glass, etc., contain few free electrons and therefore are widely used as insulators of electricity.

directions. However, when an electrical pressure (voltage) is applied across a length of wire, the free electrons in the wire give up their random motion and move or flow in one general direction. This flow of free electrons in one general direction, shown in Fig. 1.3, is called an electric current or simply current.*

Conductors and Insulators. Materials having many free electrons (Fig. 1.4), therefore, make good conductors of electricity, while materials having few free electrons (Fig. 1.5) make poor conductors. In fact, materials that have hardly any free electrons can be used to insulate electricity and are called insulators. Samples of good conductors are copper and aluminum. Samples of good insulators are glass, porcelain, rubber, polyethylene, and fiberglass.

Electric Circuit Compared with Water Circuit. An electric circuit is the path in which the electric current flows. The flow of electricity in a wire is actually the simultaneous motion of countless free electrons in one direction. It is often compared with the flow of a liquid like water. Electricity can then be said to flow in a wire as water flows in a pipe. A simple water circuit, like the one shown in Fig. 1.6, has a resemblance to a typical electric circuit, shown in Fig. 1.7. The similarity between the water circuit and the electric circuit can give one an understanding of the flow of electric currents.

Figure 1.6 illustrates water that is flowing in the pipe circuit in the direction shown by the arrows. It is evident that this current of water flows because of a pressure exerted on it. This pressure is produced by the rotary pump, often called the centrifugal pump, which is driven by a gasoline engine. A water motor is connected on the end of the pipeline, and therefore, all the water that flows around the circuit must pass through the motor. It is plain that it will cause the motor to revolve and, therefore, deliver power to the shaft and the rotating equipment connected by the shaft. Similarly, when an electric current flows in a wire, it flows because an electric pressure causes it to flow. Thus the current in Fig. 1.7 is

* By convention, however, the flow of current in a circuit is taken to be opposite in direction to that of the electrons.

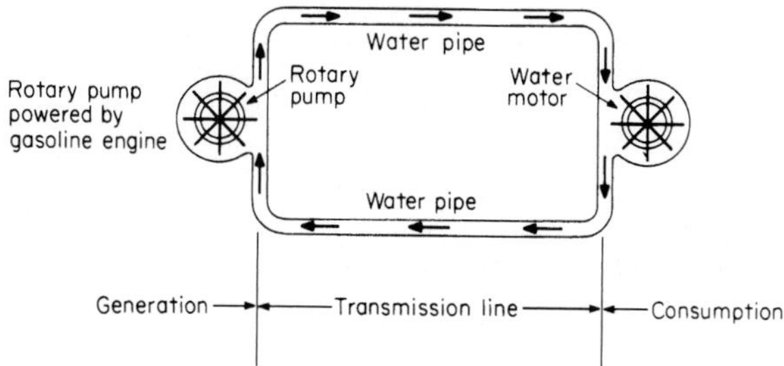

FIGURE 1.6 The water system.

made to flow because of the electric pressure produced by the dynamo, or electric generator, which is driven by a gasoline engine. As the electric current flows along the wire, it will be forced to flow through the electric motor. This motor will begin to revolve as the electricity begins to flow through it and will deliver power to the shaft and the rotating equipment connected to the shaft.

Series Circuit. An electrical circuit can be arranged in several ways, as long as the path for the electric current is closed. The simplest arrangement is the so-called series circuit. The series circuit has all the elements of the circuit connected onto each other, as illustrated in Fig. 1.8. The same current from the battery flows through all the lamps. If one of the lamps burns out, all the lamps will go out because the circuit is no longer closed.

Parallel Circuit. Another arrangement is the so-called parallel or multiple connection. Each lamp is individually connected across the battery, as shown in Fig. 1.9, instead of all the lamps being connected onto each other and then onto the battery. The lamps are now said to be in parallel with each other. If a lamp burns out in such an arrangement, the remaining lamps will continue to burn, since the path for the current through each of them is still closed.

Series-Parallel Circuit. A third arrangement is the combination of the series and the parallel circuits. Part of the circuit is in series and part in parallel, as shown in Fig. 1.10. The same current flows through each element of the series portion of the circuit. The current in the parallel part of the circuit divides, and only a portion of the current flows through each of the parallel paths.

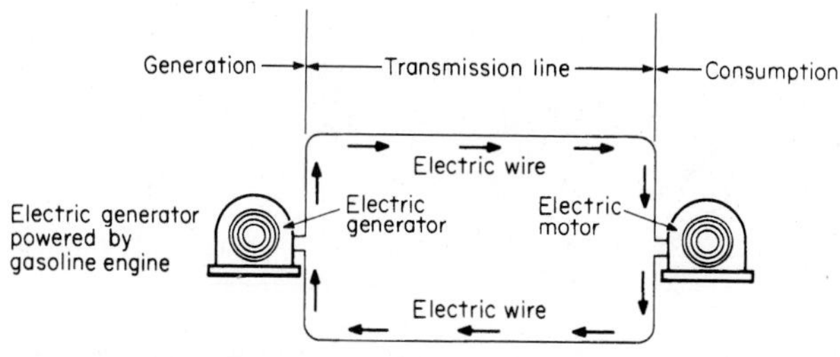

FIGURE 1.7 The electric system.

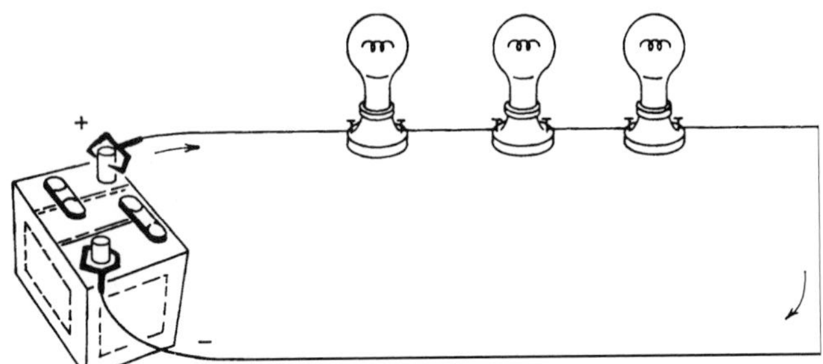

FIGURE 1.8 A series circuit. The same current flows through all the lamps.

Electric System. The circuits in Figs. 1.6 and 1.7 can be seen to consist essentially of three main divisions. The section where the pressure is produced, that is, where the engine drives the generator, is called the generator section. That part of the circuit which furnishes the path for the current from the place where it is generated to where it is used is the transmission and distribution section, and section where the electricity is used or consumed is the conversion division. conversion mean "change," and it is here that electricity is changed to the light, heat, power, or communication medium. An actual electric circuit has three parts: a generating station, a transmission and distribution line, and a customer load center. The customer load center is the place where the power is consumed. An electric circuit with the three different components is called an electric system.

The wires of the system serve to carry the electricity just as highways carry automobiles and railroad tracks carry trains. The reason one does not see the electricity moving along the wire is because it is invisible. The wires and transformers appear lifeless; however, they are very much alive and ready to do almost any work for us.

One should look on the generation, transmission, and distribution of electrical energy as one does on the manufacture, shipment, and delivery of goods. Electricity is different from a manufactured product, like shoes. The manufacturer of shoes can estimate the demand for shoes and then manufacture them in advance and put them in a warehouse. Electricity has to be manufactured or generated at the very instant when it is wanted. The customer flips a light switch to the on position or turns on the electric range and an order is flashed back through the distribution system (the retail outlet), the substation (the warehouse), the transmission line (bulk transportation), to the generator (the factory), and delivery must be made before the customer's hand releases the switch.

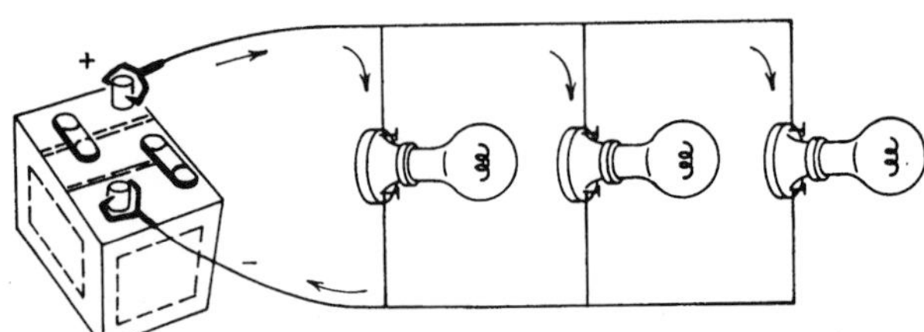

FIGURE 1.9 A parallel circuit. Each lamp is independent of the other lamps and draws its own current.

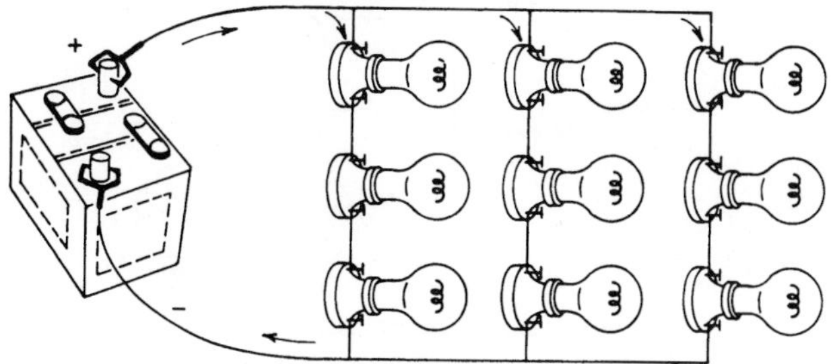

FIGURE 1.10 A series-parallel circuit. The lamps in each branch are in series, and the three branches are connected in parallel.

It is important to observe that the current path or the transmission line must have a return wire just as water must have a return pipe. The water passes out along the pipe in one side of the circuit through the water motor, where it does its work, and then returns to the rotary pump in the other pipe (Fig. 1.6). The electricity passes out along one wire to the motor, does its work, and then returns to the generator in the other wire (Fig. 1.7), a path similar to the water circuit.

Electric Current. It has been pointed out that the flow of electricity in a wire is similar to the flow of water in a pipe. When water flows in a pipe, one speaks of a current of water or a water current. Similarly, when electricity flows in a wire, it is called an electric current.

Ampere. Generally, we want to know how much water is flowing in the pipe, and we answer "10 gal per sec." In the same way, we can express how much electricity is flowing in a wire by saying "25 amps." The ampere is the unit of electric current. One can learn how much an ampere is by watching what it can do. An ordinary 60-watt, 120-volt incandescent lamp will require 1/2 amp. This means that 1/2 amp is flowing through it all the time that it is glowing.

A lamp of the type used for street lighting requires about 10 times as much current as the little house lamp. A medium-weight iron requires about 5 amps. A cooling-fan motor takes about 1 amp. A rapid-transit car motor requires about 200 amps. A good-sized factory requires several thousand amperes. The total flow of current from a large central generating station or generating plant may be as large as 20,000 amps. These figures will give one a fairly good idea of the size of an ampere and what it will do.

Ammeter. If we wish to measure the current of water flowing through a pipe, we place a meter right in the pipeline. A meter for measuring the flow of water in a pipe is called a flowmeter. When such a meter is placed in the line, water flows through it, as shown in Fig. 1.11, and the meter indicates the number of gallons per second which pass through it. It is clear that the meter must be inserted in the pipe so that the water flows through it. The number of amperes

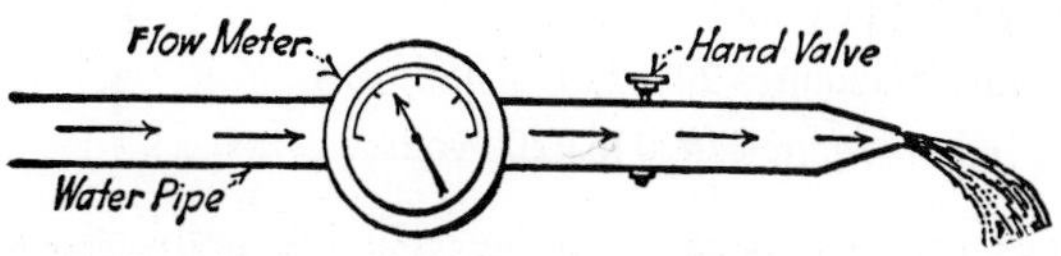

FIGURE 1.11 Flowmeter in water pipe.

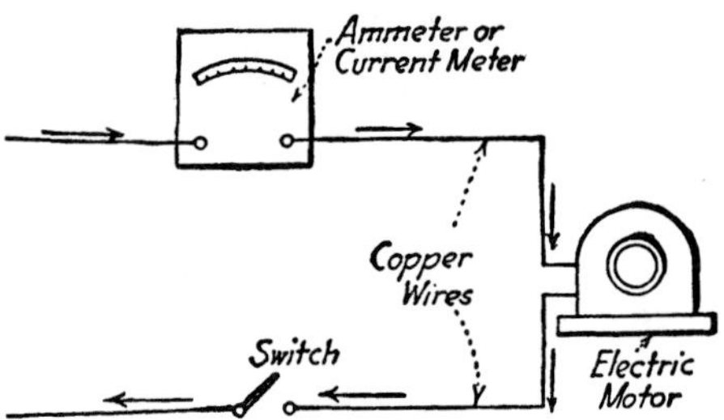

FIGURE 1.12 Ammeter in electric circuit.

of electric current flowing in a circuit can be measured by connecting a current meter or an ampere meter in the circuit, as shown in Fig. 1.12. Since such a meter is to read amperes, it is called an ampere meter or an ammeter. It should be noted that the ammeter is inserted in the line in order that all the current taken by the motor may pass through the meter.

Electric Pressure. We know that a pressure causes a current of water to flow in a pipe. Likewise, in order for current to flow in a circuit, an electric pressure must be present. Voltage is the electric force or pressure that is required. Voltage is determined by the potential difference between any two points in a circuit. Generators and batteries are devices that supply voltage to circuits.

A hand valve in the water circuit will stop the flow of water if it is closed. The water pressure would still be there, but water would not flow through the water motor or the pipe. This can be demonstrated by turning off the water faucet in the kitchen sink. The water flow ceases, but the pressure is still there. A switch placed in the electric circuit shown in Fig. 1.12 will likewise prevent the flow of current if it is in the open position. The electric pressure will still be there with the switch open if the generator is being driven by the engine. Thus there can be pressure and no current.

Volt. One must be able to talk of its strength to learn something about electrical pressure. This requires a unit with which to measure it; that unit is the volt. One volt will cause 1 ampere to flow when impressed across a 1 ohm resistor. We can learn how much a volt is by observing what it can do. We can note, for example, how much pressure or how many volts are required, in general, to force a current through a doorbell, an electric light, an iron, a washing-machine motor, a small factory motor, and a large factory motor. The most common values are as follows:

An electric doorbell requires 2 to 5 volts.

An electric iron requires 120 volts.

A washing machine motor requires 120 volts.

An electric range requires 240 volts.

An electric rapid-transit car motor requires 600 vols.

A small factory motor requires 208, 240, or 480 volts.

A large factory motor requires 2400 to 7200 volts.

These values simply mean that so many volts are required to push or force the working current through the devices or machines.

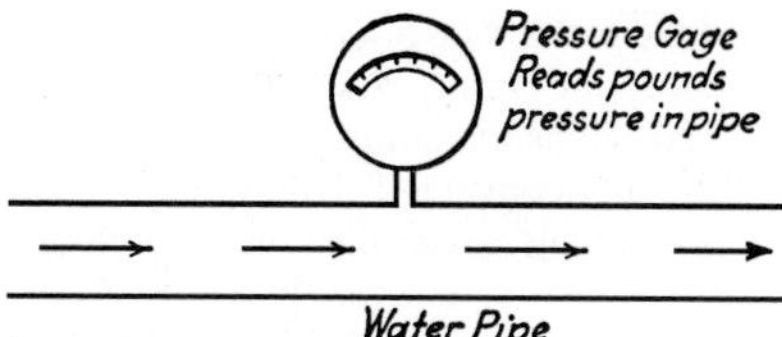

FIGURE 1.13 Water-pressure gauge.

One can gain an idea of the strength of electric pressure by observing how much of a shock is received when one puts one's hands across the two wires of a circuit. A person cannot detect or feel as little as 5 volts of electrical pressure but can feel 50 volts. At 120 volts, nearly everyone will get a very unpleasant shock, even when a very light and brief contact is made with the wires. It may prove fatal if a firm contact is made with the wires energized at 120 volts. If the hands are moist or wet and if a firm grasp is made across 120 volts, death is likely to result. All voltages should be considered dangerous and handled with great care. All higher voltages should be well guarded, and no one except an authorized person has any business getting near them. System voltages of 600 volts or less are classified as low voltages or utilization voltages in the *American National Standards Institute* (ANSI) Standard C84.1 for Electric Power Systems and Equipment—Voltage Ratings (60 Hz). System voltages of over 600 to 69,000 volts are classified as medium voltages, distribution voltages or subtransmission voltages. System voltages of over 69,000 to 230,000 volts are classified as high voltages or transmission voltages. System voltages of over 230,000 to 1 million volts are classified as extra-high voltages (EHV) or transmission voltages. Voltages of 1 million volts or more are classified as ultra-high voltages (UHV) or experimental voltages.

Voltmeter. If we wish to measure the pressure in a water circuit, all we do is tap a pressure gauge onto the pipeline, as shown in Fig. 1.13. Everyone is familiar with such a gauge. The few points to be noted are that the gauge is simply tapped on the pipeline at the point at which the pressure is wanted so that the pressure at that place can get up into the gauge and make it indicate. It is also evident that flow of water in the pipe is not disturbed by insertion of the gauge.

In the same manner, we can measure electric pressure. Electric pressure or voltage is measured utilizing the voltmeter. The voltmeter measures in volts the potential difference between two points in an electric circuit. We simply connect the two leads from a voltmeter across the line, as shown in Fig. 1.14. The current through the voltmeter will then vary

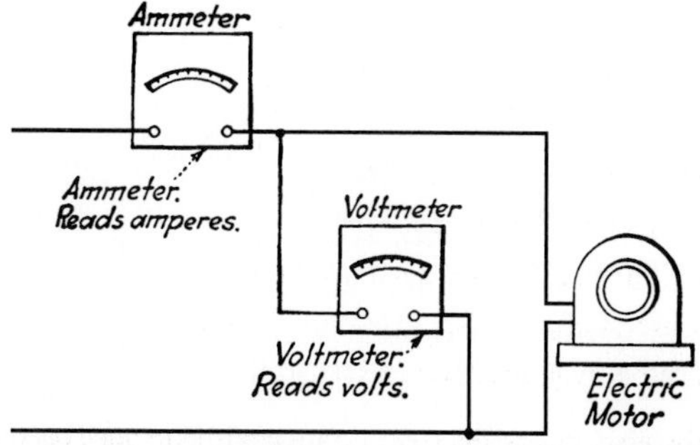

FIGURE 1.14 Ammeter and voltmeter correctly connected.

directly with the voltage, and the meter can be made to read volts. It is to be noted that the current which flows to the motor does not flow through the voltmeter. This is because the voltmeter is not a part of the circuit as the ammeter is. Figure 1.14 shows the two meters, the ammeter and voltmeter, properly connected. The ammeter reads the flow of current, and the voltmeter reads the pressure which causes the current is flow.

Waterpower. We have likened an electric current to a current of water. When a current of water flows in a pipe in a simple circuit, as shown in Fig. 1.6, power is delivered to the water motor. We know this because the water motor revolves and can do work. The power delivered depends on the amount of water flowing and the pressure under which it flows. This is self-evident, for more power will be developed if 50 gal per sec flows through the water motor than if only 25 gal per sec flows through it. Furthermore, more power will be developed with 100 lb of pressure than with only 50 lb of pressure. The power delivered in the pipeline to the motor thus depends on the amount of water flowing and on the pressure.

Electric Power. The amount of power delivered by an electric circuit to an electric motor depends on the number of amperes flowing and the number of volts of pressure in exactly the same way as in waterpower. The greater the current, the larger the number of amperes, and the greater will be the amount of power developed by it; and the greater the pressure, the more effect the current will have. The actual value of power in a direct-current circuit (not true for an alternating-current circuit) is equal to the product of volts times amperes; thus

$$\text{Power} = \text{volts} \times \text{amperes}$$

Watt. The unit of power in an electric circuit is the watt. An ordinary electric lamp, when connected to an electric circuit, as in Fig. 1.15, will draw about 150 watts from the circuit. An ordinary iron, when connected to a circuit, will draw about 550 watts of power from it. The motor shown schematically in Fig. 1.16 will draw about 5600 watts of power. It is plain that when we come to large machines, the number of watts runs up quickly. The watt is a rather small unit of power, so a larger unit equal to 1000 watts is used. The unit for 1000 is called kilo; therefore, 1000 watts is called 1 kilowatt. This unit is abbreviated kW. A kilowatt is equal to about $1\frac{1}{3}$ hp. The motor in Fig. 1.16 thus draws 5.6 kW of power, or $7\frac{1}{2}$ hp.

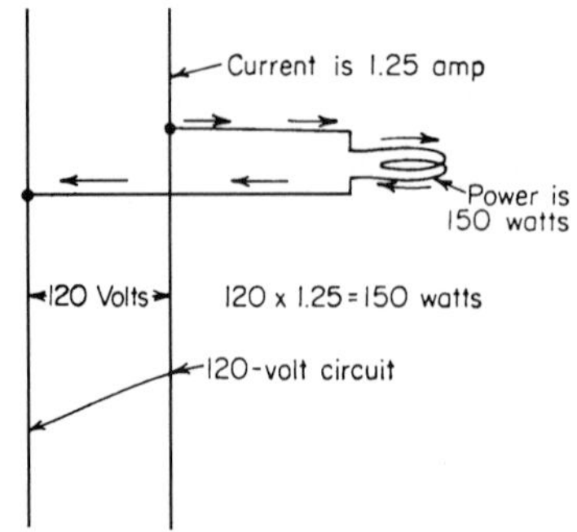

FIGURE 1.15 Electric lamp taking 1.25 amp and 150 watts from 120-volt circuit.

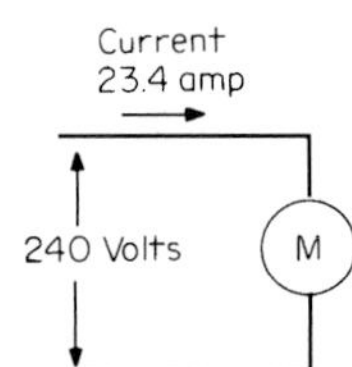

FIGURE 1.16 Motor drawing 23.4 amps of current and 5600 watts of power.

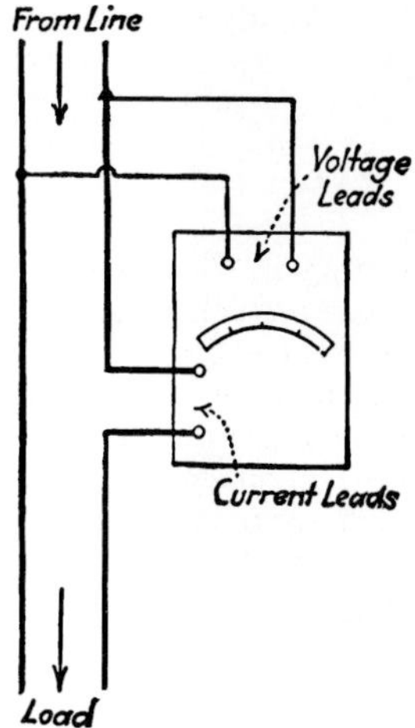

FIGURE 1.17 Wattmeter correctly connected into circuit.

Wattmeter. We have to measure electric power to know how much power any device or apparatus is drawing from the line. A wattmeter registers watts or kilowatts, and by reading it one can tell how much power any piece of apparatus is consuming. The amount of electric power delivered by a circuit depends on the amperes flowing and the volts of pressure. The meter must be connected so that the entire load current flows through it and the voltage pressure is across it. The connections are shown in Fig. 1.17. A wattmeter is essentially a combination of two instruments, an ammeter and a voltmeter. It has an ammeter coil of low resistance, which is connected into the circuit, and a voltmeter coil of high resistance, which is connected across the circuit. A wattmeter will have four terminals or binding posts, two for the current-coil leads and two for the voltage-coil leads.

Electric Energy. Electricity must act for a period of time in order to do useful work. The power expressed in watts tells how much electricity is working, and the hours express the time during which it acts. The product of these two factors gives the amount of work done.

$$\text{Power} \times \text{time} = \text{energy}$$

or

$$\text{Watts} \times \text{hours} = \text{watthours}$$

Likewise, 1000 watthours equals 1 kilowatthour which is abbreviated 1 kWh. A kilowatthour can thus be thought of as

$$1000 \text{ watts acting for 1 hour} = 1 \text{ kWh}$$

Any other combination of volts, amperes, and hours whose product is 1000 would give 1 kWh of energy.

Watthour Meter. The total amount of electrical energy consumed over a period of time, such as a day or month or year, is indicated by a watthour meter. We are all familiar with the common electric house meter. This electromechanical device with disks and dials is nothing other than an integrating watthour meter which shows how much energy the lights,

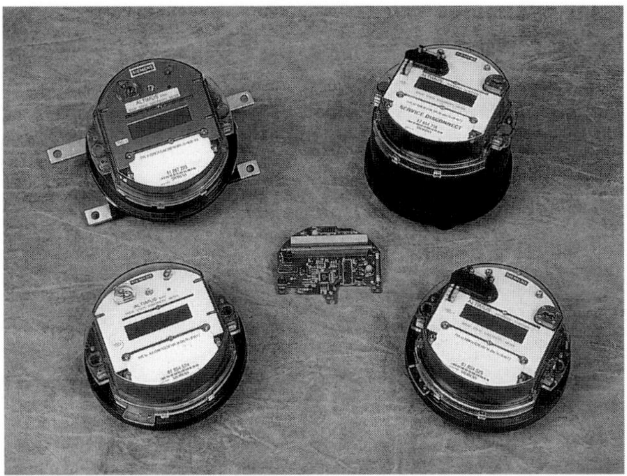

FIGURE 1.18 Typical solid state alternating-current watthour meters. (*Courtesy of Siemens Energy & Automation, Inc.*)

the television, the stereo, the iron, the toaster, the washing machine, etc., consume in the course of a month. Solid state watthour meters as in Fig. 1.18 have evolved in order to measure more than just consumption of the kilowatthour. Main printed circuit boards (Fig. 1.19 to 1.21) are installed in each meter under glass and can be programmed to provide integrations and calculations that measure peak kVAR demands (reactive power), peak kW demands, and phase angles over specified time intervals as well as recording voltage swells and voltage surges and power interruptions.

Conductors. The pipes merely furnish the path for the flowing water in the water circuit. We would say that the hole in the pipe was the path of the water if we were to be more exact. We know the water does not flow through the iron shell of the pipe. It actually passes through the opening in the pipe inside the iron shell. The hole or opening in the pipe may then be directly compared to the electric wire. The electricity, however, flows right through the hard wire. It does not need an opening. It passes right through the metal. But we find it can travel with less difficulty through some metals than through others. It travels easily through copper; therefore, we often use copper wire. Copper is therefore called a good conductor. Any substance that offers little resistance to the flow of electricity is called a good conductor.

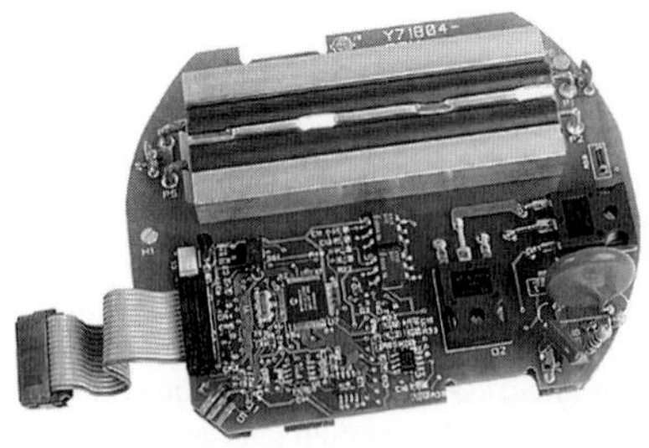

FIGURE 1.19 The main printed circuit board of a solid state meter. (*Courtesy of Siemens Energy & Automation, Inc.*)

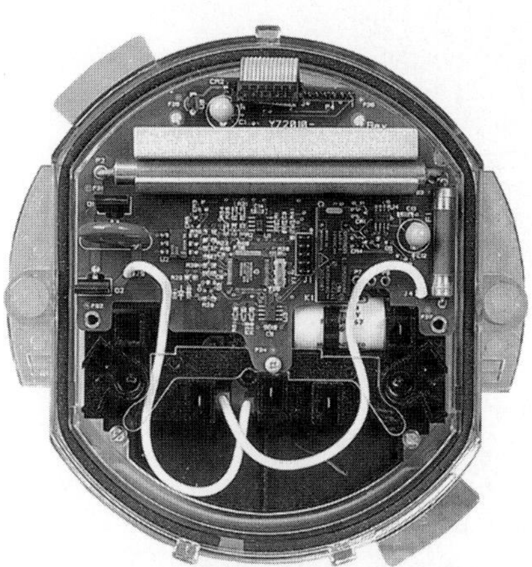

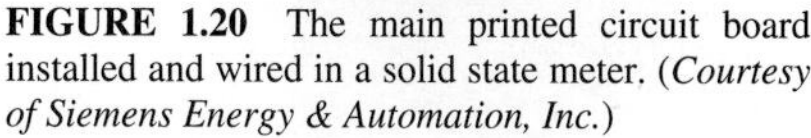

FIGURE 1.20 The main printed circuit board installed and wired in a solid state meter. (*Courtesy of Siemens Energy & Automation, Inc.*)

FIGURE 1.21 The relative position of the main printed circuit board in a solid state meter. (*Courtesy of Siemens Energy & Automation, Inc.*)

Electricity passes still more easily through silver, but silver is too expensive to be made into wires. Aluminum has been used extensively on long transmission lines because of its rather good conduction and light weight, thus making long spans between poles or towers possible. It is in general use in distribution lines as well. Iron is not a good conductor, but is used as a supporting wire in parallel with or coated by copper or aluminum.

Insulators. It was pointed out earlier that the hole in the pipes serves as the path for water. This does not mean to imply that the shell of the pipe is useless and unnecessary. It is very necessary. The shell serves to hold the water in its path. A pipe without a shell would never conduct any water. There would not be any pipe, and the water would flow everywhere. It is important to remember that the shell keeps the current of water in its path.

In an electric circuit, there must be something to keep the current from leaving the wire. The metal of the wire is its path, but there must be something to keep it from leaving the metal. For overhead installations, bare conductors are separated sufficiently from each other based upon system voltages as defined in the clearance tables of the National Electrical Safety Code (NESC). The air serves as an insulator because air has less free electrons than the wire conductor and for this reason an overhead bare metal conductor is installed on insulators made of porcelain, glass, or synthetic polymer material. For underground and some overhead applications, when the overhead conductors are desired to be placed in closer proximity to one another due to space limitations than the NESC allows, a shell is put around the wire just as in the water pipe. This shell, however, is not iron but is usually rubber, polyethylene or polyvinylchloride. The layer of rubber, polyethylene, or polyvinylchloride is called insulation, and a wire so covered is called an insulated wire. Figures 1.22 and 1.23 show conductors covered with different kinds of insulations.

Nonmetallic-sheathed cable—which is an assembly of two or more insulated conductors having an outer sheath of moisture-resistant, flame-retardant, nonmetallic material—is usually used for housing wiring where the voltage is 120 or 240 volts. Higher voltages are insulated with rubber or polyethylene.

FIGURE 1.22 Three-conductor (triplex) self-supporting cable with polyethylene insulation commonly used by electric utilities for low-voltage power-distribution secondaries and services. (*Courtesy Northwest Lineman College.*)

Wires are commonly mounted on poles, where they rest on porcelain or polymer insulators, as shown in Figs. 1.24 and 1.25. Glass, porcelain, and polymers are good insulators of electricity. Bare wires are carefully mounted on glass, "petticoat" porcelain insulators (named because of their semblance to a skirt) or polymer insulators to keep the wires from touching each other or the ground, thus preventing the electricity from leaving the wires.

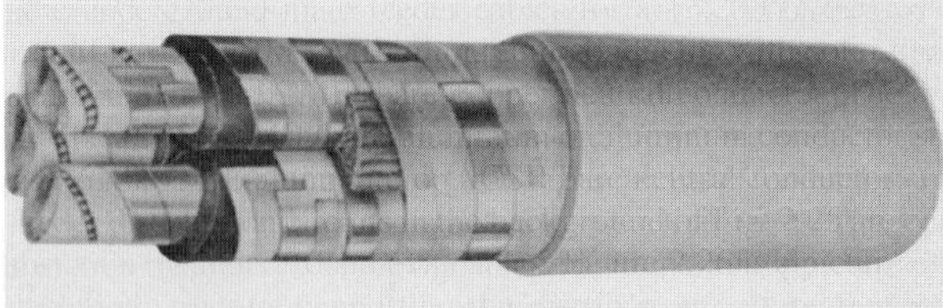

FIGURE 1.23 Three-conductor lead-sheathed paper-insulated power cable showing various layers of shielding and insulation. (*Courtesy Okonite Co.*)

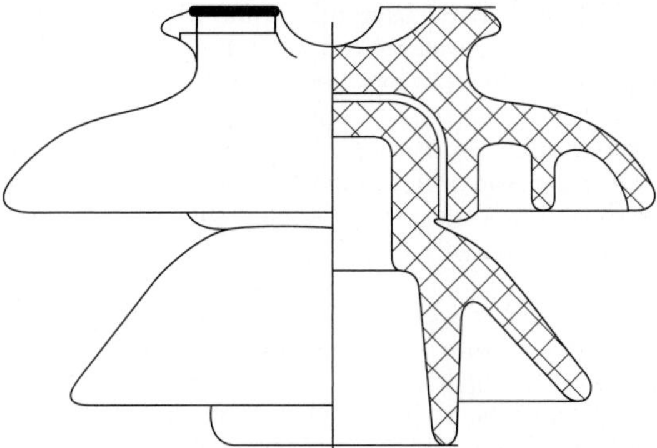

FIGURE 1.24 Two-layer porcelain-pin insulator for use in medium voltage circuits. The layers are connected together.

High-voltage transmission lines may be constructed with polymer suspension insulators. The suspension insulators are high strength, light, and durable. A polymer suspension insulator is shown in Fig. 1.26. The greater the length of the insulator, the higher the voltage it can withstand.

Resistance. One can draw up an ice cream soda faster and with less exertion through a short and large straw than through a long and small straw. The water circuit of Fig. 1.6 makes it easy to understand that less water will flow if the pipe is long and has a small opening than if it is short and has a large opening, provided the pressure is the same in both cases. This is to be expected, for in the long pipe the friction is greater because of the greater

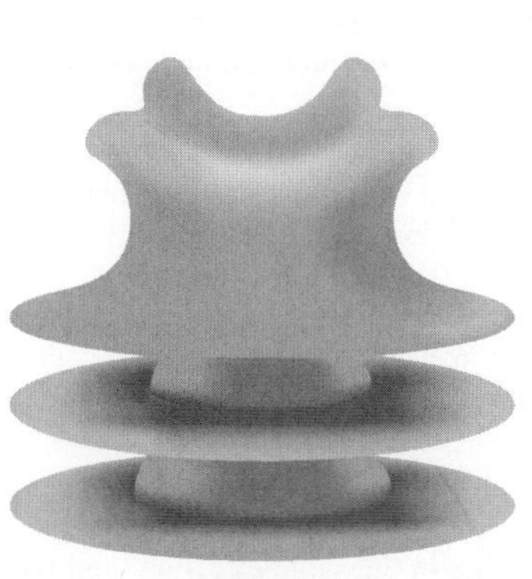

FIGURE 1.25 Polymer pole top pin insulator for use with covered or bare conductor. (*Courtesy Hendrix Wire & Cable Inc.*)

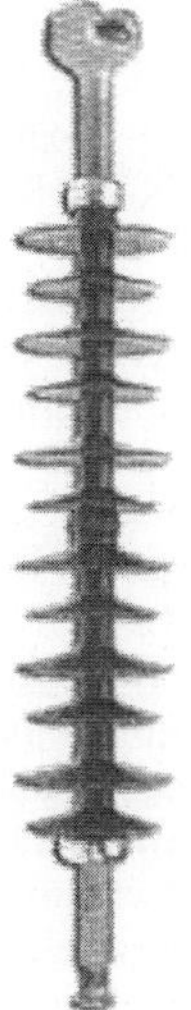

FIGURE 1.26 Polymer high strength, light weight, durable suspension insulator for use in high-voltage circuits. (*Courtesy Ohio Brass Co.*)

length, and if the pipe is small in diameter, the friction will be still greater because of the smaller space, through which the water must be forced. It may be said that a long, narrow pipe offers more resistance to the flow of water than a short, wide pipe. Likewise, an electric current flows along the path of least resistance.

An electric conductor offers a resistance to the flow of current determined by its length and its thickness or diameter. More friction must be overcome if the wire is very long than if it is short; if it is of small cross section, it will take still more effort to crowd the current through the wire.

Ohm. The unit of resistance for wire is the ohm (Ω). We can picture how much an ohm is by nothing how many feet of wire a given size takes to make an ohm of resistance. Wires used for electric purposes are supplied in regular sizes of specified diameter. The no. 10 copper wire has a diameter of about 1/10 in and has 1 ohm of resistance for each 1000 ft of length. Thus 5000 ft of this wire would have 5 ohms resistance. A wire whose cross section is one-half as large would have a resistance of 2 ohms for each 1000 ft of length.

Ohm's Law. Resistance reduces the amount of current that will flow in a circuit. The number of amperes that will flow in a circuit will not be determined completely by the voltage or pressure which causes the current to flow, but by the amount of friction or resistance in the wires. The greater the resistance of a circuit, the smaller the current will be that flows; and the smaller the resistance, the greater the current will be in a circuit with a constant voltage. This general relation between voltage, current, and resistance is commonly called Ohm's law. It is a law because it has been found to hold in every case. It is written thus:

$$\text{Current} = \frac{\text{voltage}}{\text{resistance}} \quad \text{or} \quad \text{Amperes} = \frac{\text{volts}}{\text{ohms}}$$

$$\text{Resistance} = \frac{\text{voltage}}{\text{current}} \quad \text{or} \quad \text{Ohms} = \frac{\text{volts}}{\text{amperes}}$$

$$\text{Voltage} = \text{current} \times \text{resistance} \quad \text{or} \quad \text{Volts} = \text{amperes} \times \text{ohms}$$

The law as stated above applies only to direct-current circuits, circuits in which the current continuously flows in one direction in the wire. The law is quite obvious, because it agrees with the common principle with which all of us are familiar, namely, that the result produced varies directly in amount with the magnitude of the effort or force and inversely with the resistance or opposition encountered.

Rheostats. The current in a circuit may have to be reduced or increased. The resistance in the circuit can be increased or decreased with a variable resistance that can be changed as desired. Resistance boxes, or resistors, are shown in Figs. 1.27 and 1.28. More or less resistance is put into the circuit by turning the handle of the hand-operated resistor, and the current accordingly decreases or increases. Variable resistors are commonly called rheostats and are used to control or regulate the current flowing in a circuit.

Direct Current. A water circuit has been used in which the flow of water was always in the same direction around the circuit. It is evident from Fig. 1.6 that if the engine continues to rotate in the same direction, the rotary pump will rotate in the same direction, and the current of water will continue to flow in the same direction. The electrical equivalent of this current is called a direct current. Direct current flows in the same direction in a circuit. Direct current serves a very minor role in electric distribution. The principle use of direct current is for elevators, electric furnaces, electroplating, etc.

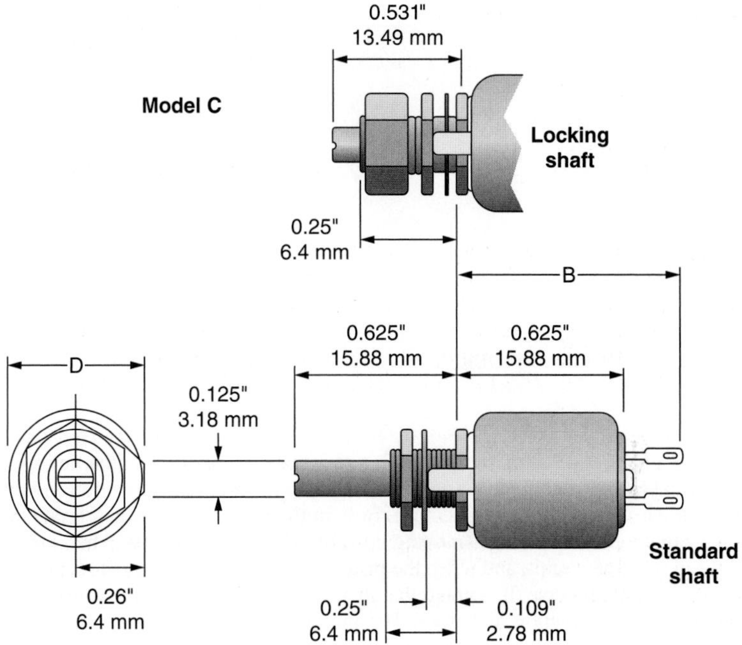

Mounting: Panels to 0.125" (3.18 mm) thick with 0.25–32 bushing and hex nut (0.063" thick)

FIGURE 1.27 Hand-operated model C rheostat used for controlling the field current in circuits. (*Courtesy Ohmite Mfg. Co.*)

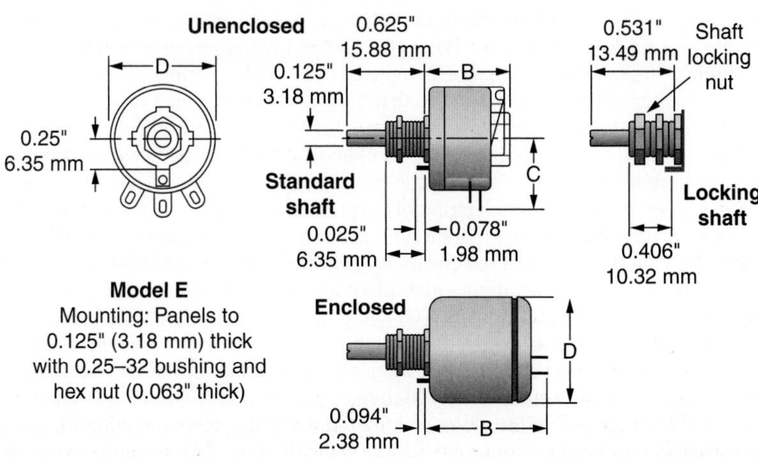

FIGURE 1.28 Hand-operated model E field rheostat. (*Courtesy Ohmite Mfg. Co.*)

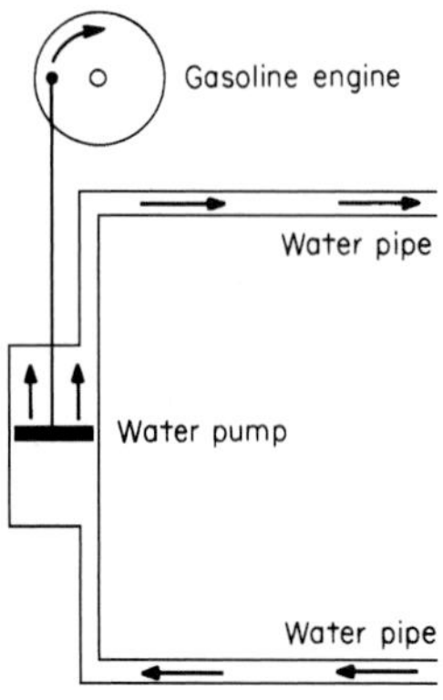

FIGURE 1.29 Alternating-current water pump.

Alternating Current. A rotary engine with a reciprocating pump, like the one shown in Fig. 1.29, will produce a current of water which will first flow in one direction and then, as the piston moves in the opposite direction, flow in the opposite direction. It can be seen in Fig. 1.29 that when the piston is moving upward, the flow of water will be as shown by the arrows in the pipe circuit, and when the piston moves downward, the direction of flow of water will have to reverse. It is noted that in such an arrangement the engine still keeps on revolving in one direction as before. The pump, however, being of a different type, causes the water current to flow in the direction for a very short time and then reverses it and makes it flow in the other direction for a very short time. This changing of direction continues all the time. The electrical equivalent of the current that flows first in one direction and then in the other direction is said to alternate in direction and is called an alternating current. An alternating current is an electric current which periodically passes through a regular succession of changing values, both positive and negative.

Cycle. The water pump illustrated in Fig. 1.29 causes the current to change its direction two times for each revolution of the engine. When the piston moves up, the water has a clockwise direction around the circuit, and when the piston moves down, it moves in a direction opposite to the hands of the clock. When the piston comes up again, the water will then move in a clockwise direction. The water current reverses its direction two times for each complete stroke of the pump. The complete stroke of the pump with its two reversals of water flow is called a cycle. A cycle is, therefore, something which repeats itself.

Frequency. Frequency is the number of complete cycles made per second. If the pump made 60 complete strokes per second, the pump could be said to have a frequency of 60 cycles per sec. Frequency of electricity is expressed in hertz (Hz), which is the unit of frequency equal to 1 cycle per sec. Almost all the electricity generated in the world is of the alternating-current type. Most of the electricity generated and consumed flows first in one direction in the wires and then in the other direction, just as water does in Fig. 1.29, and most of the alternating current is 60-Hz (60-cycle) alternating current. The currents in the house lamp, the toaster, the kitchen range, and the fan motors all flow for a very short time in one direction around the circuit and then quickly change and flow in the other direction. This continues all the time and at a very high rate. These reversals generally take place at the rate of 120 per sec or 7200 per min, making the frequency 60 per sec or 3600 per min. This is so fast that one cannot notice any flicker in the light of a house lamp. These electric currents are known as 60-Hz (60-cycle) alternating currents.

The prevailing electric frequency is 50 Hz outside North America. There is no particular advantage associated with the lower 50-Hz frequency.

Alternating-Current Generator. The discussion of electric pressure or voltage stated that a generator or dynamo produces electric pressure which causes current to flow. A description of the type of generator that produces an alternating pressure will be provided. An electric pressure which causes an alternating current to flow must alternate.

Elementary-Type Alternating-Current Generator. The simplest arrangement for generating an alternating pressure is shown in Fig. 1.30. It consists of a steel horse-shoe magnet and a bar of copper to the ends of which is attached a wire, forming the circuit shown. When this bar is moved down between the poles of the magnet so that it passes through the magnetism at the ends of the poles, an electric pressure or voltage is generated in the wire. If the bar is now raised or moved upward so that it passes through the magnetism in the other direction, a voltage will again be generated in the wire, but this time the voltage will be in the opposite direction. All that is needed to obtain an alternating voltage is a machine in which a wire moves through magnetism first in one direction and then in the other, and arrangements for connecting this moving wire to the outside circuit. Such a device is shown in Fig. 1.31 and is called an alternating-current generator. The elementary electric alternating-current generator has a wire bent into the form of a loop or coil so that it can be rotated continuously in one direction instead of having to move up and down. The two sides of the coil will alternately pass through the magnetism at the north pole and through a magnetism at the south pole. The side of the loop marked *AB* will generate a voltage in a direction from *B* to *A* (if this loop is revolved in the direction of the arrow) and the other side of the loop marked *CD* will generate a voltage in a direction from *D* to *C* at the instant shown. A current will flow if the loop is connected to a closed circuit through rings at the ends of the loop by means of sliding-contact brushes. When the loop is in the vertical position one-fourth of a revolution later, the sides of the loop are not passing through any magnetism, and during this instant there will be no voltage generated and no current flowing. When the loop has advanced one-half of a revolution from the first position, the wires are again passing through magnetism, causing a voltage to be generated and a current to flow if the circuit is closed. This time, however, the current will flow in the opposite direction, because the wires are passing different poles. An alternating current can thus be taken from the rings. The curve shown in Fig. 1.32 results if the values of generated voltages are plotted against the corresponding positions of the coil. This is the familiar sinusoidal waveform of

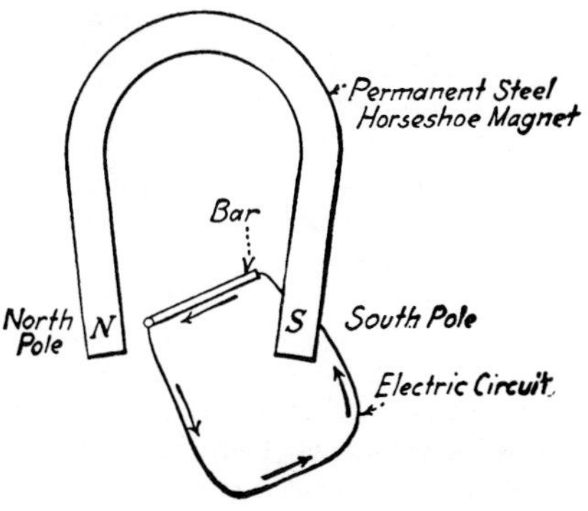

FIGURE 1.30 Elements of an alternating-current generator.

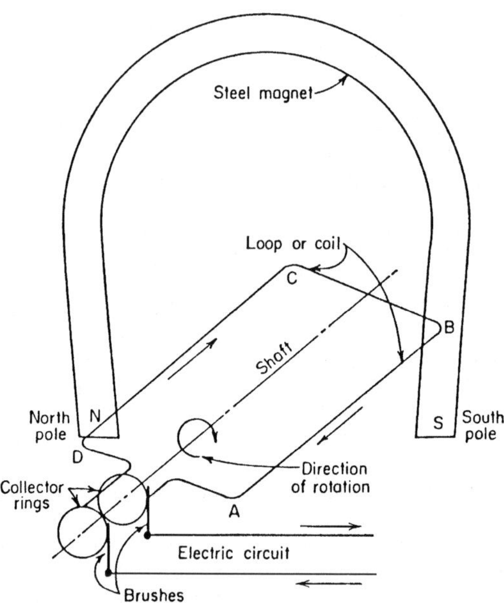

FIGURE 1.31 Elementary alternating-current generator.

an alternating voltage. Brushes sliding on the collector rings connect the loop to the electric circuit. The machine illustrated in Fig. 1.31 is a generator of alternating currents.

The voltage in the coil will reverse its direction twice for every revolution of the coil, just as the water changed its direction twice for every stroke of the pump. The frequency in

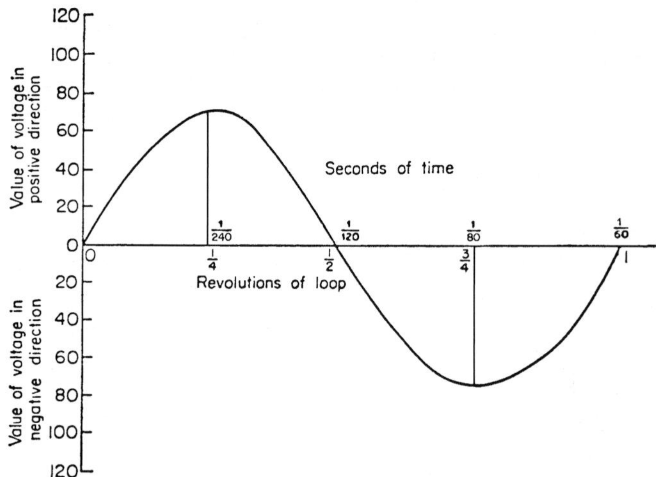

FIGURE 1.32 Alternating-voltage wave. This chart gives the positive and negative values of voltage for each position of the loop. Zero corresponds to vertical position of loop.

cycles per sec will then be the same as the revolutions of the coil per sec. If the loop revolves 60 times per sec, the frequency of the voltage will be 60 Hz. The time required for each revolution and for each cycle will be 1/60 sec.

Single-Phase Alternating-Current Generator. This generator has only two slip rings which connect to a simple circuit of two wires. A generator with only two slip rings is known as a single-phase generator, single phase meaning that the circuit has only two wires.

Two-Phase Alternating-Current Generator. A generator with another coil or loop placed on the rotating shaft so that the wires of this loop lie halfway between those of the loop already described and with the ends of the second loop brought out to another pair of slip rings will generate two voltages, one in each loop. One loop would be passing through magnetism when the other would be midway between the poles, and the voltage in one would be a maximum while the other was zero. The voltage in one loop would be ahead of the other by the time necessary for the shaft to turn through the space separating the coils or, in this case, one-fourth of a revolution (Fig. 1.33). A machine generating two voltages is called a two-phase generator. The machine would have four wires leading from it to make up the two circuits.

Three-Phase Alternating-Current Generator. Picture three coils or loops equally spaced on the shaft of the machine like the two coils of the two-phase generator. A machine built with three coils and three pairs of collector rings would obtain three alternating voltages in the three circuits, but each of these voltages would be a little ahead of the others (Fig. 1.34). The machine would be called a three-phase alternating-current generator. The cycles per sec would again be equal to the revolutions per sec if the machine were of the simple type with only two poles. The machine would have six slip rings and six wires leading from it, making up the three circuits.

It has been found that the sum of the currents at any instant in three of the six wires is zero. Some of the currents are flowing away from the generator, others are flowing toward the generator at the same instant, and the net sum of the currents in three wires is zero. This fact makes it possible to do away with three of the six wires, thus making a three-phase circuit a line with only three wires. The generator will have only three collector rings, and only three wires are connected to the machine.

Revolving-Field Alternating-Current Generator. The alternators described have stationary field poles with the conductors arranged to move past the poles. It is better construction to have the armature conductors stationary and the poles revolving, called revolving-field alternators instead of revolving-armature alternators. Figure 1.35 illustrates

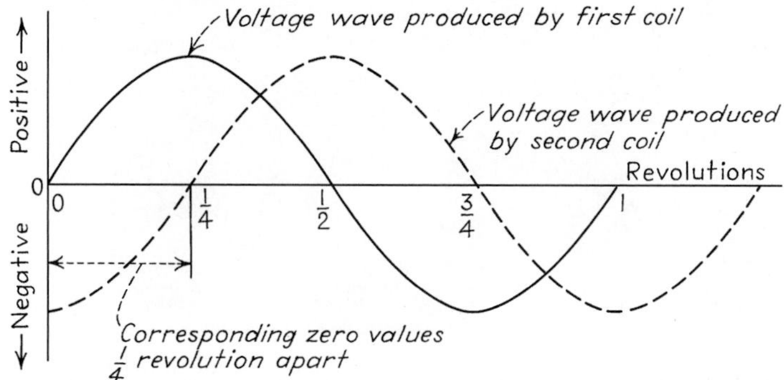

FIGURE 1.33 Voltage waves of two-phase generator.

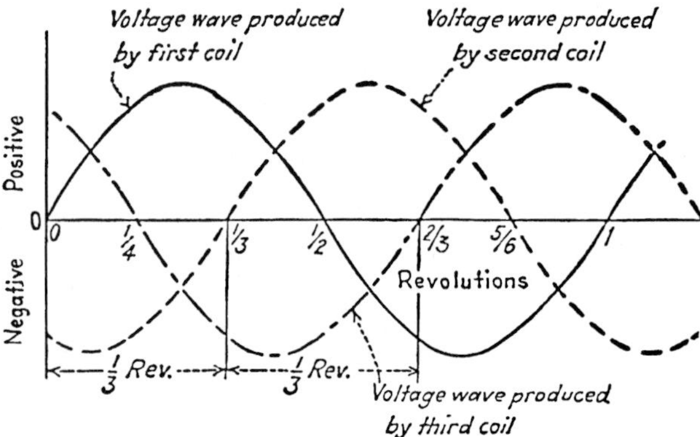

FIGURE 1.34 Voltage wave of three-phase generator. The corresponding zero values are one-third revolution apart.

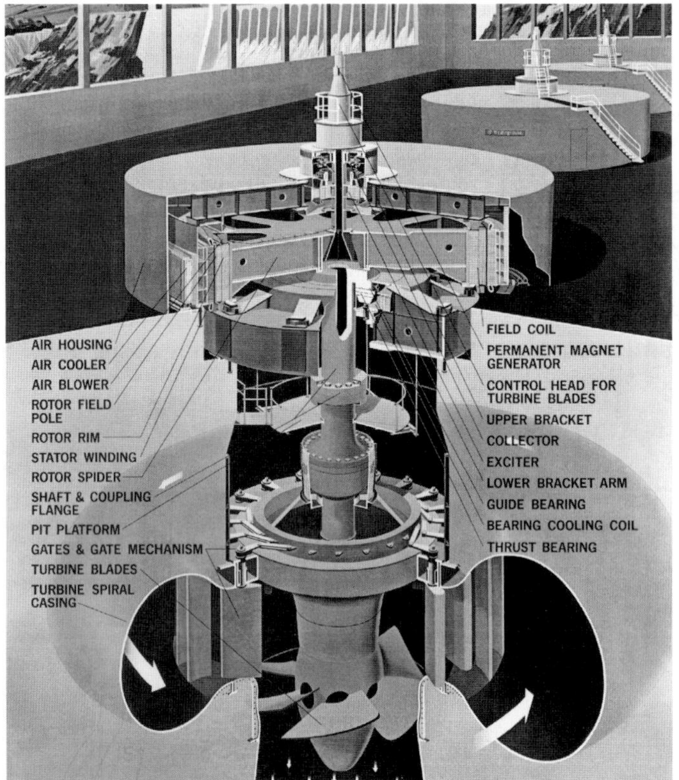

FIGURE 1.35 Three-phase revolving-field type of hydroelectric generator. Shaft is vertical and connects water turbine inside casing with revolving field on floor level. (*Courtesy Westinghouse Electric Corp.*)

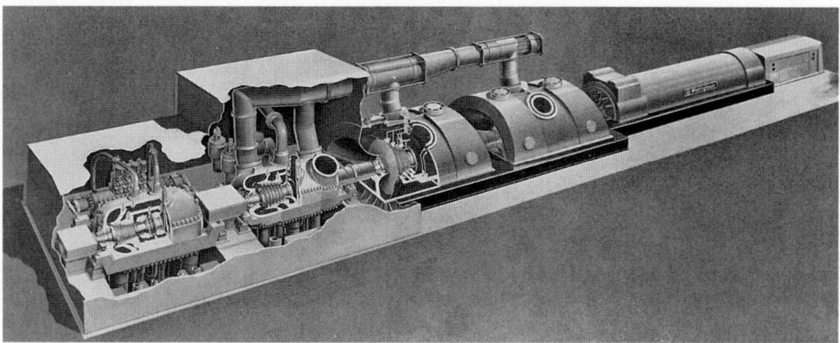

FIGURE 1.36 Cutaway view of steam turbine direct connected to three-phase, revolving-field alternating-current generator. Tandem-compound, four-flow, 3600-rpm turbine-generator unit rated 800 MW. *(Courtesy Westinghouse Electric Corp.)*

this type of construction in a hydraulic turbine-driven generator, and Fig. 1.36 illustrates it in a steam turbine-driven machine. The principal advantages of the revolving field construction are (1) the armature conductors can be more securely fastened and better insulated, and (2) the amount of material in the rotating part is considerably reduced.

Three-Phase Connections for Alternating-Current Generator. The three coils of the three-phase alternator can be connected in two ways, one to form a Y connection and the other to form a delta (Δ) connection. The Y connection has one set of corresponding ends of each of the three coils of the alternator connected together, as shown in Fig. 1.37. The name "Y connection" is taken from the appearance of the connection when shown as a diagram. The three free ends of the coils are connected to the three-phase line.

The delta connection has the three coils connected in series as shown in Fig. 1.38. The name is taken from the appearance of the connection when shown as a diagram. The line wires are connected to the junction points or the corners of the delta.

Transmission Lines. Most of the electricity used in the world is three-phase alternating-current. Long distance alternating-current transmission-line towers or poles have three-phase

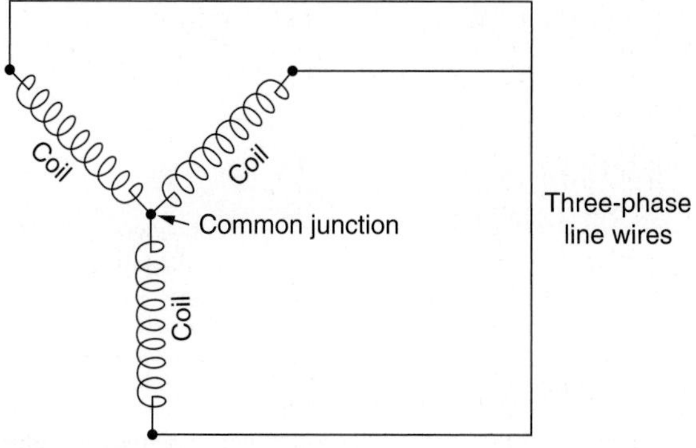

FIGURE 1.37 Y connection of three-phase alternating-current generator.

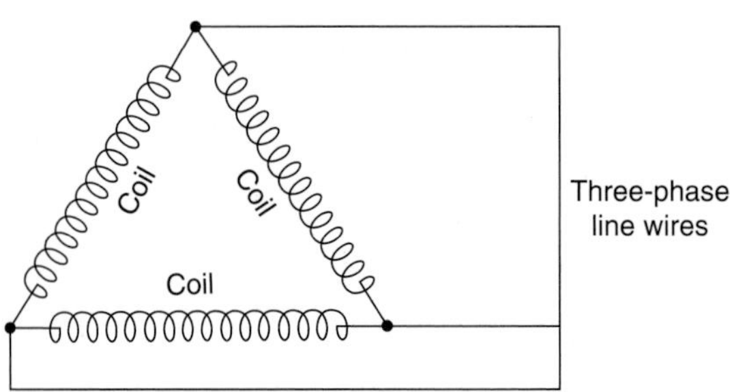

FIGURE 1.38 Delta connection of three-phase alternating-current generator.

wires and a ground or static wire for lightning protection. A typical three-phase alternating-current transmission line is shown in Fig. 1.39. In many cases, electric power companies provide lines with two sets of three wires each, as shown in Fig. 1.40. Double circuits increase the amount of power transmitted over the line, make more effective use of the right of way, and improve the probability of ensuring continuous service.

Power Factor. When an alternating voltage and the current which it causes to flow rise and fall in value together and reverse direction at the same instant, the two are said to be "in phase" and the power factor is unity. This condition is illustrated in Fig. 1.41. The same

FIGURE 1.39 Single-circuit 500,000-volt three-phase high-voltage transmission line. Note the long strings of suspension insulators needed to insulate the conductors. The insulators are installed in a V shape to maintain adequate air clearance between the conductors and the grounded steel tower by preventing the line conductors from swinging too close to the tower. The two ground wires at the top of the steel tower are used to shield the line from lightning.

FIGURE 1.40 Double-circuit high-voltage transmission line. The three conductors on each side of the steel pole constitute a single three-phase circuit. The ground wire on top of the steel pole is for protection against lightning. The steel-pole construction eliminates congestion in urban areas. Street lights may be mounted directly on the transmission-line pole as shown. (*Courtesy A. B. Chance Co.*)

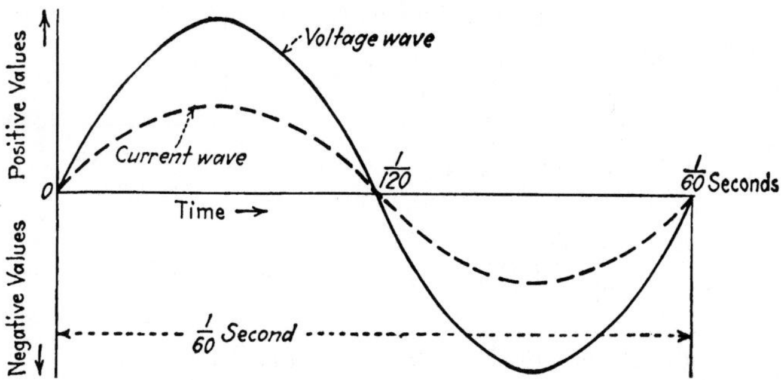

FIGURE 1.41 Voltage and current waves in phase. The power factor under these conditions is unity.

formula for power as was discussed for direct current holds for alternating current when the power factor is unity:

$$\text{Power in watts} = \text{volts} \times \text{amperes}$$

The current and voltage waves are not in phase in most cases. They do not rise and fall in value together, nor do they change direction at the same instant, but instead, the current usually lags behind the voltage. Figure 1.42 illustrates the usual condition in transmission and distribution circuits. The current and voltage are now said to be "out of phase." The current drawn by idle running induction motors, transformers, or underexcited synchronous motors lags even more than the current shown in the figure.

Occasionally, the current leads the voltage. An unloaded transmission line or an overexcited synchronous motor or a static condenser takes leading current from the line. When the current leads or lags the voltage, the power in the circuit is no longer equal to volts times amperes but is calculated from the expression

$$\text{Watts} = \text{volts} \times \text{amperes} \times \text{power factor}$$

From this,

$$\text{Power factor} = \frac{\text{watts}}{\text{volts} \times \text{amperes}}$$

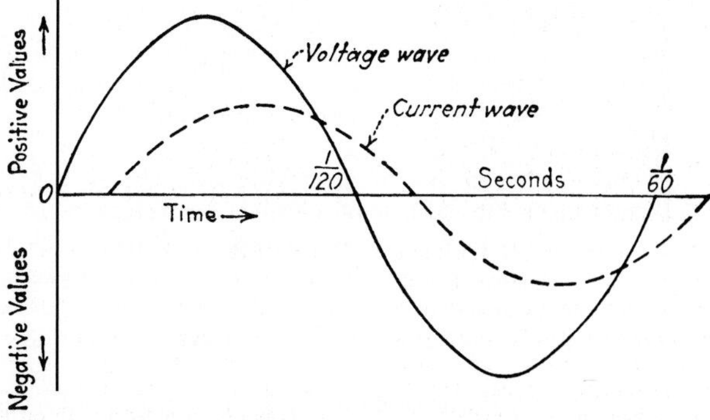

FIGURE 1.42 Current wave lagging behind the voltage wave. This is the usual condition in transmission and distribution circuits.

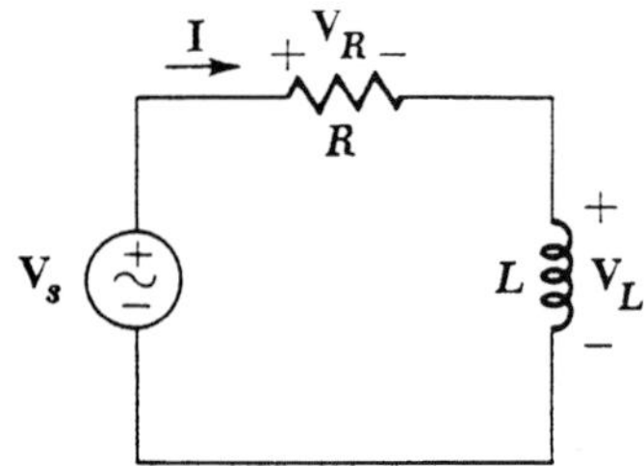

FIGURE 1.43 Electric circuit consisting of a resistive and inductive load with lagging power factor.

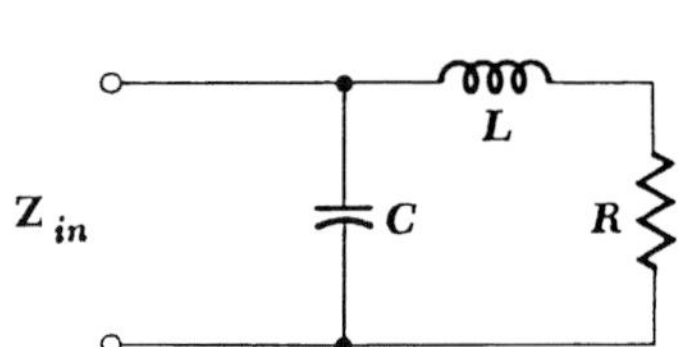

FIGURE 1.44 Electric circuit consisting of a capacitive, a resistive, and an inductive load. The power factor has been corrected to zero.

The power factor can thus be defined as the ratio of the actual power to the product of volts times amperes. The latter product is generally called volt-amperes, or apparent power. The value of the power factor depends on the amount the current leads or lags behind its voltage. When the lead or lag is large, the power factor is small, and when the lead or lag is zero, as when the current and voltage are in phase, the power factor is unity. This is the largest value that the power factor can have. The power factor is usually between 0.70 and 1.00 lagging. An average value often taken in making calculations is 0.80 lagging. Figure 1.43 is an example of an electric circuit with resistive and inductive load and a lagging power factor.

Low Power Factor. The cause of low power factor is an excessive amount of inductive effect in the electric consuming device, be it motor, transformer, or lifting magnet, etc. Induction motors, when lightly loaded, exhibit a pronounced inductive effect causing the current to lag the voltage. Idle transformers likewise have a strong tendency to lower the power factor.

Inductance. As the value of current changes in a circuit, properties in the conductor oppose the change in the current and cause a time delay prior the current reaching a steady state value. This property is known as inductance. Inductance causes the current to lag the voltage. Inductance is measured in units defined as henrys.

Capacitance. Capacitance is the direct opposite of inductance, just as heat is the opposite of cold, sweet the opposite of sour, and day the opposite of night. Capacitance is the property of a condenser, and a condenser is a combination of metal plates or foil separated from each other by an insulator such as air, paper, or rubber. The capacitance, or the capacity of the condenser to hold an electric charge, is proportional to the size of the plates and increases as the distance between the plates decreases. Capacitance causes the current to lead the voltage. Capacitance is expressed in units defined as farads.

Power-Factor Correction. One method of raising the power factor is to add capacitors to the circuit, since capacitance is the opposite of inductance and since too much inductance is the cause of low power factor. Figure 1.44 is an example of an electric circuit in which the inductive load is offset by the capacitive load and the power factor has been corrected to unity or 1.00. Capacitors are installed on poles as illustrated in Figs. 1.45 and 1.46. The capacitors can be directly connected to the circuit or switched on and off as needed.

Direct-Current Generator. We have carefully studied how an alternating current of water could be produced by the pump shown in Fig. 1.29. We observed that the current of water circulated around the pipe circuit, first in one direction and then in the opposite direction.

FIGURE 1.45 Shunt-connected pole-top capacitor bank protected by solid-material power fuses used when higher voltages, higher fault currents, and higher loads push fuse cutouts beyond their capabilities (*Courtesy of S & C Electric Co.*)

We observed that the water current reversed its direction twice for each complete stroke of the pump. All this we have carefully compared with the action of a real electric generator and found the processes are quite alike. This much is fundamental in the generation of electricity, whether it is alternating-current or direct-current electricity. To obtain alternating current, we simply fastened slip rings to the ends of the loop and connected those rings to the circuit by means of sliding brushes.

Direct-Current Water Pump. A direct current of water, which is a current that flows continually in one direction, can be obtained by adding or set of valves to our water pump, as shown in Fig. 1.47. The water will not be able to flow backward after it is once forced through the valve, and therefore, the direction of the water flow will always be the same. The piston moves up, forcing the water through the top valve, and as the piston drops, the top valve closes and the lower valve rises, allowing the water to fill the cylinder. When the piston moves upward again, the cylinder is full of water and more water will be forced into the circuit through the top valve as before.

FIGURE 1.46 Typical installation of switched pole-top cluster mounted capacitor bank protected by solid-material power fuses and surge arresters.

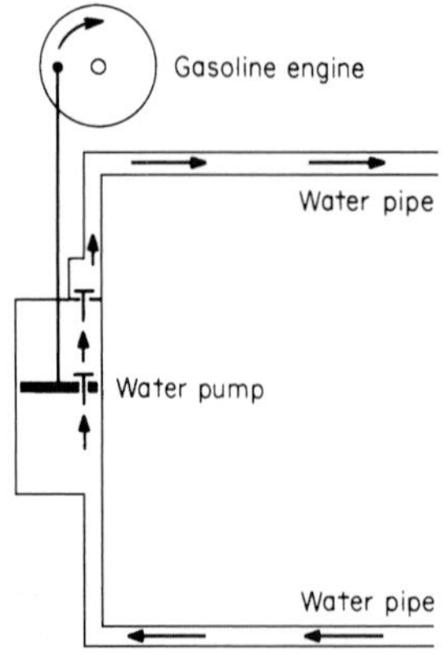

FIGURE 1.47 Direct-current water pump.

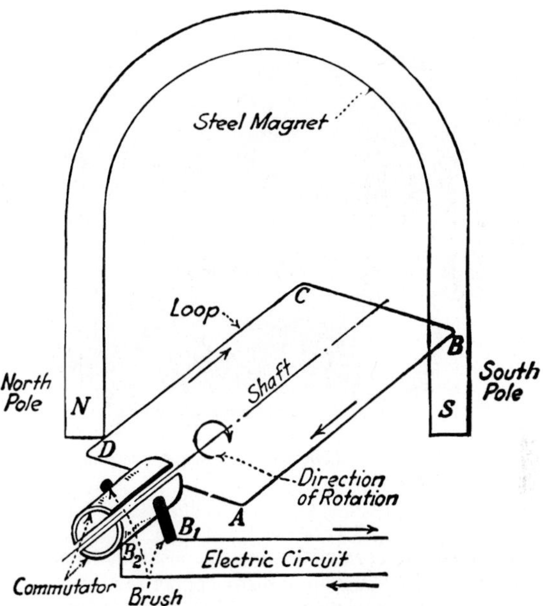

FIGURE 1.48 Elementary direct-current generator.

Elementary-Type Direct-Current Generator. A direct current of electricity can be obtained from the machine shown in Fig. 1.31 by replacing the slip rings with a commutator. The commutator is the device which changes an alternating voltage to a direct voltage. It corresponds to the valves in the hydraulic direct-current generator. The process is called commutation. The commutator consists of two bars of copper connected to the two ends of the loop or coil in the simple case. These two copper bars revolve with the shaft, as did the slip rings. These bars have brushes resting on them, which connect the machine to the electric circuit. The general arrangement is shown in Fig. 1.48.

Commutation Direct-Current Generator. From Fig. 1.48, it is clear that the commutator will first connect one side of the loop to one side of the circuit, and then as the shaft revolves, it will connect the other end of the loop to the same side of the circuit. If the brushes are correctly set, this change will take place when both sides of the loop are not generating any voltage, that is, when the coil is midway between the poles. This is the point at which the voltage in the loop changes direction. When A side of the loop is passing the south pole, the voltage is in a given direction and feeds into brush B_1, and D side of the loop feeds into brush B_2. But when the loop has revolved to its midway position, the direction of the generated voltage in the loop is about to change, and just at this instant the ends of the loop are also changed to the opposite sides of the circuit so that when the loop revolves still further, the voltage into the circuit will remain the same. The voltage that alternates in the loop is made to produce a pressure in the same direction in the outside circuit. Such a machine with its commutator is called a direct-current generator.

Commercial Direct-Current Generator. The principal parts of a commercial direct-current generator are shown in Fig. 1.49. These parts are the frame, the poles, the armature, the commutator, and the brushes. The purpose of the frame is to support the north and south

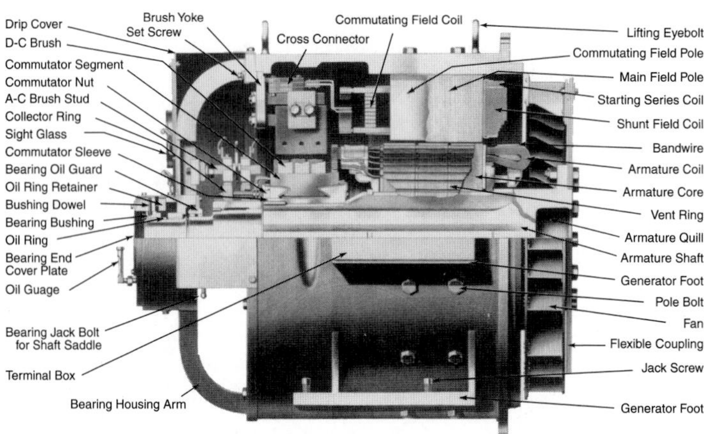

FIGURE 1.49 Sectional view of a direct-current generator showing details of construction. Generator is rated 100 kW.

poles. The armature supports the conductors which constitute the armature winding. The commutator, a split ring device made up of many copper segments, is also supported by the armature. The commutator keeps the torque on a DC motor from reversing every time the coil moves through the plan perpendicular to the magnetic field. A special brush rigging holds the brush holders and brushes. Figure 1.50 shows these parts assembled and placed in proper relation to each other.

Direct-Current Circuits. Direct-current (dc) circuits, as we have seen, require only two wires, one for the outgoing current and the other for the return current. Circuits carrying

FIGURE 1.50 Direct-current generator assembled. This machine is rated 250 kW, 250 volts, and runs at 225 rpm.

direct current will, normally, consist of only two wires. The use of direct current is generally confined to special applications, such as electroplating, battery charging, and elevator operation, or where fine speed control of industrial motors is required.

High-voltage dc circuits are used to transmit large amounts of power from remote alternating-current (ac) generating stations, often located adjacent to a large coal mine or a water reservoir, to load areas. High-voltage dc circuits are also used to transmit large quantities of electrical energy by insulated cable circuits. Alternating-current systems that are not synchronized are being interconnected with high-voltage dc circuits. High-voltage dc transmission lines originate and terminate in converter stations that change alternating current to direct current or direct current to alternating current. The converter stations use solid-state electronic devices to convert the energy from alternating current to direct current or direct current to alternating current.

Transformers. The voltages necessary to obtain economical transmission are higher than those which can be directly generated by an alternator. The voltage at which electric power is used in motors and lamps is less than that required for distribution. It is therefore necessary to raise the voltage at the generating station to the value required for transmission and to lower it at the point of consumption to the values required by the motors and lamps on the system. The voltage must be stepped up in the first case, and it must be stepped down in the latter. The transformer is the apparatus used to make these changes in ac voltage.

Hydraulic Transformer. The water circuit shown in Fig. 1.51 will make it clear how ac voltages are stepped up and down. The water flows to and fro or back and forth in the circuit, since this is an alternating circuit. As it does this, it causes the piston in *B* to move up and down. Piston *B* is connected to piston *C* by a rod pivoted at *P* so that when *B* moves up and down, *C* moves down and up. The pressure in *B* is very high and the current of water is small, just like the voltage and current in a high-voltage transmission line. The pressure in cylinder *C* will be low and the current of water large, because the size of the piston is so much larger. A mechanism such as this could be used to transform, or step down, a water circuit having a high pressure to one having a low pressure. The current on the low-pressure side would be much larger than the current on the high-pressure side.

The frequency of the current has not been changed. The only changes are the decrease in pressure and the increase in current. It should be noted that the total power delivered has not been changed, for under the discussion of power it was pointed out that the power is dependent on the product of voltage and current. The same power can be obtained from a circuit of high voltage and small current as from a circuit that has a low voltage and a large current.

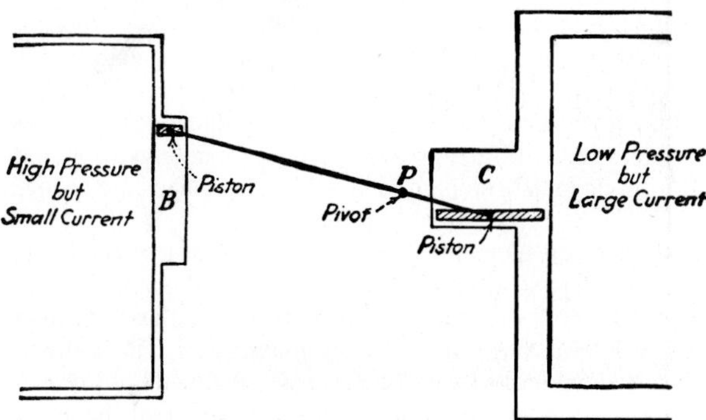

FIGURE 1.51 The hydraulic transformer.

FIGURE 1.52 Single-phase transformer with case removed showing core and windings. High-voltage winding is rated 66,000 volts, and low-voltage winding is rated 13,200 volts. Hand-operated tap changers are shown in center foreground. (*Courtesy General Electric Co.*)

Electric Transformer. An electric transformer operates in the same manner as the hydraulic transformer. The primary, or high-voltage side, of the transformer has a high electric pressure and the current is small. The secondary, or low-voltage side, of the transformer has a low pressure and the current is large. One can usually tell the high-voltage from the low-voltage side of the transformer by observing the size of the insulators or bushings on the top of the transformer case. The high-voltage side must be better insulated than the low-voltage side. Figure 1.52 shows the core and windings of a single-phase transformer removed from its tank. The high- and low-voltage leads are brought out on top of the tank.

Figure 1.53 illustrates the elementary parts of an ac electric voltage transformer. Cold-rolled silicon steel strips are so arranged as to form a closed magnetic circuit. These strips have placed on them two coils of insulated wire, one of which has many turns of small wire and the other of which has a few turns of heavy, coarse wire. The coil with many turns is the high-voltage coil and is called the primary winding in a step-down transformer. The other coil with few turns is the low-voltage coil and is called the secondary winding. Thus the winding into which current is brought is the primary, and the winding from which current is taken is the secondary. The coils correspond to the two pistons in the water transformer, and the magnetic core acts as the coupler between the two pistons.

The action in the transformer is somewhat as follows: The voltage applied to the primary causes a current to pass through the primary coil. This current creates a magnetic flux in the core. The flux in the core cuts both the primary and the secondary coils. This cutting of the primary coil creates a countervoltage in the primary coil which very nearly equals the primary voltage applied. The current in the primary at no load is only great enough to magnetize the core at no load. The secondary coil being cut by the flux will have voltage at no load, but there is no current. Now, let us apply load on the secondary. The load current in the secondary will create a counterflux in the core which reduces the magnetic flux in the core. The reduction in magnetic flux in the core reduces the primary countervoltage. The reduction in primary countervoltage increases the difference between the applied voltage and the countervoltage. More current will therefore flow into the primary, thereby increasing the magnetic flux to its former no-load value. This again raises the secondary induced voltage to its original value. All these adjustments within the transformer take place automatically and instantaneously.

It should be noted that in a modern transformer the primary and secondary windings are not placed on separate legs, as shown in the elementary diagram of Fig. 1.53, but instead, each winding is generally divided into two parts, and one-half of each winding is placed on each leg. This gives a more constant voltage with changes in load. Figure 1.54 shows a typical transformer from which the case has been removed. It shows clearly the magnetic core, the coils, and leads.

Transformer-Core Construction. The core of a transformer can be built in various shapes. The following types are in general use:

Core Type. The core is in the shape of a rectangle. Coils are placed on two legs of the core.

Shell Type. The core is rectangular in shape with a central leg in addition to the core first described. Coils are placed on the central leg.

Cruciform Type. The cruciform core is a modification of the shell type. It is the same as if two shell-type cores were set at right angles, using a common central core for the windings.

Wound-Core Type. The core is made of a ribbon of sheet steel wound in a spiral. After the core is wound and annealed, the coils are wound on the core. This type core is superseding other core types for small distribution transformers (Fig. 1.55).

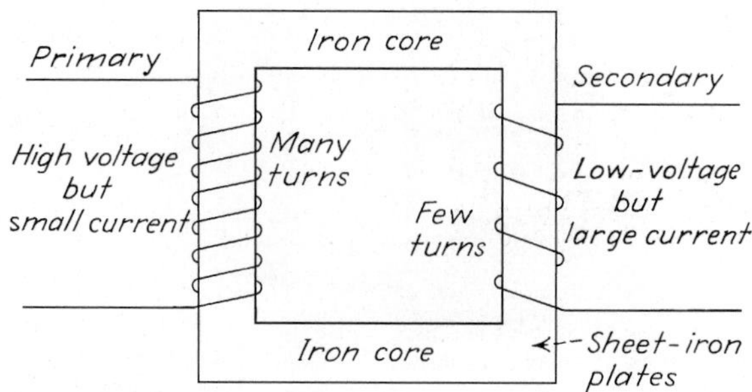

FIGURE 1.53 Electric voltage transformer.

FIGURE 1.54 Three-phase, pad-mounted transformer with case removed. Transformer is wound with taps in the high-voltage winding for a no-load tapchanger. Larger copper straps are connections to the low-voltage winding. Transformer is rated 300 kVA, 13,800Y/7960 to 208Y/120 volts.

FIGURE 1.55 Wound-core type of a single-phase distribution transformer.

Transformer–Core Material. An important consideration for a transformer is the core material because the core of an energized transformer contributes to electric system losses. Transformer losses are classified as either no-load losses or full-load losses. The no-load losses on a lightly loaded transformer are a result of the core material. The energy efficient cores are constructed of low-loss cold-rolled silicon strips or amorphous steel alloy strips.

Transformer Cooling. When the voltages of the primary and secondary windings are both low and the transformer has a small capacity, the transformer, that is, the steel core with its two windings, is placed in a metal case merely to keep out dirt and moisture. This type of transformer is designated as an air-cooled dry-type transformer. The core and coils of a high-voltage large-capacity transformer are usually placed in a tank filled with oil. The oil is a good insulator and helps to cool the windings. This type of transformer is designated as oil-cooled. It will perhaps be remembered that whenever current flows in a wire, there is a friction loss in heat. An illustration of this is found in the toaster or the iron. The same is true in a transformer; whenever currents are flowing through the primary and secondary windings, heat is given off which must be carried away so that the transformer may not become too hot and char or burn the insulation on the wire of the coils. If it becomes too hot, the transformer will be ruined, just as a person will die from running a high temperature for a long time.

More power can be safely drawn from a transformer in cold weather than in hot weather because of the lower ambient temperature. More power can be taken from a transformer for short periods of time than can be taken off continuously. A transformer which is overheated because of an overload should be replaced with a larger transformer before it becomes damaged.

Methods of Cooling Transformers. The oil in the tank is the principal cooling agent. It keeps the windings cool by carrying the heat from the coils to the surface of the tank. Many times the transformer case itself is built so that it has much more surface for cooling than if it were smooth. The metal sides of the tank are corrugated (Fig. 1.56), or pipes are connected to the top and bottom of the tank (Fig. 1.57) so that whatever oil is in the pipes is exposed on all sides. The object in either case is to increase the surface from which the heat can radiate. This is the same principle as is used in the automobile radiator.

Very large transformers may have coils of pipes placed in the top of the tank through which cold water is circulated by a pump. The cold water in the pipes cools the oil in the tank. The water pipes are placed in the top of the tank because the hottest oil is always at the top. An illustration of this method of cooling is shown in Fig. 1.58.

Transformer Temperature Limits. In general, the maximum safe temperature for such insulating materials as cotton, silk, cambric, tape, fabric, etc., is 105°C, which corresponds to 221°F. It is impossible, however, to measure the actual temperature in the transformer because the exact location of the hottest spot is not known, nor is it accessible. The temperature of the oil plus a 10 to 15° correction is usually taken as an indication of the temperature of the windings. The temperature of the oil in large power transformers is usually obtained by some form of thermometer, such as that shown in Fig. 1.57.

Distribution transformers may have devices which indicate the maximum load on the transformer. The rapid growth of loads on distribution systems makes it necessary, in many cases, to make a check of the loads on transformers in order to determine whether the transformer is still large enough to carry the load. An overload indicator, illustrated in Fig. 1.59, makes use of a bimetallic thermostat to give an indication of the winding temperature. It turns on a red light which is visible from the ground if the transformer is overloaded. A reset device makes it possible to turn the light off.

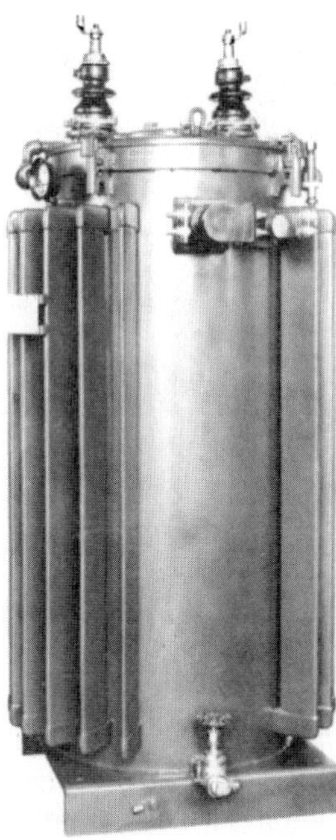

FIGURE 1.56 Single-phase transformer showing the tank corrugated to increase the cooling surface. Note oil-level gauge. Transformer is rated 250 kVA, 60 Hz, 4160Y/2400 on high-voltage side and 240/120 volts on low-voltage side. (*Courtesy General Electric Co.*)

FIGURE 1.57 Transformer tank showing pipes connected to top and bottom of tank to aid in the cooling of the oil. Oil-temperature gauge is located centrally near the top of the tank. This transformer is entirely air cooled. This three-phase oil-immersed power transformer is rated 1000 kVA, 60 Hz, 13,800 volts on high-voltage side and 480 volts on low-voltage side. (*Courtesy General Electric Co.*)

Three-Phase Transformer. A three-phase transformer is really three single-phase transformers in one case using a single combined core. The core has much the same shape as the shell-type single-phase transformer. The three-phase transformer, however, has single-phase primary and secondary windings on each leg of the core (Fig. 1.54). The common core acts to supply flux for all three phases.

A three-phase transformer weighs about two-thirds as much as the same capacity in single-phase transformers. Its biggest disadvantage lies in the fact that if one phase fails, the entire transformer must be taken out of service.

Large substation transformers employ forced-air and forced-oil cooling in order to increase the output rating. By forced circulation of the oil through the external radiators and by blowing air against the radiators, the heat is more rapidly removed from the hot oil. A typical transformer equipped with forced-air fans is shown in Fig. 1.60.

Solid-State Circuit Components. Solid-state devices are used extensively to control electric power circuits. The devices are manufactured from semiconductor materials. Semiconductor materials individually have a high resistance to the flow of electric currents and thus have been identified as semiconductors. Silicon and germanium are semiconductor materials that are commonly used to manufacture solid-state devices. A large, pure crystal of silicon has a high resistance to the flow of electric current. Silicon can be made conductive by adding other materials (referred to as doping) that either add or subtract electrons from the combination material.

FIGURE 1.58 Interior of tank of water-cooled transformer showing cooling coils in which cold water is circulated. This transformer is rated 23,333 kVA, 60 Hz, 115,000Y/66,420 volts on high-voltage side and 13,200 volts on low-voltage side. (*Courtesy-General Electric Co.*)

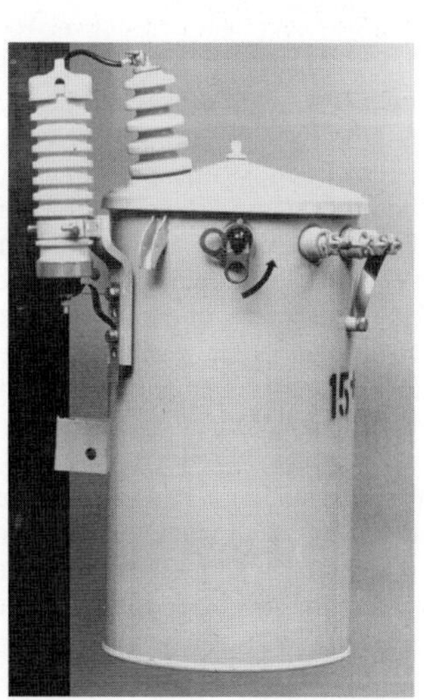

FIGURE 1.59 Self-protected pole-type distribution transformer with low-voltage circuit breaker and over-load-indicating light. Red light indicates transformer has been overloaded.

FIGURE 1.60 Cutaway view of a shelf-form substation transformer showing core, coils, internal bridge structure, bushings, and cooler with fans and oil pump mounted below. (*Courtesy Westinghouse Electric Corp.*)

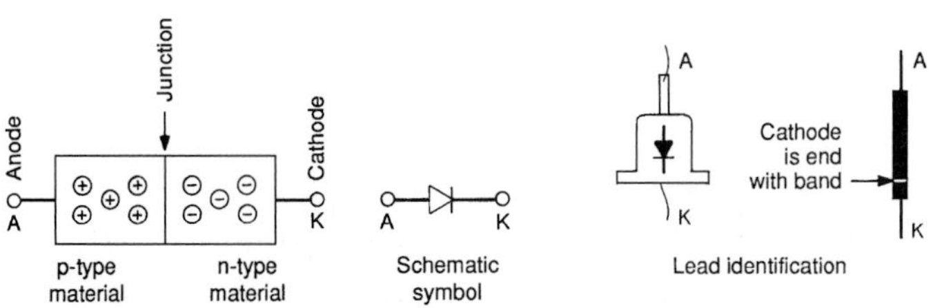

FIGURE 1.61 Diode semiconductor.

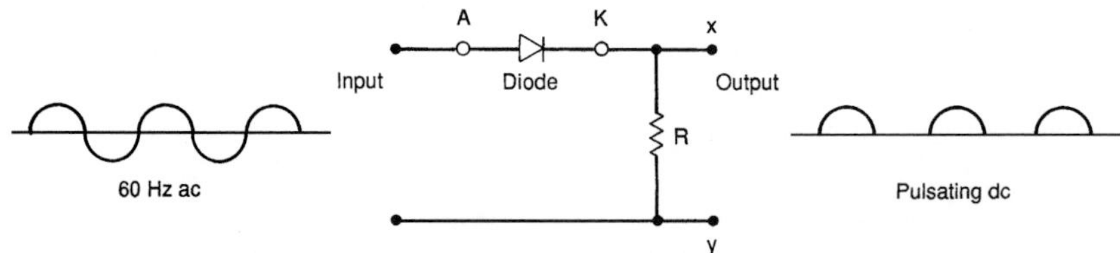

FIGURE 1.62 Diode semiconductor connected to a 60-Hz ac supply voltage with a resistor in series with it will produce a pulsating dc current through the resistor and a pulsating voltage across the resistor that can be measured from terminals *x* and *y* of the circuit.

Materials used to dope the semiconductor materials are boron, phosphorus, gallium, arsenic, and others. A silicon crystal that has been doped with phosphorus, which will donate free electrons to the material, is called an *n*-type semiconductor. An *n*-type semiconductor will have an excess number of electrons. Silicon doped with boron, which takes electrons away, is called a *p*-type semiconductor. A *p*-type semiconductor will have a deficient number of electrons. The *p*-type material with a deficient number of electrons is said to contain holes.

Diode Semiconductors. A semiconductor formed by modifying the silicon material on each end with a junction in the middle separating the *p*-type and the *n*-type materials is called a diode (Fig. 1.61).

A diode semiconductor is used in a circuit as a current rectifier. As illustrated in Fig. 1.61, the terminal of the diode on the *p*-type material is called the anode and the terminal on the *n*-type material is called the cathode. When voltage is placed across the diode semiconductor with the positive potential connected to the anode (A) and negative potential connected to the cathode (K), the holes in the *p*-type material and the electrons in the *n*-type material will both move toward the junction between the *n*- and *p*-type materials. When the impressed voltage is great enough to overcome the resistance of the barrier at the junction, the electrons will combine with the holes and flow externally from the diode, leaving the anode and returning through the cathode (Fig. 1.62).

The pulsating of the output voltage can be eliminated by adding a capacitor in the circuit connected across the output terminals of the circuit *x* and *y*. Full wave rectification can be obtained by adding a transformer on the input to the circuit with a diode connected in series with each of the transformer end terminals and a resistor connected to the cathode of each diode and the neutral terminal on the transformer. Four diodes can be used in a bridge circuit to increase the capacity of the rectified output of the circuit.*

*Additional information on semiconductors and basic solid-state conductor circuits can be obtained from *Basic Electronics: Theory and Experimentation* by Fredrick W. Hughes (Prentice Hall, 1984).

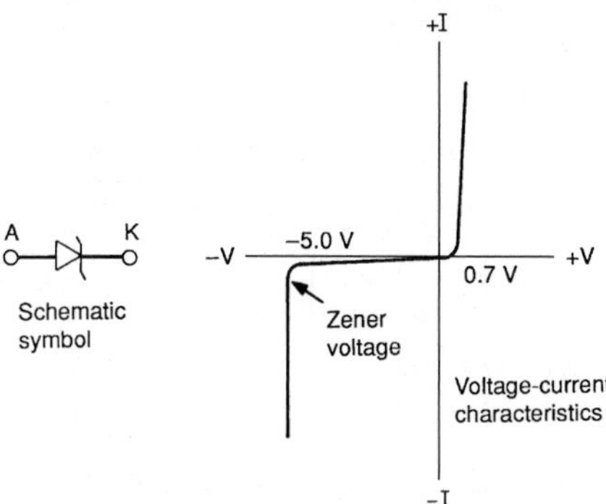

FIGURE 1.63 Zener diode symbol and operating-characteristics curve.

Zener Diode. A diode semiconductor that can be used as a voltage reference in an electric circuit is called a Zener diode. The Zener diode is manufactured by doping the silicon crystal so that when the diode is reversed bias at a voltage greater than the junction breakdown value, a large current will flow through the diode. As the current increases through the diode in the circuit, the voltage across the Zener diode within its rating will remain practically constant (Fig. 1.63).

The operating curve for the Zener diode illustrates that the voltage across the diode, when it is used in a circuit within its rating, varies from +0.7 volts when it is forward biased to –5 volts when it is negatively biased. This characteristic permits its use as a voltage reference when the proper resistance is connected in series with the diode in a circuit with other components. When the voltage applied to the diode increases beyond the breakdown value of the junction, the high current that flows in the circuit can be used to initiate operation of auxiliary components in the circuit which will correct the voltage variation and provide voltage regulation to the circuit.

Transistor Semiconductor. A semiconductor manufactured with three-elements is called a transistor (Fig. 1.64). Transistor semiconductors are constructed with two *n*-type (excess number of electrons) and one *p*-type (deficient number of electrons) or two *p*-type and one *n*-type materials. The current flowing from the emitter (E) to the collector (C) is controlled by the potential applied to the base (B). The base material is very thin.

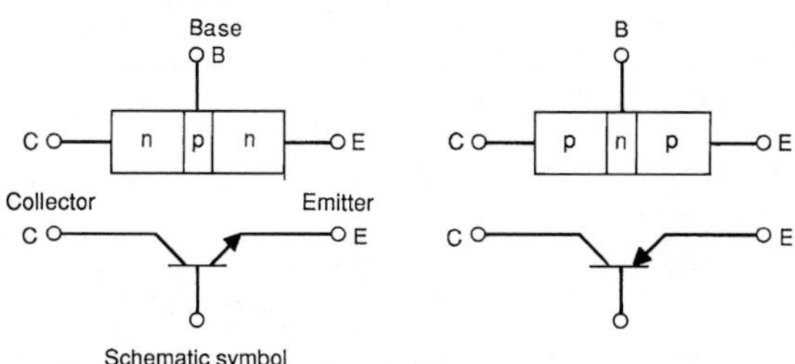

FIGURE 1.64 Transistor semiconductors.

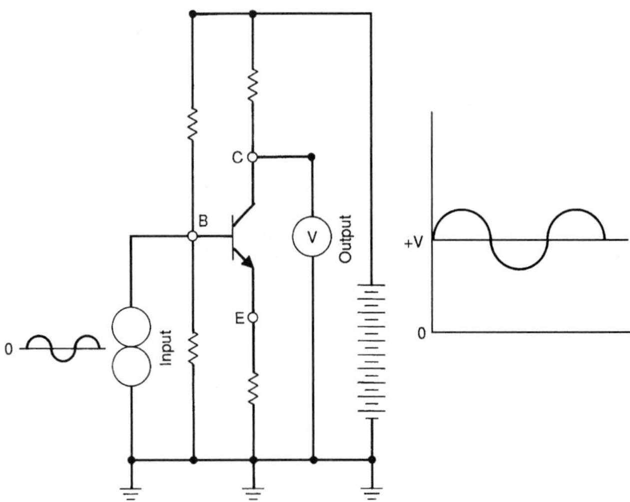

FIGURE 1.65 Schematic diagram of a circuit using a transistor to amplify a voltage.

A transistor can be used in a circuit to amplify a signal. An *npn* transistor connected in a circuit with a collector (C) reversed biased with respect to the base (B) will have a small current flow from the emitter (E) to the base and a large current flow from the emitter to the collector (Fig. 1.65).

When a small ac voltage is connected between the input terminal of the circuit and ground, the bias voltage between the base (B) and the emitter (E) will be increased and reduced accordingly. This will vary the current flow from the emitter (E) to the collector (C), which will vary the voltage above and below some positive value between the output terminal and ground. The output voltage will be in proportion to the input voltage and amplified.

Thyristor Semiconductor. A semiconductor consisting of a layer of *p*-type, a layer of *n*-type, a layer of *p*-type, and a layer of *n*-type materials with an anode (A) connection to one *p*-type material, and a layer of *n*-type materials, a gate (G) connection to one *p*-type material, and a cathode (K) connection to one *n*-type material is called a thyristor semiconductor (Fig. 1.66).

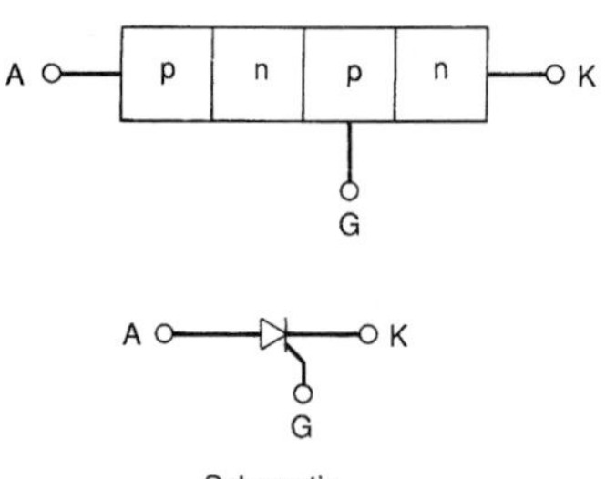

FIGURE 1.66 Thyristor semiconductor constructed with two *p*-type and two *n*-type materials.

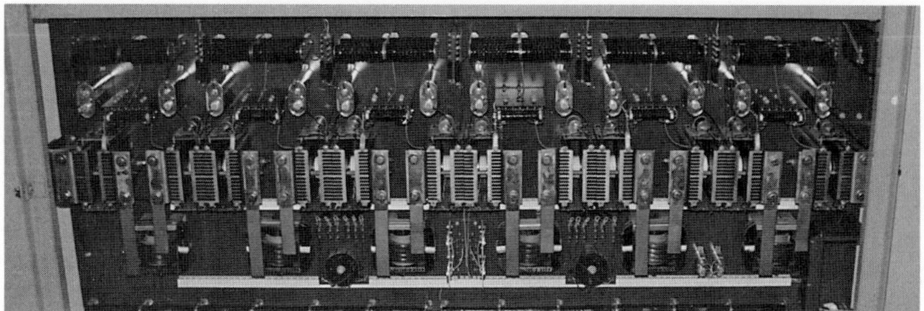

FIGURE 1.67 A picture of thyristor semiconductors and other circuit components in a valve structure for a converter station at the terminal of a high-voltage dc transmission line.

The thyristor semiconductor is used as a controlled rectifying device. The thyristor center *np* junction is reverse biased by the operating circuit. A current applied to the gate (G) terminal will reduce the voltage necessary to overcome the reverse bias of the center junction. The operation of the thyristor in a circuit designed to rectify an ac voltage can be controlled by the signal applied to the gate (G) terminal. Controlling the firing or operation of the thyristor is important when changing alternating current to direct current and reversing the operation. Thyristors are used in converter stations to convert alternating current to direct current and to convert direct current to alternating current. Thyristors are connected in series in circuits with other circuit components to operate at the high voltages necessary to carry large quantities of electric power over dc transmission lines (Fig. 1.67).

Definitions. Definitions of electrical terms used in the text are given as they appear. American National Standard definitions are listed in the *IEEE Standard Dictionary of Electrical and Electronic Terms.*[*]

[*] *IEEE Standard Dictionary of Electrical and Electronic Terms* Wiley-Interscience, New York, N.Y.

CHAPTER 2
ELECTRIC SYSTEM

ELECTRIC SYSTEM—GENERAL

The electric systems in the United States and Canada are interconnected, forming a network of electric facilities all across the continent. It is difficult to distinguish the electric system of one company from neighboring utilities. Figure 2.1 is a single-line diagram showing schematically a portion of an elementary electric system. Most electric utility companies will have many generating stations supplying a very large, complex, and intricate network of transmission and distribution circuits.

The electric system in Fig. 2.1 has nuclear, fossil-fired, and hydroelectric generating stations. The bulk of the electric energy in the United States is generated in fossil-fired steam turbine-driven generating stations using coal, natural gas, and oil fuel. Hydroelectric generating stations using water pressure to turn a turbine have been constructed at most of the sites where there is sufficient waterfall to economically generate electricity. Nuclear generating stations use the heat from an atomic chain reaction to generate steam to turn a turbine generator.

Most electric energy is generated at a voltage in a range of 13,200 to 24,000 volts. The voltage is raised to transmission levels in a transmission substation located at the generating station. Transmission substations N, F, and H, shown schematically (Fig. 2.1), serve this purpose. Transmission circuits or lines operating at 138,000 to 765,000 volts deliver the energy to the load area. Transmission substations A, B, and C (Fig. 2.1) adjacent to the load area reduce the voltage to subtransmission levels 34,500 to 161,000 volts. The subtransmission circuits or lines extend through metropolitan areas to distribution substations located in the area of the load to be served. Distribution substations 10, 11 and 12 (Fig. 2.1) reduce the voltage from subtransmission levels to distribution range usually 4,160Y/2400, 12,470Y/7200, 13,200Y/7620, 24,940Y/14,400, or 34,500Y/19,920 volts.

Distribution circuits radiate from the distribution substations to supply the customer's load. Some large industrial customers may be served directly from the transmission or subtransmission circuits. These customers will normally be metered by a high-voltage or primary-meter installation. The large industrial customer will have a substation or substations to reduce the voltage to the desired level for distribution throughout the plant area. Utility-owned distribution circuits will normally exit from a distribution substation underground and then rise and connect to overhead distribution circuits. Underground exit cables are being extended several thousand feet in some cases to minimize the overhead circuits in the vicinity of a substation. Some companies are building new distribution circuits completely underground. The dashed lines extending from substation 11(Fig. 2.1) indicate underground circuits. The solid lines indicate overhead circuits.

Branch circuits tap to the main distribution feeder circuits through high-voltage fuses commonly called distribution cutouts on overhead lines. The branch circuits may be either overhead or underground, as shown schematically. These circuits are often referred to as primaries.

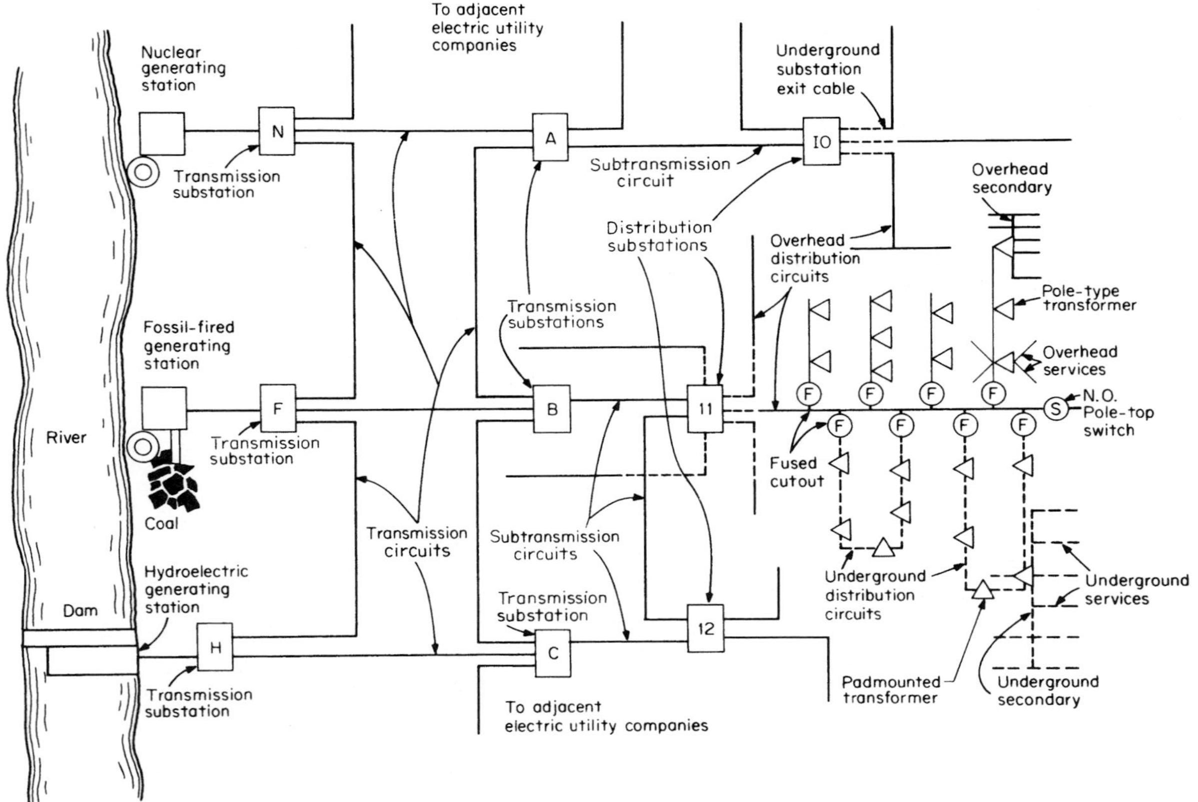

FIGURE 2.1 Single-line schematic diagram of a portion of an elementary ac electric system.

Distribution transformers connect to the branch circuits or primaries. Pole-type transformers are used on overhead circuits, and pad-mounted, or submersible transformers are used on underground circuits. The distribution transformers reduce the voltage to utilization level, usually 120/240 volts single-phase for residential service and 208Y/120 or 480Y/277 volts three-phase for commercial or industrial service. Some small commercial installations may be served by 120/240 volts single-phase. The low-voltage service to the customers may be either overhead or underground connecting directly to a distribution transformer or a secondary extending from the distribution transformers. Both types of installations are shown schematically in Fig. 2.1.

The main distribution feeder circuits often have normally open switches connecting to other main distribution feeder circuits to provide an alternative source for emergency operation. A switch serving this purpose is shown schematically in the circuit fully illustrated extending from distribution substation 11 (Fig. 2.1). Other switches may be installed in main feeder circuits to permit sectionalizing the circuit if a component of the circuit fails electrically.

Figure 2.2 is a schematic drawing illustrating a portion of an electric system pictorially. The generators in a power plant will normally be tied to the transmission lines in a transmission substation located at the generating station. Power transformers in the transmission substation raise the voltage from the generated level to the transmission level. Circuit breakers and other equipment in the transmission substation isolate defective circuits or equipment from the electric system. Substations are also used to connect the transmission lines to the distribution system. The purpose of the distribution substation is to provide equipment to protect the complete electric system from trouble on any distribution feeder. The substation is almost always equipped to isolate the feeder in trouble from the other feeders radiating from the substation. Where load and voltage conditions require it, the substation has an additional function of regulating the voltage on the feeders of the distribution system.

GENERATING STATIONS

A generating station is a plant in which the energy of coal, oil, natural gas, water, or atom is changed into electrical energy. Traditional sources of energy are not the only sources of fuel considered. Wood, rubber tires, and garbage are also being utilized successfully in some generating stations. Central generating stations have come into existence because it is inefficient and uneconomical for every user of power to generate electricity. The deregulation of the electric utility industry has significantly impacted the ownership of generating facilities. The ownership of the stations is no longer solely restricted to investor owned or cooperative electric utilities. Independent power producers (IPPs) are now contributing a greater share of new plant capacity.

Generating stations produce alternating current (ac). The frequency of the ac power generated in the United States and Canada is 60 Hz (hertz). In most other countries, the power is generated at a frequency of 50 Hz. Where direct current (dc) is required, it is usually obtained by converting alternating current to direct current.

Generating stations may be any of the following types:

1. Hydroelectric
2. Steam turbine
   a. Fossil fired
   b. Nuclear
3. Combustion turbine
4. Internal-combustion engine

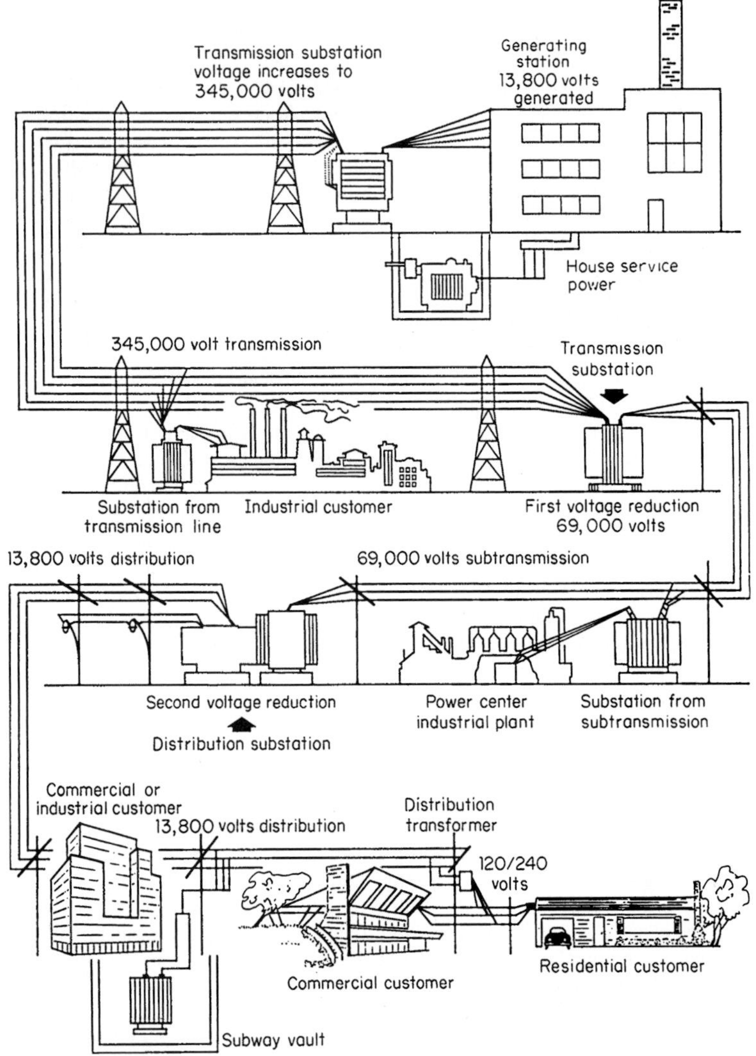

FIGURE 2.2 Portion of a typical electric system showing generating station, step-up transmission substation, transmission line, step-down transmission substation, subtransmission line, distribution substation, distribution system, and industrial, commercial, and residential customers.

5. Geothermal
6. Solar
7. Wind
8. Compressed air

Hydroelectric Generating Stations. If a generating station is a hydroelectric station, it must be located at a waterfall. This is sometimes far removed from the territory to be served. Figures 2.3 and 2.4 show the exterior and interior of typical hydroelectric plants.

FIGURE 2.3 Hydroelectric plant located on Tennessee River at Wheeler Dam. The plant contains four generating units, each rated 36,000 kVA, 13,800 volts, 60 Hz, and run at 85.7 rpm. The rotor of each generator has 84 poles. (*Courtesy General Electric Co.)*

Such plants cost little to run because they do not burn fuel, but the interest and taxes per year are higher than for steam plants because of the larger investment in a dam, land flooded by backwater, etc.

Fossil-Fired Steam-Turbine Generating Stations. Figure 2.5 is a typical fossil-fired steam turbine-driven generator power plant. The generating station contains three generating units. The three units are rated 147, 330, and 520 MW, respectively. The generating station is located near a river.

The boilers burn low-sulfur coal. The coal is mined in Wyoming and delivered to the station by 100-car trains hauling 10,000 tons. Thawing sheds, heated electrically, are used

FIGURE 2.4 Powerhouse of Bonneville hydroelectric plant located on the Columbia River in Oregon. The plant contains 10 vertical water-wheel generating units, each rated 60,000 kVA, 13,800 volts, 60 Hz. Each generator rotor has 96 poles and operates at 75 rpm. (*Courtesy General Electric Co.)*

FIGURE 2.5 Fossil-fired steam-turbine generating station with three generators. The boilers use low-sulfur coal for fuel. The plant buildings in the foreground house equipment with a capacity of 520 MW.

to thaw the coal frozen in transit during the period of cold winter weather. The train hopper cars pass over a pit where the bottoms of the cars are opened automatically and unloaded in 2 min. A conveyor belt transports the coal from the hopper pit to the coal-storage area or the silos in the plant at a rate of 3000 tons of coal per hour. The coal-storage area has a capacity of 800,000 tons.

The boiler for the 520-MW generator unit develops steam with a pressure of 2500 lb/in^2 and a temperature of 1005°F. The boiler is installed in the highest building, 240 ft, in the background of Fig. 2.5. The lower structure between the boiler enclosure and the 400-ft chimney is the electrostatic precipitator. The precipitator removes 99.2 percent of the particulates from the boiler gases before they are emitted to the atmosphere through the chimney. The electrostatic precipitator occupies an area about two-thirds the size of a football field.

The 520-MW turbine generator is located in the lower building in front of the boiler enclosure. The turbine generator (Fig. 2.6) is 18 ft high and 132 ft long. The turbine spindle has rows of blades that stem out from a shaft at the center. Steam from the boiler strikes the blades, like wind on a windmill, to spin the turbine shaft at 3600 rpm. The turbine shaft turns the generator rotor, a strong electromagnet, to create the electrical energy at 24,000 volts. Water from the river next to the generating station is used to condense the steam after it leaves the turbine. The water condensed from the steam is returned to the boiler feed pumps. The generating station is located on a 122-acre site.

Single steam-turbine generator units (Fig. 2.7) are built with a capacity of as much as 1200 MW. The majority of steam-turbine generators are smaller units with a capacity of 25, 50, 90, 125, 250, 375, 450, 525, 600, 800, or 1000 MW respectively. A typical fossil-fired steam turbine-driven generating station is shown schematically in Fig. 2.8.

Nuclear Generating Stations. Generating stations using nuclear energy are constructed similar to fossil-fired steam turbine-driven generating stations, except the boiler is replaced with a nuclear reactor. Nuclear plants do not discharge sulfur dioxide and particulates into the atmosphere like plants burning fossil fuel, thus minimizing air pollution.

FIGURE 2.6 Steam-turbine generator unit rated 520 MW. The turbine shaft and the generator shaft are coupled directly together. The turbine is in the foreground, and the generator is in the background.

Radiation, measured in millirems, is present everywhere. Natural background radiation in Iowa and Illinois along the Mississippi river is 100 to 130 millirems every year. If you lived next door to a nuclear power station, you would receive less than 5 millirems of additional radiation annually, about the same amount you would receive on a round-trip coast-to-coast jet flight.

A nuclear generating station is shown in Fig. 2.9, and a boiling-water reactor is shown in Fig. 2.10. Nuclear fission process is used to produce heat in the nuclear reactor. Fissioning occurs when a subatomic particle, known as a neutron, strikes the nucleus of an atom with sufficient force to split the nucleus. When an atom splits, energy in the form of heat is released. If enough neutrons are released, they will continue to strike and split other atoms, causing a chain reaction. Each separate fission releases only a very small quantity

FIGURE 2.7 Perspective rendering of a large steam-turbine generator unit. *(Courtesy General Electric Co.)*

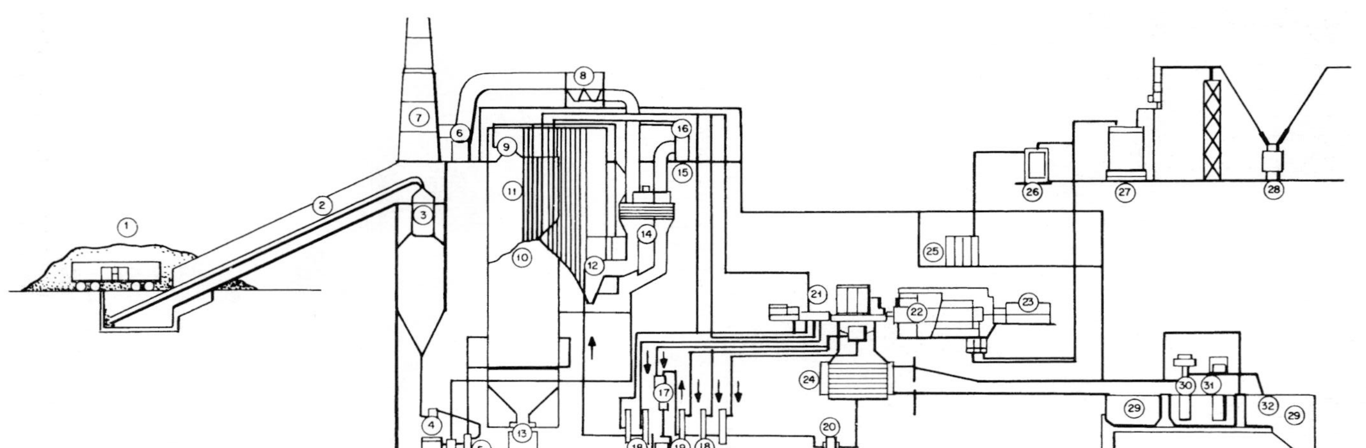

FIGURE 2.8 Fossil-fired steam-turbine generating station. (1) Reserve coal supply—300,000-ton capacity; (2) coal conveyor system—360-ton/h capacity; (3) coal bunker—1750-ton capacity; (4) coal pulverizer; (5) exhauster; (6) induced-draft fan—510,000 ft^3 /min (7) stack—reaching 346 ft above the ground; (8) electrostatic precipitator; (9) boiler drum; (10) steam generator—including furnace 133 ft high—860,000 lb of steam per hour capacity; (11) superheated steam ready to go to turbine; (12) economizer containing boiler feedwater; (13) ash sluice system; (14) air preheater; (15) forced-draft fan —340,000 ft^3 /min; (16) air intake; (17) deaerating open feedwater heater; (18) closed feedwater heaters; (19) boiler feed pump; (20) condensate pump; (21) reheat steam turbine; (22) electric generator—125,000 kW rating—138,000 kW net capability; (23) exciter; (24) steam condenser; (25) house service switchgear; (26) house service transformer; (27) 69Y/39.8-14.4 kV—main transformer—145,000 kVA; (28) circuit breaker; (29) river; (30) circulating water pumps—65,000 gal per min; (31) traveling screen; (32) debris rack.

FIGURE 2.9 Model of Carolinas Virginia Nuclear Power Associates Plant at Parr, S.C., showing cutaway view of reactor area. The nuclear reactor in which the atomic fission takes place is housed under the metal dome for protection. Heat produced by fission changes water to steam, which drives a conventional steam-turbine generator. *(Courtesy Stone and Webster Engineering Corp.)*

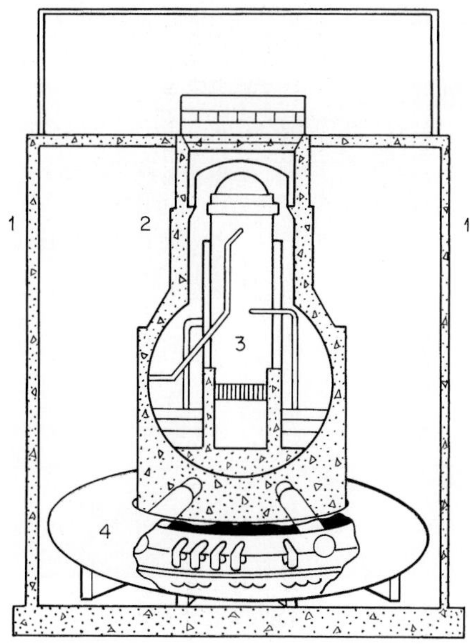

FIGURE 2.10 Boiling-water reactor (BWR). (1) Concrete reactor building, (2) steel-lined concrete drywell, (3) steel reactor vessel, (4) suppression pool.

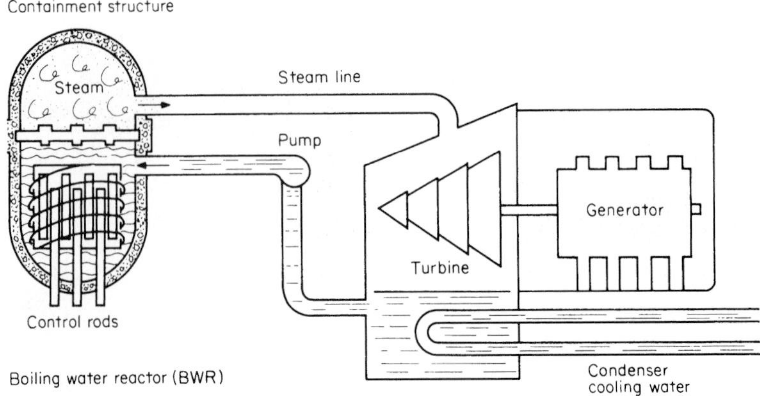

FIGURE 2.11 Schematic drawing of nuclear generating station equipment.

of heat. In a nuclear reactor, many trillions of controlled fissions take place every second. The fissioning process is maintained and controlled in the reactor. Heat from fissioning produces steam, which is piped to large turbines (see Fig. 2.11).

Everything in the world is made of atoms—the ground, your body, the food you eat. Certain materials, however, lend themselves to being split, or fissioned, more readily than others. Uranium is such a material, and at present it is the most commonly used nuclear fuel.

Combustion-Turbine Generating Stations. Generating stations using combustion turbines as prime movers for the generators can be located in an area with other industrial-type installations. Locating the generation near major electrical loads improves reliability and may reduce investment in transmission and distribution equipment.

A combustion-turbine generator facility (Figs. 2.12 and 2.13) will include a starting motor, a rotary-type compressor, combustors, ignitors, a multiple-stage turbine, an ac

FIGURE 2.12 Combustion-turbine generator installation. Units can be used for peak-load, emergency or base-load generation. High cost of fuel limits base-load operation.

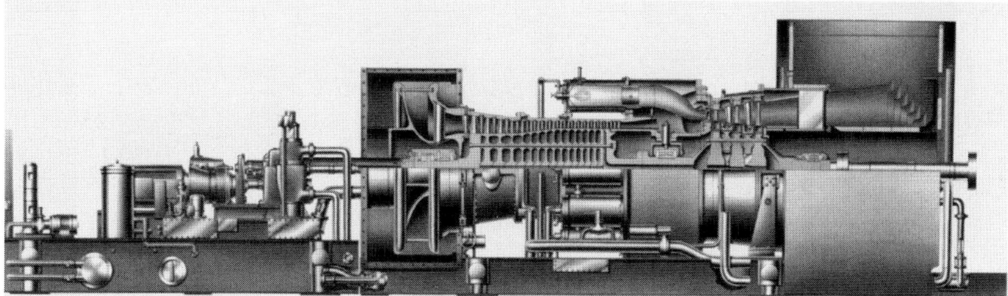

FIGURE 2.13 Sectional view of combustion turbine for a 50-MW generator. (*Courtesy General Electric Co.*)

generator, control devices, and necessary accessories. Fuel used in the turbines may be natural gas or oil in crude form or refined state.

The starting motor drives the compressor to produce pressurized air which is mixed with the fuel in the combustor and ignited. The hot gases flow through the turbine, causing it to accelerate. As the turbine power and speed increase, the starting motor is disengaged and the electric generator is synchronized to the system and connected to the load. The hot gases produced from the burning natural gas or oil expanding through the turbine produce the force to rotate the generator and produce electrical energy.

Combustion-turbine generating stations usually provide peaking power or emergency power. The capacity of the generating units is usually in the range of 15 to 95 MW. The fuel cost for combustion turbines is high, but they have the advantage of low capital cost, automatic operation, fast starting, and minimum electric transmission and distribution investment.

The hot exhaust gases from a combustion turbine can be used to produce steam in a waste-heat boiler. The steam can be used to operate a turbine generator unit, improving the efficiency of the operation and reducing the operating costs. This type of installation is called a combined cycle plant.

Internal-Combustion-Engine Generating Stations. Generating units driven by internal-combustion engines of the diesel type are usually located in remote areas, small towns, and sites requiring standby or emergency power. The engines use oil or natural gas for fuel. The diesel-engine-driven generating units have a low installed cost, can be designed for automatic operation, and can start without an ac source. The fuel cost for the units is high compared to coal- or nuclear-fired stations. Diesel-engine generating stations have a capacity in the range of 500 to 5000 kW. Construction of a diesel-engine generator unit is shown in Fig. 2.14.

Electric Generation with Renewable Energy Resources. Falling water, solar heat, wind, waves and tides, heat from below the earth's surface, agricultural waste combustion, and biomass synthetic fuel are examples of renewable energy sources that can be used to produce electricity. Hydroelectric generation, previously discussed, will continue to produce a significant amount of electricity for consumers.

Geothermal Generating Stations. Geysers located north of San Francisco produce steam that is piped to turbines to generate electricity. New sources of geothermal energy have been discovered in recent years. The geysers (hydrothermal energy) are located principally in California and neighboring states. Geopressured systems use hot water located deep in deposits of sand and sealed off from the surface of the earth by shale. Petrothermal systems

FIGURE 2.14 Diesel-engine generating station. Alternator is directly connected to the engine. The exciter is the small machine mounted on the floor in the foreground. (*Courtesy Nordberg Mfg. Co.*)

require that water or other fluids be injected into molten rock below the earth's surface and pumped out to extract the thermal energy and the dissolved methane. Geopressured and petrothermal energy is generally located throughout the United States. Tests are being conducted with methods to extract the gases from the liquids so that both may be used in a binary-cycle system to capture the geothermal heat and generate electricity.

Solar Generating Stations. The sun can be used to generate electricity by using solar cells that produce small amounts of electrical energy. The capital costs for this method of generation exceed the practical limit except for special uses, such as electrical energy for space ships. Steam can be generated by using mirrors to direct the solar energy to a boiler mounted on a tower. The steam created can be used to operate a turbine to generate electricity. The capital cost of this procedure is excessive. Solar energy can be used to heat water economically. Some homes and commercial buildings use solar panels to produce hot water for heating.

Wind Generating Equipment. Wind-propelled turbine generator units with a capacity as large as 4.0 MW are operating in the United States (Fig. 2.15). The large units are constructed with propeller rotor blades and a horizontal-axis turbine generator. The equipment is mounted on a steel tower high enough to clear the ground when the propeller blades rotate. The propeller blades vary in diameter, with some as great as 420 ft. The wind-powered generators can produce kilowatthours, reducing consumption of nonrenewable energy sources that are used to generate electricity. Intermittent operation, as a result of varying wind velocities, limits the dependability of the electric generating capacity. The ability of wind generation to reduce consumption of oil and natural gas fuels provides an economical application for the equipment.

FIGURE 2.15 Wind generator installed in a plains area on a wind farm.

Compressed-Air Generators. Air can be compressed and stored in natural caverns or underground caverns developed by mechanical methods. The caverns must be sealed to keep the air from escaping. Off-peak electrical energy is used to compress and store the air in caverns. The compressed air can be used to operate a turbine to generate electricity during electrical system peak-demand periods to reduce the base-load generating-equipment requirements.

Chemical Storage of Electrical Energy. Batteries of the lead-acid type and the nickel-cadmium type are commonly used to store small amounts of electrical energy. Zinc-chlorine batteries have been developed that can store large quantities of energy to reduce the peak-load generating requirement of an electric utility. The batteries can be installed in distribution substations close to the point where the electrical energy will be consumed. Installation of the batteries near the consumption location reduces the transmission-capacity requirement as well as the peak electrical generating requirement.

TRANSMISSION CIRCUITS

A set of conductors energized at voltages over 69,000 volts and transmitting large blocks of electrical energy over relatively long distances is called a transmission circuit or transmission line. Transmission circuits are constructed between transmission substations located at electric generating stations or switching points in the electric system. Transmission circuits may be overhead, with structures supporting conductors attached to insulators, or underground, with conductors surrounded by insulation, shielding, and a sheath or a jacket to form a cable. The distinguishing characteristics of transmission circuits are that they are operated at relatively high voltages, transmit large blocks of electrical power, and extend over considerable distances.

Overhead Transmission Circuits. Transmission lines or circuits are usually constructed for economic and capability reasons. The lines extend for long distances in rural areas. Alternating-current (ac) transmission circuits generally operate in the voltage range of 69,000 to 765,000 volts. High voltages often used for these so-called ac transmission lines are 115,000, 138,000, and 161,000 volts. Transmission circuits energized at 345, 500, or 765 kV are referred to as extra-high-voltage (EHV) lines. Experimental ac ultra-high-voltage (UHV) transmission circuits in the voltage range of 1100 to 1500 kV are being developed. A 345,000-volt ac overhead transmission circuit is shown in Fig. 2.16.

Porcelain-disk insulators in strings insulate the two bundled conductors for each phase of the three-phase line from the wood-pole structure. The small wires at the top of the structure are so-called static wires or ground wires. They are not insulated and directly connect to the metal hardware on the structure and the ground. Their purpose is to shield the line conductors from lightning.

FIGURE 2.16 A 345,000-volt three-phase bundle-conductor transmission line on a wood-pole structure. (*Courtesy Hughes Brothers Co.*)

FIGURE 2.17 Transmission line ± 250 kV dc, 456 miles long, 1000 amps, 500-MW capacity located between Minnkota Power Cooperative, Center, North Dakota, and Minnesota Power and Light Co., Duluth, Minnesota. (*Courtesy Ulteig Engineers, Inc.*)

Overhead direct-current (dc) transmission circuits are used where it is desired to transmit large amounts of power over long distances or to interconnect two ac systems that are not synchronized. The circuits are usually several hundred miles in length. A circuit in operation in the western United States extends from northern Oregon to southern California, approximately 850 miles, and operates at 400 kV.

Direct-current high-voltage (dc HV) line conductors are either positive or negative polarity. A 400-kV line could operate with one conductor 400 kV positive and one conductor 400 kV negative or ± 400 kV dc. Some of the voltages used for dc HV circuits are ± 250, ± 400, and ± 500 kV. Figure 2.17 shows an dc HV transmission-line tower.

Underground Transmission Circuits. In some areas of large cities, it is not practical or permissible to build high-voltage overhead lines. All high-voltage electric transmission and distribution circuits must, therefore, be installed underground. Underground transmission circuits may be constructed for ac or dc operation. Alternating-current underground transmission circuits are costly to install and limited in length by charging current. Circuits over 25 miles long must be compensated by using shunt reactors to absorb the capacitive charging current or the cable capacity will be used without transmitting any useful power. Alternating-current underground transmission circuits are economically attractive for transmitting large amounts of electrical energy where underground construction is required for distances in excess of 25 miles. Underground cables (Fig. 2.18) transmit ac power between substations in a large metropolitan area.

Overhead Subtransmission Circuits. Large blocks of power are transmitted from generating stations to one or more transmission substations through high-, extra-high-, or ultra-high-voltage ac lines. Intermediate-voltage lines or circuits transmit alternating current from transmission substations to subtransmission or distribution substations. These subtransmission lines carry medium-sized blocks of power and usually form a grid between substations in a metropolitan area.

Figure 2.19 illustrates a 69,000-volt three-phase subtransmission circuit constructed along an urban street to provide a source for a distribution substation. Armless-type construction is utilized to gain public acceptance. The lower circuit is a three-phase distribution feeder operating at 13,200Y/7620 volts.

FIGURE 2.18 High-voltage underground pipe-type cables installed in a tunnel. Each pipe contains three cables with paper insulation and metallic shielding surrounded and impregnated with insulating oil. The cables are capable of transmitting large quantities of three-phase ac power. (*Courtesy ABB Power T&D Company Inc.*)

Substations. A high-voltage electric system facility used to switch generators, equipment, and circuits or lines in and out of system, change ac voltages from one level to another, and/or change alternating current to direct current or direct current to alternating current is called a substation. Equipment in substations is used to connect electric generators to the system and to terminate and interconnect transmission, subtransmission, and distribution circuits. Power transformers in the substations change the voltage levels to permit interconnecting the generators and circuits forming the electric system.

Substation Types. Transmission substations are constructed at generating stations and at switching points in the transmission system (Fig. 2.1). The transmission substations near generating stations (Fig. 2.20) provide a means to connect the electric generators to the system and change the voltage to the appropriate magnitude to transmit the power the required distance. The power is usually generated at 25,000 volts ac or less and stepped up by power transformers to the transmission voltage level of 138,000, 161,000, 230,000, 345,000, 500,000, or 765,000 volts or higher. The power transformers connected to each generator and the transmission lines are switched in and out of service with circuit breakers.

The transmission substations located at switching points (Fig. 2.21) connect various circuits to the network and can serve as a source to subtransmission circuits. Power transformers installed at the substations can change the transmission system voltage from one level to another and the transmission voltage to subtransmission level. The subtransmission circuits originating at a transmission switching substation will serve as a source to distribution substations.

FIGURE 2.19 Modern 69,000-volt subtransmission line with 13,200Y/7620-volt distribution line underbuild. The picture illustrates armless-type construction utilized to enhance the appearance of the line.

FIGURE 2.20 A 345,000-volt transmission substation located adjacent to the Quad Cities 1600-MW nuclear generating station. Four 345,000-volt transmission circuits originate at the substation.

FIGURE 2.21 A transmission substation constructed at a switching point in the system. Substation is located near Iowa City, Iowa, at the interconnection points of 345,000-volt ac transmission lines constructed from Minneapolis to St. Louis and from Chicago to Omaha. Power transformers in the substation reduce the 345,000 volts to 161,000 volts to provide a source for the subtransmission system that provides power to serve Iowa City. Note microwave-reflector antenna mounted on control-house building which is used for transmitting and receiving high-frequency electric signals to control the substation equipment from a remote location.

FIGURE 2.22 Large distribution substation rated 69,000-13,200Y/7620 volts, three-phase. Step-down power transformer has a capacity of 20/26/33 MVA.

FIGURE 2.23 Distribution substation showing incoming line to disconnect switch, circuit switcher, surge arresters, power transformer, and metal-clad enclosure for feeder-circuit protective and switching equipment. (*Courtesy S&C Electric Co.*)

Distribution substations are located near the utilization point in the electric system. The power transformers in the distribution substations change the subtransmission voltage to lower levels and provide a source for the distribution circuits extending short distances to the customer's premises. Figure 2.22 illustrates a large distribution substation operating with a subtransmission voltage of 69,000 volts and a distribution circuit voltage of 13,200Y/7620 volts. Figure 2.23 is a picture of a large distribution substation supplied by a 138-kV subtransmission line. A 138-kV circuit switcher installed between the incoming subtransmission line and the power transformer is used to energize and deenergize the power transformer and to isolate the power transformer from the incoming line if a fault develops in the power transformer.

DISTRIBUTION SYSTEM

The electric facilities connecting the transmission system to the customer's equipment are called the distribution system. An electric distribution system (Fig. 2.1) consists of several parts: (1) distribution substation, (2) distribution feeder circuits, (3) switches, (4) protective equipment, (5) primary circuits, (6) distribution transformers, (7) secondaries, and (8) services.

Distribution Feeder Circuits. The conductors for a distribution feeder circuit originate at the terminals of a circuit breaker or a circuit recloser in a distribution substation. The circuits usually leave the substation underground and may be installed entirely underground. The underground cables leaving the substation are commonly called substation exit cables and

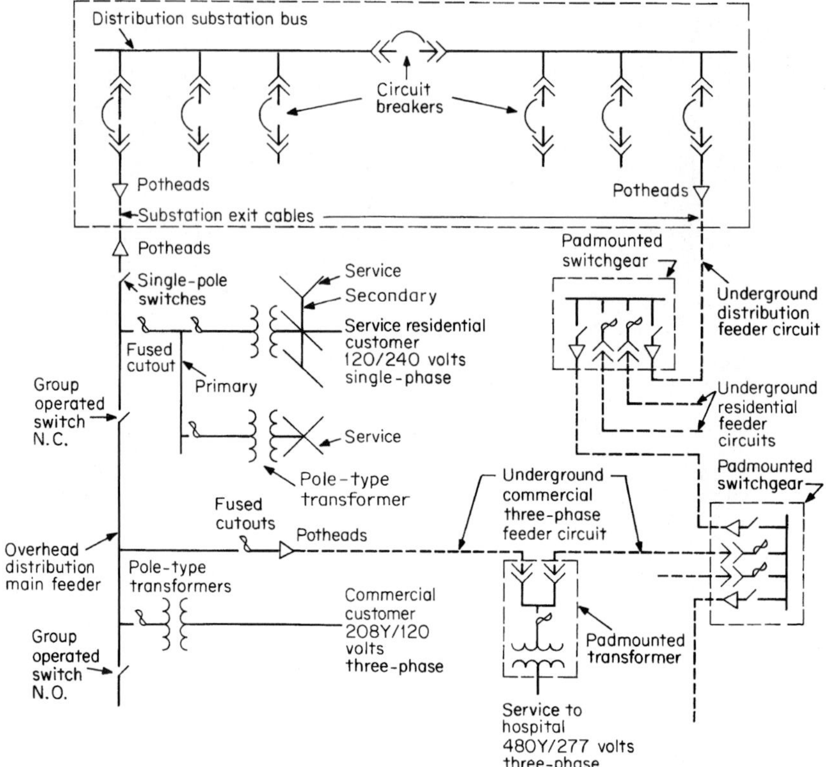

FIGURE 2.24 Single-line schematic diagram showing a portion of an electric distribution system.

normally connect to an overhead distribution feeder near the substation. Installing the distribution feeder circuits underground near the substations eliminates multiple circuits on the pole lines adjacent to the substations, thus improving the appearance of the substations and surrounding facilities.

The distribution feeder circuits leaving the substations serve as a source to branch or primary circuits and are often referred to as distribution main-feeder or express feeder circuits. Several distribution feeder circuits will originate at a distribution substation and extend in different directions to serve the customers (Fig. 2.24).

Distribution feeder circuits may have a capacity of 10 MVA or more. The substation exit cables or the underground main-feeder cables for a 13,200Y/7620-volt, three-phase feeder circuit could have a 750-kCMIL aluminum conductor with solid-dielectric insulation for each phase and a 400-kCMIL bare-copper neutral conductor. The overhead phase conductors might be a 795-kCMIL bare-aluminum or aluminum conductor steel reinforced (ACSR) with a 336.4-kCMIL aluminum or ACSR bare neutral conductor.

The main-feeder distribution circuits in the background of Fig. 2.25 are connected to the distribution substation by underground exit cables. Figure 2.26 is a picture of an open-type distribution substation structure supporting the distribution voltage-bus conductors, disconnect switches, circuit reclosers, current transformers, and lightning arresters. The substation has a transfer bus, which permits the reclosers to be removed from service for

FIGURE 2.25 Small distribution substation rated 13,200-4,160Y/2400 volts, three-phase. Step-down power transformer has a capacity of 5/6.25 MVA. All circuits entering and leaving the substation are located underground. Two 4,160Y/2400-volt distribution circuits supplied from the substation can be seen on a pole in the background of the picture.

maintenance without deenergizing a feeder circuit. The feeder circuits exit from the substation through underground cables. Figure 2.27 is a typical installation of equipment on a riser pole for substation exit cables connecting to a main-feeder or express-feeder circuit near a substation.

FIGURE 2.26 Open-type distribution substation structure containing distribution voltage main and transfer buses. Disconnect switches in structure can be opened under loads with Loadbuster tool if the current does not exceed 900 amps. (*Courtesy S&C Electric Co.*)

FIGURE 2.27 Riser pole located near a substation with power cables terminated in potheads. Lightning arresters protect the cables, and single-pole disconnect switches provide connection to overhead distribution feeder circuits.

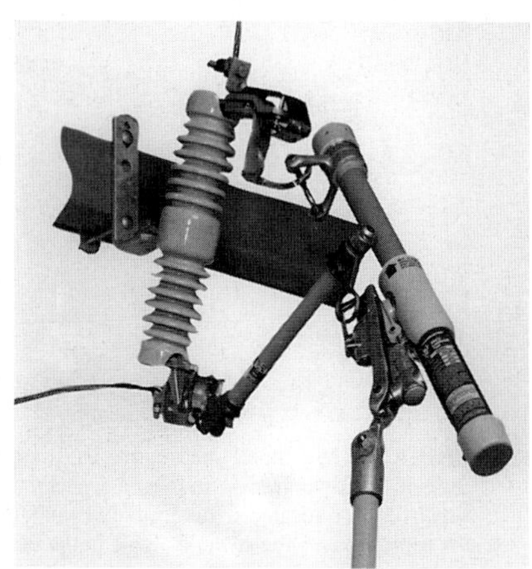

FIGURE 2.28 A fused cutout is being opened using the Loadbuster tool mounted on a universal hot-line stick. In the position shown, Loadbuster has tripped, interrupting the load current. (*Courtesy S&C Electric Co.*)

Switches. Distribution feeder circuits, whether overhead or underground, typically have switches installed at strategic locations. These can be used to redirect power flows for load balancing or for sectionalizing to permit repairing of damaged lines or equipment.

Single-pole disconnect switches are widely used whenever three-pole switching is not required. Used alone, single-pole disconnect switches without loadbreak interrupters should only be opened when deenergized. However, they can switch load currents to 900 amps when opened with Loadbuster, a portable loadbreak tool (Fig. 2.28).

Three-pole group-operated switches with loadbreak devices are generally installed at appropriate locations in distribution main-feeder circuits. They provide a means for the lineman to sectionalize the feeders or to connect the feeders to other circuits. These switches may be manually operated or power operated and installed on distribution structures or poles or in metal enclosures or vaults. Power-operated switches may be controlled from a remote location. Figure 2.29 illustrates a pole-top group-operated switch, with the contacts closed, installed in a main-feeder or express-feeder circuit.

Figure 2.30 shows a unit of pad-mounted switchgear used in an underground distribution main-feeder circuit. It contains two three-pole, 600-amp interrupter switches and two sets of three electronically controlled fuses for protection of two three-phase feeder circuits. Figure 2.31 shows a lineman bolting an incoming 750-kCMIL insulated cable to one switch terminal of the deenergized and grounded equipment. Each incoming cable is also connected to a surge arrester, one of which is visible alongside the lineman's knee. Adjacent compartment shows one of three electronic fuses.

Protective Equipment. Circuit breakers in the substations serving as a source to distribution circuits are actuated by protective relays that recognize short circuits and grounds on a circuit or line. If a circuit breaker opens, the complete distribution circuit is deenergized, and a large number of customers will have their electric service interrupted.

FIGURE 2.29 Pole-top group-operated three-phase disconnect switch installed in an overhead distribution feeder circuit. (*Courtesy ABB Power T&D Company Inc.*)

Lightning arresters and fuses are installed on distribution circuits to minimize the customer outages. Figure 2.32 is a picture of a distribution lightning arrester similar to those mounted on a pole and connected to the top conductor of the circuit in Fig. 2.33. The lightning arresters are designed to bypass electrical surges on the circuits, protecting the insulators and equipment from overvoltages. Fuses are installed in the circuits to isolate branch primary circuits and equipment from the distribution main-feeder circuit if a fault or short

FIGURE 2.30 Lineman operating interrupter switch in pad-mounted switchgear which contains two three-pole switches for sectionalizing underground main-feeder circuits and two sets of three power fuses in series with branch circuits for short-circuit protection. (*Courtesy S&C Electric Co.*)

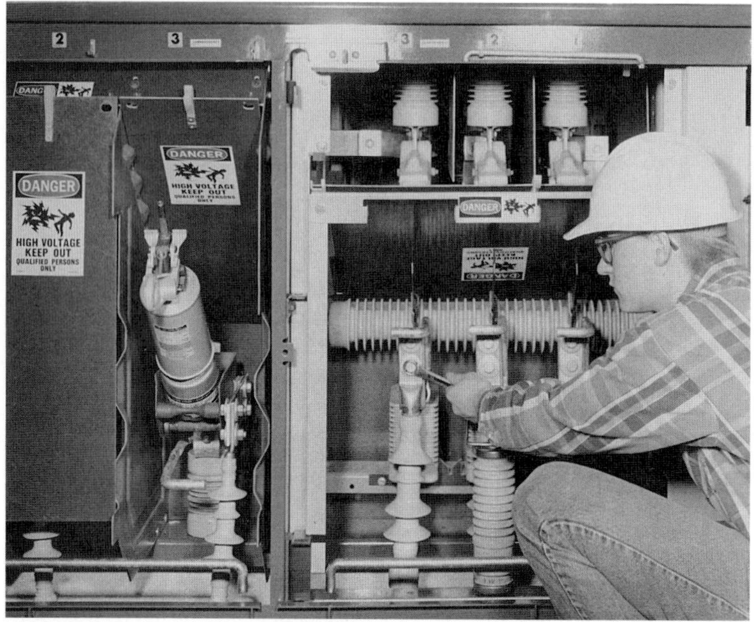

FIGURE 2.31 Installer connecting cable terminator to switch terminal in pad-mounted switchgear. View shows three-pole interrupter switch, cable terminators, surge arresters at switch terminals (one yet to be installed), grounding terminals, and insulating front barrier inserted in open gap of interrupter switch. (*Courtesy S&C Electric Co.*)

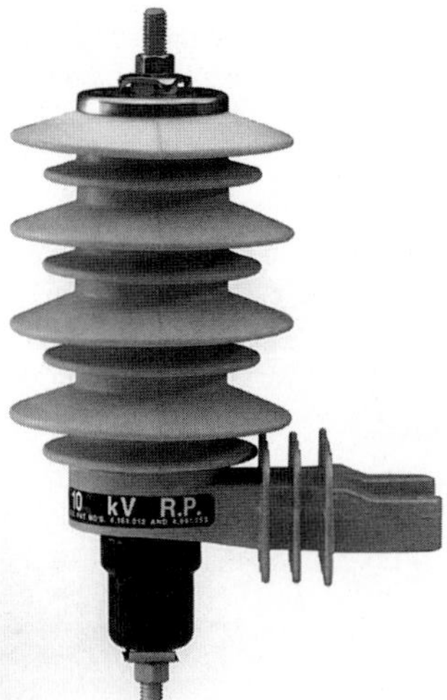

FIGURE 2.32 Distribution lightning arrester equipped with mounting bracket designed for use on overhead distribution lines and equipment. (*Courtesy Joslyn Manufacturing Corp.*)

FIGURE 2.33 A fused section of single-phase primary distribution line that is tapped through the lightning arrester prior to the connection to the fused cutout.

circuit develops on one of the taps to the circuit. It is better to have a fuse blow and interrupt service to a small number of customers than to have the substation circuit breaker open and deenergize all the customers served from the circuit.

Figure 2.34 is a picture of a fused cutout designed to be used on an overhead distribution circuit. The lineman in Fig. 2.35 is wearing a hard hat and safety glasses and is working in the proper position on the pole to open the fused cutout safely. The fused cutouts shown in Fig. 2.35 are located in the circuit at the end of an express or main feeder where three primary conductors connect to the feeder circuit.

Figure 2.36 shows two three-phase primary underground circuits originating at fuses in pad-mounted switchgear. The fuses connect to a bus in the switchgear. The bus is connected through three-pole interrupter switches to the underground distribution main-feeder circuit. The primary circuit conductors are typically smaller than the main-feeder circuit conductors.

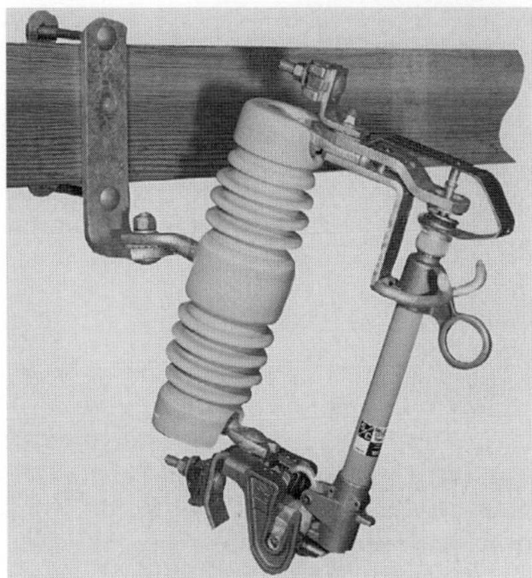

FIGURE 2.34 Crossarm-mounted fused cutout equipped with hooks for Loadbuster operation. (*Courtesy S&C Electric Co.*)

FIGURE 2.35 Fused cutout is being opened by the lineman using Loadbuster mounted on a universal hot-line tool. The lineman is deenergizing one phase of the three-phase branch circuit that is connected to the main-feeder circuit through fused cutouts. (*Courtesy Northwest Lineman College.*)

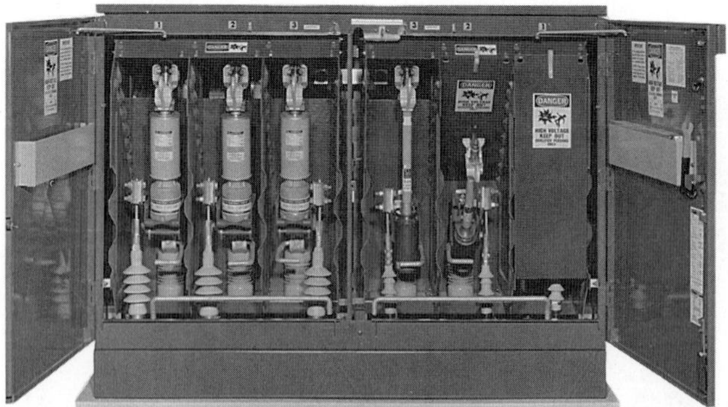

FIGURE 2.36 Pad-mounted switchgear showing fuse compartments. Left-hand compartment contains three fuses with dual-purpose front barriers removed. Right-hand compartment contains three solid-material power fuses—one dual-purpose front barrier removed, one barrier inserted in the open gap with fuse in the open position, and one barrier hanging in its normal position. (*Courtesy S&C Electric Co.*)

The fuses can carry up to 600 amps of load current. The solid-material power fuses are rated 200 amps or less. Fuses must be selected to carry anticipated loads, interrupt maximum fault currents at that location, and coordinate with upstream and downstream protective devices. The fuses connected between the bus and the underground primary-feeder cables protect the main-feeder circuit from faults on the branch circuits or equipment. Faults or short circuits usually result from cable insulation failures, frequently caused by excavating equipment hitting a cable. The fuses are operated or replaced by linemen or cablemen using hot-line tools, or "hot sticks." The fuses in the gear will coordinate properly with other fuses in the branch circuit.

Primary Circuits. All high-voltage distribution circuits greater than 600 volts up to 69,000 volts are often referred to as primaries by the lineman. The branch circuits that connect to the main- or express-feeder distribution circuits are properly called primary circuits or primaries. The overhead primary circuits normally connect to the distribution feeder circuit through fused cutouts. The primary circuits may be three-phase (Fig. 2.35) or single-phase circuits (Fig. 2.37).

Figure 2.36 shows two three-phase primary underground circuits originating at fuses in pad-mounted switchgear cubicles. The fuses connect through switches to a bus in the switchgear which is in series with an underground distribution main-feeder circuit. The underground main-feeder circuit normally connects to the bus in the switchgear through group-operated switches. The primary branch circuits serve the customers in an area surrounding the switchgear. The size of the fuses protecting the main distribution feeder circuit from a fault on the branch primary circuit and the size of the primary underground cables will limit the capacity of the circuit.

Figure 2.38 illustrates a three-phase underground primary circuit originating on a riser pole. The underground cables are supplied through fused cutouts connected to the overhead

FIGURE 2.37 Single-phase primary or branch circuits are connected to the three-phase distribution main-feeder circuit through fused cutouts. Two of the branch circuits extend away from the reader and one extends toward the reader. A street light is mounted on the pole. (*Courtesy S&C Electric Co.*)

FIGURE 2.38 Underground primary cables are terminated in combination fused cutout terminator assemblies called isolating terminators. Lightning arresters are installed on the mounting frame. The arresters are installed on the mounting frame. The arresters provide surge protection for the underground primary cables and the overhead distribution main-feeder circuit. (*Courtesy A. B. Chance Co.*)

distribution feeder circuit. The primary branch circuits provide service to small industrial, commercial, and residential customers. The small industrial and commercial customers will normally use three-phase power. The residential customers will normally be provided with single-phase service.

Distribution Transformers. The function of the distribution transformer is to reduce the voltage of the primary circuit to the utilization voltage required to supply the customer. The primary voltages vary from 34,500Y/19,920 to 4,160Y/2400 volts. A three-phase circuit with a grounded neutral source operating at 34,500Y/19,920 volts would have three high-voltage conductors and one grounded neutral conductor, a total of four conductors or wires. The voltage between the insulated or phase conductors would be 34,500 volts, and the voltage between one phase conductor and neutral, or ground, would be 19,920 volts. Similarly, a 13,200Y/7620-volt distribution feeder would have 13,200 volts between phase conductors and 7620 volts between a phase conductor and neutral, or ground.

Residential customers are normally provided 120/240 volts single-phase service. The single-phase service would have three wires. Two of the wires require insulation because they will be energized. One of the wires may be insulated or bare because it is the neutral wire that is connected to ground. The insulated wires would have 240 volts between the conductors and 120 volts between the insulated wires, or hot wires as they are often referred to by the lineman, and the ground, or neutral wire. Some small older homes may have 120-volt service which would be provided by one insulated, or hot, wire and the ground, or neutral wire. A typical single-phase transformer installed on a pole along the rear lot line of a residential area is shown in Fig. 2.39. The center low-voltage wire connected to the bushing on the side of the transformer tank in Fig. 2.39 is the neutral wire. Pad-mounted transformers are frequently used in underground residential distribution systems. Figure 2.40 shows a single-phase pad-mounted distribution transformer with the door in the closed position. In Fig. 2.41, a cableman is preparing the primary cable to install a loadbreak elbow so as to terminate the primary cable on the primary bushing of the single-phase pad-mounted transformer. The single-phase primaries normally feed through the transformer to form a loop. This permits a defective primary cable to be isolated between transformers.

FIGURE 2.39 Typical distribution-transformer installation on pole. The primary wire to which the primary winding of the transformer is connected is carried on the insulator on top of the pole. The transformer is a self-protected type complete with a primary lightning arrester, an internal fuse, and a low-voltage, or secondary, circuit breaker. The transformer is rated 7620-120/240 volts, 15 kVA.

FIGURE 2.40 Single-phase pad-mounted transformer installed on rear lot line of a residential subdivision. Transformer is rated 7620-120/240 volts, 50 kVA.

Commercial and light industrial customers normally require three-phase service. The voltage supplied is usually 480Y/277 or 208Y/120 volts. Three-phase pad-mounted transformers (Figs. 2.42 and 2.43) are used with underground primary circuits. Three-phase underground primaries are installed to form a loop to permit isolating defective underground cables without a long outage to the customer.

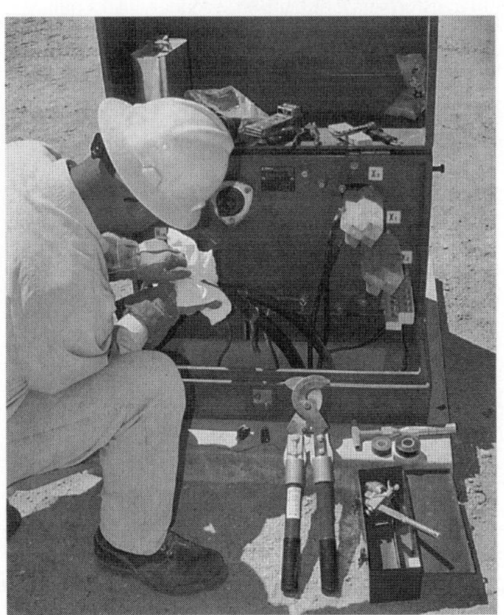

FIGURE 2.41 Cableman is installing terminations on deenergized primary cable in the single-phase pad-mounted transformer. (*Courtesy Northwest Lineman College.*)

FIGURE 2.42 Three-phase pad-mounted transformer. The transformer is rated 13,2000Y/7620-480Y/277 volts, 500 kVA. (*Courtesy ABB Power T&D Company Inc.*)

FIGURE 2.43 Three-phase pad-mounted transformer shown with the door to the primary and secondary compartments in the open position. (*Courtesy ABB Power T&D Company Inc.*)

FIGURE 2.44 Three-phase transformer bank rated 13,200Y/7620-208Y/120 volts, 150 kVA. Primary wires are connected to transformers through fused cutouts. Lightning arresters connect to primaries to provide surge protection. Low-voltage connections to transformers terminate on secondary racks mounted on pole below transformers. Quadriplex service to customer's building terminates on secondary racks.

FIGURE 2.45 Single-phase distribution transformer tapped to the three-phase primary distribution system. Two of the three phases of the primary conductor are installed on a close space bi-unit arm. Triplex secondary extends from transformer pole in line with primary. Secondary connects to residential services.

Three single-phase pole-type transformers can be cluster-mounted on a pole to provide three-phase service (Fig. 2.44). The transformers illustrated are single primary-bushing transformers connected line to ground.

Network service is provided to many businesses in areas with highly concentrated, large electric loads. Network transformers are designed for installation in a vault usually located under the sidewalk. The network transformers are designed for submersible operation. The primary underground cables terminate in a switching compartment. The primary switch provides a means to deenergize the network transformer without deenergizing the primary circuit that serves as a source to other network transformers. The secondary voltages are usually 208Y/120 or 480Y/277 volts. The secondary terminals of the network transformer connect to a network protector. The network protector is a low-voltage circuit breaker and fuse assembly complete with protective relays and control devices designed to isolate the low-voltage network from a transformer or primary cable fault. The network is supplied from several underground primary feeder circuits.

Secondaries. The conductors originating at the low-voltage secondary winding of a distribution transformer that extend along the rear lot lines, alleys, or streets past the customer's premises are called secondaries. The secondaries in residential areas are three-wire single-phase circuits (Fig. 2.44). The secondary conductors for overhead circuits are strung below the primaries and are cabled (Figs. 2.45 and 2.46) or lie in a vertical plane (Fig. 2.47). The secondaries for underground circuits are normally directly buried in the same trench with primary cables. The cables are laid in the trench without separation (Figs. 2.48 and 2.49).

Services. The wires extending from the secondary, or distribution transformer, adjacent to the customer's property to the customer's premises are called a service. An overhead triplex cable service extending to a customer's premises is shown in Fig. 2.50. The transformer

FIGURE 2.46 CSP distribution transformer mounted on pole with primary winding connected to primary, carried on vertical line post three-phase construction. The secondary mains supplied from the transformer are carried on a clevis spool mounted in a vertical plane on the pole.

FIGURE 2.47 Distribution transformer with primary connected phase to ground. The single-phase primary is supported on the insulator on top of the pole. The wires connecting to the secondary terminals of the transformer terminate on a secondary rack and connect to a triplex secondary and three triplex services. The bare conductor serves as the grounded neutral for both the primary and secondary circuits.

reduces the voltage from 7620 to 120/240 volts. The triplex service cable has two insulated aluminum conductors and a bare aluminum steel reinforced neutral wire that provides physical support for the insulated conductors. Termination of the service cables at the customer's premises is shown in Fig. 2.51. A quadriplex cable providing overhead 208Y/120-volt three-phase service to a commercial establishment is illustrated in Fig. 2.52. The underground services extend from a secondary enclosure (Fig. 2.53) or pad-mounted transformer (Fig. 2.54) to the customer's premises. The cables are installed in a trench (Fig. 2.55) or plowed in (Fig. 2.56).

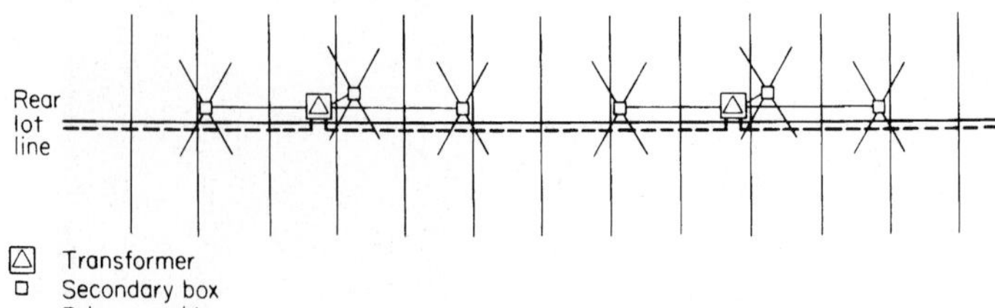

FIGURE 2.48 Plan-view schematic diagram of an underground residential electric distribution system for installation along a rear lot line. Schematic locates pad-mounted transformers, primary cable, secondary enclosures, secondary cables, and service cables.

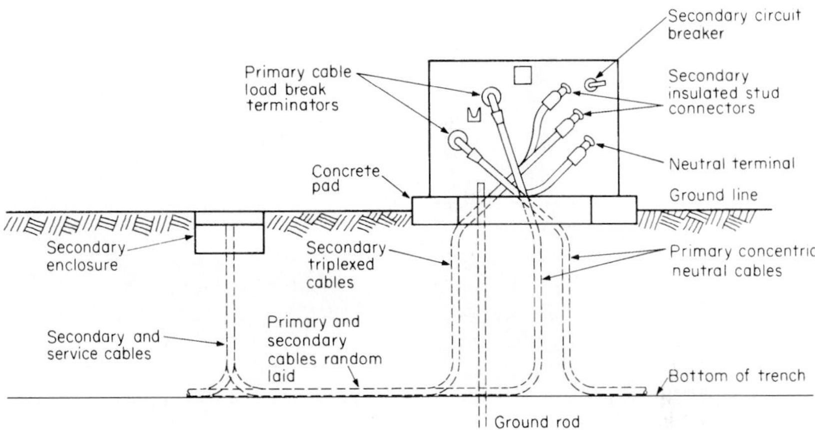

FIGURE 2.49 Sectional view locating primary and secondary cables of an underground distribution system.

FIGURE 2.50 Service wires to customer's premises terminate on pole below distribution transformer. Transformer is connected line to ground from one phase of three-phase primary circuit. Service extends perpendicular from primary circuit.

FIGURE 2.51 Cable service drop connected to residential service entrance.

FIGURE 2.52 Quadriplex service cable terminates on wooden mast adjacent to customer's service-entrance conduit. Cables enter conduit through a weatherhead.

FIGURE 2.53 Connections for underground services to the secondary in a secondary enclosure. (*Courtesy ITT Blackburn Co.*)

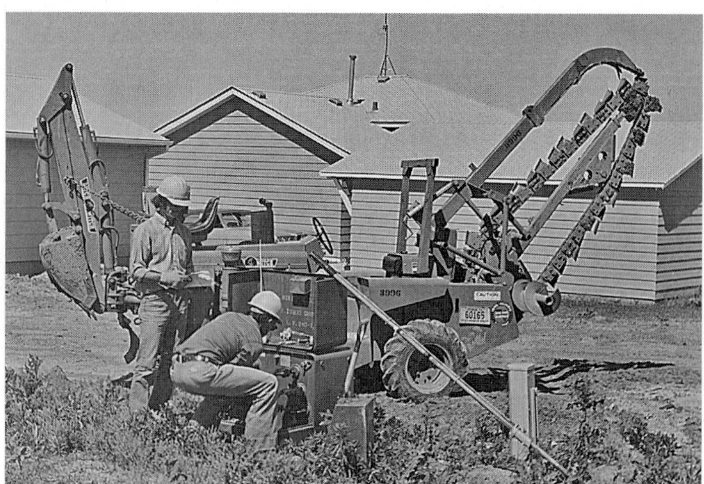

FIGURE 2.54 Lineman completing connection for underground service to a residential customer in compartment of pad-mounted transformer. Note the trencher-backhoe equipment in background of picture.

FIGURE 2.55 Cableman and helper digging trench and laying cable in trench to provide service to a new home.

FIGURE 2.56 Lineman using Ditch Witch vibrating plow to install underground service to a street light in a new subdivision. (*Courtesy The Charles Machine Works, Inc.*)

EMERGENCY GENERATION

Electric customers offering specialized services may require very reliable electric service. Customers of this type can be provided dual electric service from two distribution feeder circuits. The dual source electric service is normally supplied from one circuit with facilities to switch to the alternative circuit in case of a failure of the principal source of electrical energy. Customers are required to pay the extra cost associated with the extra electrical distribution equipment. Hospitals and similar types of customers will be equipped with emergency electric generators, in addition to dual service from the electric system. The emergency generators are designed to provide sufficient capacity to supply critical loads, such as operating rooms, emergency rooms, and minimum lighting in halls and stairways. Figure 2.57 shows an emergency generator. The emergency electric generators are designed to start automatically when the normal source of electrical energy is lost. The equipment is installed in a manner to ensure that the generator does not feed back into the electric distribution system. It is essential for the lineman and the cableman to protect themselves from a possible malfunction of the equipment which would permit a backfeed of electricity through the customer's service to the distribution facilities.

ELECTRIC SYSTEM CONTROL

The modern electric system is operated from a control center. Equipment in the control center is used to analyze the electric system operation continuously. The output of the electric generators is varied automatically to match the energy supplied with the customer's demand and the purchases and sales through interconnected electric systems. The control center is located at a strategic point in the electric-system service area. The control-center

FIGURE 2.57 Emergency electric generator rated 500 kVA, three-phase, 480Y/277 volts. Generator is driven by diesel-type engine using natural gas for fuel.

FIGURE 2.58 Control-center building designed to withstand tornado winds.

building (Fig. 2.58) is designed to protect the operating personnel and control equipment from severe weather conditions. The operating personnel responsible for the electric system use schematic diagrams drawn on a map (Fig. 2.59) mounted on a large display board to give them a picture of the entire electric system.

Instruments mounted on a panel provide essential information concerning the electric system, such as system net load, generation, net tie-line loads, area control error, and frequency. Computer monitors display schematic diagrams of the electric system, including generating-station equipment, transmission circuits, substation equipment, and distribution circuits. The monitors connect to digital computers that link up with supervisory control equipment at the generating stations and substations. Electrical outages throughout the system are continuously displayed on the schematic diagram. The electrical equipment, such as circuit breakers, is operated by action of a light pen at the appropriate location on the schematic diagram appearing on the monitor. As the circuits and generators are switched in and out of service, the changes in electrical energy flows are made automatically by the computer to correctly display the information.

Alarm logger and alphanumeric CRT equipment provide a record of electric-system operations. The displays on the alphanumeric CRT can be printed by proper inputs to the computer equipment to operate a special high-speed printer. The computers provide signals that continuously change the output of the electric generation to meet the requirements of the electric system.

The electric-system operators use two-way radio equipment to communicate with linemen and cablemen working on the electric transmission and distribution equipment (Fig. 2.60). The electric-system coordinator (Fig. 2.61) controls all switching, the placement of hold-off tags, and the installation of protective grounding equipment. Technicians assist the coordinators as needed (Fig. 2.62). Permission to install protective grounding equipment is not given until all switching has been completed to deenergize and isolate the circuit and

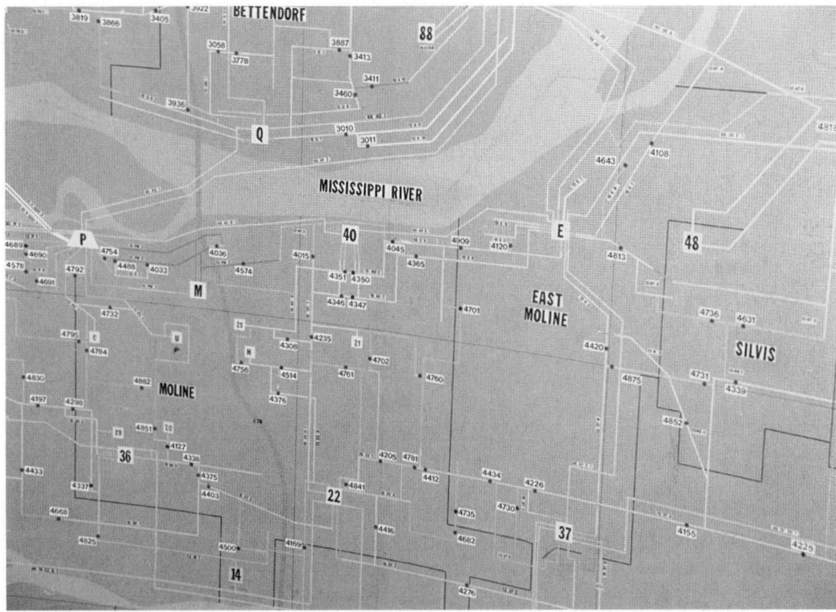

FIGURE 2.59 Map showing location of electric facilities mounted on a specially constructed support. The scale of the map is large enough to permit identification of the electric system from operating positions.

FIGURE 2.60 Distribution coordinator workstations arranged on the floor of a distribution operations control center. The distribution coordinator utilizes many computerized systems to obtain essential information on the electric distribution system operation. The workstation has several display monitors for the coordinator. (*Courtesy MidAmerican Energy Co.*)

FIGURE 2.61 Distribution coordinator at the distribution operations control-center workstation in the process of dispatching an incident for investigation by the lineman. (*Courtesy MidAmerican Energy Co.*)

FIGURE 2.62 Distribution technician assisting the distribution operations control-center distribution coordinators. The technician's workstation requires less system display monitors than that of the coordinator. (*Courtesy MidAmerican Energy Co.*)

hold-off tags have been installed. The workers installing the grounds must test the circuit to verify that it is deenergized before the grounding operation is initiated. The correct procedure for installing grounding equipment is described in chaps. 27 and 46. The workers must report to the system operator when the grounding operation is completed. When the work is completed, the lineman must obtain permission from the system operator before removing the grounding equipment from the isolated circuit or equipment. The grounds must be removed before the system operator authorizes the removal of hold-off tags and the operation of switches to reenergize the circuit. All the switching, tagging, and grounding operations must be completed in accordance with the rules in the *National Electrical Safety Code* (ANSI-C2).

Uninterruptible Power-Supply System. Electric distribution systems experience voltage dips and momentary interruptions as a result of short circuits or electrical surges. The fluctuations on the distribution system are normally caused by lightning discharges, animals bridging an insulator, or switching surges. Lightning may cause an overvoltage which results in an insulator flashover. Squirrels climb poles, get on distribution transformers, and can short out the transformer bushing. Power-system consumers may have equipment such as computers, machine tools, instruments, and other devices that require a continuous and unvarying voltage power supply. An uninterruptible power supply is used to provide a source for critical electrical loads. It must be designed and constructed specifically for the load to be served, and therefore, it is normally owned by the consumer of the electric power. Figure 2.63 is a schematic wiring diagram of an uninterruptible power supply that includes a solid-state rectifier or charger, a battery, an inverter to produce alternating current from the direct current, a solid-state or static switch, a manual transfer switch, and a motor-generator set to provide auxiliary power for the critical electric load. The solid-state rectifier filters out any variations or surges in the ac power source and keeps the battery fully charged. The battery provides power if the distribution circuit is interrupted for a short period of time. A motor-generator set would normally be used in case of a long outage of the electric distribution source circuit. The uninterruptible power supply will operate reliably to supply critical loads (Fig. 2.64).

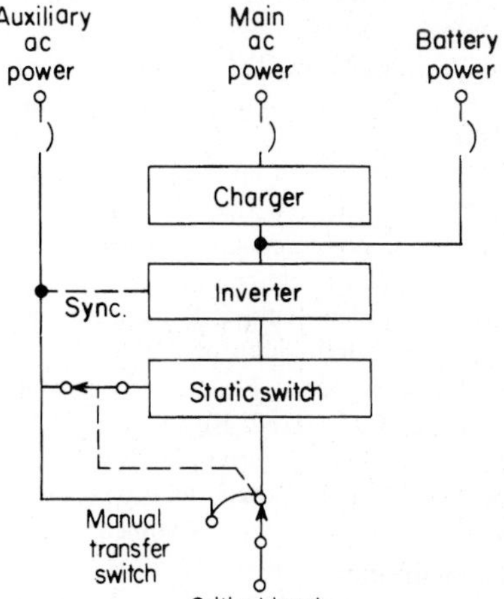

FIGURE 2.63 Schematic wiring diagram of an uninterruptible power-supply system.

FIGURE 2.64 Solid-state uninterruptible power supply complete with battery.

ELECTRIC SYSTEM—POWER QUALITY

Electric distribution companies have an obligation to provide quality power to their customers. For example, an incandescent light bulb that flickers with the start-up of a large motor is a nuisance to the customer. This problem might have been resolved by one of several methods. The methods include changing out an undersized transformer, replacing undersized secondary wiring, seeing that the loads are balanced on each phase, or ensuring that some type of reduced voltage soft starting control is installed on the motor. Electric equipment is able to tolerate a few cycles of irregular power. The distribution companies have guidelines regarding transformer sizing, allowable secondary footages, flicker limitations, and limits on allowable starting currents for motors.

With the advent of electronically controlled electrical equipment on the load side of the electric meter, a greater dependence on distortion free power has emerged. The newer electronic equipment is more sensitive to the voltage provided to the power supply. The electronic components of the devices are dependent upon a pure sine wave voltage supply.

Electric equipment or devices can be of a liner or nonlinear nature. A linear device is one in which the load resembles that of a resistor. The relationship of the current and voltage is predictable in some type of well-defined linear manner. Reactive devices also exhibit linear relationships if the frequency for the device is fixed. Reactance is either inductive or capacitive. Current lags voltage in inductive devices. Current leads voltage in capacitive devices. Linear equipment is not as affected by typical power quality problems.

Electronic equipment contains devices that are of a nonlinear nature. Nonlinear devices contain electronic components that have some inductive or capacitive characteristics that impact the relationship of voltage, current, and frequency. Electronic devices require an alternating current to direct current to alternating current conversion utilizing rectifiers and inverters. The alternating current wave-shape does not mirror that of the voltage wave-shape.

The range of tolerable voltages for computer equipment that has become the industry design guidelines for electronic equipment is known as the ITI (Information Technology Industry) Council curve (previously recognized as the CBEMA (Computer & Business Equipment Manufacturer's Association) curve). The horizontal axis of the curve is defined in cycles at 60 Hz or in seconds and the vertical axes is defined in the percentage of rated voltage. Examples of the upper design limit of the curve that should not be exceeded without voltage breakdown concerns are 300 percent rated voltage at 0.01 cycles, 200 percent rated voltage at 0.1 cycles, 130 percent rated voltage at 0.5 cycles, and 106 percent rated voltage for 1000 cycles or greater. The minimum lower limit that the equipment will tolerate without noise and voltage breakdown problems is 30 percent rated voltage at 1.0 cycle, 58 percent rated voltage at 10 cycles, 70 percent rated voltage at 100 cycles, and 87 percent at 1000 cycles. The IEEE Standard 1346—latest revision, "IEEE Recommended Practice for Evaluating Electric Power System Compatibility with Electronic Process Equipment"—provides further explanation on the significance of the curve.

According to ANSI-IEEE Standard 84.1, the electric distribution supplier should maintain steady state voltage within ±5 percent of the nominal voltage. If the nominal voltage is 120 volts, then the maximum delivered voltage is 126 volts and the minimum delivered voltage is 114 volts.

Several types of power quality problems have been defined in IEEE Standard 1159—latest revision, "IEEE Recommended Practice for Monitoring Electrical Power Quality." The problems include (a) voltage sags due to motor starting, lightning or fault clearing; (b) voltage surges or voltage swells due to lightning or high impedance fault currents; (c) flickering due to loose connections at the transformer, in the service lines extended to the meter, or in the electric meter; (d) harmonic waves in the voltage signals due to nonlinear loads such as those of adjustable speed drive motors or high intensity discharge lights; and (e) momentary interruptions due to faults on the power system causing circuit breaker tripping or relaying.

Harmonics are alternating current and voltage waveforms with frequencies that are multiples of the system frequency. If the system frequency is 60 Hz, the second harmonic is 120 Hz, the third harmonic is 180 Hz, and so on. Harmonic distortion of the voltage and current waveform has a negative effect on equipment. Harmonics are addressed in great detail in IEEE Standard 519—latest revision, "Recommended Practice and Requirements for Harmonic Control in Electric Power Systems."

Power quality investigations are required to diagnose the problem. The investigations more often than not involve the lineman. It should be noted that a high percentage of power quality problems, such as poor wiring and improper grounding, are internal to the facility and reside on the load side of the electric meter. Adherence to the NFPA 70 National Electrical Code, latest edition (NEC) Article 250 Grounding and Article 310 Conductors for General Wiring, will minimize the likelihood of internal problems. Even so, during the initial stages of the investigation, all possible power quality causes must be reviewed including all of the sources of system impedances that are encountered up to the affected device.

Equipment used to investigate the problem are as follows: true rms (root-mean-square) voltmeters and ammeters, oscilloscopes, and disturbance analyzers. More advanced diagnosis tools include harmonic analyzers, infrared detectors, gauss meters, static meters, and flicker meters.

Equipment used to mitigate the power quality problems include filters, surge suppressors, isolation transformers, power conditioners, uninterruptible power supplies (UPS), distributed generators, superconducting magnetic energy storage (SMES), mechanical transfer switches, solid state transfer switches, and fuel cells.

The *Power Quality Primer*, Barry Kennedy, McGraw-Hill Inc., and *Electrical Power Systems Quality*, Roger C. Dugan, Mark F. McGranaghan, and Wayne H. Beaty, McGraw-Hill Inc., are excellent reference books on the topic. In addition, trade journals, such as *Power Quality Assurance*, *EC&M Magazine*, and *Energy User News*, provide many timely power quality related articles.

CHAPTER 3
SUBSTATIONS

SUBSTATION TYPES

An electrical substation is an installation with electrical equipment capable of interrupting or establishing electrical circuits and changing the voltage, frequency, or other characteristics of the electrical energy flowing in the circuits.

Transmission Substations. Figure 3.1 shows an alternating-current (ac) electric transmission substation located at an appropriate point in a network of transmission lines. The substation contains equipment used to sectionalize the electric transmission system when a fault or short circuit develops on one of the circuits. The defective circuit is switched out of service automatically and deenergized, protecting the remaining portion of the transmission network from trouble. The electric transmission system is designed to permit deenergizing of any one circuit without interrupting electric service to customers.

Transmission substations located at electric generating stations (Fig. 3.2) have power transformers designed to raise the voltage from generation value, usually in the range of

FIGURE 3.1 Transmission substation with four autotransformer, 445 MVA single-phase, 500/230/34.5 kV. One transformer is a spare. Circuit breakers in the substation are used to switch circuits and power transformers in and out of service. Series reactors in the foreground limit the fault current in the 34.5-kV circuits. (*Courtesy ABB Power T&D Company Inc.*)

FIGURE 3.2 Power transformers in an ac transmission substation located adjacent to generating station. Power transformers raise the generated voltage from 25 to 765 kV. Each single-phase transformer has a capacity of 317 MVA. The three-phase bank capacity is 950 MVA. The fourth transformer serves as a spare.

13,200 to 25,000 volts three-phase, to transmission level 138,000 to 765,000 volts or higher. Circuit breakers in the transmission substation are used to switch generating and transmission circuits in and out of service.

Subtransmission Substations. Electric substations (Fig. 3.3) with equipment used to switch circuits operating at voltages in the range of 34.5 to 69 kV are referred to as subtransmission substations. Subtransmission circuits and substations are located near high-load concentrations, usually in urban areas. Subtransmission circuits supply distribution substations.

Distribution Substations. Substations located in the middle of a load area are called distribution substations. These substations may be as close together as 2 miles in densely populated areas. The substations contain power transformers that reduce the voltage from subtransmission levels to distribution level, usually in the range of 4.16Y/2.4 to 34.5Y/19.92 kV. The transformers are normally equipped to regulate the substation bus voltage. Circuit breakers or circuit reclosers in the distribution substations are installed between the low-voltage bus and the distribution circuits. The distribution circuits may vary in capacity from approximately 5 to 15 MVA. Large distribution substations may have power transformers with a capacity as large as 100 MVA and serve as a source to as many as 10 distribution circuits. If the power transformers do not have automatic tap changing underload equipment, it is usually necessary to install voltage regulators (Fig. 3.4). Distribution substations (Fig. 3.5) may be constructed near a large manufacturing plant to serve a specific industrial customer. Distribution substations are normally unattended and operate automatically or are controlled remotely by supervisory control equipment from a central control center.

FIGURE 3.3 Subtransmission ac substation equipped with oil circuit breakers, lightning arresters, disconnect switches, and potential transformers. Substation is located in the fringe area of a large city and serves as a switching center for the subtransmission circuits.

FIGURE 3.4 Portion of a distribution substation showing power-transformer distribution-voltage bushings, lightning arresters, gang-operated interrupter switch, potential transformers connected to the overhead distribution-voltage bus through solid-material power fuses and three single-phase voltage regulators connected to the bus through bypass switches. (*Courtesy S&C Electric Co.*)

FIGURE 3.5 Distribution substation located at the Nucor Steel large rolling-mill plant. The source to the substation is a 161 kV circuit supported by steel "A" frames. The power transformers are each rated 116 MVA foa, three-phase, 161 to 34.5 kV. The structures in the background are individual circuit feeders. The tall masts shown in the picture are used for lightning protection. (*Courtesy ABB Power T&D Company Inc.*)

SUBSTATION FUNCTION

An electric substation is an installation in which the following functions can be accomplished for the power system:

1. Voltage changed from one level to another level
2. Voltage regulated to compensate for system voltage changes
3. Electric transmission and distribution circuits switched into and out of the system
4. Electric power qualities flowing in the transmission and distribution circuits measured
5. Communication signals connected to the circuits
6. Lightning and switching surges eliminated from the electric system
7. Electric generators connected to the transmission and distribution system
8. Interconnections between the electric systems of various companies completed
9. Reactive kilovolt-amperes supplied to the transmission and distribution circuits and the flow of reactive kilovolt-amperes on the transmission and distribution circuits controlled
10. Alternating current converted to direct current (dc) or direct current converted to alternating current
11. Frequency changed from one value to another

SUBSTATION EQUIPMENT

Electric substations contain many types of equipment. The major components of equipment consist of the following:

Concrete foundations
Duct runs
Manholes
Steel superstructures
Bus support insulators
Suspension insulators
Lightning arresters
SF_6 circuit breakers
Oil circuit breakers
Air circuit breakers
Vacuum circuit breakers
Circuit switchers
Disconnect switches
Power transformers
Tap Changing Equipment
Coupling capacitors
Potential transformers
Current transformers
High-voltage fuses
High-voltage cables
Rectifiers
Frequency changers
Potheads
Metal-clad switchgear
Shunt reactors
Capacitors
Voltage Regulators
Synchronous condensers
Grounding transformers
Grounding resistors
Control House
Control panels
Meters
Relays
Supervisory control
Microwave
Power-line carrier
Batteries
Battery chargers
Conduits
Control wires

Power Transformers. Transformers are used to change the voltage from one level to another, to regulate the voltage level, and to control the flow of reactive kilovolt-amperes in the power system. Power transformers installed in transmission substations will normally be constructed for and operate at voltages in the range of 138,000 to 765,000 or higher volts. Figures 3.6 and 3.7 are pictures of large power transformers located in transmission substations.

Most substations will have three-phase transformers (Figs. 3.7 and 3.8). Some substations will have three single-phase transformers installed in a bank (Fig. 3.9). Often, a fourth single-phase transformer will be located at the substation to serve as a spare. If one of the transformers connected in the bank fails, the spare is energized to expedite restoration of electric service.

The capacity of the transformers will normally be in the range of 1 million to 50,000 kVA. Power transformers installed in distribution substations will normally be constructed for and operate at the lower voltages and smaller capacities. The voltages will generally be in the range of 161,000 to 4160 volts and the capacities in the range of 50,000 to 5000 kVA. Figure 3.10 is a picture of a power transformer that is part of a distribution substation.

Tap changing equipment. An option that may be added to the substation transformer is tap changing equipment. The tap changer feature allows the voltage to be adjusted in increments as system loading conditions require. The controls of the tap changing equipment are installed in a panel mounted on the side of the power transformer. Remote control of the tap changing equipment is usually installed as well. Section 19 describes in more detail the tap changing functionality.

Steel Structures. The substation's steel structures (Fig. 3.11) provide support for insulators, which are used to terminate lines and support buses that interconnect the equipment. Disconnect switches and other equipment are mounted on the steel structure (Fig. 3.12).

FIGURE 3.6 A power transformer installed in a transmission substation. The transformer is equipped with radiators for cooling the insulating oil, oil-level gauge with alarm contacts, dial hot-spot winding-temperature equipment, no-load tap changer, nitrogen cylinder and cabinet, sudden-pressure relay, bushings, and lightning arresters.

FIGURE 3.7 Large power transformer located in a transmission substation. Transformer is connected to the 138- and 345-kV buses in the substation. (*Courtesy ABB Power T&D Company Inc.*)

FIGURE 3.8 Power-transformer core and coil assembly with tap-changer underload connection cables on top and boost, or buck, transformer in background.

FIGURE 3.9 Transmission substation with three single-phase power transformers connected into a three-phase bank. (*Courtesy Westinghouse Electric Corp.*)

FIGURE 3.10 Distribution power transformer with tap-changing underload equipment in compartment in foreground. The distribution voltage connection to the transformer is completed through the bus duct on the right side of the picture to metal-enclosed switchgear.

FIGURE 3.11 The substation's steel structures provide support for insulators, which are used to terminate lines and support buses that interconnect the equipment.

FIGURE 3.12 Distribution substation with a lattice-type steel structure. A 69,000-volt subtransmission circuit in the background provides a source to the station. Underground cables connect the power transformer to the metal-clad switchgear for the distribution circuits.

Lightning Arresters. The lightning arresters protect the electric system and the substation equipment from lightning strokes and switching surges. The lightning arresters are installed in a substation near the termination of aerial circuits and close to the more valuable pieces of equipment, such as power transformers. Lightning arresters can be seen in Fig. 3.6 mounted on the power transformer adjacent to the transformer bushings. The power transformer in Fig. 3.7 has lightning arresters mounted on the radiators which are connected to both the 345- and 138-kV transformer bushings.

Lightning arresters contain semiconductor blocks, which limit the magnitude of high surge voltages, permit the large surge currents to pass harmlessly to ground, and interrupt the power-follow current after the surge is eliminated. Metal-oxide surge arresters function like a ceramic capacitor at normal line voltage, limiting the flow of current to ground. When a high voltage begins to build up across the semiconductor blocks, they provide a low-impedance path to ground, which permits the surge current to flow to ground, limiting the voltage buildup and preventing the equipment from being damaged. The semiconductor blocks in the arresters are manufactured using zinc oxide material.

Circuit Switchers. Circuit switchers employ SF_6 puffer-type interrupters for switching and protection of transformers, lines, cables, and capacitor banks and have fault-interrupting ratings suitable for use in protecting medium- to large-load transformers (Fig. 3.13). Models are available with and without integral disconnect switches. Operation of circuit switchers is initiated by manually operating a switch, by remote supervisory control equipment, or by relays that automatically sense predetermined system or equipment conditions or electrical failures (faults). The circuit switcher shown in Fig. 3.14 is used to energize and deenergize a substation capacitor bank. The substation capacitor bank is used to supply vars to the electric system and to regulate the substation voltage.

FIGURE 3.13 Distribution substation supplied from a 69-kV subtransmission circuit. The substation line terminal structure supports surge arresters. A circuit switcher is used to energize and deenergize the power transformer. The power transformer reduces the subtransmission voltage to 24,940Y/14,420 distribution voltage. (*Courtesy S&C Electric Co.*)

FIGURE 3.14 Substation capacitor bank used for power-factor correction and voltage regulation. Capacitors are individually fused. Bank switching is provided by a circuit switcher. Potential transformer connected to bank ungrounded neutral conductor provides sensing voltage for protective control devices. (*Courtesy S&C Electric Co.*)

Circuit Breakers. Oil, SF_6, air, gas, and vacuum circuit breakers are used to switch electric circuits and equipment in and out of the system. The contacts of the circuit breakers are opened and closed by mechanical linkages manufactured from insulating materials and utilizing energy from compressed air, electric magnets, or charged springs. Some of the high-voltage circuit breakers utilize compressed air to operate the contacts and interrupt the current flow when the contacts are open. Operation of the circuit breakers is initiated, utilizing dc circuits, by manually operating a switch, by remote operation of supervisory control equipment, or by relays that automatically recognize predetermined conditions or electrical failures in the system. Various types of circuit breakers can be seen in Figs. 3.15 to 3.18.

FIGURE 3.15 Three-phase high-voltage dead tank SF_6 gas power circuit breaker with vertical break group-operated disconnect switches installed as an isolator. (*Courtesy ABB Power T&D Company Inc.*)

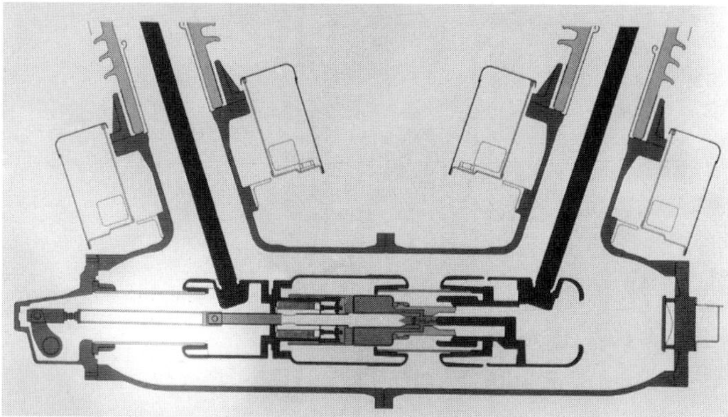

FIGURE 3.16 Cutaway view of one pole of a three-phase high-voltage dead tank SF_6 gas filled power circuit breaker showing power interrupter, contact pull rod, bushing through rods, and bushing mounted current transformers. (*Courtesy ABB Power T&D Company Inc.*)

FIGURE 3.17 SF_6 gas filled dead tank power circuit breaker with insulated composite bushings being used at an extra-high voltage (EHV) substation. (*Courtesy ABB Power T&D Company Inc.*)

FIGURE 3.18 Live tank sulfurhexafluoride (SF_6) gas filled power circuit breaker installed in an EHV substation. (*Courtesy ABB Power T&D Company Inc.*)

Disconnect Switches. High-voltage equipment is usually installed in substations with disconnect switches in series with it to facilitate isolating the equipment for maintenance. The disconnect switches are normally operated by linkages connected to a lever or gear-type handle or a motor operator for remote operation (Fig. 3.19). The switches are designed to interrupt charging current but—unless they are specially constructed—not load current. The disconnect switches shown in Fig. 3.15 on each side of the oil circuit breaker can be operated after the circuit breaker contacts are opened to visibly isolate the circuit breaker. The disconnect switches are operated by turning the crank handle, which moves the interphase rods through a gear arrangement and rotates the insulators, causing the switch blades to move and open the switch contacts.

Coupling Capacitors. Communication signals in the form of high-frequency voltages are transmitted to the transmission lines through coupling capacitors. Some of the coupling capacitors are equipped with potential devices which make it possible to measure the voltage on transmission-line circuits. The coupling-capacitor potential devices are accurate enough to be used for supplying voltages to protective relays but—unless they are specifically compensated—not accurate enough to supply voltages for meters designed for billing purposes. Figure 3.20 shows a coupling capacitor.

Potential Transformers. High voltages are measured by reducing the voltage proportionately with a potential transformer (Fig. 3.21) which has its high-voltage winding connected to the transmission or distribution circuit and its low-voltage winding connected to a meter or relay or both. Potential transformers are required to provide accurate voltages for meters used for billing industrial customers or connecting utility companies. If single-phase transformers are used, three potential transformers are generally required to measure power on a three-phase circuit.

FIGURE 3.19 Three pole high voltage gang operated center break "V" disconnect switch mounted upright on a station structure. (*Courtesy ABB Power T&D Company Inc.*)

FIGURE 3.20 Coupling-capacitor potential device designed for use on 161,000 volts. (*Courtesy General Electric Co.*)

FIGURE 3.21 Single-phase potential transformers connected between a phase conductor and ground or neutral in a 345,000-volt transmission substation. The transformers are filled with oil. (*Courtesy MidAmerican Energy Co.*)

Current Transformers. The primary winding of a current transformer is connected in series with the high-voltage conductor. The magnitude of amperes flowing in the high-voltage circuit is reduced proportionately by the ratio of the current transformer windings. The secondary winding of the current transformer is insulated from the high voltage to permit it to be connected to low-voltage metering circuits. A bushing-type current transformer is shown in the cutaway view of a SF_6 circuit breaker in Fig. 3.16. Current and potential transformers supply the intelligence for measuring power flows and the electrical inputs for the operation of protective relays associated with the transmission and distribution circuits or equipment such as power transformers.

High-Voltage Fuses. High-voltage fuses are used to protect the electrical system from faults in such equipment as potential transformers or power transformers. Figure 3.22 shows a distribution unit substation with a line terminal structure designed for appearance as well as function. A group-operated high-voltage disconnect switch and three power fuses are mounted on the terminal structure. The fuses are connected in series with the power transformer that serves as a source for the switchgear.

Metal-Clad Switchgear. Outdoor metal-clad switchgear (Fig. 3.23) provides a weatherproof housing for circuit breakers, protective relays, meters, current transformers, potential transformers, bus conductors, and other items necessary to provide electric-system requirements. Indoor switchgear must be installed in a building for protection from the elements. Figures 3.24 to 3.29 illustrate metal-clad switchgear installations and some of the equipment—including circuit breakers—normally installed in the switchgear.

FIGURE 3.22 Unit-type substation with high-voltage fuses to protect the electric system from power-transformer faults.

Shunt Reactors. When the capacitive reactance from extra-high-voltage transmission-line circuits exceeds the ability of the system to absorb the reactive kVA (kilovars), shunt reactors are installed. The shunt reactor shown in Fig. 3.30 is a three-phase, single-winding transformer which serves as an inductive reactance on the power system, thus neutralizing the capacitive reactance associated with the long extra-high-voltage transmission line. Shunt reactors are usually located in substations and connected to a transmission-line terminal through a disconnect switch.

Meters. Meters mounted on control panels are used to measure the energy flowing in a circuit or through a piece of equipment. Some of the meters are utilized to bill industrial customers and adjacent utility customers. Interstate flows of electrical energy are measured in the substations.

Relays. Protective relays installed on control panels are used to identify electrical failures on transmission and distribution circuits or in pieces of equipment, such as power

FIGURE 3.23 Outdoor metal-clad switchgear designed for appearance. (*Courtesy General Electric Co.*)

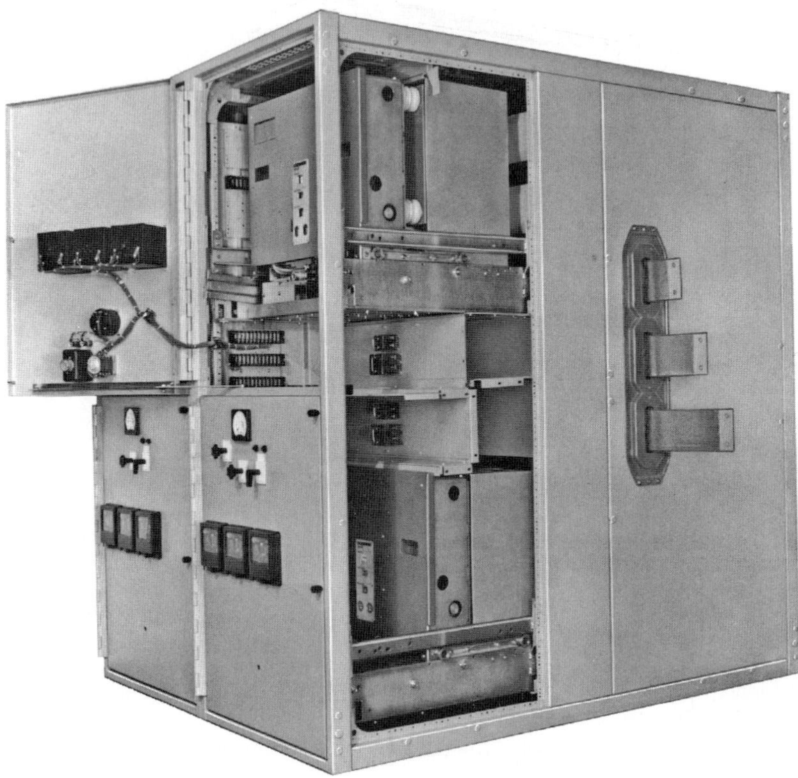

FIGURE 3.24 Indoor metal-clad switchgear with vacuum horizontal-draw-out circuit breakers. Top compartment, with side removed, has circuit breaker in the racked-out position, and bottom compartment has circuit breaker in the operating position.

FIGURE 3.25 Vacuum circuit breaker rated 15 kV removed from switchgear cubicle in Fig. 3.24. Note draw-out contacts with fingers and vacuum bottles in back of, and connected to, draw-out contacts.

FIGURE 3.26 Vacuum-bottle cutaway to show contacts and bellows that permit one contact to move without losing the vacuum.

FIGURE 3.27 Indoor metal-clad switchgear with potential compartment door open. The bottom tray with fuses and potential transformers is in the operating position, and the top tray with fuses and potential transformers is in the drawn-out position with fuses and potential transformers isolated.

transformers. Some are used to perform predetermined operations as required by system conditions. Relays automatically operate their contacts to properly identify the source of trouble and remove it from the electrical system.

Supervisory Control. The supervisory control equipment permits the remote control of substations from a system control center or other selected point of control. This equipment is used to open and close circuit breakers, operate tap changers on power transformers, supervise the position and condition of equipment, and telemeter the quantity of energy flowing in a circuit or through a piece of equipment.

Capacitors. Reactive kilovolt-amperes can be supplied to the electrical system by connecting banks of capacitors to the distribution circuits in substations or out on the distribution lines to neutralize the effect of customer inductive loads. Capacitors used in this manner help to control voltages supplied to the customer by eliminating the voltage drop in the system caused by inductive reactive loads. Capacitors in distribution substations are usually mounted in metal cubicles. The capacitors, mounted on the racks in the cubicles, are usually single-phase, single-bushing units rated 100-, 150-, 200-, 300-, or 400-kVA capacitance, 60 Hz, and a voltage consistent with the distribution system. They are connected between each of the three-phase conductors and ground. The equipment in Fig. 3.31 includes instrument transformers in the first cubicle, a three-phase vacuum switch in the second cubicle, and capacitor units in cubicles three, four, and five. Capacitor units for the higher distribution

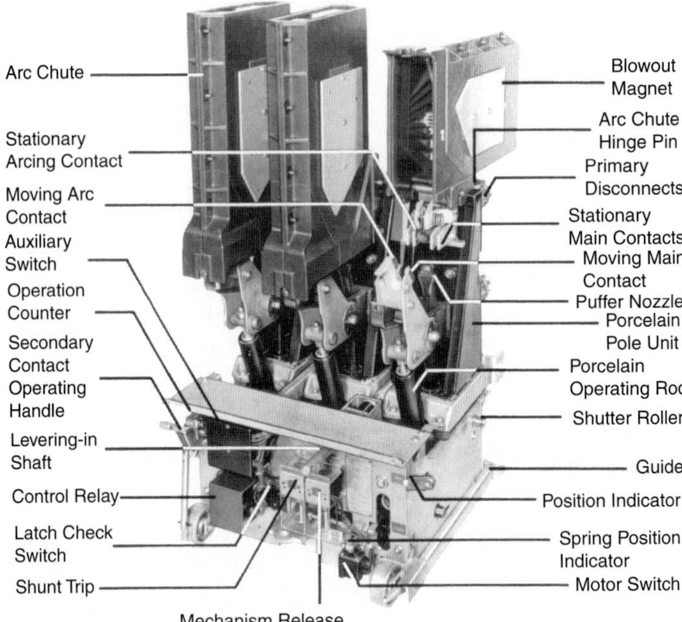

FIGURE 3.28 Horizontal-draw-out 15-kV air-magnetic circuit breaker with barriers removed and one arc chute tilted to expose contacts.

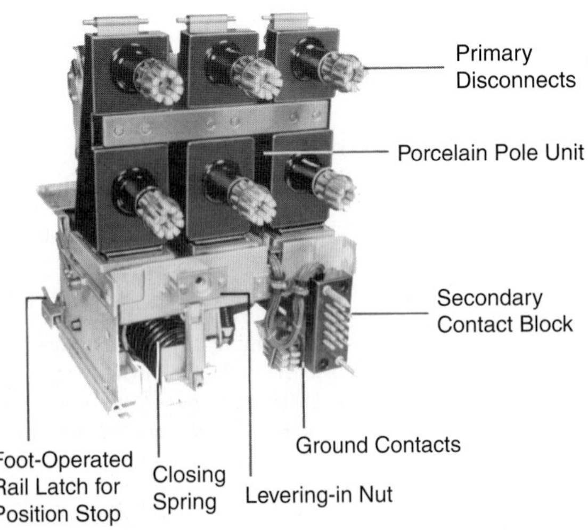

FIGURE 3.29 Back view of horizontal-draw-out 15-kV air-magnetic circuit breaker with barriers and arc chutes removed.

FIGURE 3.30 Shunt reactor installed in an extra-high-voltage substation rated 345 kV, 50 MVA. (*Courtesy MidAmerican Energy Co.*)

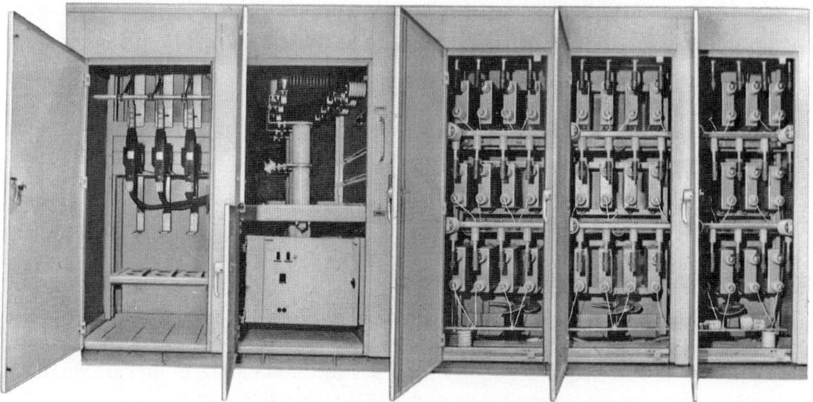

FIGURE 3.31 Metal-enclosed capacitor equipment for substation installation.

FIGURE 3.32 Control house for distribution substation constructed as a part of a decorative wall. (*Courtesy MidAmerican Energy Co.*)

voltages and those connected to conductors energized at transmission voltages are normally mounted on open-type racks (Fig. 3.14). The capacitor units are connected in series and cascaded to provide equipment that can be connected to the higher-voltage distribution and transmission systems.

Voltage Regulators. The voltage regulators maintain the system voltage on the distribution circuits. As settings on the voltage regulators are adjusted for various load conditions that occur, the desired voltage is obtained. Chapter 19 describes in more detail the voltage regulators and Chap. 40 describes voltage regulation.

Control House. The substation control house (Fig. 3.32) is used to protect the control equipment, including switchboard panels, batteries, battery chargers, supervisory control, power-line carrier, meters, and relays, from the elements. Underground and overhead conduits and control wire are installed to connect the controls of all the equipment in the substation to the control panels normally installed in the control house. Each substation usually contains several thousand feet of conduit and miles of control wire.

Control Panels. The control panels (Fig. 3.33) installed in a control house provide mountings for meters, relays, switches, indicating lights, and other control devices.

Carrier Equipment. The power-line carrier equipment provides high-frequency voltages to be used for transmitting voice communications or telemetering signals on high-voltage transmission-line circuits. In the case of voice communications, the sound frequency modulates a high-frequency signal connected to the transmission-line circuit by means of coupling capacitors. This equipment permits using the line conductors for communication, relaying, supervisory control, and metering in addition to the transmission of electric power.

Microwave Equipment. Radio signals used for point-to-point communications between substations or other power-system facilities operating in the megahertz frequencies are called microwave. The frequencies range from 952 to 13,000 MHZ or higher. Microwave radio signals are used for communication channels, protective relaying, supervisory control, and remote metering.

FIGURE 3.33 Control panels with meters, relays, and control devices mounted for operation of a substation. (*Courtesy General Electric Co.*)

Fiber optics. Special cables constructed with an ultrapure glass or plastic core of a very small diameter are capable of transmitting light by means of internal reflection. The light signal can be injected into the core by a laser diode or a light-emitting diode. The light input is capable of transmitting analog and digital signals for protective relay operations, communications, and data systems.

The fiber optic cables can be installed in conduits, direct buried in the ground, or on overhead lines in static or phase wires. The fiber optic cable can be constructed as a good insulator for electric currents and potentials. The good insulating qualities of the fiberoptic cable eliminated the need to have the cable grounded, thus making the installation on energized overhead conductors practical and economical.

The fiber optic cables can be used with optical sensors to measure current, voltage, temperature, vibration, and other quantities associated with energized electric transmission lines and substations. Fiber optic systems can eliminate the need for auxiliary devices such as current and potential transformers and the wires connecting them to instruments.

Batteries. Control batteries (Fig. 3.34) supply energy to operate circuit breakers and other equipment. It is necessary to use dc control systems with a storage battery as a source to make it possible to operate equipment during periods of system disturbances and outage. Battery chargers are used to automatically keep the batteries charged completely to provide sufficient emergency power for all necessary operations.

Rectifiers. Equipment used to convert alternating current to direct current may be installed in distribution substations if the station is adjacent to the load utilizing the direct current. Figure 3.35 is a picture of a silicon rectifier assembly.

Converter Stations. Figure 3.36 is a drawing of a converter station. Converter stations are located at the terminals of dc transmission lines. The converter stations change alternating current to direct current and invert direct current to alternating current using

FIGURE 3.34 Control battery and solid-state electronic battery charger installed in substation control house. Battery provides 125 volts for operation of control equipment.

FIGURE 3.35 Typical forced-air-cooled silicon rectifier for electrochemical service rated 10,000 amps up to 1000 volts direct current. Left-hand section contains cooling fan.

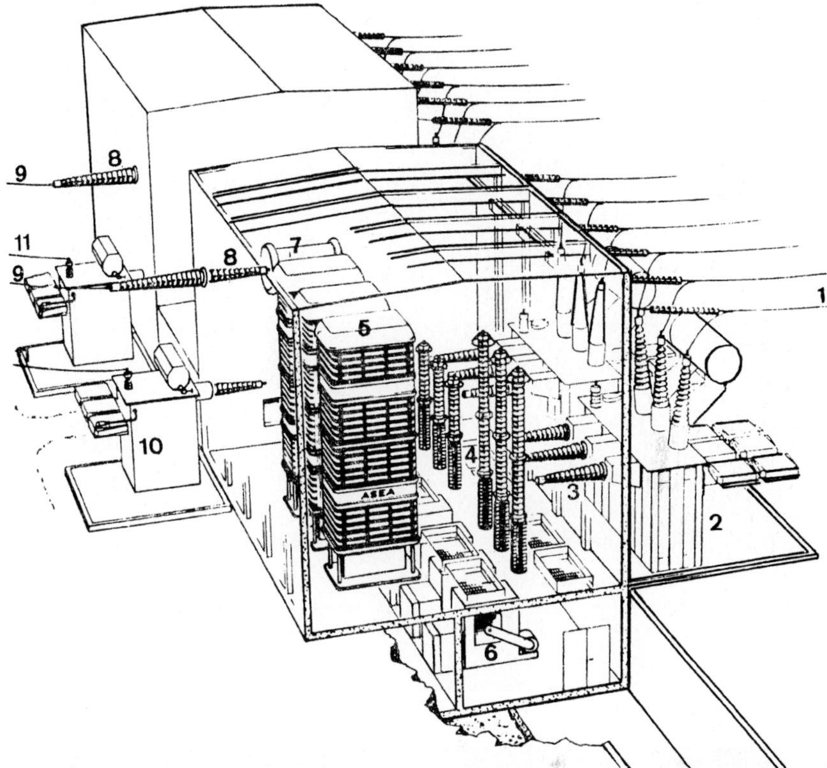

FIGURE 3.36 General arrangement of a converter station. (1) Alternating-current bus, (2) converter transformer, (3) valve side bushing of converter transformer, (4) surge arresters, (5) quadruple valves, (6) valve-cooling fans, (7) air-core reactor, (8) wall bushing, (9) outgoing dc bus work, (10) smoothing reactor, (11) outgoing-electrode-line connection. (*Source: Standard Handbook for Electrical Engineers, edited by Donald G. Fink and H. Wayne Beaty. Copyright 1978 by McGraw-Hill, Inc. Used with permission.*)

thyristor semiconductor devices. The converter stations can be located at an ac generating station, and the generator can be connected to the converter station power transformer. Converter stations are frequently located at a transmission substation to connect an ac system to a high-voltage dc transmission circuit. Two converter stations can be located adjacent to each other to connect two ac transmission systems together even though they are not synchronized.

CHAPTER 4
TRANSMISSION CIRCUITS

Alternating-current (ac) transmission is the movement of ac electrical energy from one location to another, such as from a terminal at a generating station to a substation near a load center.

Direct-current (dc) transmission is the movement of dc electrical energy from one location to another. One terminal may be at a generating station. Direct current is economically used to transmit large blocks of power long distances.

Transmission Circuit—History. The first electric generators produced direct current. These generators operated at a voltage matched to the load to be served. A Swiss engineer named Thury developed a high-voltage dc transmission system in 1899.

The invention of the ac transformer made it possible to step voltages up and down. Alternating-current generators used with transformers made it practical to locate the generators remote from the utilization area. Polyphase systems were developed because of improved efficiency over single-phase systems and reduced costs for a given quantity of power required to meet the customers' needs. The early systems generated, transmitted, and utilized two-phase ac power. The two-phase system is now obsolete. Three-phase ac systems with a frequency of 60 Hz (hertz) exist throughout the United States and Canada. The development of waterpower sites made it necessary to transmit electrical energy long distances from the generating stations to the factories located in the cities. High-voltage equipment was developed to permit the economic transmission of electrical energy from the hydroelectric generating stations to the load.

The development of transmission circuits across the continent has increased the reliability of the electric system, reduced the operating and standby reserve generating requirements, and made possible both the interchange and sale of electrical energy between utility companies, thus reducing costs and permitting the installation of large, economical generating units. Open access to lines will allow customers to negotiate their own sales independent of the local operating utility, further reducing costs and delaying the need for generation station expansion, as ideally, the power will be supplied by the areas with excess generation capacity to those areas deficient in their ability to meet demands.

The voltages used on transmission circuits have been steadily rising. A 2000-volt ac transmission line was constructed in Italy in 1886. In 1902, the maximum ac voltage rose to 80,000 volts and remained there for 5 years. In 1913, it reached 150,000 volts and stayed there for 7 years. In 1922, it reached 220,000 volts, which remained the highest voltage until 1936, when the 286,000-volt line from Hoover Dam to Los Angeles was put into operation. In 1952, a 345,000-volt grid was put into operation in Ohio (Figs. 4.1 and 4.2). Virginia Electric and Power company placed in service a 500,000-volt line 350 miles long designed to transmit 1 million kVA in 1964. American Electric Power Company energized a 765,000-volt circuit (Fig. 4.3) in 1969. American Electric Power Company has constructed approximately 2000 miles of 765,000-volt ac electric transmission lines across the states of Indiana,

FIGURE 4.1 Transmission line structure being constructed at 345 kV to provide greater stability to the electric power grid.

Ohio, Kentucky, Virginia, and adjacent areas. Experimental transmission lines have been built and tested to operate at voltages in the range of 1100 to 1500 kV.

Underground transmission circuits have been installed in large metropolitan areas and other places where overhead transmission lines are not practical. The first underground installations were developed for telegraph circuits. A cable insulated with strips of India rubber was used in 1812. Telegraph cables consisting of bare copper conductor drawn into glass tubes and joined together with sleeve joints and sealed with wax were used in 1816. Vulcanized-rubber-insulated cables proved to be successful.

Edison designed a buried system using copper rods insulated with a wrapping of jute. Two or three insulated rods were drawn into iron pipes, and a heavy bituminous compound was forced in around them. The insulated conductors were laid in 20-ft sections and joined together with specially designed tube joints from which taps could be taken if desired. The underground system was used to distribute energy for lighting in New York City. Overhead electric lines are prohibited in sections of Washington and Chicago. New York City enacted a law forcing the removal of all overhead wires from the streets in 1884.

FIGURE 4.2 Transmission-line structure supporting conductors energized at 345 kV.

FIGURE 4.3 Transmission-line tower supporting conductors energized at 765 kV. (*Courtesy American Electric Power Service Corp.*)

The short life of the early underground systems led to the development of rubber-insulated, lead-covered cables installed in ducts. Cables that were insulated with paper saturated with oil in a lead pipe were developed in England in 1890 and operated at 10,000 volts ac. There was approximately 5000 miles of underground cables constructed with copper conductor, insulated with paper saturated oil, and covered with a lead sheath operating at voltages between 6.6 and 25 kV in the United States in 1921.

Underground transmission circuits are more expensive than overhead transmission lines. The high cost of underground installations limits their installation. Factors considered in determining the necessity to invest the larger amounts of money for an underground transmission circuit include:

1. City ordinances
2. The availability and costs of rights-of-way for overhead transmission lines compared with underground transmission circuits
3. The need to provide mechanical protection for the circuits
4. The congestion, and the appearance, of the area where the circuit is to be installed.

New York City is supplied by an extensive underground transmission system with some cables operating at a voltage of 345 kV ac. Pipe-type cables (Fig. 4.4) are used extensively for underground transmission circuits. These cables operate at voltages in the range of 69 to 500 kV ac. Research is in progress on cables to operate at 765 kV ac. Charging current as a result of the capacitance between the phase conductors and their shielding conductors limits the ac power-carrying capacity of cable circuits. Uncompensated cables 26 miles long operating at 345 kV ac will have all the cable capacity absorbed by the charging current.

Direct-current transmission was initiated in Germany in 1883. The system consisted of generators operating at 2400 volts dc transmitting 5 kW over a distance of 8 km (kilometers). Transmission of electrical energy by the Thury system was developed in 1902. Generators were connected in series to transmit 150 amps at a maximum voltage of 27,000 volts. The current in the transmission line was kept constant by varying the output voltage of the

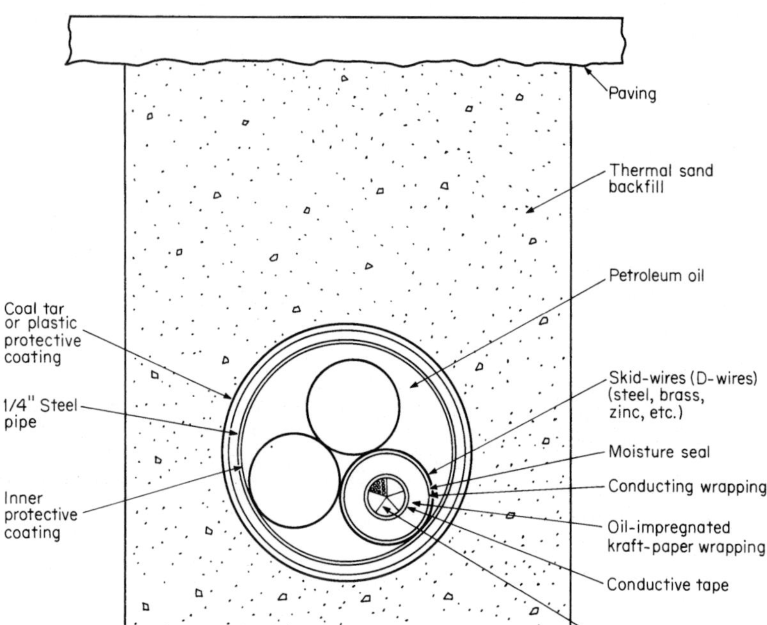

FIGURE 4.4 Cross section of a single-circuit oil-filled pipe-type cable system.

generators and switching generators in and out of service. The load was varied by bypassing and switching motors, connected in series, in and out of service.

Direct-Current Transmission. Direct-current transmission lines can consist of one conductor with a ground return, called a monopolar system, or two-conductor transmission lines with a ground return, called a bipolar system. The bipolar system operates with one conductor at a positive (+) voltage with reference to ground and one conductor at a negative (–) voltage with reference to ground. The bipolar system can be operated as a monopolar system in an emergency if one of the two line conductors is in good operating condition. Modern dc transmission lines originate and terminate at stations that convert high-voltage alternating current to direct current and invert direct current to alternating current using rectifier circuits with thyristor semiconductor elements (Fig. 4.5). The converter transformers (3) serve as the power link between ac and dc systems. The power transformers match the ac voltage to the working voltage of the thyristor valves, isolate the dc system from the ac system, and phase shift the ac supply voltage by wye and delta connections to increase the thyristor firing cycles, reducing the harmonic currents generated in the ac system and voltage ripple in the dc system. The thyristor valves (6) perform the function of converting ac voltage to dc voltage or dc voltage to ac voltage (Fig. 4.6). The valves of convertor circuits have thyristors connected in series using the number required to obtain the voltage desired for the system. Resistors and capacitors are connected to the thyristors to equalize the voltage across each of the thyristors in the valve assembly. The thyristor valves are mounted in a structure to form a 12-pulse converter group (Fig. 4.7). Valve thyristors are individually fired

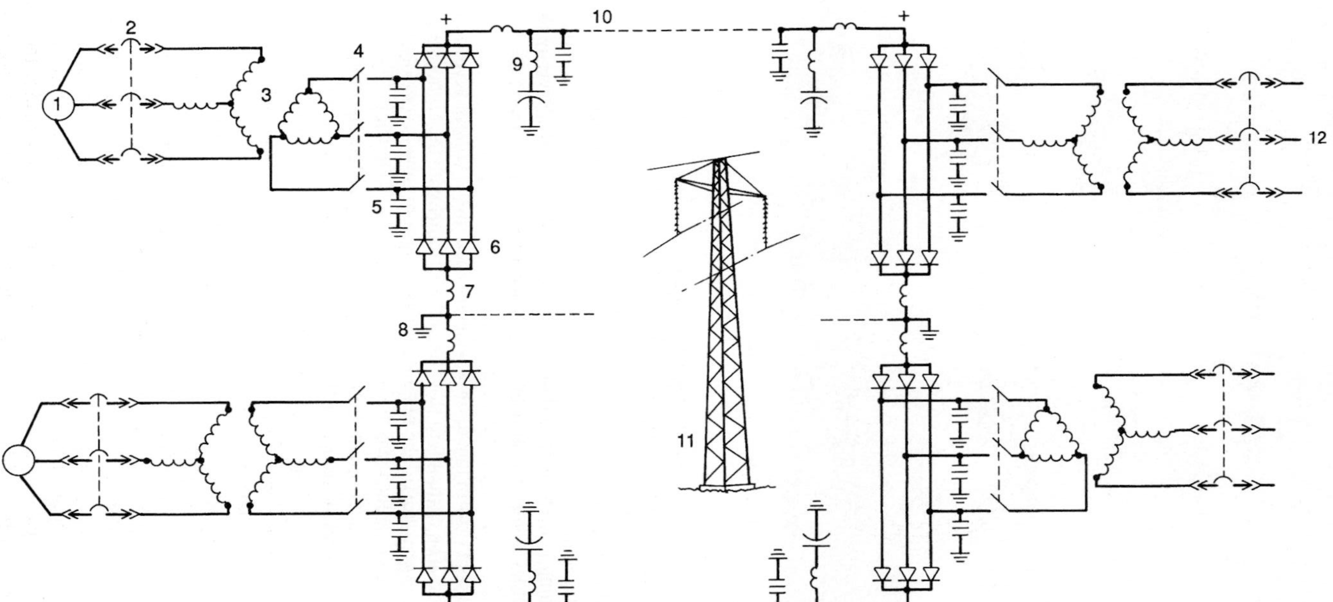

FIGURE 4.5 Schematic diagram of an ac to dc bipolar convertor station, a dc bipolar transmission line, and a dc to ac bipolar inverter station and a drawing of a typical high-voltage dc bipolar transmission-line structure supporting high-voltage conductors and static wire that serves as a ground conductor. (1)Alternating-current generator, (2) alternating-current air circuit breaker with isolating devices, (3) alternating-current power transformer, (4) gang-operated disconnect switch, (5) lightning and surge arrester, (6) thyristor valve with control connection, (7) reactor, (8) ground or neutral and ground conductor, (9) direct-current filter, (10) direct-current transmission-line conductor, (11) direct-current transmission-line structure, (12) alternating-current system.

FIGURE 4.6 Picture of a portion of a thyristor valve assembly. Capacitors and resistors that form the voltage-grading circuits can be seen.

at the proper time to provide the proper voltage. A computer develops an electronic pulse at the proper firing time. The electronic pulse is converted to a light pulse, and the light pulse is transmitted to the thyristor valves by light guides. The light pulse is reconverted to an electronic pulse that fires the thyristors. The dc filters (9) attenuate converter harmonic voltages in the dc system to avoid audiofrequency interference on telephone lines adjacent to the high-voltage dc transmission line. Alternating-current filters are installed on the ac systems by the converter equipment. The ac filter equipment is not shown in the schematic diagram. A monopolar system would have only half the equipment, shown in Fig. 4.5

The direction of current flow in the dc lines is determined by the voltage across the line terminals at each station. The current will flow from the terminal with the higher voltage to the terminal with the lower voltage. The magnitude of the current flow is controlled by the magnitude of the voltage difference and the resistance of the line conductors. Figure 4.8 is a drawing of a structure for a monopolar high-voltage transmission-line structure.

Some of the higher-voltage dc transmission lines by operational date are as follows:

Year	Voltage, kV	Country
1950	200	Russia
1954	100	Sweden
1961	±100	England-France (cross-Channel)
1962	±400	Russia
1965	±250	Sweden
1970	±400	United States
1977	±450	Canada
1978	±400	United States
1981	±500	Canada

FIGURE 4.7 Picture of a structure which contains the thyristor valve assemblies. The structure is equipped with a cooling system to remove the heat dissipated in the valve assemblies during operation.

FIGURE 4.8 Drawing of a monopolar self-supporting tangent structure with one high-voltage conductor assembly and a static wire that serves as a ground return wire.

Convertor stations constructed between 1950 and 1970 used mercury-arc valves. Stations constructed in 1970 and later use thyristor valve assemblies.

Direct-current transmission circuits are economical for installation whenever the overhead lines extend over 400 miles or underground cable circuits extend over 20 miles without a tap to the circuit or to interconnect two ac systems together that are not synchronized.

The Electric Power Research Institute (EPRI) is performing tests and completing research to develop design information for high-voltage dc lines that would be constructed to operate at voltages as high as ±1200 kV. It appears that the next higher-voltage level for dc transmission lines will be ±800 kV. EPRI and Bonneville Power Administration have published Transmission Line Reference Book-HVDC to ±600 kV. Current research and tests will provide design information for HVDC lines up to ±1200 kV to supplement the present publication.

Additional descriptive information for dc transmission systems can be found in the *Standard Handbook for Electric Engineers*, edited by Donald G. Fink and H. Wayne Beaty, published by McGraw-Hill, Inc.

Polyphase Transmission. Alternating-current voltages can be easily changed from one voltage level to another by using power transformers. Single-phase ac transmission circuits would be similar to dc circuits except for inductance and capacitance associated with ac transmission. Almost all ac transmission lines, from the smallest to the largest, are three-phase lines.

The two main reasons for building three-phase transmission lines are:

1. A single-phase two-wire transmission line operating at a certain voltage requires a certain amount of conductivity in the wires to carry the load. A two-phase four-wire line would require the same total weight of conductor. The three-phase system requires only three conductors or three-fourths of the total weight of conductor that is required by single- or two-phase systems for the same load. There is a substantial saving in conductors if a three-phase system is selected.
2. Motors operating from a three-phase line are much cheaper and more satisfactory in operation than single-phase or two-phase motors.

Transmission Voltages. Transmission lines are operated at very high voltages to deliver more power with a particular size line conductor. A three-phase ac line operating at 345,000 volts, for example, can carry 6.25 times as much power as a 138,000-volt line. This is so because the power-carrying capacity of a line varies as the square of the line voltage (kVA $= VI = V^2/R$, where $I = V/R$ and line resistance R remains constant). Raising the line voltage increases the capacity of the line by the square of the ratio between the higher and lower line voltages. Even though the voltage of the 345,000-volt line is only 2.5 times as high as that of the 138,000-volt line, 2.5 squared that is, 2.5 times 2.5 equals 6.25; therefore, the higher-voltage line can carry 6.25 times as much power as the lower-voltage line. If we assume the 138,000-volt line to be carrying a load of 80,000 kW, the 345,000-volt line can carry 6.25 times 80,000 or 500,000 kW.

In like manner, the three-phase 765,000-volt ac line of the American Electric Power System, having a voltage 5.6 times as great as the 138,000-volt line, can carry 5.6 times 5.6, or 31 times as much power; 31 times 80,000 equals 2,500,000 kW.

This tremendous increase in power-carrying capacity with the higher voltages makes it possible to send more power over the same right-of-way, thereby causing large savings in additional rights-of-way. It also reduces the number of parallel lines needed where heavy loads must be transmitted.

Figure 4.9 visually represents three high-voltage lines operating at 138,000, 345,000, and 765,000 volts, respectively, and their corresponding power-transmitting capacities of 80,000, 500,000, and 2,500,000 kW.

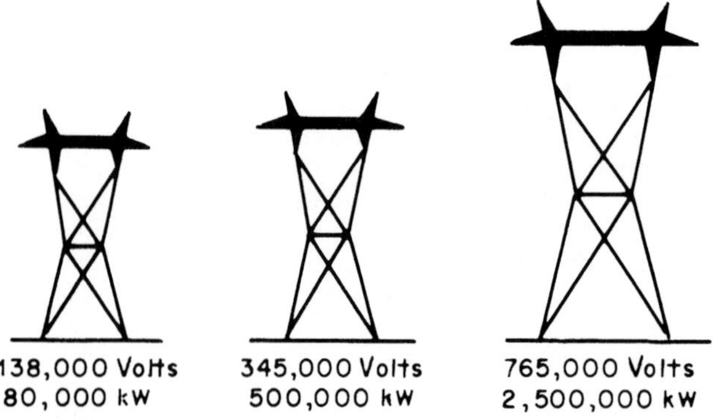

FIGURE 4.9 Comparison of power-carrying capability of three-phase ac transmission lines at different voltages. (*Courtesy Edison Electric Institute.*)

Voltage levels of ac transmission circuits vary from 34,500 to 2,250,000 volts. Voltages originally used for transmission circuits below 34,500 have been converted to distribution circuits. Circuits with voltages in the range of 34,500 to 161,000 volts have been converted to subtransmission use. In many cases, these circuits will provide high-voltage service to large industrial customers. Extra-high voltages have been used extensively for transmission circuits in the United States. The ultra-high ac and dc voltages are still experimental.

The determination of the most economical voltage to use requires careful analysis. Usually several voltages are considered. In each case the total annual operating costs, including interest, depreciation, and taxes, plus power lost in the line, are estimated for each voltage studied. If no special factors have to be considered, the voltage which gives the lowest total annual operating cost is the one selected.

The selection of the proper voltage is sometimes also governed by the voltage of existing power lines. If lines are to be connected to the other lines, as they probably will be, it is convenient to have the lines at the same voltage.

Transmission-Line Capacity. Thermal conditions of the conductors is usually the load-limiting factor of transmission circuits. Convection cooling by ambient air allows overhead conductors to carry higher currents than underground conductors. Alternating-current overhead transmission lines are limited in capacity by the thermal expansion and the resulting sag of the conductors. Short overhead transmission lines may be loaded continuously up to their thermal rating. Table 4.1 gives the thermal rating of some typical overhead ac transmission lines installed in the United States.

The current flowing in a transmission-circuit conductor can be calculated if the voltage and the power in kilovolt-amperes (kVA) or megavolt-amperes (MVA) are known. If the power is expressed in kilowatts (kW) or megawatts (MW), the power factor must be known in addition to the voltage. Kilovolt-amperes and kilowatts can be calculated if the current, voltage, and power factor are known.

The current flowing in a single-phase, two-wire circuit is given by the formula:

$$\text{Current in amperes} = \frac{\text{apparent power in kilovolt-amperes} \times 1000}{\text{line voltage in volts}}$$

or

$$\text{Current in amperes} = \frac{\text{power in kilowatts} \times 1000}{\text{line voltage in volts} \times \text{power factor}}$$

TABLE 4.1 Capacity of Typical Overhead ac Three-Phase Transmission Lines

Line voltage kV	Circuits per tower	Conductors per phase	Conductor size ACSR, kCMIL	Phase resistance Ω/mi	Ampacity A/ conductor	Thermal line rating, MVA
138	1	1	795	0.140	880	210
138	2	1	795	0.140	880	420
230	1	1	1431	0.076	1220	490
230	2	1	1431	0.076	1220	980
345	1	2	795	0.070	1750	1050
345	2	2	1590	0.035	2640	3160
500	1	2	1780	0.031	2850	2470
765	1	4	954	0.028	3890	5150

To illustrate the use of these formulas, take the example of a single-phase line carrying a load of 100,000 kVA and operating at 80,000 volts. Substituting in the first formula gives:

$$\text{Current} = \frac{100{,}000 \times 1000}{80{,}000} = 1250 \text{ amps}$$

The current just calculated is the current in each wire of the single-phase circuit.

If the kilowatts, power factor, and voltage are known instead of kilovolt-amperes and voltage, the second formula is used. For example, if an 80,000-volt circuit delivers 100,000 kW at 80 percent power factor, the current in the line will be given by:

$$\text{Current} = \frac{100{,}000 \times 1000}{80{,}000 \times 0.80} = 1562.5 \text{ amps}$$

If this load of 100,000 kW were at unity power factor, instead of 0.80, the current would be less. The kilovolt-amperes and kilowatts would be the same, and the current would be 1250 amps.

The current flowing in a three-phase circuit is given by the formula:

$$\text{Current in amperes} = \frac{\text{kilovolt-amperes} \times 1000}{1.73 \times \text{line voltage in volts}}$$

or

$$\text{Current in amperes} = \frac{\text{kilowatts} \times 1000}{1.73 \times \text{line voltage} \times \text{power factor}}$$

To illustrate the use of this formula, take the example of a three-phase line carrying 100,000 kVA of load and operating at 138,000 volts. To find the current flowing in each line conductor, substitute in the formula. Thus:

$$\text{Current} = \frac{100{,}000 \times 1000}{1.73 \times 138{,}000} = 418.9 \text{ amps}$$

If the amperes at unity power factor are known and it is desired to know what the current in amperes would be at any other power factor, the procedure is to divide the current at unity power factor by the power factor at which the current is desired. Thus:

$$\text{Current at any desired power factor} = \frac{\text{current at unity power factor}}{\text{desired power factor}}$$

To illustrate, if the current at unity power factor is 100 amps, the current at 0.8 power factor is:

$$\text{Current at 0.8 power factor} = \frac{100}{0.80} = 125 \text{ amps}$$

The formulas and equations presented are summarized and arranged for ready reference in Table 4.2. It might be pointed out as a matter of interest that power factor does not have to be considered in dc calculations because there is no lag or lead of current in such circuits. In ac circuit calculations, the power factor is important. If its value is not known for a given circuit or line, the value 0.80 is often assumed as a basis for calculation. It will also be noted that the factor 1.73 is used with line voltage for all three-phase calculations.

Transmission-Cable Capacity. Solid-dielectric cables are used extensively for subtransmission and transmission underground circuits. The insulating compounds used most extensively are crosslinked polyethylene (XLPE) and ethylene propylene rubber (EPR).

A typical extruded solid-dielectric power-transmission cable is shown in Fig. 4.10. The cable consists of a stranded copper or aluminum conductor covered by a thin extruded layer of semiconducting (carbon-black-filled) polyethylene. The semi-conducting conductor shield provides a smooth conductor surface for the insulation and excludes air from the

TABLE 4.2 Summary of Circuit Calculations

To find	Direct current	Alternating current: Single-phase	Alternating current: Three-phase
Amperes when kilovolt-amperes and voltage are known	$\frac{\text{Kilovolt-amperes} \times 1000}{\text{Volts}}$	$\frac{\text{Kilovolt-amperes} \times 1000}{\text{Volts}}$	$\frac{\text{Kilovolt-amperes} \times 1000}{1.73 \times \text{volts}}$
Kilovolt-amperes when amperes and voltage are known	$\frac{\text{Amperes} \times \text{volts}}{1000}$	$\frac{\text{Amperes} \times \text{volts}}{1000}$	$\frac{1.73 \times \text{volts} \times \text{amperes}}{1000}$
Amperes when kilowatts, volts, and power factors are known	$\frac{\text{Kilowatts} \times 1000}{\text{Volts}}$	$\frac{\text{Kilowatts} \times 1000}{\text{Volts} \times \text{power factor}}$	$\frac{\text{Kilowatts} \times 1000}{1.73 \times \text{volts} \times \text{power factor}}$
Kilowatts when amperes, volts, and power factor are known	$\frac{\text{Amperes} \times \text{volts}}{1000}$	$\frac{\text{Amperes} \times \text{volts} \times \text{power factor}}{1000}$	$\frac{1.73 \times \text{amperes} \times \text{volts} \times \text{power factor}}{1000}$

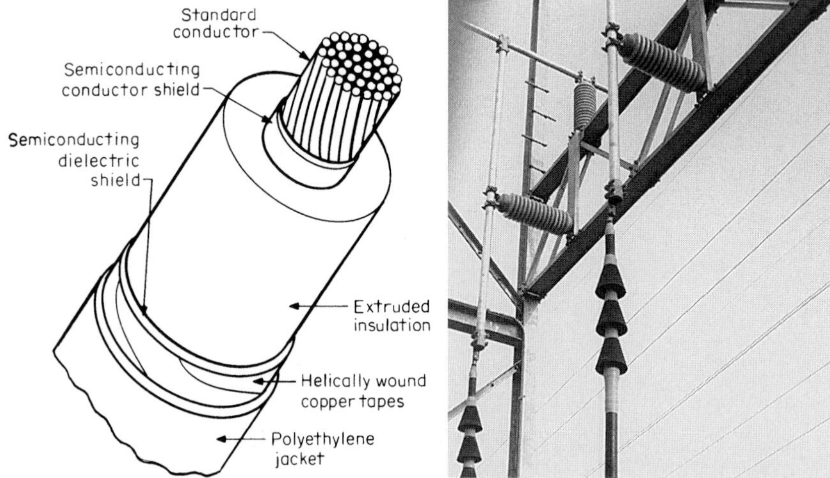

FIGURE 4.10 Construction of solid-dielectric extruded-insulation, high-voltage transmission cable.

FIGURE 4.11 Solid-dielectric subtransmission circuit cables operating at 69 kV ac terminated in a substation. The cable terminations consist of tape-formed stress cones with rubber rain shields above the stress cones.

interface. The insulation is extruded over the conductor shield. A thin semiconducting polyethylene shield is extruded over the insulation. The semiconducting shields over the conductor and the insulation prevent electric discharges, or corona, that would damage the insulation. A metallic layer on top of the dielectric shield provides a good conductor for the charging current, preventing damage to the semiconducting shield from the heat generated by the charging current. The metallic shield may be copper tapes, concentric wires, tubular corrugated copper, copper wires embedded in the semiconducting insulation shield, or a lead sheath. The metallic shield is usually protected by a polyethylene jacket. Extruded-rubber-insulated solid-dielectric cables operating at 69 kV ac are pictured in Fig. 4.11. Figure 4.12 is a cross-sectional view of the same cables.

Submarine cables constructed with solid-dielectric insulations have been successful. Figure 4.13 illustrates at three-conductor solid-dielectric cable. Figures 4.14 and 4.15 show submarine cables being installed by cablemen and linemen.

Extruded solid-dielectric cables, with a 2000-kCMIL copper conductor installed direct-buried with the three cables in a flat configuration with one three-phase circuit per trench would have the following capacity:

Voltage, kV	Apparent power, MVA
138	289
230	470
345	700

High-pressure oil-filled (HPOF) paper-insulated cables in which the three-phase conductors are enclosed in a steel pipe containing oil are also used for underground-cable transmission circuits.

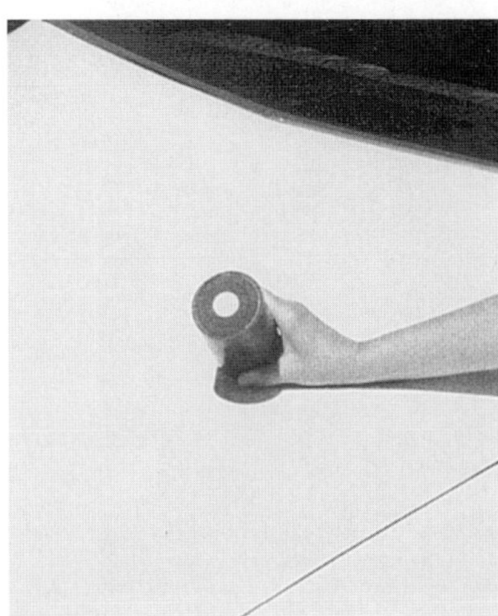

FIGURE 4.12 Cross-sectional view of 750-kCMIL, 69-kV, solid-dielectric, extruded-insulation, copper-tape-shielded, polyethylene-jacketed cable.

FIGURE 4.13 Three-conductor solid- dielectric insulated cable for submarine installation. The three insulated conductors are covered with jute and protected by polyethylene-covered steel wires. (*Courtesy Okonite Co.*)

FIGURE 4.14 Reels of submarine transmission cables mounted on a barge to permit installation under water. (*Courtesy Okonite Co.*)

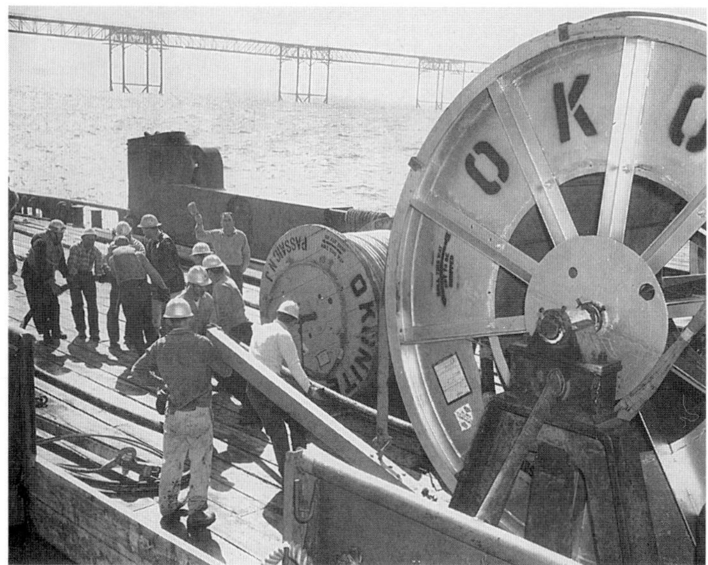

FIGURE 4.15 Submarine cable in picture is being unreeled and placed in the water. (*Courtesy Okonite Co.*)

A cross-sectional view of a typical pipe-type cable installation is detailed in Fig. 4.4. The cables are terminated in potheads. Each phase conductor is installed in a nonmetalic pipe (see Fig. 4.16) to a junction point called a trifurcator where the three-phase cables enter a common pipe. Shipping lengths of cable will normally be approximately 2500 ft for

FIGURE 4.16 Cablemen installing kraft-paper-insulated copper-tape-shielded cable in pipes at a terminal in a substation. The trifurcator, or junction device, is located below the terminal structure back of the reels. (*Courtesy Anaconda Wire and Cable Co.*)

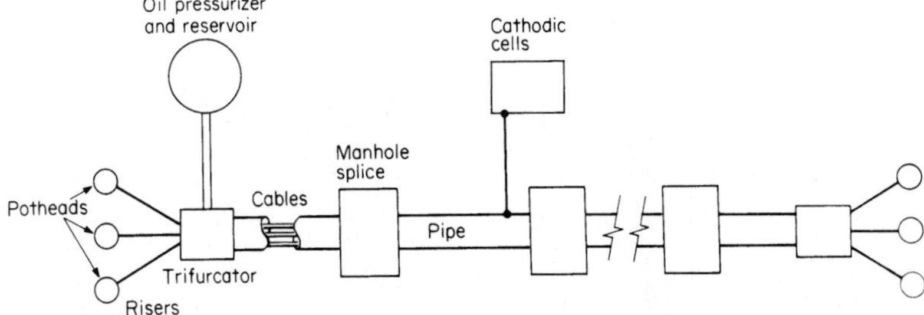

FIGURE 4.17 Schematic drawing of a pipe-type cable transmission circuit.

138 and 230 kV and 1750 ft for 345 and 500 kV. The cables are spliced in manholes. The steel pipe is usually direct-buried between the manholes. Oil pressure is maintained in the steel pipe with the use of pumps and reservoirs. The steel pipe is protected from corrosion by cathodic protection equipment. Alarms are installed on the system to detect low oil pressure. A schematic diagram of a pipe-type cable system appears in Fig. 4.17. The transmission circuit may consist of two cables installed in parallel. The two pipes can be placed in the bottom of the same trench separated as far as possible to minimize the ambient temperature and the installation cost. The dual facilities increases the reliability.

The pressurized oil can be pumped through the pipe for one cable circuit and returned through the parallel cable-circuit pipe. The capacity of oil-filled pipe-cable systems can be increased significantly by circulating the oil past the cable at several feet per second and cooling it externally by means of an oil-to-air heat exchanger (see Table 4.3).

TABLE 4.3 Capacities of Three-Phase (3 ϕ) Oil-Filled Paper-Insulated Pipe-Type Cables

	2000-kCMIL conductor				2500-kCMIL conductor
Voltage, kV	138	230	345	500	765
Pipesize, nominal, in.	8	10	10	12	14
Insulation thickness, in.	0.505	0.835	1.025	1.25	1.56
Naturally cooled cables					
Capacity, MVA	224	344	453	571	645
Dielectric losses, watts/cond. ft	0.733	1.39	2.71	3.99	8.70
Resistive losses, watts/cond. ft	7.72	7.02	5.49	4.40	2.30
Pipe temperature, °C	62.0	59.2	58.4	57.1	64.7
Total 3ϕ reactive losses per cable mile, MVAR	4.7	8.9	17.3	31.8	69.4
Forced-cooled cables					
Capacity,MVA	426	688	941	1352	2137
Dielectric losses, watts/cond. ft	0.733	1.39	2.71	3.99	8.7
Resistive losses, watts/cond. ft	28.7	28.4	24.2	25.2	24.9
Pipe temperature, °C	50.2	37.8	38.8	32.4	29.1
Total 3ϕ reactive losses per cable mile, MVAR	4.7	8.9	17.3	31.8	69.4

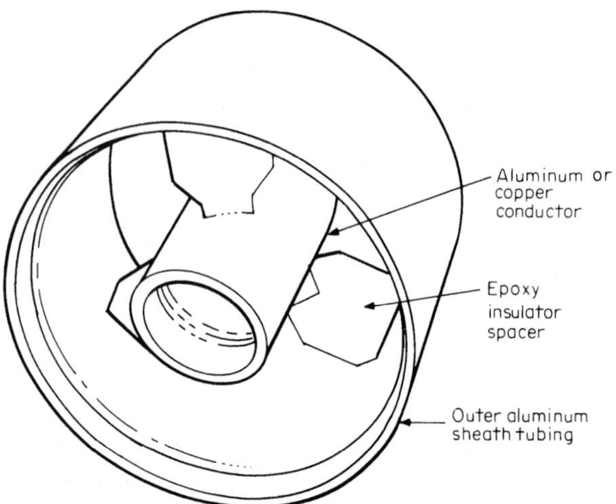

FIGURE 4.18 Cross-sectional view of gas-spacer cable construction.

Compressed-gas-insulated or gas-spacer cables using sulfur hexaflouride (SF_6) gas as the insulating medium are used for high-capacity insulated transmission circuits. The gas-spacer cable transmission system is an ac three-conductor isolated-phase design. The conductor can be copper or aluminum tube with a diameter of approximately 3 to 10 in, depending on the current-carrying capacity required and the operating voltage. The conductor is surrounded by an outer sheath of extruded-aluminum tubing, the diameter determined by the space required to provide insulation for the operating voltage. These diameters vary from 8.5 in at 138 kV up to 20 in at 500 kV and are approximately 30 in at 1000 kV.

The conductor is centered and maintained concentric with the outer enclosure by insulators installed in the enclosure at intervals along the cable (Fig. 4.18). The optimum insulating SF_6 gas pressure is approximately 50 psig. The cable assemblies are manufactured in approximately 40-ft lengths. The conductors are spliced together by plug-in type terminals or welding. The outer enclosures are welded together to form a continuous leak-proof pipe. The cables can be installed on top of the ground, overhead on supporting structures, or direct-buried. Direct-buried circuits would require a trench 4 to 7 ft wide, depending on the operating voltage. Thermal sand is used for a back-fill material. Figure 4.19 shows a surface installation protected by concrete walls. The cables are terminated in potheads consisting of an extension of the outer sheath, the inner conductor, and a porcelain insulator. The capacities and charging currents of typical gas-spacer cable installations are itemized in Table 4.4.

Self-contained cables with a central duct (Fig. 4.20) using pressurized oil for forced cooling have been used in some locations for underground transmission circuits. The cables can be direct-buried or installed in ducts. The capacity of the circuit is limited severely when these cables are installed in ducts as a result of the inability to dissipate the heat generated in the cable.

Direct-current underground transmission-cable circuits have advantages over ac cables since they do not have induction losses or a continuous flow of charging current. Solid-dielectric extruded-polyethylene-insulated cables can be used on voltages up to 300 kV. Pipe-type high-pressure oil-filled cables with paper-tape insulation can be used for circuits with a voltage of ±600 kV. Self-contained oil-filled cables will operate satisfactorily for voltages of ±300 kV.

FIGURE 4.19 Gas-insulated (SF_6) cables rated 3800 amps at 230 kV. Two transmission-cable circuits are located adjacent to each other for economical installation.

Direct-current cable insulation must be thick enough to provide a safety factor of 3 to 5 with respect to dc insulation breakdown because of surges. The surges may be caused by malfunction of conversion equipment, operation of switches, accidental polarity reversal, or lighting strokes on overhead lines connecting to the underground cables.

Table 4.5 itemizes the capacity of high-pressure oil-filled pipe-type 2000-kCMIL conductor cable operating at various dc voltages. The high cost of converter stations limits the use of underground dc transmission circuits.

Transmission-Line Environmental Considerations. Underground transmission circuits would eliminate most of the environmental problems associated with electric power transmission. The large investment associated with underground cable installations and technical limitations to the length of ac underground cable circuits prevent using underground installations to solve all the environmental problems.

Rights-of-Way. Transmission-line environmental considerations include visual impact, natural land conditions, electric field effects, land use, corona generation and associated radio and television interference or audible noise, and the limiting of the voltages induced in objects located in the transmission-line right-of-way. Rights-of-way for transmission lines should be selected with the purpose of minimizing the visual impact to the

TABLE 4.4 Single-Circuit Three-Phase ac Gas-Spacer Cable Characteristics

Operating voltage, kV	Capacity, MVA	Charging current amps/mi	Uncompensated maximum length, mi
138	400	4.0	440
230	700	5.3	350
345	1200	7.4	295
500	2200	12.6	250
765	4200	20.0	
1000	6500	25.0	

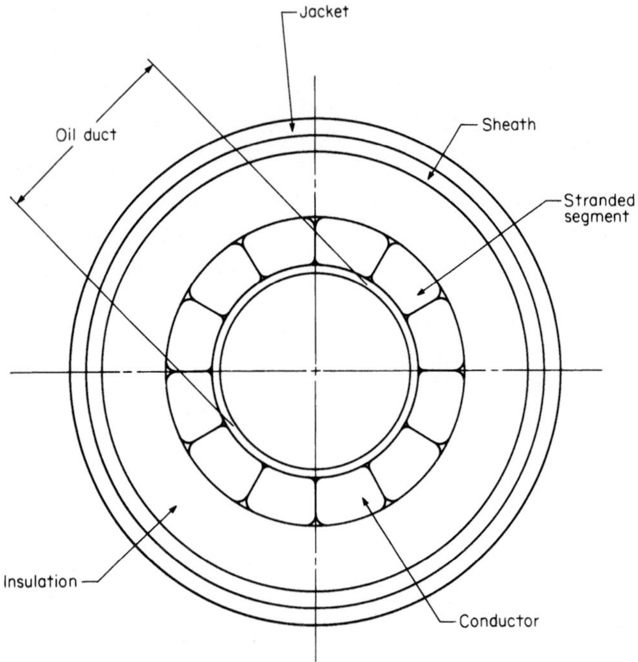

FIGURE 4.20 Cross-sectional view of a self-contained cable used for underground transmission circuits.

public. Existing rights-of-way should be given priority as the locations for additions to existing transmission facilities.

The first step to be taken prior to the design or construction of any line is to conduct a survey and make a map of the territory where the line is to be constructed. Aerial photography is usually used to survey the land in rural areas for transmission lines. The final location of the line is, of course, not known, and therefore, a wide strip of land must be included on the map. The main points between which the line is to be built will be known. The intervening territory should be laid out on a scale large enough to show clearly all division lines, towns, roads, streams, hills, ridges, railroads, bridges, buildings, and existing communications and power lines.

TABLE 4.5 Data for 2000-kCMIL Conductor High-Pressure Oil-Filled Pipe-Type Underground Cable Installation Operating at Various dc Voltages

	Voltage dc, kV			
	±180	±300	±450	±655
Single-conductor loss, kW/mi	42.7	40.7	39.5	33.8
Maximum dc current, amps	1120	1095	1030	1000
Average electric stress, kV/in	356	360	440	525
Capacity, MVA	403	657	927	1307

With the map completed, the following principles should be used as guides in selecting the exact route:

1. *Select the Shortest Route Practicable.* The shortest line naturally is the most economical, other things being equal. This means that the line should be a straight line between the terminals of the line, for a straight line is the shortest distance between any two points.
2. *Parallel Streets and Highways as Much as Possible.* This makes the line readily accessible both for construction and for inspection and maintenance. The hauling and delivery of materials will be greatly facilitated if this can be done.
3. *Follow Section Lines.* Transmission lines located in rural areas should follow section lines. Doing this causes less damage to farmers' property and, therefore, makes it possible to purchase the right-of-way more economically. Paralleling railroads is desirable for the same reason, because the farms have already been cut, and therefore, the additional damage is negligible.
4. *Route in Direction of Possible Future Loads.* If there is a possibility of adding power loads, the route selected should be such as to come as near as possible to such locations, provided that the additional cost is not excessive.
5. *Avoid Crossing Hills, Ridges, Swamps, and Bottom Lands.* Hills and ridges subject the line to lightning and storms. Swamps and bottom lands subject the line to floods. Furthermore, delivery of material as well as the construction of the line becomes difficult. Extra guying and cribbing are often necessary in swamps.
6. *Coordinate with Communication Lines.* High-voltage transmission lines can cause disturbances or interferences in communications lines by induction and therefore may require transposition of the power conductors. This adds to the cost of the line. In some cases it is advisable to relocate or place the communications circuits underground. Distribution lines are often constructed on the same poles with communications circuits and/or installed underground in the same trench or conduit system.
7. *Preserve the Environment.* Line routes should be selected which are the least visible from scenic locations or other areas where people pass or congregate. A simple test of visibility can be carried out by touring roads and points close to a proposed right-of-way and noting the visual impact of the line prior to final selection.

Width of Right-of-Way. When the route of the line has been selected, permission must be obtained to construct the line or land must be purchased. This is known as the right-of-way. In this connection, it should be pointed out that it is well to select alternate tentative routes so that if difficulties arise in obtaining permits on any one route or the prices demanded become inflated, another route can be chosen. High-voltage lines usually require a continuous strip of land for the right-of-way. Table 4.6 gives a general idea of the widths of rights-of-way necessary for different lines. The width of the right-of-way will vary with the type of construction, the conductor horizontal spacing, the voltage of the line, the terrain, and the area needed for construction and maintenance of the line.

The reason for acquiring a right-of-way much wider than the actual space occupied is to prevent tall trees from being blown into the line and to prevent damage from forest fires. Sometimes it is possible to purchase a narrower right-of-way and obtain permission to clear back the remaining distance beyond the edge of the right-of-way.

Electric Field Effects. Electric field effects of transmission lines are not hazardous. Lineman regularly work on energized extra-high-voltage (EHV) and lower-voltage transmission lines using hot-line tools and bare-hand methods with no ill effects. A study of 10 linemen who regularly worked on energized 345-kV transmission lines by personnel of

TABLE 4.6 Approximate Range of Widths of Rights-of-Way Required for Various Types of Overhead Lines

	Feet
Pole lines	
Single circuit	15–50
Double circuit	30–75
H-frame	60–250
Tower lines	
Single circuit	50–250
Double circuit	75–250

John Hopkins University in Baltimore, Maryland, over a period of 10 years determined that the linemen were in excellent health.

Magnetic Effects. A voltage can be induced on a person or conductor under an overhead energized electric transmission-line conductor. The voltage is induced by the magnetic field developed by the current flow in the line conductors. The induced voltage will be insignificant at all times under dc lines and under normal conditions under ac lines. High fault current on an ac transmission line may increase the induced voltage on a conducting object under the line. The fault current on dc lines is limited by a converter station, which limits induced voltages.

Electric Fields. The magnitude of electric fields at ground level under all over-head high-voltage transmission lines is very low. Static charges on farm fences directly under the lines are usually drained to ground through the metal or wood posts or contacts with grass, weeds, or the ground. Vehicles, such as a tractor or combine, directly under an EHV line may become charged, and a farmer bridging the rubber tires of the vehicle might feel a tingle if the vehicle discharged through his body to ground. The effect could be annoying, but it is not electrically dangerous, as lineman can verify. The annoying shock may startle a person, causing a fall or other act that could result in a personal injury. Vehicles operated under EHV transmission lines or facilities such as fences installed under EHV lines should be grounded to prevent annoying shocks and acts that might lead to injuries. All metal buildings or structures should be grounded whether they are built near a transmission-line right-of-way or not. Grounding the building will eliminate any static charge.

Linemen use special conductive clothing or electrostatic shields while working on energized ultra-high-voltage (UHV) transmission lines. It is necessary that special precautions be taken when UHV transmission lines are designed and constructed to protect the public from hazardous voltages. The major characteristics affecting a transmission line's electrostatic conditions are voltage, conductor type, phase configuration, phase spacing, and conductor height.

Corona. Ionization of the air around energized high-voltage transmission-line conductors, when the voltage gradient exceeds the air withstand limits, establishes corona conditions. Corona is visible at night as a luminous glow close to the conductors. Corona and resulting radio-interference voltages and audible noise can be controlled by proper transmission-line design and construction precautions. The use of bundled conductors and appropriate hardware on ac EHV and UHV lines keeps corona to acceptable limits if the equipment is installed correctly. Bundling of conductors and selection of conductor hardware will have minimal effect on dc lines because of the space charge surrounding the conductors. The line conductors must be kept clean and free from damage that results in rough or pointed projections. Grease on a conductor can cause the generation of corona. Corona on dc lines is generally caused by the same factors discussed for ac lines. However, corona is generally less on dc lines and does not increase as much during bad weather compared with conditions on ac lines.

Radio interference on high-voltage dc transmission lines is caused by the pulses created by the ignition of the thyristor valves in the converter stations, corona on the line, and partial electric discharge over the insulators. The pulses generated in a converter station can be controlled by shielding and proper design and construction of the thyristor valve housing. Proper design, construction, and maintenance will keep radio interference caused by corona and partial electric discharge over the insulators of both ac and dc high-voltage lines within allowable limits.

Television transmissions are above the 50-MHz frequency. Interference from electric transmission lines above the 50-MHz frequency is very small and does not normally cause problems.

Induced voltages on objects in the right-of-way of ac UHV transmission lines can be controlled by the height of the towers or by shielding. The right-of-way for the transmission line can be purchased and fenced to prevent access to the land. Induced voltages are not a problem on high-voltage dc lines. It is important that transmission lines be designed to minimize the cost and protect the safety of the public.

Individuals opposed to the construction of electric transmission lines have intervened in hearings on the citing of high-voltage lines and have received considerable publicity. The interveners have demonstrated that the weak electric fields under extra-high-voltage ac electric transmission lines will light fluorescent tubes. Electric fields generated by an automobile's ignition system or by a television receiver can make a fluorescent tube glow. A silk cloth rubbed along the surface of a fluorescent tube will cause the lamp to glow. Electric appliances found in the home produce electric fields, allowable by the American National Standards Institute publications, that are greater than those found under high-voltage lines. Research completed by the Electric Power Research Institute, the U.S. Department of Energy, the Minnesota Department of Health, Iowa State University, and other institutions has not identified biologic effects from electric fields that are detrimental to the public.

CHAPTER 5
DISTRIBUTION CIRCUITS

Distribution is that system component which delivers the energy from the generators, or the transmission system, to the customers. It includes the substations that reduce the high voltage of the transmission system to a level suitable for distribution and the circuits that radiate from the substation to the customers (Fig. 5.1).

The distribution circuits that carry power from the substations to the local load areas are known as express feeders, distribution main feeders, or primary circuits and generally operate at voltages between 2400 and 34,500 volts. The loading of distribution circuits varies with customer load density, type of load supplied, and conditions peculiar to the area served. The nominal capacity of distribution circuits operating at different voltage levels is itemized in Table 5.1.

The distribution circuits may be overhead or underground depending on the load density and the physical conditions of the particular area to be served. The vast majority are constructed overhead on wood poles (Figs. 5.2 and 5.3) that may support communication facilities and street lights. Street lighting is an important part of the distribution system.

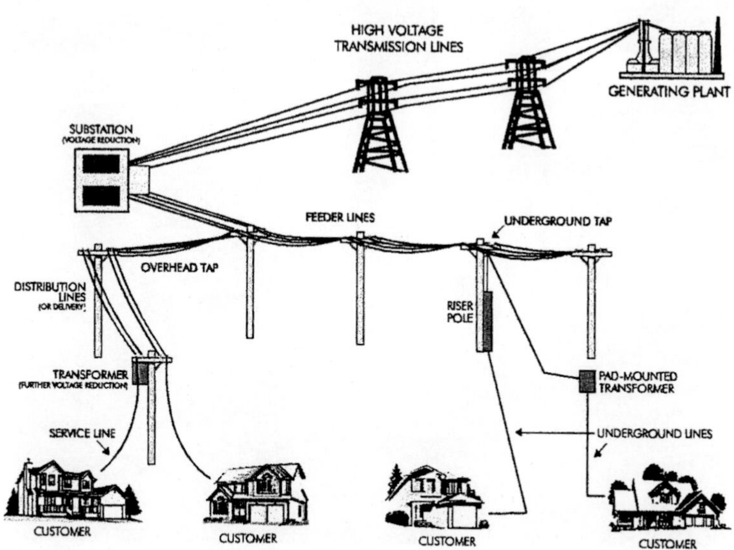

FIGURE 5.1 Schematic diagram of a distribution system. The distribution circuits may be overhead or underground. The transformers may be pole-type or pad-mounted.

TABLE 5.1 Distribution Feeder Capacities of Distribution Circuits

Voltage class, kV	Capacity, MVA	
	Average	Maximum
4–5	2.5	3.5
12–15	7.5	15.0
20–25	12.0	20.0
30–35	18.0	30.0

FIGURE 5.2 Detroit Edison distribution construction crews in the process of upgrading a new distribution line. (*Courtesy Altec Manufacturing Co.*)

FIGURE 5.3 Linemen in the process of transferring transmission and distribution facilities on a pole replacement project. The 13.8 kV line is under-built of the 161 kV subtransmission line. (*Courtesy Altec Manufacturing Co.*)

Distribution transformers are installed in the vicinity of each customer to reduce the voltage of the primary circuit to 120/240 volts or other utilization voltage required by the customer. Secondary circuits and services carry the power at utilization voltage from the distribution transformers to the customers.

Distribution circuits originate in a distribution substation and include the conductors and equipment necessary to distribute electricity to the customer. The distribution substation may be located at a generating station or at a point remote from the generation and connected to the generation by transmission and subtransmission circuits. Distribution substations are located near the customers utilizing the electricity to ensure good voltage and reliable service.

Distribution circuits can have a major impact on the customer's voltage and service continuity. The voltage must be maintained within a range that permits the customer's equipment to perform satisfactorily and within limits specified by regulatory agencies. If the customer's base voltage is 120 volts, then the minimum normal service voltage would be 114 volts and the maximum normal service voltage would be 126 volts. It is good practice to maintain the voltage as close to the base voltage as practical and to limit the range of the voltage variations from base. The maximum utilization voltage would normally be 126 volts, and the minimum utilization voltage would normally be 110 volts, with limited periods of operation as low as 106 volts and as high as 127 volts.

It is normal to operate the distribution system at higher voltages during peak-load periods to compensate for voltage drop in the distribution equipment and the customer's wiring system. It is important that distribution-circuit equipment be designed and installed in a manner to prevent objectionable light flicker as a result of momentary variations in voltage level.

The electric distribution system and distribution circuits are described in general in Chap. 2.

Three-Phase, Three-Wire Distribution Circuits. Alternating-current distribution-system feeders are commonly operated as three-phase, three-wire or three-phase, four-wire circuits. A three-phase circuit can be obtained from either a Y or a delta (Δ) connection of the three coils in the ac generator. Similarly, in distribution systems, the Y or delta connection is obtained by the proper connection of the secondary coils of the transformers in the substation.

Figure 5.4 shows a three-phase, three-wire distribution feeder together with the Y and the delta connections of the secondary windings of the distribution substation step-down transformers. The transformer secondary connections are shown in the conventional and practical way of representation. It should be pointed out here that the Y connection is generally used on distribution feeders when the feeder is operated as a four-wire circuit. The voltage between any two of the three wires for either method of connection is the same, that is, the voltages between conductors A and B, B and C, and C and A are equal. In the delta connection, this voltage between line wires is the same as the transformer-coil voltage, but in the Y connection, the voltage between line conductors is the square root of 3 times the transformer-coil voltage.

Three-Phase, Four-Wire Distribution Circuits. The three-phase, four-wire feeder circuit is in general use, especially in the larger cities. This circuit can be obtained from a Y connection of the distribution substation transformer secondary. Figure 5.5 shows this type

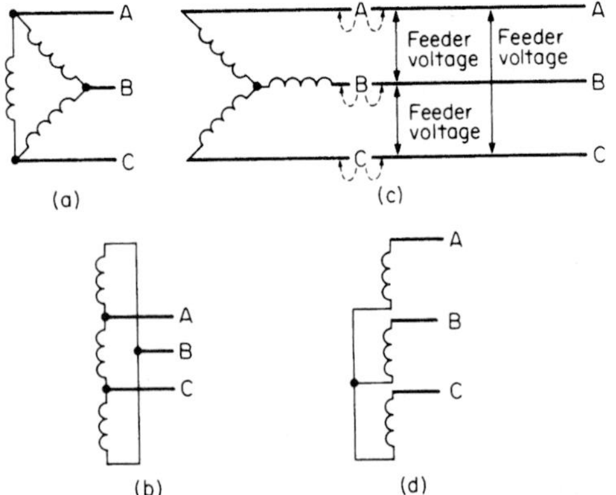

FIGURE 5.4 Three-phase, three-wire distribution circuit. (*a*) Conventional and (*b*) practical way of picturing delta connection of transformer secondaries. (*c*) Conventional and (*d*) practical way of picturing Y connection of transformer secondaries.

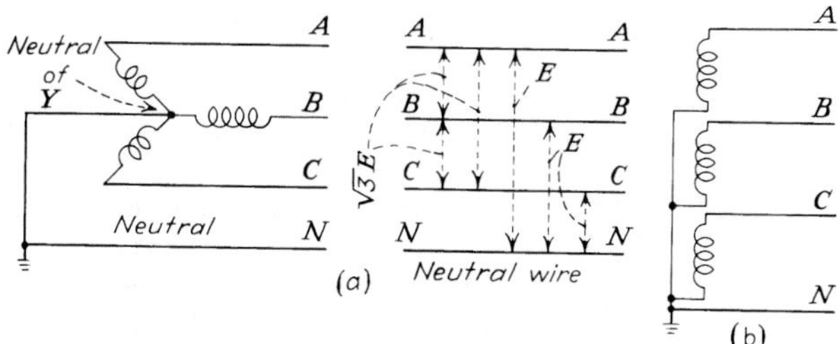

FIGURE 5.5 Three-phase, four-wire distribution circuit. (*a*) Conventional and (*b*) practical way of picturing Y connection of distribution substation transformer secondary with neutral brought out.

of circuit. The fourth wire is called the neutral wire because it is connected to the neutral point of the Y connection.

It will be noted that the voltage between any two of the line wires is no longer equal to *E*, the transformer secondary-coil voltage, but instead is the square root of 3 × *E*. The voltage *E* still exists between any line wire and neutral, but the voltage between any two line wires is the square root of 3 × *E* volts. If the voltage *E* between any wire and neutral is 7620 volts, then the voltage between any two-phase wires is equal to the square root of 3 × *E*, or 1.732 × *E*, which, in this example, would be 1.732 × 7620, or 13,200 volts. Single-phase voltages equal to 7620 volts would be available by using one-phase wire and the neutral conductor which is grounded.

The advantages of the four-wire feeder over the three-wire feeder are increased power-carrying capacity and better voltage regulation. Since the line voltage in the four-wire system is the square root of 3 or 1.73 times that of the three-wire system, its power-carrying capacity for the same current in the wires will also be 1.73 times as great, or for the same power transmitted, the current in the four-wire system will only be 58 percent as large. Moreover, the voltage regulation is also greatly improved. Many large cities have changed over to this system of feeders in order to get the benefit of this increased feeder capacity. Merely stringing the fourth wire and reconnecting the transformer secondaries from delta to Y make possible an increase in load-carrying capacity of 73 percent.

In earlier years, 2400 volts delta, 4160Y/2400, and 4800 volts delta primary distribution-circuit voltage levels were in wide use, amounting to approximately 60 percent of the systems in 1955. In 1969, usage of distribution voltages below 5000 volts was down to about 12 percent. The next higher levels, 12,470Y/7200, 13,200Y/7620, and 13,800Y/7960 volts, are now the most prevalent. Higher primary distribution voltages are in use, including 24,500Y/14,400 and 34,500Y/19,920 volts.

From an application standpoint, there are a number of factors to be considered. Factors favorable to the higher levels include the steady increase in consumer kilowatthours, the permissible use of smaller conductors, better voltage regulation, and the availability of existing industrial feeders or subtransmission circuits at levels that are suitable, or could be made suitable, for the start of the higher-voltage distribution circuits. Some negative factors are the existence of a widespread use of the present levels and the possibility of an alternative solution of using more substations to supply present voltage levels as loads increase. At the increased voltage levels, more customers would be interrupted in case of system faults.

Single-Phase Primary Circuits. A single-phase circuit is generally obtained from a three-phase circuit. It could be obtained from a single-phase ac generator, but such machines are very rare today. Figures 5.6 and 5.7 illustrate how single-phase primaries are obtained from three-phase, three-wire and three-phase, four-wire feeders. These single-phase

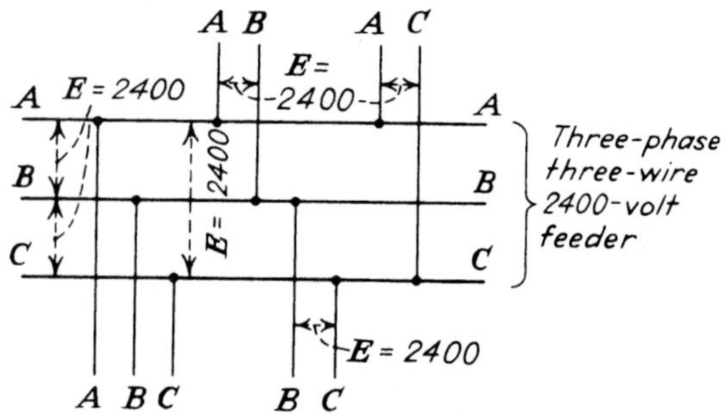

FIGURE 5.6 ABC is a 2400-volt three-phase primary feeder, and AB, BC, and AC are single-phase 2400-volt primaries obtained by tapping the 2400-volt three-phase, three-wire feeder.

primaries are tapped to the feeder circuit. By tapping two line wires in the three-wire feeder, the voltage of the primary will naturally be the same as the voltage across the line wires, 2400 volts in this example. By tapping from line wire to neutral in the four-wire feeder, the voltage of the primary will also be 2400 volts in this example.

An attempt is always made to keep the loads on the single-phase primaries connected to the three phases of the three-phase line balanced so that the currents in the wires of the three-phase feeder also will be balanced.

Single-Phase Secondary Circuits. Secondary circuits, for residential customers, are generally single-phase, three-wire circuits, although they may be only two-wire circuits. The secondaries originate from the secondary winding of the distribution transformer. Figure 5.8 shows the single-phase, two-wire primary connecting to the primary winding of a single-phase transformer. The secondary winding of a single-phase distribution transformer is nearly always made up in two coils which can be connected either in series or in parallel, as shown. They are most often connected in series. Each coil has induced in it 120 volts so that when they are connected in series, the voltage across the outside wires forms a 240-volt, single-phase, two-wire circuit. When the neutral is used, either outside wire and the neutral

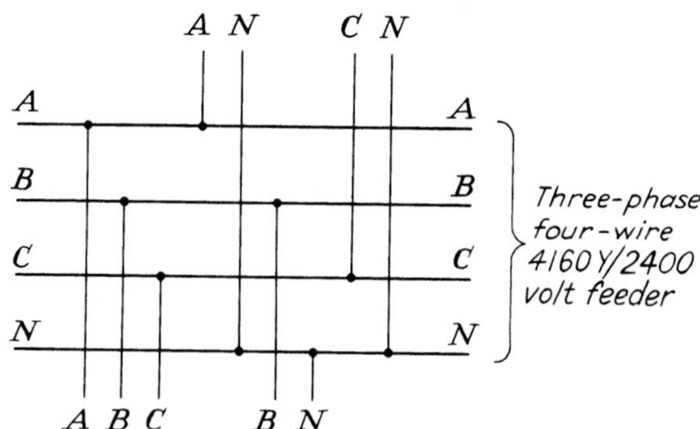

FIGURE 5.7 ABC is a 4160Y/2400-volt primary feeder obtained by tapping the three-phase wires, and AN, BN, and CN are single-phase 2400-volt primaries obtained by tapping a phase wire and neutral of the three-phase, four-wire 4160Y/2400-volt feeder.

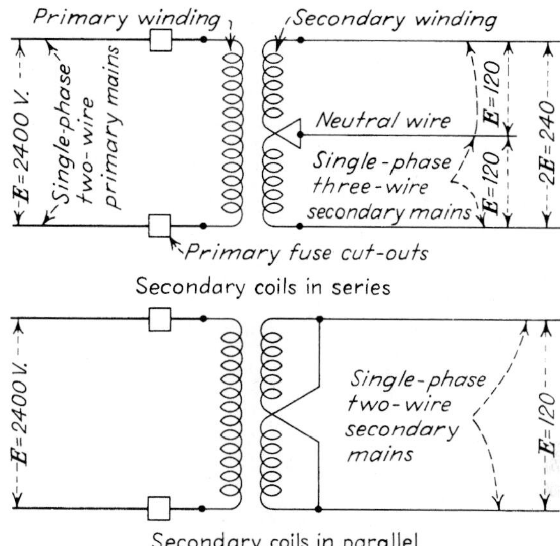

FIGURE 5.8 Standard connections of common step-down single-phase distribution transformer.

form a 120-volt single-phase circuit. A consumer using lamps and small household appliances may have a two-wire service connected to the premises as the load would be small and would be at 120 volts. If the consumer uses an electric range for cooking, the service connection would consist of all three wires, because most electric ranges operate on 120 and 240 volts. Any consumer having a single-phase motor larger than 1/2 hp would likely operate it from the two outside wires of the secondary. Using higher voltage draws less current and keeps the line loss and the voltage drop at a minimum. Any office building or store using single-phase service would also take the three-wire service.

Three-Phase Secondary Circuits. Commercial and light industrial customers usually require three-phase service to operate three-phase motors. Three-phase motors are more efficient and less expensive than single-phase motors. Large single-phase motors cause a sudden drop in the voltage when they start which appears as a voltage flicker that affects incandescent lights and distorts television pictures. Three-phase motors draw less current than single-phase motors when they start, thus minimizing voltage flicker. Installation of three-phase service helps to keep the current in each phase wire of the distribution circuit equal. The voltages normally used for three-phase secondary circuits are 208Y/120 volts four-wire, 240 volts three-wire, 480Y/277 volts four-wire, and 480 volts three-wire.

Figure 5.9 illustrates the proper connections for single-phase pole-type transformers to obtain 208Y/120-volt service for a three-phase customer from a four-wire primary-feeder circuit. The primary and the secondary transformer neutral connections should be connected to the distribution feeder circuit neutral and to ground. Transformer connections for a 480Y/277-volt three-phase, four-wire service would be similar to those illustrated in Fig. 5.9.

Three-phase, four-wire secondary voltages of 208Y/120 or 480Y/277 volts can be obtained by proper selection of pole-type transformers installed as shown in Fig. 5.10. Armless-type construction is used for the three-phase, four-wire primary feeder circuit to enhance the appearance of the facilities. Using single-primary-bushing transformers with one side of the transformer winding connected to the transformer tank simplifies the installation

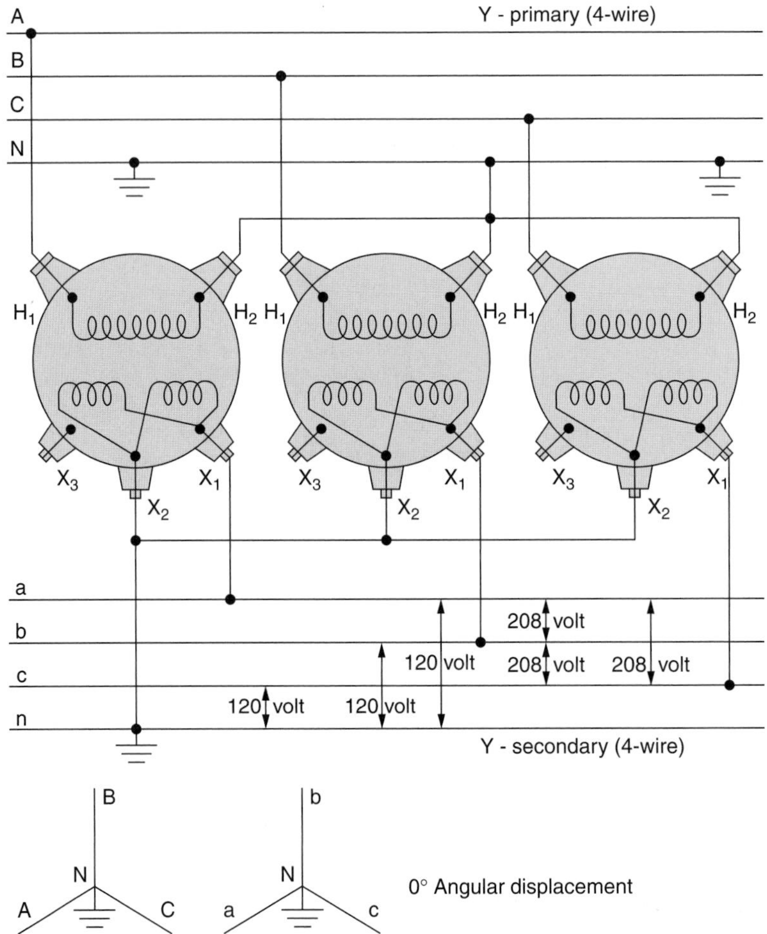

FIGURE 5.9 Three-phase Y-Y pole-type transformer connections used to obtain 208Y/120 volts three-phase from a three-phase, four-wire primary. If single-primary-bushing transformers are used, the H_2 terminal of the transformer primary winding would be connected to the tank of the transformer.

and minimizes the cost. The tank of each transformer is connected to the primary neutral wire and ground.

Figure 5.11 illustrates the proper connections for single-phase pole-type transformers to obtain 240/120-volt single-phase service from a 240-volt service for a three-phase customer from a three-wire primary-feeder circuit. The transformer providing the source for the single-phase load carries two-thirds of the 240/120-volt single-phase load and one-third of the 240-volt three-phase load. The other two transformers carry one-third of the single-phase and two-thirds of three-phase loads. The transformers must have matched voltage ratios and impedances. Only one of the transformer's secondary bushings can be grounded. Grounding more than one transformer secondary bushing will cause a short circuit.

Figure 5.12 pictures a three-phase delta-delta transformer installation providing three-phase, three-wire service. It is good practice to ground one of the transformer secondary

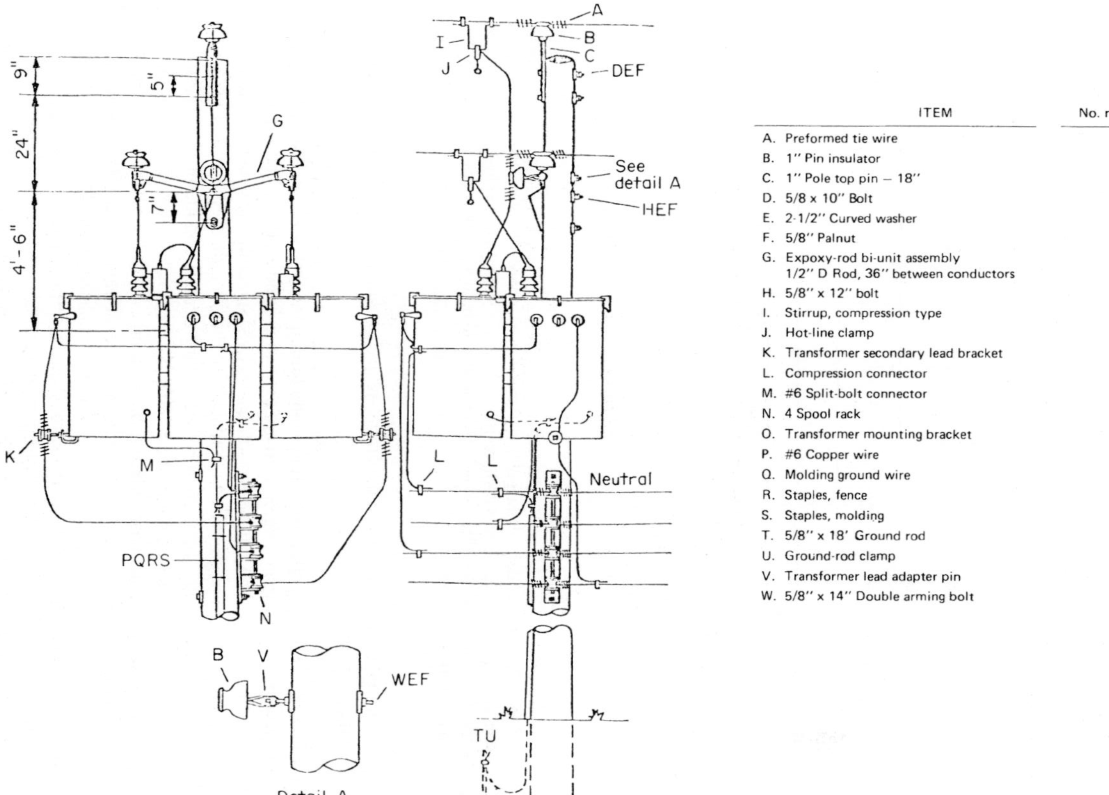

ITEM	No. required
A. Preformed tie wire	7
B. 1'' Pin insulator	4
C. 1'' Pole top pin – 18''	1
D. 5/8 x 10'' Bolt	2
E. 2-1/2'' Curved washer	8
F. 5/8'' Palnut	8
G. Expoxy-rod bi-unit assembly 1/2'' D Rod, 36'' between conductors	1
H. 5/8'' x 12'' bolt	4
I. Stirrup, compression type	3
J. Hot-line clamp	3
K. Transformer secondary lead bracket	2
L. Compression connector	5
M. #6 Split-bolt connector	3
N. 4 Spool rack	1
O. Transformer mounting bracket	1
P. #6 Copper wire	5#
Q. Molding ground wire	30'
R. Staples, fence	10
S. Staples, molding	12
T. 5/8'' x 18' Ground rod	1
U. Ground-rod clamp	1
V. Transformer lead adapter pin	1
W. 5/8'' x 14'' Double arming bolt	1

FIGURE 5.10 Three-phase Y-Y pole-type single-primary-bushing transformer installation details showing connections to provide three-phase, four-wire low-voltage service from a three-phase, four-wire primary. Dimensions in illustration are shown for distribution system operating at a primary voltage of 13,200Y/7620 volts.

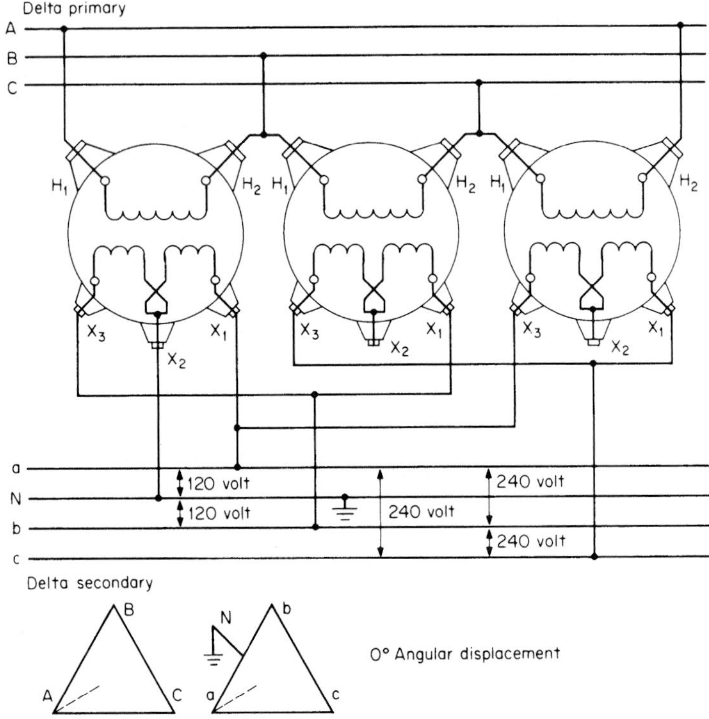

FIGURE 5.11 Three-phase delta-delta pole-type transformer connection used to obtain 240/120-volt single-phase service from a 240-volt three-phase service and a three-phase, three-wire primary. The transformer with the secondary connection to bushing X_2 provides a source for single-phase 120/240-volt service.

bushings, as illustrated in Fig. 5.11, even if a single-phase service is not desired, to limit the secondary voltage to ground.

Neutral Conductor. The secondary-neutral conductors in the three-phase, four-wire system and in the single-phase, three-wire system are always grounded, that is, connected to a low-resistance earth terminal. This can be accomplished by connecting the neutral conductor to a ground rod driven deep enough to be in contact with moist earth, as shown in Fig. 5.10 and discussed and illustrated in Chap. 26. The secondary-neutral conductor also must be grounded near the customer's service entrance. This is often accomplished by making a firm connection with water mains embedded in the ground. When this is not sufficient or available, grounds are driven on the customer's premise or metallic grids or plates are embedded in moist earth and an electrical connection is made with them. When only one ground connection is made to the primary-neutral, the neutral conductor is called unigrounded. General practice is to ground the primary-neutral conductor at numerous points along its length; it is then called multigrounded. The National Electrical Safety Code requires that multigrounded primary-neutral conductors be connected to ground at a minimum of four locations per mile and at each piece of equipment connected to the primary and the neutral conductors along the line. When both primary and secondary circuits are Y-connected four-wire circuits and both circuits run along the same route, a single neutral conductor can serve both circuits. Such an arrangement is called a common-neutral system.

FIGURE 5.12 Three single-phase transformers connected to a 13,800-volt three-phase, three-wire primary circuit to provide three-phase, three-wire 480-volt service to a small manufacturing plant. Installation includes fused cutouts and lightning arresters to protect the installation. Current transformers and potential transformers installed above the distribution transformers provide currents and voltages for primary metering.

FIGURE 5.13 Three-phase, four-wire primary-feeder circuit, armless-type construction to enhance the appearance of the distribution circuit. (*Courtesy A. B. Chance Co.*)

In the case of the single-phase secondary, either two- or three-wire, the neutral conductor is grounded at numerous locations, including the consumer's service entrance. The reason for grounding these circuits is primarily for safety.

Distribution Environmental Considerations. The compatibility of distribution systems with the environments in which they are located has become an increasingly important consideration. The location and appearance of both substations and overhead lines have become primary criteria for the planning and design of distribution systems. A strong trend to use underground rather than overhead construction for distribution lines has developed.

Techniques have been developed for improving the appearance of overhead lines by the selection of materials, structural shapes, and colors which are in harmony with the environment. The improved overhead lines should meet appearance objectives in many situations (Fig. 5.13).

The adverse visual impact of the existing overhead lines can be reduced to a significant extent by selective conversion of existing overhead distribution lines to underground programs of limited magnitude. The optimal environmental improvement per dollar expended for conversions can be obtained by selecting overhead lines that have a particularly adverse impact because of congestion of facilities, location, or exposure to public view.

Efforts to make distribution facilities more compatible with their environments by choice of location, consideration for appearance in the design of visible facilities, and the use of underground construction for new lines where economically feasible should be promoted and encouraged by all.

Harmonic Voltages. Voltages with frequencies which are a multiple of the base frequency are referred to as harmonic voltages. These harmonic voltages in distribution circuits can be originated by overvoltages which cause distribution transformers to be saturated, rotating

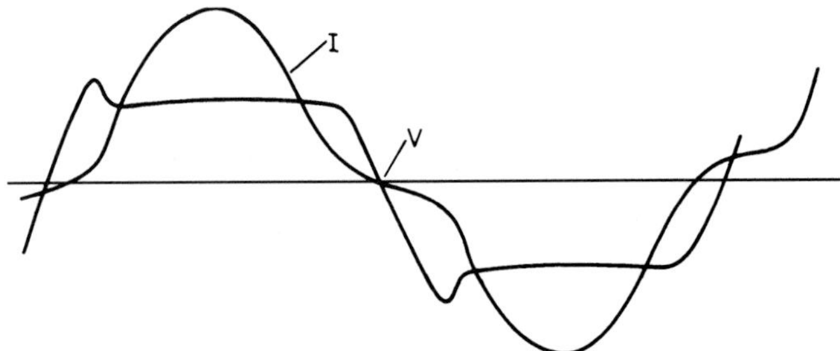

FIGURE 5.14 Plot of voltages and currents in distribution circuit distorted by harmonics generated by an arc furnace.

machinery that is not properly constructed or operated, arc furnaces, arc-welding equipment, and solid-state devices such as thyristor-controlled equipment, diode bridges, rectifiers, inverters, and static-arc sources. The installation of customer-owned generating equipment that is not properly manufactured, like wind generators operating in parallel with the distribution circuits, and the proliferation of thyristor-controlled devices have greatly increased the harmonic voltages on the distribution system and the harmonic currents that flow from the voltage sources (Fig. 5.14). Harmonics in the distribution system can damage equipment such as capacitor banks, motors, transformers, and loadbreak devices. Protective relays may malfunction, fuses may melt erroneously, and inductive interference may develop in communication circuits from the harmonic voltages and currents. Harmonics can increase power losses significantly in the distribution circuits, causing a major increase in unaccounted-for energy and operating costs.

Harmonics should be eliminated at the source by modification of the equipment or the installation of filtering equipment to block its path to the distribution circuit.

Electrical Interference. High-frequency voltages and currents can be generated by the customer's equipment connected to distribution circuits, by arcing between components of the circuit, and by corona. The electrical interference may interfere with operation of radios or television sets. Radio interference is more common than television interference because of the lower electrical frequencies.

The best method of eliminating electrical interference is to prevent it by constructing the distribution lines properly. Loose connections, defective insulators, inadequate clearance between high-voltage and ground conductors, and contamination on insulators may cause arcing. Corona occurs on high-voltage transmission circuits more frequently than on distribution circuits. Contaminated distribution-line insulators may permit arcing to be generated during fog or high-humidity conditions.

The customer's equipment, such as thyristor-controlled devices, defective thermostat contacts, and heating devices with a broken wire such as a heating pad or heat tapes, is a common source of electrical interference. Electrical interference from power-line sources will generate a buzzing sound in radio and television audio outputs. Horizontal bands of lines and spots that move up slowly on the television screen can be caused by electrical interference. Arcing-type electrical interference normally causes a raspy intermittent buzz in the audio output and two bands on the television screen that become wider and denser as the intensity of the interference increases.

Observation and information from the public can be helpful in locating the electrical interference. A radio receiver and/or a meter that measures field-strength intensity in the audio and very high frequency (VHF) range can be used to good advantage. Special receivers and instruments with a directional antenna are available for locating electrical interference sources.

Neutral to Earth Voltages. Neutral to earth voltage is the voltage measured from the electrical system neutral and/or any structure bonded to the neutral conductor and remote earth which can be established with a ground rod driven in the earth at a location isolated from the neutral system grounds. The surface of the customer's electrical utilization equipment and other metallic surfaces may have voltages impressed on them that can be felt and measured between the metallic surface and a remote ground source. These voltages are usually caused by defective insulation of the customer's appliances or wiring. Unbalanced loads connected to the 120-volt circuits in the customer's premises may create voltages on metallic surfaces, if the electrical equipment grounding wire is not isolated from the neutral conductor beyond the customer's distribution panelboard, as required by the *National Electrical Code*. When the neutral conductor is connected to the equipment grounding wire, the voltage drop on the neutral conductor will be impressed on the equipment grounding wire and all metallic surfaces connected to it. The primary-neutral conductor on a single-phase electric distribution line normally has a small voltage on it that can be measured with a voltmeter connected to the primary neutral conductor and a remote ground rod. This voltage results from a portion of the return current from the customer's loads flowing through the impedance of the neutral conductor. Adequate size neutral conductors and multiple ground rods connected to the neutral can minimize the level of voltage on the primary-neutral conductor. The voltage on the primary neutral will appear on the customer's neutral conductor, since they are normally connected at the distribution transformer location.

Stray Voltage. Stray voltage has been defined as low-level voltage present across points (for example, drinking cup to rear hooves), which will cause a current to flow through an animal when the animal simultaneously comes into contact with the points. Stray voltage can be measured by connecting the leads of a voltmeter from the drinking cup or waterline in the barn or milking pallor to a 4-in.-square plate on the floor of the barn. The surface of the floor of the barn must be moist, and pressure must be applied to the copper plate. A 500-ohm resistor is placed across the voltmeter leads to simulate the resistance of the cow.

Isolation. Isolation of the primary distribution system neutral conductor from the secondary neutral conductor will prevent the primary neutral voltage drop from contributing to stray voltage. However, this will not eliminate stray voltage created by the customer's wiring or equipment. Isolation is permitted by the *National Electrical Safety Code* as a temporary means of reducing stray voltage. Isolation is undesirable from a safety standpoint because the customer loses the benefits of the low primary ground resistance in parallel with the customer's grounding system. If the isolating device fails, it probably will not be detected for an extended period of time. Failure of the isolating device could permit a high voltage to appear on the customer's wiring, if the distribution transformer is damaged by

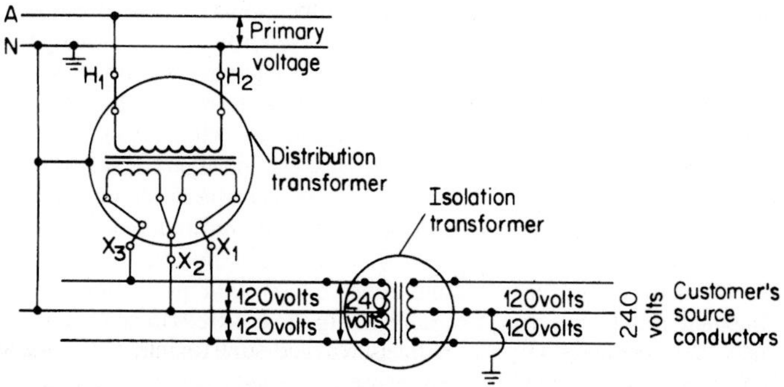

FIGURE 5.15 Schematic diagram for an isolation transformer installation.

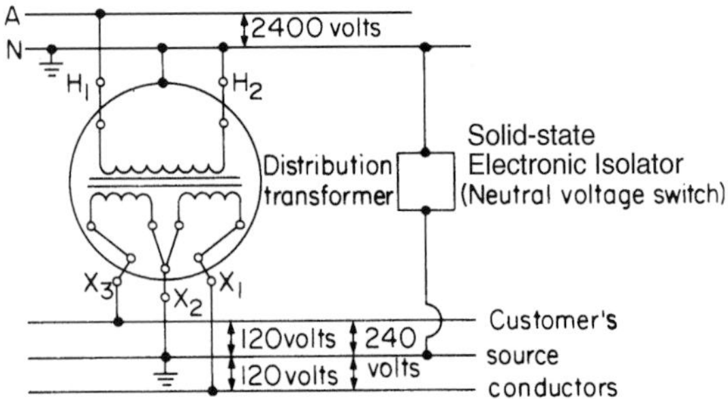

FIGURE 5.16 Schematic diagram for installation of a solid-state electronic isolation device.

lightning or fails as a result of some other problem such as an overload. Isolation increases the cost of the facilities providing service.

Isolation can be accomplished by installing an isolation transformer with a 1:1 ratio on the secondary of the distribution transformer near the customer's distribution panelboard (Fig. 5.15). The isolation transformer must have adequate capacity for the load on the customer's circuits. An isolation transformer will not eliminate stray voltages if the customer's wiring or equipment is defective.

Isolation can be accomplished with a device containing solid-state electronic components connected between the primary and secondary neutral conductors near the distribution transformer. The primary and secondary neutral conductors must each be connected by insulated wire to a ground rod. The ground rods must be separated by 12 ft or more. The solid-state electronic isolation device becomes conducting if a voltage develops between the primary and secondary neutral conductors that is greater than the isolator's designed isolation level, thus limiting the high voltage that could appear on the low-voltage secondary circuits (Fig. 5.16).

An electronic grounding system has been developed to eliminate neutral-to-earth potentials found to be present in milking parlors of dairy farms. The system can be used for other installations with similar characteristics. The electronic grounding system measures the potential between earth and neutral at the power utilization location. The measured voltage is amplified, and a current proportional to the measured voltage is supplied to remote earth, which eliminates the potential between earth and neutral. The electronic grounding system is described fully in a paper entitled "Computation, Measurement and Mitigation of Neutral-to-Earth Potentials on Electric-Distribution Systems," by William K. Dick and David F. Winter, in IEEE Transactions on Power Delivery, Vol. PWRD-2, No. 2, April 1987.

The best way to prevent problems as a result of the neutral-to-remote earth voltages in a milking parlor on a dairy farm is to construct the milking-parlor facilities properly with an equal-potential plane. The equal-potential plane can be established by reinforcing the concrete floor with a 6 × 6-in reinforcing mesh connected at 3-ft intervals with no. 4 AWG or larger copper bonding wire, which is connected to the ground rod at the milking-building electric-distribution panel. All metal-conductive equipment that the cows can touch in the milking building, such as metal posts, feeding troughs, curbs, grates, etc., must be connected together with the bonding conductor and connected to the steel-reinforcing wire mesh in the concrete floor.

It is difficult to retrofit the concrete floor for animal confinement areas to establish an equipotential plane. However, if the electrical wiring is installed as specified by the *National Electrical Code*, it is easy and inexpensive to provide protection for the animals in the confinement area from stray voltage. Painting the concrete floor of the animal confinement areas with a proper paint for application to concrete will provide a moisture seal and insulate the animals from remote earth. A solvent-based sealer formulated with acrylic and silicone resins is recommended. The concrete should receive two coats of the sealer with a skid-resistant additive placed in the second coat.

CHAPTER 6
CONSTRUCTION SPECIFICATIONS

Many transmission and distribution lines are built by contracting companies employing linemen and cablemen. A contract is normally completed between the electric utility company and the contracting company to specify the responsibilities for completing the project. Construction specifications are included in the contract. These specifications are used by linemen and cablemen installing the facilities.

CONSTRUCTION CONTRACTS

The contract describes the work to be completed, such as the following: Construct a 161-kV wood-pole H-frame transmission line to be owned by the utility company and to extend from Riverside Generating Station of the company in Pleasant Valley Township, Scott County, Iowa to the Valley Substation, owned by the company, south of the town of Walcott in Blue Grass Township, Scott County, Iowa.

The scope of the project is described in the contract and would be similar to the following: The contractor agrees to provide all necessary materials, tools, equipment, labor, and supervision and to pay all related costs listed in the contract necessary for the proper construction and completion of the transmission line described and specified in "Specifications for the construction of a 161-kV transmission line from Riverside Generating Station to Valley Substation" dated October 24, 1980, which specifications, including all exhibits thereto attached, are made a part of the contract. A hold harmless clause, a schedule for commencing and completing the work, and provisions for payment are usually included in the contract.

SPECIFICATIONS

General specifications would normally consist of the following:

Definitions

Items furnished by contractor

Materials furnished by others

Changed conditions

Changes in work

Inspection and testing of work

Correction of work
Guarantee
Contractor's right to terminate contract
Assignment
Separate contracts
Connection to work of others
Progress of work
Losses to be borne by contractor
Delays
Indemnity against loss of livestock and crop and fence damage
Indemnity against claims by subcontractors or miscellaneous third parties
Performance by owner of contractor's work
Owner's right to terminate contract
Subcontracts
Performance bond
Liens
Patents and royalties
Government labor regulations
Indemnity against loss due to damage or accidents
Labor
Progress payments
Final payment
Payments and payments withheld
Engineer
Drawings and specifications
Drawings furnished by the contractor
Schedule of work
Overtime
Use of premises
Accidents
Fire prevention
Contractor's liability insurance
Protection of and responsibility for work property
Safety and accident prevention
Right-of-way, easements, and permits
Public regulations, permits, and laws
Cleaning up
Acceptance
Contractor's responsibility

Special Conditions. Such conditions itemized in the specifications would probably consist of these items:

Definitions
Scope
Facilities at site
Standards
Inspection of the site
Surveys
Right-of-way
Cooperation with other contractors
Access to the work
Installing gates and maintaining fences
Underground structures
Damages
Loss or damage to livestock
Arrangement with owners of adjacent property
Roads
Use of explosives
Communication equipment
Warning to contractor's employees
Safety precautions
Equal opportunity
Insurance
Withholding of state income tax
Plan and profile, route map, and construction drawings
Special provisions
Cost accounting code
Commencement and completion of work

Technical Specifications. Such specifications for land clearing would cover the following:

Scope of Work. The work to be performed under the terms of this specification shall consist of clearing the owner's right-of-way for a new 161-kV transmission line. The land to be cleared will be from the Riverside Generating Station in Pleasant Valley Township, Scott County, Iowa, to the Valley Substation, south of the town of Walcott in Blue Grass Township, Scott County, Iowa. The width of the right-of-way is 150 ft.

Conditions for Clearing. The contractor shall clear the right-of-way of all obstructions that will interfere with the operation of the electric transmission line. The clearing to be performed, as a part of these specifications, shall be done in strict accordance with the contract documents. The actual work involved consists of clearing a strip of land 75 ft wide on each side of the centerline of the transmission line, by cutting and/or trimming of all trees and brush within the right-of-way limits. The work includes the removal or trimming of trees which the engineer may classify as "danger trees."

Danger Trees. The contractor must have written permission from the engineer before any danger trees are cut. The trees to be cut shall be marked with a distinctive blaze, by personnel furnished by the contractor, in accord with the instructions of the engineer.

Clearing in developed areas, such as those planted for orchards or other shade, fruit, or ornamental trees, shall be conducted in accordance with directions given by the engineer. The contractor shall promptly notify the engineer whenever any objections to the trimming or felling of any trees or to the performance of any other work on the land are made by the landowner.

Clearing. All trees, brush, stumps, and other inflammable material, except grass and weeds, shall be removed unless the contractor is otherwise instructed by the engineer. All trees and brush shall be cut 3 in or less from the ground line (measured on the uphill side). All stumps must be cut so that passage of trucks and tractors will not be hindered.

All trees located within highway rights-of-way and that are to be removed shall be cut at the ground line or as near to the ground line as possible, in accordance with highway commission practices. The extent of clearing of brush, hedges, and the like within 150 ft of any public road shall be determined by the engineer. Wherever the right-of-way crosses orchards, parks, or indicated special areas, all clearing shall be done specifically as directed by the engineer.

Disposal. The contractor shall dispose of all trees and brush by either chipping and spreading or hauling away unless otherwise specified in the special provisions or as otherwise directed by the engineer. Disposal of any limbs or brush shall be conducted in conformance with state and local laws and applicable regulations and shall be accomplished in such a manner as not to create a hazard or a nuisance. The contractor shall obtain all necessary permits. No damage of any nature shall be inflicted by the contractor upon adjoining property by unwarranted entry or disposal upon said property. All logs and brush shall be the property of the contractor unless otherwise specified in the "special provisions."

Stump Removal. Any stumps that interfere with installation of the structure shall be completely removed and disposed of by the contractor at no extra cost to the owner.

Seeding Right-of-Way. In wooded areas where the contractor may deem it necessary to bulldoze a road for construction purposes or in uncultivated areas where the ground is denuded of vegetation due to construction operations, the denuded areas shall be seeded with a mixture of 2 lbs of Alsike clover and 6 lbs of White Dutch clover per acre. The seed shall be furnished by the contractor and shall be well raked in to ensure germination. Areas where germination fails shall be reseeded at the contractor's expense.

Maintaining Fences during Clearing. Where the removal of trees requires the removal of fencing along the owner's right-of-way, the contractor shall take steps to provide temporary fencing as directed by the engineer. Such temporary fencing shall remain the property of the contractor and shall be removed when permanent fencing is installed or when directed by the engineer. During disposal and/or removal of trees, the contractor shall keep the fences in good repair where the work has affected their condition and shall keep all gates and temporary openings closed except when in actual use. The contractor shall be responsible for any damage caused by failure to keep fences in good repair or gates closed. On completion of the work, the contractor shall restore to their previous condition all fences that have been damaged during clearing operations. This includes the removal of gates installed for the contractor's use if the landowner wishes those gates removed.

Treatment. Not less than 2 weeks before cutting, all trees and brush shall be given an adequate basal spray with a chemical that meets with the approval of the engineer. Treatment prior to cutting is required to allow absorption of the chemical by the tree and roots in order to prevent regrowth. This chemical shall be furnished by the contractor.

The basal spray shall saturate the bottom 2 ft of the tree or brush. It is important that the entire circumference be treated. Also, it is important that the ground line and all exposed roots be thoroughly saturated. Some species, such as box elder, willow, and osage orange, are very susceptible to chemical treatment, while others, particularly ash and linden, are tolerant. Because of the difference of susceptibility, some species will require heavier applications than others to gain good kill. Among these are ash, linden, maple, hickory, and oak.

In the event that the owner is not able to secure the right-of-way to do the work in time to allow the contractor to basal spray all trees and brush prior to cutting, all stumps shall be saturated thoroughly with chemical, especially at the ground line and on all exposed roots. Spraying after cutting may be done only when approved by the engineer. Care shall be used to prevent any chemical from contacting any foliage or farm land off the owner's right-of-way. Chemical spraying shall be prohibited on any state or federal conservation lands. Such lands are marked on the plan and profile sheets.

Staking of Right-of-Way. The owner will stake the right-of-way limits in advance of the contractor's clearing operations. If for any reason the stakes are removed or lost, the contractor shall inform the engineer who will arrange to have the stakes reset.

Engineering Specifications. Such specifications would include the following:

Local Conditions. The contractor must check all local conditions affecting the work and shall make a thorough examination of the route of the line, plans, specifications, and premises in order to be entirely familiar with the details and construction of the installation.

Workmanship. All work shall be executed in a neat and skillful manner as specified or detailed in these specifications and/or drawings as listed herein and in accordance with best construction practice.

Drawings. The character of the line, location of line, and details of various structures are shown on drawings listed herein. These plans and drawings are intended to be complete and final and shall be followed as closely as possible. Dimensions as shown on drawings are not guaranteed and contractor should check their accuracy before proceeding with the work.

Pole Locations. Stakes showing the location of each pole will be set by the utility company. These stakes will be marked with a structure number as shown on drawings. The size of poles required, class, height, and other data for each structure number will be shown on supplementary tabulation sheets to be furnished to the contractor by the company.

Depth of Pole Setting. All poles are to meet American National Standards Institute specifications: western red cedar, Pentrex butt-treated with full-length lifespan treatment or full-length pressure-treated fir, and shall be set in accordance with the following table except where directed otherwise in writing by the engineer:

	Depth of setting	
Height of pole, ft	In earth	In rock
50	7′0″	5′0″
55	7′6″	5′6″
60	8′0″	6′0″
65	8′6″	6′6″
70	9′0″	7′0″
75	9′6″	7′6″

Poles shall be set to stand perpendicular and in exact alignment when the line is completed unless otherwise specifically called for. Pole-setting depths shall have a tolerance of not more than ± 3 in except as otherwise directed in writing by the engineer. Poles set partly in earth and partly in rock shall be set to the depth shown for earth.

At locations where the poles of any one structure are located at different ground elevations, the bottoms of the poles shall be leveled by deepening the hole on the higher ground, the hole on the lower ground having the depth shown above. In no case shall any pole be set to any depth such that the top of the Pentrex treatment is less than 9 in. above ground level at the pole. In special cases it may be necessary for the contractor to excavate the shelf as indicated on the drawings in order to meet the above requirements. In special cases a longer pole may be required on the downhill side which shall be set at least as deep as shown in the table.

Pole Selection. The contractor shall select heavier poles for longer spans and angle points. The supplementary tabulation shows the class pole to be used for each structure. In addition, on this type of construction, the contractor shall match, as nearly as possible, the ground line and top dimensions of both poles in each structure. In possible, circumference at ground line of the two poles in each fixture shall not differ more than 2 in. A contractor who finds it impossible, because of variation in sizes of poles received, to meet this requirement shall secure the written approval of the engineer before setting poles.

Pole Setting. Holes shall be dug of sufficient size to permit free insertion of tamping bar on all sides of poles after poles are set. In backfilling, tampers shall continually tamp in earth until the hole is completely filled. No earth shall be added until that already in place is solid and tight. After hole is filled, excess earth shall be piled up and packed firmly around pole.

Line poles shall be set to stand perpendicular when line is completed. Poles at corners and angles shall be set with a rake of approximately 1/4 in. for each foot of height of pole above ground. No change from this specification shall be made except upon the written direction of the engineer. H-type fixtures are all double-armed. Poles shall be set with gains at right angles to line, as no reaming or extra drilling of holes will be permitted to get arm bolts inserted. Poles shall be set not over 2 in. more or less than the specified distance apart.

Installation of Arms and Braces. Arms on the two-pole fixtures are to be completely assembled. Poles are to be gained with bolt holes drilled so that the assembled double arm may be spread and slipped over top of pole and bolted in place. This may be done, at contractor's option, either before or after poles are set. Poles are to be drilled for shield-wire ridge iron, which the contractor will bolt in place. Arms must be level after installation. This leveling must be done by leveling the bottom of pole holes. It will not be permissible to level arms by drilling additional holes in poles. Poles are to be furnished with top X-brace bolt holes drilled. The X braces shall be installed so as to maintain the vertical alignment of the structures without causing undue permanent stresses in either poles or X braces. In some cases, the company may require the contractor to bore and gain certain poles, in which case contractor will be allowed a unit price shown in the quotation. All field gains and holes shall have brush or spray preservative treatment.

Special Structures. Poles for three-pole or special structures are to be furnished roofed only, with the contractor to do all boring and gaining. Structures are to be erected in accordance with drawings as furnished for each structure. All field gains and holes shall be brush or spray treated with preservative.

Anchors and Guys. Galvanized-steel guy-anchor assemblies are to be used for all heavy guying. Screw anchors are to be used where specifically indicated on drawings or where approved by the engineer. Sizes of anchor rods, assemblies, screw anchors, and guy wire are shown on drawings. Anchors shall be installed to the specified depth; anchor rods are to lie in the direction of strain. Thimble-eye guy nuts or guy-anchor assemblies will be installed. Guys shall be installed before wire is pulled and shall be pulled up to sufficient tension to pull pole over slightly. After wire is pulled, where two or more guys are used on one pole, they shall be inspected and equalized if necessary to see that each guy takes its proper portion of the strain. Prefabricated guy grips or two-bolt guy clamps shall be used for holding wire, and the loose end of the guy wire shall be fastened with a guy-wire clip.

Eight-foot guy guards will be installed on all anchor guys. The following numbers of guy clamps are to be used:

3/8″ EHS guy 3 clamps

1/2″ EHS guy 4 clamps

Insulators. Tangent line structures will be insulated with a string of ten 15,000# 10-in. ball-and-socket disk insulators, as shown by drawings; dead-end structures with a string of twelve 25,000# 10-in. clevis-type disk insulators. All cotter keys shall be bent sufficiently to prevent loss by vibration. All insulators shall be carefully handled to avoid chipping or breaking, and no chipped, cracked, or broken insulators shall be assembled on the line. It shall be the contractor's responsibility to inspect all insulators as taken from their crates and to replace all defective units.

Splices. Splices or joints may be made in the conductors and shield wires where necessary, except at crossings over railroads, important highways, and communication lines, and shall be at least 50 ft from the nearest pole or structure. All splices and dead ends shall be made up in strict accordance with the drawings and instructions furnished by the company, using steel-core aluminum sleeves and clamps and a hydraulic press. Particular care shall be taken by the contractor in making up all splices or joints in order to see that the proper procedure is followed. The company reserves the right to inspect any or all joints and to reject those which do not meet with the approval of the engineer. All joints shall be made by one foreman and a crew especially trained for this work.

Stringing Conductors and Shield Wires. Conductors and shield wires shall be handled with care and must not be trampled on or driven over. They shall be continuously inspected during installation, and any cuts, kinks, and other injured portions shall be cut out and the wire spliced, said splices to be made strictly in accordance with the company's specifications and drawings. Conductors and shield wires may be strung out with reels in a stationary position with conductors supported on the structures in free-running stringing blocks or paying off from reels towed along the line. Conductors or shield wires shall not be dragged out on the ground or dragged over the arms of the structures or over fences or foreign objects which may damage conductors. Each reel shall be equipped with a suitable braking device to keep the conductors or shield wires always under tension and clear of the ground.

Sagging Conductors and Shield Wires. The power-line conductors and the overhead shield wires shall be strung in accordance with sag and tension charts provided by the company. These sag charts are initial-stringing sag charts, and no conductors or overhead shield wire shall be prestressed beyond that required by customary practice. The air temperature at the time of sagging shall be determined with a certified etched-glass thermometer. All conductor grips, come-alongs, stringing blocks, and pulley wheels shall be of a design which will prevent damaging the conductors or shield wires by kinking, scouring, or unduly bending them during stringing operations.

Sags shall be accurately measured in several spans approximately equal to the ruling span during stringing and checked against the sag charts for the prevailing air temperature. Sagging shall not be done during periods of high wind which, in the opinion of the engineer, might prevent accurate sag measurements. A tolerance of ± 1/2 in. sag per 100 ft of span length, based on the sag and tension charts will be permitted provided that all conductors in the same span assume the same sag and satisfactory clearances are obtained. The engineer will compute and check the sag at all points to be checked, and the contractor shall furnish the necessary crew for signaling and climbing purposes.

The methods for checking sag and the points at which the checks are to be made shall be agreed upon between the engineer and the contractor, the intent of these specifications being that the engineer shall be assured, by means of a sufficient and reasonable number of checks, that proper clearances are obtained at all points, that the proper tensions are being obtained, and that the general appearance of the line shall be satisfactory. The contractor shall make adequate provision to prevent pulling any structure out of line during the stringing and sagging operations.

Dead Ends. All dead ends of the power conductors will be made with Aluminum Company of America three-piece dead-end clamps. These three pieces are catalog no. 1744.2 aluminum compression dead-end body, catalog no. 772.1 compression steel clevis end, and catalog no. A1763.2 aluminum compression jumper terminal. The shield wires will be dead ended by means of guy thimble or roller and three Tu-base clamps as shown on the drawings.

Wire Attachment. After sagging and dead ending conductors and shield wire, the contractor shall straighten any poles that may have been pulled out of plumb in the stringing and sagging operation. Conductors and shield wires shall then be securely clamped in their proper suspension clamps.

Pole Grounding. All H-frame poles will have a no. 4 bare-copper ground wire installed as shown on the drawing attached as an exhibit. Contractor shall securely staple 15 ft of no. 4 bare-copper wire forming a flat spiral on the butts of all of the above poles, the free end to be stapled along the ground section and extending up 12 in above the ground line. A solderless clamp connector shall be installed 6 in. below ground level as shown on the drawings, connecting the butt section of the ground wire with the down lead from the pole top. The down lead shall be connected, as shown on the drawings, to a 5/8-in. × 12-ft Copperweld ground rod, and the two ground rods of each structure shall be connected by a no. 4 copper wire buried not less than 12 in. below ground. Eight feet of wood molding shall be installed to protect the ground wire and shall extend 6 in below grade. Ground molding shall be stapled to pole with four staples.

Exhibits. Exhibits attached to the specifications would include the following: (Figs. 6.1–6.6).

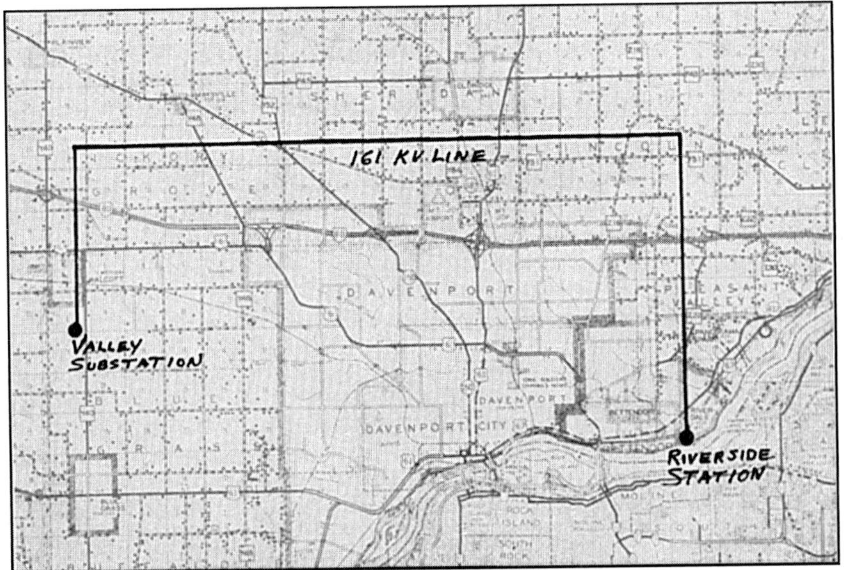

FIGURE 6.1 Map showing route of 161-kV Riverside-Valley line.

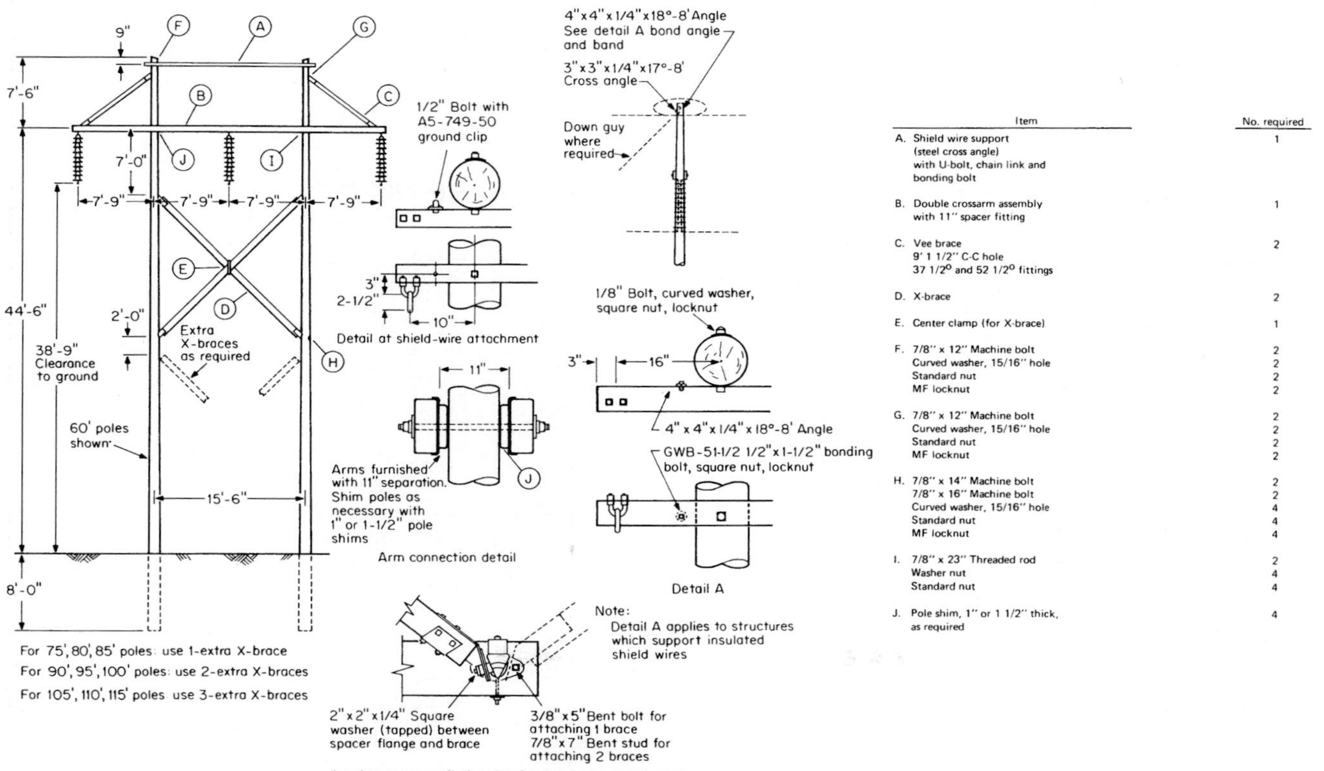

Item	No. required
A. Shield wire support (steel cross angle) with U-bolt, chain link and bonding bolt	1
B. Double crossarm assembly with 11'' spacer fitting	1
C. Vee brace 9' 1 1/2'' C-C hole 37 1/2° and 52 1/2° fittings	2
D. X-brace	2
E. Center clamp (for X-brace)	1
F. 7/8'' x 12'' Machine bolt Curved washer, 15/16'' hole Standard nut MF locknut	2 2 2 2
G. 7/8'' x 12'' Machine bolt Curved washer, 15/16'' hole Standard nut MF locknut	2 2 2 2
H. 7/8'' x 14'' Machine bolt 7/8'' x 16'' Machine bolt Curved washer, 15/16'' hole Standard nut MF locknut	2 2 4 4 4
I. 7/8'' x 23'' Threaded rod Washer nut Standard nut	2 4 4
J. Pole shim, 1'' or 1 1/2'' thick, as required	4

FIGURE 6.2 Tangent H-frame structure.

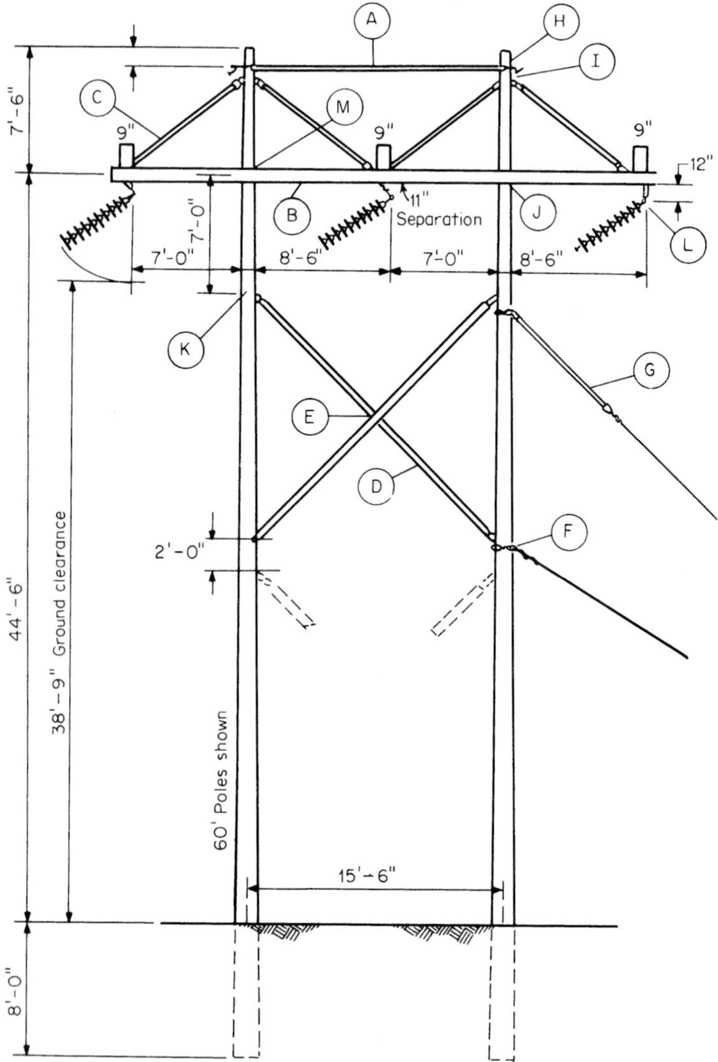

FIGURE 6.3 Angle structure (1/2° to 6°).

Item	No. required
A. Shield wire support (steel cross angle) with U-bolt, chain link, and bonding bolt	1
B. Double crossarm assembly with 11" spacer fitting	1
C. Vee brace 9' - 1 1/2" C-C hole 37 1/2° and 52 1/2° fittings	4
D. X-brace	2
E. Center clamp (for X-brace)	1
F. Pole band	1 Set (4 Sections) 1 Set (4 Sections)
Stud bolt, 7/8" x 8"	8
Links	2 Sets
Guy roller, 15/16" hole	1
7/8" x 3" Machine bolt/MF locknut	1
G. Fiberglass strain insulator	1
H. 7/8" x 12" Machine bolt	2
Curved washer, 15/16" hole	2
Standard nut	2
MF locknut	2
I. 7/8" x 12" Machine bolt	2
Standard nut	2
MF locknut	2
J. 7/8" x 23" Threaded rod	2
Washer nut	4
Standard nut	4
K. 7/8" x 14" Machine bolt	2
7/8" x 16" Machine bolt	2
Curved washer, 15/16" hole	4
Standard nut	4
MF locknut	4
L. Swinging angle bracket	3
M. Pole shim, 1" or 1 1/2" thick, as required	4

FIGURE 6.3 (*Continued*)

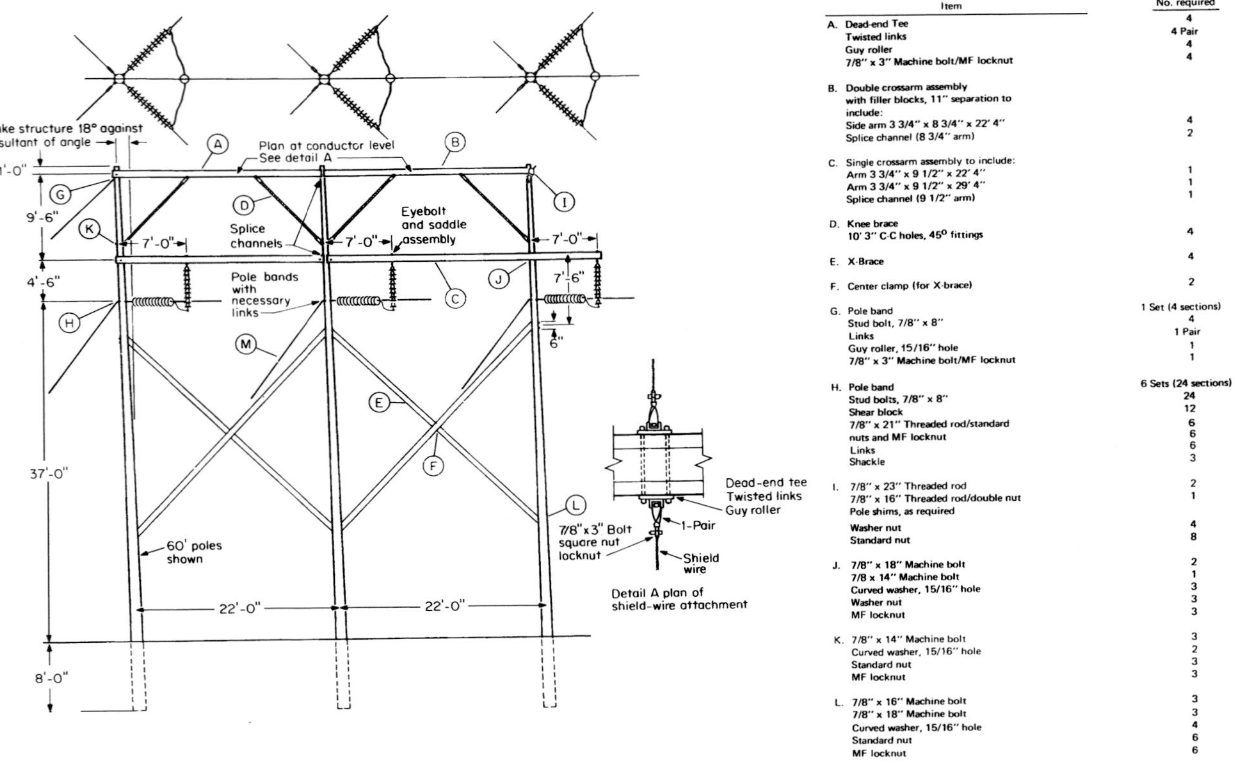

Item	No. required
A. Dead-end Tee	4
Twisted links	4 Pair
Guy roller	4
7/8" x 3" Machine bolt/MF locknut	4
B. Double crossarm assembly with filler blocks, 11" separation to include:	
Side arm 3 3/4" x 8 3/4" x 22' 4"	4
Splice channel (8 3/4" arm)	2
C. Single crossarm assembly to include:	
Arm 3 3/4" x 9 1/2" x 22' 4"	1
Arm 3 3/4" x 9 1/2" x 29' 4"	1
Splice channel (9 1/2" arm)	1
D. Knee brace 10' 3" C-C holes, 45° fittings	4
E. X-Brace	4
F. Center clamp (for X-brace)	2
G. Pole band	1 Set (4 sections)
Stud bolt, 7/8" x 8"	4
Links	1 Pair
Guy roller, 15/16" hole	1
7/8" x 3" Machine bolt/MF locknut	1
H. Pole band	6 Sets (24 sections)
Stud bolts, 7/8" x 8"	24
Shear block	12
7/8" x 21" Threaded rod/standard	6
nuts and MF locknut	6
Links	6
Shackle	3
I. 7/8" x 23" Threaded rod	2
7/8" x 16" Threaded rod/double nut	1
Pole shims, as required	
Washer nut	4
Standard nut	8
J. 7/8" x 18" Machine bolt	2
7/8 x 14" Machine bolt	1
Curved washer, 15/16" hole	3
Washer nut	3
MF locknut	3
K. 7/8" x 14" Machine bolt	3
Curved washer, 15/16" hole	2
Standard nut	3
MF locknut	3
L. 7/8" x 16" Machine bolt	3
7/8" x 18" Machine bolt	3
Curved washer, 15/16" hole	4
Standard nut	6
MF locknut	6
M. Fiberglass strain insulator	6

FIGURE 6.4 Three-pole angle and dead-end structure (60° to 90°).

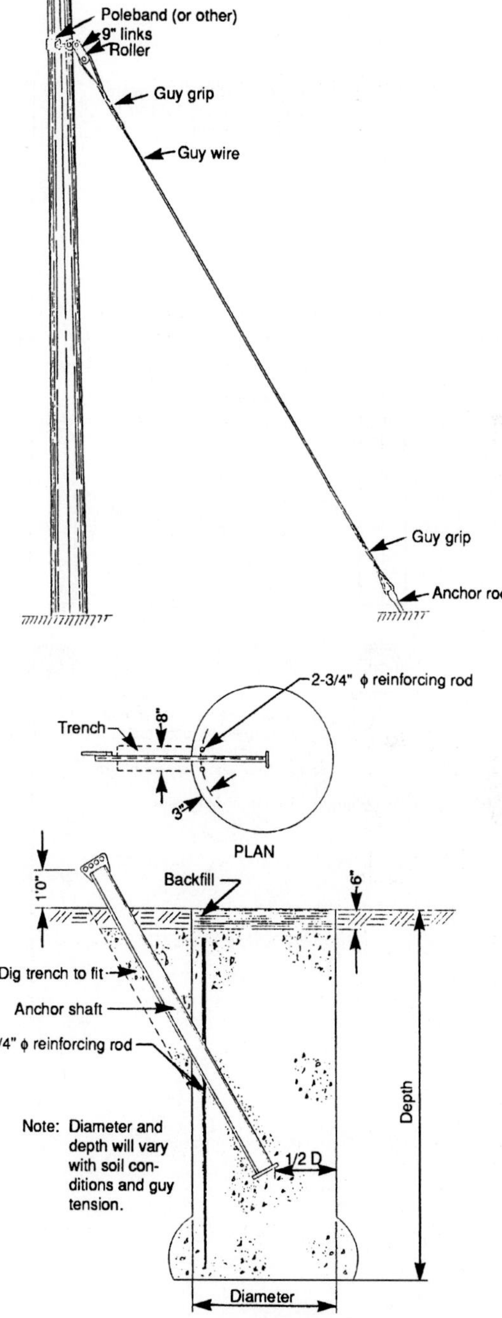

FIGURE 6.5 Down-guy and anchor installation details.

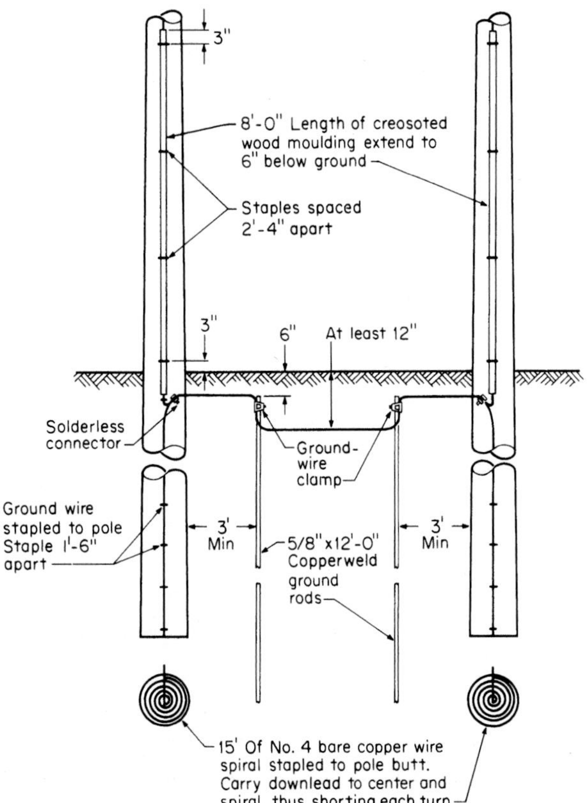

FIGURE 6.6 Pole-grounding installation details.

CHAPTER 7
WOOD-POLE STRUCTURES

Structures supporting electric lines must be designed to support conductors, insulators, and shield wires in a manner that provides adequate electrical clearances and ample mechanical strength as specified in the *National Electrical Safety Code* (ANSI C2), published by the Institute of Electrical and Electronics Engineers, Inc. Safe clearances must be maintained when the conductor temperature is elevated as a result of large currents flowing in a circuit and when the conductors are ice-coated and strong winds are blowing. Common types of structures include wood poles, reinforced-concrete poles, steel poles, steel towers, and aluminum towers. The total cost of the installed facilities normally determines the type of structure used. Lines built with wood-pole structures have generally proved to be the most economical where adequate physical and electrical requirements could be satisfied. The American National Standards Institute's publication entitled *Specifications and Dimensions for Wood Poles* (ANSI 05.1) provides additional detailed information for wood-pole structures.

Wood-Pole Types. Southern yellow pine, western red cedar, and Douglas fir are the most commonly used species of trees used for wood poles in the United States. Western larch trees are used for some poles in the western states. Jack pine poles are frequently used in Canada. The type of poles used is determined by the physical requirements of the poles needed to construct the line and the cost of the poles as a result of the shipping expense from the location where the species is grown. The suppliers of the poles usually shave and reduce the diameter of the timber to improve the appearance of the poles.

Wood-Pole Preservative Treatment. The poles are given a preservative treatment to prevent deterioration. Southern yellow pine poles are treated under pressure because of the thick sap wood. Western red cedar poles are normally treated without pressure. Douglas fir poles can be treated with or without pressure. The treatment of poles has been specified by the American Wood Preservers' Association. Experience has shown that the proper treatment of the wood can double the service life of the poles.

Pentachlorophenol, and chromated copper arsenates (CCA) are used commonly to provide a preservative treatment of wood poles.

The Electric Power Research Institute (EPRI) has determined that poles can be treated with ammoniacal copper fatty acid economically. Ammoniacal copper fatty acid treatment is as effective as pentachlorophenol treatment for poles. Poles treated with ammoniacal copper fatty acid have a light color and are easier to climb because the fatty acid softens the wood. The ammoniacal copper fatty acid treatment of poles is acceptable environmentally because it does not contain materials that are toxic to mammals.

The American Wood Preservers' Association has established standards for copper naphthenate treatment of wood poles. Copper naphthenate treatment has environmental advantages.

TABLE 7.1 Wood-Pole Classifications, Dimensions, and Breaking Strengths

		Class								
		1			2			3		
		M.H.B.L.*								
		4500			3700			3000		
		Minimum circumference at top, in								
		27			25			23		
		Minimum circumference, in at ground line†								
Length, ft	Ground line distance from butt, ft	WC	SP DF	WL	WC	SP DF	WL	WC	SP DF	WL
25	5	37.4	33.8	33.2	34.9	31.7	31.2	32.9	29.7	29.2
30	5.5	40.2	36.6	35.6	37.7	34.1	33.6	35.2	32.1	31.1
35	6	42.5	39.0	38.0	40.0	36.5	35.5	37.5	34.0	33.0
40	6	45.0	41.0	40.0	42.5	38.5	37.5	39.5	36.0	35.0
45	6.5	47.3	42.9	41.9	44.3	40.4	39.4	41.3	37.4	36.9
50	7	49.1	44.7	43.8	46.1	41.8	40.8	43.1	38.8	38.3
55	7.5	50.9	46.1	45.2	47.9	43.1	42.2	44.4	40.1	39.7
60	8	52.7	47.5	46.6	49.7	44.5	43.6	45.7	41.5	40.6
65	8.5	54.0	48.9	48.0	50.5	45.9	45.5	47.0	42.9	42.0
70	9	55.4	50.2	49.4	51.9	47.2	46.4	48.4	44.8	43.4
75	9.5	56.7	51.6	50.8	53.2	48.1	47.3	49.7	45.1	44.3
80	10	58.0	53.0	51.7	54.5	49.5	48.7	50.5	46.0	45.2
85	10.5	59.3	53.9	53.1	55.3	50.4	49.6	51.3	46.9	46.1
90	11	60.6	54.7	54.0	56.6	51.7	50.5	52.6	47.7	47.5
95	11	61.6	55.7	55.5	57.6	52.7	52.0			
100	11	63.1	57.2	56.5	59.1	53.7	53.0			
105	12	63.7	58.0	57.2	59.7	54.5	53.7			
110	12	65.2	59.0	58.2	60.7	55.5	54.7			

*Minimum horizontal breaking load applied 2 ft below top of the pole.
†This circumference at ground line for measurement in the field.
Legend: WC = western red cedar, SP = southern yellow pine, DF = douglas fir, WL = western larch.
Source: Commonwealth Edison Co.

Pole Classification. Wood poles are commonly classified by length, top circumference, and circumference measured 6 ft from the butt end. The lengths vary in 5-ft steps, and the circumference at the top varies in 2-in. steps. Thus we have lengths, in feet, of 25, 30, 35, 40, etc., and minimum top circumferences, in inches, of 15, 17, 19, 21, etc.

The circumference measured 6 ft from the butt end determines to which class, numbered from 1 to 10, a pole of a given length and top circumference belongs. The classification from 1 to 10 determines the strength to resist loads applied 2 ft from the top of the pole.

TABLE 7.1 Wood-Pole Classifications, Dimensions, and Breaking Strengths (*Continued*)

Class											
4			5			6			7		
M.H.B.L.*											
2400			1900			1500			1200		
Minimum circumference at top, in											
21			19			17			15		
Minimum circumference, in a ground line†											
WC	SP DF	WL	WC	SP DF	WL	WC	SP DF	WL	WC	SP DF	WL
30.4	27.7	26.7	28.4	25.7	24.7	25.9	23.2	23.2	24.4	21.7	21.2
32.7	29.6	29.1	30.2	27.6	26.6	28.2	25.1	24.6	26.2	23.6	23.1
34.5	31.5	31.0	32.0	29.0	28.5	30.0	27.0	26.5	27.5	25.0	24.5
36.5	33.5	32.5	34.0	31.0	30.0	31.5	28.5	28.0			
38.3	34.9	33.9	35.8	32.4	31.4	32.8	29.9	28.9			
39.6	36.3	35.3	37.1	33.8	32.8						
41.4	37.6	36.7									
42.7	38.5	38.1									
44.0	39.9	39.0									
44.9	40.7	40.4									

The American National Standards Institute's publication entitled *Specifications and Dimensions for Wood Poles* (ANSI 05.1) provides much technical data for wood utility poles. The circumference of western red cedar, southern yellow pine, Douglas fir, and western larch poles at the top and at the ground line for various classes and lengths are outlined in Table 7.1. The weights of commonly used wood poles are outlined in Table 7.2.

Vertical pole loading capacities are shown in Table 7.3, which gives the maximum weight that may be installed on poles of different heights and classes. Weights were calculated for three-cluster or crossarm-mounted transformers. Weights in the table would require adjustment for other material mounting configurations. Values in the table are based upon kiln-dried poles with fiber stresses for wood poles, published by the American

TABLE 7.2 Weights of Wood Poles That Are Used Extensively

Length of pole, ft	Species*	Class			
		Average weight, lb per pole			
		1	2	4	5
30	WC	880	750	540	440
	DF	1110	930	690	600
	SP	1279	1082	784	660
	WL	955	800	595	515
35	WC	1055	880	660	570
	DF	1435	1260	875	770
	SP	1568	1343	1004	862
	WL	1235	1085	750	660
40	WC	1320	1145	790	705
	DF	1760	1560	1120	920
	SP	1884	1623	1219	1059
	WL	1515	1340	965	790
45	WC	1585	1365	1010	880
	DF	2070	1845	1350	1125
	SP	2223	1911	1444	1274
	WL	1780	1585	1160	965
50	WC	1760	1585	1230	1145
	DF	2500	2150	1600	1300
	SP	2585	2214	1687	1494
	WL	2150	1850	1375	1120
55	WC	2025	1760	1410	1410
	DF	2860	2475	1815	1540
	SP	2993	2567	1934	1719
	WL	2460	2130	1560	1325
60	WC	2290	1935	1670	
	DF	3360	2820	2040	1740
	SP	3451	2943	2186	1953
	WL	2890	2425	1755	
65	WC	2815	2200	1935	
	DF	3835	3250	2340	2015
	SP	4015	3341	2457	2237
	WL	3300	2795	2010	
70	WC	3170	2640	2290	
	DF	4340	3640	2590	2240
	SP	4620	3781	2732	2488
75	WC	3695	3170	2640	
	DF	4800	4050	2925	
	SP	5198	4235	3021	
80	WC	4400	3695	3080	
	DF	5240	4400	3200	
	SP	6400	5170	3615	

*WC = western red cedar, DF = Douglas fir, WL = western larch, SP = southern pine.
Source: Commonwealth Edison Co.

TABLE 7.2 Weights of Wood Poles That Are Used Extensively (*Continued*)

Length of pole, ft	Species*	Class / Average weight, lb per pole: 1	2	4	5
85	WC	4840	3960		
	DF	5780	4930		
	SP	7200	5745		
90	WC	5810	4930		
	DF	6300	5400		
	SP	8140	6405		
95	WC	6750	5950		
	DF	6935	5985		
100	WC	7500	6550		
	DF	7600	6700		

National Standards Institute, degraded to 75 percent of values found by standard tests to allow for variation in fiber strengths of poles. The calculations, applying a factor of safety of 5 to 1, provide for two 250-lb linemen and their equipment on the pole and for stress caused by hoisting cables.

Operating Voltages. Wood-pole structures are commonly used for most distribution-line structures and are frequently used for subtransmission lines operating at voltages of 34.5 through 161 kV and for transmission lines operating at 138 through 230 kV. Some extra-high-voltage (EHV) transmission lines operating at 345 and 500 kV are constructed with wood-pole structures.

Distribution Structures. Distribution-circuit structures normally consist of single wood poles using line-post or pin-type insulators with fiberglass support brackets or crossarms. Construction with line-post insulators or fiberglass support brackets with pin-type insulators is used to create a pleasing appearance for overhead electric distribution systems (Figs. 7.1 and 7.2). Construction details and materials required for a three-phase, four-wire 13.2Y/7.6-kV electric distribution line built on wood-pole structures with fiberglass insulator support brackets to obtain close space construction are shown in Fig. 7.3.

Many companies are converting 34.5-kV ac transmission circuits to 34.5Y/19.9-kV three-phase, four-wire distribution circuits (Figs. 7.4 and 7.5). Three-phase or single-phase lines are tapped to the converted circuits to supply distribution customers. The lines tapped to the 34.5Y/19.9-kV circuits in many cases previously operated at a lower primary voltage. It was necessary to reinsulate, install distribution transformers with the proper voltage rating, and rebuild the lower-voltage lines to operate at the higher voltage. The higher voltage of 34.5Y/19.9 kV has the advantage of operating at a small amount of current to supply large customer loads. The high distribution voltage with a small amount of current reduces the required conductor size and has the advantages of low electrical losses and reduced voltage drop on the lines.

TABLE 7.3 Maximum Weight That May Be Installed on Wood Poles

Pole length, ft		Mounting distance from pole top to uppermost attachment, ft	Pole class — Maximum allowable weight, lb: 1	2	4	5
30		1	4,650	3,500	1,700	730
		4	6,700	5,200	2,730	1,270
35		1	3,800	2,900	1,450	655
		4	5,240	4,000	2,070	970
		5	5,800	4,480	2,350	1,100
		8	7,900	6,200	3,360	1,630
40		1	3,200	2,300	1,100	590
		4	4,250	3,150	1,570	890
		5	4,670	3,500	1,700	1,020
		8	6,200	4,750	2,550	1,520
45		1	2,560	1,890	835	465
		4	3,400	2,550	1,170	670
		5	3,760	2,810	1,310	750
	①	8	4,950	3,780	1,830	1,070
		13	8,000	6,200	3,200	1,950
		16	10,600	8,350	4,500	2,820
		17	11,600	9,120	5,050	3,200
50		1	2,260	1,340	660	
		4	2,950	1,800	910	
		5	3,200	2,000	1,000	
		8	4,160	2,650	1,380	
	①	13	6,600	4,250	2,350	
		16	8,600	5,650	3,220	
		17	9,400	6,200	3,660	
		20	12,300	8,400	4,900	
55		13	5,750	3,600		
	①	16	7,300	4,700		
		17	7,900	5,100		
		20	10,200	6,700		
60		13	4,960	2,850		
	①	16	6,400	3,750		
		17	6,900	4,100		
		20	8,900	5,300		
65		13	3,750	2,550		
	①	16	4,850	3,300		
		17	5,260	3,600		
		20	6,800	4,700		

① Mounting heights are provided for equipment located as underbuild on 34.5-kV lines. If higher poles are required for clearance (i.e., railroad crossings), the preferred mounting height for equipment is nearer the ground to reduce the stress on the pole.

Source: Commonwealth Edison Co.

FIGURE 7.1 Distribution circuit rated 13.2Y/7.6 kV, three-phase, constructed on a wood pole with a fiberglass bracket supporting pin-type insulators. Single-phase circuit, supplied from main-feeder circuit, through fused-disconnect switch runs perpendicular to main circuit.

FIGURE 7.2 Armless-type construction for 4.16Y/2.4-kV and 13.2Y/7.6-kV distribution-circuit operation.

Subtransmission Structures. Alternating current (ac) subtransmission lines operating at voltages of 138 kV and below can be built on single wood-pole structures.

Wishbone-type construction on a single-pole structure is popular for use on 69-, 115-, and 138-kV subtransmission lines. This type of construction provides a symmetrical conductor arrangement which has proved to be economical to build and reliable to operate. Figure 7.6 pictures a single-pole structure with wishbone-type construction operating at 69kV. Subtransmission structures have been simplified to make them as inconspicuous as possible, to improve the aesthetic appearance, and to preserve the environment. This type of construction minimizes the width of right-of-way needed and reduces the number of trees that must be removed.

The structures may use light-colored wood poles and gray-colored post-type insulators (Fig. 7.7). Figure 7.8 is a drawing of a wood-pole tangent structure for a 69-kV three-phase circuit using horizontal post-type insulators with a triangular conductor configuration for installation in a straight portion of the line. Figure 7.9 is a drawing of a 69-kV-circuit single-pole structure designed for installation where it is necessary for the line to change direction. If a line is to be built in an area where appearance is not critical, such as an industrial location, crossarms are frequently used. The use of crossarms reduces the required pole height and permits the addition of a second circuit on the same wood-pole structures if the electrical system requires it at some future time. Figure 7.10 is a drawing of a typical 69-kV-circuit tangent structure with crossarms and provisions for a future second circuit. Figure 7.11 is a drawing of a 7° to 30° angle structure.

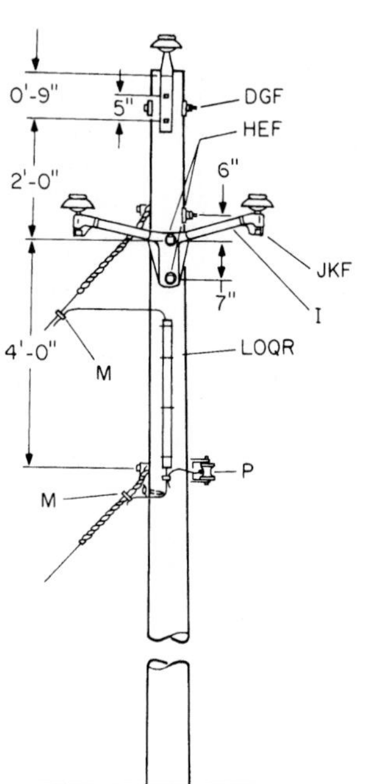

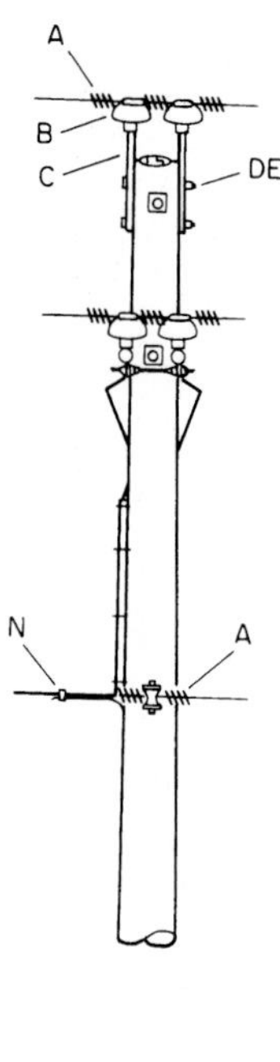

Item	No. required
A. Preformed tie wire size required	7
B. 1" Pin insulator	6
C. 1" Pole-top pin	2
D. 5/8" x 10" Bolt	3
E. 5/8" Round washer	2
F. 5/8" Palnut	9
G. 2 1/2" Curved washer	1
H. 5/8" x 12" Bolt	2
I. Two-phase brackets 48" spacing	2
J. 5/8" x 14" Double arming bolt	2
K. 2 1/4" Square washer	8
L. #6 Copper wire	1#
M. Split-bolt connector, size required	3
N. Compression connector, size required	1
O. Molding, ground wire	5′
P. Single-clevis insulator	1
Q. Staples, molding	6
R. Staples, fence	6

FIGURE 7.3 Three-phase 13.2Y/7.6-kV armless-type wood-pole structure (angle pole 6° to 20°).

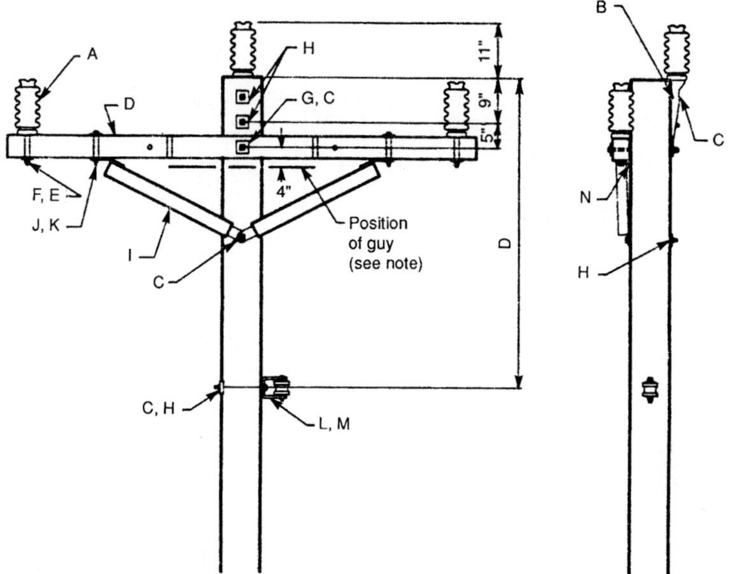

Conductor clearances		
Span length (ft)	Pole length (ft)	Neutral location D
200 & less	All	4'0"
201–270	35	6'0"
271–310	40 & greater	7'6"

NOTE: Guy for angles greater than 1°.

Item	Description
A	Insulator, 35 kV, post-type, clamp top.
B	Bracket for post-type insulator.
C	Bolt, 3/4" diameter, length required.
D	Crossarm, wood, 10' long, 4 pin.
E	Washer, round for 5/8" diameter bolt.
F	Bolt, stud, 5/8" diameter, 7" long for line post insulator.
G	Washer, square, flat for 3/4" diameter bolt.
H	Washer, square, curved for 3/4" diameter bolt.
I	Brace, crossarm, wood.
J	Bolt, carriage, 1/2" diameter, 6" long.
K	Washer, round for 1/2" diameter bolt.
L	Insulator, spool, 3" diameter.
M	Clevis for spool insulator.
N	Gain 3" X 4".

FIGURE 7.4 Drawing of a 34.5Y/19.9-kV three-phase tangent distribution-circuit structure for installation in a straight portion of the line or where any angle in the line is small.

Transmission Structures. Two-pole wood structures commonly referred to as H-type structures are frequently used for three-phase ac transmission circuits operating at voltages of 138 through 230 kV. Most subtransmission circuits operating in the voltage range of 34.5 through 115 kV are constructed on single wood-pole structures; however, a few are installed on two-pole H-type wood structures. Extra-high-voltage circuits operating at 345 kV are frequently constructed on two-pole wood structures. A few 500-kV installations have been constructed with wood-pole structures. The wood-pole structures are aesthetically acceptable in wooded areas and can be installed economically in many locations.

Figure 7.12 shows a typical tangent H-type two-pole wood structure commonly used for 161-kV three-phase ac circuits. The structure is usually built on 65-ft poles with some variation due to terrain and has $15^1/_2$-ft clearance between conductors. A complete structure with crossarms and insulators ready for erection will weigh from 5 to 6 tons. A three-pole,

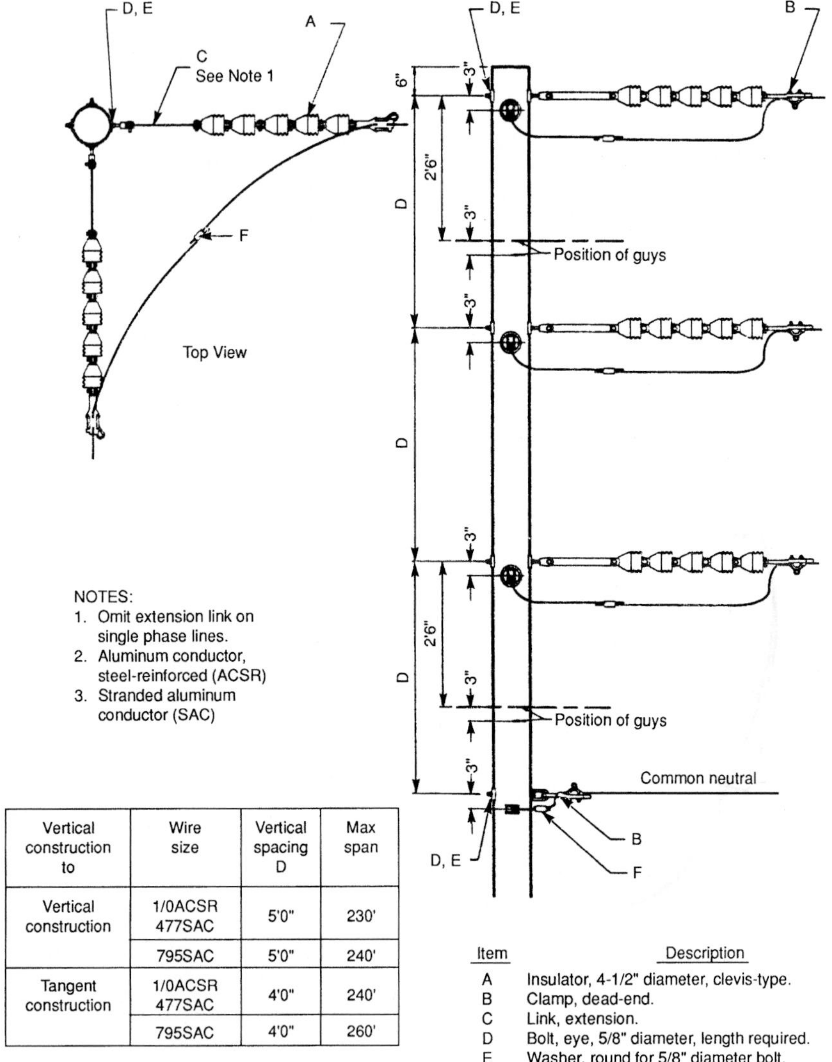

Vertical construction to	Wire size	Vertical spacing D	Max span
Vertical construction	1/0ACSR 477SAC	5'0"	230'
	795SAC	5'0"	240'
Tangent construction	1/0ACSR 477SAC	4'0"	240'
	795SAC	4'0"	260'

FIGURE 7.5 Drawing of a 34.5Y/19.9-kV three-phase angle distribution-circuit structure.

60° to 90° angle, dead-end structure can be seen in the background (Fig. 7.12). Figure 7.13 is a drawing of a two-pole H-type structure for a 161-kV line. Figure 7.14 is a drawing of a three-pole H-type angle structure for a 161-kV line.

Figure 7.15 is a drawing for a double K-type two-pole tangent structure for an EHV ac 345-kV transmission line. A typical line for this use could have 7/16-in extra-high-strength steel shield wires and two 795-kCMIL 26/7 aluminum conductor steel reinforced (ACSR) bundled conductors per phase, with spans of 1000 ft between structures. Structure safety

FIGURE 7.6 Wood-pole line using wooden crossarms arranged to provide symmetrical arrangement of the line conductors. (*Courtesy of Hughes Brothers.*)

FIGURE 7.7 Post-type insulator in use in horizontal position. Note the absence of crossarm and braces and narrow right-of-way required. (*Courtesy Lapp Insulator Co.*)

factors with $^1/_2$ in. of radial ice would be 4 for vertical and transverse loads. The structure vertical safety factor for $1^1/_2$ in. of radial ice would be 1.5, and the structure safety factor for 100 mph winds and bare conductors would be 2.3.

Wood-Substitute Poles. The clear cutting of timber instead of selective harvesting has decreased the supply of wood poles available for use in constructing electric lines. Solid-wood poles are becoming more difficult to obtain and more expensive. Substitute poles made of concrete, steel, aluminum, and fiberglass are described in Chap. 8. Laminated wood poles have been made by gluing layers of wood together under pressure. This process makes it possible to use smaller trees to manufacture large poles, thus increasing the availability. Laminated poles have been used satisfactorily for special structures and streetlight poles. The poles are either square or rectangular in cross section, making it difficult to use existing pole-line hardware. The poles have been more expensive than solid-wood poles.

EPRI has developed a composite-wood pole using a composite-wood material developed by the Institute of Wood Research. The composite-wood material is made from elongated wood flakes treated with a preservative, aligned, and bonded together with a synthetic adhesive. The preservative treatment prevents biological deterioration. The number of laminations depends on the size of the pole and the stress placed on the poles. The poles are octagonal with a hollow core (Fig. 7.16). The pole dimensions are similar to those for the solid-wood poles they are replacing. The poles can be manufactured in lengths up to

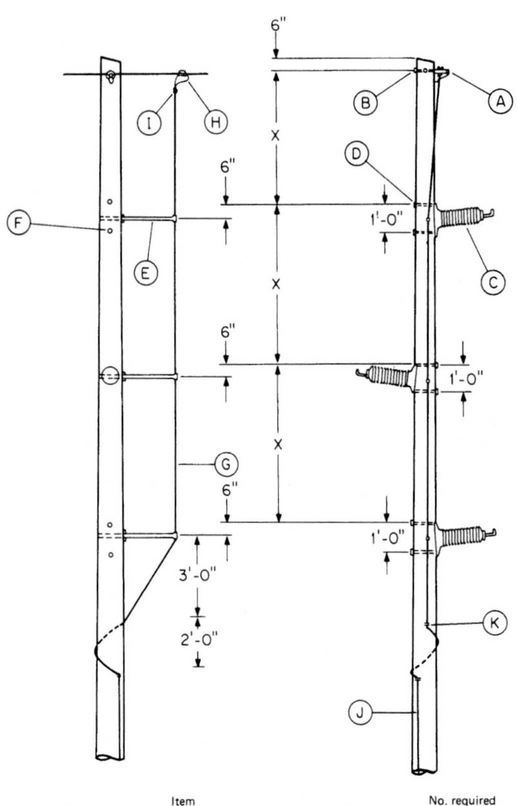

Item	No. required
A. Ground wire bracket	1
1/2" x 4" Lag screw	1
B. 5/8" x 14" mach. bolt	1
3" Square curved washer/5/8" bolt hole	1
Standard nut	1
Locknut	1
C. 88 kV Horizontal post insulator, gray color	3
D. 3/4" x 12" Machine bolt	2
3/4" x 14" Machine bolt	3
3/4" x 16" Machine bolt	1
3" Square curved washer/3/4" bolt hole	6
Standard nut	6
MF Locknut	6
E. Fiber glass downlead bracket	3
F. 5/8" x 12" Machine bolt	3
Standard nut	3
2 1/4" Square Washer/5/8" bolt hole	6
G. 6A Copperweld groundwire	
H. Two-bolt connector	1
I. Split-bolt Connector	1
J. Ground-wire molding, 1/2" groove	
K. Bronze staple, for 1/2" plastic molding	

Note: Phase spacing (dimension X) varies with span lengths, conductor size, conductor sag, and midspan clearances.

FIGURE 7.8 A 69-kV single-circuit armless-type tangent structure.

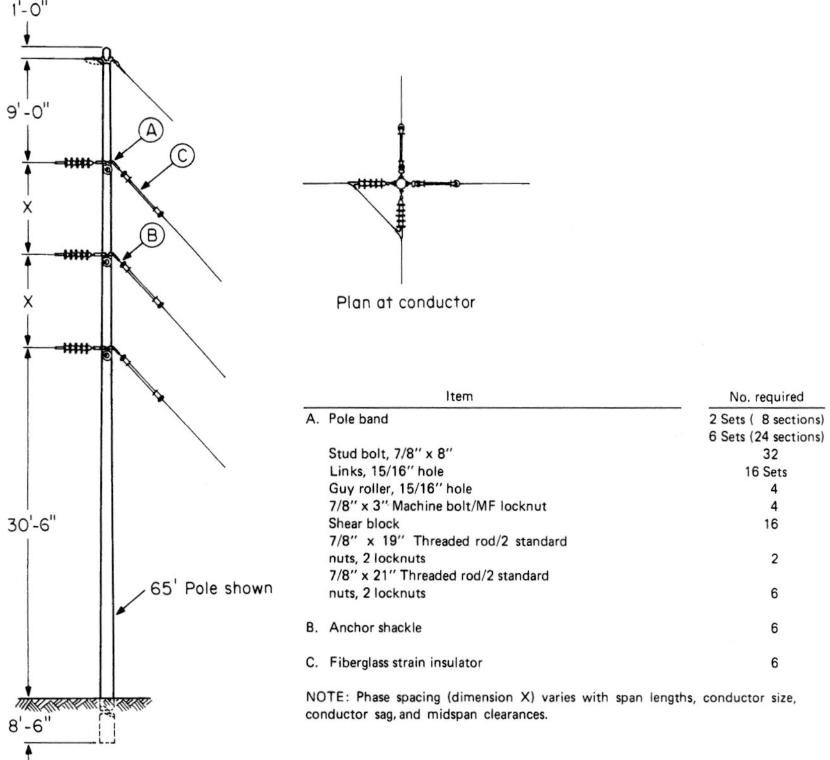

Item	No. required
A. Pole band	2 Sets (8 sections) 6 Sets (24 sections)
Stud bolt, 7/8" x 8"	32
Links, 15/16" hole	16 Sets
Guy roller, 15/16" hole	4
7/8" x 3" Machine bolt/MF locknut	4
Shear block	16
7/8" x 19" Threaded rod/2 standard nuts, 2 locknuts	2
7/8" x 21" Threaded rod/2 standard nuts, 2 locknuts	6
B. Anchor shackle	6
C. Fiberglass strain insulator	6

NOTE: Phase spacing (dimension X) varies with span lengths, conductor size, conductor sag, and midspan clearances.

FIGURE 7.9 A 69-kV single-circuit dead-end structure (45° to 90° angle in line).

100 ft. The composite-wood poles can be climbed without problems; they are lightweight, smooth, clean, and straight; and they have a satisfactory dielectric strength. The poles should have a long life as demonstrated by accelerated aging tests. EPRI studies indicate the composite poles should prove to be economical as a replacement for longer-length solid-wood poles.

Wood-Poles Fires. Fires on wood poles and crossarms can be initiated by leakage currents. Insulators on wood-pole lines can become contaminated by dust or chemicals in the air during dry weather. Light rain, fog, or wet snow can moisten the insulators, the crossarms, and the poles, causing leakage currents to flow to ground at the base of the pole or a ground wire on the pole. A dry area on the assembly, such as the space between the pole and the crossarm, can impede the flow of leakage current, establishing a voltage across the dry area which may approach the voltage between the line and ground. A small air gap may exist between the pole, the crossarm, and the through bolt securing the crossarm to the pole. An arc can be established across this air gap if the dielectric strength of the insulation of the air in the gap is exceeded. High humidity will reduce the dielectric strength of the air. The arc established across the air gap will ignite the wood in the dry area between the pole and the

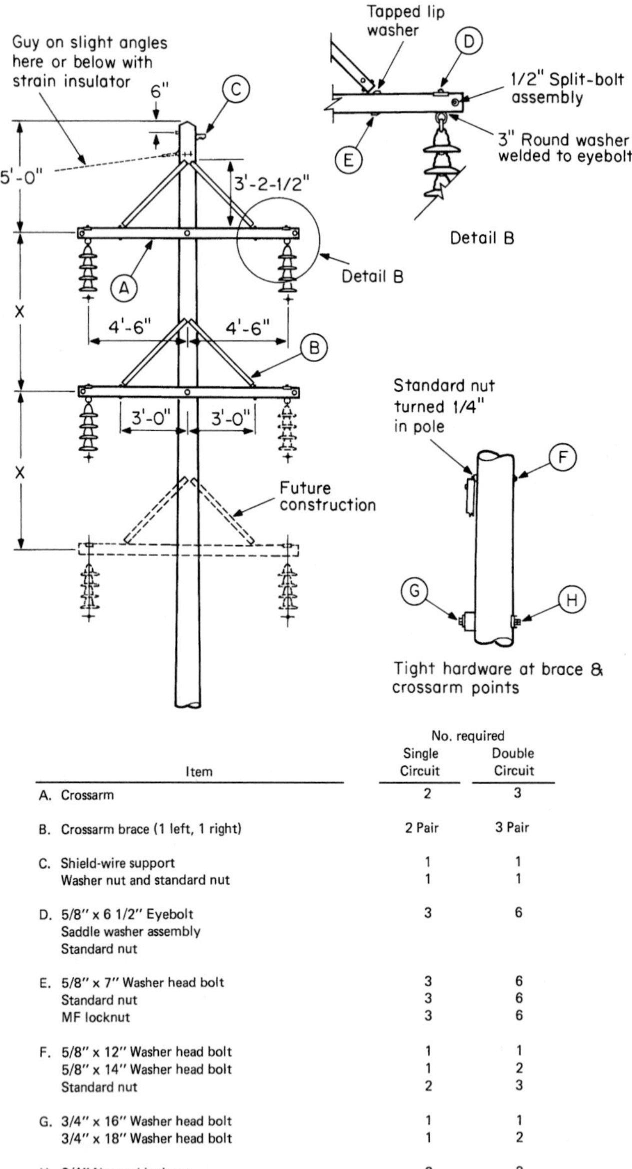

Item	No. required Single Circuit	Double Circuit
A. Crossarm	2	3
B. Crossarm brace (1 left, 1 right)	2 Pair	3 Pair
C. Shield-wire support	1	1
Washer nut and standard nut	1	1
D. 5/8" x 6 1/2" Eyebolt	3	6
Saddle washer assembly		
Standard nut		
E. 5/8" x 7" Washer head bolt	3	6
Standard nut	3	6
MF locknut	3	6
F. 5/8" x 12" Washer head bolt	1	1
5/8" x 14" Washer head bolt	1	2
Standard nut	2	3
G. 3/4" x 16" Washer head bolt	1	1
3/4" x 18" Washer head bolt	1	2
H. 3/4" Nut and locknut	2	3
Curved guide washer staked	2	3

NOTE: Phase spacing (dimension X) varies with span lengths, conductor size, conductor sag, and midspan clearances. Spacing will be determined in accordance with appropriate design standard.

FIGURE 7.10 A 69-kV double-circuit crossarm construction tangent structure (0° to 30° angle).

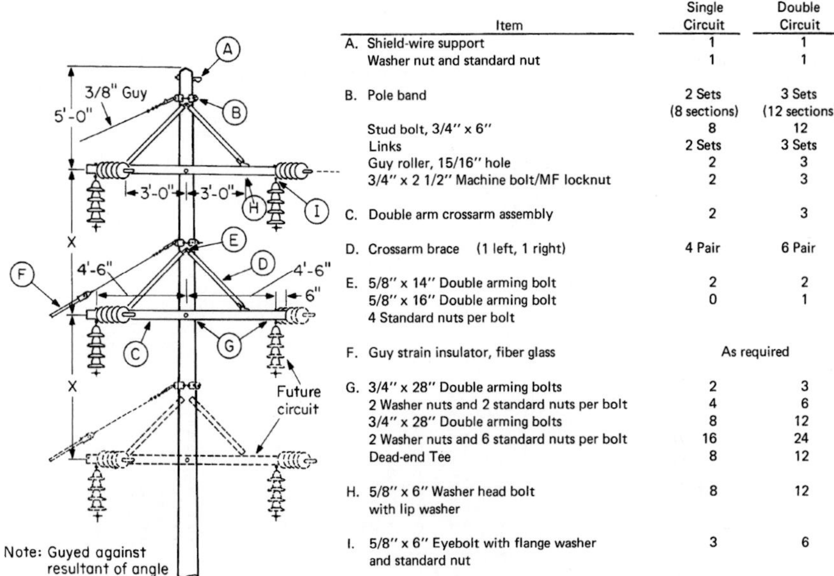

Item	No. required Single Circuit	No. required Double Circuit
A. Shield-wire support	1	1
Washer nut and standard nut	1	1
B. Pole band	2 Sets (8 sections)	3 Sets (12 sections)
Stud bolt, 3/4" x 6"	8	12
Links	2 Sets	3 Sets
Guy roller, 15/16" hole	2	3
3/4" x 2 1/2" Machine bolt/MF locknut	2	3
C. Double arm crossarm assembly	2	3
D. Crossarm brace (1 left, 1 right)	4 Pair	6 Pair
E. 5/8" x 14" Double arming bolt	2	2
5/8" x 16" Double arming bolt	0	1
4 Standard nuts per bolt		
F. Guy strain insulator, fiber glass	As required	
G. 3/4" x 28" Double arming bolts	2	3
2 Washer nuts and 2 standard nuts per bolt	4	6
3/4" x 28" Double arming bolts	8	12
2 Washer nuts and 6 standard nuts per bolt	16	24
Dead-end Tee	8	12
H. 5/8" x 6" Washer head bolt with lip washer	8	12
I. 5/8" x 6" Eyebolt with flange washer and standard nut	3	6

NOTE: Phase spacing (dimension X) varies with span lengths, conductor size, conductor sag, and midspan clearances. Spacing will be determined in accordance with appropriate design standard.

FIGURE 7.11 A 69-kV double-circuit crossarm construction for a 7° to 30° angle structure.

FIGURE 7.12 Two poles combined to make a H-frame structure. (*Courtesy Hughes Brothers.*)

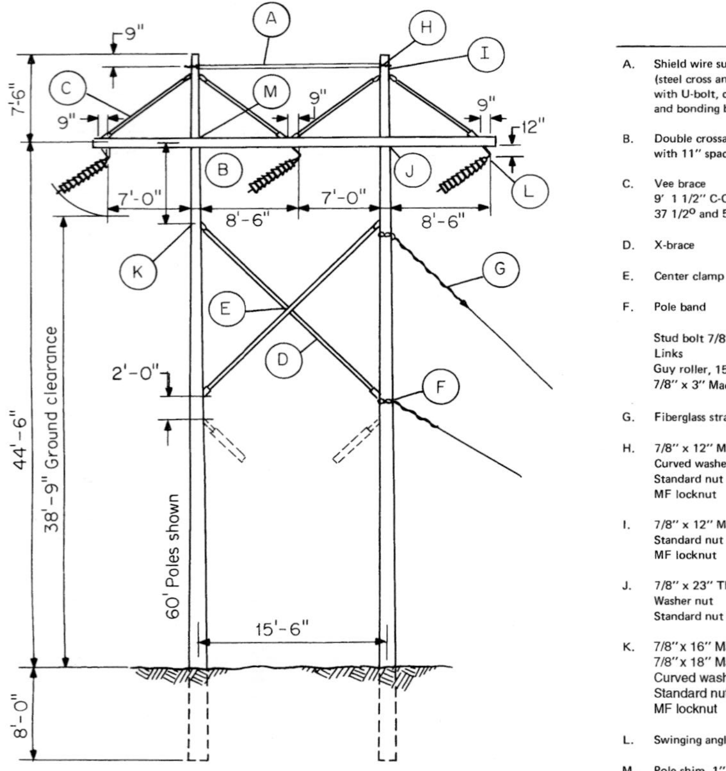

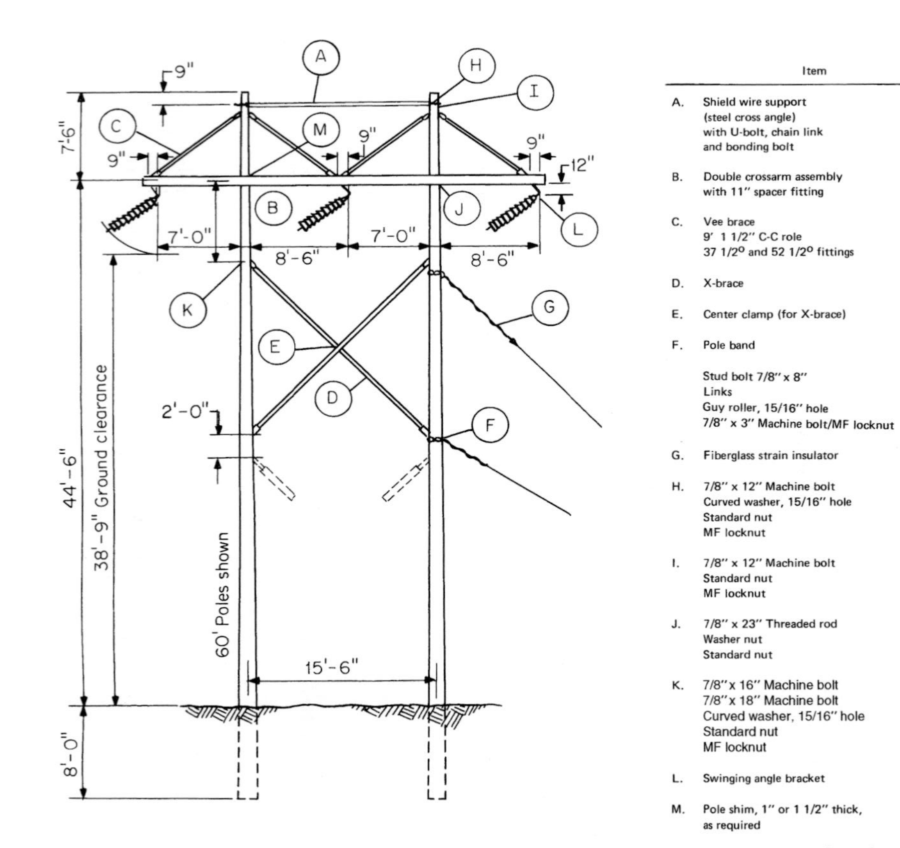

	Item	No. required
A.	Shield wire support (steel cross angle) with U-bolt, chain link and bonding bolt	1
B.	Double crossarm assembly with 11" spacer fitting	1
C.	Vee brace 9' 1 1/2" C-C role 37 1/2° and 52 1/2° fittings	4
D.	X-brace	2
E.	Center clamp (for X-brace)	1
F.	Pole band	1 Set (4 sections) 1 Set (4 sections)
	Stud bolt 7/8" x 8"	8
	Links	2 Sets
	Guy roller, 15/16" hole	1
	7/8" x 3" Machine bolt/MF locknut	1
G.	Fiberglass strain insulator	1
H.	7/8" x 12" Machine bolt Curved washer, 15/16" hole Standard nut MF locknut	2 2 2 2
I.	7/8" x 12" Machine bolt Standard nut MF locknut	2 2 2
J.	7/8" x 23" Threaded rod Washer nut Standard nut	2 4 4
K.	7/8" x 16" Machine bolt 7/8" x 18" Machine bolt Curved washer, 15/16" hole Standard nut MF locknut	2 2 4 4 4
L.	Swinging angle bracket	3
M.	Pole shim, 1" or 1 1/2" thick, as required	4

FIGURE 7.13 A 161-kV H-frame structure for installation where the line has an angle of $1^1/_2°$ to $6°$.

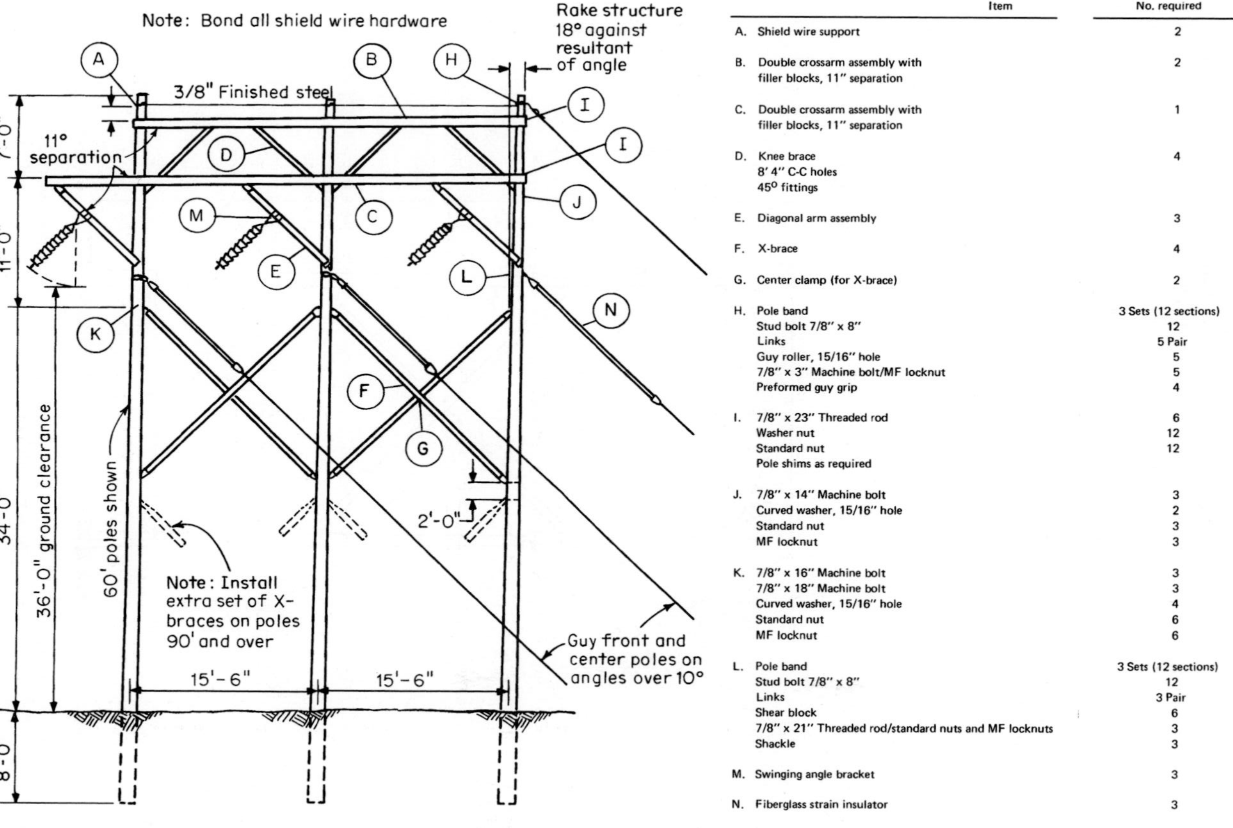

Item	No. required
A. Shield wire support	2
B. Double crossarm assembly with filler blocks, 11" separation	2
C. Double crossarm assembly with filler blocks, 11" separation	1
D. Knee brace	4
8' 4" C-C holes	
45° fittings	
E. Diagonal arm assembly	3
F. X-brace	4
G. Center clamp (for X-brace)	2
H. Pole band	3 Sets (12 sections)
Stud bolt 7/8" x 8"	12
Links	5 Pair
Guy roller, 15/16" hole	5
7/8" x 3" Machine bolt/MF locknut	5
Preformed guy grip	4
I. 7/8" x 23" Threaded rod	6
Washer nut	12
Standard nut	12
Pole shims as required	
J. 7/8" x 14" Machine bolt	3
Curved washer, 15/16" hole	2
Standard nut	3
MF locknut	3
K. 7/8" x 16" Machine bolt	3
7/8" x 18" Machine bolt	3
Curved washer, 15/16" hole	4
Standard nut	6
MF locknut	6
L. Pole band	3 Sets (12 sections)
Stud bolt 7/8" x 8"	12
Links	3 Pair
Shear block	6
7/8" x 21" Threaded rod/standard nuts and MF locknuts	3
Shackle	3
M. Swinging angle bracket	3
N. Fiberglass strain insulator	3

FIGURE 7.14 A 161-kV-line three-pole 6° to $17^{1}/_{2}^{\circ}$ angle structure.

crossarm. The leakage currents will help maintain the arc and the fire after it is ignited (Fig. 7.17).

Pole fires can be extinguished with a dry-chemical fire extinguisher designed for use on class C fires (electrical fires) followed by the use of a pressurized-water fire extinguisher on the burnt area after the burning has been minimized. The lineman must be careful to keep the water from the extinguisher clear of the high-voltage circuit while using it to cool the embers to prevent reignition.

Pole fires in locations where experience indicates they may develop can be prevented by cleaning the insulators and/or eliminating air gaps where arcing might develop. Pole-line hardware can be grounded, bridging the air gaps where ignition could occur. Metal shunts can be installed across the dry areas on the wood poles where fires might be started.

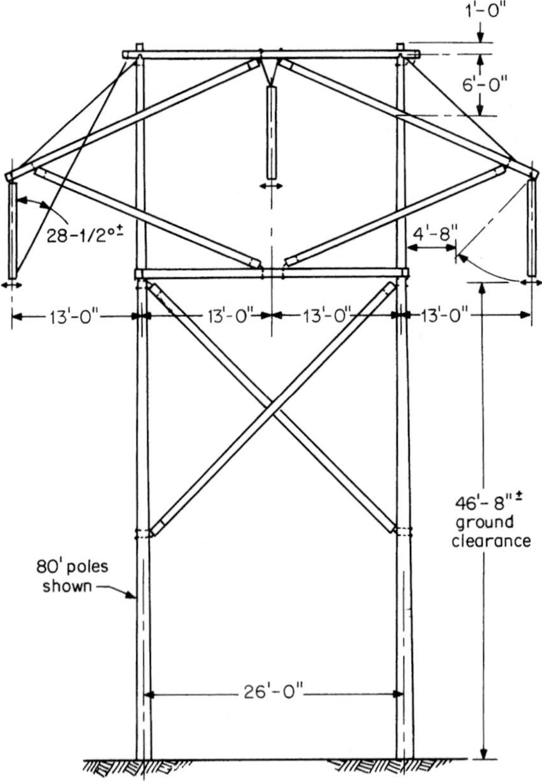

FIGURE 7.15 A 345-kV EHV ac transmission-line tangent structure. (*Courtesy Hughes Brothers.*)

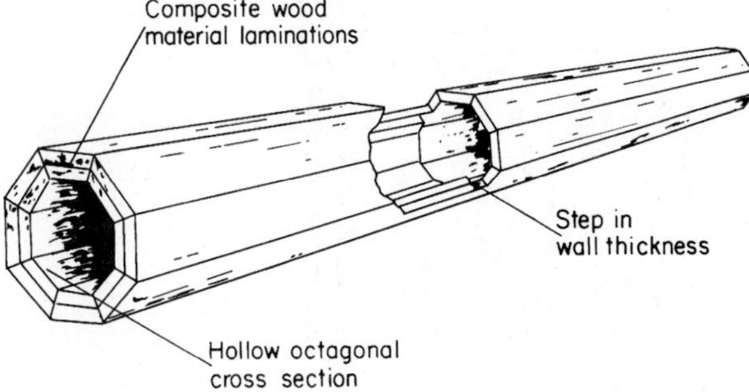

FIGURE 7.16 Sketch of EPRI-developed composite-wood material laminated pole.

FIGURE 7.17 Wood-pole fire between pole and crossarm around the through bolt.

CHAPTER 8

ALUMINUM, CONCRETE, FIBERGLASS, AND STEEL STRUCTURES

Environmental considerations have increased the use of aluminum, concrete, fiberglass, and steel to support electric transmission and distribution circuits. Wood-pole structures have generally proved to be the most economical for distribution and transmission lines operating at voltages through 345 kV. The design of structures to make them compatible with the environment has resulted in the use of materials and colors to comport with the natural surroundings. Aluminum, concrete, fiberglass, and steel structures, properly used, may reduce the widths of rights-of-ways required for distribution and transmission lines.

Aluminum Structures. Ornamental street lighting is one of the most common applications of aluminum poles (Fig. 8.1). The hollow-tubular street-lighting poles give a pleasing appearance. The poles are lightweight, making them easy to handle and install. The hollow pole facilitates the installation of the low-voltage cables supplying the lamps from underground electric distribution circuits.

Aluminum poles can be used for subtransmission circuits installed along urban streets (Fig. 8.2). Transmission lines operating at voltages through 500 kV have been constructed

FIGURE 8.1 Ornamental street-lighting poles made of aluminum.

FIGURE 8.2 Octagonal aluminum transmission-line pole.

FIGURE 8.3 Linemen working from bucket truck on conductors supported by aluminum H-frame structures.

FIGURE 8.4 Linemen assembling and preparing V-shaped guyed-aluminum towers to be airlifted by helicopter to the installation site. (*Courtesy L. E. Meyers Co.*)

with H-frame (Fig. 8.3) and guyed-aluminum towers (Fig. 8.4). Guyed-aluminum towers have also been fabricated in Y and V shapes. The towers can be completely assembled on the ground at a convenient location and airlifted to the site for erection using a helicopter. The structures can be airlifted, complete with insulators, stringing sheaves, tag lines for the conductor pulling, and guys. Airlifting the structures to the installation site can save time and money in inaccessible locations such as swamps, forest, or mountainous areas. Aluminum towers and poles do not need to be painted for protection, which minimizes the maintenance required.

Concrete Poles. Poles manufactured with concrete are used for street lighting and for distribution and transmission lines. Concrete poles have been used to improve the aesthetics of overhead circuits and are preferred where the life of wooden poles is unduly shortened by decay and woodpeckers. Compared to wood poles, concrete poles have disadvantages of being more expensive, lower in insulation level, more difficult to climb, heavier to handle, and more difficult to drill when necessary to install equipment. Concrete has advantages, compared to wood, of long life and availability on demand.

Hollow-type concrete poles are made by putting the concrete materials and steel reinforcing rods into a cylinder of the desired length and taper. Revolving the cylinder in a lathe-like machine for a period of 10 to 15 min then forces the concrete materials to the outside, thereby forming the hollow pole when the concrete cures. The hollow-type pole is lighter than the solid-type pole and provides a means for making connections through the pole to underground cables or services. The hollow-type poles are commonly used for ornamental street-lighting installations served from underground electric circuits (Fig. 8.5.)

The hollow-spun prestressed-concrete poles have a high-density concrete shell completely encasing a reinforcing cage containing prestressed high-tensile steel wires. Prestressing produces poles with a high strength-to-weight ratio that can be used for distribution and transmission lines. Table 8.1 lists typical data for hollow-spun prestressed-concrete poles equivalent to class 1 wood poles.

Figure 8.6 illustrates the method used to transport long prestressed-concrete poles, being used to construct a transmission line, to the site where they are to be installed. The poles are

FIGURE 8.5 Hollow-type concrete street-lighting pole.

TABLE 8.1 Characteristics of Hollow-Spun Prestressed-Concrete Poles Equivalent to Class1 Wood Poles

Pole length, ft					
Nominal	Actual	Above ground	Butt diameter, in	Approx. weight, lb	Moment capacity, ft-kips
30	32.8	27.8	14.57	2300	73
40	39.4	33.4	14.57	2600	88
45	45.9	39.4	15.75	3300	105
50	52.5	45.5	16.93	3900	122
60	59	51	18.11	4700	138
65	65.6	57.1	19.29	5500	155
70	72.2	63.2	20.47	6300	172
80	78.7	68.7	21.65	7200	188
85	85.3	74.8	22.83	8200	205
90	91.8	80.8	24.01	9200	222
100	98.4	86.4	26.38	10400	237
105	105	92.5	27.56	11600	255
110	111.5	98.5	28.74	12800	271
115	118.1	104.6	28.74	14000	289

Notes: 1. Equivalents are based on using a 2.5 overload factor for prestressed concrete and a 4.0 overload factor for wood.
2. Pole taper is 0.18 in/ft.
3. Poles longer than 90 ft are usually two pieces.
4. Moment capacity is ultimate moment required at ground line.
5. Ft-kips is 1000 ft - lb.

Source: PowerSpan, Division of Valmont Industries, Inc.

FIGURE 8.6 Hollow-spun prestressed-concrete poles loaded on truck parked at the edge of the right-of-way where a 161-kV transmission line is to be constructed. (*Courtesy Interstate Power Co.*)

unloaded from a transport truck with a crane (Fig. 8.7). The tall poles required are obtained by splicing two sections together. Galvanized-steel davit arms are usually bolted to the poles with a bracket using holes provided in the pole by the manufacturer. Suspension insulators can be fastened to the davit arms, and a crane can be used to set the pole and the assembled equipment attached to it in an augered hole lined with a split-steel casing (Fig. 8.8).

Concrete can be poured around the pole after it has been set and centered in the hole lined with a split-steel casing. The steel casing is removed with the assistance of a crane (Fig. 8.9). Figure 8.10 is a picture of the base of a concrete pole and its concrete foundation

FIGURE 8.7 Concrete pole has been unloaded from a transport truck with a crane. The crane is supporting the pole with the aid of slings and a cradle while the sections of the pole are bolted together and the equipment is attached by the workmen. (*Courtesy Interstate Power Co.*)

FIGURE 8.8 Concrete-pole assembly is being raised to set it in the hole that has been prepared for it with a crane and the aid of a Caterpillar tractor with an A frame installed on it. Note the split-steel casing next to the pole, which can be used to line the hole in the ground for the pole. (*Courtesy Interstate Power Co.*)

FIGURE 8.9 Crane and Caterpillar tractor with an A frame are being used to remove split-steel casing after concrete has been poured into hole around pole. (*Courtesy Interstate Power Co.*)

FIGURE 8.10 Base of a completed concrete-pole and foundation installation. A ground wire has been installed inside the hollow concrete pole and brought out at the base of the pole. The ground wire will be connected to a ground rod to be installed in undisturbed earth near the pole and the exposed extension of the steel reinforcing rods. The reinforcing rods in the pole and in the concrete foundation are connected together. (*Courtesy Interstate Power Co.*)

after concrete has set up and the forms have been removed. A finished concrete-pole structure with conductors and downguys in place is shown in Fig. 8.11. The top circuit is a 161-kV three-phase ac transmission line. The bottom, underbuilt circuit is a 69-kV three-phase ac transmission line.

Square-tapered prestressed-concrete poles are constructed by placing the stressed reinforcing material in a form and pouring the form full of concrete. The poles may have a center void developed by use of a cardboard or plastic tube. A copper wire is usually cast in the pole for grounding purposes. A prestressed-concrete 50-ft distribution pole has been found to deflect 2 ft before cracks appeared and 6 ft before failure. Characteristics of square-tapered prestressed-concrete poles are itemized in Table 8.2.

Large numbers of prestressed-concrete poles have been used in Florida for street-lighting, electric-distribution, subtransmission, and transmission circuits. Figure 8.12 pictures an ornamental square-tapered concrete street-lighting pole. An overhead 34.5Y/19.9-kV three-phase distribution circuit utilizing square-tapered prestressed-concrete poles is illustrated in Fig. 8.13. The pole shown supports a streetlight and three single-pole switches. The installation is clean and neat, satisfying visual environmental requirements.

Florida Power and Light Company has constructed 138-kV transmission lines using line-post insulators on single square-tapered prestressed-concrete poles. Two-pole H-type structures have been used for double-circuit 240-kV transmission lines.

Composite concrete and steel telescoping poles manufactured with a series of nested steel tubes fabricated from high-strength steel and pumped full of concrete can be used to support transmission circuits. Pumping concrete into the pole raises it to its full height, ranging up to 150 ft. The assembled telescoping structures are manufactured to be transported to the installation site on a standard trailer. The steel-reinforced concrete foundation is constructed with a tube to accept the telescoping pole. The telescoped pole is installed on

FIGURE 8.11 A precast hollow concrete pole with conductors for a 161-kV transmission line and underbuilt 69-kV subtransmission line. (*Courtesy Interstate Power Co.*)

the base with crossarms, insulator strings, stringing sheaves, and tag lines in place. The pole is pumped full of concrete, using a commercially available pump and transit-mix concrete. A 100-ft pole can be pumped to its full height in less than 15 min.

Fiberglass Poles. Fiberglass is immune to freezing, rotting, the pecking of birds, and damage from nails, all of which are injurious to wood poles. Fiberglass is an excellent insulator, even when wet from rain. Poles made with fiberglass do not require painting. Figure 8.14 shows a fiberglass pole installed in a residential area for a streetlight supplied from an underground distribution system.

Linemen use bucket trucks to eliminate the need for pole steps when fiberglass poles are located in areas accessible to trucks. Fiberglass poles are too expensive to be used extensively for overhead distribution circuits, but fiberglass rods have been used for insulator

TABLE 8.2 Tabulation of Square-Tapered Prestressed-Concrete-Pole Characteristics

Length, ft	Nominal square pole top, in	Minimum breaking strength, lb	Minimum load with no cracks, lb	Repetitive testing load, lb	Minimum crack closure after repetitive loading, lb
25	$4^1/_2$	1200	510	850	255
35	5	1300	600	900	300
45	6	2600	1000	2400	900
60	9	6000	2800	4200	1200

FIGURE 8.12 Square-tapered ornamental street-lighting pole in a residential area with an underground electric distribution system.

FIGURE 8.13 Prestressed-concrete pole supports three-phase 34.5Y/19.9-kV distribution line with single-pole sectionalizing switches and a streetlight. (*Courtesy S&C Electric Co.*)

supports (Fig. 8.15), eliminating the need for crossarms. Armless construction using fiberglass components is used extensively to enhance the appearance of overhead lines. Fiberglass components are reasonably priced and economically acceptable where the appearance of the facilities is important.

FIGURE 8.14 Fiberglass street-lighting pole.

FIGURE 8.15 Armless-type distribution line with fiberglass assembly mounted on wood pole to support porcelain insulators.

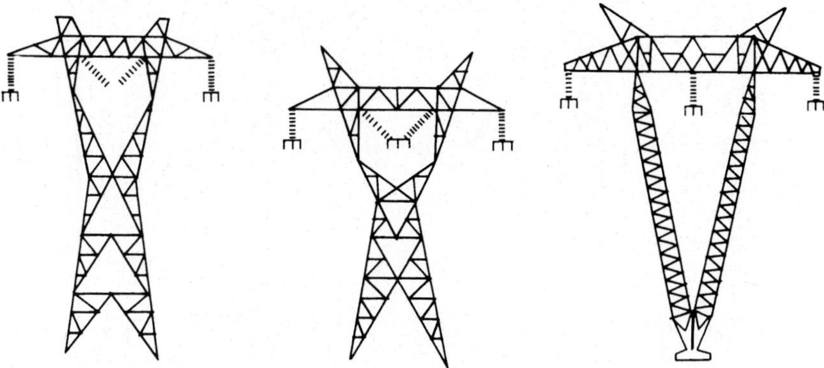

FIGURE 8.16 Standard tower designs.

Steel Structures. Steel towers have been used extensively to support subtransmission-line and transmission-line conductors. Standard types of steel towers are shown in Fig. 8.16. Towers of this type are required for 765-kV three-phase ac transmission lines (Fig. 8.17). Specially designed steel structures have been created to blend with the natural beauty of the landscape (Fig. 8.18). The size of transmission-line towers should be kept to the minimum feasible to preserve the environment and minimize the conflict with other use of the land. Tubular-steel poles have been used to support the subtransmission-line and transmission-line conductors energized at voltages of 345 kV and below (Figs. 8.19 and 8.20). Weathered galvanized-steel structures blend with the sky and reduce the silhouette appearance. Galvanized steel eliminates the need to paint the towers for several years, reducing maintenance costs. Steel is free from damage by woodpeckers and bacteria.

FIGURE 8.17 A 765-kV transmission-line angle tower in foreground and V-type guyed tangent tower in background. (*Courtesy American Electric Power Co.*)

FIGURE 8.18 Aesthetically designed transmission-line structures.

Steel transmission-line towers are of either the rigid or the flexible type. The rigid tower is firm in all directions, whereas the flexible tower is free to deflect in the direction of the line. Rigid towers, in order to be rigid, must have three or more legs, but the flexible construction consists of only two legs. Figure 8.21 shows the four-legged rigid-type tower. Rigid towers can resist the same strain in all directions. Flexible towers (Fig. 8.22) can resist strain only transversely or across the line and have very little strength in the direction of the line. They will give or deflect if there is any unbalance in conductor pull; in fact, they depend largely on the conductors to hold them in position.

Flexible H-frame steel structures are usually more expensive than wood-pole structures. The steel H-frame structures have many advantages that have increased their use. A 345-kV three-phase ac transmission line constructed across agricultural land in a heavy loading district can have a ruling span of 1400 ft compared to 850 ft for wood-pole structures when

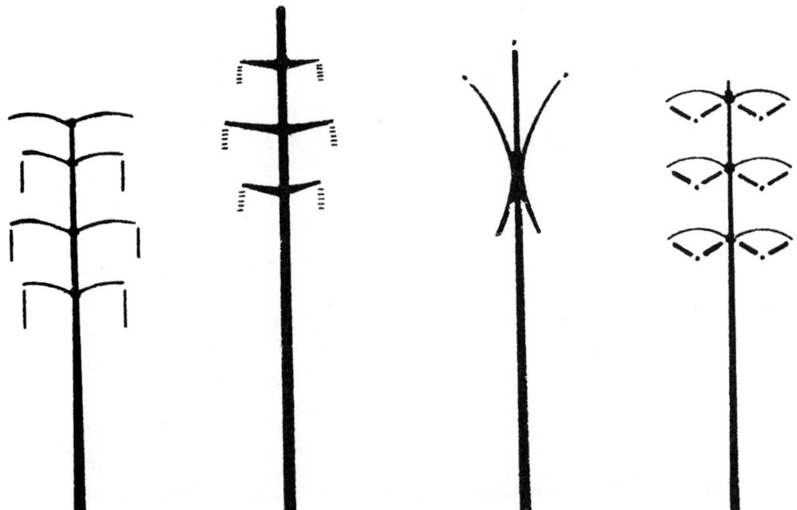

FIGURE 8.19 Inconspicuous steel transmission-structure designs.

FIGURE 8.20 Steel pole supporting double-circuit 161-kV transmission line.

FIGURE 8.21 Four-legged rigid-type 230-kV transmission tower.

the conductor is 954-kCMIL 45/7 ACSR two-conductor bundle. The number of structures required is reduced considerably, minimizing the conflicts with cultivation of the land and reducing the cost of the right-of-way.

The cost of structure assembly of steel H-frame structures is estimated to be approximately 50 percent of that for standard rigid-latticework steel towers for this type of circuit; it is also less than that of wood-pole structures as a result of the reduced number of structures. The smaller number of structures minimizes the number of insulators and the line hardware required and reduces the labor costs to string, sag, and clip in the conductors. The total cost for the construction of this type of line is approximately the same for rigid and flexible steel towers.

FIGURE 8.22 Linemen performing hot-stick maintenance work on H-frame two-pole steel flexible transmission-line tower operating at 230 kV ac. (*Courtesy A. B. Chance Co.*)

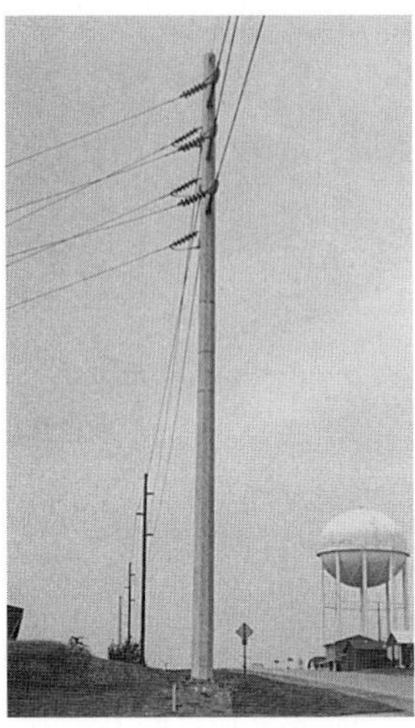

FIGURE 8.23 Galvanized-steel pole used with wood poles in a 69-kV three-phase subtransmission line to eliminate the need for guys and anchors.

Steel poles can be used to establish a rigid structure in subtransmission and distribution lines to eliminate the need for guys and anchors (Fig. 8.23). Steel poles are frequently installed for angle structures in lines with wood poles used for tangent structures. Elimination of the guys and anchors improves the appearance of the facilities, enhances the environment, and solves right-of-way problems in urban areas. The use of line-post insulators and armless-type construction minimizes the silhouette established.

CHAPTER 9
DISTRIBUTION AUTOMATION

Electric systems are becoming more remotely monitored controlled, and automated to improve reliability and power flow. Vendors are introducing communication equipment that retrofits existing isolation devices to control the isolating equipment. The new equipment allows the Control Room personnel (described in Chap. 2) to utilize the supervisory controlled and data acquisition (SCADA) system to not only monitor the system but also to perform remote site switching or isolating if needed. Personal computers and personal computer software applications have aided the advancement of distribution automation. A communication system must be in place to transmit the signals via the SCADA system to the equipment.

MONITORING

Past and present distribution automation projects are aimed toward system monitoring. Monitors are placed on the distribution system to sense momentary and continuing power outages, under and over voltages, and excessive fault currents. In some instances the status device provides feedback to the SCADA system. The majority of devices simply log or display status and must be field verified to obtain system information.

Circuit fault indicators shown in Fig. 9.1 may be installed in either overhead or underground applications. An example of an overhead application is the utilization of the device just beyond a three-phase recloser positioned as shown in Fig. 9.2. In the event of a single-phase line-to-ground fault, the recloser (reclosers are referenced in Chap. 18) will lock out after progressing through the desired sequence of trips and recloses. The indicator will sense fault current and remain in the faulted position. It is important to note that the fault indicators installed behind automatic reclosing devices must have the inrush current restraint option or else the non-faulted unit or units will falsely indicate faults. The lineman dispatched to troubleshoot the outage verifies which of the three phases experienced the fault as shown on the face of the indicator and then traces the faulted phase or phases out until he determines the cause of the fault.

Underground fault indicators are beneficial in large underground systems that require trouble-shooting. The benefits are briefly described in Chap. 34. Figure 9.3 diagrams a large underground system that has several indicators, and a cable fault on the system has occurred. The lineman troubleshooting a faulted underground section verifies that fault current has flowed through the indicator and proceeds to the next location. The fault will be in the cable section between the last indicator flagged with a fault and the first indicator that has not registered a fault. Fault indicators installed on concentric neutral or tape shield neutral cable must be installed in the area with the neutral either completely removed or doubled back or else the device will not function properly.

Event recorders are available for many distribution devices. Capacitor (capacitors are referenced in Chap. 1) controllers may be monitored with devices as shown in Fig. 9.4. The

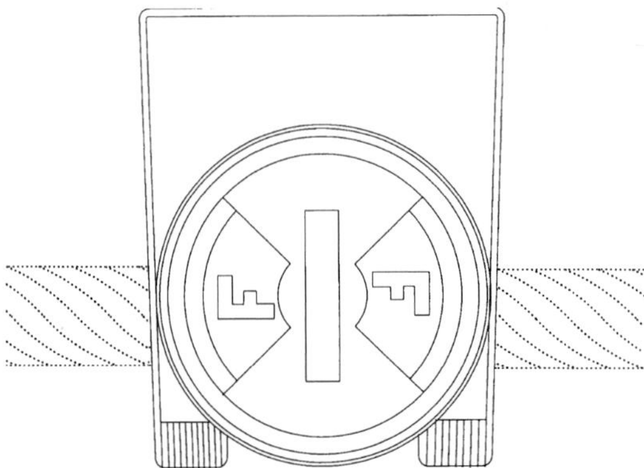

FIGURE 9.1 Circuit fault indicators are installed in either overhead or underground systems, and the indicators are beneficial in speeding the diagnosis to locate system problems. (*Courtesy MidAmerican Energy Co.*)

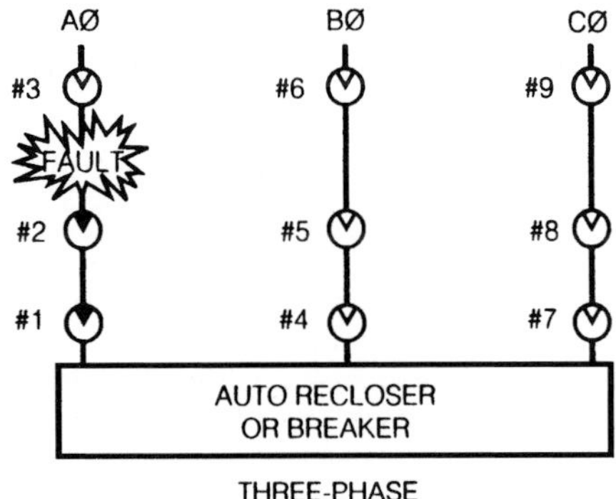

FIGURE 9.2 Fault indicator positioned on an overhead line beyond a three-phase recloser. Based on the display of the indicator, the lineman will be able to more rapidly track the cause of the fault and attempt to restore electric service to the customer without electric service. (*Courtesy Cooper Power Systems, Inc.*)

monitor records voltage and temperature data. In the event of a system disturbance or fault, the recorder stores and logs this information. The information may be either retrieved with a laptop computer taken to the worksite, with a telecommunications link to the SCADA system, or with a telecommunications link to a personal computer in the office.

Another advantage of monitoring devices is their ability to provide warning indications of malfunctioning equipment prior to the equipment being called on to operate. For example, monitors installed on the capacitors indicates status points or error conditions as shown in Fig. 9.5. If the error condition existed, the equipment may be maintained as soon as the error message appears and the capacitor will be available as needed.

Conventional voltage sensing controls should be served whenever possible from a dedicated or a lightly loaded transformer. The controls have over- and under-voltage overrides as well as voltage recording capabilities. If it is connected to the secondary of a transformer serving other customers, the controls will be directly affected by those customer's load

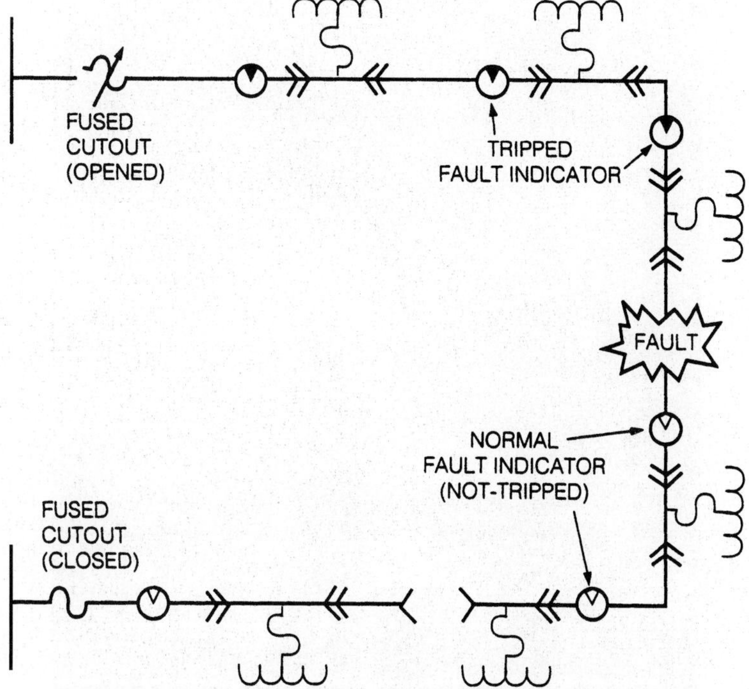

FIGURE 9.3 Schematic of an underground looped distribution system with fault indicators located to aid the lineworker to observe the indicator target to isolate the problem area. Locators may be installed with visible target mounted flush on the door of the pad-mounted gear or installed so that gear must be opened to observe target. (*Courtesy Cooper Power Systems, Inc.*)

instead of the main line voltage, and the voltage data will be corrupted. A "fixed load" transformer is manufactured specifically for this purpose.

AUTOMATED SWITCHING, SECTIONALIZING, AND FAULT ISOLATION

Three-phase, group-operated disconnect switches are ideal isolation devices offering the opportunity to automate the distribution system. Hydraulically powered or motorized operators may be installed in the handle mechanism to control the operation of the switch as shown in Figs. 9.6 and 9.7. Monitors mounted on the base of the switch blade contact mechanism provide system status signal to a remote terminal unit (RTU). A radio link to the SCADA system communicates the RTU information and alerts the control room personnel of a change in system status. The control room personnel may then send a signal back to the RTU. The RTU and operator are linked via plug connectors. The operating mechanism changes the status of the switch.

Pad-mounted switchgear equipment (pad-mounted switchgear equipment is referenced in Chap. 17) with two incoming sources may be controlled such that if power is lost on the incoming primary source, the back-up primary feed will be switched to serve the load. Thus an automatic load transfer has occurred. The switching transfer will not occur if the control device senses a fault on the load side of the line.

FIGURE 9.4 Lineman interrogating capacitor bank event recorder installed on pole at a height out of reach of typical vandalism. The information is being downloaded into a laptop personal computer that is not shown. (*Courtesy EnergyLine Systems, Inc.*)

In the completely automated disconnect switch distribution system, the system senses a fault has occurred and the isolating devices are activated without manual intervention to reconfigure the system to isolate the least amount of customers affected by the fault from the rest of the system. The customers without electric power will have service restored once the cause of the fault on the isolated section is investigated.

AUTOMATED VAR CONTROL

Capacitor Switching. A switched capacitor bank is energized or connected to the line on a given schedule. It is sometimes switched in a number of steps, because of its size. The schedule is determined by time, temperature, circuit voltage, or other considerations. It is switched on and off to meet circuit requirements. The switching can be done manually or automatically. A nonswitched capacitor bank is connected to the line at all times and is disconnected only for maintenance purposes.

```
EnergyLine Prog. Cap. Control - Basic Version
Phys. Location:              C-746 13-47-1, Rt 84 & 14 St, Rapid City, IL
Program ID:             C2RP15X1
Upload Time             10/27/94  2:38:11 pm
P.C. Upload Time              2:42:58 pm 10/27/94
Bank Status at upload ...
Voltage:                  123.4
Temperature:                       62
Cap. Bank State:             Out of Circuit/Switch Open
Control Mode:                Automatic Control
Voltage Override Status:                  None
Error Conditions:            All Sensors O.K.
Auto. Control Strategy:            Temperature w. Voltage Override
Softw. Man. Oper.:                 Disabled
Softw. Man. Oper. Requested:              Open Switch
Softw. Remote/Local Ena.:                 Remote
Hardw. Man. Override Sw.:                 Auto
Hardw. Remote/Local Sw.:                  Local
Low and High Voltage Overrides:                   118.0   126.0
Bank Voltage Rise Plus Hysteresis:                  3.8
High and Low Comm. Addr.:                 00              00
Summer and Winter Control Strategies:                Temperature                      Temperature
Start., End Months of Summer:             May                   September
Daily Operating Schedules ...
Sunday                 Saturday                           8      00       23      00
(n.a.)                 (n.a.)                             0      00        0      00
Automatic Holidays:                Enabled
Holidays by date ...
January                       1
July                          4
December                     25
(unused)               --
(unused)               --
(unused)               --
(unused)               --
(unused)               --
(unused)               --
(unused)               --
(unused)               --
(unused)               --
(unused)               --
(unused)               --
(unused)               --
High and Low Temp. Setpts:                      85      82
Temp. Change Time Threshold:              10 Min
Under/Over Voltage Override Time Threshold:                60 Sec
Max. Auto. Control Cycles/Day:                        4
Cap. Bank Motor Pulse Time:                7 Sec
Cap. Bank Minimum Sw. Voltage:                    100.3
Daylight Sav. Time Auto. Changeover:                 Enabled
Total Sw.Cycles Since Installation:                217
Daily Low, High Volts/Temps, and Cycles Follow ...
  101.5    123.6       52       71       1
  117.4    122.6       50       65       1
  116.4    124.5       49       66       1
  117.4    124.6       52       67       1
  116.8    124.9       51       72       1
  117.4    124.5       52       78       1
  116.1    124.0       56       80       1
  117.0    124.3       46       56       1
  108.4    122.9       39       64       4
  115.9    124.3       35       65       1
  117.2    124.3       35       69       1
  117.2    124.0       37       71       1
```

FIGURE 9.5 Sample sheet of capacitor bank controller event recorder. (*Courtesy MidAmerican Energy Co.*)

FIGURE 9.6 Automated three-phase, group-operated switch. Note the current transformers mounted on the top of the rack at the same level as the disconnect blades. (*Courtesy Kearney-National Company*)

Capacitor Connections. Figure 9.8 shows a connection diagram for a typical three-phase capacitor bank. The oil switch indicated can be a three-pole device or three single-pole devices that can be controlled automatically to switch the bank in or out to control power factor or to regulate system voltage (Fig. 9.9). Figure 9.10 shows a capacitor application using

FIGURE 9.7 DC motor driven remote operator unit controlling the handle of a three-phase, group-operated disconnect switch. (*Courtesy Kearney-National Company*)

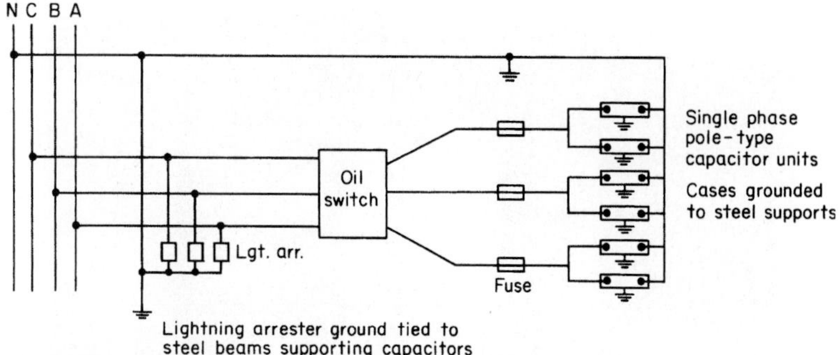

FIGURE 9.8 Typical connection diagram of three-phase capacitor bank.

FIGURE 9.9 Lineman is adjusting automatic control devices for a capacitor-bank installation. Upper control on panel senses loss of individual capacitor units and initiates bank switching before damaging overvoltages occur. Lower control senses line voltage and initiates switching for voltage regulation. (*Courtesy S&C Electric Co.*)

FIGURE 9.10 Pole-top microprocessor controlled capacitor bank. (*Courtesy Maysteel Electric Utility Products & Enclosures*)

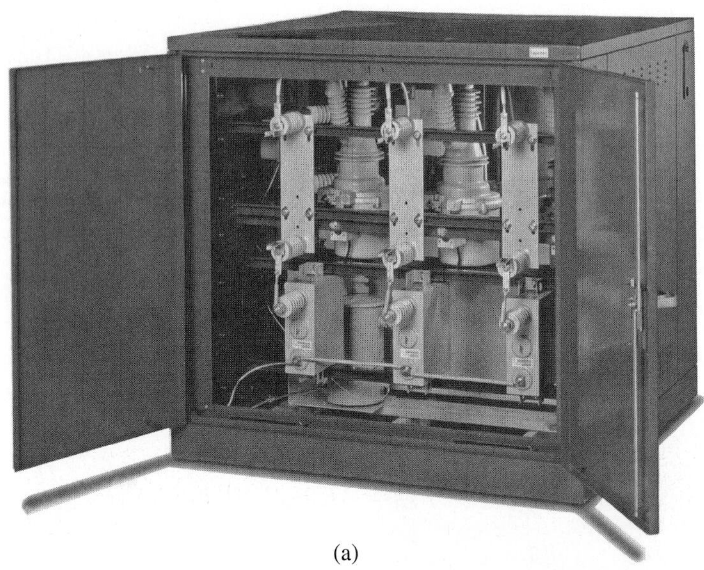

(a)

FIGURE 9.11 Pad-mount microprocessor controlled capacitor bank (a) front panel removed to display the remote controlled switches and capacitor units. (b) dead-front and clear see-through barriers installed. (*Courtesy Maysteel Electric Utility Products & Enclosures*)

remote controlled capacitor switches and Figs. 9.11 (a) and (b) show a three-phase pad-mount capacitor bank utilizing remote controlled capacitor switches.

Capacitor banks are turned on and off by oil or vacuum switches. The switches may be activated based upon a clock setting, voltage, or radio controls. The clock settings are established

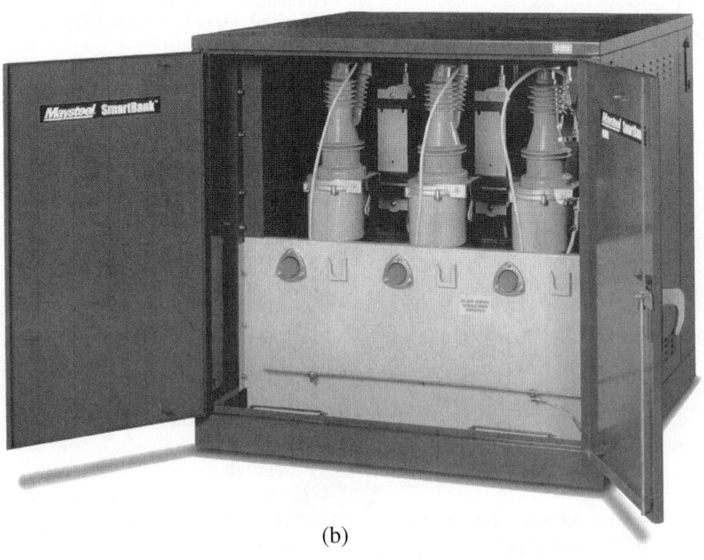

(b)

FIGURE 9.11 (*Continued*)

FIGURE 9.12 Static Var compensator on the manufacturer's loading dock waiting for shipment to the customer. (*Courtesy of Power Quality Systems.*)

based upon engineering system studies. The clocks may be adjusted in the spring and fall for standard time zone changes. Capacitor banks may be brought on line based upon the voltage on the circuit. Sensors monitor the voltage of adjacent customers and supply the desired reactive power flow required to compensate the system. Radio controlled banks are activated based upon MW, MVAR, and MVA readings in the SCADA system. A radio signal is sent to the switch to alter the status of the capacitor bank.

Static Var Compensators. The static var compensator is a device utilized when a simple electric distribution system capacitor will not meet customer power quality acceptance. Due to harmonic and switching transient issues associated with capacitors, static var compensators are utilized for power factor correction. The static var compensator (Figs. 9.12 to 9.16) stabilizes line and system voltage instantaneously as the reactive component of customer load varies.

800/900 MHz RADIO SYSTEMS

Upgraded communications systems are replacing limited low-band radio systems. The high-band systems provide users greater flexibility, clearer reception, and more channel options.

Users of low-band systems had limited range and had to compete for air time. Limited range users often had communications that were walked on or cut out by other users not realizing the channel was being utilized. The high-band systems have provided users increased flexibility as the users may be divided into call groups. Each call group is assigned a designation on the radio. Unlike the low-band system in which only limited frequency and corresponding channels were available, the call groups access the radio system and the systems hunt to find an open channel among the several channels available. Thus, the users have freer access to a radio channel to communicate vital information.

FIGURE 9.13 The Static Var compensator has arrived at the utility service center pending installation at the job site. (*Courtesy of Power Quality Systems.*)

The 800/900 MHz system requires multiple antenna sites that are linked utilizing fiber-optic or microwave communication systems. The linking of antennas provides simultaneous radio conversation broadcasting, thus linking wide areas of the communications operating territory. Distance limitations of the low-band systems were on the order of 50 to 60 miles. The distance limitation of the high-band communications is bound only by the positioning of antenna sites.

FIGURE 9.14 The Static Var compensator is being installed on the distribution system one pole upstream of the customer service entrance. (*Courtesy of Power Quality Systems.*)

FIGURE 9.15 Three Static Var compensators installed on the distribution system to perform voltage regulation, flicker control, and power factor correction. (*Courtesy of Power Quality Systems.*)

FIGURE 9.16 Platform mounted static Var compensator installation. (*Courtesy of Power Quality Systems.*)

PERSONAL COMPUTER APPLICATIONS

System data may also be gathered and monitored on personal computers. Computer software vendors are marketing programs to improve electric system reliability and performance. Examples of software programs available are transformer load management, protective device coordination, and outage trouble response. The data is then utilized to study the system to determine required corrective action.

Laptop personal computers are becoming a part of the lineman/cableman's work tools. Laptops are stationed in the truck. Outage orders are dispatched to the truck via wireless communication from the central office mainframe computer to the truck-mounted computer terminal. The lineman/cableman proceeds to each order noting specifics that provide estimated restoration times for the customer service representatives to relay to all customers calling in reference to that specific order. Upon completion of the order, the lineman/cableman indicates causes of the outage and denotes a completed status and then proceeds to the next order in his assigned responsibility area.

FIBER-OPTIC POSSIBILITIES

Fiber-optic systems offer unlimited potential to distribution companies to provide interactive two-way communication service to their customers. The companies may implement new functions to improve system restoration, such as immediate detection of customer outages, fault detection and immediate system isolation, and improved power quality.

Communication systems have been greatly enhanced by the development and advancement of fiber optic cable. The fiber optic cable system allows greater amounts of communication data to be passed via light waves versus an electrical signal sent on a traditional twisted paired copper wire system (one wire to send the signal, the other to receive the signal). Another advantage of fiber optic systems is that the quality of service is not affected by the outside influence of electromagnetic fields.

Utilities are installing fiber optic communication systems in a variety of ways that also incorporate the benefits of fiber optic technology. Some fiber optic systems are being installed as a part of the overhead transmission shield wire. The shield wire cable utilized is called optical ground wire (OPGW) (Fig. 9.17). Other overhead fiber optic systems are being installed in the traditional communication zone as defined by the *National Electrical Safety Code* (NESC) on the distribution pole. All dielectric, self-supporting (ADSS) cables are used for this application. All dielectric cables by definition do not carry current, radiate energy, or produce heat. Finally, in cases of underground systems, fiber cables are being installed in high-density polyethylene (HDPE) conduits.

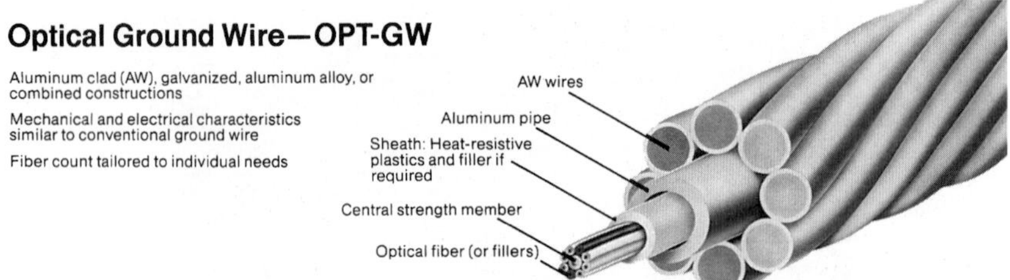

FIGURE 9.17 Optical ground wire (OPGW) cable utilized as the shield wire of a transmission line. (*Courtesy of Alcoa Fujikura Ltd.*)

Each fiber conductor in the cable transmits beams of light from the source end to the receiving end. The fiber conductor is made of glass, silica, or plastic. A thin coating is added to the fiber conductor to keep the light contained. Fiber optic sensors convert the light signals to electric signals for processing.

Of course if the utility installs the fiber optic system, it must design, construct, operate, and maintain the system. The new challenges of the fiber optic system present the lineman opportunities to learn new skills, such as splicing, fusing, and terminating the fiber. New safety precautions unique to the cable are required. Lasers, glass scraps, and chemical cleaners are all items for the lineman to be aware of.

POWERLINE CARRIER

Powerline carrier systems are in place to provide channel capabilities for transmission line relaying (refer to carrier equipment in Chap. 3). Utilization of the powerline carrier communication medium for distribution automation purposes is being explored. The amount of data transmitted per second is slow and voice communication is limited, yet this communication medium performs well over short distances. The powerline carrier transmitters and receivers are capacitively coupled to the transmission line. Reliability is a not a concern as back-up power is provided when the main power source is out to keep the system in operation.

AUTOMATED METER READING AND METER DISCONNECTS

The ability to remotely read and control electric meters is now a reality. The solid-state electric meters have the ability to store data and upon demand communicate the stored data to devices at remote sites. The meters may be controlled to either turn on or turn off power. Meters may be programmed to limit consumption to specified amounts to take advantage of demand side management incentive rate programs.

CHAPTER 10
EMERGENCY SYSTEM RESTORATION

Customers expect reliable electric service due to their ever-increasing dependence upon and utilization of electric equipment and electronic devices. Lifestyle conveniences are associated with reliable electric service. Electric service interruptions negatively affect reliability. Seldom is the lineman more visible to the community than during the repair of service interruptions to the electric transmission and distribution system. The service interruptions occur for a variety of reasons. However, emergency restoration situations caused by storms generated the most publicity.

Storms vary in terms of magnitude and severity. Electric system damage may be localized or widespread. Storms may affect several hundred customers or several million customers. Storm damage clean-up and electric system restoration may require only local labor, equipment, and materials for repair or the storm may require both local and mutual assistance emergency manpower response, equipment, and materials. Storms are caused by winds, lightning, ice, or other naturally occurring weather events. There are several recent examples of the devastating damage and destruction created by intense storms throughout the Untied States and Canada. The 2005 hurricanes, Katrina and Rita, that devastated New Orleans, Louisiana and Gulf coastal regions of Alabama and Mississippi, the series of 2004 Florida hurricanes, Charley, Frances, Ivan, and Jeanne, and the 1998 ice storm in Quebec and the Northeast United States are just a few examples.

The rescue and recovery of human life remains top priority in electrical emergency restoration situations. Successful electric system restoration depends upon recognizing and assessing the magnitude of the situation, directing resources safely and efficiently, acquiring materials, and keeping everyone informed as to the details of the situation. While the exact date and time of emergency electric system situations cannot be typically predicted, they can be prepared for and thus dealt with in a methodical manner in order to minimize service disruption to customers.

The lineman's performance is critical in the restoration process. Sixteen hour work days on a continuous basis are required. The lineman relies upon many factors in order to be successful. This chapter will delve into several of the aspects of the restoration process required in emergency conditions.

EMERGENCY CONDITIONS

Linemen are expected to respond to electrical system emergencies caused by storms (Ice (Fig. 10.1), tornados (Figs. 10.2 and 10.3), hurricanes (Figs. 10.4 and 10.5)), sabotage or terrorism (both to equipment and computer systems), or catastrophic equipment failure

FIGURE 10.1 Conductor of a 161-kV transmission line with ice measuring 5 in. in diameter. The devastation resulted from the ice and wind conditions. (*Courtesy L. E. Meyers Co.*)

(fires or explosions). Other types of emergencies include serious injuries to employees or to the public (including downed wires or downed poles) or environmental emergencies (oil spills and fuel leaks). Storm damage is typically more widespread and thus requires more flexibility in terms of the response to the event as opposed to the other emergencies that are more localized in nature.

FIGURE 10.2 Winds in excess of 90 mph damaged this 345,000-volt transmission line structure.

FIGURE 10.3 Another view of wind damage to a transmission line structure.

Business continuity plans, disaster recovery plans, mutual assistance plans, storm plans, and general switching procedures are established to ensure that organizations are prepared for the unexpected. The plans and procedures include strategies to be followed to resume business activities. Frequently, the plans and procedures are reviewed and practiced with employees for clarity and understanding. Mock storms utilizing past storm data are scheduled.

EMERGENCY SWITCHING PROCEDURES

Critical to the lineman is the assurance that upstream protective devices are controlled such that a portion of the de-energized system will not become energized unexpectedly. At all times, the lineman expects the system to be properly monitored and controlled as well as trouble orders efficiently dispatched. All conductors are treated as energized unless confirmed as isolated and grounded.

If mutual assistance emergency response linemen from other companies are involved, the review of the general switching procedures occurs. Topics of discussion include hot line warning procedures, clearance procedures, and hold order procedures. Safety rules are also addressed.

FIGURE 10.4 Structure with transmission and distribution facilities damaged in a hurricane.

STORM PLAN

Storm plans provide individuals guidance, direction, and expectations regarding defined roles that each person performs in order to maximize the utilization of everyone involved in the recovery effort. Plans emphasize restoring electric service as safely and quickly as possible. Public health and safety as well as environmental concerns are considered. Coordination with city, state, and federal authorities are addressed. While the plans are well defined, flexibility to adapt to each unique case is permitted.

Many individuals with varying backgrounds and past experiences are utilized to support the lineman for storm recovery. People are needed to:

- Answer phones
- Assess storm damage
- Supervise crews
- Dispatch orders
- Investigate wire down orders
- Order materials
- Support computer systems
- Assist and guide mutual assistance crews
- Consult industrial customers
- Perform system planning analysis studies

Coordination of these tasks is the responsibility of storm directors, managers, and supervisors.

Storm plans address logistics concerning mutual assistance emergency response manpower, tree trimmers, fleet, materials, lodging, meals, and communications (phones, computers, radios,

FIGURE 10.5 Distribution facility damaged in a hurricane.

fax machines, and printers). The first few hours of response are critical as the situation is properly assessed to ensure appropriate action.

Assessment. Initial assessment data includes:

- Number of locked out circuits
- Number of trouble orders
- Number of customers out
- Number of broken poles
- Number of wires down calls
- Number of police/fire personnel standing by calls
- Number of tree damage calls
- Number of spans of downed conductor
- Number of damaged transformers
- The amount of local manpower available to assist
- Forecasted weather conditions
- Input from other emergency agencies

FIGURE 10.6 Emergency response tree trimmers enroute to assist in restoration work due to hurricane damage. (*Courtesy Asplundh Tree Expert Co.*)

Based upon the initial assessment of the situation, accurately determining sufficient staffing requirements related to field crews and how to fill the staffing needs are determined. In many cases, manpower is supplemented with mutual assistance emergency response crews (Figs. 10.6 and 10.7). Manpower needs include:

- Linemen
- Servicemen
- Groundmen
- Substation electricians
- Metermen
- Tree trimmers
- Mutual assistance emergency response crews

The tree trimmers are required to clear debris from the facilities. In some cases, the tree trimmers are sent out to some areas in advance of the linemen and in other instance, the tree trimmers are sent out with the linemen (Fig. 10.8).

FIGURE 10.7 Tree trimmers responding to request for mutual assistance. (*Courtesy Asplundh Tree Expert Co.*)

FIGURE 10.8 Tree crew on-site and repairing ice storm damage (*Courtesy Asplundh Tree Expert Co.*)

Logistics involved with mutual assistance emergency response crews involves staging areas for assignments, work and rest hours, lodging, meals, refueling of vehicles, first aid treatment facilities, and laundry.

A various assortment of fleet are considered for utilization.

- Single-bucket aerial devices
- Double-bucket aerial devices
- Material handlers
- Digger derricks
- Dump trucks
- Pick-up trucks
- All terrain vehicles
- Cranes
- Trenchers
- Back-hoes

Working height requirements of the aerial devices range from 25 to over 150 in.

Material needs are addressed. Adequate reserves of poles, conductor, cross-arms, and connectors are typically on-hand. However, as the assessment of the situation continues, material requirements become more defined and arrangements are made so that repair crews are well-stocked. Customers are not interested in excuses if the power is not restored due to a shortage of a given piece of material.

Material needs include standard sizes and sufficient quantities of:

- Poles
- Overhead transformers
- Overhead conductors, both primary and secondary

- Cross-arms, both in-line and dead-end arms
- Braces
- Insulators
- Steel pins and pole-top pins
- Top ties and spool ties
- Dead-end clamps, hot line clamps and bails, and service wedge clamps
- Machine bolts, oval-eye bolts, and double-arming bolts
- Washers and nuts
- Cut-outs
- Arresters
- Fuse links
- Compression connectors
- Sleeves and splices
- Guy wires and associated guying materials

Not to be forgotten in the assessment portion of the recovery operation is the need of the environmental service division and of the legal department. Having both of these areas plugged into the effort pays dividends in the future.

Repair. After all of the information has been processed, the recovery moves from an assessment stage to one of repair. People and systems (including computer systems and radio and data communication systems) are in place to support the lineman. Decisions are made regarding the staffing of the office support teams. Round the clock coverage occurs, both for the office staff and for the lineman. The lineman responds to prioritized trouble orders. Orders are prioritized based upon public health and safety issues. Critical transmission and substation infrastructure issues receive the next level of attention followed by locked out circuits. To isolate the affected areas in order to restore service, the lineman is directed utilizing switching procedures.

Job briefing conferences are conducted at defined intervals, typically at the start of each designated shift to review current status of customer interruptions, goals for the period, work assignments, and safety reminders.

As the recovery effort unfolds, it is important to communicate both internally and externally to address how things are going. Predetermined arrangements are worked out with the public officials and the press so that press briefings share critical announcements with everyone. Press announcements inform the public as to the status of the recovery process. Electrical safety reminders and tips addressing downed or fallen wires, proper utilization of stand-by generators, and operating specialized equipment are provided as well. In addition, some customers have specials individual issues that require immediate resolution.

Poststorm Assessment. Poststorm assessments provide feedback regarding processes that require revising. The poststorm assessment involves representatives from multiple areas in order to provide varied viewpoints. Success of the recovery operation is measured after the storm, in terms of how quickly and safely the system was restored to normal and by the amount of complaints received by the host utility or local media.

Mock Storms. Successful recovery plans are well-developed and practiced prior to actual implementation. The practice sessions require all of the individuals involved in the response plan available to practice and utilize all processes and equipment identified so that everyone is familiar with them.

OUTAGE MANAGEMENT SYSTEMS

The outage management systems (OMS) is a valuable tool in terms of analyzing all of the outage data and projecting abnormal protective devices based upon customer calls or protective device status information provided from intelligent electronic devices (IEDs). The OMS serves as an outage data collector. The system provides dispatchers with targeted outage, crew, restoration, and network information (Fig. 10.9). Outage data collected includes start date and time, location, and descriptions of the specific events.

OMS displays the information in both text and pictorial form (Figs. 10.10 and 10.11). The data are sorted by state, district, service center, substation, circuit, and customer. Statistical snapshots of the information provide information as to the scope of the situation.

The OMS contains maps based upon connectivity models of the entire distribution system. The maps are in several formats; single-line maps, individual circuit maps, district maps, layered informational maps, and combinations of most of the desired features are available. The hierarchy of the distribution system exists in the connectivity model. Protective devices and system isolation points are modeled. Network model changes completed in the field are stored and displayed. Each and every customer meter is associated with a transformer. The transformers are in turn associated to line fuses. The line fuses are associated with higher-level protective devices such as a recloser or a set of mainline fuses. The highest level of upstream protective devices is typically the circuit breaker.

As customer calls are received and entered into OMS, the system groups the calls based upon the connectivity tree of the electric system, diagnoses the information, and targets the most likely protective devices in the electric system where the assessment needs to begin.

The work orders are assigned in a prioritized methodology to restoration crews. The crew status is maintained on the orders.

Crew statuses logged include:

- Assigned
- Dispatched
- En route
- Arrived on site

FIGURE 10.9 Delivery system expert utilizing the OMS to analyze outage information to respond to system interruptions.

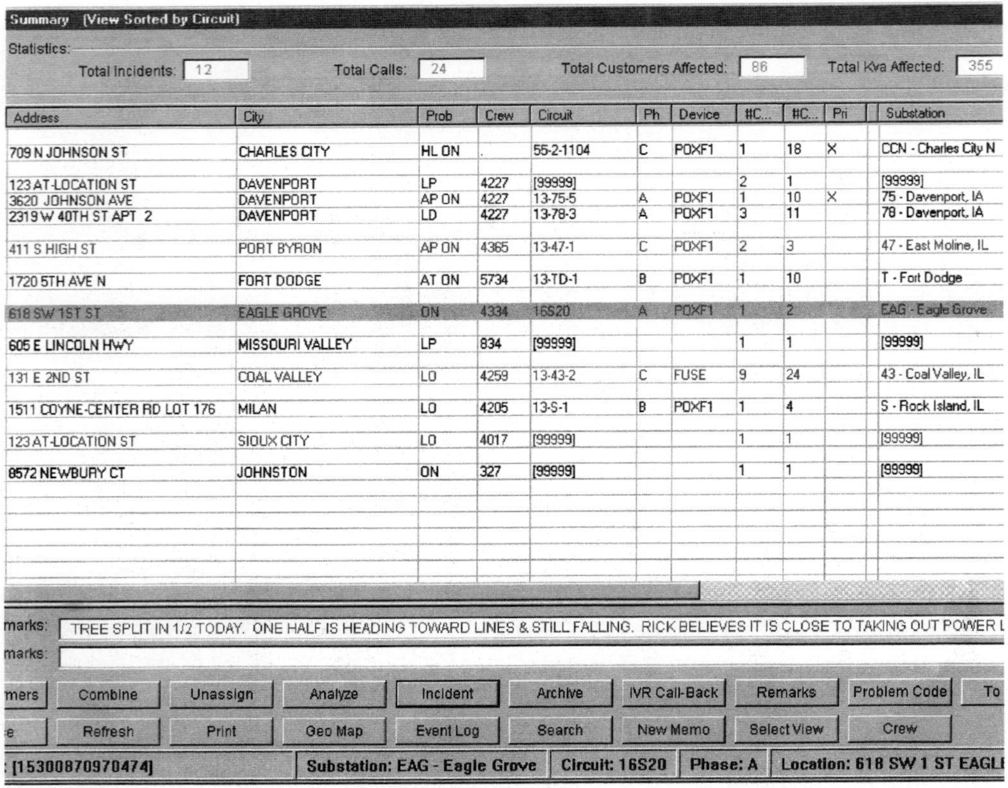

Summary [View Sorted by Circuit]

Statistics:

Total Incidents: 12 Total Calls: 24 Total Customers Affected: 86 Total Kva Affected: 355

Address	City	Prob	Crew	Circuit	Ph	Device	#C...	#C...	Pri	Substation
709 N JOHNSON ST	CHARLES CITY	HL ON	.	55-2-1104	C	POXF1	1	18	X	CCN - Charles City N
123 AT-LOCATION ST	DAVENPORT	LP	4227	[99999]			2	1		[99999]
3620 JOHNSON AVE	DAVENPORT	AP ON	4227	13-75-5	A	POXF1	1	10	X	75 - Davenport, IA
2319 W 40TH ST APT 2	DAVENPORT	LD	4227	13-78-3	A	POXF1	3	11		78 - Davenport, IA
411 S HIGH ST	PORT BYRON	AP ON	4365	13-47-1	C	POXF1	2	3		47 - East Moline, IL
1720 5TH AVE N	FORT DODGE	AT ON	5734	13-TD-1	B	POXF1	1	10		T - Fort Dodge
618 SW 1ST ST	EAGLE GROVE	ON	4334	16S20	A	POXF1	1	2		EAG - Eagle Grove
605 E LINCOLN HWY	MISSOURI VALLEY	LP	834	[99999]			1	1		[99999]
131 E 2ND ST	COAL VALLEY	LO	4259	13-43-2	C	FUSE	9	24		43 - Coal Valley, IL
1511 COYNE-CENTER RD LOT 176	MILAN	LO	4205	13-S-1	B	POXF1	1	4		S - Rock Island, IL
123 AT-LOCATION ST	SIOUX CITY	LO	4017	[99999]			1	1		[99999]
8572 NEWBURY CT	JOHNSTON	ON	327	[99999]			1	1		[99999]

marks: TREE SPLIT IN 1/2 TODAY. ONE HALF IS HEADING TOWARD LINES & STILL FALLING. RICK BELIEVES IT IS CLOSE TO TAKING OUT POWER L

marks:

mers | Combine | Unassign | Analyze | Incident | Archive | IVR Call-Back | Remarks | Problem Code | To

e | Refresh | Print | Geo Map | Event Log | Search | New Memo | Select View | Crew

: [15300870970474] | Substation: EAG - Eagle Grove | Circuit: 16S20 | Phase: A | Location: 618 SW 1 ST EAGL

FIGURE 10.10 Incident summary screen of known problems within the service territory. (*Courtesy CGI Information Systems and Management Consultants Inc.*)

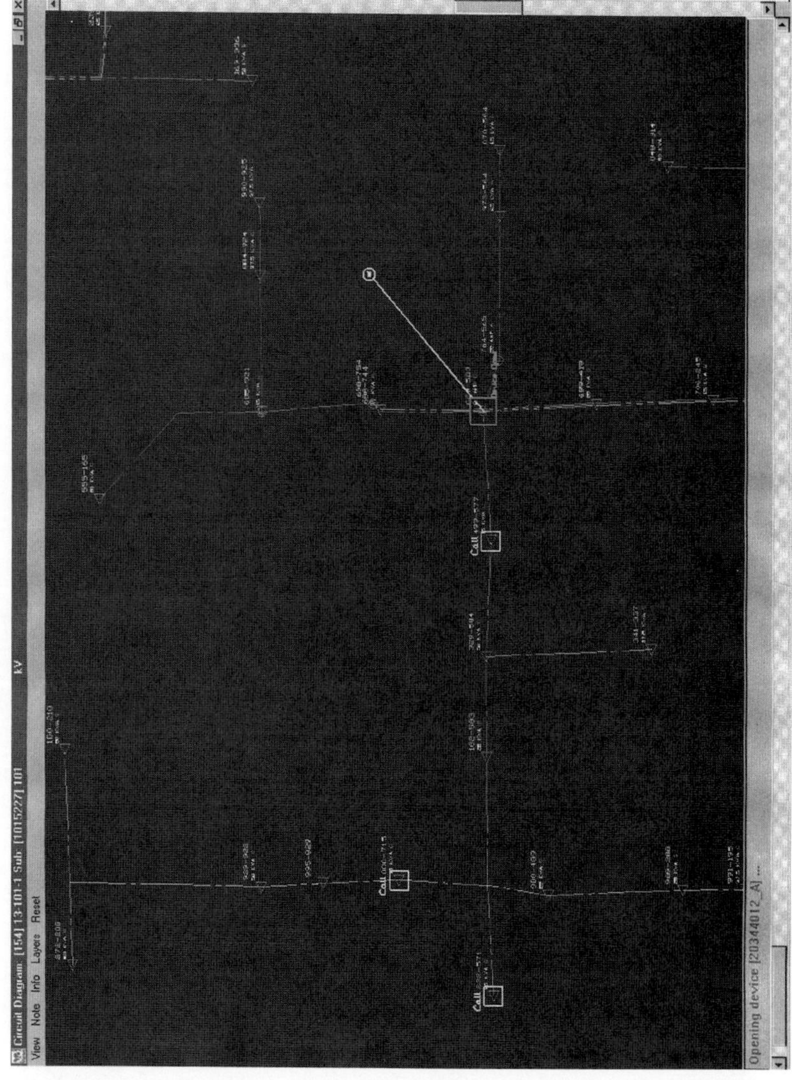

FIGURE 10.11 Geographical view of an area that has experienced an interruption of service. (*Courtesy CGI Information Systems and Management Consultants Inc.*)

- Estimated time to repair
- Service restoration, and
- Job completed

Crew management is accomplished using the OMS. Tracking of employee hours is logged in the system.

OMS is capable of creating switch plans and tracking clearances utilized in the isolation of damaged facilities while returning unaffected areas to normal.

RADIO COMMUNICATIONS SYSTEMS

In order for orders to be properly dispatched and switching procedures to be followed, reliable communication networks are established for both company and mutual assistance emergency response crews. In some cases, company personnel (guides) are assigned to each crew. In other cases, mobile phones or hand-held devices are utilized. In order to ensure safety for everyone involved, radio procedures and guidelines are reviewed and incorporated.

CONCLUSION

Rather than being reactive, a proactive approach to system emergencies will minimize disruption of service to the customer. Customers expect well-coordinated and expertly managed disaster recovery operations. In most cases, due to the dedication of the lineman and of the support teams, exceptional response is what the customers receive.

CHAPTER 11
UNLOADING, HAULING, ERECTING, AND SETTING POLES

Unloading Wood Poles. Poles are shipped on railroad flatcars (Fig. 11.1) or flatbed semi-truck trailers. Great care must be exercised while unloading poles from the flatcar to avoid any possibility of an accident. Before starting the unloading operation, the railroad car brakes should be set to prevent the car from moving. The wheels should be chocked as an added precaution. Flags and barricades should be set up to safeguard the area. Mechanized equipment reduces the time required and the number of workers needed to unload the poles safely (Fig. 11.2).

The lineman should inspect the stakes holding the poles to be sure the poles are secure before cutting the banding wires. The banding wires should be cut from the end of the poles cautiously by the lineman on the ground. After the banding wires are cut, equipment such as the Cary-Lift loader (Fig. 11.2) can be used to safely remove several poles at a time and place them on the pole storage pile (Fig. 11.3).

If mechanized equipment is not available, wood poles can be unloaded from a railroad flatcar with a winch line on a line truck or with a mechanical hoist. After the railroad car brakes have been set, the wheels choked, and the area barricaded, the lineman should restrain the poles on the railroad car by placing chains around the piles of poles near each end of the railroad car and secure each chain with a hoist.

When the poles have been secured, the lineman should install additional stakes in the empty pockets on the side of the railroad car opposite the area to receive the poles. A long V sling should be attached to the railroad car approximately 8 ft from the ends of the poles on the unloading side (Fig. 11.4) and extended over the poles to the winch of the line truck or a mechanical hoist (Fig. 11.5). Unloading skids should be secured to the side of the railroad car and extended to the ground, or pole pile, to permit the poles to roll from the car when they are released (Figs. 11.6 and 11.7). Steel rails, or timbers, can be used for this purpose.

The lineman should check the chains and the V sling to be sure the poles are secure and then, starting in the center and working to each end of the car, remove the stakes and cut the banding wires on the unloading side of the railroad car. When all personnel are in the clear, the chain hoists should be operated to provide slack in the chains. When the chains are slack, the V sling should be released slowly allowing the poles to roll off the railroad car onto the ground.

The remaining poles are rolled off the car with a cant hook. The lineman using the cant hook stands at the end of the poles and remains alert in a position to move out the danger if the poles on the car shift position unexpectedly. After all the poles have been removed from the railroad car, the skids and chocks are removed and the area cleaned up, disposing of the banding wires and stakes.

FIGURE 11.1 Railroad carload of wood poles. Stakes and binding wires hold the poles in place for shipment.

Roofing Poles. Most poles are completely impregnated with wood preservatives, eliminating the need to roof them. If poles require roofing, the work is normally completed in the pole yard, where the poles can be readily handled with mechanized equipment.

The roofing operation consists of forming a roof on the top of the pole. The roof can be simple slant cut as in Fig. 11.8(*a*) or a gable roof as in Fig. 11.8(*b*). Either type prevents water, snow, or ice from collecting on the top of the pole and causing decay. If poles are completely impregnated with wood preservatives, a flat roof is adequate, as in Fig. 11.8(*c*).

To cut a gable roof, the pole is supported in a raised position. The pole is then turned so that the heaviest sag or curve is nearest to the ground. This is done by having one person turn the pole at the butt end with a cant hook while another sights along the pole and decides when the pole is in the right position. By putting the pole in this position, the direction of the roof will be same as the pole line. When this pole is erected in the line, it appears straight. Figure 11.9 is a sketch of a pole provided with roof and gains and gives the correct relation of these to the crook or bend of the pole. The side of the pole which has the longest concave sweep between the ground line and the point of the load is the face of the

FIGURE 11.2 Cary-Lift loader being used by lineman to unload poles from the flatcar. The hydraulically controlled clamping device holds the poles firmly in place on the forks. (*Courtesy Pettibone Corp.*)

FIGURE 11.3 Cary-Lift loader being used by lineman to place wood poles in storage crib. (*Courtesy Pettibone Corp.*)

pole, as illustrated in Fig. 11.9. The pole supplier normally marks the pole length and class on the face side of the pole with a branding iron. The face of the pole should be determined before the pole is set and before a crossarm or fiberglass insulator supports and insulators are installed. The face of the pole may be different from the side of the pole branded by the supplier as a result of the way the pole was stored in a pole pile.

The gable roof is made by sawing the two sides of the pole top (Fig. 11.10) at an angle of 45° with the pole. One way of obtaining the 45° angle is to measure along the length of the pole a distance equal to one-half the pole-top diameter and to saw accordingly. This roof will shed the rain and prevent any accumulation of water, thereby decreasing possible decay.

Pole Gains. A gain is the notch cut into a pole into which the crossarm is placed when mounted. Gains are cut to provide a flat surface to help maintain the arms in alignment.

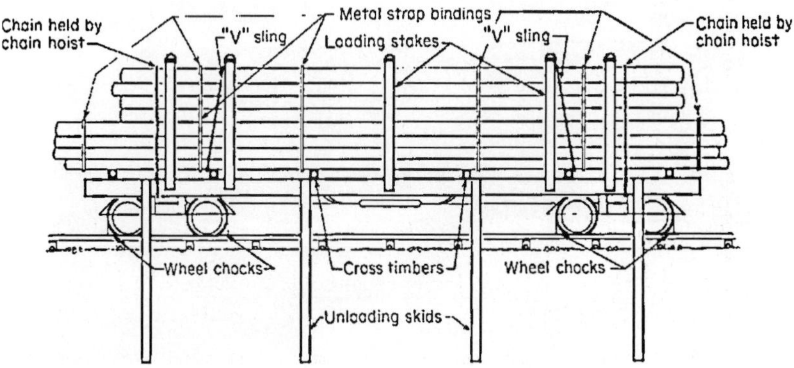

FIGURE 11.4 Railroad car of poles viewed from unloading side after preparations for unloading have been completed.

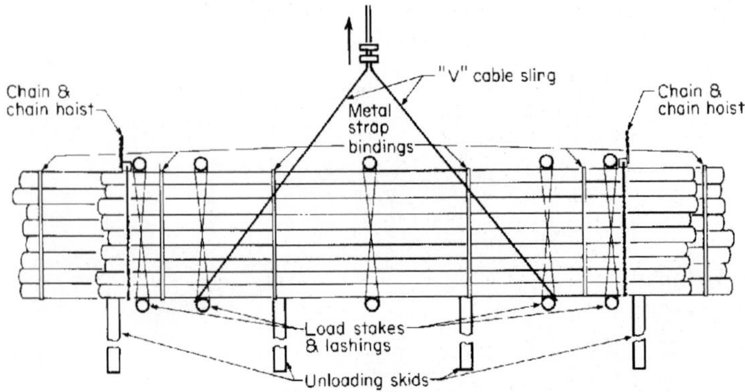

FIGURE 11.5 Railroad car of poles (top view) after preparations for unloading have been completed.

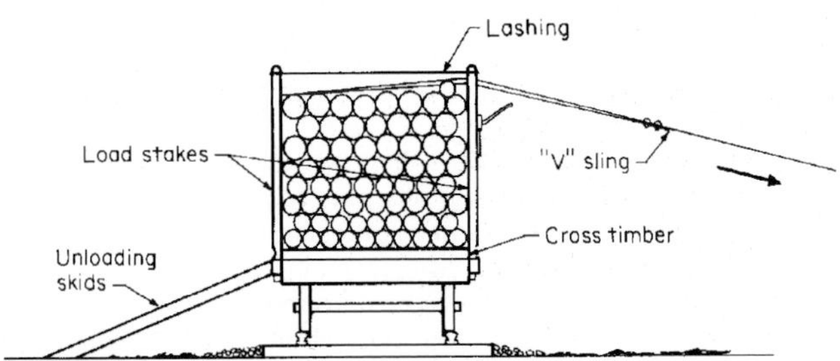

FIGURE 11.6 Railroad car of poles (end view) after preparations for unloading have been completed.

FIGURE 11.7 Steel rails secured to the railroad car and extended to the pole pile permit the poles to roll from the car.

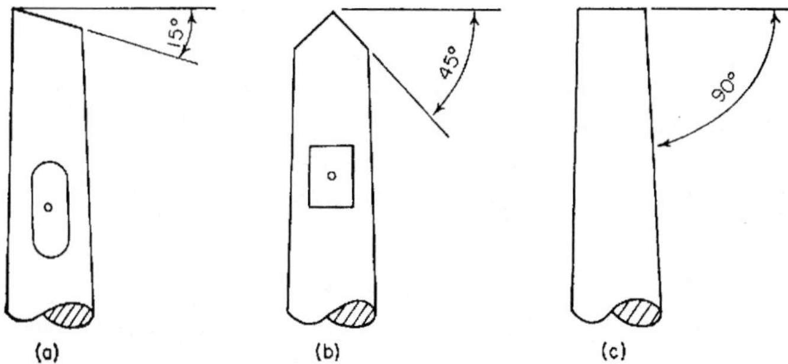

FIGURE 11.8 Three types of pole roofs. (*a*) Slant, (*b*) gable, (*c*) flat.

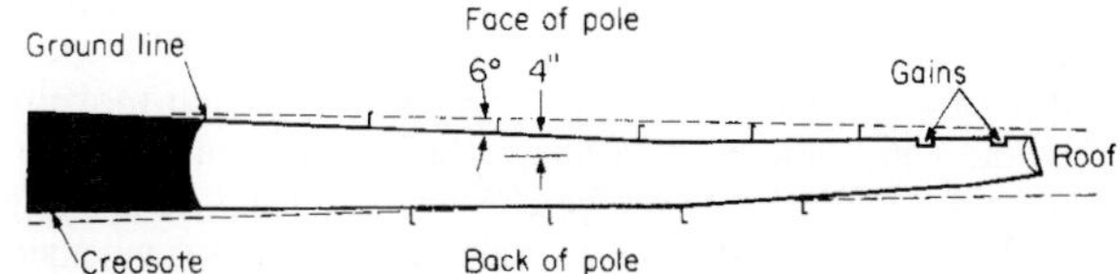

FIGURE 11.9 Sketch of completely framed pole showing relation of parts to curvature of pole.

To cut a gain, one must first saw along the upper and lower edges of the gain with a handsaw and then chisel out a round portion, thereby making a flat recess. It is good practice to hollow the gains out slightly in the center to ensure a snug fit of the crossarm, thereby preventing its rocking from side to side.

The first crossarm gain is usually cut 12 in from the top of the pole. The succeeding gains are generally spaced 24, 30, or 36 in apart. The spacing selected depends largely on whether the pole is to carry buck arms or not. When buck arms must be provided, these are placed between the line arms at right angles. In order to permit the use of 28-in crossarm braces, the spacing between the line arm and buck arm should be at least 18 in. Gains should be square with the axis of the pole and about 1/2 in deep. The height depends on the height of the crossarm. The crossarm should fit snugly.

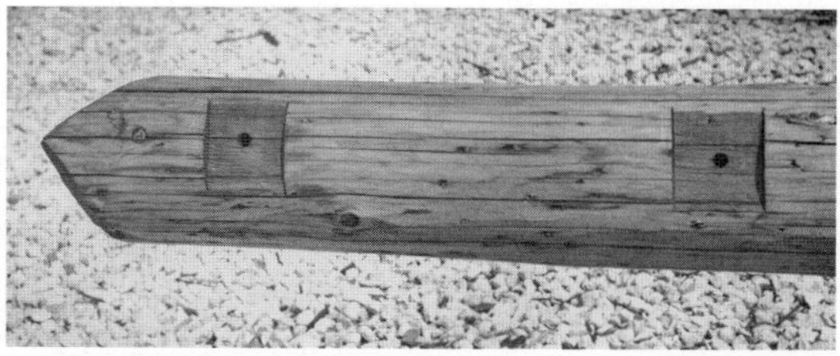

FIGURE 11.10 Pole top showing pole roof cut at 45° and two gains with crossarm through-bolt holes. (*Courtesy Wisconsin Electric Power Co.*)

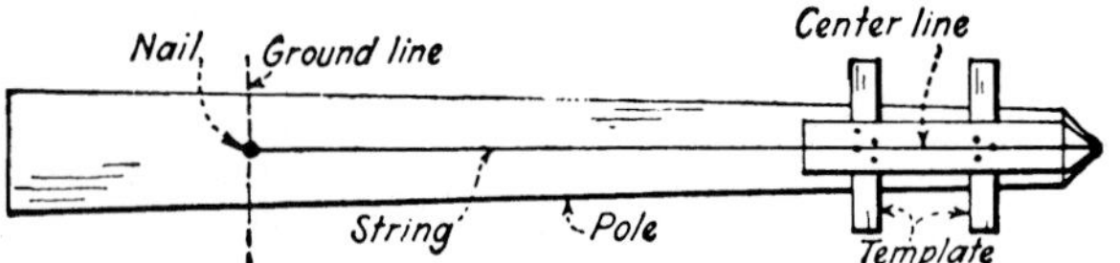

FIGURE 11.11 Template for use in marking gains.

If several gains are to be cut, a gaining template (Fig. 11.11) is a great convenience. In using the template, the point of the roof of the template is placed over the roof of the pole and the template is shifted until its centerline lies in line with a string stretched from the point of the roof to the center of the butt indicated by a nail at the ground line. Then the outlines of the gains are marked with a sharp tool by marking along the side pieces of the template.

The vertical spacing of the crossarms, and consequently the gains, depends on the nature of the circuits and the voltages of the circuits carried by the arms. The specifications and standards for the type of construction to be used will show the details for gaining the poles.

One of the difficulties encountered in cutting more than one gain on a pole is to keep the flat surfaces of the gains parallel. This can be checked by resting or hanging tee squares, one in each gain (Fig. 11.12). By sighting over them it is easy to see whether the gains are properly leveled. The square also can be used to advantage in getting buck arms at right angles with the line gains.

The next operation consists of boring the holes for the crossarms. The size of the hole for the standard crossarm through bolt is 11/16 in. This is for the 5/8- or 10/16-in through bolt. This operation is generally performed with an electrically or hydraulically driven power drill.

A scheme often used for locating the center of the crossarm bolt hole is to draw two diagonal pencil lines across the gain (Fig. 11.13). The intersection of these lines determines the center of the bolt hole. In case the pole surface is not smooth, it is best not to guess at the pole center but to measure for the vertical center.

Another scheme which takes less time is the use of a template. This consists of a short length of crossarm in which the hole has been bored. This is placed centrally over the pole in the gain, and the center is marked with a punch.

Wherever necessary, poles are painted. The desirability of painting is generally determined by appearance. Poles used in streets of large cities may be shaved and painted to improve appearance. Some companies make a practice of painting the roof and the gains with one or more coats of approved paint. Poles can be obtained from a supplier in colors or a natural light color for installation where appearance is particularly important.

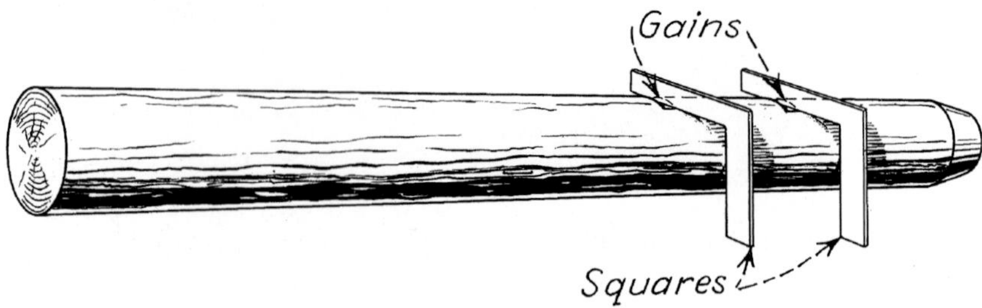

FIGURE 11.12 Method of determining whether gains are cut parallel. Test is made by sighting over edges of squares. If gains are parallel, the edges of the tee squares will be in line.

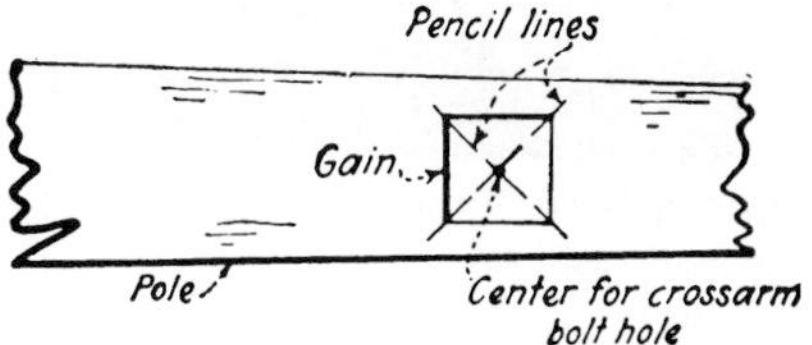

FIGURE 11.13 Scheme for determining center for crossarm bolt hole. Two diagonal pencil lines are drawn as shown. The intersection of these lines is the bolt-hole center.

FIGURE 11.14 Pole on trailer connected to the back of line truck. Flag on the end of the pole is displayed to warn the public. (*Courtesy MidAmerican Energy Co.*)

Hauling Poles. Loading of poles and hauling them from pole storage to the work site are hazardous to the lineman handling the poles and to the public. Moving poles short distances in the pole yard can be completed efficiently with mechanized equipment. Poles are usually loaded on a trailer, and the trailer is pulled by a line truck to the work site (Fig. 11.14). The trailer is placed near the pile for convenience in loading the poles. The brakes on the trailer should be set and the wheels chocked while the poles are loaded on the trailer with mechanized equipment if available or with the boom of the line truck (Fig. 11.15).

FIGURE 11.15 Pole is being lifted out of pile for loading on trailer with boom of line truck. Notice lineman controlling pole with cant hook. (*Courtesy A. B. Chance Co.*)

FIGURE 11.16 Drawbar with eye attached to pole.

FIGURE 11.17 Drawing eye coupled to truck.

FIGURE 11.18 Pole secured to trailer with chain and load binder.

If one of the poles is to be used as the tongue for the trailer, the pole selected should be of sufficient size and strength to carry the load. The drawbar is chained to the tongue pole (Fig. 11.16). The eye of the drawbar is engaged in the pintle hook on the back of the truck, and safety chains on the truck are fastened to the tongue pole and drawbar (Fig. 11.17). The poles should be blocked into position as they are loaded on the trailer to keep them from shifting. The lineman must be careful not to overload the trailer or balance the load of poles so that the load is tongue heavy.

When all the poles have been loaded, they must be secured with cables or chains or both to keep the load from shifting in transit (Figs. 11.18 and 11.19). Figure 11.20 shows the overall view of a pole on a trailer complete with rigging. Flags should be installed on the end of the poles during daylight hours and warning lights should be used after dark. The lineman must take care to protect the public while hauling poles. The driver should signal turns and stops early to warn other drivers while hauling poles on a highway. The poles may swing into adjacent traffic lanes when short turns are necessary. It is desirable for a lineman to use flags to warn the public if the area is congested.

When the terrain is rugged and access roads are not available, poles can be transported by helicopter (Fig. 11.21). The specialized helicopters are designed to lift and carry the

FIGURE 11.19 Rear view of trailer. Pole is secured with load-binder cable.

FIGURE 11.20 Overall view of trailer. (*Courtesy Pacific Gas & Electric Co.*)

heavy structures. Structures are completely or partially assembled, depending upon weight, and then delivered to the job site. Pole holes are dug or foundation bases installed and correspondingly identified so that each structure is delivered to the correct location and lowered to the ground by the helicopter pilot to complete the installation. A two-pole structure can be transported by helicopter as shown in Fig. 11.22.

Setting poles. The specifications and drawings give the location and size of each pole to be set. The pole locations are usually identified by stakes before the construction work is

FIGURE 11.21 Transporting a fully framed pole by helicopter over rugged terrain. A 1-in manila rope sling 8 ft long was attached to each pole top with a spliced eye at the outer end for rigging to electrically operated sling hook suspended 4 ft below the helicopter. Each pole was numbered and keyed to a map. Pole holes were correspondingly identified by a 2-ft^2 cardboard. (*Courtesy Pacific Gas & Electric Co.*)

FIGURE 11.22 Helicopter transporting preassembled wooden-pole H-frame structure across the Roanoke River in Virginia. (*Courtesy American Electric Power System.*)

FIGURE 11.23 Mechanical pole-hole digger mounted on a truck equipped with four-wheel drive. Unit is complete with motor, hydraulic drive system, auger, and turntable to facilitate proper location and alignment of hole for pole. (*Courtesy Northwest Lineman College.*)

started. If one of the stakes has been removed or destroyed, the proper location can probably be determined by measuring and aligning from adjacent stakes. If several stakes have been lost or destroyed, it may be necessary to ask the foreman or engineer for assistance in resetting the stakes for the poles.

The holes for the poles are normally dug by a mechanical digger, sometimes referred to as an earth-boring machine. The mechanical digger consists of an auger, or helix, operated by a hydraulic system. The mechanical digger used to dig holes for large transmission-line poles is a complete unit normally mounted on a flatbed truck (Fig. 11.23). The casing around the auger operating shaft can be equipped with a pulley, for a winch line and pole grabber, to make it possible to set poles up to 65 ft long, which are commonly used for distribution lines, with the mechanical digger unit.

The standard way of digging pole holes for the smaller poles used for distribution lines is to use a derrick- or boom-mounted auger (Fig. 11-24). An auger attached to the boom of a corner-mounted derrick line truck is easily aligned to dig the pole hole in the proper location. A line truck, equipped with an auger, increases the versatility of the work that can be performed by the line crew.

Pole-Hole Diameter. The diameter of the hole is determined by the size of the pole to be set. The hole should be large enough to allow plenty of space on each side of the pole for the free use of tamping equipment, which requires at least 3 in all around the pole. The diameter of the hole at the top should not be greater than at the bottom, but rather it should be larger at the bottom than at the top. A slight increase at the bottom is often necessary to allow for the shifting of the pole in lining-in and to accommodate the larger diameter of the pole at the butt.

Pole-Hole Depth. The depth of setting is determined by the length of the pole and by the holding power of the soil or earth. The required average setting depths have been determined by experience. The recommended depths of setting in soil and rock are given in Table 11.1 and Fig. 11.25 for various pole lengths from 25 to 135 ft. It will be noted that a pole set in rock need not be set as deep as a pole set in soil.

FIGURE 11.24 Boom- or derrick-mounted auger. The auger is supported by the line-truck boom and is operated by the truck's hydraulic system.

TABLE 11.1 Recommended Pole-Setting Depths in Soil and Rock for Various Lengths of Wood Poles

Length of pole, ft	Setting depth in soil, ft	Setting depth in rock, ft
25	5.0	3.5
30	5.5	3.5
35	6.0	4.0
40	6.0	4.0
45	6.5	4.5
50	7.0	4.5
55	7.5	5.0
60	8.0	5.0
65	8.5	6.0
70	9.0	6.0
75	9.5	6.0
80	10.0	6.5
85	10.5	7.0
90	11.0	7.5
95	11.0	7.5
100	11.0	7.5
105	12.0	8.0
110	12.0	8.0
115	12.0	8.0
120	13.0	8.5
125	13.0	8.5
130	13.0	8.5
135	14.0	9.0

Source: Commonwealth Edison Company.

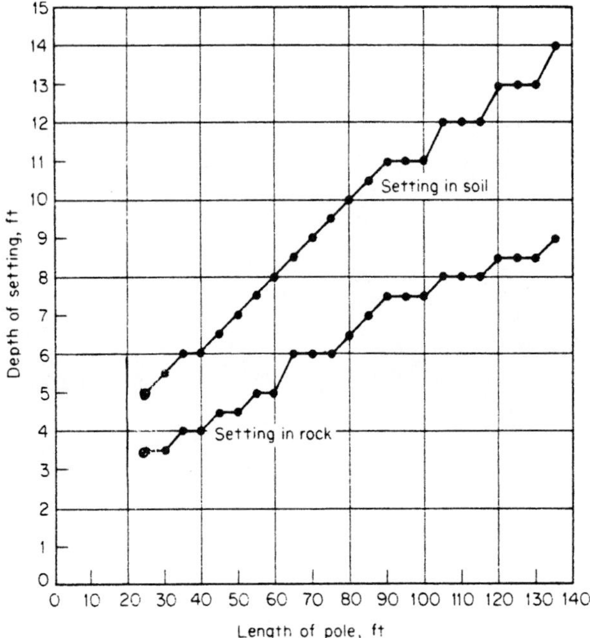

FIGURE 11.25 Recommended pole-setting depths in soil and rock.

A rule often followed for determining the setting depth in soil is to take 10 percent of the pole length and add 2 ft, with a minimum of 5 ft. The setting depth of a 40-ft pole would thus be:

$$10\% \text{ of } 40 = 4\text{ft} + 2\text{ft} + 2\text{ft} = 6\text{ft}$$

This matches the table. The pole-hole depths in Table 11.1 are for poles to be set in the straight portion of the line. Holes for poles that are to be set at angle or corner points in the line should be at least 1/2 ft deeper. If soil is soft or marshy, dig the pole holes deeper. Holes for poles set on a slope should be 1/2 ft deeper than normal, measured from the lowest point on the ground.

Crushed rock is often used for backfill around transmission-line poles and distribution poles requiring extra stability (Fig. 11.26). The rock must be well tamped to obtain the desired strength.

Poles set in sandy or swampy ground should be set considerably deeper or be supported by guys, braces, or cribbing. One form of cribbing uses a barrel into which the pole is set. The barrel is then filled with concrete or small stones to make the pole secure.

When exceptional stability is required of a pole, an artificial foundation of concrete may be placed around the base (Fig. 11.27). This concrete filling should extend approximately 1 ft from the pole on all sides and should be carried 1/2 ft above the ground level. The top should be beveled to shed the rain. A good mix to use is 1 to 2½ to 5, that is, 1 part cement, 2½ parts sand, and 5 parts broken stone or gravel. Enough water should be added to make the mixture flow freely and take its place without tamping. The pole must be firmly braced in position, and the bracing must be left in position until the concrete is set. No line work should be done on the pole within a week after the concrete is poured.

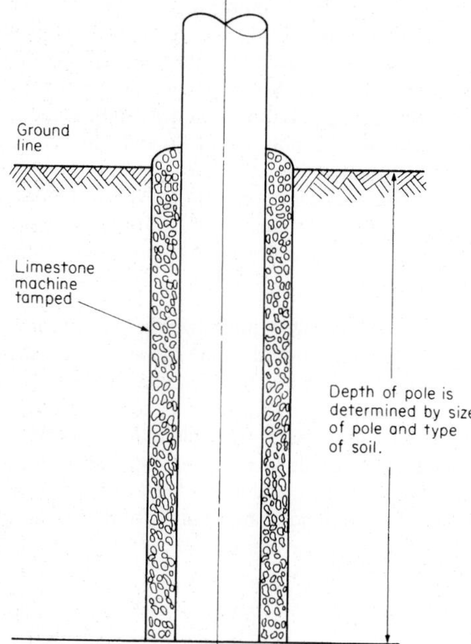

FIGURE 11.26 Crushed rock used as backfill around pole to provide stability.

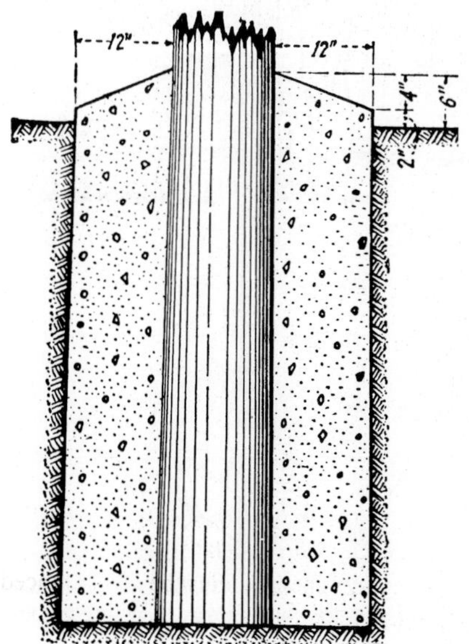

FIGURE 11.27 Concrete foundation for wooden pole. The mixture should be made thin enough that it does not require tamping. (*Courtesy Texas Power and Light Co.*)

Poles set on curves, on corners, or at points of extra strain should be set at least 6 in deeper than the values given in Table 11-1. The side strain, however, should always be taken up with terminal, side, or line guys, leaving the pole to carry the vertical load only.

Poles subjected to slight side pulls can be kept erect by the use of pole keys. Some companies use treated timber to key poles. Expandable steel-fabricated steel keys have proved to have longer life and to be more economical to install. Street lights with a long mast arm supporting the light fixture, very small angles in the line, and service drops at right angles to the line produce slight side pulls on a pole. When the poles are set in firm soil, one key placed 1 ft below ground surface on the strain side of the pole is sufficient. In loose soil, however, a second key placed on the opposite side at the butt of the pole is required (Figs. 11.28 and 11.29).

Pole-Hole Digging. Mechanical diggers minimize the physical effort, and time, required to dig holes for poles. The lineman should locate the site where the pole is to be set and determine the best route to get the truck to the location. Special precautions must be taken if the hole is to be dug near an energized circuit. If adequate clearance cannot be maintained, protective equipment must be used to insulate the energized conductors to prevent accidental contact. If adequate protection cannot be obtained from existing conductors, the circuit should be deenergized, isolated, and grounded before the work is started. The truck should be located properly to permit accurate location of the auger.

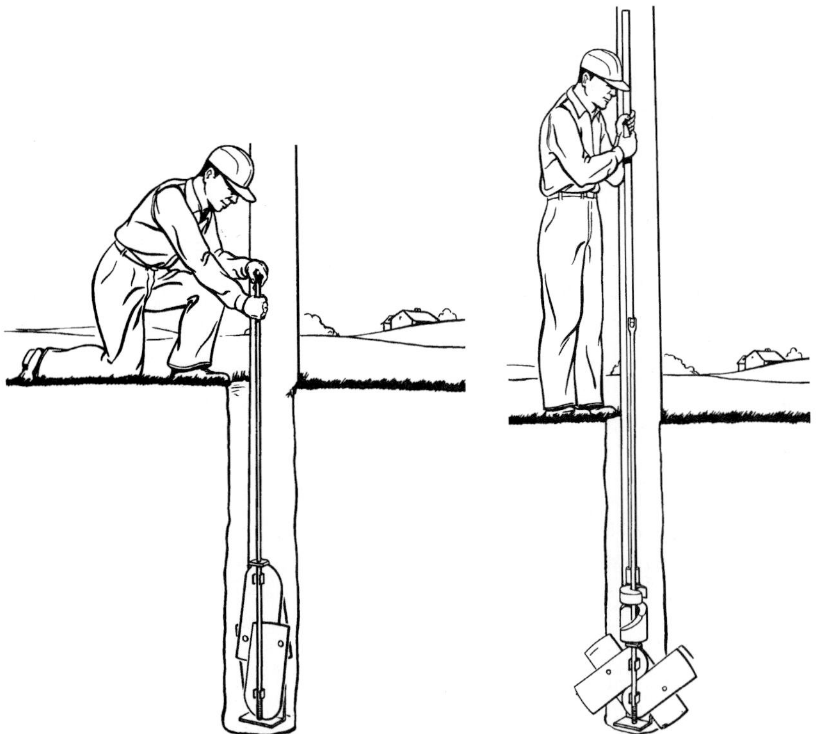

FIGURE 11.28 Expandable steel pole key has been placed in the hole adjacent to the butt of the pole by the workman on the side of the pole opposite the overhead strain. (*Courtesy A. B. Chance Co.*)

FIGURE 11.29 Workman expands steel pole key with an expanding and tamping bar. (*Courtesy A. B. Chance Co.*)

FIGURE 11.30 Power pole-hole borer mounted on rear of truck and tilted forward into traveling position. Power for auger is obtained from special engine. (*Courtesy Northwest Lineman College.*)

It may be necessary to start angle-structure holes by hand. A post-hole-type auger, a digging bar, a 5-ft shovel, a 9-ft shovel, and a 9-ft spoon should be available. The power digging equipment should be inspected before the auger is operated. The lineman should ensure that all personnel maintain adequate clearance to prevent clothing from being caught by the auger or other mechanical equipment.

When pole holes have been dug along paved streets, near farm buildings, or on any premises used by the public and poles are not set at once, the holes should be completely covered with a plank and dirt, and cordoned off with lighted barricades.

Pole-Hole Power Borer. Pole holes are dug with a power earth borer. This greatly reduces the physical effort required and speeds up the work. A power borer is very similar to the familiar wood brace except that it is much larger and is power driven; Fig. 11.30 shows a typical power borer mounted on the rear of a truck body. The borer can be tilted up while moving between work sites.

When a hole is to be dug, the truck is driven over the pole location and placed so that the borer is directly over the pole stake. The truck outriggers are lowered to stabilize the vehicle. The table supporting the earth-borer motor and mechanism is moved, or rotated, to place the point of the auger over the stake when it is lowered (Fig. 11.31). The boring equipment must be positioned to maintain proper alignment while the pole hole is dug. Then the operator lowers the auger (Fig. 11.32) while it rotates until it has taken a full "bite." Then the auger is raised until it is above ground and away from the hole and cleaned of the soil (Figs. 11.33 to 11.35). The auger is lowered again, and another bite is taken. Again the auger is raised and cleaned. This is repeated until the desired depth of hole has been reached.

The completed hole will have a ridge of loose dirt completely around it. The loose dirt should be moved away from the edge of the hole to prevent it from falling into the hole. All losse dirt should be removed from the bottom of the hole by use of hand tools. A post-hole digger, a long-handled shovel, and a spoon are usually required (Figs. 11.38 and 11.39).

Holes of different diameters are obtained by the use of augers of different diameters. Augers having different diameters can be mounted on a digging bar. The change from one to the other can be made in a few minutes. The 9-in size, the smallest, is generally used for digging guy-anchor holes. The 16-, 24-, 36-, and 48-in sizes are carried on the truck and are used for different sizes of poles that must be set.

FIGURE 11.31 Pole-hole power borer in position to start digging. Truck outriggers are down, and point of auger is touching ground at location established by a stake. (*Courtesy Northwest Lineman College.*)

Pole-Hole Derrick Auger. Derrick-mounted power pole-hole augers (Fig. 11.36) are commonly used for digging distribution-pole holes. This type of power-hole digger increases the flexibility of the distribution-line crew in utilizing the vehicle and equipment efficiently. The procedures and precautions are similar to those described for the pole-hole power boring equipment. The derrick-mounted auger can be used in locations several feet from the truck by extending the derrick or boom, on which it is mounted. The auger is driven by the hydraulic system that operates the boom. The point of the auger is lowered to the ground at the pole location, usually identified by a stake. It is important to locate the auger accurately. The lineman must be sure the boom moves easily to keep the pole hole plumb and to prevent damage to the equipment. The auger should be started to rotate slowly to prevent tilting.

The auger rotation should be stopped when the auger worm is full of dirt to avoid damage to the equipment. The rotation of the auger must be stopped before the auger worm gets stuck. The dirt is removed from the hole by raising the derrick and reversing the rotation. If the derrick does not raise the auger, reverse the rotation of the unit to free it from the earth. A short-handled shovel can be used to remove the dirt from the auger.

Pole-Hole Digging in Rock. The rock-cutting drill (Fig. 11.37) is mounted on the boom of a standard digger-derrick line truck. The drill cuts a hole in the rock at the rate of 16 to 20 in (41 to 51 cm) an hour. The drill cuts a circular hole in the rock, leaving a core. The

FIGURE 11.32 Auger lowered into hole for another bite of ground. Note loose dirt around hole. (*Courtesy Northwest Lineman College.*)

rock dust is vacuumed out of the hole by the mechanism as the rock is penetrated. The rock core remaining after the drill reaches the proper depth for the pole to be set is broken at the base with a hydraulic wedge and removed in one piece. The rock drill can eliminate the need to use explosives to form a hole in rock for a pole. The use of explosives is hazardous and requires special training to ensure their safe use.

Pole-hole Digging by Hand. Digging holes with shovel and spoon is resorted to when it is impossible to use a power digger. The shovel is a straight-bladed tool built as a spade (Fig. 11.38), but the spoon has the blade at an angle (Fig. 11.39). The shovel is used to loosen the soil, and the spoon is used to remove soil from the hole.

An experienced hole digger will dig the hole largely without the aid of the spoon under ordinary conditions of soil and moisture. A certain amount of practice is required to do this. One must know just how to strike the soil with the shovel so that the soil will break loose from its position and yet stick to the blade of the shovel. In case the soil is very dry, the shovel can be used to loosen the soil and the spoon used to lift the loose soil from the hole.

Sometimes a bar must be used to break the earth loose if there is frost, rock, or hard clay. One end of the bar should be sharpened to a blunt chisel point and the other end to a round point similar to the sharpened end of a pencil. Such a digging bar is shown in Fig. 11.40.

The usual procedure to be followed is to outline the hole around the stake. The diameter of this circle should be at least 6 in greater than the diameter of the pole the butt. For the

FIGURE 11.33 Auger raised with full bite of ground. (*Courtesy Northwest Lineman College.*)

FIGURE 11.34 Auger has been raised and is being moved away from hole. (*Courtesy MidAmerican Energy Co.*)

FIGURE 11.35 Auger has been positioned away from hole and dirt is being knocked from the auger. (*Courtesy Northwest Lineman College.*)

first 2 or 3 ft of digging, it is advisable to use a short-handled shovel (Fig. 11.41), as it is more convenient to hold. The spoon is not necessary as the earth can be lifted out with the shovel. For the remainder of the hole the long-handled shovel and spoon are used in the usual manner. Where quicksand or swamp soil is encountered and cave-ins occur, a large bottomless barrel is lowered into the hole as it is dug. As the hole is dug deeper, the barrel or drum is forced down, thus preventing the side of the hole from falling in (Fig. 11.42). The drum will also provide a greater bearing surface for the pole when it is set and tamped in the inside of the drum.

Setting Poles with Derrick. The most common method of setting poles is with a truck derrick. The truck should be equipped with a derrick (boom), winch line, and pole grabber (claws). Hand tools required include a cant hook, pike poles, and rope lines. Protective equipment required includes rubber gloves, hard hats, safety glasses, and polyethylene pole covers, or epoxy-rod pole guards.

The line truck should be positioned appropriately to dig the hole, if a boom-mounted auger is to be used, and to erect the pole (Fig. 11.43). The crew should be informed how the work is to be performed. The hazards should be identified so that all personnel understand the safe way of performing the work. The truck outriggers must be positioned in the proper manner to stabilize the vehicle when the pole is raised. Wood poles are heavy and

FIGURE 11.36 Drilling for pole with power auger mounted on boom of truck. (*Courtesy Northwest Lineman College.*)

could tip a line truck over if proper precautions are not taken. The weights of common sizes and types of wood poles are itemized in Table 11.2.

If the pole is to be set near an energized line, the pole should be covered with protective equipment with proper insulation for the voltage of the energized circuit. The winch line is attached to the pole slightly above the midpoint so that the butt end outweighs the top end (Fig. 11.44). The truck derrick is used to move the pole from the pole trailer to the proper position for setting in the hole. The lineman may have to balance the pole and guide it to the proper position (Fig. 11.45). The lineman should not stand or pass under the pole while it is suspended by the winch cable and derrick. The pole is gradually raised until it is in a near vertical position and the butt end is clear of the ground. Linemen wearing rubber gloves guide the pole as it is raised (Fig. 11.46).

The crew leader, or foreman, directs the operation, using standard hand signals for line work. As the pole reaches vertical position, the pole claws are used to grab the pole and stabilize it (Fig. 11.47). When the pole claws have grabbed the pole, the derrick and winch line are manipulated as directed by the crew leader utilizing a plumb bob to set the pole in the hole (Figs. 11.48, 11.49, and 11.50). If the pole is to be set in an existing energized line, it may be necessary to hold the energized conductors apart to have adequate clearance. This can be done with hot-line tools and rope lines, as shown in Fig. 11.51.

After the pole is set in the hole and properly aligned in accordance with the specifications, the hole around the pole is backfilled and tamped (Figs. 11.52 and 11.53). After the pole is secured by backfilling and tamping, lineman removes the temporary protective

FIGURE 11.37 Drilling pole hole with rock-cutting drill mounted on boom of line truck.

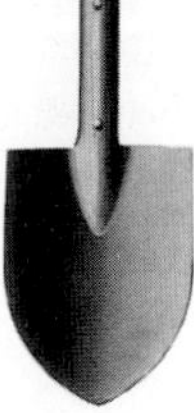

FIGURE 11.38 Straight shovel used to loosen soil in digging a pole hole. (*Courtesy Leach Co.*)

covering used to set the pole and installs the equipment on the pole necessary to complete the project (Fig. 11.54).

Transmission structures are usually assembled on the ground complete with insulators, hardware, conductor stringing blocks, and small rope lines (Fig. 11.55).

The structures are set in place with caterpillar-mounted power derricks or mobile cranes (Fig. 11.56).

Setting Poles with Helicopter. Erecting single-pole lines and transmission structures using conventional construction methods in mountainous, swampy, muddy, or snow and

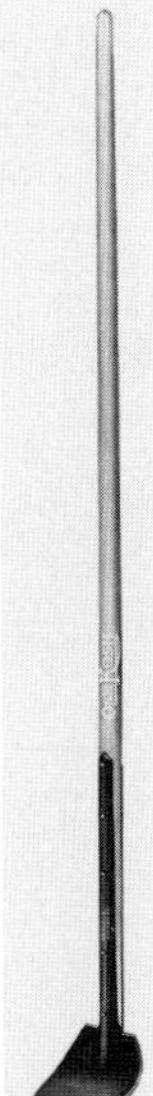

FIGURE 11.39 Spoon used to lift soil out of a pole hole while digging. (*Courtesy Leach Co.*)

FIGURE 11.40 Digging bar used in hard or stony soil. (*Courtesy Leach Co.*)

ice conditions is difficult, slow, and very costly (Fig. 11.57). Single poles are often transported by helicopter to mountainous locations where land transportation is difficult or practically impossible. The poles are normally airlifted with crossarm, insulators, and hardware in place (Fig. 11.58). When the poles arrive at the site, they can usually be lowered into the pole hole directly. Instead of unloading the pole, the helicopter hovers over the assigned pole hole and slowly lowers the pole into it. One or two crewman are usually needed to guide the butt of the pole into the hole.

Transmission structures can be assembled on the ground at a convenient marshalling yard (Figs. 11.59 and 11.60).

The helicopter hovers over the assembled transmission structure at the marshaling yard while the ground crew attaches the structure to the cables extended from a winch installed on the helicopter. Communications are executed by using two-way radios and standard hand

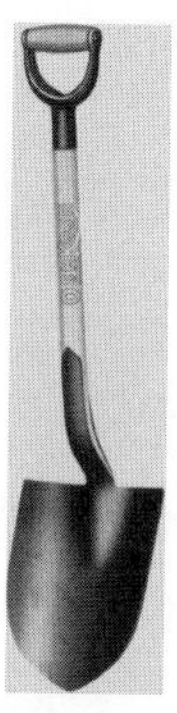

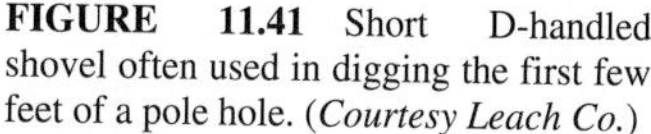

FIGURE 11.41 Short D-handled shovel often used in digging the first few feet of a pole hole. (*Courtesy Leach Co.*)

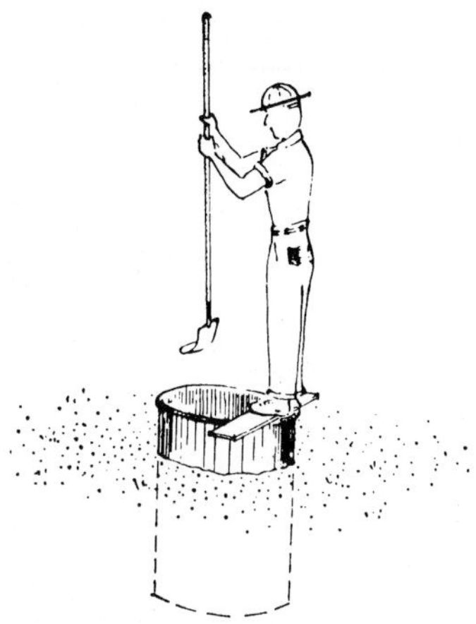

FIGURE 11.42 Lineman using bottomless drum to dig a pole hole in sand or swampy soil. The drum is forced down as the hole is dug, thereby preventing the sides of the hole from caving in. Drum also helps to hold pole erect because of greater bearing surface. (*Courtesy Electric Lineman, Series 100.*)

FIGURE 11.43 Line truck with poles to be set on trailer behind truck. Crew leader is holding tailboard conference to discuss the hazards and to give directions for completing the job. (*Courtesy Northwest Lineman College.*)

TABLE 11.2 Weights of Wood Poles That Are Used Extensively

Length of pole, ft	Species*	Class / Average weight, lb per pole 1	2	4	5
30	WC	880	750	540	440
	DF	1110	930	690	600
	SP	1279	1082	784	660
	WL	955	800	595	515
35	WC	1055	880	660	570
	DF	1435	1260	875	770
	SP	1568	1343	1004	862
	WL	1235	1085	750	660
40	WC	1320	1145	790	705
	DF	1760	1560	1120	920
	SP	1884	1623	1219	1059
	WL	1515	1340	965	790
45	WC	1585	1365	1010	880
	DF	2070	1845	1350	1125
	SP	2223	1911	1444	1274
	WL	1780	1585	1160	965
50	WC	1760	1585	1230	1145
	DF	2500	2150	1600	1300
	SP	2585	2214	1687	1494
	WL	2150	1850	1375	1120
55	WC	2025	1760	1410	1410
	DF	2860	2475	1815	1540
	SP	2993	2567	1934	1719
	WL	2460	2130	1560	1325
60	WC	2290	1935	1670	
	DF	3360	2820	2040	1740
	SP	3451	2943	2186	1953
	WL	2890	2425	1755	
65	WC	2815	2200	1935	
	DF	3835	3250	2340	2015
	SP	4015	3341	2457	2237
	WL	3300	2795	2010	
70	WC	3170	2640	2290	
	DF	4340	3640	2590	2240
	SP	4620	3781	2732	2488
75	WC	3695	3170	2640	
	DF	4800	4050	2925	
	SP	5198	4235	3021	
80	WC	4400	3695	3080	
	DF	5240	4400	3200	
	SP	6400	5170	3615	
85	WC	4840	3960		
	DF	5780	4930		
	SP	7200	5745		

TABLE 11.2 Weights of Wood Poles That Are Used Extensively (*Continued*)

Length of pole,ft	Species*	Class Average weight, lb per pole 1	2	4	5
90	WC	5810	4930		
	DF	6300	5400		
	SP	8140	6405		
95	WC	6750	5950		
	DF	6935	5985		
100	WC	7500	6550		
	DF	7600	6700		

*WC = western red cedar, DF = Douglas fir, WL = western larch, SP = southern pine.
Source: Commonwealth Edison Co.

signals for line work. The helicopter airlifts the structure to the site where it is to be installed. Figure 11.61 shows an H-frame structure being lowered into its holes. Figure 11.62 shows a K-frame transmission structure being set in predrilled holes after being transported 15 miles by helicopter. The helicopter pilot is directed from the ground by two-way radio. Sometimes considerable saving in time and money can be realized by the use of a helicopter for transport and for erection.

Setting Poles Along Rear Lot Lines. Powered dollies equipped with hydraulically operated derrick and auger are available for installing poles or pad-mounted transformers in

FIGURE 11.44 Pole is insulated with polyethylene pole covers designed for use when setting a pole near 13.2Y/7.62-kV three-phase circuit. Winch line is attached to pole so that the pole is a little butt-heavy. (*Courtesy Northwest Lineman College.*)

FIGURE 11.45 Linemen guiding and balancing pole while it is supported by derrick and winch line. (*Courtesy Northwest Lineman College.*)

FIGURE 11.46 Lineman, wearing rubber gloves and protectors, guiding pole as it is raised. Top of pole is insulated with polyethylene pole covers. (*Courtesy Northwest Lineman College.*)

FIGURE 11.47 Pole claws are used to grab and stabilize pole. (*Courtesy Northwest Lineman College.*)

FIGURE 11.48 Crew leader uses standard hand signal and plumb bob to direct lineman operating derrick to set pole in hole. (*Courtesy Northwest Lineman College.*)

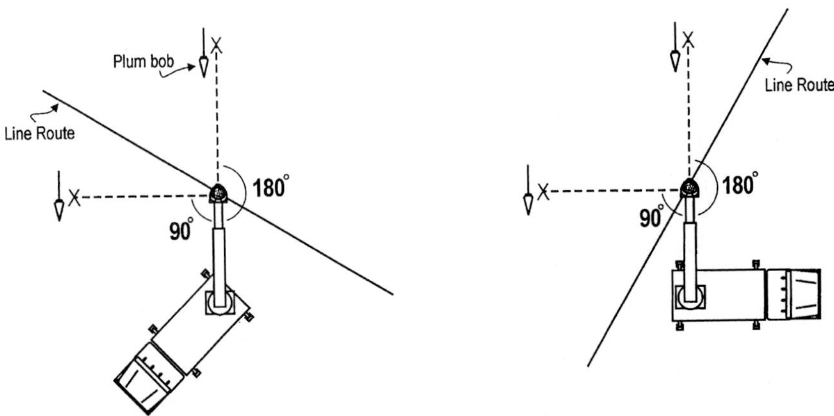

FIGURE 11.49 Position of lineman when using plumb bob. Lineman stands at 90° to 180° from derrick regardless of line route and vehicle location. This ensures that straightening the pole in one direction has no effect when aligning the pole in the other direction. (*Courtesy Northwest Lineman College.*)

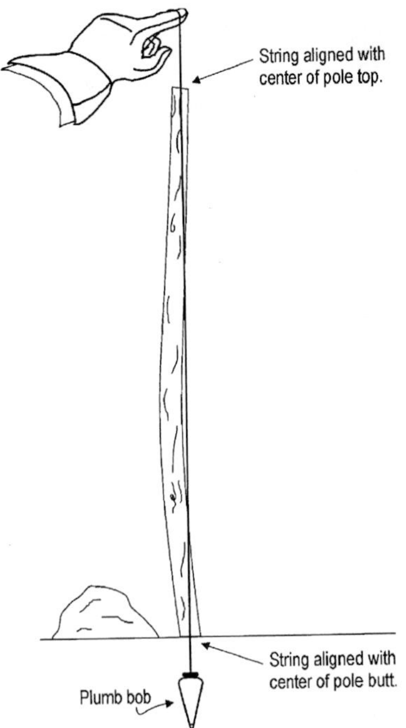

FIGURE 11.50 Straightening a pole with a plumb bob. String is first aligned with center of pole butt. Equipment operator is then directed to move derrick as needed to align center of pole top with the plumb bob string. (*Courtesy Northwest Lineman College.*)

FIGURE 11.51 Energized conductors held in the clear with roller link-stick hot-line tools and rope lines. (*Courtesy A. B. Chance Co.*)

FIGURE 11.52 Lineman and groundman backfilling and tamping dirt in hole around pole. (*Courtesy Northwest Lineman College.*)

FIGURE 11.53 Hydraulic-powered tamper packing dirt around pole to keep pole in place. (*Courtesy Pitman Manufacturing Co.*)

FIGURE 11.54 Lineman in bucket mounted on top of boom attaching equipment and conductor to pole after it is set. Note protective equipment on the lineman and the energized conductor. (*Courtesy Altec Industries.*)

FIGURE 11.55 Lineman assembling H-frame transmission-line structure on the ground to replace structure destroyed by ice storm. Note damaged structure and conductors laying on the ground. (*Courtesy Nebraska Public Power District.*)

FIGURE 11.56 Raising assembled H frame for 138,000-volt line with power boom mounted on caterpillar-tractor. Lifting yoke is used to equalize lift. (*Courtesy American Electric Power System.*)

FIGURE 11.57 Lineman setting poles singly for 161-kV transmission-line structure in rough terrain. (*Courtesy L. E. Meyers Co.*)

FIGURE 11.58 Setting pole by helicopter. The fully framed pole, weighing about 1100 lb, is delivered by a Sikorsky S-58 helicopter and set into hole. Only one lineman is needed to guide pole into hole. A 4-mile average round trip over a 1500-ft elevation change took only $8^1/_2$ min. (*Courtesy Pacific Gas and Electric Co.*)

FIGURE 11.59 Two-pole transmission structure assembled on airport concrete apron. (*Courtesy L. E. Myers Co.*)

FIGURE 11.60 Assembled two-pole transmission structures laid out on airport concrete apron in rotation so that when the helicopter picks them up they are headed in the proper direction. Poles and timbers for additional structures are located adjacent to the assembled structures. (*Courtesy L. E. Myers Co.*)

FIGURE 11.61 Setting H-frame structure by helicopter. Two linemen guide each pole into hole. (*Courtesy American Electric Power System.*)

FIGURE 11.62 Helicopter used to transport and erect K-frame transmission structure to rebuild 345-kV transmission line destroyed by ice storm. (*Courtesy Nebraska Public Power District.*)

unaccessible locations (Fig. 11.63). The poles are hauled to the site on a trailer pulled by a line truck (Fig. 11.64). The boom of the line truck is used to unload the pole from the trailer and place it on the two-wheeled dolly. The two-wheeled dolly with pole is pulled to the work site by the powered dolly (Figs. 11.65 and 11.66). The hydraulically powered auger, supported by the hydraulically operated derrick on the powered dolly, is used to dig the hole for the distribution-line pole (Fig. 11.67). The auger is removed from the boom, and the derrick is attached to the pole, using a nylon sling to set the pole (Fig. 11.68). The pole is raised by the power unit and guided to the hole where it is to be set by the linemen and groundmen (Fig. 11.69).The installation is completed in the same manner as previously

FIGURE 11.63 Power-operated dolly on trailer for transport to job site.

FIGURE 11.64 Line truck with poles on trailer located along street when poles are to be installed on a rear lot line.

FIGURE 11.65 Lineman operates power unit to pull pole on pole-transport dolly.

FIGURE 11.66 The equipment has turf-type tires to prevent lawn damage.

FIGURE 11.67 Power-driven auger used to dig pole hole along rear lot line of homes in residential area.

described. The power-operate dolly eliminates the need to set a distribution pole manually, thus saving effort and manpower and reducing the cost of installation. The customer's lawn is protected from damage that would result if the line truck was driven to the rear lot-line location.

Setting Poles Manually. The piking method is the oldest method of raising poles. It is a manual method. It gets its name from the tool called a "pike pole" (Fig. 11.70), employed after the pole is lifted beyond the reach of the crew raising the pole. The piking method is used only where a derrick cannot be brought in. The use of power-operated boom is a faster and more economical method as only a few workers are needed.

The size of the piking crew depends upon length and the weight of the pole to be raised. It varies from 5 for a 25-ft pole to 10 for a 50-ft pole, as shown in Table 11.3. The size of the crew required is not absolutely fixed, but the numbers indicated in Table 11.3 are the size of crews generally employed. As will be noted, one of the crew is a jennyman, one is stationed at the butt of the pole near the hole, and the balance of the crew are called the "pikers."

The first step in the procedure of raising a pole by the piking method is to lay the butt end of the pole over the hole against a bump board or bar, as shown in Fig. 11.71. The use of the board or bar protects the walls of the hole and prevents them from being caved in by the butt of the pole as the pole is raised. The upper end of the pole is raised and placed on the pole support (Fig. 11.72). Pole supports are made in various forms, two of which are shown in Fig. 11.73. The main duty of the man at the butt is to keep the pole from rolling. This is done by means of a cant hook, illustrated in Fig. 11.74. The men stand side by side on both sides at the top end of the pole (Fig. 11.75). They then push toward each other and up by use of their arms, with the jennyman catching the weight between lifts. They move along the pole until the pole is high enough to permit the use of pikes. The fourth step (Fig. 11.76) is to punch the pikes into the pole and prepare to raise the pole. As the pole is raised, the man carries the jenny forward, always ready to support the pole if need be.

FIGURE 11.68 Power-dolly derrick used to lift pole from transport dolly.

FIGURE 11.69 Pole has been raised by boom and is being placed in hole prepared for it.

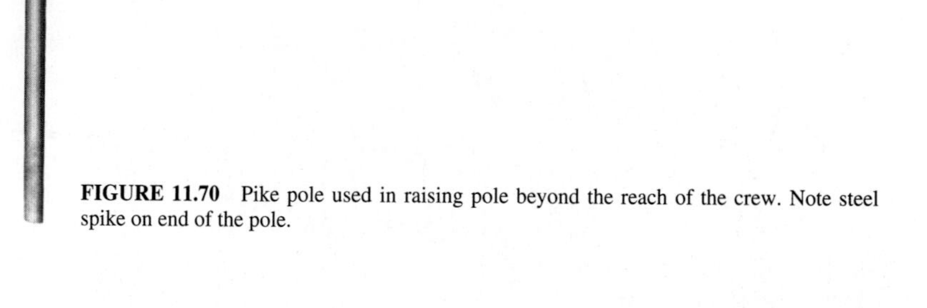

FIGURE 11.70 Pike pole used in raising pole beyond the reach of the crew. Note steel spike on end of the pole.

The raising continues until "high pike!" is called by one of the men. This means that his pike is too high to be effective if raised any farther. The low man brings his pike down first, because he does not have to lower his pike through the rest of the pikes, but has a clear path. The other men follow in order until all the pikes are lowered. The pole is raised in this manner until it drops into the hole. Figure 11.77 shows the pole almost raised and ready to slide into the hole. Figure 11.77 shows very clearly the duty of each one of the crew. The five

TABLE 11.3 Average Size of Crew Required to Manually Raise Poles of Different Lengths

Pole length, ft	Size of crew	Number of pikers	Number of Jennymen	Number of men at butt
25	5	3	1	1
30	6	4	1	1
35	7	5	1	1
40	8	6	1	1
45	10	8	1	1
50	10	8	1	1

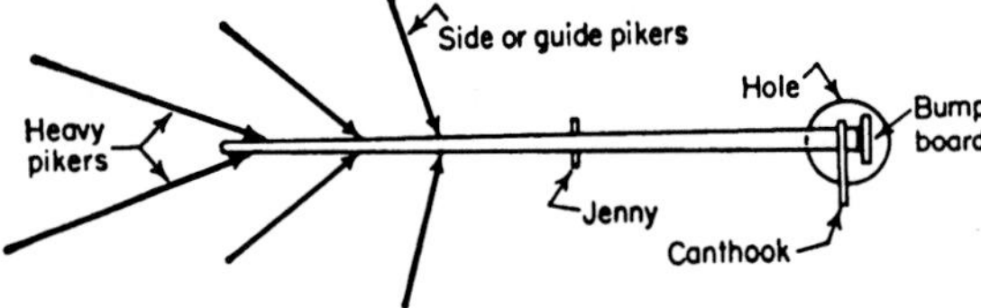

FIGURE 11.71 Plan view of pole-piking method of pole raising.

FIGURE 11.72 Second step in raising a pole. Pole is being lifted from ground into jenny.

FIGURE 11.73 Two forms of pole supports. The one on the left shown open is called the jenny, and the one on the right is known as the mule.

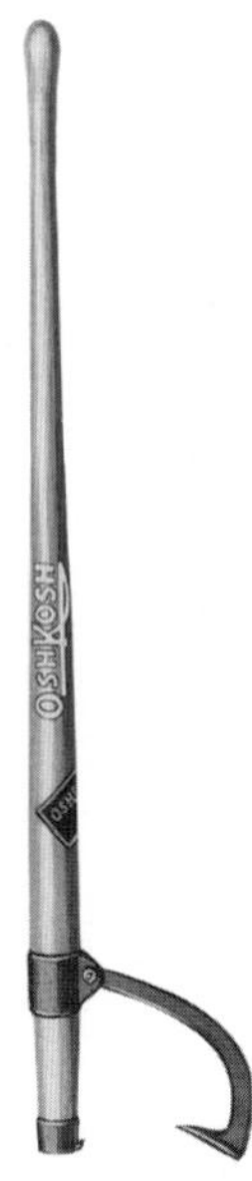

FIGURE 11.74 Cant hook used by man at butt of the pole to keep it from rolling and to turn it if necessary.

pikers raise the pole, the jennyman keeps the jenny always snugly under the pole, and the buttman guides the bottom of the pole and keeps the pole from rolling.

It will be noted that the pikers are not directly underneath the pole. At least one should be well out on each side to guide the pole as well as to help lift it.

FIGURE 11.75 Third step in raising a pole. Pikers have raised onto jenny and are now raising the pole as high as they can without use of pike poles.

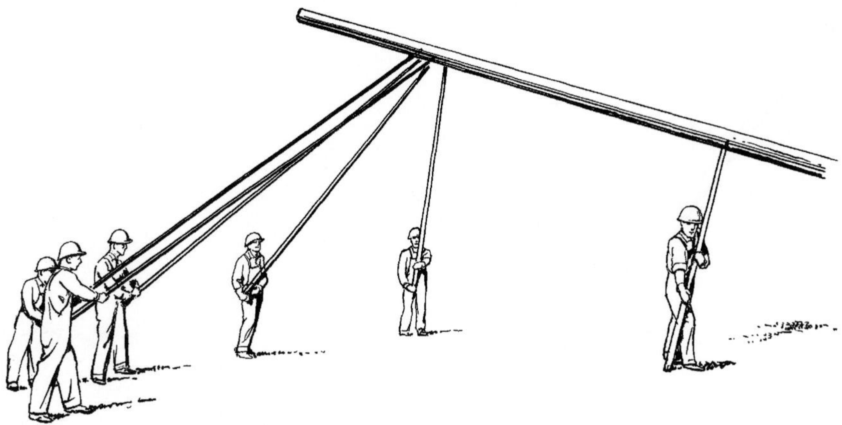

FIGURE 11.76 Fourth step in raising pole. Pikers have placed pike poles against the pole and are raising it into vertical position. Note how pikers are spread out.

Another thing worth noting is the manner of holding the pike pole after the pole is partly raised. It is held in the palm of the hand with the other hand underneath, both arms extended downward. If the pike pole is held in this manner, the worker is in a comfortable position, and if the weight of the pole is suddenly thrown on the pike pole, he is in a good position for grounding the pike pole or moving into the clear. If the pole is very long, the pike pole can be supported on the shoulder as shown in Fig. 11.77. The pike still rests in the palm of the piker's hand. In this way the piker can exert a strong push on the pike.

If the pole to be raised is a very large, heavy pole and the crew is small, it is well to "trench" the hole, that is, to cut a ditch back from the hole. The pole is then placed in this trench. This allows the pole to begin to slide into the hole earlier than if the pole lay flush on the ground. Furthermore, it allows the weight of the lower end of the pole to balance a portion of the weight of the pole above the point of the trench on which the pole is resting.

Facing the Poles. Facing the poles means turning the poles in the proper direction after they have been placed in the ground. The side of the pole which has the longest concave sweep between the ground line and the top is the face side (Fig. 11.78). Poles are normally branded on the face side by the supplier. The pole's curvature should be checked with respect to the brand after it is stored and before a gain is cut for a crossarm or line hardware is installed. The crossarms and other material must be in the proper position to permit the installation of the conductors.

On straight lines it is customary to set adjacent poles with the gains facing in opposite directions (Fig. 11-79). This system is sometimes called gain-to-gain or back-to-back. It provides for the maximum strength in the line.

Crossarms on poles on each side of the center of a curve should face the center of the curve (Fig. 11.80). The pole at the center of the curve is often equipped with double arms.

Poles next to angles should face the angle, as shown in Fig. 11.81. In case of large angles, the angle pole is generally double-armed or dead-ended.

Poles next to corners should face the corner (Fig. 11.82). In case of urban distribution, the arms on the corner pole are generally double arms, as illustrated.

Poles on steep grades should face up the grade, as shown in Fig. 11.83.

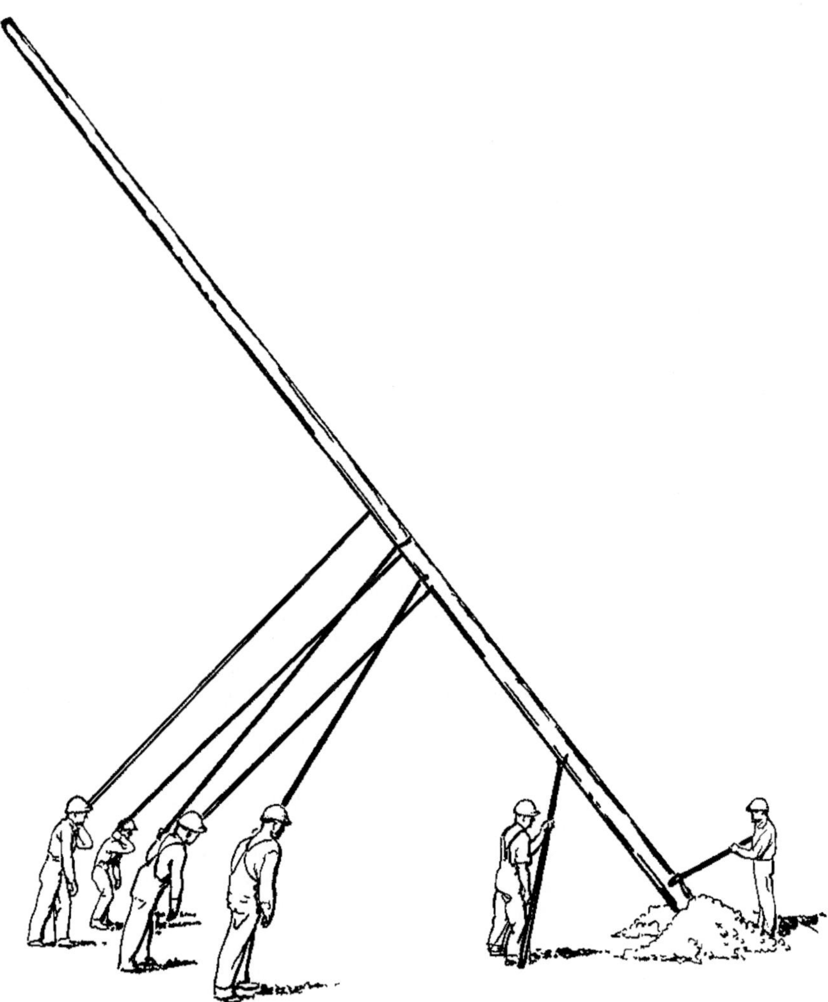

FIGURE 11.77 Last step in raising a pole. All pikes have been lowered. Pole is about to slide into hole. Note jennyman and buttman with cant hook. Also note manner of pushing on pike poles.

Poles next to street crossings should face the street (Fig. 11.84). The poles at the intersections are generally double armed as illustrated.

Poles next to poles supporting long spans should face away from the long span, thereby being in a better position to carry part of the load (Fig. 11.85). The poles on each side of a long span are generally of special construction.

Poles immediately preceding end poles shall face toward the end of the line (Fig. 11.86). The end pole, being a terminal pole, is usually of special construction. Sometimes it is good to have the last two poles face the terminal pole.

Poles set at line terminals, curves, corners, and other points of abnormal stress should be given a slight rake against the direction of the pull. This should be sufficient to allow for

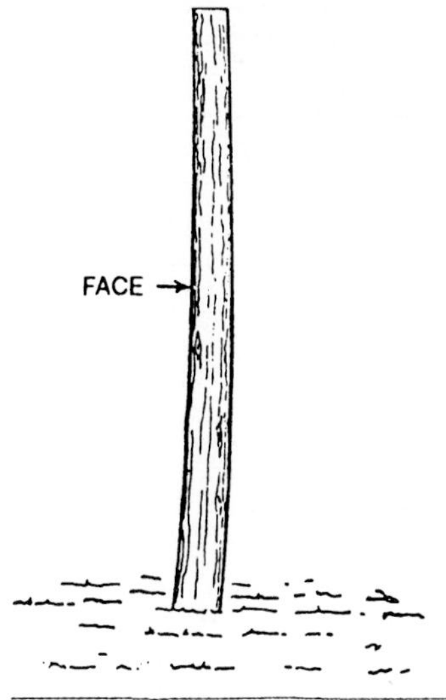

FIGURE 11.78 Concave side of the pole is the face. The brand markings should have been applied on the face of the pole by the supplier.

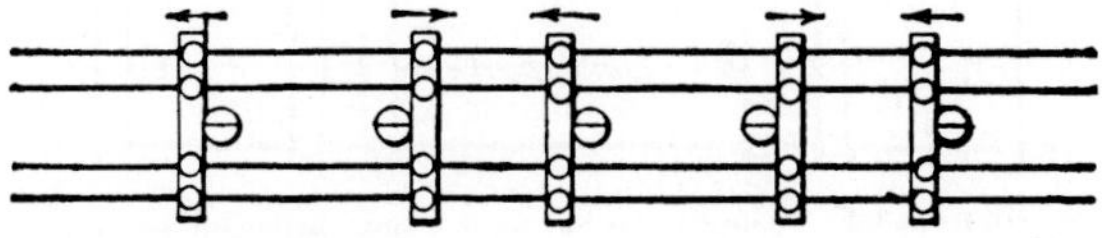

FIGURE 11.79 Facing crossarms on straight lines. Every second pole has crossarm facing in the same direction.

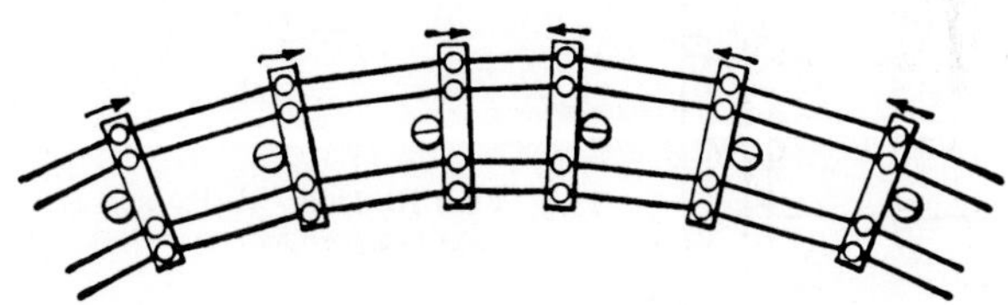

FIGURE 11.80 Facing crossarms on curve. Note that the three poles on each side of the center of the curve face the curve.

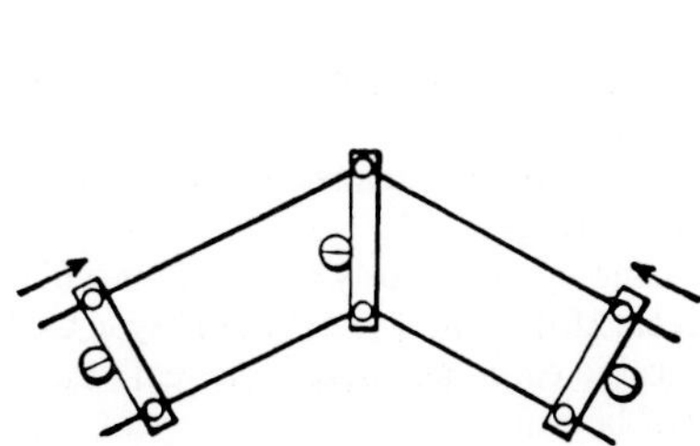

FIGURE 11.81 Facing crossarms at an angle. Adjacent poles face the angle.

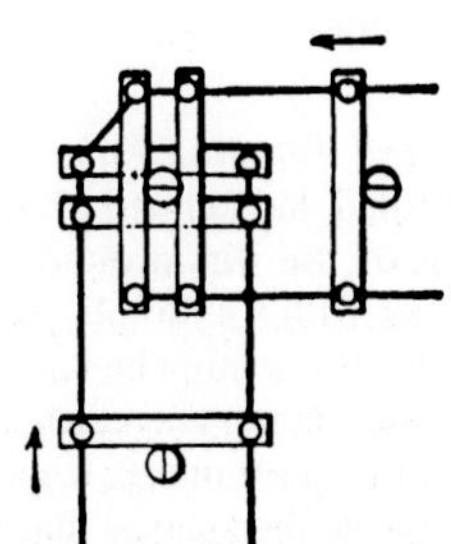

FIGURE 11.82 Facing crossarms at a corner. Adjacent poles face the corner.

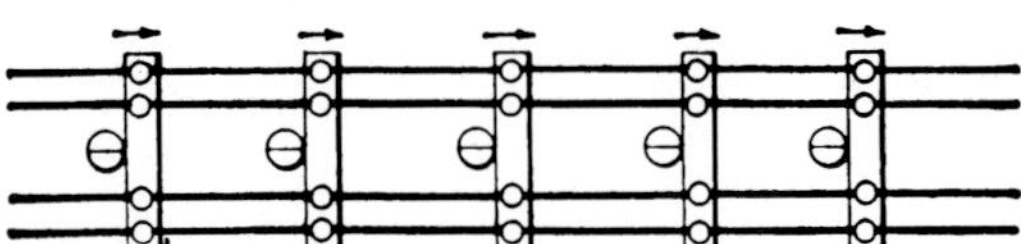

FIGURE 11.83 Facing poles on steep grade. All poles face upgrade.

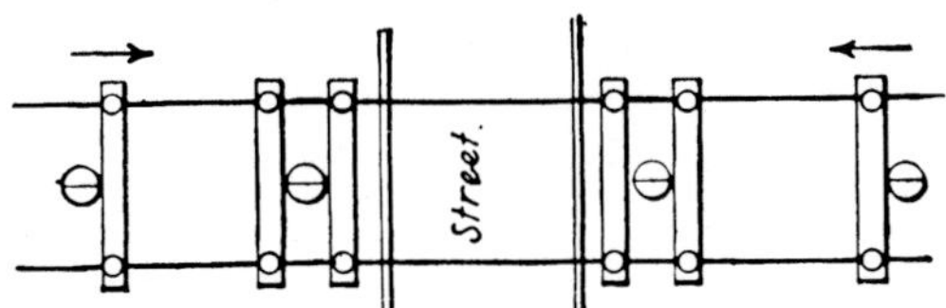

FIGURE 11.84 Facing poles at a crossing. Adjacent poles face the crossing.

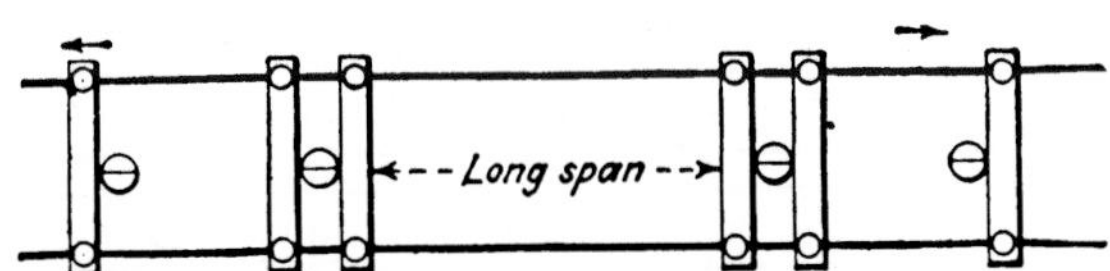

FIGURE 11.85 Facing poles on long spans. Adjacent poles face away from long span.

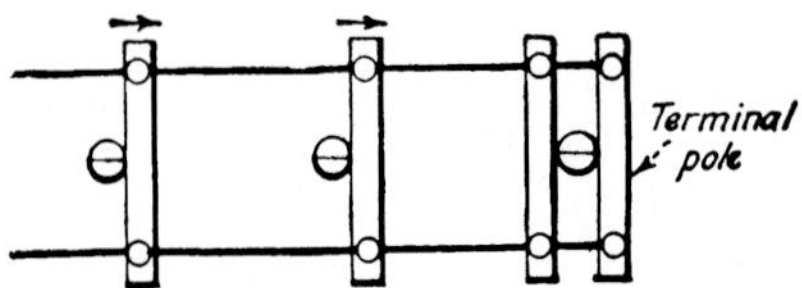

FIGURE 11.86 Facing poles at terminals. Adjacent poles face terminal.

the change in pole position caused by the continued pull of the load and the normal creepage of the anchor.

Backfilling and Tamping Pole Hole. After the pole is properly faced and lined in, the hole is backfilled and tamped (Fig. 11.87).

Tamping of the dirt around the pole is completed manually with a tamping bar (Fig. 11.88) or with a hydraulic-powered tamper (Fig. 11.53).

Too much stress cannot be laid on proper tamping, as a poorly tamped hole will not hold the pole in alignment. The earth should be backfilled slowly and each layer thoroughly tamped until the tamp makes a solid sound as the earth is struck. In general, the tamping should be done so thoroughly that no dirt need be hauled away.

If small stones or gravel are readily available, these should be used in backfilling, as a better foundation can thus be obtained. Care should be taken that plenty of earth is tamped

FIGURE 11.87 Crew backfilling and tamping.

in to fill the spaces between the stones. The backfill should be piled well up around the base of the pole to allow for settling. An examination should be made 1 or 2 months later and backfilling added if it has settled below the ground level.

Pulling Poles. The boom of the line truck should not be used to pull poles out of the ground. The strain placed on the boom, if it is used to pull poles, will exceed the capacity of the boom; and if the boom is rocked to free the pole, the boom can be seriously damaged. A hydraulic pole puller powered by the truck hydraulic system or a portable hydraulic pump driven electrically or with a gasoline engine can be used successfully (Fig. 11.89). The hydraulic pole puller consists of a hydraulic cylinder, a swivel chain hook assembly, a baseplate, 6 ft of 1/2-in alloy chain, and carrying handles. The puller will lift the pole approximately 16 in. The line truck boom is used to secure the pole while it is being pulled. If the pole is near energized conductors, it should be covered as shown in Fig. 11.51. The personnel guiding the pole should wear rubber gloves. When the hydraulic cylinder reaches the maximum position, the chain around the pole is released, the cylinder is lowered, the chain is refastened, and the pole is lifted farther. This process is continued until the pole is free from the earth. The boom of the line truck is used to lower the pole to the ground. The groundmen and/or the linemen guide the pole while it is being lowered.

FIGURE 11.88 Wood and steel tamper bars used in pole setting.

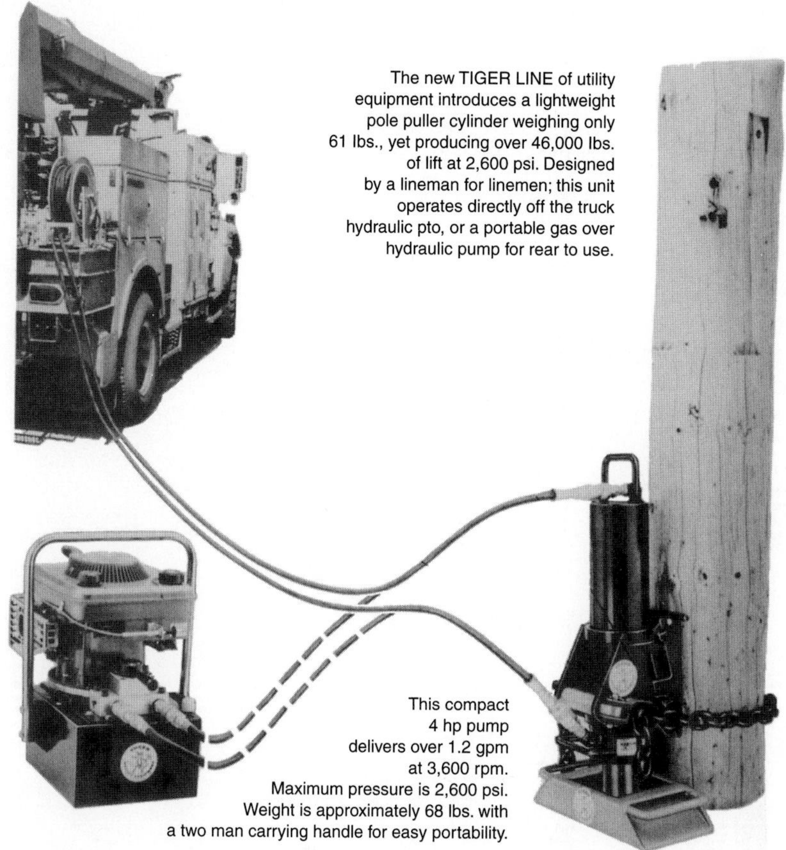

FIGURE 11.89 Pole puller in position to raise pole. The pole puller can receive its power from the line truck's hydraulic system or a portable power-driven hydraulic pump. (*Courtesy Thiermann Industries, Inc.*)

CHAPTER 12
GUYING POLES

A *guy* is a brace or cable fastened to the pole to strengthen it and keep it in position. Guys are used wherever the wires tend to pull the pole out of its normal position and to sustain the line during abnormal loads caused by sleet, wind, and cold. Guys counteract the unbalanced forces imposed on the poles by dead-ending conductors; changing conductor sizes, types, and tensions; or angles in the transmission or distribution line. The guy should be considered as counteracting the horizontal component of the forces with the pole or supporting structure as a strut resisting the vertical component of the forces.

Anchor or Down Guys. An *anchor* or *down guy* consists of a wire running from the attachment near the top of the pole to a rod and anchor installed in the ground (Fig. 12.1). This type of guy is preferable if field conditions permit its installation since it transfers the unbalanced force on a pole or structure to the earth without intermediate supports.

An anchor or down guy used at the ends of pole lines in order to counterbalance the pull of the line conductors is called a *terminal guy* (Fig. 12.2). All 90° corners in the line are considered as dead ends. They should be guyed the same as terminal poles except that there will be two guys, one for the pull of the conductors in each direction (Fig. 12.3).

Where the line makes an angle, a side pull is produced on the pole. Side guys should be installed to balance the side pull (Figs. 12.4 and 12.5). Figure 12.6 shows side guys installed in the line as it changes from one side of the highway to the other.

Where a branch line takes off from the main line, an unbalanced side pull is produced. A side guy should be placed on the pole directly opposite to the pull of the branch line. Anchor or down guys are installed at regular intervals in transmission lines that extend long distances in one direction to protect the line from excessive damage as a result of broken conductors (Fig. 12.7). Guys installed to protect the facilities and limit the damage, if a conductor breaks, are called *line guys* or *storm guys*.

An anchor with a horizontal strut at a height above the sidewalk to clear pedestrains on the sidewalk is referred to as a sidewalk guy. Figure 12.8 illustrates two sidewalk guys installed at one pole to serve as branch guys.

Span Guy. A span or overhead guy consists of a guy wire installed from the top of a pole to the top of an adjacent pole to remove the strain from the line conductors. The overhead or span guy transfers the strain on a pole to another structure. This may be to another line pole or to a stub pole on which there is no energized equipment. A span guy is always installed to extend from the strain pole to the same or lower level on the next line pole.

Head Guy. A guy wire running from the top of a pole to a point below the top of the adjacent pole is called a *head guy* (Fig. 12.9). Lines on steep hills are normally constructed with head guys to counteract the downhill strain of the line.

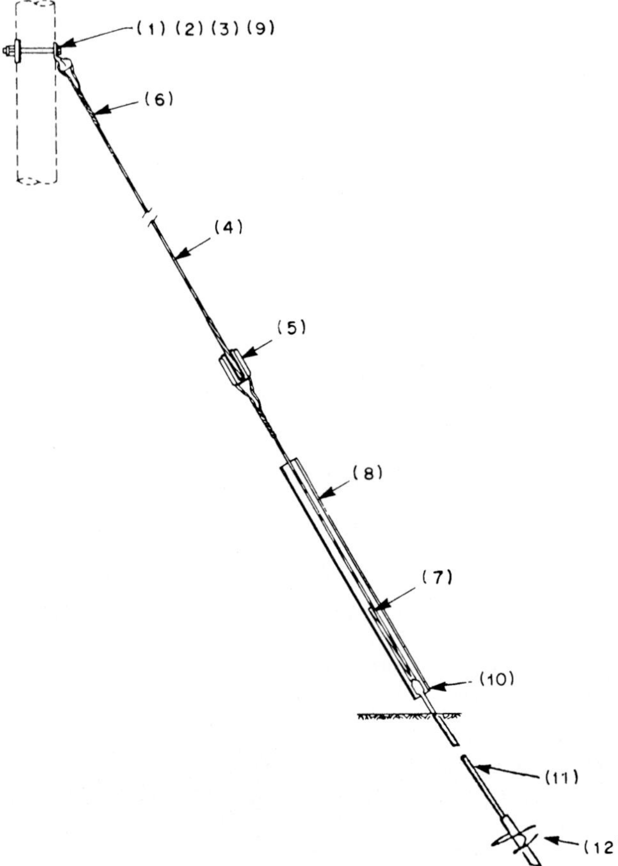

FIGURE 12.1 Anchor or down guy assembly. (1) Galvanized machine bolt with nut, (2) locknut, (3) square curved galvanized washer, (4) galvanized steel guy wire, (5) porcelain guy strain insulator, (6) prefabricated guy dead-end grip, (7) prefabricated guy dead-end grip, (8) plastic guy guard, (9) angle thimbleye, (10) eyenut, (11) steel anchor rod, (12) power-installed screw anchor.

Electric lines crossing railroad tracks must be reinforced by the use of double arms and head guys as shown in Fig. 12.10. These guys run from the top of the pole adjacent to the track to a point about halfway down on the next pole.

Arm Guys. A guy wire running from one side of a crossarm to the next pole is called an *arm guy*. Arm guys are used to counteract the forces on crossarms which have more wires dead-ended on one side than on the other (Fig. 12.11).

Stub Guys. A guy wire installed between a line pole and a stub pole on which there is no energized equipment is called a *stub guy* (Fig. 12.12). An anchor or down guy is used to secure the stub pole. This type of guy is often installed to obtain adequate clearance for guy

FIGURE 12.2 Guy wires installed on terminal or end pole. Guy wires are connected to system neutral conductor to provide a low-resistance ground, eliminating the need for a guy insulator.

FIGURE 12.3 Terminal guys installed at corner of transmission line. Entire pull of line conductors has to be carried by two sets of three guy wires. (*Courtesy Ioslyn Manufacturing and Supply Co.*)

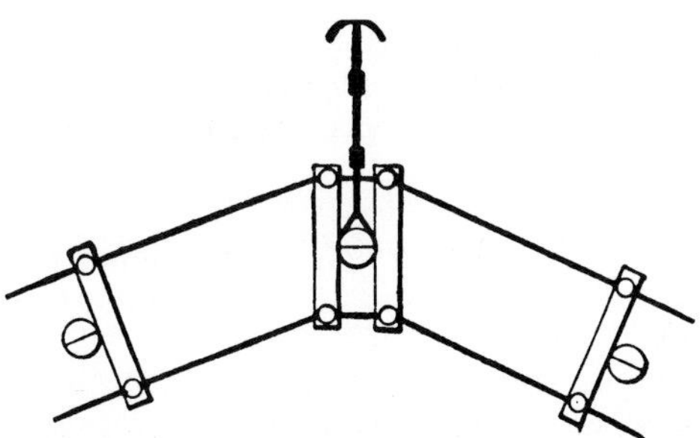

FIGURE 12.4 Guy installed at angle in line. The guy opposes the side pull of the line conductors.

FIGURE 12.5 Side guys counterbalancing the side pull of the line conductors due to a branch line extending perpendicular from the single-phase primary line. A fiberglass strain insulator is installed in series with the down guy at the point where the guy connects to the top of the pole. The fiberglass insulator isolates the down guy from the energized wires and provides clearance from the grounded guy for a lineman working on the primary conductor.

wires extending across streets or highways (Fig. 12.13). When transmission or distribution lines parallel to streets or roads must be guyed toward the street or road, the necessary clearance can be obtained by the use of a pole stub on the opposite side.

Push Guys. A pole used as a brace to a line pole is often referred to as a *push guy* (Fig. 12.14). A push brace or guy is used where it is impossible to use an anchor or down guys. For example, when it is impossible to obtain sufficient right of way for a down guy, a push brace can usually be installed (Fig. 12.15).

Guy Wire. The wire or cable normally used in a down guy is seven-strand galvanized steel wire or seven-strand alumoweld wire (Fig. 12.16). Alumoweld wire consists of steel wire strands coated with a layer of aluminum to prevent corrosion.

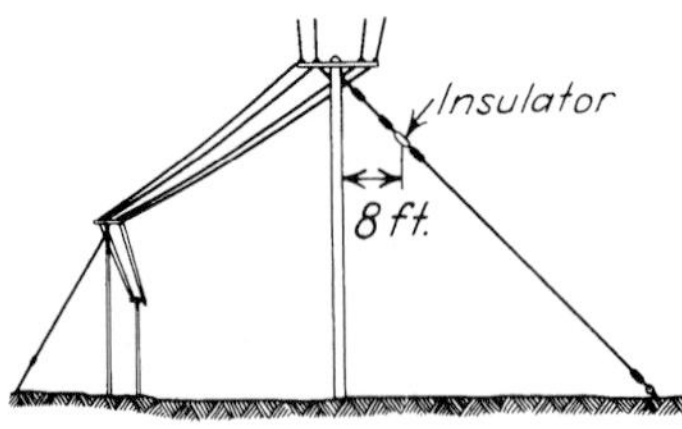

FIGURE 12.6 Down guys installed in line to balance side pulls caused by angles in the line.

FIGURE 12.7 Line guys installed on three-phase transmission line. Note that poles are guyed in both directions, primarily installed to protect line from broken conductors.

FIGURE 12.8 Vertical or "sidewalk" guy completely installed with strain insulator and guy guard. This type of guy is used when the anchor must be set within 5 feet of the pole. The horizontal strut shown consists of a 2 inch galvanized iron pipe and should have a length equal to the distance between the anchor and the pole. This type of guy is often called a sidewalk guy because the pole can be set on one side of a sidewalk and the anchor on the other. *(Courtesy Wisconsin Electric Power Co.)*

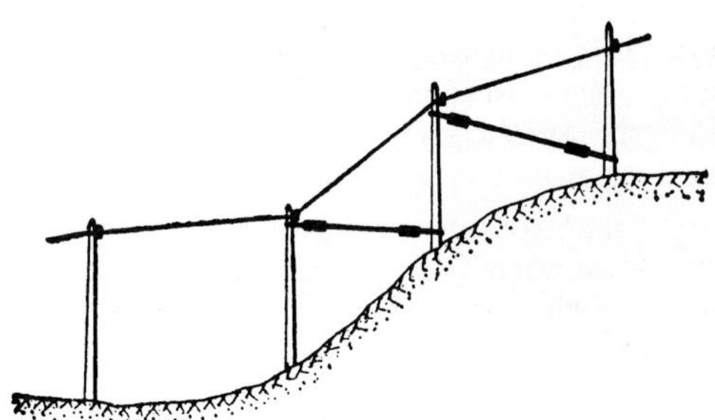

FIGURE 12.9 Head guys installed on steep grade.

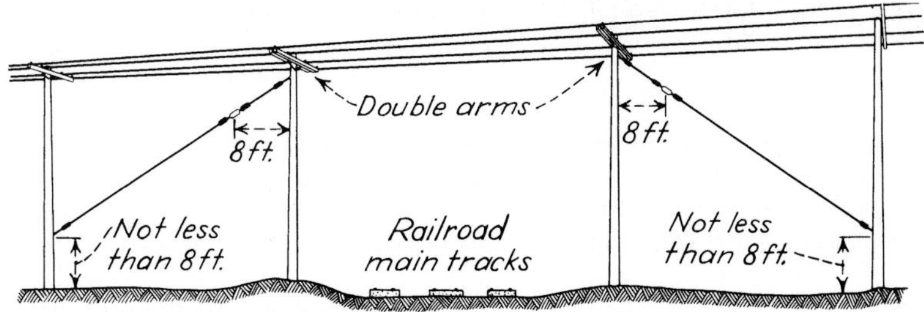

FIGURE 12.10 Line reinforcement by use of double arms and head guys at a railroad crossing.

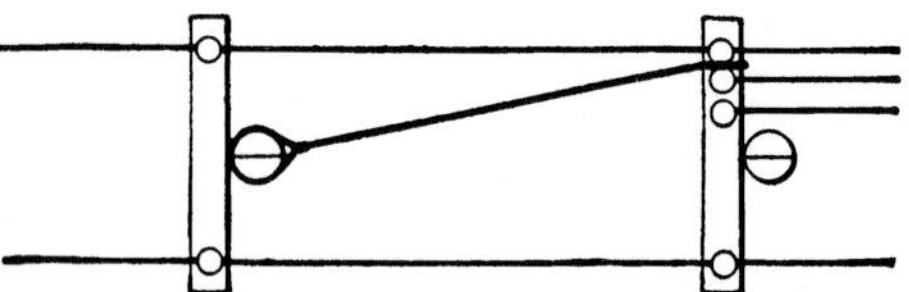

FIGURE 12.11 Arm guy. Guy opposes unbalanced pull on crossarm.

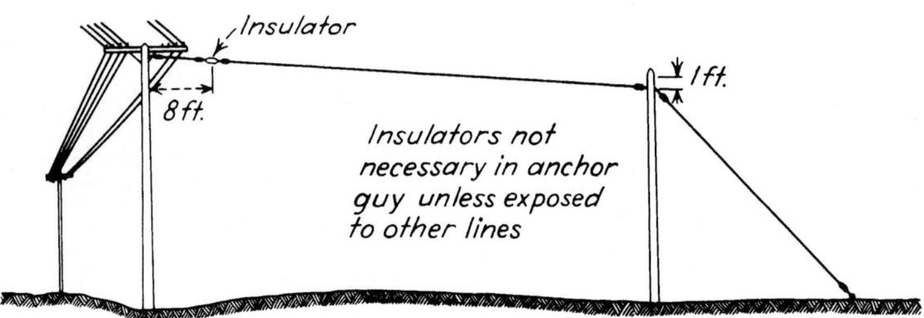

FIGURE 12.12 Use of pole stub to secure clearance over highway. Pole stub is then guyed in usual manner.

FIGURE 12.13 Stub guy used to obtain clearance of guy wire over highway. (*Courtesy Copperweld Steel Co.*)

Guy wire is used in various sizes with diameters from $^1/_4$ to $1^1/_4$ inches. The breaking or ultimate strength of the various types of guy wire is given in Table 12.1. Alumoweld guy cable, seven-strand No. 8 wire, has a minimum breaking strength of 15,930 pounds, and seven-strand No. 5 wire has a minimum breaking strength of 27,030 pounds.

Anchors. The log anchor is the oldest type of anchor used to counteract the unbalanced forces on electric transmission and distribution line structures. The log anchor is frequently

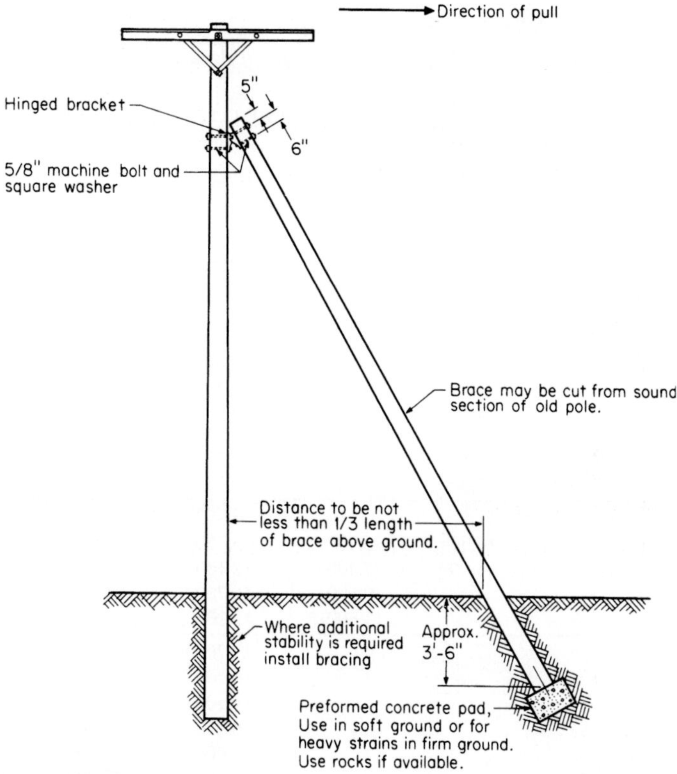

FIGURE 12.14 Push brace or guy.

FIGURE 12.15 Single-pole brace used in place of stub guy.

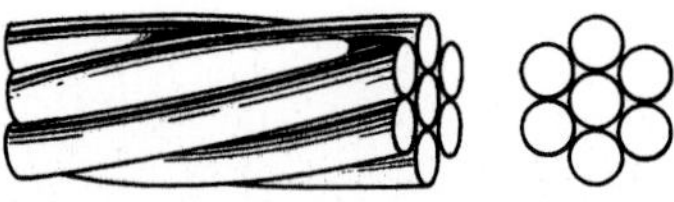

FIGURE 12.16 Seven-strand steel guy cable. (*Courtesy Joslyn Hfg and Supply Co.*)

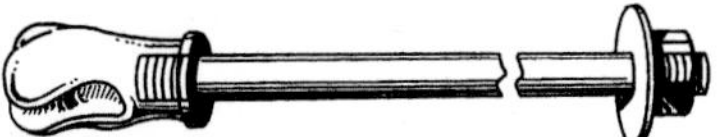

FIGURE 12.17 Guy anchor rod used with log anchor. (*Courtesy Joslyn Hfg and Supply Co.*)

TABLE 12.1 Guy Strand Sizes and Strengths

		Minimum breaking strength of strand, lb				
Number of wires in strand	Diameter, in	Utilities grade	Common grade	Siemens-Martin grade	High-strength grade	Extra-High-strength grade
7	9/32	4,600	2,570	4,250	6,400	8,950
7	5/16	6,000	3,200	5,350	8,000	11,200
7	3/8	11,500	4,250	6,950	10,800	15,400
7	7/16	18,500	5,700	9,350	14,500	20,800
7	1/2	25,000	7,400	12,100	18,800	26,900
7	9/16		9,600	15,700	24,500	35,000
7	5/8		11,600	19,100	29,600	42,400
19	1/2			12,700	19,100	26,700
19	9/16			16,100	24,100	33,700
19	5/8			18,100	28,100	40,200
19	3/4			26,200	40,800	58,300
19	7/8			35,900	55,800	79,700
19	1			47,000	73,200	104,500
37	1			46,200	71,900	102,700
37	1 1/8			58,900	91,600	130,800
37	1 1/4			73,000	113,600	162,200

referred to as a *dead man*. This nickname comes from the gravelike dimensions of the excavation necessary to install the log anchor. The cost of the excavation necessary to install the log anchor in the ground limits its use. Figures 12.17 and 12.18 illustrate the component parts of a log anchor assembly.

Manufactured anchors are easier, quicker, and less expensive to install than log anchors. The common types of manufactured anchors consist of the expansion anchor (Fig. 12.19), the screw-type anchor (Fig. 12.20), and the "never-creep" or plate-type anchor (Fig. 12.21).

Log or manufactured anchors cannot be installed to secure a down guy assembly in an area where rock formations are close to the surface of the ground. If rock is located close to the surface, expanding rock-type anchors are used (Fig. 12.22). Rock anchors, properly installed, will develop holding power equal to the full strength of the anchor rod.

A log dead-man anchor 8 ft long with a diameter of 9 in. installed 6 ft deep in loose or sandy-type soil, with an angle of pull for the guy wire and rod assembly equal to 45°, should have a holding power of approximately 28,000 lb. This holding power can be determined by calculating the weight of the soil that must be lifted to extract the log from the earth. Loose sandy-gravel-type soil weighs approximately 100 lb per cubic foot (ft^3). The volume

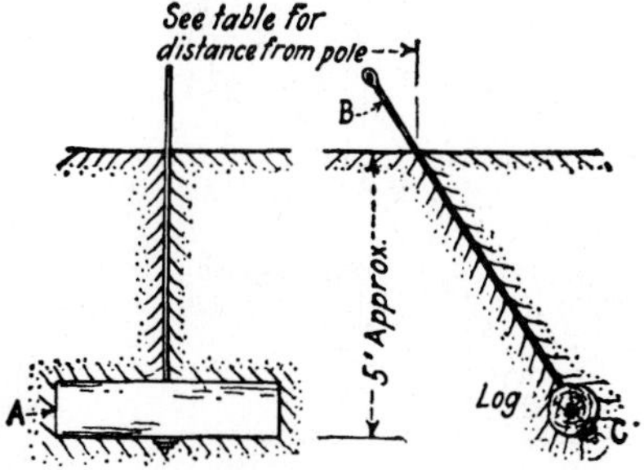

Size of pole, ft	Minimum distance from pole, ft	Material Loc.	No.	Description
30 and 40	15	A	1	12″ × 6′—0″log
45 and 50	20	B	1	Anchor rod
55 and 60	25	C	1	Anchor-rod washer

FIGURE 12.18 Views showing log anchor installed. Tables list materials required as well as distance from pole at which anchor should be placed for various heights of poles.

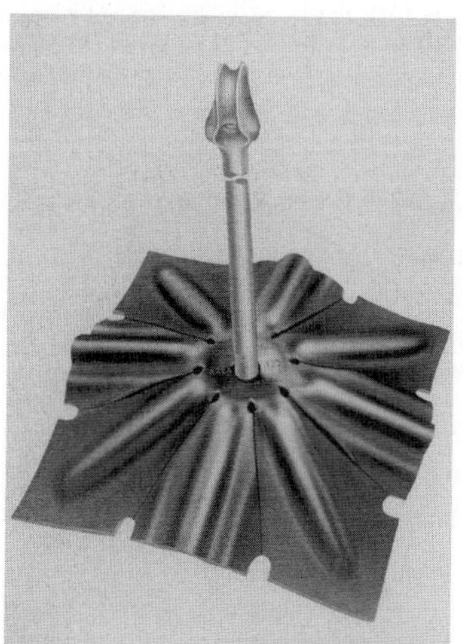

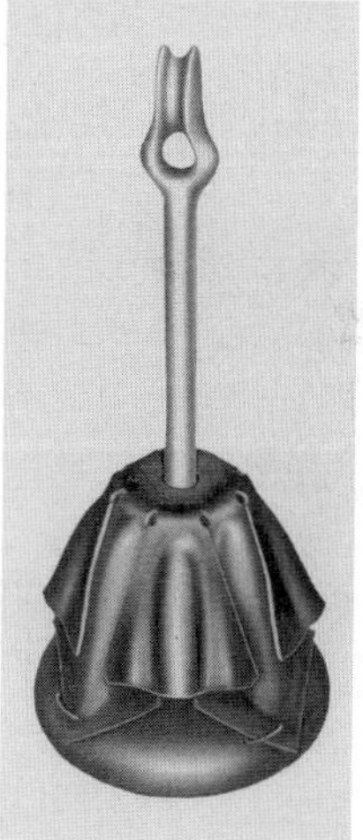

FIGURE 12.19 Expansion anchor. (*a*) Expanded position and (*b*) closed position. The anchor is inserted in the hole in the closed position. Then the top of the anchor is struck several blows with a ram to cause the leaves to expand into solid undisturbed earth at the bottom of the hole.

of the soil of the fill above the anchor, and in line with the pull, is approximately 280 ft^3. Tests performed by the American Electric Power Co. on this type of anchor, when installed in wet loam and clay-type soil, determined that a 3/4-in anchor rod failed with a pull of 27,000 lb without disturbing the log in the earth. The holding strength of the different types of manufactured anchors is itemized in Table 12.2.

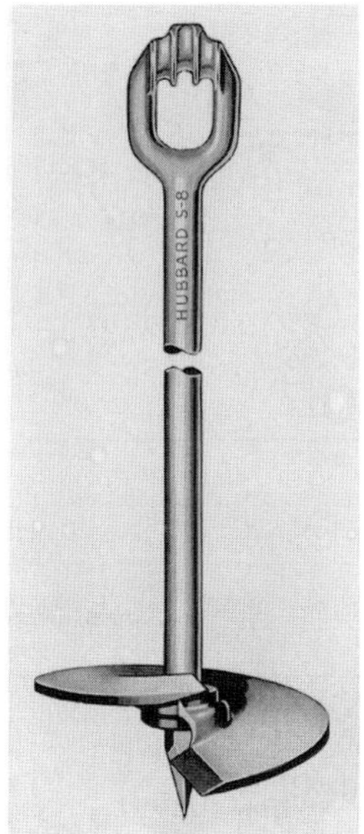

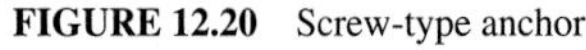

FIGURE 12.20 Screw-type anchor.

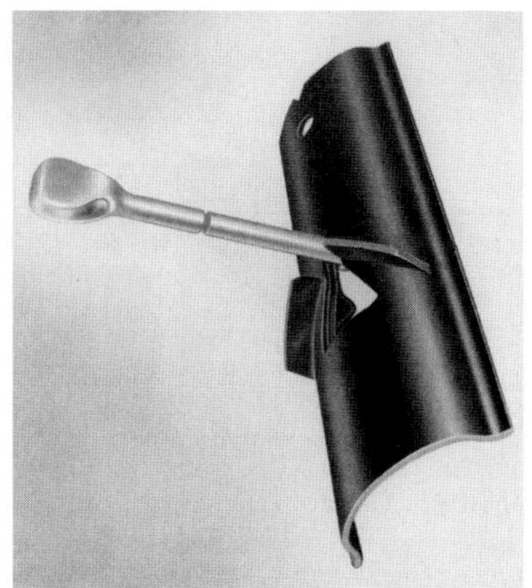

FIGURE 12.21 View showing never-creep or plate-type anchor installed. *(Courtesy A. B. Chance Co.)*

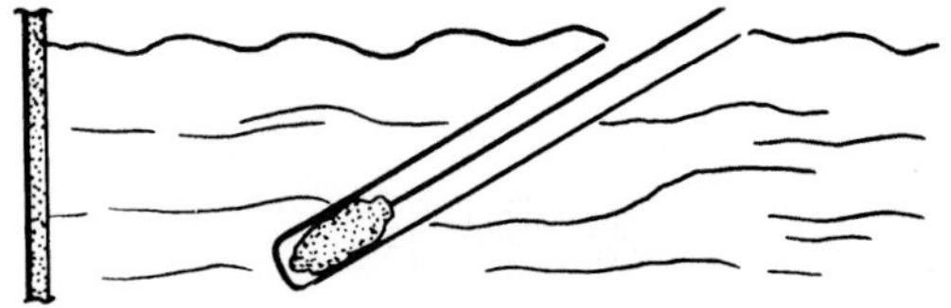

FIGURE 12.22 Rock anchor expanded in hole drilled in the rock. The greater the strain, the more firmly it is wedged in the rock.

TABLE 12.2 Holding Power of Commonly Used Manufactured-Type Anchors

Anchor		Holding strength, lb	
Type	Size, in	Poor soil	Average soil
Expansion	8	10,000	17,000
Expansion	10	12,000	21,000
Expansion	12	16,000	26,500
Screw or helix*	8	6,000	15,000
Screw or helix	11 5/16	9,500	15,000
Screw or helix	32	10,000	23,000
Screw or helix	34	15,500	27,000
Never-creep		21,000	34,000
Cross-plate		18,000	30,000

*Screw or helix anchors power-installed.

TABLE 12.3 Anchor Rod Strength

Anchor rod nominal diameter, in	Ultimate tensile strength, lb/sq. in	Yield point, lb/sq. in	Full rod section		Threaded section	
			Ultimate load, lb	Yield load, lb	Ultimate load, lb	Yield load, lb
1/2	70,000	46,500	13,700	9,120	12,750	8,450
5/8	70,000	46,500	20,400	13,500	18,700	12,400
3/4	70,000	46,500	29,600	19,600	27,100	18,000
1	70,000	46,500	53,200	35,300	42,400	28,100
1 high strength	90,000	59,500	68,200	45,100	54,500	36,000
1 1/4	70,000	46,500	85,800	57,000	67,800	45,000
1 1/4 high strength	90,000	59,500	110,000	73,000	87,200	57,600

Anchor Rods. The anchor rods serve as the connecting link between the anchor and the guy cable (Fig. 12.17). The rod must have an ultimate strength equal to, or greater than, that required by the down guy assembly. The strength of commonly used anchor rods is itemized in Table 12.3. Anchor rods vary in diameter from 1/2 to 1 1/4 in and in length from 3 1/2 to 12 ft. Power-installed 5/8-in D anchor rods are normally 6 to 8 ft long. If a greater length is needed, a 3 1/2-ft rod can be coupled to the longer rod to reach a greater depth (Fig. 12.23).

Figure 12.24 illustrates a helix anchor. Several helixes are stacked on a 1 1/2-in sq steel shaft. Each helix acts essentially as a separate anchor for increased holding power. Anchor rods and extension rods for power-installed helix anchors are fabricated with a connecting device designed to fit over the 1 1/2-in sq drive hub (Fig. 12.25). Figure 12.26 illustrates a power-installed swamp anchor. Extra-heavy galvanized steel pipe is used as an anchor rod to obtain the proper depth. The pipe is manufactured in standard 21-ft lengths. The pipe is cut to the desired length or joined with couplings, if necessary. Commonly used sizes are

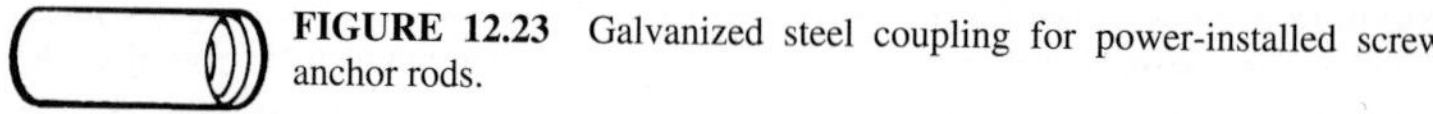

FIGURE 12.23 Galvanized steel coupling for power-installed screw anchor rods.

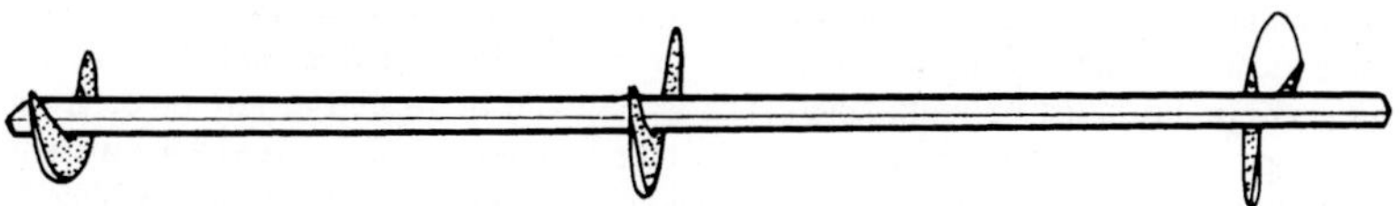

FIGURE 12.24 Helix anchor designed for heavy guy loading.

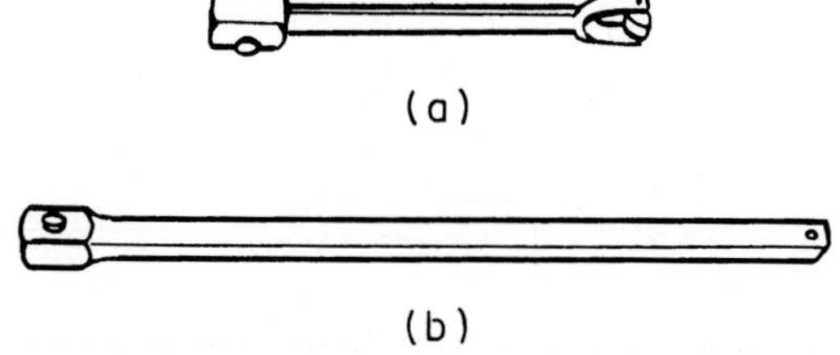

FIGURE 12.25 (*a*) Helix anchor guy rod and (*b*) extension.

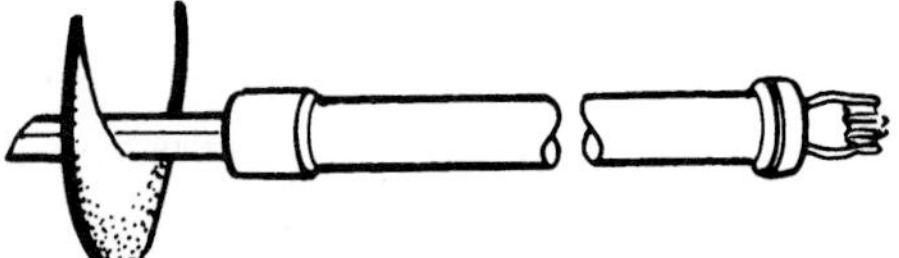

FIGURE 12.26 Swamp anchor to be used in swampy areas where greater depth than normal is necessary.

FIGURE 12.27 Lineman installing threaded eyenut on anchor rod. *(Courtesy A. B. Chance Co.)*

$1\frac{1}{2}$-in-diameter pipe with a 10-in swamp anchor and 2-in-diameter pipe with a 15-in swamp anchor. Galvanized steel eyenuts are installed on power-installed screw anchor rods to provide a connecting device for the guy-strand cable (Fig. 12.27).

Guy Insulators. The *National Electrical Safety Code* requires that ungrounded guys, which are attached to supporting structures that carry open-supply conductors of more than 300 volts, or guys that are exposed to such conductors, be insulated. Otherwise, the guys must be effectively grounded. It may be desirable to install insulators in guys that are effectively grounded to maintain adequate electrical clearances and provide safe working space for the lineman. Linemen must recognize the grounded guy. If work is to be done on a grounded guy assembly, electrical continuity must be maintained. The guy should be treated as any other grounded device. Proper protective equipment must be installed on the guy in the area of the work while the lineman is working on or near energized conductors. Ungrounded guys are insulated with porcelain, wood, fiberglass, or other materials of suitable mechanical and electrical properties.

A porcelain guy insulator usually consists of a porcelain piece pierced with two holes, at right angles to each other, through which the guy wires are looped (Fig. 12.28).

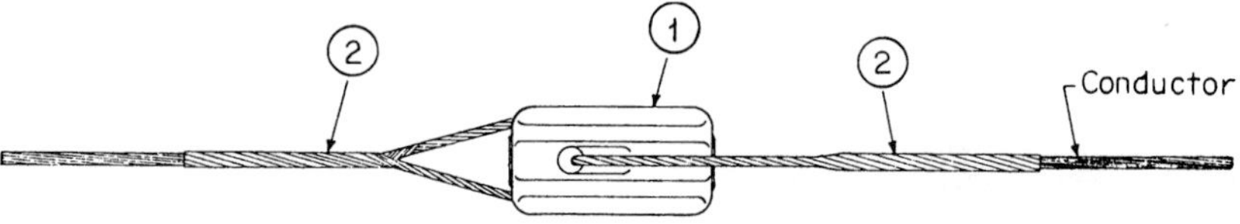

FIGURE 12.28 Porcelain guy insulator assembly. (1) Insulator porcelain guy strain, (2) guy grip.

FIGURE 12.29 Wood strain insulators equipped with arcing horns inserted in down guys. (*Courtesy Copperweld Steel Co.*)

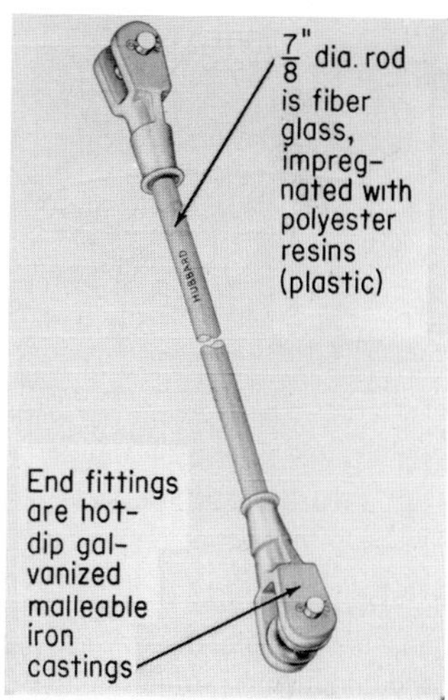

FIGURE 12.30 Fiberglass strain insulator. Fiberglass rod is impregnated with polyester resins.

Since this puts the porcelain between the loops under compression, it makes for great strength.

When higher values of insulation are needed in guy, wood or fiberglass strain insulators are used. Long wood strain insulators require arcing horns (as shown in Fig. 12.29) to bypass lightning strokes around the wood. The use of wood strain insulators on newly constructed pole lines has been discontinued. Fiberglass strain insulators (Figs. 12.30 and 12.31) are impervious to moisture. They can withstand a direct stroke of lightning to the pole structure without bursting. For the same reason, they do not require arcing horns to bypass the stroke. The fiberglass insulator takes up less airspace in the vicinity of the pole.

FIGURE 12.31 Epoxiglass strain insulator installed in down guys. (*Courtesy A. B. Chance Co.*)

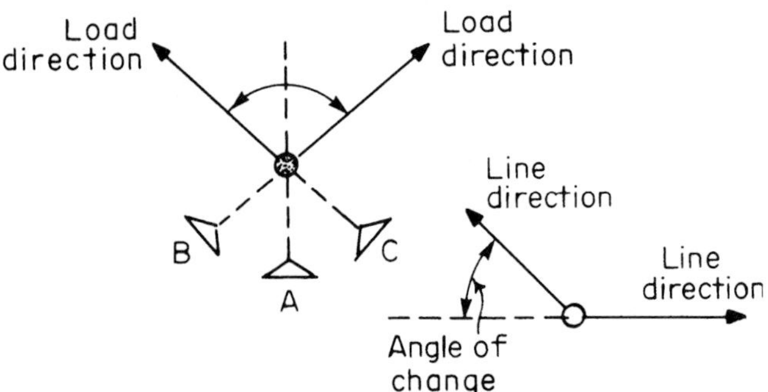

FIGURE 12.32 Proper location of guys and line angle as illustrated.

Guy Size Determination.* Factors to be considered in guying pole lines are the weight of the conductor or cable, the size and weight of crossarms and insulators, wind pressures on poles and conductors, strain due to the contour of the earth, line curvatures, pole heights, and dead-end loads, plus the vertical load due to sleet and ice. To reduce unbalanced stresses to a minimum, correct angling and positioning of guy wires are essential. Where obstructions make it impossible to locate a single guy in line with the load or pull, two or more guys can be installed with their resultant guying effect in line with the load.

Where lines make an abrupt change in direction, the guy anchor is normally placed so that it bisects the angle formed by the two conductors, as shown by anchor A in Fig. 12.32. Under heavy load conditions, it may be necessary to use two anchors, each dead-ending a leg of the line load, as shown by anchors B and C. Long straight spans require occasional side and end guys to compensate for heavy icing and crosswind on conductors and poles.

These, and all other factors that might make it advisable to use guys, should be carefully considered in initial designs for line construction.

To compute the load on the guy, the line load must be determined first. When the line is dead-ended, the line load can be calculated by multiplying the ultimate breaking strength of the conductor used S by the number of conductors N.

For example, if three 1/0 aluminum-conductor steel-reinforced (ACSR) conductors are dead-ended on a pole, the line load will be 12,840 lb:

$$S \times N = \text{line load}$$

$$4280 \times 3 = 12{,}840$$

The ultimate breaking strength of the conductor used is found for the conductor size as shown in Tables 12.4 through 12.10.

To determine the line load to be guyed on a single anchor where the line changes direction, multiply the ultimate breaking strength of the conductor used S by the number of conductors used N by the multiplication factor for the angle change of line direction M, as listed in Table 12.11. For example, if the pole to be guyed carries three 1/0 ACSR conductors with an angle change of 90°, the line load will be 18,156 lb:

$$S \times N \times M = \text{line load}$$

$$4280 \times 3 \times 1.414 = 18{,}156$$

* Material under this heading is from Encyclopedia of Anchoring. Copyright A. B. Chance Co. Used by permission.

TABLE 12.4 Hard-Drawn Copper Wire, Solid

Size AWG*	Diameter, in	Breaking strength, lb
10	0.102	529
8	0.129	826
6	0.162	1,280
4	0.204	1,970
2	0.258	3,003
1	0.289	3,688
1/0	0.325	4,517
2/0	0.365	5,519
3/0	0.410	6,722
4/0	0.460	8,143

*AWG = American wire guage.

TABLE 12.5 Stranded Hard-Drawn Copper Wire

Size, AWG or cir mils	No. of strands	Bore. diameter, in	Breaking strength,lb
6	7	0.184	1,288
4	7	0.232	1,938
2	7	0.292	3,045
1	7	0.328	3,804
1/0	7	0.368	4,752
2/0	7	0.414	5,926
3/0	7	0.464	7,366
4/0	7	0.522	9,154
4/0	19	0.528	9,617
250	19	0.574	11,360
300	19	0.628	13,510
350	19	0.679	15,590
400	19	0.726	17,810
450	19	0.770	19,750
500	19	0.811	21,950
500	37	0.813	22,500
600	37	0.891	27,020
700	61	0.964	31,820
750	61	0.998	34,090
800	61	1.031	36,360
900	61	1.094	40,520
1,000	61	1.152	45,030
1,250	91	1.289	56,280
1,500	91	1.412	67,540
1,750	127	1.526	78,800
2,000	127	1.632	90,050
2,500	127	1.824	111,300
3,000	169	1.998	134,400

TABLE 12.6 Copperweld Copper, Three-Wire

Type A		
Conductor	Cable diameter, in	Breaking load, lb
2A	0.366	5876
3A	0.326	4810
4A	0.29	3938
5A	0.258	3193
5D	0.31	6035
6A	0.23	2585
6D	0.276	4942
7A	0.223	2754
7D	0.246	4022
8A	0.199	2233
8D	0.219	3256
9½D	0.174	1743

TABLE 12.7 Copperweld Stranded

		Breaking strength, lb	
Size	Diameter, in	High-strength	Extra high-strength
13/19(19 No. 6)	0.810	45,830	55,530
21/32(19 No. 8)	0.642	31,040	37,690
5/8(7 No. 4)	0.613	24,780	29,430
1/2(7 No. 6)	0.486	16,890	20,460
3/8(7 No. 8)	0.385	11,440	13,890
5/16(7 No. 10)	0.306	7,758	9,196
3 No. 6	0.349	7,639	9,754
3 No. 8	0.277	5,174	6,282
3 No. 10	0.220	3,509	4,160
3 No. 12	0.174		

If separate anchors are installed, each in line with a leg of the line, consider the leg as dead ends.

The manner of measuring the height and lead of a guy on sloping ground and of a stub pole guy is illustrated in Fig. 12.33. On sloping ground the lead is the horizontal distance from the guy to the pole, and the height is the distance above this horizontal line, as shown in Fig. 12.33a. On a pole stub guy the lead is the horizontal distance from the pole stub to the line pole, and the height is the vertical distance above this horizontal line, as shown in Fig. 12.33b. The tension in the guy wire for a given line-conductor load depends on the distance the anchor is installed from the base of the pole or the guy angle (Fig. 12.34).

After the line load is known, the chart in Fig. 12.35 is used as a quick reference for determining the load on the guy at different angles of pull. To use the chart, determine the line

TABLE 12.8 Aluminum Stranded

Size, cir mils or AWG	No. of strands	Diameter, in	Ultimate strength, lb
6	7	0.184	528
4	7	0.232	826
3	7	0.260	1,022
2	7	0.292	1,266
1	7	0.328	1,537
1/0	7	0.368	1,865
2/0	7	0.414	2,350
3/0	7	0.464	2,845
4/0	7	0.522	3,590
266,800	7	0.586	4,525
266,800	19	0.593	4,800
336,400	19	0.666	5,940
397,500	19	0.724	6,880
477,000	19	0.793	8,090
477,000	37	0.795	8,600
556,500	19	0.856	9,440
556,500	37	0.858	9,830
636,000	37	0.918	11,240
715,500	37	0.974	12,640
715,500	61	0.975	13,150
795,000	37	1.026	13,770

TABLE 12.9 Aluminum Stranded

Size, cir mils or AWG	No. of strands	Diameter, in	Ultimate strength, lb
795,000	61	1.028	14,330
874,500	37	1.077	14,830
874,500	61	1.078	15,760
954,000	37	1.124	16,180
954,000	61	1.126	16,860
1,033,500	37	1.170	17,530
1,033,500	61	1.172	18,260
1,113,000	61	1.216	19,660
1,272,000	61	1.300	22,000
1,431,000	61	1.379	24,300
1,590,000	61	1.424	27,000
1,590,000	91	1.454	28,100

TABLE 12.10 ACSR

Size, cir mils or AWG	No. of strands	Diameter, in	Ultimate strength, lb
6	6 × 1	0.198	1,170
6	6 × 1	0.223	1,490
4	6 × 1	0.250	1,830
4	7 × 1	0.257	2,288
3	6 × 1	0.281	2,250
2	6 × 1	0.316	2,790
2	7 × 1	0.325	3,525
1	6 × 1	0.355	3,480
1/0	6 × 1	0.398	4,280
2/0	6 × 1	0.447	5,345
3/0	6 × 1	0.502	6,675
4/0	6 × 1	0.563	8,420
266,800	18 × 1	0.609	7,100
266,800	6 × 7	0.633	9,645
266,800	26 × 7	0.642	11,250
300,000	26 × 7	0.680	12,650
336,400	18 × 1	0.684	8,950
336,400	26 × 7	0.721	14,050
336,400	30 × 7	0.741	17,040
397,500	18 × 1	0.743	10,400
397,500	26 × 7	0.783	16,190
397,500	30 × 7	0.806	19,980
477,000	18 × 1	0.814	12,300
477,000	24 × 7	0.846	17,200
477,000	26 × 7	0.858	19,430
477,000	30 × 7	0.883	23,300
556,500	26 × 7	0.914	19,850
556,500	30 × 7	0.927	22,400
556,500	30 × 7	0.953	27,200
605,000	24 × 7	0.953	21,500
605,000	26 × 7	0.966	24,100
605,000	30 × 19	0.994	30,000
636,000	24 × 7	0.977	22,600
636,000	26 × 7	0.990	25,000
636,000	30 × 19	1.019	31,500
666,600	24 × 7	1.000	23,700
715,500	54 × 7	1.036	26,300
715,500	26 × 7	1.051	28,100
715,500	30 × 19	1.081	34,600
795,000	54 × 7	1.093	28,500
795,000	26 × 7	1.108	31,200
795,000	30 × 19	1.140	38,400
874,500	54 × 7	1.146	31,400
900,000	54 × 7	1.162	32,300
954,000	54 × 7	1.196	34,200
1,033,500	54 × 7	1.246	37,100
1,113,000	54 × 19	1.292	40,200
1,272,000	54 × 19	1.382	44,800
1,431,000	54 × 19	1.465	50,400
1,590,000	54 × 19	1.545	56,000

TABLE 12.11 Line-Angle Change Multiplication Factor

Angle change of line direction, deg	Multiplication factor M
15	0.262
30	0.518
45	0.766
60	1.000
75	1.218
90	1.414

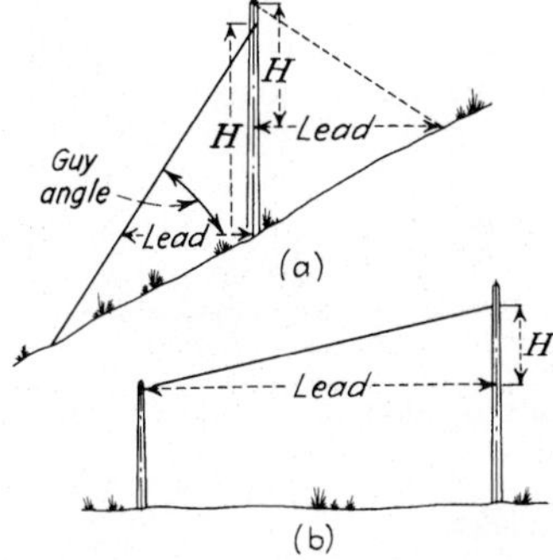

FIGURE 12.33 Methods of measuring height and lead to determine guy angle. (*a*) Sloping ground and (*b*) stub guy.

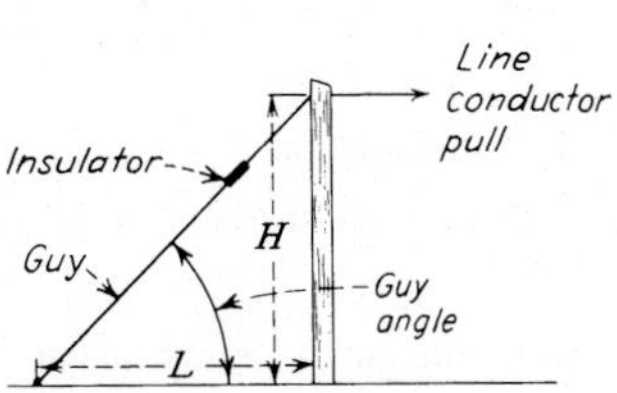

FIGURE 12.34 Sketch of guy installation where H is height of guy attachment to pole, L is lead or distance from pole to anchor rod.

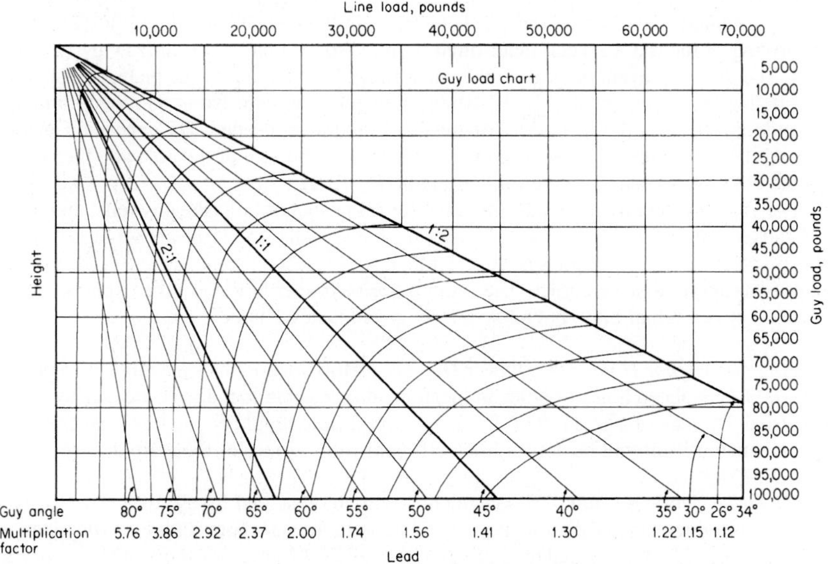

FIGURE 12.35 Chart for determining guy load if line load and guy angle are known. (*Courtesy A. B. Chance Co.*)

load, and using the figures across the top of the chart, follow the vertical and curved line until it intersects the line indicating the angle of the proposed guy. From this point, follow the horizontal line across the chart to the right-hand side. The figure at this point will be the guy load in pounds. For example, assuming a line load of 10,000 lb and a guy angle of 60°, follow the line at 10,000 down the chart, following its curve, until it intersects the line indicating the 60° angle. Reading across the chart to the right, the figure obtained is 20,000 lb, which is the guy load. An alternative method is to multiply the factor at the bottom of the chart by the line load. For example, the multiplication factor for a 60° angle is 2.00, and the line load is 10,000 lb; 2.00 multiplied by 10,000 lb equals 20,000 lb.

In using the breaking strength of conductor, it should be considered that a conductor properly sagged (according to the NESC) will not exceed 60 percent of its breaking strength when fully loaded. This automatically allows a safety factor of $1^2/_3$, However, additional safety factors will be required on important crossings especially over highways, railroads, or rivers, where safety factors of 2 and 3 are generally used. After the guy load has been found, select an anchor with holding power in the soil class, allowing for the desired safety factor. The anchor, rod, and guy strand must be selected with proper size and strength to coordinate with the anchor maintaining the desired safety factor (see Tables 12.1 through 12.3).

Guy Construction. The installation of a guy divides itself naturally into the five steps outlined:

1. Digging in the anchor
2. Inserting the strain insulator in the guy wire or grounding the guy wire or both
3. Fastening the guy wire to the pole
4. Tightening the guy wire and fastening it to the anchor
5. Mounting the guy-wire guard

Digging in the Anchor. The manner of digging in the guy anchor depends on the type of anchor. See Fig. 12.36 for various types of guy anchors in general use. The screw-type anchor is screwed into the earth in the manner shown in Fig. 12.37.

The patented "never-creep" anchor is installed in the manner shown in Fig. 12.38. For the driving of the rod a special maul fitted with oil-soaked hickory inserts (Fig. 12.39) is used. The inserts prevent damage to the anchor rod. For hanging of the plate an installing bar fitted with a special hook (Fig. 12.40) is employed. The same bar is also used for tamping and for retrieving the anchor in case the guy is no longer needed. The expansion anchor (Fig. 12.19) is installed by inserting the anchor into the bottom of the hole and then causing the blades to expand into the solid ground (Fig. 12.41).

The anchor is an important part of a line, for if the anchor fails, the guy fails, and if the guy fails, the pole fails, and if the pole fails, the corner fails, and if the corner fails the line fails. In other words, the chain is no stronger than its weakest link, and the line is no stronger than its weakest angle or corner. The importance of proper has led to the expression, "A line well guyed is half built, but a line poorly guyed is never built."

Digging the Anchor Hole with a Power Digger. Most anchor holes are dug with a power digger. Only isolated holes or holes in locations inaccessible to a digger are dug by hand. The usual diameter of a hole is 9 in. The length of the hole should be such, as a general rule, that the anchor itself will be not less than 6 ft below the surface of the ground.

Power-Drive Screw Anchor. Screw anchors may be installed by use of a power drive. Tools required for the installation of power installed screw anchors (PISA) are depicted in Fig. 12.42 and 12.43. A special tubular wrench (see Fig. 12.44) slipped over the anchor rod

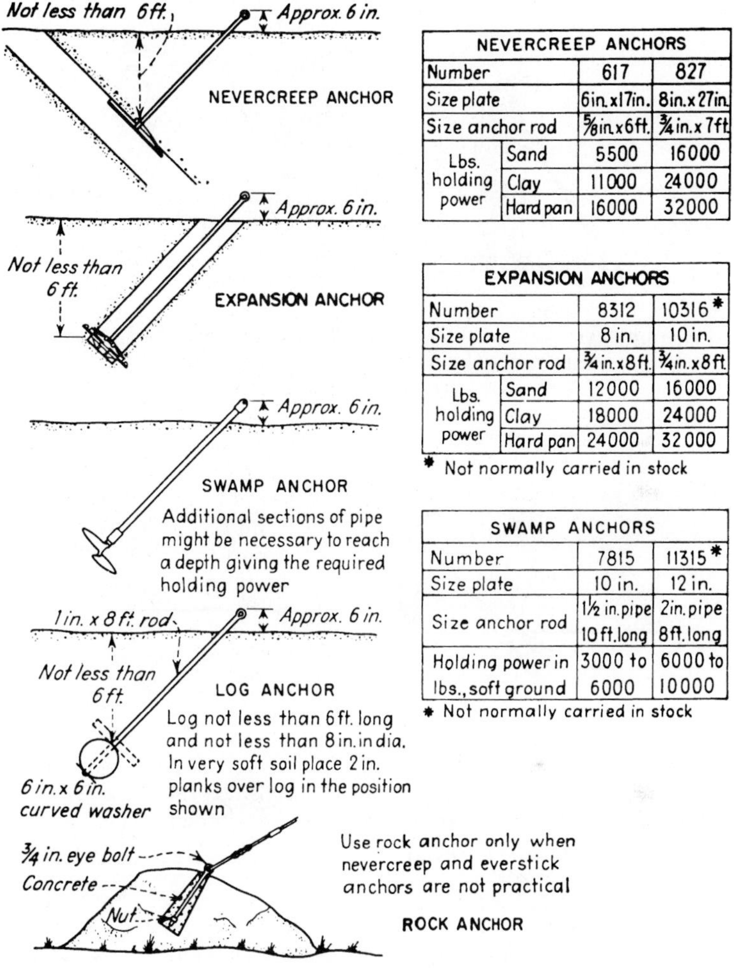

NEVERCREEP ANCHORS			
Number		617	827
Size plate		6in.x17in.	8in.x27in.
Size anchor rod		⅝in.x6ft.	¾in.x7ft.
Lbs. holding power	Sand	5500	16000
	Clay	11000	24000
	Hard pan	16000	32000

EXPANSION ANCHORS			
Number		8312	10316*
Size plate		8 in.	10 in.
Size anchor rod		¾in.x8ft.	¾in.x8ft.
Lbs. holding power	Sand	12000	16000
	Clay	18000	24000
	Hard pan	24000	32000

* Not normally carried in stock

SWAMP ANCHORS		
Number	7815	11315*
Size plate	10 in.	12 in.
Size anchor rod	1½ in. pipe 10 ft. long	2 in. pipe 8 ft. long
Holding power in lbs., soft ground	3000 to 6000	6000 to 10000

* Not normally carried in stock

FIGURE 12.36 Various types of guy anchors in general use.

transmits the torque from the power drive directly to the hub of he anchor wing. Since no torque is transmitted through the anchor rod, the rod has to be only heavy enough to withstand the pull of the guy. Moreover, a larger-diameter wing can be readily installed in this manner, thus providing greater holding power.

The steps in the installation procedure are as follows:

1. Remove the eye from the anchor rod.
2. Slide a tubular wrench over the anchor rod to engage the square hub shank of the helix. The wrench is held in position by spring-loaded dogs and a hex nut (see Fig. 12.45).
3. Attach the wrench to the driving bar of the power digger (Fig. 12.46).
4. Drive the anchor down with the power digger, which feeds as well as screws the anchor into the earth. The feed should match the natural penetration of the helix so that a

FIGURE 12.37 Steps in installation of manual screw-type anchor. (*a*) Digging small hole with bar and (*b*) screwing in anchor.

FIGURE 12.38 Steps in the installation of the never-creep anchor. Locate the spot desired for the anchor rod, measure from that point back from the pole the length of the rod, and start the hole at this point. A boring-machine auger of almost any size can be used. (*a*) Bore the hole as nearly at right angles to the line of strain as conditions will allow. Then proceed with steps *b*, *c*, *d*, and *e* as shown. (*Courtesy A. B. Chance Co.*)

FIGURE 12.39 Driving the rod for a never-creep anchor. Lineman is using a special maul fitted with hickory inserts to prevent damage to the rod. (*Courtesy A. B. Chance Co.*)

FIGURE 12.40 Lineman lowering the never-creep anchor plate with special installing bar. Bar has hook which engages plate. The same bar can also be used to tamp soil into the hole and to retrieve plate if necessary. (*Courtesy A. B. Chance Co.*)

FIGURE 12.41 Lineman using ram to strike blows at anchor to cause it to expand. Four or five blows are usually required. The anchor is fully expanded when the hollow sound of the first few blows changes to a dull thud. The ram is 10 ft long, has a diameter $2^1/_2$ in, has a heavy collar welded to each end, and weighs 35 lb. (*Courtesy Wisconsin Electric Power Co.*)

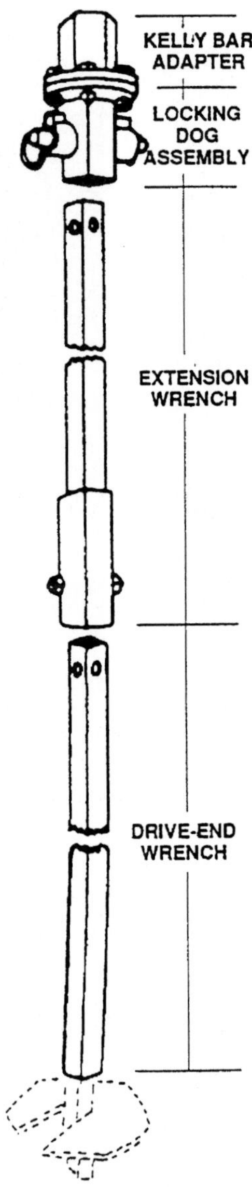

FIGURE 12.42 PISA anchor accessories. (*Courtesy A. B. Chance Co.*)

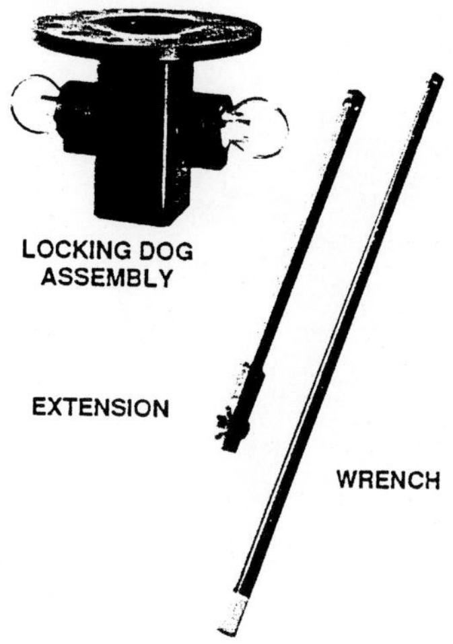

FIGURE 12.43 Additional PISA anchor accessories. (*Courtesy A. B. Chance Co.*)

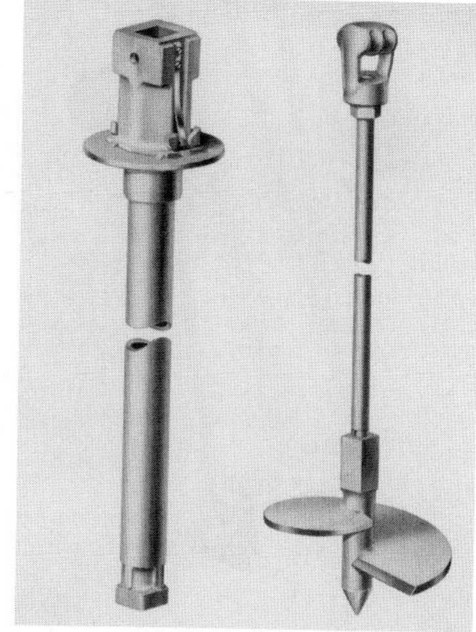

FIGURE 12.44 Special tubular wrench used with power drive to rotate screw anchor. Turning torque is applied on square hub shank next to helix.

minimum of earth is disturbed. Driving at proper rpm and feed eliminates churning the earth, which often results from hand turning, as the anchor then must be drawn down by the ability of the soil to resist crumbling. Churning is similar to stripping a thread (see Figs. 12.47 and 12.48).

5. Release the spring-loaded dogs and remove the wrench from the anchor rod (Fig. 12.49).

6. Replace the eye on the anchor rod (Fig. 12.50).

FIGURE 12.45 Tubular wrench is slid over anchor rod. Square hub of anchor is engaged by square opening in wrench. Spring-loaded dogs on wrench slip over hex nut to lock in position. (*Courtesy A. B. Chance Co.*)

FIGURE 12.46 Wrench being fastened to driving bar of power drive. A bolt securely locks wrench in place. (*Courtesy A. B. Chance Co.*)

Besides eliminating the hard work of installation when done manually, the use of power saves time. The average time that is required for actual installation is about 5 min.

Guy Assembly. The tools needed to assemble a guy include bolt cutters, wrenches, pliers, hoist, wire grips, and pulling eye for preformed grips. A typical guy assembly is illustrated in Fig. 12.51. The material required to connect the anchor rod to the pole includes guy wire, hardware, insulators, and preformed guy-wire grips. The lineman assembling a guy must determine the length of the guy wire needed to cut the wire accordingly. All guy wire, whether galvanized, copperweld, or alumoweld has steel core strands; the wire is springy and difficult to handle. Before a guy wire is cut, it is good practice to wrap tape around the

FIGURE 12.47 Installing screw anchor by means of power drive. Power is used to rotate the screw as well as feed the anchor into the earth. The anchor is driven into the ground until locking dogs on the wrench are just above ground level. (*Courtesy A. B. Chance Co.*)

FIGURE 12.48 Anchor almost completely in place. Note ratchet on driving bar which is used to advance the anchor according to the pitch of the helix. (*Courtesy A. B. Chance Co.*)

wire on each side of the place where the cut is to be made. This will keep the strands from fraying or unlaying after the cut is made. When the guy wire is cut, hold both sides of the cut firmly so that the free ends will not fly into your face or into a live conductor. A good procedure is to place both feet on the guy wire and make the cut between them. After the cut has been made, each end can then be carefully released.

When the guy wire has been cut to the proper length, the insulators should be installed in accordance with the specifications, if they are required, using prefabricated guy-wire grips (Fig. 12.52) or three-bolt clamps (Fig. 12.53). An insulator properly assembled in series with the guy strand is shown in Fig. 12.28. The guy wire is attached to the pole by the lineman, using

FIGURE 12.49 Releasing wrench from anchor by lifting the dogs clear of the hex nut on the anchor. The wrench can then be pulled free and uncoupled from the drive shaft. (*Courtesy A. B. Chance Co.*)

FIGURE 12.50 Replacing nut on end of anchor rod. Anchor is now ready for hooking up guy wire. (*Courtesy A. B. Chance Co.*)

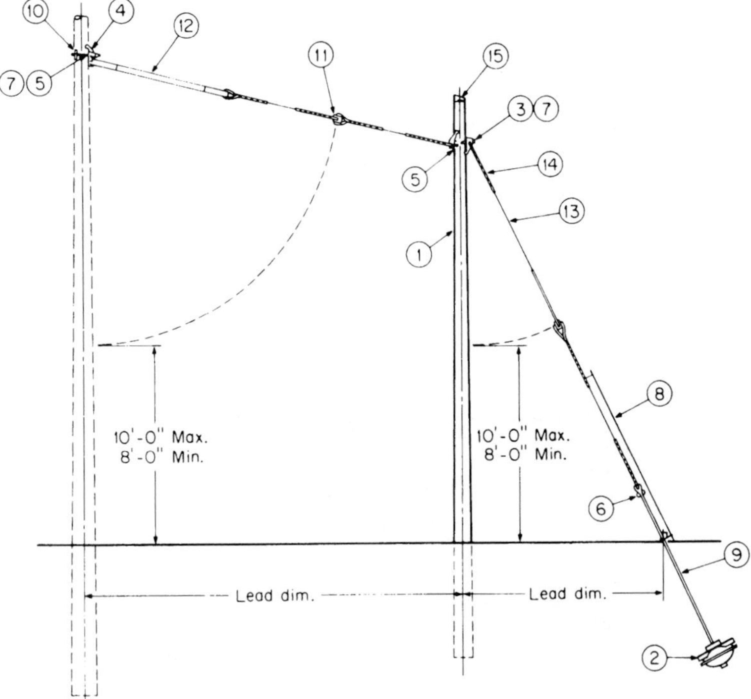

FIGURE 12.51 Stub guy and down guy assembly. (1) Pole stub; (2) anchor, expansion; (3) guy attachment, thimbleeye; (4) guy attachment, eye type; (5) bolt, standard machine, with standard nut; (6) clamp, guy bond, for 3/4-in rod, twin eye; (7) locknut, 3/4-in galvanized; (8) protector, guy; (9) rod, twin eye, 3/4-in × 9 ft, with two standard nuts; (10) washer, 4-in × 4-in square curved; (11) insulator, strain, porcelain; (12) insulator, strain, fiberglass guy; (13) wire, guy; (14) grip, guy wire; (15) pole topper.

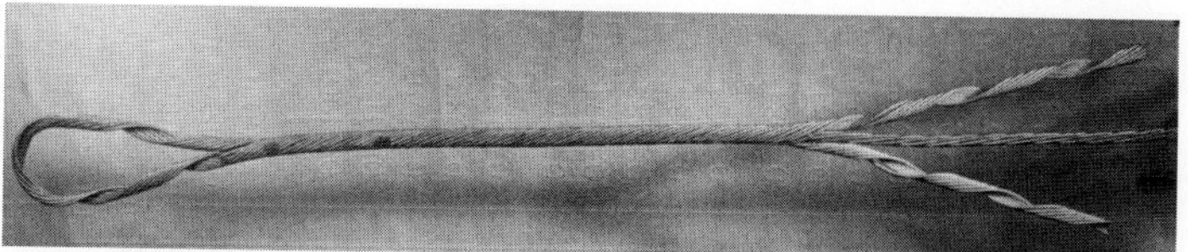

FIGURE 12.52 Prefabricated guy grip partially installed on end of guy strand. (*Courtesy Preformed Line Products Co.*)

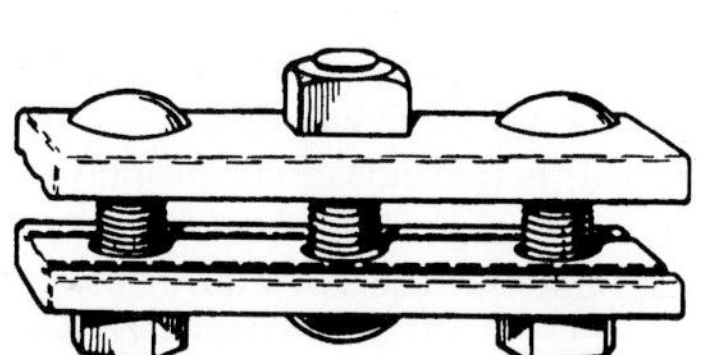

FIGURE 12.53 Three-bolt guy-wire clamp. (*Courtesy A. B. Chance Co.*)

FIGURE 12.54 Lineman has secured span guy with three-bolt clamps and down guy with prefabricated grip. Thimbleeye nuts are installed on through bolt, and pole is protected with curved washers. (*Courtesy Copperweld Steel Co.*)

proper procedures. Figure 12.54 illustrates guys secured with three-pole clamps and prefabricated grips. The guy wire can be fastened to the pole by wrapping the turns of the wire around the pole and clamping the free end of the wire. Strain plates are used with guy hooks to provide a bearing surface for the guy strand to prevent damage to the wood fibers of the pole. The guy hooks are used to keep the guy strand from slipping down on the pole (Fig. 12.55). The

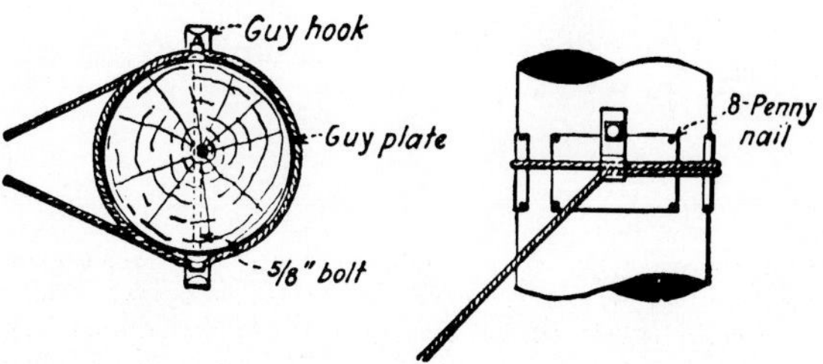

FIGURE 12.55 Method of fastening guy wire to pole. (*Courtesy Hughes Brothers*)

FIGURE 12.56 Lineman pulling up on anchor guy. Note the Coffing hoist or chain jack and the two come-alongs being used. It replaces, and is easier to use than, the ordinary block and tackle. It is operated with a back-and-forth pump-handle motion of the handle. The main advantage is that it locks itself in place. It will not slack off when the hand is taken off the handle. (*Courtesy Wisconsin Electric Power Co.*)

guy strands are clamped together with three-bolt clamps. The end of the guy strand extends approximately 1 foot beyond the clamps and is clipped to the main guy wire.

Guy wires should be placed on the poles and "pulled" before the line conductors are placed on the pole. If the line wires were placed first, the poles would be pulled out of position.

A guy wire is pulled up or tightened from the ground by means of a hoist and approved grips, as shown in Figs. 12.56 and 12.57. These grips (Fig. 12.58) are also called *come-alongs*.

FIGURE 12.57 Cutaway view of never-creep anchor installation, showing lineman tightening up on guy wire with hoist attached to guy wire with come-alongs. (*Courtesy A. B. Chance Co.*)

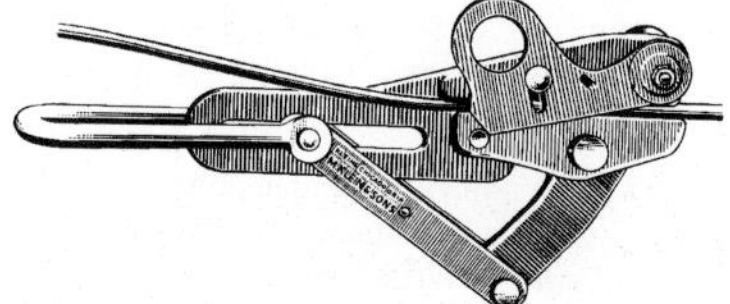

FIGURE 12.58 Improved "Chicago" steel wire grip or come-along. (*Courtesy Mathias Klein and Sons.*)

FIGURE 12.59 Lineman tightening bolts on anchor guy clamp. Enough end is left on the guy wire beyond the clamp for attachment of a come-along for later retightening. (*Courtesy Wisconsin Electric Power Co.*)

Come-alongs or wire eccentrics will grip a straight wire at any point without slipping. The guy is drawn up until the pole is pulled over slightly toward the guy.

In threading the guy wire through the anchor rod, a thimble should be used to protect the guy wire from a sharp bend. This is shown in Fig. 12.59.

Prefabricated guy dead-end grips can be used to terminate the guy strand to the eye on the anchor rod (see Fig. 12.60). The guy strand is pulled taut with a hoist by the lineman, then it is wrapped with tape and cut to the proper length. The prefabricated grip is threaded through the eye on the anchor rod, and the legs of the grip are wrapped around the guy

FIGURE 12.60 Prefabricated guy dead-end partially installed. The legs of the grip must be completely wrapped around the guy strand, and the ends must be snapped in place before the hoist is removed. Snapping the ends of the grip into place ensures that the grip will have rated holding strength and prevents the grip from unwrapping. (*Courtesy Preformed Line Products Co.*)

FIGURE 12.61 Pole is held by a down guy. Lineman is properly sagging low-voltage secondary wires with a block and tackle. (*Courtesy Wisconsin Electric Power Co.*)

strand. The ends of the legs of the grip are snapped into position, and the hoist is removed, completing the installation. Guy-grip dead-ends may be removed and reapplied 3 times within 3 months after installation, if necessary, to retension the guys. Guy-grip dead-ends that were removed after their original installation should not be reused.

Figure 12.61 shows the manner of tensioning low-voltage secondary wires on a pole held in place by a head guy. Note that the lineman on the pole does not do any direct pulling on the rope. This should always be done by someone on the ground.

The lineman is pulling on the fall line to his right with his left hand and on the same rope on the other side of the sheave to his left with his right hand. His right and left hands are

FIGURE 12.62 Showing use of come-along, dynamometer, and chain jack in pulling down guy to proper tension. (*Courtesy Coffing Hoist Co.*)

FIGURE 12.63 Typical metal guy guard or protector made of galvanized steel.

FIGURE 12.64 Lineman installing plastic guy guard. (*Courtesy MidAmerican Energy Co.*)

pulling in opposite directions; hence no extra strain is placed on his "hooks," and there is less danger of breakouts. A *foul* was thrown into the set of blocks before the operation was started. This can be seen in Fig. 12.61 just above the lineman's left forearm. This enables the lineman to cinch the blocks to hold the strain. When the lineman cannot pull the guy wire tight enough by himself, the helper on the ground also pulls on the fall line. In this case the fall line is passed over the lineman's safety strap, and the helper sets himself about one-half span away before he starts to pull. In this case, however, the foul cannot be used to cinch the strain. The helper must keep pulling on the rope until the lineman secures the wire.

Figure 12.62 shows the use of a dynamometer in pulling a down guy to the proper tension.

Guys should not be installed where they will interfere with traffic. Figure 12.63 shows a typical metal guy protector. The ground end of anchor guys exposed to pedestrian traffic must be provided with a substantial and conspicuous marker not less than 8 feet long. Figure 12.64 shows a lineman installing a plastic guard. The guard should be light-colored so that it will be visible at night. The guard should extend from a point near the ground to 8 feet above the ground. Plastic guy guards installed on guy wires are shown in Fig. 12.65.

Pole keys can be used to prevent poles with an unbalanced load from leaning. Street lights installed on poles often cause the poles to lean as a result of the side strain, especially if the pole is set in soft ground. The pole key (Fig. 12.66) is lowered into the pole hole alongside the pole, and expanded with a ram. The anchor rod used to lower the pole key into the pole hole is unscrewed and removed (Fig. 12.67). The pole hole is filled with earth and thoroughly tamped. The finished installation prevents the pole from leaning due to an unbalanced load if the pole key has been properly installed. A treated-wood key can be installed just below the ground surface on the opposite side of the pole from the metal key to provide additional stability (Fig. 12.68).

FIGURE 12.65 Plastic guy guards installed on guy wires.

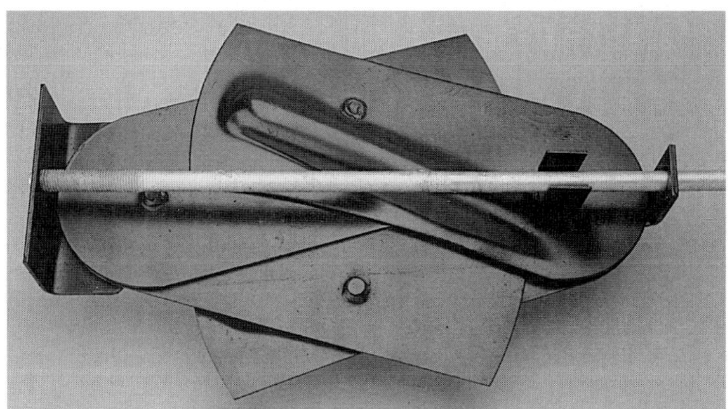

FIGURE 12.66 An expansion metal pole key. (*Courtesy A. B. Chance Co.*)

FIGURE 12.67 Groundman unscrewing and removing the anchor rod used to lower the key into the pole hole prior to filling the hole with dirt. (*Courtesy A. B. Chance Co.*)

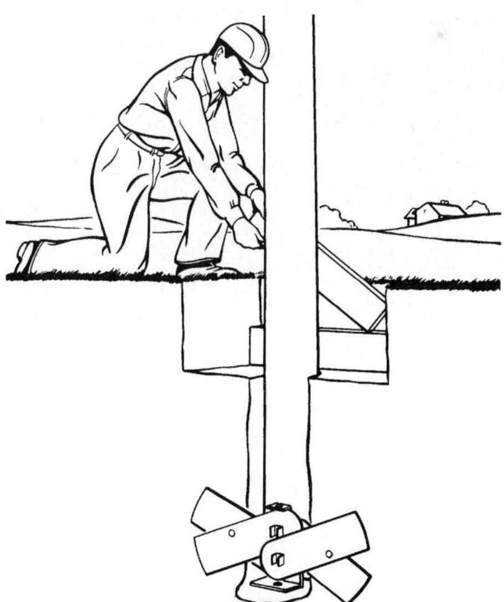

FIGURE 12.68 Pole key has been expanded. Wood key is being installed adjacent to the pole just below the surface of the ground by a groundman. (*Courtesy A. B. Chance Co.*)

CHAPTER 13
INSULATORS, CROSSARMS, AND CONDUCTOR SUPPORTS

Overhead conductors are supported on the distribution pole using insulators, crossarms, or conductor supports. The *National Electrical Safety Code* (NESC) establishes the criteria to be followed in the design, construction, and operation of the electric distribution system. The requirements for insulators, crossarms, and conductor supports varies depending upon the grade of construction as defined in the Code.

Insulators. An insulator prevents the flow of an electric current. The insulator supports electrical conductors. The function of an insulator is to separate the line conductors from the pole or tower. Insulators are designed to withstand electrical, mechanical, and environmental stresses. Insulators are fabricated from porcelain, glass, fiberglass, and polymers.

Porcelain insulators are manufactured from clay. Special clays are selected and mixed mechanically until a plastic-like compound is produced. The clay is then placed in molds to form the insulators. The molds are placed in an oven to dry the clay. When the clay is partially dry, the mold is removed and the drying process is completed. When the insulator is dry, it is dipped in a glazing solution and fired in a kiln. The glaze colors the insulator and provides a glossy surface. This makes the insulator surface self-cleaning.

Large porcelain insulators are made up of several shapes cemented together. Care must be taken when cementing the insulators together to prevent a chemical reaction on the metal parts, causing cement growth. Cement growth can cause stresses on the porcelain great enough to crack the porcelain.

Glass insulators are made from sand, soda, ash, and lime. The materials are properly mixed and melted in an oven until a clear-liquid plastic-like material is produced. The plastic-like compound is placed in a mold and allowed to cool. After the glass insulator is cool, it is placed in an annealing oven.

Fiberglass insulators are manufactured with rods of fiberglass treated with epoxy resins. Rubber-like compounds are applied to the rods to fabricate suspension, dead-end, and post-type insulators.

Polymer insulators are formed by using various sizes of silica bound together chemically with a resin. The compound's composition is approximately 90 percent silica. Polymer insulators are replacing porcelain, glass, and epoxy insulators. Polymer insulators have excellent mechanical and dielectric strength. The insulators are lightweight, virtually nonbreakable and much easier to handle than any of the other types of insulators. Polymer insulators are less costly than porcelain, glass, or epoxy insulators because of the ease of manufacture. Field tests and current installation success of polymer insulators have proven that they are durable and will provide many years of service in an adverse atmosphere.

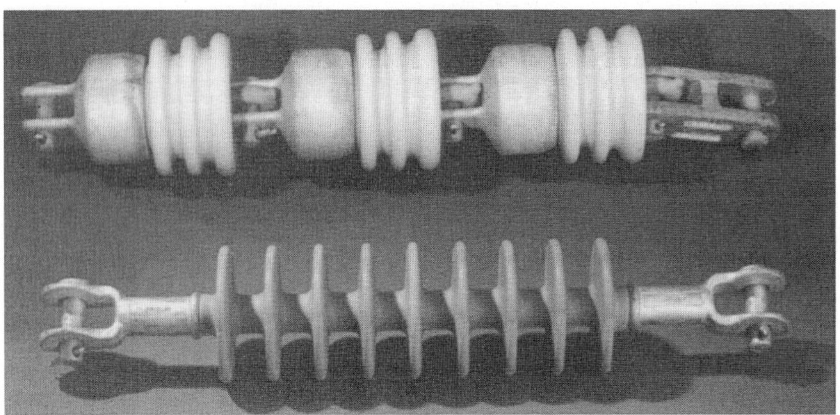

FIGURE 13.1 Porcelain dead-end bells and one-piece polymer dead-end insulators for a typical 25-kV class distribution system with a design BIL of 110 kV. (*Courtesy A. B. Chance Co.*)

The galvanized end fitting of the polymer insulator is crimped to the insulator's fiberglass rod. Polymer material covers the rod. The greatest amount of stress occurs at the interface of the end fitting and the rod.

Conductor Dead-end Insulators. The primary conductor is terminated at the crossarm, at the conductor support, or on the pole using a conductor dead-end shoe and porcelain dead-end bells or a conductor dead-end shoe and a one-piece polymer dead-end insulator. Both types of dead-end insulators are shown in Fig. 13.1. The length and size of the insulator required is dependent upon the voltage class of the distribution system and the basic insulation level for the system.

Pin-Type Insulators. The pin insulator gets its name from the fact that it is supported on a pin. The pin holds the insulator, and the insulator has the conductor tied to it. Pin insulators are made of glass, porcelain, or polymers. The porcelain insulator (Fig. 13.2) is a one-piece insulator when used on 15 kV or less class voltage lines but consists of two, three, or four layers cemented together to form a rigid unit when used on higher-voltage lines. It is usually one piece for voltages below 23,000 volts, two pieces for voltages from 23,000 to 46,000 volts, three pieces for voltages from 46,000 to 69,000 volts, and four pieces for voltages from 69,000 to 88,000 volts. The use of several layers helps to spill the rain and provides a long, dry arc-over path. These layers flare out at the bottom into a bell shape. Sometimes these layers are called petticoats because of their appearance, and pin insulators of this multilayer construction are called petticoat insulators. Figures 13.3 shows a typical multilayer petticoat insulator.

Pin insulators are usually used on distribution lines and seldom used on transmission lines. The glass pin insulator is principally used on low-voltage circuits. The porcelain pin insulator is used on main-feeder circuits and primaries.

The polymer pin insulator has applications with spacer cable or in areas subject to vandalism. When using the polymer insulator with a bare conductor, a bare-tie wire should be used (Fig. 13.4). When using the polymer tie-top insulators with covered conductors, a covered-tie wire is required to avoid degradation of the conductor covering (Fig. 13.5).

The smaller insulators are threaded to fit on a 1-inch pin, and the larger units are threaded to fit on a $1^3/_8$-inch pin. When insulators are to be mounted on steel pins, a thimble, threaded to fit the pin, is cemented into the porcelain.

FIGURE 13.2 Porcelain top-groove pin insulators mounted in position on double crossarm. Note use of steel insulator pins and crossarm spreader bolt.

FIGURE 13.3 A 23,000-volt two-layer porcelain pin insulator. (*Courtesy Ohio Brass Co.*)

HPI insulator with bare conductor and bare tie wire

FIGURE 13.4 Polymer pin insulator with bare-tie wire for use with bare conductor. (*Courtesy Hendrix Wire & Cable Inc.*)

Insulator pins. The function of an insulator pin is to hold the insulator mounted on it in position. The insulator pins are made of iron or steel. Steel pins are in general use. A steel pin provided with nylon threads is depicted in Fig. 13.6. The details of a typical steel pin are shown in Fig. 13.7. The pin is usually provided with lead threads that prevent any localized pressure on the insulator. The insulator pin has a broad base that rests squarely on the crossarm, as shown in Figs. 13.2. and 13.8

The spacing of the pins is generally suited to the voltage of the circuit in accordance with the NESC. The spacing should provide sufficient working space for the lineman. The end pins are generally spaced 4 in from the end of the arm.

Metal pins that clamp the crossarm are mounted by clamping them on the arm at the desired spacings. Since these pins do not require pin holes, their use avoids weakening the crossarm.

Crossarms and conductor supports. Overhead conductors rest on insulators supported by crossarms or conductor supports (Fig. 13.8). The crossarms and conductors supports are designed to support the load that they will bear plus appropriate overload capacity factors due to stresses by ice or wind. An NESC table contains the dimensions of a crossarm section of selected Southern Pine or Douglas Fir woods. The conductor support is a more compact style of close space construction utilizing apitong wood. The apitong supports are 48 inches in length (Fig. 13.9).

Single arms are used on straight lines where no excessive strain needs to be provided for. As already mentioned in Chap. 11, every other crossarm faces in the same direction.

Double arms are frequently used at line terminals, at corners, at angles, or at other points where there is an excessive strain (Fig. 13.10). Where lines cross telephone circuits or

HPI insulator with covered conductor and covered tie wire

FIGURE 13.5 Polymer pin insulator with covered-tie wire for use with covered conductor. (*Courtesy Hendrix Wire & Cable Inc.*)

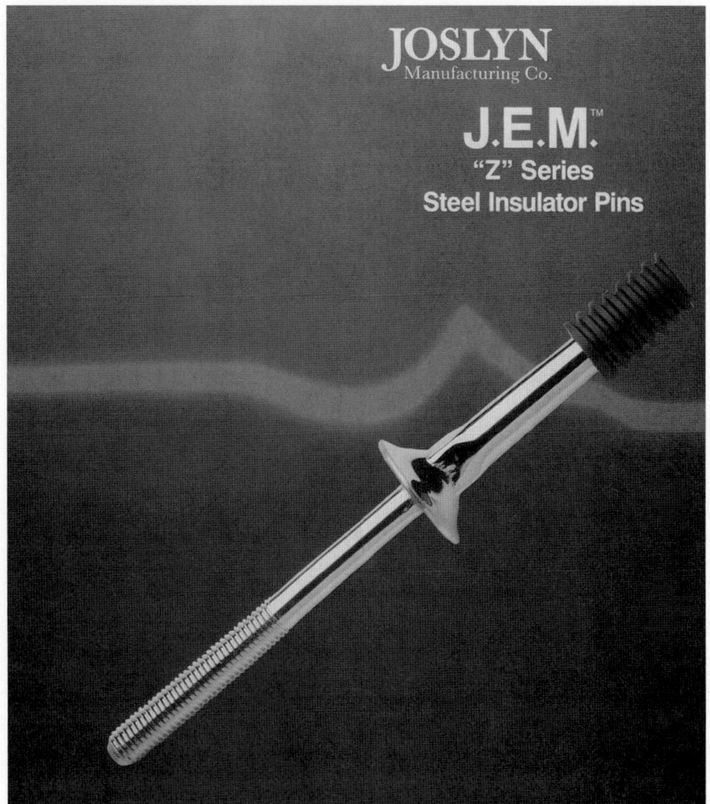

FIGURE 13.6 Steel pin with nylon threads. (*Courtesy Joslyn Mfg. and Supply Co.*)

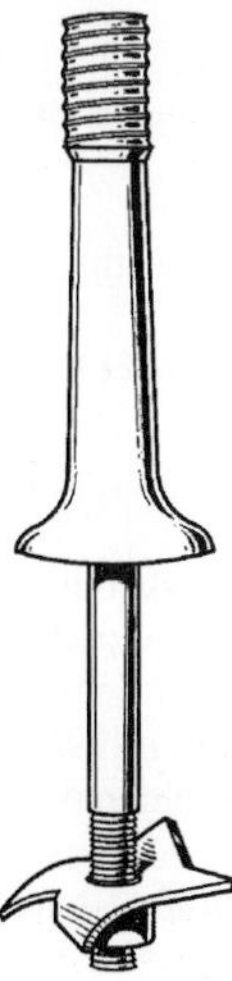

FIGURE 13.7 Steel pin provided with lead threads.

FIGURE 13.8 Assembled view of wood crossarm, steel insulator pin, porcelain pin insulator, line conductor, and tie wire.

railroad crossings, double arms may be used, as more than ordinary safety is required at such points.

Buck arms may be used at corners and at points where branch circuits are taken off at a right angle to the main line (Fig. 13.11).

Side arms are used in alleys or other locations where it is necessary to clear buildings, etc. (Fig. 13.12).

Fiberglass crossarms can be used at points where the distribution lines dead-end or at locations where extra safety is necessary, instead of double-wood crossarm (Fig. 13.13). The fiberglass crossarms are gray and have a small cross-sectional area that minimizes the mass at the top of the pole. The fiberglass crossarms are often installed where the appearance of a pole line is important. The bracket between the pole and the fiberglass crossarm eliminates the need for braces.

Galvanized steel crossarms may be used in place of double-wood crossarms of fiberglass crossarms to obtain the safety factor required (Fig. 13.14).

FIGURE 13.9 Picture of a distribution pole line with conductor support construction. Conductor support construction is used to enhance the appearance of the pole line due to a more compact style of close space construction utilizing apitong wood. The apitong supports are 48 inches in length.

Crossarm braces are used to give strength and rigidity to the crossarm. Metal crossarm braces are made of either galvanized flat bar or light angle iron. The size used varies with the size of the arm and the weight of the conductors. The usual flat-strap brace for ordinary distribution work (Fig. 13.15) is 38 in long and 1/4 by 1¼ in. One end is attached to the crossarm by means of a carriage bolt and the other to the pole by means of a lag screw. One brace extends to each side of the arm. Angle-iron braces are made in one piece and bent into the shape of a V, as shown in Fig. 13.16. These braces are fitted to the bottom of the crossarm instead of the side as is the flat type.

Wooden crossarm braces are used extensively for medium voltages. Their use increases the insulation of the line. A set of wooden braces is illustrated in Fig. 13.17, and their use is illustrated in Figs. 13.9, 13.10, and 13.11.

Where crossarms extend to one side of the pole only, as is sometimes necessary in alley construction, special braces are required. The brace must be longer and more rigid. A single brace made of galvanized angle iron 1¾ by 1¾ by 3/16 in and 5 to 7 ft long is generally used for distribution lines and is shown in Figs. 13.12 and 13.18. The brace is provided with a step near the middle for use of the lineman.

A variety of bolts and screws are used in line construction, and only the common ones are illustrated. Figure 13.19 shows a typical crossarm through bolt used for fastening the crossarm to the pole. Its common size is 5/8 by 12 to 16 in. A bolt for attaching two crossarms, one on each side of the pole, is shown in Fig. 13.20. Figure 13.21 illustrates a carriage bolt used for attaching the brace to the crossarm. The typical bolt for this purpose has a diameter of 3/8 in and is 4, 4½, or 5 in long. Figure 13.22 shows a lag screw used to fasten the crossarm braces to the pole. The recommended size is 1/2 by 4 in. Lag screws are supposed to be screwed into place either into a small bored hole or after being started

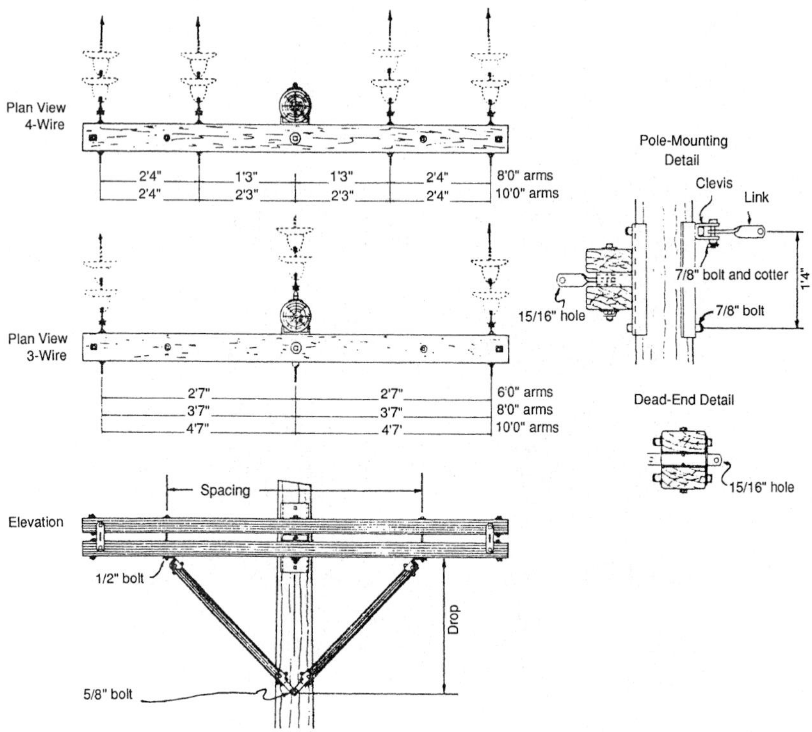

FIGURE 13.10 Double-crossarm construction details. The steel bracket installed between the crossarms and the pole eliminates the need to cut a gain on the pole for the crossarms. (*Courtesy Hughes Brothers.*)

by hammering. They should be screwed into place to keep the threads from injuring the fibers of the wood.

The complete assembly of the pieces of pole hardware described above is shown in Fig. 13.23. This figure also lists the standard parts and illustrates their correct use.

Crossarms may be mounted before or after the pole is erected; the practice varies. If the pole is located where it is accessible from a bucket truck and if the pole has not been preframed and drilled, the lineman frames and drills the pole while working from an insulated bucket (Fig. 13.24). The pin-type insulators are installed on the crossarm by a groundman or lineman working on the ground (Fig. 13.25). The groundman hands the crossarm, complete with pins, insulators and crossarm braces, to the lineman in the bucket of the truck (Fig. 13.26). The lineman installs the crossarm on the pole, using a through bolt while working from the insulated bucket on the truck (Fig. 13.27).

In case the crossarm is mounted after the pole is in place and the pole is located in an area inaccessible to a bucket truck, the lineman climbs the pole, carrying his small tools (pliers, knife, and connectors), hand line, hammer, and lag wrench, and fastens the pulley of the hand line to the pole. The crossarm is then pulled up by the groundman, as illustrated in Fig. 13.28. Various methods of tying the hand line to the crossarm are shown in Fig. 13.29.

The crossarm is attached to the pole (Fig. 13.30) by means of a 5/8-in galvanized machine bolt driven from the back of the pole. A $2\frac{1}{4}$-in square washer is placed under the head and the nut of the bolt, and then the bolt is drawn up.

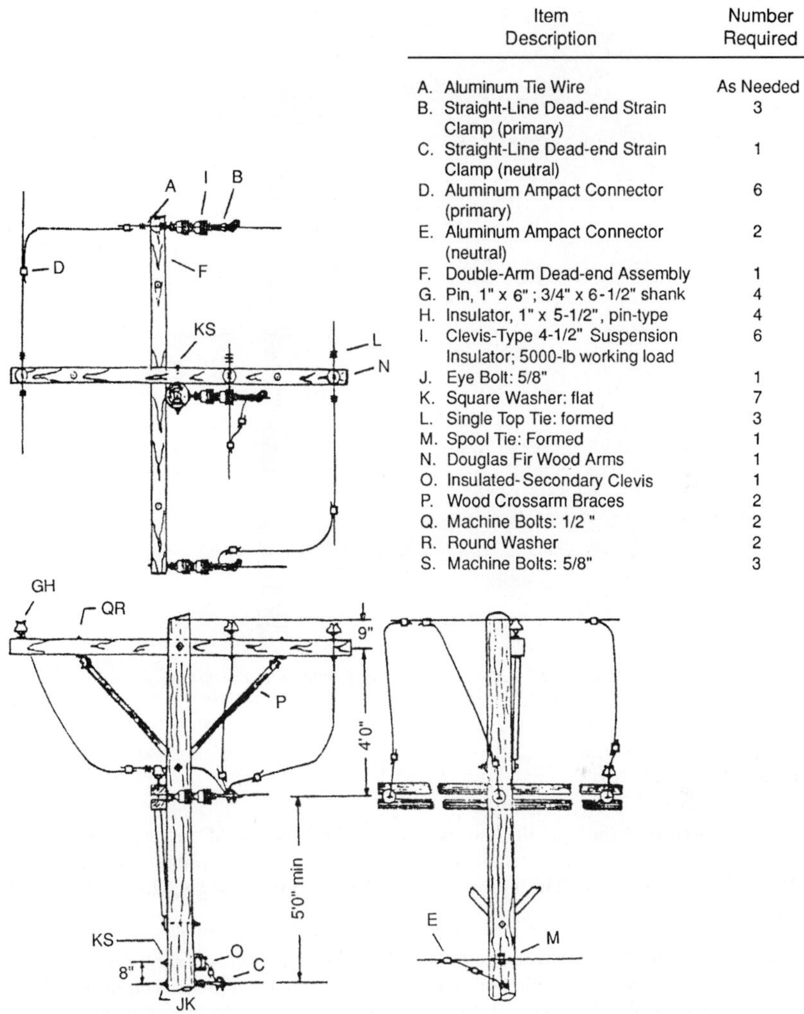

Item Description	Number Required
A. Aluminum Tie Wire	As Needed
B. Straight-Line Dead-end Strain Clamp (primary)	3
C. Straight-Line Dead-end Strain Clamp (neutral)	1
D. Aluminum Ampact Connector (primary)	6
E. Aluminum Ampact Connector (neutral)	2
F. Double-Arm Dead-end Assembly	1
G. Pin, 1" x 6" ; 3/4" x 6-1/2" shank	4
H. Insulator, 1" x 5-1/2", pin-type	4
I. Clevis-Type 4-1/2" Suspension Insulator; 5000-lb working load	6
J. Eye Bolt: 5/8"	1
K. Square Washer: flat	7
L. Single Top Tie: formed	3
M. Spool Tie: Formed	1
N. Douglas Fir Wood Arms	1
O. Insulated-Secondary Clevis	1
P. Wood Crossarm Braces	2
Q. Machine Bolts: 1/2 "	2
R. Round Washer	2
S. Machine Bolts: 5/8"	3

FIGURE 13.11 Three-phase, 13,200Y/7620-volt distribution line constructed with 10-ft crossarms. Double crossarm is used to support conductors for a tap to the main circuit on a tangent pole in the line. A down guy would be installed to counter the pull of the tapped-line conductors.

After the through bolt is in place, the crossarm braces are attached. It is necessary first to sight the crossarm, that is, make sure it is horizontal. In case more conductors are to be carried on one side than on the other, that side can be higher, as it will settle later. The braces are usually $1\frac{1}{4}$ by $\frac{1}{4}$ by 28 in. Sometimes the braces are bolted to the crossarm before the arm is pulled up. The end of the brace attached to the crossarm is held in place by means of a 3/8- by 4-in galvanized carriage bolt, and the ends at the pole are fastened by means of a $\frac{1}{2}$- by 5-in lag screw. The braces are attached on the back of the crossarm (see Fig. 13.31). When the standard 28-in braces are used, the hole in the crossarm should be located 19 in from the middle of the crossarm.

FIGURE 13.12 Alley-arm brace used with side arm to clear obstructions. Note pole step on brace for use by lineman.

The appearance of the pole line can be improved by the use of armless-type construction. Pin-type insulators can be installed on epoxy rods to provide proper clearances without the use of crossarms. The epoxy rods are available in single, double, and triple assemblies (Fig. 13.32).

Post-Type Insulators. Post-type insulators are used on distribution, subtransmission, and transmission lines and are installed on wood, concrete, and steel poles. The line-post insulators are manufactured for vertical or horizontal mounting. The line-post insulators are usually manufactured as one-piece solid porcelain units (Fig. 13.33) or fiberglass epoxy-covered rods with metal end fittings and rubber weather sheds (Fig. 13.34). In some instances the fiberglass rod is combined with a porcelain insulator unit (Fig. 13.35). The insulators are fabricated with a mounting base for curved or flat surfaces, and the top is designed for tying the conductor to the insulator or fitted with a clamp designed to hold the conductor.

Line-post insulators designed for vertical mounting are mounted on crossarms (Fig. 13.36). This type of construction is often used for long-span rural distribution circuits. Figure 13.37 illustrates a distribution circuit constructed with porcelain horizontal line-type post insulators. Armless construction using post-type insulators permits the construction of subtransmission and transmission lines on narrow rights-of-way and along city streets. A 69-kV subtransmission line constructed with porcelain line-post insulators with conductor clamp top is pictured in Fig. 13.38. A 230-kV transmission line constructed with steel poles and line-post-type insulators is illustrated in Fig. 13.39.

FIGURE 13.13 Fiberglass crossarms are used on distribution pole of 13,200Y/7620-volt three-phase line with a three-phase primary circuit tapped to the conductor. The line is constructed with close spacing to meet environmental requirements. Solid blade disconnect isolation switches are also installed. (*Courtesy A. B. Chance Co.*)

Post-type insulators have been used to construct circuits that operate at voltages through 345 kV. The higher transmission voltages require more than one segment of porcelain to be bolted together (Fig. 13.40). Lines constructed with post-type insulators are clean-looking and aesthetically acceptable to the public (Fig. 13.39).

Suspension Insulators. The suspension insulator, as its name implies, is suspended from the crossarm and has the line conductor fastened to the lower end (Fig. 13.41). The suspension-type insulator was developed when voltages were increased above 44,000 volts.

FIGURE 13.14 Galvanized steel insulator supports are used for 13,200Y/7620-volt three-phase line constructed with close spacing for appearance reasons. A galvanized steel crossarm is used to support dead-ended conductors tapped to main circuit. (*Courtesy Sherman & Reilly, Inc.*)

FIGURE 13.15 Standard flat-strap crossarm brace.

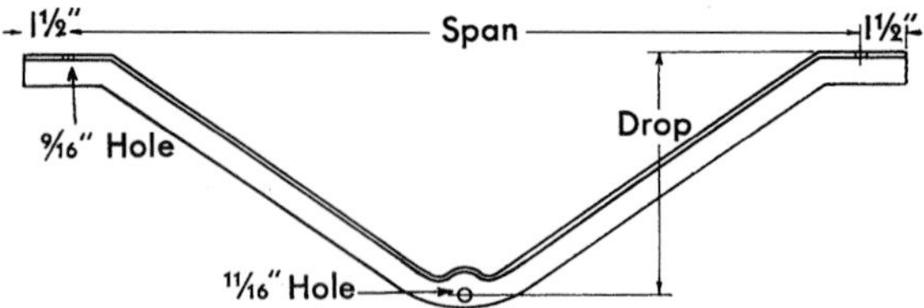

FIGURE 13.16 A V-shaped angle-iron crossarm brace. (*Courtesy Joslyn Mfg. and Supply Co.*)

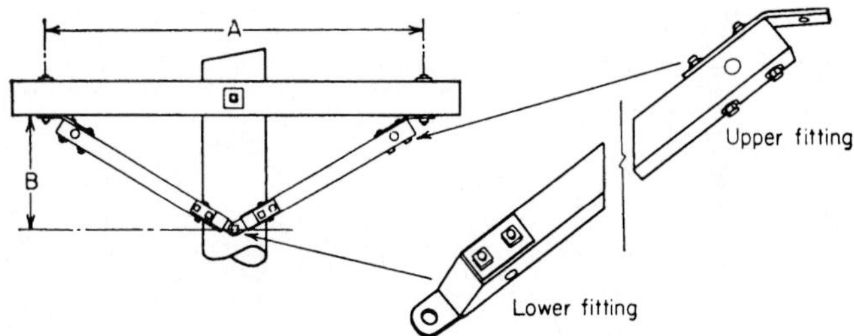

FIGURE 13.17 Wooden crossarm brace used to increase insulation properties of line. (*Courtesy Joslyn Mfg. and Supply Co.*)

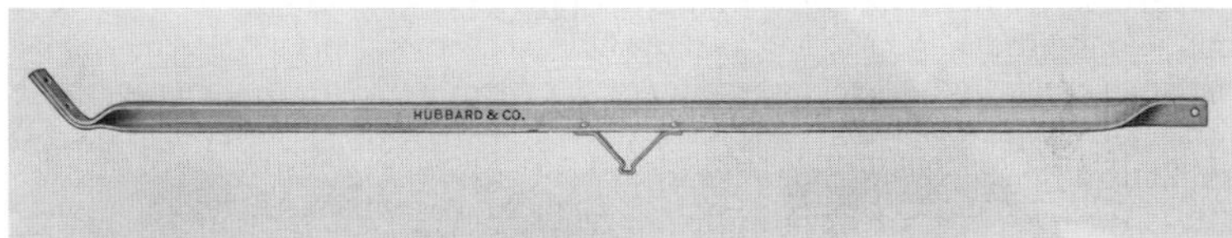

FIGURE 13.18 Alley-arm brace.

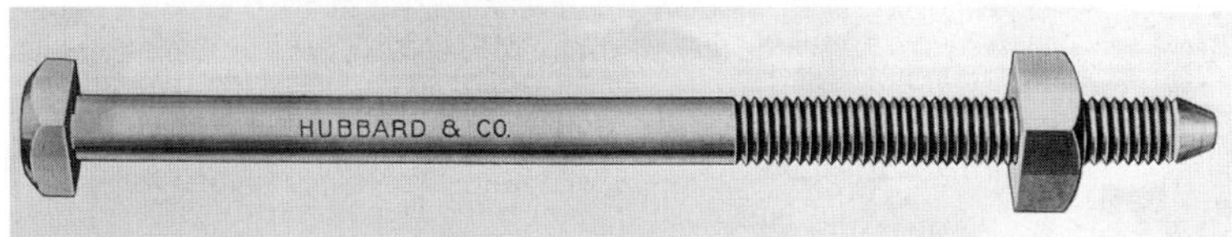

FIGURE 13.19 Crossarm bolt.

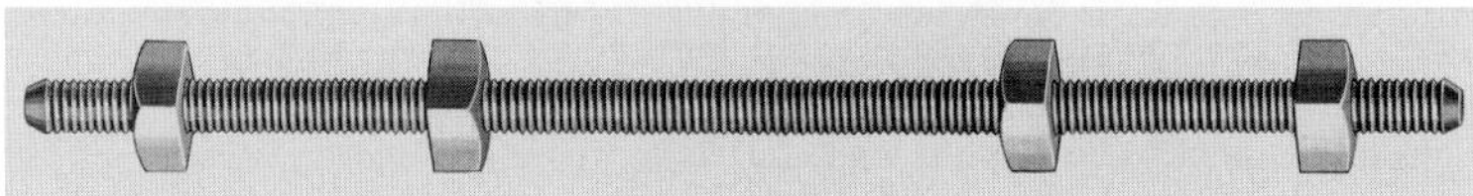

FIGURE 13.20 Bolt for double crossarms.

FIGURE 13.21 Carriage bolt for fastening brace to crossarm.

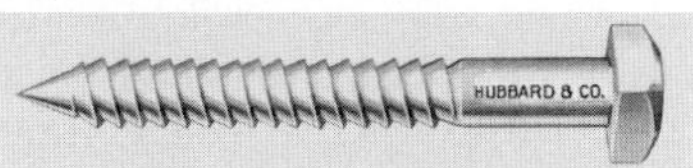

FIGURE 13.22 Lag screw for fastening crossarm brace to pole.

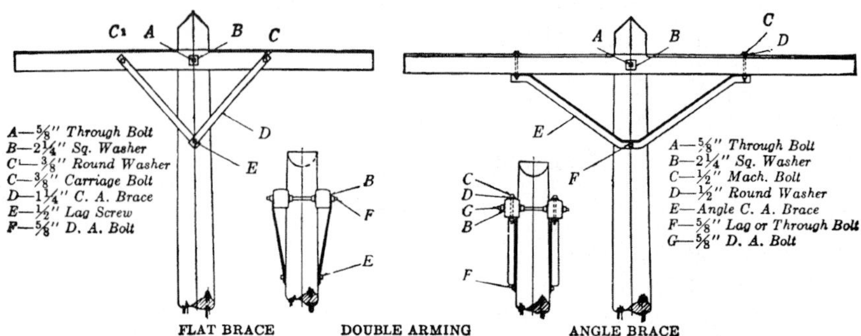

FIGURE 13.23 Sketches showing standard sizes and use of bolts, lags, and washers in single- and double-crossarm assembly.

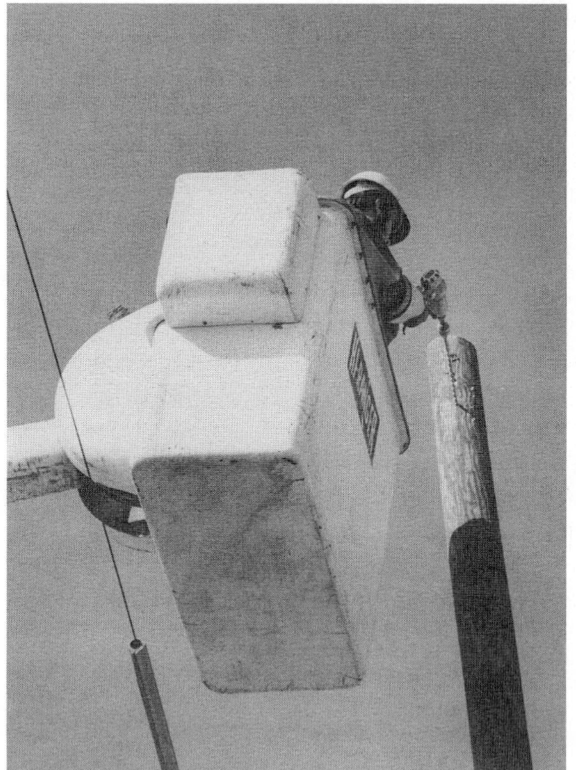

FIGURE 13.24 Lineman drills hole in pole, using power drill operated by truck hydraulic system. (*Courtesy MidAmerican Energy Co.*)

FIGURE 13.25 Lineman and groundman installing pin-type insulator with steel pin on crossarm. (*Courtesy Northwest Lineman College.*)

FIGURE 13.26 Lineman in bucket of truck takes crossarm completely assembled from groundman in preparation to raise it to proper height for installation. (*Courtesy MidAmerican Energy Co.*)

FIGURE 13.27 Lineman completing installation of crossarm. Pin-type insulator for center-phase conductor is supported by special pole-top steel pin fabricated so that it can be bolted to pole. (*Courtesy Northwest Lineman College.*)

FIGURE 13.28 Groundman hoisting crossarm to lineman on pole. Note that groundman stands to one side so that he is out of danger in case any material or tool should drop. The lineman has just taken hold of the crossarm and is about to remove the hand-line hitch from its upper end. The groundman is holding the hand line taut to take up the weight of the crossarm. (*Courtesy Wisconsin Electric Power Co.*)

At higher voltages the pin insulator becomes quite heavy, and it is difficult to obtain sufficient mechanical strength in the pin to support the insulator. The suspension-type insulator is manufactured from porcelain (Fig. 13.42), glass (Fig. 13.43), and epoxy glass rod and rubber materials (Fig. 13.44).Transmission (Fig. 13.41), subtransmission (Fig. 13.45), and distribution (Fig. 13.44) circuits are constructed with suspension insulators. Economics, voltage of the circuit, width of right-of-way, size of the conductor, length of the span, and clearances required all enter into the decision regarding the type and material of the insulators to be used.

The porcelain insulator unit consists essentially of two metal pieces insulated from one another with porcelain. First the porcelain is cemented into the metal cap or the top of the

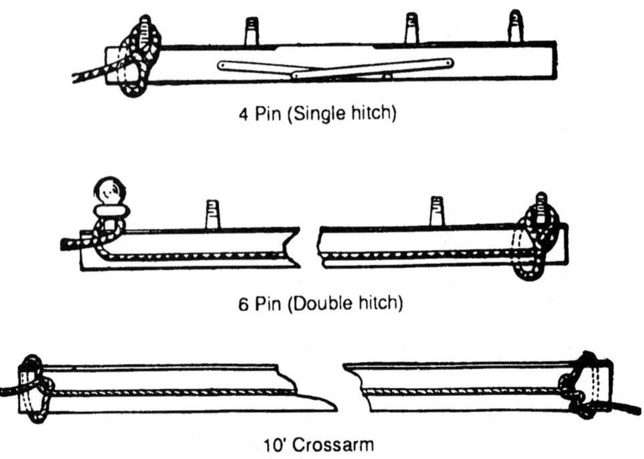

FIGURE 13.29 Methods of tying hand line to crossarm for the purpose of raising.

unit, and then the metal pin on the bottom of the unit is cemented into the porcelain. A typical cemented-type suspension insulator is shown in Fig. 13.46.

The diameter of the porcelain disks varies from 3⅜ to 11½ in or more. The 10-in disk is perhaps the most common size. The number of units to use in a string depends largely on the voltage of the line (Fig. 13.47). Other factors, such as climate, type of construction, and degree of reliability required, also enter into consideration. Table 13.1 gives a general idea of the usual number of units employed for the various standard distribution and transmission voltages.

Strain insulators are used where a pull must be carried as well as insulation provided. Such places occur wherever a line is dead-ended—at corners, at sharp curves, or at extra long spans, as at river crossings or in mountainous country. In such places the insulator not only must be a good insulator electrically, but also must have sufficient mechanical strength to counterbalance the forces due to the tension of the line conductors.

Strain insulators are built in the same way as suspension insulators except that they are made stronger mechanically. Furthermore, when a single string is not able to withstand the pull, two or more strings are arranged in multiple (Fig. 13.48).

Glass suspension insulators have been used on the circuits in many cases. Glass is used more commonly in countries other than the United States. Glass insulators can be used interchangeably with porcelain insulators wherever their strength and voltage characteristics are adequate.

The development of extra-high-voltage (EHV) bundle-conductor transmission circuits resulted in heavy and cumbersome installations when porcelain suspension insulators were used. The polymer suspension insulator is used to address the issue. The polymer suspension insulator consists of a long epoxy-fiberglass rod of extremely high mechanical and dielectric strength with polymer weather sheds covering the rod. A heavy silicone grease is used as an interface between the rod and the weather sheds. Forged-steel end attachments

FIGURE 13.30 Lineman placing the crossarm on the through bolt. Notice the "assist" given to the lineman by the helper. The hand line is still attached to the trailing end of the crossarm, and the helper is keeping it taut to take up some of the weight. The through bolt was put into the pole previously with the threaded end toward the lineman. The bolt has a diameter of 5/8 in, and the hole in the pole has a diameter of 11/16 in. If the bolt fits loosely, the lineman usually bends it slightly near its head end by hitting it with a hammer when it is part way in. This jams the bolt in the hole so that it will not push out when the lineman pushes the crossarm on. Where through-bolt strength is very important, the through bolt is installed so that the head end is toward the lineman. This places the weaker threaded end in the pole opposite the crossarm where the strain is least. (*Courtesy Wisconsin Electric Power Co.*)

FIGURE 13.31 Lineman fastening crossarm braces to the pole with a 1/2- by 5-in lag bolt. The lag bolt has a "drive" thread; the edge of the thread toward the end of the bolt is beveled. The bolt is not driven in full length. It is left about 1/4 in off tight position. It is finally seated by three or four turns with a lag wrench. Note the working position of the lineman. He has set himself such that he can get a good two-handed free swing at the lag. His safety strap is just long enough to place him the correct distance away. He does not have to "reach"; neither does he have to "choke up." (*Courtesy Wisconsin Electric Power Co.*)

FIGURE 13.32 Pin-type insulators mounted on epoxy rod single- and bi-unit assemblies to eliminate use of crossarm. Eyes on bi-unit assembly are for mounting stringing equipment. (*Courtesy A. B. Chance Co.*)

are swaged to the rod under pressure. The insulators are manufactured with maximum design tension ratings of 10,000, 20,000, 40,000, and 80,000 lb. A single polymer suspension insulator can replace four parallel strings of porcelain or glass suspension insulators. The synthetic-material insulators weigh only 5 to 10 percent of the weight of the suspension insulators they replace.

A 500-kV polymer insulator weighs 40 lb, and a 765-kV polymer insulator weighs 65 lb. The synthetic-material insulators are less susceptible to mechanical damage. Structures can be assembled on the ground and erected without danger of destroying the insulator. The polymer insulators have a high ratio of strength to weight, resistance to vandalism (gunfire), a life expectancy equal to that of porcelain, freedom from tracking and deterioration caused by surface leakage currents in contaminated environments, and a low cost.

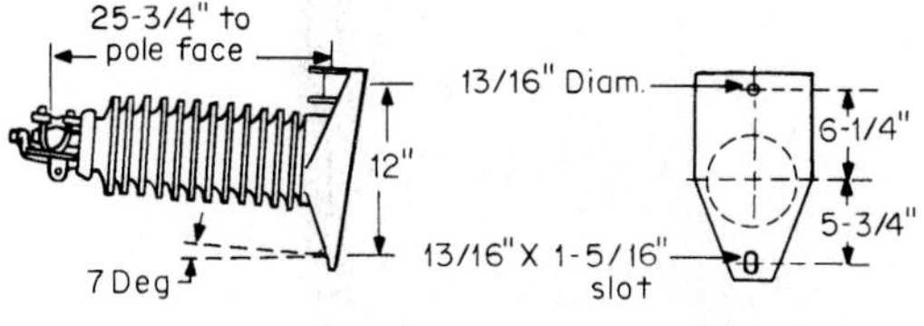

Characteristics				
Typical application			kV	69
Flash-over voltage	Impulse critical 1.2 x 50 mu-sec wave	Positive	kV	330
		Negative	kV	425
	60 Hz	Dry	kV	200
		Wet	kV	180
Standard glaze color			Sky tone	
Radio influence voltage	Test voltage to ground		kV	44
	Maximum RIV at 1000 kHz		Micro-volts	200
Leakage distance			Inches	53
Dry arcing distance			Inches	19-1/4
Cantilever			Pounds	2800

FIGURE 13.33 An 88-kV porcelain horizontal post insulator.

Spacer Cable Spacers. In some instances, bundled, covered, primary conductor is installed. The covered cable is often referred to as "spacer" cable. The messenger supported, bundled configuration has a few advantages, most notably:

- Less tree trimming issues with less tree contacts exposure
- Less animal contact exposure
- Ability to install multiple circuits in congested areas

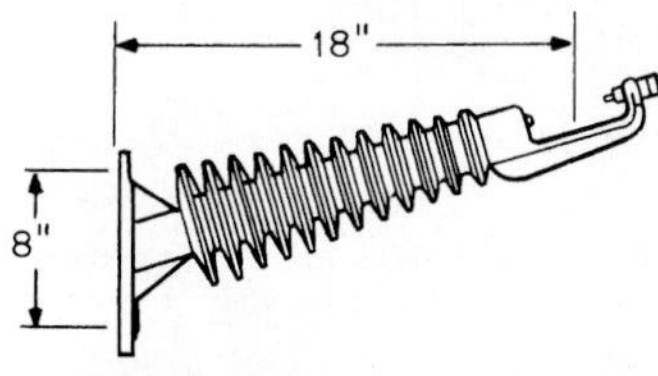

Mechanical characteristics (ultimate ratings)	
Vertical cantilever, lbs	1,500
Longitudinal cantilever, lbs	1,000
Tensile, lbs	4,000
Compression, lbs	1,500
Electrical characteristics	
60 Hz flashover, dry, kV	140
60 Hz flashover, wet, kV	120
Impulse flashover, positive, kV	210
Impulse flashover, negative, kV	320

FIGURE 13.34 A 15-kV tapered-skirt standoff insulator.

FIGURE 13.35 Fiberglass rod and pin insulator used to support top conductor and fiberglass rod and porcelain assemblies to support side conductors. This arrangement permitted increasing the voltage on the circuit from 4.16 to 13.2 kV without replacing the pole. (*Courtesy A. B. Chance Co.*)

FIGURE 13.36 Post-type insulator in use in vertical position. Note special supporting hardware used in place of conventional pin. (*Courtesy Lapp Insulator Co.*)

FIGURE 13.37 A 13.2Y/7.6-kV three-phase four-wire distribution circuit constructed with sky-gray line-post insulators to enhance the appearance. Phase conductors are 477-mcm ACSR. Neutral conductor, 4/0 AWG ACSR is supported by porcelain spool-type insulator fastened to pole with a metal bracket. Crew is holding tailgate safety meeting before starting to work on assignment.

FIGURE 13.38 Post-type insulator in use in horizontal position. Note narrow right-of-way required.

FIGURE 13.39 A 230-kV transmission line located along a city street. Line uses porcelain post-type insulators for the transmission circuit and the underbuilt distribution circuit on steel poles.

- Narrower right-of-way requirements, and
- Less need for overhang easements due to the compactness of the installation.

A variety of spacer cable system construction applications are depicted in Figs. 13.49 through 13.51.

The spacer cable spacers support the conductors and maintain the phase balancing. The spacers are installed in 30 in. intervals. The spacers provide strength and flexibility. The spacers are a gray, track resistant, high-density polyethylene with long leakage distance (Fig. 13.52 and 13.53). A self washing shed design is featured.

Low-Voltage Insulators. Porcelain spool insulators are used for secondary-voltage circuits operating at voltages below 600 volts. Open-wire secondaries of bare conductor or conductor covered with weather-proof material are usually supported by porcelain spool insulators mounted on secondary steel racks (Fig. 13.54).

Porcelain spool insulators supported by steel clevises are often used to support neutral conductors for three-phase, four-wire high-voltage distribution circuits or the bare neutral conductor of insulated twisted-multicable secondary circuits (Fig. 13.55). Porcelain insulators reinforced by a one-piece pressed-steel housing with attached galvanized steel screws, often referred to as house knobs, are available for supporting low-voltage circuits.

Strain insulators used in guy wires are described and illustrated in Chap. 12, "Guying Poles."

Insulator Washing. Insulators can be cleaned by hand-wiping and washing. Insulators, hand-wiped or newly installed in areas with an atmosphere that might cause insulator contamination, are usually sprayed with a silicone gel to extend the cleaning cycle and to make it easier to remove contaminants from the insulator's surface. Contaminated insulators may cause excessive leakage current to flow, causing pole fires or insulator flashover and interruptions to electric service.

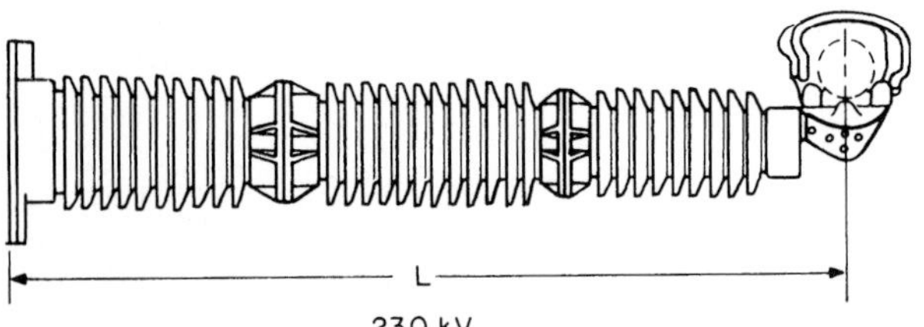

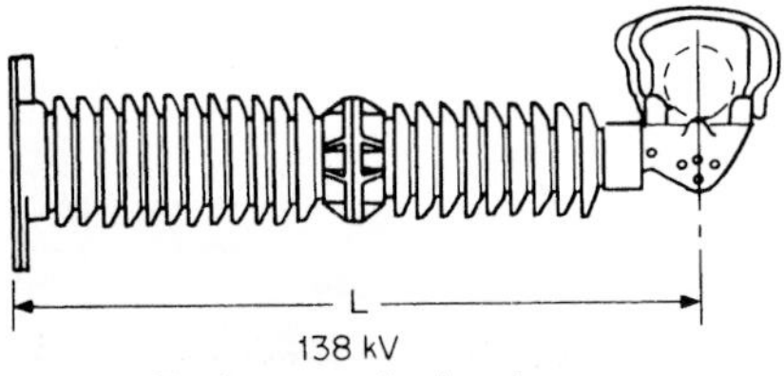

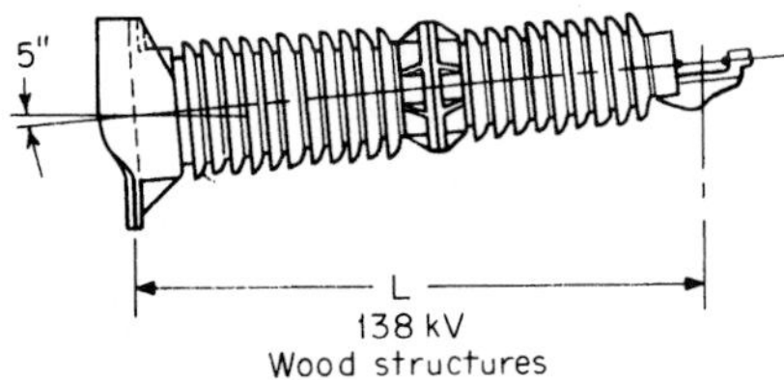

Line Voltage	Lapp Catalog no.	Unit length (L), inches	Ultimate strength	
			Cantilever, lb	Tension, lb
138 kV	70148	53.0	2800	5,000
	79135	58.0	4000	10,000
	301543	56.75	2800	5,000
	305071	60.50	4000	10,000
230 kV	301372	101.50	3500	10,000
	90974	89.0	5800	10,000

FIGURE 13.40 Porcelain horizontal line-post insulators. (*Courtesy Lapp Insulator Co.*)

Insulators can be washed safely, while energized, if proper precautions are taken (Figs. 13.56 and 13.57). The water must have a minimum resistance of 1000 ohms per cubic centimeter. Normally water obtained from city water systems will be satisfactory. The hose nozzle must be grounded, normally to the tower, static wire, or neutral conductor. The water pressure must be maintained high enough to prevent a solid stream of water flowing from the hose nozzle. The hose nozzle must be kept at a safe distance from the energized conductors (Table 13.2).

The washing of energized insulators should start on the bottom and work up to prevent dirty water from falling on contaminated insulators. Overspray should be avoided to prevent inadequate water volume from contacting insulators and causing flashover.

Insulators Used in Pad-Mounted Switchgear. Pad-mounted switchgear used with underground circuits requires the use of space-saving insulators to keep the equipment compact.

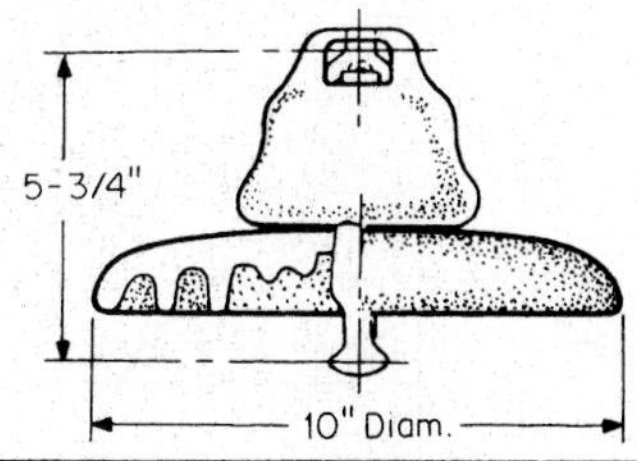

Characteristics				
ANSI class (Std. C.29.2-1971)				52.8
Flash-over voltage	Impulse critical 1.2 x 50 mu-sec wave	Positive	kV	125
		Negative	kV	130
	60 Hz	Dry	kV	80
		Wet	kV	50
Low-frequency puncture voltage			kV	110
Radio influence voltage	Test voltage to ground		kV	10
	Maximum RIV at 1000 kHz		Micro-volts	50
Leakage distance			Inches	13
Dry arcing distance			Inches	7.75
Section length			Inches	5.75
Porcelain disc diameter			Inches	10
Strength rating	M & E rating		Pounds	40,000
	ANSI M & E category		Pounds	36,000
Recommended maximum sustained load			Pounds	20,000
Impact strength			Inch Pounds	90
Color			Skytone	

FIGURE 13.42 Porcelain suspension insulator, ball-and-socket type.

FIGURE 13.41 Single-circuit 345,000-volt three-phase high-voltage transmission line. Note the long string of insulators needed to insulate the conductors. Also note two ground wires at extreme top of the tower. These are used to shield the line from lightning.

Cycloaliphatic epoxy-resin insulators, sold under the tradename Cypoxy, are specially formulated to provide the electrical and mechanical strength characteristics detailed in industry standards for porcelain insulators.

Cypoxy insulators and insulating components have given trouble-free service in a wide range of environments including pad-mounted switchgear. The ability to mold Cypoxy into intricate shapes with light weight and high strength has made it possible to develop space-saving switch designs not possible with porcelain insulation. These insulators are non-tracking, self-scouring, and nonweathering. Various ways in which this insulating material can be used are illustrated in Fig. 13.58.

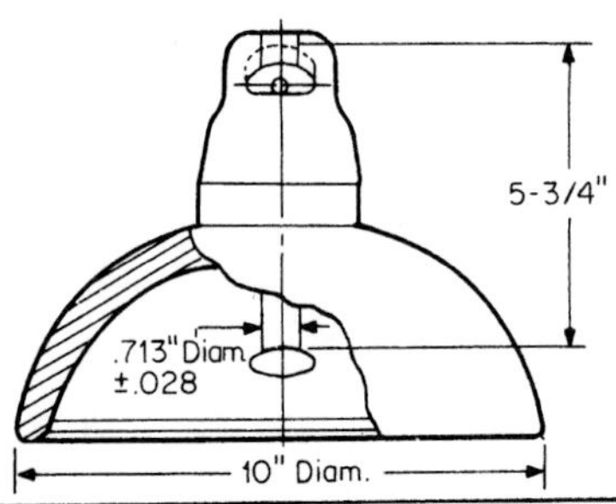

Characteristics				
ANSI class (Std. C29.2-1962)				52-5
Flash-over voltage	Impulse	Positive	kV	125
		Negative	kV	125
	Low frequency	Dry	kV	70
		Wet	kV	55
Low-frequency puncture voltage			kV	130
Radio influence voltage	Test voltage to ground		kV	10
	Maximum RIV at 1000 kHz		Micro-volts	50
Leakage distance			Inches	11-1/2
Impact strength			Inches Pounds	400
Routine proof test			Pounds	15,000
Strength ratings	M & E rating		Pounds	30,000
	Maximum sustained load		Pounds	16,500

FIGURE 13.43 Toughened glass suspension insulator, ball-and-socket type.

FIGURE 13.44 Epoxy glass rod, with rubber weather sheds, suspension insulators used on vertical corner of a distribution circuit. (*Courtesy A. B. Chance Co.*)

FIGURE 13.45 A 69-kV subtransmission circuit with porcelain suspension insulators. Wishbone wooden crossarms provide triangular spacing. (*Courtesy MidAmerican Energy Co.*)

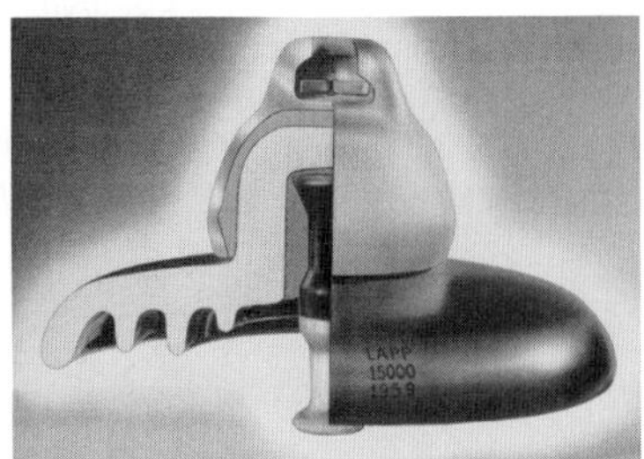

FIGURE 13.46 Sectional view of cemented-type ball-and-socket porcelain suspension insulator. (*Courtesy Lapp Insulator Co.*)

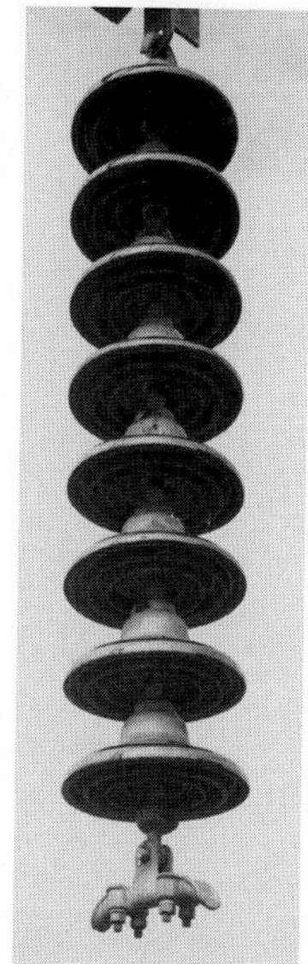

FIGURE 13.47 A string of eight suspension insulator units suspended from a crossarm. Note conductor clamp at bottom of the string.

TABLE 13.1 Approximate Number of 10-in Porcelain or Glass Disk Insulators Required in Suspension String for Various Line Voltages

Line voltage, volts	Number of suspension units required in string	Line voltage, volts	Number of suspension units required in string
13,200	2	138,000	8, 9, or 10
23,000	2 or 3	154,000	9, 10, or 11
34,500	2 or 3	230,000	12 to 16
69,000	4 or 5	345,000	18 to 20
88,000	5 or 6	500,000	24 to 28
110,000	6, 7, or 8	765,000	30 to 35

FIGURE 13.48 Two strings of suspension insulators used in multiple to serve as strain insulators at 230-kV dead-end tower.

FIGURE 13.49 Double-circuit three-phase spacer cable installation. (*Courtesy Hendrix Wire & Cable Inc.*)

FIGURE 13.50 A 25-kV single-phase distribution transformer connected to the three-phase spacer cable installation. (*Courtesy Hendrix Wire & Cable Inc.*)

FIGURE 13.51 Mid-span spacer installed in the three-phase spacer cable installation. (*Courtesy Hendrix Wire & Cable Inc.*)

FIGURE 13.52 Spacer cable spacer. (*Courtesy Hendrix Wire & Cable Co.*)

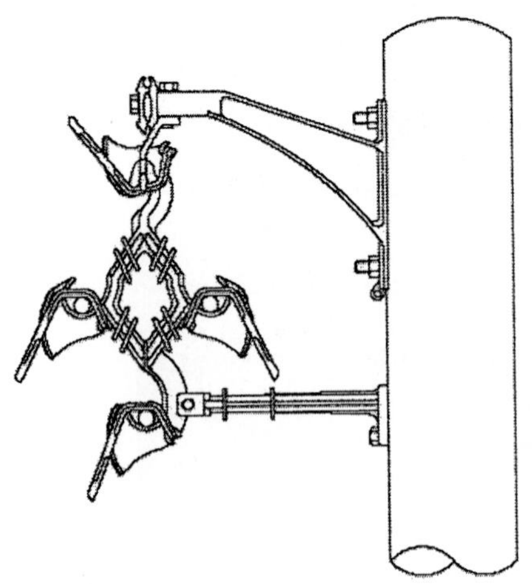

FIGURE 13.53 Spacer cable spacer installed and hanging from the pole mounted bracket. The spacers are a gray, track resistant, high-density polyethylene with long leakage distance. (*Courtesy Hendrix Wire & Cable Co.*)

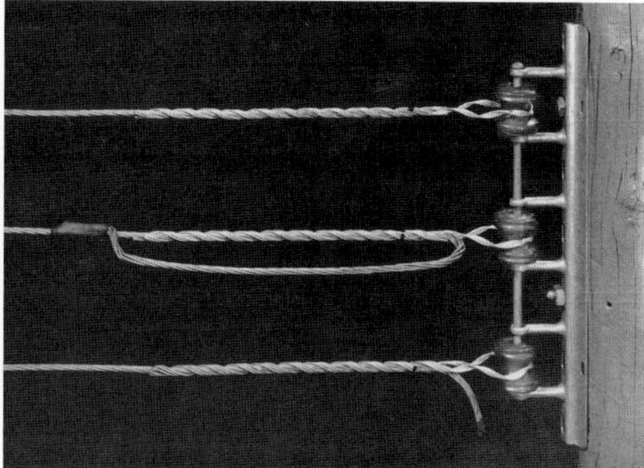

FIGURE 13.54 Open-wire bare secondary wires supported by porcelain spool insulators on a galvanized steel rack. Wires are secured to insulators with preformed distribution grips. (*Courtesy Preformed Line Products Co.*)

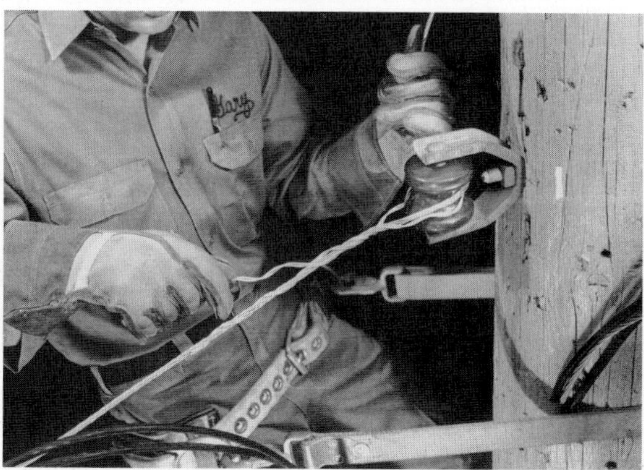

FIGURE 13.55 Lineman securing bare neutral conductor to porcelain spool insulator fastened to pole with a galvanized steel clevis by a through bolt. Conductor is being secured with a preformed grip. (*Courtesy Preformed Line Products Co.*)

FIGURE 13.56 Lineman in helicopter positioned to begin spraying energized substation insulator with silicone gel.

FIGURE 13.57 From a helicopter, the lineman is washing insulators on energized transmission-line structure to remove contamination.

TABLE 13.2 Standard Minimum Length of Stream Distances When "Hot Washing" with a Minimum Dynamic Pressure of 550 psi at the Nozzle for Various Voltages

Voltage (phase-to-phase), kV	Length of stream distance, ft	
	15/64- or 1/4-in nozzle	5/16-in nozzle
4–12	7	10
13–23	10	13
24–70	12	15
71–115	15	18
230	15	20
500	20	20

Source: Pacific Gas and Electric Company.

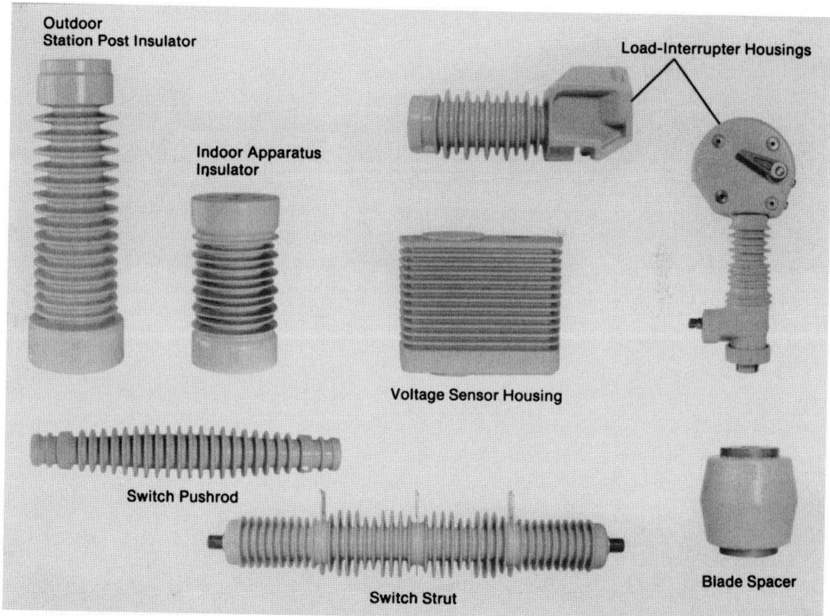

FIGURE 13.58 Cycloaliphatic epoxy-resin insulators are used for the applications illustrated. (*Courtesy S&C Electric Co.*)

CHAPTER 14
LINE CONDUCTORS

The wires and cables over which electric energy is transmitted are made of copper, aluminum, steel, or a combination of copper and steel or aluminum and steel. A conductor is a material that readily permits the flow of an electric current. Most metals can be used for conductors in an electric circuit. Materials, other than those mentioned, that conduct electricity are not generally used to make wires and cables because of economic or physical reasons. Gold, silver, platinum, nickel, zinc, tungsten, molybdenum, boron, cobalt, cadmium, beryllium, magnesium, silicon, and the like may be used for special wire applications or combined with other materials to improve their characteristics for forming wires.

Copper Conductors. Copper is a commonly used line conductor. It conducts electric current very readily, ranking next to silver. It is very plentiful in nature, and therefore its cost is comparatively low. It can be easily spliced.

Three kinds of copper wire are in use: hard-drawn copper, medium-hard-drawn copper, and annealed copper, also called *soft-drawn*. Copper wire is hard-drawn as it comes from the drawing die. To obtain soft or annealed copper wire, the hard-drawn wire is heated to a red heat to soften it.

For overhead line purposes, hard-drawn copper wire is preferable on account of its greater strength. Annealing, or softening, it reduces the tensile strength of the wire from about 55,000 to 35,000 lb/in^2. Because of this, it is not good practice to have any soldered splices when using hard-drawn wire, as the soldering anneals the wire near the joint, thereby reducing its strength. Joints in hard-drawn wire should, therefore, be made with splicing sleeves. Annealed or soft-drawn copper wire is used for ground wires and special applications where it is necessary to bend and shape the conductor. Medium-hard-drawn copper is used for distribution especially for wire sizes smaller than No. 2. Soft copper wire has a yield point less than one-half that of medium-hard copper and hence stretches permanently with a correspondingly lighter loading of ice and wind.

Aluminum Conductors. Aluminum is widely used for distribution and transmission-line conductors. Its conductivity, however, is only about two-thirds that of copper. Compared with a copper wire of the same physical size, aluminum wire has 60 percent of the conductivity, 45 percent of the tensile strength, and 33 percent of the weight. The aluminum wire must be 100/66 = 1.66 times as large as the copper wire in cross section to have the same conductivity. An aluminum wire of this size will have 75 percent of the tensile strength and 55 percent of the weight of the equivalent copper conductor.

Figure 14.1 shows cross sections of copper, aluminum, and aluminum-conductor steel-reinforced (ACSR) conductors which have equal conductivity or, what amounts to the same thing, equal current-carrying capacities.

The tensile strength of aluminum is only about one-half that of hard-drawn copper, namely, 27,000 lb/in^2. But since the cross section of an aluminum wire must be about twice

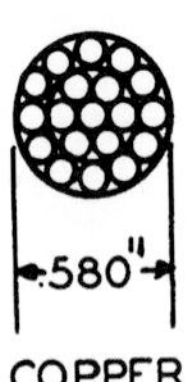

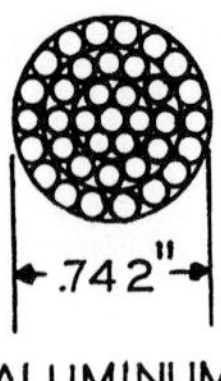

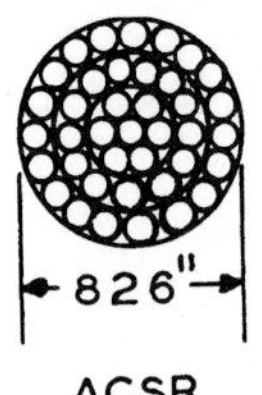

FIGURE 14.1 Relative cross sections of copper, aluminum, and ACSR conductors having equal current-carrying capacities.

that of copper to have the same conductivity, the actual breaking strength of the aluminum conductor is about the same as that of copper. The weight per foot is still only 55 percent as great, as pointed out above, in spite of being twice as large in cross section. This fact makes aluminum conductors preferred in many cases, because the lighter weight permits longer spans and, therefore, fewer towers and insulators. This is a decided advantage where long spans are common. The larger size helps to keep corona loss down.

When an aluminum conductor is stranded, the central strand is often made of steel, which serves to reinforce the cable. Such reinforcement gives great strength for the weight of conductor. Reinforced aluminum cable called ACSR is therefore especially suited for long spans.

Steel Conductors. Steel wire is used to a limited extent where minimum construction expense is desired. Steel wire, because of its high tensile strength of 160,000 lb/in^2, permits relatively long spans, therefore requires few supports. Bare steel wire, however, rusts rapidly and is therefore very short-lived. Steel is also a poor conductor compared with copper, being only about 10 to 15 percent as good. Galvanized steel conductors are commonly used for shield or static wires in transmission and subtransmission lines. Most guy wires are galvanized steel stranded cables.

Copperweld Steel Conductors. The disadvantages of short life and low conductivity of steel led to the development of the copperweld steel conductor. In this conductor a coating of copper is securely welded to the outside of the steel wire. The copper acts as a protective coating to the steel wire, thus giving the conductor the same life as if it were made of solid copper. At the same time, the layer of copper greatly increases the conductivity of the steel conductor, while the steel gives it great strength. This combination produces a very satisfactory yet inexpensive line conductor. Its chief field of application is for overhead primary conductors in rural lines, for guy wires, and for overhead ground wires.

The conductivity of copperweld conductors can be raised to any desired percentage, depending on the thickness of the copper layer. The usual values of conductivity of wires as manufactured are 30 and 40 percent. Figure 14.2 shows a stranded copperweld conductor.

Alumoweld Steel Conductors. Steel wire can be covered with aluminum to prevent the steel from rusting as well as to improve its conductivity. Figure 14.3 shows such an aluminum-covered stranded steel cable used for guys. By its use the guys can be expected to last as long as modern treated wooden poles.

Figure 14.4 shows an ACSR conductor in which one conductor is a steel strand covered with a welded layer of aluminum and the other conductors are aluminum strands. The layer

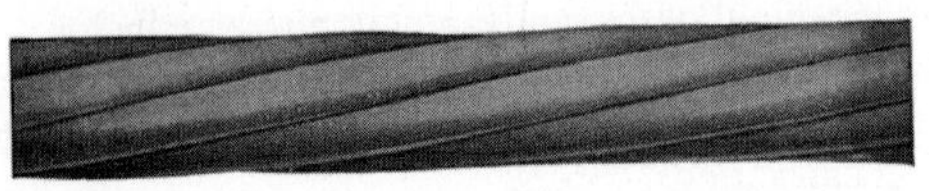
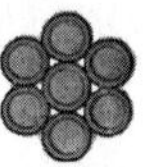

FIGURE 14.2 Stranded copperweld conductor. (*Courtesy Copperweld Steel Co.*)

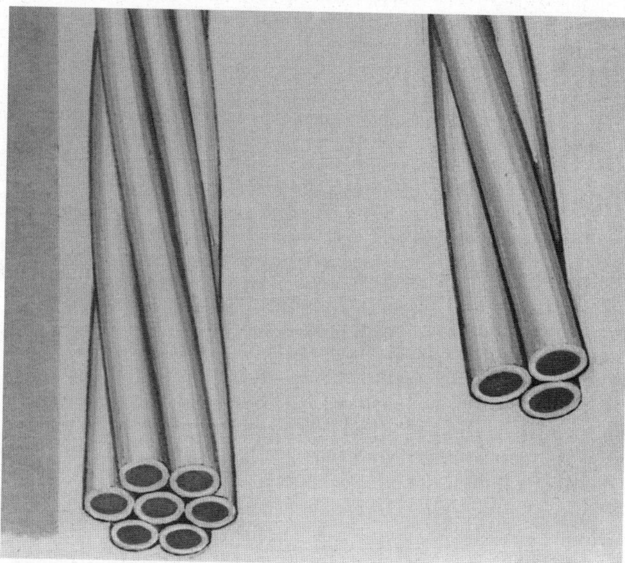

FIGURE 14.3 Alumoweld guy strand. Each strand is covered with a welded layer of aluminum to protect the steel core from rusting. (*Courtesy Copperweld Steel Co.*)

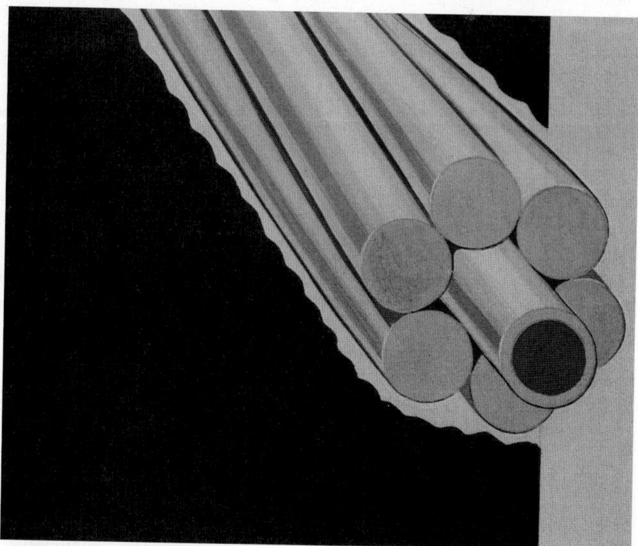

FIGURE 14.4 ACSR cable with steel strand covered with layer of welded aluminum. The layer of aluminum on the steel strand prevents electrolytic corrosion between strands, prevents rusting of the steel strand, and increases the conductivity of the steel strand by a substantial amount. (*Courtesy Copperweld Steel Co.*)

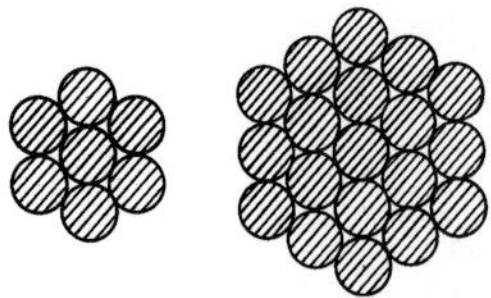

FIGURE 14.5 Sketches showing arrangement of wires in 7- and 19-strand conductors. The wires are arranged in concentric layers about a central core.

of aluminum increased the conductivity of the steel strand in addition to preventing it from rusting. All the strands of the cable, the alumoweld strand, and the aluminum strands thus have the same life.

Conductor Classes. Conductors are classified as solid or stranded. A *solid conductor*, as the name implies, is a single conductor of solid circular section. A *stranded conductor* is composed of a group of wires made into a single conductor. A stranded conductor is used where the solid conductor is too large and not flexible enough to be handled readily. Large solid conductors are easily injured by bending. The size No. 0 wire is the approximate dividing line between solid and stranded conductors. The strands in the stranded conductor are usually arranged in concentric layers about a central core (Fig. 14.5). The smallest number of wires in a stranded conductor is 3. The next number of strands is 7, then 19, 37, 61, 91, 126, etc. Figure 14.6 illustrates a typical stranded conductor. Both copper and aluminum are thus stranded.

Conductor Covering. Conductors on overhead transmission lines are bare conductors, that is, they are not covered. Conductors on overhead distribution circuits, below 5000 volts, are usually covered. The common covering is triple-braid weatherproof cotton, neoprene, or polyethylene. The triple-braid cotton is saturated with a black moistureproof compound. Figure 14.7 shows a conductor and two layers of braid. This covering, of course, is not sufficient to withstand the voltage at which the line is operating, and the conductors must, therefore, be mounted on insulators. In fact, the wires should always be treated as though they were bare.

Wire Sizes. Wire sizes are ordinarily expressed by numbers. There are, however, several different numbering methods, so that in specifying a wire size by number it is also necessary to state which wire gauge or numbering method is used. The most used wire gauge in the United States is the *Brown and Sharpe* gauge, also called *American wire gauge* (AWG).

American Wire Gauge. The American standard wire gauge is shown in Fig. 14.8. This cut is full size, so that the widths of the openings on the rim of the gauge correspond to the diameters of the wires whose numbers stand opposite the openings. The actual size of these

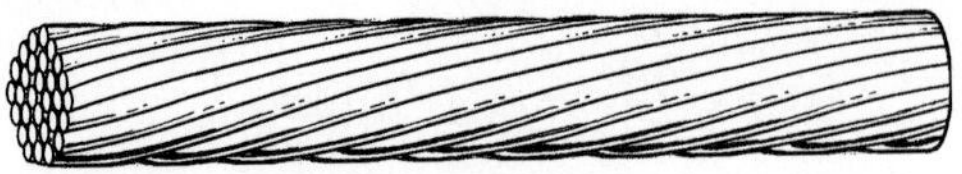

FIGURE 14.6 Typical stranded cable.

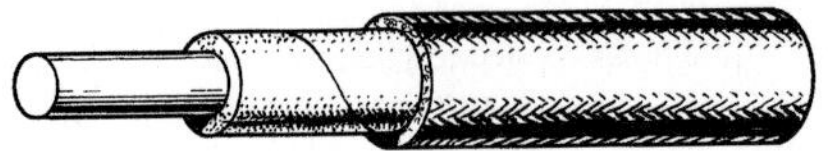

FIGURE 14.7 Solid conductor covered with two layers of braided cotton and saturated with a weatherproofing compound.

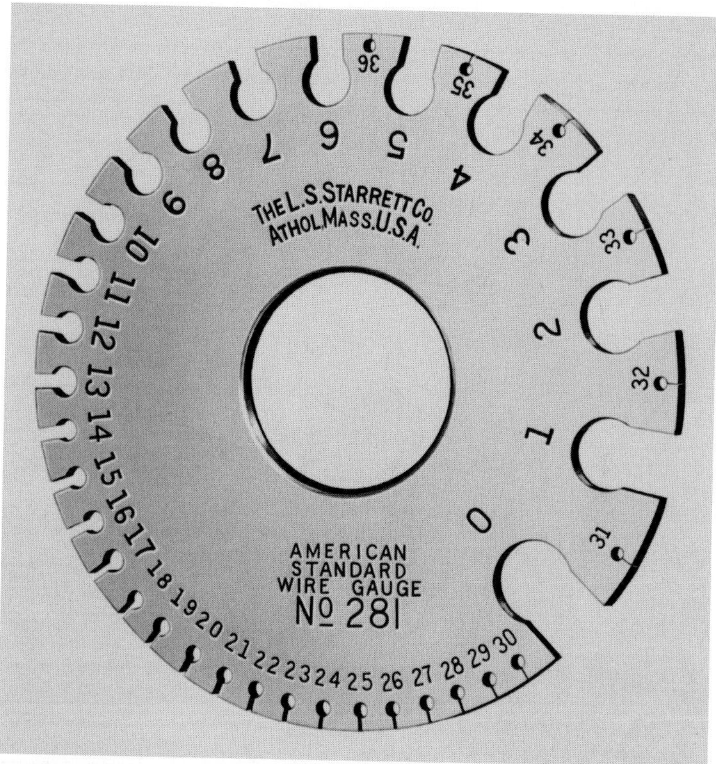

FIGURE 14.8 American standard wire gauge. Cut is full size. Widths of openings on rim correspond to diameters of wires whose numbers are opposite the openings. (*Courtesy L. S. Starrett Co.*)

wire numbers is perhaps better illustrated by Table 14.1. This table gives, for wire sizes from No. 0000 to No. 8, the diameter of the wire in inches and a full-size end and side view of the wire corresponding to each number of the gauge. This table should make it easy to get an idea of the actual physical size of some of the wires commonly used in distribution work. Note that the greater the number of the wire, the smaller the wire is. That came about by numbering the wire by the number of steps required in the wire-drawing process. The smaller the wire, the greater the number of steps required to draw it.

Descriptive details for solid copper wire are itemized in Table 14.2. Note in Table 14.2 that with every increase of 3 in the gauge number, the cross section and weight are halved and the resistance is doubled. An increase of 10 in the gauge number increases the resistance 10 times and cuts the weight and cross section to one-tenth. Thus, for example, the No. 5 wire is three sizes smaller than the No. 2 wire. Its cross section is one-half that of the No. 2, for 33,090 is approximately one-half of 66,360, and its resistance is twice that of the No. 2, for 0.3260 is twice 0.1625. Furthermore, the resistance of the No. 12 is 10 times that of the No. 2 for 1.65 is nearly 10 times 0.1625. This general relation between gauge numbers is good to remember. Properties of solid aluminum wire are listed in Table 14.3.

Circular Mil. The circular mil is the unit of cross-sectional area customarily used in designating the area of wires. A *circular mil* is the area contained in a circle having a diameter

TABLE 14.1 American Wire Sizes (Bare Conductor)

Gauge number	Diameter, in	Full-size end view	Full-size side view
8	0.1285		
7	0.1443		
6	0.162		
5	0.1819		
4	0.2043		
3	0.2294		
2	0.2576	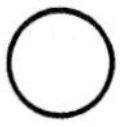	
1	0.2893	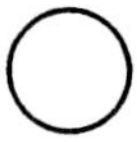	
0	0.3249	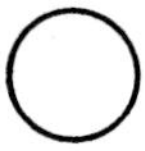	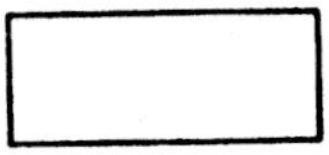
00	0.3648	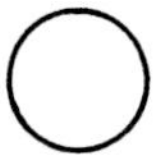	
000	0.4096	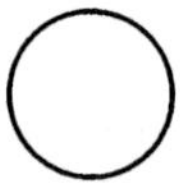	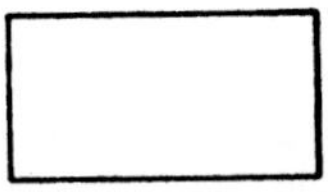
0000	0.460	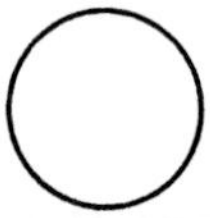	

TABLE 14.2 Copper Wire—Weight, Breaking Strength, DC Resistance
(*Based on ASTM Specifications B1, B2 and B3*)

		Area		Weight		Hard		Medium		Soft	
Size, AWG	Diameter, in	Cir mils	in^2	Lb per 1,000 ft	Lb per mile	Breaking strength, minimum, lb	DC resistance at 20°C (68°F) maximum,* ohms per 1000 ft	Breaking strength, minimum, lb	DC resistance at 20°C (68°F) maximum,* ohms per 1000 ft	Breaking strength, maximum,† lb	DC resistance at 20°C (68°F) maximum,* ohms per 1000 ft
4/0	0.4600	211,600	0.1662	640.5	3,382	8,143	0.05045	6,980	0.05019	5,983	0.04901
3/0	0.4096	167,800	0.1318	507.8	2,681	6,720	0.06362	5,666	0.06330	4,744	0.06182
2/0	0.3648	133,100	0.1045	402.8	2,127	5,519	0.08021	4,599	0.07980	3,763	0.07793
1/0	0.3249	105,600	0.08291	319.5	1,687	4,518	0.1022	3,731	0.1016	2,985	0.09825
1	0.2893	83,690	0.06573	253.3	1,338	3,688	0.1289	3,024	0.1282	2,432	0.1239
2	0.2576	66,360	0.05212	200.9	1,061	3,002	0.1625	2,450	0.1617	1,928	0.1563
3	0.2294	52,620	0.04133	159.3	841.1	2,439	0.2050	1,984	0.2039	1,529	0.1971
4	0.2043	41,740	0.03278	126.3	667.1	1,970	0.2584	1,584	0.2571	1,213	0.2485
5	0.1819	33,090	0.02599	100.2	528.8	1,590	0.3260	1,265	0.3243	961.5	0.3135
6	0.1620	26,240	0.02061	79.44	419.4	1,280	0.4110	1,010	0.4088	762.6	0.3952
7	0.1443	20,820	0.01635	63.03	332.8	1,030	0.5180	806.7	0.5153	605.1	0.4981
8	0.1285	16,510	0.01297	49.98	263.9	826.1	0.6532	644.0	0.6498	479.8	0.6281
9	0.1144	13,090	0.01028	39.61	209.2	660.9	0.8241	513.9	0.8199	380.3	0.7925
10	0.1019	10,380	0.008155	31.43	166.0	529.3	1.039	410.5	1.033	314.0	0.9988
11	0.0907	8,230	0.00646	24.9	131	423	1.31	327	1.30	249	1.26
12	0.0808	6,530	0.00513	19.8	104	337	1.65	262	1.64	197	1.59
13	0.0720	5,180	0.00407	15.7	82.9	268	2.08	209	2.07	157	2.00
14	0.0641	4,110	0.00323	12.4	65.7	214	2.63	167	2.61	124	2.52
15	0.0571	3,260	0.00256	9.87	52.1	170	3.31	133	3.29	98.6	3.18
16	0.0508	2,580	0.00203	7.81	41.2	135	4.18	106	4.16	78.0	4.02

*Based on nominal diameter and ASTM resistivities.
†No requirements for tensile strength are specified in ASTM B3. Values given here based on Anaconda data.
Source: *Standard Handbook for Electrical Engineers*, Donald G. Fink, Editor-in-Chief, H. Wayne Beaty, Associate Editor. Copyright 1978 by McGraw-Hill, Inc. Used with permission.

TABLE 14.3 Aluminum Wire—Dimensions, Weight, DC Resistance (*Based on ASTM Specifications B230, B262, and B323*)

Conductor size, AWG	Diam. at 20°C (68°F), mils	Area at 20°C (68°F)		DC resistance at 20°C (68°F),* ohms per 1,000 ft	Weight at 20°C (68°F),† lb		Length at 20°C (68°F), ft per ohm
		Cir mils	in²		Per 1,000 ft	per ohm	
2	257.6	66,360	0.05212	0.2562	61.07	238.4	3903
3	229.4	52,620	0.04133	0.3231	48.43	149.9	3095
4	204.3	41,740	0.03278	0.4074	38.41	94.30	2455
5	181.9	33,090	0.02599	0.5139	30.45	59.26	1946
6	162.0	26,240	0.02061	0.6479	24.15	37.28	1544
7	144.3	20,820	0.01635	0.8165	19.16	23.47	1225
8	128.5	16,510	0.01297	1.030	15.20	14.76	971.2
9	114.4	13,090	0.01028	1.299	12.04	9.272	769.7
10	101.9	10,380	0.008155	1.637	9.556	5.836	610.7
11	90.7	8,230	0.00646	2.07	7.57	3.66	484
12	80.8	6,530	0.00513	2.60	6.01	2.31	384
13	72.0	5,180	0.00407	3.28	4.77	1.45	305
14	64.1	4,110	0.00323	4.14	3.78	0.914	242
15	57.1	3,260	0.00256	5.21	3.00	0.575	192
16	50.8	2,580	0.00203	6.59	2.38	0.361	152
17	45.3	2,050	0.00161	8.29	1.89	0.228	121
18	40.3	1,620	0.00128	10.5	1.49	0.143	95.5
19	35.9	1,290	0.00101	13.2	1.19	0.0899	75.8
20	32.0	1,020	0.000804	16.6	0.942	0.0568	60.2
21	28.5	812	0.000638	20.9	0.748	0.0357	47.8
22	25.3	640	0.000503	26.6	0.589	0.0222	37.6
23	22.6	511	0.000401	33.3	0.470	0.0141	30.0
24	20.1	404	0.000317	42.1	0.372	0.00884	23.8
25	17.9	320	0.000252	53.1	0.295	0.00556	18.8
26	15.9	253	0.000199	67.3	0.233	0.00346	14.9
27	14.2	202	0.000158	84.3	0.186	0.00220	11.9
28	12.6	159	0.000125	107	0.146	0.00136	9.34
29	11.3	128	0.000100	133	0.118	0.000883	7.51
30	10.0	100	0.0000785	170	0.0920	0.000541	5.88

*Conductivity = 61.0% IACS.
†Density = 2.703 g per cu cm (0.09765 lb per cu in).
Source: *Standard Handbook for Electrical Engineers*, Donald G. Fink, Editor-in-Chief, H. Wayne Beaty, Associate Editor. Copyright 1978 by McGraw-Hill, Inc. Used with permission.

of 1/1000 in (Fig. 14.9). A mil is a thousandth of an inch. The use of the circular mil as the unit of area came about because of the custom of using a square as the unit of area in measuring the areas of squares or rectangles. Naturally, when the areas of circles had to be measured, a circle having unit diameter suggested itself as the natural unit of measurement.

The circular mils of cross section in a wire of any diameter are obtained by multiplying the diameter in thousandths of an inch by itself. Thus the number of circular mils of cross section in the No. 10 wire is $102 \times 102 = 10{,}404$. The circular-mil cross section of the No. 2 wire is $257.6 \times 257.6 = 66{,}370$.

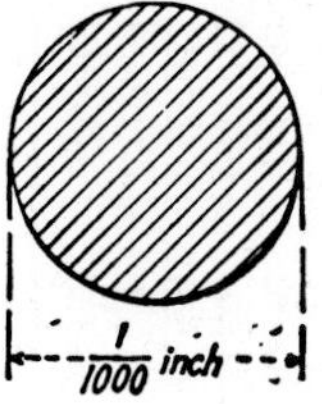

FIGURE 14.9 Area of circle whose diameter is 1/1000 in is 1 cir mil.

If one can remember the relations pointed out above and the data for the No. 10 copper wire, one can build up the remainder of the table any time. The approximate data for the No. 10 copper wire are easily remembered because its resistance per 1000 ft is 1 ohm, its diameter is 1/10 in, and its circular-mil cross section is 10,000. Its weight per 1000 ft is 31.4 lb. With these figures as a basis, one can calculate the data for the Nos. 7, 4, 1, 13, 16, 19, etc., wire numbers by using the factors 2 and 1/2, remembering that for every increase in wire number of three sizes, the cross section and weight are halved and the resistance is doubled. In other words, the larger the number of the wire, the smaller the wire.

To obtain the data for wires one number larger, the factors are 1.25 and 0.80, and for two numbers larger they are 1.60 and 0.625. Thus the data for the No. 11 wire are obtained by multiplying the resistance of the No. 10 wire by 1.25, that is 1.0×1.25 equals 1.25, and dividing the cross-sectional area of 10,000 by 1.25, which equals 8000. The results obtained in this way are, of course, approximate, but they are close enough to give one a fair idea of the wire.

Stranded Conductor Data. The classes of concentric lay stranded cable for various applications are given in Table 14.4. Descriptive data for copper stranded conductors are itemized in Table 14.5. Table 14.6 provides data for copper-clad steel-copper conductors.

Table 14.7 lists properties of aluminum stranded conductors. ACSR cable descriptive data are itemized in Table 14.8. Galvanized steel cable properties are listed in Table 14.9.

TABLE 14.4 Aluminum Cable—Stranding Classes, Uses
(*Based on ASTM Specification B231*)

Construction	Class	Application
Concentric lay	AA	For bare conductors usually used in overhead lines
	A	For conductors to be covered with weather-resistant (weatherproof), slow-burning materials and for bare conductors where greater flexibility than is afforded by Class AA is required. Conductors intended for further fabrication into tree wire or to be insulated and laid helically with or around aluminum or ACSR messengers shall be regarded as Class A conductors with respect to direction of lay only
	B	For conductors to be insulated with various materials such as polyethylene, rubber, paper, varnished cloth, etc., and for the conductors indicated under Class A where greater flexibility is required
	C, D	For conductors where greater flexibility is required than is provided by Class B conductors

Source: *Standard Handbook for Electrical Engineers*, Donald G. Fink, Editor-in-Chief, H. Wayne Beaty, Associate Editor. Copyright 1978 by McGraw-Hill, Inc. Used with permission.

TABLE 14.5 Copper Cable, Classes AA, A, B—Weight, Breaking Strength, DC Resistance
(*Based on ASTM Specifications B1, B2, B3, and B8*)

Conductor size, kCMIL or AWG	No. of wires (ASTM stranding class)	Wire diameter, in	Conductor diameter, in	Conductor area, in^2	Conductor weight, lb		Hard		Medium		Soft	
					Per 1,000 ft	Per mile	Breaking strength, minimum,* lb	DC resistance at 20°C (68°F), ohms per 1,000 ft	Breaking strength, minimum,* lb	DC resistance at 20°C (68°F), ohms per 1,000 ft	Breaking strength, maximum,† lb	DC resistance at 20°C (68°F), ohms per 1,000 ft
5,000	169 (A)	0.1720	2.580	3.927	15,890	83,910	216,300	0.002265	172,000	0.002253	145,300	0.002178
5,000	217 (B)	0.1518	2.581	3.927	15,890	83,910	219,500	0.002265	173,200	0.002253	145,300	0.002178
4,500	169 (A)	0.1632	2.448	3.534	14,300	75,520	197,200	0.002517	154,800	0.002504	130,800	0.002420
4,500	217 (B)	0.1440	2.448	3.534	14,300	75,520	200,400	0.002517	156,900	0.002504	130,800	0.002420
4,000	169 (A)	0.1538	2.307	3.142	12,590	66,490	175,600	0.002804	138,500	0.002790	116,200	0.002697
4,000	217 (B)	0.1358	2.309	3.142	12,590	66,490	178,100	0.002804	139,500	0.002790	116,200	0.002697
3,500	127 (A)	0.1660	2.158	2.749	11,020	58,180	153,400	0.003205	120,400	0.003188	101,700	0.003082
3,500	169 (B)	0.1439	2.159	2.749	11,020	58,180	155,900	0.003205	122,000	0.003188	101,700	0.003082
3,000	127 (A)	0.1537	1.998	2.356	9,353	49,390	131,700	0.003703	103,900	0.003684	87,180	0.003561
3,000	169 (B)	0.1332	1.998	2.356	9,353	49,390	134,400	0.003703	104,600	0.003684	87,180	0.003561
2,500	91 (A)	0.1657	1.823	1.963	7,794	41,150	109,600	0.004444	85,990	0.004421	72,650	0.004273
2,500	127 (B)	0.1403	1.824	1.963	7,794	41,150	111,300	0.004444	87,170	0.004421	72,650	0.004273
2,000	91 (A)	0.1482	1.630	1.571	6,175	32,600	87,790	0.005501	69,270	0.005472	58,120	0.005289
2,000	127 (B)	0.1255	1.632	1.571	6,175	32,600	90,050	0.005501	70,210	0.005472	58,120	0.005289
1,750	91 (A)	0.1387	1.526	1.374	5,403	28,530	77,930	0.006286	61,020	0.006254	50,850	0.006045
1,750	127 (B)	0.1174	1.526	1.374	5,403	28,530	78,800	0.006286	61,430	0.006254	50,850	0.006045
1,500	61 (A)	0.1568	1.411	1.178	4,631	24,450	65,840	0.007334	51,950	0.007296	43,590	0.007052
1,500	91 (B)	0.1284	1.412	1.178	4,631	24,450	67,540	0.007334	52,650	0.007296	43,590	0.007052
1,250	61 (A)	0.1431	1.288	0.9817	3,859	20,380	55,670	0.008801	43,590	0.008755	36,320	0.008463
1,250	91 (B)	0.1172	1.289	0.9817	3,859	20,380	56,280	0.008801	43,880	0.008755	36,320	0.008463

Con-ductor size, kCMIL or AWG	No. of wires (ASTM stranding class)	Wire diameter, in	Conductor diameter, in	Conductor area, in^2	Conductor weight, lb		Hard		Medium		Soft	
					Per 1,000 ft	Per mile	Breaking strength, minimum,* lb	DC resistance at 20°C (68°F), ohms per 1,000 ft	Breaking strength, minimum,* lb	DC resistance at 20°C (68°F), ohms per 1,000 ft	Breaking strength, maximum,† lb	DC resistance at 20°C (68°F), ohms per 1,000 ft
1,000	37 (AA)	0.1644	1.151	0.7854	3,088	16,300	43,830	0.01100	34,400	0.01094	29,060	0.01058
1,000	61 (A–B)	0.1280	1.152	0.7854	3,088	16,300	45,030	0.01100	35,100	0.01094	29,060	0.01058
900	37 (AA)	0.1560	1.092	0.7069	2,779	14,670	39,510	0.01222	31,170	0.01216	26,150	0.01175
900	61 (A–B)	0.1215	1.094	0.7069	2,779	14,670	40,520	0.01222	31,590	0.01216	26,150	0.01175
850	37 (AA)	0.1516	1.061	0.6676	2,624	13,860	37,310	0.01294	29,440	0.01288	24,700	0.01245
850	61 (A–B)	0.1180	1.062	0.6676	2,624	13,860	38,270	0.01294	29,840	0.01288	24,700	0.01245
800	37 (AA)	0.1470	1.029	0.6283	2,470	13,040	35,120	0.01375	27,710	0.01368	23,250	0.01322
800	61 (A–B)	0.1145	1.031	0.6283	2,470	13,040	36,360	0.01375	28,270	0.01368	23,250	0.01322
750	37 (AA)	0.1424	0.997	0.5890	2,316	12,230	33,400	0.01467	26,150	0.01459	21,790	0.01410
750	61 (A–B)	0.1109	0.998	0.5890	2,316	12,230	34,090	0.01467	26,510	0.01459	21,790	0.01410
700	37 (AA)	0.1375	0.963	0.5498	2,161	11,410	31,170	0.01572	24,410	0.01563	20,340	0.01511
700	61 (A–B)	0.1071	0.964	0.5498	2,161	11,410	31,820	0.01572	24,740	0.01563	20,340	0.01511
650	37 (AA)	0.1325	0.928	0.5105	2,007	10,600	29,130	0.01692	22,670	0.01684	18,890	0.01627
650	61 (A–B)	0.1032	0.929	0.5105	2,007	10,600	29,770	0.01692	22,970	0.01684	18,890	0.01627
600	37 (AA–A)	0.1273	0.891	0.4712	1,853	9,781	27,020	0.01834	21,060	0.01824	17,440	0.01763
600	61 (B)	0.0992	0.893	0.4712	1,853	9,781	27,530	0.01834	21,350	0.01824	18,140	0.01763
550	37 (AA-A)	0.1219	0.853	0.4320	1,698	8,966	24,760	0.02000	19,310	0.01990	15,980	0.01923
550	61 (B)	0.0950	0.855	0.4320	1,698	8,966	25,230	0.02000	19,570	0.01990	16,630	0.01923
500	19 (AA)	0.1622	0.811	0.3927	1,544	8,151	21,950	0.02200	17,320	0.02189	14,530	0.02116
500	37 (A–B)	0.1162	0.813	0.3927	1,544	8,151	22,510	0.02200	17,550	0.02189	14,530	0.02116

Note: See p. 14.14 for footnotes.

TABLE 14.5 Copper Cable, Classes AA, A, B—Weight, Breaking Strength, DC Resistance (*Continued*)
(*Based on ASTM Specifications B1, B2, B3, and B8*)

Con-ductor size, kCMIL or AWG	No. of wires (ASTM stranding class)	Wire diameter, in	Conductor diameter, in	Conductor area, in²	Conductor weight, lb		Hard		Medium		Soft	
					Per 1,000 ft	Per mile	Breaking strength, minimum,* lb	DC resistance at 20°C (68°F), ohms per 1,000 ft	Breaking strength, minimum,* lb	DC resistance at 20°C (68°F), ohms per 1,000 ft	Breaking strength, maximum,† lb	DC resistance at 20°C (68°F), ohms per 1,000 ft
450	19 (AA)	0.1539	0.770	0.3534	1,389	7,336	19,750	0.02445	15,590	0.02432	13,080	0.02351
450	37 (A–B)	0.1103	0.772	0.3534	1,389	7,336	20,450	0.02445	15,900	0.02432	13,080	0.02351
400	19 (AA–A)	0.1451	0.726	0.3142	1,235	6,521	17,810	0.02750	13,950	0.02736	11,620	0.02645
400	37 (B)	0.1040	0.728	0.3142	1,235	6,521	18,320	0.02750	14,140	0.02736	11,620	0.02645
350	12 (AA)	0.1708	0.710	0.2749	1,081	5,706	15,140	0.03143	12,040	0.03127	10,170	0.03022
350	19 (A)	0.1357	0.679	0.2749	1,081	5,706	15,590	0.03143	12,200	0.03127	10,170	0.03022
350	37 (B)	0.0973	0.681	0.2749	1,081	5,706	16,060	0.03143	12,450	0.03127	10,580	0.03022
300	12 (AA)	0.1581	0.657	0.2356	926.3	4,891	13,170	0.03667	10,390	0.03648	8,718	0.03526
300	19 (A)	0.1257	0.629	0.2356	926.3	4,891	13,510	0.03667	10,530	0.03648	8,718	0.03526
300	37 (B)	0.0900	0.630	0.2356	926.3	4,891	13,870	0.03667	10,740	0.03648	9,071	0.03526
250	12 (AA)	0.1443	0.600	0.1963	771.9	4,076	11,130	0.04400	8,717	0.04378	7,265	0.04231
250	19 (A)	0.1147	0.574	0.1963	771.9	4,076	11,360	0.04400	8,836	0.04378	7,265	0.04231
250	37 (B)	0.0822	0.575	0.1963	771.9	4,076	11,560	0.04400	8,952	0.04378	7,559	0.04231
4/0	7 (AA–A)	0.1739	0.522	0.1662	653.3	3,450	9,154	0.05199	7,278	0.05172	6,149	0.04999
4/0	12—	0.1328	0.522	0.1662	653.3	3,450	9,483	0.05199	7,378	0.05172	6,149	0.04999
4/0	19 (B)	0.1055	0.528	0.1662	653.3	3,450	9,617	0.05199	7,479	0.05172	6,149	0.04999
3/0	7 (AA–A)	0.1548	0.464	0.1318	518.1	2,736	7,366	0.06556	5,812	0.06522	4,876	0.06304
3/0	12—	0.1183	0.492	0.1318	518.1	2,736	7,556	0.06556	5,890	0.06522	4,876	0.06304
3/0	19 (B)	0.0940	0.470	0.1318	518.1	2,736	7,698	0.06556	5,970	0.06522	5,074	0.06304
2/0	7 (AA–A)	0.1379	0.414	0.1045	410.9	2,169	5,926	0.08267	4,640	0.08224	3,867	0.07949

Conductor size, kCMIL or AWG	No. of wires (ASTM stranding class)	Wire diameter, in	Conductor diameter, in	Conductor area, in^2	Conductor weight, lb		Hard		Medium		Soft	
					Per 1,000 ft	Per mile	Breaking strength, minimum,* lb	DC resistance at 20°C (68°F), ohms per 1,000 ft	Breaking strength, minimum,* lb	DC resistance at 20°C (68°F), ohms per 1,000 ft	Breaking strength, maximum,† lb	DC resistance at 20°C (68°F), ohms per 1,000 ft
2/0	12—	0.1053	0.438	0.1045	410.9	2,169	6,048	0.08267	4,703	0.08224	3,867	0.07949
2/0	19 (B)	0.0837	0.419	0.1045	410.9	2,169	6,152	0.08267	4,765	0.08224	4,024	0.07949
1/0	7 (AA–A)	0.1228	0.368	0.08289	325.8	1,720	4,752	0.1042	3,705	0.1037	3,067	0.1002
1/0	12—	0.0938	0.390	0.08289	325.8	1,720	4,841	0.1042	3,755	0.1037	3,191	0.1002
1/0	19 (B)	0.0745	0.373	0.08289	325.8	1,720	4,901	0.1042	3,805	0.1037	3,191	0.1002
1	3 (AA)	0.1670	0.360	0.06573	255.9	1,351	3,621	0.1302	2,879	0.1295	2,432	0.1252
1	7 (A)	0.1093	0.328	0.06573	258.4	1,364	3,804	0.1314	2,958	0.1308	2,432	0.1264
1	19 (B)	0.0664	0.332	0.06573	258.4	1,364	3,899	0.1314	3,037	0.1308	2,531	0.1264
2	3 (AA)	0.1487	0.320	0.05213	202.9	1,071	2,913	0.1641	2,299	0.1633	1,929	0.1578
2	7 (A–B)	0.0974	0.292	0.05213	204.9	1,082	3,045	0.1657	2,361	0.1649	2,007	0.1594
3	3 (AA)	0.1325	0.285	0.04134	160.9	849.6	2,359	0.2070	1,835	0.2059	1,530	0.1990
3	7 (A–B)	0.0867	0.260	0.04134	162.5	858.0	2,433	0.2090	1,885	0.2079	1,592	0.2010
4	3 (AA)	0.1180	0.254	0.03278	127.6	673.8	1,879	0.2610	1,465	0.2596	1,213	0.2509
4	7 (A–B)	0.0772	0.232	0.03278	128.9	680.5	1,938	0.2636	1,505	0.2622	1,262	0.2534
5	7 (B)	0.0688	0.206	0.02600	102.2	539.6	1,542	0.3323	1,201	0.3306	1,001	0.3196
6	7 (B)	0.0612	0.184	0.02062	81.05	427.9	1,288	0.4191	958.6	0.4169	793.8	0.4030
7	7 (B)	0.0545	0.164	0.01635	64.28	339.4	977.1	0.5284	765.2	0.5257	629.5	0.5081
8	7 (B)	0.0486	0.146	0.01297	50.97	269.1	777.2	0.6663	610.7	0.6629	499.2	0.6408
9	7 (B)	0.0432	0.130	0.01028	40.42	213.4	618.2	0.8402	487.4	0.8359	395.9	0.8080
10	7 (B)	0.0385	0.116	0.008155	32.06	169.3	491.7	1.060	388.9	1.054	314.0	1.019

Note: see p. 14.14 for footnotes.

TABLE 14.5 Copper Cable, Classes AA, A, B—Weight, Breaking Strength, DC Resistance (*Continued*)
(*Based on ASTM Specifications B1, B2, B3, and B8*)

					Conductor weight, lb		Hard		Medium		Soft	
Conductor size, kCMIL or AWG	No. of wires (ASTM stranding class)	Wire diameter, in	Conductor diameter, in	Conductor area, in²	Per 1,000 ft	Per mile	Breaking strength, minimum,* lb	DC resistance at 20°C (68°F), ohms per 1,000 ft	Breaking strength, minimum,* lb	DC resistance at 20°C (68°F), ohms per 1,000 ft	Breaking strength, maximum,† lb	DC resistance at 20°C (68°F), ohms per 1,000 ft
12	7 (B)	0.0305	0.0915	0.005129	20.16	106.5	311.1	1.685	247.7	1.676	197.5	1.620
14	7 (B)	0.0242	0.0726	0.003225	12.68	66.95	197.1	2.679	157.7	2.665	124.2	2.576
16	7 (B)	0.0192	0.0576	0.002028	7.974	42.10	124.7	4.259	100.4	4.237	81.14	4.096
18	7 (B)	0.0152	0.0456	0.001276	5.015	26.48	78.99	6.773	63.91	6.738	51.03	6.513
20	7 (B)	0.0121	0.0363	0.0008023	3.154	16.65	50.04	10.77	40.67	10.71	32.09	10.36
ASTM designation	B8						B1 & B8		B2 & B8		B3 & B8	

*No. 10 AWG and smaller, based on Anaconda data.
†No requirements for tensile strength are specified in ASTM B3. Values given here based on Anaconda data.

Weight and Resistance

Stranding class	Conductor size, kCMIL or AWG	Increment of resistance and weight, %
AA	4−1	1
	1/0−1000	2
A, B, C, D	200 and under	2
	Over 2000 − 3000	3
	Over 3000 − 4000	4
	Over 4000 − 5000	5

Source: *Standard Handbook for Electrical Engineers*, Donald G. Fink, Editor-in-Chief. H. Wayne Beaty, Associate Editor. Copyright 1978 by McGraw-Hill, Inc. Used with permission.

Resistance
(ASTM requirements)

Temper	Conductivity at 20°C (68°F), IACS, %	Resistivity at 20°C (68°F), ohms (mile, lb)
Hard	96.16	910.15
Medium	96.66	905.44
Soft	100	875.20

The resistance in this table are trade maximums and are higher than the average values for commercial cable.

TABLE 14.6 Copper-Clad Steel-Copper Cable—Weight, Breaking Strength, DC Resistance (*Based on ASTM Specification B229*)

Hard-drawn copper equivalent,* kCMIL or AWG	Conductor type	Conductor stranding: EHS 30% copper-clad wires, No.	Conductor stranding: EHS 30% copper-clad wires, Diam., in	Conductor stranding: Hard-drawn copper wires, No.	Conductor stranding: Hard-drawn copper wires, Diam., in	Conductor diam., in	Conductor area, in^2	Conductor weight, lb: Per 1,000 ft	Conductor weight, lb: Per mile	Breaking strength min., lb	DC resistance at 20°C (68°F). ohms/1,000 ft
350	E	7	0.1576	12	0.1576	0.788	0.3706	1,403	7,409	32,420	0.03143
350	EK	4	0.1470	15	0.1470	0.735	0.3225	1,238	6,536	23,850	0.03143
300	E	7	0.1459	12	0.1459	0.729	0.3177	1,203	6,351	27,770	0.03667
300	EK	4	0.1361	15	0.1361	0.680	0.2764	1,061	5,602	20,960	0.03667
250	E	7	0.1332	12	0.1332	0.666	0.2648	1,002	5,292	23,920	0.04400
250	EK	4	0.1242	15	0.1242	0.621	0.2302	884.2	4,669	17,840	0.04400
4/0	E	7	0.1225	12	0.1225	0.613	0.2239	848.3	4,479	20,730	0.05199
4/0	EK	4	0.1143	15	0.1143	0.571	0.1905	748.4	3,951	15,370	0.05199
4/0	F	1	0.1833	6	0.1833	0.550	0.1847	710.2	3,750	12,290	0.05199
3/0	E	7	0.1091	12	0.1091	0.545	0.1776	672.7	3,552	16,800	0.06556
3/0	EK	4	0.1018	15	0.1018	0.509	0.1546	593.5	3,134	12,370	0.06556
3/0	F	1	0.1632	6	0.1632	0.490	0.1464	563.2	2,974	9,980	0.06556
2/0	F	1	0.1454	6	0.1454	0.436	0.1162	446.8	2,359	8,094	0.08265
1/0	F	1	0.1294	6	0.1294	0.388	0.09206	354.1	1,870	6,536	0.1043
1	F	1	0.1153	6	0.1153	0.346	0.07309	280.9	1,483	5,266	0.1315
2†	A	1	0.1699	2	0.1699	0.366	0.06801	256.8	1,356	5,876	0.1658
2	F	1	0.1026	6	0.1026	0.308	0.05787	222.8	1,176	4,233	0.1658
4†	A	1	0.1347	2	0.1347	0.290	0.04275	161.5	852	3,938	0.2636
6†	A	1	0.1068	2	0.1068	0.230	0.02688	101.6	536.3	2,585	0.4150
8†	A	1	0.1127	2	0.07969	0.199	0.01995	74.27	392.2	2,233	0.6598

*Area of hard-drawn copper cable having the same dc resistance as that of the composite cable.
†Sizes commonly used for rural distribution.
Source: *Standard Handbook for Electrical Engineers*, Donald G. Fink, Editor-in-Chief, H. Wayne Beaty, Associate Editor. Copyright 1978 by McGraw-Hill, Inc. Used with permission.

Conductor Selection. The size of a line conductor depends on a number of factors. The most important are the line voltage, the amount of power to be transmitted, and mechanical strength. The line voltage is a very important factor, for it will be recalled that for a given amount of power the greater the voltage, the less the current. That is the reason for using as high a line voltage as possible. The current in turn determines the size of the conductor.

Equally important is the magnitude of the load to be transmitted. For a given voltage, the larger the load, the larger the current. That a conductor must also have sufficient mechanical strength to carry its own weight and any load due to sleet or wind is obvious if the line is to give reliable and uninterrupted service. Other factors are length of line, power factor of the load, length of span, etc. Table 14.10 gives the current-carrying capacities of copper and aluminum wires for the wire sizes used frequently in outdoor circuits. These values of current were obtained by experiment and are of such magnitude that the heat produced by them in the

TABLE 14.7 Aluminum Conductor—Physical Characteristics EC-H19, Classes AA and A

Cable code word	Conductor size		Current-carrying capacity,* amps	Stranding		Conductor diam., in	Rated strength, lb	Nominal weight, lb†	
	Cir mils or AWG	in²		Class	No. × diam. of wires, in			Per 1,000 ft	Per mile
Peachbell	6	0.0206	95	A	7 × 0.0612	0.184	563	24.6	130
Rose	4	0.0328	130	A	7 × 0.0772	0.232	881	39.2	207
Iris	2	0.0522	175	AA, A	7 × 0.0974	0.292	1,350	62.3	329
Pansy	1	0.0657	200	AA, A	7 × 0.1093	0.328	1,640	78.5	414
Poppy	1/0	0.0829	235	AA, A	7 × 0.1228	0.368	1,990	99.1	523
Aster	2/0	0.1045	270	AA, A	7 × 0.1379	0.414	2,510	124.9	659
Phlox	3/0	0.1317	315	AA, A	7 × 0.1548	0.464	3,040	157.5	832
Oxlip	4/0	0.1663	365	AA, A	7 × 0.1739	0.522	3,830	198.7	1,049
Sneezewort	250,000	0.1964	405	AA	7 × 0.1890	0.567	4,520	234.7	1,239
Valerian	250,000	0.1963	405	A	19 × 0.1147	0.574	4,660	234.6	1,239
Daisy	266,800	0.2097	420	AA	7 × 0.1953	0.586	4,830	250.6	1,323
Laurel	266,800	0.2095	425	A	19 × 0.1185	0.593	4,970	250.4	1,322
Peony	300,000	0.2358	455	A	19 × 0.1257	0.629	5,480	281.8	1,488
Tulip	336,400	0.2644	495	A	19 × 0.1331	0.666	6,150	316.0	1,668
Daffodil	350,000	0.2748	506	A	19 × 0.1357	0.679	6,390	328.4	1,734
Canna	397,500	0.3124	550	AA, A	19 × 0.1447	0.724	7,110	373.4	1,972
Goldentuft	450,000	0.3534	545	AA	19 × 0.1539	0.770	7,890	422.4	2,230
Cosmos	477,000	0.3744	615	AA	19 × 0.1584	0.793	8,360	447.5	2,363
Syringa	477,000	0.3743	615	A	37 × 0.1135	0.795	8,690	447.4	2,362
Zinnia	500,000	0.3926	635	AA	19 × 0.1622	0.811	8,760	469.2	2,477
Hyacinth	500,000	0.3924	635	A	37 × 0.1162	0.813	9,110	469.0	2,476
Dahlia	556,500	0.4369	680	AA	19 × 0.1711	0.856	9,750	522.1	2,757
Mistletoe	556,500	0.4368	680	AA, A	37 × 0.1226	0.858	9,940	522.0	2,756
Meadowsweet	600,000	0.4709	715	AA, A	37 × 0.1273	0.891	10,700	562.8	2,972
Orchid	636,000	0.4995	745	AA, A	37 × 0.1311	0.918	11,400	596.9	3,152
Heuchera	650,000	0.5102	755	AA	37 × 0.1325	0.928	11,600	609.8	3,220

TABLE 14.7 Aluminum Conductor—Physical Characteristics EC-H19, Classes AA and A *(Continued)*

Cable code word	Conductor size		Current-carrying capacity,* amps	Stranding		Conductor diam., in	Rated strength, lb	Nominal weight, lb†	
	Cir mils or AWG	in^2		Class	No. × diam. of wires, in			Per 1,000 ft	Per mile
Verbena	700,000	0.5494	790	AA	37 × 0.1375	0.963	12,500	656.6	3,467
Flag	700,000	0.5495	790	A	61 × 0.1071	0.964	12,900	656.8	3,468
Violet	715,500	0.5622	800	AA	37 × 0.1391	0.974	12,800	672.0	3,548
Nasturtium	715,500	0.5619	800	A	61 × 0.1083	0.975	13,100	671.6	3,546
Petunia	750,000	0.5892	825	AA	37 × 0.1424	0.997	13,100	704.3	3,719
Cattail	750,000	0.5892	825	A	61 × 0.1109	0.998	13,500	704.2	3,718
Arbutus	795,000	0.6245	855	AA	37 × 0.1466	1.026	13,900	746.4	3,941
Lilac	795,000	0.6248	855	A	61 × 0.1142	1.028	14,300	746.7	3,943
Cockscomb	900,000	0.7072	925	AA	37 × 0.1560	1.092	15,400	845.2	4,463
Snapdragon	900,000	0.7072	925	A	61 × 0.1215	1.094	15,900	845.3	4,463
Magnolia	954,000	0.7495	960	AA	37 × 0.1606	1.124	16,400	895.8	4,730
Goldenrod	954,000	0.7498	960	A	61 × 0.1251	1.126	16,900	896.1	4,731
Hawkweed	1,000,000	0.7854	990	AA	37 × 0.1644	1.151	17,200	938.7	4,956
Camellia	1,000,000	0.7849	990	A	61 × 0.1280	1.152	17,700	938.2	4,954
Bluebell	1,033,500	0.8124	1015	AA	37 × 0.1672	1.170	17,700	970.9	5,126
Larkspur	1,033,500	0.8122	1015	A	61 × 0.1302	1.172	18,300	970.6	5,125
Marigold	1,113,000	0.8744	1040	AA, A	61 × 0.1351	1.216	19,700	1,045	5,518
Hawthorn	1,192,500	0.9363	1085	AA, A	61 × 0.1398	1.258	21,100	1,119	5,908
Narcissuus	1,272,000	0.999	1130	AA, A	61 × 0.1444	1.300	22,000	1,194	6,304
Columbine	1,351,500	1.062	1175	AA, A	61 × 0.1489	1.340	23,400	1,269	6,700
Carnation	1,431,000	1.124	1220	AA, A	61 × 0.1532	1.379	24,300	1,344	7,096
Gladiolus	1,510,500	1.187	1265	AA, A	61 × 0.1574	1.417	25,600	1,419	7,492
Coreopsis	1,590,000	1.250	1305	AA	61 × 0.1615	1.454	27,000	1,493	7,883
Jessamine	1,750,000	1.375	1385	AA	61 × 0.1694	1.525	29,700	1,643	8,675
Cowslip	2,000,000	1.570	1500	A	91 × 0.1482	1.630	34,200	1,876	9,911
Sagebrush	2,250,000	1.766	1600	A	91 × 0.1572	1.729	37,700	2,132	11,257
Lupine	2,500,000	1.962	1700	A	91 × 0.1657	1.823	41,800	2,368	12,503

Note: See p. 14.18 for footnotes.

Cable code word	Conductor size		Current-carrying capacity,* amps	Stranding		Conductor diam., in	Rated strength, lb	Nominal weight, lb†	
	Cir mils or AWG	in^2		Class	No. × diam. of wires, in			Per 1,000 ft	Per mile
Bitterroot	2,750,000	2.159	1795	A	91 × 0.1738	1.912	46,100	2,606	13,760
Trillium	3,000,000	2.356	1885	A	127 × 0.1537	1.996	50,300	2,844	15,016
Bluebonnet	3,500,000	2.749	2035	A	127 × 0.1660	2.158	58,700	3,350	17,688

Class of stranding. The class of stranding must be specified on all orders. Class AA stranding is usually specified for bare conductors used on overhead lines. Class A stranding is usually specified for conductors to be covered with weather-resistant (weatherproof) materials and for bare conductors where greater flexibility than afforded by Class AA is required.

Lay. The direction of lay of the outside layer of wires with Class AA and Class A stranding will be right hand unless otherwise specified.

*Ampacity for conductor temperature rise of 40°C over 40°C ambient with a 2 ft/s crosswind and an emissivity factor of 0.5 without sun.

†Nominal conductor weights are based on ASTM standard stranding increments. Actual weights will vary with lay lengths. Invoicing will be based on actual weights.

Source: *Standard Handbook for Electrical Engineers*, Donald G. Fink, Editor-in-Chief, H. Wayne Beaty, Associate Editor. Copyright 1978 by McGraw-Hill, Inc. Used with permission.

TABLE 14.8 ACSR Conductor—Physical Characteristics

ACSR													Rated strength, lb			
	Cross section								Nominal weight, lb†			Zinc-coated core				
	Aluminum			Current-carrying capacity,* amps	Stranding no. and diam of strand, in		Diameter, in		Per 1,000 ft							
Code word	cmils or AWG	in²	Total, in²		Aluminum	Steel	Complete cond.	Steel core	Total	Al	Steel	Standard weight coating	Class B coating	Class C coating	Aluminum-coated core
Turkey	6	0.0206	0.0240	95	6 × 0.0661	1 × 0.0661	0.198	0.0661	36.1	24.5	11.6	1,190	1,160	1,120	1,120
Swan	4	0.0328	0.0382	130	6 × 0.0834	1 × 0.0834	0.250	0.0834	57.4	39.0	18.4	1,860	1,810	1,760	1,760
Swanatc	4	0.0328	0.0411	130	7 × 0.0772	1 × 0.1029	0.257	0.1029	67.0	39.0	28.0	2,360	2,280	2,200	2,160
Sparrow	2	0.0522	0.0608	175	6 × 0.1052	1 × 0.1052	0.316	0.1052	91.3	62.0	29.3	2,850	2,760	2,680	2,640
Sparatc	2	0.0522	0.0654	175	7 × 0.0974	1 × 0.1299	0.325	0.1299	106.7	62.0	44.7	3,640	3,510	3,390	3,260
Robin	1	0.0657	0.0767	200	6 × 0.1181	1 × 0.1181	0.355	0.1182	115.1	78.2	36.9	3,550	3,450	3,340	3,290
Raven	1/0	0.0830	0.0968	230	6 × 0.1327	1 × 0.1327	0.398	0.1327	145.3	98.7	46.6	4,380	4,250	4,120	3,980
Quail	2/0	0.1046	0.1221	265	6 × 0.1490	1 × 0.1490	0.447	0.1490	183.2	124.4	58.8	5,310	5,130	5,050	4,720
Pigeon	3/0	0.1317	0.1537	310	6 × 0.1672	1 × 0.1672	0.502	0.1672	230.8	156.7	74.1	6,620	6,410	6,300	5,880
Penguin	4/0	0.1662	0.1939	350	6 × 0.1878	1 × 0.1878	0.563	0.1878	291.1	197.7	93.4	8,350	8,080	7,950	7,420
Waxwing	266,800	0.2094	0.2210	430	18 × 0.1217	1 × 0.1217	0.609	0.1217	289.5	250.3	39.2	6,880	6,770	6,650	6,540
Owl	266,800	0.2096	0.2368	410	6 × 0.2109	7 × 0.0703	0.633	0.2109	342.4	250.5	91.9	9,680	9,420	9,160	9,160
Partridge	266,800	0.2095	0.2436	440	26 × 0.1013	7 × 0.0788	0.642	0.2364	367.3	251.7	115.6	11,300	11,000	10,600	10,640
Merlin	336,400	0.2642	0.2789	500	18 × 0.1367	1 × 0.1367	0.684	0.1367	365.2	315.7	49.5	8,680	8,540	8,400	8,260
Linnet	336,400	0.2640	0.3070	510	26 × 0.1137	7 × 0.0884	0.720	0.2652	462.5	317.0	145.5	14,100	13,700	13,300	13,300
Oriole	336,400	0.2642	0.3259	515	30 × 0.1059	7 × 0.1059	0.741	0.3177	527.1	318.1	209.0	17,300	16,700	16,200	15,900
Chickadee	397,500	0.3121	0.3295	555	18 × 0.1486	1 × 0.1486	0.743	0.1486	431.6	373.1	58.5	9,940	9,780	9,690	9,530
Brant	397,500	0.3122	0.3527	565	24 × 0.1287	7 × 0.0858	0.772	0.2574	512.1	375.0	137.1	14,600	14,300	13,900	13,900
Ibis	397,500	0.3119	0.3627	570	26 × 0.1236	7 × 0.0961	0.783	0.2883	546.6	374.7	171.9	16,300	15,800	15,300	15,100
Lark	397,500	0.3121	0.3849	575	30 × 0.1151	7 × 0.1151	0.806	0.3453	622.7	375.8	246.9	20,300	19,600	18,900	18,600
Pelican	477,000	0.3747	0.3955	625	18 × 0.1628	1 × 0.1628	0.814	0.1628	518.0	447.8	70.2	11,800	11,600	11,500	11,100
Flicker	477,000	0.3747	0.4233	635	24 × 0.1410	1 × 0.1628	0.846	0.2820	614.6	450.1	164.5	17,200	16,700	16,200	16,000
Hawk	477,000	0.3744	0.4354	640	26 × 0.1354	7 × 0.1053	0.858	0.3159	656.0	449.6	206.4	19,500	18,900	18,400	18,100
Hen	477,000	0.3747	0.4621	645	30 × 0.1261	7 × 0.1261	0.883	0.3783	747.4	451.1	296.3	23,800	23,000	22,100	21,300
Osprey	556,500	0.4369	0.4612	690	18 × 0.1758	1 × 0.1758	0.879	0.1758	604.1	522.2	81.9	13,700	13,500	13,400	12,900
Parakeet	556,500	0.4372	0.4938	700	24 × 0.1523	7 × 0.1015	0.914	0.3045	716.9	525.1	191.8	19,800	19,300	18,700	18,500
Dove	556,500	0.4371	0.5083	710	26 × 0.1463	7 × 0.1138	0.927	0.3414	766.0	524.9	241.1	22,600	21,900	21,200	20,900

ACSR													Rated strength, lb			
	Cross section								Nominal weight, lb†			Zinc-coated core				
	Aluminum			Current-carrying capacity,*	Stranding no. and diam of strand, in		Diameter, in		Per 1,000 ft							
Code word	cmils or AWG	in²	Total, in²	amps	Aluminum	Steel	Complete cond.	Steel core	Total	Al	Steel	Standard weight coating	Class B coating	Class C coating	Aluminum-coated core	
Eagle	556,500	0.4371	0.5391	710	30 × 0.1362	7 × 0.1362	0.953	0.4086	871.9	526.2	345.7	27,800	26,800	25,800	24,800	
Peacock	605,000	0.4753	0.5370	740	24 × 0.1588	7 × 0.1059	0.953	0.318	779.7	570.9	208.8	21,600	21,000	20,400	20,100	
Squab	605,000	0.4749	0.5522	745	26 × 0.1525	7 × 0.1186	0.966	0.356	832.3	570.4	261.9	24,300	23,600	22,800	22,500	
Teal	605,000	0.4751	0.5834	750	30 × 0.1420	19 × 0.0852	0.994	0.426	939.5	572.0	367.5	30,000	29,000	28,000	28,000	
Swift	636,000	0.4994	0.5133	745	36 × 0.1329	1 × 0.1329	0.930	0.1329	643.7	596.9	46.8	13,800	13,600	13,500	13,400	
Kingbird	636,000	0.4997	0.5275	750	18 × 0.1880	1 × 0.1880	0.940	0.1880	690.8	597.2	93.6	15,700	15,400	15,300	14,800	
Rook	636,000	0.4996	0.5643	765	24 × 0.1628	7 × 0.1085	0.977	0.326	819.2	600.0	219.2	22,600	22,000	21,400	21,100	
Grosbeak	636,000	0.4995	0.5808	775	26 × 0.1564	7 × 0.1216	0.990	0.365	875.2	599.9	275.3	25,200	24,400	23,600	22,900	
Egret	636,000	0.4995	0.6135	775	30 × 0.1456	19 × 0.0874	1.019	0.437	988.2	601.4	386.8	31,500	30,500	29,400	29,400	
	653,900	0.5136	0.5321	760	18 × 0.1906	3 × 0.0885	0.953	0.1906	676.2	613.8	62.4	14,800	14,700	14,500	14,500	
Flamingo	666,600	0.5238	0.5917	790	24 × 0.1667	7 × 0.1111	1.000	0.333	858.9	629.1	229.8	23,700	23,100	22,400	22,100	
Gannet	666,600	0.5234	0.6086	795	26 × 0.1601	7 × 0.1245	1.014	0.373	917.3	628.7	288.6	26,400	25,600	24,800	24,000	
Starling	715,500	0.5620	0.6535	835	26 × 0.1659	7 × 0.1290	1.051	0.387	984.8	675.0	309.8	28,400	27,500	26,600	25,700	
Redwing	715,500	0.5617	0.6896	840	30 × 0.1544	19 × 0.0926	1.081	0.463	1,110	676	434	34,600	33,400	32,700	31,600	
Coot	795,000	0.6243	0.6416	860	36 × 0.1486	1 × 0.1486	1.040	0.1486	804.7	746.2	58.5	16,800	16,600	16,500	16,300	
Tern	795,000	0.6242	0.6674	875	45 × 0.1329	7 × 0.0886	1.063	0.266	895.8	749.7	146.1	22,100	21,700	21,200	21,200	
Cuckoo	795,000	0.6244	0.7053	885	24 × 0.1820	7 × 0.1213	1.092	0.364	1,024	750	274	27,900	27,100	26,400	25,600	
Condor	795,000	0.6240	0.7049	885	54 × 0.1213	7 × 0.1213	1.093	0.364	1,024	750	274	28,200	27,400	26,600	25,800	
Drake	795,000	0.6247	0.7264	890	26 × 0.1749	7 × 0.1360	1.108	0.408	1,094	750	344	31,500	30,500	29,600	28,600	
Mallard	795,000	0.6245	0.7669	900	30 × 0.1628	19 × 0.0977	1.140	0.489	1,235	752	483	38,400	37,100	35,800	35,100	
Ruddy	900,000	0.7066	0.7555	945	45 × 0.1414	7 × 0.0943	1.131	0.283	1,015	849	166	24,400	24,000	23,500	23,300	
Canary	900,000	0.7068	0.7984	955	54 × 0.1291	7 × 0.1291	1.162	0.387	1,159	849	310	31,900	31,000	30,200	29,300	

*Ampacity for conductor temperature rise of 40°C over 40°C ambient with a 2 ft/s crosswind and an emissivity factor of 0.5 without sun.

†Nominal conductor weights are based on ASTM standard stranding increments. Actual weights will vary within standard tolerances for wire diameters and lay lengths. Invoicing will be based on actual weights.

Source: *Standard Handbook for Electrical Engineers*, Donald G. Fink, Editor-In-Chief, H. Wayne Beaty, Associate Editor. Copyright 1978 by McGraw-Hill, Inc. Used with permission.

TABLE 14.9 Galvanized Steel Strand—Dimensions, Weight, Breaking Strength
(Based on ASTM Specifications A363 and A475)

Strand diameter, in		Stranding		Strand area, in^2	Strand weight, lb per 1,000 ft	Breaking strength, minimum, lb							
						Utilities grade*							
Nominal	Actual	No. of wires	Diameter of coated wires, in			1	2	3	4	Common	Siemens-Martin	High strength	Extra high strength
1 1/4	1.253	37	0.179	0.9311	3248	.....	.....	.....	.....	44,600	73,000	113,600	162,200
1 1/8	1.127	37	0.161	0.7533	2691	.....	.....	.....	.....	36,000	58,900	91,600	130,800
1	1.001	37	0.143	0.5942	2057	.....	.....	.....	.....	28,300	46,200	71,900	102,700
1	1.000	19	0.200	0.5969	2073	.....	.....	.....	.....	28,700	47,000	73,200	104,500
7/8	0.885	19	0.177	0.4675	1581	.....	.....	.....	.....	21,900	35,900	55,800	79,700
3/4	0.750	19	0.150	0.3358	1155	.....	.....	.....	.....	16,000	26,200	40,800	58,300
5/8	0.625	19	0.125	0.2332	796	.....	.....	.....	.....	11,000	18,100	28,100	40,200
5/8	0.621	7	0.207	0.2356	813	.....	.....	.....	.....	11,600	19,100	29,600	42,400
9/16	0.565	19	0.113	0.1905	637	.....	.....	.....	.....	9,640	16,100	24,100	33,700
9/16	0.564	7	0.188	0.1943	671	.....	.....	.....	.....	9,600	15,700	24,500	35,000
1/2	0.500	19	0.100	0.1492	504	.....	.....	.....	.....	7,620	12,700	19,100	26,700
1/2	0.495	7	0.165	0.1497	517	.....	.....	.....	25,000	7,400	12,100	18,800	26,900
7/16	0.435	7	0.145	0.1156	399	.....	.....	.....	18,000	5,700	9,350	14,500	20,800
3/8	0.360	7	0.120	0.07917	273	.....	.....	.....	11,500	4,250	6,950	10,800	15,400
3/8	0.356	3	0.165	0.06415	220.3	.....	.....	8,500					

TABLE 14.9 Galvanized Steel Strand—Dimensions, Weight, Breaking Strength (*Continued*)
(*Based on ASTM Specifications A363 and A475*)

Strand diameter, in		Stranding				Breaking strength, minimum, lb							
						Utilities grade*							
Nominal	Actual	No. of wires	Diameter of coated wires, in	Strand area, in²	Strand weight, lb per 1,000 ft	1	2	3	4	Common	Siemens-Martin	High strength	Extra high strength
5/16	0.327	7	0.109	0.06532	225	6,000							
5/16	0.312	7	0.104	0.05946	205	.....	.....	.....	.....	3,200	5,350	8,000	11,200
5/16	0.312	3	0.145	0.04954	170.6	.....	.....	6,500					
9/32	0.279	7	0.093	0.04755	164	4,600	.....	.....	.....	2,570	4,250	6,400	8,950
1/4	0.240	7	0.080	0.03519	121	.....	.....	.....	.....	1,900	3,150	4,750	6,650
1/4	0.259	3	0.120	0.03393	116.7	.....	3,150	4,500					
7/32	0.216	7	0.072	0.02850	98.3	.....	.....	.....	.....	1,540	2,560	3,850	5,400
3/16	0.195	7	0.065	0.02323	80.3	2,400							
3/16	0.186	7	0.062	0.02113	72.9	.....	.....	.....	.....	1,150	1,900	2,850	3,990
5/32	0.156	7	0.052	0.01487	51.3	.....	.....	.....	.....	870	1,470	2,140	2,940
1/8	0.123	7	0.041	0.00924	31.8	.....	.....	.....	.....	540	910	1,330	1,830
Elongation in 24 in:						%							
....	......	......	......	....	....	10	8	5	4	10	8	5	4

*Used principally by communication and power and light industries.
Note: Sizes and grades in boldface type are those most commonly used and readily available.
Source: *Standard Handbook for Electrical Engineers*, Donald G. Fink, Editor-In-Chief, H. Wayne Beaty, Associate Editor, Copyright 1978 by McGraw-Hill, Inc. Used with permission.

TABLE 14.10 Current-Carrying Capacities of Copper and Aluminum Conductors When Used Outdoors

Copper		Aluminum	
Cir mils or AWG	Amps	Cir mils or AWG	Amps
		3,364,000	2150
		3,156,000	1580
		1,780,000	1460
1,000,000	1300	1,590,000	1380
900,000	1220	1,431,000	1300
800,000	1130	1,272,000	1210
700,000	1040	1,113,000	1110
600,000	940	954,000	1000
500,000	840	795,000	897
400,000	730	636,000	776
300,000	610	477,000	646
No. 4/0	480	336,400	514
No. 3/0	420	266,800	443
No. 2/0	360	No. 4/0	381
No. 1/0	310	No. 3/0	328
No. 1	270	No. 2/0	283
No. 2	230	No. 1/0	244
No. 3	200	No. 1	209
No. 4	180	No. 2	180

conductor will bring it up to a temperature of 75°C (167°F). The air is assumed to have a temperature of 25°C, and a steady wind of 1.4 miles per hour (2 ft per sec) is blowing over the conductors. The frequency is 60 Hz. If ability to carry current were the only requirement, one could easily pick from this table the right size of conductor to be used for any given current.

While the values of current shown for the different conductors could be carried in emergency operation, they are far too high for normal loading. Normal values would be approximately one-half of the values shown. Furthermore, current-carrying capacity is not the only consideration in determining conductor size. Power loss and voltage regulation are equally important considerations.

If no other factor has to be considered, a good rule to follow for copper conductors is to allow 1000 cir mils per amp. For example, a current of 300 amps would thus require a copper conductor having 300,000 cir mils of area.

The conductor, besides requiring adequate current-carrying capacity, must be of sufficient size and strength to support itself and any additional load due to ice, sleet, and wind. The weight of the conductor for a given span can be easily computed, but the magnitude of the additional load due to sleet and wind cannot be accurately determined. For this purpose the Bureau of Standards of the U. S. government has divided the entire area of the United States into *loading districts*, as shown in Fig. 14.10. The three loading districts are known as heavy, medium, and light.

In the heavy loading district, the resultant loading on the conductor is assumed to be that due to weight of the conductor plus the added weight of a layer of ice with a radial thickness of 1/2 in, combined with a transverse horizontal wind pressure of 4 lb/ft^2 on the projected area of the ice-covered conductor at 0°F.

In the medium loading districts, the loading is assumed to be that caused by 1/4 in of radial thickness of ice combined with a transverse wind pressure of 4 lb/ft^2 at 15°F.

In the light loading districts, the loading assumed is that produced without any ice on the conductor but with a wind pressure of 9 lb/ft^2 at 30°F.

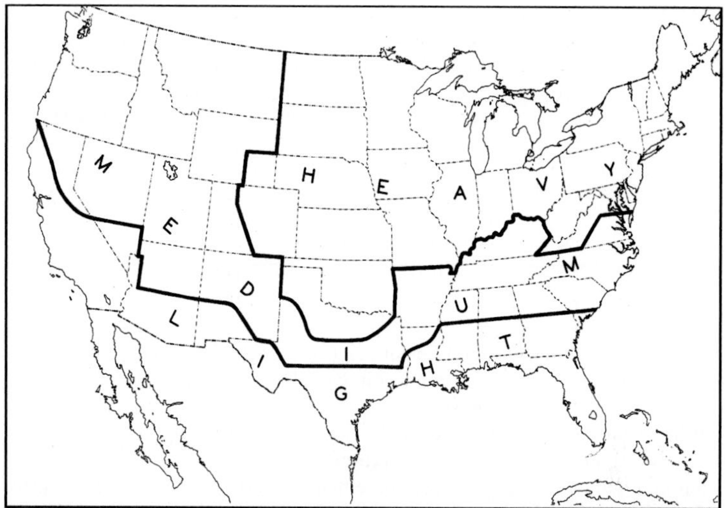

FIGURE 14.10 General loading map showing the territorial division of the United States with respect to loading of overhead lines. (*Taken from National Electrical Safety Code*, 6th ed.)

To these resultant values a constant in pounds per foot as shown in Table 14.11 is added. As the map shows, the light loading district consists of the extreme South and a part of the Southwest; the medium loading district covers a band through the South and the entire West; and the heavy loading district includes the north central and northeastern states.

The following is the statement on conductor loading as specified by the *National Electrical Safety Code*: "The loading on conductors shall be assumed to be the resultant loading per foot equivalent to the vertical load per foot of the conductor, ice-covered where specified, combined with the transverse loading per foot due to a transverse horizontal wind pressure upon the projected area of the conductor, ice-covered where specified, to which equivalent resultant shall be added a constant." Table 14.11 gives the values for ice, wind, temperature, and constants which shall be used to determine the conductor loading.

In addition to the ice load, the side push due to wind on the conductor and ice must be taken into account, as mentioned above. After taking everything that causes tension on the conductor into account, namely, weight of the conductor itself, the weight of the probable ice load, the transverse force due to probable high winds, and the shortening of the conductor and consequent increase of tension due to extreme cold, the engineer multiplies the resultant pull by 2 and selects the conductor that will stand that force before breaking. This gives a factor of safety of 2 under the assumed conditions.

The electrical losses of a transmission line are equal to the current in amperes squared, multiplied by the resistance of the conductors.

$$\text{Losses (watts)} = I^2 \text{ (amps)} \times R \text{ (ohms)}$$

The current required for a given quantity of power decreases as the voltage increases.

$$P \text{ (watts)} = I \text{ (amps)} \times V \text{ (volts)}$$

It is much more effective to reduce the current than to decrease the resistance of the conductors to permit the transmission of energy economically. If the losses are too great, the conductor will overheat and fail mechanically.

TABLE 14.11 Conductor Loading

	Loading district		
	Heavy	Medium	Light
Radial thickness of ice, in	0.50	0.25	0
Horizontal wind pressure, lb/ft^2	4	4	9
Temperature, °F	0	+15	+30
Constant to be added to the resultant for all conductors, lb/ft:	0.30	0.20	0.05

The need to transmit large quantities of electric energy over circuits for long distances has led to the extensive use of extra-high-voltage (EHV) and experimental development of ultra-high-voltage (UHV) transmission lines. The selection of conductors for EHV and UHV transmission lines is influenced by line losses, current-carrying capacity, conductor material costs, and environmental considerations.

Aluminum conductor with steel reinforcement is normally selected for transmission-line conductors because of the good current-carrying capacity and minimum conductor and structure cost. Steel reinforcement of the aluminum conductor does not affect the electrical characteristics of the conductor appreciably. The ACSR conductor will permit higher current densities without loss of the strength since the sag of the conductors is limited by the steel strands as the aluminum expands. Limiting the sag of the aluminum conductor permits the use of longer spans and shorter and fewer supporting structures.

Steel-supported aluminum conductor (SSAC) is constructed like ACSR and is the same in appearance except that the aluminum wires are factory-annealed to provide a very low yield strength. When tension is applied to SSAC, the aluminum elongates, forcing a steel core to carry the conductor load. SSAC can operate continuously at high temperatures without any detriment to mechanical properties. SSAC will sag significantly less at high temperatures than will other types of conductors when maximum tension under ice and wind loading is the same. SSAC has a high capability for the damping of mechanical oscillations, such as aeolian vibration, which is caused by the wind's blowing across the conductors.

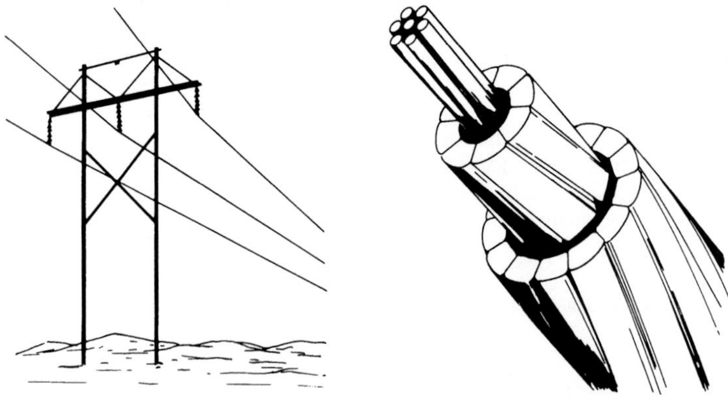

FIGURE 14.11 Construction details of aluminum self-damping conductor. (*Courtesy Alcoa Conductor Products Company Division of Aluminum Company of America.*)

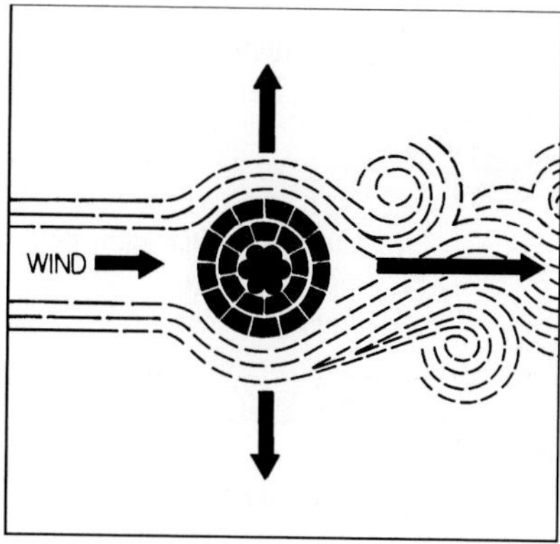

FIGURE 14.12 Diagram of forces on aluminum self-damping conductor as a result of wind. (*Courtesy Alcoa Conductor Products Company Division of Aluminum Company of America.*)

Aluminum self-damping conductor (SDC) provides a means of controlling aeolian vibration (Fig. 14.11). The conductor is manufactured with trapezoid-shaped aluminum strands surrounding a stranded steel core. The spaces between the conductor layers allow the conductor components to move, generating friction and impact forces which absorb energy and counter aeolian vibration (Fig. 14.12). The use of self-damping aluminum conductor permits higher stringing tensions, longer spans, and/or shorter towers. The aluminum self-damping conductor can be installed in accordance with the procedures for ACSR; however, the control of aeolian vibration permits conductor tensions to be increased to the maximum value specified by the *National Electrical Safety Code*.

Environmental considerations involve visual impact and voltage gradient. Steel reinforcement of the conductor results in shorter structures, or fewer structures, or both, thus improving the appearance of the facilities. The voltage gradient is an electrical characteristic determined by the physical size, arrangement, and surface condition of the conductors. The voltage gradient determines corona loss, radio and TV interference, and audible noise level. The voltage gradient should be minimized. This is accomplished by increasing phase-to-phase spacing and the size and/or number of conductors per phase. The increasing cost of installing generation equipment requires the use of larger conductors to reduce the cost of supplying the losses. This has resulted in the selection of conductors with high ampere capacity, lower voltage gradients, and higher supporting structure costs.

The proper conductor selection is a compromise between economics associated with the cost of electric generation, conductor material, and supporting structures; mechanical strength requirements; current-carrying capacity required; and environmental impact.

Vibration Dampers. Line conductors vibrate as a result of aeolian vibration. The vibration has been found to vary from 3 to 150 Hz. Vibration dampers are installed on the conductors to control the vibration and prevent conductor damage. The Stockbridge-type dampers shown in many of the transmission-line pictures throughout the handbook have been used for many years. The elastomeric dampers developed by Alcoa are smaller, lighter, and more efficient than Stockbridge dampers (Figs. 14.13 and 14.14). The elastomeric damper can protect long span lengths from vibration without causing corona on EHV transmission lines.

Conductor Spacers. Transmission lines with bundled conductors, that is, two or more conductors per phase, use spacers to keep the conductors from wrapping together. Alcoa spacer dampers are designed with damping mechanisms capable of reducing vibration by

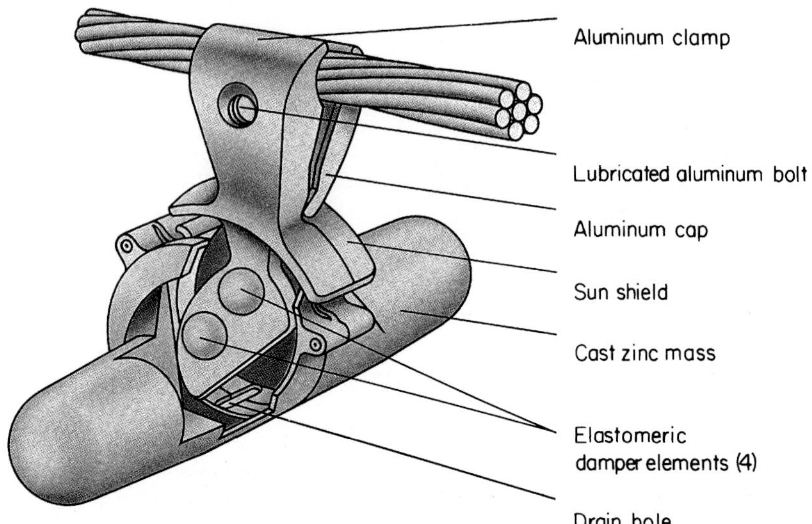

FIGURE 14.13 Cutaway view of the elastomeric damper for use on static or ground wires. (*Courtesy Alcoa Conductor Products Company Division of Aluminum Company of America.*)

converting the mechanical energy of the vibrating conductors into heat in the damping element (Fig. 14.15).

The spacer dampers required to prevent conductor clashing are normally adequate to prevent aeolian vibration.

Conductor Reels. All types and sizes of wire are shipped on reels normally. Limited amounts of small wire can be shipped in coils wrapped with paper. Figure 14.16 illustrates reel construction and size for all-aluminum and ACSR wire.

FIGURE 14.14 Elastomeric damper used on transmission-line phase conductors. (*Courtesy Alcoa Conductor Products Company Division of Aluminum Company of America.*)

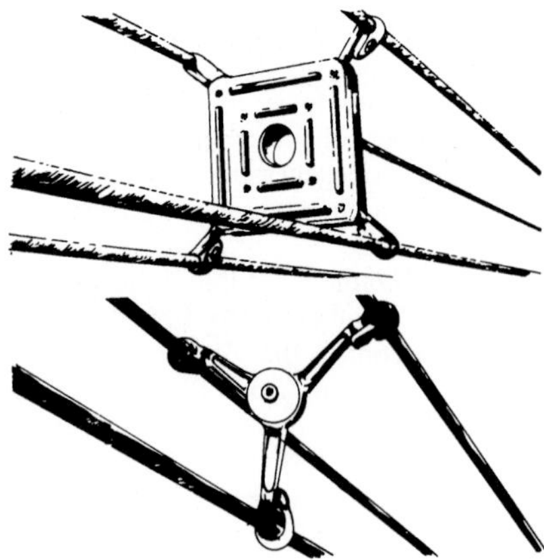

FIGURE 14.15 Transmission-line bundle conductor spacers with vibration damping mechanisms. (*Courtesy Alcoa Conductor Products Company Division of Aluminum Company of America.*)

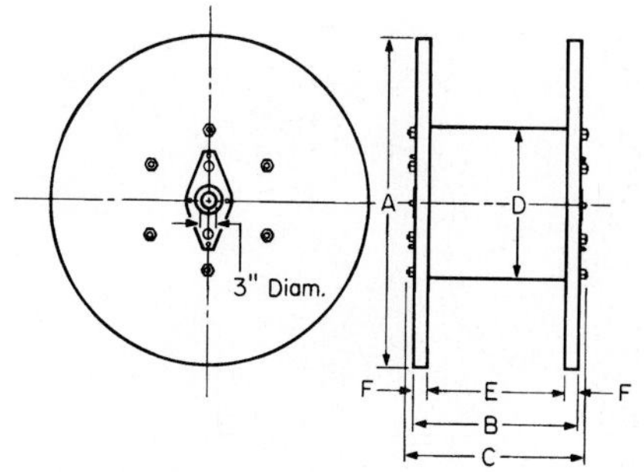

Approximate dimensions, in								
Diameter						Lagging thickness	Approximate weight, lb	
A	B	C	D	E	F		With lags	Without lags
69	43½	46	30	38	2¾	1¾	900	—
69	32½	35	30	27	2¾	1¾	720	—
65	32½	35	30	27	2¾	1¾	650	—
58	32½	35	30	27	2¾	1¾	540	—
52	31½	33	23½	26	2¾	1¾	420	—
46	31½	33	23½	26	2¾	1¾	380	—
44	22½	24½	22½	18	2¼	1¾	205	140
40	22½	24½	22½	18	2¼	1¾	185	125
36	22½	24½	22½	18	2¼	1¾	—	100

FIGURE 14.16 Reel for all-aluminum and ACSR cables.

Fiber-Optic Cables. Overhead ground and phase wires can be fabricated with optical fibers as an integral part of the conductor. Phase conductors on high-voltage lines require the use of an insulator constructed with optical fibers in the core.

Ground or static wire with the optical fibers fabricated in the conductor is the method frequently used to establish a fiber-optic communication system (Fig. 14.17). The optical ground wire has three main components in the completed assembly: the optical fiber cable assembly, an aluminum pipe, and strand members consisting of aluminum-clad, alloy, or galvanized steel wires.

The optical fiber cable assemblies can be classified as either tight buffered or loose buffered. The tight-buffered cable has the fibers stranded in close contact around a relatively stiff central strength member. All expansions and contractions resulting from external forces and temperature changes are directly supported by the central strength member. The central strength member protects the optical fibers from tensile stress and strain and contraction of the fiber coating material. The loose-buffered cable has the fibers placed in a spacer or tube to make the movement of the whole cable and the optical fibers independent of each other. The aluminum pipe around the optical fibers provides protection while the conductor is being strung and keeps the optical fibers dry while the assembly is in service. Table 14.12 provides data for standard ground or static wire and fiber-optic ground wire (OPT-GW).

Aluminum Conductor Composite Core (ACCC) Cable. ACCC conductor (Fig. 14.18) has been introduced to the industry as a solution to address transmission constraint issues.

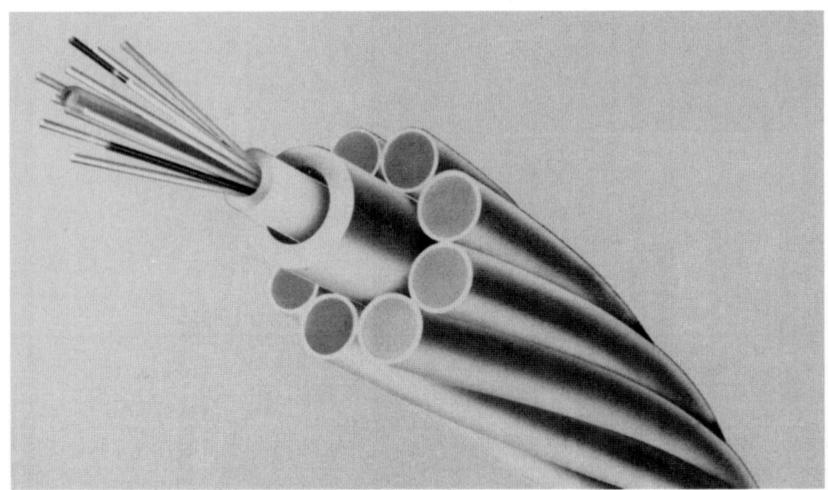

FIGURE 14.17 Alumaweld ground or static wire with fiber-optic cable placed in the core. (*Courtesy Aluminum Company of America.*)

TABLE 14.12 Physical Characteristics of Standard 1/2-in Extra-High-Strength Ground Wire and Equivalent Fiber-Optic Ground Wire

	Standard ground wire reference	OPT-GW equivalent	
		Diameter	Strength
Size, reference	1/2 in EHS	75 mm^2 AW	99 mm^2 AW
Nominal diameter, in	0.495	0.5	0.55
Stranding (no. × diam, in)	7 × 0.165	8 × 0.1358	28 × 0.0835
Aluminum pipe, OD/ID, in	—	$\frac{0.2283}{0.1339}$	$\frac{0.2165}{0.126}$
Area, sq in			
Strand	0.1497	0.1159	0.1533
Pipe		0.026	0.0244
Total	0.1497	0.1419	0.1777
Rated strength, lbf	26,900	20,380	27,310
Weight, lb/1000 ft			
Strand	517	334.42	443.7
Pipe		30.53	28.6
Fiber		10.1	10.1
Total	517	375	482.4
Strength-to-weight ratio	52	54.5	56.6

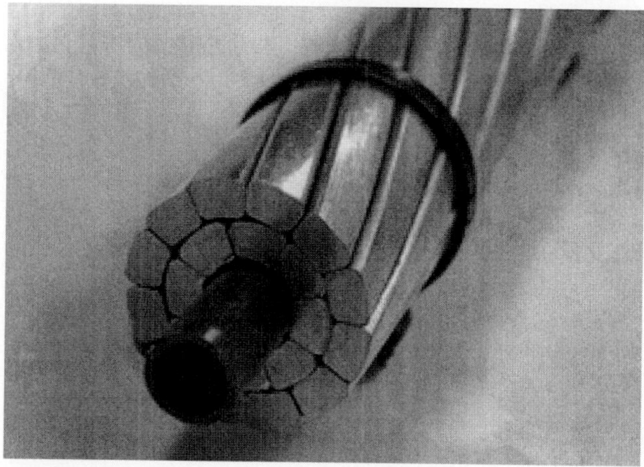

FIGURE 14.18 Aluminum Conductor Composite Core conductor. *(Courtesy Composite Technology Corporation)*

The cables utilize composite metallic materials with properties that are of lighter weight but with increased ampacities capabilities. Initial tests have demonstrated that the ACCC conductor operates with cooler temperature performance and lower sag characteristics. As the technology proves itself in trial installations, greater industry acceptance and utilization will occur.

CHAPTER 15
DISTRIBUTION TRANSFORMERS

The purpose of a distribution transformer is to reduce the primary voltage of the electric distribution system to the utilization voltage serving the customer. A distribution transformer is a static device constructed with two or more windings used to transfer alternating-current electric power by electromagnetic induction from one circuit to another at the same frequency but with different values of voltage and current.

Figure 15.1 shows distribution transformers in stock at an electric utility company service building. The distribution transformers available for use for various applications, as shown, include pole-type (Figs. 15.2 and 15.3), pad-mounted (Fig. 15.4), vault or network type (Fig. 15.5), and submersible (Fig. 15.6).

The distribution transformer in Fig. 15.2 is self-protected. It is equipped with a lightning arrester, a weak-link or protective-link expulsion-type fuse (installed under oil in the transformer tank), a secondary circuit breaker, and a warning light. The transformer primary bushing conductor is connected to one phase of the three-phase primary circuit through a partial-range current-limiting fuse. The transformer tank is grounded and connected to the

FIGURE 15.1 Electric utility distribution storage yard. Forklift trucks are used to load transformers on line trucks. Storage area is covered with concrete to provide accessibility and protect transformers.

FIGURE 15.2 Typical pole-type distribution transformer installation with the transformer bolted directly to the pole. The unit is equipped with a surge arrester, a low-voltage circuit breaker, and an overload warning light. A partial-range current-limiting fuse is mounted on the primary bushing, connecting it in series with the high-voltage winding to prevent a violent failure of the transformer if an internal fault develops.

primary and secondary common-neutral ground wire. The self-protected transformer contains a core and coils, a primary fuse mounted on the bottom of the primary bushing, a secondary terminal block, and a low-voltage circuit breaker.

Convential overhead transformers are also available. The key component distinguishing the conventional transformer from the self-protected transformer is the lack of internal fusing in the conventional model. An external fuse/cutout combination is mounted between the distribution primary and the conventional transformer. A conventional unit is depicted in Fig. 15.3.

Pad-mounted transformers are used with underground systems. Three-phase pad-mounted transformers are used for commercial installations, as illustrated in Fig. 15.4, and single-phase pad-mounted transformers are used for underground residential installations, as described and shown in Chap. 34. Vault-type distribution transformers are installed for commercial customers where adequate space is not available for pad-mounted transformers. The vault-type transformer (Fig. 15.5) may be installed in a vault under a sidewalk or in a building. They are often used in underground electric network areas.

Submersible single-phase distribution transformers (Fig. 15.6) are used in some underground systems installed in residential areas, as described in Chap. 34. Corrosion problems and the high cost of installation have minimized the use of submersible transformers.

Distribution Transformer Operation. The basic principle of operation of a transformer is explained in Chap. 1. A schematic drawing of a single-phase distribution transformer appears in Fig. 15.7. The single-phase distribution transformer consists of a primary winding and a secondary winding wound on a laminated steel core. If the load is disconnected from the secondary winding of the transformer and a high voltage is applied to the primary winding of the transformer, a magnetizing current will flow in the primary winding. If we assume the resistance of the primary winding is small, which is usually true, this current is limited by the countervoltage of self-induction induced in the highly inductive primary

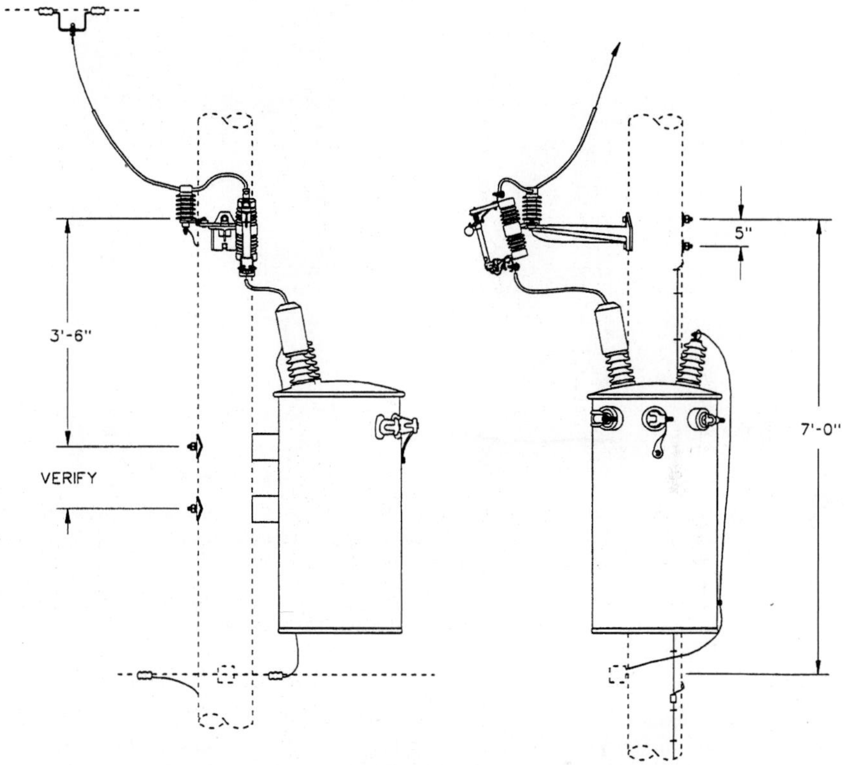

FIGURE 15.3 Two view construction standard detail of a conventional transformer to be installed on the distribution system. Note the fused cutout and lighting arrestor combination external to the transformer. (*Courtesy MidAmerican Energy Co.*)

FIGURE 15.4 Three-phase pad-mounted transformer installed in shopping center parking lot. Transformer reduces voltage from 13,200Y/7620 to 208Y/120 volts. Primary and secondary terminals of transformer are connected to underground cables. (*Courtesy ABB Power T&D Company Inc.*)

FIGURE 15.5 Vault-type distribution transformer 1000 kVA, three-phase 13,200 GrdY/7620 to 208Y/120 volts. High-voltage terminals H_1, H_2, and H_3 are designed for elbow-type cable connections, and low-voltage terminals X_1, X_2, and X_3 are spade-type for bolted connections to cable lugs. (*Courtesy ABB Power T&D Company Inc.*)

FIGURE 15.6 Submersible distribution transformer strapped to a pallet for shipment. Cables are for low-voltage connections. High-voltage connections are made with elbow cable connectors to bushing inserts installed in wells shown on top of the transformer. (*Courtesy ABB Power T&D Company Inc.*)

winding. The windings of the transformer are constructed with sufficient turns in each winding to limit the no-load or exciting current and produce a countervoltage approximately equal to the applied voltage. The exciting current magnetizes, or produces a magnetic flux in the steel transformer core. The magnetic flux reverses each half cycle as a result of the alternating voltage applied to the primary winding. The magnetic flux produced cuts the turns of the primary and secondary windings. This action induces a countervoltage in the primary winding and produces a voltage in the secondary winding. The voltages induced in each turn of the primary and secondary winding coils will be approximately equal, and the voltage induced in each winding will be equal to the voltage per turn multiplied by the number of turns.

Voltage V = volts per turn × number of turns N
Voltage primary winding = V_p
Voltage secondary winding = V_s
Volts per turn = V_t

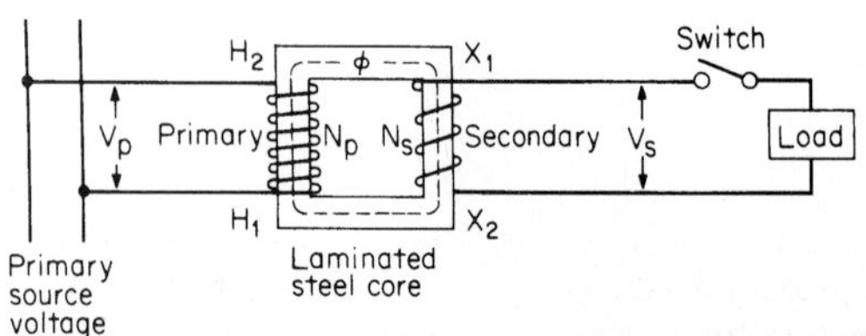

FIGURE 15.7 Schematic drawing of single-phase distribution transformer.

Number of turns of primary winding = N_p
Number of turns of secondary winding = N_s

Therefore, the voltage of the primary winding is equal to the applied voltage and equal to the number of turns in the primary winding multiplied by the volts per turn.

$$V_p = V_t \times N_p$$

Likewise the voltage of the secondary winding will equal the volts per turn multiplied by the number of turns in the secondary winding.

$$V_s = V_t \times N_s$$

If we use basic algebraic operations and divide both sides of an equation by the same number, we can obtain the following equation:

$$\frac{V_p}{N_p} = \frac{V_t \times \cancel{N_p}}{\cancel{N_p}}$$

$$\frac{V_p}{N_p} = V_t$$

and

$$\frac{V_s}{N_s} = \frac{V_t \times \cancel{N_s}}{\cancel{N_s}}$$

$$\frac{V_s}{N_s} = V_t$$

Quantities equal to the same quantity are equal, so

$$\frac{V_p}{N_p} = \frac{V_s}{N_s}$$

Both of the above quantities are equal to V_t. If we multiply both sides of the equation by the same quantity N_p, we obtain

$$\frac{V_p}{N_p} \times N_p = N_p \times \frac{V_s}{N_s}$$

or

$$V_p = \frac{N_p}{N_s} \times V_s$$

Likewise,

$$V_s = \frac{N_s}{N_p} \times V_p$$

Therefore, if we know the source voltage, or primary voltage, and the number of turns in the primary and secondary windings, we can calculate the secondary voltage. For example, if a distribution transformer has a primary winding with 7620 turns and a secondary winding of 120 turns and the source voltage equals 7620 volts, the secondary voltage equals 120 volts.

$$V_s = V_p \times \frac{N_s}{N_p}$$

$$= \cancel{7620} \times \frac{120}{\cancel{7620}} \text{volts}$$
$$= 120 \text{volts}$$

When a load is connected to the secondary winding of the transformer, a current I_s will flow in the secondary winding of the transformer. This current is equal to the secondary voltage divided by the impedance of the load Z.

$$I_s = \frac{V_s}{Z}$$

Lenz' law determined that any current that flows as a result of an induced voltage will flow in a direction to oppose the action that causes the voltage to be induced. Therefore, the secondary current, or load current, in the distribution transformer secondary winding will flow in a direction to reduce the magnetizing action of the exciting current or the primary current. This action reduces the magnetic flux in the laminated steel core momentarily and, therefore, reduces the countervoltage in the primary winding. Reducing the countervoltage in the primary winding causes a larger primary current to flow, restoring the magnetic flux in the core to its original value. The losses in a distribution transformer are small, and therefore, for all practical purposes, the voltamperes of the load, or secondary winding, equal the voltamperes of the source, or primary winding.

$$V_p \times I_p = V_s \times I_s$$

If we divide both sides of this equation by the same quantity V_p, we obtain

$$\frac{\cancel{V_p} \times I_p}{\cancel{V_p}} = \frac{V_s \times I_s}{V_p}$$

or

$$I_p = \frac{V_s}{V_p} \times I_s$$

Therefore, if the primary source voltage of a transformer is 7620 volts, the secondary voltage is 120 volts, and the secondary current, as a result of the load impedance, is 10 amps, then the primary current is 0.16 amp.

$$I_p = \frac{120}{7620} \times 10 = 0.16 \text{ amp}$$

A three-phase transformer is basically three single-phase transformers in a single tank in most cases. The windings of the three transformers are normally wound on a single multilegged laminated steel core.

Amorphous Metal. Some distribution transformers are being manufactured with amorphous metal cores. The amorphous metal reduces the core loss in a distribution transformer, thus increasing the transformer efficiency. The core losses of distribution transformers built with an amorphous metal core are approximately 60 percent less than the core losses of distribution transformers with a laminated steel core.

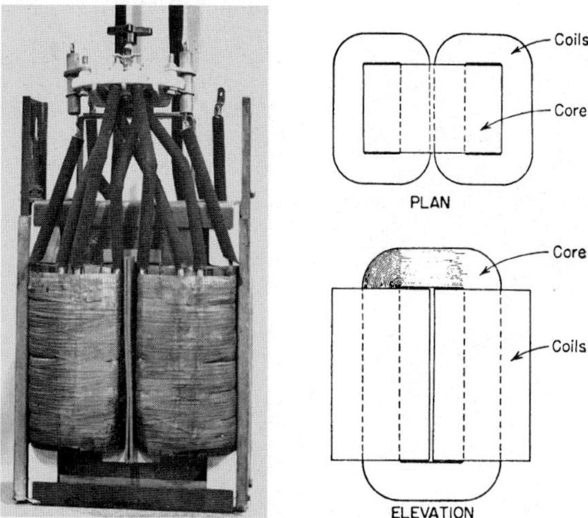

FIGURE 15.8 Views illustrating core-type transformer construction. (*Courtesy ABB Power T&D Company Inc.*)

Distribution Transformer Construction. Two types of construction are the *core type* and the *shell type*. In the core type, the core is in the form of a rectangular frame with the coils placed on the two vertical sides (Fig. 15.8). The coils are cylindrical and relatively long. They are divided, part of each primary and secondary being on each of the two vertical legs. In the shell type (Fig. 15.9), the core surrounds the coils, instead of the coils surrounding the core. The coils in the shell type are generally flat instead of cylindrical, and the primary and secondary coils are alternated (Fig. 15.10). The core-and-coil assembly of the distribution

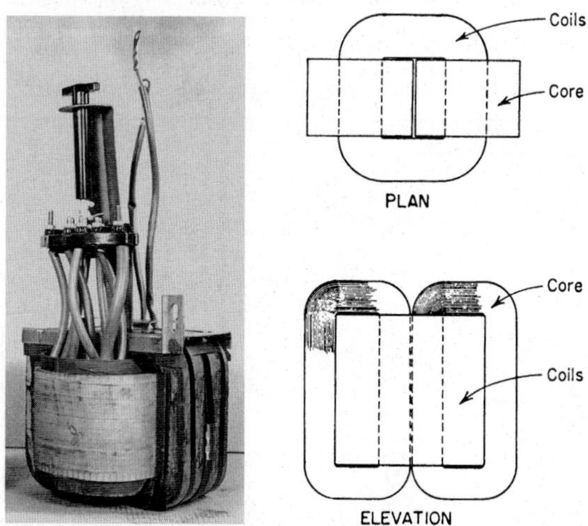

FIGURE 15.9 Views illustrating shell-type transformer construction. (*Courtesy ABB Power T&D Company Inc.*)

FIGURE 15.10 Windings of a shell-type distribution transformer cut to show cross section of winding conductors. High-voltage winding is located between the secondary winding on the inner and outer surfaces. The high-voltage winding has small conductors and a large number of turns. The secondary winding has large rectangular conductors and a small number of turns. (*Courtesy ABB Power T&D Company Inc.*)

transformer is completely immersed in insulating oil or other insulating liquids to keep the operating temperatures low and to provide additional insulation for the windings.

Distribution Transformer Ratings. The capacity of a distribution transformer is determined by the amount of current it can carry continuously at rated voltage without exceeding the design temperature. The transformers are rated in kilovolt-amperes (kVA) since the capacity is limited by the load current which is proportional to the kVA regardless of the power factor. The standard kVA ratings are itemized in Table 15.1.

Distribution Transformer Connections. Single-phase distribution transformers are manufactured with one or two primary bushings. The single-primary-bushing transformers can only be used on grounded wye systems. The two-primary-bushing transformers can be used on three-wire delta systems or four-wire wye systems if they are properly connected. Figure 15.11 illustrates schematically the connections of a single-phase transformer

TABLE 15.1 Standard Ratings of Distribution Transformers, kVA

Overhead type		Pad-mounted type	
Single-phase	Three-phase	Single-phase	Three-phase
5	15	25	75
10	30	$37^1/_2$	$112^1/_2$
15	45	50	150
25	75	75	225
$37^1/_2$	$112^1/_2$	100	300
50	150	167	500
75	225		750
100	300		1000
167	500		1500
250			2000
333			2500
500			

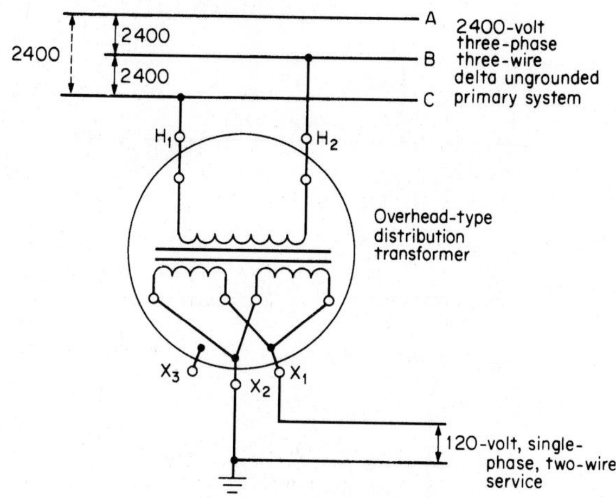

FIGURE 15.11 Single-phase overhead or pole-type distribution transformer connection for 120-volt two-wire secondary service. Transformer secondary coils are connected in parallel.

to a three-phase 2400-volt three-wire ungrounded delta primary voltage system to obtain 120-volt single-phase two-wire secondary service. The connections for similar systems operating at other primary distribution voltages such as 4800, 7200, 13,200, and 34,400 would be identical.

Figure 15.12 illustrates the proper connections for a single-phase transformer to a three-phase three-wire ungrounded delta primary voltage system to obtain 120/240-volt single-phase three-wire service. Normally the wire connected to the center low-voltage bushing will be connected to ground. Grounding the wire connecting to the center bushing limits the

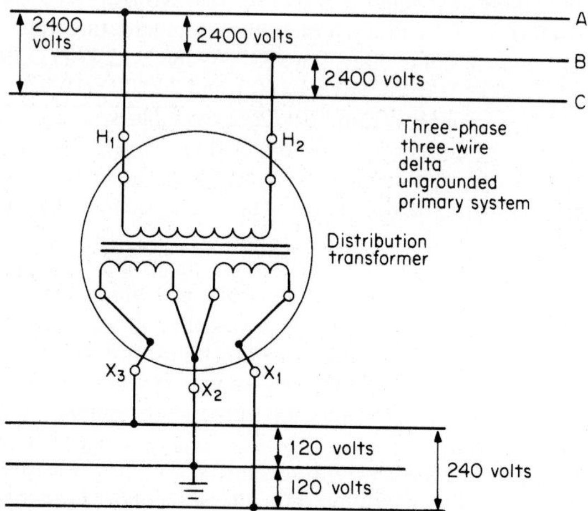

FIGURE 15.12 Single-phase distribution transformer connected to give 120/240-volt three-wire single-phase service. Transformer secondary coils are connected in series.

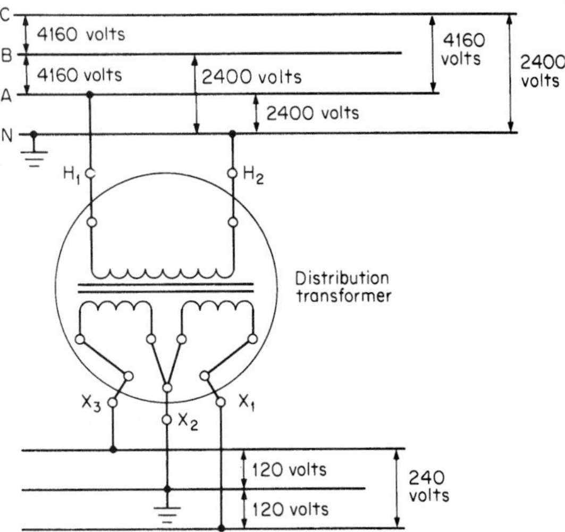

FIGURE 15.13 Single-phase distribution transformer connected to provide 120/240-volt three-wire single-phase service. Primary winding is connected line to neutral or ground.

voltage above ground to 120 volts even though the wires connecting to the outside secondary bushings have 240 volts between them.

Figure 15.13 illustrates schematically the single-phase distribution transformer connections to a three-phase four-wire wye grounded neutral primary system rated 4160Y/2400 volts to obtain 120/240-volt single-phase secondary service. The three-phase four-wire wye grounded neutral system has voltages between phases equal to the phase or line to neutral voltage multiplied by 1.73. In Fig. 15.13 the primary system line to neutral voltage is 2400 volts, and the voltage between phases is 1.73×2400, or 4160 volts. This system is designated as a 4160Y/2400-volt system. Other standard three-phase four-wire wye grounded neutral primary system voltages are 8320Y/4800, 12,470Y/7200, 13,200Y/7620, 13,800Y/7970, 22,860Y/13,200, 24,940Y/14,400, and 34,500Y/19,920.

The transformer connections to obtain 120, 120/240, or 240-volt secondary service are normally completed inside the tank of the transformer (Fig. 15.14). The transformer nameplate provides information necessary to complete connections. The voltages should be measured when the transformer is energized, to ensure that the connections are correct before the load is connected to the transformer.

Single-phase pole or overhead-type transformers can be used to obtain three-phase secondary service (Fig. 15.15). There are four normal connections: the delta-delta (Δ-Δ), the wye-wye (Y-Y), the delta-wye (Δ-Y), and the wye-delta (Y-Δ). Figure 15.16 illustrates the proper connections for three single-phase distribution transformers connected to a three-phase three-wire ungrounded delta primary-voltage system to obtain three-phase three-wire delta secondary service. The illustration is for a 2400-volt primary system and 240-volt secondary service. Other similar systems operating at other primary voltages would be connected the same way. Single-phase transformers with secondary windings constructed for voltages of 240/480 could be used to obtain 480-volt three-phase secondary service.

Figure 15.17 illustrates the proper connections for three single-phase distribution transformers connected to a three-phase four-wire grounded neutral wye primary-voltage system to obtain 208Y/120- or 480Y/277-volt secondary service. Figure 15.18 illustrates the

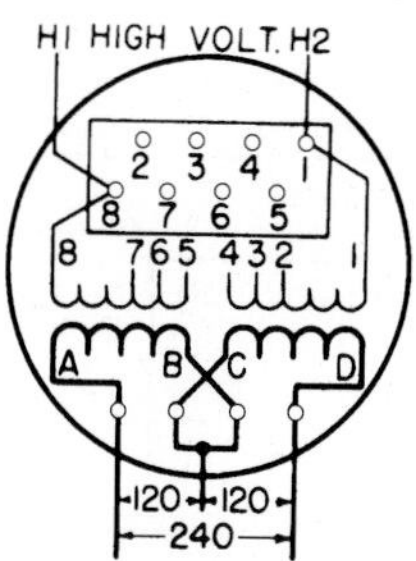

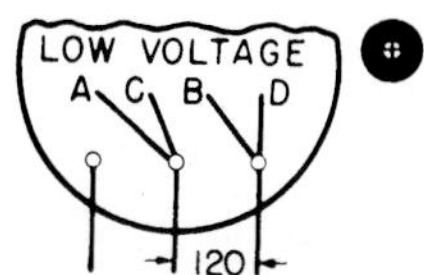

FIGURE 15.14 The left view shows connections for 120/240-volt three-wire service. For 240-volt two-wire service, the neutral bushing conductor is not connected to the jumper between terminals *B* and *C* inside the transformer tank. For 120-volt two-wire service, the connections are made as shown in the right view. All connections are completed inside the transformer tank. (*Courtesy ABB Power T&D Company Inc.*)

proper connections for three single-phase distribution transformers connected to a three-phase three-wire ungrounded neutral delta primary-voltage system to obtain 208Y/120- or 480Y/277-volt secondary service.

Transformers must be constructed with the proper windings for the primary-voltage system and the desired secondary voltage. Properly manufactured distribution transformers can be connected wye-delta if it is desired to obtain three-wire three-phase secondary voltages from a three-phase four-wire grounded neutral wye primary-voltage system.

The standard secondary system three-phase voltages are 208Y/120, 240, 480Y/277, and 480 volts.

If one of the transformers from a delta-connected bank is removed, the remaining two are said to be open-delta-connected. This can be done, and the remaining two transformers will still transform the voltages in all three phases and supply power to all three phases of the secondary mains. The proper connections, using two transformers, to obtain three-phase

FIGURE 15.15 Cluster-mounted bank of transformers bolted directly to the pole and protected by surge arresters and partial-range current-limiting fuses. The current-limiting fuses are from two different manufacturers, which gives one fuse a different appearance. The transformers are self-protected with an internal weak-link fuse in series with each of the primary windings and a secondary circuit breaker in series with each of the secondary windings. The transformers are connected to a 13,200Y/7620-volt primary circuit and provide 208Y/120-volt three-phase service for the customer.

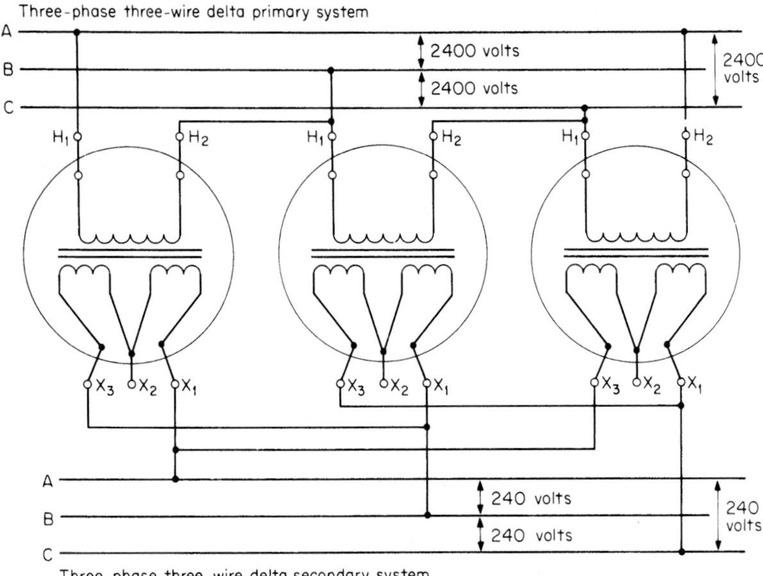

FIGURE 15.16 Three single-phase distribution transformers connected delta-delta (Δ-Δ).

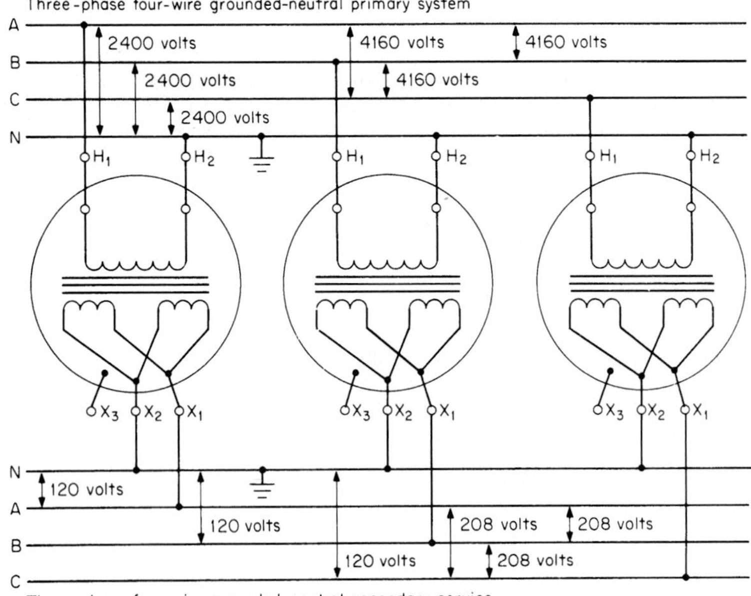

FIGURE 15.17 Three single-phase distribution transformers connected wye-wye (Y-Y). Proper voltage windings would be used with other primary voltages. Secondary windings constructed for 277 volts would provide 480Y/277-volt three-phase secondary service.

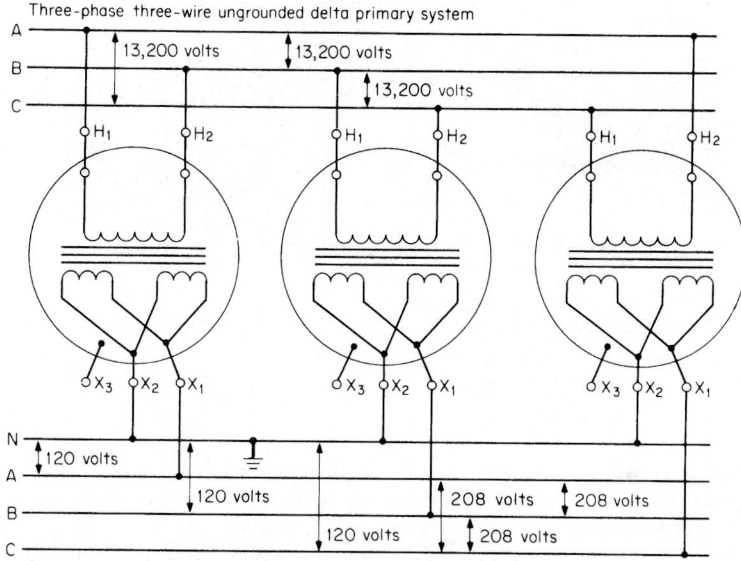

FIGURE 15.18 Three single-phase distribution transformers connected delta-wye (Δ-Y). Windings must be rated for the primary system and the desired secondary voltage.

service, for a delta primary circuit are shown in Fig. 15.19. The capacity of the two transformers is now, however, only 58 percent instead of 66⅔ percent of what it would appear to be with two transformers.

The open-delta connection is often used where an increase in load is anticipated. The third unit is added when the load grows to the point at which it exceeds the capacity of the two transformers. Furthermore, if one transformer of a three-phase bank should become defective, the defective transformer can be removed and the remaining two transformers continue to render service to at least part of the load.

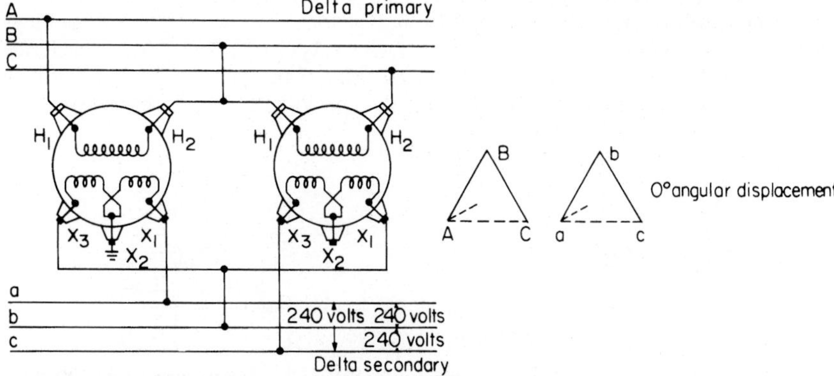

FIGURE 15.19 Two single-phase transformers connected into a bank to transform power from three-phase primary voltage to 240 volts three-phase. Primaries and secondaries are open-delta connected.

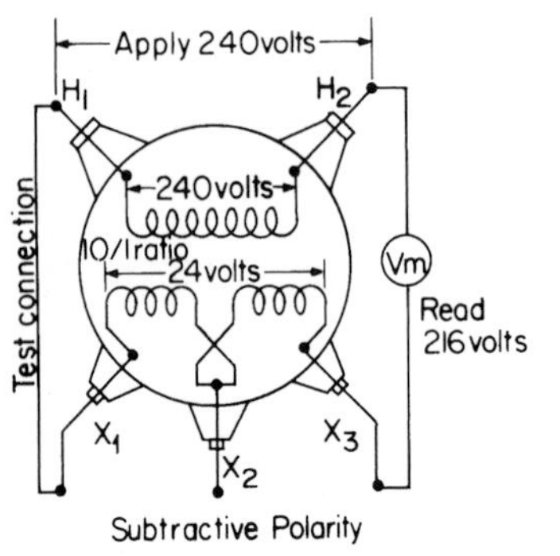

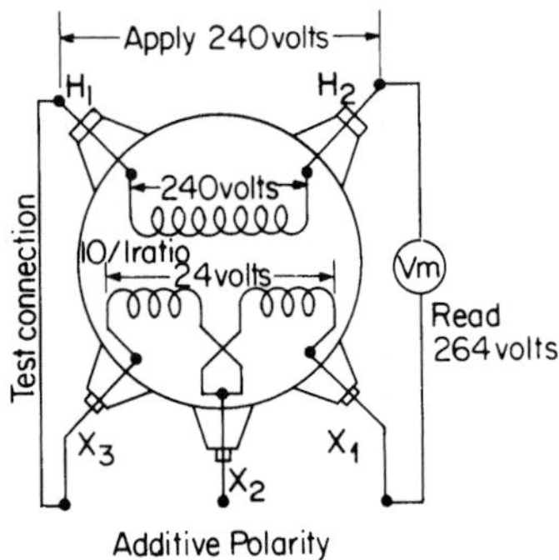

FIGURE 15.20 Polarity tests.

Distribution Transformer Polarity. In connecting transformers in parallel or in a three-phase bank, it is important to know the polarity of the transformer terminals or leads. In building transformers in the factory, the ends of the windings can be so connected to the leads extending out through the case that the flow of current in the secondary terminal with respect to the corresponding primary terminal is in the same direction or in the opposite direction. When the current flows in the same direction in the two adjacent primary and secondary terminals, the polarity of the transformer is said to be *subtractive*, and when current flows in opposite directions, the polarity is said to be *additive*.

Polarity may be further explained as follows: Imagine a single-phase transformer having two high-voltage and three low-voltage external terminals. Connect one high-voltage terminal to the adjacent low-voltage terminal, and apply a test voltage across the two high-voltage terminals. Then if the voltage across the unconnected high-voltage and low-voltage terminals is less than the voltage applied across the high-voltage terminals, the polarity is *subtractive;* while if it is greater than the voltage applied across the high-voltage terminals, the polarity is *additive* (Fig. 15.20).

From the foregoing, it is apparent that when the voltage indicated on the voltmeter is greater than the impressed voltage, it must be the sum of the primary and secondary voltages, and the sense of the two windings must be in opposite directions, as in Fig. 15.21. Likewise, when the voltage read on the voltmeter is less than the impressed voltage, the voltage must be the difference, as illustrated in Fig. 15.22. When the terminal markings are arranged in the same numerical order, H_1H_2 and X_1X_2 or H_2H_1 and X_2X_1, on each side of the transformer, the polarity of each winding is the same (subtractive). If either is in reverse order, H_2H_1 and X_1X_2 or H_1H_2 and X_2X_1, their polarities are opposite (additive). The

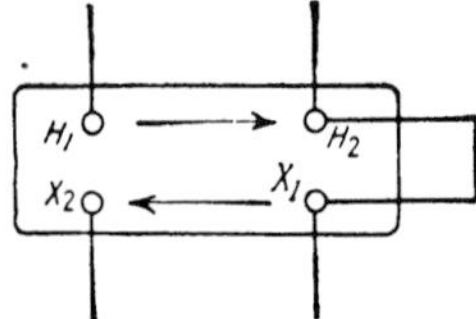

FIGURE 15.21 Sketch showing polarity markings and directions of voltages when polarity is additive.

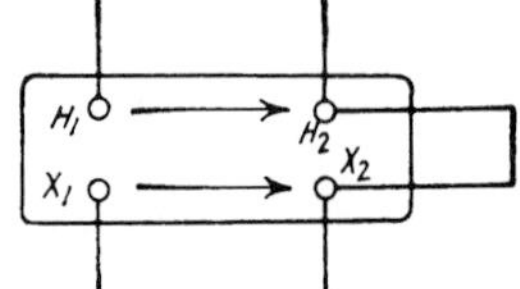

FIGURE 15.22 Sketch showing polarity markings and directions of voltages when polarity is subtractive.

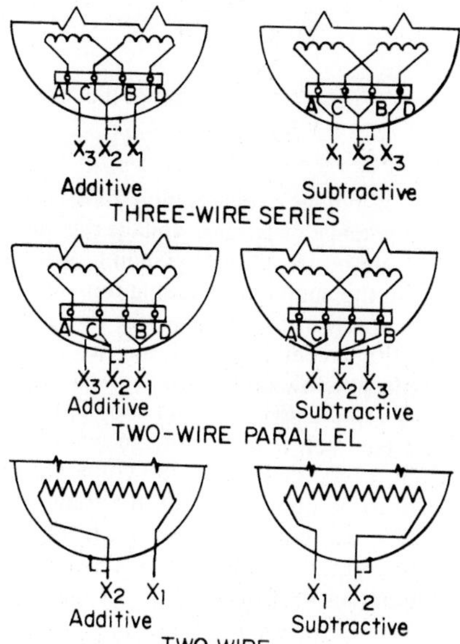

FIGURE 15.23 Standard low-voltage connections for distribution transformer.

nameplate of a transformer should always indicate the polarity of the transformer to which it is attached.

Additive polarity is standard for all single-phase distribution transformers 200 kVA and below having high-voltage ratings of 9000 volts and below.

Subtractive polarity is standard for all single-phase distribution transformers above 200 kVA irrespective of voltage rating.

Subtractive polarity is standard for all single-phase transformers 200 kVA and below having high-voltage ratings above 9000 volts.

Standard low-voltage terminal designations are illustrated in Fig. 15.23.

Paralleling Single-Phase Distribution Transformers. If greater capacity is desired, two transformers of the same or different kVA ratings may be connected in parallel. Single-phase transformers of either additive or subtractive polarity may be paralleled successfully if connected as shown in Fig. 15.24 and if the following conditions are met:

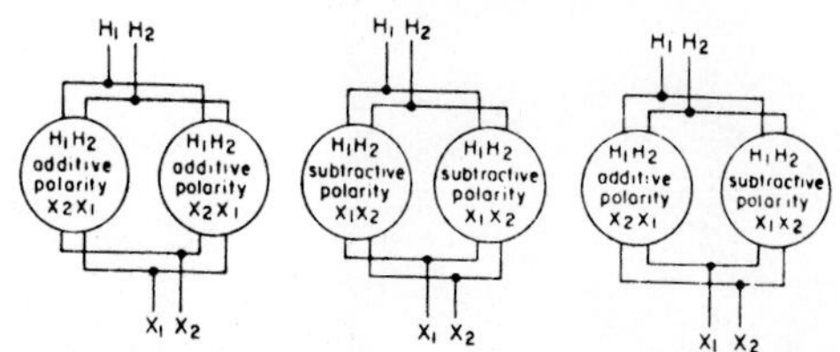

FIGURE 15.24 Schematic diagram showing proper connections for single-phase distribution transformers connected in parallel.

1. Voltage ratings are identical.
2. Tap settings are identical.
3. Percentage of impedance of one is between 92½ and 107½ percent of the other.
4. Frequency ratings are identical.

Distribution Transformer Tapped Windings. Windings of distribution transformers can be manufactured with various taps to broaden the range of primary voltages suitable for a given installation. The standard taps available are on the primary winding of the transformer and are designed to provide variations in the number of winding turns in steps of 2½ percent of the rated voltage. The distribution transformers may be designed with four 2½ percent taps above rated voltage, two 2½ percent taps above and two 2½ percent taps below rated voltage, or four 2½ percent taps below rated voltage. A standard distribution transformer with a primary winding rated 2400 volts and supplied with two 2½ percent taps above and two 2½ percent taps below rated voltage would be constructed with primary winding taps rated 2520, 2460, 2340, and 2280 volts. Tap changers on distribution transformers are designed to be operated while the transformer is deenergized. The operating handle (Fig. 15.25) can be located outside of the tank or inside the tank above the oil level. The transformer nameplate provides information concerning the taps available on the distribution transformer.

Distribution Transformer Standards. The standards are published by American National Standards Institute (ANSI). ANSI Standard C57.12.20 covers "Requirements for Overhead-Type Distribution Transformers, 67000 volts and below; 500 kVA and smaller." ANSI Standard C57.12.25 covers "Requirements for pad-mounted compartmental-type Single Phase Distribution Transformers with Separable Insulated High-Voltage Connectors, High-Voltage, 34,500 GrdY/19,920 Volts and Below; Low-Voltage, 240/120; 167 kVA and Smaller." ANSI Standard C57.12.26 covers "Requirements for Pad-Mounted Compartmental-Type Self-Cooled, Three-Phase Distribution Transformers for Use with Separable Insulated High-Voltage connectors, High-Voltage, 24,940 GrdY/14,400 Volts and Below; 2500 kVA and Smaller."

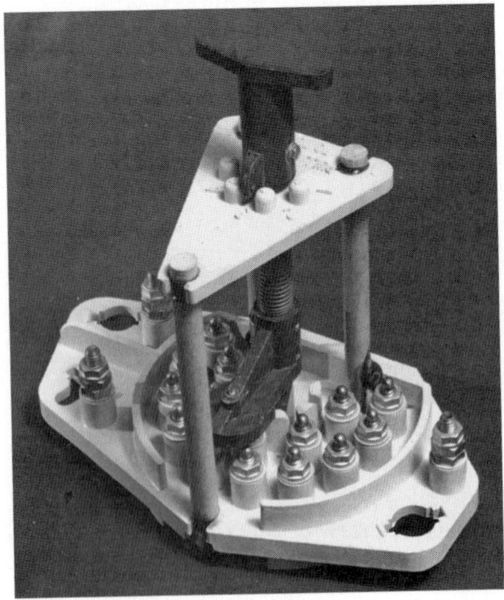

FIGURE 15.25 No-load changer for distribution transformer. (*Courtesy ABB Power T&D Company Inc.*)

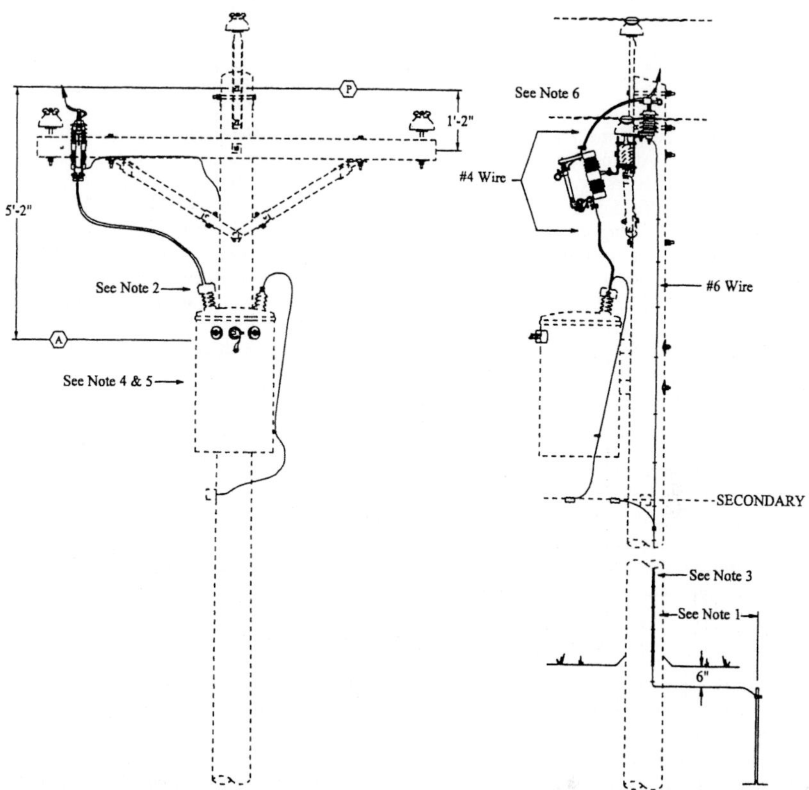

FIGURE 15.26 Diagram of a conventional single-phase, overhead distribution transformer. (*Courtesy MidAmerican Energy Co.*)

A typical specification for conventional single-phase overhead distribution transformers (Fig. 15.26) would read as follows:

> This standard covers the electrical characteristics and mechanical features of single phase, 60 Hz, mineral oil immersed, self-cooled, conventional, overhead type distribution transformers rated: 15, 25, 50, 75, 100, and 167 kVA; with high voltage: 7200/12,470Y, 7620/13,200Y, or 7970/13,800Y volts; and low voltage: 120/240 volts; with no taps. The unit shall be of "conventional" transformer design. There shall be no integral fuse, secondary breaker, or overvoltage protection.
>
> All requirements shall be in accordance with the latest revision of ANSI C57.12.00 and ANSI C57.12.20, except where modified by this standard.
>
> The transformer shall be bar coded in accordance with the proposed EEI "Bar Coding Distribution Transformers: A Proposed EEI Guideline." The manufacturer and serial number shall be bar coded on the nameplate. The serial number, manufacturer, and the Company Item ID (if given) shall be bar coded on a temporary label.
>
> The transformer shall meet interchangeability dimensions for single position mounting as shown in Figs. 1 and 9 of ANSI C57.12.20.
>
> The two primary bushings shall be wet process porcelain cover-mounted. The primary connectors shall be tin-plated bronze of the eye-bolt type. The H1 bushing is to be supplied with a wildlife protector and an eyebolt connector equipped with a covered handknob tightening device. The eyebolt connector shall accept a 5/16 in diameter knurled stud terminal with the

wildlife protector in place. The secondary bushings shall be wet process porcelain or epoxy, mounted on the tank wall. Secondary connectors shall be per ANSI C57.12.20, Section 6.1.2, except terminals on 75 kVA 120/240-volt units shall accommodate up to 1000 kcmil-61 stranded conductor.

The cover of the transformer shall be coated for wildlife protection. The coating shall have a rating of at least 10-kV AC rms. The insulating finish shall be capable of withstanding 10 kV at a 200 volt per sec rate of rise, tested per ASTM D149-87, using electrode type 3 (1/4 in cylindrical rods).

The transformer tank and cover shall be designed to withstand an internal pressure of 20 psig without tank rupture or bushing or cover displacement, and shall withstand an internal pressure of 7 psig without permanent distortion. The transformer tank shall retain its shape when the cover is removed.

Support lugs for single-position mounting shall be provided and shall be designed to provide a static-load safety factor of 5. Lifting lugs shall be provided, with location and safety factor as specified in ANSI C57.12.20. The tank and cover finish shall be light gray No. 70 (Munsell 5BG7.0/0.4). The interior shall be coated with a durable protective finish from the top to at least below the top liquid level.

The transformer shall be equipped with a tank ground per ANSI C57.12.20, Paragraph 6.5.4.1 with eyebolt to accept an AWG 6 solid copper ground wire; a tank ground for ground strap connection; and a ground strap, properly sized to carry load and fault current, connected between the secondary neutral terminal and the tank ground. The grounding strap shall be supplied with a connection hole at each end. "J" hook or slotted connection points shall not be allowed.

The transformer shall be equipped with neoprene or Buna-N gaskets on all primary and secondary bushings and tank fittings.

A Qualitrol no. 202-032-01 automatic pressure relief device, with operation indication, or approved equivalent; or a cover assembly designed for relief of excess pressure in accordance with ANSI C57.12.20, Section 6.2.7.2, shall be supplied on the transformer.

Final core and coil assembly shall result in a rigid assembly that maintains full mechanical, electrical, and dimensional stability under fault conditions.

The coil insulation system shall consist of electrical kraft paper layer insulation coated on both sides with thermosetting adhesive. Curing of the coils shall cause a permanent bond layer to layer and turn to turn.

The internal primary leads shall have sufficient length to permit cover removal and lead disconnection from the high voltage bushings. These leads shall be adequately anchored to the coil assembly to minimize the risk of damaging the coil connection during cover removal.

The internal secondary leads shall be identified by appropriate markings, embossed or stamped on a visible portion of the secondary lead near the end connected to the bushing. These leads shall have sufficient length to permit series or parallel connection of the secondary windings. If the internal secondary leads are soft aluminum, a hard aluminum-alloy tab shall be welded to the bushing end of the lead for connection to the bushing.

The transformer should be designed so that the voltage regulation shall not exceed 2.5 percent at full load and 0.8 power factor for all transformers covered in this specification.

All transformers shall be filled with Type I insulating oil meeting the following requirements:

a. All applicable requirements of ASTM D 3487, Paragraph 82a.
b. Minimum breakdown value of 28 kV per ASTM D 1816, Paragraph 84a using a 0.4 in gap.
c. The PCB content shall be less than 2 parts per million which shall be noted upon the nameplate of each unit. In addition, the manufacturer shall provide and install a "No PCB" identification on the tank wall, near the bottom of the tank centered below the low-voltage bushings. The decal shall be not less than 1 in by 2 in in size and shall indicate "No PCB's" in blue lettering on a white background, and be sufficiently durable to remain legible for the life of the transformer.

A typical specification for self-protected single-phase overhead distribution transformers (Figs. 15.2 and 15.3) would read as follows:

Distribution transformers; overhead type; single-phase; 25-kVA capacity; 13,200 GrdY/7620-120/240 volts; 95-kV BIL; self-protected, with primary weak-link fuse and secondary circuit breaker; red overload warning light; oil-insulated, self-cooled (OISC); no taps; 60 Hz; one surge arrester; one primary bushing, with wildlife protector and eyebolt connector equipped with an insulated hand knob-tightening device designed to accept a 5/16-in-diameter knurled stud terminal with wildlife protector in place; tank ground, with eyebolt designed to accept No. 6 solid copper ground wire; a tank ground terminal for a ground strap connection; a ground strap, properly sized to carry load and fault current, connected between the secondary neutral terminal and the tank ground terminal; cover coated to provide 10-kVAC rms protection from wildlife; 65°C rise rating, sky gray in color, Qualitrol No. 202-032-01 Automatic Pressure Relief Device, with operation indication, or approved equivalent; and standard accessories.

The primary weak-link fuse, secondary circuit breaker, and red overload warning light shall be coordinated and compatible with the nameplate ratings.

The arrester shall be a 9/10-kV, direct-connected lead-type distribution class surge arrester (in accordance with the latest revision of ANSI Standard C62.2), with ground lead disconnector, and wildlife protector. The arrester primary lead shall have a minimum length of 24 in with a weatherproof jacket, for connection to the line side of a current-limiting fuse to be installed on the primary bushing in the filed.

All requirements shall be in accordance with the latest revision of ANSI Standard C57.12.20 except where modified by this specification.

The transformer shall be bar coded in accordance with the proposed EEI "Bar Coding Distribution Transformers: A Proposed EEI Guideline." The manufacturer and serial number shall be bar coded on the nameplate. The serial number, manufacturer, and the Company Item ID (if given) shall be bar coded on a temporary label.

The transformer shall meet interchangeability dimensions for single position mounting as shown in Figs. 1 and 9 of ANSI C57.12.20.

All transformers shall be filled with Type I insulating oil meeting the following requirements:

a. All applicable requirements of ASTM D 3487, Paragraph 82a.
b. Minimum breakdown value of 28 kV per ASTM D 1816, Paragraph 84a using a 0.4 in gap.
c. The PCB content shall be less than 2 parts per million which shall be noted upon the nameplate of each unit. In addition, the manufacturer shall provide and install a "No PCB" identification on the tank wall, near the bottom of the tank centered below the low-voltage bushings. The decal shall be not less than 1 in by 2 in in size and shall indicate "No PCB's" in blue lettering on a white background, and be sufficiently durable to remain legible for the life of the transformer.

A typical specification for a single-phase pad-mounted distribution transformer (Figs. 15.27 and 15.28) would read as follows:

Distribution transformers; low profile; dead front; pad-mounted; single-phase; 50-kVA capacity; 13,200 GrdY/7620-120/240 volts; OISC; 95-kV BIL; 60 Hz; 65°C rise rating; and no taps on the primary winding. All requirements shall be in accordance with the latest revisions of ANSI Standard C57.12.25, except where modified by this specification.

The transformer shall be equipped with two primary bolted-on bushing wells rated 8.3 kV and 200 A in accordance with the latest revision of ANSI/IEEE Standard 386. The bushing wells shall be equipped with 8.3/14.4-kV 200-A load break bushing inserts to accommodate load break elbows for sectionalizing cable in a loop-feed primary system.

The secondary terminals are to be equipped with bolted standard copper-threaded studs only.

A ground pad shall be provided on the outer surface of the tank. The low-voltage neutral shall have a fully insulated bushing. A removable ground strap sized for the rating of the transformer shall be provided and connected between the neutral bushing and a ground terminal on the tank of the transformer.

The high-voltage and low-voltage ground pads inside the transformer compartment shall be equipped with eyebolt connectors to accept 1/0 bare copper wire.

FIGURE 15.27 Single-phase, pad-mounted distribution transformer. (*Courtesy A. B. Chance Co.*)

FIGURE 15.28 Single-phase, pad-mounted distribution transformer with cover open. (*Courtesy ABB Power T&D Company Inc.*)

The transformer shall be equipped with an RTE Bay-O-Net, or approved equivalent, externally replaceable overload sensing primary fuse for coordinated overload and secondary fault protection coordinated and in series with an 8.3-kV partial-range current-limiting fuse. The current-limiting fuse shall be sized so that it will not operate on secondary short circuits or transformer overloads. The current-limiting fuse shall be installed under oil inside the transformer tank. Each load-sensing and current-limiting fuse device shall be capable of withstanding 15-kV system recovery voltage across an open fuse.

The transformer shall be equipped with an oil drip shield directly below the Bay-O-Net fuse holder to prevent oil dripping on the primary elbows or bushings during removal of the fuse.

The transformer housing shall be designed to be tamper-resistant to prevent unauthorized access.

The transformer shall be bar coded in accordance with the proposed EEI "Bar Coding Distribution Transformers: A Proposed EEI Guideline." The manufacturer and serial number shall be bar coded on the nameplate. The serial number, manufacturer, and the Company Item ID (if given) shall be bar coded on a temporary label.

All transformers shall be filled with Type I insulating oil meeting the following requirements:

a. All applicable requirements of ASTM D 3487, Paragraph 82a.

b. Minimum breakdown value of 28 kV per ASTM D 1816, Paragraph 84a using a 0.4 in gap.

c. The PCB content shall be less than 2 parts per million which shall be noted upon the nameplate of each unit. In addition, the manufacturer shall provide and install a "No PCB" identification on the tank wall, near the bottom of the tank centered below the low-voltage bushings. The decal shall be not less than 1 in by 2 in in size and shall indicate "No PCB's" in blue lettering on a white background, and be sufficiently durable to remain legible for the life of the transformer.

A typical specification for a three-phase pad-mounted distribution transformer (Fig. 15.4) would read as follows:

Distribution transformers; three-phase;dead front; 300-kVA capacity; high-voltage windings rated 13,800 GrdY/7970 volts; low-voltage windings rated 208 GrdY/120 volts; the high-voltage winding shall have four $2^1/_2$ percent taps below rated voltage; and the no-load tap-changer switch is to be located in the high-voltage compartment.

All requirements shall be in accordance with the latest revision of ANSI Standard C57.12.26, except where changed by this specification.

The transformer housing shall be designed to be tamper-resistant to prevent unauthorized access in accordance with the latest revision to the Western Underground Committee Guide 2.13, entitled "Security for Padmounted Equipment Enclosures."

The transformer shall have specific dimensions for Type B transformers for loop-feed systems as defined in the latest revision of ANSI Standard C57.12.26. The transformer primary bushing wells and parking stands shall be located as shown in Fig. 6 of the Standard.

The primary and secondary windings of the transformer shall be wound on a five-legged core and connected GrdY-GrdY.

The transformer shall be equipped with six primary bolted-on bushing wells rated 8.3 kV and 200 A in accordance with the latest revision of the ANSI/IEEE Standard 386. The bushing wells shall be equipped with 8.3/14.4-kV 200-A load break bushing inserts to accommodate load break elbows for sectionalizing cables in a loop-feed primary system.

The transformer secondary bushings are to be equipped with copper studs complete with bronze spade terminals.

The low-voltage neutral shall have a fully insulated bushing. A removable ground strap sized for the rating of the transformer shall be provided and connected between the neutral bushing terminal and a ground terminal in the compartment on the tank of the transformer.

The high- and low-voltage compartments shall each have a ground terminal on the transformer tank equipped with an eyebolt connector designed to accept a 1/0 bare copper wire. A ground pad shall be provided on the external surface of the transformer tank.

The high- and low-voltage terminals for the transformer windings shall be separated by a metal barrier.

The transformer shall be equipped with RTE Bay-O-Net, or approved equivalent, externally replaceable overload-sensing primary fuses, for coordinated overload and secondary fault protection, coordinated and in series with 8.3-kV partial-range current-limiting fuses. The current-limiting fuses shall not operate on secondary short circuits or transformer overloads. The current-limiting fuses shall be installed under oil inside the transformer tank. Each load-sensing and current-limiting fuse device shall be capable of withstanding 15-kV system recovery voltage across an open fuse.

The transformer shall be equipped with oil drip shields directly below the Bay-O-Net fuse holders, to prevent oil dripping on the primary elbows or bushings during removal of the fuses.

The transformer tank shall have a removable cover or handhole of the tamper-resistant design.

A Qualitrol no. 202-032-01 automatic pressure relief device with an operation indication, or approved equivalent, shall be supplied on the transformer.

The transformer tank shall be equipped with a 1-in oil drain valve with a built-in oil sampling device.

The transformer shall be bar coded in accordance with the proposed EEI "Bar Coding Distribution Transformers: A Proposed EEI Guideline." The manufacturer and serial number shall be bar coded on the nameplate. The serial number, manufacturer, and the Company Item ID (if given) shall be bar coded on a temporary label.

All transformers shall be filled with Type I insulating oil meeting the following requirements:

a. All applicable requirements of ASTM D 3487, Paragraph 82a.

b. Minimum breakdown value of 28 kV per ASTM D 1816, Paragraph 84a using a 0.4 in gap.

c. The PCB content shall be less than 2 parts per million which shall be noted upon the nameplate of each unit. In addition, the manufacturer shall provide and install a "No PCB" identification on the tank wall, near the bottom of the tank centered below the low-voltage bushings. The decal shall be not less than 1 in by 2 in in size and shall indicate "No PCB's" in blue lettering on a white background, and be sufficiently durable to remain legible for the life of the transformer.

The above specifications can be changed to meet the kVA capacity and voltages required. Changes in the high-voltage requirements also require revisions in the related items to obtain the proper insulation and surge-withstand capabilities.

Safety Precautions for Distribution Transformers. Distribution transformers must be protected from violent explosions. When a distribution transformer is removed from service as a result of an automatic operation of the transformer protective devices, the transformer should be tested to be sure it does not have an internal fault before it is reenergized by a lineman. Lineman have closed a fused cutout on a distribution transformer, which had an internal fault, and have been injured by the explosion and oil fire that resulted. An internal protective link fuse, commonly installed in self-protected distribution transformers, has a limited interrupting rating and cannot be depended on to operate in a manner to prevent catastrophic failure of the transformer. The interrupting rating of a fused cutout, in series with the primary winding of the transformer with or without an internal weak-link fuse, may exceed the available fault current from the electric system and fail to prevent the explosion of the transformer as a result of an internal arcing fault. The operation of the cutout, with an expulsion fuse, may take one-half cycle, permitting internal arcing under oil in the transformer, which can cause a violent explosion and fire.

Current-Limiting Fuses. A current-limiting fuse can be installed in series with the primary winding of a distribution transformer to prevent violent transformer failures as a result of an internal arcing fault. The current-limiting fuse can be installed external to the transformer or inside the tank under oil. The current-limiting fuse is usually installed externally on the bushing of a pole-mounted distribution transformer (Fig. 15.29) and internally under oil in a pad-mounted transformer. The current-limiting fuse installed in series with the primary winding of the distribution transformer must be the proper size to coordinate with the other transformer protective devices. Current-limiting fuses are discussed in Chap. 17, "Fuses."

FIGURE 15.29 Lineman is installing a partial-range current-limiting fuse on the terminal on the primary bushing of a self-protected overhead-type distribution transformer. Notice that the wire from the surge arrester is long enough to connect to the top of the fuse. The transformer assembly will be raised and mounted on a pole, in the vicinity, when all the equipment has been installed on the transformer.

CHAPTER 16
LIGHTNING AND SURGE PROTECTION

Lightning. Lightning flashes are currents of electricity flowing from one cloud to another or between the cloud and earth. They always flow by the path of least resistance, just as water takes the path of least resistance when it flows down the hillside. Such paths down the hillside are not generally straight but, on the contrary, are very crooked. The same is true of the path taken by a bolt of lightning; this fact explains the zigzag path of the flash. Trees, high buildings, towers, transmission and distribution lines, and poles are paths of less resistance than air, and therefore lightning often strikes them and flows through them to ground, with damaging results (Fig. 16.1).

Overhead lines have charges of electricity induced on them when charged clouds pass over them. Induced charges trapped on electric circuits can create abnormal voltages on the transmission and distribution lines when lightning strokes travel to ground, discharging the

FIGURE 16.1 The damaging results to the underground electric system due to a nearby lightning strike on the overhead distribution system.

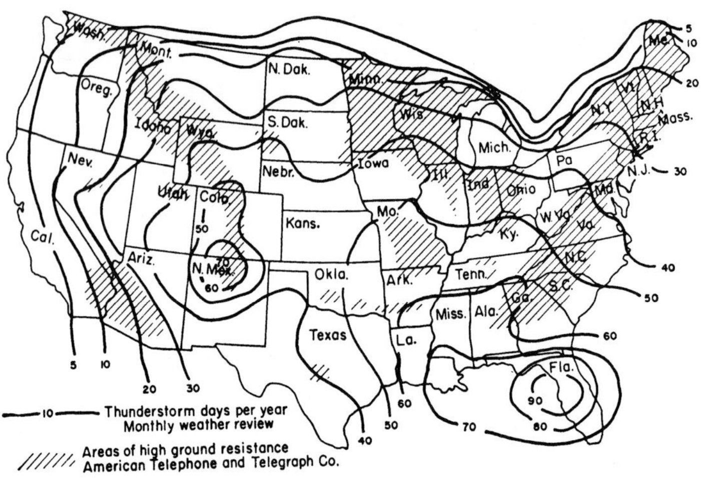

FIGURE 16.2 Number of electrical storm days per year. Lines are marked with the average number of lightning-producing storm days occurring in each area per year. (*Courtesy American Telephone and Telegraph Co.*)

clouds. The abnormal voltages placed on the line will travel along the line until dissipated by surge arrestor operation, attenuation, and/or leakage or by failure of insulators, transformers, or other apparatus connected to the line.

Lightning-producing storms are very common. It is estimated that 44,000 thunderstorms rage daily over the earth, producing 100 flashes of lightning every second. Figure 16.2 shows graphically the approximate number of lightning-producing storm days in various parts of the continental United States during a year. In some areas the number is as low as 5 days, and in other parts it is as high as 90 days. Thunderstorms bring rain, strong winds, tornadoes, and lightning. All these can cause major damage to electric facilities.

Surge. Large overvoltages that develop suddenly on electric transmission and distribution circuits are referred to as surges or transients. The surges can be caused by direct lightning strokes or induced charges as a result of lightning strokes to ground, as described, or by circuit-switching operations as well as the operation of devices connected to the lines. The large overvoltages may damage electrical equipment connected to the lines and line insulators, even though they are transient or quick to disappear.

Shield Wire. A ground wire, or static wire, mounted above the other line conductors and strung along the line, the same as other line conductors, can provide protection or shielding to the circuit from direct lightning strokes (Figs. 16.3 and 16.4). The shield wire on an overhead line may be a galvanized steel, copper-covered steel, or aluminum-covered steel cable.

The shield wire on lines supported by metal towers is grounded by a connection to each tower, with a metal clamp which fastens it to the tower and the ground conductors and ground rods installed at the base of the tower.

The shield wire on lines supported by wood poles is connected to ground at frequent intervals by a ground conductor extending from the metal clamp, which supports the shield wire at the top of the pole, to a ground rod adjacent to the pole just below the ground surface. The grounds may be installed on each pole or may be as far apart as every fifth pole. If the pole line has a common-neutral ground wire on the support structures for a distribution circuit, the ground wire must be connected to the common-neutral conductor on the structure. The *National Electrical Safety Code* requires that the common-neutral for primary and

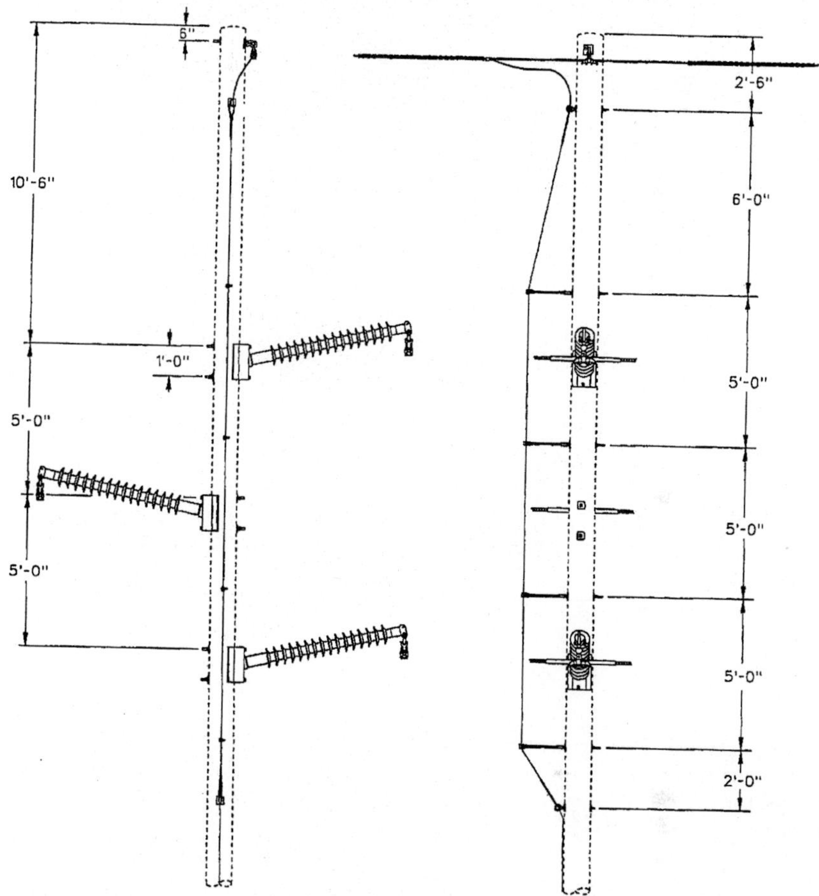

FIGURE 16.3 The shield wire for this 161-kV tangent transmission line structure is connected to ground by a wire that extends down the pole and connects to the common-neutral conductor below the phase wires and to a ground rod near the base of the pole. (*Courtesy MidAmerican Energy Co.*)

secondary circuits be grounded at each piece of equipment connected to the line and have at least four connections to ground per mile.

A properly installed shield wire is highly effective, but its effectiveness depends on low ground resistance. The shield wire virtually brings the ground or earth potential above the transmission and/or distribution line, and hence the stress on the line and equipment insulation due to lightning direct strokes is greatly reduced.

Lightning or Surge Arresters. American National Standards Institute's *Surge Arresters for Alternating-Current Power Circuits* (ANSI/IEEE C62.1) provides specifications for surge arresters. ANSI C62.2, *Guide for the Application of Valve-Type Surge Arresters for Alternating-Current Systems*, provides proper methods to apply them. A surge arrester is a device that prevents high voltages, which would damage equipment, from building up on a circuit, by providing a low-impedance path to ground for the current from lightning or transient voltages, and then restoring normal circuit conditions.

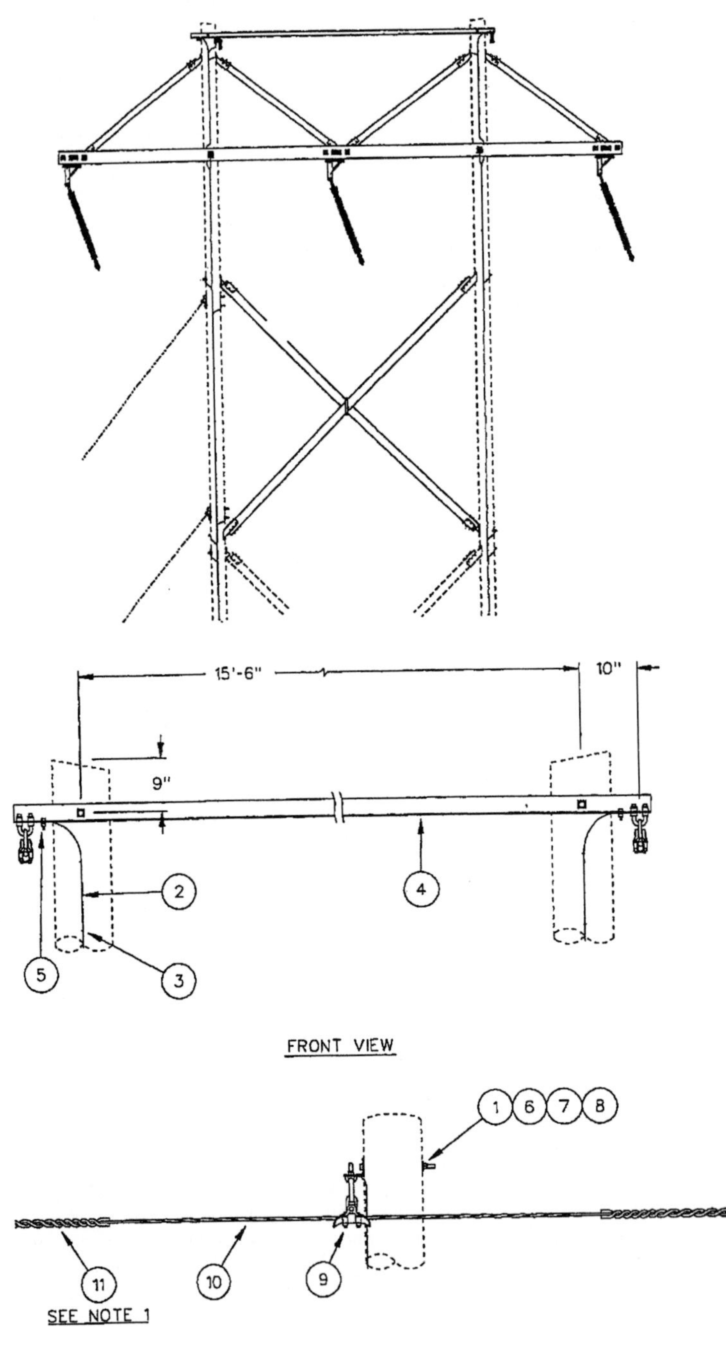

FIGURE 16.4 A 161-kV two-pole transmission line is protected with an overhead shield wire. (*Courtesy MidAmerican Energy Co.*)

Surge arresters perform a function on the electric system similar to that of a safety valve on a steam boiler. A safety valve on a boiler relieves high pressure by blowing off steam until the high pressure is reduced to normal. When the pressure is reduced to normal, the safety valve closes again and is ready for the next abnormal condition. When a high voltage (greater than the normal-line voltage) exists on the line, the arrester immediately furnishes a path to ground and thus limits and drains off the excess voltage. Furthermore, when excess voltage is relieved, the action of the arrester must prevent any further flow of power current. The function of a surge arrester is, therefore, twofold-first, to provide a point in the circuit at which the overvoltage impulse can pass to earth without damage to line insulators, transformers, or other connected equipment and, second, to prevent any follow-up power current from flowing to ground.

Elementary Surge Arrester. The elementary form of an arrester is a simple horn gap connected in series with a resistance, as shown in Fig. 16.5. A horn gap is the space that is formed between the two sides of a V when the lower point of the V is removed. One reason for providing the gap is to prevent current leakage when the voltage on the line is normal. If there was no surge arrester protecting the motor shown in Fig. 16.5, the path of least resistance would likely be through the windings of the machine, to its frame, and then to ground. This, of course, would damage the motor by puncturing its insulation. If a surge arrester is connected from the line to the earth, however, it is likely that the current from the overvoltage would find it easier to flow to ground by way of the arrester and thus leaves the motor undamaged. While the high-voltage charge is being drained off, the current flow to ground is limited by the resistance in series with the horn gap. When the discharge has passed, the heat of the arc between the V gap creates an upward draft of air, which tends to carry the arc upward with it until it is stretched out to such a length that it breaks. The magnetic field on the inside of the V helps to force the arc upward. Thus it is clear that all this may be done without interrupting the load circuit.

It should be pointed out, however, that this elementary form of arrester is rarely used in practice. The general principle of operation, however, is correct.

Classes of Valve-Type Surge Arresters. Four classes of valve-type surge arresters are manufactured for alternating-current systems: secondary (Fig. 16.6), distribution, intermediate,

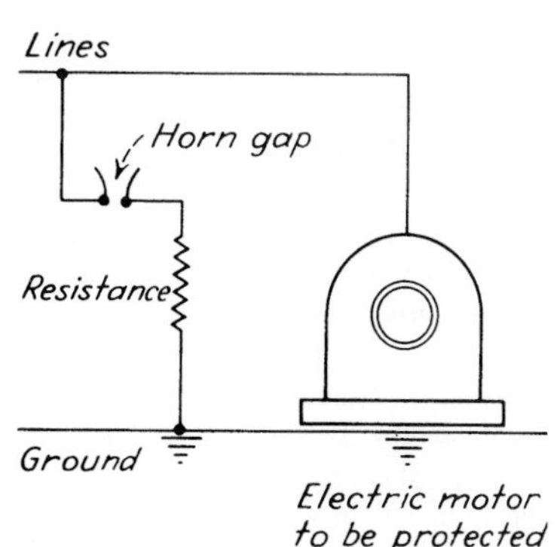

FIGURE 16.5 Elementary surge arrester consisting of horn gap and resistance.

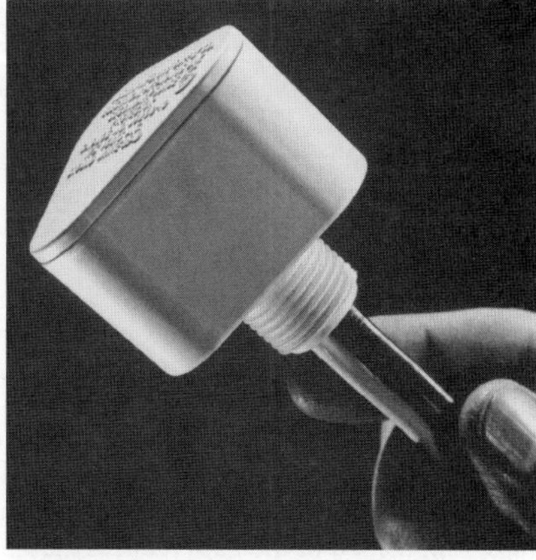

FIGURE 16.6 Secondary valve-type surge arrester. Arrester can be connected to three-wire 120/240-volt service at the service entrance or mounted on the lightning distribution panel board and connected to the main cable terminations. (*Courtesy Joslyn Mfg. and Supply Co.*)

and station. These four classes of surge arresters have similar components, but are manufactured with different characteristics. Secondary arresters are manufactured with voltage ratings of 175 and 650 volts. Distribution surge arresters for the standard voltage distribution systems are 3, 6, 9, 10, 12, 15, 18, 21, 25, 27, and 30 kV. Intermediate arresters are manufactured with voltage ratings of 3 kV through 120 kV. Station arresters are manufactured with voltage ratings of 3 kV and above. These four different classifications of arresters have different overvoltage conducting characteristics, current discharge capabilities, and maximum surge discharge voltages. The cost of surge arresters increases as the capacities are increased.

Secondary arresters are used on service and other low-voltage alternating-current circuits. Distribution arresters are used on primary distribution systems to protect insulators, distribution transformers, power cables, and other equipment. Intermediate-type surge arresters are often used on substation exit cables and other locations in the distribution system needing a high level of lightning and surge protection. Station-type surge arresters are used in substations and generating stations to provide a high level of surge protection for the major pieces of equipment. Surge voltages are generated by operating switches and other equipment connected to the electric transmission system as well as by lightning.

The surge arresters with silicon-carbide-valve blocks have gaps in a sealed porcelain casing and grading resistors or capacitors to control the sparkover voltages of the gaps. The silicon-carbide-valve blocks function as a nonlinear resistor that passes large surge currents and limits the power follow current.

When a surge voltage of high magnitude develops on a circuit protected with surge arresters with silicon-carbide-valve blocks, the series gaps in the arresters will spark over, permitting the nonlinear resistor valve block to pass the surge current to ground. The surge voltage will be limited by the voltage drop across the arrester and the connecting wires ($V = I \times Z$). The resistance of the silicon carbide valve blocks, while conducting the large surge current, is very low, keeping the voltage across the surge arrester low. This keeps the voltage across the equipment, connected in parallel with the surge arrester, below the value that would damage the insulation of the equipment. After the surge current has been bypassed to ground, power follow current will flow through the surge arrester to ground. The nonlinear resistor silicon-carbide-valve block develops a high resistance to the 60-Hz current, preventing a short circuit to ground on the line. When the instantaneous value of the 60-Hz power follow current goes through zero, the arrester gaps interrupt the current, returning the circuit conditions to normal.

The metal-oxide-varistor (MOV) surge arrester is now the industry standard (Fig. 16.7 through Fig. 16.12). First generation MOV surge arresters, constructed with zinc-oxide-valve blocks, do not need gaps. The zinc-oxide-valve blocks in an arrester effectively insulate the electrical conductors from ground at normal electric system operating voltages, limiting the leakage current flow through the arrester to a small value. When a surge voltage develops on the electric system protected by metal-oxide surge arresters, the arrester valve blocks conduct large values of current to ground, limiting the $I \times R$ voltage drop across the arrester. If the connections to the arrester have a low impedance, the stress on the electric system is kept below the *basic insulation level* (BIL) or critical insulation level of the equipment. The zinc-oxide-valve blocks or disks start conducting sharply at a precise voltage level and cease to conduct when the voltage falls below the same level. The arrester zinc-oxide-valve blocks have the characteristics of a nonlinear resistance.

The metal-oxide gapless arresters are smaller and lighter than valve-type arresters with gaps. Elimination of the gaps reduces the frequency of failure of arresters as a result of surface contamination, which can cause arrester gap spark over. Metal-oxide valve-type arresters without gaps act quickly to limit the surge voltage, improving the electric system protection.

The present generation of MOV surge arresters produced by some manufacturers have been designed with a combination of metal oxide and gapped arrester technologies. For the gapped MOV arrester, during steady state conditions, the line voltage is shared by both the gap structure and the MOVs. When surges occur, the gaps spark over, leaving only the

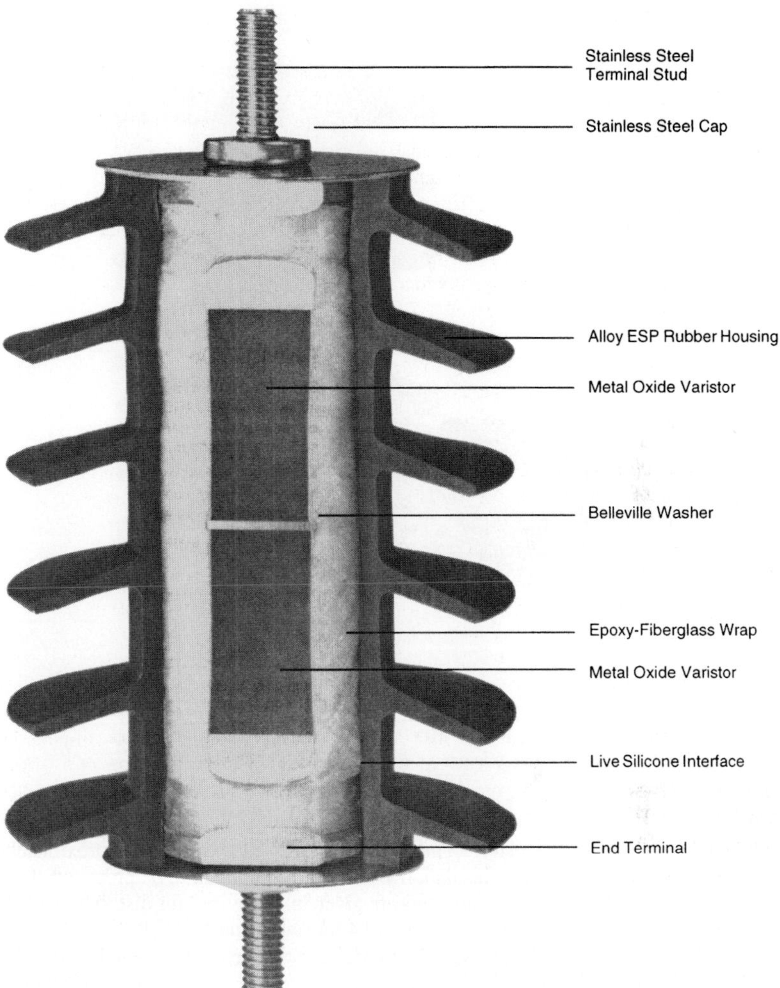

FIGURE 16.7 Sectional view of metal-oxide-varistor (MOV) valve distribution-class surge arrester with zinc-oxide-valve blocks encased in a polymer to prevent sharp flying porcelain fragments if the arrester should fail. (*Courtesy Ohio Brass Co.*)

MOVs in the circuit. The result is a lower discharge voltage and increased arrester front-of-wave (FOW) sparkover voltage withstand ability than exhibited by the gapless MOV surge arrester.

Surge Arrester Application. Application of surge arresters requires knowledge of the maximum phase-to ground system voltage, maximum discharge voltage of the arrester, and basic insulation level (BIL) of the equipment to be protected. The BIL for distribution systems is provided in Table 16.1. The basic-impulse insulation level is the crest value of voltage of a standard full-impulse voltage wave that the insulation will withstand without failure. The surge arrester must be selected so that it will not conduct current as a result of maximum

FIGURE 16.8 A 8.4 kV MCOV/10 kV rated overhead distribution MOV surge arrester. (*Courtesy Ohio Brass Co.*)

system operating voltage and will have a maximum-surge discharge voltage approximately 20 percent less than the basic impulse insulation level of the equipment to be protected. Typical descriptive data for valve-type surge arresters are given in Table 16.2 for distribution class, in Table 16.3 for intermediate class, and in Table 16.4 for station class.

The metal-oxide valve arresters with zinc oxide blocks have a very small leakage current at normal system line-to-ground voltage. The zinc oxide disks go into conduction sharply at a precise voltage level and cease to conduct when the voltage falls below this level. Valve arresters with either zinc oxide or silicon-carbide-valve blocks with the same rating will provide adequate protection for the line and its equipment with approximately the same stress on the protected facilities as a result of the arrester discharge voltage. However, the uniformity of the operation of the zinc oxide disks without the need for series gaps permits the selection of a lower-rated voltage arrester, which will provide better protection for the line and its equipment. When a metal-oxide arrester with zinc oxide blocks is chosen, it is necessary to be sure that the arrester's maximum power-frequency continuous operating voltage (MCOV) should not be exceeded. Typical descriptive data for MOV arresters is provided in Tables 16.5 through 16.8. Table 16.9 lists the commonly applied voltage ratings of distribution surge arresters. The MOV arresters are defined as (a) normal duty, (b) heavy duty, (c) riser pole, and (d) station class. Note the difference in the characteristics in the normal and heavy duty arresters described in Tables 16.5 and 16.6. Specifically, the normal duty distribution arrester IEEE Standard C62.11 design and test requirements are: (a) Low current, long duration; 20 surges with 2000 μs duration and 75A

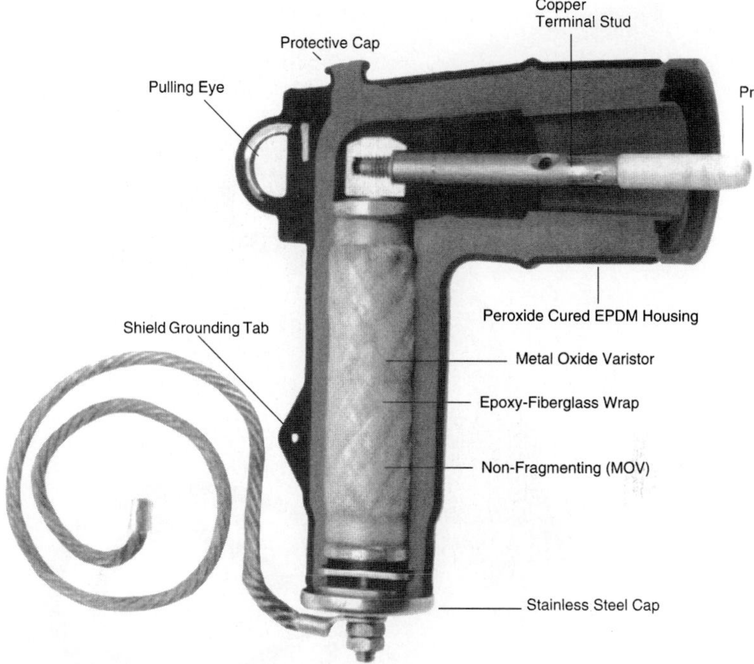

FIGURE 16.9 Sectional view of a dead-front MOV elbow surge arrester utilized in an underground distribution system. *(Courtesy Ohio Brass Co.)*

magnitude, (b) High current, short duration; 2 surges with 4/10 μs duration and 65-kA magnitude, and (c) Duty cycle, 22 5-kA discharges. The heavy duty distribution arrester IEEE Standard C62.11 design and test requirements are: (a) Low current, long duration; 20 surges with 2000 μs duration and 250-A magnitude, (b) High current, short duration; 2 surges with 4/10 μs duration and 100-kA magnitude, and (c) Duty cycle, 20 10-kA discharges and 2 40-kA discharges.

A few surge arrester application examples are as follows:

Surge Arrestor Selection Criteria

Maximum phase-to ground system voltage

Maximum discharge voltage

Maximum continuous operating voltage (MCOV)

Basic impulse level of the equipment to be protected

Insulation Protective Margin Equation

$$\text{Protective margin} = \left[\left(\frac{\text{Insulation withstand}}{V_{\max}}\right) - 1\right] \times 100$$

The equation is taken from IEEE Standard C62.2, *Application Guide for Surge Arresters*. The minimum recommended margin is 20%.

FIGURE 16.10 A 8.4 kV MCOV/10 kV rated underground distribution dead-front MOV elbow surge arrester. (*Courtesy Ohio Brass Co.*)

Example: Determine the margin of protection for both the normal duty and heavy duty MOV arrester on a 15-kV class, 13,200/7620 volt, four wire, overhead system utilizing the 20 kA 8/20kV discharge voltage.

Solution: From Table 16.1, the 15-kV equipment class BIL is 95, from Table 16.9, a 10-kV rated arrestor is utilized. The corresponding maximum discharge voltage is 41.5 for the normal duty arrester (Table 16.5) and 37.5 for the heavy duty arrester (Table 16.6).

$$\text{Protective margin} = \left[\left(\frac{\text{Insulation withstand}}{V_{\text{max}}}\right) - 1\right] \times 100$$

$$\text{Normal duty arrester protective margin} = [(95/41.5)-1] \times 100 = 129\%$$

$$\text{Heavy duty arrester protective margin} = [(95/37.5)-1] \times 100 = 153\%$$

Not included are the effects of faster rates of current rise on the discharge voltage, reduced insulation levels due to various factors, and line and ground leads.

Example: Determine the margin of protection for a 15-kV underground system utilizing only a riser pole arrester and then utilizing both a riser pole arrester and an elbow arrester at the open point. Assume arrester lead length to be 5 ft and utilize the 20 kA 8/20kV discharge voltage.

Solution:

$$\text{Protective margin} = \left[\left(\frac{\text{Insulation withstand}}{V_{\text{max}}}\right) - 1\right] \times 100$$

FIGURE 16.11 MOV station surge arrester utilized on a substation feeder primary riser pole.

FIGURE 16.12 Station-class metal-oxide surge arresters installed in a substation. (*Courtesy Ohio Brass Co.*)

TABLE 16.1 Basic Insulation Levels for Distribution Systems

Voltage	Basic Insulation Level (BIL)
5 kV	70
15 kV	95
25 kV	125
35 kV	150

Protective margin calculation—Arrester only at the riser pole

In the case of utilizing only an arrester at the riser pole, V_{max} consists of both the arrester discharge voltage and the arrester lead length calculated at 1.6 kV per foot. V_{max} is doubled due to the reflection of the wave at the open point in the underground system.

V_{max} = 2 × (discharge voltage of the arrester + voltage drop due to arrester lead length)

From Table 16.7, the 10 kV, 20 kA, 8/20 maximum discharge voltage is 29.8. The voltage drop due to a 5 ft lead length is 5 × 1.6 or 8.0 kV.

$$V_{max} = 2 \times (29.8 + 8.0) = 75.6\text{kV}$$

$$\text{Protective margin} = \left[\left(\frac{\text{Insulation withstand}}{V_{max}}\right) - 1\right] \times 100$$

$$\text{Protective margin} = [(95/75.6) - 1] \times 100 = 25.7\%$$

Protective margin calculation—Arrester at both the riser pole and the open point

In the case of utilizing an arrester at the riser pole and at the open point of the underground system, V_{max} consists of the arrester discharge voltage, the arrester lead length calculated at 1.6 kV per foot and some percentage of the front of wave voltage of the open point arrester. (The percentage is typically between 15 and 40. For this calculation we will use a value of 30%.)

V_{max} = (discharge voltage of the arrester + arrester lead length voltage drop + 0.3 × open point front of wave (FOW) voltage)

From Table 16.7, the 10 kV, 20 kA, 8/20 maximum discharge voltage is 29.8. The voltage drop due to a 5 ft lead length is 5 × 1.6 for 8.0 kV. Finally from Table 16.8, the open point FOW voltage of a 10-kV arrester is 38.7. For purposes of the calculation, 30% of 38.7 equals 11.61.

$$V_{max} = (29.8 + 8.0 + 11.61) = 49.41\text{ kV}$$

$$\text{Protective margin} = \left[\left(\frac{\text{Insulation withstand}}{V_{max}}\right) - 1\right] \times 100$$

$$\text{Protective margin} = [(95/49.41) - 1] \times 100 = 92.3\%$$

TABLE 16.2 Typical Application Data for Silicon Carbide Distribution-Class, Nonlinear Resistor Valve-Type Lightning Arrestors with Gaps (Electrical Characteristics)

Arrester rating max. line to ground, kV rms	Minimum 60-Hz sparkover, kV rms	Maximum $1\frac{1}{2} \times 40$ wave kV crest	Maximum impulse sparkover ANSI front-of-wave kV crest, maximum	Discharge voltage in kV crest for discharge currents of 8×20 microsecond waveshape with following maximum crest amplitudes				
				1500-amp kV crest, maximum	5000-amp kV crest, maximum	10,000-amp kV crest, maximum	20,000-amp kV crest, maximum	65,000-amp kV crest, maximum
3	6	14	17	10	12.4	13.8	15.5	19
6	11	26	33	19	23	26	29	35
9	18	38	47	30	36.5	41	46	56
10	18	38	48	30	36.5	41	46	56
12	23.5	49	60	38	46	52	58	71
15	27	50	75	45	55	62	70	85
18	33	58	90	54	66	74	83	103

TABLE 16.3 Typical Application Data for Silicon Carbide Intermediate-Class, Nonlinear Resistor Valve-Type Lightning Arrestors with Gaps (Electrical Characteristics)

Arrester rating kV rms maximum valve-off or maximum reseal rating	Maximum circuit voltage phase to phase, kV rms		Maximum front-of-wave impulse sparkover, kV crest	Maximum 100% impulse sparkover 1.2 × 50 wave, kV crest	Minimum 60-Hz sparkover, kV rms	Maximum discharge voltage with discharge current, 8 × 20 wave, kV crest				
	Ungrounded neutral 100% arrester	Effectively grounded neutral 80% arrester				1.5 kA	3 kA	5 kA	10 kA	20 kA
3	3	3.75	11	11	4.5	5.2	6	6.6	7.5	8.7
4.5	4.5	5.63	16	15	6.8	7.8	9	9.9	11.3	13.1
6	6	7.50	21	19	9	10.4	11.9	13.2	15	17.4
7.5	7.5	9.38	26	23.5	11.3	13	14.9	16.5	18.8	21.8
9	9	11.25	31	27.5	13.5	15.6	17.9	19.8	22.5	26.1
10	10	12.50	35	31	15	17.5	20.0	22.0	25.0	29.0
12	12	15.00	40	35.5	18	20.8	23.8	26.4	30	34.8
15	15	18.75	50	43.5	22.5	25.9	29.7	32.9	37.5	43.5
18	18	22.50	59	51.5	27	31.1	35.7	39.5	45	52.2
21	21	26.25	68	59	31.5	36.3	41.6	46.1	52.5	60.9
24	24	30.00	78	67	36	41.5	47.6	52.7	60.0	69.6
27	27	33.75	88	75	40.5	46.7	53.5	59.2	67.5	78.3
30	30	37.50	97	81	45	51.8	59.4	65.8	75	87
36	36	45.00	116	95	54	62.2	71.3	79	90	104.4
39	39	48.75	126	102	58.5	67.4	77.3	85.5	97.5	113.1
45	45	56.25	144	116	67.5	77.7	89.1	98.7	112.5	130.5
48	48	60.0	154	123	72	82.9	95.1	105.3	120	139.2
60	60	75.00	190	153	90	103.6	119	131.2	150	174
72	72	90.00	228	180	108	124.3	142.6	158	180	208.8
78	78	97.50	245	195	117	134.7	154.5	171	185	226.2
84	84	105.00	262	209	126	145	166.4	184.2	210	243.6
90	90	112.50	282	223	135	155.4	178.2	197.3	225	261
96	96	120.00	300	236	144	165.7	190.1	210.5	240	278.4
108	108	135.00	335	263	162	186.5	213.9	237	270	313.1
120	120	150.00	370	290	180	207.2	238	263	300	347.9

TABLE 16.4 Typical Application Data for Silicon Carbide Station-Class, Nonlinear Resistor Valve-Type Lightning Arrestors with Gaps (Electrical Characteristics)

Arrester rating, maximum permissible line-to-ground kV rms (maximum valve-off or maximum reseal rating)	Minimum 60-Hz sparkover, kV rms	Maximum switching surge sparkover, kV crest	Maximum 100% $1\frac{1}{2} \times 40$ impulse sparkover, kV crest	Maximum impulse sparkover ASA front-of-wave, kV crest	Maximum discharge voltage for discharge currents of 8×20 microsecond waveshape with the following crest amplitudes					
					1500 amp, kV crest	3000 amp, kV crest	5000 amp, kV crest	10,000 amp, kV crest	20,000 amp, kV crest	40,000 amp, kV crest
3	5	8	8	12	7	8	8.5	9	10	11.5
6	11	17	17	24	15	16	17	19	20	23
9	16	25	24	35	21	23	24	26	28	31.5
12	22	34	32	45	28	30	32	35	38	42
15	27	42	40	55	35	38	40	44	47	52.5
21	36	56	55	72	47	52	55	60	65	73
24	45	70	65	90	56	61	65	71	76	84
30	54	85	80	105	69	75	80	87	94	105
36	66	104	96	125	83	90	96	105	113	126
39	72	113	104	130	90	98	104	114	123	137
48	90	141	130	155	112	122	130	142	153	168
60	108	169	160	190	137	150	160	174	189	210
72	132	206	195	230	167	184	195	212	230	252
90	160	242	228	271	206	226	240	262	283	315
96	175	257	247	294	222	242	258	280	304	336
108	195	294	266	332	244	264	282	316	333	378
120	220	323	304	370	275	301	320	350	378	420

TABLE 16.5 Normal Duty Polymer Housed Metal-Oxide Valve Distribution Class Surge Arrester Electrical Characteristics

Rated voltage, kV	MCOV, kV	0.5 μsec 5 kA maximum IR, kV	500 A switching surge maximum IR, kV	8/20 maximum discharge voltage, kV					
				1.5 kA	3 kA	5 kA	10 kA	20 kA	40 kA
3	2.55	12.5	8.5	9.8	10.3	11.0	12.3	14.3	18.5
6	5.1	25.0	17.0	19.5	20.5	22.0	24.5	28.5	37.0
9	7.65	33.5	23.0	26.0	28.0	30.0	33.0	39.0	50.5
10	8.4	36.0	24.0	27.0	29.5	31.5	36.0	41.5	53.0
12	10.2	50.0	34.0	39.0	41.0	44.0	49.0	57.0	74.0
15	12.7	58.5	40.0	45.5	48.5	52.0	57.5	67.5	87.5
18	15.3	67.0	46.0	52.0	56.0	60.0	66.0	78.0	101.0
21	17.0	73.0	49.0	55.0	60.0	64.0	73.0	84.0	107.0
24	19.5	92.0	63.0	71.5	76.5	82.0	90.5	106.5	138.0
27	22.0	100.5	69.0	78.0	84.0	90.0	99.0	117.0	151.5
30	24.4	108.0	72.0	81.0	88.5	94.5	108.0	124.5	159.0
36	29.0	134.0	92.0	104.0	112.0	120.0	132.0	156.0	202.0

Source: Ohio Brass Co.

TABLE 16.6 Heavy Duty Polymer Housed Metal-Oxide Valve Distribution Class Surge Arrester Electrical Characteristics

Rated voltage, kV	MCOV, kV	0.5 μsec 10 kA maximum IR, kV	500 A switching surge maximum IR, kV	8/20 maximum discharge voltage, kV					
				1.5 kA	3 kA	5 kA	10 kA	20 kA	40 kA
3	2.55	12.5	8.0	9.5	10.0	10.5	11.0	13.0	15.3
6	5.1	25.0	16.0	19.0	20.0	21.0	22.0	26.0	30.5
9	7.65	34.0	22.5	24.5	26.0	27.5	30.0	35.0	41.0
10	8.4	36.5	23.5	26.0	28.0	29.5	32.0	37.5	43.5
12	10.2	43.5	28.2	38.0	32.9	34.8	38.5	43.8	51.5
15	12.7	54.2	35.0	38.4	41.0	43.4	48.0	54.6	64.2
18	15.3	65.0	42.1	46.0	49.1	52.0	57.5	65.4	76.9
21	17.0	69.5	44.9	49.2	52.5	55.7	61.5	69.9	82.2
24	19.5	87.0	56.4	61.6	65.8	69.6	77.0	87.6	103.0
27	22.0	97.7	63.2	69.2	73.9	78.2	86.5	98.4	115.7
30	24.4	108.4	70.0	76.8	82.0	86.8	96.0	109.2	128.4
36	29.0	130.0	84.2	92.0	98.2	104.0	115.0	130.8	153.8

Source: Ohio Brass Co.

TABLE 16.7 Riser Pole Polymer Housed Metal-Oxide Valve Distribution Class Surge Arrester Electrical Characteristics

Duty cycle voltage rating, kV	MCOV, kV	0.5 μsec 10 kA maximum IR, kV	500 A switching surge maximum IR, kV	8/20 maximum discharge voltage, kV					
				1.5 kA	3 kA	5 kA	10 kA	20 kA	40 kA
3	2.55	8.7	5.8	6.5	7.0	7.4	8.1	9.0	10.6
6	5.1	17.4	11.7	13.0	14.0	14.7	16.2	18.1	21.1
9	7.65	25.7	17.5	19.3	21.0	21.9	24.0	27.0	31.6
10	8.4	28.5	19.2	21.2	23.0	24.0	26.5	29.8	34.8
12	10.2	34.8	23.3	25.9	28.0	29.4	32.3	36.2	42.2
15	12.7	43.1	29.1	32.3	35.0	36.6	40.2	45.1	52.7
18	15.3	51.4	34.9	38.6	41.9	43.8	48.0	54.0	63.2
21	17	57.6	38.7	42.8	46.4	48.6	53.6	60.2	70.5
24	19.5	68.8	46.6	51.6	55.9	58.5	64.2	72.1	84.3
27	22	77.1	52.4	57.9	62.9	65.7	72.0	81.0	94.8
30	24.4	85.5	57.6	63.5	69.0	72.0	79.5	89.4	104.4
36	29	102.8	69.8	77.2	83.8	87.6	96.0	108.0	126.6

Source: Ohio Brass Co.

TABLE 16.8 Deadfront Elbow Arrester Electrical Characteristics

Voltage rating (kV rms)	MCOV (kV rms)	Equivalent F.O.W.* (kV crest)	Maximum discharge voltage (kV crest) using an 8/20 μs current wave				
			1.5 kA	3 kA	5 kA	10 kA	20 kA
3	2.55	12.9	10.6	11.2	11.7	13.0	15.7
6	5.1	25.8	21.3	22.3	23.5	25.9	31.5
10	8.4	38.7	31.9	33.5	35.2	38.9	47.2
12	10.2	51.6	42.5	44.8	46.9	52.0	62.9
18	15.3	71.1	58.5	61.4	64.6	71.3	86.2
24	19.5	96.8	79.7	84.0	87.9	97.5	117.9
27	22.0	110.0	90.0	94.9	99.7	110.1	133.7

*The equivalent front-of-wave is the maximum discharge voltage for a 5 kA impulse current wave which produces a voltage wave cresting in 0.5 μs.

Source: Joslyn Manufacturing Co.

TABLE 16.9 Commonly Applied Voltage Ratings of Arresters

System Voltage (kV rms)		Recommended Arrester Rating per IEEE C62.22 (kV rms)		
Nominal	Maximum	Four-Wire Wye Multi-Grounded Neutral	Three-Wire Wye Solidly Grounded Neutral	Delta and Ungrounded Wye
2.4	2.54	—	—	3
4.16Y/2.4	4.4Y/2.54	3	6	6
4.16	4.4	—	—	6
4.8	5.08	—	—	6
6.9	7.26	—	—	9
8.32Y/4.8	8.8Y/5.08	6	9	—
12.0Y/6.93	12.7Y/7.33	9	12	—
12.47Y/7.2	13.2Y/7.62	9	15	—
13.2Y/7.62	13.97Y/8.07	10	15	—
13.8Y/7.97	14.52Y/8.38	10	15	—
13.8	14.52	—	—	18
20.78Y/12.0	22Y/12.7	15	21	—
22.86Y/13.2	24.2Y/13.87	18	24	—
23	24.34	—	—	30
24.94Y/14.4	26.4Y/15.24	18	27	—
27.6Y/15.93	29.3Y/16.89	21	30	—
34.5Y/19.92	36.5Y/21.08	27	36	—
46Y/26.6	48.3Y/28	36	—	—

Protector Tubes. Another device used to protect overhead transmission lines from lightning discharges is the protector tube. Basically, the protector tube consists of a fiber tube with an electrode in each end. It is designed so that the impulse voltage breakdown through the tube is lower than that of the line insulation to be protected. On a transmission tower the tube is often mounted below the conductor, as shown in Fig. 16.13. The upper electrode is connected to an arc-shaped horn located at the proper distance below the conductor. The arc-shaped horn maintains a constant length of external gap between the upper electrode and the line conductor, even though the insulator string swings from side to side. The lower electrode to be solidly grounded. An installation in which the protector tubes hang downward beside the insulator strings is shown in Fig. 16.14.

When lightning strikes the line, external and internal series gaps, break down instead of flashing over the insulator string because the tube has lower breakdown voltage. After breakdown of the gaps, the power follow current volatilizes a small layer of the fiber wall, and the gas given off mixes in the arc to help deionize the space between the electrodes. A pressure is built up in the tube, and the hot gases are discharged through the lower electrode, which is hollow. If the deionizing action is sufficiently strong and if the voltage does not build up too rapidly across the tube, the arc will go out at a current zero and will not be reestablished. While the tube is discharging it is a good conductor, but after the arc is extinguished, it becomes a good insulator again.

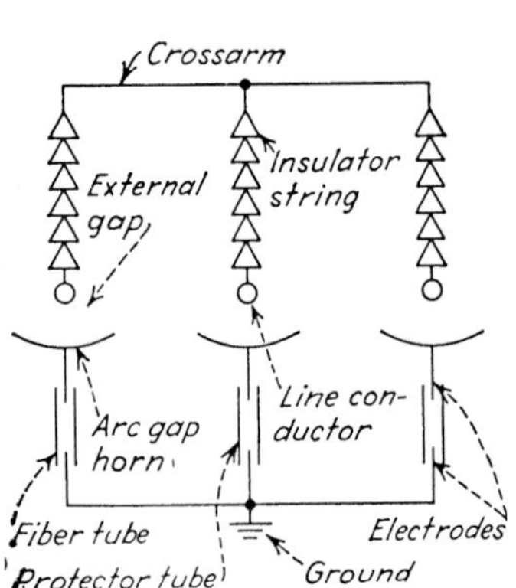

FIGURE 16.13 Diagrammatic representation of protector tube installation on a three-phase tower line.

FIGURE 16.14 Installation of expulsion protective gaps on a 115,000-volt three-phase transmission line. Tubes are mounted on one side of insulator string instead of below conductor.

CHAPTER 17
FUSES

Fuses are relatively inexpensive protection devices connected into circuits to open the circuit and de-energize the apparatus to prevent or limit damage due to an overload or short circuit. Fuses are used to protect the electric system or source from interruption or damage when a short circuit or overload occurs on a circuit beyond the fuse. A fuse is an intentionally weakened spot in an electric circuit. It utilizes an element made of silver, tin, lead, copper, or an alloy such as tin-lead, which will melt at a predetermined current maintained for a predetermined length of time. When the current through such a metal becomes excessive, the resistance offered by the metal to the flow of current develops enough heat to melt the metal, thereby opening the circuit before abnormal current can damage the electric source or any connected apparatus.

Fuses are usually enclosed to prevent the molten metal from flying and doing damage or causing a fire. Enclosing the fuse aids in quenching the arc. The melting of the fuse is often accompanied by a puff of smoke and vaporized metal. This action is sometimes referred to as the *blowing* of the fuse.

Fuses can be broadly classified into low- and high-voltage fuses. Low-voltage fuses are of the plug or cartridge type. High-voltage fuses commonly used on an electric distribution system are the expulsion, open-link, current limiting, liquid, and boric acid types.

Plug-Type Fuses. Low-voltage plug-type fuses can be used for residential services and for ordinary lighting branch circuits on panel boards. The fuse consists of a small cup of solid insulating material within which the fusible wire connects the center contact with the outer metal screw shell, as shown in Fig. 17.1. The plug is the same size as the ordinary incandescent-lamp base.

Cartridge Fuse. Cartridge fuses are used on circuits rated 600 volts and below. Fuses are manufactured differently for use on 240- and 480-volt circuits. A fuse of the proper voltage and current rating should be used for the circuit application. In the cartridge fuse the fusible element is contained within a completely enclosed insulating tube. This tube is called the *cartridge*, because of its resemblance to an ordinary shotgun cartridge. Cartridge fuses in which the element is not replaceable after the fuse has blown are called *nonrenewable fuses*. Cartridge fuses in which the fusible element may be readily replaced by the

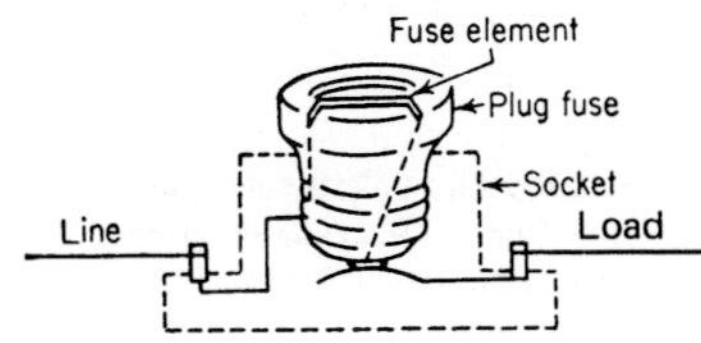

FIGURE 17.1 Plug fuse screwed in place in socket. Fusible element is contained in an insulated cup. Cup has transparent cover to permit inspection of fusible element.

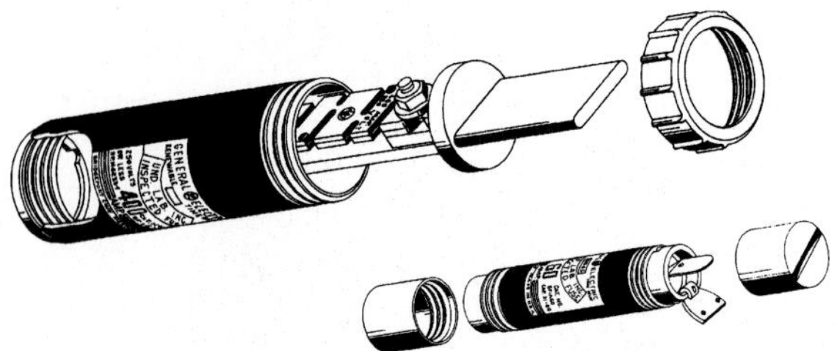

FIGURE 17.2 Renewable cartridge-type fuse. Only the blade shown in the inside of the tube need be replaced in case the fuse blows. (*Courtesy General Electric Co.*)

user with suitable renewal elements supplied by the manufacturer are called *renewable fuses* (Fig. 17.2). The cartridge or tube may thus be used repeatedly with the new renewal elements. In this way, a considerable saving is effected, especially in the larger fuse sizes.

Distribution Cutouts. A distribution cutout provides a high-voltage mounting for the fuse element used to protect the distribution system or the equipment connected to it (Fig. 17.3). Distribution cutouts are used with installations of transformers, capacitors, cable circuits, and sectionalizing points on overhead circuits (Fig. 17.4). Enclosed, open, and open-link cutouts are used for different distribution-circuit applications. Cutouts normally use an expulsion fuse. An expulsion fuse operates to isolate a fault or overload from a circuit as a result of the arc from the fault current eroding the tube of the fuse holder, producing a gas that blasts the arc out through the fuse tube vent or vents.

The mechanical differences between enclosed, open, and open-link cutouts are in their external appearance and method of operation. Enclosed cutouts have terminals, fuse clips, and fuse holders mounted completely within an insulating enclosure. Open cutouts have these parts completely exposed as the name indicates. Open-link cutouts have no integral fuse holder; the arc-confining tube for the cutouts is incorporated in the fuse link.

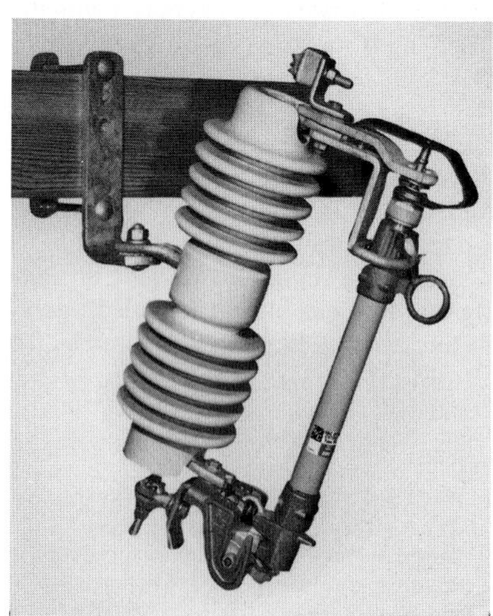

FIGURE 17.3 Distribution cutout with expulsion fuse mounted. Fuse is in the closed position, completing circuit between top and bottom terminals. (*Courtesy S&C Electric Co.*)

FIGURE 17.4 Distribution cutouts installed in transition point from an overhead electric distribution system to a three-phase underground system. The underground primary cable system is protected by the fused cutouts.

The construction of the cutout fuse holder provides for a dropout operation. Fuses are manufactured with an expendable cap.

Enclosed Distribution Cutout. An enclosed distribution fuse cutout is one in which the fuse clips and fuse holder are mounted completely within an enclosure. A typical enclosed cutout, as shown in Fig. 17.5 and Fig. 17.6, has a porcelain housing and a hinged door supporting the fuse holder. The fuse holder is a hollow vulcanized-fiber expulsion tube. The fuse link is placed inside the tube and connects with the upper and lower line terminals when the door is closed. When the fuse blows or melts because of excessive current passing through it, the resultant arc attacks the walls of the fiber tube, producing a gas which blows out the arc. The melting of the fusible element of some cutouts causes the door to drop open, signaling to the lineman that the fuse has blown.

Each time the fuse blows, a small amount of the vulcanized fiber of the expulsion tube is eroded away. The larger the value of the current interrupted, the more material is

FIGURE 17.5 Enclosed primary cutout. (*Courtesy ABB Power T&D Company Inc.*)

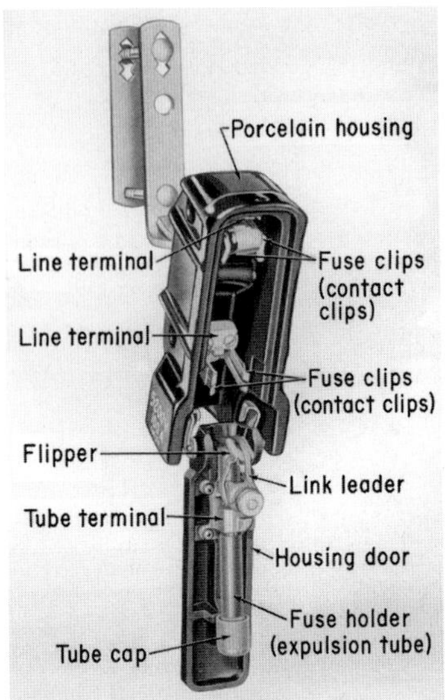

FIGURE 17.6 Enclosed primary cutout showing door and cartridge assembly.

FIGURE 17.7 Open porcelain insulator body distribution cutout with arc chute designed for load break operation. (*Courtesy ABB Power T&D Company Inc.*)

consumed. In general, a hundred or more operations of average current values can be successfully performed before the cutout fails.

Enclosed cutouts can be arranged to indicate when the fuse link has blown by dropping the fuse holder. The enclosed cutout is designed and manufactured for operation on distribution circuits of 7200 volts and below. The standard current ratings of the cutouts are 50, 100, or 200 amps.

Open Distribution Cutout. Open-type cutouts are similar to the enclosed types except that the housing is omitted (Fig. 17.7). The open-type cutout is designed and manufactured for all distribution-system voltages. The open-type is made for 100- or 200- amp operation. Some cutouts can be uprated from 100 to 200 amps by using a fuse tube with dual venting action rated for 200-amp operation. The open-type cutouts are manufactured with a porcelain insulator body (Fig. 17.7) or a polymer insulator body (Fig. 17.8). The polymer insulator body is much lighter and thus easier to handle during installation.

Open-Link Distribution Cutout. This type of cutout differs from the open cutout in that it does not employ the fiber expulsion tube. The fuse link is supported by spring terminal contacts, as shown in Fig. 17.9. An arc-confining tube surrounds the fusible element of the link. During fault clearing, the spring contacts provide link separation and arc stretching. The arc-confining tube furnishes the necessary expulsion action for circuit interruption. The open-link primary cutout cannot interrupt large values of fault current, and therefore its use is limited.

Primary Fuse Links. A primary fuse link consists of the button, upper terminal, fusible element, lower terminal, leader, and sheath, as shown in Fig. 17.10. The button is the upper terminal and the leader is the lower terminal. Fuse links for open-link cutouts are similar except that they have pull rings at each end, as shown in Fig. 17.11. In either case, the sheath aids in the interruption of low-value faults, and it provides protection against damage during handling.

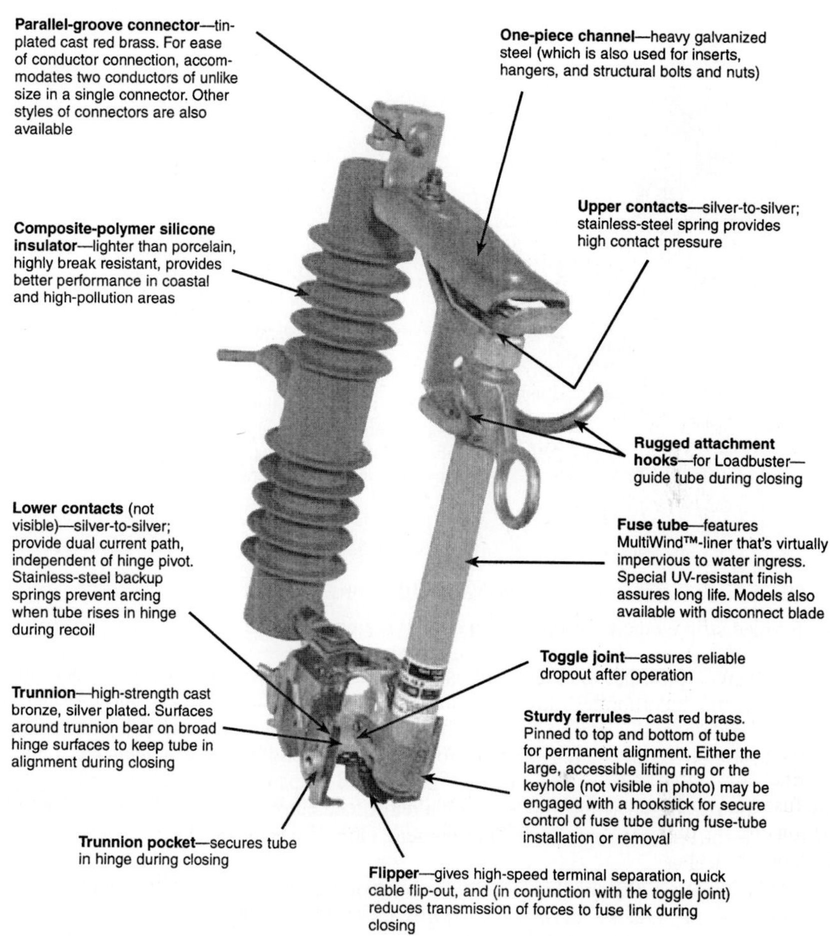

FIGURE 17.8 Open polymer insulator body cutout. (*Courtesy S&C Electric Co.*)

Standards specify the size of the fuse holder into which the link must fit freely. Links rated 1 to 50 amps must fit into a 5/6-in-diameter holder, 60- to 100-amp links must fit into a 7/17-in-diameter holder, and 125- to 200-amp links must fit into a 3/4-in-diameter holder. Links must withstand a 10-lb pull while carrying no load current, but are generally given a 25-lb test.

Fusible elements are made in a wide variety of designs, one group of which is shown in Fig. 17.12. Most silver-element fuse links utilize the helically coiled construction shown in Fig. 17.10. This construction permits the fusible element to absorb, without damage, vibration as well as thermal shock due to current surges and heating and cooling throughout the daily load cycle. The fusible elements shown in Fig. 17.12 are made in four designs:

1. High-surge dual elements for links rated 1 to 8 amps
2. Wire element for links rated 5 to 20 amps

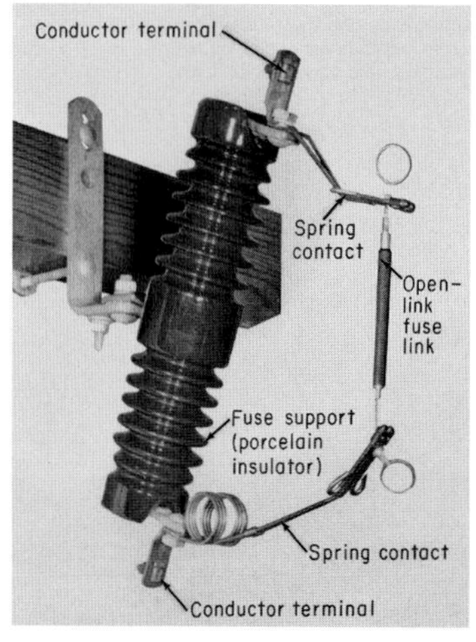

FIGURE 17.9 Open-link primary cutout showing spring terminal contacts and fuse link enclosed in an arc-confining tube. During fault clearing, the spring contacts provide link separation and arc stretching. (*Courtesy S&C Electric Co.*)

3. Die-cast tin element for links rated 25 to 100 amps
4. Formed strip element for links rated over 100 amps

Link designs are further distinguished as strain or spring type. The fusible element of a strain-type link incorporates a high-resistance strain wire for strengthening purposes, while the fusible element of a spring-type link has, in addition to the strain wire, a pretensioned spring to stretch the arc, thereby aiding circuit clearing after the fusible element has melted.

The high-surge-element link of fuses rated 1 to 8 amps (Fig. 17.12*a*) is a spring-type link having two heater wires in series with a reinforced eutectic alloy solder joint. The solder joint is the central component of the element. One wire is fastened to the shank, and the other to the leader.

Links rated 5 to 20 amps (Fig. 17.12*b*) are also of the spring type, consisting of a fusible tin wire and a separate strain wire. Both wires are soldered to the shank and leader.

Links rated 25 to 100 amps (Fig. 17.12*c* and *d*) are the strain type with die-cast elements, whereas links rated over 100 amps (Fig. 17.12*e*) are the strain type with formed strip elements. The strip element is lap-soldered to the shank and the leader, and the strain wire is securely fastened to the shank and leader.

Fuse Link Operation. When a fault occurs, the fusible element is melted by the excessive current, and an arc forms across the open gap. The arc is temporarily sustained in a conducting path of gaseous ionized arc products. Gas pressure builds rapidly, and this pressure, acting in conjunction with a spring-loaded flipper at the lower end of the fuse tube, rapidly ejects the fuse link leader, lengthening and cooling the arc. For low values of fault current, the arc acts on the fuse link sheath, generating considerable amounts of deionizing gases. When the current passes through the next zero value, as it changes the direction of flow, the arc is interrupted (Fig. 17.13). As the voltage increases again across the opening in the fuse link, the arc attempts to reestablish itself. A restrike, however, is prevented by the deionizing gases which will have rebuilt the dielectric strength of the open gap. For large values of fault current, the sheath is rapidly destroyed, and the arc erodes fiber from the inner wall of the cutout tube, generating large amounts of deionizing gas. During the fault interruption process, the cutout expels large amounts of gas under very high pressure

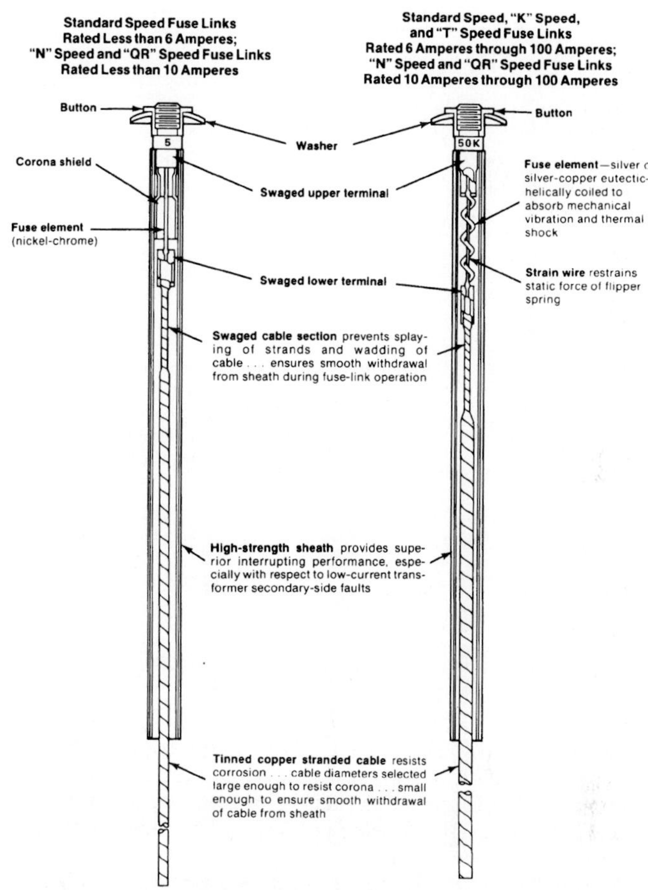

FIGURE 17.10 Typical primary cutout fuse link showing component parts: button, upper terminal, fuse element, lower terminal, cable (leader), and sheath. Helically coiled fuse element is typical of silver-element fuse links. (*Courtesy S&C Electric Co.*)

as well as some fuse link particles (Fig. 17.5). There may also be a very loud noise reported by individuals living in or near the vicinity. Accordingly, when one is working in the vicinity of a fuse cutout, care should be taken to stay clear of the exhaust path. In addition, when closing a cutout, it is good practice to look down, away from the discharge path, since there is always the possibility of closing into a short circuit (Fig. 17.14).

Time-Current Curves. A *time-current curve* is a curve that is plotted between the magnitude of a fault current and the time required for the fuse link to open the circuit. It is obvious that the greater the current, the faster the fuse melts and the shorter the time required for it to blow. Figure 17.15 is a typical time-current curve for a 10K fuse link (normal current rating 10 amps). Thus, a current of 10 amps would not cause the fuse to blow; 20 amps, which is twice normal, would require 300 sec or 5 min; 30 amps would require 3 sec; and 100 amps would require about 0.15 sec. These time-current curves are useful in providing required coordination of fuses, reclosers, and sectionalizers.

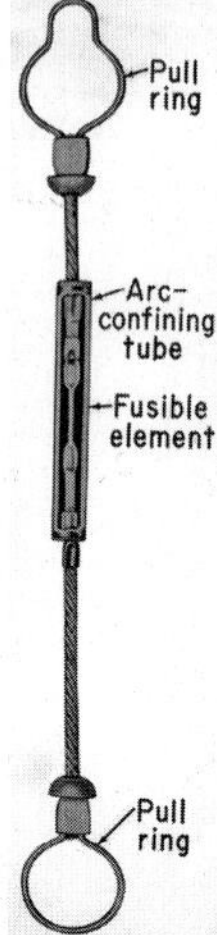

FIGURE 17.11 Open primary cutout fuse link provided with pull rings for handling with hot-line tools. Hooks are inserted in rings for placement in cutout. Cutaway shows fusible element inside arc-confining tube.

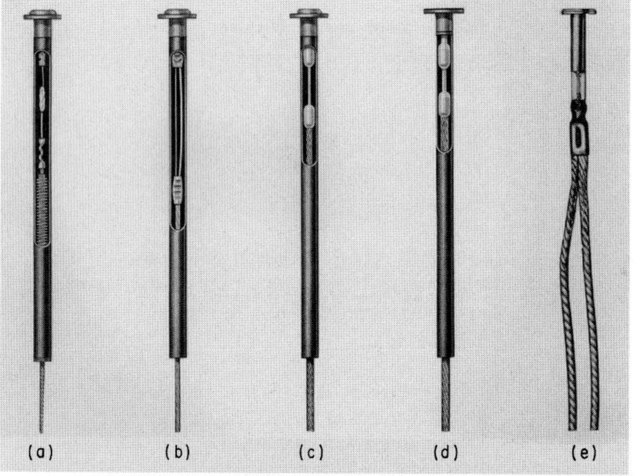

FIGURE 17.12 Primary cutout fuse links having different fusible elements. (*a*) High-surge dual element (1 to 8 amps), (*b*) wire element (5 to 20 amps), (*c*) die-cast tin element (25 to 50 amps), (*d*) die-cast tin element (65 to 100 amps), (*e*) formed strip element (140 to 200 amps).

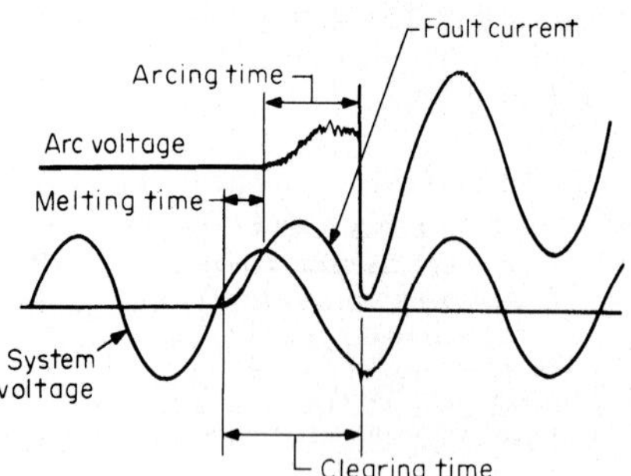

FIGURE 17.13 Diagram of voltages, current, and timing reference recorded with an oscillograph to show fuse operation.

FIGURE 17.14 Lineman preparing to close cutout. Note that lineman is wearing appropriate protective equipment, is positioned to close the fuse cutout with a final vigorous thrust while looking down and away, and is located well away from the cutout exhaust path, recognizing that the cutout will operate if closed into a fault. (*Courtesy Northwest Lineman College.*)

Fuse links are identified by their amperage ratings and letter designations such as K, T, N, H, and QR. Links of the same current rating have about the same 300-sec points on the time-current curve, but they have widely varying shapes of time-current curves, depending on the type of metal used for the fusible element and the link's letter designation. For example, Fig. 17.16 shows the time-current curves for a number of fuse links that could be used to protect a specific capacitor bank. This plot illustrates graphically why it is important to

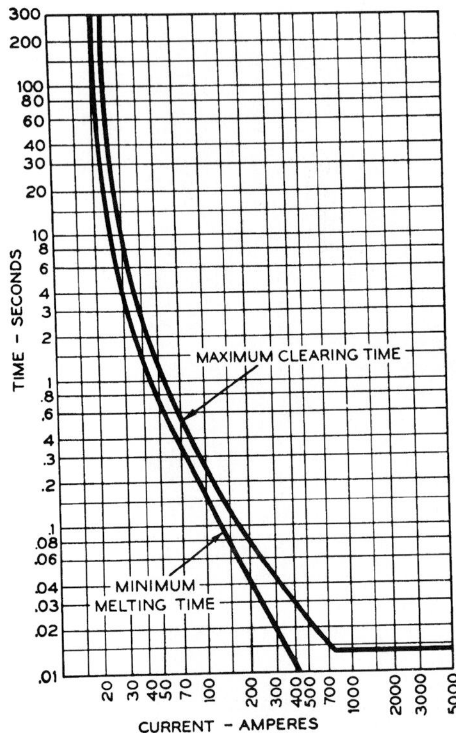

FIGURE 17.15 Typical time-current curves for 10K fuse link. Fault currents are plotted along the abscissa, and time to blow is along the ordinate.

replace a blown fuse link with the same type link or one that is called for in the system fusing schedule. Failure to do so may result in miscoordination with other protective devices and needless electric system outages and damage to equipment. Careless storage and handling of fuse links may also result in incorrect operation.

Fuse links designated as K (kwick) and T (tardy) with the same current rating have the same 300-sec points, but have slightly different time-current curves, as illustrated in Figs. 17.16 and 17.17. It can be seen in both figures that the K link operates faster than the T link on the higher-current end of the curves by an amount of time equal to the vertical difference between the two curves for a given fault current. Figure 17.17 illustrates that for an overcurrent of 100 amps the 15K fuse will operate in 0.5 sec while the 15T fuse will require 1.5 sec, a difference of 1.0 sec.

Recommended Size of Primary Fuse. Table 17.1 gives the recommended size of primary fuse to use with different transformer voltage and kilovoltampere ratings. The table also gives the normal full-load primary current rating of the transformer.

It is general practice not to protect distribution transformers against small overloads. To do so would cause unnecessary blowing of fuses and frequent interruption of the service, both of which are undesirable. It is therefore customary to provide fuses which have a higher current rating than the current rating of the transformer. This is clearly shown by a comparison of the normal current-rating column with the fuse-rating column in Table 17.1.

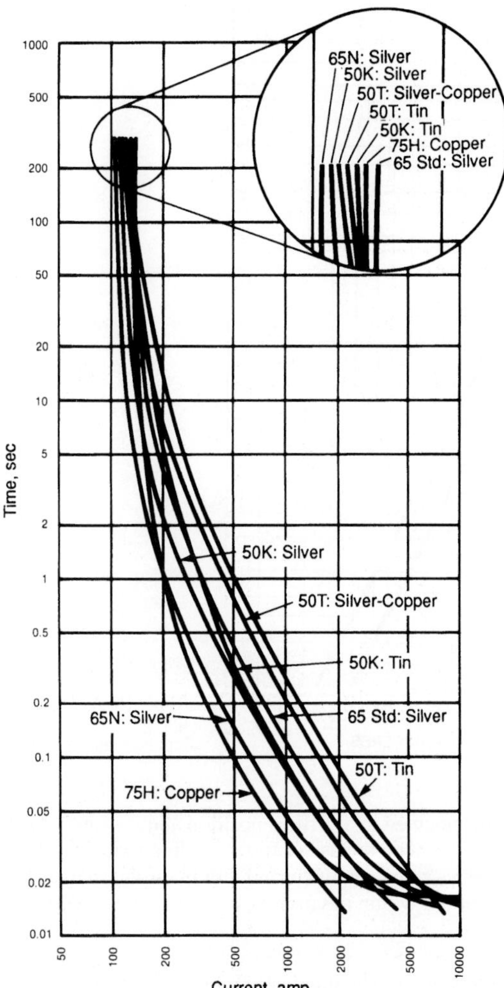

FIGURE 17.16 Total clearing time-current characteristic curves for fuse links recommended for a 1200-kvar grounded wye-connected capacitor bank rated 13.8-kV three-phase, with two 7.97-kV, 200-Kvar capacitor units per phase. (*Courtesy S&C Electric Co.*)

Liquid Fuse. This fuse consists of a glass tube of proper length and diameter, at one end of which a fusible element is mounted. The fusible element is held in tension by means of a helically coiled spring secured at the bottom end, the fusible element being attached to the upper end. The tube is filled with an arc-quenching liquid like oil which has a high dielectric strength. When fault current melts the fusible element, the spring contracts very suddenly and opens a gap in the tube proportionate to the voltage rating of the fuse. As the arc is drawn down into the liquid, the liquid extinguishes or quenches it (see Fig. 17.18).

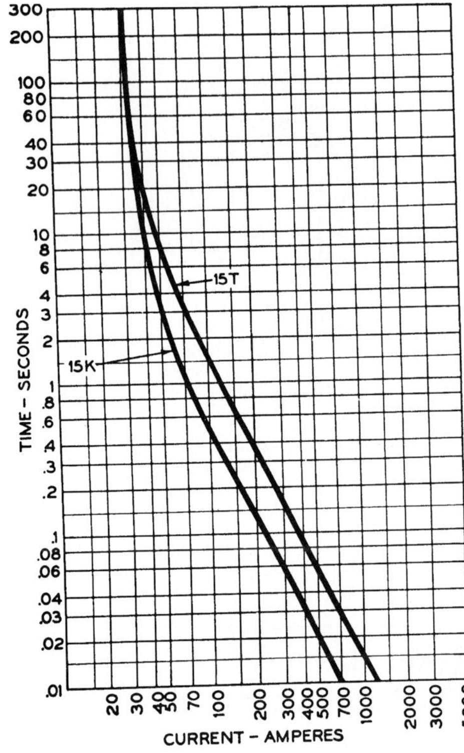

FIGURE 17.17 Curves showing melting times of fast (K) links and slow (T) links having the same 15-amp rating.

The fuse should be mounted in a vertical position and is usually held in spring clips, so that it is readily removable with the use of fuse-handling tools.

In case of an excessive short-circuit current, the arc produced may cause enough heat to vaporize some of the liquid within the fuse. A vent cap is provided which yields to relieve the internal pressure.

Solid-Material Power Fuse. The interrupting medium in this fuse is compressed boric acid. It is pressed into cylindrical blocks with a hole in the center of the block forming the bore. When an arc is drawn in the boric acid tube, the generated gas consists principally of steam formed from the water of crystallization of the boric acid. Steam is less readily ionized than organic gases and thus helps to maintain high dielectric strength around the fuse under operating conditions. There is no flame discharge when the fuse operates (see Figs. 17.19 and 17.20).

Because steam can be condensed, a copper-mesh condenser can be provided for indoor installations.

To replace a blown fuse, the fittings are removed from the blown fuse tube and clamped on a new fuse unit.

Boric acid fuses are made in both fixed and dropout forms. Figure 17.21 illustrates an outdoor dropout switch-hook-operated type. Boric acid fuses are used in pad-mounted switchgear to isolate cable and equipment faults on underground circuits (Fig. 17.22). The operating characteristics of the boric acid fuse meet the requirements necessary to maintain the dielectric strength of the atmosphere in metalclad switchgear.

TABLE 17.1 Suggested Transformer Fusing Schedule (Protection Between 200 and 300 percent Rated Load)*
(Link sizes are for EEI-NEMA K or T fuse links except for the H links noted.)

	2,400 Δ				4,160Y/2,400		4,800 Δ				8,320Y/4,800	
	Figures 1 and 2		Figure 3		Figures 4, 5, and 6		Figures 1 and 2		Figure 3		Figures 4, 5, and 6	
Transformer size, kVA	Rated amp	Link rating	Rated amp	Link rating	Rated amp	Link rating	Rated amp	Link rating	Rated amp	Link rating	Rated amp	Link rating
3	1.25	2H	2.16	3H	1.25	2H	0.625	1H†	1.08	1H	0.625	1H†
5	2.08	3H	3.61	5H	2.08	3H	1.042	1H	1.805	3H	1.042	1H
10	4.17	6	7.22	10	4.17	6	2.083	3H	3.61	5H	2.083	3H
15	6.25	8	10.8	12	6.25	8	3.125	5H	5.42	6	3.125	5H
25	10.42	12	18.05	25	10.42	12	5.21	6	9.01	12	5.21	6
37.5	15.63	20	27.05	30	15.63	20	7.81	10	13.5	15	7.81	10
50	20.8	25	36.1	50	20.8	25	10.42	12	18.05	25	10.42	12
75	31.25	40	54.2	65	31.25	40	15.63	20	27.05	30	15.63	20
100	41.67	50	72.2	80	41.67	50	20.83	25	36.1	50	20.83	25
167	69.4	80	119.0	140	69.4	80	34.7	40	60.1	80	34.7	40
250	104.2	140	180.5	200	104.2	140	52.1	60	90.1	100	52.1	60
333	138.8	140	238.0		138.8	140	69.4	80	120.1	140	69.4	80
500	208.3	200	361.0		208.3	200	104.2	140	180.5	200	104.2	140

*Reprinted with permission from "Distribution System Protection and Apparatus Co-ordination," published by Line Material Industries.
†Since this is the smallest link available and does not protect for 300 percent load, secondary protection is desirable.

TABLE 17.1 Suggested Transformer Fusing Schedule (Protection Between 200 and 300 percent Rated Load)*(*Continued*)

	7,200 Δ				12,470Y/7,200		13,200Y/7,620		12,000 Δ			
	Figures 1 and 2		Figure 3		Figures 4, 5, and 6		Figures 4, 5, and 6		Figures 1 and 2		Figure 3	
Transformer size, kVA	Rated amp	Link rating	Rated amp	Link rating	Rated amp	Link rating	Rated amp	Link rating	Rated amp	Link rating	Rated amp	Link rating
3	0.416	1H†	0.722	1H†	0.416	1H†	0.394	1H†	0.250	1H†	0.432	1H†
5	0.694	1H†	1.201	1H	0.694	1H†	0.656	1H†	0.417	1H†	0.722	1H†
10	1.389	2H	2.4	5H	1.389	2H	1.312	2H	0.833	1H†	1.44	2H
15	2.083	3H	3.61	5H	2.083	3H	1.97	3H	1.25	1H	2.16	3H
25	3.47	5H	5.94	8	3.47	5H	3.28	5H	2.083	3H	3.61	5H
37.5	5.21	6	9.01	12	5.21	6	4.92	6	3.125	5H	5.42	6
50	6.49	8	12.01	15	6.94	8	6.56	8	4.17	6	7.22	10
75	10.42	12	18.05	25	10.42	12	9.84	12	6.25	8	10.8	12
100	13.89	15	24.0	30	13.89	15	13.12	15	8.33	10	14.44	15
167	23.2	30	40.1	50	23.2	30	21.8	25	13.87	15	23.8	30
250	34.73	40	59.4	80	34.73	40	32.8	40	20.83	25	36.1	50
333	46.3	50	80.2	100	46.3	50	43.7	50	27.75	30	47.5	65
500	69.4	80	120.1	140	69.4	80	65.6	80	41.67	50	72.2	80

*Reprinted with permission from "Distribution System Protection and Apparatus Co-ordination," published by Line Material Industries.
†Since this is the smallest link available and does not protect for 300 percent load, secondary protection is desirable.

TABLE 17.1 Suggested Transformer Fusing Schedule (Protection Between 200 and 300 percent Rated Load)*(*Continued*)

	13,200 Δ				14,400 Δ				24,900Y/ 14,400	
	Figures 1 and 2		Figure 3		Figures 1 and 2		Figure 3		Figures 4, 5, and 6	
Transformer size, kVA	Rated amp	Link rating	Rated amp	Link rating	Rated amp	Link rating	Rated amp	Link rating	Rated amp	Link rating
3	0.227	1H†	0.394	1H†	0.208	1H†	0.361	1H†	0.208	1H†
5	0.379	1H†	0.656	1H†	0.347	1H†	0.594	1H†	0.374	1H†
10	0.757	1H†	1.312	2H	0.694	1H†	1.20	2H	0.694	1H†
15	1.14	1H	1.97	3H	1.04	1H	1.80	3H	1.04	1H
25	1.89	3H	3.28	5H	1.74	2H	3.01	5H	1.74	2H
37.5	2.84	5H	4.92	6	2.61	3H	4.52	6	2.61	3H
50	3.79	6	6.56	8	3.47	5H	5.94	8	3.47	5H
75	5.68	6	9.84	12	5.21	6	9.01	12	5.21	6
100	7.57	8	13.12	15	6.94	8	12.01	15	6.94	8
167	12.62	15	21.8	25	11.6	12	20.1	25	11.6	12
250	18.94	25	32.8	40	17.4	20	30.1	40	17.4	20
333	25.23	30	43.7	50	23.1	30	40.0	50	23.1	30
500	37.88	50	65.6	80	34.7	40	60.0	80	34.7	40

*Reprinted with permission from "Distribution System Protection and Apparatus Co-ordination," published by Line Material Industries.
†Since this is the smallest link available and does not protect for 300 percent load, secondary protection is desirable.

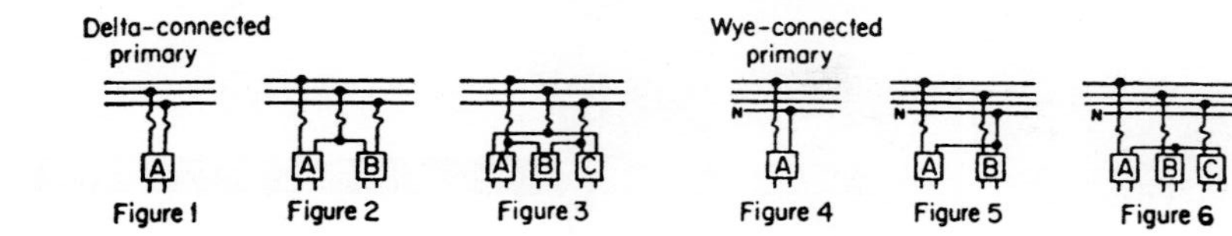

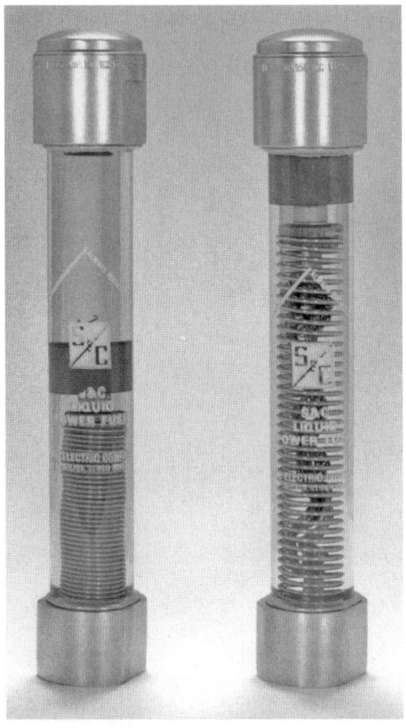

FIGURE 17.18 Two views of liquid fuse showing (right) fuse before being blown and (left) fuse after being blown. (*Courtesy S&C Electric Co.*)

FIGURE 17.19 External view of 69-kV, 300-amp solid-material power fuse. Power fuses of this type are available in ratings to 138 kV and are generally preferred protective devices for small- to medium-size load transformers in utilities and industrial substations. (*Courtesy S&C Electric Co.*)

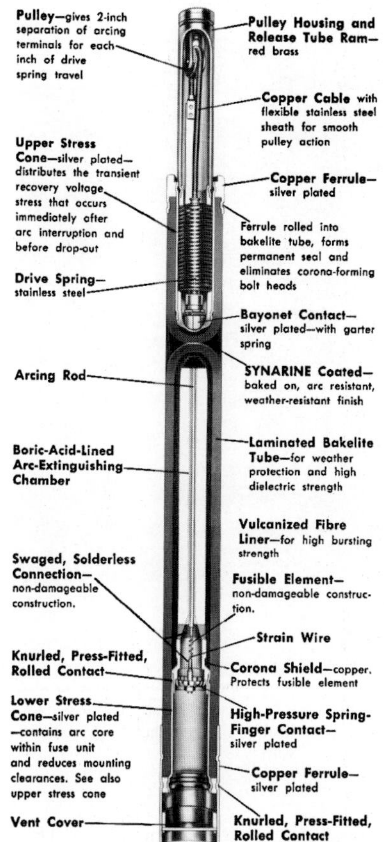

FIGURE 17.20 Internal view of solid-material power fuse showing construction details. Note boric-acid-lined arc-extinguishing chamber. (*Courtesy S&C Electric Co.*)

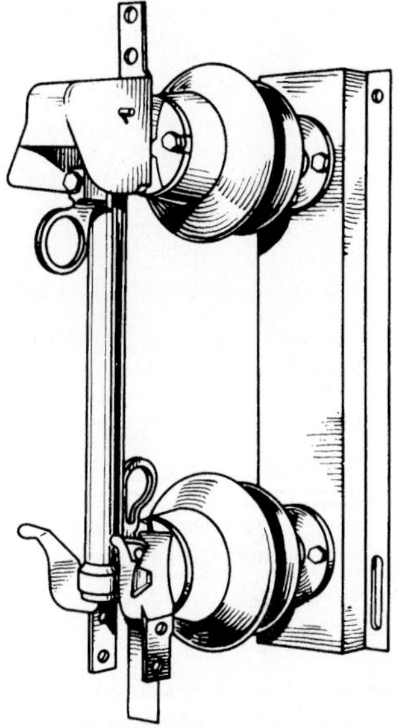

FIGURE 17.21 Solid-material power fuse designed to drop out when blown. Fuse is operated with customary switch hook. (*Courtesy ABB Power T&D Company Inc.*)

Current-Limiting Fuse. The *current-limiting fuse* (CLF) is a nonexpulsion fuse consisting of a silver fuse element wound on a central core enclosed in a tube filled with high-purity silica sand. Under fault conditions the silver element melts and establishes an arc. The fuse element is burned back by the arc. The heat of the arc melts the adjacent sand to produce a glasslike substance known as *fulgurite.* The rapid heat absorption of the sand, together with the rapid burn back of the fuse element, acts to insert a high resistance into the arc path that prevents the current from reaching its natural crest. This produces the fuse's current-limiting action and lowers the energy let-through during faults. In addition, the current-limiting fuse has an interrupting rating in the 50,000-amp symmetric range and a less than ½-cycle clearing time. The current-limiting fuse is produced in two versions: the backup or partial-range and the general-purpose or full-range.

The backup CLF is constructed as described with the central core generally being of ceramic construction. No gas is produced during fault clearing; therefore, the fuse can be

FIGURE 17.22 Open-door view, pad-mounted switchgear equipped with 400-amp solid-material power fuses, two three-pole interrupter switches, metering transformers, and meters. Cables and cable terminators for underground circuits have not been installed in unit shown. (*Courtesy S&C Electric Co.*)

hermetically sealed, making it suitable for submersion in oil-filled apparatus. This type of fuse is not capable of interrupting low-magnitude fault currents. It will only interrupt a range of fault currents from stated minimum interrupting rating (usually 3 to 7 times the continuous current rating) to the maximum interrupting rating. Currents below this range but above the continuous current rating of the fuse will melt the fuse element but will not have enough energy to cause interruption. These currents must be interrupted by a series-connected interrupting device (Fig. 17.23). If the fuse is exposed to low-fault currents, the continued arcing will cause the fuse to burn up or explode.

The general-purpose CLF adds modifications to the basic CLF design to make the fuse capable of clearing any current greater than the continuous rating and less than the maximum interrupting rating of the fuse (Fig. 17.24). In this fuse, part of the fuse element is constructed from a solder alloy with a melting temperature one-third of the 960° melting temperature of the silver portion of the fuse element. The alloy portion is known as the *M spot*. The central core of the fuse is constructed from a material which evolves gas when heated by the element melting. The lowered melting temperature and the gas-evolving core make the general-purpose CLF capable of clearing the lower range of currents that the backup CLF cannot clear. High-current operation is the same.

Current-limiting fuses are expensive compared to the cost of expulsion fuses. The use of the partial-range current-limiting fuse in series with an expulsion fuse minimizes the operating costs since the expulsion fuse will operate for the most common faults that occur most frequently. Figure 17.25 illustrates an expulsion fuse in series with a partial-range

FIGURE 17.23 Partial-range current-limiting fuse is installed on top of distribution cutout with expulsion fuse. Current-limiting fuse will operate for high-fault-current insulation failures in the distribution transformer. Operating characteristics of the current-limiting fuse will prevent destruction of the transformer and the lid from blowing off if a fault develops inside the transformer tank. Expulsion fuse will operate for low primary current or secondary faults. (*Courtesy A. B. Chance Co.*)

CLF mounted on a metal sheet, as it would be installed in a pad-mounted transformer or pad-mounted switchgear used with underground circuits.

Current-limiting fuses applied as illustrated in Figs. 17.23 through 17.25 and properly sized can prevent catastrophic failure of a transformer. Catastrophic failure of a distribution transformer can result from a high-current arc formed under oil. The failure can result in rupture of the transformer tank and an oil fire that is dangerous to the public and to linemen working near the transformer. When a lineman reenergizes a transformer that has been removed from service as a result of operation of the transformer protective equipment, it is important that special precautions be taken to protect him. The transformer should be tested and/or energized through a current-limiting fuse to prevent a high-fault current from being generated under oil in the transformer.

Electronic Fuses. The *electronic fuse* integrates state-of-the-art electronics with an advanced-design high-current fuse. The electronic componentry provides current sensing, time-current characteristics, and control power for the fuse. High-speed interruption of fault currents to 40,000 amps symmetric is provided by the high-current fuse section (Fig. 17.26). The fuse is rated 600 amps continuous with voltage ratings of 4.16 through 25 kV.

The electronic fuse has a control module and an interrupting module (Fig. 17.27). The control module has an integral toroidal current transformer, which provides line current sensing and control power for the electronic circuitry. The current transformer also provides the energy to operate the interrupting module in the event of a fault. Electrical output from the current transformer is processed by the electronics located inside the factory-sealed cast-aluminum control module housing, which serves as both a path for continuous current and a Faraday cage to shield sensing circuits against interference from external electric fields. When a fault occurs, the electronics within the control module initiates a “trip” signal in accordance with electronically derived time-current characteristics. The signal to “trip” is delivered to the interrupting module through a low-resistance contact. The interrupting module of the fuse assembly has a centrally positioned main current section which carries load current during normal operating conditions. The main current section in the interrupting module is rapidly opened when a trip signal is received from the control

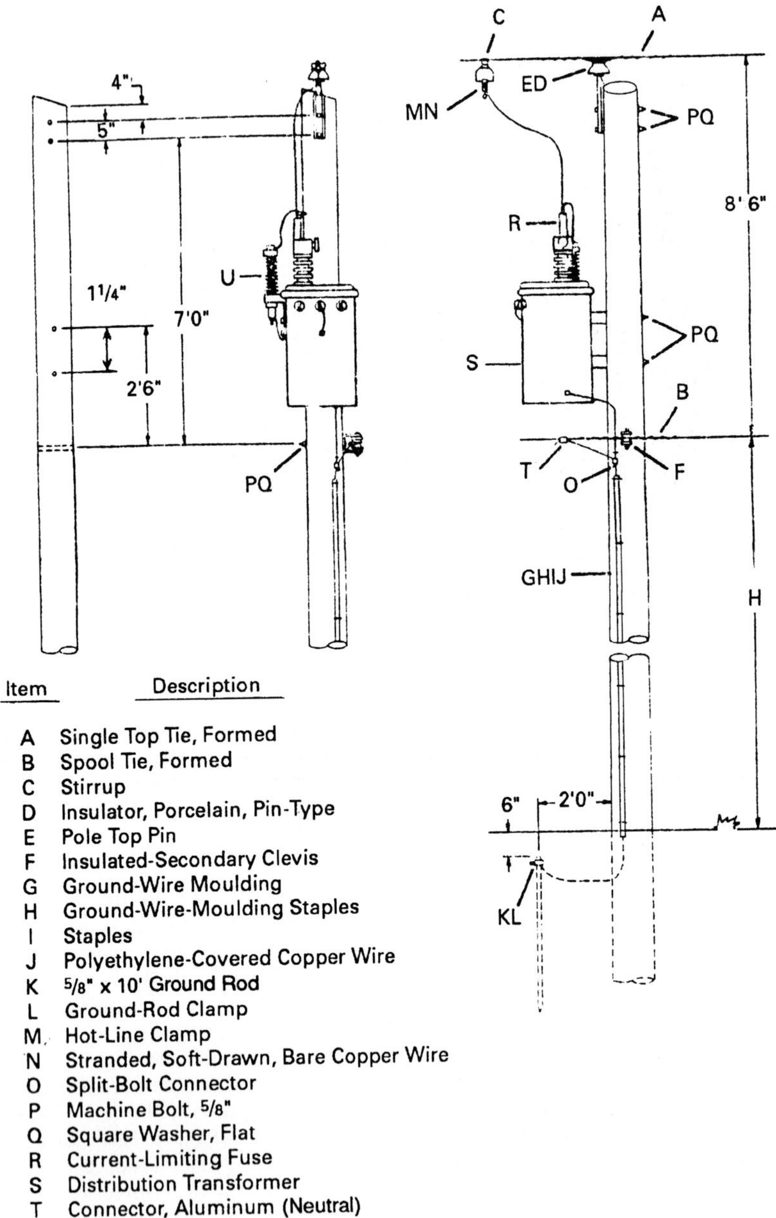

FIGURE 17.24 Distribution transformer installation on 7620-volt single-phase primary with a current-limiting fuse and surge arrester. A general-purpose or full-range current-limiting fuse would be used with a conventional distribution transformer, and a partial-range current-limiting fuse would be used with a self-protected distribution transformer, which has an expulsion fuse inside the transformer tank. Dimension H varies with the length of the pole.

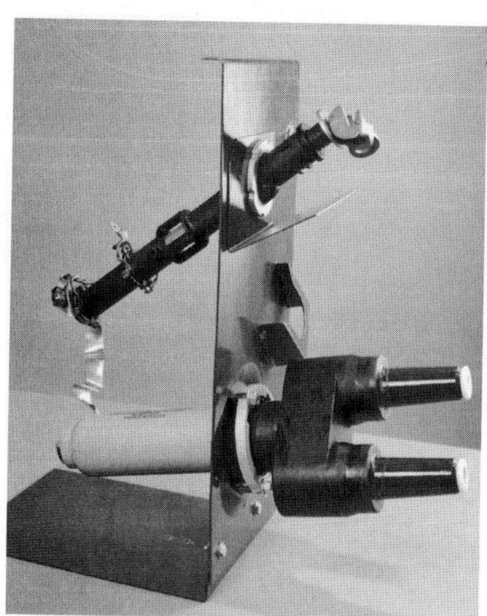

FIGURE 17.25 Partial-range current-limiting fuse connected in series with draw-out type of expulsion fuse. Both fuses would be installed in the tank of the pad-mounted transformer or switching equipment. Draw-out feature facilities replacement of expulsion fuse which is designed to operate for most common electrical faults. (*Courtesy A. B. Chance Co.*)

FIGURE 17.26 Electronic fuse with interrupting module on top and control module connected to and mounted below the interrupting module. Live switching up to 400 amps is provided by interrupter mounted on assembly above fuse unit. (*Courtesy S&C Electric Co.*)

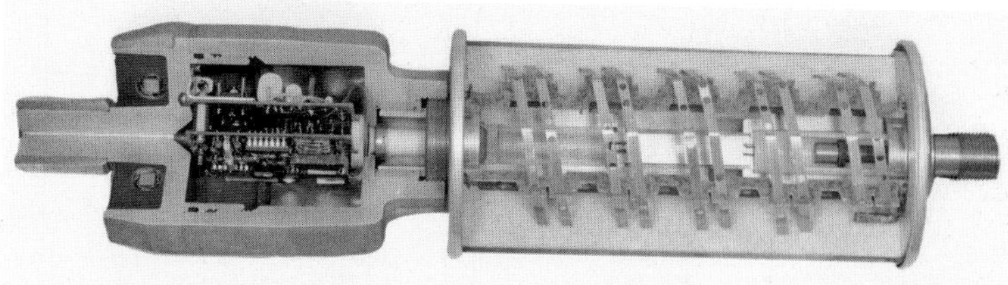

FIGURE 17.27 Cutaway view of electronic fuse. Toroidal current transformer, far left in control module, provides line current sensing and control power for electronics in the control module. Tripping signal from control module activates a gas-generating power cartridge in the interrupting module, causing the main current path to be separated at two points. Fault current is then diverted through the sand-embedded fusible elements. Fault current is interrupted in a current-limiting mode. (*Courtesy S&C Electric Co.*)

module by the action of a gas-generating power cartridge and associated insulating piston. The fault current is shunted into the circuit-interrupting section and the coaxially wound fusible elements, as a result of the opening of the main current path through the interrupting module. The fault current flowing through the fusible element causes the element to open, and the silica sand in the interrupting module limits the magnitude of the energy generated as a result of the short circuit.

The electronic fuses equipped with inverse-curve type of time-current characteristics are ideally suited for service-entrance protection and coordination because these fuses incorporate unique time-current characteristics designed for coordination with source-side overcurrent relays and in-plant load-side feeder fuses. When used for service-entrance protection and coordination, the electronic fuses provide a buffer to prevent in-plant problems on the customer's system from affecting service on the utility company system. In addition, the electronic fuses provide protection for the switchgear bus against damage from high-current bus faults, as well as backup protection for the load-side feeder fuses (Figs. 17.28 and 17.29).

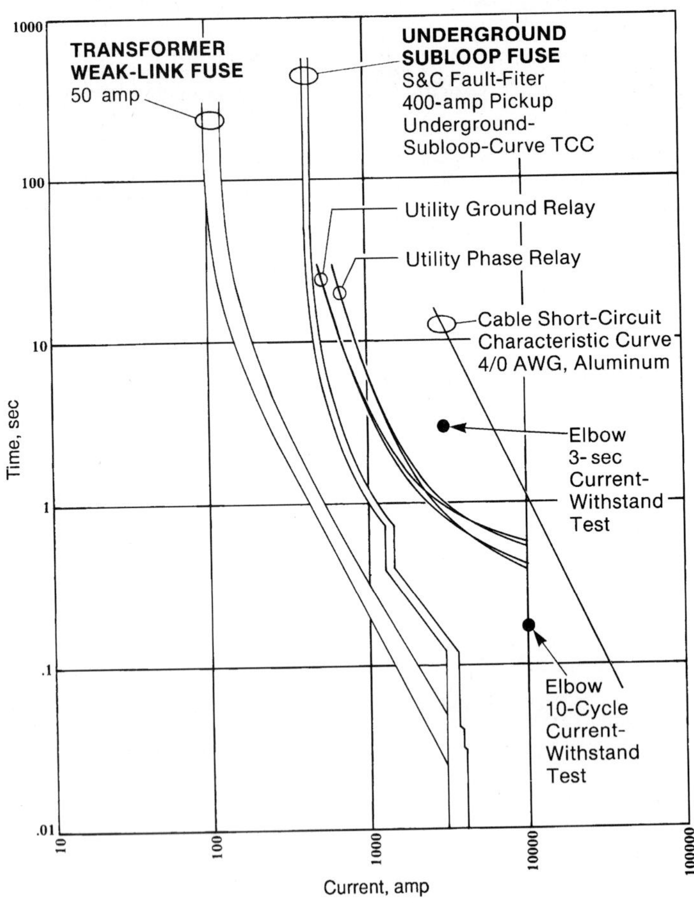

FIGURE 17.28 Fault-Filter fuses with electronic control for an underground-sub-loop type of time-current curve (TCC) to provide fault protection for transformers, elbow connectors, and underground cable. Coordination is readily achieved with both the transformer's weak-link fuses and the utility's substation circuit breakers. (*Courtesy S&C Electric Co.*)

FIGURE 17.29 Open-door view, incoming bay of metal-enclosed switchgear. Incoming cables are connected to terminals of three-pole interrupter switch. Fault-Filter electronic fuses protect the utility system from downstream faults and provide backup protection for power fuses in feeder sections. (*Courtesy S&C Electric Co.*)

Protective Overcurrent Coordination. The coordination of overcurrent protective devices involves their selection and use in such a manner that temporary faults will be quickly removed and permanent faults will be restricted to the smallest section of the system possible.

The locations of the protective devices are known as *coordinating points*. Coordinating points are usually established at the substation, at positions along the feeder, in branch lines off the feeder, and on the primary side of distribution transformers. In the single-line diagram of Fig. 17.30, representing an elementary distribution system, points *A, B, C, E, F, H,* and *I* are coordinating points. Each device must be selected so that it can carry the normal line or load current and will respond properly to an excessive fault current. Device *A* in the substation nearest to the power source is normally a circuit breaker; devices *B, H, I, E,* and *F* are fused cutouts or power fuses; and device *C* could be a circuit recloser. Devices *A* and *C* could be adjusted to open before the fuses melt, to clear temporary faults and to reclose to restore service to all customers. Devices *A* and *C* are adjusted to open after time delay if the first operation fails to clear the fault. A permanent fault as shown in Fig. 17.30 would be isolated or cleared from the source by fuse *E* after device *C* failed to clear the fault when it opened and reclosed. After device *E* has isolated the branch with the fault, normal load current will continue to flow in the remainder of the system.

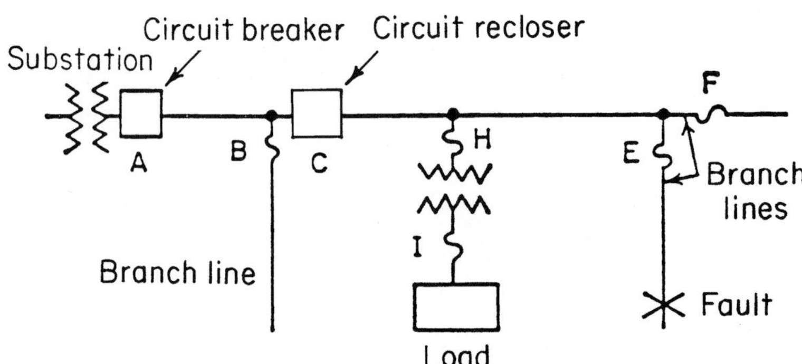

FIGURE 17.30 One-line diagram of an elementary distribution system. Overcurrent protective devices are shown as *A, B, C, E, F, H,* and *I.* Fault is shown as *X.* Device *E* should disconnect the branch line and permit the remainder of the system to operate.

The minimum fault current on each circuit must be determined and compared with the minimum operating current of the protective device, to be sure the protective device will operate to clear the fault. The minimum fault current calculations must allow for a high-impedance fault that may occur if an energized wire is lying on the ground. The maximum fault current must be determined to ensure that each protective device has adequate interrupting capacity to de-energize the faulted circuit without a catastrophic failure of the protective equipment. These conditions are necessary to protect the linemen and cablemen working on the circuit and the public, who may be near the circuit and its equipment, as specified by the *National Electrical Safety Code*.

Additional detailed information on fuses can be obtained from the ANSI C37.40, *Service Conditions and Definitions for High Voltage Fuses, Distribution Enclosed Single-Pole Air Switches, Fuse Disconnecting Switches and Accessories;* ANSI C37.42, *Specifications for Distribution Cutouts and Fuse Links;* ANSI C37.46 *Specifications for Power Fuses and Fuse Disconnecting Switches;* and ANSI C37.47, *Specifications for Distribution Fuse Disconnecting Switches, Fuse Supports and Current-Limiting Fuses.*

CHAPTER 18
SWITCHES

A switch is used to disconnect or close circuits that may be energized. If the circuit conducts current when it is operated, special devices need to be installed on the switch contacts to interrupt or establish the current flow. High-voltage switches are operated remotely with a mechanism or directly with a hot-line tool, called a *switch stick*. The operating mechanism may be manually operated at the location of the switch by a lineman or by control signals initiated by remote devices to an electric, hydraulic, or pneumatic operating mechanism, at the switch location, from a remote point.

Switches may be divided into four general classes:

1. Air switches
  a. Circuit breaker
  b. Air break
  c. Disconnect
2. Oil switches
  a. Oil circuit breaker
  b. Oil circuit recloser
    (1) Sectionalizer
3. Vacuum switches
  a. Vacuum circuit breaker
  b. Vacuum recloser
4. Sulfur hexafluoride gas (SF_6) switches
  a. Circuit breaker
  b. Circuit switcher
  c. Recloser

The following standards provide additional information on switches:

ANSI/IEEE C37.30, *Definitions and Requirements for High Voltage Air Switches, Insulators and Bus Supports*

ANSI C37.35, *Application, Installation, Operation, and Maintenance of High-Voltage Air Disconnecting and Load Interrupter Switches*

ANSI C37.45, *Specifications for Distribution Enclosed Single-Pole Air Switches*

ANSI/IEEE C37.60, *Requirements for Overhead, Pad-Mounted, Dry Vault, and Submersible Automatic Circuit Reclosers and Fault Interrupters for AC Systems*

ANSI/IEEE C37.61, *Guide for the Application, Operation and Maintenance of Automatic Circuit Reclosers*

Air Switches. As their names imply, air switches are switches whose contacts are opened in air. Air switches are further classified as air circuit breakers, air break switches, and disconnects.

Circuit Breaker. A device used to complete, maintain, and interrupt currents flowing in a circuit under normal or faulted conditions is called a *circuit breaker*. The circuit breaker has a mechanism that mechanically, hydraulically, or pneumatically operates the circuit breaker contacts. Insulating oil, air, compressed air, vacuum, or sulfur hexaflouride gas is used as an arc-interrupting medium and a dielectric to insulate the contacts after the arc is interrupted. If it is desired to open the circuit automatically during overload or short circuit, the circuit breaker is equipped with a tripping mechanism to accomplish this. Circuit breakers are thus normally used where control of the circuit as well as protection from overload, short circuit, etc., is desired, such as at generating stations and substations.

Most of the different types of circuit breakers are illustrated in Chap. 3, "Substations."

Air Break Switch. The air break switch can have both blade and stationary contacts equipped with arcing horns as shown in Fig. 18.1. These horns are pieces of metal between which the arc forms when a circuit-carrying current is opened. As the switch opens, these horns spread farther and farther apart, thereby lengthening the arc until it finally breaks. That such arcs do form is convincingly shown in Fig. 18.2, which shows the arcs formed in air when a three-phase group-operated air break switch was opened at the source end of a 50-mile 132,000-volt transmission line. The line was without load, but 20 amps of charging current was flowing into the line. The arcs measured 123 ft in length, while the spacing of the phases was only 16 ft. Favorable oblique winds kept the arcs from short-circuiting across phases. Wind also has a cooling effect which aids arc extinction.

Air break switches are usually mounted on substation structures or on poles and are operated from the ground level (Fig. 18.3). In a three-phase circuit all three switches, one in each phase, are opened and closed together as a "group," as the system is called. The switches can be operated by a handle connected to the rod extending from the switch to the base of the pole. Many air break switches are operated by mechanized equipment connected to the switch-operating rod from a remote location. The automation of distribution circuits has resulted in the installation of many mechanized operators for key air break switches so that they can be remotely controlled from a central operations center.

Switches for underground distribution circuits are usually installed in pad-mounted switchgear (Fig 18.4). The switches are operated with the cabinet doors closed to provide protection for the lineman or cableman (Fig. 18.5).

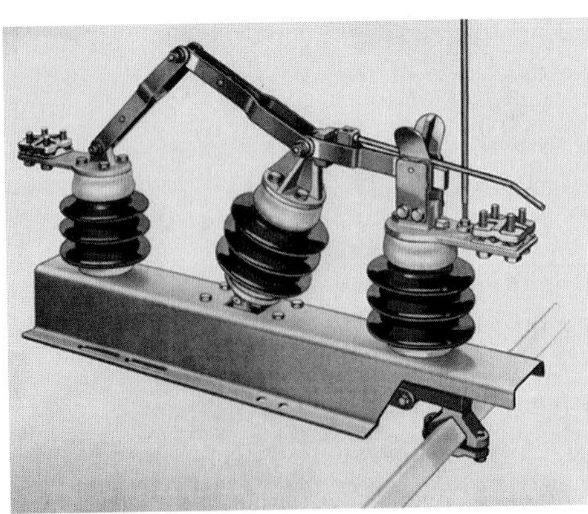

FIGURE 18.1 Single pole of high-voltage horn gap air break switch. Switch is in closed position. Switch is opened by tilting middle insulator to left. Operating bar is shown below. Long metal rods constitute horn gap. (*Courtesy James R. Kearney Corp.*)

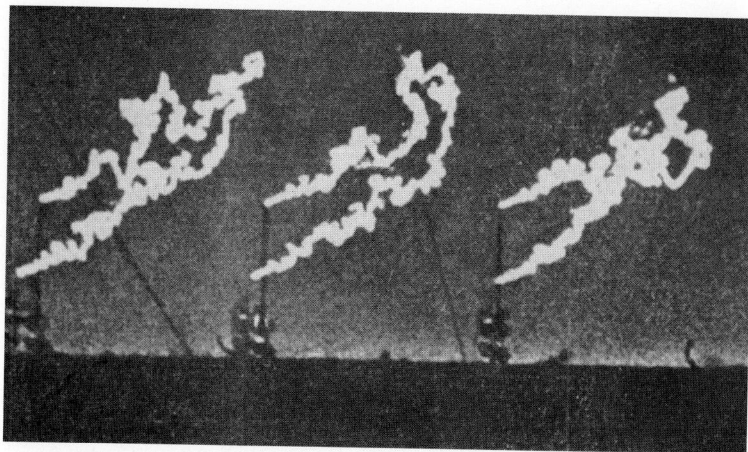

FIGURE 18.2 Arcs formed as a three-phase group-operated switch is opened at the source end of a 132-kV 50-mile unloaded transmission line. Arcs measured 123 ft in length. Spacing between phases was 16 ft. Charging current interrupted was 20 amps in each phase.

Customers with essential loads may require dual electric sources with automatic transfer from one source to the other, if one circuit is taken out of service. The switchgear illustrated in Fig. 18.6 is equipped with switches in series with each underground source circuit and a bus-tie switch. These switches are power-operated and controlled by an automatic control device for source transfer. The underground load feeder circuits are connected to the switchgear bus through power fuses and air break switches with interrupter contacts to permit energized operation while carrying load. The bus-tie switch will permit the load to be served from either source. The power source circuits are monitored by an automatic control device which in turn controls and sequences the power operation of the incoming circuits and bus-tie switches.

FIGURE 18.3 Group-operated three-pole air break switch with arc interrupters. The switch is in the open position. The switch is installed at a normally open point between two distribution main feeder circuits. *(Courtesy S & C Electric Co.)*

FIGURE 18.4 Dead-front pad-mounted switchgear unit utilized in underground distribution systems. The switchgear unit features include cable termination bushings, switches, and fuses. The unit is contained within the enclosure and is packaged to minimize electrical exposure during switching and fusing operations. *(Courtesy Federal Pacific)*

FIGURE 18.5 Lineman operating one of two three-pole interrupter switches connected in series with the main underground distribution circuit in the pad-mounted switchgear, which also contains two sets of three solid-material power fuses through which circuits tapped to the main circuit are served. *(Courtesy S&C Electric Co.)*

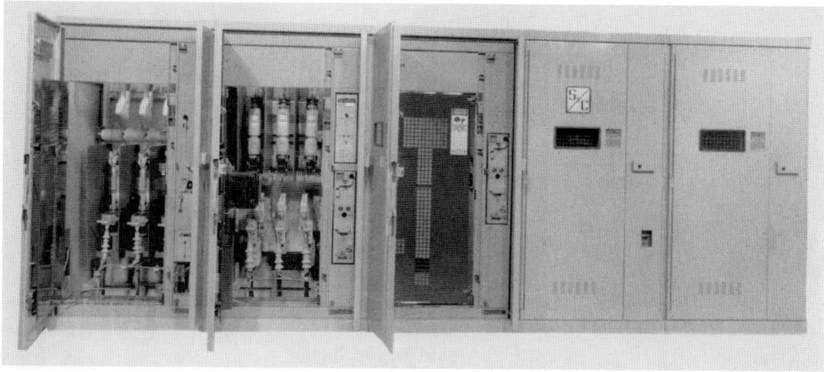

FIGURE 18.6 View of five-bay lineup of metal-enclosed switchgear. Feeder bay, left, contains three-pole interrupter switch, power fuses for feeder circuit short-circuit protection, and open-phase detector to protect load from single phasing due to blown fuse. Incoming bay, second from left, contains three-pole power-operated interrupter switch, Fault-Fiter electronic fuses, switch operator, and control device for automatic source transfer. Center bay contains normally open tie switch. Two right-hand bays mirror two left-hand bays. (*Courtesy S & C Electric Co.*)

Upon loss of service from either of the two power sources—for a preset time sufficient to confirm permanent loss of service—the interrupter switch associated with the deenergized source automatically opens and the bus-tie switch closes to restore service to the affected feeders. When the normal power source is returned to service the affected feeder bus is retransferred to its normal power source, either automatically or manually depending on the operating mode which has been selected. Overcurrent relays are used to block closing of the bus-tie switch in the event of a bus fault. Paralleling of power sources is precluded by mechanical and electrical interlocks. The equipment can be manufactured and equipped to permit operation of the switches from a remote control point, if that type of operation is desired, instead of the automatic mode.

Disconnect Switch. A *disconnect switch* is an air break switch not equipped with arcing horns or other load break devices. The disconnect switch cannot be opened until the circuit in which it is connected is interrupted by some other means, such as a "load buster" attached to a hot-line tool, illustrated and discussed in Chap. 2, or a circuit breaker, illustrated and discussed in Chap. 3. If a disconnect switch is opened while current is flowing in the line, an arc is likely to be drawn between the blade and stationary contact which might easily jump across to the other conductor or to some grounded metal and cause a short circuit. The hot arc could also melt part of the metal, thereby damaging the switch.

Disconnect switches are used to complete a connection to or isolate the following:

1. Two energized transmission or distribution lines
2. Transmission or distribution lines from substation equipment
3. Substation equipment
4. A distribution feeder circuit and a branch circuit

Disconnect switches are frequently used to isolate a line or a piece of apparatus such as a circuit breaker to complete maintenance work. In most circumstances, it is necessary to test the equipment for high voltage and, if proved deenergized, to ground it before the maintenance work is performed.

Oil Switch. An *oil switch* is a high-voltage switch whose contacts are opened and closed in oil. The switch is actually immersed in an oil bath, contained in a steel tank, as shown in

FIGURE 18.7 Exposed view of contacts of a three-phase oil switch. These contacts are immersed in oil, which insulates the phases and assists in quenching the arcs formed when the switch contacts are opened.

Fig. 18.7. The reason for placing high-voltage switches in oil is that the oil may help to break the circuit when the switch is opened. With high voltages, a separation of the switch contacts does not always break the current flow, because an electric arc forms between the contacts. If the contacts are opened in oil, however, the oil will help to quench the arc since oil is an insulator. Furthermore, if an arc should form in the oil, it will evaporate part of the oil, because of its high temperature, and so will partially fill the interrupters surrounding the switch contacts with vaporized oil. This vapor develops a pressure in the interrupters which assists in quenching or breaking the arc by elongating the arc.

The three lines of a three-phase circuit can be opened and closed by a single oil switch. If the voltage is not extremely high, the three poles of the switch are generally in the same tank (Fig. 18.8), but if the voltage of the line is high, the three poles of the switch are placed in separate oil tanks. The poles are placed in separate tanks to make it impossible for an arc

FIGURE 18.8 Three-phase oil circuit recloser for use on circuits operating at 15 kV or below. Operating mechanism can be operated with a switch stick to switch circuits on and off while energized carrying load.

to form between any two phases when the switch is opened or closed. An arc between phases would be a short circuit across the line and would probably blow up the tank.

When an oil switch is to open the circuit automatically because of overload or short circuit, it is provided with a trip coil. This trip coil consists of a coil of wire and a movable plunger. In low-voltage circuits carrying small currents, this coil is connected in series with the line. When the current exceeds its permissible value, the coil pulls up its plunger. The plunger trips the mechanism, and a sprig opens the switch suddenly.

In high-voltage circuits or in circuits carrying large currents, a current transformer is connected into the line and the secondary leads from this transformer supply the current to the trip coil of the oil switch (see Fig. 18.9). Since there is a fixed ratio between primary and secondary currents of the current transformer, the coil can be adjusted to trip at any predetermined value of current in the line.

The use of the current transformer on such circuits serves the double purpose of providing a small current for operating the tripping coil and of insulating the coil from the high voltage of the line.

Oil Circuit Recloser. An oil circuit recloser is a type of oil switch designed to interrupt and reclose an alternating-current circuit automatically (Fig. 18.10). It can be made to repeat this cycle several times. Reclosers are designed for use on single-phase circuits (Fig. 18.11) or on three-phase circuits (Fig. 18.12).

A recloser opens the circuit in case of fault as would a fuse or circuit breaker. The recloser, however, recloses the circuit after a predetermined time (for hydraulically controlled reclosers about 2 sec). If the fault persists, the recloser operates a predetermined number of times (1 to 4) and "locks out," after which it must be manually reset before it can be closed again. If, however, the fault was of a temporary nature and cleared before lockout, the recloser would reset itself and be ready for another full sequence of operations.

Temporary faults arise from wires swinging together when improperly sagged, from tree branches falling into the line, from lightning surges causing temporary flashover of line insulators, and from animals on the conductors short-circuiting the insulators.

A recloser is unlike a fuse link because it distinguishes a temporary from a permanent fault. A fuse link interrupts temporary and permanent faults alike. Reclosers give temporary faults repeated chances, usually four, to clear or be cleared by a subordinate device like a fuse or sectionalizer. If the fault is not cleared after four operations, the recloser recognizes it as a permanent fault and operates to lock out leaving the line open.

A recloser can be magnetically operated by a solenoid connected in series with the line. Minimum trip current is usually twice the normal load current rating of the recloser coil. The operations are performed by a hydraulic mechanism and a mechanical linkage system.

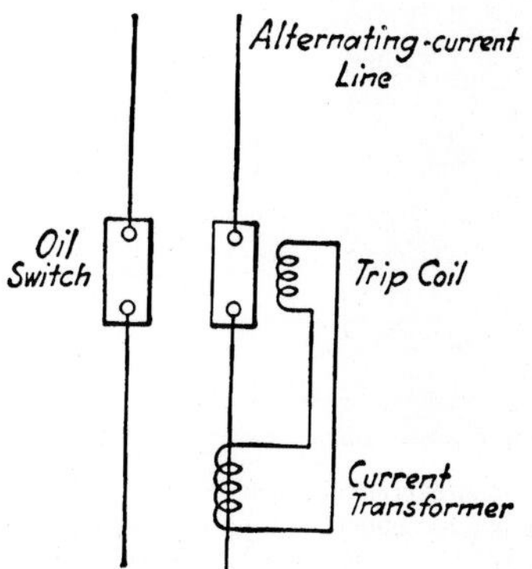

FIGURE 18.9 Current transformer used to supply current to trip coil of oil switch.

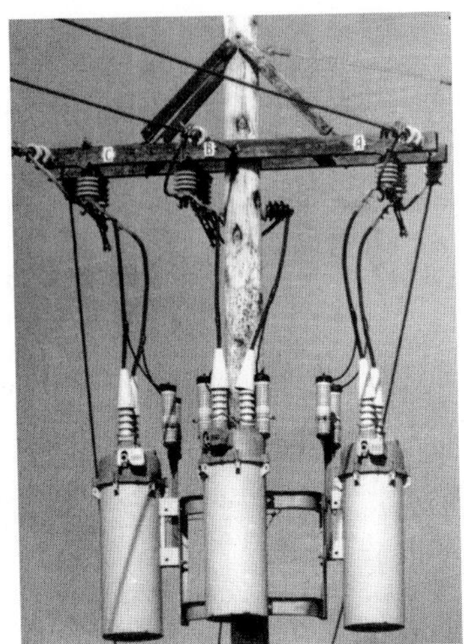

FIGURE 18.10 Three single-phase oil circuit reclosers installed in 13,200Y/7620-volt three-phase line. Reclosers have a current rating of 560 amps and interrupting rating of 8000 amps. Also note installation of lightning arresters.

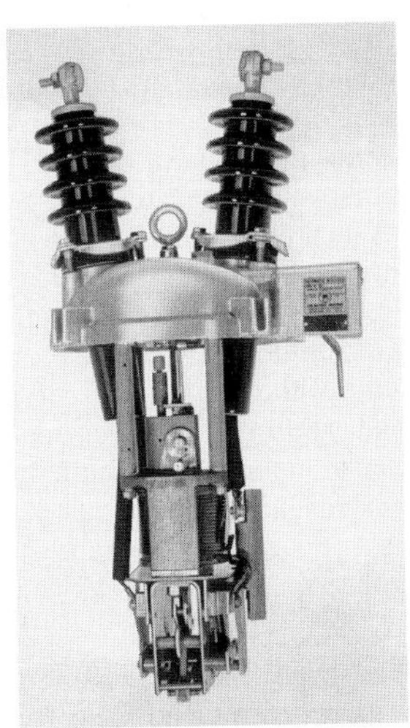

FIGURE 18.11 Single-phase oil circuit recloser removed from oil tank. Recloser shown has a maximum voltage rating of 14.4 kV and a maximum current-interrupting rating of 1250 amps.

FIGURE 18.12 Oil circuit recloser with tank removed. Recloser is rated 15 kV, three-phase, 560 amps continuous and 16,000 amps interrupting capacity.

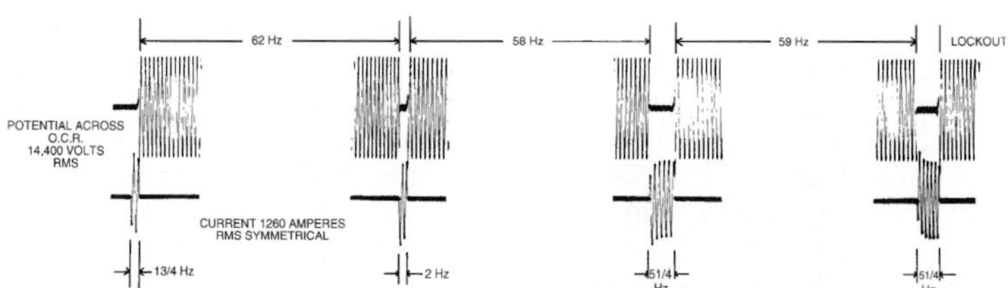

FIGURE 18.13 Oscillogram of single-phase recloser operation interrupting 1260 amps at 14.4 kV. Recloser is set for two fast reclosures of approximately 2 Hz and two delayed reclosures of approximately 6 Hz.

When the fault current reaches twice the normal line current, the increased magnetic field pulls the plunger down into the coil. As the plunger moves downward, the lower end trips the contact assembly to open the contacts and break the circuit. As soon as the contacts are open, there is no more current in the coil to hold them open, so a spring closes the mechanism and reenergizes the line.

This and succeeding operations are depicted on the oscillogram in Fig. 18.13 as it cycles through to lockout. Note that the line is held open for approximately 60 cycles between reclosures. This provides time for sectionalizers to operate while no current is flowing in the line. During lockout the contacts are held open until the recloser is reset manually. If a temporary fault clears before the recloser locks out, all mechanical operations cease and the recloser becomes ready to cycle over again when the next fault occurs.

Examination of the oscillogram (Fig. 18.13) shows that the first two openings of the recloser are faster than the last two. The first two openings occur in about 2 Hz each, whereas the last two occur in about 5 Hz each. This is arranged so that in case the fault takes place on a branch line or tap protected with a fused cutout, the recloser will try to remove the fault before the fuse has time to melt. If the recloser does not clear the fault after two trials, it will give the fuse time to blow on the longer 5-Hz openings and disconnect the faulted branch. The fact that the fault did not clear with two openings indicates that it was of a permanent nature and could be removed only by disconnecting the defective branch line.

The objective of recloser and fuse coordination is to eliminate outages due to temporary faults and, in the case of permanent faults, to restrict outages to a minimum number of customers for the shortest possible time. This requires satisfying the following conditions whenever possible:

1. Fuses should be protected by a recloser fast trip to clear transient faults.
2. Reclosers and/or fuses in series should be sized such that the source device operates before the load device over the entire range of fault currents.
3. Fuses should be sized as large as fault current and source side protective devices will permit in order to avoid blowing fuses on overload.

In determining time-current characteristic (TCC) curves, tests are conducted at 25°C ambient without any preloading. In accordance with National Electrical Manufacturers Association (NEMA) standards, minimum melting curves are determined by taking average test values minus the manufacturer's tolerance, while total clearing curves represent average melting time plus the manufacturer's tolerance plus the arcing time.

Since fuse links are thermal devices, it should be evident that preloading and/or ambient temperatures higher than 25°C will result in links blowing faster than indicated on the published curves. To account for operating variables, it is recommended that 75 percent of the minimum melting curve be used in coordination work. It is generally agreed that for tin fuse links, this 25 percent margin consists of 10 percent for preloading, 5 percent for extraordinary ambient temperatures, and 10 percent to prevent damaging the fuse link. The concept of damaging a fuse link provides an explanation for fuses blowing on "nice, sunny days for no apparent reason." Fuse link damage is caused by a fault which lasts long enough for the link to reach its melting temperature, but not long enough for the necessary heat of fusion to be added to completely melt the link. When the link cools, the physical dimensions have changed and consequently the time-current characteristics have been altered.

Recloser curves are plotted to average values with manufacturing tolerances resulting in variations of ±10 percent in current or time, whichever is greater. The hydraulic fluid used provides consistent timing for ambient temperatures above 0°C. Below this, operations may be slightly slower, but coordination with fuses will be maintained since they are also affected by the low ambient temperature.

To examine fuse-recloser coordination, assume the recloser is the source side device and the fuse is the load side or protecting device. The curve shown in Fig. 18.14 will illustrate coordination between a 35-amp type H recloser and a 15T fuse link. Four operations to lockout may be selected using a maximum of two curves. If fast tripping is desired, select at least one operation on the A curve with the remainder on the B or C curve. The C curve is chosen since it allows more room for coordination. To determine the number of fast trips, the following should be considered. First, the purpose of a fast trip is to prevent transient faults from becoming permanent ones and also to prevent fuse blowing due to transient faults. Since the momentary interruptions resulting from using two fast trips should not be

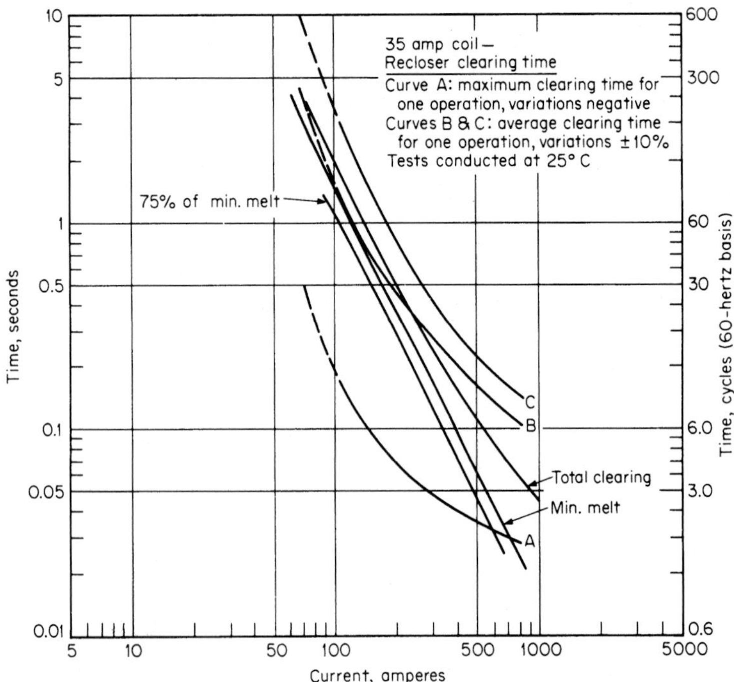

FIGURE 18.14 Coordination curves for 35-amp recloser and 15T fuse link.

objectionable, it should be used to provide better fuse protection. By comparing the maximum clearing time for a recloser fast trip to 75 percent of the fuse melting time, it is evident that a 35-amp type H recloser can protect a 15T fuse with one fast trip for faults up to 550 amps. For two fast trips it would obviously be too conservative to compare twice the maximum clearing time with 75 percent of the fuse minimum melting time since the reclosing interval has a cooling effect on the fuse. For a reclosing interval of 1 to 2 sec, the cooling factor can be handled by comparing twice the average clearing time for a fast trip to 75 percent of the fuse minimum melting time. Thus, for two fast trips, a 35-amp type H recloser can protect a 15T fuse for faults up to 350 amps.

Thus, the fault duty at the fuse location determines the degrees of protection afforded by the recloser fast trips.

Sectionalizer. A *sectionalizer* is another type of oil switch designed to isolate faults on distribution circuits in conjunction with reclosers. Sectionalizers are usually installed on taps or branches off main lines. While in appearance (Fig. 18.15) a sectionalizer is similar to a recloser, it should not be confused with it because it does not interrupt a fault current. In fact, a sectionalizer waits until the recloser has opened the line and then sectionalizes the faulty line while the line is still open and no current flowing. It will be remembered from Fig. 18.13 that the recloser holds the line open for about 1-sec intervals (60 Hz) between reclosures. It is during these periods that the sectionalizer functions.

When a fault occurs beyond the sectionalizer, the recloser will operate. If the fault is of a permanent nature, the sectionalizer will count the number of operations of the recloser and trip and lock itself out after a predetermined number, usually three, of recloser operations. The recloser continues on its fourth operation and restores service up to the sectionalizer. A sectionalizer must therefore always be backed up by a recloser of proper size.

The sectionalizer illustrated in Fig. 18.16 consists of a set of contacts, a spring-controlled trip, and a solenoid operating coil also connected in series with the line. All are immersed in a bath of insulating oil and housed in a tank.

FIGURE 18.15 Typical external view of single-phase sectionalizer.

FIGURE 18.16 Internal view of single-phase sectionalizer shown in Fig. 18.15.

FIGURE 18.17 Three-phase vacuum recloser installed to protect an electric distribution circuit from failure of an underground cable tap. The distribution circuit operates at 13,200Y/7620 volts.

FIGURE 18.18 Vacuum recloser with tank removed. Recloser is rated 15.5 kV, three-phase, 560 amps continuous, 12,000 amps interrupting capacity.

Vacuum Reclosers. *Vacuum reclosers* operate in the same manner as oil circuit reclosers and can be used in the electric distribution system in place of oil circuit reclosers. A vacuum recloser installation is illustrated in Figs. 18.17, 18.18, and 18.19. Vacuum medium has an outstanding dielectric strength and develops rapid recovery of the dielectric strength in the arc path between the separating contacts following a current zero (Fig. 18.20). The vacuum interrupter limits the arc time to a short period, usually interrupting the current at the first current zero. The contact travel is short, minimizing the energy requirements of the operating mechanism.

The vacuum reclosers will operate for long periods without maintenance. Contact erosion occurs each time the recloser operates. The amount of contact erosion is directly related to the current magnitude when the interruptions occur. When the contacts have eroded approximately $^1/_8$ in, the vacuum interrupter should be replaced. The recloser mechanism should be kept clean and checked for proper operation at regular intervals. The insulators should be kept clean and inspected for damage when other routine maintenance is performed.

FIGURE 18.19 External view of vacuum bottle with operating mechanism removed from vacuum recloser.

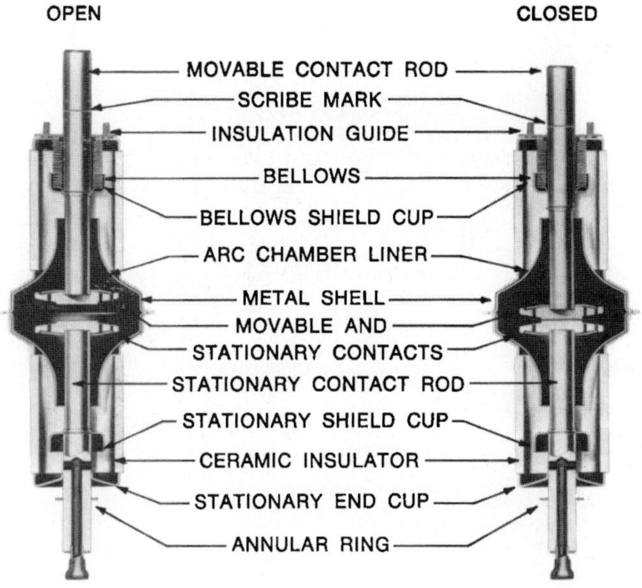

FIGURE 18.20 Vacuum interrupter.

Sulfur Hexafluoride Gas (SF_6) Switches. Circuit breakers, circuit switches, and reclosers are manufactured to use SF_6 gas as an insulating and interrupting medium. SF_6 gas has proved to be a very efficient insulating and interrupting medium, permitting a reduction in the physical size of the equipment. Circuit breakers and circuit switches that use SF_6 as an interrupting medium are illustrated and discussed in Chap. 3. The SF_6 recloser is constructed to operate on all distribution system voltages through 35 kV. The operation and construction of the SF_6 recloser are similar to those of reclosers using oil as an insulating and interrupting medium, except the contacts rotate instead of operating vertically. The recloser uses an electronic control system similar to the controls for the reclosers that use vacuum and oil as an interrupting medium.

Electronic Controlled Reclosers. In the hydraulically controlled three-phase reclosers it is normally necessary to "untank" the unit to change the characteristics, such as number of fast and slow trips, total number of operations, time-current trip characteristics, or the reclose times. However, in the electronically controlled reclosers, changing of all the functions of the control is accomplished by easy replacement of minimum trip resistors and time-current plugs, as shown in Fig. 18.21. A much finer and repeatable control function is obtained since the temperature of the oil does not have to be taken into account. The electronic-type control is used on reclosers using oil, vacuum, and SF_6 gas as an interrupting material.

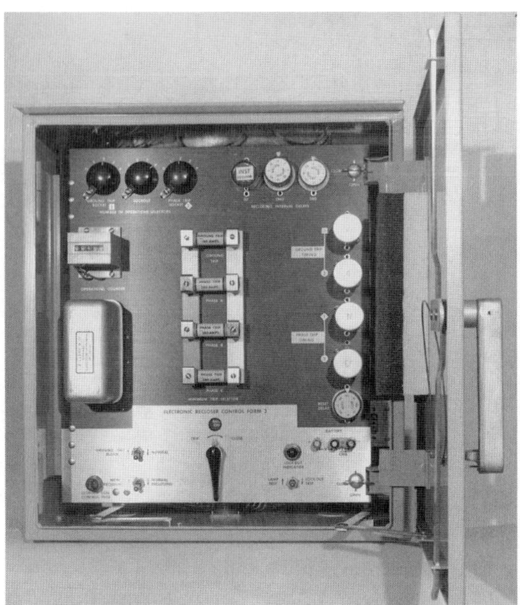

FIGURE 18.21 Electronic controls for recloser. Operating characteristics can be selected by switches and interchangeable components on front of panel.

CHAPTER 19
TAP-CHANGING TRANSFORMERS AND VOLTAGE REGULATORS

System voltages fluctuate due to the ever-changing load consumption characteristics of customers. Voltage at the customer's premises must be maintained within a range that will permit the customer load to operate properly. The *American National Standard for Electric Power Systems and Equipment—Voltage Ratings (60 Hz)* ANSI C84.1 provides detailed information on standard nominal system voltages and voltage ranges. This standard specifies that 120-volt alternating-current nominal voltage must be maintained between 114 and 126 volts (+/−5 percent). There are several options available to utilize in order to stay within the desired utilization voltage range and compensate for the system load changes. The options include balancing load, increasing feeder size, applying tap-changing transformer equipment, installing capacitors, or installing voltage regulators. The concept of voltage regulation is discussed in more detail in Chap. 40. It was stated in Chap. 3 that if the substation power transformer is not equipped with load-tap-changing equipment, it is usually necessary to install voltage regulators. Voltage regulators are used to vary the source voltage to the customer by the proper amount to keep the voltage within the limits desired. Voltage regulators maintain a reasonably constant voltage at the point of utilization. Voltage regulation equipment may be installed in the substation or on a pole out on the distribution circuit.

Tap-Changing Transformers. Distribution substation power transformers or distribution line transformers may be equipped with *tap-changing* equipment. The distribution substation transformers (Fig. 19.1) are normally three-phase with delta-connected primary windings and wye-connected secondary windings to provide a source for three-phase four-wire grounded-neutral alternating-current distribution feeder circuits (Fig. 19.2). The windings of the transformer illustrated are constructed to reduce the subtransmission voltage to the distribution-system primary voltage level. The three distribution voltage windings of the transformer have taps near the neutral, or common, connection. The taps in each winding connect to switch contacts, making it possible to change the number of turns in the transformer secondary windings to vary the substation distribution feeder-bus voltage as desired.

The tap-changer switches in the low-voltage windings are constructed to permit their operation underload without interrupting the circuits. The high-voltage tap changer is designed for deenergized operation. The tap changers in the high-voltage windings are set to the proper position before the transformer is energized to coordinate with the subtransmission-system voltage level at the substation. The rated voltage for each tap in the low-voltage winding is tabulated in the chart shown in Fig. 19.2. The ratio between high- and low-voltage windings is determined by the rated voltage of the taps at which the transformer operates. For example,

FIGURE 19.1 Distribution substation transformer with tap-changing underload equipment. Tap-changing switches are located in upper compartment on side of transformer, and control equipment is located in lower compartment.

if the high-voltage winding is set on no-load tap position 3, rated 69,000 volts, and the low-voltage winding is set on underload tap position lower 4, rated 13,455 volts, the ratio of the windings will be 69,000/13,455 or 5.13/1. If the subtransmission voltage is 68,500 volts, the distribution primary voltage will be 68,500/5.13, or 13,353Y/7710 volts, for the transformer tap position stated.

The transformer tap changer that operates underload can be controlled manually by an operator using control switches or automatically by voltage-sensitive devices. The automatic regulating equipment can adjust the primary distribution voltage for various distribution feeder line loadings and associated voltage drop with the aid of line-drop compensator circuits. Voltage regulation and the control circuits for a transformer tap changer are described in other chapters. If the distribution primary feeder circuits are short, the substation power transformer tap-changing equipment may be adequate to regulate the voltage. Power transformers with tap-changing underload equipment normally have a capacity to vary the voltage ±10 percent from rated voltage.

Feeder Voltage Regulators. The function of a feeder voltage regulator is to maintain constant voltage on an alternating-current primary distribution feeder circuit with variations in load. *American National Standard Requirements, Terminology, and Test Code for Step-Voltage and Induction Regulators* ANSI/IEEE C57.15 is the industry guideline. The feeder voltage regulator can be three-phase (Fig. 19.3) or single-phase (Fig. 19.4). If the substation power transformer is not manufactured with tap-changing underload equipment, or the

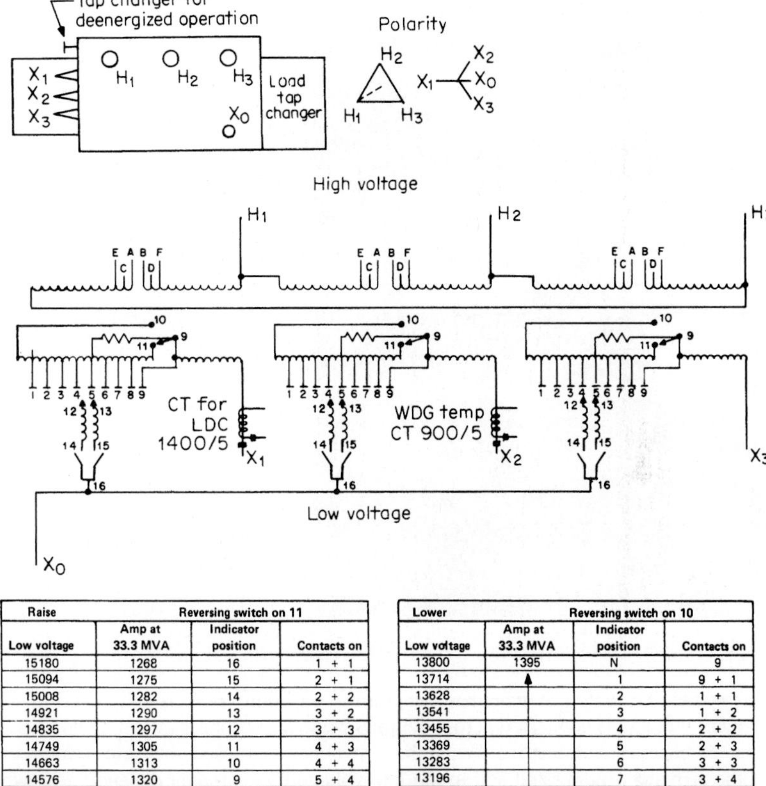

Raise	Reversing switch on 11		
Low voltage	Amp at 33.3 MVA	Indicator position	Contacts on
15180	1268	16	1 + 1
15094	1275	15	2 + 1
15008	1282	14	2 + 2
14921	1290	13	3 + 2
14835	1297	12	3 + 3
14749	1305	11	4 + 3
14663	1313	10	4 + 4
14576	1320	9	5 + 4
14490	1328	8	5 + 5
14404	1336	7	6 + 5
14318	1344	6	6 + 6
14231	1352	5	7 + 6
14145	1361	4	7 + 7
14059	1369	3	8 + 7
13973	1377	2	8 + 8
13886	1386	1	9 + 8
13800	1395	N	9

Lower	Reversing switch on 10		
Low voltage	Amp at 33.3 MVA	Indicator position	Contacts on
13800	1395	N	9
13714	↑	1	9 + 1
13628		2	1 + 1
13541		3	1 + 2
13455		4	2 + 2
13369		5	2 + 3
13283		6	3 + 3
13196		7	3 + 4
13110		8	4 + 4
13024		9	4 + 5
12938		10	5 + 5
12851		11	5 + 6
12765		12	6 + 6
12679		13	6 + 7
12593		14	7 + 7
12506	↓	15	7 + 8
12420	1395	16	8 + 8

FIGURE 19.2 Schematic diagram of three-phase distribution substation power transformer windings. Transformer is rated 20/26.7/33.3 mVA high-voltage winding 69,000 Δ volts, low-voltage winding 13,800Y/7970 volts.

primary distribution feeder circuit is long and heavily loaded, a voltage regulator will be installed to maintain proper voltage (Fig. 19.5).

A three-phase voltage regulator basically consists of three single-phase regulators installed in a single tank. Three single-phase regulators are often used to control the voltage on a three-phase distribution circuit. One single-phase voltage regulator will be used on long single-phase taps to the main three-phase distribution circuit (Fig. 19.6).

The single-phase feeder voltage regulator, or step voltage regulator, is essentially a transformer having a low-voltage secondary winding connected in series with the line and arranged so that the number of turns in the winding can be varied. The voltage is changed by changing the number of turns in the secondary winding. This is accomplished by means of a

FIGURE 19.3 Three-phase step-type primary distribution feeder circuit voltage regulator installed in a distribution substation. Regulator is rated 13,800Y/7970 volts. Switches in structure permit regulator to be bypassed and isolated.

rotary tap-changing switch. A schematic diagram of the connections is shown in Fig. 19.7. The primary winding is connected across the line, while the secondary winding with its taps is connected in series with the line. When the *A* switches are closed, the series winding will extend the primary winding and add to the line voltage in an amount depending on which transfer switches are contacted by *x* and *y*. When the *B* switches are closed, the voltage in the series winding will be opposite in direction to the primary and so reduce the load voltage.

Regulators generally provide for a variation in load from 10 percent below to 10 percent above normal line voltage. This is usually accomplished in 32 steps of $^5/_8$ percent each. The tap-changing switch is motor-driven and is immersed in oil. The reversing switch provides the 10 percent regulation on either side of the neutral position.

The use of vacuum switches while changing taps on the voltage regulators has reduced maintenance. A tap changer, using a vacuum interrupter, can operate a million times while carrying large current before it is necessary to complete major maintenance.

Power thyristors are being installed in some step voltage regulators to electronically switch the number of turns in the winding to completely eliminate arcing in the insulating oil. The use of *silicon controlled rectifiers* (SCRs), which permit an arcless changing of the taps in the voltage regulator, eliminates the need for periodic maintenance of the equipment.

Single-phase step voltage regulators are manufactured in pad-mounted tanks with provision for cable elbow connectors providing dead-front construction (Fig. 19.8).

The control transformer illustrated in Fig. 19.9 provides a voltage source for the regulator control circuits. The control transformer is an autotransformer that supplies voltages above and below ground to the control circuits (Fig. 19.10).

Two diode rectifiers are connected to the center-tapped supply; one supplies power to the circuit, and the other a sensing voltage through a resistive divider.

Reference voltages are generated from a Zener reference and a resistive divider. These are applied to one input each of two detectors (comparators). The difference in the two reference

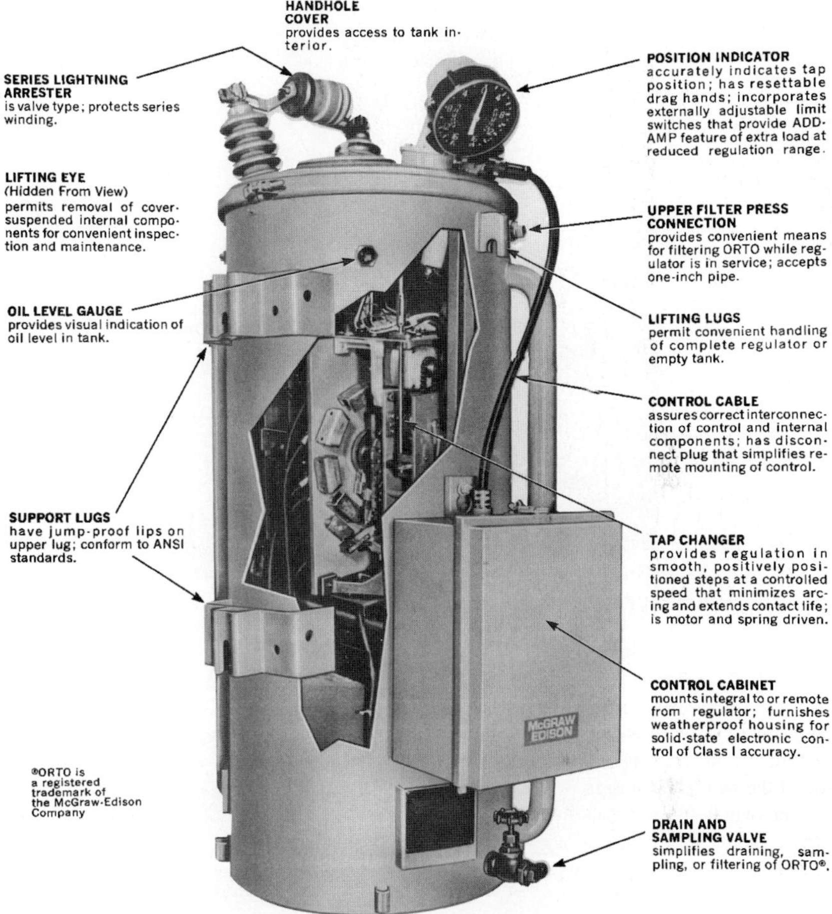

FIGURE 19.4 Cutaway view of a typical single-phase pole-type distribution feeder voltage regulator. Regulator is rated 167 kVA.

voltages is the basis of the bandwidth of the control. The sensed voltage is filtered and applied to the other input of both detectors.

One output of each of the detectors goes low if the sensed voltage is out of band on the low side. This initiates the timer and turns on the OUT-BAND light-emitting diode (LED). The other output of the two detectors goes high when the voltage is out of band on the high side. This is ANDED with the output of the timer which goes high after the predetermined delay and the associated Triac fires.

When the control selector switch is in the automatic position, the tap-changer motor (Fig. 19.11) is connected to the control output Triacs. When the selector switch is in either the manual-raise or manual-lower position, the tap-changer motor is disconnected from the automatic circuit and the proper motor circuit is connected to ground to drive the tap changer in the direction selected.

FIGURE 19.5 Three-phase step-type primary distribution feeder circuit voltage regulator installed in a long circuit remote from a substation.

FIGURE 19.6 Single-phase pole-type distribution feeder voltage regulator installed in a distribution circuit.

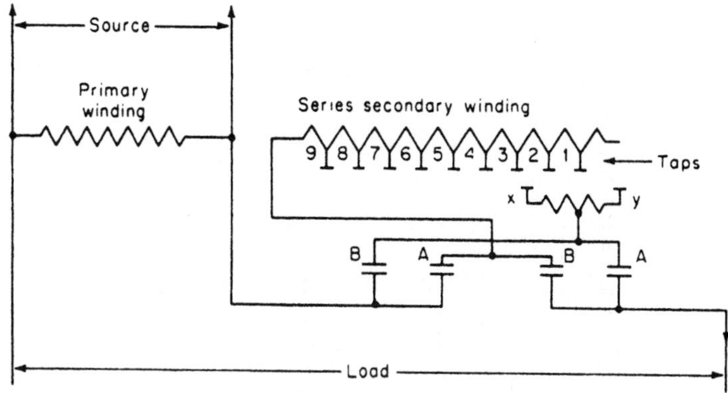

FIGURE 19.7 Connection diagram of step voltage regulator. *A* and *B* are reversing switches, and 1 to 9 are transfer switches which contact *x* and *y*. When *A* switches are closed, the line voltage is raised, and when *B* switches are closed, the line voltage is lowered.

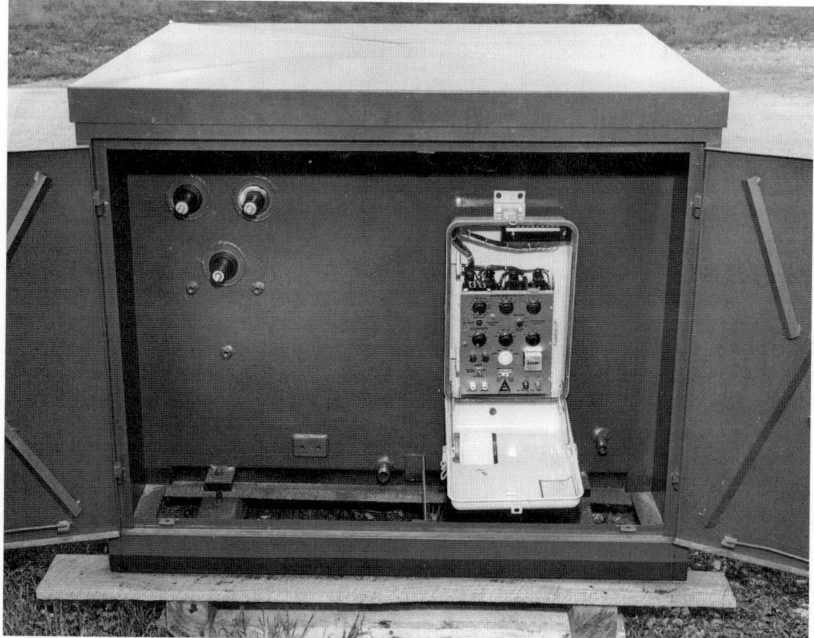

FIGURE 19.8 Pad-mounted primary distribution feeder voltage regulator with compartment door open. High-voltage cables connect to the regulator with elbow connectors. Voltage-sensitive control device case lid is open, displaying voltage control dials.

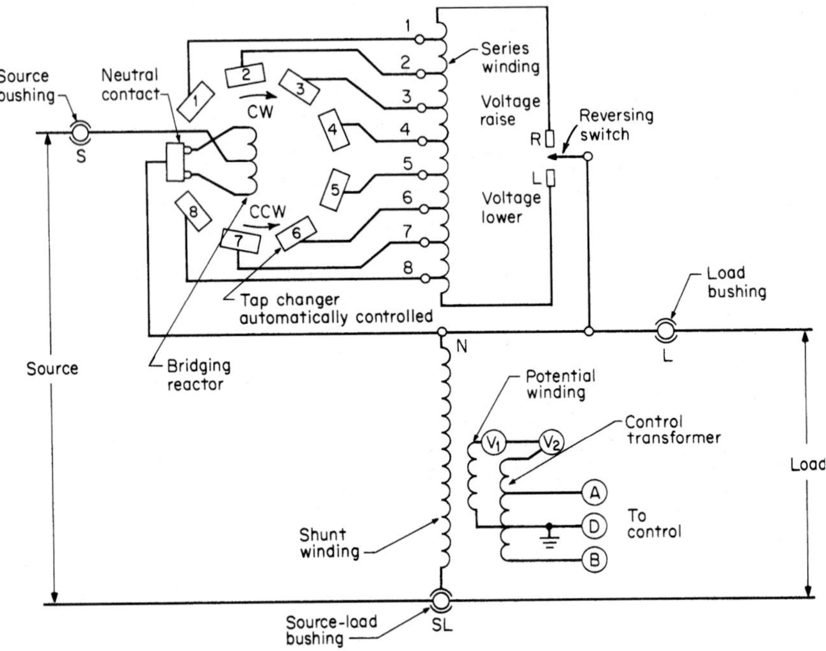

FIGURE 19.9 Single-phase regulator power circuit schematic diagram.

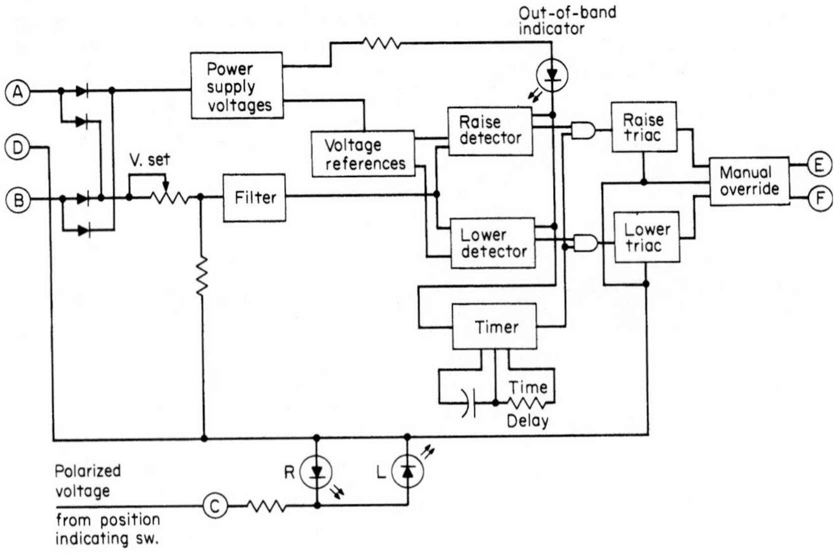

FIGURE 19.10 Regulator control circuit schematic diagram.

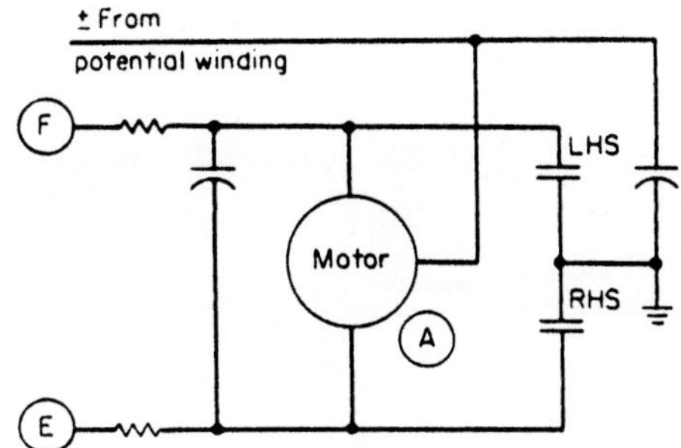

FIGURE 19.11 Regulator motor circuit schematic diagram.

In automatic operation, the voltage from the potential winding is compared with the voltage-level setting on the control. If the voltage falls outside the band, the time-delay sequence is initiated. At the end of the delay time (30 sec is standard), the proper Triac is fired, driving the tap changer in the direction that will correct the voltage. The block diagram of the control (Fig. 19.10), in the conjunction with the motor circuit diagram (Fig. 19.11), shows this sequence of operation.

Time delay, which can be altered by changing a resistor, is set by an *R*/*C* timing circuit. A capacitive discharge circuit rapidly discharges the capacitor if the input to the timer goes high, resetting the timer. Thus, the timer is quickly reset when voltage goes back in band. Time delay must be continuous without resetting before a tap change is accomplished.

A holding switch—either LHS or RHS (Fig. 19.11)—seals in shortly after the motor starts to run, thus ensuring the completion of the tap change. Resistors in the Triac return from the motor limit Triac current to an acceptable level under all conditions. The Triacs only have to initiate a tap change, at which time the motor torque requirement of the tap changer is low. The holding switch takes over almost immediately after the start of a tap change and carries the motor current for the higher torque required to complete the operation.

A pair of microswitches (Fig. 19.12) are actuated by the tap-changer reversing switch. When the reversing-switch actuating segment is centered (the tap changer is in the neutral position), the potential voltage is fed through a series connection of the normally closed contacts of the microswitches to the neutral lamp mounted on the tank wall. As the tap changer moves off the neutral position, the reversing-switch segment actuates one of the microswitches (which microswitch depends on whether the segment is moving in a lower

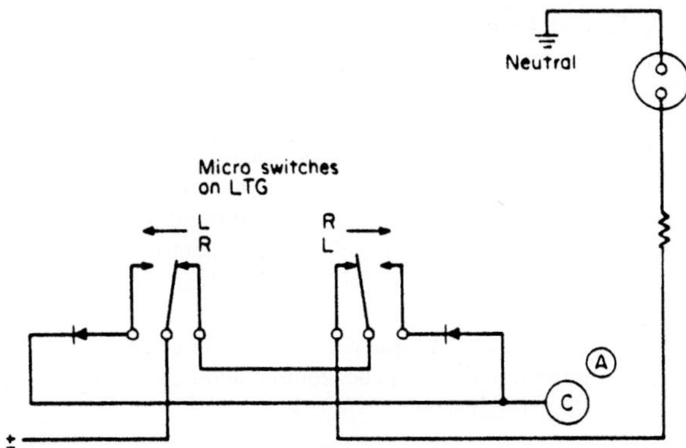

FIGURE 19.12 Regulator indicator circuit schematic diagram.

FIGURE 19.13 Voltage regulator bypass switches mounted on substation steel structure above voltage regulators. Switches bypass voltage regulators and interrupt magnetizing current in the correct sequence . . . all with one pull of a hookstick. (*Courtesy S&C Electric Co.*)

or raise direction). On one side, the supply is half-wave-rectified by a diode to form positive pulses and, on the other side, to form negative pulses. These positive and negative pulses are fed through a current-limiting resistor to a pair of parallel reverse-connected LEDs: Positive pulses actuate one; negative pulses actuate the other. Thus, one LED—designated L—remains on as long as the tap changer is in the lower position. The other LED—designated R—remains on as long as the tap changer is in the raise position.

Voltage Regulator Bypass Switches. Bypass switches for voltage regulators are shown in Fig. 19.13. The voltage regulators must be in the neutral position to operate the bypass switches with the circuit energized. When the voltage regulator is in the neutral position, the series winding output voltage is zero. When the lineman operates the bypass switch to deenergize the voltage regulator, the switch contacts short-circuit the series winding of the voltage regulator to maintain the continuity of the feeder circuit and then in sequence open the contacts in series with the primary winding, interrupting the magnetizing current. If the switches are operated without the voltage regulator in the neutral position, zero voltage output from the series winding, a short circuit will develop, creating a dangerous situation.

Summary. Tap-changing equipment and voltage regulators are widely used within the industry to maintain voltage ranges within acceptable levels. Chapter 40 explains in more detail the concept of voltage regulation.

CHAPTER 20
TRANSMISSION TOWER ERECTION

The general procedure for erecting transmission towers is similar to that for poles, but towers present more problems. The towers require foundations; they are higher and heavier and are therefore more difficult to erect; the conductors are larger and the spans longer, making wire stringing a more difficult job.

Only the general procedure of the construction process will be given. Illustrations will be used to show the various steps. The discussions will be limited to a description of the operations shown in the illustrations.

The order of the operations in tower erection may be briefly outlined as follows:

1. Clearing right-of-way for the line
2. Installing tower footings
3. Grounding tower base
4. Erecting transmission towers
5. Insulator installation

These operations will be illustrated in the pages that follow, and comments will be made on the operations performed.

Clearing Right-of-Way. The right-of-way is cleared of all obstructions that interfere with the operation of the electric transmission line. In scenic and residential areas, clearing of natural vegetation is limited. Trees, shrubs, grass, and topsoil that are not cleared are protected from damage during the construction of the tower line. At road crossings, or other special locations of high visibility, right-of-way strips through forest and timber areas are cleared with varying alignment to comport with the topography of the terrain. Where rights-of-way enter dense timber from a meadow or other clearing, trees are feathered in at the entrance of the timber for a distance of 150 to 200 yards. Small trees and plants are used for transition from natural ground cover to larger areas (Fig. 20.1).

A strip of land is cleared on each side of the centerline of the transmission line by cutting and/or trimming the trees and brush. All "danger trees," trees considered to be hazardous to the transmission line, are removed (Fig. 20.2). All trees, brush, stumps, and other inflammable material, except grass and weeds, are removed from the right-of-way. All trees and brush are cut 3 in or less from the ground line so that the passage of trucks and tractors will not be hindered. The trees and brush cut are disposed of by chipping and spreading, burning, or hauling away. Disposal of the debris by burning, or otherwise, is accomplished in accordance with state and local laws and regulations without creating a hazard or a nuisance.

FIGURE 20.1 Right-of-way cleared for transmission line. The contour of the land and the environment must be taken into consideration by the use of selective clearing techniques. (*Courtesy Asplundh Tree Expert Co.*)

FIGURE 20.2 High-voltage alternating-current transmission line with right-of-way cleared. (*Courtesy A. B. Chance Co.*)

The right-of-way is treated with a chemical spray to retard the growth of brush or trees that if left untreated could endanger the operation of the transmission line (Fig. 20.3).

Installing Tower Footings. Tower sites are graded in accordance with the specifications (Fig. 20.4). Usually the slope of the grade is not greater than 3:1. All topsoil is removed prior to grading the tower location. After the tower construction has been completed and the footings backfilled, the topsoil is replaced. Any excess graded material is removed from the right-of-way.

FIGURE 20.3 Right-of-way is sprayed with chemicals to control unwanted woody growth. Chemicals and methods must be tailored to the season of the year and the varying conditions of the area to meet ecological needs. (*Courtesy Asplundh Tree Expert Co.*)

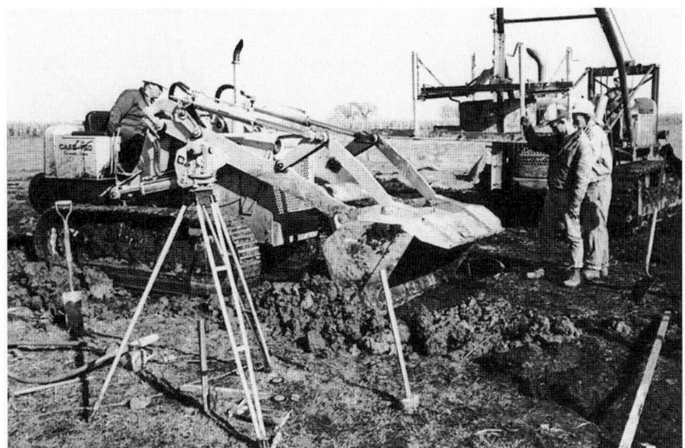

FIGURE 20.4 Workmen grading tower site with a track-driven tractor. Note transit used to properly locate footing. (*Courtesy L. E. Myers Co.*)

The excavation is made according to the dimensions and depths shown on the drawings, allowing sufficient space for proper construction of forms and installation of caissons as required (Figs. 20.5 through 20.10).

Excavations may require the use of equipment such as a wellpoint for dewatering the excavation. The excavations are adequately braced and shored to guard against movement or settlement of adjacent structures, utilities, roadways, or railroad facilities.

Reinforcing bars and embedded stub angles are installed properly. The reinforcing steel is placed and firmly wired before the concrete pouring is started (Fig. 20.11). Exposed reinforcement steel for bonding future construction is protected from corrosion. Reinforcing steel ties, or spacers, are 1 in or more from the finished surface of the exterior or exposed portion of the footings. All accessories in contact with the formwork, or the soil, are galvanized steel, plastic, or some other corrosion-resistant material.

Piles are driven to the depth specified in the foundation drawing. Driving is done with fixed leads that hold the pile firmly in position and alignment. Suitable anvils are used to prevent excessive damage to the piles. Driving of the piles is continuous without intermission until the pile has been driven to the specified depth. The tops of the piles are cut off true and level at the proper elevation.

Concrete for the footings is thoroughly mixed and in a uniform workable state when placed. Concrete is placed before the initial set has occurred, which is accomplished by pouring the concrete in the forms within $1^1/_4$ hours or before the drum of agitating transport equipment has revolved 300 rev (Fig. 20.12). Before placing concrete, the earth foundation is compacted. Granular soil is thoroughly moistened by intermittent sprinkling. The surface, however, is not muddy or frozen when the concrete is poured. Rock surfaces on which concrete is placed is thoroughly cleaned of dirt, debris, and disintegrated material by a high-velocity jet of air or water. All standing water is removed from the depressions in the area where the concrete is to be placed. The steel reinforcing rods, embedded steel, and forms are cleaned of all loose rust scale, paint, mud, dirt, or dried mortar. The forms are oiled before the concrete is poured.

The concrete is conveyed with the proper consistency to the place of final deposit as rapidly as practicable by methods which will prevent segregation, loss of ingredients, rehandling, or premature coating of reinforcing steel or forms (Fig. 20.13). The concrete is consolidated by use of high-frequency internal vibrations and hand spading and rodding while it is poured. The tops of all foundations are troweled smooth, and all irregular projections are removed when the forms have been removed. Any exposed reinforcing steel,

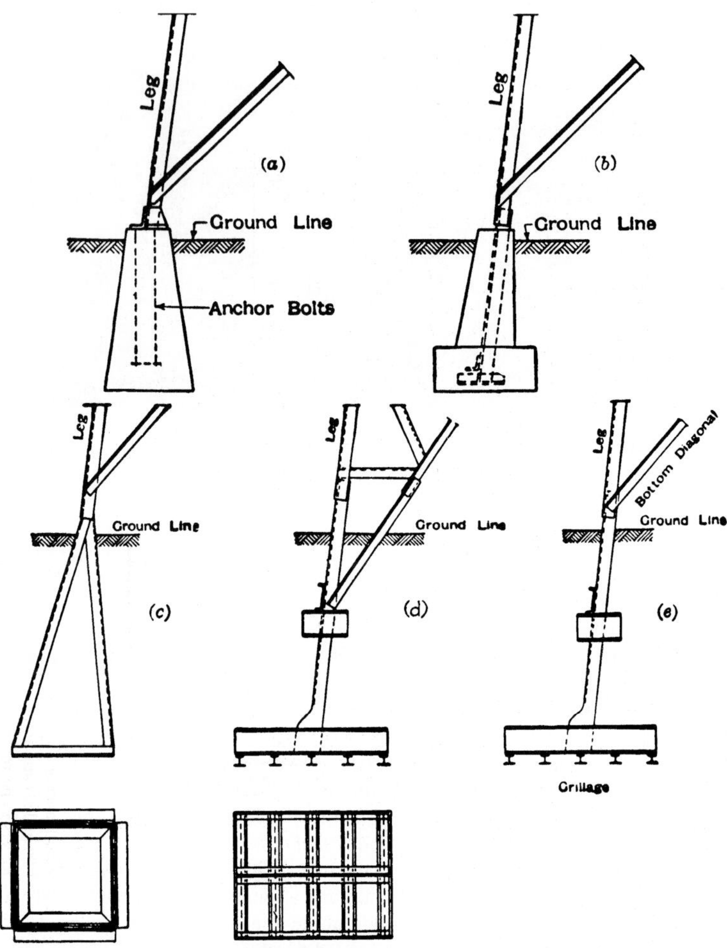

FIGURE 20.5 Various types of tower anchors. Since towers are not massive, footings must be provided to anchor them firmly in the ground. Anchors support the tower and prevent it from blowing over when the lines are subjected to sleet and wind. The anchors shown in (*a*) and (*b*) consist of tapering masses of concrete in which anchor bolts or stub angles are embedded. The anchors shown in (*c*), (*d*), and (*e*) are merely extensions of the tower structure embedded in the earth. These extensions are enlarged so that a large bearing surface is presented. (*Courtesy American Bridge Co.*)

broken corners, or edges are cleaned and painted with an epoxy-resin bonding compound (Fig. 20.14). The concrete foundations are properly cured. Curing of the concrete usually requires seven consecutive days, or more, with the temperature of the air in contact with the concrete at 50°F or greater, before the steel for the tower is attached. Backfilling of dirt around the completed tower footing is completed as soon as practicable after the concrete work is finished. Care is taken to exclude large lumps of dirt from the backfill. The backfilling is completed evenly on all sides and properly tamped.

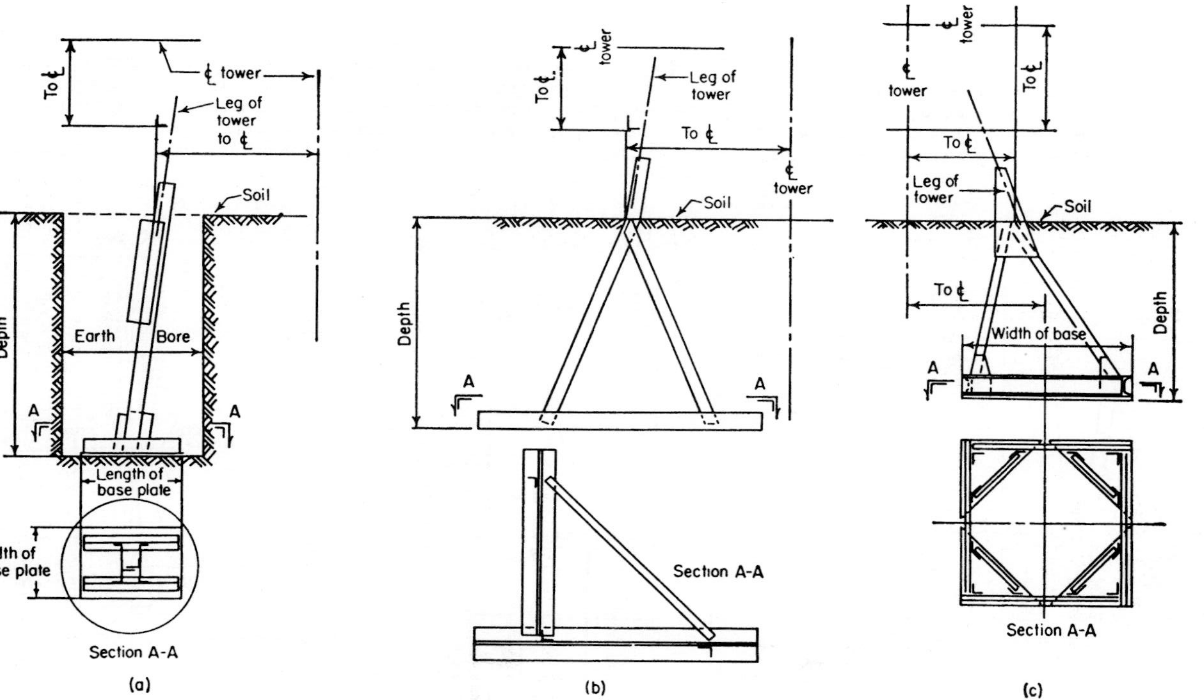

FIGURE 20.6 Tower anchoring systems. (*a*) Single-post earth anchor, used with tangent and small-angle towers in line and at sites where earth-augering equipment can economically and feasibly be employed. (*b*) Triangled earth anchor, used for normal soil conditions with line and angle towers. (*c*) Quadruped earth anchor, used with dead-end, line, and heavy-angle towers in normal soil conditions. The soil-bearing capacity of this anchor is doubled when it is encased in a concrete footing. (*Courtesy Blaw-Knox Co.*)

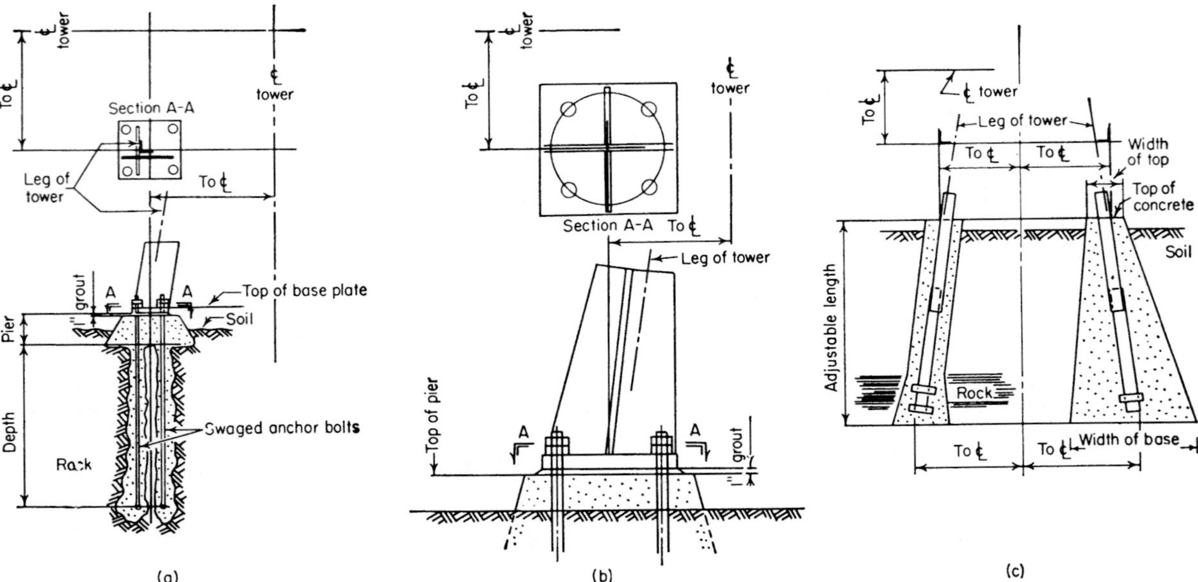

FIGURE 20.7 Tower anchoring systems. (*a*) Rock anchor, used where welded base assembly is preferred and at sites where elevation of subsurface rock is known before erection. Insertion of the anchor's narrow shaft in drilled rock eliminates rock removal by blasting or other costly means. Swedged anchor bolts in the shaft increase the bond between steel and grout. (*b*) Welded base assembly, an easy-to-erect anchor for use where structural loads require the strength of a welded unit in good soil conditions. (*c*) Adjustable rock and concrete anchor, used for applications where economical concrete anchorage is desirable and where rock may be encountered. (*Courtesy Blaw-Knox Co.*)

FIGURE 20.8 Backhoe is being used to dig hole for tower footing. The backhoe digs to a depth of $7\,^1/_2$ ft, has a reach of 15 ft from the centerline of the axle, and a swing of 180°. Large models are available to dig deeper holes for tower foundations. (*Courtesy Ditch Witch.*)

Grounding Tower Base. All steel towers are grounded, usually by means of ground rods. Some tower bases are actually extended to buried steel footers which ground them, while others are connected to buried copper or steel wires. The usual ground rod is a 5/8- or 3/4-in-diameter 8- to 10-ft-long copperweld rod. The rods are driven into undisturbed earth near the tower footing, one rod near each leg. Each tower leg is connected to its own ground rod. If the tower is located near a substation, the four ground rods are often connected together with a heavy copper wire which runs around the tower in a shallow trench. Often the tower

FIGURE 20.9 Large auger is being used to dig footing hole for steel tower footing. (*Courtesy L. E. Myers Co.*)

FIGURE 20.10 Lining up tower footings with a template. The template, which has the same dimensions as the lower end of the tower, must be used to give correct spacing and slope to the anchors which will be embedded in the concrete. The anchors are bolted to this template and are held in their proper position until the concrete is poured and has hardened. Complete template and all four concrete forms are in place, ready for the concrete. A tower anchor or stub is bolted to each corner of the template. Note the ends of the template sides extending beyond the limits of the tower footings at the corners. This extra length is used for lining up towers with a larger base than the one shown. (*Courtesy Wisconsin Electric Power Co.*)

FIGURE 20.11 Reinforcing steel in place and wired. Concrete footing is partially poured.

FIGURE 20.12 Concrete is being poured into forms for transmission-line tower footing from transport-agitating equipment. (*Courtesy L. E. Myers Co.*)

FIGURE 20.13 Helicopter delivering concrete to hopper. A 1/3-yd^3 2400-lb bucket was used for delivery. Unloading was done by suspending bucket over hopper and releasing concrete with a ground line. Several lifts per tower were needed for the footings. (*Courtesy Southern California Edison Co.*)

FIGURE 20.14 Finished concrete foundation with bottom section of steel transmission-line pole set in position.

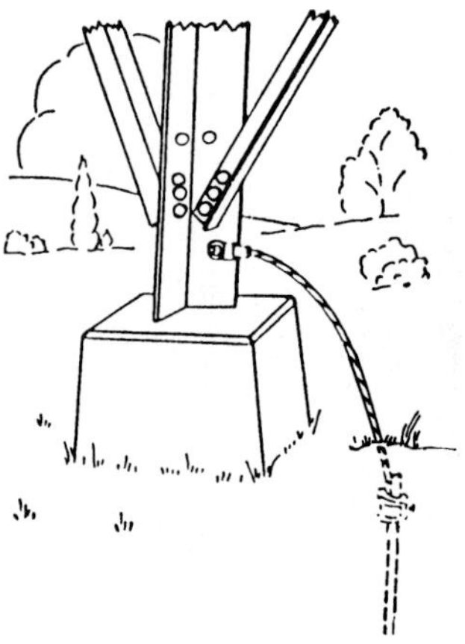

FIGURE 20.15 Ground rod connected to tower leg. (*Courtesy Copperweld Steel Co.*)

ground is also connected to the substation ground by means of the same heavy copper wire. The connection to the tower consists of a heavy clamp bolted to the leg above the ground surface, where it can be removed for testing purposes if need be (Fig. 20.15).

Erecting Transmission Towers. The lineman exercises care in unloading, delivering, handling, and erecting towers so as not to damage the finish (Fig. 20.16). Tower members

FIGURE 20.16 Delivering members of steel transmission tower by helicopter. Made-up bundles were tightly banded for slinging below helicopter. All loads were carried horizontally on a single hook having pilot-controlled mechanical and electric releases and an automatic underload release. (*Courtesy Southern California Edison Co.*)

FIGURE 20.17 Horizontal assembly of the base of a high-voltage transmission-line tower. The first side or panel of the tower has been assembled. Note one anchor or stub in the center foreground. With this type of assembly and modern methods of tower erection, the tower can be assembled on level land, even though it is some distance from the tower footing. Later a crane will pick up the completed tower, carry it to its location, and set it on its foundation. (*Courtesy Wisconsin Electric Power Co.*)

stored pending erection are sorted and neatly piled and supported by suitable blocking (Fig. 20.17). Tower members are not used as unloading or loading skids. Steel erection drawings and structure lists provide the guidelines necessary to assemble the towers. The towers are erected by assembling them in sections on the ground and hoisting the successive sections into place (Figs. 20.18 through 20.24). If the towers are erected by assembling in sections, 50 percent or more of the bolts are in place on each section before another section is started.

The piece-by-piece method of tower erection is similar to that used in erecting steel buildings or other permanent structures. The method is used when the towers are large and heavy and when the ground is rough.

FIGURE 20.18 Linemen are using flatbed truck with A frame to assemble section of transmission-line tower for a 345-kV alternating-current circuit. (*Courtesy L. E. Myers Co.*)

FIGURE 20.19 A 132,000-volt transmission-line tower assembly. This is the top of the tower, here shown being assembled separately. It will be bolted to the bottom part of the tower after it has been assembled. Note steel in the right background being sorted and placed for the assembly of the bottom part of the tower. (*Courtesy Wisconsin Electric Power Co.*)

FIGURE 20.20 A 132-000-volt transmission-line tower almost completely assembled. Note that the bottom part is being bolted to the subassembly of the top part. This tower is almost complete and ready for erection. The dark square sections of steel shown in the top assembly are the mounts for the insulator strings which carry the line conductors. (*Courtesy Wisconsin Electric Power Co.*)

FIGURE 20.21 A completely assembled 132,000-volt transmission-line tower ready for erection. The crane is about to pick the tower off the ground and carry it to its footing. This is a far cry from the old gin pole method of erection, where the tower had to be hinged to its footings while being erected. With this method of erection, a level piece of ground can be picked for the tower assembly even though it is a short distance away from the tower base. (*Courtesy Wisconsin Electric Power Co.*)

FIGURE 20.22 A 132,000-volt transmission-line tower being set on its footings. The first leg is in its approximately correct position. From this point on, the crane operator must operate his crane to get the other three legs correctly placed. After that, it is just a matter of bolting the tower legs to their footings. (*Courtesy Wisconsin Electric Power Co.*)

In assembling a tower in place, light members are often simply lifted into place. Sometimes one of the corner legs is used for raising the other members. For heavier towers, a small boom is rigged on one of the tower legs for hoisting purposes.

When the tower assembly is complete, all bolts are drawn up tight to specified torque. Palnuts are placed on all bolts and tightened with a special palnut wrench.

The development of the helicopter has made possible its use for transport of poles, towers, and line materials for erection of poles and towers. It is especially suited for transport if the terrain is rugged, access roads are not available, or the right-of-way has not been cleared. Figures 20.25 to 20.31 illustrate the helicopter's suitability in the transport and erection of towers in sections or V-type transmission-line towers fully assembled.

Steel pole H-frame structures, Dreyfuss-type Y-shaped steel structures (Fig. 20.32), and single steel pole structures (Fig. 20.33) are used to improve the aesthetic effects and minimize the right-of-way required. Steel poles eliminate the need for down guys. Single steel poles are used to support transmission-line circuits along city streets. The steel poles are

FIGURE 20.23 Erecting lofty tower by use of two crawler cranes. First tower was completely assembled on ground. Then two cranes double-team to erect tower by the "up-ending" method. (*Courtesy Bethlehem Steel Co.*)

FIGURE 20.24 An extra-high-voltage transmission-line tower completely assembled with insulators and conductor-stringing dollies in place has been transported to the installation site by a helicopter. Linemen are fastening the tower legs to the foundation anchor bolts while the helicopter supports the tower. (*Courtesy Columbia Helicopters, Inc.*)

FIGURE 20.25 Helicopter used to erect transmission tower in sections. Bottom section 40 ft high and weighing 2000 lb was lifted onto the stubs. Upper section was lifted in place by use of guide shoes on upper and lower assemblies. Sikorsky S-58 helicopter has a lifting capability of about 4000 lb. This varies with altitude and temperature. (*Courtesy Southern California Edison Co.*)

FIGURE 20.26 Helicopter lifting an aluminum V-shaped two-legged tower in the marshaling yard where it was fabricated to transport it to the line location. Towers are 50 to 95 ft high and weigh upward of 3900 lb. (*Courtesy American Electric Power System.*)

fabricated in a factory and shipped to the site in sections. The foundations are constructed in the same manner as those described for towers. Linemen assemble the steel pole towers in the field and erect structures with the use of cranes and mechanized equipment (Figs. 20.34 through 20.40). Some steel transmission-line poles are fabricated with telescoping construction. The steel poles are shipped to the site, bolted to the concrete foundation, and raised by pumping the interior of the pole full of concrete, using a high-pressure pump. The poles are erected very fast. The concrete reinforces the steel assembly when it has cured.

Insulator Installation. Insulators are handled with care to prevent chipping or cracking the porcelain or glass or damaging epoxy assemblies. Excessive bending strain on the pin shanks or caps is avoided. A cradle, or similar device, is used to lift porcelain or glass insulators. Insulators that are damaged are discarded. The insulators are kept clean and free from grass, twigs, dirt, or other foreign matter. All cotter pins are properly and adequately spread before the insulator strings are attached to the transmission-line structures. The suspension assemblies on all conductors of the line are placed so that all nuts face the center of the structure and cotter pins are parallel to the line with the points of the pins facing inward or toward the space between the conductors. Installation of insulators is illustrated in Figs. 20.41 and 20.42.

FIGURE 20.27 Helicopter transporting completely assembled tower including guy wires to line location. (*Courtesy American Electric Power system.*)

FIGURE 20.28 Helicopter lowering V-shaped aluminum tower onto its single-pin foundation. Helicopter hovers overhead while tower is being securely guyed. (*Courtesy American Electric Power System.*)

FIGURE 20.29 While the helicopter holds the tower in place, members of the ground crew grab each of four guy lines, race to their preset respective anchor points, and secure the tower. (*Courtesy American Electric Power System.*)

FIGURE 20.30 Ground crew connecting and tightening the tower guys. As soon as the guy lines are pulled taut, the helicopter disengages and returns to the marshaling yard for another tower. (*Courtesy American Electric Power System.*)

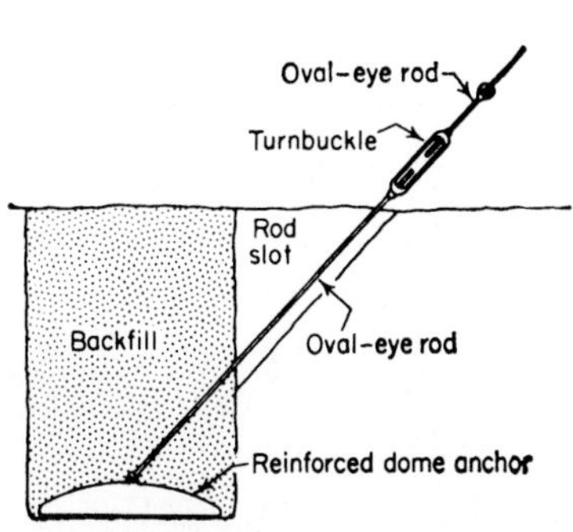

FIGURE 20.31 Typical anchor assembly used to anchor the guys on V-shaped two-legged aluminum towers. A single-center compression foundation, 1 by 2 ft, supports the tower. The anchors range in size from 4 to 6 ft in diameter. (*Courtesy A. B. Chance Co.*)

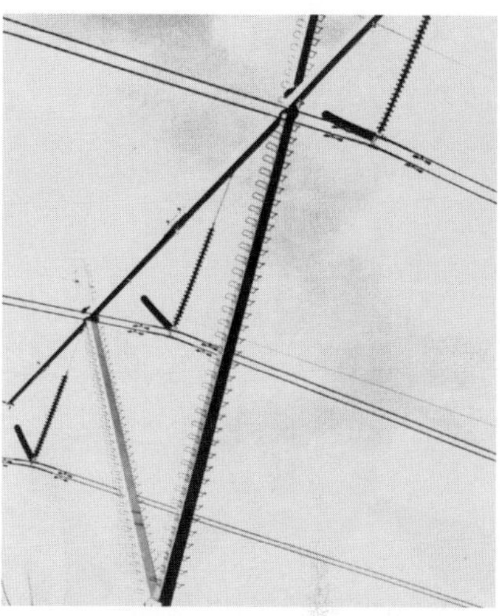

FIGURE 20.32 Dreyfuss-type Y-shaped steel pole structure used for 500-kV ac transmission line. Note steel wings welded to tower members for lineman to use for climbing tower. (*Courtesy A. B. Chance Co.*)

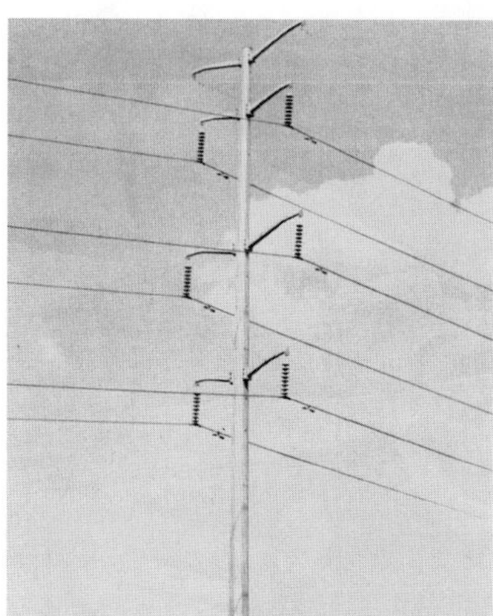

FIGURE 20.33 Single-steel pole structure on edge of golf course along railroad right-of-way. Structures support double-circuit 161-kV ac transmission. (*Courtesy A. B. Chance Co.*)

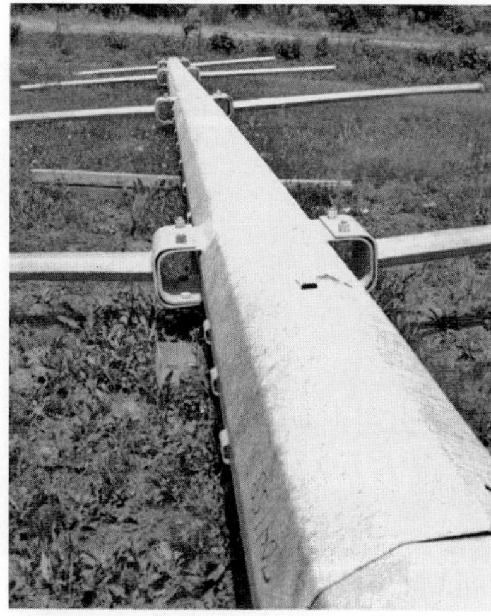

FIGURE 20.34 Single steel pole transmission structure assembled on ground prior to installation. Note brackets on side of pole to facilitate the installation of steps for climbing.

FIGURE 20.35 Bottom section of steel pole has been installed and bolted to concrete foundation. Crane is in position to lift top section of pole.

FIGURE 20.36 Crane starts lift of top section of steel pole for double-circuit 161-kV transmission. Lineman directs crane operator.

FIGURE 20.37 Top section of pole is guided into position by linemen working from a bucket truck. Note insulators were attached to the transmission-line structure before it was raised from the ground.

FIGURE 20.38 Lineman in bucket guides top section of steel pole down over bottom section to provide proper joint overlap.

FIGURE 20.39 Steel transmission pole assembly is completed. Top arms on pole without insulators will support static wires for lightning protection.

FIGURE 20.40 Lineman has installed removable steps on steel transmission-line pole. Lineman is descending pole after removing crane hitch from top of pole.

FIGURE 20.41 Lifting completely assembled 115,000-volt tower onto its footings by means of a mobile power crane. The crane has a 60-ft boom. Note that the tower is being erected with the insulator strings attached to crossarms. They have to be securely fastened, however, to avoid swinging against the tower and breaking the porcelain on the insulator disks. (*Courtesy Virginia Electric and Power Co.*)

FIGURE 20.42 Crane lifting steel transmission-line pole into position. Insulators are secured to steel arms in a manner to restrain them and prevent breakage. (*Courtesy L. E. Myers Co.*)

CHAPTER 21
STRINGING LINE CONDUCTORS

The installation of conductors on poles and towers must be accomplished in a manner to provide an electric circuit that will operate reliably and not endanger the public. Conductors with surface scratches or defects energized at high voltages will have corona losses and generate radio-interference voltages that will be transmitted through the atmosphere.

Slack Conductor Stringing. Slack conductor stringing is limited to short lengths of line operating at low voltages utilizing conductors with a weatherproof covering or installations where scratches on the surface of the conductor are not important. Scratches on ground or neutral conductors will not cause problems if the strength of the wire is not impaired.

The reels of wire can be mounted on a vehicle in such a manner that the reels are free to rotate. The ends of the conductors are fastened to a pole, tower footing, or other fixed object. The vehicle is then slowly propelled along the route of the line, allowing the conductors to unwind as the reels are moved forward. In this method the conductors are not dragged over the ground, causing them to become scratched or damaged. They are simply laid out on the ground without being pulled over it. The method, however, cannot be used on one-circuit tower lines or on X-braced H-frame lines because the center or middle conductor must be placed over the internal framework of the tower or over the X braces of the H frames. This obviously is impossible if the reels are on the lead end of the conductor.

If the reels on which the conductors are wound remain in a fixed location and are raised off the ground or supported in their carriages in such a way that they are free to rotate, the conductors can then be pulled out, thereby rotating the reels and unwinding the conductors, as illustrated in Fig. 21.1.

The conductor should never be payed out from a nonrotating reel or coil, as each turn removed gives the conductor a complete twist which may cause kinks or other damage.

Unreeling conductors from stationary reels provides two possibilities, one to simply draw the conductors forward, sliding or dragging them over the ground, and the other to keep the conductors suspended in the air in tension so that they will not touch the ground. When the conductors are pulled forward, sliding over the ground, they are apt to become scratched and nicked, especially if the ground is rough and covered with surface stone. Aluminum conductors are more easily injured in this manner than copper.

Tension Conductor Stringing. The process of installing overhead transmission- or distribution-line conductors in a manner which keeps the conductors off the ground, clear of vehicular traffic and other structures that might damage the conductors, and clear of energized circuits is called *tension stringing*. Tension stringing, to reconductor existing energized lines, provides a means to keep the new conductors under control and prevent them from contacting the energized conductors. The new conductors being installed are effectively grounded for the protection of the workmen. Transmission-line conductor installation requires the linemen to install conductor stringing blocks on the transmission-line insulator

FIGURE 21.1 Reel of triplexed low-voltage secondary cable. Axle in reel is supported by winch line on boom of line truck. Axle has a swivel joint, permitting reel to rotate easily. Lineman is restraining rotation as cable is laid out. Cable and other line material were transported to job site in material trailer behind the truck.

strings to permit pulling the conductors in under tension (Figs. 21.2 and 21.3). The conductor stringing blocks are built for the installation of one to four conductors for each point on the tower (Fig. 21.4). The sheaves on the stringing blocks are usually lined with a conductive-type neoprene, or urethane, to protect the phase conductors which are usually aluminum or aluminum and steel (ACSR). The blocks used for static, or ground, wires are not lined since the static wires are usually galvanized steel, copperweld, or alumoweld, which are not easily damaged.

The conductive neoprene lining of the stringing blocks permits the conductors to be effectively grounded by a jumper from the stringing block to a ground wire on the metal tower structure. Effectively grounded stringing blocks will eliminate induced, static, or impulse voltages that might be present during construction. The stringing blocks, often called *travelers*, can be equipped with outriggers to permit the installation of the pilot line with a helicopter.

When linemen install the stringing blocks on the insulators, they normally place a lightweight rope called a *finger line* over the traveler (Fig. 21.5). The finger line must be long enough to reach the ground on both sides of the traveler. Finger lines are used by the linemen to pull the pilot line through the traveler from the ground. If the pilot line is installed with a helicopter, finger lines are not used. Pilot lines are lightweight rope used to pull pulling ropes through the travelers, as shown in Fig. 21.2. A pilot-line winder (Fig. 21.6) is used to provide the power necessary to pull the conductor pulling rope from the reel on the take-up reel stand through the travelers. Tension is kept on the pulling rope, often referred to as the *bull line*, to prevent interference with objects on the ground.

The conductor reels, tensioners, and pulling machines must be in line before the conductor pulling is started. The tensioner, often called a *bull-wheel* tensioner (Fig. 21.7), and the bull-wheel puller (Fig. 21.8) should be set up as near to midspan as possible (Fig. 21.9). The slope of the conductors between the equipment and the stringing blocks at the first tower must not be steeper than 5 horizontal to 1 vertical.

The transmission-line conductor retarding bull wheels should have a minimum of five turns of the conductor over the bull-wheel and have multiple grooves lined with neoprene or other approved nonmetallic resilient material so that the conductor will cushion into the lining to prevent flattening or otherwise damaging the conductors as they are payed out

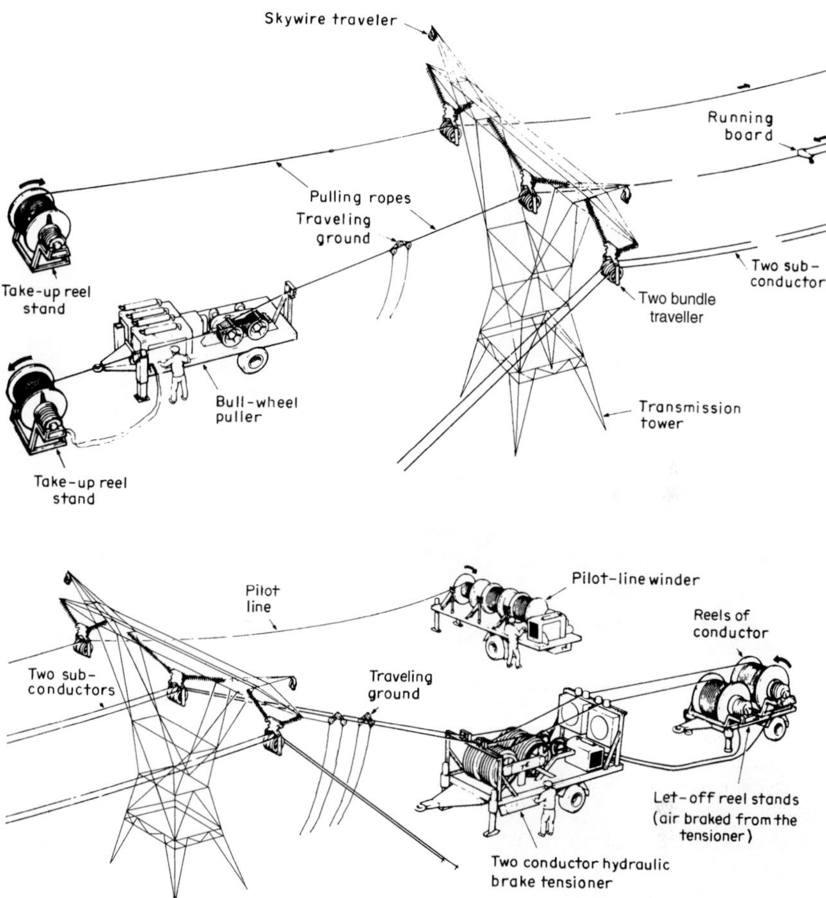

FIGURE 21.2 Typical equipment setup for tension stringing a two-conductor bundle transmission line. (*Courtesy Timberland Equipment Ltd.*)

(Fig. 21.10). The diameter of the bull-wheel grooves must not be less than the conductor diameter, plus 25 percent. The bull-wheels should have a minimum bottom grove ("root") diameter of 25 times the conductor diameter.

The conductors should be pulled directly from the cable reel onto the tensioners and into the stringing blocks without touching the ground. The conductor pulling tension should not exceed 70 percent of the conductor sagging tension. Care must be taken at all times to ensure that the conductors do not become kinked, twisted, abraded, or damaged and that foreign matter does not become deposited on them. Dependable communication must be maintained between the linemen operating the pulling equipment and the tensioning equipment and observers at intermediate points at all times during the wire-stringing operations (Fig. 21.11).

Bundled conductors are pulled simultaneously by one pulling line with the use of a unidirectional articulated running board (Fig. 21.12). Special bundle-conductor-type stringing blocks are required. For stringing two conductors (see Fig. 21.13) the block consists of two

FIGURE 21.3 Take-up reel stand in combination with a bull-wheel puller.

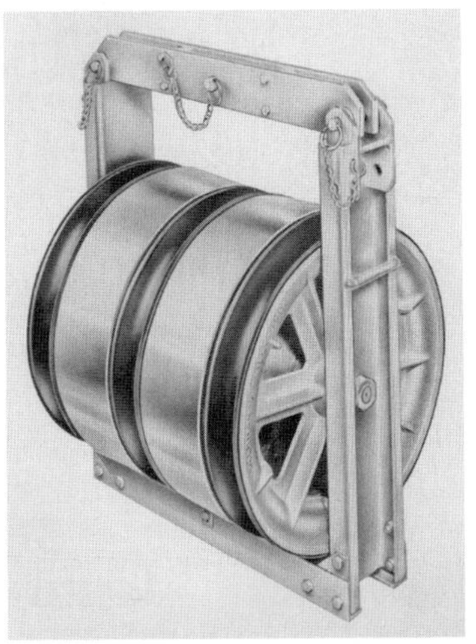

FIGURE 21.4 Three-conductor bundle stringing block. (*Courtesy Sherman & Reilly, Inc.*)

FIGURE 21.5 Finger line looped through traveler extends to the ground. (*Courtesy American Electric Power Co.*)

FIGURE 21.6 Linemen using a pilot-line winder to pull conductor pulling rope through travelers. Note frame of winder is grounded. (*Courtesy Sherman & Reilly, Inc.*)

sheaves and one drum, all of which turn independently of one another. The two outer sheaves support the conductors, and the center portion of the drum carries the pulling line. The drum is designed to accept the running board. For stringing a bundle of three conductors, a block with three sheaves and a two-part drum are used, as shown in Fig. 21.4. The pulling line rides in the center groove as does the center conductor, shown in Fig. 21.14. The sheave spacing corresponds to the final spacing of the bundled conductor.

The tensioner for a two-conductor bundle consists of two pairs of bull wheels, one pair for each conductor. The bull wheels are geared together in such a way that a small auxiliary engine can be used to even up the conductors if one should sag more than the other. The engine is also used to air-cool the brakes and to reverse the bull wheels slowly, if this should become necessary.

When a bundle of four conductors is strung, it is good to string the bottom two conductors first and leave them in soft sag. The top conductors are then strung over them.

FIGURE 21.7 Bull-wheel tensioner and conductor reels set up to pull three conductors per phase. The three conductors are called *subconductors* and form a three-conductor bundle per phase. (*Courtesy Sherman & Reilly, Inc.*)

FIGURE 21.8 Linemen setting up bull-wheel pullers prior to starting the conductor pulling operation. (*Courtesy L. E. Myers Co.*)

The conductors are grounded while they are being installed. Traveling grounds can be used at the tensioner and puller sites. The use of travelers with a conductive-type neoprene lining permits grounding the conductors at each structure. The grounding cable must be large enough to adequately conduct fault current to ground without fusing.

When the conductors being installed reach the pulling equipment, they must be secured to prevent them from developing slack, which might let them touch the ground (Fig. 21.15).

FIGURE 21.9 Tensioners and reels of conductor set up for installing conductors on transmission line in background. (*Courtesy L. E. Myers Co.*)

FIGURE 21.10 Lineman operating bull-wheel tensioner used for two subconductors per phase. (*Courtesy L. E. Myers Co.*)

FIGURE 21.11 Lineman with two-way radio directing conductor-pulling operation. (*Courtesy L. E. Myers Co.*)

New distribution-circuit conductors are installed by using tension stringing methods, as described, except that the equipment for pulling and tensioning the conductors is smaller and lighter (Figs. 21.16 through 21.18). Distribution circuits are shorter in distance than most transmission lines, and the conductors are usually smaller with a limit of one conductor per phase. Finger lines can be used to pull the pilot line through the stringing blocks (Fig. 21.19). However, care must be taken to tie the finger lines off high enough on the pole to keep them out of reach of the public.

FIGURE 21.12 Articulated running board designed for use in stringing bundled conductors. Vertical links hold running board in horizontal position for correct entrance into stringing block. (*Courtesy Sherman & Reilly, Inc.*)

FIGURE 21.13 Pulling board about to be drawn over two-conductor stringing block. Central drum carries pulling line, and two outer sheaves carry the two-conductor bundle. The wedge-shaped running board ensures block centering and placing conductors in grooves. (*Courtesy Sherman & Reilly, Inc.*)

FIGURE 21.14 Three-conductor bundle block supported on suspension insulator string. Block has three sheaves. Center sheave carries the pulling line. After running board passes the block, the third conductor follows the pulling line on the center sheave. (*Courtesy Sherman & Reilly, Inc.*)

FIGURE 21.15 Conductors secured temporarily with conductor clamps and steel cables anchored to frame of bull-wheel pullers. (*Courtesy L. E. Myers Co.*)

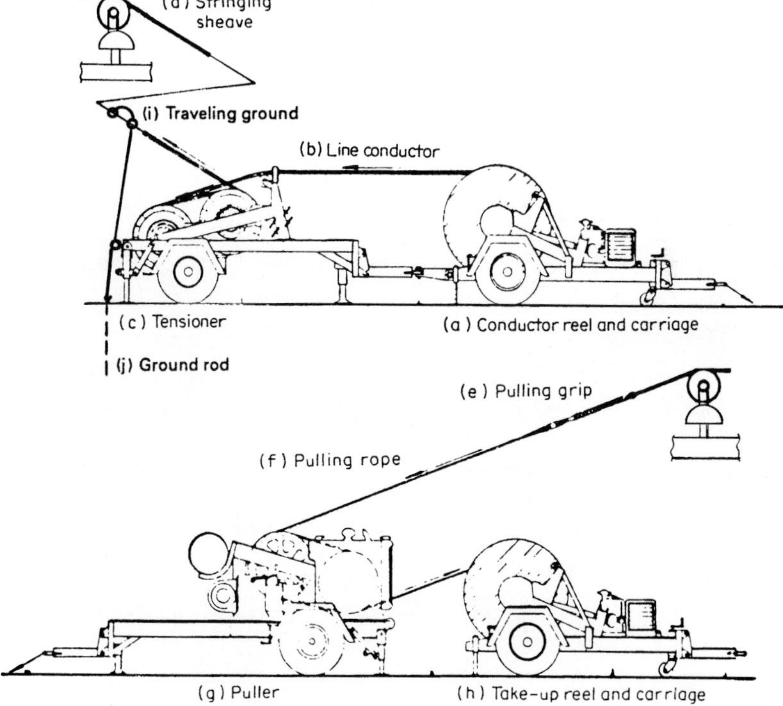

FIGURE 21.16 Diagrammatic sketch showing arrangement of tension stringing equipment to install new line conductor. (*a*) Conductor reel and carriage, (*b*) line conductor, (*c*) tensioner, (*d*) stringing sheave, (*e*) pulling grip, (*f*) pulling rope, (*g*) puller, (*h*) take-up reel and carriage, (*i*) traveling ground, (*j*) ground rod. The conductor being installed is grounded on the reel, by traveling grounds and by grounds connected to the metal stringing blocks. (*Courtesy Peterson Engineering Co., Inc.*)

FIGURE 21.17 Tension stringing of line conductors with four-reel payout tensioner operated in conjunction with four-reel take-up puller. Each reel is provided with a brake to produce tension in the conductor. Lineman kneeling is using a walkie-talkie two-way radio to communicate with the lineman at the puller, in order to coordinate pulling and tensioning. (*Courtesy Ameren UE.*)

FIGURE 21.18 Tension stringing of line conductors with four-reel rope take-up puller equipped with automatic level winder. Separate gasoline engine drives take-up reels. Each reel can hold 3000 ft of $^3/_4$-in rope. Tension stringing suspends conductors above ground in each span and controls the new conductor to prevent accidental contacts from adjacent energized wires. The new conductor is effectively grounded to protect the workmen. (*Courtesy Ameren UE.*)

FIGURE 21.19 Finger lines have been installed in stringing blocks for a new distribution circuit. Pilot line has been pulled through block for static wire. Lineman is checking equipment on first pole before pulling continues. Normally pilot line will be pulled through stringing blocks with finger lines from the ground. (*Courtesy A. B. Chance Co.*)

Tension stringing procedures are used successfully to reconductor distribution lines. The energized conductors must be kept isolated from the new conductors being installed. The linemen often use hot-line-extension arms (Figs. 21.20 through 21.22) to provide clearance for the new conductors. The new conductors being installed adjacent to the energized wires are effectively grounded and controlled to prevent them from being energized by accidental contact with energized wires.

FIGURE 21.20 Groundman has taken hot-line-extension arm from material and equipment trailer. The hot-extension arm will be sent up pole with hand line to lineman.

FIGURE 21.21 Lineman has removed energized conductor covered with insulated-line hose from insulator on crossarm. Hot-line-extension arm is being installed on crossarm. Energized conductor will be secured in the clear by clamp on hot-extension arm. Note stringing block clamped to crossarm with finger line in place.

FIGURE 21.22 All three energized phase conductors on pole in foreground have been set out in the clear on hot-line-extension arms. Stringing blocks for phase conductors are installed on crossarm. Stringing block for neutral-ground wire is clamped to pole below crossarm. The new conductors being installed will be grounded on the reels, by traveling grounds and grounded stringing blocks. The multigrounded neutral conductor is the best source for the ground connections.

Pilot lines can be installed for distribution-line tension stringing by using the *spider system.* Pilot-line controllers are installed on the lead pole by a lineman approximately 15 to 20 ft above the ground. A pilot-line storage reel is hung on each controller (Fig. 21.23). The pilot line comes off the storage reel over an arm that releases the brake on the pilot-line controller when the rope is pulled and engages the brake when the rope is slack. This eliminates "backwinding" and holds tension during the installation of the pilot line.

All pilot lines can be carried up each pole and placed in the stringing blocks when the blocks are placed on the pole. As this is accomplished, the pilot lines can be tied up and left in the clear of any obstructions below, such as tree limbs, transformers, and street lights.

The pulling line is positioned at the opposite end of the spider controller and storage reels and is tied to the end of the pilot line. The partially empty storage reel is then mounted on a pole-mounted winch or a line truck winch. The pole-mounted winch is powered hydraulically or electrically. The line truck's hydraulic system can be used to supply power for the pole-mounted winch.

The winch turns the reel and rewinds the pilot line, thus hauling in the pulling line under tension and eliminating the need for linemen to walk through while carrying the pulling line over each structure.

Fast, efficient, safe, and field-tested, the spider system is used by many utilities. The pilot line is used to pull the pulling line, or bull rope, through the stringing blocks. The pulling rope is connected to the conductor with a conductor grip (Fig. 21.24).

The conductors are effectively grounded while they are being installed (Fig. 21.25). The grounding cable connecting to the running ground must be large enough to safely conduct

FIGURE 21.23 Spider system of installing pilot line for tension stringing of distribution-circuit conductors. Reels are mounted high enough on pole to prevent pedestrian contact and overnight vandalism. (*Courtesy Sherman & Reilly, Inc.*)

FIGURE 21.24 Attaching the Kellum grip on a 795,000-cir mil ACSR cable. This grip is used for pulling the cable over the stringing sheaves on the structures. When the grip is in place, it is taped down to the conductor to keep it from pulling back and releasing. (*Courtesy Northwest Lineman College.*)

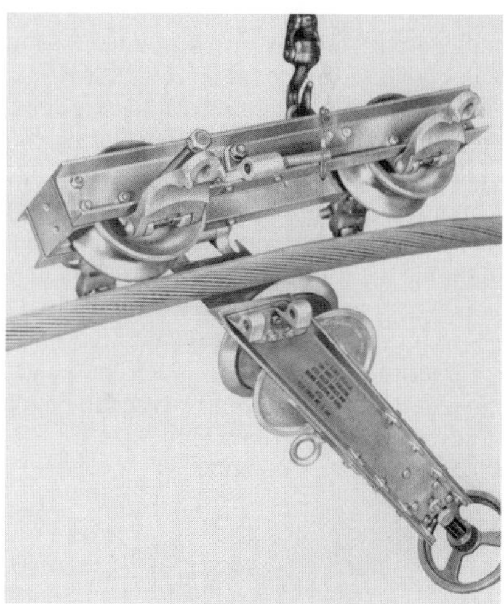

FIGURE 21.25 Running ground device designed to maintain positive contact with the conductor being strung regardless of tension. (*Courtesy Sherman & Reilly, Inc.*)

FIGURE 21.26 Pole-mounted hydraulically powered four-drum puller being used to install new distribution-line conductors. The hydraulic hoses shown connect to the line truck's hydraulic system to obtain operating power.

FIGURE 21.27 In case conductors have to be strung over a live line, guard poles which hold the wires above the live line are used. Shown here are two guard poles carrying three line conductors and two ground wires of a 132,000-volt line over a 4800-volt single-phase distribution feeder. (*Courtesy Wisconsin Electric Power Co.*)

fault current to the point of ground attachment. The multigrounded neutral conductor provides the best source for a ground, if it is available.

Equipment similar to the spider system described is available that can be used to install small distribution conductors under tension without pulling in a larger pulling rope.

The equipment shown in Fig. 21.26 has ropes and power available to install some distribution conductors under tension. The ropes can be placed in the stringing blocks while the lineman is mounting the block, or finger lines can be installed in the blocks to pull the stringing rope through the blocks at a later time. The conductor tension pulling equipment is powered by the line truck's hydraulic system. The equipment can be pole-mounted as shown or can be installed on a trailer.

If distribution- or transmission-line conductors are to be strung over highways or energized circuits, guard poles should be used to protect the public and the workmen (Fig. 21.27).

CHAPTER 22
SAGGING LINE CONDUCTORS

Sagging operations begin as soon as the conductor stringing is completed, to establish the proper conductor tension for the conditions at the time the work is performed. The correct conductor tension and sag for various sag or control spans are detailed in the construction specifications or standards to provide proper clearances. The weather conditions, including temperature and wind velocity, must be taken into consideration to complete the process. The proper conductor tension can be obtained by establishing the proper sag in the control spans. Sighting methods using a surveyor's transit or targets can be used to measure the sag in the conductor. The tension in the conductor can be measured with a dynamometer or determined by timing mechanical waves induced into the conductor.

All conductor grips, come-alongs, stringing blocks, and pulley wheels used in the sagging operations must be of a design which will not damage the conductors by kinking, scouring, or unduly bending them. Sag sections must be selected before the operations are started. The lengths of sag sections are limited by dead-ends, angle structures, and the terrain. The sag sections are normally not longer than $4^1/_2$ miles.

Sag sections must have control spans. The control spans will be used to measure the sag for each sag section. If the control spans are properly sagged, the conductors in the sag section should all be at the proper tension. The control spans should be approximately the length of the ruling span. Ruling spans are a calculated dead-end span length assumed to have the same conductor tension as all other spans in the sag section, even though they vary in length. Control spans should be located away from the dead-end sections and large line angles in a level portion of the line, if possible. If only one span is used for checking the sag, it should be approximately in the middle of the section being pulled to tension. Where two or more spans are used for checking the sag, they should be located approximately equidistant from each other and from each end of the sag section. The sag of spans on both sides of points in the line where the grade varies by more than 10° should be checked.

The use of more than one sag control span in each sag section will eliminate problems due to cumulative sheave effect and human error. One control span is permissible in short sag sections of five spans or less. A minimum of two control spans should be used in sag sections of five to eight spans in length. A minimum of three control spans should be used in sag sections of nine or more spans. Sag sections are normally limited to approximately 20 spans. The control spans selected should be checked to be sure there are no major discrepancies between the actual span lengths and those on the plan and profile sheets included with the specifications.

All wires on new lines being sagged under deenergized conditions should be grounded (Figs. 22.1 and 22.2). When a bundle of three or four conductors is sagged, the top conductors are sagged first and bottom conductors are sagged last. The tension in the bottom conductors should be about 3 percent less than those in the top conductors. All wires must be pulled up to the proper sag without exceeding the necessary tension. The wires must not

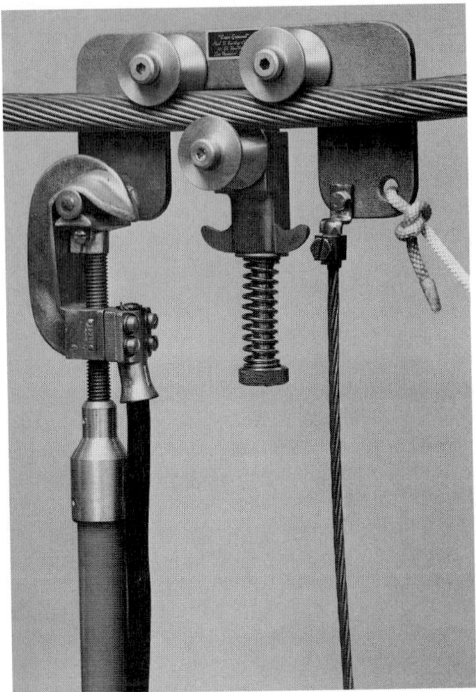

FIGURE 22.1 Ground device installed on moving conductor while sagging and stringing operations are completed. Grounding cable is attached with hot-line tool. Rope secured to pole or tower prevents ground device from moving with conductor.

be pulled up to a tension greater than the sagging tension. The conductors must be sagged to the initial sag specifications.

The temperature of the conductor must be determined with a thermometer that is normally inserted into a piece of conductor approximately 30 in long and suspended at the approximate elevation of the wire to be sagged. Two thermometers should be used at locations near sag spans in the sag section. The thermometers should be in position for a minimum of 15 min before the temperature is recorded to determine the proper wire sag. After the conductors have been properly sagged, intermediate spans should be inspected to

FIGURE 22.2 Ground device installed on conductor for a field demonstration.

FIGURE 22.3 Close-up view of towerman as he observes the wire sagging and keeps the supervisor informed of conditions at cable sheaves. (*Courtesy American Electric Power System.*)

FIGURE 22.4 Line conductors pulled up close to correct sag and snubbed off to earth anchors. The snub must be made far enough out from the tower to keep the downpull on the crossarms to a safe value. Lineman shown is using coffing hoist with wire grips to pull conductors up to exact sag. The conductors are 795,000-cir-mil ACSR cables and have been strung over 14-in stringing sheaves. The desired sag is equivalent to a tension of 7000 lb. (*Courtesy Detroit Edison Co.*)

ensure that the sags are uniform and correct. The wires are sagged from the tensioner end of the sag section to the puller end.

The foreman initiates and supervises the sag operations, maintaining communication with the linemen at each sag span and other points under observation in the sag section between the tensioner end and the pulling end (Figs. 22.3 and 22.4). The sagging operations cannot be completed on days with strong winds because of the conductor uplift from wind pressure. The aerodynamic effect of the wind on the conductors can cause sagging errors.

Special precautions must be taken to sag distribution conductors that are near energized circuits or are energized (Figs. 22.5 through 22.12).

FIGURE 22.5 Schematic diagram of distribution line being pulled up to proper sag. Note use of free-running sheave block at every point of support.

FIGURE 22.6 Distribution-phase conductor in stringing block mounted on insulated epoxy stand-off bracket in position for sagging operation. (*Courtesy Sherman & Reilly, Inc.*)

FIGURE 22.7 Distribution-circuit conductors in stringing blocks mounted on line-post insulators. Circuit can be energized in this position while sagging operations are completed if proper precautions are taken. (*Courtesy Sherman & Reilly, Inc.*)

FIGURE 22.8 Rope basket employed to guard against unstrung line conductors dropping low or falling onto street. (*Courtesy Union Electric Co.*)

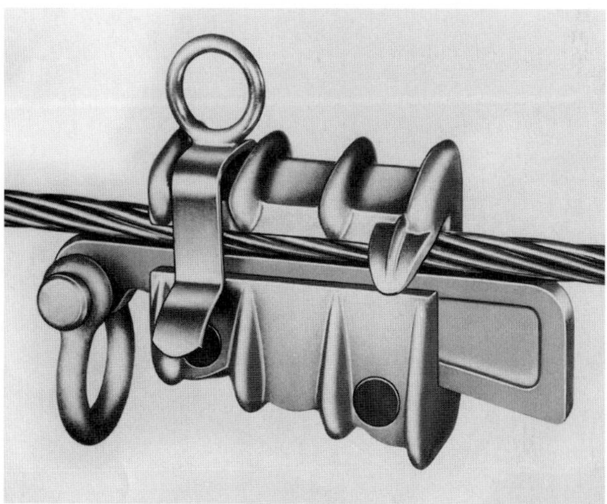

FIGURE 22.9 Come-along grip depends on the wedge action to grip conductor. The greater the pull on the wedge, the harder the conductor is forced against the steel grip insert in the body. Come-alongs can be installed in live-line work with a tie stick, using the lifting ring provided on the top. (*Courtesy James R. Kearney Corp.*)

FIGURE 22.10 Pulling grip or come-along being installed on an energized circuit using a hotline tool. Note manner in which wire is held by grip when tension is applied to linkage by attaching a hoist to the eye of the linkage. (*Courtesy A. B. Chance Co.*)

FIGURE 22.11 Lineman pulling up on energized conductor with hoist and grip or come-along to properly sag energized conductors. Personal and line protective equipment is used while working energized circuit from insulated-bucket truck.

FIGURE 22.12 Lineman sagging energized-distribution conductors. Note linemen wear rubber gloves and sleeves and perform work on an insulated platform. All conductors except the one being sagged, with the use of rope blocks, are covered with protective equipment. (*Courtesy A. B. Chance Co.*)

Sagging Line with Transit. The method of determining conductor sag used commonly by linemen is to climb one structure and place targets at the specified sag level for the conditions that prevail. The lineman then climbs the structure on the opposite end of the control span and mounts a surveyor's transit, or an especially designed sag scope (Fig. 22.13 and 22.14), at a point equal to the sag distance below the conductor support. The sag scope with a variable mount can be easily leveled under a variety of field conditions. The clamping device for steel, aluminum, or wood-pole structures permits adjustments from 0 to 3 in. The lineman then sights through the transit and observes conductor sag. When the conductor is at the proper level, he tells the foreman supervising the operations to prevent overstressing of the conductor (Fig. 22.15). If the structures on each end of the control span are not at the same height or not located on level ground, special adjustments must be made in the target and transit locations to compensate for the differences in elevation.

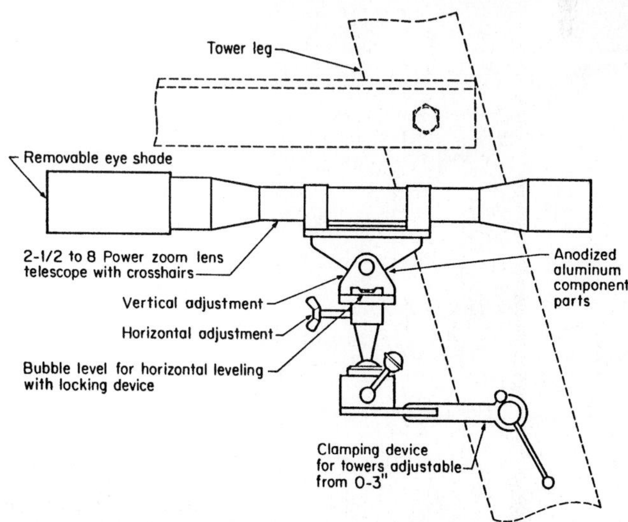

FIGURE 22.13 Sag scope with zoom capabilities for wire sagging. (*Courtesy Hi-line Industries, Inc.*)

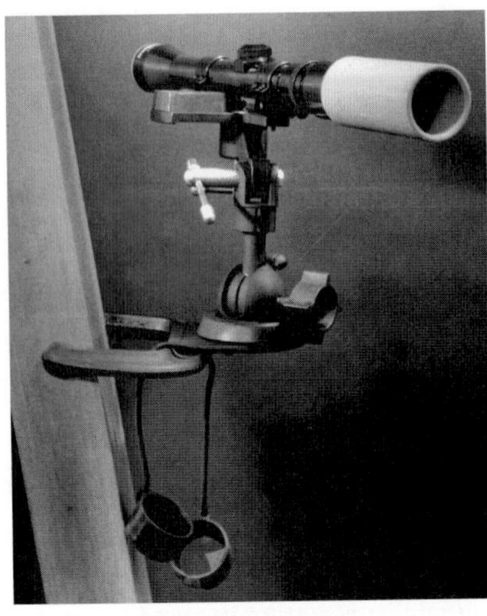

FIGURE 22.14 Sag scope with 3- to 9-power zoom telescope cross hairs for precision sighting and a two-way bubble level. (*Courtesy Sagline Corp.*)

In the case of an H-frame line, the line conductors are on one or the other side of the poles. A lineman stationed on one pole and looking to the corresponding pole a span length away cannot include the lowest point of the conductor in his line of sight. To get around this difficulty, a transit is securely fastened to the pole at a distance equal to the desired sag below the conductor support. Then the transit is leveled. To observe the sag, the transit is sighted at the conductor at midspan and then is swung around until in line with the pole a span length away. The sag is then observed, and the line conductor is drawn up to the specified distance.

FIGURE 22.15 Lineman sighting through a transit telescope at a target on the next structure while a line conductor is being sagged. (*Courtesy Wisconsin Electric Power Co.*)

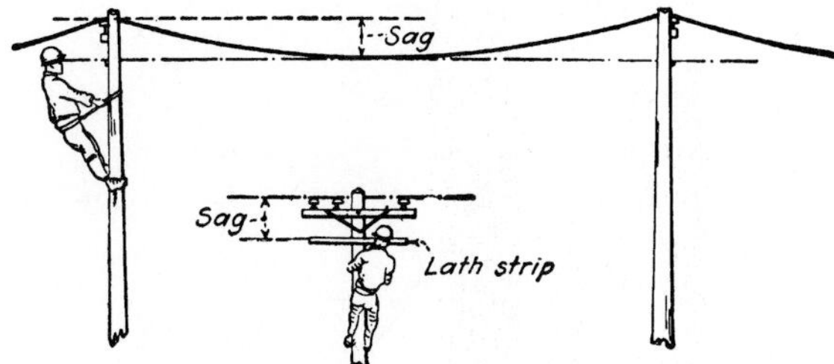

FIGURE 22.16 Sketch showing lineman sighting for sag by means of two lath strips nailed to poles at proper distance below the conductor when resting on the insulator.

A simple and accurate method of measuring the sag is by use of targets placed on the poles below the crossarm, as shown in Fig. 22.16. The targets may be a light strip of wood, like a lath, nailed to the pole at a distance below the conductor when resting on the insulator equal to the desired sag. The lineman sights from one lath to the next. The tension on the conductor is then increased until the lowest part of the conductor in the span coincides with the lineman's line of sight. Careful attention must be paid to the temperature at the time of sagging in of the conductor as the sag varies considerably with temperature. This is illustrated in Fig. 22.17, where typical values of sag are shown for the three values of temperature. The reason, of course, that the sag is greater at the higher temperature is because of the expansion of the metal in the conductor, causing it to be elongated.

The method of sighting for sag gives very satisfactory results when the sag is not less than 6 in and when the visibility is good. If the visibility is not good, it is often helpful to provide a suitable target at the distant pole for visual contrast with the conductor. This may be a piece of wood or metal approximately 2 ft square, painted white with a horizontal black line about 1 in wide across the center. For the sighting end, a device may be used, having a sighting slot in place of the black line (Fig. 22.18). The distant target and the slot are set below the support according to the desired sag, and the conductor sag is varied until the low point of the conductor is in line with the slot and the black line.

Sag Measurement by Timing. Accurate sagging by sighting is difficult if the spans are long or if the sags are quite small. Furthermore, if the poles used in sighting are not at the

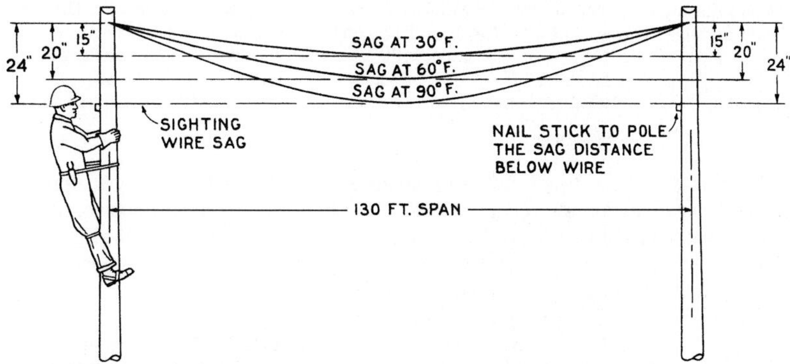

FIGURE 22.17 Sketch showing how sag increases with rise in temperature also illustrates the correct manner of sighting for sag.

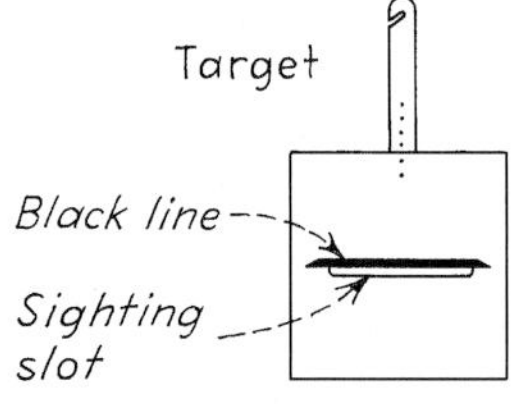

FIGURE 22.18 Sighting target about 2 ft square and painted white. Black line is about 1 in wide. Sighting slot is also about 1 in wide.

FIGURE 22.19 Sag Measurement by timing. The rope over the wire is used for applying an impulse to the conductor which creates a wave that travels along the wire. The lineman on the ground is holding a stopwatch in one hand and the rope in the other. (*Courtesy Wisconsin Electric Power Co.*)

same level, it is even more difficult. A mechanical wave, instigated near one support, will travel to the next support, be reflected in reverse phase, and pass back and fourth repeatedly between the supports. Although the wave is attenuated continuously, the time required for the wave to return is independent of the span length or the size or type of conductor. The sag of the conductor can be determined by timing the wave returns with a standard stopwatch and converting the time measured to sag from the equation

$$\text{Sag}\,D = 48.3(t/2n)^2$$

where D is in inches, t is time in seconds, and n is the number of return waves counted.

The number of returns depends primarily upon the conductor weight or size and the length of the span being sagged. For long spans and/or larger conductor sizes, the wave energy imparted to the conductor will dissipate more rapidly, and the number of return waves which can be counted accurately is less than for lightweight conductors and/or short spans.

Attach a rope to the conductor about 3 ft from the support. Then give the rope a sharp jerk, as in Fig. 22.19, and at the same time start a stopwatch or note the reading of the second hand on your (pocket) watch. Striking the conductor causes a wave or ripple to travel along the conductor until it reaches the next pole. When it reaches the next pole, it is reflected and returns to the pole where you are stationed. Here it is also reflected, and thus starts its second round trip. This will continue until the energy of the blow has been expended. The time in seconds required for the wave to return to the near support corresponds to a definite sag which can be calculated or read from prepared tables or graphs. The time is independent of the span length or the size or type of the conductor.

Observe the travel of this wave, and count the number of returns until the wave has returned 3, 5, 10, or 15 times. Upon the arrival of the third, fifth, tenth, or fifteenth return wave read the stopwatch or your (pocket) watch and note the time elapsed in seconds. A sufficient number of tests should be made until at least three identical readings are obtained. Then refer to Fig. 22.20 and read the corresponding sag opposite the time in seconds. If this is not the desired sag, change the pull on the conductor and repeat the timing operation.

The choice of the number of return waves depends principally upon the span length and the size of the conductor. For long spans and large conductors, the number of return waves which can be accurately counted is less than for short spans and small conductors. The distance of the "jerk" line from the support of a conductor will largely determine the force of the impulse given to the conductor as well as the ease with which the lineman with the stopwatch can feel the returning waves. A distance of 3 ft or more is commonly used. The largest number of return waves which can be observed should be used as this minimizes errors in recording the time.

The return wave may be felt by a man on the pole by placing a finger lightly on the conductor, if the line is de-energized and effectively grounded. Or readings may be made from

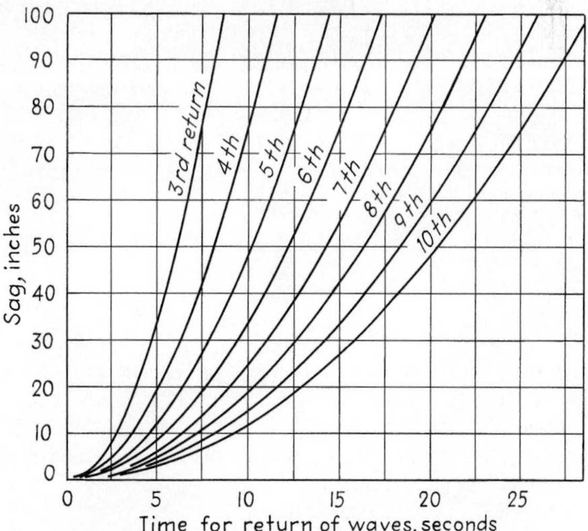

FIGURE 22.20 Graphs giving relation between time in seconds and sag in inches for third to tenth return of wave. (*Courtesy Copper Wire Engineering Association.*)

the ground by throwing a light, dry, and clean polypropylene rope over the conductor about 3 ft from the support, as shown in Fig. 22.19. This light rope may also be used to give the impulse initiating the wave. When the line is "hot," that is, energized, the lineman handling the rope must wear rubber gloves while handling the clean and dry polypropylene rope, which must be constructed and maintained properly for use on energized lines.

Care must be taken not to count 1 when the impulse is given to the line, but to count 1 on the first return of the wave. In other words count, "hit," 1, 2, 3, etc.

This sagging method is most satisfactory if the line is not in motion. Any vibration of the line, such as might be caused by working on it or by a strong wind, makes it difficult to determine the exact time of the return wave. The conductor should rest on the crossarm or in the stringing block to reflect the wave, although it need not be tied in.

Unisagwatch. The watch (Fig. 22.21) developed by Philip C. Evans takes the place of the stopwatch and sag-time tables and curves. The Unisagwatch carries logarithmic scales calibrated in feet and decimals for both third and fifth returns of a mechanical wave, providing direct readings of sags up to 100 ft. The watch is calibrated for an accuracy of 1 percent and has dials for reading sags in feet or meters. Sagging with this instrument eliminates the need for conversion tables and setting and sighting targets. The Unisagwatch is suitable for use in poor visibility conditions due to weather or in a selectively cleared right-of-way. The sag of an energized conductor may be readily checked by an observer on the ground.

FIGURE 22.21 Unisagwatch calibrated to determine wire sag regardless of span length, temperature, composition, size or weight of conductor. (*Courtesy Sagline Corp.*)

FIGURE 22.22 Traction dynamometer used to pull up line conductors to desired tension. Dynamometer has one hand for continuous load indication and the other for maximum tension indication. (*Courtesy John Chatillon & Sons.*)

Sagging Line with Dynamometer. When conductors are to be strung with unusually small sags, the measuring of the sag may not be as accurate or convenient as measuring the conductor tension. This is done by means of a dynamometer which is inserted in the pulling equipment. Such a dynamometer is shown in Fig. 22.22, and the manner of using it is shown in Fig. 22.23. One of the pointers indicates the pull at all times, and the other will remain at the point of maximum load after the tension is released. The pull is increased until

FIGURE 22.23 Traction dynamometer used with block and tackle in pulling line conductor to proper tension. (*Courtesy John Chatillon & Sons.*)

the tension is obtained for the desired conductor type and size, span length, loading district, and prevailing temperature.

Any convenient span in the section of line may be used for measuring the tension, but care should be taken that the tension is uniform in all spans. After stringing, a common practice is to allow the conductor to remain untied for several hours before final measurement and tying. On a long section of line it is advisable to check the sag or tension at several spans to ensure uniformity of the tension in the line. Figure 22.24 illustrates a tension dynamometer equipped with a swivel hook designed for a maximum conductor tension of 20,000 lb.

A shunt dynamometer can be used to measure conductor tension without being in series with the conductor. The operating principle of the shunt dynamometer is based on the relation of the tension in the conductor to the force necessary to displace it in a direction perpendicular to the axes of tension. The use of the shunt dynamometer has been limited because of the cost of the instrument and the complications associated with its successful application.

Clipping In Conductors. Conductors and shield wires should be clipped in as soon as practical after the sagging is completed. In all cases the conductors and static wires, or ground wires, should be clipped in within 2 days after the sagging is completed (Fig. 22.25). Conductors should be lifted from the stringing sheaves by using a standard suspension clamp or a large plate hook. Rope slings should not be used. The inside diameter of the plate-hook seat should be at least 1/16 in greater than the diameter of the conductor to be lifted.

The center longitudinal portion of the plate-hook seat must be bent down at substantially the same curvature as found on the suspension clamp. The inside surfaces of the plate-hook must be free of all burrs and rough edges. The conductors must be attached to the insulator

FIGURE 22.24 Dynamometer equipped with a swivel hook to prevent conductor twisting. (*Courtesy John Chatillon & Sons.*)

assemblies at all suspension and dead-end points in accordance with the specifications or the standards for the line under construction. Armor rods and vibration dampers are usually specified for all high-voltage transmission lines. The dampers should be placed on the conductors as soon as the conductors are clipped in, to avoid the risk of conductor damage as a result of critical wind conditions.

FIGURE 22.25 Hoisting work platform to linemen who are to apply armor rods, attach corona shield, and clamp conductor to insulator string. (*Courtesy American Electric Power System.*)

Any conductor supported above ground in long spans under relatively high tension is subject to vibration. Such vibration is caused by the passage of air over the conductor at right angles to the line. As the air blows over the conductor, eddy currents are set up on the leeward side. These eddy currents oscillate, first in one direction and then in the other. When the frequency of these eddies corresponds to some natural frequency of the line conductor, a tendency to vibrate exists. Changes in velocity or direction of wind dampen the original vibrations and set up new ones. It is for this reason that a steady gentle breeze may set up far more destructive vibrations than high winds of unsteady force and direction.

Under certain conditions these vibrations may build up an amplitude which produces alternating stresses large enough to cause fatigue failure. The failure usually takes place at the point of support, very seldom at a dead-end or at a joint. Failure may not be experienced over a period of years, or it may develop in a few months. When a vibration passes through a wire, the conductor tends to conform to the waveform of the vibration. When this vibration encounters some obstacle such as a supporting clamp, it may be reflected back on itself with consequent increase in destructive stresses. More destructive still is the effect of two vibrations of similarly directed bending tendencies meeting at opposite ends of a common supporting clamp.

Vibration fatigue of conductors may be prevented either by reinforcing the conductor at points of stress to resist the effects of repeated bending or by reducing the vibration to negligible amplitudes with damping devices. One form of damper known as the *Stockbridge*

FIGURE 22.26 The Stockbridge damper clamped to transmission-line conductor to reduce conductor vibration by changing natural frequency of line conductor. (*Courtesy Aluminum Company of America.*)

damper is illustrated in Fig. 22.26. It consists of a resiliently supported weight with a suitable clamp for attaching it to the conductor. The action of the damper is to absorb continuously the energy of the vibrations started by winds of all kinds and to prevent these vibrations from building up to damaging proportions. The damper is attached a suitable distance ahead and back on the line from various types of conductor supports, dead-ends, etc.

Distribution-circuit conductors are fastened to insulators by the use of either wire ties or clamps. Wire ties are used in distribution systems where the pin insulator is employed. Insulator clamps are used on some high-voltage pin-insulator lines and entirely on suspension insulators.

Conductors should occupy such a position on the insulator as will produce minimum strain on the tie wire. The function of the tie wire is only to hold the conductor in place on the insulator, leaving the insulator and pin to take the strain of the conductor.

In straight-line work the best practice is to use a top-groove or *saddle-back* insulator. All these insulators carry grooves on the side as well. When the conductor is placed in the top groove (Fig. 22.27*a*), the tie wire serves only to keep the conductor from slipping out. The groove practically relieves the tie wire of any strain and allows the entire strain to be placed on the insulator and pin. When side-groove insulators are used in straight-line work, the wires should be placed as shown in Fig. 22.27*b*. The conductors on the pins nearest to the pole are placed on the side away from the pole. This is to increase the climbing space for the lineman. All other wires are placed on the pole side of the insulator. This prevents the conductors on the end pins from falling in case their tie wires break.

On corners and angles where the wires are not dead-ended, the conductors should be placed on the outside of the insulators (Fig. 22.28) on the far side of the pole and on the inside of the insulators on the near side of the pole, irrespective of the type of insulator employed. This puts the conductor pull against the insulator instead of away from the insulator.

Distribution-circuit conductors are usually secured to the insulators with prefabricated ties (see Figs. 22.29 through 22.36). The ties are packaged and identified by the manufacturer to facilitate their correct application. The prefabricated ties can be installed on energized circuits with proper protective equipment in place to insulate the tie from grounded equipment or other energized phase conductors. Hot-line tools or rubber gloves and rubber sleeves should be used, depending on the voltage applied to the conductor.

Ties used for tying conductors to pin-type insulators should be relatively simple and easy to apply. They should bind the line conductor securely to the insulator and should reinforce the conductor on both sides of the insulator. Any looseness between the conductor, tie wire, and insulator will result in chafing and injury to the conductor. The hand-wrapped

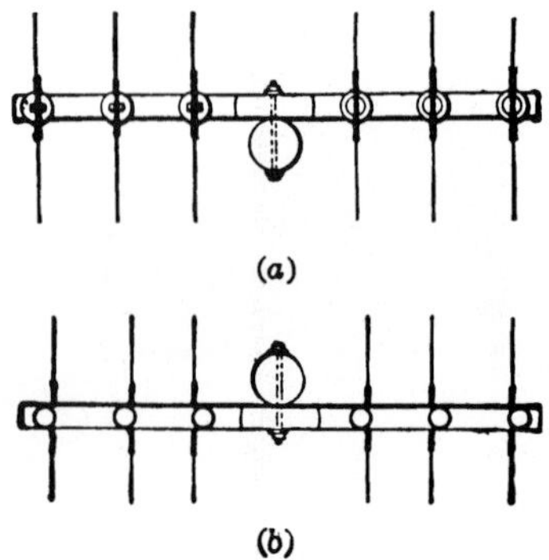

FIGURE 22.27 Position of wires on insulator in straight-line work for (*a*) top-groove insulators, and (*b*) side-groove insulators. Top-groove insulators are generally used with conductors size 0 and larger and side-groove insulators with size 2 and smaller.

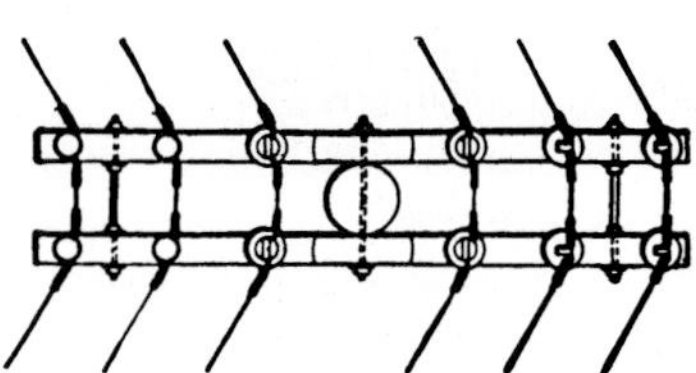

FIGURE 22.28 Position of wires on both top- and side-groove insulators at angles. All conductors are placed on side of insulator so that pull is against insulator.

ties illustrated in Figs. 22.37 through 22.40 have been widely used and when properly applied give excellent service. These ties should be applied by hand without the use of pliers, and care should be taken to use the proper length and size of fully annealed tie wire specified for each conductor.

In general, the tie wire should be the same kind of wire as the line wire. If the line wire is a bare conductor, the tie wire should also be bare; if the line conductor is covered, the tie wire should also be covered. Copper tie wires should be used with copper line conductors and aluminum tie wires with aluminum line conductors. The tie wires, however, should always be made of soft-annealed wire as the hard-drawn tie wire would be too brittle and cannot be wrapped snugly. A hard tie wire might also injure the line conductor.

A tie wire should never be used the second time.

Good practice is to use No. 6 tie wire for line conductors of sizes No. 4 and smaller, No. 4 tie wire for line conductors No. 1 to No. 4, and No. 2 tie wire for line conductors No. 0 and larger, as shown in Table 22.1.

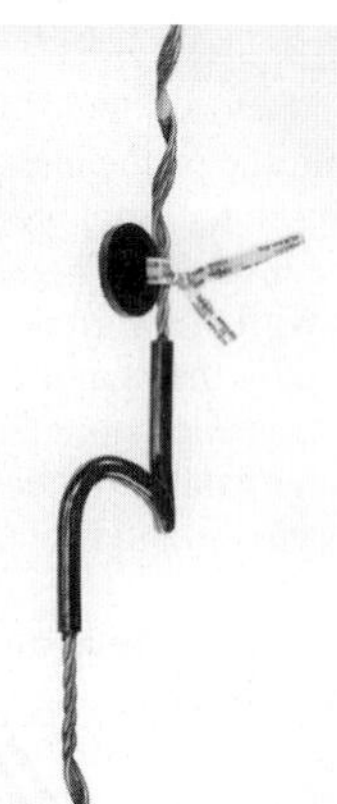

FIGURE 22.29 Prefabricated conductor tie kit for top-tie insulator. (*Courtesy Preformed Line Products Co.*)

FIGURE 22.30 Installation of prefabricated conductor tie on deenergized circuit. The neoprene pad is centered in the insulator groove between the conductor and the insulator. The center of the tie is positioned over the insulator so that both legs are parallel to the conductor. (*Courtesy Preformed Line Products Co.*)

The length of the tie wire varies. The approximate lengths for various ranges of conductor sizes for some of the common ties are given in Table 22.2.

Rules of good tying practice are as follows:

1. Use only fully annealed tie wire.
2. Use a size of tie wire which can be readily handled, yet one which provides adequate strength.
3. Use a length of tie wire sufficient for making the complete tie, including an end allowance for gripping with the hands. The extra length should be cut from each end after the tie is completed.

FIGURE 22.31 The prefabricated tie is rotated around the insulator so that one leg is around the neck of the insulator wrapped onto the conductor and the end snapped in place. (*Courtesy Preformed Line Products Co.*)

FIGURE 22.32 Completed application of prefabricated tie with both legs wrapped around the conductor and the ends of the tie snapped in place. (*Courtesy Preformed Line Products Co.*)

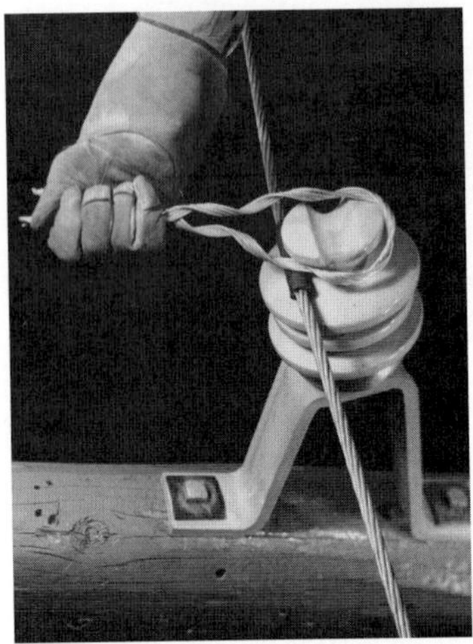

FIGURE 22.33 Prefabricated conductor tie for application as a side tie. Neoprene pad is placed around the conductor to prevent conductor contact with insulator. The legs of the tie are squeezed together to enlarge the loop, enabling it to be pushed over the head of the insulator. (*Courtesy Preformed Line Products Co.*)

FIGURE 22.34 The side tie is positioned around the neck of the insulator so that the conductor is between the legs of the tie. (*Courtesy Preformed Line Products Co.*)

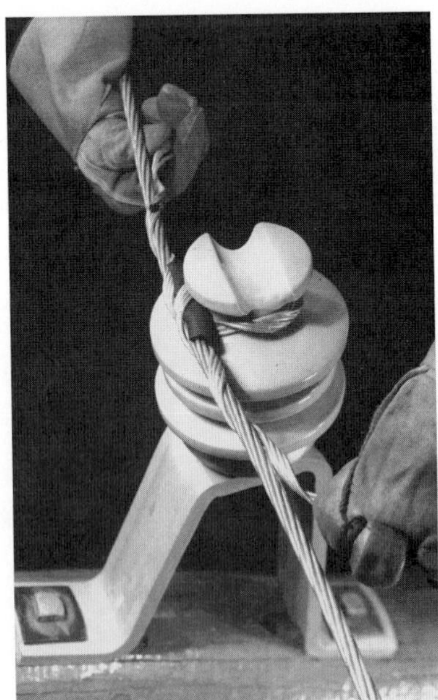

FIGURE 22.35 The legs of the side tie are pulled firmly and wrapped around the conductor. (*Courtesy Preformed Line Products Co.*)

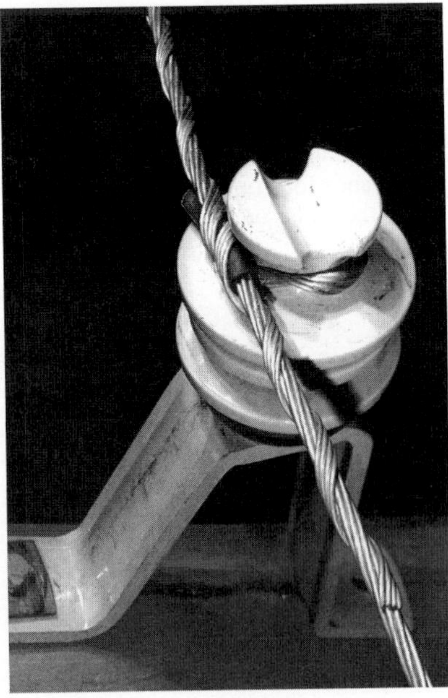

FIGURE 22.36 Prefabricated conductor side-tie installation completed. When both legs are completely wrapped around the conductor, the ends of the legs must be snapped into position against the conductor. (*Courtesy Preformed Line Products Co.*)

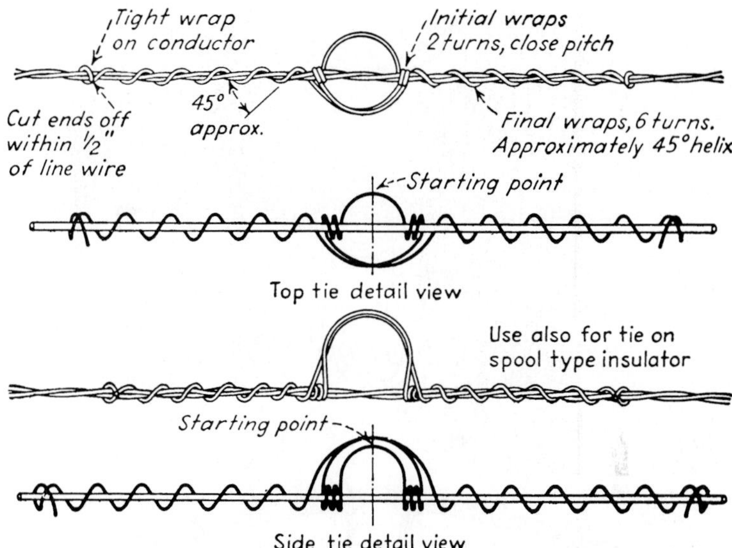

FIGURE 22.37 Single-pin-type insulator tie for copper conductors. (*Courtesy Copper Wire Engineering Association.*)

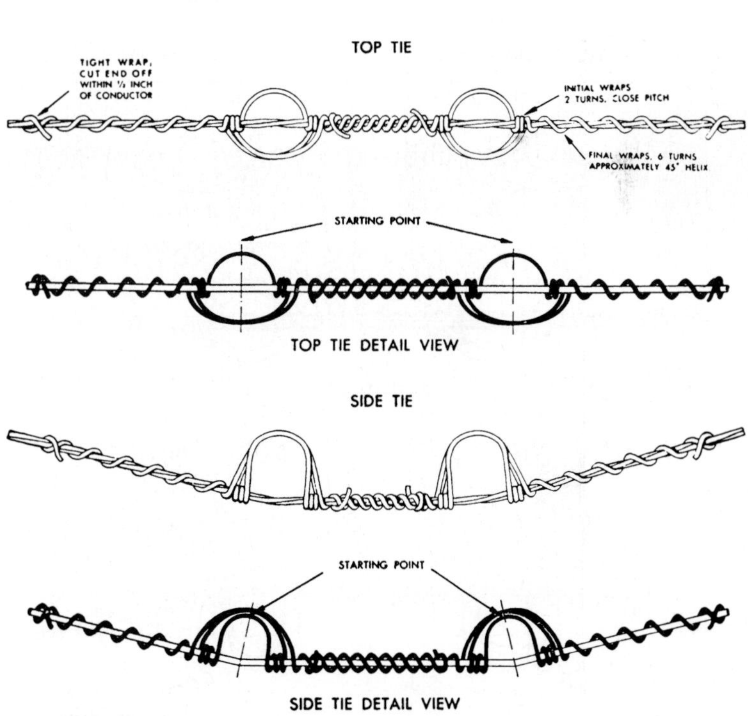

FIGURE 22.38 Double-pin-type insulator tie for copper conductors. (*Courtesy Copper Wire Engineering Association.*)

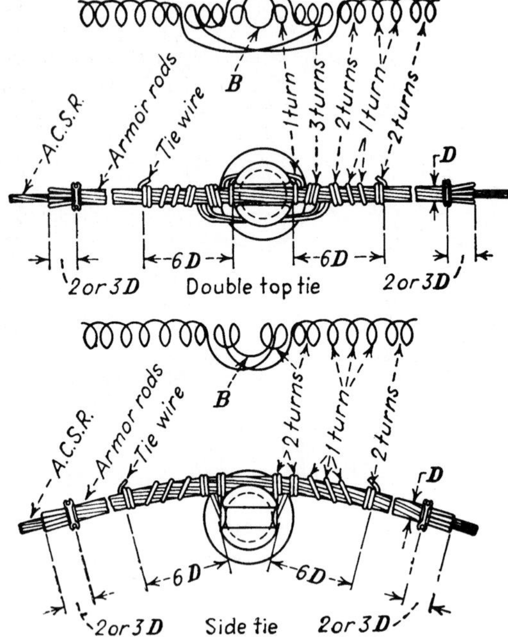

FIGURE 22.39 Single-pin-type insulator tie for ACSR and aluminum cables equipped with armor rods. (*Courtesy Aluminum Company of America.*)

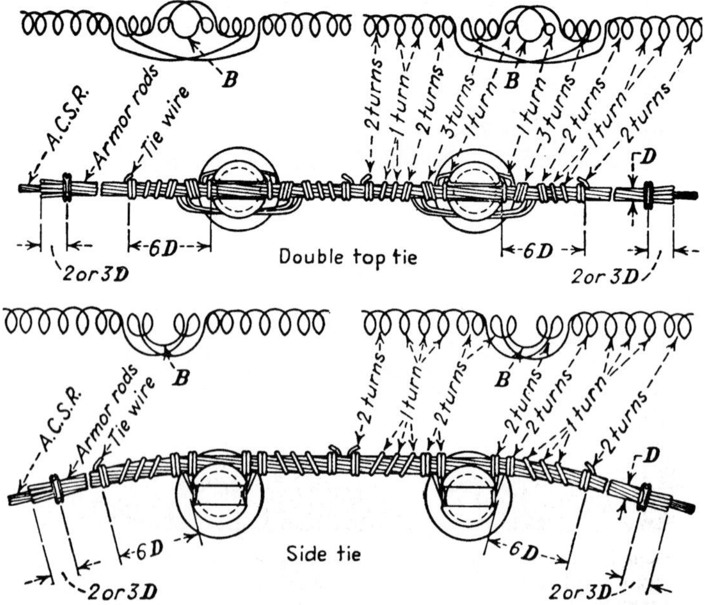

FIGURE 22.40 Double-pin-type insulator tie for ACSR aluminum cables equipped with armor rods. (*Courtesy Aluminum Company of America.*)

TABLE 22.1 Size of Tie Wire to Use with Various Sizes of Line Conductors

Size of line conductor, AWG	Size of tie wire AWG
No. 4 and smaller	No. 6
No. 1 to No. 4	No. 4
No. 0 and larger	No. 2

4. A good tie should

a. Provide a secure binding between line wire, insulator, and tie wire

b. Have positive contacts between the line wire and the tie wire so as to avoid any chafing contacts

c. Reinforce the line wire in the vicinity of the insulator

5. Apply without the use of pliers.

6. Avoid nicking the line wire.

7. Do not use a tie wire which has been previously used.

8. Do not use hard-drawn wires for tying or fire-burned wire which is usually either only partially annealed or injured by overheating.

Steps in Making Ties. The simplest tie is the Western Union tie. It is universally used in distribution systems and is made as follows:

Place the middle of the tie wire on one side of the insulator and the cable on the other side, both bearing in the side groove. Each end of the tie wire is then brought around the neck of the insulator, but underneath the cable, then up and around the cable, and served closely for about seven wraps.

The looped Western Union is quite similar except that the start is made on the cable side and the two ends are wound around the insulator once before being wrapped.

Armor Rods. Distribution-line conductors can be reinforced with armor rods, as illustrated in Figs. 22.41 and 22.42. This armor consists of a spiral of rods surrounding the conductor for a short distance (Fig. 22.43). The attachment of the conductor to its support is made in the middle of this armored length (Fig. 22.44). Since this provides a cable of much larger diameter than the conductor itself, the resistance to bending is greatly increased. This

TABLE 22.2 Approximate Length of Tie Wires Required for Different Types of Ties

Type of tie	Length of tie wire, in
Western Union	{ No. 0000 bare cable, 54 2,000,000-cir-mil cable, 87
Bridle	{ No. 0000 bare cable, 54 2,000,000-cir-mil cable, 87
Horseshoe	All sizes, 39
Armored Western Union	All sizes, 102
Amored top	All sizes, 114
Stirrup	All sizes, two pieces-each 54

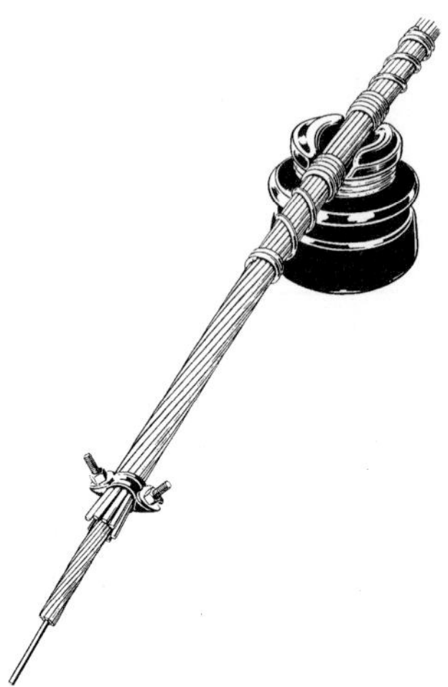

FIGURE 22.41 An ACSR line conductor cable reinforced with armor rods. These armor rods reduce vibration and protect against chafing and flashover at point of support. Increased conductor size at support provides better grip for tie wire and also reduces danger of cable slippage. (*Courtesy Aluminum Company of America.*)

FIGURE 22.42 Typical single-phase rural line illustrating the use of armor rods to minimize the effects of conductor vibration. (*Courtesy Ohio Brass Co.*)

FIGURE 22.43 Lineman applying armor rods to ACSR conductor cable. (*Courtesy L. E. Myers Co.*)

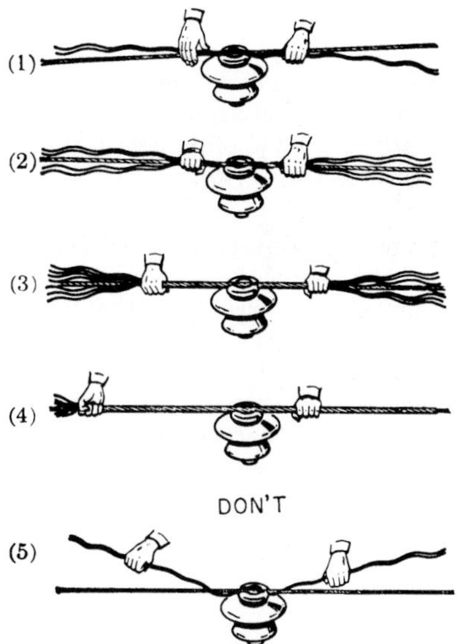

FIGURE 22.44 Steps in the application of preformed armor rods. (1) Center preformed rods (maximum two or three) at support of suspension point and wrap on one or two turns on both sides of center. Distortion will be avoided by not applying too many rods at one time. (2) Follow the same procedure with additional rods until all rods are applied at center for several pitches. (3) After all rods are applied several turns at center, the whole group can be rotated until the set is completely installed. (4) Always rotate the rods from center out toward end. They easily fit into position. (5) Never grip preformed rods too far from the point of application, as distortion will usually result. (*Courtesy Preformed Lined Products Co.*)

FIGURE 22.45 Linemen securing energized conductor to clamp-top line-post insulator. Note protective equipment used on conductors. Linemen wear rubber gloves, rubber sleeves, hard hats, and safety glasses. Work is being performed from an insulated-bucket truck. (*Courtesy A. B. Chance Co.*)

FIGURE 22.46 Linemen cutting in dead-end on an energized circuit. Note bolt-type dead-end clamps attached to bell insulators to be used to dead-end the conductors. (*Courtesy A. B. Chance Co.*)

not only reduces the stress by distributing the bending but also strengthens the cable in the region of maximum bending stress.

Clamp-top line-post insulators can be used on distribution circuits to eliminate tying the conductor to the insulator (Fig. 22.45). The conductor can be protected with armor rods if necessary before it is placed in conductor clamp. Dead-end clamps of the bolted, or compression, type are used to dead-end distribution conductors (Fig. 22.46).

Spacer Installation. Bundled-conductor transmission lines use spacers to maintain uniform distance between the conductors for each phase. The number of spacers to be installed at equal intervals in each span will be detailed in the specifications for the line. Spacers should be installed immediately after clipping-in operations are completed in each pull to minimize conductor damage due to contacts between the subconductors. The bolted spacers should be installed with the conductor clamp free to rotate on one conductor without binding, and then the second clamp should be bolted loosely on the second conductor. After the spacer is in place properly, the lineman should hold the clamp firmly in place and tighten all bolts to the proper tension.

Corona Shield Installation. Extra- and ultra-high-voltage transmission lines require corona shields at each point of conductor support. The corona shields are normally installed after the conductors are clipped in and spacers installed.

CHAPTER 23
JOINING LINE CONDUCTORS

Transmission- and distribution-line conductors must be joined together with full-tension splices if the conductors are under tension. Bolted connectors can be used to join electric conductors at locations where the conductors are slack, such as between conductor dead-ends. Figures 23.1 and 23.2 illustrate compression tools used for joining and dead-ending electric distribution conductors.

Compression-, internally fired-, implosive-, and automatic-type splices are used to obtain the strength and electrical conductivity needed to join together conductors in tension. The number of locations where the conductors are joined should be kept to a minimum. All splices in transmission-line conductors should be made at least 50 ft or more from the suspension or dead-end points. Splices should not be made in spans crossing over railroads, rivers, canals, or interstate highways. It is advisable to avoid splices in spans crossing over communications circuits or electric transmission and distribution lines, if possible.

The splices must be applied in accordance with the transmission-line specifications or distribution standards and in accordance with the splice manufacturer's instructions. The conductors of a transmission line requiring a full-tension splice should be laid out straight for a distance of 40 ft each way from the joining point and straightened at the ends to prepare the conductors for splicing. The conductors should be supported and if necessary laid on polyethylene sheets to keep them clean if laid along the ground in preparation for splicing.

Compression Splice for ACSR Conductor. A compression splice for an aluminum conductor with steel reinforcing (ACSR) usually consists of two separate splices, one for the steel core and the other for the overall cable (Fig. 23.3). The complete splice, therefore, consists of a small compression slice for the central steel core to establish the tensile strength of the steel core of the spliced conductors and a much larger aluminum compression splice over the outside of the two joined cables to establish good conductivity for the spliced aluminum strands of wire. The plugs are for sealing the holes in the aluminum sleeve after the paste filler has been injected.

The steps for making a compression splice on an ACSR conductor are as follows (Fig. 23.4):

Caution. Before proceeding, make sure that the bores in the two sleeves are perfectly clean.

1. A compression connector shall be used only for the size of conductor for which it is specified. Neither the connector nor the conductor shall be altered to attempt installation of a fitting on a conductor for which it is not designed.
2. Select the correct press and dies for the conductor and compression connector (Figs. 23.5, 23.6, and 23.7).
3. Slip the aluminum compression sleeve over one cable end and back it out of the way along the cable.

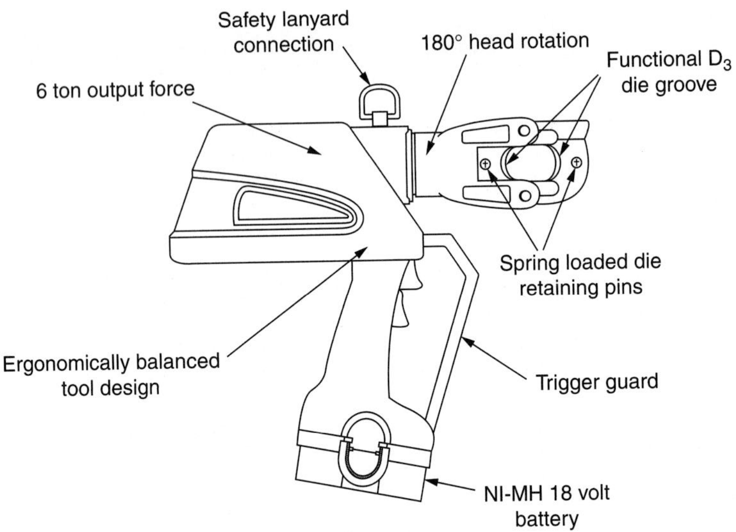

FIGURE 23.1 Sketch of a battery actuated compression tool used to join and terminate electric conductors. The tool utilizes standard compression dies or the tool may be used with a die-less tool head. (*Courtesy Burndy Corp.*)

4. Cut off the aluminum strands from each cable end, exposing the steel core for a distance of a little more than half the length of the steel compression sleeve (Figs. 23.8 and 23.9). Be careful not to nick the steel core. Before cutting, secure the aluminum strands of the cable just behind the cut.
5. Insert the steel-core ends into the steel-core compression sleeve, making sure that the ends are jammed against the stop in the middle of the sleeve (Fig. 23.10).

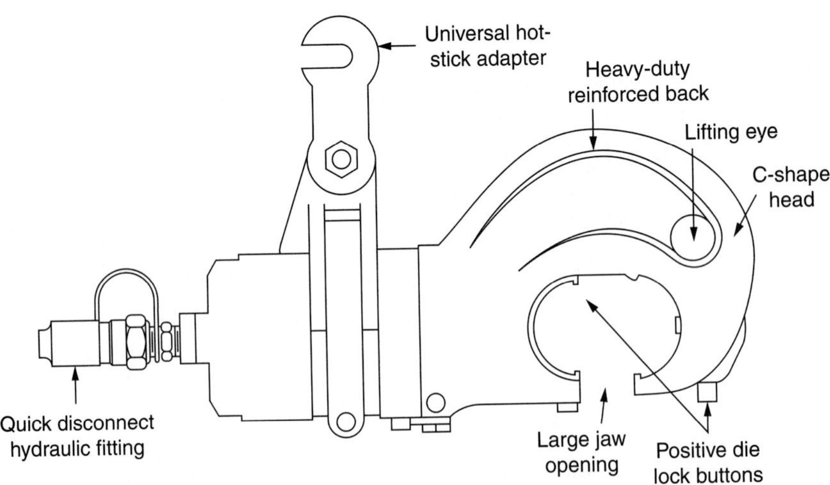

FIGURE 23.2 Sketch of a hydraulically operated tool used to join and terminate electric conductors. (*Courtesy Burndy Corp.*)

FIGURE 23.3 Parts of compression splice for ACSR conductor. (*Top*) Sleeve for steel core of conductor, flanked by plugs for sealing holes in the large aluminum-alloy sleeve, after paste filler has been injected. (*Bottom*) Aluminum-alloy splicing sleeve for complete cable. (*Courtesy James R. Kearney Corp.*)

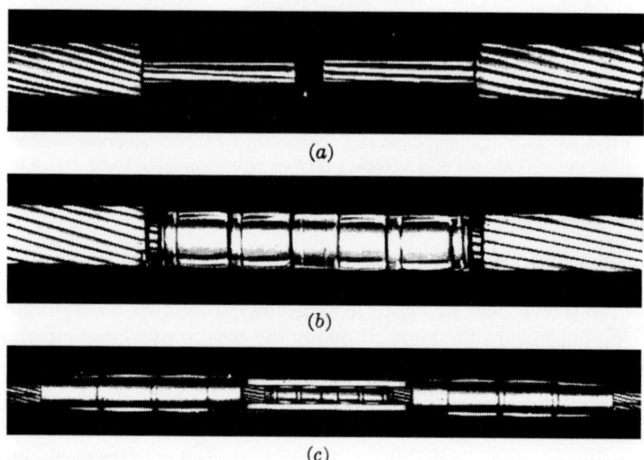

(*a*)

(*b*)

(*c*)

FIGURE 23.4 Several steps for completing a compression splice on an ACSR conductor as described in the text are shown (*a*) *Step 4*: Aluminum strands cut back a distance equal to one-half the length of the steel sleeve, (*b*) *Step 6*: Completed compressed sleeve on the steel core, (*c*) *Step 9*: Completed compression splice on ACSR conductor. Cutaway view shows inner-steel-core compressed sleeve. (*Courtesy Burndy Corp.*)

FIGURE 23.5 An electrically operated hydraulic oil pump used to provide power to operate a compression tool to splice conductors. The pump is designed to be used to splice deenergized conductors on or near the ground. (*Courtesy MidAmerican Energy Co.*)

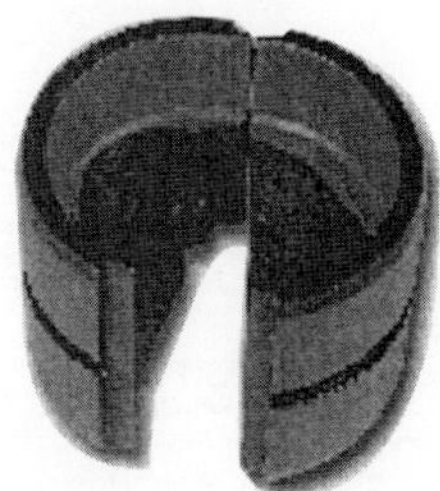

FIGURE 23.6 Example of a die used in a compression tool to compress splicing sleeve or other connecters of conductors.

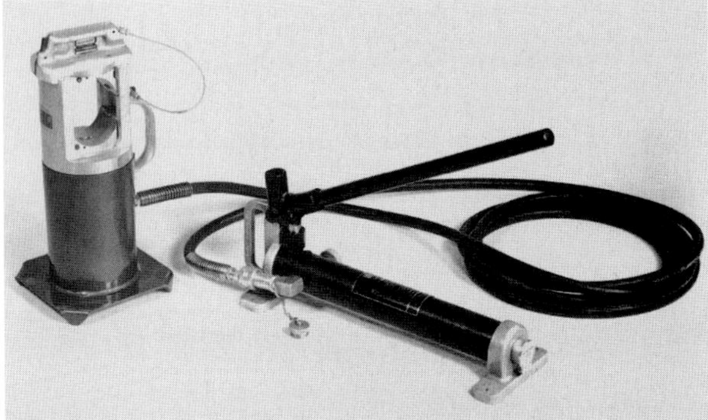

FIGURE 23.7 Compression tool operated by foot, electric, or pneudraulic power pump capable of developing 60 tons of compression force. It can be used to install compression sleeves on conductors as large as 1,780,000 cir mils and weighs 38 lb. (*Courtesy Burndy Corp.*)

6. Compress the sleeve over its entire length, making the first compression at the center and working out toward the ends, allowing the dies to overlap their previous position (Figs. 23.11 and 23.12).

7. Measure from the center of the steel-core splice along the aluminum wire strands the distance of one-half the length of the aluminum splicing sleeve and mark the cable surface. This will facilitate centering the sleeve when it is slipped over the steel-core splice. Brush the aluminum conductor surface with a wire brush to a bright finish. Coat the brushed conductor with an inhibitor-type compound (Fig. 23.13).

8. a. Slip the aluminum splicing sleeve up over the steel-core splice to the mark placed on the outside surface of the conductor, centering the sleeve over the steel-core splice (Fig. 23.14).

FIGURE 23.8 Lineman cutting conductor with cable cutter in preparation for completing splice. The conductor to be inserted in the splice is held with a clamp to prevent the wire strands from unraveling. (*Courtesy L. E. Myers Co.*)

FIGURE 23.9 Lineman is holding conductor that has been prepared for splicing. Aluminum strands have been cut back, exposing the steel core of the conductor. The aluminum strands have been secured with tape to prevent them from unraveling. The ends of the aluminum wires have been cleaned with a wire brush to remove oxidation from the conductor, giving the surface a polished finish. The steel core of the conductor has been inserted into one end of the steel-core splicing sleeve. Note traveling ground on conductor extends up to transmission-line structure. Note tension-stringing equipment in background of picture. (*Courtesy MidAmerican Energy Co.*)

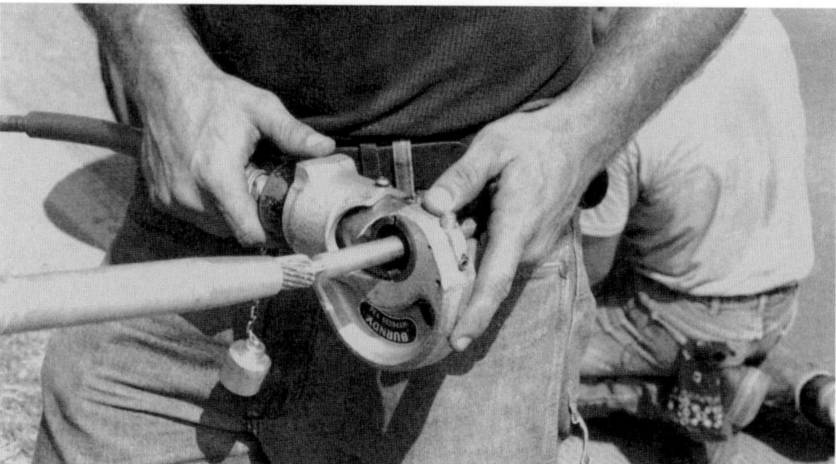

FIGURE 23.10 Lineman starting to compress splicing sleeve on steel core of a conductor with hydraulically powered compression tool. Outer splicing sleeve has been placed on the conductor so that it can be pulled back over the compression-spliced steel core after operation is completed. (*Courtesy MidAmerican Energy Co.*)

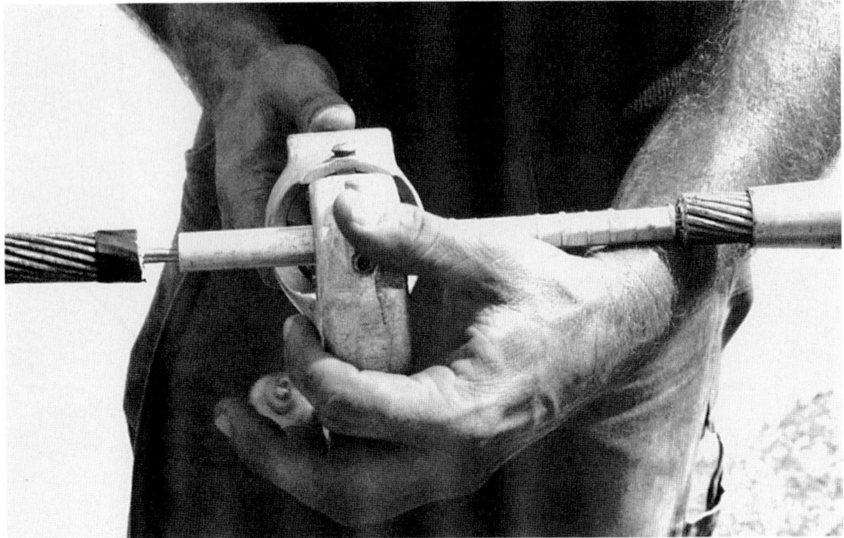

FIGURE 23.11 Lineman completing compression splice of steel core of ACSR conductor. Compression of the splice starts in the middle and proceeds to the end of the sleeve. (*Courtesy MidAmerican Energy Co.*)

FIGURE 23.12 Steel-core compression splice has been completed. Lineman is cleaning ends of aluminum conductor strands after removing tape that secured the strands. The outer sleeve will be moved so that it is centered over the spliced steel core. (*Courtesy MidAmerican Energy Co.*)

FIGURE 23.13 Lineman is brushing the conductor to be spliced with an inhibitor compound before moving the aluminum splicing sleeve into the proper position to start the compression. (*Courtesy MidAmerican Energy Co.*)

  b. Using a caulking gun equipped with a tapered nozzle, inject filler paste through the holes provided in the aluminum splicing sleeve until the paste is visible at the ends of the sleeve. The preferred filler paste contains zinc chromate.
  c. Insert the plugs in the filler holes and hammer them firmly in place.

9. Compress the aluminum sleeve with a hydraulically powered compression tool. Make the first two compressions with the inner edges of the dies matching the positions stenciled on the aluminum sleeve. Make additional compressions advancing to the ends, allowing the dies to always overlap the previous position (Figs. 23.15, through 23.18).

FIGURE 23.14 Lineman has centered the aluminum splicing sleeve over the compressed-core splicing sleeve. (*Courtesy MidAmerican Energy Co.*)

FIGURE 23.15 Linemen are starting to compress outer-aluminum splicing sleeve with hydraulically powered compression tool. (*Courtesy MidAmerican Energy Co.*)

FIGURE 23.16 One end of the aluminum splicing sleeve has been compressed. (*Courtesy MidAmerican Energy Co.*)

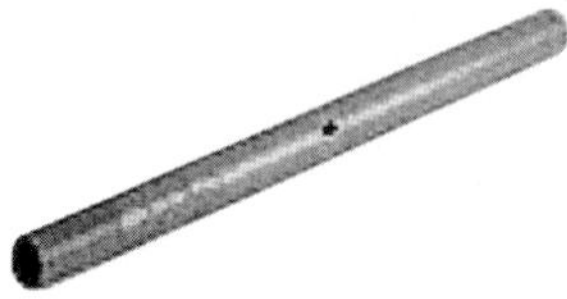

FIGURE 23.17 A 41 in. full tension two-piece full tension compression splice for use on a 1,275,000-cir-mil expanded ACSR cable having an outside diameter of 1.6 in. This size cable is installed on transmission lines. (*Courtesy Burndy Corp.*)

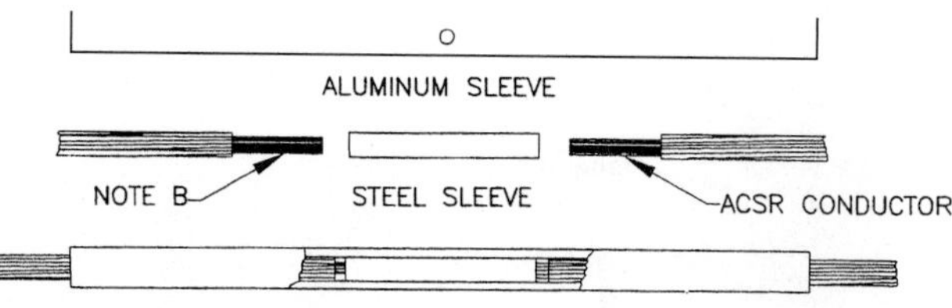

CUT-AWAY VIEW OF ASSEMBLY

OPERATIONS: METHOD OF APPLYING TWO PIECE SLEEVE FOR ACSR LINE SPLICE (15)

(1) SLIP THE ALUMINUM SLEEVE BACK OUT OF THE WAY OVER ONE END OF THE CONDUCTOR.

(2) USING A TUBE CUTTER OR A HACK-SAW, CUT OFF THE ALUMINUM STRANDS, EXPOSING THE STEEL CORE FOR A DISTANCE OF 1/2 INCH MORE THAN HALF THE LENGTH OF THE STEEL SLEEVE. USE CARE NOT TO NICK THE STEEL CORE.

(3) INSERT THE STEEL CORE INTO THE STEEL SLEEVE, BRINGING THE ENDS EXACTLY TO THE CENTER.

(4) MAKE THE REQUIRED NUMBER OF COMPRESSIONS ON THE STEEL SLEEVE.

(5) WRAP TWO TURNS OF TAPE AROUND THE END OF THE CONDUCTOR OPPOSITE THE SLEEVE WITH THE EDGE OF THE TAPE EXACTLY ONE-HALF THE LENGTH OF THE ALUMINUM SLEEVE FROM THE CENTER OF THE COMPRESSED STEEL SLEEVE.

(6) BRUSH THE ALUMINUM CONDUCTOR WITH A WIRE BRUSH TO A BRIGHT FINISH.

(7) COAT THE BRUSHED CONDUCTOR WITH AN APPROVED CORROSION INHIBITING COMPOUND.

(8) SLIP THE ALUMINUM SLEEVE OVER THE STEEL JOINT WITH THE END TOUCHING THE TAPE INSTALLED IN STEP (5).

(9) USING A CAULKING GUN, INJECT APPROVED CORROSION INHIBITING COMPOUND THROUGH HOLES PROVIDED IN THE ALUMINUM SLEEVE UNTIL THE SPACE BETWEEN THE ALUMINUM SLEEVE AND COMPRESSED STEEL SLEEVE IS COMPLETELY FILLED.

(10) INSERT ALUMINUM PLUGS IN THE FILLER HOLES AND HAMMER FIRMLY IN PLACE.

(11) MAKE THE REQUIRED NUMBER OF COMPRESSIONS ON THE ALUMINUM SLEEVE, AS SPECIFIED ON THE STANDARD, BEGINNING AT THE MARK NEAREST THE CENTER AND WORKING TOWARD THE ENDS OF THE SLEEVE.

FIGURE 23.18 Sketch and instructions for the installation of a two-piece full tension compression splice for use on ACSR transmission line cable. (*Courtesy WE Energy Co.*)

Compression sleeves are manufactured for transmission-line ACSR conductors that contain a gripping unit internal to the aluminum-alloy splicing sleeve. The conductor is prepared for the splice in the same manner as described above. The conductor is inserted into the splicing sleeve so that the steel cores of the conductors that are being spliced are in the gripping unit. The gripping unit establishes the continuity of the tension strength of the steel core of the conductors being spliced. The splice is completed in the same manner as described above for compression of the outer aluminum conductor sleeve. Only one die is required in the compression tool. The two-step operation in which a sleeve is compressed on the steel core before compressing the outer-conductor splicing sleeve is avoided.

Compression Splices for Distribution Conductors. Single-sleeve full-tension splices are normally installed for aluminum, aluminum-alloy, ACSR, and copper distribution-line conductors (Fig. 23.19). Compression splices for smaller distribution-line conductors can be made with lightweight hand-powered compression tools like those shown in Figs. 23.20, through 23.22. Power-driven and hand-operated pump-type compression tools can be used for the larger distribution-line splices. Smaller hydraulically powered compression tools using line-truck hydraulic-tool circuits can be used to splice both transmission- and distribution-line conductors. Distribution-line conductors can be spliced and tapped while energized if the truck is properly equipped with an insulating boom and the lineman wears the proper protective equipment and follows procedures for working on energized conductors.

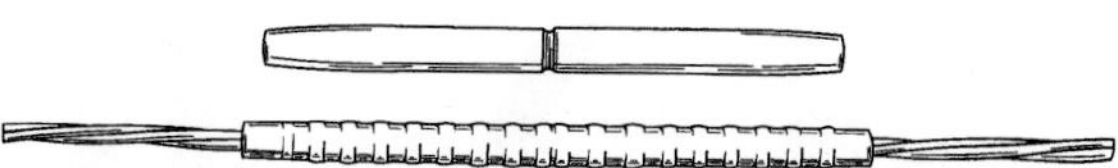

FIGURE 23.19 Compression-type splicing sleeve used for distribution-line conductors. Upper view shows sleeve before compression, and lower view shows sleeve fully compressed. Markings on sleeve are produced by die used in compression. (*Courtesy Coppperweld Steel Co.*)

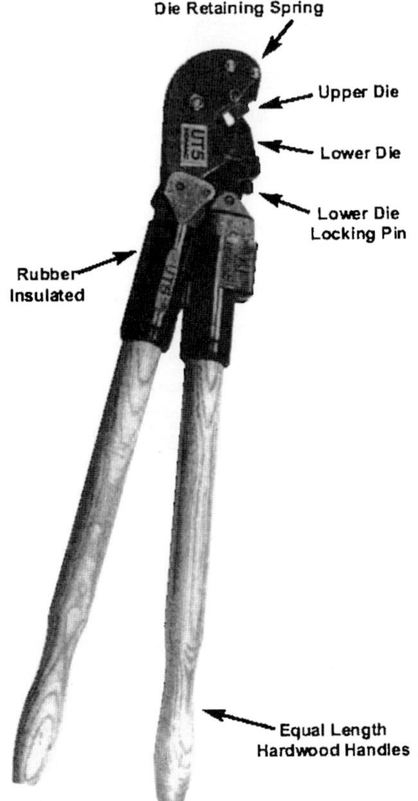

FIGURE 23.20 Hand-operated compression tool modeled after bolt cutter uses principle of lever to develop force needed to compress sleeve onto conductor. (*Courtesy Burndy Corp.*)

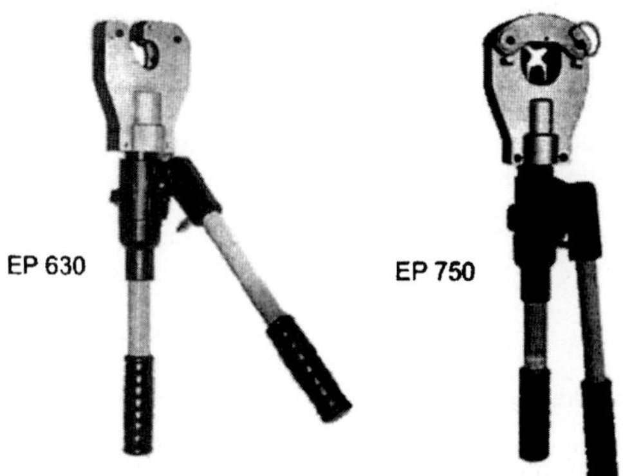

FIGURE 23.21 Hand-operated die-less compression tool. (*Courtesy Homac Manufacturing Co.*)

FIGURE 23.22 Hand-operated crimping tool for the single indent copper conductor splices. (*Courtesy Homac Manufacturing Co.*)

Single-sleeve compression splices should be completed in the following manner:

1. Select the proper compression splicing sleeve for the type and size of distribution-line conductor to be spliced.
2. Use the proper compression tools and dies for the sleeve and the wire.
3. Clean the conductor thoroughly to remove corrosion and film. Aluminum conductors should be brushed with a wire brush until the wire has a bright finish.
4. Cover the conductor with an inhibitor compound.
5. Cut off the ends of the conductors to be joined with a tube cutter or a hack saw to obtain flush ends. The splice should be made at least 12 in from the edge of the nearest tie wire, armor rod, or dead-end clamp on the distribution conductor.
6. Support the conductors to be joined in a straight line on each side of the joint to keep the splice straight. The wires should be straight for a distance of 10 ft or more if possible.
7. Wrap two turns of tape around one conductor exactly one-half the length of the sleeve from the end of the conductor if the sleeve has no center stop.
8. If the sleeve is not completely filled with inhibitor compound, apply additional compound to the conductor so that the sleeve will be completely filled when installed.
9. Slip the ends of the conductors into the sleeve with one end of the sleeve touching the tape and the conductor ends touching inside the sleeve or against the stop if one is provided in the sleeve.
10. Make the required number of compressions on the sleeve beginning at the center and working toward the ends of the sleeve.

Compression tools can be used with the proper accessories in a manner similar to the procedures described for splicing conductors to dead-end conductors, complete taps, and install terminals or lugs (Figs. 23.23 through 23.33).

Secondary and service conductors can be spliced with bare or insulated compression connectors, using a lightweight hand-operated compression tool (Fig. 23.34). The tools are designed so that the compression cycle must be completed before the tool can be removed, ensuring proper compression of the connector. The hand tool is used to install compression streetlight tap connectors and pin terminals (Fig. 23.35). Large-size service conductors are spliced with compression connectors using hydraulically operated tools powered by a truck hydraulic system or hand-operated or ac-dc electric-driven hydraulic-fluid pump, previously illustrated. The head of the hydraulically operated tool weighs approximately 15 lb and can be equipped with a hot-stick adapter for use on energized lines. The hoses connecting the hydraulic-fluid pump to the tool head must be insulated for energized work. Special hydraulic fluid with a high dielectric strength must be used to operate the tool. The tools operate with a hydraulic-fluid pressure of 10,000 lb/in^2. The lineman must protect the equipment from damage to ensure safe operation. Specific dies are needed for different size conductors.

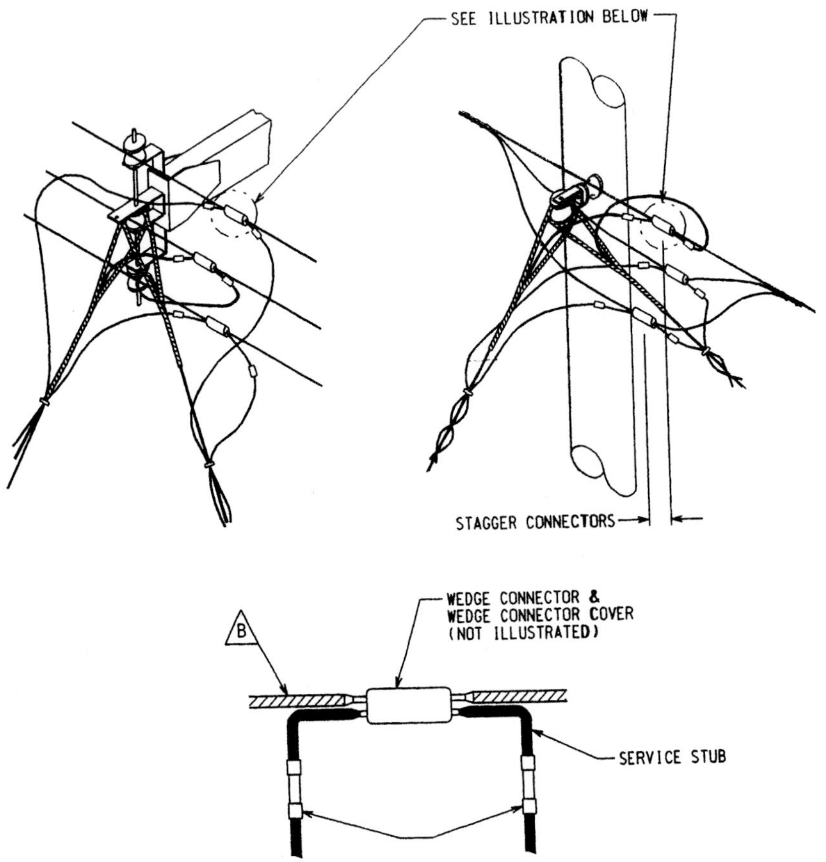

FIGURE 23.23 Sketch of a wedge connection for secondary service wire to house service connection. (*Courtesy WE Energy Co.*)

Internally Fired Connector System. Internally fired connectors can be used for both transmission and distribution lines. The connectors are made for joining stranded aluminum, aluminum-alloy, and aluminum-steel composite conductors. The housing of the connectors is made of high-tempered aluminum alloy, tapered at the ends where the conductors enter (Fig. 23.36). An impact tool contains a specially designed cartridge that is loaded with a fast-burning propellant charge. Igniting the charge creates instantaneous high pressure in the cartridge. This pressure drives cylindrical sets of wedge-shaped serrated aluminum jaws (into which the conductor ends have been inserted) at high velocity into the tapered ends of the housing. The jaws clamp and lock the conductor ends in position, providing the required holding strength and establishing a low resistant current path across the housing.

The C-member is installed by the lineman hitting the impact tool end with a hammer (Fig. 23.37).

The impact tool is convenient for lineman to use to install propelled tap connectors (Fig. 23.38). The tool is loaded with a specially designed cooler-coded shell that contains a small propellant charge. The tap connector consists of a tapered C-shaped member and a

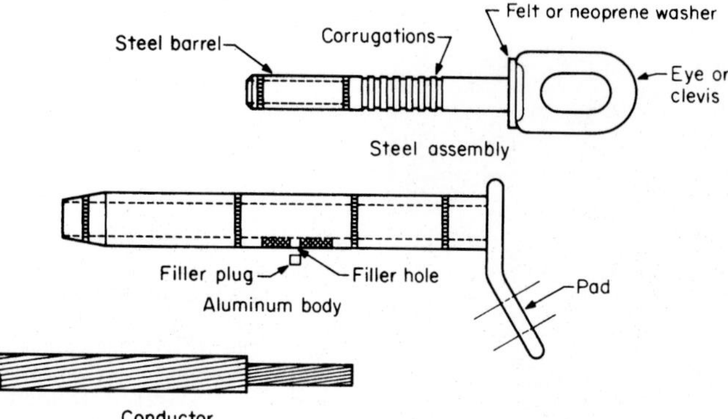

Conductor

PROCEDURE:

A. Prepare the conductor and dead end as stated in procedure for splicing ACSR conductors.
B. Slide aluminum body over conductor until sufficient working length protrudes from pad end.
C. Cut back aluminum strands a distance equal to the depth of wire bore in steel barrel plus 1/2 inch. Do not nick steel strand. File off all burs or sharp edges from ends of steel and aluminum strands.
D. Insert steel strands in steel barrel to full depth.
E. Select die size for compressing steel barrel. The die size number on die and die size number for steel barrel marked on steel assembly must be the same.
F. Make the required number of compressions on steel barrel beginning at knurl adjacent to corrugations and working toward mouth of barrel.
G. Coat steel assembly with inhibitor compound up to felt or neopreme washer.
H. Slide aluminum body over conductor and steel assembly until pad end butts solidly against felt or neopreme washer. Washer should butt against shoulder of clevis or eye.
I. Align eye or clevis of the steel assembly so that the pad of the aluminum body will be in the proper position and the eye or clevis will line up properly with the attaching hardware.
J. Select die size for compressing aluminum body. The die size number or die and die size number for aluminum marked on aluminum body must be the same.
K. Hold aluminum body in position against felt or neopreme washer and compress onto the steel assembly by making the required number of compressions beginning at the knurl mark nearest pad and working to the second knurl mark from pad end. No compressions shall be made between the second and third knurl mark from the pad end.
L. Inject Inhibitor compound into filler hole until compound is visible at the end of aluminum body. Insert and drive filler plug into hole and peen edge of hole over top surface of plug.
M. Compress aluminum body onto the conductor by making the required number of compressions beginning at third knurl from pad end and working toward outer end.
N. Finish the installation to provide the compressed portion of the connector with a smooth uniform appearance. Remove any flashing with file or emery cloth and crocus cloth. Wash the connector and adjacent wire with solvent to remove all inhibitor compound or other possible causes of corona.

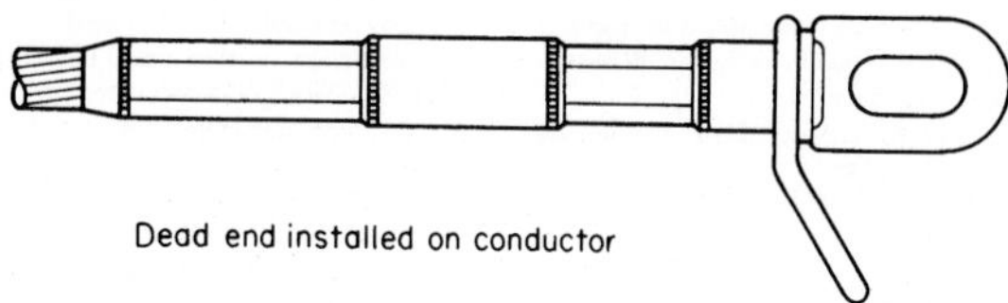

Dead end installed on conductor

FIGURE 23.24 Compression dead end for ACSR conductor.

wedge (Fig. 23.39). The lineman hooks the C-member over the main and tap conductors and places the wedge in position. The joint can be completed on energized conductors using hot sticks or with rubber glove and sleeve procedures (Fig. 23.40).

The lineman clamps the tool over the tap and detonates the propellant in the tool by striking the end of the tool with a hammer. The tool drives the wedge at high velocity between the main conductor and the tap wire within the support of the C member. This action spreads the C member. As it attempts to revert to its original shape, the C member exercises a permanent retentive force on the electrical contact areas. The locking tap ensures that the wedge remains in position even under severe service conditions (Fig. 23.41).

FIGURE 23.25 Conductor dead-ended with compression-type connector. Conductor is 795,000 cir mil ACSR. Line constructed using vertical conductor arrangement on steel poles. Jumper between dead-ended conductors is terminated with a lug-type compression connector. Jumper terminal lug is bolted to pad on compression dead-end connector. Circuit operates at 161 kV. Linemen are preparing to replace a defective insulator.

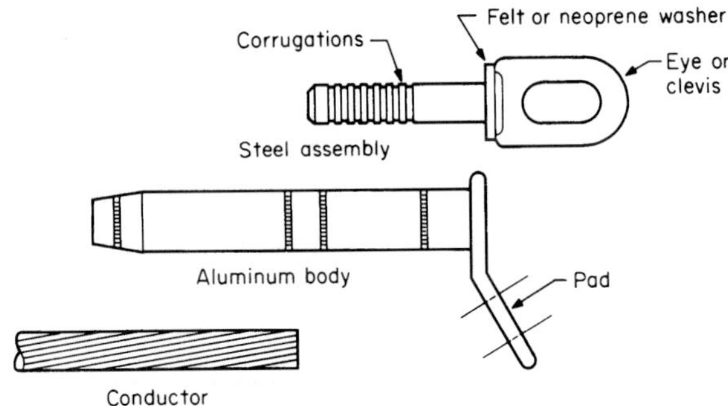

PROCEDURE:

A. Prepare the conductor and dead end as stated in procedures for splicing distribution conductors.

B. Coat steel assembly with inhibitor compound up to felt or neoprene washer.

C. Slide aluminum body over steel assembly until pad end butts solidly against felt or neoprene washer. Washer should butt against shoulder of clevis or eye.

D. Align eye or clevis of the steel assembly so that the pad of the aluminum body will be in the proper position and the eye or clevis will line up properly with the attaching hardware.

E. Select die size for compressing aluminum body. The die size number on die and die number for aluminum marked on aluminum body must be the same.

F. Hold aluminum body in position against felt washer and compress onto the steel assembly by making the required number of compressions beginning at the knurl mark nearest pad and working to the second knurl mark from pad end. No compressions shall be made between the second and third knurl mark from the pad end.

G. Insert conductor to the full depth of the bore and mark conductor at end of body. Check by measuring.

H. Inject sufficient inhibitor compound into the aluminum body to ensure that excess compound will be visible when body is completely compressed.

I. Insert clean, brushed end of the conductor into the body to the mark on the conductor.

J. Compress aluminum body onto the conductor. Using same die as used in step E, by making the required number of compressions beginning at third knurl mark from pad and working toward outer end.

K. Finish the installation to provide the compressed portion of the connector with a smooth uniform appearance. Remove any flashing with file or emery cloth and crocus cloth. Wash the connector and adjacent wire with solvent to remove all inhibitor compound or other possible causes of corona.

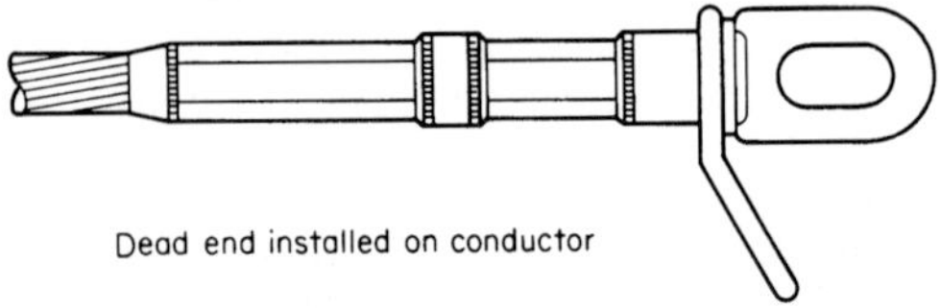

FIGURE 23.26 Compression dead end for aluminum and aluminum-alloy conductors.

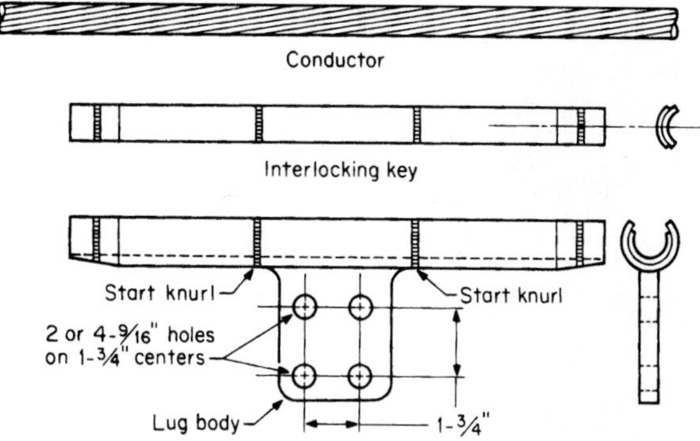

PROCEDURE:

A. Prepare the conductor and tee lug as stated in procedure for splicing distribution conductors.

B. Coat the clean, brushed conductor and inner surface of key and body with sufficient compound so that the groove will be completely filled.

C. Wire sizes smaller than 477 MCM do not have interlocking key.

D. Insert clean, brushed conductor and interlocking key into groove.

E. Select correct die size for tee lugs. The die size number on die and die size number marked on tee lug must be the same.

F. Make the required number of compressions for tee lugs. The center portion which is over the pad is not compressed. Make initial compression beyond "start" knurl on one end of the barrel of tee lug. Make second compression beyond "start" knurl on opposite end and complete compressions to the end of the barrel of tee lug. Then complete compressions on starting end.

G. Clean contact areas of pad and connecting fitting. Coat with inhibitor compound. Wire brush through compound and assemble joint without removing compound.

H. Finish the installation to provide the compressed portion of the connector with a smooth uniform appearance. Remove any flashing with file or emery cloth and crocus cloth. Wash the connector and adjacent wire with solvent to remove all inhibitor compound or other possible causes of corona.

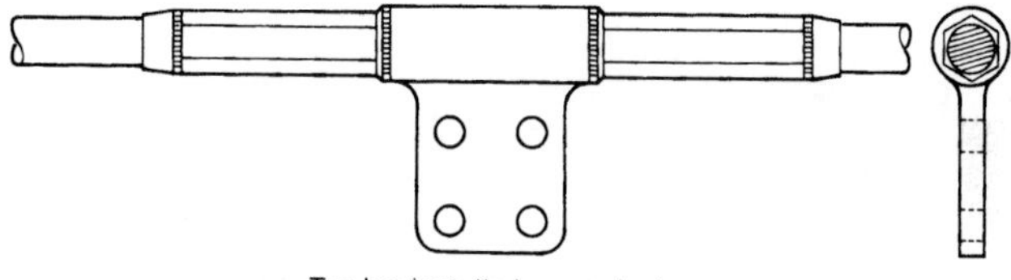

Tee lug installed on conductor

FIGURE 23.27 Compression tee-lug connector for installation on aluminum conductor.

The impact tool can be used to remove the taps. The propellant reduces the effort needed to remove the tap. The main conductor is free of connector bodies after the tap has been removed.

Implosive Compression Connectors. Implosive connectors are manufactured for splicing and terminating all transmission-line conductors. The implosive compression splice for ACSR conductor consists of a steel sleeve, an inner aluminum sleeve, and an outer aluminum sleeve on which the implosive charge is mounted (Fig. 23.42). The inner aluminum sleeve is used instead of filler compound to fill the central void in the splice. The conductor is prepared for splicing in the same manner as described for a hydraulically compressed splice. The implosive charge is detonated by a No. 8 strength blasting cap, initiated by a safety fuse or electrical means (Fig. 23.43). The implosion-compressed splice has a strength equal to 95 percent of the conductor's breaking strength, meeting the requirements for a full-tension splice. The use of a hydraulic press and its accessories is avoided. The finished splice is corrosion-resistant and corona-free (Fig. 23.44). The splices are completed

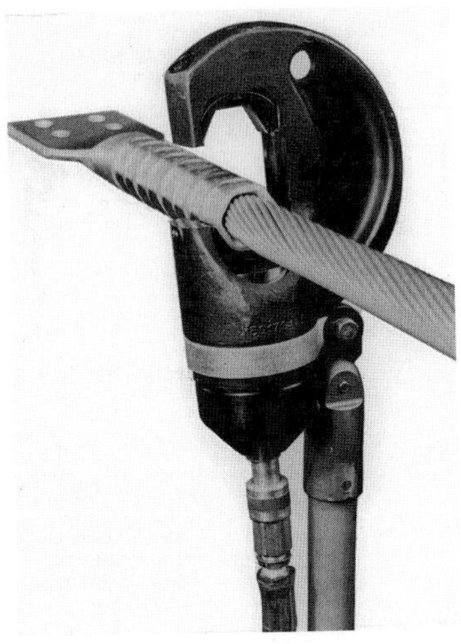

FIGURE 23.28 Compression straight-lug terminal installed on large aluminum conductor with bucket-truck hydraulic tool circuit or hand-operated hydraulic pump providing force to operate compression tool on energized conductors. Hydraulic hose and fluid have a high dielectric. Electric pump may be used to provide force to operate compression tool on deenergized conductors. (*Courtesy Somerset Division of Homac Manufacturing Co.*)

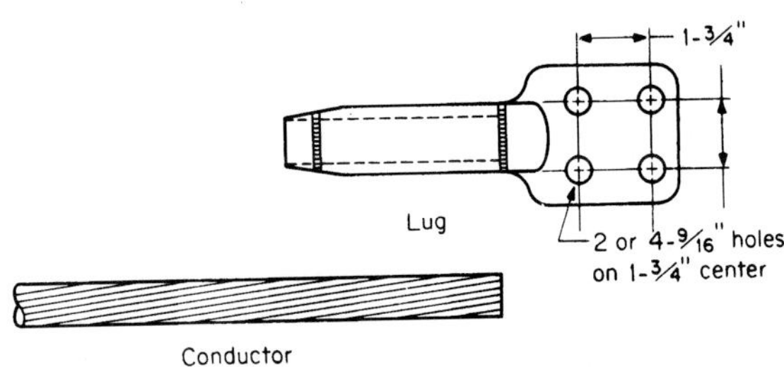

PROCEDURE:

A. Prepare the conductor and lug as stated in procedures for splicing distribution conductors.

B. If lug is prefilled with inhibitor compound, mark conductor end with pencil or crayon at a distance equal to length from barrel end of connector to the knurl at pad end plus 3/8 inch.

C. If lug is not prefilled with inhibitor compound, insert conductor to the full depth of the bore and mark conductor with pencil or crayon at end of barrel. Remove conductor after marking. Inject sufficient inhibitor compound into the barrel to ensure that excess compound will be visible when barrel is completely compressed.

D. Insert clean, brushed end of the conductor into the barrel to the mark on the conductor.

E. Select correct die size for straight lugs. The die size number on die and die size number marked on lug must be the same.

F. Make the required number of compressions for straight lugs. Beginning at pad end and working to outer end.

G. Clean contact areas of pad and connecting fittings. Coat with inhibitor compound. Wire brush through compound and assemble joint without removing compound.

H. Finish the installation to provide the compressed portion of the connector with a smooth uniform appearance. Remove any flashing with file or emery cloth and crocus cloth. Wash the connector and adjacent wire with solvent to remove all inhibitor compound or other possible causes of corona.

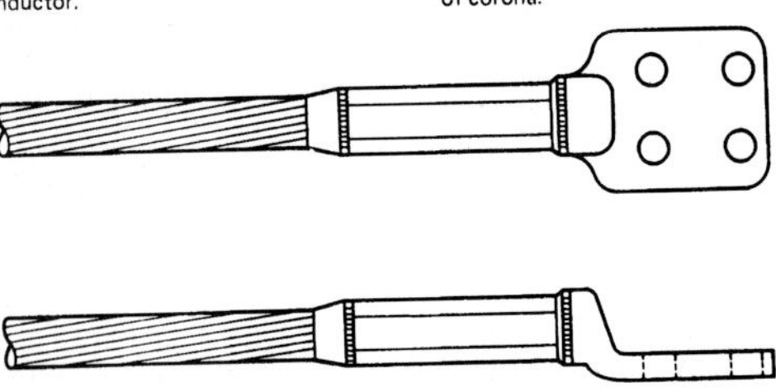

FIGURE 23.29 Compression straight-lug terminal for aluminum conductor.

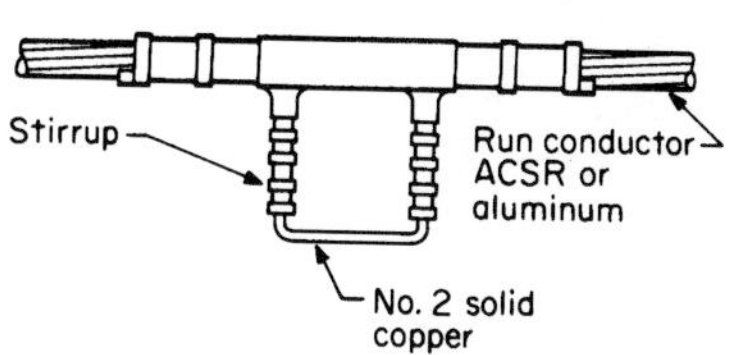

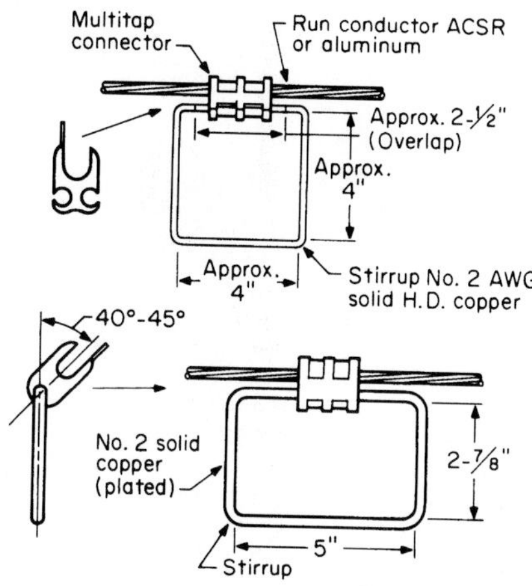

Notes:

1. Clean all conductors thoroughly with soft wire brush or with No. 70 emery cloth.
2. After the cleaning operation, coat the conductors with corrosion-inhibiting grease before installing connector. Apply grease not later than one hour after the cleaning operation. If connectors are factory-packed with grease, be sure any supplementary grease used on run or tap conductors is similar or compatible.
3. Be sure connector contact surfaces are clean. (Consider connectors factory-packed with grease to have clean contact surface when received.)
4. Always place copper conductor below aluminum conductor.
5. Install connectors directly on line conductor where possible instead of over armor rods.
6. Use compression tool with proper dies to install connector.

FIGURE 23.30 Compression stirrup connector. Stirrup permits use of hot-line taps on distribution line without damaging line conductor.

FIGURE 23.31 Compression stirrup installed on aluminum distribution line. Hot-line tap installed on stirrup permits readily removable connection for distribution transformer. (*Courtesy Sherman & Reilly, Inc.*)

FIGURE 23.32 Compression-type tap connector. Connector can be used to make a nontension tap to primary or secondary conductors with all combinations of aluminum to aluminum or aluminum to copper conductors. (*Courtesy James R. Kearney Co.*)

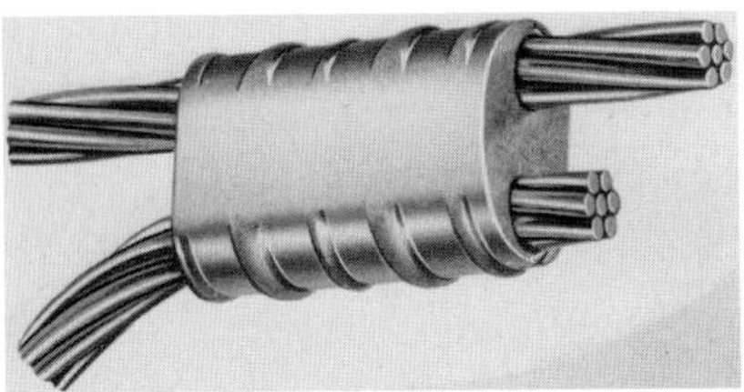

FIGURE 23.33 Compression-type tap connector installed to make a copper-conductor connection to an aluminum primary conductor. Conductors and connector must be cleaned with a wire brush and coated with inhibitor compound before connector is compressed. Aluminum conductor must be kept above the copper conductor to prevent copper salts from washing down onto aluminum conductor. (*Courtesy James R. Kearney Co.*)

without birdcaging the conductor and without the "banana" look sometimes obtained with a hydraulically compressed splice. The finished splice can be pulled through the stringing blocks without bending and without damaging the conductor. The manufacture specifies the maximum line angle for various stringing tensions when the splice is to be pulled through the stringing blocks.

Automatic Tension Splice. This type of joint is made with a single-bore sleeve that has internal gripping jaws at each end (Fig. 23.45).

Automatic line splices may be used in any span where the tension during installation exceeds 15 percent of the rated breaking strength of the conductor. It is necessary to force

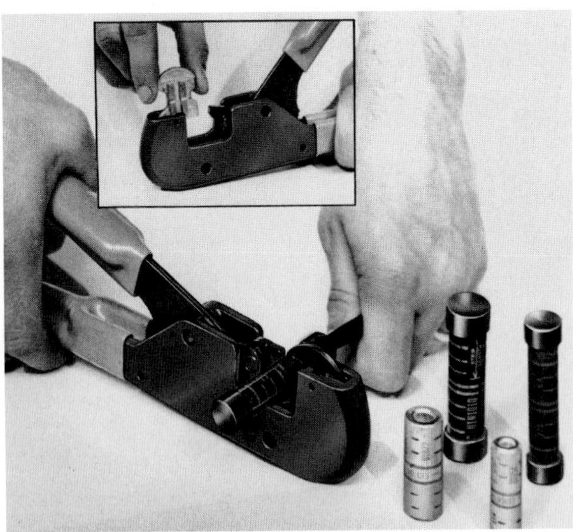

FIGURE 23.34 Compression tool and self-insulated and bare connectors used for splicing stranded conductor sizes no. 10 through no. 1/0 aluminum and ACSR normally used for secondary and service conductors. (*Courtesy Somerset Division of Homac Manufacturing Co.*)

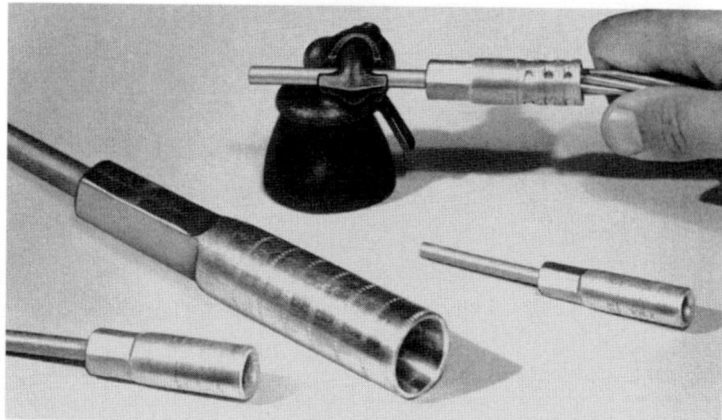

FIGURE 23.35 Compression-pin terminals and completed pin terminal joint used to connect aluminum conductors to copper alloy equipment clamps designed for use with copper. Cold-flow problems of aluminum conductors in copper clamps are eliminated. (*Courtesy Somerset Division of Homac Manufacturing Co.*)

the teeth of the jaws into the conductor to obtain reliable gripping. This should be accomplished by imposing several severe jerks on the conductor or by exerting enough pulling tension on the conductor to obtain 15 percent of the rated breaking strength. Automatic line splices *cannot be used* in taps or jumpers.

Splices should not be made in spans crossing railroads and preferably not in adjacent spans that are dependent on for withstanding the longitudinal load of the crossing conductors. Splices should not be installed within 12 in of a tie wire or armor rod.

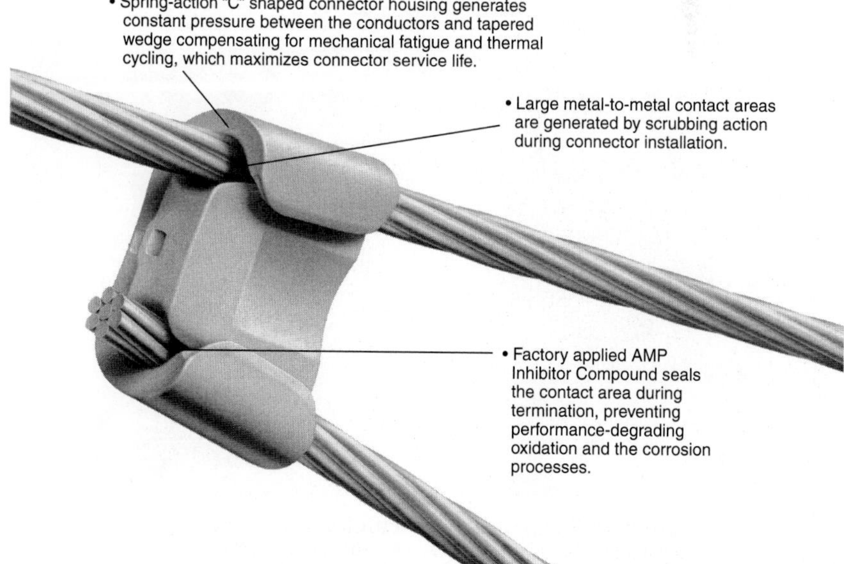

FIGURE 23.36 Spring-action "C" shaped connector. (*Courtesy Amp., Inc.*)

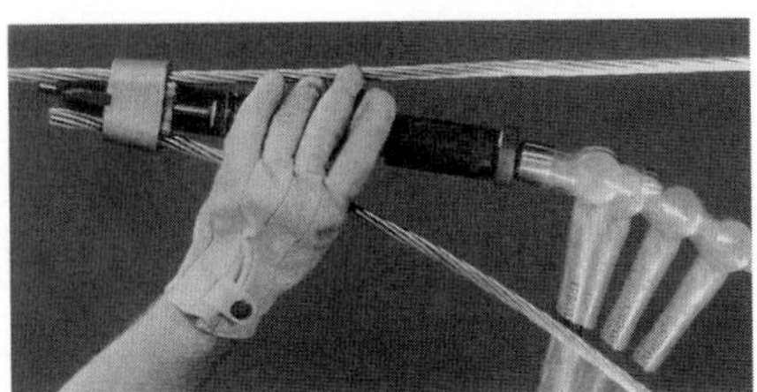

FIGURE 23.37 Taps utilize Wedge Pressure Technology. Using the impact tool, a wedge is driven into a specially shaped C-member. During installation, several key aspects of utility connector performance are addressed: the connector-to-conductor surfaces are abraded; the conductive surfaces are maximized; an oxidation inhibitor is applied to the surfaces; and spring tension is applied within the connection. (*Courtesy Amp., Inc.*)

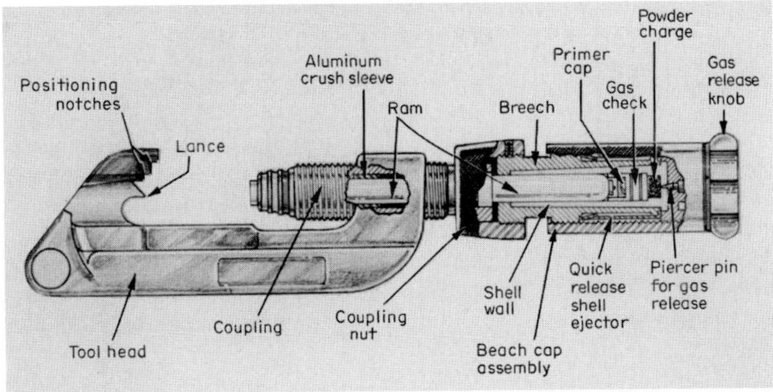

FIGURE 23.38 Detail and sectional view of impact tool. (*Courtesy Amp., Inc.*)

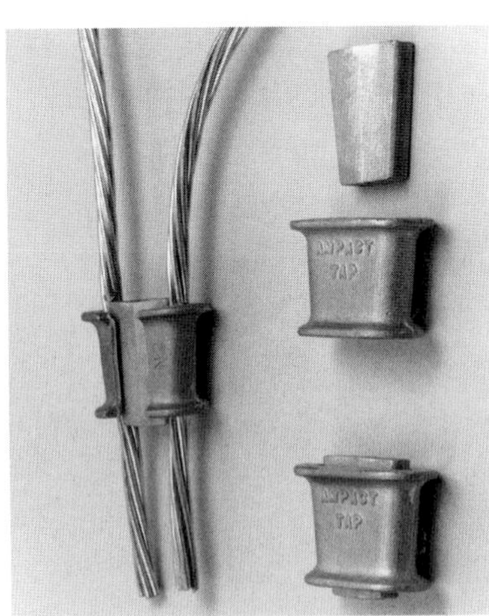

FIGURE 23.39 Impact tap component parts, assembly, and completed tap joint. (*Courtesy Amp., Inc.*)

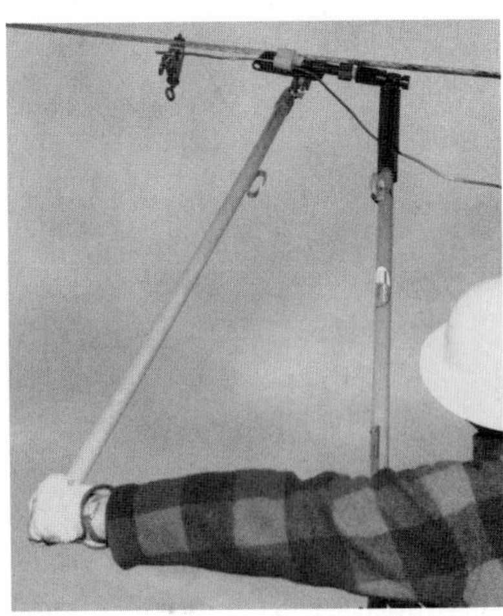

FIGURE 23.40 Lineman installing impact tap with hot-line tools. Impact tool is in position to complete the connection. A piggyback hot-line clamp is used to hold the tap conductor in pace while regular tap is completed. (*Courtesy Amp., Inc.*)

FIGURE 23.41 Dead-end pole with jumpers connected between conductors with impact taps. (*Courtesy Amp., Inc.*)

FIGURE 23.42 Full-tension implosive joint for ACSR conductor. Connector consists of steel sleeve, an aluminum-alloy filler tube, and an aluminum-alloy sleeve. (*Courtesy C-I-L, Inc.*)

FIGURE 23.43 Detonation and compression of implosive connector on an ACSR conductor.

FIGURE 23.44 Completed implosive compression splice for an ACSR conductor.

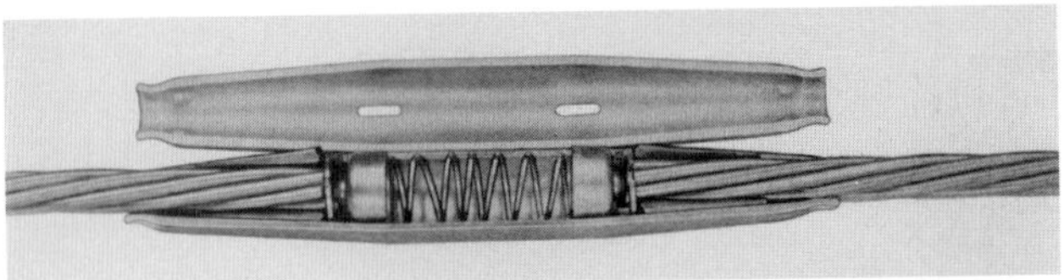

FIGURE 23.45 Automatic tension splice for use on aluminum, aluminum-alloy, and copper conductors. Ends of conductors are surrounded by free-floating self-aligning chucks. The spring serves to position the chucks against the conductors. (*Courtesy Reliable Electric Co.*)

Automatic line splices must remain free of dirt, which would impair the proper operation of the internal jaws. Splices that have lost their original protective equipment, such as plastic end inserts, must be carefully inspected for the presence of dirt and other contaminants that would impair the operation of the jaws.

During installation, automatic line splices must not be dragged through dirt, mud, ice, or water and must remain free of fouling until the conductor reaches its final installed tension. Any splice with a deformed or dented barrel cannot be used. The preceding conditions interfere with the proper seating of the internal jaws.

The making of a good splice with an automatic line splice requires following these important procedures:

1. Choose the proper size and type of splice for the conductor size and material.
2. Clean the conductor with a wire brush and remove all burrs but *do not grease* (automatic splices are pregreased with a material that is grit-free. The presence of grit prevents proper jaw operation).
3. Be sure the conductor strands are in normal lay and not deformed.

FIGURE 23.46 Lineman working in insulated bucket using rubber gloves, rubber sleeves, safety glasses, and insulated hard hat to install bolted parallel-groove clamp to complete jumper between energized dead-ended conductors. Temporary insulated jumper with hot-line taps is used to bypass regular connection being completed. Conductors are terminated with bolted-type dead-end clamps.

4. Insert the conductor with one sharp, continuous thrust, forcing it in as far as possible.

   a. Aluminum splices are provided with a pilot cup for retaining and guiding the strands into jaws of the chuck. This pilot cup must remain in place and will be driven into the interior of the splice until it strikes the center barrier.

   b. Copper splices are provided with slotted openings which allow visible examination to determine if the conductors are properly inserted into the jaws.

5. It is extremely important to set the jaws into the conductor by applying one or more momentary jerks to the conductor, forcing the jaws to bite into the conductor.

When an attempt is made to withdraw the conductors, the jaws clamp down on the conductors because of a taper in the bore of the sleeve and on the jaws. Tension causes a wedging action which increases with the pull applied on the conductor. A loss of tension could cause the jaws to release their grip and allow the conductor to drop out. This type of splice is therefore used where the wires are kept in tension. It is also used where service must be restored quickly. It is especially suited for live-line splicing.

Bolted Joints. Conductors can be joined together with bolted conductors if the wires are not under tension (Fig. 23.46). Bolted connectors are used successfully on both copper and aluminum conductors. The conductors must be cut to the proper length, cleaned, and coated with an inhibitor to obtain a good connection. Bolted parallel-groove clamps, U-bolt clamps, tap lugs, and dead-end clamps are manufactured for use on a range of conductor sizes. The lineman must use care to obtain the proper tension when tightening the bolts on a bolted connector.

CHAPTER 24
LIVE-LINE MAINTENANCE WITH HOT-LINE TOOLS

The trend in electric-system management is to do more and more of the testing, repair, and maintenance work while the lines are live (energized). The reason for this is to reduce the number of times the service is interrupted, or the number of outages, as interruptions in service are called. Live-line maintenance has been made possible by the development of special tools and procedures for such work.

Under live-line maintenance, a great variety of work is included. The most common of these live-line operations are as follows:

1. Replacing insulators
   a. Pin
   b. Suspension
   c. Strain
2. Replacing crossarms
3. Replacing poles
4. Tapping a hot line
5. Cutting slack in or out
6. Splicing conductors
7. Installing vibration dampers
8. Installing armor rods
9. Phasing conductors

It will not be the object here to give detailed instruction covering the procedure of every hot-line-tool live-line maintenance job. This would be impossible because of the great variety of jobs and types of line construction and special conditions. This section is therefore intended to give the reader only a general conception of the methods and procedures employed in the most fundamental operations of live-line maintenance work with hot sticks.

The Institute of Electrical and Electronics Engineers, Inc. (IEEE), issued IEEE Standard 516 in 1995, entitled *IEEE Guide for Maintenance Methods on Energized Power-Lines*, and this standard contains the following information:

1. Minimum tool-insulation distance ac energized-line work
2. Minimum tool-insulation distance dc energized-line work
3. Altitude correction factor
4. Minimum approach distance ac energized-line work air insulation

5. Minimum approach distance dc energized-line work
6. Physical aspects of energized-line work
   a. Electric fields
   b. Field intensities in the work area
      (1) Shielding and fields
      (2) Types of shielding
   c. Effects of fields

The guide should be reviewed for the information listed and other safety precautions required to complete live-line maintenance with hot-line tools.

The maintenance of high-voltage lines while energized, or "hot," may appear on first thought to be hazardous, especially when compared with working on dead, or deenergized lines. Actually, however, the work is completed safely when the lineman is continually conscious of the fact that the lines are "hot" and of the need to be careful to follow the correct procedures. Furthermore, there is no possibility of the line being "hot" when it was thought to be dead, as there is when working on deenergized lines. Also there is no possibility of confusion with live duplicate circuits, for in live-line work, every conductor is worked "hot" (energized) and every operation is planned and performed accordingly.

Tools. The insulated-pole portions of the hot-line tools are made from fiberglass or wood. Wood was used originally before fiberglass became available. Fiberglass has proven to be the best material for insulated-pole tools.

If the hot stick pole is made of wood, it must be kept dry and clean. Scratches, scars, or other abrasions on the stock will damage the wood surface and permit moisture and dirt to enter the fibers of the wood. Moisture and dirt will form conducting paths along the hot stick that will render it hazardous for use.

If the pole is made of fiberglass, it will not absorb moisture and therefore does not have to be dried periodically. Fiberglass is not affected by creosote, gasoline, oil, grease, or most solvents. The smooth pole does not have any grooves to trap dirt or moisture. Fiberglass poles are stronger and a better insulator than wood. They can also be molded in bright colors for improved visibility. However, fiberglass insulated-pole tools should be treated carefully and kept clean and dry. Any material that becomes contaminated and/or wet loses its insulating qualities.

Hot-line tools are of great variety and those most generally used can be classified as follows (Figs. 24.1 and 24.2):

1. Wire tongs
2. Wire-tong supports or saddles
3. Insulated tension links
4. Auxiliary arms
5. Strain carriers
6. Tie sticks
7. Platforms
8. Insulated hoods
9. Insulated tools

The *IEEE Guide on Terminology for Tools and Equipment to be Used in Live-Line Working* (ANSI/IEEE Standard 935-1989) provides detailed information on live-line tool nomenclature.

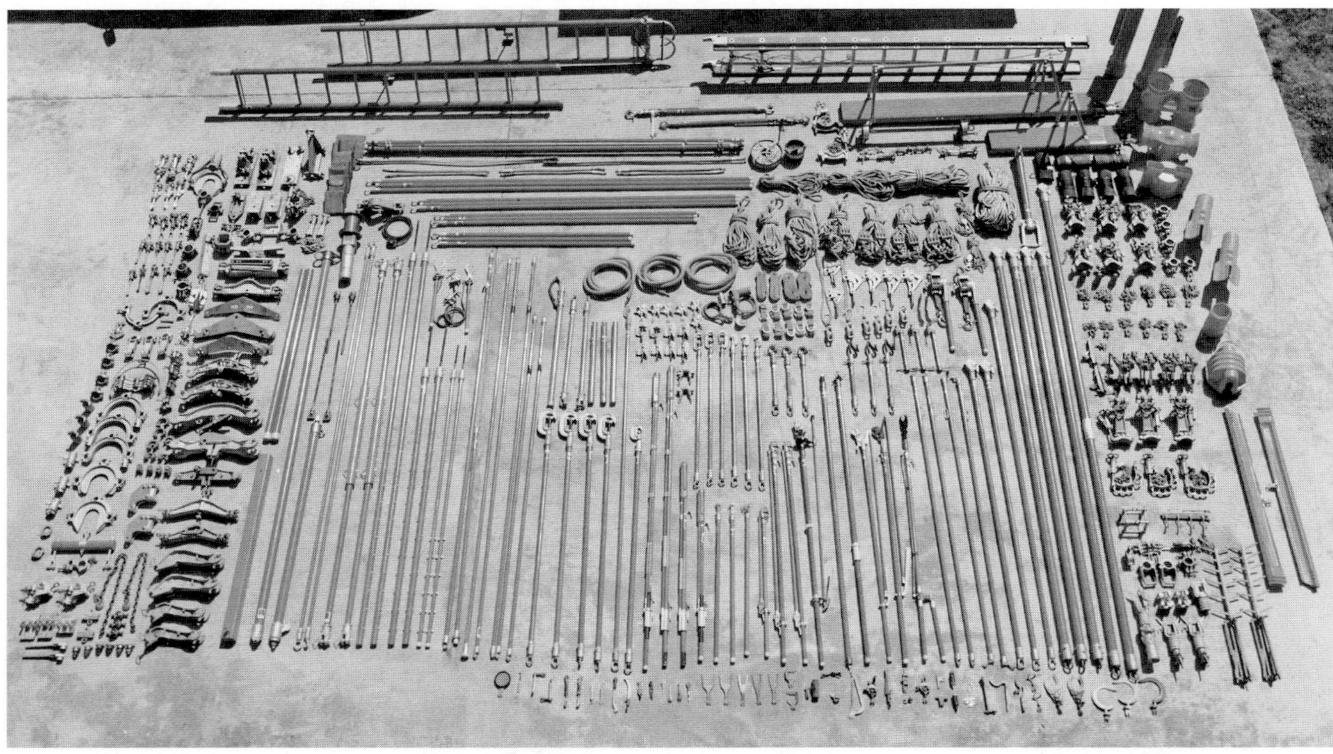

FIGURE 24.1 A display of hot-line tools used for live-line maintenance of distribution and transmission circuits operating at voltages up to 765 kV. (*Courtesy A. B. Chance Co.*)

FIGURE 24.2 Special hot-line tool trailer. Note elevated cover. Tools are for live-line work on distribution and transmission lines. (*Courtesy A. B. Chance Co.*)

Wire Tongs. A wire tong is a slender insulated pole provided with a hook clamp on one end and a swivel on the other end (see Fig. 24.3). Tongs are used primarily in holding live conductors. They are sometimes referred to as lifting sticks or holding sticks. The hook clamp is opened or closed by turning the pole. A live conductor is clamped by placing the hook over the conductor and turning the pole until the jaws of the hook are completely closed. The swivel clevis on the other end is for the application of the lifting tackle.

The wire tong is used to hold the conductor away from the point where work is to be done such as when replacing insulators, crossarms, or transformers. The simplest case is illustrated in Fig. 24.4, where the wire tongs are positioned to hold the live-line conductor to one side. A more complicated operation is illustrated in Fig. 24.5, where a number of wire tongs are used to hold the three conductors of a three-phase line away from the pole.

Wire-Tong Supports. A wire-tong support serves the special purpose of guiding and holding a wire tong in a fixed position after the conductor has been moved. A typical tong support is shown in Fig. 24.6. The support consists essentially of a base or saddle fastened by a chain to the pole or tower. A swivel clamp is mounted on the saddle. The wire tong slides between the jaws of this clamp, while the swivel action permits movement of the tong in all directions. Wing nuts are used to tighten the jaw of the clamp onto the tong.

A multiple form of tong support is illustrated in Fig. 24.7. When a number of saddles are needed on the same side of a pole or tower, a slide type of wing-tong support may be used. This support consists of a round rod secured to the pole at each end by means of a chain. On the rod are placed the tong holders. These holders slide up and down the rod freely but can be clamped in any position. One to six tongs can be supported from one fastening. The rod is available in lengths up to 6 ft.

Insulated Tension Link. An insulated tension link is a strain insulator with a conductor clamp on one end and a swivel on the other, as shown in Fig. 24.8. It is commonly used between the line conductor and the rope tackle when pulling conductors in the clear of structures when changing insulators, crossarms, or poles.

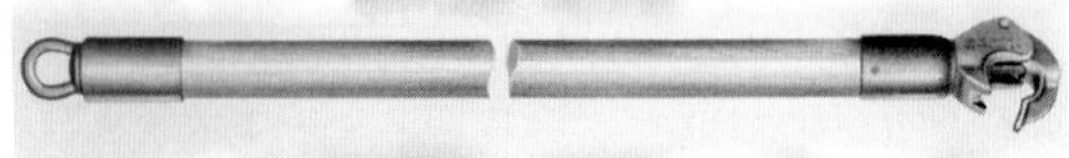

FIGURE 24.3 Wire tong consisting of pole, conductor clamp, and swivel. (*Courtesy A. B. Chance Co.*)

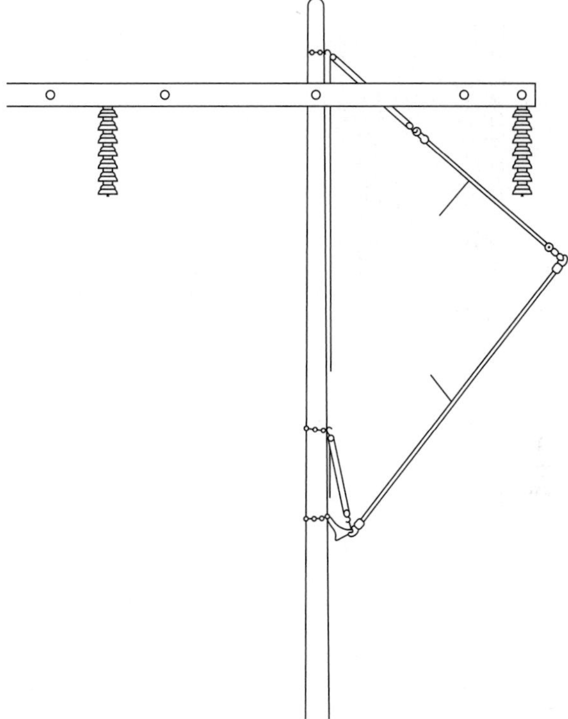

FIGURE 24.4 Sketch of two wire tongs positioned to hold live conductor on one side so as to create needed working space. (*Courtesy A. B. Chance Co.*)

Figure 24.9 shows two tension links in use when changing post insulators. The wire tongs hold the conductor at the desired distance, and the links keep the tongs from swaying toward the pole.

Auxiliary Arm. An auxiliary arm is an arm that is used to support the line conductors temporarily while the insulators or crossarms are being replaced. It may be used to raise and support the conductors vertically, or it may be used to support one or more conductors to one side horizontally. An auxiliary arm is especially useful where it is desired to install a pole of greater height or where low-voltage lines must be cleared preparatory to working on higher-voltage lines above them. Clamps are provided on the top of the arm to hold the conductors securely.

Strain Carrier. The strain carrier is used to take the load or strain from suspension insulators while the insulators are changed. Two types of strain carriers are in use, one for light loads up to about 2000 lb and one for heavy loads up to about 10,000 lb.

The light-duty strain carrier consists of two rocker arms hinged on a bar (Fig. 24.10). By bringing together the two rocker arms, the insulator string is compressed, and the insulator units may be removed and replaced.

The heavy-duty strain carrier consists of two parallel insulating tension rods which are set astride the insulators. The strain is relieved by taking up screws on the ends of the rods,

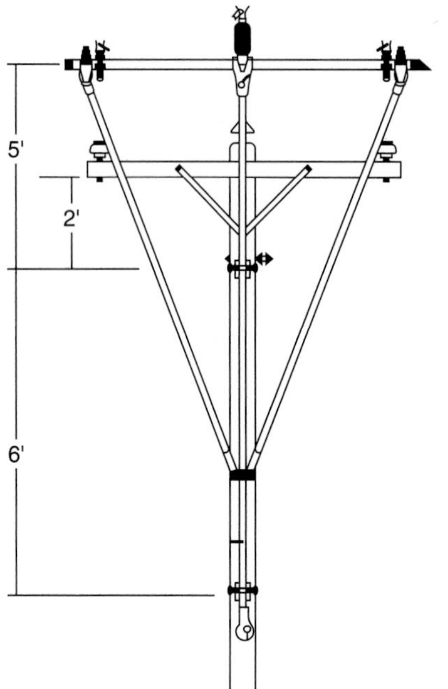

FIGURE 24.5 Sketch of wire tong cluster used to hold each line conductor away from the pole. (*Courtesy A. B. Chance Co.*)

thereby compressing the insulator string. The insulator string is supported in a cradle consisting of two insulated rods below the insulator string. The strain carrier may be used on either pole or tower construction.

Tie Sticks. Tie sticks are used to fasten or unfasten the tie wire which holds the conductor to the pin-type insulator (see Fig. 24.11). Also, tie sticks are usually equipped with a small double hook called the *tie assistant*. It is fastened to the tie stick a short distance from the end. It has several uses, such as prying loose ends of old tie wires that may be wrapped close to the conductor, holding the conductor down in the insulator groove while being tied in, or as a means of hanging the tie stick on the conductor when not in use.

Platform. The platform is small and collapsible and may be secured to a pole or tower. Figure 24.12 shows a small utility platform, and Fig. 24.13 shows a more elaborate type

FIGURE 24.6 Typical wire-tong base for anchoring tong to pole. (*Courtesy James R. Kearney Corp.*)

FIGURE 24.7 Multiple wire-tong support. Each support can be clamped in any position along the rod. (*Courtesy A. B. Chance Co.*)

designed for hot-line work. This platform is equipped with a hand railing. The lineman secures his safety trap to the hand rail instead of the pole. Its use enables the lineman to reach farther out from a pole or tower and thus perform the work with greater reach and ease.

Insulated Hoods. Insulated hoods are available to place over conductors and insulators to add to the safety of lineman doing hot-stick work (see Figs. 24.14 and 24.15).

Insulated Tools. Many other tools can be adapted to hot-line work by mounting them on the ends of insulated poles. Among these are pliers, wire cutters, tree trimmers, cotter-key pullers, screwdrivers, fuse pullers, saws, wrenches, brushes, and clamps. Some of these are illustrated in Fig. 24.16.

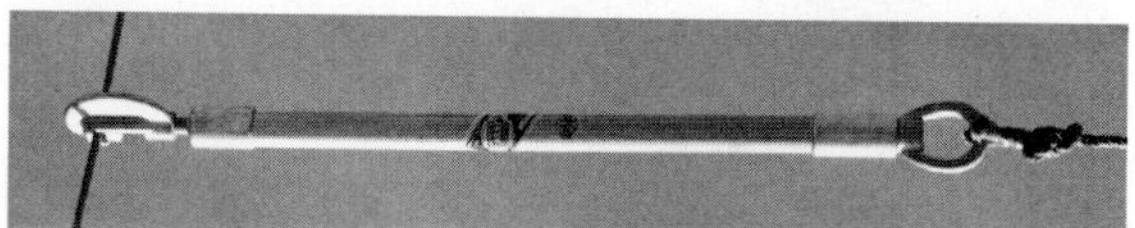

FIGURE 24.8 Insulated tension link used to spread line conductors for pole installations and removals, pulling conductors in the clear of structures for maintenance work, and handling extremely heavy conductors on H-frame and tower lines. (*Courtesy James R. Kearney Corp.*)

FIGURE 24.9 Two conductors supported by wire tongs, with two insulated tension links being used in series with ropes to maintain the wires at a proper distance from the pole to complete the maintenance work. (*Courtesy James R. Kearney Corp.*)

Required Clearance. Table 24.1 gives the minimum safe working distances from the conductors or from the hot end of sticks to the lineman as specified in the Occupational Safety and Health Act (OSHA) Standards. For 345 to 362 kV, 500 to 552 kV, and 700 to 765 kV, the minimum working distance and the minimum clear hot-stick distance may be reduced provided that such distances are not less than the shortest distance between the energized part and a grounded surface. The IEEE *Guide for Maintenance Methods on Energized Power-Lines* provides additional information on safe working distances that takes into account switching-surge voltages and the altitude at the location where the work is being performed.

Changing Pin Insulators. The principal parts of a pole line are the conductor, insulator, crossarm, and pole. Since the conductor hardly ever fails, live-line maintenance resolves itself into replacing insulators, crossarms, and poles. To replace one of these requires that

FIGURE 24.10 Light-duty strain carrier of the rocker-arm type. Illustration shows strain carrier mounted on a corner pole and used to relieve the insulator string of conductor pull. (*Courtesy James R. Kearney Co.*)

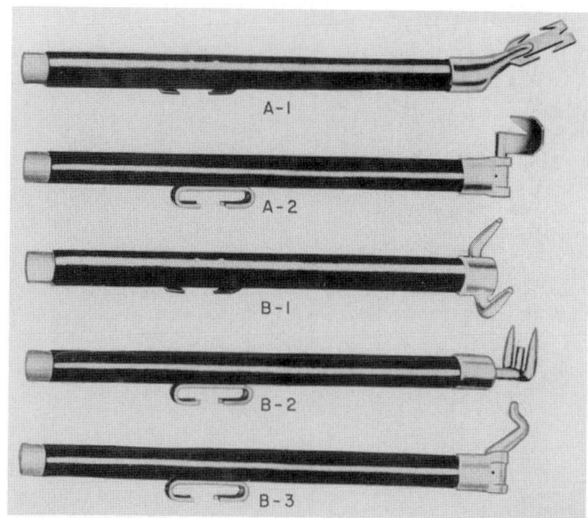

FIGURE 24.11 Tie sticks for use on live lines. (A) Blade type: (1) fixed, (2) rotary. (B) Prong type: (1) two-prong, (2) three-prong, (3) rotary-prong.

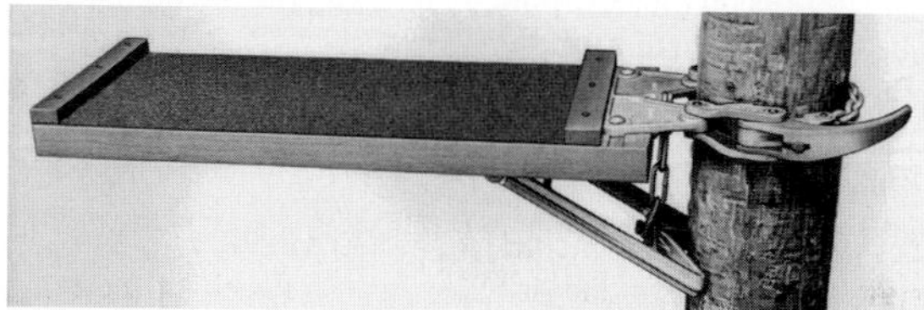

FIGURE 24.12 Small utility platform. (*Courtesy A. B. Chance Co.*)

FIGURE 24.13 Insulated platform designed for live-line maintenance work. (*Courtesy James R. Kearney Co.*)

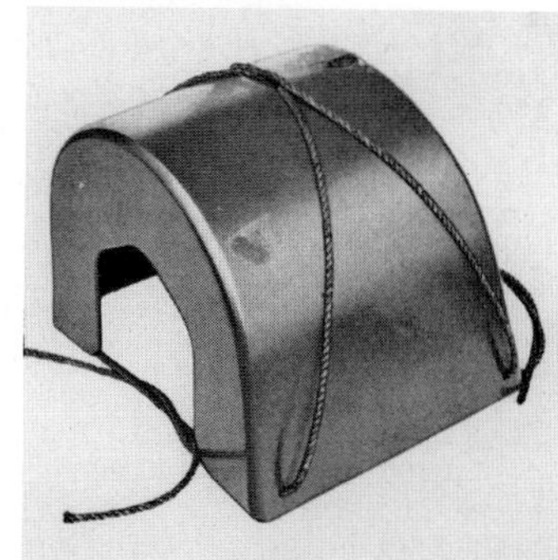

FIGURE 24.14 Insulated insulator cover. (*Courtesy James R. Kearney Co.*)

FIGURE 24.15 Polyethylene insulator hood, conductor covers, and crossarm guard in use on live-line maintenance work on 34.5-kV line. (*Courtesy A. B. Chance Co.*)

the conductors be untied from the insulators, removed, and held in the clear. When so held, the insulators or crossarms, or both, can be removed and replaced. If the pole is to be replaced also, the best practice is to set the new pole first and then to mount all wire tongs and saddles and other tools on the new pole. In this way the old pole, which usually is weak, will not have to be specially guyed for this operation.

The steps in the process are preformed in the following order:

1. Fasten wire tongs to conductor
2. Unite conductor from insulator

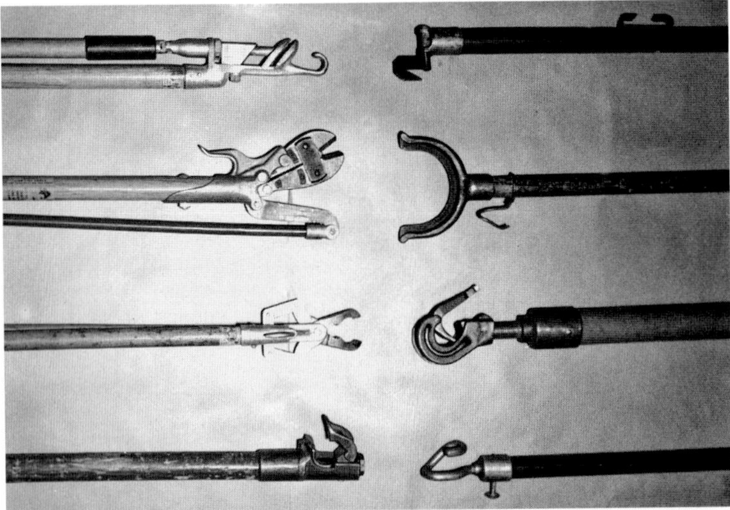

FIGURE 24.16 Heads of hot-line tools used in holding off and repairing lines. Hot-line tools must be inspected and tested periodically for safe use. (*Courtesy American Electric Power System.*)

TABLE 24.1 Hot-Line Tool Maintenance Work Clearances

	Minimum clearance	
Line voltage, volts	Feet	Inches
1,100 to 15,000	2	1
15,100 to 36,000	2	4
36,100 to 46,000	2	7
46,100 to 72,500	3	0
72,600 to 121,000	3	2
138,000 to 145,000	3	7
161,000 to 169,000	4	0
230,000 to 242,000	5	3
345,000 to 362,000	8	6
500,000 to 550,000	11	3
765,000 to 800,000	14	11

3. Move conductor into clear
4. Remove old insulator, or crossarm, or both
5. Mount new insulator, or crossarm, or both
6. Return conductor to insulator
7. Tie in conductor
8. Remove wire tongs or auxiliary arm

Moving Line Conductors. Three methods for moving line conductors into the clear are available. These are:

1. Wire-tong method
2. Auxiliary-arm method
3. Combination of wire-tong and auxiliary-arm methods

Wire-Tong Method. In the wire-tong method, the conductor is moved and supported by the use of two wire tongs, one being used for lifting, called the *lifting tong* or *stick*, and the other for holding, called the *holding tong* or stick (see Fig. 24.17). The lifting stick is clamped onto the conductor to be moved, and its saddle is fastened to the pole in such a position that several feet of swivel end of the lifting stick extend through the saddle. The saddle clamp is then tightened onto the stick by turning the wing nut. Next, a set of blocks is attached between the saddle and the swivel on the end of the lifting stick. This block and tackle will later be used to lift and move the conductor out of reach.

Another saddle for the holding stick is fastened to the pole just below the crossarm braces. The holding stick is now clamped onto the conductor and placed in the top saddle.

The conductor may now be untied from the insulator by means of the insulated tie sticks. After the conductor is untied, it is ready to be pushed out of reach. To move the conductor, pull up on the set of blocks, thereby raising the conductor off the insulator. If the outer conductor is being moved, the distance raised should be about 10 in. If the center conductor is being moved, it should be raised enough to clear the end of the crossarm. When the conductor has been moved outward to a safe working distance, the saddle clamps on both sticks are tightened firmly. The other outside conductor is moved next, and the inside or center conductor is moved last, since it must be moved above one of the outer conductors. The saddles for the sticks supporting the outer conductors may be placed on opposite sides of

FIGURE 24.17 Two wire tongs being used to hold live conductor on a rural line in the clear while lineman is preparing to replace the pole-top pin. The longer lower tong is the lifting tong and supports the conductor; the shorter upper tong is the holding tong and is used to hold the conductor firmly, at the proper distance from the pole. (*Courtesy American Electric Power System.*)

FIGURE 24.18 Three sets of wire tongs holding three live conductors firmly in the clear while a new pole and crossarm are being installed. The tongs are all supported on the new pole. (*Courtesy American Electric Power System.*)

the poles, while a saddle extension may be used for the saddles supporting the middle conductor. Figure 24.18 shows three live conductors of a three-phase, four-wire circuit moved in the clear by means of wire tongs preparatory to changing the pole and crossarm.

After the insulators, or crossarms, or both have been replaced, the above procedure is reversed and the conductors are brought back in. As the conductors are returned, they are tied in, using the special tie sticks.

For heavier construction such as the wishbone, H-frame, or at angles in the line, the same procedure may be followed except that an insulated link stick with blocks may be used to help move the conductor away from the insulator. The top holding stick should still be used because it helps to stabilize the conductor on the line stick.

Auxiliary-Crossarm Method. Where construction permits, the work of moving conductors may be simplified by the use of the auxiliary crossarm. The auxiliary crossarm may be used to lift all conductors from a crossarm vertically at one time, or it may be used to hold two or more conductors out on one side of the pole.

Mast Arm. When the auxiliary arm is used as a mast arm, that is, in lifting the conductors up from a crossarm, two saddles are mounted on the pole. The first saddle is mounted just beneath the crossarm braces. The assembled mast arm is then placed in position in this saddle and pushed up against the conductors and clamped into position. The lower saddle is then mounted on the pole below the point where the bracing is fastened to the mast arm, and the main lifting pole is clamped in position. Saddle extensions should be used if necessary.

The conductors are then untied from the insulators with the tie sticks and placed in the holders on the mast arm provided for that purpose. When all the conductors have been transferred, the whole mast-crossarm assembly is lifted up until the conductors are in the clear.

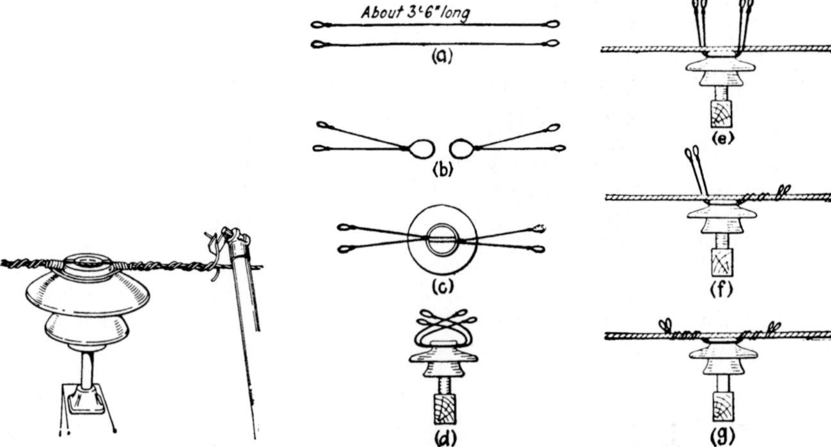

FIGURE 24.19 Removing tie wire with tie stick. (*Courtesy A. B. Chance Co.*)

FIGURE 24.20 Steps in tying in conductors by use of live-line maintenance tools. Tie wires are attached to insulator before insulator is screwed in place on pin.

If the line makes an angle greater than 5°, it is advisable to side guy the mast arm using an insulated link stick and a rope.

Side Arm. When the auxiliary arm is used as a side arm, it should be placed on the two-conductor side of the pole. One end of the side arm is mounted on a saddle at a point just below the crossarm braces. The conductors are then transferred to the holders on the side arm provided for that purpose with the use of lifting and holding tongs. The outside conductor is moved first. A block and tackle fastened to the upper end of the lifting tongs is also used to assist in moving the conductor.

The third conductor on the opposite side of the pole is moved and supported with wire tongs.

Removing the Tie Wire. The tie wire is untied by starting to unwind the end of the tie wire with the tie-wire blade on the tie stick, as shown in Fig. 24.19. If the tie wire has no end which can be easily reached with the blade of the tie-wire stick, an end is made by cutting the tie wire at some convenient point with a hack saw. The tie wire is then unwound by pulling the end of the tie wire if it points away from the pole. If the end of the tie wire points toward the pole, it is started by pushing up with the tie-wire stick. If it has a long wrapping, it should be cut with the hack saw as soon as the untied portion becomes difficult to handle. If it gets too long, there is danger of its touching the other conductor and shorting the line.

During the process of removing old tie wires and tying in, the lineman should not be allowed to sit on the crossarm or put his legs or arms through the braces but should be made to stand with his full weight in the climbing hooks with his feet as close together as possible, and he should be instructed to lean back in his safety belt and keep his hands off the pole, crossarms, and crossarm braces. If the will do this, he will benefit from the insulation properties of the pole and the crossarm.

Tying in Conductor. Particular mention should be made of the procedure employed in tying the conductor onto the new insulator. Special provision is made for this by providing the new insulator with special tie wires before the insulator is screwed in place. Figure 24.20*d* shows the insulator in place with the tie wires fastened to the insulator. Figure 24.20*a*, *b*, and *c* shows the details of the tie wires and the method of fastening them to the insulator. Each piece of the tie wire is placed in the neck of the insulator and bent until the ends meet (both ends being the same length). They are then given a couple of twists close to the neck of the insulator and allowed to extend horizontally in the same direction as the top groove of the insulator, one pair of tie-wire ends extending out in the opposite direction for the other side of the insulator, as shown in Fig. 24.20*c*. These ends are now bent over the top of the insulator, as shown in Fig. 24.20*d*, so that in screwing the insulator on the pin the tie-wire ends will not come in contact with any part of the live line. The insulator is then sent up on the pole and screwed onto the pin until it is tight and with the top groove of the insulator parallel to the line conductor. The line conductor is then placed in the top groove of the insulator, care being taken that both ends of the tie wire come on the same side of the line conductor and that the two pairs of tie-wire ends come on opposite sides of the line conductor. When in this position, the insulator is ready to be tied in. The tie stick is used for this purpose. The man making the tie should be well below the crossarm. The tie stick is passed up on the far side of the line conductor from the pair of tie-wire ends on that side of the insulator. One horn of the tie stick is put through the loop in the end of the tie wire nearest the insulator. This end is then pulled downward. The same is then done with the other end of the tie wire. When both ends are pointing downward, the end of the tie wire which was first pulled down is pushed upward with a stiff punch on the other side of the line conductor; the other tie wire is likewise pushed up. By repeating this operation, both ends of the tie wire are thus served up to the loops in their ends. These loops can then be removed if desired, although there is no reason for removing them, by cutting them with a hack saw and breaking them off or by twisting them with a tie stick until they break off close to the line conductor. Obviously, the same method can be used in tying the line conductor in the neck of the insulator as will be necessary on angles. In the case of double arms, half the above tie is made on one insulator and half on the other insulator.

This is by no means the only tie which can be applied; any desired form of tie can be made. This particular tie is chosen here only because it lends itself to a clear and accurate description and is perhaps the easiest tie to learn.

The sketches in Fig. 24.21 illustrate the most common types of hot-line ties in use. Note that both wires on one side of an insulator are turned in the same direction and that both wires on the other side of the insulator are turned in the opposite direction. They are started from the top and bottom of the conductor as in B and D (Fig. 24.21) or from opposite sides of the conductor as in A and C. This provides a self-snubbing action which prevents the conductor from rolling and assists in holding the conductor tightly against the cap of the insulator.

Note that eyes are shown on all ties. A lightweight stick with prong-type head and this type of tie (Fig. 24.22) make a very satisfactory working combination. Some linemen prefer having the tie without eyes. In this case, the blade-type head is more convenient (Fig. 24.23). The use of eyes is especially helpful in untying as it makes it unnecessary to pry up the ends of the tie wire.

A handy tie stick is one with the blade-type head on one end and the prong-type head on the other.

Prefabricated ties manufactured to fit specific insulators and conductors provide a fast, firm, and economical system for securing wires to pin- and post-type insulators. Hot-stick application of a prefabricated tie for a conductor installed on top of a pin-type insulator is illustrated in Figs. 24.24 through 24.31. Application of a prefabricated tie for a conductor installed on side groove of a horizontal line post insulator with hot sticks is shown in Figs. 24.32 through 24.36.

Removing Conductors from Clamp-Top Insulators. In case the pin insulator is of the clamp-top type, removal of the conductor from the insulator is greatly simplified. Likewise, fastening of the conductor is also made easy.

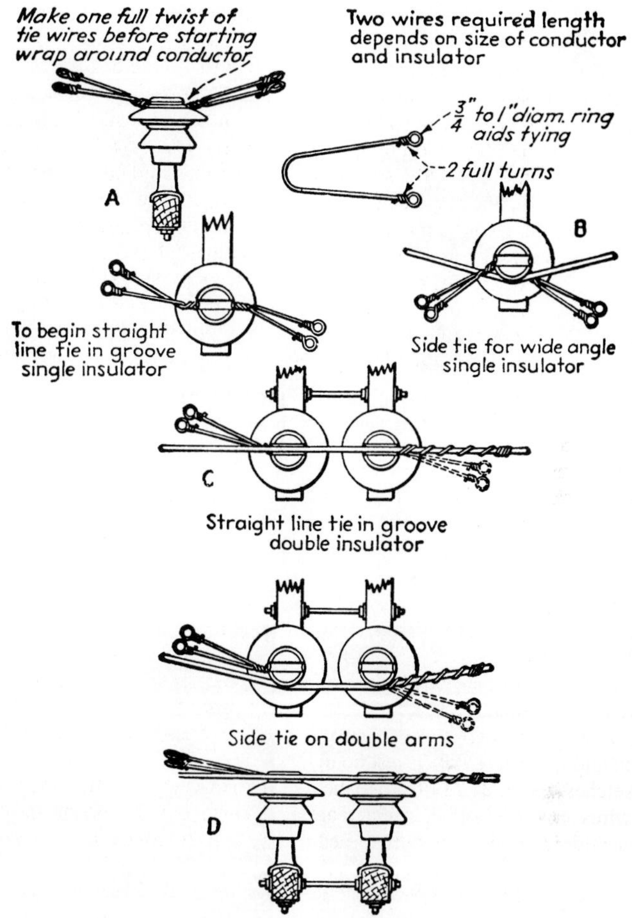

FIGURE 24.21 Most common types of hot-line ties. (*Courtesy A. B. Chance Co.*)

The clamp-top insulator, illustrated in Fig. 24.37, is similar to other pin insulators except that it has a bolted clamp cemented to the top. Two carriage bolts hold the keeper piece over the conductor.

To remove the conductor, therefore, it is only necessary to unscrew the two nuts. This is done by means of a special ratchet wrench and socket secured to the end of an insulated handle. Figure 24.38 shows this operation being performed, and Fig. 24.39 is a close-up view of the ratchet wrench and socket for loosening the nuts on the clamp-top bolts. Before this is done, the lifting and holding tools are attached to the conductor and secured to the pole.

After the nuts are loosened and backed off about 5/8 in, the conductor is raised slightly, about 1/2 in, to take its weight off the clamp seat. The clamp is then lifted from the supporting pintles with a special wire cradle mounted on a hot stick and slid along the conductor away from the insulator. A close-up view of this process is shown in Fig. 24.40, and a view of the complete operation is shown in Fig. 24.41. With the clamp slid away from the

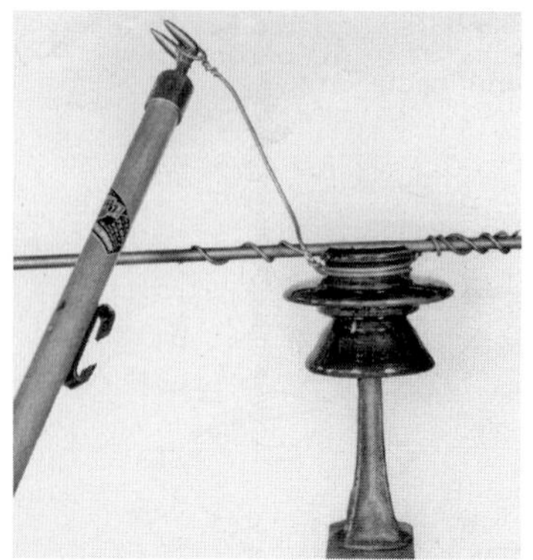

FIGURE 24.22 Tying in a conductor using tie wires provided with eyelets and a prong tying tie stick. (*Courtesy James R. Kearney Corp.*)

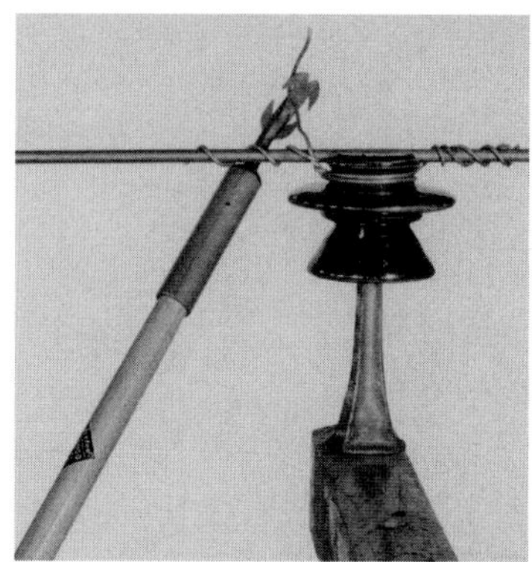

FIGURE 24.23 Tying in line conductor with blade-type tie stick. (*Courtesy James R. Kearney Corp.*)

FIGURE 24.24 Lineman is installing Top-Tie to secure energized conductor to top of pin-type insulator. The pole is covered with insulated pole guard to provide safe working space. Lineman places unmarked leg of tie in a clamp stick one pitch in from the end and places tie on insulator and conductor. (*Courtesy A. B. Chance Co.*)

FIGURE 24.25 Lineman uses hot stick with tie loop to wrap the marked tie leg to completion while holding the tie loop snug against the indicator with clamp stick. (*Courtesy A. B. Chance Co.*)

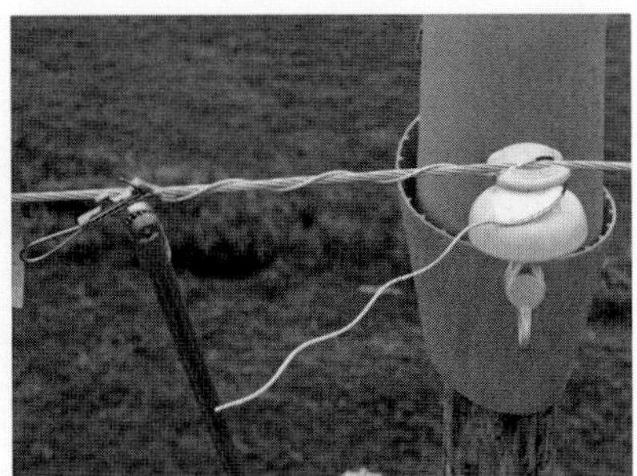

FIGURE 24.26 Lineman snaps marked leg of tie securely into place with tie loop on hot stick. (*Courtesy A. B. Chance Co.*)

FIGURE 24.27 Lineman slips the second tie of the set between the wrapped and unwrapped legs of the first tie using clamp stick on the marked tie leg. The marked leg must remain on top of the conductor parallel to the previously installed leg. The tie wires must not cross. A clamp stick is used to hold conductor in position on top of insulator. (*Courtesy A. B. Chance Co.*)

FIGURE 24.28 Lineman removes clamp stick from marked leg and places it on the unmarked leg, holding it while marked leg is wrapped to completion with tie loop. (*Courtesy A. B. Chance Co.*)

FIGURE 24.29 Lineman applies unmarked leg of second prefabricated tie wire with tie loop. (*Courtesy A. B. Chance Co.*)

FIGURE 24.30 Lineman wraps remaining unapplied leg of first tie around energized conductor. (*Courtesy A. B. Chance Co.*)

FIGURE 24.31 Lineman has completed installation of prefabricated tie securing conductor to insulator. Clamp stick holding conductor can now be removed. (*Courtesy A. B. Chance Co.*)

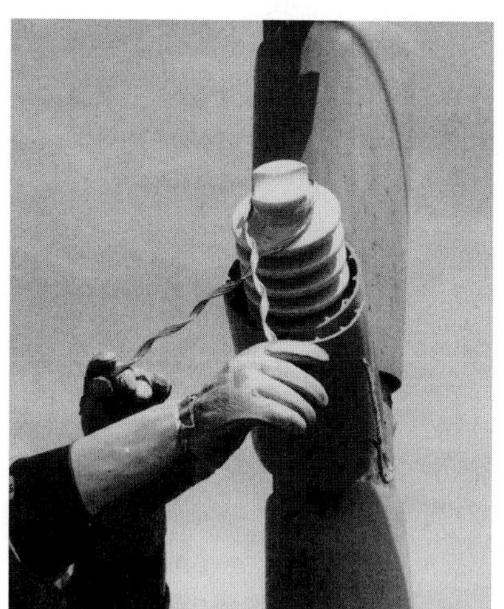

FIGURE 24.32 Lineman places tie over insulator side groove before energized conductor is lowered onto insulator side groove. Note plastic cover-up material on insulator support and pole. (*Courtesy A. B. Chance Co.*)

FIGURE 24.33 Lineman rotates tie on insulator with tie loop on hot stick so legs of tie point skyward. The conductor supported by a clamp stick is lowered onto insulator side groove with conductor between tie legs. (*Courtesy A. B. Chance Co.*)

FIGURE 24.34 Lineman applies left leg to one pitch with tie loop on hot stick and holds tie while applying the right-hand leg. (*Courtesy A. B. Chance Co.*)

FIGURE 24.35 Lineman completes application of left leg of tie after making sure right leg is snapped into place. (*Courtesy A. B. Chance Co.*)

FIGURE 24.36 Lineman has completed application of prefabricated tie. Clamp stick supporting conductor and temporary protective equipment can now be removed. (*Courtesy A. B. Chance Co.*)

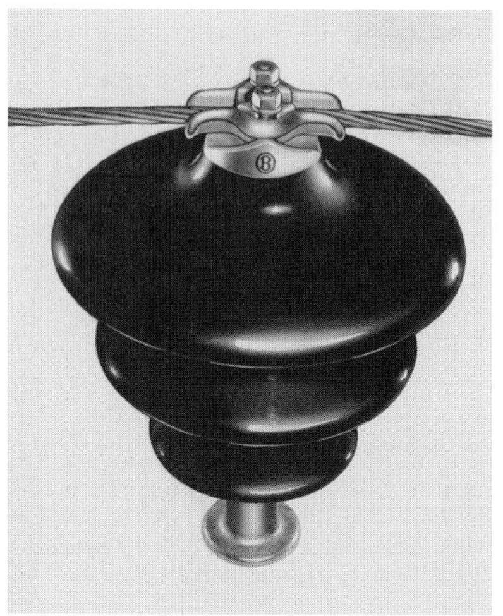

FIGURE 24.37 Clamp-top type of pin insulator using a clamp instead of the usual tie wires to hold the conductor in position. (*Courtesy Ohio Brass Co.*)

FIGURE 24.38 Unscrewing the two nuts that hold the keeper piece over the conductor. Note that the lifting and holding tools are already attached to the conductor and secured to the pole. (*Courtesy James R. Kearney Corp.*)

FIGURE 24.39 Ratchet wrench and socket for loosening the nuts on clamp-top bolts. (*Courtesy James R. Kearney Corp.*)

FIGURE 24.40 Cradle for lifting the clamp, a suitably formed wire attached to a standard hot stick. (*Courtesy James R. Kearney Corp.*)

FIGURE 24.41 Lifting the clamp from the supporting pintles with the wire cradle. (*Courtesy James R. Kearney Corp.*)

FIGURE 24.42 With the clamp slid away from the clamp-top seat, the hot conductor can be moved away from the insulator as shown. (*Courtesy James R. Kearney Corp.*)

clamp-top seat, the hot conductor is free and can be moved to any desired position to clear the insulator. The completed operation is shown in Fig. 24.42.

For attaching the conductor to the insulator, the reverse procedure is followed.

CHANGING SUSPENSION AND STRAIN INSULATORS

The principal parts of a steel-tower line are the steel tower, steel crossarm, insulator string, and conductor. Suspension insulators are installed on tangent (in-line) towers. At crossings, angles, and dead-ends, strain insulators are used instead of suspension insulators. The most frequently performed live-line maintenance work on tower lines will be the replacement of suspension or strain insulators. The steel towers, steel crossarms, and conductors seldom fail.

In order to replace any defective units in the insulator string, the weight of the conductor must be taken off the string. This is accomplished by the use of the strain carrier. By drawing up on the strain carrier, the insulator string is relieved of the conductor load and can then be disconnected from the conductor clamp. The lineman is then free to remove and replace one or more of the insulator disks. When this is accomplished, the conductor load is transferred back to the insulator string and the strain carrier is removed. These steps can be listed as follows:

1. Attach strain carrier
2. Draw up on strain carrier
3. Disconnect insulator string
4. Replace defective insulator units
5. Reconnect insulator string

6. Release tension from strain carrier

7. Remove strain carrier

Clearing Working Space. Before proceeding with the work, the lineman should make sure that sufficient working space is available and that the required clearance from live conductors can be maintained. If protective gaps are in the way, they should be pushed out of the way or removed temporarily. Jumpers around dead-end strings also should be pushed out of the way and held away by use of a holding stick in the regular manner.

REPLACING SUSPENSION INSULATORS

Wire-Tong Method. In many cases, especially on light-pole lines, the suspension insulators may be replaced with the use of wire tongs and other hot-line tools.

The procedure is similar to that used with pin-type insulators. The conductor and clamp are secured with a lifting tong. The assembly is then pushed away from the pole or tower a distance of about 6 in. The lifting tong is then secured in its saddle clamp. The holding tong is fastened to the conductor, and the conductor is then pulled back to the pole or tower. This will cause the conductor to be raised slightly and thus transfer the conductor load onto the lifting tong. The insulator string will now hang limply. The insulator string may now be disconnected from its clamp and the line conductor moved farther out of the way. A link stick may be used in pulling the conductor away or in pulling it back, but the holding tong should always be used as it will help to stabilize the conductor in its position away from the pole or tower. After the defective insulator units are replaced, the conductor is pulled back in and the string is reconnected to the conductor clamp. All tongs and other tools are then removed in reverse order of their installation.

Strain-Carrier Method. The strain carrier is placed in position vertically adjacent to the insulators. As the suspension string hangs vertically, the strain carrier also hangs vertically. A cradle for supporting the insulator string will usually not be required. The fixed end of the carrier is hooked onto the conductor on opposite sides of the clamp. The adjustable or take-up end is fastened to the crossarm, allowing enough slack in the tightening arrangement for taking the strain off the insulator string.

After the strain is taken off the insulator string, the insulators hang limply and may be disconnected from the strain clamp. This requires pulling out a cotter key or pin. This is performed with a cotter-key puller, which consists of a special cotter-key pulling hook fastened to the end of another hot stick. If the cotter key is hard to pull out, the conductor load on the string should be further reduced by additional tightening of the strain carrier.

With the insulator string now free from the clamp, the string can be removed and the defective insulator units replaced. It has been found preferable in most cases to remove the entire string of insulators rather than to replace the insulator disks singly. This is especially so when the insulator string is short, consisting of only two or three disks. The insulator string may be removed by using a shepherd hook, fork stick, lifting yoke, sling, or other suitable device.

After the insulators are replaced, the string is reconnected to the clamp. The cotter key is installed with a special cotter-key replacer. The cotter key is spread by means of a screwdriver fastened to the other end of the cotter-key-replacer hot stick. The strain is now transferred to the insulator string by letting off on the strain carrier. The carrier and other tools should then be removed in the reverse order of their installation.

REPLACING STRAIN INSULATORS

The strain carrier is first placed in position as with suspension insulators. Since the strain insulator strings are somewhat horizontal, the carrier can be slid over the insulators out to the line conductor. The fixed end of the carrier is attached behind the strain clamp on the span side of the clamp. The adjustable or take-up end of the carrier is fastened to the pole, crossarm, or tower, allowing enough slack in the tightening arrangement for taking the strain off the insulator string. The cradle for supporting the slack string of insulators is then placed in position by hanging the adjustable hook over the conductor and fastening the other end to the crossarm. When the insulator string is short, that is, consists of only two or three disks, the supporting cradle need not be used, as the insulators can then be handled by means of a yoke arrangement on the end of a hot stick.

The next step is to take the conductor load off the insulators. This is accomplished by drawing up on the strain carrier. In the buck-saw type of strain carrier, this is done by pulling together the ends of the two hinged members using a set of blocks. In the cradle type of strain carrier, the conductor load is transferred to the carrier by taking up on the jackscrew at the rear of the two tension members.

After the strain is taken off the strain insulator string, the string may be disconnected from the clamp. The defective insulator disks may now be replaced and the complete string reconnected to the clamp. The conductor load is then returned to the strain insulator string by letting off on the strain carrier. All hot-line tools should now be removed in reverse order of their placement.

PULLING SLACK

String Insulator. The same general procedure is followed when pulling or removing slack from a conductor when the line is energized as is used when the line is deenergized. When the lines are energized, properly insulated tools and insulated links must, of course, be used. The order of work is as follows: The pulling grip or come-along is attached to the conductor with a tie stick, holding stick, or other hot stick. An insulated link stick is next attached to the pulling eye of the come-along, pole, crossarm, or tower. A set of blocks or a hoist is placed between the end of the link stick and the pole, crossarm, tower, or conductor (Fig. 24.43). The conductor is then secured by taking up on the set of blocks or hoist. After the conductor is thus secured, the clamp holding the conductor to the insulator string is loosened by unscrewing the nuts on the holding bolts by means of a socket wrench using hot-line procedures. The slack is then taken up by pulling further on the blocks or the hoist. When sufficient slack has been removed, the conductor clamp is tightened. If all the unwanted slack cannot be pulled up in one operation, the foregoing procedure is repeated until the desired conductor sag is obtained. When sufficient slack has been pulled up, the blocks or hoist, insulated stick, and pulling grip are removed. The pulling grip is loosened by tripping the releasing lock with a tie stick or other hot stick.

Pin Insulator. The same procedure is followed when pulling slack in a line supported on pin insulators. Instead of unscrewing the nuts on the conductor clamp, the tie wires holding the conductor to the insulator are loosened but not removed. Then the slack is pulled up and held while the insulator ties are remade.

CUTTING OUT CONDUCTOR

If a short length of conductor needs to be cut out of the line, the slack can be pulled in both directions from the same pole or tower. The extra conductor may then be cut out and the conductor spliced and then refastened to the insulators.

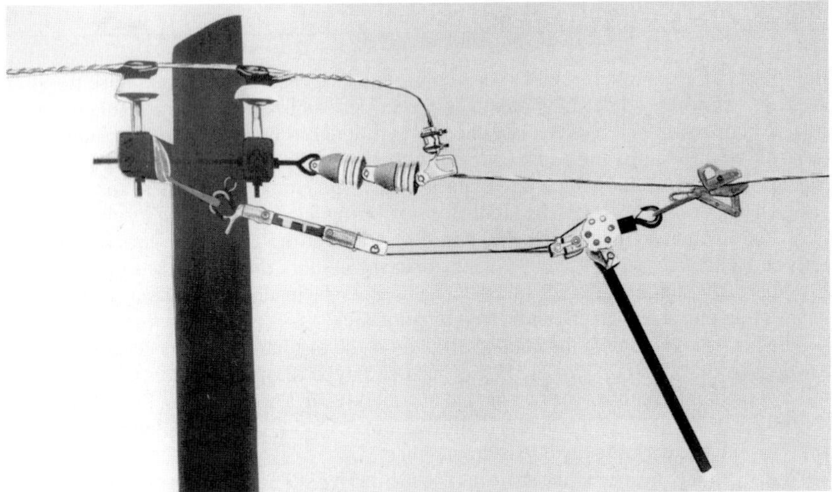

FIGURE 24.43 A hoist with a nylon strap and an insulated handle is connected to an insulated link stick so that slack can be pulled out of a 15-kV distribution-line conductor. (*Courtesy A. B. Chance Co.*)

In order not to interrupt the flow of current in the line when the conductor is cut, a jumper is placed around the section where the conductor is to be cut out. However, before the ends of the jumper are tapped onto the conductor, the jumper must be firmly held away from the lineman by means of a holding tong or stick securely fastened to the pole or tower. An insulated link stick with rope also may be used for this purpose. The ends of the jumper are then clamped onto the live conductor. The desired length of conductor is cut out of the line, and the ends are spliced. After splicing is completed, the jumper is removed.

If the line conductor is a stranded cable, extreme care must be taken to prevent the cable from unraveling and allowing the wires to get out of control.

SPLICING LIVE CONDUCTORS

The procedure for splicing a hot conductor with live-line tools is as follows: The conductors to be spliced are pulled tight with blocks, using link sticks on the come-alongs for insulation. The conductors are connected together with a temporary jumper to bypass the current while the splice is completed. The ends of the conductors are held in one position with holding sticks. The splicing connector is next fastened to the end of one conductor. Then the end of the other conductor is brought over and placed in the connector. The connector is now firmly fastened, compressed, or fired, completing the splice. Insulated tools, of course, are used for all the operations.

TAPPING LIVE LINE

Taps are easily made to live connectors by the use of appropriate equipment. A stirrup is installed on the conductor to be tapped, as discussed and illustrated in Chap. 23. The conductor to be tapped to the live line is fastened to the body of a special hot-line tapping clamp

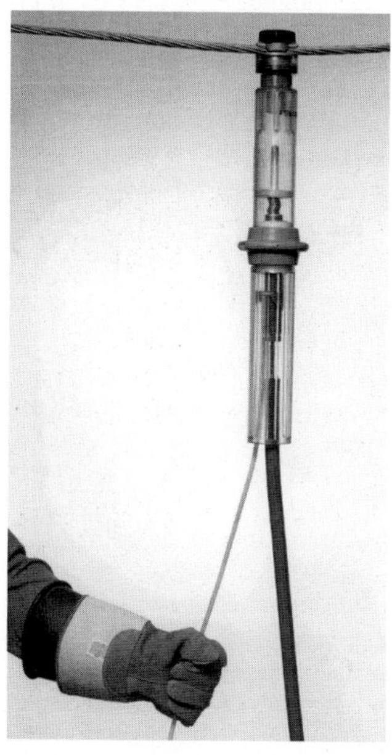

FIGURE 24.44 Load-pickup tool designed for 15-kV, 200-amp operation. The tool has a one-way mechanism design which precludes its use to interrupt load current. (*Courtesy A. B. Chance Co.*)

and is then held with a hot stick and hooked over the live-line conductor stirrup. The clamp is then tightened by turning the screw head of the clamp using the hot stick.

Hot-line clamps are not designed to pick up load current or fault current. If there is load on the line being energized or if there is any doubt as to the condition of the insulation of the line being energized, the line should be energized by a switch designed to energize circuits, carry load current, and withstand the forces associated with fault current.

Load-pickup tools are available to energize circuits operating at voltages of 15 kV or below, which will carry 200 amps of load current (Fig. 24.44). Once the load has been energized successfully through the load-pickup tool, the conductor is connected to the main circuit with the hot-line clamp using a hot stick, and the load-pickup tool is removed.

APPLYING VIBRATION DAMPERS

Vibration dampers of different types may be applied to lines while energized using live-line tools for their installation.

A special tool simplifies the installation of the Stockbridge type of damper on live conductors (Fig. 24.45). The damper is placed on the head of the tool, as shown in Fig. 24.46. A spring holds the damper clamp in place whether it is open or closed. With the damper clamp in the open position, the damper is placed on the live conductor. The locking nut on the bottom of the clamp, which has previously been placed in a socket, is tightened by turning the hot stick. The tool is removed by pulling down on the hot stick, compressing the spring, and turning the stick a quarter turn to engage the socket in a locked position on a projecting lug on the head of the tool. The tool may then be easily lifted away. A ring is provided on the side of this tool for attaching the link stick and hand line or blocks for aid in handling heavy dampers of this type.

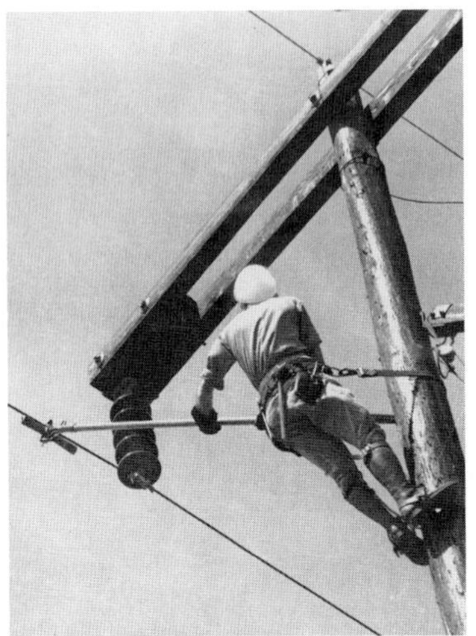

FIGURE 24.45 Applying Stockbridge vibration damper to a live line with a special tool. (*Courtesy A. B. Chance Co.*)

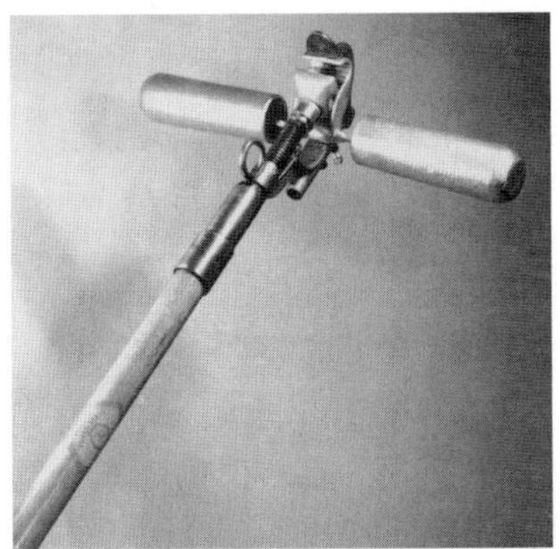

FIGURE 24.46 Vibration damper placed in a special tool preparatory to fastening a live-line conductor. (*Courtesy A. B. Chance Co.*)

APPLYING ARMOR RODS

In order to install armor rods, it is first necessary to remove the conductor from the insulator in the usual manner, and move it with wire tongs to a suitable working position. The conductors are supported on either a clamp or a holding stick. The armor rods are then applied with a special tool (Fig. 24.47). The tool consists essentially of two split wheels into which different size dies are set. The armor rods are placed into these dies, and the assembly is placed around the conductor by means of holding or clamping sticks. The wheels or

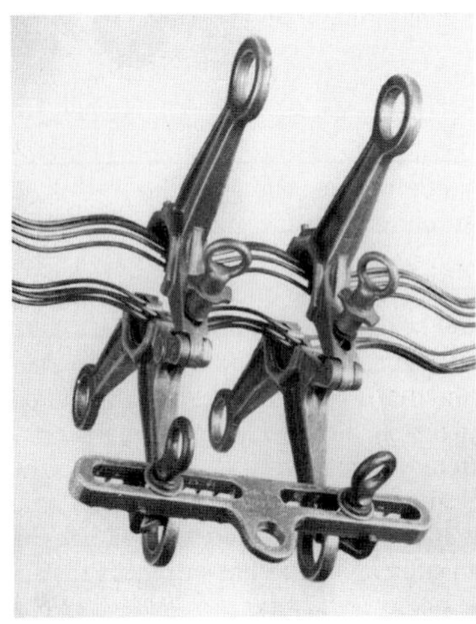

FIGURE 24.47 Special armor-rod wrenches and wrench holder used to apply armor rods while line is live. (*Courtesy A. B. Chance Co.*)

holders are rotated in the proper direction depending on the lay of the wire, one holder moving clockwise and the other counterclockwise. Rotation of the holders is accomplished by means of hooks fitted onto hot sticks engaging lugs on the holders. At the same time the holders are rotated, they are moved apart, starting near the center of the armor rods and moving toward the ends. The rotary movement of the holders is continued until the armor rods are completely and tightly twisted into place. The holders are then pushed off the rods and removed by means of clamp sticks or other hot sticks.

Armor clips are then inserted on each end of the armor rods and clamped into place by tightening the bolt nuts. Special spring sockets mounted on the ends of hot sticks are used to tighten the nuts, the spring being used to thrust the saddle into place as soon as the two halves of the clip are in alignment.

PHASING OUT

Phasing out is absolutely necessary when a new line is to be paralleled with another line, new or old, and after repairs or changes have been made on either of two lines which have previously operated in parallel. It is necessary in the latter case because of the possibility of interchanging conductors when making repairs or changes.

The process of phasing out consists of determining whether the phases of a given line or apparatus correspond with the phases of another line with which it is to operate in parallel. This problem arises most frequently at sectionalizing switches. The voltage across corresponding lines or phases should be zero. Therefore, to determine if zero voltage exists across corresponding lines, it is necessary to read the voltage. This is done with a phasing-out voltmeter.

Phasing-Out Voltmeter. A phase tester for use on 15-kV circuits is illustrated in Fig. 24.48. The tester consists of two high-resistance units connected in series with a milliammeter that is calibrated to read voltage in kilovolts mounted on Epoxiglas housings. The resistors and instrument are connected in series with 20 ft of 15-kV insulated flexible cable coiled on a reel. Universal insulated hot sticks are fastened to the resistor housing to provide the lineman with isolation from the instrument. Only the necessary length of cable should be unreeled to avoid contact with the insulated cable or the cable with a ground that could distort the reading. Resistor units can be attached to the basic phase tester to extend the voltage range of the instrument (Fig. 24.49).

Phasing-out Procedure. Circuits to be connected together or connected in parallel must first be checked for proper phasing to prevent short circuit. The phase tester should be connected between all phases of each line and/or between each phase and ground of both lines to be sure that both lines are energized and that the tester is indicating properly. The tester is then connected between one phase of one line and a phase of the other line to be sure the circuits are properly coupled. If the circuits are properly coupled, the voltmeter will have reading equal to line-to-line voltage. If both lines are energized and coupled together, the tester is connected from a phase of one line to each phase of the other line in turn to determine the corresponding phases of each line. The conductors that have zero voltage between them are in phase and can be connected together for parallel operation. Figure 24.50 shows the linemen reading the voltage on a phase tester connected between a phase conductor of one line and a phase conductor of another line. The tester should be removed from the lines as soon as the readings are completed.

Since the operation of high-voltage paralleling is usually done through a circuit breaker, the customary procedure in phasing out is to open the disconnects on one side of the circuit

FIGURE 24.48 Lineman holding 15-kV phase tester. Wire hooks on resistor units provide a contact surface. Universal hot sticks are bolted to the two units before the tester is used by the lineman. (*Courtesy MidAmerican Energy Co.*)

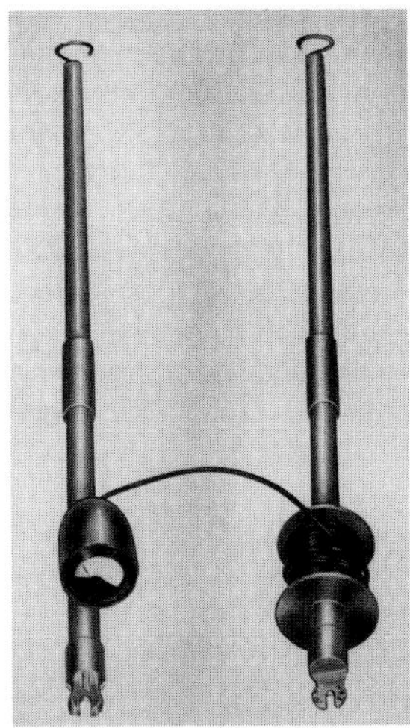

FIGURE 24.49 Resistors have been attached to a basic 15-kV phase tester to increase the voltage range of the instrument to 75 kV. (*Courtesy A. B. Chance Co.*)

breaker and then close the circuit breaker so that the corresponding blades and contact jaws of the open disconnects are alive from the two sources that are to be paralleled. The instrument is checked for proper operation by connecting it between phases on each side of the open disconnect switch. This procedure verifies that both circuits to be phased out are energized if the instrument works correctly. The phasing-out voltmeter is then used to detect the presence of any difference in potential between a blade and the corresponding contact jaw of each disconnect.

FIGURE 24.50 The tester units are attached to hot sticks. The contact hooks are placed over the conductors, and the cable is kept taut. (*Courtesy A. B. Chance Co.*)

If the voltmeter gives no indication as the two leads are successively applied to each pair of the three disconnect terminals, then no potential exists between each jaw and blade. This shows that the correct phases are connected to each of the two parts of the respective disconnects. A voltage read between the jaw and hinge of one of the disconnect switches indicates that the circuits are out of phase. The proper phasing can be determined by checking between the jaw of one switch and the hinge of the other switches. The conductors with zero voltage between them have the same phase relationship. The phase conductors must be connected to provide the proper phase relationship before the circuits are paralleled. If no voltage is read across the open disconnects, the high-voltage voltmeter should be connected from the jaw of one switch to the blade of an adjacent switch, verifying that the circuits are similar. Ungrounded neutral or three-wire, three-phase circuits without adequate capacitive coupling will require the use of two high-voltage voltmeters simultaneously on adjacent phases to obtain a proper reading.

Similar tests can be completed for proper phasing with a low-voltage meter if potential transformers or potential devices are connected to the lines on both sides of the separation between circuits. The voltmeter leads are connected to the low-voltage terminals of the potential devices in the same sequence as described for the high-voltage tester on energized primary lines. Transmission-line circuits are usually phased out in substations using the low-voltage terminals of potential devices installed in the substation for metering and relaying circuits.

Insulator Testing. All live-line testing requires a knowledge of the voltage distribution over the units of a suspension string or across the shells of a multipart pin insulator. Unfortunately, this distribution is not uniform, which would make the voltage across all units the same. Instead, the distribution is quite uneven, that is, the voltage across the unit nearest the line conductor is several times as great as the voltage across the unit nearest the crossarm or support. In some cases, the voltage across the unit nearest the conductor will be five times as much as across the unit next to the support.

Now, if for any reason one or more units becomes defective, the distribution curve becomes distorted.

Therefore, if the voltage across each unit of a string can be conveniently measured, defective units can be located easily. Any unit whose voltage reading is lower than the reading for a like unit on a string in good condition would show that the former unit is not withstanding its share of the voltage drop and is therefore defective. Actually, if a unit is very defective, the voltage drop across it will approach zero, as the broken-down insulation tends to act as a conductor. Ordinarily, if a reading is less than 60 percent of the average reading, the unit is probably defective.

Insulator Tester. One make of tester employing a meter for reading the voltage distribution is illustrated in Fig. 24.51. The meter on which the voltage is read is mounted on the end of the insulated hot stick. Although this meter actually is an ammeter, being actuated by a condenser current, its readings are proportional to the voltage; therefore, it may be looked on as a voltmeter.

In the case of disk-type insulators in a string or insulators in a stack, the two prongs are touched to the metal portions on both sides of the insulator, and readings are taken on the meter. Since the two prongs actually are the leads from the voltmeter, the readings give the voltage drop across the insulator. These readings are then compared with those for the corresponding units on a good insulator string. If any of the readings are much lower, the units are defective.

In the case of multilayer pin-type insulators, the prong is inserted between the porcelain layers until it makes contact with the cement. This prong is one lead from the voltmeter. The other lead from the meter connects to a condenser made of several metal tubes located

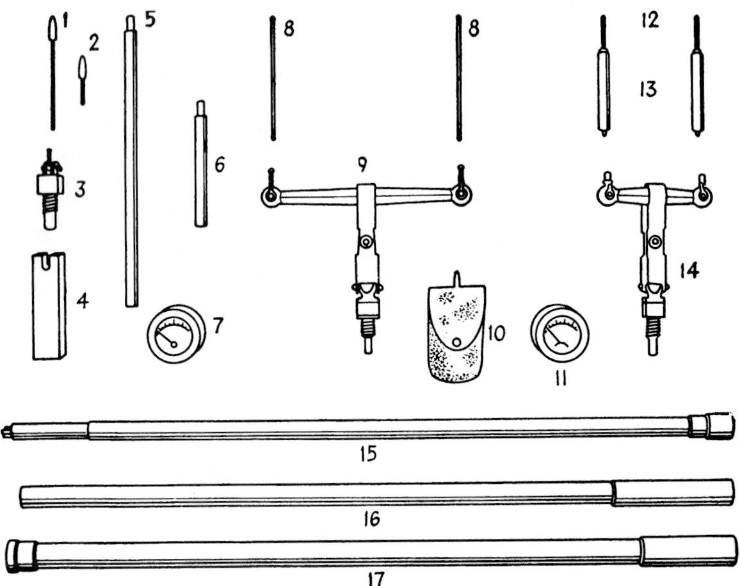

FIGURE 24.51 Parts of meter-type live-line insulator tester.

in the hollow of the stick. The metal tube is one plate of the condenser, and the earth is the other plate. The condenser is made variable by providing two metal tubes of different length. The longer one is used at all times unless the reading goes off the scale. Then the shorter one is used. On the other hand, if the reading is too small and needs to be increased, both tubes may be joined and used.

On a two-piece insulator, a reading of the upper part will give the condition without testing the bottom part. If the upper part is bad, the meter reading will drop. If the lower part is bad, the meter reading on the upper part will be above normal.

This type of insulator tester is made with different type heads and a variety of contact tips to fit the many types of insulators in use. The holding stick is furnished in three sections for convenience in reaching the insulator.

The phasing-out high-voltage voltmeter can be used to complete the tests, following the same procedure as described for the insulator tester, if the voltages are within the range of the voltmeter capacity.

A tool has been developed to easily test insulator strings as shown in Fig. 24.52. The tool is lightweight and easy for the lineman to handle with a hot-stick as it is slid along the insulator string.

Hot-Line-Tool Care. The outside surface of the insulated stick portion of hot-line tools must be kept clean and dry. The trailers used to store and transport hot-line tools are usually equipped with electric heaters to reduce the humidity and prevent condensation. The surface of the sticks should be inspected before they are used. Tools that are battered, scratched, or dirty should be maintained before they are used. Hot-line tools are factory-tested by the manufacturer at 100,000 volts per foot. Many companies have laboratories for testing protective equipment and hot-line tools. The insulated stick portion of the hot-line tools is normally tested at 50,000 volts per foot after delivery to the customer. The tools can be tested in the field using a portable hot-stick tester (Fig. 24.53).

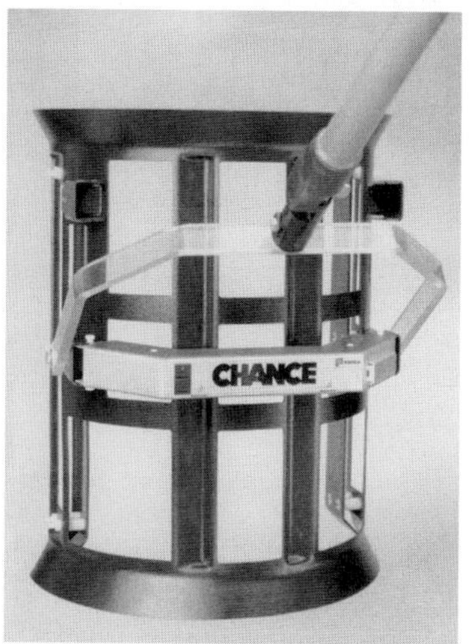

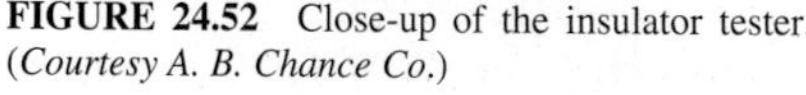

FIGURE 24.52 Close-up of the insulator tester. (*Courtesy A. B. Chance Co.*)

FIGURE 24.53 Portable hot-stick tester. (*Courtesy A. B. Chance Co.*)

The tester is energized from a 120-volt ac source. The meter is set to zero without a stick in it while energized to eliminate the effect of leakage current. A test bar is used to ensure proper operation (Fig. 24.54). The tester puts 1800 volts across the two center electrodes with the two outer electrodes grounded. The outer electrodes shield the inner electrodes to prevent the lineman using the tester from affecting the meter reading while the tester is moved along the surface of the insulated portion of the hot-line tool (Fig. 24.55).

FIGURE 24.54 Test bar being used to check operation of tester. (*Courtesy A. B. Chance Co.*)

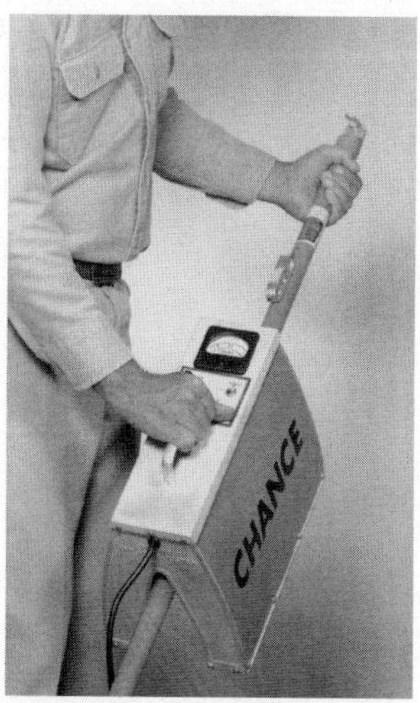

FIGURE 24.55 Lineman using portable tester to check insulation of hot-line tool. Tester can be used to check all kinds of insulated sticks regardless of the material. (*Courtesy A. B. Chance Co.*)

FIGURE 24.56 Tomcat being used to change an insulator string on a 345-kV energized transmission line. Tomcat is operated by a lineman sitting in the van adjacent to the transmission-line structure.

Robotic Hot-Line-Tool Live-Line Maintenance. Tomcat, a remote-controlled manipulator arm, is being used to complete live-line maintenance (Fig. 24.56). Tomcat (teleoperator for operations, maintenance, and construction using advanced technology) was developed by Philadelphia Electric Company and the Electric Power Research Institute. Tomcat is

FIGURE 24.57 Tomcat being used to install strain carrier adjacent to energized 345-kV insulator string.

FIGURE 24.58 Tomcat has completed installation of strain carrier and disconnected the insulator string, and the insulators are being lowered to the ground by a groundman. New insulators will be raised to the structure by the groundman, and Tomcat will be used to complete the replacement.

operated by a lineman from a position in a van like the one shown in Fig. 24.56. The lineman uses closed-circuit television and a master-slave control to perform hot-line maintenance work. Fiberoptic cable between the manipulator and the equipment in the van provides the communication link and electrical insulation from the energized line. Figure 24.57 shows Tomcat being used to install a strain carrier on an energized 345-kV line prior to changing out a string of suspension insulators. Figure 24.58 pictures the insulator string being lowered to the ground by a lineman or groundman after Tomcat has completed the installation of the strain carrier and disconnected the insulator string from the line and its support on the structure. Tomcat can be programmed to complete certain operations automatically, which permits it to function robotically.

CHAPTER 25
LIVE-LINE MAINTENANCE FROM INSULATED AERIAL PLATFORMS

The *IEEE Guide for Maintenance Methods on Energized Power-Lines* (IEEE Standard 516) provides detailed information concerning work by linemen from insulated aerial platforms. The procedures discussed in the standard should be followed while performing work on energized power lines.

Insulated Aerial Lifts. The American National Standard entitled, *Vehicle-Mounted Elevating and Rotating Devices* (ANSI Standard A92-2) provides information on insulated aerial devices commonly used by linemen and cablemen. Mobile extension ladders have long been useful for tree trimming and for the replacement of burned-out lamps in street lighting fixtures. The insulated aerial platform has replaced the insulated extension ladder for most operations. The mobile insulated aerial platform has an insulated section in the upper portion of the boom assembly and an insulated insert in the lower portion of the lifting boom (Fig. 25.1). The fiberglass-insulated section of the upper portion of the boom provides protection from ground for the lineman and the insert near the mounting mechanism for the boom provides isolation for a groundman if the metal in the boom assembly accidentally contacts an energized conductor. The mobile insulated aerial platform is an improved device for street-lighting maintenance and a safer platform for tree trimming near energized lines, but its greatest field of usefulness is for live-line maintenance (Figs. 25.2 and 25.3). It can serve as a platform from which to perform live-line operations in three ways:

1. Using rubber gloves on distribution lines
2. Using conventional hot-line tools on distribution and transmission lines
3. Using bare hands on medium- and high-voltage transmission lines

For tree trimming or replacement of street lamps, the platform usually consists of a single bucket or basket with room for only one lineman, while for live-line maintenance, single and double buckets are employed.

Use of the aerial platform is limited only by accessibility to the work area near the overhead lines. As access is sometimes impossible, the aerial platform can only supplement regular line crews using conventional pole climbing procedures.

Description of Insulated Aerial Lift. The lifting-boom pedestal is placed behind the truck cab. Outriggers from the pedestal base are used to stabilize the boom when it is extended. The boom and baskets are positioned over the cab while traveling. The upper member of the boom and the baskets are made of fiberglass to provide the required insulation

FIGURE 25.1 Aerial lift trucks set-up for a field exhibition. The vehicles have the outriggers extended to stabilize the vehicle. The single bucket aerial device is extended away from the truck. The insulated sleeves in the upper and lower portions of the lift device boom are visible.

FIGURE 25.2 Two lineman in a double-bucket truck preparing for a field demonstration.

FIGURE 25.3 Two lineman working in a double-bucket truck in a substation.

from ground. A fiberglass section may be installed in the lower boom to provide protection for groundmen. All movements of the lift are controlled by the lineman in the basket. Mobile aerial platforms, baskets, or bucket trucks used for live-line bare-handed work should be tested and specially certified for the voltage of the line to be maintained. The illustrations to follow depict the various features mentioned.

Maintenance with Rubber Gloves. Live-line maintenance from an insulated platform with rubber gloves is limited to the distribution voltages. Some electric companies work on energized lines with voltages through 20 kV to ground or 34.5 kV phase-to-phase; their employees use rubber gloves and sleeves from an insulated aerial bucket or an insulated platform clamped to the structure (Fig. 25.4). It is good practice to test the mobile insulated aerial platform or bucket truck in the field each day before the work is started and each time the assembly is to be used on a higher distribution voltage or if there is some question about the insulating conditions. The insulating arm of the bucket truck can be tested by raising the bucket with all personnel at ground level until the bucket contacts a conductor energized at the voltage to be worked. Contact should be maintained for 3 minutes or more. If the insulation of the assembly does not give indication of excessive leakage current or electrical failure, the work can proceed. Figure 25.5 shows two linemen in an insulated aerial double basket on a 15-kV line using rubber gloves and sleeves. An insulated boom extension on the aerial-lift device is being used to support the conductor while the insulator is being replaced. Aerial-lift platform accessories improve the safety and the efficiency of the linemen performing energized-line maintenance work. Other operations that are usually completed with rubber gloves and rubber sleeves are installing transformers, switches, and cutouts, replacing insulators, repairing conductors, and inspecting equipment. Work on distribution circuits operating at 12,500Y/7200 volts or higher requires the use of insulated aerial devices or insulated platforms to supplement the rubber gloves and sleeves. Rubber gloves and sleeves must be manufactured and rated for a voltage greater than the distribution voltage to be worked. Distribution voltages lower than 12,500Y/7200 volts may be worked using rubber gloves while the lineman's gaffs are engaged in wood poles.

FIGURE 25.4 Lineman working from an insulated platform mounted on the pole. (*Courtesy A. B. Chance Co.*)

FIGURE 25.5 Rubber-glove work demonstrated on 15-kV line installing a conductor jumper.

Maintenance with Hot-Line Tools. Live-line maintenance from an insulated aerial platform using hot-line tools is more conveniently and more rapidly done than when working from the pole or tower. The ease of reaching elevated positions and the reduction in the required protective covering, as well as the reduced number of linemen required, favor this method. Sometimes a combination of linemen working both from the platform and the pole or tower is preferred.

Maintenance Using Bare-Hand Method. The insulation of mobile aerial devices and fiberglass ladders to be used for bare-hand live-line work should be tested each time they are used in the field, before the electrical work is started. The devices can be tested by contacting an energized conductor to be worked on with the device and reading leakage current to ground at the ground location. Contact should be maintained for 3 min. The leakage-current readings should remain substantially constant. If the leakage current increases or decreases by more than 20 percent from the initial reading, a longer test period should be used and the cause of the variation should be identified and corrected before the bare-hand electrical work is started. ANSI Standard A92.2 provides detailed testing information. When live-line work is to be done bare-handed from an insulated aerial platform, the lineman in the basket must be at the same potential as the live conductor on which the repair is to be made. This is accomplished by bonding the metal-mesh lining of the basket to the live-line conductor. The mesh lining, also referred to as *basket shielding*, is brought to line potential by means of a bonding lead, one end of which is permanently attached to the shielding while the other is clamped onto the energized conductor as the basket approaches it. A short hot stick is used to fasten the metal clamp. One lead is employed for each lineman. Since the two leads are electrically interconnected, the second lead serves as a safety measure in case one is accidentally knocked off. If both were removed accidentally, the lineman would receive a charging-current shock each time he touched or removed his hand from the energized conductor.

The linemen, wearing conductive suits if working on EHV or UHV lines, are also “bonded” to the shielding by means of a spring-type ankle clamp or by the wearing of conductive-soled shoes.

While the linemen and their shielding are thus bonded to the energized conductor, the linemen are maintained at conductor potential. They can move around in the baskets and touch the conductor and any metal accessories which are at the same potential without having any physical awareness of the fact that they are charged at line potential. As long as the lineman is at the same potential as the conductor he is working on, and as long as he is insulated from ground, no current can flow from conductor to man. The lineman can work bare-handed on the conductor or on any metal fittings at the same potential as the conductor without experiencing any discomfort. The rule is, “Touch only that which you are clipped to.” Work is completed safely on transmission lines using proper equipment and procedures.

In many locations it is impossible to position an insulated bucket truck at the work site. Bare-hand live-line work can be performed by linemen in special conducting suits working from special insulated ladders (Figs. 25.6 through 25.8). The conducting suit is connected to the energized conductors.

Touching live conductors with bare hands when insulated from ground is similar to the contact birds make when resting on live conductors, except that the birds would call it the barefeet method. The birds apparently experience no uncomfortable effects or they would leave the conductor. Of course, the birds are thoroughly insulated from ground, since they require no auxiliary support.

Live-line maintenance is completed on EHV circuits constructed with steel towers. The procedures for performing work on 345-kV and 765-kV energized lines are similar. Linemen performing bare-hand maintenance on energized EHV circuits wear a specially designed conductive suit. The conductive suit has a steel mesh that equalizes the charge around a

FIGURE 25.6 Lineman being positioned from the aerial lift bucket truck to work from insulated ladders on a 345-kV transmission line.

lineman's body as he approaches the energized conductor from an insulated platform. The metal-mesh suit acts like a Faraday cage, shielding the lineman from the electric field. The lineman contacts the energized line conductor with a wire or wand that connects to the conductive suit. The wand draws an arc as it approaches the energized conductor. The connection between the wand and the conducting suit bypasses the charging current, eliminating any sensation the lineman would feel if he contacted the energized conductor directly with his bare hand before being clipped to the conductor. A crane with an insulated tension link stick

FIGURE 25.7 Lineman setting up to work from insulated ladders on a 345-kV transmission line.

FIGURE 25.8 Lineman working from insulated ladders on a 345-kV transmission line.

in series with the winch line on the crane is used to support the energized bus conductor while the bus support is replaced.

Live-line bare-hand maintenance is frequently performed with the aid of a helicopter (Fig. 25.9). The lineman is carried to the energized line conductor while sitting on a platform suspended below the helicopter. When the lineman approaches the energized line, he connects the metal platform below the helicopter and his conductive suit to the energized line conductor on which the work is to be performed. The lineman uses a portable two-way radio to communicate with the helicopter pilot. The helicopter makes it possible to perform work in areas where the transmission line is inaccessible for an insulated bucket truck.

FIGURE 25.9 Lineman replacing conductor spacer on 765-kV line using bare-hand techniques while being supported by a helicopter. The lineman wears a conductive suit that is clipped to the energized conductors. (*Courtesy American Electric Power Company, Inc. and Haverfield Helicopters*)

Protective Gaps. Linemen performing live-line maintenance work on EHV and UHV transmission lines may be working in the vicinity of insulator strings with a small number of insulators shorted out. The insulators may be shorted out because they are defective or as a result of the procedures necessary to complete the work. The dielectric strength of the air surrounding the insulators being worked on or between the lineman and the energized conductor or between the lineman in contact with the energized conductor and the transmission tower depends on the air humidity, temperature, density, and pressure and the shape of the electrodes separated by the air gap.

Protective gaps can be installed in the vicinity of the work being performed, usually on an adjacent tower, to protect the linemen performing bare-hand or hot-line-tool work. The gap distances are specified by the organization responsible for the facilities and the work being performed. The linemen performing the work then always maintain an approach distance greater than the gap distance to ensure their safety. The protective gaps ensure that a surge voltage on the energized transmission line will always cause a flashover at the temporary protective-gap installation instead of at the location where the linemen are performing the work.

CHAPTER 26
GROUNDING

Proper grounding for electrical systems is very important for the safety of the linemen, cablemen, groundmen, and the public. Correct operation of protective devices is dependent on the grounding installation. The *National Electrical Safety Code* (ANSI-C2 Standard) specifies the grounding methods for electrical supply and communications facilities. *The National Electrical Code* (NFPA-70 Standard) specifies the grounding methods for all facilities that are within or on public or private buildings or other structures.

The grounded static wires for transmission lines and the grounded neutral wire for primary distribution circuits originate at substations. The static wires for transmission lines will normally originate and terminate at a substation. The substation may be located at a generating station.

The substation grounding system governs the proper functioning of the whole grounding installation and provides the means by which grounding currents are conducted to remote earth. It is very important that the substation ground have a low-ground resistance, adequate current-carrying capacity, and safety features for personnel. The substation grounding system normally consists of buried conductors and driven ground rods interconnected to form a continuous-grid network. The surface of the substation is usually covered with crushed rock or concrete to control the potential gradient when large currents are discharged to ground and to increase the contact resistance to the feet of personnel in the substation. Substation equipment should be connected to the ground grid with large conductors to minimize the grounding resistance and limit the potential between the equipment and the ground surface to a safe value under all conditions. All substation fences should be constructed inside the ground grid and connected to the grid frequently to protect the public and workmen.

Ground Resistance. The electrical resistance of the earth is largely determined by the chemical ingredients of the soil and the amount of moisture present. Measurements of ground resistances completed by the Bureau of Standards are summarized in Table 26.1.

The resistance of the soils tested varied from 2 to 3000 ohms. The type of soil, the chemical ingredients, and the moisture level surrounding an electrode determine the resistance, as illustrated in Fig. 26.1. If the soil has uniform resistivity, the greatest resistance is in the area immediately surrounding the electrode which has the smallest cross section of soil at right angles to the flow of current through the soil. The area of the current path 8 to 10 ft from the electrode is so large that the resistance is negligible compared to the area immediately surrounding the ground rod. Measurements show that 90 percent of the total resistance surrounding an electrode is generally within a radius of 6 to 10 ft. A variation of a few percent in moisture will make a great difference in the effectiveness of a ground connection made with electrodes of a given size for moisture contents less than approximately 20 percent. Experimental tests made with red-clay soil indicated that with only 10 percent moisture content, the resistivity was over 30 times that of the same soil having a moisture content of about 20 percent (Fig. 26.2).

TABLE 26.1 Resistance of Different Types of Soil

Soil	Resistance, ohms		
	Average	Minimum	Maximum
Fills and ground containing more or less refuse such as ashes, cinders, and brine waste	14	3.5	41
Clay, shale, adobe, gumbo, loam, and slightly sandy loam with no stones or gravel	24	2.0	98
Clay, adobe, gumbo, and loam mixed with varying proportions of sand, gravel, and stones	93	6.0	800
Sand, stones, or gravel with little or no clay or loam	554	35	2700

Bureau of Standards Technologic Paper 108.

The soil resistivity also varies greatly with the temperature (Fig. 26.3). The water in the soil freezes when the temperature is below 32°F, causing a large increase in the resistance of a ground connection. Grounding electrodes which do not extend below the frost line where the soil freezes will have a large variation in resistance as the seasons change. The depth of the ground electrode is important to the electrical performance. Driven ground rods should be long enough to reach the permanent moisture level of the soil (Fig. 26.4). Soil is seldom of uniform resistivity at different depths. The first few feet of soil near the surface normally has a relatively high resistance because it is subject to changes in moisture content as a result of rainfall.

The ground rod (Fig. 26.5) should be driven deep enough to be in perpetually moist earth. In most localities, a depth of 8 to 10 ft is usually necessary to meet this requirement. A ground rod that penetrates perpetually moist earth will have a fairly low resistance. Low values of resistance are desirable and essential for the proper functioning of lightning arresters. The ground resistance can be measured as the ground rod is driven, as shown in Fig. 26.5. The ohmmeter shown resting on a cable reel gives a continuous indication of the ground resistance as the rod is driven into the earth. The resistance of grounds on lines in service can be measured easily with a clamp-on ground-resistance tester (Fig. 26.6). The ground-wire molding must be removed to gain access to the ground wire near its connection to the ground rod.

If when the rod is driven to its full length it is found that the resistance is not well below 25 ohms, resort can be had to one of the two following procedures:

1. Extend the length of the rod
2. Drive additional rods

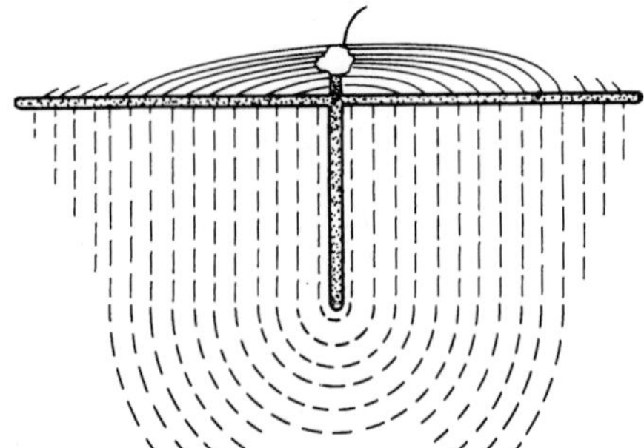

FIGURE 26.1 Resistance of earth surrounding an electrode. This may be pictured as successive shells of earth of equal thickness. With increased distance from the electrode, the earth shells have greater area and therefore lower resistance. *(Courtesy Copperweld Steel Co.)*

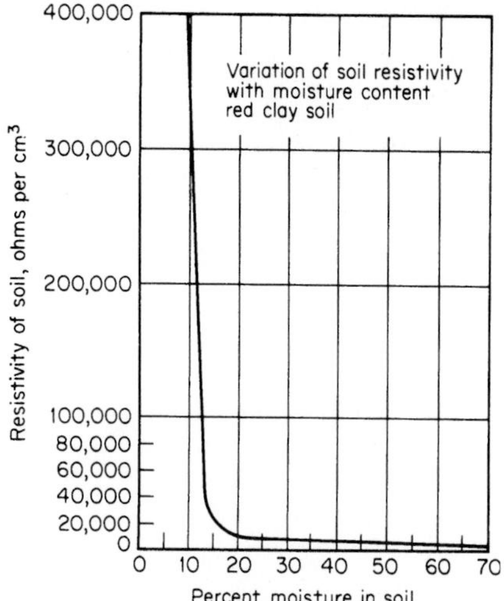

FIGURE 26.2 Variation of soil resistivity with moisture content. (*Courtesy Copperweld Steel Co.*)

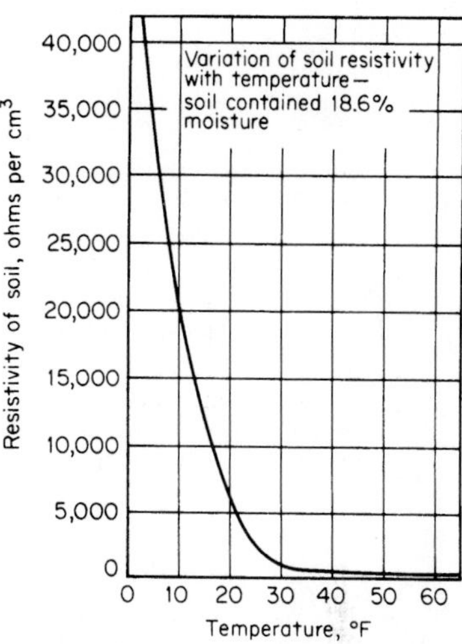

FIGURE 26.3 Variation of soil resistivity with temperature. (*Courtesy Copperweld Steel Co.*)

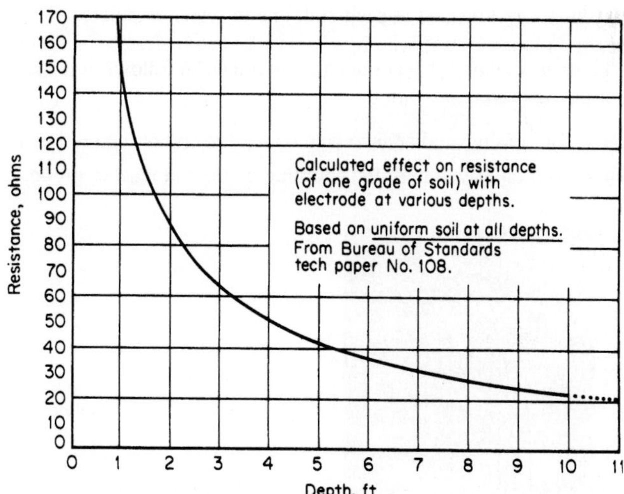

FIGURE 26.4 Chart showing the relation between depth and resistance for a soil having a uniform moisture content at all depths. In the usual field condition, deeper soils have a higher moisture content and the advantage of depth is more pronounced. (*Courtesy Copperweld Steel Co.*)

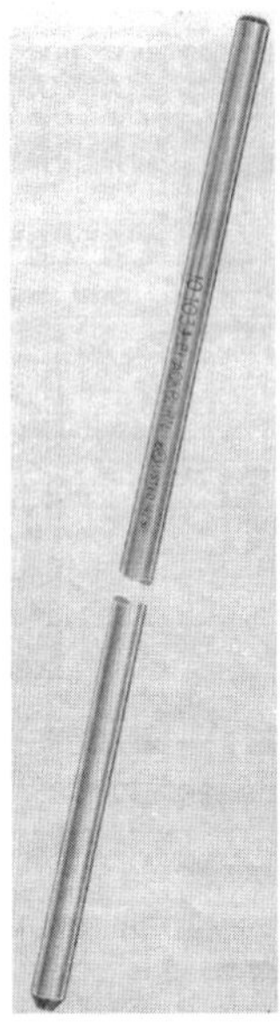

FIGURE 26.5 Copper bonded steel ground rod. (*Courtesy of Blackburn.*)

A sectional ground rod is one that makes possible adding section upon section in order that the rod can be driven deeper into more favorable moist soil, thereby lowering its ground resistance. Figure 26.7 shows the parts of a typical sectional ground rod and gives details of its construction. The sections are joined by a heavy tapped-bronze coupling. A special impact-resisting steel driving stud is screwed into the top section to take the hammer blows while driving. It protects the threading for subsequent attachment of additional sections of rod if needed. The sections are threaded at each end, one of which is pointed. The usual section length is 8, 10, or 12 ft.

The installation procedure for sectional ground rods is as follows. It is the same whether hand driven or power driven.

1. The coupling and driving stud are screwed on the top end of the sectional rod.
2. This assembly is then driven until the coupling on the rod reaches ground level.

FIGURE 26.6 Lineman measuring ground resistance of ground on distribution circuit without disconnecting ground connection.

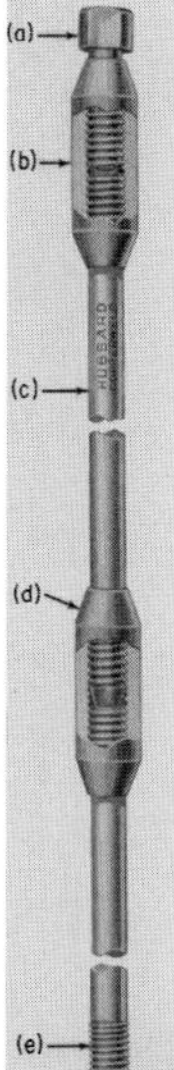

FIGURE 26.7 Typical sectional ground rod showing driving stud, couplings, rod sections, and driving point. (*a*) Removable driving stud of special steel takes the blows, protects the thread. (*b*) Cutaway view showing coupling and driving stud installed. (*c*) The steel core gives strength and rigidity for driving. The molten-welded copper covering provides high conductivity, permanent protection against corrosion. (*d*) A strong union and tight contact—easily made as successive sections are joined by the heavy bronze coupling. (*e*) Bottom end of sectional rods are pointed for easier installation. Rods are threaded on both ends. (*Courtesy Copperweld Steel Co.*)

3. The driving stud is then removed.
4. The lower end of the second sectional rod is then screwed into the coupling.
5. Another coupling is screwed onto the top end of the second sectional rod.
6. The driving stud is screwed into the second coupling.
7. Driving is now continued until the second section reaches ground level.
8. This sequence is repeated until resistance measurements indicate that the desired value of resistance has been obtained.

The other means of securing lowered ground resistance is by use of additional rods. When two or more driven rods are well spaced from each other and connected together, they provide parallel paths to earth. They become, in effect, resistances in parallel or multiple and tend to follow the law of metallic parallel resistances. Figure 26.8 shows one, two, three, and four ground rods installed at each pole, respectively. The rods are connected together with wire no smaller than that used to connect to the top of the pole.

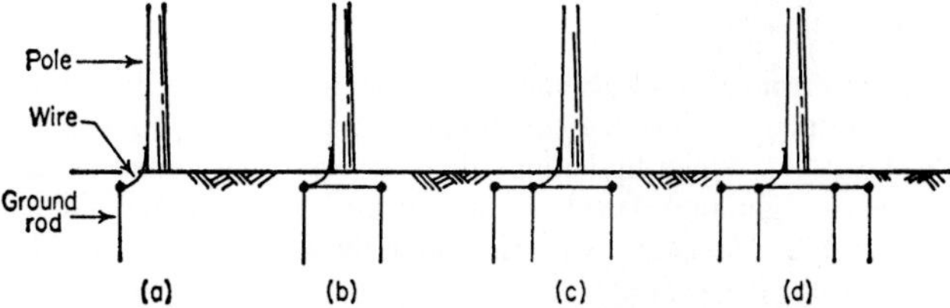

FIGURE 26.8 Sketches showing one ground rod (*a*) and two, three, and four ground rods at poles (*b*), (*c*), and (*d*), respectively. Rods at each pole are connected together electrically.

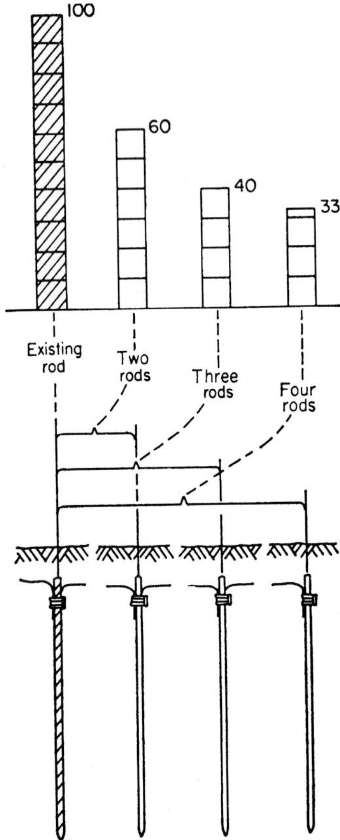

FIGURE 26.9 Installations of multiple ground rods. Upper graphs show approximate extent to which ground resistance is reduced by the use of two, three, or four rods connected in multiple.

If the rods were widely separated, two rods would have 50 percent of the resistance of one rod, but because they are usually driven only 6 ft or more from each other, the resistance of two rods is about 60 percent of that of one rod. Three rods, instead of having 33 percent of one rod, have on an average about 40 percent. Four rods, instead of having 25 percent of one rod, have about 33 percent. These values are shown graphically in Fig. 26.9.

Multiple rods are commonly used for arrester grounds and for station and substation grounds. In addition to lowering the resistance, they provide higher current-carrying capacity and can thus handle larger fault currents.

Transmission-Line Grounds. High-voltage electric transmission lines are designed and constructed to withstand the effects of lightning with a minimum amount of damage and interruption of operation. When lightning strikes an overhead ground or static wire on a transmission line, the lightning current is conducted to ground through the metal tower or the ground wire installed along the pole. The top of the structure is raised in potential to a value determined by the magnitude of the lightning current and the surge impedance of the ground connection. If the impulse resistance of the ground connection is high, this potential may be many thousands of volts. If the potential exceeds the insulation level of the

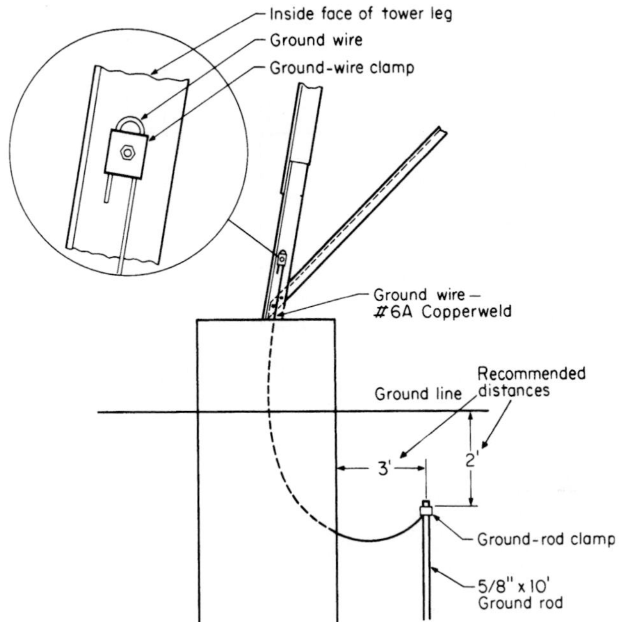

FIGURE 26.10 Ground wire connects ground rod to metal tower.

equipment, flashover will result, causing a power arc which will initiate the operation of protective relays and removal of the line from service. If the transmission structure is well grounded and proper coordination exists between the ground resistance and the conductor insulation, flashover can usually be avoided. The transmission-line grounds are installed in accordance with the specifications provided. Ground rods are usually used to obtain a low-ground resistance (Figs. 26.10 through 26.13).

A pole-butt grounding plate (Fig. 26.14) or a butt coil can be used on wood-pole structures. A butt coil is a spiral coil of bare copper wire placed at the bottom of a pole, as shown in Fig. 26.15. The wire of the coil continues up the pole as the ground-wire lead. Sometimes it is wound around the lower end of the pole butt a few times as a helix (see Fig. 26.16) to increase the amount of wire surface in contact with the earth. Butt coils should have enough turns to make good contact with the earth. A coil having seven turns and 13 ft or more of wire would provide satisfactory ground contact. Number 6 or larger AWG soft-drawn or annealed copper wire should be used. To prevent the coil from acting as a choke coil, the ground lead from the innermost turn is stapled to all the other turns.

If the soil has a high resistance, a grounding system called a counterpoise may be necessary. The counterpoise for an overhead transmission line consists of a special grounding terminal which reduces the surge impedance of the ground connection and increases the coupling between the ground wire and the conductors. Counterpoises are normally installed for transmission-line structures located in areas with sandy soil or rock close to the surface. The types of counterpoises used are the continuous, or parallel, type and the radial, or crow-foot, type (Fig. 26.17). The continuous, or parallel, counterpoise consists of one or more conductors buried under the transmission line for its entire length or under sections with high-resistance soils. The counterpoise wires are connected to the overhead ground, or

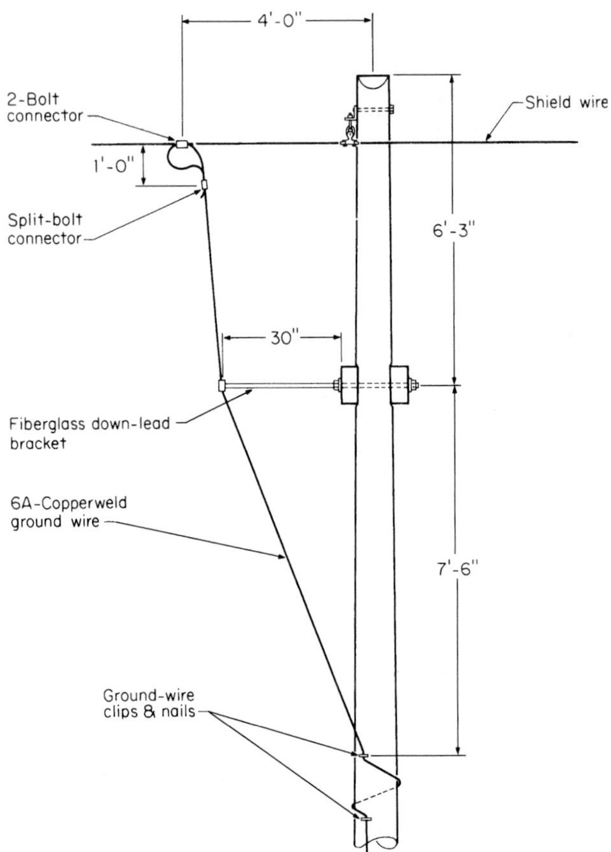

FIGURE 26.11 Shield-wire connection to ground wire on wood pole for transmission line.

static wire, at all supporting structures. The radial-type counterpoise consists of a number of wires extending radially from the tower legs. The number and length of the wires will depend on the tower location and the soil conditions. The continuous counterpoise wires are usually installed with a cable plow at a depth of 18 in. or more (Fig. 26.18). The wires should be deep enough that they are not disturbed by cultivation of the land.

Distribution-Circuit Grounds. A multigrounded neutral conductor for a primary distribution line is always connected to the substation grounding system where the circuit originates and to all grounds along the length of the circuit. The primary neutral conductor must be continuous over the entire length of the circuit and should be grounded at intervals not to exceed $^1/_4$ mile. The primary neutral conductor can serve as a secondary neutral conductor. If separate primary and secondary neutral conductors are installed, the conductors should be connected together if the primary neutral conductor is effectively grounded. All equipment cases and tanks should be grounded (Figs. 26.19 and 26.20). Lightning-arrester ground terminals should be connected to the common neutral if available and to a separate

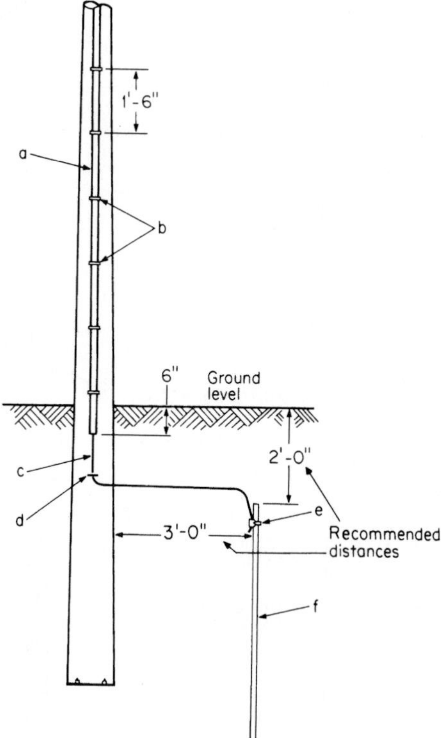

FIGURE 26.12 Ground-wire connection to ground rod for wood-pole transmission-line structure (*a*) Ground-wire molding; (*b*) staples, molding; (*c*) 6A Copperweld wire; (*d*) staples, fence; (*e*) ground-rod clamp; (*f*) 5/8-in × 10-ft ground rod.

ground rod installed as close as is practical. Switch handles, streetlight fixtures, down guys, and transformer secondaries must be properly grounded (Fig. 26.21). The common conductor on the transformer secondary for a single-phase 120/240-volt service must be grounded. A three-phase delta- or open-delta-connected transformer secondary for 240- or 480-volt service would have the center tap of one of the transformer secondary terminals grounded. If the center tap of one of the transformer secondaries is not available, one phase conductor is normally grounded. The neutral wire for three-phase, four-wire, Y-connected secondaries for 208Y/120-volt or 480Y/277-volt service is grounded (Fig. 26.22). If the primary circuit has an effectively grounded neutral conductor, the primary and secondary neutrals are connected together as well as to ground. Many times the primary distribution-line grounded neutral conductor is used for the secondary grounded neutral conductor and is called a common grounded neutral conductor. The neutral conductor of the service to the customer must be connected to ground on the customer's premises at the service-entrance equipment. A metal water pipe provides a good ground. Care must be taken to avoid plastic, cement-lined, or earthenware piping. Driven grounds may be installed near the customer's service entrance to provide the required ground connection.

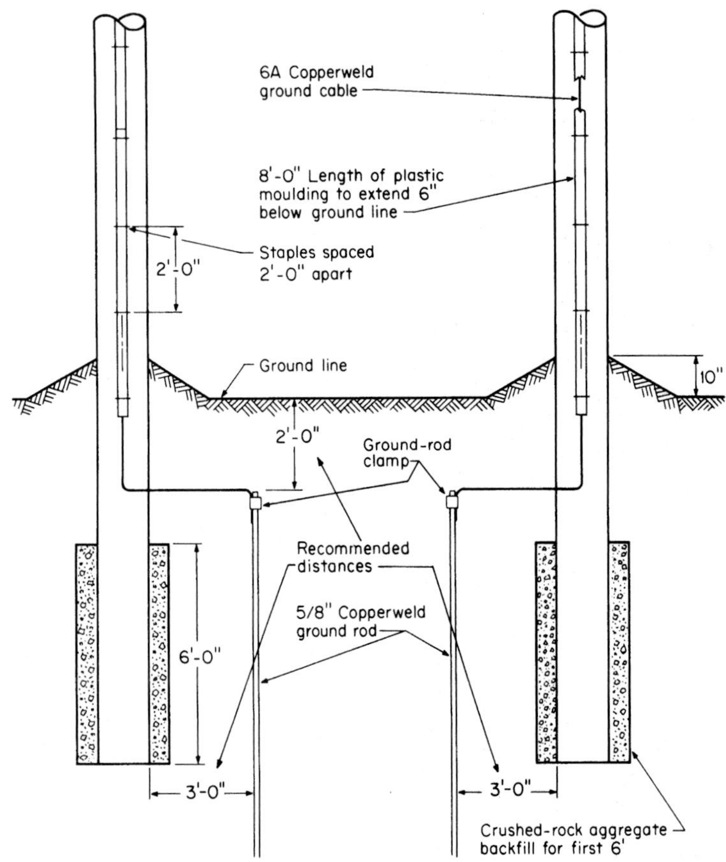

FIGURE 26.13 Ground-wire connections to ground rods for wood-pole H-frame transmission-line structure.

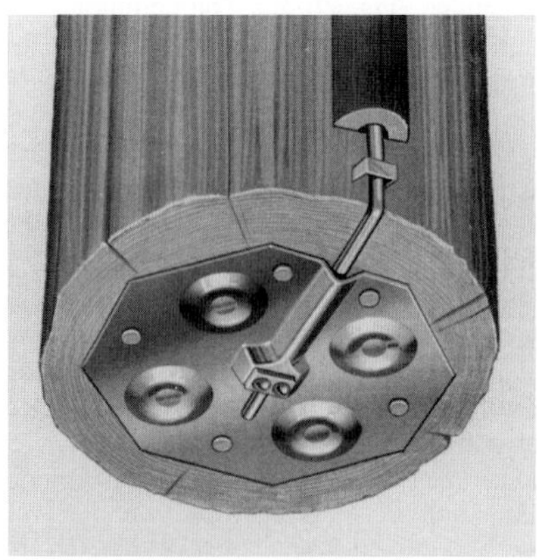

FIGURE 26.14 Pole-butt grounding plate attached to pole of a transmission or distribution line before the pole is set. The grounding plate is made from copper plate with moisture-retaining cups to maintain a good ground. (*Courtesy Homac Manufacturing Co.*)

FIGURE 26.15 Butt coil in position on butt of pole. Entire weight of pole helps to maintain contact with earth below pole. The spiral coil is stapled to the pole butt. Wire from inner spiral is stapled to all turns as it crosses them to shunt out the turns.

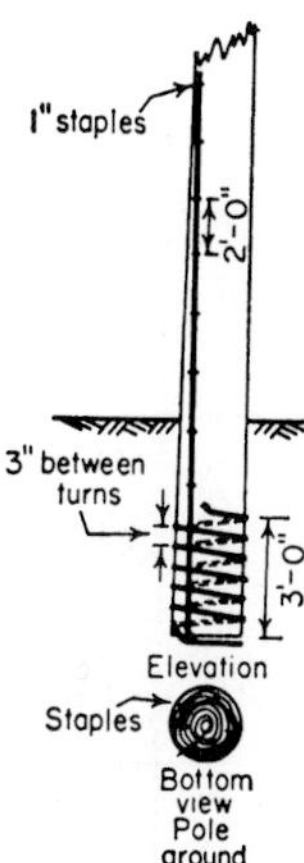

FIGURE 26.16 Views of butt-coil ground. Elevation shows six twin helix with 3 in between turns. Bottom view shows spiral pancake coil of seven turns.

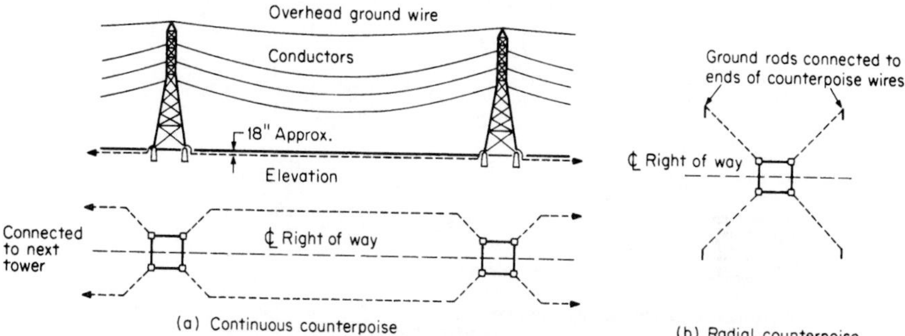

FIGURE 26.17 Two general arrangements of counterpoise installations. (*Courtesy Copperweld Steel Co.*)

FIGURE 26.18 Lineman using vibrating plow to install continuous counterpoise wire. (*Courtesy The Charles Machine Works, Inc.*)

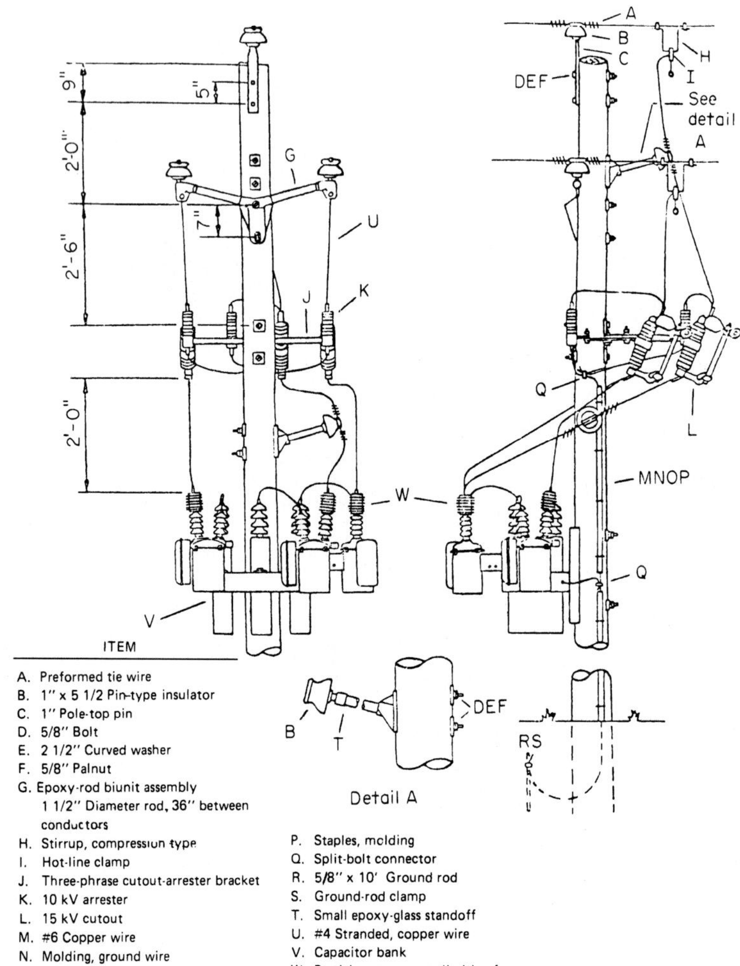

FIGURE 26.19 Capacitor bank installation for 13.8-kV circuit with grounding details illustrated.

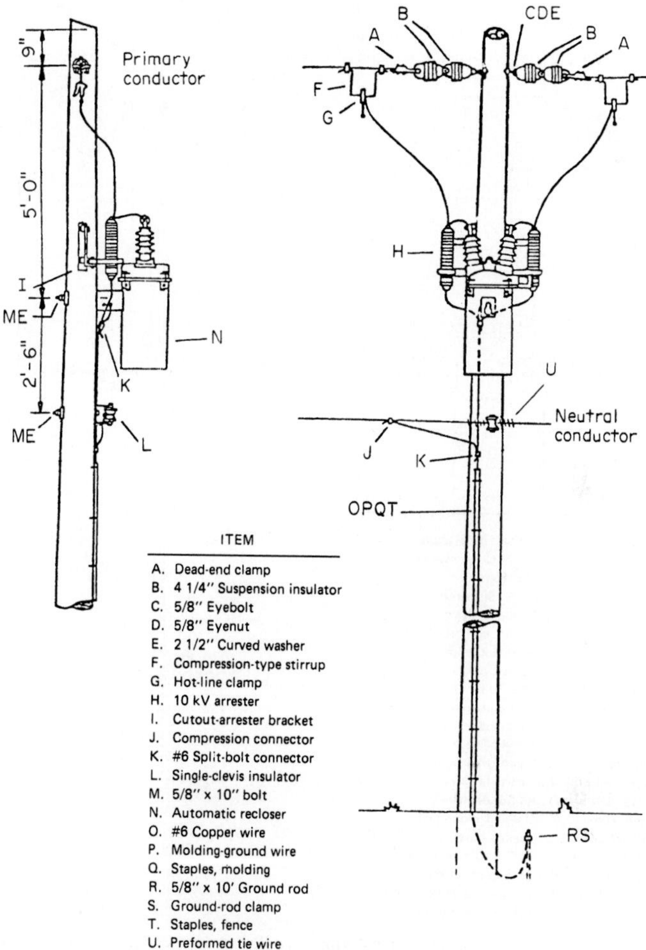

FIGURE 26.20 Single-phase 15-kV rural distribution automatic recloser installation with grounding details illustrated.

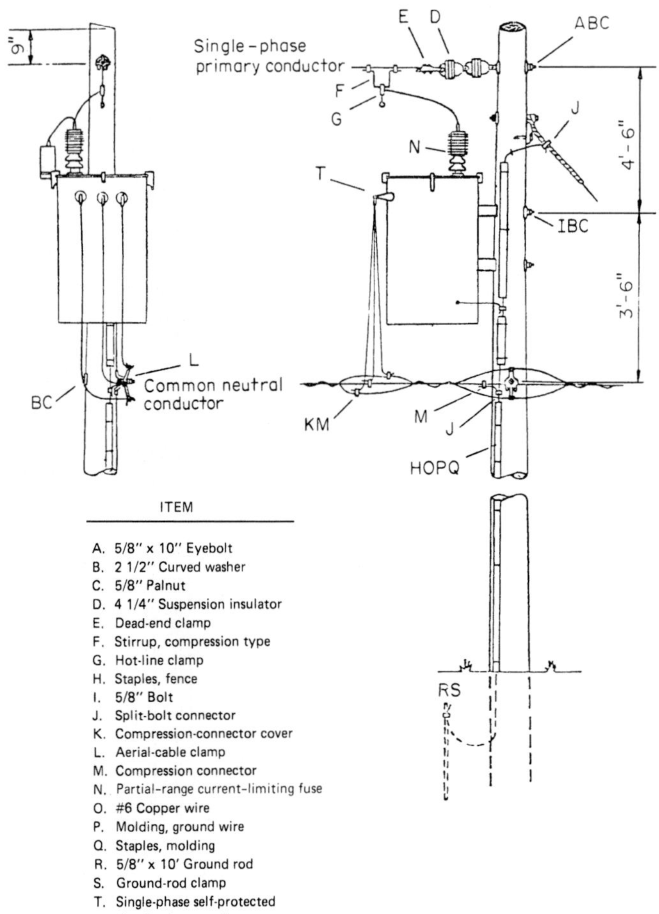

FIGURE 26.21 Distribution transformer installed on dead-end pole for 15-kV line with grounding details illustrated.

Installing Driven Grounds. Since the most common ground in distribution systems is the driven ground, its installation will be briefly described. A galvanized steel rod, stainless-steel rod, or copperweld rod is driven into the ground beside the pole at a distance of 2 ft, as shown in Fig. 26.23. The rod should be driven until the top end is 2 ft below the ground level. After the rod is driven in place, connection is made with the ground wire from the pole. The actual connection to the ground rod is usually made with a heavy bronze clamp. The clamp is placed over the ground-rod end and the ground-wire end. The size of the ground wire should not be less than no. 6 AWG copper wire. Number 4 wire is generally used. When the setscrew in the clamp is tightened, the ground wire is squeezed against the ground rod. Several precautions are taken in making the actual connection to minimize the possibility of the ground wire being pulled out by frost action or packing of the ground above it. The ground wire may be brought up through the clamp from the bottom. It is then

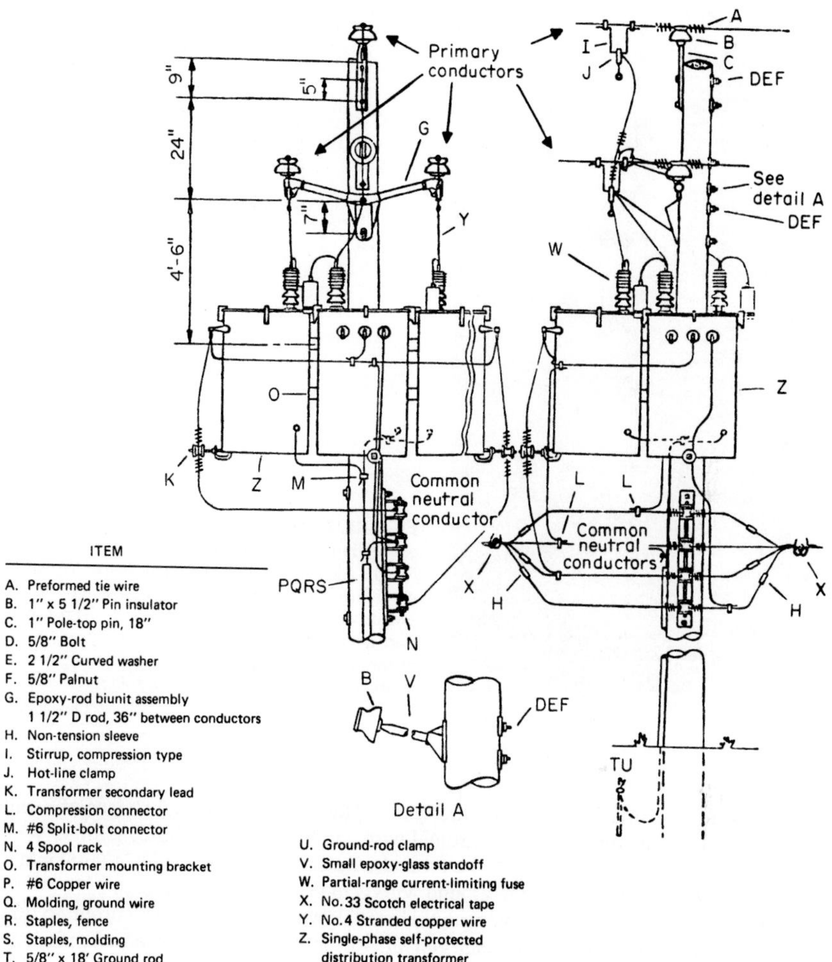

FIGURE 26.22 Three-phase Y-Y 15-kV distribution transformer bank with quadruplex secondaries.

bent over the clamp so that it cannot pull out. The wire is trained loosely so that there is plenty of slack between the ground-wire molding and the ground rod. This connection must be made before the lightning arresters are completely connected up on the pole. If this were the last connection—the one that completes the lightning-arrester circuit to ground—the lineman could get a high-voltage electric shock if one of the lightning arresters is defective. The last connection is always made at the lightning arrester by the lineman, either to its top or its bottom terminal.

Ground-Wire Molding. The molding protects the wire as well as the lineman. The molding should extend well below the surface of the ground so that in case the ground rod bakes

FIGURE 26.23 Lineman helper driving the ground connection. The rod is copperweld, has a diameter of 5/8 in. and is 8 ft long. Lineman helper is using a ground-rod driver which slips over the end of the rod. The driver must be operated with short, sharp strokes. Care must be taken not to lift the driver above the head of the rod and have it slide off, as this might cause injury. (*Courtesy Bashlin Industries. Inc*).

out, that is, fails to make contact with moist earth, there will not be any chance of a child or animal coming in contact with the ground wire and the ground at the same time. When the installation is finally covered up, the ground wire should be completely out of sight.

Another reason for covering the ground wire is to protect persons in case of lightning-arrester trouble. If a lightning arrester should break down and allow power current to flow, the ground wire will become "hot" and remain so for an indefinite period of time if the resistance of the ground connection is high. Anyone touching the wire while it is energized could get an electric shock.

CHAPTER 27
PROTECTIVE GROUNDS

Protection of the lineman is most important when a transmission or distribution line or a portion of a line is removed from service to be worked on using de-energized procedures. Precautions must be taken to be sure the line is de-energized before the work is started and remains de-energized until the work is completed. The same precautions apply to new lines when construction has progressed to the point where they can be energized from any source.

The installation of protective grounds and short-circuiting leads at the work site protect against the hazards of static charges on the line, induced voltages, and accidental energizing of the line.

When a de-energized line and an energized line are parallel to each other, the de-energized line may pick up a static charge from the energized line because of the proximity of the lines. The amount of this static voltage "picked up" on the de-energized line depends on the length of the parallel, weather conditions, and many other variable factors. However, it could be hazardous, and precautions must be taken to protect against it by grounding the line at the location where the work is to be completed. This will drain any static voltage to ground and protect the workman from this potential hazard.

When a de-energized line parallels an energized line carrying load, the de-energized line may have a voltage induced on it in the same manner as the secondary of a transformer. If the de-energized line is grounded at a location remote from where the work is being done, this induced voltage will be present at the work location. Grounding the line at the work location will eliminate danger from induced voltage.

Grounding and short-circuiting protect against the hazard of the line becoming energized from either accidental closing in of the line or accidental contact with an energized line which crosses or is adjacent to the de-energized line.

The procedures established to control the operation of equipment in an electrical system practically prevent the accidental energizing of a transmission or distribution line. Hold-off tagging procedures have proven to be very effective. If a circuit should be inadvertently energized, the grounds and short circuits on the line will cause the protective relays to initiate tripping of the circuit breaker at the source end of the energized line in a fraction of a second and de-energize the hot line. During this short interval of time, the grounds and short circuits on the line being worked on will protect the workmen (Fig. 27.1). If it isn't grounded, it isn't dead!

Testing before Grounding. The lineman must test each line conductor to be grounded to be sure it is de-energized (dead) before the protective grounds are installed. The circuit can be tested with a high-voltage voltmeter such as is described in Chap. 24 and illustrated in Fig. 27.20. Special testing equipment that produces an audible sound is available for detecting high voltage before applying grounding equipment as shown in Figs. 27.2 and 27.3. A high-voltage voltmeter can be used to check for static or induced voltage if the lineman is unable to positively determine that the line is safe for grounding. The high-voltage voltmeter

FIGURE 27.1 Grounding cluster installed on four-wire, three-phase line. Line conductor mounted on side of pole is the grounded neutral conductor. The grounding conductors create a low-impedance short circuit around the lineman's work area. (*Courtesy A. B. Chance Co.*)

connected between ground and conductor energized with a static charge will give an indication of voltage and then may gradually indicate that voltage has reduced to zero, as the static charge is drained off to ground through the high-voltage voltmeter.

Protective Ground Installation. After the testing is completed, the protective grounds should be installed in the following sequence:

1. Connect one end of the grounding conductor to an established ground. The grounded neutral conductor on a distribution line and the grounded static wire on a transmission line are the best grounded sources. A driven ground rod at the work site connected to the neutral conductor or static wire provides additional protection. If the work is being performed on a wood-pole structure, the installation of a cluster block on the pole below the work area connected to the grounding conductor, as shown in Fig. 27.1, will ensure the establishment of an equal-potential zone on the pole for the lineman performing the work.

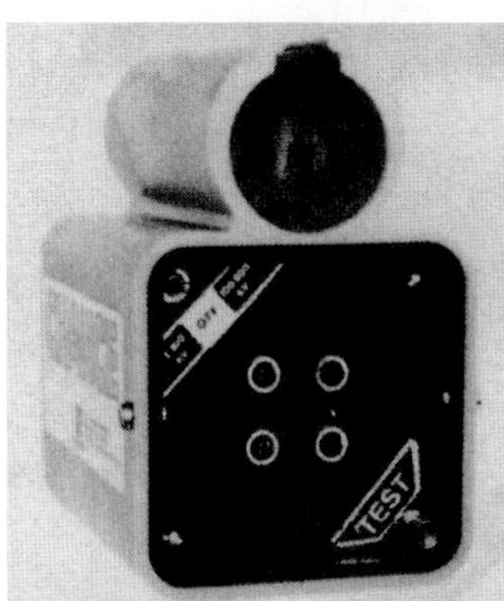

FIGURE 27.2 Tool utilized for testing potential on an energized circuit. (*Courtesy A. B. Chance Co.*)

FIGURE 27.3 Lineman using hot-line tool to detect energized conductor. (*Courtesy A. B. Chance Co.*)

2. Connect the other end of the grounding conductor, using a hot-line tool, to the bottom conductor on vertical construction or the closest conductor on horizontal construction (Fig. 27.4).
3. Install grounds or jumpers from a grounded conductor to the ungrounded conductors in sequence until all conductors are grounded and short-circuited together.

When the work is completed and the protective grounds are to be removed, always remove the grounds from the line conductors in reverse sequence, removing the connection to ground last.

When a connection is made between a phase conductor and ground, the grounding lead, connection to the earth, and the earth itself become a part of the electric circuit. All parts of the grounding circuit must be adequate to provide protection to the workmen under severe conditions (Figs. 27.5 through 27.7).

Protective Ground Requirements. An inadequate or poorly installed ground can be a safety hazard. It can give a lineman a false sense of security without actually protecting him under severe conditions. The protection that is provided by a ground is as good (or as bad) as the care that is taken to make sure it is installed properly. The requirements of a good ground are as follows:

1. A low-resistance path to earth
2. Clean connections
3. Tight connections
4. Connections made to proper points
5. Adequate current-carrying capacity of grounding equipment

FIGURE 27.4 Lineman completing the installation of protective grounds using a hot-line tool. Grounding operation was completed by connecting the cluster block mounted on the pole below the work area to a driven ground rod. Then a grounding jumper was connected from the cluster block to one phase wire. Jumpering was completed in sequence from bottom to top, providing the lineman with a zone of protection as the protective grounds are installed. The grounded static wire at the top of the pole provides an effective ground source. (*Courtesy A. B. Chance Co.*)

A good available ground (earth connection) is the system neutral; however, it usually is not available on a transmission line. However, if there is a common neutral at the location where the line is to be grounded, it must be connected to the grounding system on the structure so that all grounds in the immediate vicinity are connected together. If the structure is metal, the neutral should be bonded to the steel, either with a permanent connection or a temporary jumper. If the structure is a wood pole that has a ground-wire connection to the static wire, the neutral should be permanently connected to the ground wire. If there is no permanent connection, a temporary jumper should be installed. If there is no ground wire on the pole, the common neutral should be temporarily bonded to whatever temporary grounding connection is used to ground the line.

Static wires are connected to earth at many points along the line and provide a low-resistance path for grounding circuits. These are normally available for use as the ground-end connection of the grounding leads. If the line has a static wire, it is usually grounded at each pole. A ground wire runs down the wood pole(s) and connects to the ground rod. All metal poles and towers are adequately grounded when they are installed and serve as a good ground-connection point, provided the connection between the metal structure and the

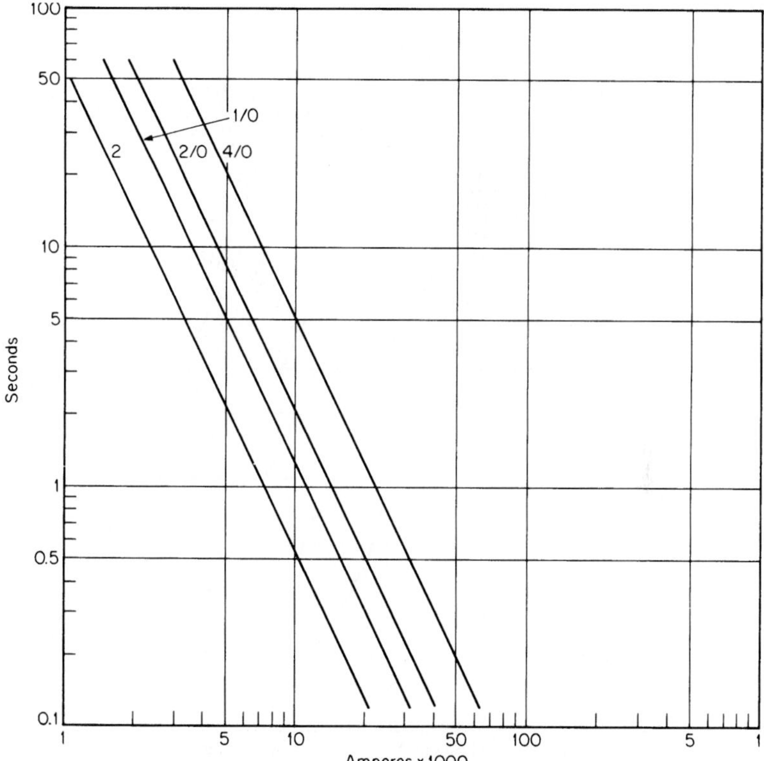

FIGURE 27.5 Suggested maximum allowable fault currents for copper grounding cables. Total current-on time, for minimum off time between consecutive current-on periods-reclosures. These values exceed IPCEA recommendations for cable installations by 1.91. (Based on tests of 10- to 60-cycle duration, 30°C ambient.) (*Courtesy A. B. Chance Co.*)

ground rod has not been cut or disconnected. If there is an exposed lead from the structure to the ground, it should be checked for continuity.

If a neutral conductor or static wire grounded at the structure is not available, use any good metallic object, such as an anchor rod or a ground rod, which obviously extends several feet into the ground as a ground for the protective ground source. If no such ground is available, the lineman must provide a ground by driving a ground rod or screwing down a temporary anchor until it is firm and in contact with moist soil.

Protective Ground Locations. A line is grounded and short-circuited whenever it is to be worked on using de-energized procedures. It is grounded before the work is started and must remain so until all work is completed.

Special devices are used to permit the grounding of conductors during stringing operations (Fig. 27.8). De-energized lines are grounded for the protection of all the men working on the line. The protective grounds are installed from ground in a manner to short-circuit the conductors so that the lineman and everything in the working area is at equal potential. This

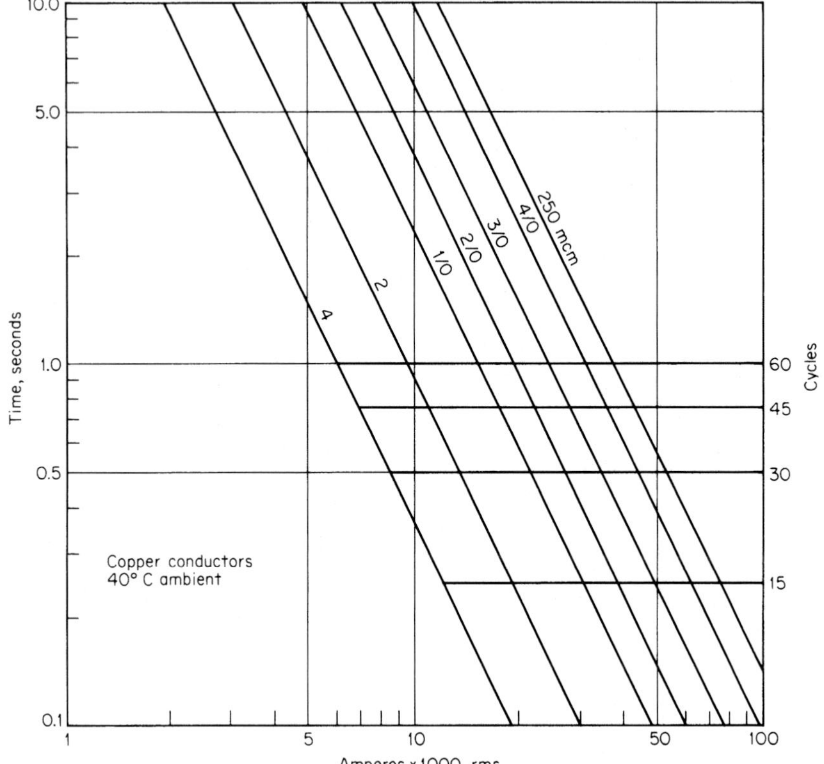

FIGURE 27.6 Fusing current time for copper conductors, 40°C ambient temperature. (*Courtesy A. B. Chance Co.*)

is the "man-shorted-out" concept, since the grounding, short-circuiting, and bonding leads will carry any current which may appear because of potential differences in the work area.

When a line is to be worked on de-energized, the line must be grounded and short-circuited at the work location, even though the work location is within sight of the disconnecting means used to deenergize the line.

When work is to be done at more than one location in a line section, the line section being worked on must be grounded and short-circuited at one location, and only the conductor(s) being worked on must be grounded at the work location.

When a conductor is to be cut or opened, it must be grounded on both sides of the open or grounded on one side of the open and bridged with a jumper cable across the point to be opened (Fig. 27.9).

A cluster block can be installed on a wood-pole structure to facilitate the ground connections and to help maintain the equal-potential area on the pole (Figs. 27.10 through 27.12).

Protective grounds installed on a steel transmission-line structure to connect conductors to ground and together are illustrated in Fig. 27.13.

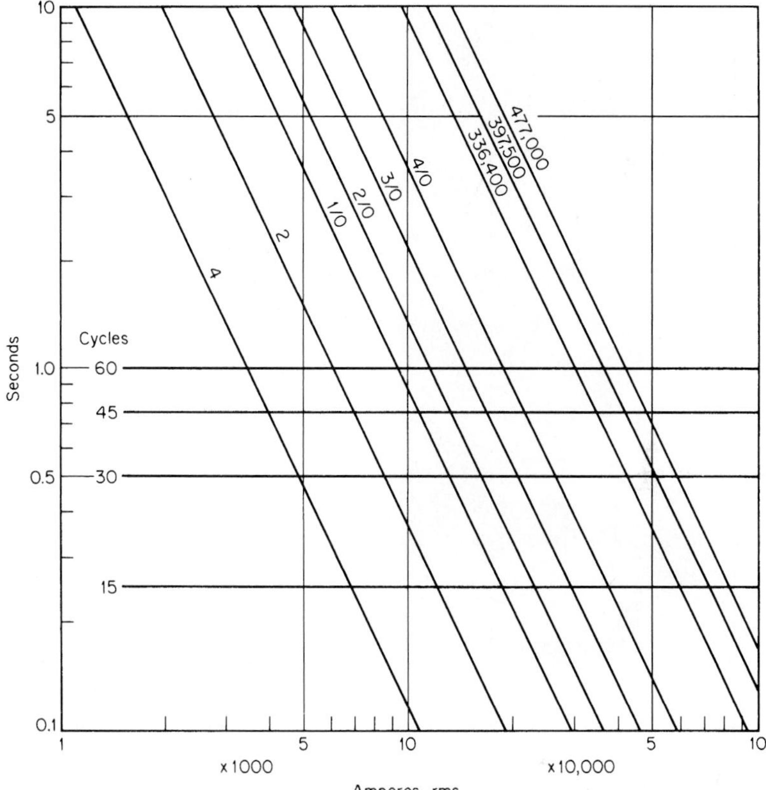

FIGURE 27.7 Fusing currents, aluminum conductor, 40°C ambient temperature. (*Courtesy A. B. Chance Co.*)

Underground-System Protective Grounds. Underground transmission and distribution cable conductors are not readily accessible for grounding. Transmission cables originating and terminating in substations can normally be grounded with portable protective grounding equipment at the cable termination points in the same manner as described for overhead transmission circuits. If the cable termination points are not accessible, special equipment manufactured and provided for grounding the circuit must be used. Underground distribution cables that originate in substation switchgear to supply complete underground circuits must be tested and grounded with special equipment provided at the substation before work is performed on the cable conductors. Underground equipment installed in vaults may be equipped with oil switches which can be used to isolate and ground the cable conductors. Underground-cable circuits originating at a riser pole in an overhead line can be grounded at the riser pole with portable protective grounding equipment (Figs. 27.14 through 27.16). Underground cable circuits originating in pad-mounted switchgear illustrated in Fig. 27.17 can be grounded at the switchgear near the cable termination. The switches in the switchgear must be open to isolate the cable circuit. Opening the switch provides a visible break in the circuit. A high-voltage voltmeter can be used by the lineman to be sure

FIGURE 27.8 Running ground connection designed to keep the conductor grounded as stringing operations are in process. A rope is used to secure the connector in place. (*Courtesy Sherman & Reilly, Inc.*)

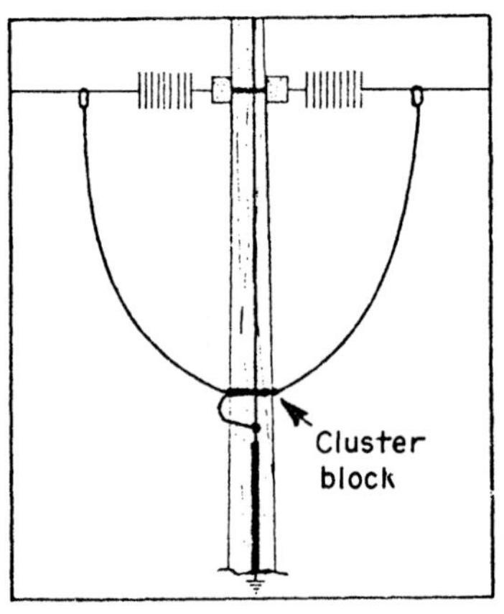

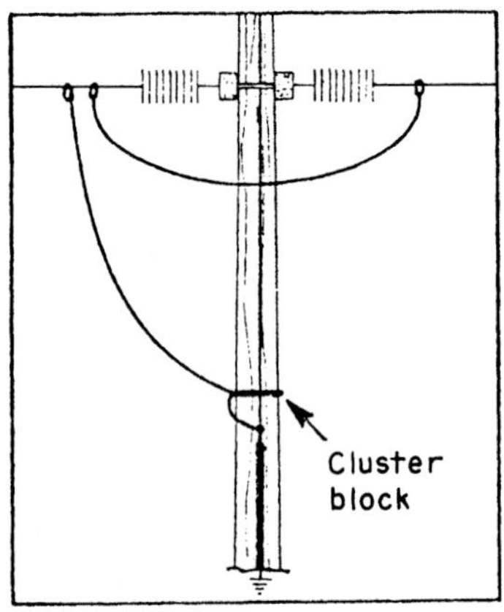

FIGURE 27.9 Method for grounding and bridging conductors across an open point is illustrated.

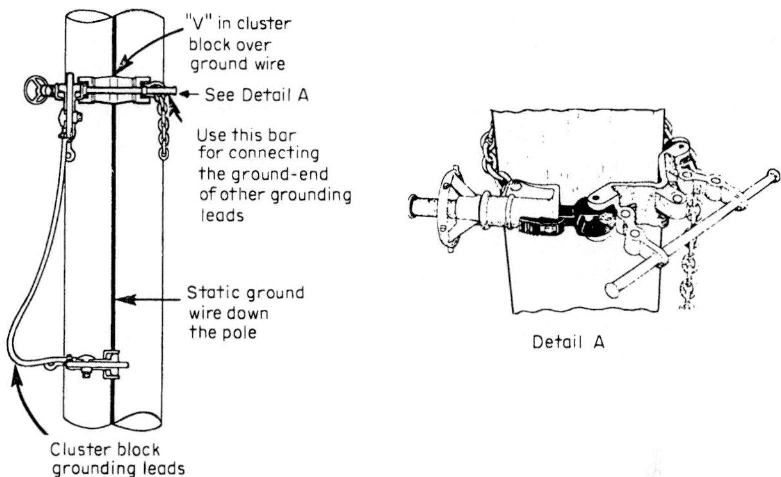

FIGURE 27.10 Metal cluster block equipped with 4-ft chain binder. The connection bar will accommodate four ground-lead clamps.

the circuit does not have a feedback from the remote end, proving that the cables are de-energized. A special conductor is available above the cable terminations for the application of portable grounding devices in the switchgear.

Underground-cable conductors terminating in pad-mounted transformers can be isolated, tested, and grounded at the transformer location. The lineman opens the pad-mounted transformer-compartment door and installs a feed-through device with a hot-line Grip-All clamp stick to prepare for the isolating and grounding operation (Fig. 27.18). The cable to be worked on must be isolated to separate it from a source of energy at the remote end. The lineman or cableman identifies the cable to be worked on at the pad-mount transformer with the markings on the cable and maps of the circuit. When the proper cable has been identified, the elbow cable terminator can be separated from the bushing on the transformer with

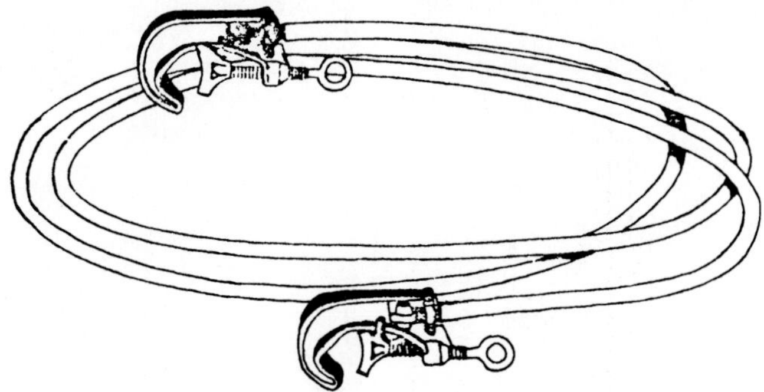

FIGURE 27.11 Grounding and short-circuiting jumper equipped with C-type clamps designed for operation with hot-line tool. Jumper cable is 1/0 copper conductor or larger with a large number of strands to provide flexibility.

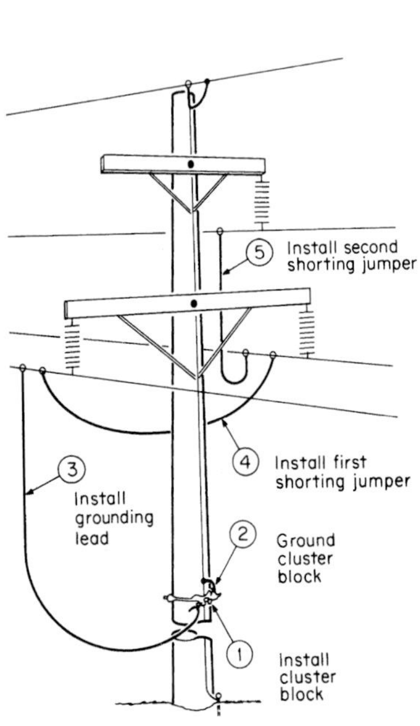

FIGURE 27.12 Protective grounds installed on conductors of a single-circuit wood-pole subtransmission line.

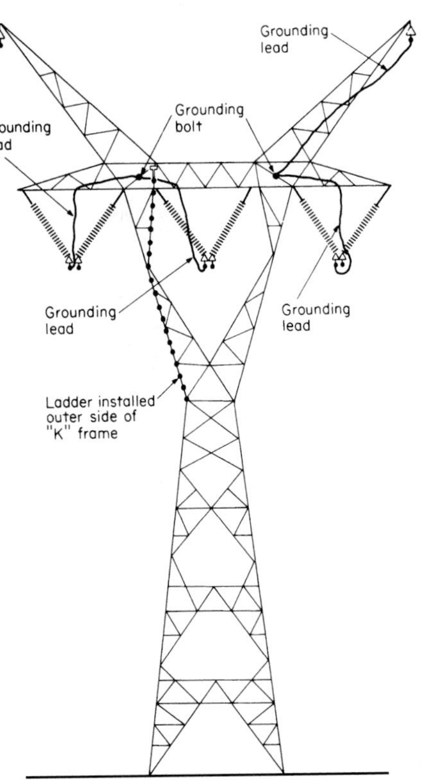

FIGURE 27.13 Protective grounds installed on conductors of a single-circuit transmission line constructed on steel towers.

FIGURE 27.14 Lineman testing cable circuit on underground-circuit riser pole with high-voltage voltmeter prior to installation of protective grounds. (*Courtesy MidAmerican Energy Co.*)

FIGURE 27.15 Lineman installing grounding jumper to common neutral conductor. (*Courtesy MidAmerican Energy Co.*)

FIGURE 27.16 Lineman completing grounding operation of cable riser circuit. Lineman uses hot stick to connect grounding lead to conductor attached to cable pothead. All three cutouts between the distribution line and the cable potheads are open. The grounding jumpers connect each pothead to the neutral conductor grounding, short-circuiting all three underground cables. (*Courtesy MidAmerican Energy Co.*)

FIGURE 27.17 Lineman has installed grounding jumpers with special grounding clamps on three-pole interrupter switch in pad-mounted switchgear. Note that insulating front barrier is inserted in open gap of three-pole interrupter switch to isolate incoming cables from bus and upper switch contacts. (*Courtesy S&C Electric Co.*)

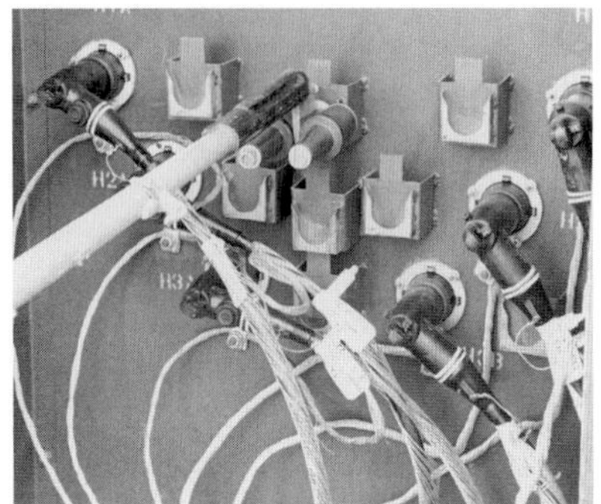

FIGURE 27.18 Using hot-line tool to mount feed-through device on bracket in high-voltage compartment of pad-mounted transformer. (*Courtesy MidAmerican Energy Co.*)

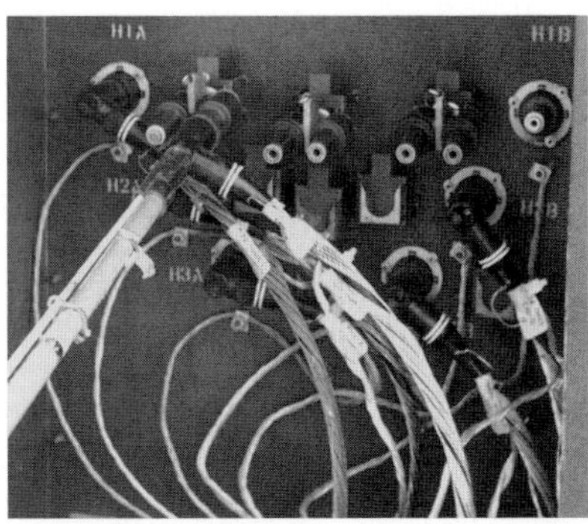

FIGURE 27.19 Removing cable to be grounded from bushing with hot-line tool hooked to elbow-connector terminating cable and installing elbow connector on feed-through device. (*Courtesy MidAmerican Energy Co.*)

the Grip-All clamp stick. The lineman then installs the elbow on the bushing on the feed-through device with the hot stick (Fig. 27.19). A high-voltage tester equipped with a special connection can be used to test the isolated cable through the second bushing on the feed-through device (Fig. 27.20). If the cable is found to be de-energized, the lineman can connect a special grounding conductor to the ground wires in the pad-mounted transformer compartment and place it on the bushing of the feed-through device with a Grip-All clamp stick (Figs. 27.21 through 27.23). The cable is now properly grounded so that the necessary work can be performed safely.

The *IEEE Guide for Protective Grounding of Power Lines* (IEEE Standard 1048) provides additional information about the installation of protective grounds.

FIGURE 27.20 High-voltage voltmeter with special probe for testing cable circuit through feed-through device. (*Courtesy MidAmerican Energy Co.*)

FIGURE 27.21 Lineman testing cable circuit with high-voltage voltmeter through feed-through device. Note one lead of the high-voltage voltmeter is connected to grounding conductor in pad-mounted transformer compartment. (*Courtesy MidAmerican Energy Co.*)

FIGURE 27.22 Connecting special grounding jumper with elbow connector on one end and C clamp on the other end to grounded neutral cable in pad-mounted transformer compartment. C clamp is connected to ground lead. (*Courtesy MidAmerican Energy Co.*)

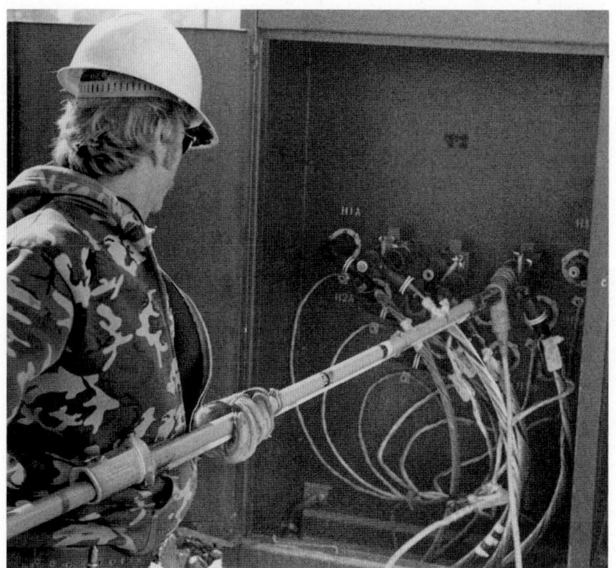

FIGURE 27.23 Lineman installs elbow connector of grounding jumper on feed-through-device bushing, grounding cable. (*Courtesy MidAmerican Energy Co.*)

CHAPTER 28
STREET LIGHTING

Streetlight systems have developed gradually from candles in the seventeenth century to gas lighting in the eighteenth century and electric lighting in the twentieth century. The first electric lights were electric arc lamps installed in Paris, France. The development of the incandescent lamp, initiated by Thomas Edison, greatly improved street lighting. The modern streetlight system reduces traffic hazards, helps to prevent crime, and enhances the beauty of our surroundings. The incandescent streetlight lamp that was commonly used in the early part of the twentieth century is gradually being replaced by electric discharge lamps of the fluorescent, mercury-vapor, metal-halide, and high-pressure-sodium types. The mercury-vapor lamp has had wide use because of its good efficiency, long life, and good light maintenance. Metal-halide and high-pressure-sodium lamps have a greater light output per watt of electric power input than the mercury-vapor lamps. High-pressure-sodium lamps are currently the most prominent light source for new street-lighting installations. The American National Standards Institute's publication, entitled *Practice for Roadway Lighting* (ANSI/IES RP8), covers street-lighting design for installations in the United States.

Street-Lighting Terms

Light. A form of radiant energy. Measurements are based on a unit of luminous intensity equal to the light emitted by a "standard candle" in a horizontal direction.

Lamp. A source of light.

Luminaire. The device which directs, controls, or modifies the light produced by a lamp. It consists of a light source and all necessary mechanical, electrical, and decorative parts.

Candlepower. The amount of light that will illuminate a surface 1 ft distant from the light source to an intensity of 1 footcandle (fc).

Lumen. The unit of luminous flux. It is equal to the flux on a unit surface, all points of which are at unit distance from a uniform point source of one candle. A uniform 1-candlepower source of light emits a total of 12.57 lumens.

Footcandle. The unit of illumination produced on a surface, all points of which are 1 ft distant from a uniform point source of 1 candle, or the illumination of a surface 1 ft^2 in area on which a flux of 1 lumen is uniformly distributed.

Ballast. An auxiliary device used with vapor lamps, on multiple circuits, to provide proper operating characteristics. It limits the current through the lamp and also may transform voltage.

Mast Arm. An attachment for a pole used to support a luminaire.

Part of Street-Lighting System. A street-lighting system consists of the following:

1. Circuit
   - **a.** Multiple or parallel
   - **b.** Series

2. Light-fixture support
    a. Mast arm
    b. Post
3. Luminaire
    a. Incandescent
    b. Electric discharge
        (1) Mercury vapor
        (2) Metal halide
        (3) Low-pressure sodium vapor
        (4) High-pressure sodium vapor
        (5) Fluorescent
4. Light control
    a. Reflector
    b. Refractor
    c. Diffuser
5. Time control
    a. Manual
    b. Automatic time switch
    c. Light-sensitive relay
6. Circuit switching
    a. Pilot wire
    b. Cascading

Multiple-Circuit Streetlights. The lamp for a multiple-circuit streetlight fixture is manufactured to operate on a constant line voltage of 120, 208, 240, 277, or 480 volts. The incandescent-type lamps operate directly from the line-voltage supply. Electric-discharge-type lamps require a ballast designed for the particular type of lamp to provide starting voltage, sustaining voltage during lamp warmup period, and operating voltage. The ballast for an electric-discharge lamp must provide a relatively high voltage to start the lamp and control the current in the lamp during the warmup period. The impedance of the lamp decreases as the arc in the lamp gets hot. Electric-discharge lamps take longer to reach full brilliance in cold weather as a result of the warmup requirement. The lamps or ballasts for use on a multiple-type circuit are connected across a constant-voltage source usually obtained from the secondary of a distribution transformer (Figs. 28.1 and 28.2). In a multiple-circuit setup the lamps are connected in parallel across the circuit. The distribution transformer delivers a constant low voltage, usually 120 volts, to the circuit. All lamps therefore have the same voltage impressed across them.

A multiple-circuit setup requires that both wires be brought to each lamp. This makes each lamp independent of the others, that is, any lamp can be switched on or off or burn out without affecting the other lamps.

Ornamental streetlight fixtures installed in commercial areas or residential subdivisions with underground distribution lines are usually connected in multiple. The secondary voltages used for multiple streetlight circuits are safer for the lineman and the public than the higher operating voltages used for series-circuit streetlights.

Series-Circuit Streetlights. In a series circuit, all the lamps are connected in series. The same current, therefore, flows through all the lamps, which requires a high voltage to maintain the proper current when a large number of lamps are installed in a circuit. If a series circuit develops an open point in the circuit, the constant-current transformer that energizes the

FIGURE 28.1 Multiple-circuit streetlight installed on a wood pole in an area served by an overhead electrical distribution system.

circuit develops a high output voltage, creating a hazard, particularly if the open circuit develops as a result of a tree limb falling on the wire, severing the wire so that it falls to the ground. Very few series streetlight circuits remain in service, and therefore, they will not be discussed further.

Light-Fixture Support. The streetlight fixture can be supported above the street in two different ways:

1. Supported from a bracket attached to a post (mast arm)
2. Supported on top of a post

Streetlights installed in areas served with an overhead electrical distribution system usually are mounted on the poles supporting the overhead lines (Figs. 28.3, 28.4, and 28.5). The length of the luminaire support or mast arm will determine the location of the luminaire with respect to the edge of the street. The luminaire should be located from 0 to 6 ft out over the street surface for roadways less than 40 ft wide and from 6 to 12 ft out over the street surface for roadways more than 40 ft wide. A 3-ft mast arm is usually used for a light installed on a pole located along the edge of an alley. However, 6-ft mast arms are normally installed on poles located along a street in a residential area with streets less than 40 ft wide, and 12-ft mast arms are normally installed on poles along a street in a commercial or industrial area where the streets are more than 40 ft wide.

If the lighting circuit is placed underground along the street, the most convenient support for the lighting fixture is an ornamental pole or post. The supply circuit for the lamp enters the pole or post underground and passes up inside the hollow post or pole. The light fixture can be mounted directly on the top of the post, as illustrated in Fig. 28.6, or supported

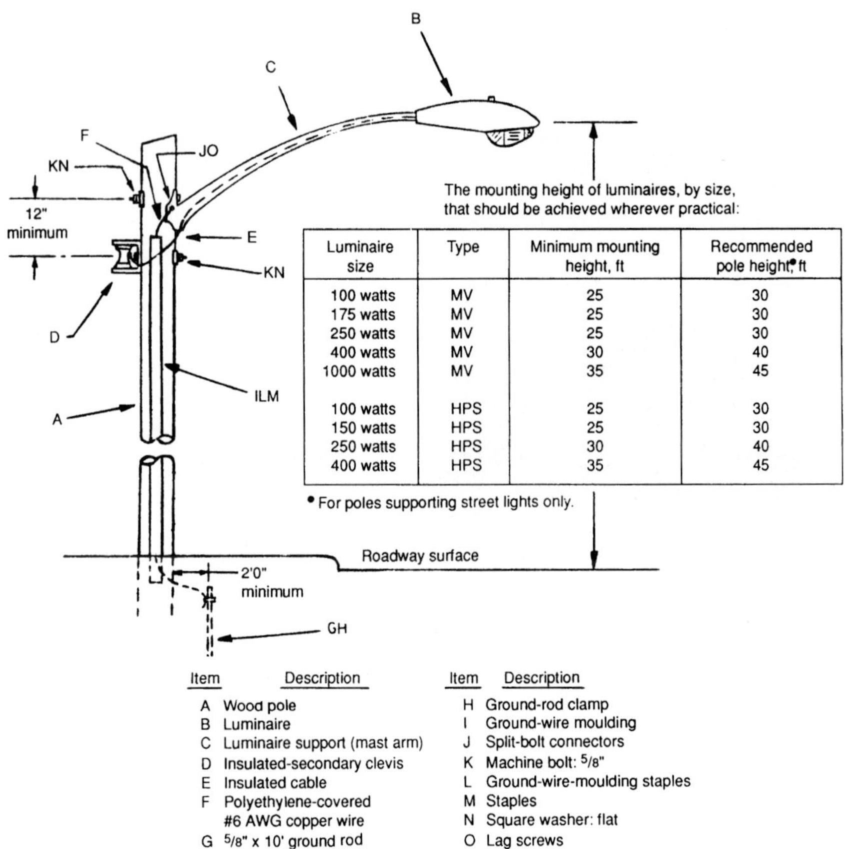

Luminaire size	Type	Minimum mounting height, ft	Recommended pole height,* ft
100 watts	MV	25	30
175 watts	MV	25	30
250 watts	MV	25	30
400 watts	MV	30	40
1000 watts	MV	35	45
100 watts	HPS	25	30
150 watts	HPS	25	30
250 watts	HPS	30	40
400 watts	HPS	35	45

FIGURE 28.2 Construction details for streetlight installed on a wood pole for installation in an area served by an overhead electrical distribution system.

by a bracket or mast arm out over the street, as shown in Figs. 28.7 and 28.8. Construction details for various types of ornamental streetlight installations served by underground supply circuits are shown in Figs. 28.9 through 28.11.

Area lighting has become common to enhance security and to permit nighttime sports activities. Figure 28.12 is an illustration of an ornamental pole and luminaire installation used to provide light for a parking lot. Figures 28.13 and 28.14 show the installation of luminaires with high-pressure-sodium lamps mounted on wood poles to light tennis courts.

Streetlight Grounding. All luminaire supports (mast arms) and luminaires installed on wood poles should be grounded to the neutral conductor, if available, or to a driven ground rod located near the base of the pole and connected to the mast arm by a polyethylene-covered no. 6 AWG copper wire. All metal poles, concrete poles, and metal U-guards and conduits must be grounded to a minimum height of 8 ft above ground, as required by the *National Electrical Safety Code* (ANSI C-2).

FIGURE 28.3 Multiple-circuit streetlight with a 6-ft mast arm installed on a wood pole supporting a 13.2Y/7.62-kV distribution circuit.

FIGURE 28.4 Multiple-circuit streetlight with a 12-ft mast arm installed on a wood pole supporting a 13.2Y/7.62-kV distribution circuit.

Streetlight Lamps. Lamps used for street lighting are of two principal types:

1. Incandescent

2. Electric discharge:

- **a.** Mercury vapor
- **b.** Metal halide
- **c.** Low-pressure sodium vapor
- **d.** High-pressure sodium vapor
- **e.** Fluorescent

Incandescent Lamp. In an incandescent lamp, a wire filament is sealed in a glass bulb from which the air is extracted. Electric current flowing through the filament heats it to incandescence and produces light. It is necessary to heat the filament to about 5000°F to produce full brilliance. The metal tungsten has a melting point of 6120°F, well above the operating temperature, and is therefore well suited for use in incandescent lamps. The absence of oxygen prevents the filament from burning up. A mixture of inert gases is often put into the bulb. These gases carry any tungsten that evaporates from the filament to the top of the bulb. A typical incandescent lamp, the PS 25, rated 2500 lumens, is shown in Fig. 28.15.

Lamps are rated as to light output in lumens. A *lumen* is defined as one unit of light output. Incandescent lamps used on multiple circuits are operated at 120 volts and are rated in watts input as well as lumens output. Table 28.1 lists the incandescent lamps most commonly used.

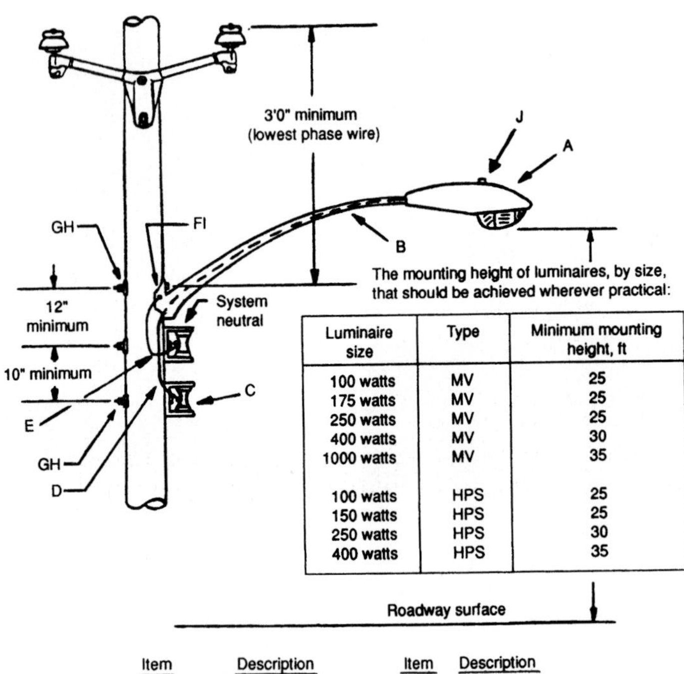

Luminaire size	Type	Minimum mounting height, ft
100 watts	MV	25
175 watts	MV	25
250 watts	MV	25
400 watts	MV	30
1000 watts	MV	35
100 watts	HPS	25
150 watts	HPS	25
250 watts	HPS	30
400 watts	HPS	35

Item	Description	Item	Description
A	Luminaire	F	Split-bolt connectors
B	Luminaire support (mast arm)	G	Machine bolt: 5/8"
C	Insulated-secondary clevis	H	Square washer: flat
D	Insulated cable	I	Lag screws
E	Polyethelyne-covered #6 AWG copper wire	J	Photoelectric control

FIGURE 28.5 Multiple-circuit streetlight installation on wood pole supporting 13.2Y/7.62-kV distribution circuit.

FIGURE 28.6 Streetlight fixture mounted on top of an ornamental pole. Power is supplied to the lamp from a buried cable that extends up through the hollow center of the pole to the luminaire.

FIGURE 28.7 Streetlight fixture supported from pole. Power is supplied from underground circuit.

FIGURE 28.8 Streetlight luminaire attached to ornamental pole. Power is supplied from underground circuit.

Electric-Discharge Lamps. In electric-discharge lamps, electrons flow between electrodes through an ionized gas or vapor. All gaseous conduction lamps have a negative-resistance characteristic. This means that the lamp resistance decreases as the lamp heats up. As the resistance decreases, the current increases. In fact, the current will increase indefinitely unless a current-limiting device is provided. All gaseous conduction lamps, therefore, have current limiters called ballasts.

The main advantage of a gaseous-discharge lamp is that it gives out more light than an incandescent lamp of comparable rating. Mercury and sodium are the most economical chemicals used in electric-discharge lamps.

Vapor Lamps. The mercury-vapor lamp contains a small quantity of argon gas, which permits a starting discharge between the special starting electrode and one of the main electrodes. During the starting interval, the liquid mercury gradually vaporizes and eventually the full lamp current flows between the main electrodes, causing the mercury vapor to give off a considerable amount of light that is extremely rich in blue, green, and yellow colors.

Where more normal color appearance of people or objects is important, improvement is obtained by the use of mercury lamps having fluorescent materials coated on the inside of the outer bulbs. The phosphors convert some of the invisible ultraviolet energy into visible light, in colors that supplement those from the mercury-vapor discharge.

A typical mercury lamp is shown in Fig. 28.16, and the essential parts of its construction are indicated in Fig. 28.17. Note the main electrodes and the starting electrode, the starting resistor, the inner arc tube, and the "mogul" screw base. A typical 400-watt mercury lamp is rated at 135 volts, and 3.2 amps and in its clear design gives out an average of 20,500 lumens initially when operated in a vertical position. When it is operated horizontally, its light output is slightly reduced, but this burning position is commonly used for street lighting because it permits a more effective distribution of light from the fixture. The mercury-vapor lamp has an output of approximately 50 lumens per watt and a life expectancy of more than 24,000 burning hours.

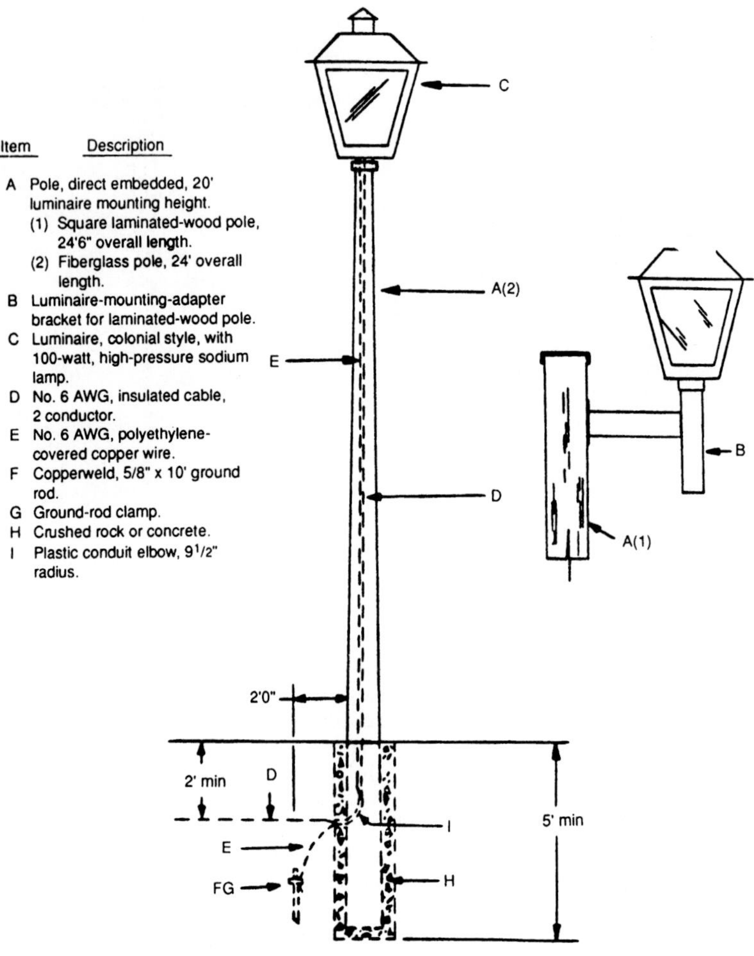

FIGURE 28.9 Ornamental, direct-embedded post streetlight installation supplied from underground wiring. Luminaire has a high-pressure-sodium lamp.

The metal-halide lamp is similar in construction and operation to the mercury-vapor lamp. The electrodes in the lamp are modified to permit use of the metallic halide compounds. Inert gases are placed within the arc tube and halogen compounds are introduced into the mercury, increasing the efficiency of the lamp. Ballasts must be constructed to produce a starting voltage higher than that normally utilized for mercury-vapor lamps. The lamp has an output of approximately 75 lumens per watt and a life of approximately 15,000 burning hours.

The low-pressure sodium-vapor lamp operates in a similar manner electrically. However, its light is limited to an intense yellow-orange color. This lamp type has little present-day use in the United States.

The high-pressure sodium-vapor lamp operates at high pressure and temperature. The lamp is elliptical and smaller than a mercury-vapor lamp of equal lumen output. The

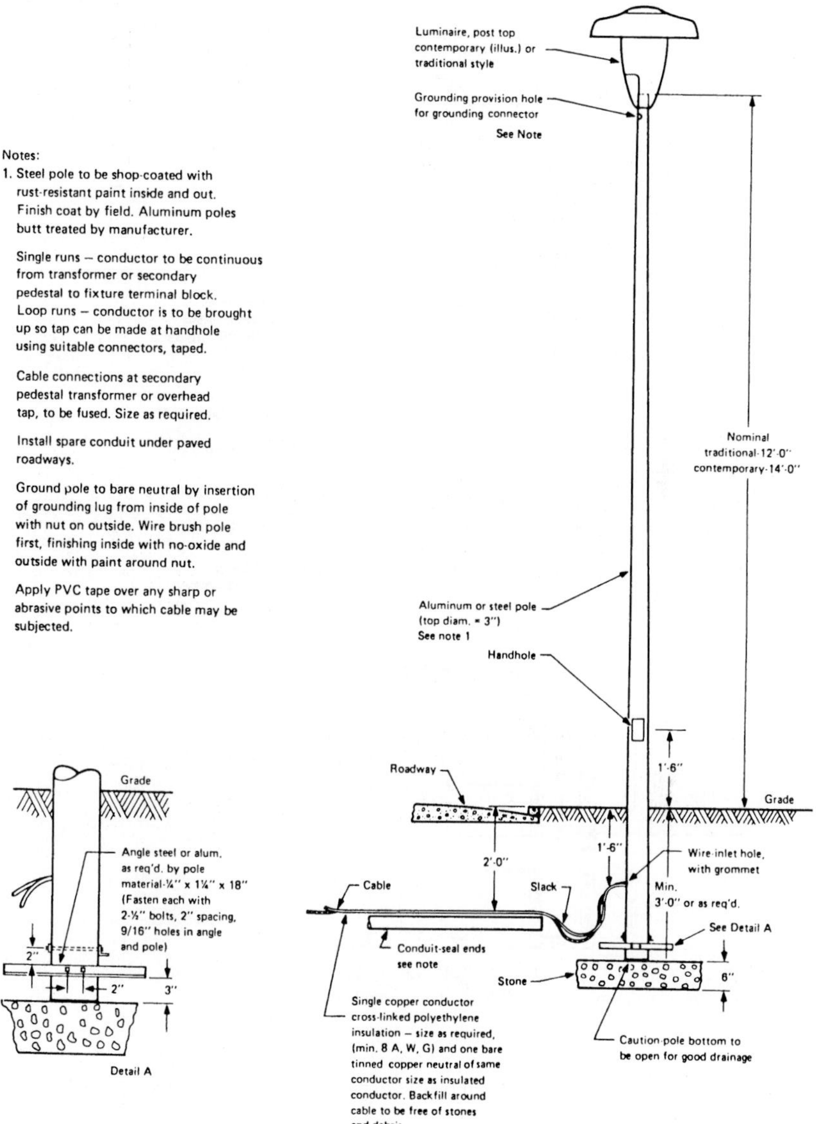

FIGURE 28.10 Multiple-circuit streetlight installation with post-top luminaire and underground supply. (*Courtesy American Electric Power System.*)

lamp is constructed with a ceramic arc tube utilizing sodium and other metallic traces and is contained by a borosilicate glass envelope (Fig. 28.18). High voltage for starting the lamp is generated by high-impulse electronic equipment. The ballast is designed specifically for the high-pressure-sodium lamp and is not interchanged with ballasts for other electric-discharge lamps. The high-pressure sodium-vapor lamp has an output of

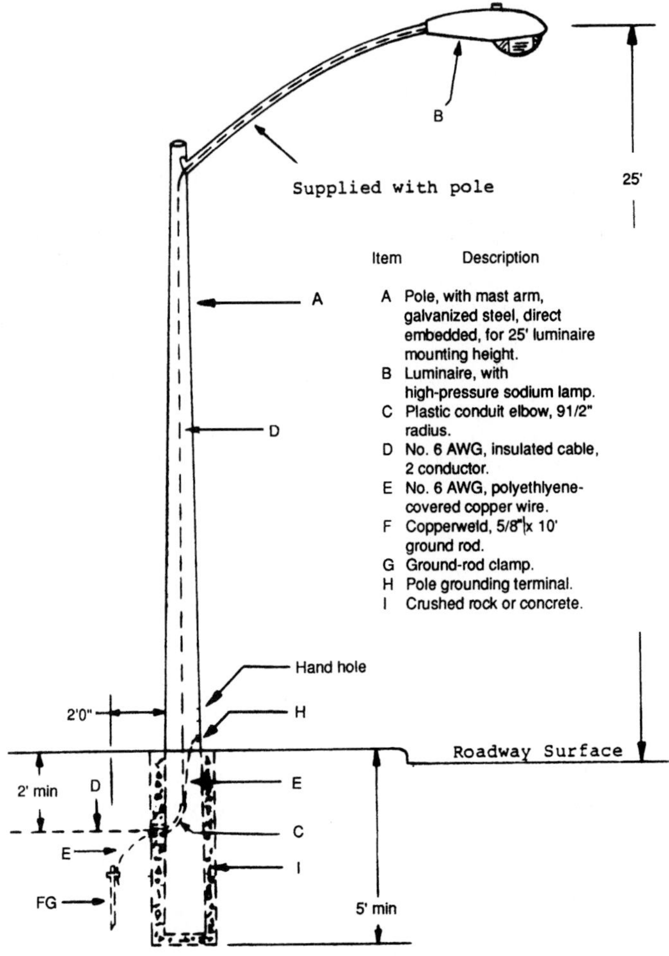

FIGURE 28.11 Installation details for streetlight luminaire mounted on galvanized-steel pole with power supplied from underground circuit.

approximately 100 lumens per watt and a life expectancy of approximately 24,000 burning hours.

Fluorescent Lamp. The fluorescent lamp is built in the shape of a long, slender glass tube with an electrode at each end. See Fig. 28.19 for details of construction.

In the fluorescent lamp, the flow of current takes place between electrodes through a mercury vapor as in a mercury-vapor lamp. The inside of the glass tube, however, is coated with a thin visible layer of fluorescent material called phosphor. This material will glow and give off visible light when struck by the electrons flowing through the mercury vapor between electrodes. The light produced is thus a secondary effect of the current flow. A "cool white" color is usually preferred, although other colors can be obtained by the use of other phosphors.

FIGURE 28.12 Ornamental pole and luminaire installed in parking lot of a hospital. High-pressure-sodium lamps are installed in the luminaires.

FIGURE 28.13 High-pressure-sodium lamps are installed in luminaires mounted on wood poles to light tennis courts.

FIGURE 28.14 Installation details of luminaires installed on wood poles for area lighting.

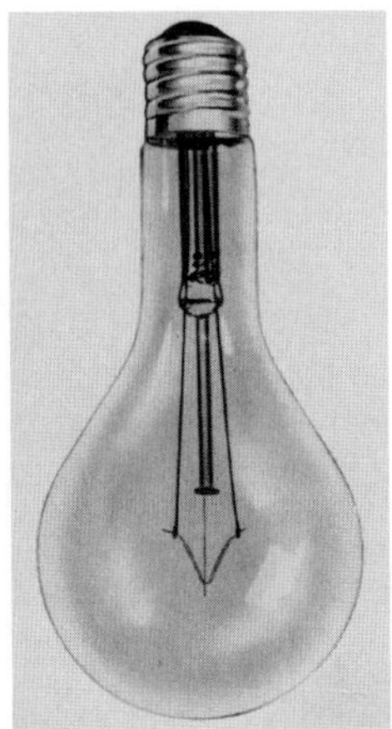

FIGURE 28.15 Typical incandescent lamp rated 2500 lumens. (*Courtesy General Electric Co.*)

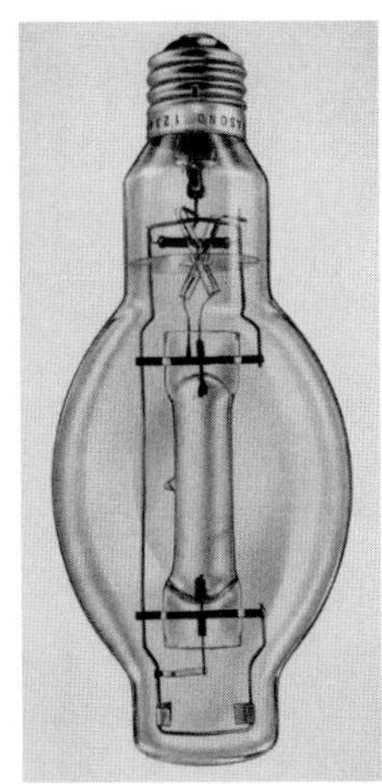

FIGURE 28.16 Typical mercury-vapor lamp rated 400 watts. (*Courtesy General Electric Co.*)

A small measured amount (about a drop) of mercury placed in the arc tube permits the lamp to operate, and a ballast in the form of a choke coil in the circuit limits the current. Lamps require starters to provide a temporary heating circuit. This circuit opens after a few seconds' operation.

The fluorescent lamp has an output of approximately 70 lumens per watt and a life expectancy of 18,000 or more burning hours.

Streetlight Control. Bare lamps emit light in many directions. Light in the skyward direction is largely wasted, and crosswise light on streets is also partly wasted.

Streetlight fixtures (luminaires) at street intersections can be designed to direct their light equally in four directions, and luminaires along the sides of streets should emit light in two principal directions. Such light-distribution patterns are illustrated in Figs. 28.20 and 28.21. Figure 28.20*a* shows uniform distribution in all directions, and Fig. 28.20*b* shows a light pattern in four main directions. Figure 28.21 shows light distribution in two principal directions along the street.

The required control of the light emitted from the lamp is obtained by the use of reflectors, refractors, and diffusing globes or combinations thereof.

TABLE 28.1 Ratings of Incandescent Lamps in Common Use

Bulb shape	Volts	Watts	Initial lumens	Nominal lumens	Rated life, h
PS 25	120	189	2,920	2,500	3000
PS 35	120	295	4,950	4,000	3000
PS 40	120	405	6,900	6,000	3000
PS 40	120	620	10,800	10,000	3000

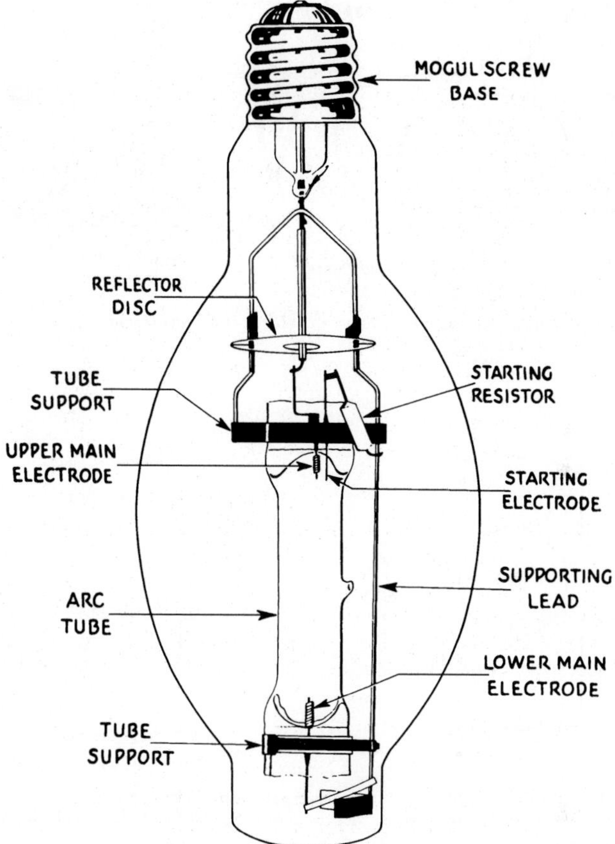

FIGURE 28.17 Details of the mercury-vapor lamp shown in Fig. 28.16. (*Courtesy General Electric Co.*)

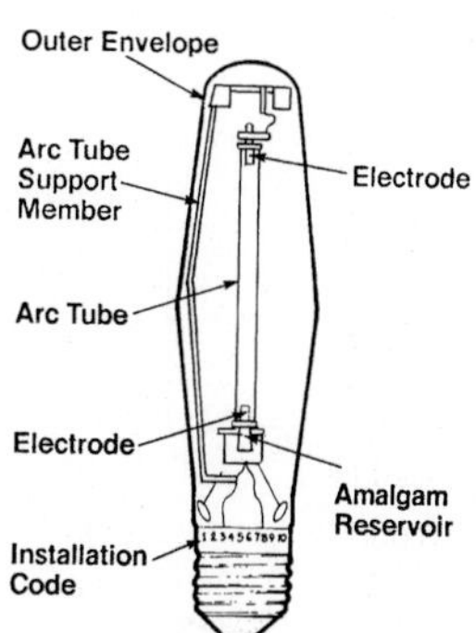

FIGURE 28.18 Details of a high-pressure-sodium lamp. (*Courtesy ITT Outdoor Lighting.*)

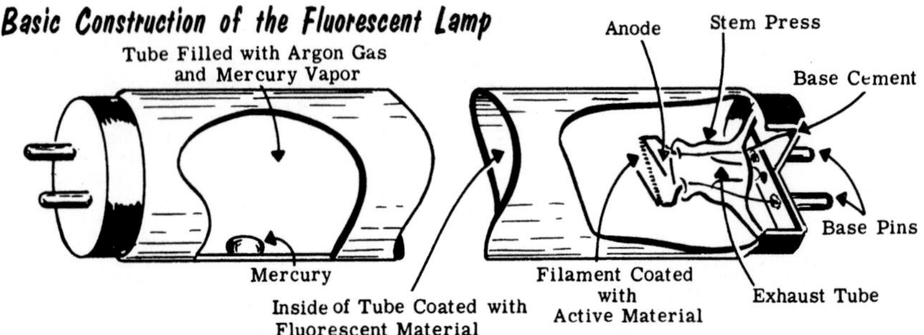

FIGURE 28.19 Details of construction of fluorescent lamp. (*Courtesy General Electric Co.*)

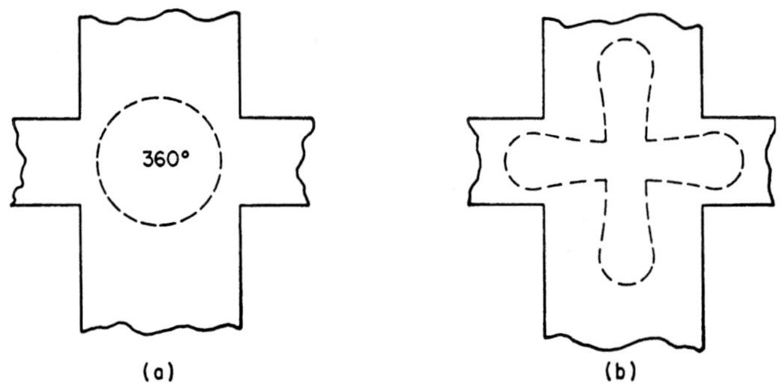

FIGURE 28.20 Light-distribution patterns at street intersection. (*a*) Uniform, (*b*) four-way.

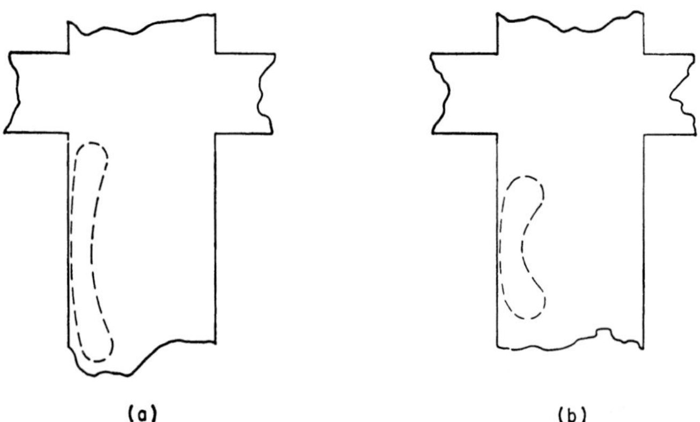

FIGURE 28.21 Light-distribution patterns along (*a*) street and (*b*) Two-way.

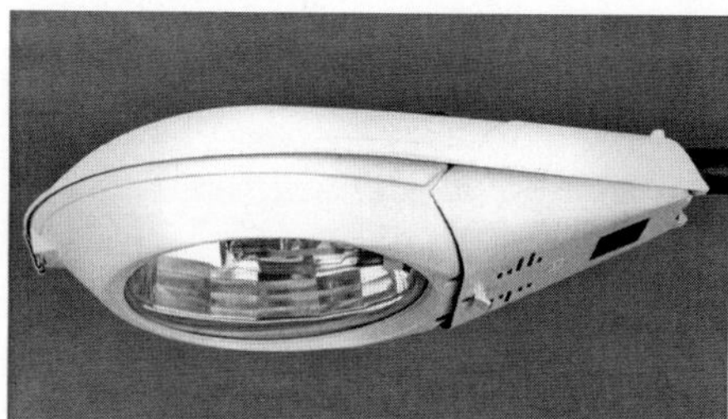

FIGURE 28.22 Streetlight luminaire with lamp and reflector back of lamp shown. Refractor has been removed from luminaire. (*Courtesy General Electric Co.*)

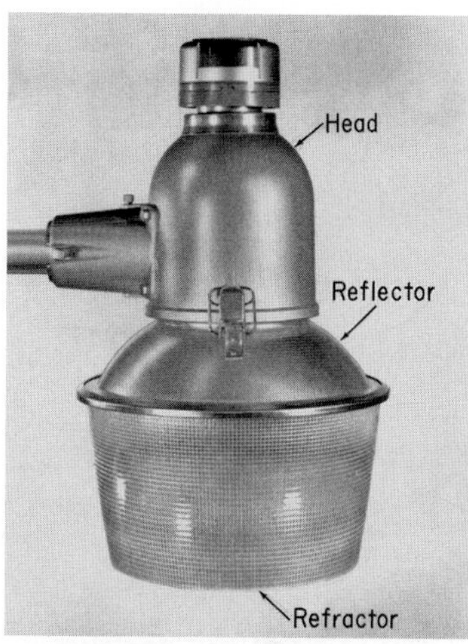

FIGURE 28.23 Luminaire using reflector and refractor for light control. Note photoelectric control mounted on top of luminaire. (*Courtesy General Electric Co.*)

A reflector is a surface which stops the light ray and redirects it in the desired direction. Figure 28.22 shows a typical light fixture equipped with a reflector. Such a reflector will stop the light rays shining upward and redirect them downward to the ground. Reflectors may be made of porcelain, metal, or silvered glass. They are made in various shapes to give different types of light distribution.

A refractor is made of a transparent material, usually glass or acrylic plastic, which passes light rays but bends or deflects them in the desired direction. The enclosing globe is actually a series of optical prisms which bend the light rays. A bowl refractor is illustrated in Fig. 28.23. It is used with pendant luminaires, while dome refractors are used on upright post luminaires.

Refractors must be positioned correctly to get the desired light distribution in the horizontal plane. They are adjusted vertically with respect to the lamp filament.

Sometimes the enclosing globe is made to diffuse the light, that is, scatter the light in many directions. Diffusing the light reduces the glare by spreading the apparent light over a larger area at lower brightness. The emitted light from the lamp so lights up the globe that the globe itself appears to be the light source.

Special types of light fixtures often combine the several methods of light control in the same fixture. Such a combination fixture is shown in Fig. 28.24.

Methods of Time Control. As the purpose of street lighting is to illuminate streets and intersections when natural lighting from the sun is not sufficient, the lights should be switched on as darkness approaches and off when daylight returns. This can be accomplished either manually or automatically by the following methods:

1. Manual
2. Clock-operated time switch
3. Photoelectric control
4. Cadmium-sulfide-cell control

FIGURE 28.24 Luminaire mounted on ornamental streetlight pole in a residential area with underground distribution utilizing reflector and diffusing bowl for light control.

Manual. This method requires that some member of the operating personnel be given the specific task of closing the lighting circuits at dusk and opening them at dawn, as well as at other daylight hours when unusual weather conditions indicate the need for artificial light. On series circuits, the switch on the primary side of the constant-current transformer is operated. On multiple circuits fed by transformers whose loads are street lighting only, the primary side of the transformer is frequently switched. However, if the transformer supplies power to other loads in addition to street lighting, then the street-lighting secondary is switched.

Automatic Clock Control. Clock-operated time switches can be

1. Motor driven
2. Spring driven, hand rewound
3. Spring driven, motor rewound

In case of a power interruption, the motor-driven time switch stops and has to be reset to the correct time manually. The spring-driven hand-rewound clock has to be rewound regularly. The spring-driven motor-rewound clock rewinds itself automatically. The last arrangement is the most satisfactory when clock-driven time switches are employed.

Time switches are usually equipped with astronomical dials which automatically change the turn-on and turn-off times to conform to the seasonal changes of sunrise and sunset. This gradual change is reflected in the shape of a cam which is driven by a small electric motor. The cam allows the contacts to be made or broken in accordance with the shape of the cam. The dial is adjusted at the factory for the locality in which the time switch is to be used. Figure 28.25 shows a typical time switch equipped with an astronomical dial.

Photoelectric-Cell Control. A photoelectric-cell-controlled relay may be employed to control street-lighting circuits by turning on the lights whenever the natural illumination falls below a given level. The photoelectric cell activates a relay which switches the lights on or off.

Cadmium-Sulfide-Cell Control. The cadmium-sulfide cell is commonly used for automatic street-lighting circuit control. This device can be mounted on the luminaire itself, as illustrated in Fig. 28.23, thereby eliminating the need of control circuits wherever secondary voltage is available.

FIGURE 28.25 Motor-driven time switch with astronomical dial. (*Courtesy General Electric Co.*)

A diagram of the basic circuit for this type of control is shown in Fig. 28.26. Its operation is based on the variable resistance of the cadmium-sulfide cell rather than on the small current-generating characteristic of the photoelectric tube. The cadmium-sulfide cell has a very high resistance (about 15,000,000 ohms) when dark, but when exposed to sunlight, it has a relatively low resistance, less than 1000 ohms. The cell is placed in series with the holding coil of the relay. Thus in the daytime the resistance of the cell is low, and sufficient current flows through the relay coil to hold the contacts open and keep the lights off. As evening approaches and the resistance of the cell begins to increase owing to the decreasing light level, the current of the holding coil decreases. When the natural light level reaches the control setting, the current of the holding coil lets the relay contacts close and the streetlight is energized.

The next morning, as the light level begins to increase, the resistance of the cell again is reduced. When the light level reaches the control setting, the current in the relay holding coil becomes large enough to cause the relay to open the contacts and de-energize the light. The relay holds these contacts open again throughout the day until the following evening.

The controls are manufactured to fail on, as described, or fail off: the streetlight remains off if the control fails. The fail-on control has the advantage of the streetlight being on in daytime, indicating that the control has failed and that there is no interruption of streetlight output at night. The fail-off control has the advantage of eliminating daytime burning of the streetlight, thus conserving energy.

A typical cadmium-sulfide-cell control unit is shown in Fig. 28.27. The cell is located directly behind the window in the cover.

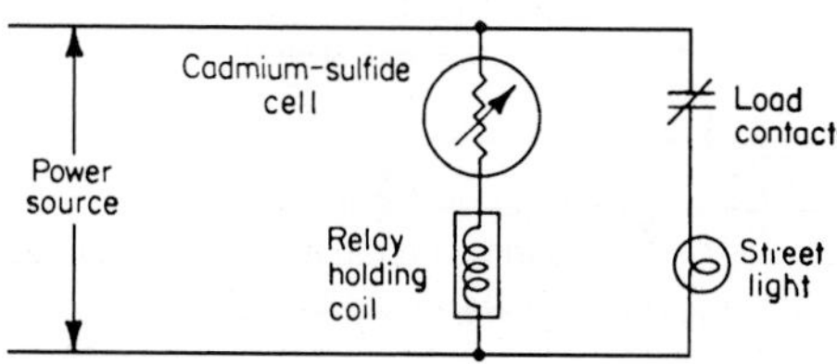

FIGURE 28.26 Cadmium-sulfide-cell control circuit.

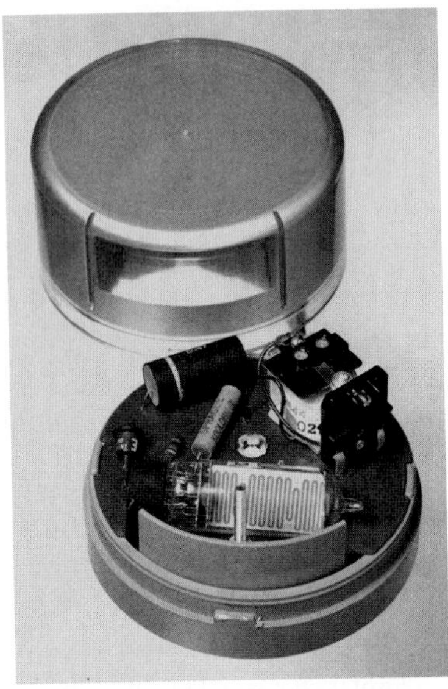

FIGURE 28.27 Assembled view of cadmium-sulfide cell streetlight control unit. Cadmium-sulfide cell is in lower control location directly behind window in cover. (*Courtesy General Electric Co.*)

Switching Several Circuits. Two methods are in use to switch multiple lighting circuits, fed from different secondaries or transformers, on and off. They are:

1. Pilot wire
2. Cascading

In the first method, a pilot wire operates a relay in each lighting circuit. The pilot wire can be either energized or de-energized at night. If the pilot wire is energized at night when lighting is needed, the relay must be of the circuit-closing type, as shown in Fig. 28.28. Energizing the pilot wire energizes the relays, and the relays close the lighting circuits. If the pilot wire is de-energized at night, the relays must be of the circuit-opening type. Normally, the relay contacts are closed. De-energizing the pilot wire reduces the pull on the

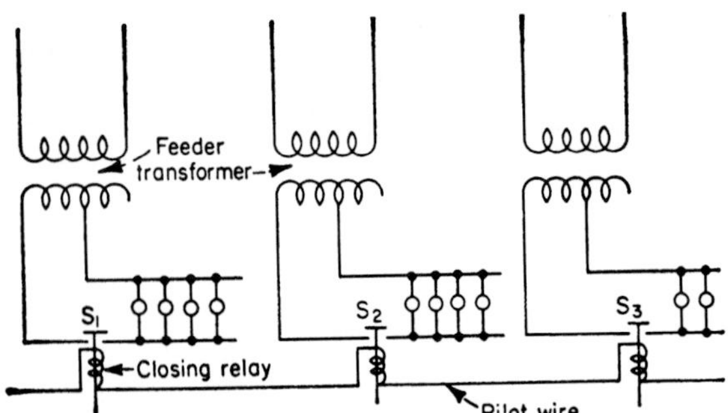

FIGURE 28.28 Pilot-wire control of multiple lighting circuits supplied from several feeder transformers.

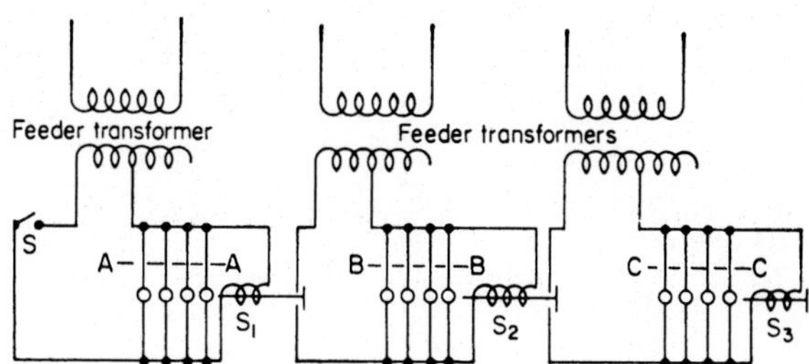

FIGURE 28.29 Cascade control of multiple lighting circuits supplied from several feeder transformers.

relays; the armatures fall back and in so doing close the lighting circuits. This method has the advantage that should a defect occur in the pilot wire or relays during the daytime, the lights will come on. The burning lamps indicate immediately that a fault has developed, and repairs can be made during the daylight hours.

Switching lighting circuits on and off by the cascade method means that the first circuit, upon being energized, activates a relay which in turn energizes the second circuit. Another relay activated by the second circuit energizes a third circuit and so on, as illustrated in Fig. 28.29. The only extra wiring required is from the end of one circuit to the relay of the next circuit. Figure 28.29 shows that when circuit A is energized, the relay S_1 for circuit B is activated, energizing circuit B. This is turn causes the relay S_2 for circuit C to operate, which energizes circuit C. The word *cascade* means to put one after the other.

Design of Street-Lighting Installations. For information on the design of street-lighting systems, the reader is encouraged to refer to the *Practice for Roadway Lighting* (ANSI/IES RP8), published by the American National Standards Institute (ANSI) under the sponsorship of the Illuminating Engineering Society. This provides information on the design of street-lighting systems and the measurement of their performance. Copies may be obtained from the Illuminating Engineering Society, 345 East 47th Street, New York, N.Y. 10017.

CHAPTER 29
UNDERGROUND SYSTEM

Underground transmission and distribution are installed when:

1. Space is not available for overhead lines, as in the congested downtown areas of large cities.
2. The hazard of high-voltage overhead lines is too great, as in heavily built-up areas of large cities.
3. The appearance of numerous heavy overhead lines would be unsightly, as in dense downtown areas or in new or redeveloped districts of cities.
4. Community ordinances, administrative codes, or franchise agreements require the installation of underground electric facilities in defined areas.

Underground lines cost more than the equivalent overhead lines. Underground lines are therefore used only when necessity demands them, as stated above.

Parts of Underground System. An underground system may consist of six parts:

1. Conduits or ducts
2. Manholes
3. Cables
4. Transformer vaults
5. Risers
6. Transformers

The first three parts listed above are illustrated in Fig. 29.1. The ducts are the hollow tubes in the conduit runs connecting the manholes, one with another. The cable is the electrical power circuit placed in the conduit through which power flows, and the manhole is the chamber where cable ends are spliced together to make a continuous circuit. Transformer vaults (Fig. 29.2) are underground rooms in which the power transformers, network protectors, voltage regulators, circuit breakers, meters, etc. are housed. Cables terminate in transformer vaults or customers' substations or connect with overhead lines in potheads, as illustrated in Fig. 29.3. The last termination is called a riser.

Conduits

Types. The hollow tubes or ducts running from manhole to manhole can be made of plastic polyvinyl chloride (PVC), fiberglass, fiber, tile, concrete, or steel. Plastic and fiberglass are currently used in most cases. Plastic PVC conduit is available for direct-burial installation or encasement in concrete.

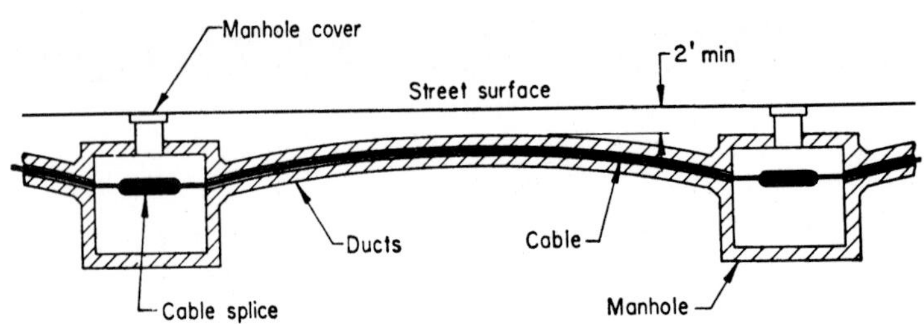

FIGURE 29.1 Principal parts of an underground conduit system for underground electrical power transmission and distribution.

Sizes. Ducts are made in varying sizes with inside diameters normally from 2 to 6 in. The size of duct selected depends on the size of the cable to be drawn into it or to be installed in the future. In general, when a single cable is installed in a duct, the diameter of the duct should be at least $^1/_2$ to $^3/_4$ in greater than the diameter of the cable. When more than one cable is drawn into one duct, the diameter of the duct should be at least $^1/_2$ to $^3/_4$ in greater than the diameter of the circle enclosing all cables. The duct opening should always be large enough to make the pulling in of the cable as easy as possible.

Duct runs should preferably be short, straight runs between manholes. This facilitates pulling in the cable. When this is impossible, as in going around obstacles, the bends should be laid out with the greatest possible radius of curvature, thereby avoiding sharp curves.

Duct runs should be slightly graded, that is, they should slope gently toward one or both manholes, thus allowing any water that may seep in to drain to one of the manholes. Figure 29.4 illustrates a duct run on level ground. It has a double slope, that is, half the run drains to each manhole. Good practice requires that the slope be no less than 3 in for every 100 ft of length. Figures 29.5 and 29.6 illustrate the manner of providing proper grade on moderate and steep slopes. In these cases, the drainage is all in one direction.

In general, the top of a duct run should not be less than 2 ft below the street surface, as indicated in the diagrams.

Manholes

A manhole is the opening in the underground duct system which houses cable splices and which cablemen enter to pull in cable and to make splices and tests. The manhole is therefore often called a splicing chamber or cable vault.

Manholes are generally built of reinforced concrete or brick, and the covers are made of steel. The opening leading from the street to the manhole chamber is called the chimney or throat. An opening having a minimum diameter of 32 in is recommended. This size opening is large enough for a man to enter on a ladder and also to pass the equipment needed for splicing and testing.

Manholes are usually located at street intersections so that the ducts can be run in all four directions. The distance between manholes seldom exceeds 500 ft, as this is about as long a cable as can be pulled into the duct. In long blocks, additional manholes are placed in the middle of the block.

Two-Way Manholes. The shape and size of manholes depend on the number and size of cables to be accommodated and the number of directions in which ducts leave the manhole. The simplest shape of manhole is illustrated in Fig. 29.7, where ducts and cables enter and

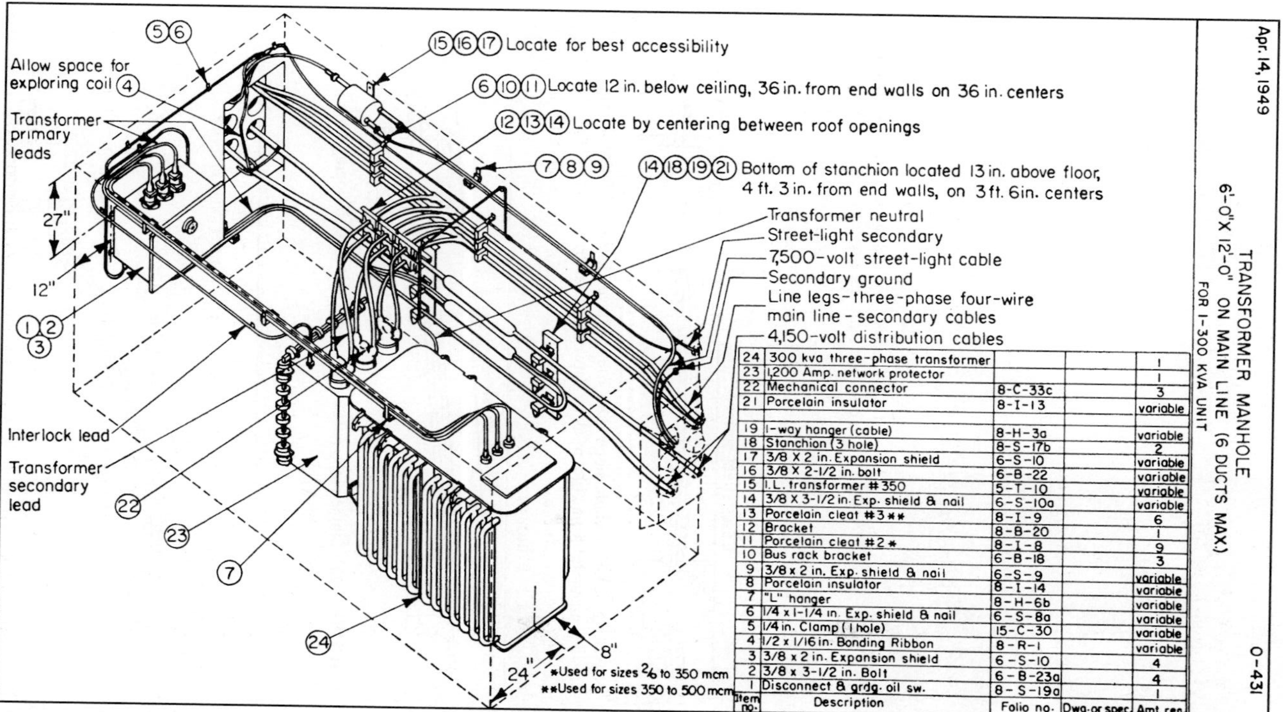

Item no.	Description	Folio no.	Dwg. or spec.	Amt. req.
24	300 kva three-phase transformer			1
23	1,200 Amp. network protector			1
22	Mechanical connector	8-C-33c		3
21	Porcelain insulator	8-I-13		variable
19	1-way hanger (cable)	8-H-3a		variable
18	Stanchion (3 hole)	8-S-17b		2
17	3/8 X 2 in. Expansion shield	6-S-10		variable
16	3/8 X 2-1/2 in. bolt	6-B-22		variable
15	I.L. transformer #350	5-T-10		variable
14	3/8 X 3-1/2 in. Exp. shield & nail	6-S-10a		variable
13	Porcelain cleat #3**	8-I-9		6
12	Bracket	8-B-20		1
11	Porcelain cleat #2*	8-I-8		9
10	Bus rack bracket	6-B-18		3
9	3/8 x 2 in. Exp. shield & nail	6-S-9		variable
8	Porcelain insulator	8-I-14		variable
7	"L" hanger	8-H-6b		variable
6	1/4 x 1-1/4 in. Exp. shield & nail	6-S-8a		variable
5	1/4 in. Clamp (1 hole)	15-C-30		variable
4	1/2 x 1/16 in. Bonding Ribbon	8-R-1		variable
3	3/8 x 2 in. Expansion shield	6-S-10		4
2	3/8 x 3-1/2 in. Bolt	6-B-23a		4
1	Disconnect & grdg. oil sw.	8-S-19a		1

FIGURE 29.2 Typical transformer vault. (*Courtesy Public Service Electric and Gas Company of New Jersey.*)

FIGURE 29.3 Typical cable riser with potheads, fused cutouts, and lightning arresters. (*Courtesy S & C Electric Co.*)

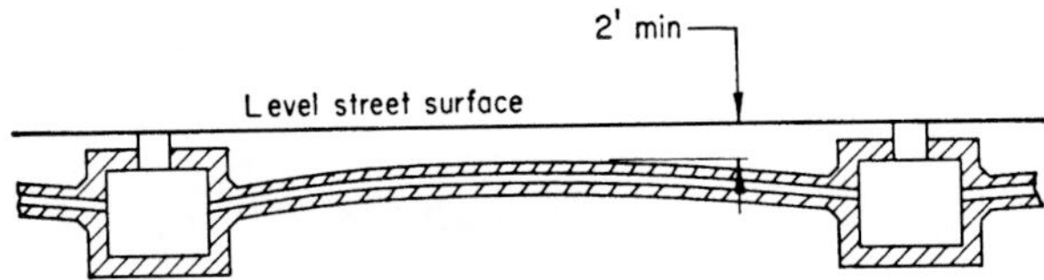

FIGURE 29.4 Double-slope duct run on level ground. Water drains toward manholes from point midway between manholes.

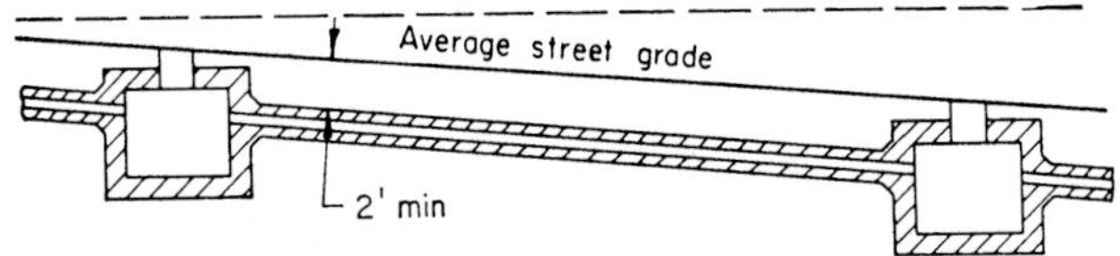

FIGURE 29.5 Single-slope duct run between manholes on sloping street.

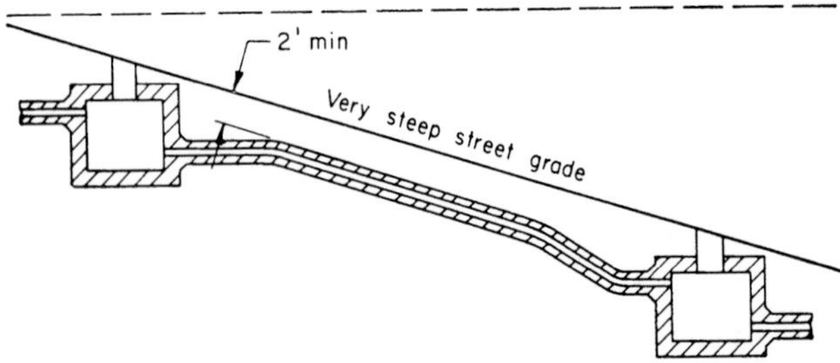

FIGURE 29.6 Single-slope duct run on street with steep grade. Note that duct run leaves at bottom of one manhole and enters near top of next manhole.

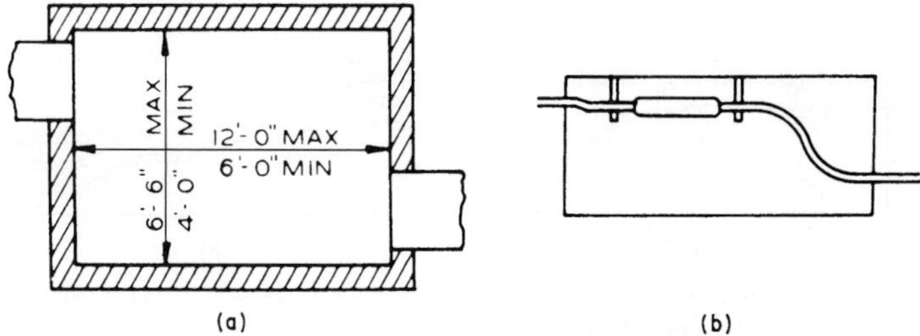

FIGURE 29.7 Plan views of two-way manhole. Note how cable is racked around sides of manhole with double-reverse curve and recommended dimensions shown.

leave in only two directions. Such a manhole is called a two-way manhole. Note in Fig. 29.7*b* that the cables do not pass through the manhole in a straight line but instead are offset, that is, they enter and leave at opposite corners. This is to provide opportunity to rack the cables around the sides of the manhole. Cables expand and contract in length with changes in temperature caused by changes in electrical power flow through the cable. The most practical way to take care of this expansion and contraction is to provide a large reverse curve in the cable as it passes through the manhole. This reverse curve, consisting of two large-radius 90° bends, enables the cable to take up the expansion movements, thereby reducing to minimum the danger of damaging the cable.

A dimensioned side view of such a two-way manhole is given in Fig. 29.8. Note that the inside vertical height should not normally be less than 7 ft or more than 8 ft. These dimensions allow for the racking of the cables with their splices around the side of the manhole and sufficient space and headroom for the splicer to work.

In case ducts and cables do not enter and leave the manhole at the same level, the manhole design would appear as in Fig. 29.9. Note the change in level of the cables as they pass through the manhole in the side view and the reverse curve of the cables in the plan view.

A two-way 90° turning manhole with cables at the same level would appear as in Fig. 29.10. Only one curve is needed here to allow for cable expansion. Such a manhole is usually called a corner manhole.

Three- and Four-Way Manholes. Manholes designed for three-way duct and cable lines are illustrated in Fig. 29.11, and manholes designed for four-way lines are illustrated in Fig. 29.12. The headroom in each case need not exceed 8 ft but should not be less than 7 ft.

Manhole Drainage. Drainage should be provided in manholes to keep them free of water. Oftentimes a "dry well" is resorted to. This requires that a part of the manhole floor be left open, the dirt excavated to a depth of a foot or two, and the hole filled with stones. This permits the water to seep away gradually. In case a connection to the city storm sewer

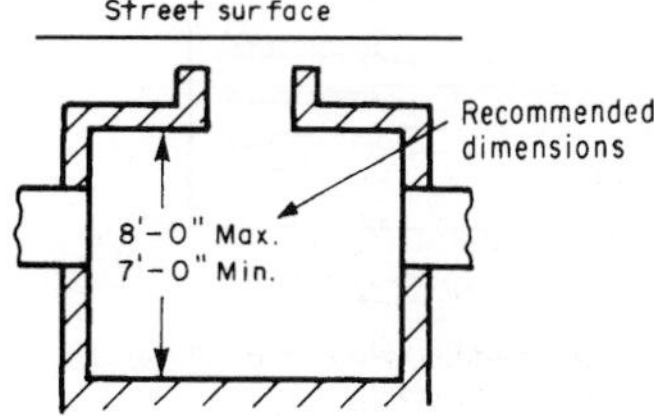

FIGURE 29.8 Side view of two-way manhole. Ducts enter and leave at same level.

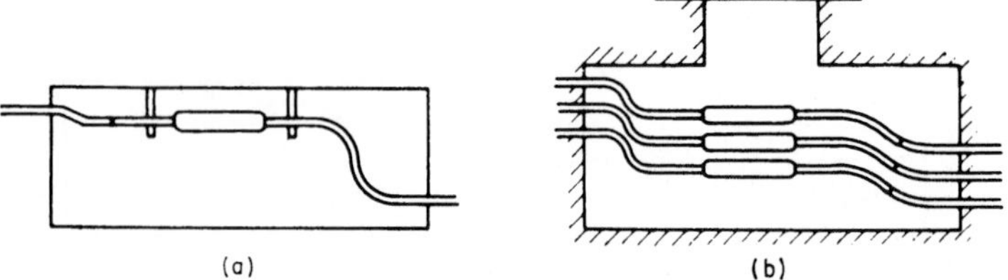

FIGURE 29.9 Plan view (*a*) and side view (*b*) of manhole when ducts and cables enter and leave manhole at different levels.

line can be made conveniently, this is often done. Figure 29.13 shows a cross section or side view of a typical sewer connection. In areas along waterfronts where the soil is so moist or the water table so high that water will not drain out of a manhole, the only alternative remaining is to build the manhole walls and floor of waterproofed concrete. In addition, the concrete is usually painted with a waterproof paint. In extreme cases, an automatic pump is installed which removes the water from the sump below floor level.

Distribution Manholes. When only a few trunk ducts are required for primary feeders to serve only the local distribution system, smaller manholes are satisfactory. The cables are usually smaller in diameter and therefore more easily racked. The dimensions for a two-way distribution manhole are given in Fig. 29.14. This manhole could accommodate a maximum of six trunk ducts. Note that the headroom is reduced to 7 ft. Most distribution manholes are made with precast concrete. The precast-concrete manholes are made at a remote location and hauled to the site and set in place. This method of construction for small manholes has proven to be economical.

A similar distribution manhole providing for primary and secondary circuits radiating in four directions at a street intersection is shown in Fig. 29.15.

The same type of cable manhole may be used for secondary distribution at locations where it is necessary to accommodate several circuits. This occurs frequently in congested areas adjacent to buildings requiring large and multiple service from either or both the primary and secondary circuits.

Handholes. When small splicing chambers are necessary on lateral two-way duct lines, so-called handholes are used. They should be restricted to one-conductor cables only because the cable makes a complete loop once around in the hole, as shown in Fig. 29.16. The handhole cover should be large enough to permit working from above the handhole.

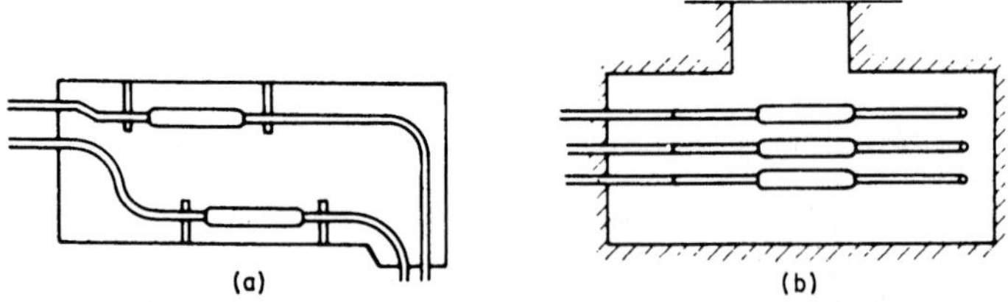

FIGURE 29.10 Plan view (*a*) and side view (*b*) of two-way manhole when ducts and cables make a 90° turn. Such a manhole is usually called a corner manhole.

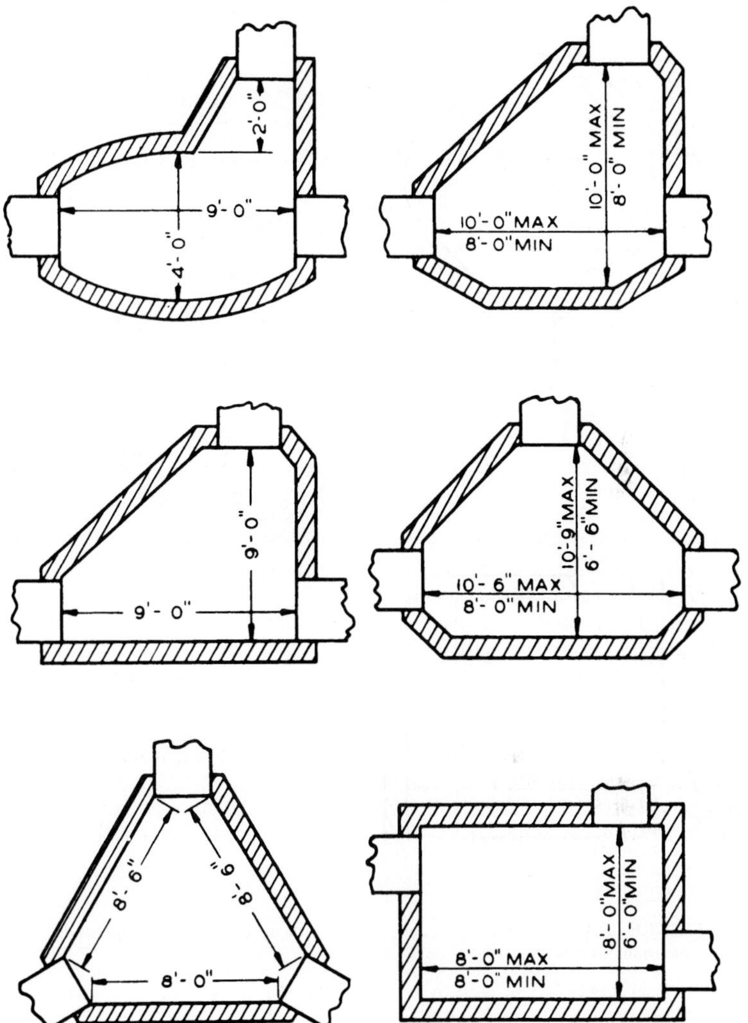

FIGURE 29.11 Possible variations of three-way manhole with recommended dimensions shown.

Cable

On systems operating above 2 kV to ground, the design of the conductors or cables installed in nonmetallic conduit should consider the need for an effectively grounded shield and/or sheath. Bending during handling, installation, and operation should be controlled to avoid damage to any of the components of the supply cable. Pulling tensions and sidewall pressures of cable should be limited to avoid damage to the supply cable. Manufacturer's recommendations may be used as a guide. Ducts should be cleaned of foreign material which could damage the supply cable during pulling operations. Cable lubricants should not be

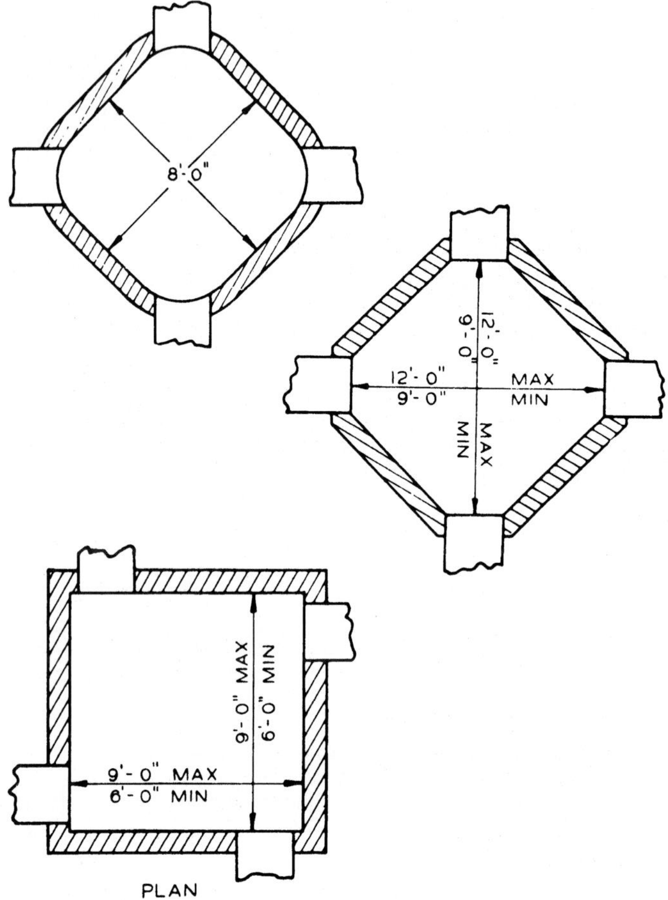

FIGURE 29.12 Possible variations of four-way manhole with recommended dimensions shown.

detrimental to cable or conduit systems. On slopes or vertical runs, consideration should be given to restraining cables to prevent downhill movement. Supply, control, and communications cables should not be installed in the same duct unless the cables are maintained or operated by the same utility.

Cable supports should be designed to withstand both live and static loading and should be compatible with the environment. Supports should be provided to maintain specified separation between cables. Horizontal runs of supply cables should be supported at least 3 in above the floor or be suitably protected. This rule does not apply to grounding or bonding conductors. The installation should allow cable movement without destructive concentration of stresses. The cable should remain on supports during operation. Adequate working space should be provided. Where cable and/or equipment is to be installed in a joint-use manhole or vault, it should be done only with the concurrence of all parties concerned. Supply and communication cables should be racked from separate walls. Crossings should

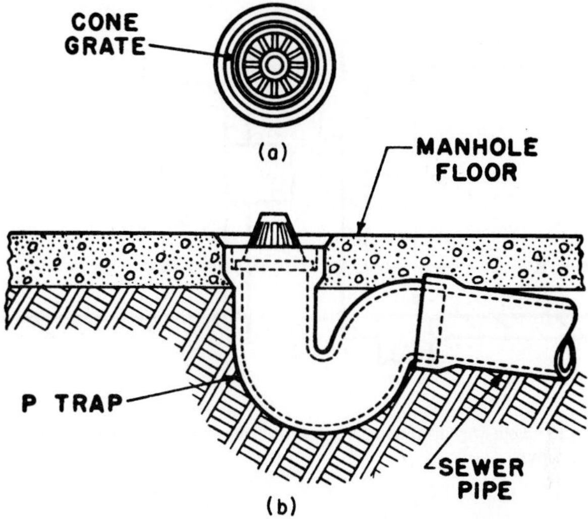

FIGURE 29.13 Manhole drainage provided with sewer connection. (*a*) Plan view; (*b*) elevation.

be avoided. Where supply and communications cables must be racked from the same wall, the supply cables should be racked below the communication cables. Supply and communications facilities should be installed to permit access to either without moving the other. Clearances should be maintained as specified in Table 29.1.

These separations do not apply to grounding conductors, and they may be reduced by mutual agreement between the parties concerned when suitable barriers or guards are installed. Cables should be permanently identified by tags or otherwise at each manhole or other access opening of the conduit system. This requirement does not apply where the position of a cable, in conjunction with diagrams or maps supplied to workmen, gives sufficient identification. All identification should be of a corrosion-resistant material suitable for the environment. All identification should be of such quality and located so as to be readable with auxiliary lighting. Where cables in a manhole are maintained or operated by

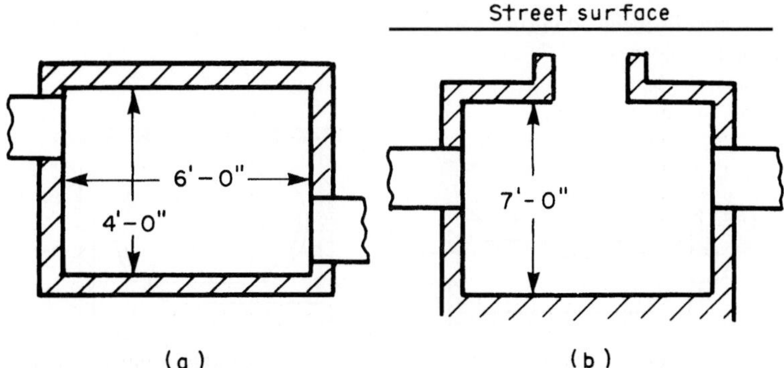

FIGURE 29.14 Two-way distribution manhole with recommended dimensions shown. (*a*) Plan view; (*b*) elevation.

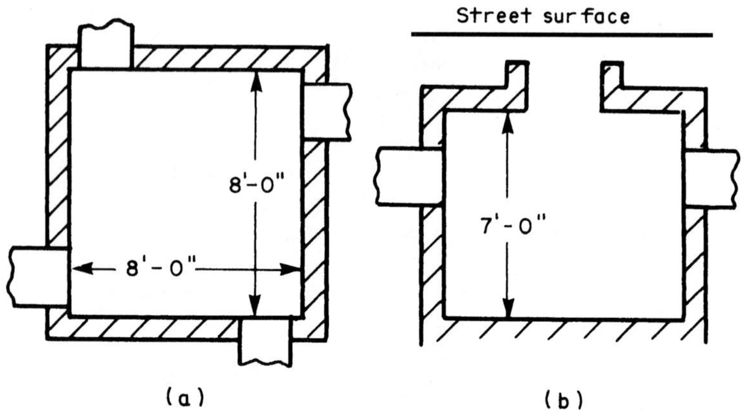

FIGURE 29.15 Four-way distribution manhole with recommended dimensions shown. (*a*) Plan view; (*b*) elevation.

different utilities or are of supply and communication usage, they should be permanently marked as to company and/or type of use.

Insulation shielding of cable and joints shall be effectively grounded. Cable sheaths or shields which are connected to ground at a manhole shall be bonded or connected to a common ground. Bonding and grounding leads shall be of a corrosion-resistant material suitable for the environment or suitably protected.

Although fireproofing is not a requirement, it may be provided in accordance with each utility's normal service reliability practice to provide protection from external fire.

Special circuits operating at voltages in excess of 400 volts to ground and used for supplying power solely to communications equipment may be included in communications cables under the following conditions:

1. Such cables shall have a conductive sheath or shields which shall be effectively grounded, and each such circuit shall be carried on conductors which are individually enclosed with an effectively grounded shield.
2. All circuits in such cables shall be owned or operated by one party and shall be maintained only by qualified personnel.
3. Supply circuits included in such cables shall be terminated at points accessible only to qualified employees.

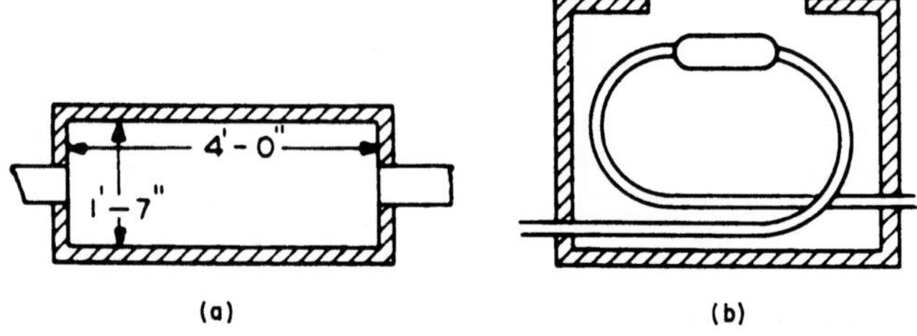

FIGURE 29.16 Distribution handhole with recommended dimensions shown. (*a*) Plan view; (*b*) elevation.

TABLE 29.1 Minimum Separation Between Supply and Communications Facilities in Joint-Use Manholes and Vaults

Phase-to-phase supply voltage	Inches surface to surface
0 to 15,000	6
15,001 to 50,000	9
50,001 to 120,000	12
120,001 and above	24

4. Communications circuits brought out of such cables, if they do not terminate in a repeater station or terminal office, shall be protected or arranged so that in event of a failure within the cable the voltage on the communications circuit will not exceed 400 volts to ground.
5. Terminal apparatus for the power supply shall be so arranged that live parts are inaccessible when such supply circuits are energized.
6. Such cables shall be identified, and the identification shall meet the pertinent requirements.

The requirements do not apply to supply circuits of 550 volts or less which carry power not in excess of 3200 watts.

Cable Construction. Power cables are constructed in many different ways. Cables are generally classified as nonshielded or shielded cables. Nonshielded cables normally consist of a conductor, conductor-strand shielding, insulation, and a jacket. The use of nonshielded power cables is limited to the lower primary voltages, which normally do not exceed 7200 volts. Shielded power cables are generally manufactured with a conductor, conductor-strand shielding, insulation, semiconducting insulation shielding, metallic insulation shielding, and a sheath. If the sheath is metallic, it may serve as the metallic insulation shielding and be covered with a nonmetallic jacket to protect the sheath (Fig. 29.17). Shielded cables are commonly used on circuits operating at 4,160Y/2400 volts or higher (Fig. 29.18).

Multiconductor cables used for three-phase power circuits consist of three single conductor cables with a common sheath and/or jacket (Fig. 29.19).

Cable Conductors. Copper and aluminum are commonly used conductor materials. The conductors of large power cables have a stranded conductor to make a cable as flexible as possible, thus facilitating bending. Power cables with a conductor size of no. 2 AWG and smaller may use a solid conductor. Copper conductor has a high conductivity and is easily worked and handled. Aluminum conductor is lighter in weight and is commonly used to reduce the initial installation cost. Aluminum conductors must be larger than copper conductors to supply a given load because of the lower conductivity. Aluminum will "cold flow" if improper connectors are used. The surface of aluminum conductors oxidizes quickly and can be damaged by galvanic action when touching other metals with moisture present. The manufacturer's instructions must be followed carefully to prevent a defective splice or termination of aluminum conductors.

The stranded conductors have several shapes that are used for various power-cable constructions (Fig. 29.20).

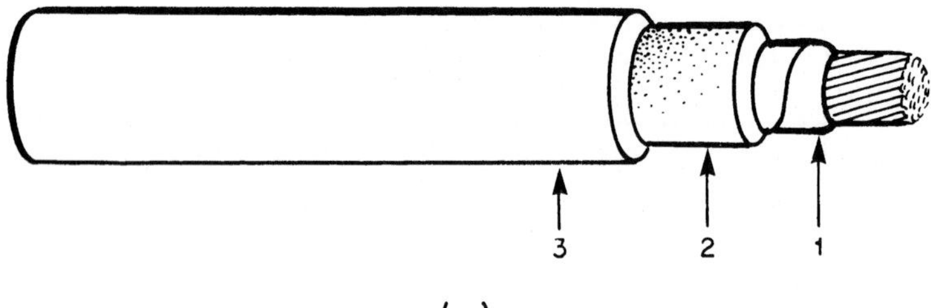

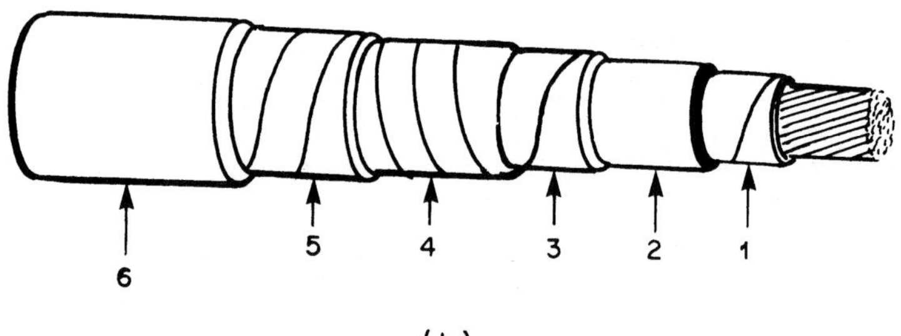

FIGURE 29.17 Cable construction. (*a*) Nonshielded cables: (1) strand shielding, (2) insulation, (3) cable jacket. (*b*) Shielded power cable: (1) strand shielding, (2) insulation, (3) semiconducting material, (4) metallic shielding, (5) bedding tape, (6) cable jacket.

FIGURE 29.18 Single-conductor power cable. (*Courtesy The Okonite Co.*)

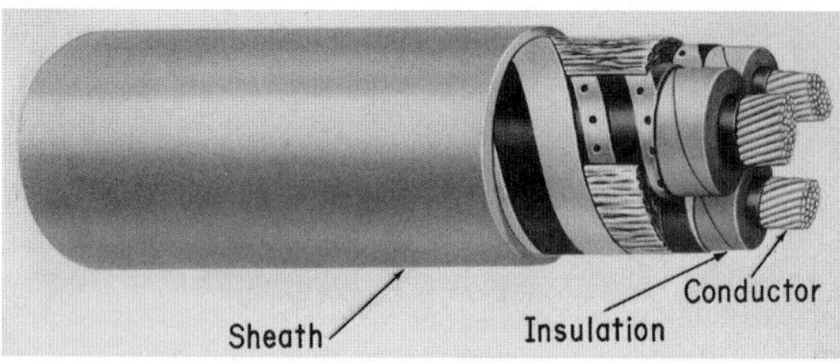

FIGURE 29.19 Typical three-conductor power cable.

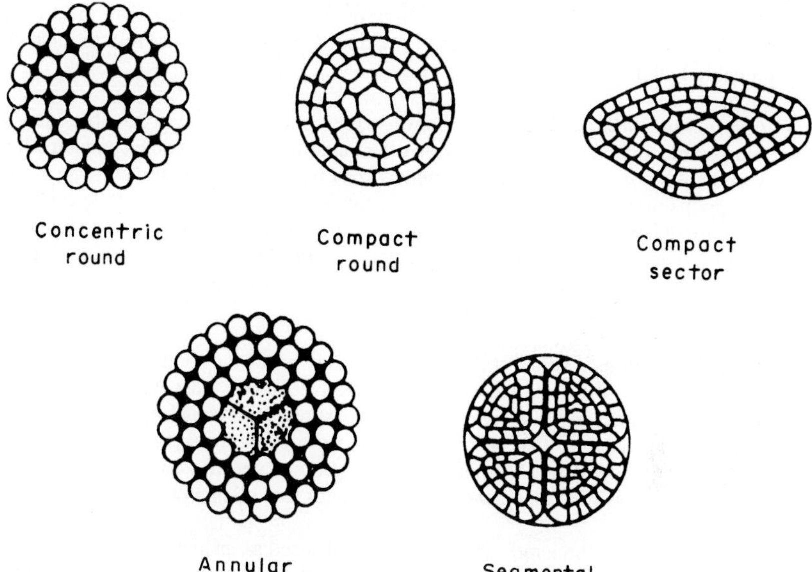

FIGURE 29.20 Diagrams of common shapes of stranded conductors.

Concentric-round-stranded conductors have wires of the same diameter. When six wires of the same size are helically wrapped around a single wire, they lie tightly to the inner wire. This progression of six additional wires in each layer will produce a conductor of any diameter.

The compact-round-stranded conductor has a smaller diameter than the concentric-round conductor but is less flexible. The smaller diameter is obtained by laying all the strands in the same direction and then rolling each layer to eliminate the spaces.

A compact-sector-stranded conductor is used to obtain a smaller-diameter cable. It is normally furnished with impregnated-paper or vanished-cambric insulation. Sector conductors are stiffer and more difficult to splice. However, they are cheaper and lighter in weight (Fig. 29.21).

Annular conductors are usually composed of wires stranded concentrically around a ropelike cord. This conductor is generally used for sizes above 1000 kCMIL to help combat the "skin effect" problem. It has been shown that alternating current flows more densely near the outer surface of a conductor; therefore, increasing the surface area increases the current carrying capacity.

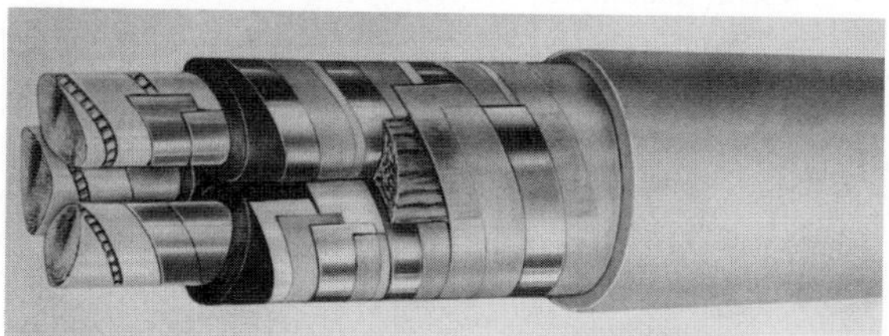

FIGURE 29.21 Power cable with sector-shaped conductors showing various layers of shielding and insulation. (*Courtesy The Okonite Co.*)

A segmental conductor is composed of three or four segments which are electrically separated from one another. This reduces the "skin effect" problem and results in small-diameter cable with high current-carrying ability.

Cable-Conductor Sealing. The space between the conductor strands in cables can be filled with a semiconducting compound during the manufacture of the cable conductor. The filling compound is designed to prevent water from penetrating the stranded-cable conductor during manufacture, storage, installation, splicing, and terminating of the cable.

Water in the interstices of the cable conductor is one of the principal causes of electrochemical trees that cause premature failure of polymer-insulated cables. Compound filling of the spaces between conductor stranding during manufacture of the conductor eliminates the possibility of water being trapped in the conductor.

Cable-Strand Shielding. Strand shielding is applied between the conductor and the cable insulation. It is made from materials such as conductive fibrous tapes, conducting paints, conducting rubber, and graphite compounds. The strand shielding should be securely anchored to the dielectric so that no ionizable air can exist between them if the cable is sharply bent. A semiconducting tape has proven that it can perform this function in addition to providing an easily stripped and clean conductor for splicing or terminating.

Cable Insulation. The three most common materials used as insulation for power cables are (1) taped insulation, such as impregnated paper and varnished cambric; (2) rubber or rubberlike synthetic compounds, such as butyl rubber, silicone rubber, and oil-based rubber; and (3) thermoplastics, such as PVC, polyethylene, and cross-linked polyethylenes. The insulation best suited for a specific cable depends on many factors such as operating voltage, current load, ambient temperature, type of installation (underground, ducts, direct buried, etc.), and type of outer covering.

Impregnated-paper insulation is used for cables operating at 5 kV and above where low dielectric loss, low power factor, and low ionization are important considerations. It consists of multiple layers of paper tape helically wrapped around the conductor to be insulated. The total wall of paper is then vacuum dried and impregnated with an insulating compound. The impregnating compound generally is a pure mineral oil or oil blended to obtain higher viscosity. Impregnated-paper insulation has good electrical properties (high dielectric strength, low power factor, and low dielectric loss), has good frequency stability, and is thinner than rubber or varnished-cambric insulation for cables of the same voltage rating. Impregnated-paper insulation is susceptible to deterioration due to water and must have a lead cover, which makes the cable difficult and expensive to terminate and splice.

Varnished-cambric insulation is commonly used on cables ranging between 600 and 23,000 volts. It is closely woven cotton cloth, both sides of which have been coated with several layers of insulating varnish. These tapes are wrapped helically around each conductor. Between the layers, a viscous, moisture-repelling, nondrying slipper compound is used. Varnished-cambric insulation has moderately high dielectric and impulse strength; is ozone, corona, and oil resistant; and is used extensively in transformers and oil switches. Varnished-cambric insulation tends to have a high power factor and low ionization level, and it must be covered by a lead sheath for moisture protection if used on a cable in a nonprotected atmosphere.

Rubber insulation is generally used in cables from 600 to 15,000 volts. Rubber insulations are made from either synthetic rubber, natural rubber, or butyl synthetic rubber. The compounds may contain as little as 20 percent or as high as 90 percent rubber. The balance of the compound consists of suitable organic or inorganic materials which facilitate manufacture and develop desirable properties not possessed by rubber alone. The rubber insulations are rated anywhere from 60° to 130°C maximum.

Oil-base rubber insulation is the oldest type of the rubber compounds. It consists of a vulcanized vegetable oil and bitumen base which is compounded with other ingredients to form an ozone-resistant insulation. This type of insulation carries a 75°C conductor temperature rating and is furnished on power cables for operation up to 35 kV.

Butyl synthetic rubber insulation has molecules that have been tailor-made to meet the requirement of the highest-grade vulcanized insulation for high-voltage applications. It is basically a heat- and ozone-resisting rubber. Butyl rubber has excellent resistance to moisture, excellent dielectric strength and insulation resistance, and low dielectric loss factor. Butyl dielectrics are rated at 90°C up to 2000 volts, 85°C up to 15,000 volts, and 80°C up to 28,000 volts by the Insulated Power Cable Engineers Association (IPCEA).

Silicone rubber is a synthetic compound that has excellent heat, moisture, and ozone resistance. Cables insulated with silicone rubber can be operated at temperatures up to 125°C for power applications and are flexible down to –54°C. The tensile strength, elongation, and abrasion resistance of silicone rubber at ordinary temperatures are not as great as those of organic rubber compounds. Silicone rubber is extremely resistant to flame in the sense that when exposed to direct flame it will gradually burn and leave an ash which is electrically nonconducting, and if contained by a glass braid, it will serve as an insulator until the emergency is over and the wiring can be replaced. It is more expensive than most other types of cable insulation.

Ethylene-propylene rubber-base (EPR) insulations are vulcanizable compounds which can be made with one of the EPR elastomer gums developed by several chemical companies. Cables made with EPR insulations show great promise of combining in one dielectric a moisture resistance approaching that of polyethylene, the dielectric-strength retention of oil base, and the heat and form stability of butyl insulations. These insulations are rated at 125°C for emergency or hot-spot operation and a normal operating temperature of at least 90°C.

The Association of Edison Illuminating Companies has written *Specifications for Ethylene Propylene Rubber Insulated Shielded Power Cables Rated 5 Through 69 kV.**

PVC is the thermoplastic most commonly used for low-voltage power and control cables. Its power factor and dielectric loss limit its use to low voltages, 600 volts or less, for general-purpose wiring. From the standpoint of performance, PVC differs from the vulcanized rubberlike compounds in its sensitivity to temperature and melting when in high heat. Thermoplastic building wiring insulation is well known as type THW or TW.

Polyethylene cable insulation contains relatively small amounts of material other than the base polymer. It may contain coloring materials to protect if from exposure to sunlight and weather, and it should contain a small amount of carbon black. Polyethylene has a low power factor and dielectric loss, high dielectric and impulse strength, excellent moisture resistance, and high impact strength and abrasion resistance.

Cross-linked polyethylene cable insulation has compounds consisting of the polyethylene resin, fillers, antioxidants, and vulcanizing agents. It has most of the electrical advantages of polyethylene plus much better mechanical properties such as superior resistance to heat deformation and environmental stress cracking. It should be noted that the insulation thickness used on these cables is thin as that of polyethylene power cables and in many cases much thinner as a result of its exceptional electric strength. Most cross-linked polyethylenes are rated at 90°C at 15 kV with a 100 hour per year overload temperature of 130°C for 5 years of its life. For the same voltage rating, the insulation thickness of the rubber type is generally thickest, cross-linked polyethylene the thinnest, and straight polyethylene somewhere between these two. The specific thicknesses for rubber and polyethylene are listed in the NEMA-IPCEA manual.

---

*Association of Edison Illuminating Companies, 51 East 42nd Street, New York, N.Y. 10017

The Association of Edison Illuminating Companies has written *Specifications for Thermoplastic and Crosslinked Polyethylene Insulated Shielded Power Cables Rated 5 through 46 kV* and *Specifications for Crosslinked Polyethylene Insulated Shielded Power Cable Rated 69 Through 138 kV*. Experimental cross-linked polyethylene insulated cables have been developed for operation at 230 and 345 kV.

Cable Insulation Shielding. Insulation shielding confines the dielectric field within the cable to obtain symmetrical radial distribution of voltage stress within the dielectric. The shielding protects cables connected to overhead lines or otherwise subject to induced potentials and limits radio interference. If the shield is grounded, it reduces the hazard of shock and provides a ground return in case of a cable failure.

Conducting nonmetallic tape is used directly over the insulation of shielded power conductors to provide more intimate contact between the insulation and metallic shield. It is at least 0.0025 in thick and consists of carbon-black-filled cloth, rubber, or neoprene. Extruded semiconducting polyethylene is used over the insulation of concentric neutral underground cables. The extruded jackets are superior to tapes in both durability and moisture-penetration resistance.

Insulation shieldings are either metal, tape, braid, or wire. In general, rubber-covered cable with a synthetic jacket should be shielded when applied to systems over 3000 volts. Requirements for shielding are given in the following industrial specifications: *Solid Type Impregnated-Paper-Insulated Lead-Covered Cables*, issued by AEIC, and *Wire and Cable with Rubber, Rubber-like and Thermoplastic Insulations*, issued by IPCEA and NEMA. The metallic tape is generally 5 mils thick and is applied over the conducting nonmetallic tape.

Cable Bedding Tapes. Bedding tapes are generally used in a cable to add mechanical strength. They are located in various places by different manufacturers-under the sheath, under the shielding, and between the conductor and insulation. If one is not sure whether a tape is bedding or conductive, it should always be considered to be semiconductive and handled accordingly.

Cable Filler Tapes. Filler tapes of suitable materials generally are used in the interstices of cables, where necessary, to give the completed assembly a substantially circular cross section. Fillers are normally either treated hemp rope, synthetic rubber compound, or thermoplastic material.

Cable Jackets, Sheaths, or Coverings. The jacket or sheath over the cable insulation is a protective covering which provides mechanical protection; in some degree it is used for electrical reasons (Fig. 29.22). Sheaths are made from basically four different materials-fibrous, rubber and rubberlike, thermoplastic, and metallic.

Fibrous jackets made with cotton braid provide a relatively low-cost, flexible covering with good mechanical strength. They are the most extensively used type of protective

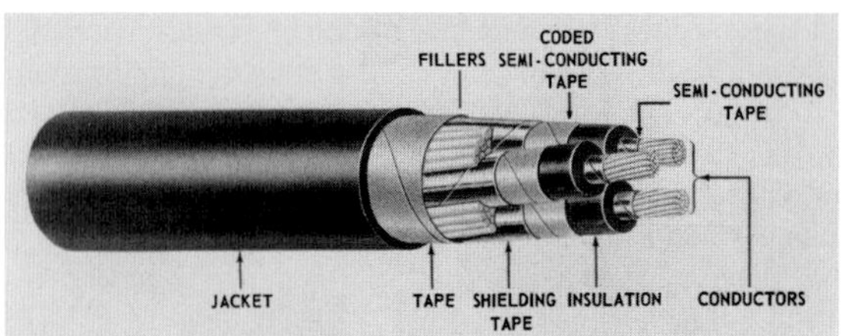

FIGURE 29.22 Three-conductor rubber-insulated shielded cable with polyvinyl chloride jacket.

covering for electrical wires and cables installed in a dry location. Various materials, such as compounds of petroleum asphalt and stearin pitch, are used for saturating and finishing cotton braids to provide moisture resistance and nonflammability. A final coating of paraffin wax yields a nontacky finish that facilitates pulling the cables into conduits.

Jute wrapped on the outside of cable acts as a cushion and binder, protects galvanizing from abrasion during installation, and acts as a structure to hold intact corrosion-resisting asphalt coatings. The most common use of jute in the cable industry is on underground and submarine cables. Jute is usually spiral wrapped rather than woven. It has been used to provide a bedding under metallic coverings.

Basket-weave glass-fiber braids are often used in preference to cotton braids, particularly where better resistance to deterioration at elevated temperatures is desirable. They are sometimes used for underground cable because of the good stability of the glass fiber under wet service conditions.

Synthetic rubber is a good heavy-duty jacket for aerial, duct, and buried installations. Two types of jackets commonly used are neoprene and Buna-N. The use of Buna-N has been limited to special service conditions, such as extremely low temperature locations or where maximum oil resistance is desired. Its strength and flexibility at low temperatures and oil resistance are excellent. Neoprene has excellent weather resistance and good aging and mechanical properties. Black neoprene properly compounded will not deteriorate on long exposure to sunlight and weather or in wet locations, such as underground. Heavy-duty portable cords and cables are frequently made with neoprene-compounded sheaths. Neoprene is frequently used in place of cotton braids and other fibrous coverings.

PVC has excellent resistance to acids, alkalies, cutting, flame, moisture, oils, ozone, sunlight, and weathering. It is a good heavy-duty jacket when installed under conditions that eliminate harmful plastic flow. It is not as flexible as neoprene and cannot be used at extremely low temperatures. PVC compound jackets are often used over lead and interlocked armor sheaths for protection against oils, gasoline, acids, and alkalies.

Polyethylene is quite stable physically throughout its temperature range and resistant to many destructive chemicals and oils up to at least 70°C. Excellent moisture resistance makes polyethylene suitable as a nonmetallic sheath in damp locations and for aerial installation. In thin sections polyethylene is weak in abrasion resistance and often deforms at high temperatures. It will also cold flow under pressure.

Nylon is a thermoplastic of the polyamide type. A thin-layer film of nylon is sometimes extruded over a wall of polyethylene insulation to supplement the physical properties of the latter. Nylon is not subject to cold flow and therefore can be used as a protective covering. Nylon is mechanically strong and is not affected by oils.

Metallic sheaths provide mechanical protection to the cable and may also function as a current return. Lead sheaths are used with rubber-, varnished-cambric-, and paper-insulated cable as mechanical protection and to prevent entrance of moisture or loss of insulation impregnant (Fig. 29.23).

FIGURE 29.23 Three-conductor varnished-cambric-insulated, shielded, lead-sheathed cable. (*Courtesy The Okonite Co.*)

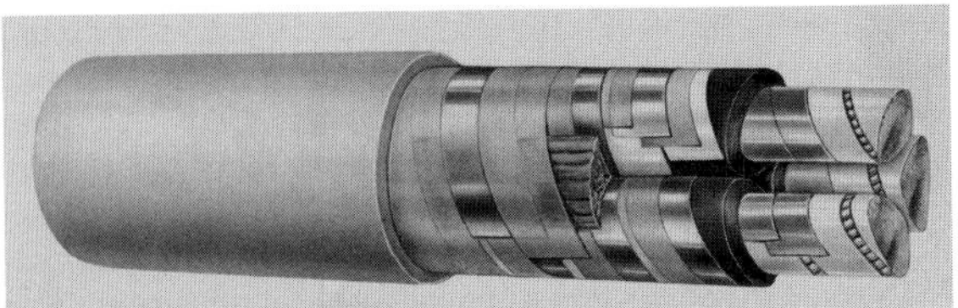

FIGURE 29.24 Three-conductor lead-sheathed, paper-insulated power cable showing various layers of shielding and insulation. (*Courtesy The Okonite Co.*)

Lead is particularly suited to the purpose since it may be applied by extrusion presses in long, continuous lengths. It forms a flexible covering, allowing the cable to be bent easily during handling and installation. Experience has shown that an intact lead sheath is the most effective protection for insulation against moisture, weather, and oxidation (Fig. 29.24).

The physical properties of aluminum are quite good. It is lightweight and capable of withstanding high internal pressures. It is not as ductile a metal as lead and therefore requires a larger bending radius. It is also necessary to protect it against chemical corrosion, particularly if it is drawn into ducts or laid directly on the ground. A thermoplastic or neoprene jacket is sometimes applied over the aluminum to eliminate this corrosion. Aluminum has a higher ac loss in the sheath as compared to lead; this will influence the effective ac resistance of the conductors when calculating the current-carrying capacity of the cable.

Steel armors for cables consist of flat-metal-tape armor, interlocked-metal-tape armor, steel-wire armor, or basket-weave armor. The metallic armor is usually applied over a jute bedding which is applied over a fibrous covering, jacket, or lead sheath for mechanical strength and mechanical outer protection. Single-conductor cables, when armored with steel, are subject to the magnetic properties of the steel. This condition affects the reactance of the circuit and causes heat losses in the armor, which, in turn, affect the current-carrying capacity of the cable.

Rubber Cable Specification. This specification covers a typical 15-kV single conductor, rubber-insulated, shielded, jacketed power cable with semicompressed concentric-stranded aluminum conductor. Application of this cable is for 60-Hz Y-connected system with grounded neutral operating at 13,200Y/7620 volts.

The cable shall be suitable for installing in ducts or direct burial in the earth in wet or dry locations.

Cable shall be furnished with a 750-kCMIL stranded aluminum conductor.

Maximum conductor temperature shall be 90°C normal, 130°C emergency, and 250°C short-circuit.

Conductor: Alloy 1350 aluminum. Stranding shall be class B stranded per ASTM B231 and semicompressed.

Insulation System: The insulation shall be heat, moisture, ozone, corona, and impact resistant thermosetting-rubber-based elastomeric compound. The insulation shall be nonblack in color. When examined under 15 power magnification, there shall be no inclusions, contaminants, or voids. The insulation shall consist of a conductor energy-suppression layer, a primary wall of insulation, and an outer energy-suppression layer.

Conductor Energy-Suppression Layer: The conductor energy-suppression layer shall be extruded over the stranded conductor with good concentricity and shall be a black insulating material which is compatible with the conductor and the primary insulation. The extruded energy-suppression layer shall have a specific inductive capacitance of 10 or greater. In addition to its energy-suppression function, this layer shall provide stress relief at the interface with the primary insulation. The energy-suppression layer shall be easily removed from the conductor with the use of commonly available tools and shall have a minimum average thickness of 0.027 in and a minimum thickness at any part of the cable of 0.024 in when measured over the top of the strands. The outer surface of the extruded stress-suppression layer shall be cylindrical, smooth, and free of significant irregularities.

Primary Insulation: The primary insulation shall be a thermosetting rubber-based elastomeric compound and shall be nonblack in color. The primary insulation shall be extruded directly over the conductor energy-suppression layer with good concentricity and shall be compatible with the conductor energy-suppression layer. The outer surface of the extruded primary insulation shall be cylindrical, smooth, and free of significant irregularities.

Insulation Energy-Suppression Layer: The insulation energy-suppression layer shall be extruded directly over the primary insulation with good concentricity and shall be a black insulating material compatible with the primary insulation and adjacent components of the insulation shielding system. This extruded stress-suppression layer shall have a specific inductive capacitance of 10 or greater. The extruded energy-suppression layer shall have a minimum average thickness of 0.040 in and a minimum thickness at any part of the cable of 0.035 in when measured over the outer surface of the insulation. The outer surface of the extruded layer shall be cylindrical, smooth, and free of significant irregularities.

Insulation Thickness: The minimum average thickness of the insulation system shall be 0.225 in. The inner and outer energy-suppression layers are to be included with the primary insulation when measuring the insulation-system thickness since they are insulating in nature.

Insulation Shielding System: The cable insulation shielding system shall consist of a nonmagnetic metallic conducting shielding tape and an underlying nonmetallic semiconducting cloth tape.

> The underlying semiconducting tape shall consist of cloth coated, impregnated, or calendared with a carbon-black semiconducting compound and shall be compatible with the insulation system and the metallic shielding tape. The semiconducting tape shall be applied smoothly and tightly, and completely cover the insulation system with a minimum 10 percent lap on itself. The tape shall have an average thickness of 0.006 in and be legibly identified as being semiconducting by surface printing throughout the length of the cable.
>
> The metallic shielding tape shall be 0.005-in-thick copper. The shielding tape shall be helically applied, smoothly and tightly, and completely cover the underlying semiconducting tape with a 20 percent lap upon itself. The shielding tape shall be compatible with the semiconducting underlying tape and the jacket which will be extruded over the shielding tape. The metallic tape shall be free from burrs, wrinkles, or contaminates and shall be made electrically continuous by welding, soldering, or brazing.

Jacket: The jacket shall be PVC extruded over the metallic shield. The minimum average thickness shall be 0.080 in, and the minimum thickness at any point on the cable shall not be less than 80 percent of the minimum average.

Identification: The outer surface of each jacketed cable shall be durably marked by sharply defined surface printing. The surface marking shall be as follows:

Manufacturer's name

Conductor size

Conductor material

Rated voltage

Year of manufacture

Symbol for supply cable (NESC C2-350.G)

Cable-Diameter Dimensional Control: The overall diameter of the 15-kV insulated cable shall not exceed 1.67 in.

Production Tests: The following production tests on each length of cable shall be performed:

Conductor Tests: The conductor shall be tested in accordance with ASTM B-231 and IPCEA Standard S-19 (NEMA WC-3, Section 6.3).

Conductor Energy-Suppression Layer: The integrity of the conductor energy-suppression layer shall be continuously monitored prior to or during the extrusion of the primary insulation using an electrode system completely covering the full circumference of the conductor energy-suppression layer at a test voltage of 2 kV dc.

Insulation Tests

Preliminary Tank Test: Each individual length of insulated conductor, prior to the application of any further coverings, shall be immersed in a water tank for a minimum period of 24 hours. At the end of 16 hours (minimum), each length of insulated conductor shall pass a dc voltage withstand test applied for 5 min, the applied dc voltage to be 70 kV.

At the end of 24 hours (minimum), each length of insulated conductor shall pass an ac voltage withstand test applied for 5 min, the applied ac voltage to be 35 kV. Each length of insulated conductor following the 5-min ac voltage test shall have an insulation resistance, corrected to 15.6°C (60°F), of not less than the value of R, as calculated from the following formula (after electrification for 1 minute):

$$R = K \log_{10} D/d$$

R = insulation resistance in megohms/1000 ft

K = 21,000 megohms/1000 ft

D = diameter over insulation

d = conductor diameter

Final Test on Shipping Reel: Each length of completed cable shall pass an ac voltage withstand test applied for 1 min, the applied ac voltage to be 35 kV. Each length of completed cable, following the 1-min ac voltage test, shall have an insulation resistance not less than that specified above.

Dimensional Verification

Each length of insulated conductor shall be examined to verify that the conductor is the correct size.

Each length of insulated conductor shall be measured to verify that the insulation wall thickness is of the correct dimension.

Each length of completed cable shall be measured to verify that the jacket wall thickness is of the correct dimension.

Conductor and Shield Continuity: Each length of completed cable shall be tested for conductor and shield continuity.

Test Reports: Certified test reports of all cable will be required, one copy of test reports to be sent to the ordering party.

Shipping Lengths: The shipping lengths will be specified on the inquiry and purchase order. Unless otherwise specified on the purchase order, the cable shall be furnished in one continuous length per reel.

Reels: Reels shall be heavy-duty wooden reels (nonreturnable, if possible) properly inspected and cleaned to ensure that no nails, splinters, etc. may be present and possibly damage the cable. Each end of the cable shall be firmly secured to the reel with nonmetallic rope. Wire strand is not acceptable for securing cable to the reel. Deckboard or plywood covering secured with steel banding is required over the cable.

Cable Sealing: Each end of the cable shall be sealed hermetically before shipment using either heat-shrinkable polyethylene caps or hand taping, whichever the manufacturer recommends. Metal clamps which may damage pulling equipment are not acceptable. The end seals must be adequate to withstand the rigors of pulling in.

Preparation for transit: Reels must be securely blocked in position so they will not shift during transit. The manufacturer shall assume responsibility for any damage incurred during transit which is the result of improper preparation for shipment.

Reel Marking: Reel markings shall be permanently stenciled on one flange of the wooden reel. Such markings shall consist of the purchaser's requirements as indicated elsewhere in this specification, but shall consist, as a minimum, of a description of the cable, purchaser's order number, manufacturer's order number, date of shipment, footage, and reel number. An arrow indicating direction in which reel should be rolled also shall be shown.

Cable Terminations. As pointed out in the beginning of this section, underground transmission is resorted to only when necessary because of congestion in downtown areas or where overhead lines detract from the appearance of the community. However, it is possible that once the cables have passed underground through some crowded area, they may change to overhead lines. This is arranged either by continuing the cable as an aerial cable supported from a messenger wire or by connecting the cable to an open overhead line through a riser and pothead.

Aerial Cable. An aerial cable is illustrated in Fig. 29.25. The cable is supported at close intervals by rings attached to a tightly strung steel cable called a messenger. Instead of rings being used, the cable can be supported by means of a continuous wire wrapped around both the cable and the messenger in spiral fashion with a spinning machine.

Cable Riser. When it is desired to connect an underground cable to an overhead line, the cable is brought out of a manhole to a pole through a run of conduit which is usually

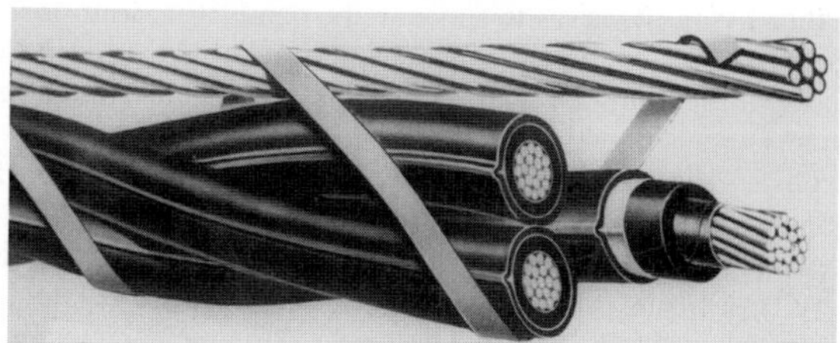

FIGURE 29.25 Three-conductor rubber-insulated, shielded, self-supporting aerial cable. (*Courtesy The Okonite Co.*)

FIGURE 29.26 Single-conductor polymer terminators on three phase feeder riser pole.

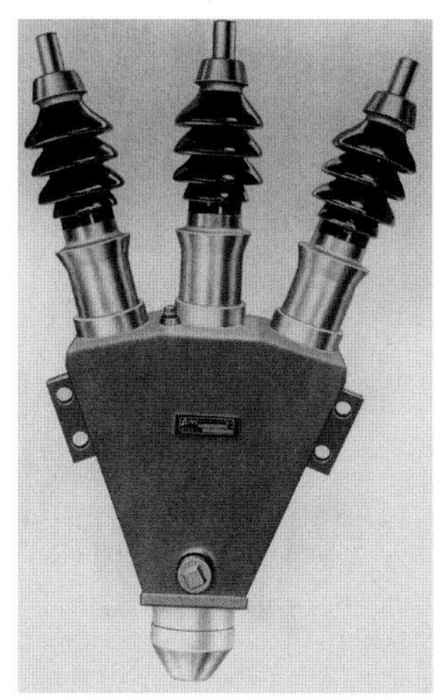

FIGURE 29.27 Typical three-conductor pothead. (*Courtesy G. & W. Electric Specialty Co.*)

extended 10 ft or more up the pole in metallic pipe or plastic duct for mechanical protection. If the cable is low voltage, it terminates in a so-called weather head, and if it is a high-voltage cable, it terminates in a so-called pothead. The whole assembly is called a riser. Figure 29.3 shows such an assembly consisting of the conduit attached to the pole and the cable coming through it and terminating in the pothead. When the cables are single-conductor cables, they are fanned out just below the crossarm and terminate in single-conductor polymer terminators like that shown in Fig. 29.26. When the cable is a multiconductor cable, it terminates in a multiconductor pothead like that shown in Fig. 29.27. From the pothead the conductors spread out into the overhead line. The pothead is supported by means of bolts fastened to the pole or crossarm.

Weather Head. A weather head is a simple underground-overhead fitting which provides a cap or roof for the vertical conduit to prevent rain from entering it. Its use is restricted to connecting underground secondary cables to overhead secondary lines or service wires.

Pothead. A pothead (Fig. 29.27) provides for the connection between an underground cable and an overhead wire. It serves to separate the conductors from their close position to one another in the cable to the much wider separation in the overhead line without permitting arcing between them. It also seals off the end of the cable so that moisture cannot enter the insulation of the cable.

The pothead consists essentially of a pot, from which it gets its names, and a cover. The pot has a hole in the bottom through which the cable enters. The pot is attached to the sheath of the cable by a wiped joint or clamp. The conductors of the cable are made to flare out to give the necessary separation. The conductors of the cable are connected to terminals encased in porcelain attached to the cover. These projections of porcelain have "petticoats" on them the same as pin insulators and naturally serve the same purpose. The overhead-line conductors are connected to these terminals.

FIGURE 29.28 Pipe-type cable on reels before being pulled into the pipe. (*Courtesy Anaconda Wire and Cable Co.*)

Pipe-Type Cable. Underground transmission circuits with voltages in the range of 69,000 to 500,000 volts are usually installed utilizing pipe-type cable. The cable consists of a copper or aluminum conductor, conductor shielding, paper-tape insulation, and insulation shielding (see Fig. 29.28). Three conductors are pulled into the pipe to form a three-phase circuit. The pipe is installed direct buried or in a tunnel. Manholes are installed for splices. Figure 29.29 shows the three insulated conductors in the pipe with a cableman completing a splice in a manhole. The pipe is coated to protect it from electrolysis damage. Cathodic protection is normally installed to protect the cable pipe. The pipe is filled with insulating oil and the oil is continually pumped through the pipe to provide a cooling medium and maintain a high dielectric strength between the energized cables and ground. The cables are terminated in potheads, as illustrated in Fig. 29.30.

Pipe-type cables are also described and illustrated in Chap. 4. Experimental pipe-type cables have been developed for operation at 765 kV ac and ±600 kV dc.

Compressed-Gas Cable. Rigid isolated-phase compressed-gas cables using sulfur hexafluoride (SF_6) gas that can operate at voltages up to 1000 kV ac are described and illustrated in Chap. 4. Three-conductor rigid compressed-gas cable has been developed for operation at 345 kV ac. The three-phase cable costs less to fabricate and install than the isolated-phase

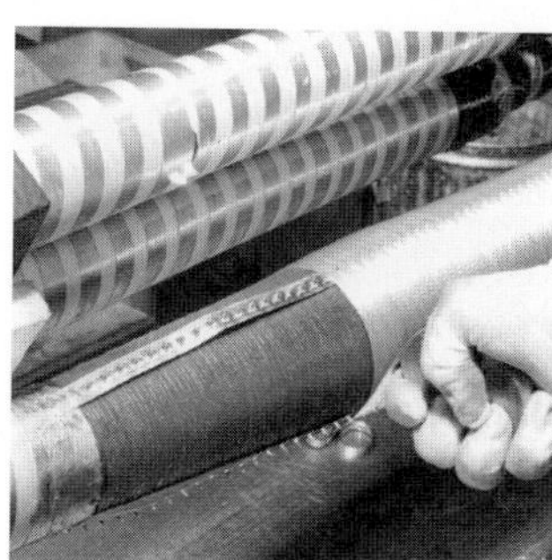

FIGURE 29.29 Cableman completing splice of pipe-type cable in manhole. (*Courtesy Anaconda Wire and Cable Co.*)

FIGURE 29.30 Cableman and lineman terminating pipe-type cable with potheads. The porcelain insulator is in the process of being placed over the cable termination for one phase. (*Courtesy Anaconda Wire and Cable Co.*)

compressed-gas cable. Experimental flexible compressed-gas cables for operation at 345 kV ac have been fabricated. The flexible cable is easier to handle and ship, requires less excavation, and can be routed around obstacles when it is installed. It is estimated that the installed cost of flexible compressed-gas cables will be 25 percent less than the current costs for fabricating and installing rigid compressed-gas cables.

Transformers

Vault-type, pad-mounted, submersible, and direct-buried transformers are used in underground systems. Figure 29.2 illustrates the installation of a vault-type transformer complete with a network protector and a primary oil switch. The primary and secondary cables connecting to the transformer are shown. Transformers are discussed in Chap. 15.

CHAPTER 30
LAYING CONDUIT

A conduit is a structure containing one or more ducts. A duct is a single enclosed runway (hole) for conductors or cables. Conduit systems should be located so as to be subject to the fewest disturbances practical. Conduit systems extending parallel to other subsurface structures should not be located directly over or under other subsurface structures. If this is not practical, the rules on clearances should be followed. Conduit alignment should be such that there are no protrusions which would be harmful to the cable. When bends are required, the minimum radius shall be sufficiently large to prevent damage to cable being installed in the conduit. The maximum change of direction in any plane between lengths of straight, rigid conduit without the use of bends should be limited to 5°.

Routes through the unstable soils such as mud or shifting soil or through highly corrosive soils should be avoided. If construction is required in these soils, the conduit should be constructed in such a manner as to minimize movement and/or corrosion. When conduit must be installed longitudinally under the roadway, it should be installed in the shoulder or, to the extent practical, within the limits of one lane of traffic. The conduit system shall be located so as to minimize the possibility of damage by traffic. It should be located to provide safe access for inspection or maintenance of both the structure and the conduit system. The top of a conduit system should be located not less than 36 in below the top of the rails of a street railway or 60 in below the top of the rails of a railroad.

Where unusual conditions exist or where proposed construction would interfere with existing installations, a greater depth than specified above may be required. Where this is impractical or for other reasons, this clearance may be reduced by agreement between the parties concerned. In no case, however, shall the top of the conduit or any conduit protection extend higher than the bottom of the ballast section which is subject to working or cleaning. At crossings under railroads, manholes, handholes, and vaults should not, where practical, be located in the roadbed. Submarine crossings should be routed and/or installed so they will be protected from erosion by tidal action or currents. They should not be located where ships normally anchor.

The clearance between a conduit system and other underground structures paralleling it should be as large as necessary to permit maintenance of the system without damage to the paralleling structures. A conduit which crosses over another subsurface structure shall have a minimum clearance sufficient to prevent damage to either structure (Fig. 30.1). These clearances should be determined by the parties involved. When conduit crosses a manhole, vault, or subway tunnel roof, it may be supported directly on the roof with the concurrence of all parties involved. Conduit systems to be occupied by communications conductors shall be separated from conduit systems to be used for supply systems by 3 in of concrete, 4 in of masonry, or 12 in of well-tamped earth. Lesser separations may be used where the parties concur. If conditions require a conduit to be installed parallel to and directly over a sanitary or storm sewer, it may be done provided both parties are in agreement as to the method.

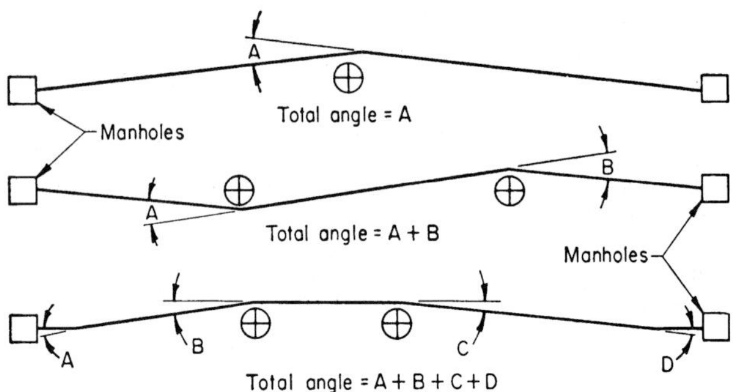

FIGURE 30.1 Typical outlines of duct runs showing manner of avoiding obstructions ⊕. Angles *A, B, C,* and *D* are degree deflections from straight line.

Where a conduit run crosses a sewer, it shall be designed to have suitable support on each side of the sewer to avoid transferring any direct load onto the sewer. Conduit should be installed as far as practical from a water main in order to protect if from being undermined if the main breaks. Conduit which crosses over a water main shall be designed to have suitable support on each side as required to avoid transferring any direct loads onto the main. Conduit should have sufficient clearance from fuel lines to permit the use of pipe-maintenance equipment. Conduit and fuel lines shall not enter the same manhole. Conduit should be so installed as to prevent detrimental heat transfer between the steam and conduit systems.

The bottom of the trench should be undisturbed, tamped, or relatively smooth earth. Where the excavation is in rock, the conduit should be laid on a protective layer of clean, tamped backfill. All backfill should be free of materials that may damage the conduit system. Backfill within 6 in of the conduit should be free of solid material greater in maximum dimension than 4 in or with sharp edges likely to damage it. The balance of backfill should be free of solid material greater in maximum dimension than 8 in. Backfill material should be adequately compacted.

Duct material shall be corrosion resistant and suitable for the intended environment. Duct material and/or the construction of the conduit shall be designed so that a cable fault in one duct would not damage the conduit to such an extent that it would cause damage to cables in adjacent ducts. The conduit system shall be designed to withstand external forces to which it may be subjected by the surface loading, except that impact loading may be reduced one-third for each foot of cover, so no impact loading need be considered when cover is 3 ft or more. The internal finish of the duct shall be free of sharp edges or burrs which could damage supply cable.

Conduit, including terminations and bends, should be suitably restrained by backfill, concrete envelope, anchors, or other means to maintain its design position under stress of installation procedures, cable-pulling operations, and other conditions such as settling and hydraulic or frost uplift (Figs. 30.2 and 30.3). Ducts shall be joined in such a way as to prevent solid matter from entering the conduit line. Joints shall form a sufficiently continuous smooth interior surface between joining duct sections that supply cable will not be damaged when pulled past the joint. When conditions are such that externally coated pipe is required, the coating shall be corrosion resistant and should be inspected and/or tested to see that it is continuous and intact before backfilling is begun. Precautions shall be taken to prevent damage to the coating when backfilling.

Conduit installed through a building wall shall have internal and external seals intended to prevent the entrance of gas into the building insofar as practical. The use of seals may be

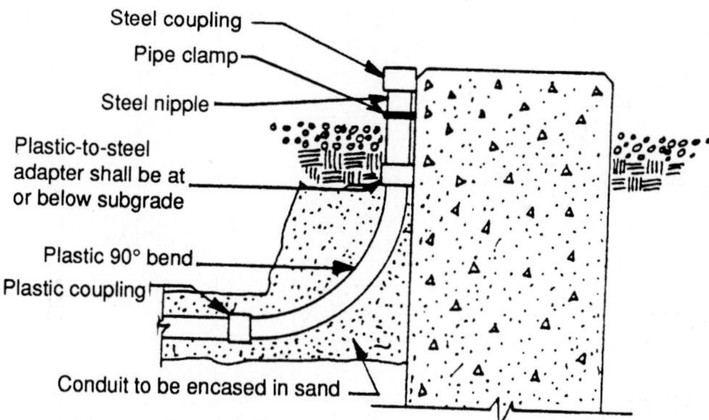

FIGURE 30.2 Typical conduit-riser construction details for a conduit system buried directly in the earth.

supplemented by gas-venting devices in order to minimize building up of positive gas pressures in the conduit. Conduit installed in bridges shall be such as to allow for expansion and contraction of the bridge. Conduits passing through a bridge abutment should be installed so as to avoid or resist any shear due to soil settlement. Conduit of conductive material installed on bridges shall be effectively grounded. Conduit should be installed on compacted soil or otherwise supported when entering a manhole to prevent shear stress on the conduit at the point of manhole entrance.

Laying Out Line. In general, a conduit run should be as straight as possible. When this is impossible due to obstructions, the run is made up of combinations of straight sections and curved sections. The combinations of straight sections and curves can follow the typical outlines shown in Fig. 30.1 depending on the number and position of the obstructions to be avoided. After the outline for a particular set of conditions has been made, the total angular displacement is determined by adding the degrees at angles *A*, *B*, *C*, and *D* as illustrated in Fig. 30.1; thus:

$$\text{Total angle} = A° + B° + C° + D°$$

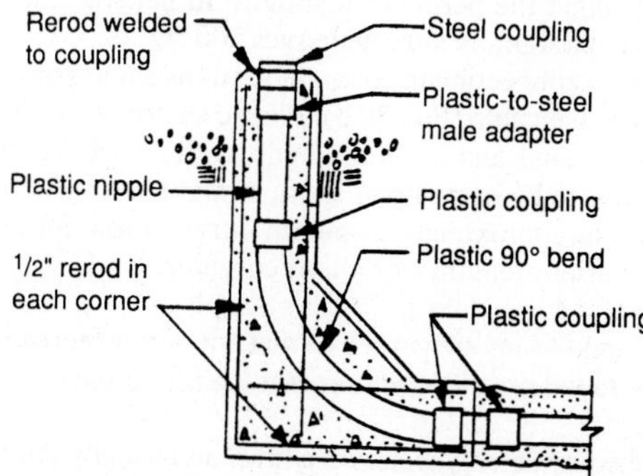

FIGURE 30.3 Typical conduit-riser construction details for a conduit system encased in concrete. The terminals of some conduit systems buried directly in the earth are secured by encasing the ducts in concrete.

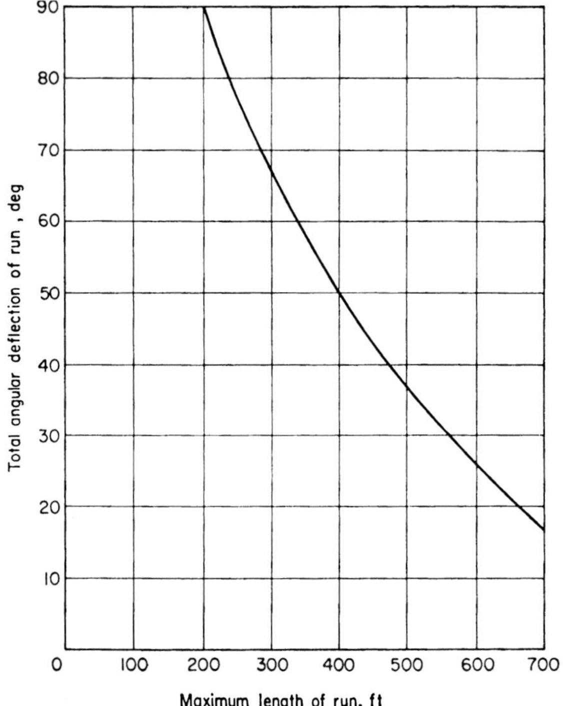

FIGURE 30.4 Chart showing relation between total angular deflection of duct and permissible maximum length of run.

If there is a displacement in a vertical plane also, this is included in the summation of the total angle. This total angle limits the length of the run between manholes, as shown in Fig. 30.4. The greater the total angle, the shorter the permissible duct length, and the smaller the angle, the longer the permissible length. In general, the maximum length of duct, even when entirely straight, is not much over 500 ft.

The radius of the curve connecting the straight sections should be as large as possible to facilitate the pulling in of the cable (Fig. 30.5). Plastic or polyvinyl chloride (PVC) duct is available with preformed bends and in short lengths with a small amount of curvature.

If the curved section is to be made up of short, straight sections, these straight sections should be short enough to approximate a smooth curve. Table 30.1 shows the preferred lengths of sections for various lengths of radii of curvature.

Grading the Line. Conduit runs are graded so that any water seeping into the conduit will drain into the manhole. Good practice requires that the minimum grade be no less than 3 in for every 100 ft of run.

Grades are established by an engineer with a level and usually marked with wood grade stakes driven 5 ft apart on the center line of the bottom of the conduit trench. The top of the grade stakes indicates the top surface of the bottom duct sheathing.

The methods of grading are double slope and single slope. Figure 29.4 illustrates the double-slope grade. This is the preferred method in the level areas as it reduces the distance any seep water needs to drain. The duct run is highest midway between manholes, coming

FIGURE 30.5 View showing a conduit run making a turn with a large radius of curvature. Note use of spacers to keep ducts in correct alignment.

TABLE 30.1 Preferred Lengths of Sections

Length of radius of curvature, ft	Length of short straight sections of conduit, in
10 to 20	6
20 to 30	9
30 to 150	18
150 and over	36

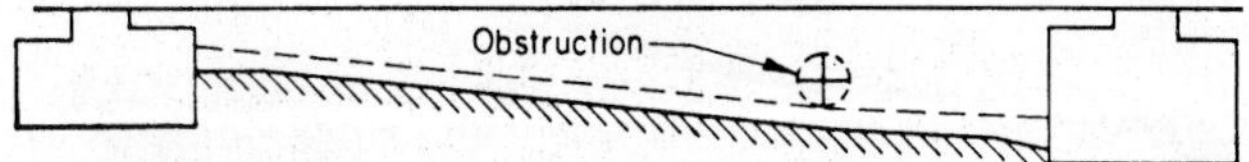

FIGURE 30.6 Manner of grading conduit run in order to pass under obstruction.

to within 2 ft of the street surface, and grades off in each direction. Figure 29.5 illustrates single-slope grading. The entire conduit run is graded in only one direction. This is the logical method when the area slopes from one manhole to the other. Sometimes obstructions also prevent the use of double-slope grading, as illustrated in Fig. 30.6. Here the single-slope grade is increased in order to have the conduit pass under the obstruction.

Arrangements of Ducts. The National Electrical Manufacturers Association (NEMA)* publication entitled *User's Manual for the Installation of Underground Plastic Duct* (Bulletin No. TCB2) discusses the procedures in detail. Ducts can be arranged in the trench in various ways, as shown in Fig. 30.7, depending on the number of ducts actually needed and the number of spares to be provided for future load growth. A rule often followed is to double the number initially filled with cable. This would therefore provide for a 100 percent increase in load.

Since cables carrying electric power become heated from the electrical losses in the conductors, they should be so installed that cooling can take place by radiation. When a large number of cables are close together, the ability to radiate heat is greatly reduced. Hence the

*National Electrical Manufacturers Association, 155 East 44th Street, New York, N.Y. 10017.

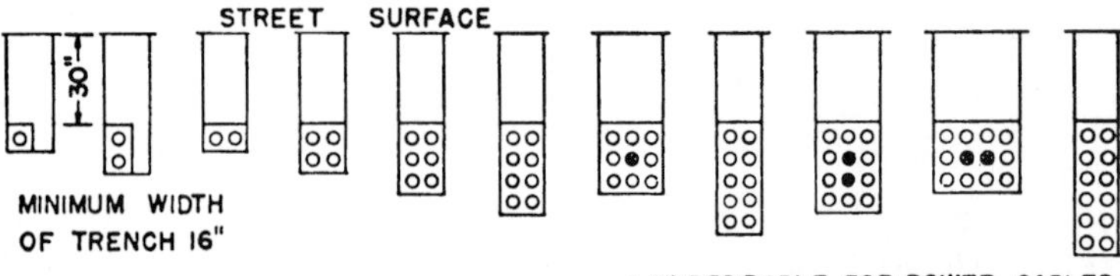

FIGURE 30.7 Various arrangements of ducts in a conduit run. Interior ducts are sometimes called dead ducts because they do not radiate heat so well as outside ducts. They are therefore generally used for control cables.

number of cables in a duct run should be kept low. A recommended maximum of ducts in a single run is nine. When a greater number is used, the heat radiation per cable becomes greatly reduced, and the total power that can be carried in the conduit run is only slightly increased.

Excavating the Trench. The trench for the conduit run is first marked off. If a digging machine is to be used (Fig. 30.8), only one side of the trench need be marked. Cuts in a paved street must conform to the shape of the required excavation. The width of the cut is made equal to the overall width of the conduit run plus necessary clearance for the digging equipment. The sides of the excavation are shored where necessary. The trench is dug true to the marked line with perpendicular sides. The sides are trimmed smooth to provide for a uniform sheath of concrete around the ducts if they are to be encased. The entire trench is opened between manholes or termination before any duct is laid in order to determine if any obstructions exist that must be bypassed.

FIGURE 30.8 Backhoe being used to dig trench for conduit run. (*Courtesy John Deere Co.*)

When the required depth of the trench is known for all points along its length, the grade may be established. The bottom of the trench should be relatively smooth, undisturbed earth, well-tamped earth, or sand. Trenches which have been dug too deep at any point are partially refilled and tamped solid.

Directional Boring. In some instances, an open cut trench is not an option. Perhaps the conduit system must cross under a paved roadway, a body of water, or an environmentally sensitive area. In these instances, directional boring techniques are utilized to complete the conduit installation. Boring is accomplished utilizing directional boring construction equipment (Fig. 30.9). The equipment controls a guided boring "missle-shaped" tool that is attached to the lead end of a drill rod. The 10-ft sections of drill rod are connected together as the tool is extended beneath the ground's surface (Fig. 30.10), from the starting point to the end point without disturbing the ground-line between the start and end points. This technique minimizes the clean up required over the course of the conduit route. High-pressure fluid jets on the head of the directional boring tool provide the equipment operator the ability to control the device as it moves along underground. Once the device has completed its desired route, backreaming tools are utilized to solidify the tunnel or place the conduit system. The soil conditions dictate the best directional boring technique to use. In loose soils, it may be necessary to utilize drilling fluids such as bentonite to hold the bored hole to reduce the occurrence of collapsed soil until the conduit system is installed. The bore lengths range in distance from 75 ft to distances as great as 1200 ft. Prior to beginning directional boring, it is important to determine and locate any known underground facilities or obstructions that are installed and must be avoided.

Cutting the Duct. Plastic and fiberglass ducts can be cut with a hand saw, power saw, or rotary cutter. The duct must be cut square. Remove the burrs left by sawing and the ridges

FIGURE 30.9 Directional boring equipment used to bore a hole from the start point to the end point without disturbing the ground-line in between these two points. (*Courtesy Vermeer Manufacturing Co.*)

FIGURE 30.10 The design of the directional boring equipment is compact with a narrow profile. Note the 10-ft sections of drill rod on the side of the equipment that are utilized to extend the boring tool. (*Courtesy Vermeer Manufacturing Co.*)

left by a rotary cutter with a file or a knife. Sharp edges on the outside or inside of the duct must be removed by beveling the edge of the duct. If the burrs and ridges are not removed, a cable may be damaged when it is pulled into the duct. If burrs and ridges are not removed, a defective joint may result.

Joining the Duct. Surfaces of the duct to be joined must be clean and free from dirt, foreign materials, and moisture. Clean the outside surface of the dust spigot end for the depth of the socket and the inside surface of the socket with a clean, dry cloth.

Fiberglass ducts are manufactured with a bell or socket on one end which is equipped with a gasket. The spigot end of one duct is pushed into the bell end of the next duct. The gasket in the bell end of the fiberglass duct provides the seal.

Plastic (PVC) duct joints are sealed with a cement. Solvent cements are specified for each thermoplastic duct material and should not be interchanged between duct materials. Combination cements are available for joining ducts and fittings of different materials. The solvents used in the cements are flammable. Fumes from the solvents can be harmful if adequate ventilation is not provided. Forced ventilating equipment should be used while cementing ducts in manholes, vaults, or enclosed areas. Open flames, smoking, and exposed heating elements must be avoided while applying the cement to the ducts. The cement should be applied with a brush or applicator to easily and rapidly coat the surfaces of the ducts to be joined. While the spigot surface of the duct is still wet with cement, insert the duct spigot with a slight twisting motion into the socket until it bottoms home at the socket shoulder. The bonding of the duct joint begins immediately and should be completed within 15 sec. After assembly, wipe any excess cement from the duct at the end of the socket to prevent excess cement from dripping on and possibly weakening a lower duct.

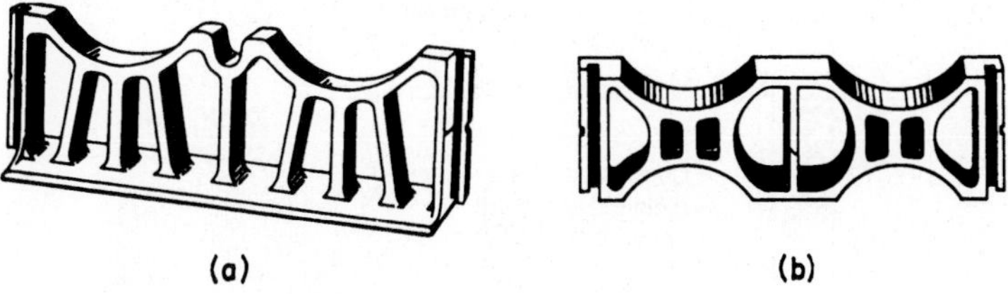

FIGURE 30.11 Plastic base spacer (*a*) and intermediate spacer (*b*).

Laying Ducts. The laying of ducts is begun midway between manholes where duct spacing is standard and then working toward the manholes where the ducts may flare out. If the duct run is located between a termination such as a riser (Fig. 30.2) and a manhole, the laying of ducts is started at the termination and laid toward the manhole. Manufactured plastic spacers are normally used to support the ducts, to give them proper spacing for cooling, and to secure the ducts so that they will not float or move in any direction (Fig. 30.11). If the ducts are to be encased in concrete, the manufactured spacers provide the proper clearance from the bottom of the trench and the proper spacing between ducts for encasement (Fig. 30.12). The use of bricks or wood is not recommended because these materials may deform the duct wall. The spacers are normally placed approximately 5 ft apart to provide adequate support during the placement of concrete or backfilling of the trench with soil. The bottom layer of ducts is placed between the terminals of the conduit run, which may be any combination of manholes, handholes, vaults, or risers, before the second layer is started. It is recommended that the end ducts that do not terminate in a riser be full length. The ducts should terminate in a manhole, handhole, or vault with end bells that are grouted in place or manufactured duct terminators which are designed for installation into the walls to create a watertight duct-entry system (Fig. 30.13). The spacers normally provide for anchoring the ducts into position after all have been installed. Some manufactured spacers provide vertical openings on either side of the spacers of the duct assembly through which reinforcing rods can be passed and driven into the trench floor. The reinforcing rods will

FIGURE 30.12 Assembled bank of ducts being installed. Note use of spacers to hold ducts in proper position.

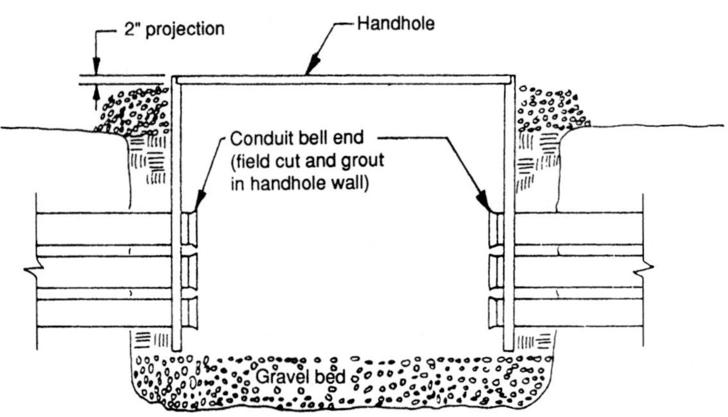

FIGURE 30.13 Conduit run terminated in a handhole. The end bells provide an outlet to the handhole with rounded edges to prevent cable damage.

secure the duct assembly while concrete is being poured for encasement or while earth is placed around the ducts. If the duct assembly cannot be secured with reinforcing rods, a trench jack may be placed directly over the spacer location and wedges used between the trench jacks and the spacers to hold the ducts in place while the concrete is poured or the earth placed in the trench. The open ends of the ducts should be closed with plastic plugs to prevent dirt, animals, or other objects from entering and becoming lodged. This should be done whenever the work is halted for any length of time, even for overnight. Figure 30.14 is a cross-sectional view of ducts laid in a trench in preparation for backfilling with earth or encasement with concrete.

Burial of Ducts in Earth. Plastic (PVC) duct is manufactured in different wall thicknesses and strengths. The extra-strength plastic ducts can be directly buried in the ground in many locations, eliminating the need for and cost of concrete encasements. The ANSI/NEMA TC8 publication entitled *Extra-Strength PVC Plastic Utilities Duct for Underground Installation* provides detailed information on ducts. The extra-strength plastic ducts are available with inside diameter measurements of 1 through 6 in. The ducts are available in type EB for encased burial in concrete and type DB for direct burial without encasement. A suitable backfill properly placed is critical for a satisfactory direct-buried duct

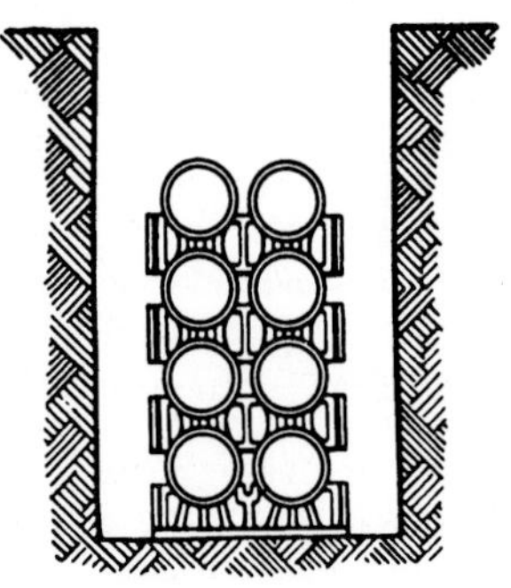

FIGURE 30.14 Ducts resting on spacers installed in a trench.

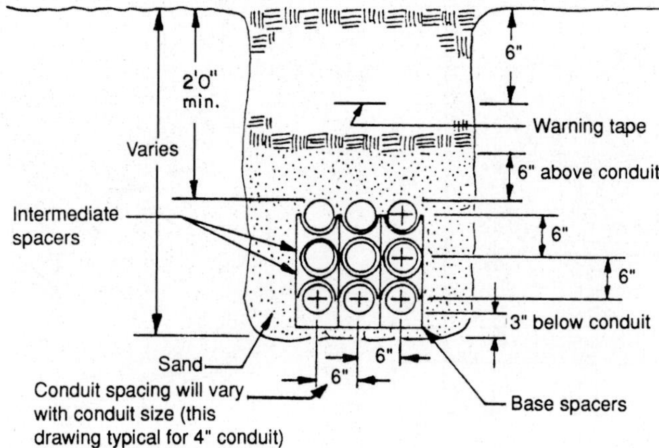

FIGURE 30.15 Cross-sectional view of a completed conduit installation with ducts directly buried in the earth.

installation. A stable backfill material should be selected. Sand and small gravel are the best materials. Sand and gravel mixed with silts or clays in which the sand and gravel amount to at least 50 percent of the mixture are satisfactory. Rock-free sandy loam and small crushed quarry stone are suitable. Soils of highly organic content, identified by odor or spongy feel, or highly plastic clays are not suitable for backfilling. Each soil layer should be compacted to 85 to 90 percent of its maximum density. Care should be taken when compacting to avoid excessive tamping in order to prevent shifting or distorting of the ducts. Vibratory methods are the preferred method of compacting sand and gravel. The fill should be in a nearly saturated condition while the compaction is in progress. After the initial cover of select backfill over the duct bank has been completed, the remaining trench should be backfilled to grade with sufficient depth to obtain the required density when tamped (Fig. 30.15).

Encasement of Ducts in Concrete. The concrete envelope surrounding an encased duct structure is usually 3 in thick on the top, bottom, and sides. The concrete mixture used is generally one part cement, three parts clean sand, and five parts broken stone or gravel. The maximum size aggregate should be $^3/_8$ in and the slump should be 7 to 9 in to ensure proper distribution of the concrete around the ducts. The concrete should be well mixed and not too wet. Very wet concrete has a tendency to float the ducts and change the alignment. Care should be taken to limit the fall of the concrete to a minimum height from the chute to the top tier of ducts to minimize flotation effects. It is recommended that the pouring of the concrete start at the point the ducts are terminated in a manhole or at another end of the conduit run where the ducts are securely restrained. The concrete is poured around the ducts as soon as possible after they have been placed to protect them from mechanical damage. Continuous spading is necessary to ensure a flow of concrete between and under the individual ducts. A long, flat tool or spatula worked carefully up and down between each vertical line of ducts will help eliminate voids. Powerdriven tampers should not be used for this purpose. The pouring is continued to a height of 3 in above the top layer of ducts. If a trench jack and wedges were used to secure the ducts, the wedges should be removed before the concrete sets and the voids left by them filled with concrete and spaded. Figure 30.16 shows a conduit bank after pouring of the concrete is completed.

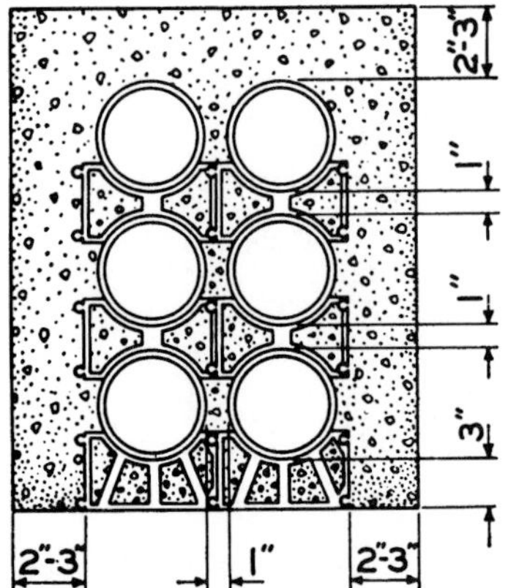

FIGURE 30.16 Cross-sectional view of a completed conduit run. The spacers and ducts are all encased in a concrete envelope.

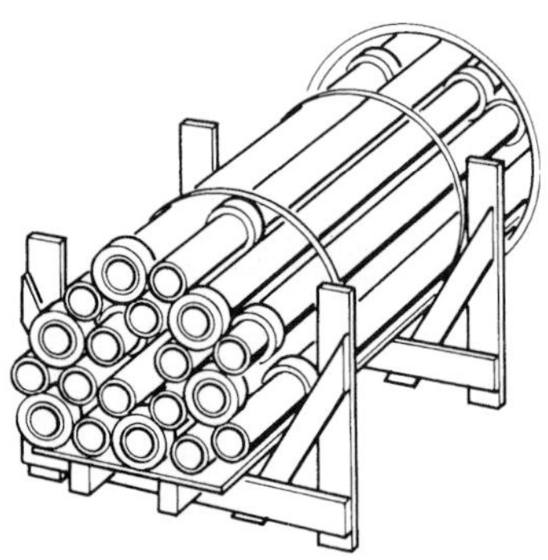

FIGURE 30.17 Cradle holds plastic ducts in proper formation while circular-type spacers are placed over each duct prior to banding. Cradle also keeps duct formation lined up with casing entrance. (*Courtesy National Electrical Manufacturers Assoc.*)

After the concrete has taken its initial set, the trench is backfilled. The dirt should be placed in the trench in 6- to 12-in layers, and each layer carefully tamped. In case a pavement was cut, this is then replaced.

Steel encasement of Ducts. Plastic (PVC) ducts can be installed in driven or bored steel casings (Figs. 30.17 through 30.19). Steel casings are frequently used when crossing under highways and railroads. A steel casing may be used to support a conduit run in soft soil provided proper bearing support is available at the pipe terminals. The steel casing used should be strong enough not to bend between supports. The extra-strength ducts normally used for direct burial should be used inside the steel casing.

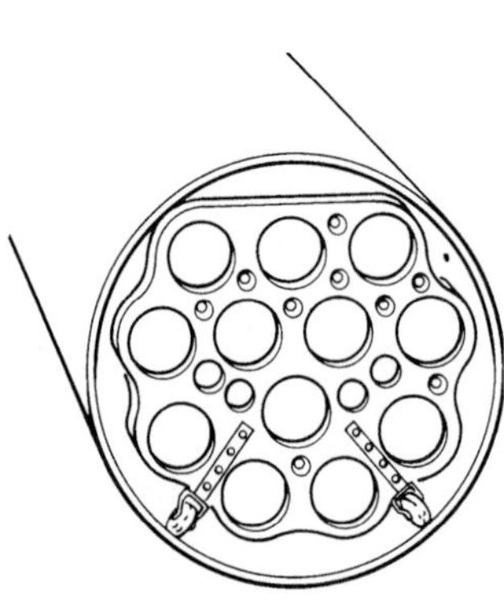

FIGURE 30.18 Custom bore spacer with small openings for wire guides on each side. Wires prevent rotation of duct bank as it advances through the casing.

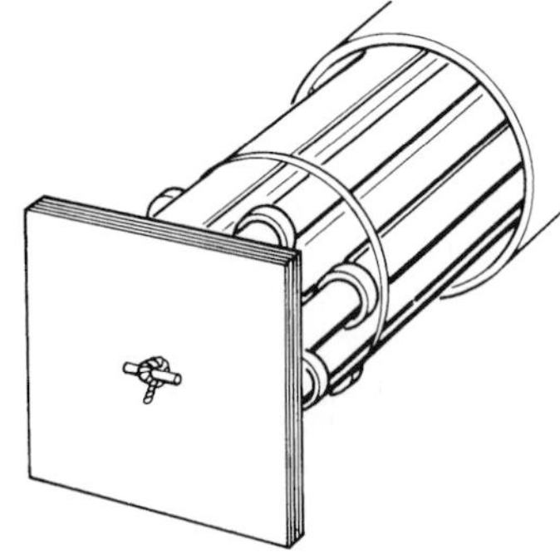

FIGURE 30.19 Push plate, of either wood or metal, covers entire duct bundle and spreads load pressure evenly on the bundle. Winch line is threaded through center duct as well as rear push plate so that formation can be pulled and pushed into the casing. (*Courtesy National Electrical Manufacturers Assoc.*)

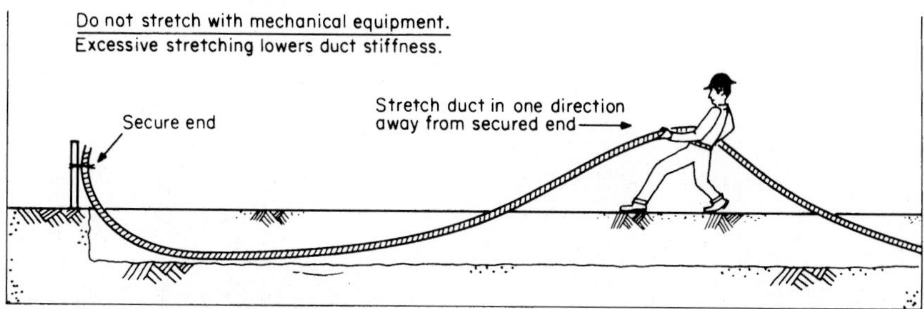

FIGURE 30.20 Lineman installing corrugated-plastic duct in trench. Duct is unreeled and placed in trench. (*Courtesy National Electrical Manufacturers Assoc.*)

The space between the ducts and the steel casing must be filled to provide a means of dissipating the heat in the power cables installed in the ducts. A cement grout used to fill the voids in the casing contains an additive called Elastizell or Mearlcrete which reduces the grout density to approximately 75 lb/ft^3. The cement grout should have about an 11-in slump, which is fluid enough to allow low hydraulic pumping pressure. The steel casing should be filled from both ends, and the hoses should be withdrawn as back-pressure develops while the grout is being pumped. Ducts in a cradle, as illustrated in Fig. 30.17, are properly separated to permit the installation of the cement grout. Special care must be taken not to deform the ducts while installing them in the steel casing and while pumping in the grout to fill the casing.

Flexible-Plastic Ducts. Plastic (PVC) utility duct is available in smooth-walled and corrugated form, packaged in coils or on reels. ANSI/NEMA Publication No. TC7 provides detailed information on smooth-walled, coilable polyethylene electrical plastic duct. The smooth-walled, coilable plastic ducts are available with inside-diameter dimensions of $^1/_2$ through 3 in. ANSI/NEMA Publication No. TC5 provides detailed information on corrugated-polyolefin coilable plastic utilities duct. The corrugated-plastic ducts are available with inside-diameter dimensions of $1^1/_4$ through 5 in. The manufacturer's instructions should be followed for proper installation of the flexible-plastic ducts. Duct may be unreeled directly into the trench for installation. The flexibility of the ducts permits their installation around corners without the use of prefabricated bends or the splicing of short sections of duct (Fig. 30.20). The flexible ducts can be installed with plowing equipment if the soil is appropriate.

CHAPTER 31
MANHOLE CONSTRUCTION

Manholes, handholes, and vaults shall be designed to sustain all expected loads which may be imposed on the structure. The horizontal and/or vertical design loads shall consist of dead load, live load, equipment load, impact load due to water table or frost, and any other load expected to be imposed on and/or occur adjacent to the structure. The structure shall sustain the combination of vertical and lateral loading that produces the maximum shear and bending moments in the structure.

In roadway areas, the live load shall consist of the weight of the moving tractor-semi-trailer truck illustrated in Fig. 31.1. The vehicle wheel load shall be considered applied to an area as indicated in Fig. 31.2. In case of multilane pavements, the structure shall sustain the combination of loadings which result in vertical and lateral structure loadings which produce the maximum shear and bending moments in the structure. Loads imposed by equipment used in road construction may exceed loads to which the completed road may be subjected.

In designing structures not subject to vehicular loading, the minimum live load shall be 300 lb/ft^2. Live loads shall be increased by 30 percent for impact. When hydraulic, frost, or other uplift will be encountered, the structure shall be either of sufficient weight or so restrained as to withstand this force. The weight of equipment installed in the structure is not to be considered as part of the structure weight. Where pulling-iron facilities are furnished, they should be installed with a factor of safety of 2 based on the expected load to be applied to the pulling iron.

A clear working space sufficient for performing the necessary work shall be maintained. The horizontal dimensions of the clear working space shall not be less than 3 ft. The vertical dimensions shall not be less than 6 ft, except in manholes, where the opening is within 1 ft horizontally of the adjacent interior side wall of the manhole. Where one boundary of

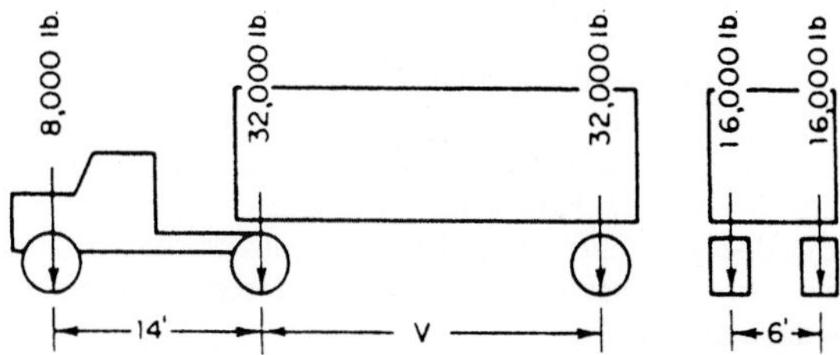

Variable spacing - 14 ft. to 30 ft. inclusive

Spacing to be used is that which results in vertical and lateral structure loading which produces the maximum shears and bending moments in the structure

FIGURE 31.1 Vehicle loading requirements for manhole construction.

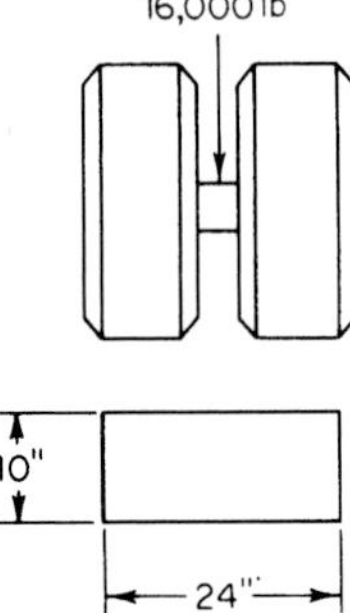

FIGURE 31.2 Vehicle wheel loading requirements for manhole construction.

the working space is an unoccupied wall and the opposite boundary consists of cables only, the horizontal working space between these boundaries may be reduced to 30 in. In manholes containing only communications cables and/or equipment, one horizontal dimension of the working space may be reduced to not less than 2 ft provided the other horizontal dimension is increased so that the sum of the two dimensions is at least 6 ft.

Round access openings in a manhole containing supply cables shall be not less than 26 in. in diameter. Round access openings in any manhole containing communication cables only, or manholes containing supply cables and having a fixed ladder which does not obstruct the opening, shall be not less than 24 in. in diameter. Rectangular access openings should have dimensions not less than 26 by 22 in. Openings shall be free of protrusions which will injure personnel or prevent quick egress.

When not being worked in, manholes and handholes shall be securely closed by covers of sufficient weight or proper design that they cannot be easily removed without tools. Covers should be suitably designed or restrained so that they cannot fall into manholes or protrude into manholes sufficiently far to contact cable or equipment. Strength of covers and their supporting structures shall be at least sufficient to sustain the applicable loads.

Vault or manhole openings should be located so that safe access can be provided. When in the highway, they shall be located outside the paved roadway when practical. They should be located outside the area of street intersections and crosswalks whenever practical to reduce the traffic hazards to personnel working at these locations. Personnel access openings in vaults or manholes should be located so that they are not directly over the cable or equipment. Where these openings interfere with curbs, etc., they can be located over the cable if one of the following is provided: a conspicuous warning sign, a protective barrier over the cable, or a fixed ladder. In vaults, other types of openings may be located over equipment to facilitate work on this equipment.

Where accessible to the public, access doors to utility tunnels and vaults shall be locked unless qualified persons are in attendance to prevent entry by unqualified persons. Such doors shall be designed so that a person on the inside may exit when the door is locked from the outside. This rule does not apply where the only means of locking is by padlock and the latching system is so arranged that the padlock can be closed on the latching system to prevent locking from the outside.

Fixed ladders shall be corrosion resistant.

Where drainage is into sewers, suitable traps or other means should be provided to prevent entrance of sewer gas into manholes, vaults, or tunnels.

Adequate ventilation to open air shall be provided for manholes, vaults, and tunnels having an opening into enclosed areas used by the public. Where such enclosures house transformers, switches, regulators, etc., the ventilating system shall be cleaned at necessary

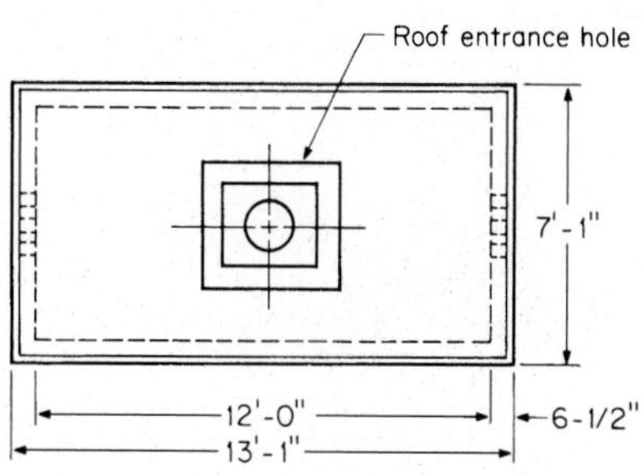

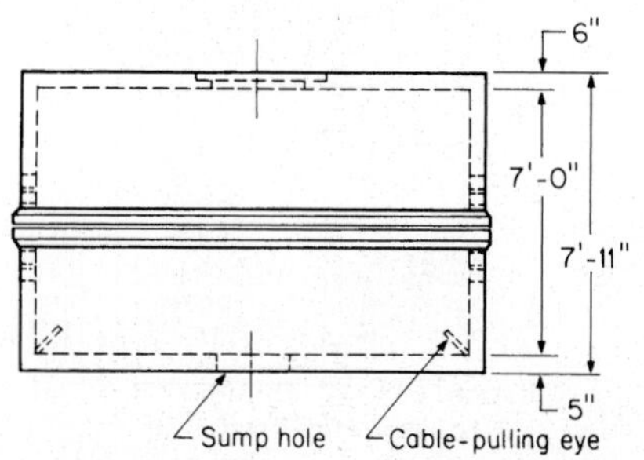

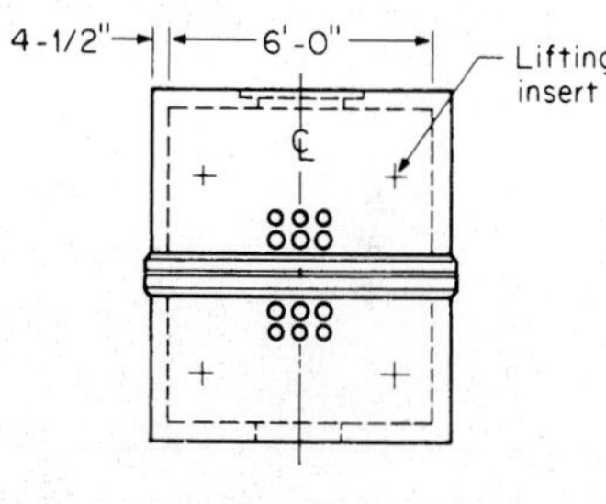

FIGURE 31.3 Straight-type precast cable manhole. Approximate weights top section 14,800 lb and bottom section 14,650 lb. Manhole is designed for use with 12-duct conduit run. (*Courtesy Commonwealth Edison Co.*)

intervals. This does not apply to enclosed areas under water or in other locations where it is impractical to comply.

Supply cables and equipment should be installed or guarded in such a manner as to avoid damage by objects falling or being pushed through the grating.

Manhole and handhole covers should have an identifying mark which will indicate ownership or type of utility.

Description. Manholes are usually made of brick, concrete blocks, reinforced concrete, precast concrete, precast plastic, fiberglass, or a combination of these materials. The manhole cover is made of steel. As a rule, manholes are constructed in place. However, sometimes it is more economical and convenient to precast the manhole and transport it to the location in the street. This procedure eliminates the field parking of materials, pouring of concrete, etc., in crowded streets and avenues.

Precast manholes can be fabricated in various configurations such as a straight manhole, an L manhole, and T manhole (Figs. 31.3 through 31.5). Large manholes are usually precast in sections or in pieces (Fig. 31.6). Manholes for direct-buried plastic conduit systems are often fabricated with precast concrete, plastic, or fiberglass pipe with a minimum internal diameter of 60 in. and a minimum depth of 7 ft (Figs. 31.7 and 31.8). The plastic conduits entering the manhole are terminated with an end bell and grouted if precast concrete pipe is used for the manhole. The conduits joined to a plastic or fiberglass manhole must have a watertight connection normally completed with mechanical devices. Precast concrete pipe can be used in an area subject to vehicular traffic by setting it on a 6-in. reinforced poured concrete base with a 6-in. tile 1 ft in length to serve as a sump. Circular manholes in an area not subject to vehicular traffic can be set on an 18-in. layer of crushed rock. All manholes installed in an area subject to

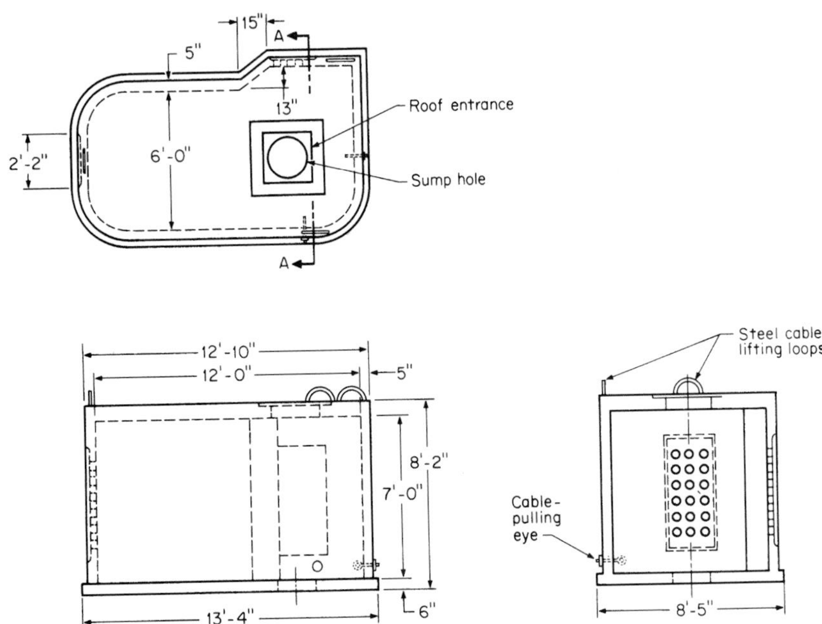

FIGURE 31.4 An L-type precast cable manhole. Approximate weight 30,000 lb. Manhole is designed for use with 18-duct conduit run. (*Courtesy Commonwealth Edison Co.*)

vehicular traffic must be able to support a minimum load of 12,500 lb/ft^2. Manholes installed in areas not subject to vehicular traffic must support a minimum load of 300 lb/ft^2.

The area at the manhole installation site must be cleared of obstructions such as gas pipes and water mains. Mechanized equipment such as a backhoe can be used to excavate the area for the manhole. The bottom of the excavation should be firm, undisturbed, or compacted earth, leveled without any large rocks or obstructions that would prevent the manhole from setting properly (Fig. 31.9). The excavation should be shored to protect the worker preparing the excavated area for the manhole installation. The precast manhole can be moved to the site on a low-boy trailer. A mobile crane is normally used to lift the manhole from a trailer and place it in the excavated area (Figs. 31.10 through 31.12).

The hole around the precast manhole must be backfilled and thoroughly compacted. Sand or the excavated material, if it is fine and dry, can be used for backfilling.

Manhole frame and cover are installed after the precast manhole is in place. Precast manhole necks can be used to extend the manhole opening to the proper level for the surrounding area (Fig. 31.13).

As manholes differ in size the complexity owing to variations in the number of ducts entering and leaving; whether one-, two-, three-, or four-way; size of cables; length of splices; horizontal offsets of conduits; etc., only general principles of construction can be outlined.

Floors. Manhole floors are built of poured concrete approximately 6 in. thick and then surfaced with cement mortar about ½ in. thick. The mixture for the concrete is usually:

One part portland cement

Three parts clean sand

Five parts broken stone or gravel

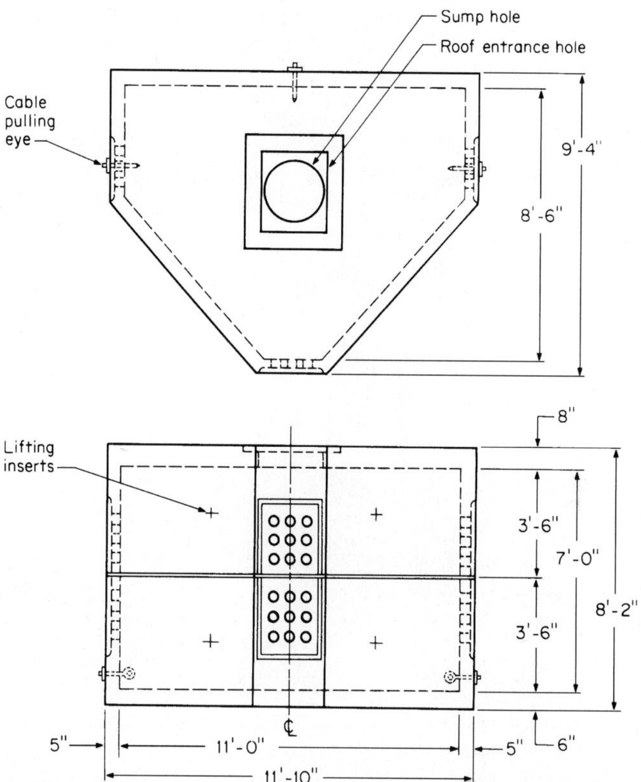

FIGURE 31.5 A T-type precast cable manhole. Approximate weights top section 15,065 lb and bottom section 13,235 lb. Manhole is designed for use with 18-duct conduit run. (*Courtesy Commonwealth Edison Co.*)

The mixture for the cement mortar usually is:

One part portland cement

Three parts clean sand

The cement mortar may be placed after the walls have been completed or after the roof has been constructed.

The floor should slope toward the sewer drain or sump with a grade of about 1 in in 6 ft.

Walls. Walls are commonly built of poured concrete, concrete blocks, or bricks. The thickness of the wall varies from 8 in. to 2 ft depending on the height and length of the wall. It also depends on whether the soil is firm or fluid. Brick walls are usually half again as thick as concrete walls. Concrete walls may be reinforced if necessary. A typical brick wall is shown in Fig. 31.14.

The conduit runs should terminate at the interior side of the manhole wall in conduit bells. A conduit bell (Fig. 31.15) is a flared opening that makes it easier to pull the cable and eliminates the sharp corner that might injure the cable sheath as the cable is pulled in. It also protects the cable sheath as the cable moves in and out owing to load cycling. If the

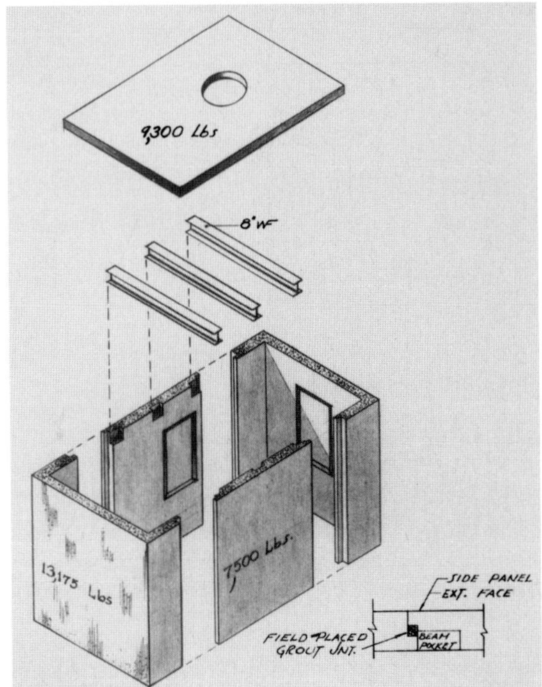

FIGURE 31.6 Precast sections of concrete manhole. (*Courtesy Eugene Water and Electric Board.*)

conduit run has not been installed at the time the manhole wall is being erected, an opening should be left in the wall for the conduit bells. The hole can be closed later, permitting correct alignment of bells and ducts.

Pulling eyes (Fig. 31.16) are usually embedded in the walls opposite the conduit runs to facilitate the pulling in of the cables.

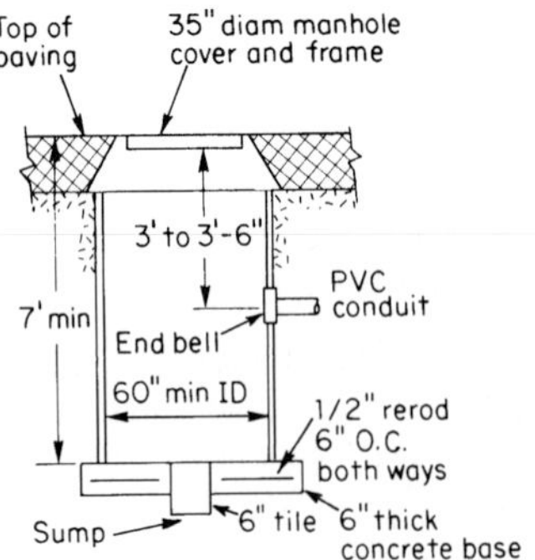

FIGURE 31.7 Concrete manhole for use in areas subject to vehicular traffic.

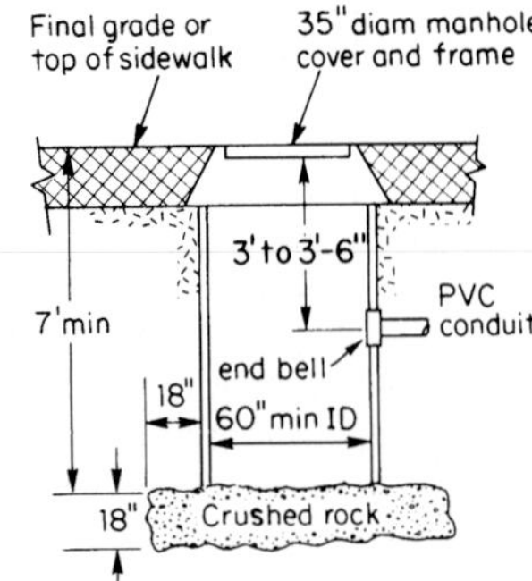

FIGURE 31.8 Fiberglass, plastic, or concrete manhole for use in areas not subject to vehicular traffic.

FIGURE 31.9 Workers checking bottom of excavation to be sure it is level before setting precast manhole.

FIGURE 31.10 Placing precast sections in position in excavated manhole. (*Courtesy Eugene Water and Electric Board.*)

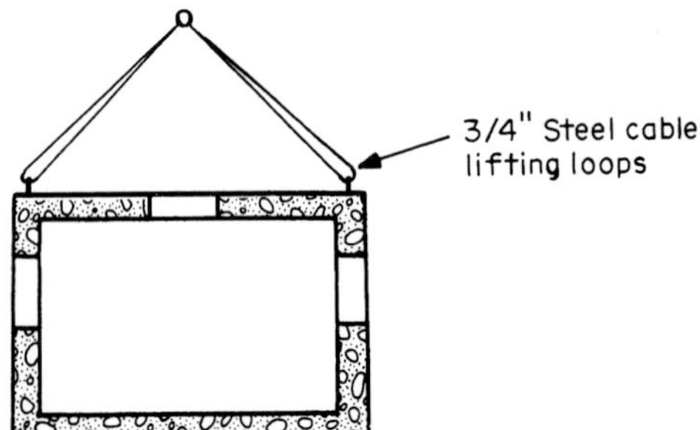

FIGURE 31.11 Typical method of lifting single-piece precast manhole to place it in excavated area.

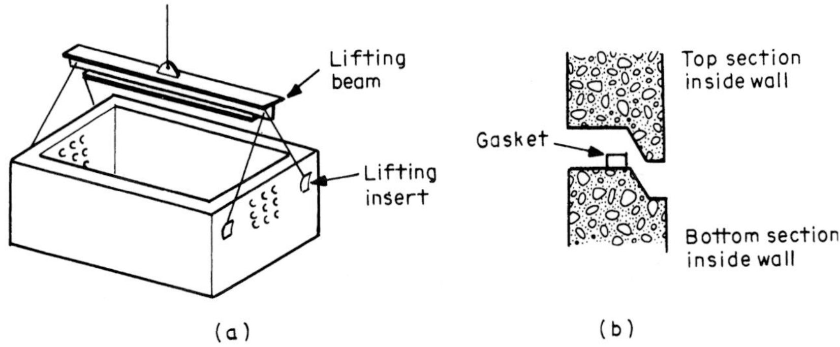

FIGURE 31.12 (*a*) Typical method of lifting sectional precast manhole to place it in excavated area. (*b*) Detail of joint gasket installation.

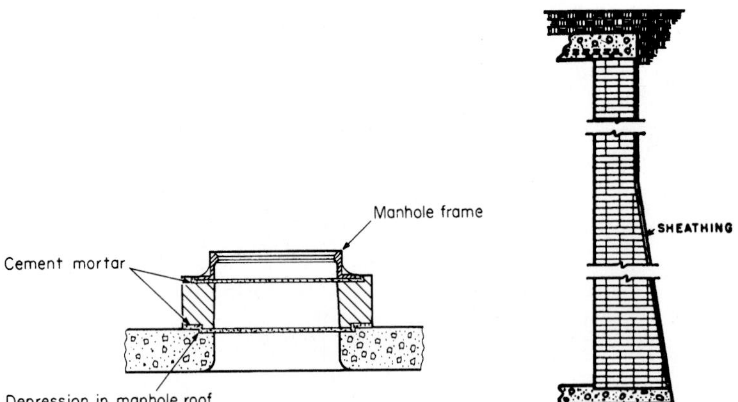

FIGURE 31.13 Manhole frame and cover in place on precast manhole. A precast concrete manhole neck is used to extend opening proper distance so that cover will be flush with surface of terrain.

FIGURE 31.14 Typical brick-wall construction of manhole. Note poured-concrete floor and roof.

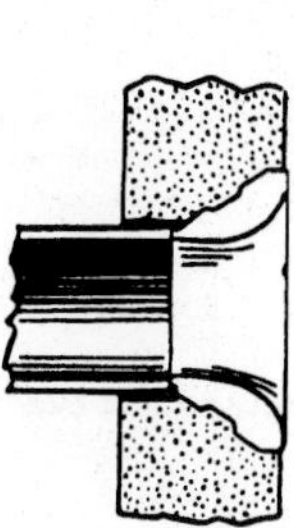

FIGURE 31.15 Typical plastic duct bell used to terminate duct in manhole wall.

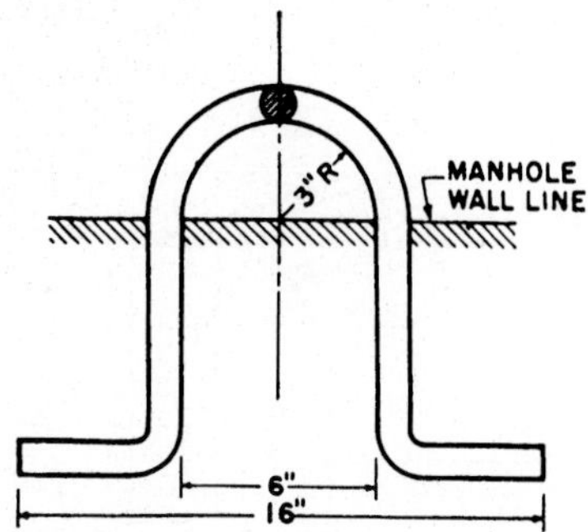

FIGURE 31.16 Typical galvanized wrought-iron pulling eye built into manhole wall opposite duct runs.

Roofs. Manhole roofs are usually constructed of reinforced concrete. If, however, it is necessary to restore traffic on the street as quickly as possible, a roof of steel beams and brick is used. This eliminates the delay in the time required for the concrete to set. A top-view picture with the reinforcing bars in place is shown in Fig. 31.17.

The concrete mixture for the roof is usually:

One part portland cement

Two parts clean sand

Four parts 3/4 in. crushed stone or gravel

FIGURE 31.17 Manhole constructed of reinforced concrete showing reinforcing rods before pouring of concrete. Note that conduit bank at manhole is being cast integrally with manhole walls. (*Courtesy Edison Electric Institute.*)

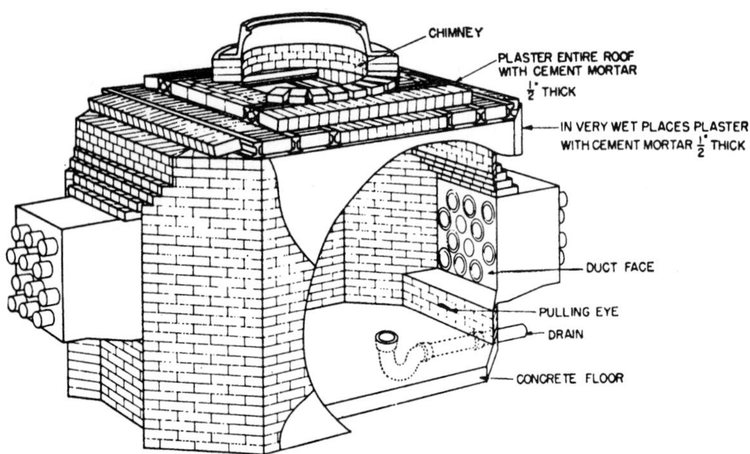

FIGURE 31.18 Cutaway view of typical brick manhole showing concrete floor, brick walls and roof, chimney, two-way duct runs, 12 ducts, duct bells, pulling eye, and drain. (*Courtesy Edison Electric Institute.*)

The roof opening, or throat or chimney as it is sometimes called, must be located so all is clear directly below. The opening may be circular or rectangular in shape, depending on the type of cover frame used. A removable metal cover is used to close the opening.

A completed manhole is shown in the cutaway sketch of Fig. 31.18. The manhole is constructed with concrete floor and brick walls and roof. The sketch also shows the concrete-encased conduit run and the circular opening in the roof from which the manhole derives its name.

FIGURE 31.19 Installation of a 750-kVA oil-filled network transformer rated 14,000-480Y/277 volts in a vault. (*Courtesy General Electric Co.*)

FIGURE 31.20 Linemen placing network transformer in vault constructed below sidewalk in a business area. (*Courtesy General Electric Co.*)

Vaults. Large manholes are often referred to as vaults. Vaults are used for transformer and network protector installation. The construction of a vault is similar to that of the manhole and must meet the same construction specifications. Figures 31.19 and 31.20 show transformers being installed in vaults by linemen.

CHAPTER 32
PULLING CABLE

Installing Cable Racks. Before cables are pulled into ducts, provision for supporting the cable ends on racks in the manholes is made. Allowing the cable ends to lie on the manhole floor enables them to be stepped on or damaged by falling objects. The rack support channel is fastened to the manhole wall with expansion bolts. Holes for the expansion bolts are drilled 3 in. deep. Safety goggles are worn when concrete or brick is being drilled.

The spacing between racks adjacent to the proposed cable splice is usually 3 ft. In general, a cable should be supported at points 6 in. from each end of the splice. Additional racks are spaced about 3 to $4^1/_2$ ft apart depending on the nature of the cable bends.

Selecting the duct. Before the start of cable installation, the duct to be occupied should be selected throughout the entire length of the conduit run. As far as possible, the same relative position in the duct bank should be maintained. In general, the longer cables should be installed in lower ducts and the shorter cables in upper ducts. In this way the vacant spaces will be left in the upper ducts, where they can be utilized later on without disturbing the cables already installed.

Automated Cable-Pulling Equipment. Portable cable-pulling equipment or its accessories and a winch truck can be used to clean a duct and pull the steel cable contained on the cable puller equipment or the winch line on a truck through the duct simultaneously (Figs. 32.1 through 32.3). The hose from an air compressor is attached to the air inlet on the portable cable-pulling equipment or to the air control valve assembly that can be mounted on a winch truck. The steel cable on the truck or the pulling equipment is fastened to the double-disk birdie and inserted in the duct to be used (Figs. 32.4 and 32.5). A duct-sealing head attached to the outlet air hose is fastened around the steel cable, inserted into the duct on top of the birdie, and locked in place to provide an air seal (Figs. 32.6 and 32.7). The air control valve is operated so that the compressed air pressure forces the double-disk birdie through the duct (Fig. 32.8). The birdie cleans any dirt out of the duct and pulls the winch line through the duct. When the birdie reaches the end of the duct, the air seal is broken, and the travel of the birdie and the steel winch cable stops. The steel cable is now available to be used to pull in the electric power cable. If the duct has an obstruction in it, the workmen may have to use manual methods to clear the ducts and to clean them.

Rodding the Duct. *Rodding* is the term used to describe the insertion of a flexible fiberglass or steel rod into a duct to clear or dislodge obstructions. Rodding equipment is used if the automated cable equipment is not successful or if the automated equipment is unavailable. Figure 32.9 shows a steel rod on portable reels. The size of the rod is such as to make it stiff enough for pushing. Coupling of rods is accomplished by use of a leader and pickup fittings on the ends, shown in Fig. 32.10.

FIGURE 32.1 Equipment utilized for ensuring the duct is clear of obstructions.

Cleaning the Duct. Before the cables are installed, each duct must be checked to see that it is clean and free from all obstructions. If the automated procedures are not successful, rodding equipment can be used to pull a cable through the duct as previously described. The duct can then be cleaned with a mandrel, a circular wire brush, and a swab attached to the cable. The mandrel (Fig. 32.11) is about 1/4 to 3/8 in. smaller in diameter than the duct. A common mandrel is a steel tube rounded and fitted with a pulling eye at each end. It will remove serious obstructions such as concrete which has dripped through a joint by working it back and forth until the obstruction is cleared. The circular wire brush has the same diameter as the duct and is used to remove small pebbles or bits of concrete, etc. The swab is made by stuffing waste into a holding grip.

FIGURE 32.2 Duct cleaning equipment being prepared for a field demonstration.

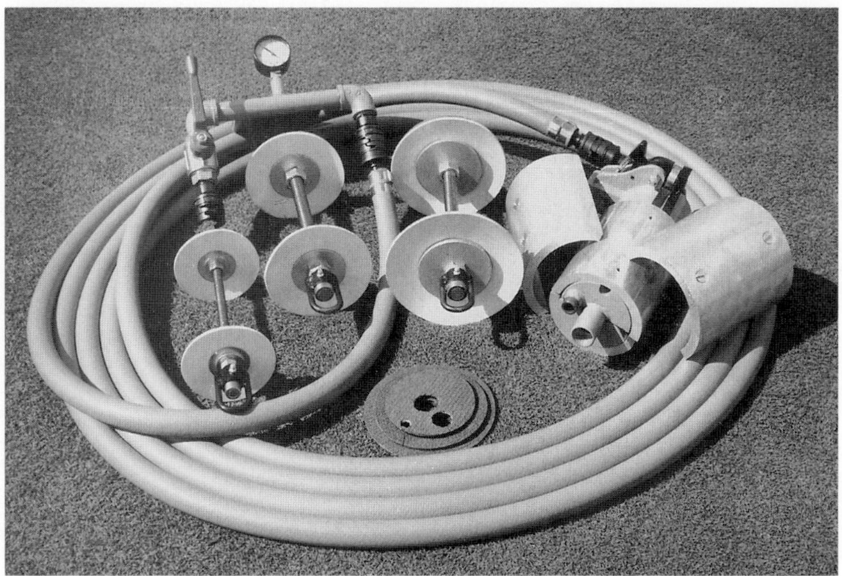

FIGURE 32.3 Air hose and accessories used with winch truck or portable cable-pulling equipment for automated duct cleaning and cable pulling. (*Courtesy Sherman & Reilly, Inc.*)

Selecting the Pulling-In Manhole. The selection of the manhole from which the cable is to be pulled is important. Cables can be pulled on the first pull only as far as the pulling rig will permit (Fig. 32.12). When the pulling eye in the cable end or on the basket grip reaches the sheave, the pulling must stop. The length of cable then in the manhole may not be sufficient to reach the cable rack on which the splice will be supported. Furthermore, additional cable length is needed to train the cable around the walls to the center of the splice on the rack. The shape of the manhole and the position of the splice on the wall will determine the amount of cable length needed. In a manhole where the ducts

FIGURE 32.4 Double-disk birdie with winch line of cable-pulling equipment is being inserted into a duct. The double-disk birdie will clean the duct and pull the winch line to the far end of the duct so that the winch line can be used to pull the power cable being installed through the duct. (*Courtesy Sherman & Reilly, Inc.*)

FIGURE 32.5 Double-disk birdie with winch line of cable-pulling equipment is being inserted into a duct for a field demonstration.

come in at one corner and leave at the diagonally opposite corner, the splice is always closer to one duct than to the other (Fig. 32.13). When there is considerable difference in duct elevations, the length of cable from the splice to one entrance may be less at one end than at the other. The rule, therefore, is that the manhole requiring the shorter cable length in the manhole to the center of the splice should be the pulling end, as required length of cable can be left at the feeding end.

Manhole Riggins. The rigging used to pull in cable varies with the type of manhole, length of cable, and the conditions prevailing on the street.

FIGURE 32.6 The duct-sealing head with air hose attachment is partially inserted into the duct. The winch line attached to the double birdie is threaded through the head. The head will be locked into place to provide an air seal on the end of the duct. The double birdie will pull the winch line through the duct when compressed air is supplied to the head. (*Courtesy Sherman & Reilly, Inc.*)

The simplest rigging arrangement possible is a pulling eye that has been embedded in the manhole wall (Fig. 32.14). Here the pulling eye is directly opposite to the duct, and the pulling block holds the pulling rope in line with the duct. In case the embedded pulling eye is located much below the duct level, a chain or rope of proper length is inserted between the pulling block and the pulling eye as shown in Fig. 32.15. This is the usual situation in a multi-way manhole, where ducts enter the manhole on all sides and the eye must be placed below the duct level. A sheave in the manhole opening is often used in deep manholes to guide the pulling rope to the pulling winch on the street as shown.

When no embedded pulling eyes are available, the sheaves are supported on a so-called sheave stand. It consists of two timbers, channel irons, or I beams bolted together with a space between them for the sheaves. Holes through the two parallel members (Fig. 32.16) permit placing the sheaves at the desired locations.

Attaching Pulling Rope. The pulling rope is attached to the cable by means of a woven cable grip or by means of a clevis or eye.

The principle of a woven cable grip, or *basket* grip, as it is sometimes called, shown in Fig. 32.17, is that the wires, which are in the form of an open weave, will contract laterally and thus hold fast to the cable sheath when a longitudinal pull is applied. Since the pull is thus applied to the sheath, this attachment is satisfactory only if the sheath is tight and if the pull is light.

FIGURE 32.7 The duct-sealing head with air hose attachment is partially inserted into the duct for the field demonstration.

When the sheath is loose or the pull heavy or there is no sufficient clearance in the duct for the wire mesh surrounding the cable, the pulling rope must be attached to a clevis or eye. The eye in turn is fastened to the conductors in the cable as well as to the sheath surrounding the cable (Fig. 32.18). Such eyes are usually attached at the cable factory.

Pulling eyes can be attached in the field by first folding back about 4 in of sheath and removing the insulation. Then the conductor strands are wrapped around the lug of the eye for sweating. The sheath is then turned back and sealed with solder by wiping to the conductors and lug. The steps in this operation are illustrated in Fig. 32.19.

Since the pulling rope has a tendency to twist during the pull, a swivel (Fig. 32.20) should be inserted between the pulling rope and the cable.

FIGURE 32.8 The duct-sealing head with air hose attachment is fully inserted into the duct and in the process of clearing obstructions and cleaning the duct as it travels through the duct system.

Feeding Tube. To prevent injury to the cable by scraping on the manhole frame or at the duct opening or in passing over other cables, a feeding tube is sometimes used. This guiding tube is usually made of flexible steel tubing about 4 in in diameter with a flare at the reel end and a removable nozzle fitting at the duct end so that it can be adapted to various size ducts. The tubes vary in length from 10 to 20 ft. Figure 32.21 shows typical feeding tubes.

Handling Cable. A common cause of underground system outages is damaged cable resulting from improper handling and installation. It is important that extreme care be exercised while handling and installing the cable. Minor physical damage to the cable jacket or installation can lead to premature failure.

FIGURE 32.9 Flexible steel rod on reels. (*Courtesy Superior Switchboard and Devices Division.*)

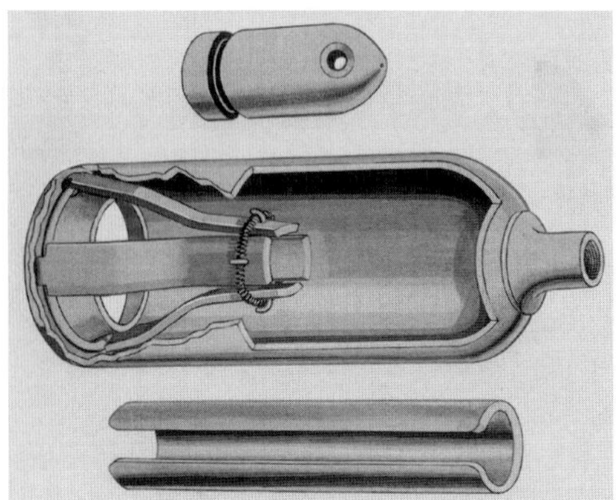

FIGURE 32.10 Coupling fixtures for flexible steel rod. (*Courtesy Superior Switchboard and Devices Division.*)

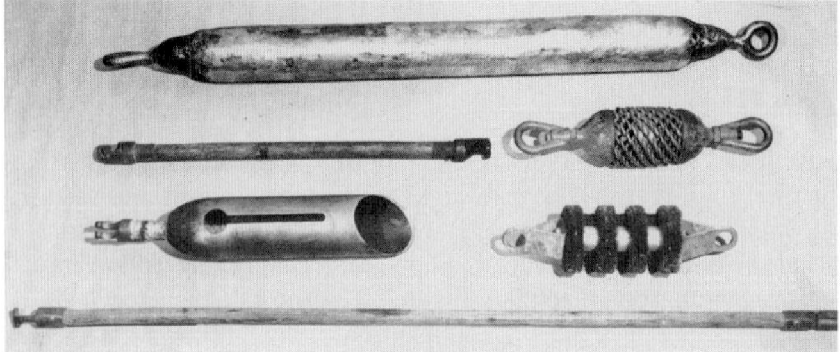

FIGURE 32.11 Tools used for duct cleaning and testing. Note mandrel, wire brush, and swab. (*Courtesy Edison Electric Institute.*)

FIGURE 32.12 Lineman on pole has threaded steel winch cable, which was drawn through the riser conduit by a birdie and compressed air, through a pulling block mounted on the pole above the duct. The steel cable will be threaded through a pulling block at the base of the pole. The steel cable is fastened to a winch on a line truck that will be used for power to pull the electric power cable through the duct.

Cable packaged on reels should arrive on the site suitably protected against damage. Reels should be upright and with means to prevent cable unrolling or unwrapping. The lineman or cableman should inspect the cable for obvious and hidden damage. If the reels or reel wrap is damaged, the inspector should use extra caution. The cable must arrive at the site with the ends sealed to prevent moisture from entering the cable. The cable must be protected from cold weather prior to installation. If the temperature is less than 14°F, the cable reels should be warmed for 24 hours prior to the starting of the cable-pulling operation.

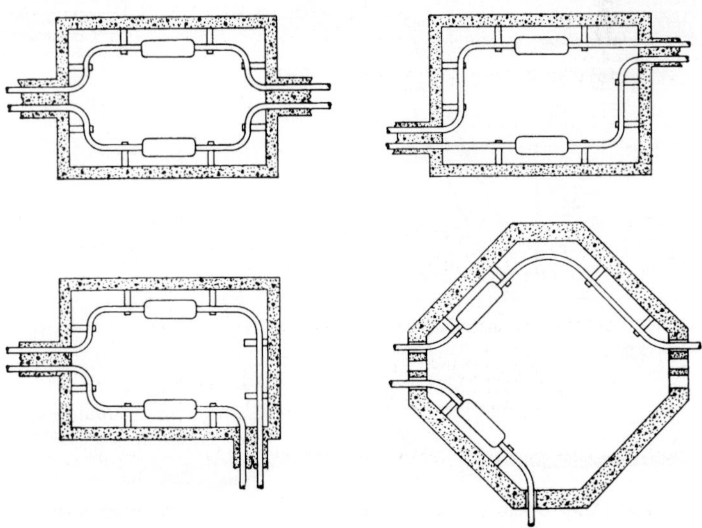

FIGURE 32.13 Types of cable training in manholes of different shapes. (*Courtesy Edison Electric Institute.*)

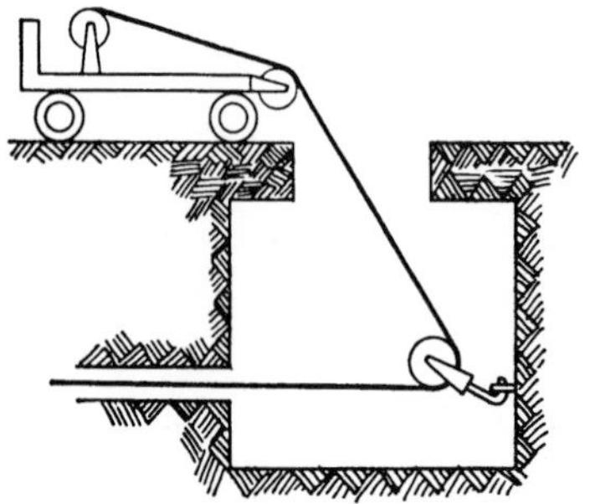

FIGURE 32.14 Pulling cable by use of eye embedded in manhole wall opposite duct. Pulling block holds pulling rope in line with duct.

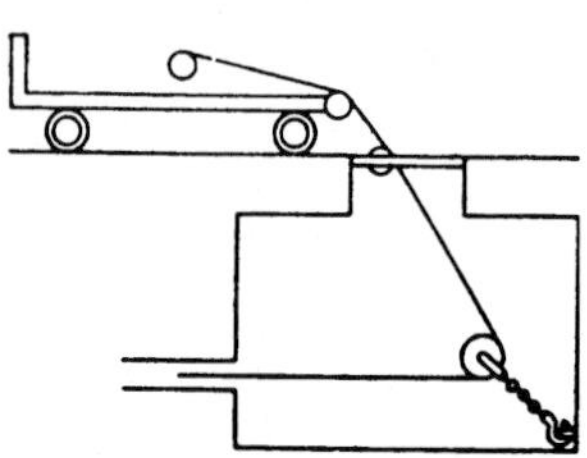

FIGURE 32.15 Pulling eye located below duct level requires use of short rope or chain between pulling block and pulling eye.

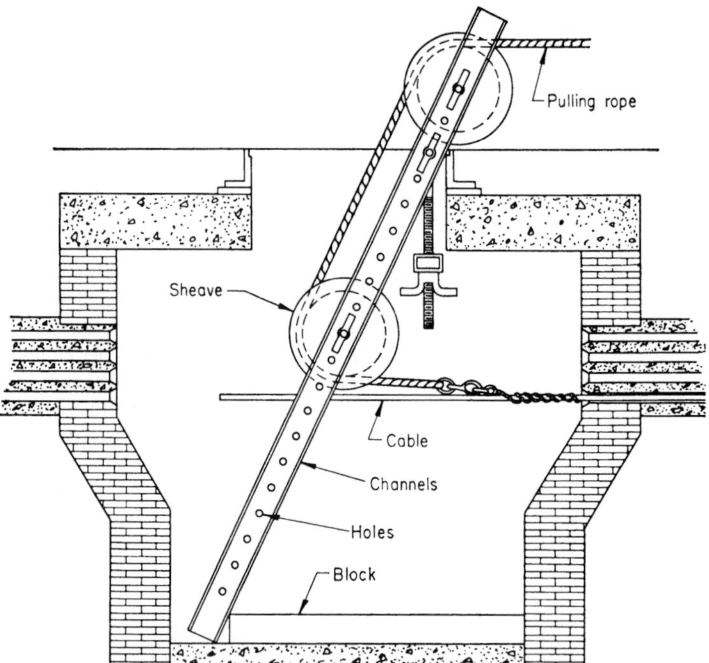

FIGURE 32.16 Double-sheave stand made of two parallel channel irons. Holes in channel members permit placing sheaves at proper levels.

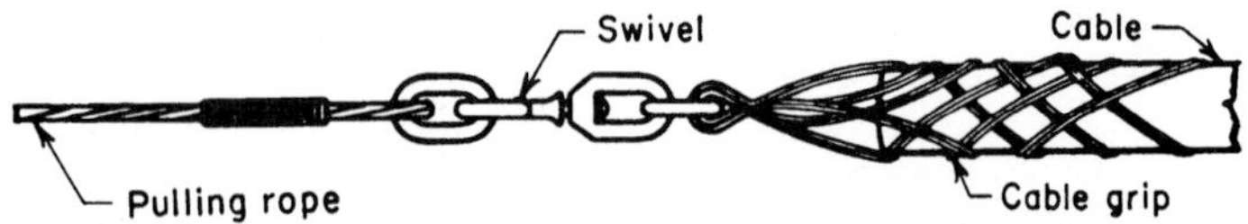

FIGURE 32.17 Woven cable grip, sometimes called *basket* grip. Note swivel inserted between pulling rope and cable grip.

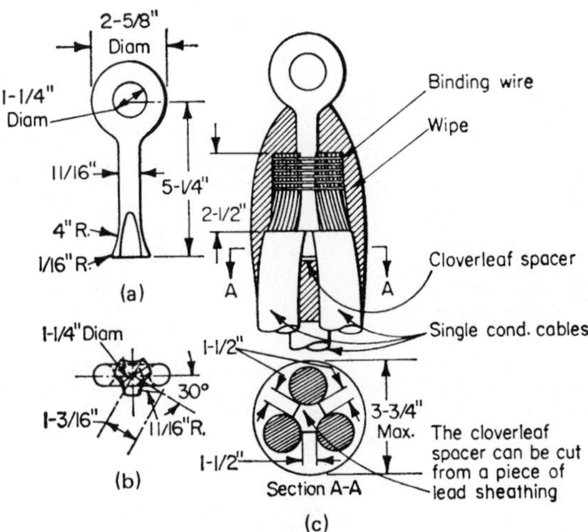

FIGURE 32.18 Pulling eyes attached to cable end. Conductors are sweated onto lug of eye, and sheath is wiped onto conductors and lug. (*a*) Clevis type, (*b*) eyebolt type, (*c*) eyebolt type used with triplexed cables. (*Courtesy Edison Electric Institute.*)

The lineman or cableman unloading the cable at the installation site should use a cradle to support the reel flanges or a shaft through the arbor hole to lift the reels. If a forklift is used by the workmen, the forks must lift the cable reel at 90° to the flanges and be long enough to make complete lifting contact with both flanges. The cable surface and/or the protective wrap must be protected from damage by the lift forks or other equipment. The cable is often placed on a trailer designed for handling cable at the service center and transported to the installation site. The cable can be installed while the reel is mounted on the trailer, minimizing the handling of the cable and the danger of damage while it is being installed (Fig. 32.22).

Cable Lubrication. To protect the cable from excessive tension during pulling in, the cable is lubricated. This reduces the friction between cable and duct walls, especially on curves in the duct run, by as much as 70 percent.

The lubricants for lead-sheath cables are greases, oils, soapstone, etc. The lubricant is applied with a paddle or brush just before the cable enters the feeding tube. A coating about 1/16 in thick is ample and will amount to 6 to 8 lb per 100 ft on 3-in cable.

No lubricant should be applied to the first and last 5 ft of cable for convenience and cleanliness in splicing.

Pulling the Cable. The cable is drawn into the duct by means of a winch or capstan. The winch is usually mounted on a truck or a portable cable puller located near the manhole or the riser conduit at the pulling end.

The reel of cable must be properly placed at the feeding end to cause minimum flexing of the cable. It should always be located on the side of the manhole toward which the cable is to be pulled. This avoids putting a reverse bend in the cable and makes it unwind easily.

The end of the cable is then threaded into the feeding tube extending from above the edge of the manhole down to the end of the duct. The pulling grip is now slipped over the

THE END OF THE CABLE IS CUT SQUARE.
THE LEAD SHEATH IS MARKED AT APPROXIMATELY 2" AND 4" FROM THE END.
THE LEAD IS STRIPPED OFF UP TO THE 2" MARK.

THE LEAD BETWEEN THE 2" AND 4" MARK IS CUT IN SCALLOPS WITH EVERY OTHER ONE REMOVED.
THE SCALLOPS ARE LEFT ABOUT 1 1/2" LONG AND 3/16" WIDE AT THE TIP.
THIS SECTION IS REMOVED.

THESE SCALLOPS ARE THEN BENT BACK OVER THE LEAD.
THE INSULATION IS CLEARED EVEN WITH THE EDGE OF THE SHEATH.

ON SINGLE CONDUCTOR CABLES, OPEN THE CONDUCTOR AND INSERT SHANK.
ON SINGLE CONDUCTOR CABLES WITH ROPE CORE, REMOVE ROPE CORE AFTER OPENING THE CONDUCTOR.
ON MULTIPLE CONDUCTOR CABLES, THE CONDUCTORS ARE SPREAD ENOUGH TO ALLOW THE SHANK TO ENTER.
THE CONDUCTORS ARE THEN HAMMERED OVER THE SHANK AND BOUND AGAINST IT WITH WIRE.
TIN CONDUCTORS AND SHANK.

PULL BACK THE SCALLOPS AND HAMMER THEM DOWN AGAINST THE CONDUCTORS.

WIPE PULLING-EYE, SCALLOPS AND CONDUCTOR, INTO A SMOOTH JOINT.

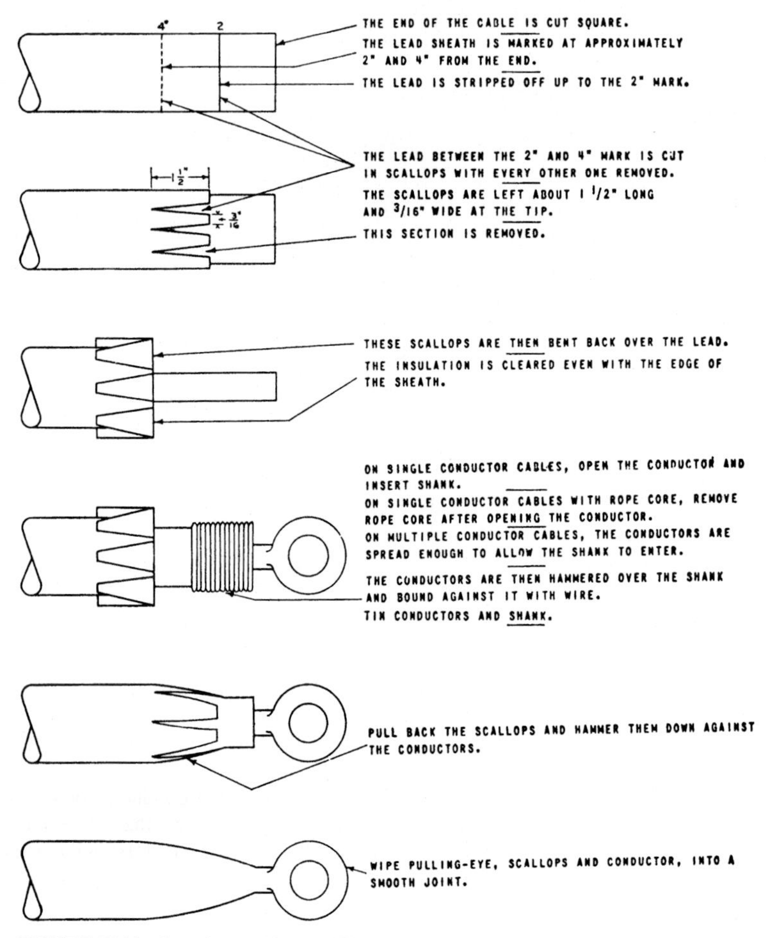

FIGURE 32.19 Steps in attaching a pulling eye. (*Courtesy Edison Electric Institute.*)

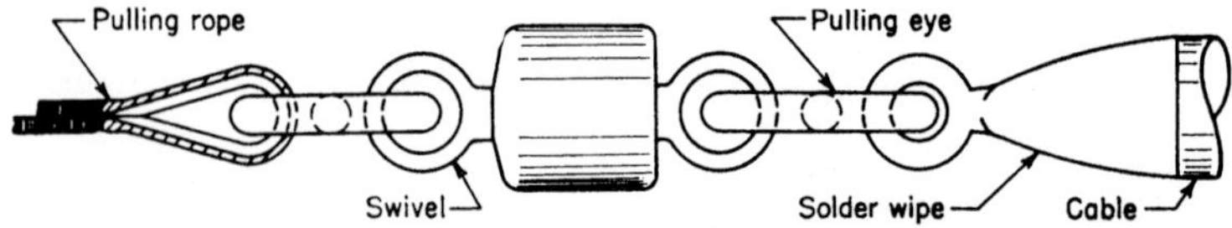

FIGURE 32.20 Swivel inserted between pulling rope and cable to prevent twisting of cable. (*Courtesy Edison Electric Institute.*)

FIGURE 32.21 Typical flexible-cable feeding tubes used to guide cable into duct and prevent abrasion. (*Courtesy Edison Electric Institute.*)

end of the cable, the swivel is attached, and finally the pulling rope is attached. In case a pulling eye is provided on the cable, the swivel and pulling rope are attached to it.

The cable is inspected and greased as it enters the funnel of the feeding tube. Lead-covered cable must be completely covered with lubricant. On cables having a rubber jacket on the outside, a mixture of soapstone and water or a pulling compound specified by the cable manufacturer is applied to the jacket as the cable is drawn in.

Workers stationed at the cable reel regulate the amount of slack in the cable so that it passes freely into the tube without being loose on the reel and without scraping on the manhole frame.

A man should be stationed in the manhole at the pulling end to signal a stop to prevent pulling the cable over the sheaves. Another should be stationed at the top of the manhole to relay signals to the winch operator.

Excessive pulling tensions that exceed the elastic limit of the conductor or stress the extruded portions of a cable will cause permanent damage to the cable that will eventually cause the cable to fail. The maximum pulling tension should not exceed 0.008 lb/cir mil of

FIGURE 32.22 Cable mounted on specially designed trailer for transportation and handling of the cable while it is being installed. Cable has been attached to steel winch line. Lineman is guiding cable over pulley mounted on a steel duct protruding out of the ground in an opening in a concrete pad for a pad-mounted transformer.

copper conductor or 0.005 lb/cir mil of aluminum conductor, when the pulling eye is attached to the cable conductor, with the maximum tension to be limited to 5000 lb for single-conductor cable and 6000 lb for multi-conductor cables. If a basket is placed over the surface of the cable for pulling, the maximum tension should not exceed 1000 lb. Sidewall pressure on the insulation or sheath of the cable at a bend in the duct should not exceed 300 lb per foot of radius of the duct bend.

The speed at which the cable is drawn into the duct varies with conditions and cable size. It is good practice, however, to pull the cable slowly. In a straight duct a single cable can be pulled in at about 60 ft/min. If a higher speed is used, there is difficulty in greasing properly and in inspecting the cable. An average speed of about 40 ft/min, or 2/3 ft/sec, is preferable when a single cable is being pulled. When two or more cables are pulled into one duct, the speed should be further reduced to about 20 ft/min. A slow speed will prevent the possibility of crossing the conductors as they enter the duct.

The savings in time obtained with higher speeds is negligible, as the time required for drawing in a 400-ft length at 40 ft/min is only 10 min. This is a small part of the total time needed to prepare for the pulling-in operation.

Slack Pulling. When the cable end has been drawn up to the manhole rigging, the first pulling operation must stop. If not enough cable length is then available in the manhole to reach to the center of the splice as the cable is trained around the wall, additional pulling is necessary. To obtain this additional slack in the cable, a new cable grip must be applied to the cable the required number of feet from the cable end. A special *basket* grip (Fig. 32.23) is used. This grip is open at both ends and can be slipped over the cable to the desired position. Care must be used to avoid injuring the cable.

When the cable has been pulled in as far as necessary, the forward end is placed on the cable supports. No attempt, however, should be made to train the cable into its final position. This will be done by the splicing crew. A check should be made to see that the seal at the end of the cable is still watertight. At the feeding end the cable is cut off with enough slack left to train it properly around the manhole walls. The end of the cable is then also placed on its supports, and the end of the cable is sealed. The end of the cable remaining on the reel is also sealed.

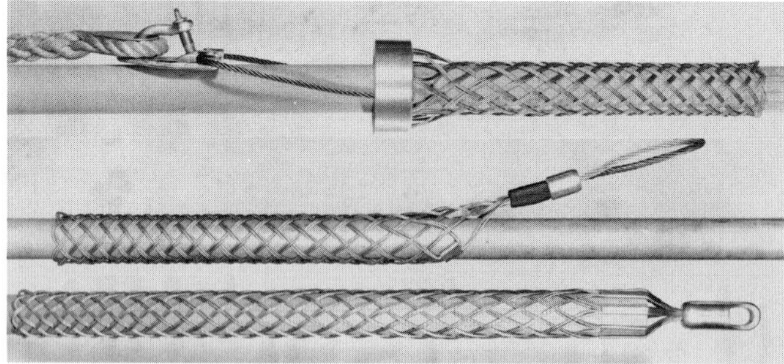

FIGURE 32.23 Special basket grips applied to cable to pull up more slack at pulling end. (*Courtesy Edison Electric Institute.*)

CHAPTER 33
SPLICING CABLE

Multiple lengths of cable are connected together to form a continuous length. These connections are called joints or splices. If underground ducts are used, the splices are normally completed in a manhole. When a cable is cut, in preparation for splicing or for any other reason, it must be protected from moisture and dirt. Cable ends exposed to the atmosphere will collect moisture and contamination and so should be properly sealed at all times except during the period when the splice is being completed or the cable termination is being installed. Cable ends should be thoroughly inspected before they are trained into final position for splicing.

If the splices are to be completed in a manhole, the cables and the splices should be supported on racks mounted on the manhole wall as illustrated in Chap. 29. The joint or splice is supported between two brackets located on the central portion of the manhole wall. Because of expansion and contraction of the cables, caused by changes in cable temperature, it is necessary to provide reverse curves or offset bends in the portion of the cable in the manhole. These reverse curves enable the cable to take up the expansion and contraction movements without cracking or buckling of the insulation or lead sheath. Therefore, the training of cables on the manhole wall is not only for making a neat arrangement in the manhole but more importantly for providing the expansion space needed to absorb the movements of the cable.

In general, the bending radius R (see Fig. 33.1) of the reverse curves in lead-covered cables must be no less than 8 times the overall diameter of the cable. Thus, a 3-in cable should have a minimum bending radius of 24 in or more, and a 4-in cable should have a minimum bending radius of 32 in or more.

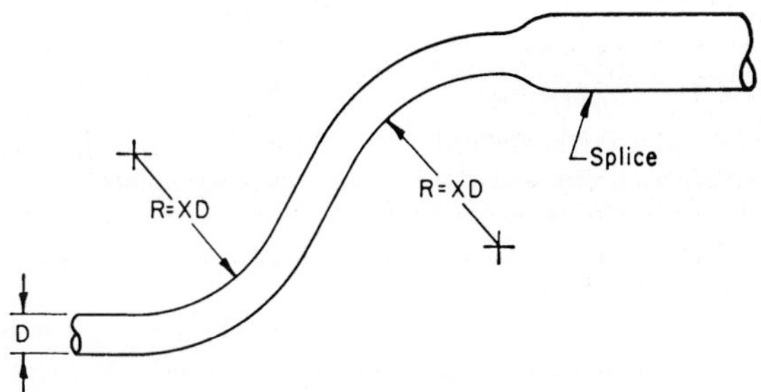

FIGURE 33.1 Sketch showing minimum bending radius on reverse curves of cable as it is trained around wall in manhole; D is cable diameter, X is number of times D, and R is bending radius.

The range of values of R in terms of cable diameter varies from 5 to 12 and depends on the size and number of conductors and whether it is a polyethylene-, thermoplastic-, rubber-, varnished-cambric-, or paper-insulated cable, as well as on the rated voltage of the cable.

At least 6 in of straight cable should extend beyond each end of the splice to provide space for resting on the saddles of the supporting racks. The clips holding the saddles on the bracket should be spread enough to permit the saddle to slide freely along the bracket.

All cables and joints should be so racked in the manhole that they are not directly under the manhole cover.

The exact makeup of a splice depends on the specific cable construction, that is, whether the cable is a single- or multiple-conductor cable; whether the insulation is rubber, polyethylene, thermoplastic, varnished cambric, or impregnated paper; or whether the cable insulation has a conducting shield; etc. The general installation procedure is similar to those described in this chapter. The cable manufacturer's specific instructions must be followed in all instances. The worker must maintain tools in good condition and keep them clean and dry. The working area and the area of the cable being worked on should be protected from moisture with a rubber blanket or other waterproof material. The cable splicer or lineman performing the work must maintain clean and dry hands.

Splices vary greatly with type and voltage of cable, so that drawings showing the essential dimensions for each type and voltage of cable splice are required. Figure 33.2 shows such a drawing. These drawings will give for each particular joint the length of lead sheath to be stripped back from the cable end, the amount of conductor insulation to be cut back so as to give the proper stepping or penciling, the diameter to which the individual conductor insulation should be built up, and the correct completed outside diameter of the insulation. The cable splicer must have the drawings and materials for the particular joint to be made on hand before starting.

The cable ends to be spliced are trained into position with ends overlapped. The bending radius is made as large as, or larger than, required. Paper- and varnished-cambric-insulated cables should not be bent when the temperature is below 14°F. In cold weather, cables must be heated so that they are warm all the way through before they are subjected to bending.

Cables are then marked at the centerline of the splice. Cables are cut off squarely at this mark with a hacksaw or cable cutter. The new cable ends should now butt together squarely. Cables should not be retrained after the ends are cut off because the conductors and the conductor strands will slide to unequal lengths.

Cable Jacket and Sheath Removal. Cables normally have a jacket or sheath or both. The most common cable jackets are metallic, rubber, or thermoplastic. It is necessary to remove the jacket the prescribed distance. The sheath of the cable should be cleaned to keep the shavings and contamination from coming in contact with the exposed cable insulation before the sheath is removed.

Lead cable sheaths can be cleaned by scraping with a shave hook or rasp. The entire surface must appear shiny and be free of all oil, cable-pulling compounds, and contamination.

The lead sleeve used to cover and protect the splice is cleaned inside and slid over one of the cable ends far enough to be out of the way while the splice is being made. Before this is done, each end of the sleeve should be thoroughly scraped with a rasp or steel brush for about 3 in. The scraped portion should then be coated with stearine flux to retard oxidation.

With a chipping knife mark the cable sheath where the ring cut is to be made according to the drawing. The length of sheath to be removed from each cable end is usually $1^1/_2$ in less than one-half the total length of the lead sleeve. Score the sheath at the ring cut by tapping the chipping knife lightly with a hammer around the sheath at the position marked (see Fig. 33.3). Cut about halfway through the lead. Slit the lead sheath lengthwise from

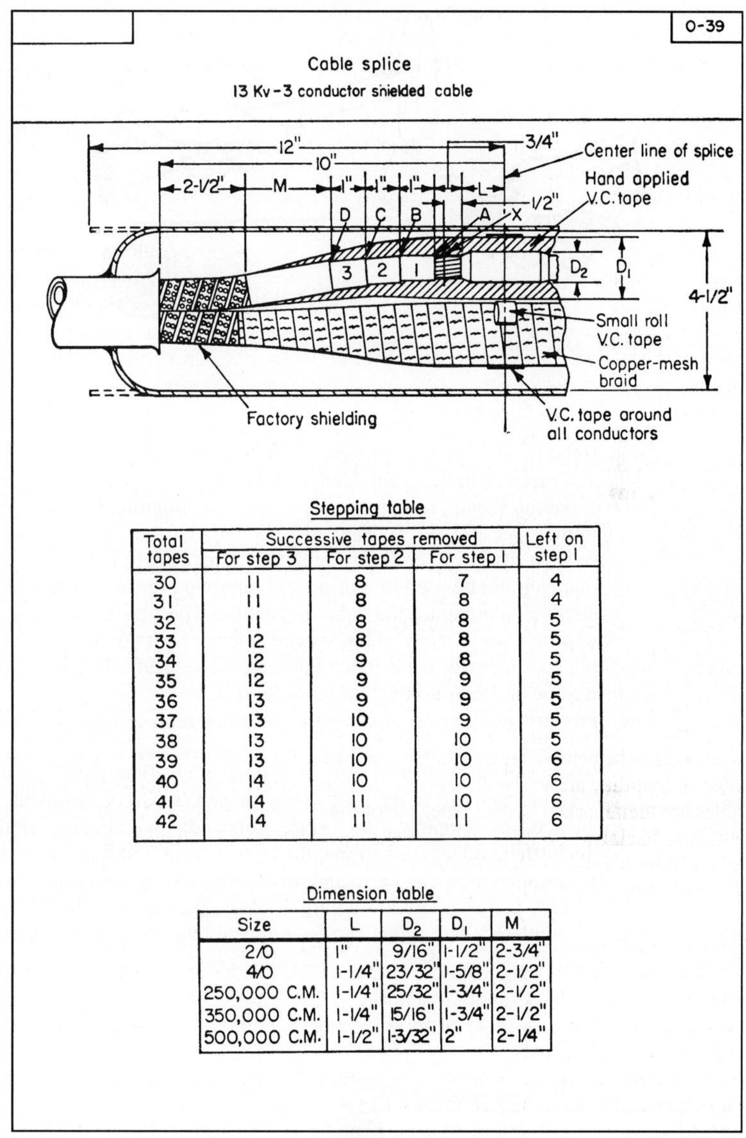

Stepping table

Total tapes	Successive tapes removed			Left on step 1
	For step 3	For step 2	For step 1	
30	11	8	7	4
31	11	8	8	4
32	11	8	8	5
33	12	8	8	5
34	12	9	8	5
35	12	9	9	5
36	13	9	9	5
37	13	10	9	5
38	13	10	10	5
39	13	10	10	6
40	14	10	10	6
41	14	11	10	6
42	14	11	11	6

Dimension table

Size	L	D_2	D_1	M
2/0	1"	9/16"	1-1/2"	2-3/4"
4/0	1-1/4"	23/32"	1-5/8"	2-1/2"
250,000 C.M.	1-1/4"	25/32"	1-3/4"	2-1/2"
350,000 C.M.	1-1/4"	15/16"	1-3/4"	2-1/2"
500,000 C.M.	1-1/2"	1-3/32"	2"	2-1/4"

FIGURE 33.2 Drawing of 13-kV three-conductor shielded cable splice. Accompanying table gives all essential dimensions for this particular splice. (*Courtesy Public Service Electric and Gas Co. of New Jersey.*)

FIGURE 33.3 Making ring cut with chipping knife.

the end of the cable to the score with the chipping knife held on a slant so that it passes between the insulation or metallic braid, if shielded, and the sheath as the knife progresses (see Fig. 33.4). To remove the sheath, loosen the edge along the longitudinal cut from the insulation with the hammer (see Fig. 33.5). Then grip the edge of the sheath with pliers and tear off at the score with a twisting motion, thereby forming a slight bell in the end of the lead sheath. Great care should be used in scoring and removing the lead sheath to prevent cutting into or otherwise damaging the cable insulation. The new ends of the lead sheath are belled out approximately $^1/_4$ in with a special fiber or hardwood tool. The tool is driven under the sheath parallel to the cable to raise the end of the sheath. This will provide space for a binder tape to be carried into the bell for at least $^1/_4$ in.

Shielded cable ends are not belled, since the shielding tape is extended to the stress cone.

Rubber or thermoplastic jackets or sheaths must be cleaned of all contamination such as wax, cable-pulling compound, and dirt in the area where the jacket is to be removed. This type of jacket can be cleaned by scraping with a knife, nonconductive abrasive cloth, or rasp. The jacket must be buffed and cleaned so that the tape or resin will bond and form a moisture tight seal. The cleaned area can be temporarily protected by covering it with a layer of vinyl tape. This tape picks up any loose particles from the surface when the inner portion of the splice is completed. The jacket can be removed by making a circular score

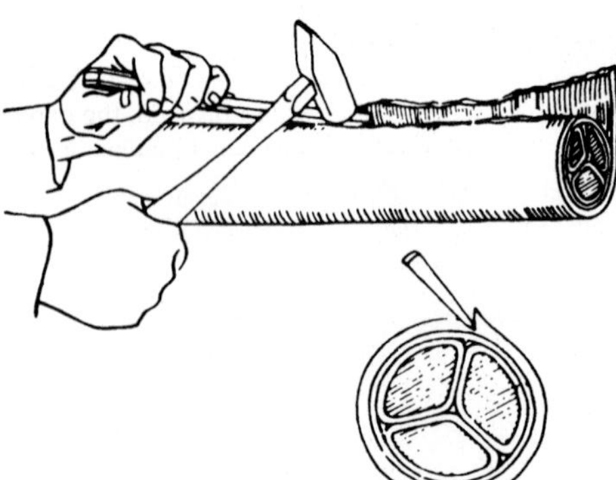

FIGURE 33.4 Making longitudinal cut in sheath. Chipping knife is held at an angle.

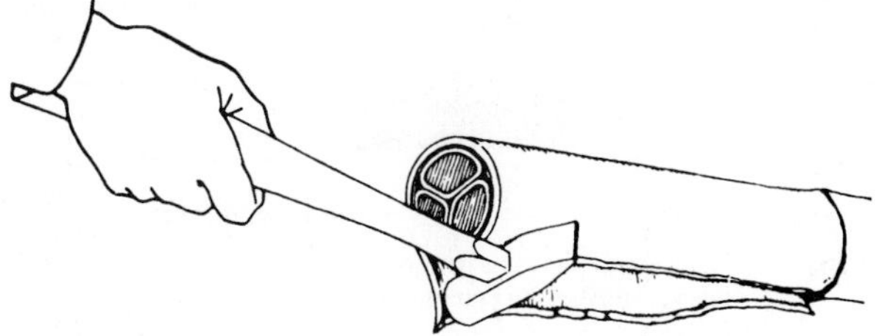

FIGURE 33.5 Loosening cable sheath from insulation.

approximately half-way through its thickness and then making a longitudinal cut from the end of the cable back to the score. The cable metallic shielding or insulation should never be cut during this operation. All bedding tapes should be removed back to the jacket.

Cable Metallic Shielding Removal. The metallic shielding should be terminated as evenly and smoothly as possible. A glass cloth tape can be wrapped around the shielding at the required extension beyond the cable sheath. This area can be tinned and held in place with solder. A ground strap or braid may be attached during the soldering process. It is important to be careful not to overheat the cable insulation during the soldering process. Acid core solder or acid flux cannot be used. A special aluminum flux must be used when soldering aluminum shielding.

Metallic shielding can be removed by lightly holding a sharp knife against it at the point of removal and then pulling the shielding away from the cable against the blade of the knife. The shielding tapes should be rolled back at the terminating edge to eliminate sharp points.

Removal of Cable Semiconducting Material. All traces of semiconducting material must be removed from the exposed cable insulation back to within approximately $^1/_4$ in of the metallic shielding. The extension of the semiconducting material must not be overlapped when the insulation is built up in the cable splice area. The semiconducting material residue can be removed by scraping with a knife, a nonconductive abrasive cloth, or a rasp. The method of cleaning depends on the types of cable insulation.

Rubber insulations are best cleaned by first scraping with a knife or a rasp and then buffing with a nonconducting abrasive cloth. The semiconducting material normally strips rather cleanly from polyethylene or cross-linked polyethylene cables; therefore, they may only need to be buffed with an abrasive cloth.

The use of solvents for cleaning the cable insulation is a rather controversial subject. Improper solvents or poor workmanship may leave a conductive residue on the insulation, or the solvent may run under the shielding and semiconducting layer. Hidden solvents may cause the shielding tapes to crack or deteriorate. If solvents are to be used, the cable manufacturer's instructions must be followed carefully. Paper- or varnished-cambric-tape-insulated cables may have a black conducting or metallic paper tape under the shielding. These tapes must be removed to within approximately $^1/_4$ in of the edge of the insulation shielding.

Cable Insulation Removal. The conductor must not be nicked or cut while removing the cable insulation. Cotton-type ties are made around the conductor insulation back of the trimming marks as shown on the manufacturer's drawing. The factory insulation is then removed on each conductor from the cable end for a length equal to one-half the connector length plus $^1/_2$ in. Some drawings specify a square-end insulation trim, and others a sloping trim. The two cable ends will now appear as in Fig. 33.6.

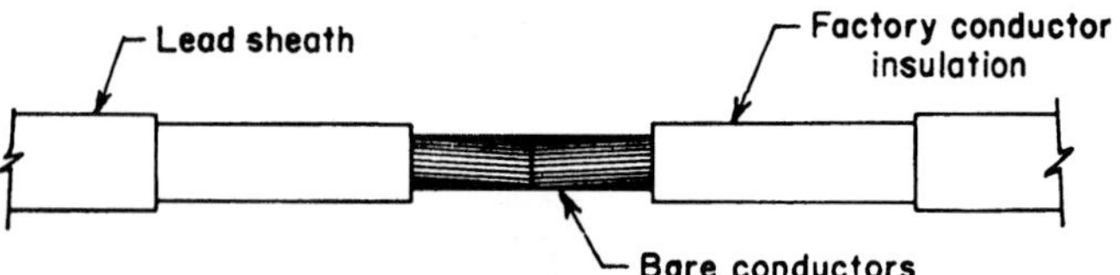

FIGURE 33.6 Cable ends showing bare conductors after factory insulation has been removed. Note squarely fitting butt joint of conductors.

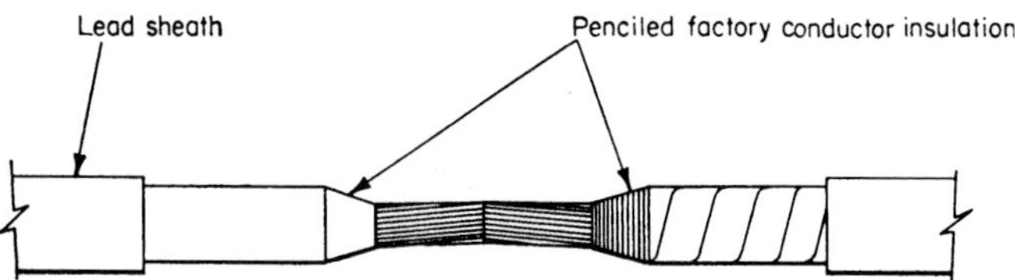

FIGURE 33.7 Cable splice showing factory conductor insulation after being penciled.

To avoid a direct radial creepage path from the conductor outward, the ends of the factory insulation adjacent to the connector are *penciled*, or *stepped*.

On rubber, polyethylene, and varnished-cambric cables the factory insulation is usually penciled as shown in Fig. 33.7. Penciling insulation is similar to sharpening a pencil. Cuts are made at an angle, leaving the insulation tapered. A sharp knife should be used to avoid ragged edges. Keep the diagonal cut in a round and uniform cone shape (Fig. 33.8). After the pencil is formed, smooth with flint or nonconducting abrasive cloth. A mechanical penciling tool may be used to remove and taper the insulation of rubber, polyethylene, and cross-linked polyethylene insulation. It is very important to obtain a smooth and uniform pencil.

On paper-insulated cables, "step" the insulation as shown in Fig. 33.9. Loop a steel piano wire with weights on the ends around the insulated conductor and tear off the layers of paper tape. Start with the step farthest from connector. Tear tapes by count, progessing toward the connector as the required number of tapes is removed for each step. Temporarily bind the edges of paper tape in each step with saturated flax twine. The exposed insulation is covered with a muslin wrapping for protection.

Cable Paper-Insulation Tests. Moisture tests should be made on one or more fillers, several layers of paper insulation from each conductor, and several layers of belt insulation, from all cables rated above 5 kV. The test samples should be placed in a pail of paraffin

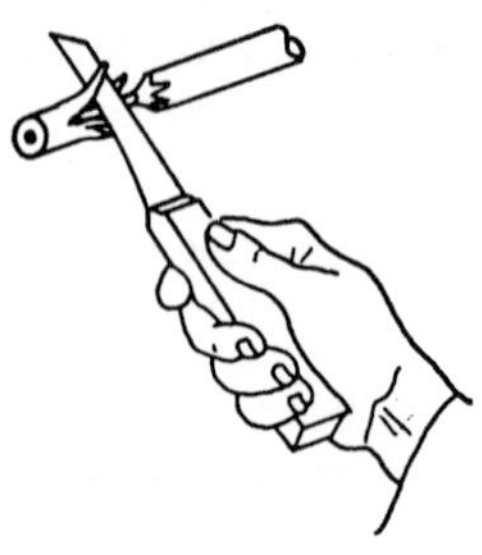

FIGURE 33.8 Penciling conductor insulation.

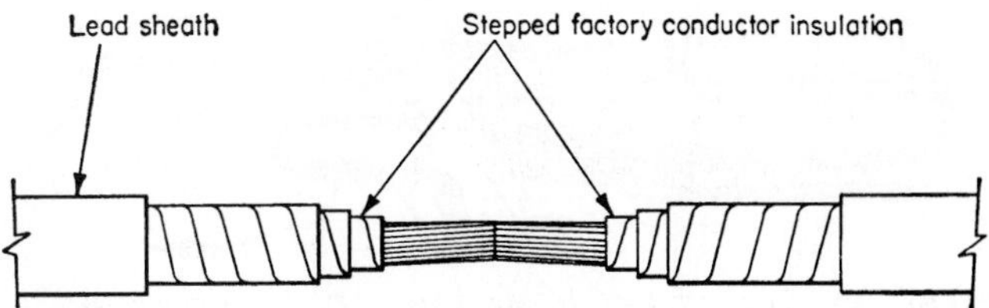

FIGURE 33.9 Cable splice showing factory conductor insulation after being stepped.

which is maintained at 300°F. If frothing or bubbling occurs at the surface of the paraffin within a few seconds, the insulation is wet and the condition should be reported before proceeding any further.

Cable Splice Connector. The proper choice of a connector is important to completing a reliable splice. A smooth tubular connector, such as a crimp or compression sleeve connector, or a split-soldered connector is used for most splices and is necessary for cables operating at voltages higher than 5 kV. Compression-type connectors must be used on thermoplastic-insulated cables to prevent damage to the insulation by heat. The conductor must be cleaned prior to applying the connector.

Apply stearine flux to the exposed surface of the strands if a solder-type connector is to be used. Ordinarily, solder flux in paste form is applied with a brush. If flux in stock form is used, melt it by holding it against a hot ladle and permit flux to drip over surface of conductor. Then open the slot of the connector (Fig. 33.10) sufficiently to pass over the conductors, and slip the connector over the strands of one conductor (see Fig. 33.11). Then insert the end of the other conductor into the connector until it butts the end of the first conductor squarely. Work the connector until it is centered over the ends of both conductors with the slot on top. Tightly compress the connector around the conductor with pliers.

To make the soldered connection, use two solder ladles. With one ladle containing hot solder held above the connector and the other dry ladle held underneath to catch the overflow, pour solder over the ends of the connector and into the slot until the strands and connector are heated enough to make the solder flow freely through the cable strands and out the ends of the connector (see Fig. 33.12). Return excess solder to the solder pot. Now hold the stock of soldering flux against the heated connector, and allow the flux to melt and run into the slot and down through the ends of the conductor. Continue to alternate pouring solder and applying flux until the connector is tinned to a bright, shiny cast. Allow the solder to cool in the ladle, and pour until the solder becomes plastic and fills the slot. While the connector is still hot, smooth off burrs with dry cotton tape. While the solder is cooling, polish the connector with a loop of 1-in cotton tape drawn rapidly back and forth over it. The splice will now appear as in Fig. 33.13.

The compression connector is applied with the proper tool in the manner specified by the manufacturer. The number of compressions or crimps varies with the type of tool used and the size of the conductor. Hydraulic compression tools are used on the larger conductors.

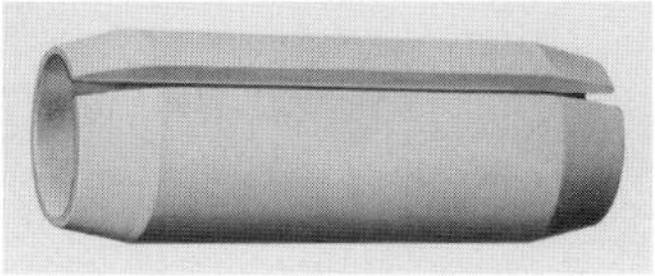

FIGURE 33.10 Typical split tinned copper cable connector. (*Courtesy Reliable Electric Co.*)

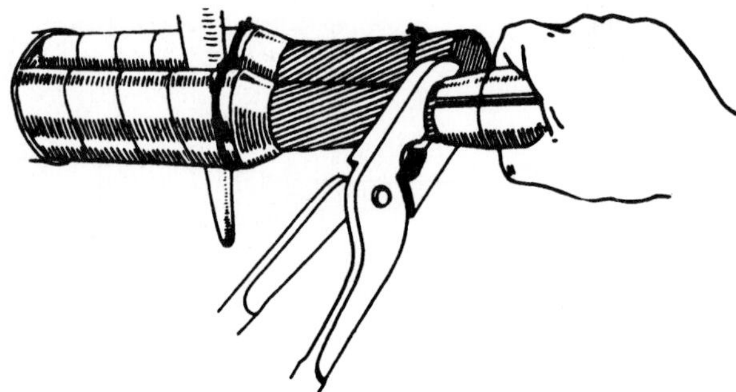

FIGURE 33.11 Slipping split connector over conductor strands. Strands are held in the round with string and special eagle-claw pliers.

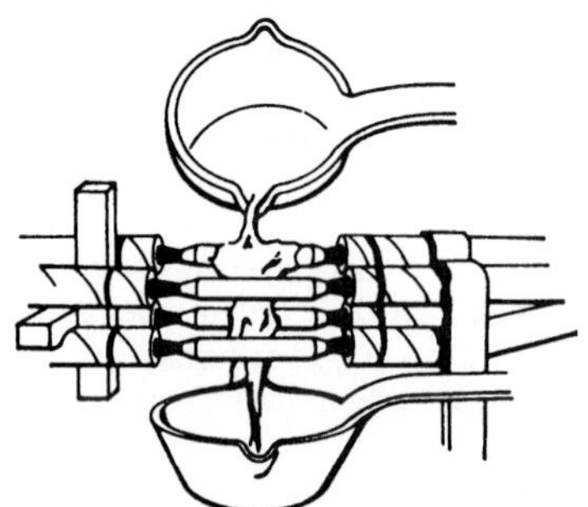

FIGURE 33.12 Pouring solder over ends of conductors and into connector slot.

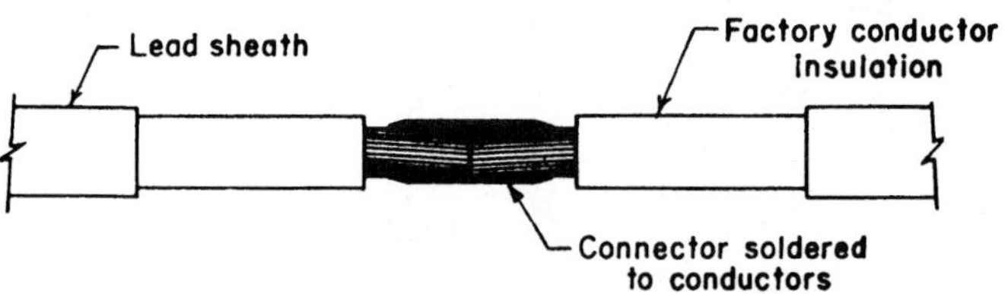

FIGURE 33.13 Cable splice showing connector soldered to conductors.

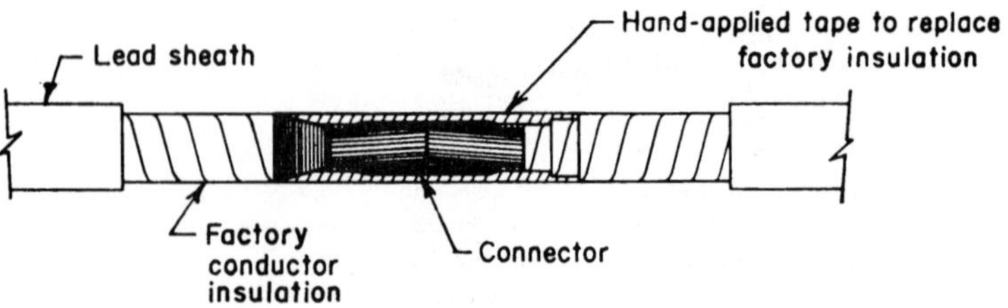

FIGURE 33.14 Cable splice after factory conductor insulation has been replaced with hand-applied tape.

A semiconducting tape is normally applied over the compression connector and the exposed conductor. The semiconducting tape eliminates problems that might develop from the irregularities due to indentations on the compression connector.

Installing Taped-Cable Insulation. The temporary protective cloth wrappings are removed. Then wrap rubber tape on rubber-insulated cables and varnished-cambric tape on varnished-cambric- or paper-insulated cables. Unless otherwise specified, the tape shall be applied with a half lap. One layer half-lapped is then equal to two thicknesses.

First fill in the space at both ends between the connector and the trimmed insulation. Then insulate over the connector and the pencils or steps to the diameter of the factory-applied insulation. The splice will now appear as shown in Fig. 33.14. Now continue tapping over the entire exposed cable. The first layer of tape applied over the factory insulation should be wrapped in the same direction as the factory insulation and should extend back into the crotch as far as possible. In the case of high-voltage paper cable, one outer layer of factory-applied tape insulation is removed before the hand-applied taping is started. All varnished-cloth tape should be drawn up as tightly as possible without breaking the varnish over the fabric. All tape must be applied smoothly and carefully to expel all air and avoid wrinkles.

When rubber tape or rubberlike tape is applied, the surface should be cleaned with the specified solvent and coated with rubber cement. Rubber tape should be applied with uniform and sufficient tension to reduce its original width approximately one-third. Rubber tape should be protected with friction tape and asphaltum paint or with a layer of weather-resistant tape as specified.

The joint specifications usually provide for approximately a 75 percent greater thickness of hand-applied insulation than is found in the factory insulation.

Figure 33.15 shows the manner of wrapping the tape. Figure 33.16 shows the three conductors of a three-conductor cable in various stages of the taping operation. The lower conductor is completely taped, the middle conductor has the space at the ends of the connector filled, and the upper conductor is without any insulation replaced.

At the completion of the taping operation, lock the last turn in place with a tape tie. This is made by passing the end of the tape under the preceding turn and pulling the tape tight (see Fig. 33.17). Cut off excess tape. A completely taped splice is shown in Fig. 33.18.

Prefabricated Cable Splice Installation. Prefabricated cable splices are normally used on underground distribution cable installations with rubber, ethylene propylene rubber, polyethylene, or cross-linked polyethylene insulation (Fig. 33.19). The splices may be direct-buried or placed in enclosures to permit access. The prefabricated splice consists of a metallic connector, a connector shield, insulation, and an insulation-shielding jacket.

FIGURE 33.15 Applying $^{1}/_{2}$-in tape half-lapped over conductor strands. Note connector and penciled insulation on lower conductor, completely wrapped middle conductor, and wrapping in progress on upper conductor.

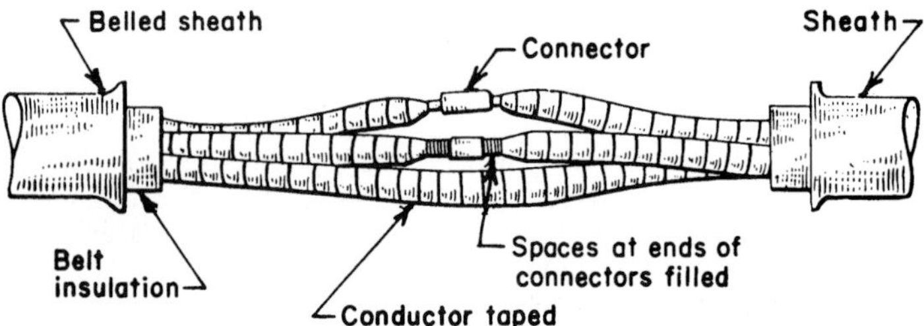

FIGURE 33.16 Several stages in the splicing of a three-conductor cable. Upper conductor shows connector in place, middle conductor has spaces filled between connector and penciled conductor insulation, and lower conductor is completely taped.

FIGURE 33.17 Wrapping ended in tape tie.

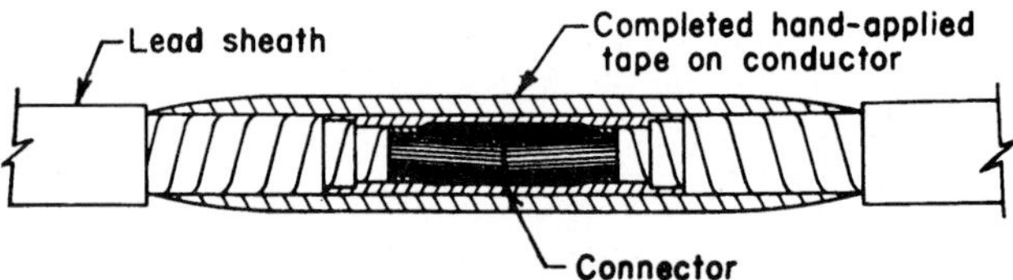

FIGURE 33.18 Cable splice after completing the taping of the conductor: On rubber-insulated cables this should be equal to $1^1/_2$ times factory insulation, and on varnished-cambric- or paper insulated cables, it should be equal to 2 times factory insulation.

FIGURE 33.19 Prefabricated splice installed on concentric neutral cross-linked polyethylene insulated UD cable. The cable and splice will be buried in the ground. (*Courtesy AMP, Inc.*)

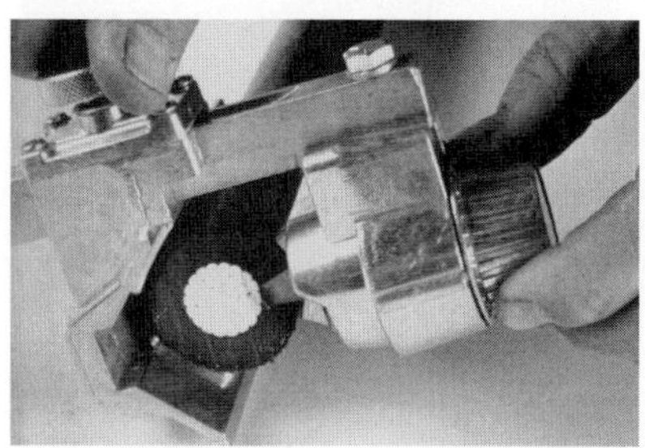

FIGURE 33.20 Solid-dielectric cable insulation stripper being set to the correct depth by the lineman. (*Courtesy Speed Systems.*)

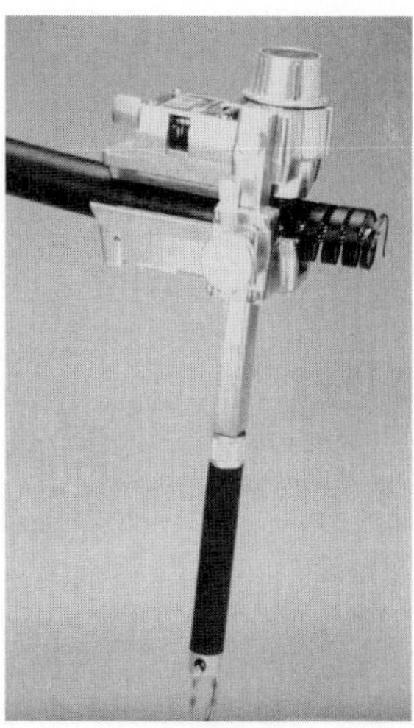

FIGURE 33.21 Solid-dielectric cable insulation partially cut for removal. (*Courtesy Speed Systems.*)

Cables are prepared in the manner previously described for other splicing methods. Special tools are available to prepare the cable ends for the prefabricated splice. (Figs. 33.20 through 33.26). Apply one half-lapped layer of vinyl tape over the semiconductive jacket of the cables. Apply silicone grease over cable insulation and vinyl tape. Slide splice end caps onto each cable until against concentric wires. Slide splice body onto one of the cables. Place cable conductors into connector and crimp the connector with the proper tool in the manner specified by the manufacturer (Fig. 33.27). The splice body is slid into final position over cable connector, leaving a small area of insulation exposed at both ends of the splice body (Fig. 33.28). The cable splicer or lineman removes the vinyl tape applied to the semiconductive jacket prior to completing the splice assembly. The end caps are slid onto the splice body, completing the semiconductive shielding jacket over the cable and splice insulation. The concentric neutral

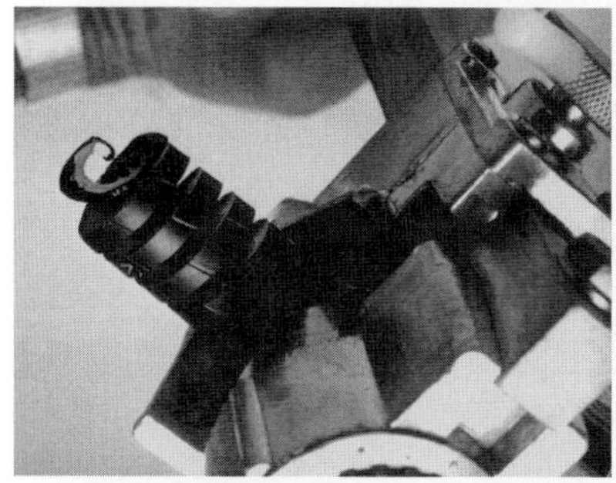

FIGURE 33.22 Solid-dielectric cable insulation being removed by special tool. (*Courtesy Speed Systems.*)

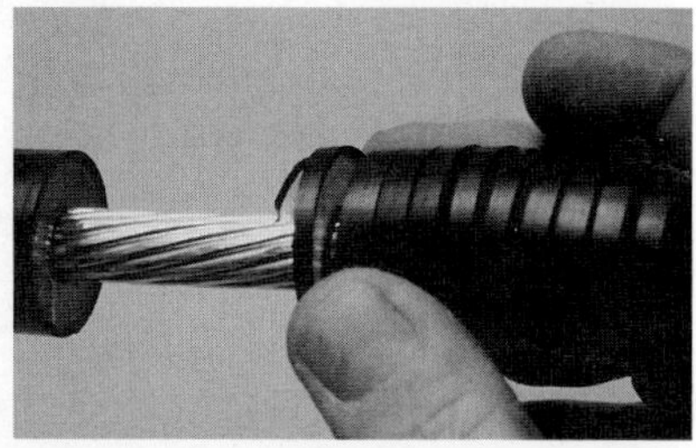

FIGURE 33.23 Solid-dielectric cable semiconducting jacket, insulation, and semiconducting conductor shield being removed after being cut by special tool. (*Courtesy Speed Systems.*)

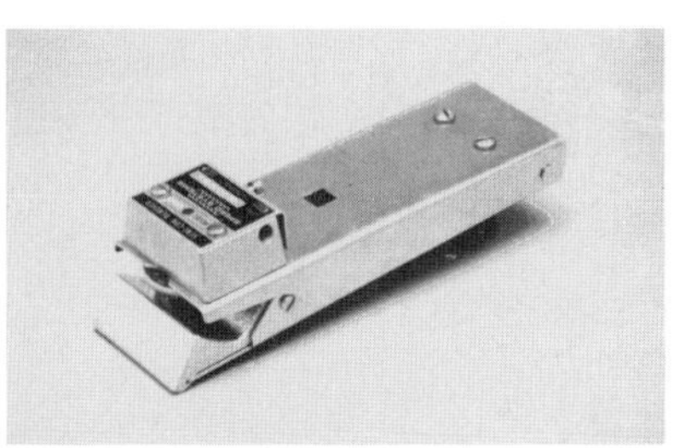

FIGURE 33.24 Special tool manufactured to score semiconducting insulation jacket of solid-dielectric insulated cable. (*Courtesy Speed Systems.*)

FIGURE 33.25 Cable jacket scoring tool being used to score semiconducting jacket extruded over solid-dielectric cable insulation. (*Courtesy Speed Systems.*)

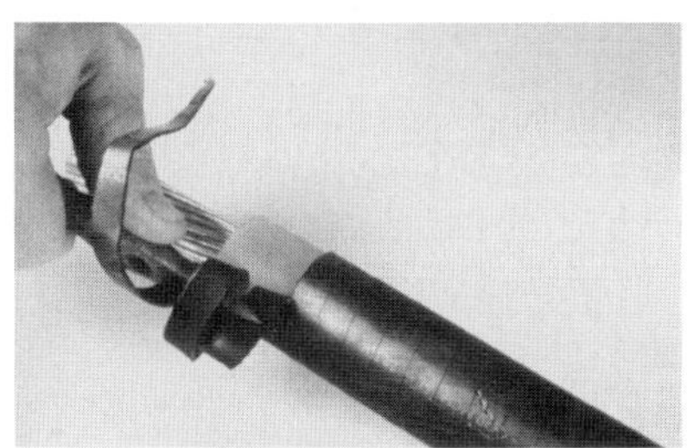

FIGURE 33.26 Cable semiconducting jacket shielding being removed from cable after being scored without damaging the solid-dielectric cable insulation. (*Courtesy Speed Systems.*)

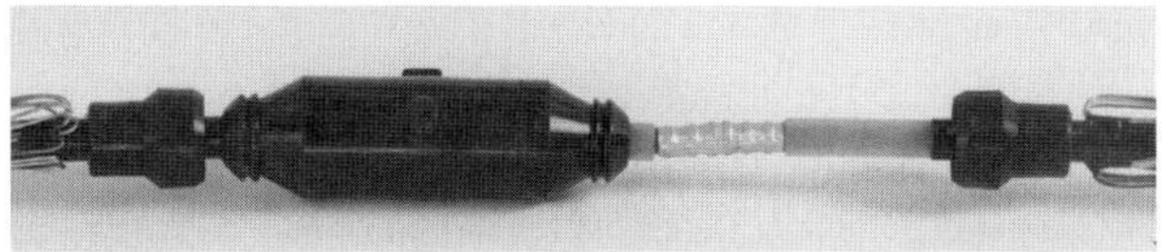

FIGURE 33.27 Prefabricated splice on solid-dielectric insulated, concentric neutral cable. Conductor connector has been crimped. (*Courtesy 3M Co.*)

FIGURE 33.28 Cable splicer sliding body of splice over connector. (*Courtesy 3M Co.*)

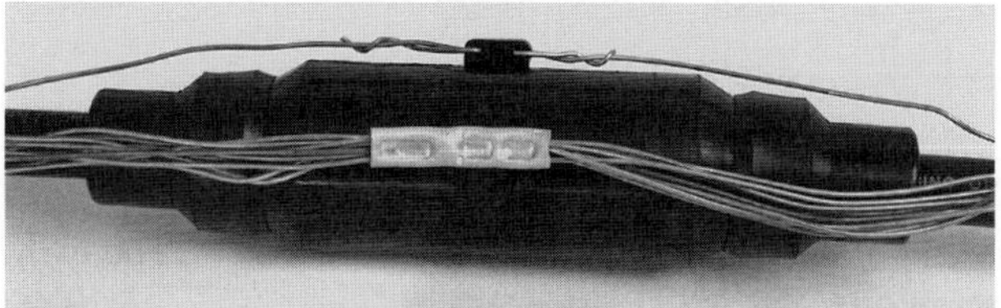

FIGURE 33.29 Completed installation of prefabricated solid-dielectric insulated concentric neutral cable splice. (*Courtesy 3M Co.*)

wires are extended over the splice and connected together with a compression-type connector. The concentric wires are connected to the grounding eyes on the splice body to ensure that the splice jacket is grounded (Fig. 33.29). The insulation shield on a prefabricated splice normally consisting of a conductive jacket retains the electric field completely within the spliced cable insulation. The electric field is properly distributed to eliminate any highly concentrated electric stress areas (Figs. 33.30 and 33.31).

Applying Cable Shielding Tape. Single-conductor and multiple-conductor cables rated at 13 kV and above have shielding tapes wrapped around the outside of the insulation of each conductor. The multiple-conductor cables have an additional shielding tape wrapped around the outside of the three conductors directly under the lead sheath, as shown in Fig. 33.32. Splices made on these cables must have the shielding continued over the applied insulation of each conductor.

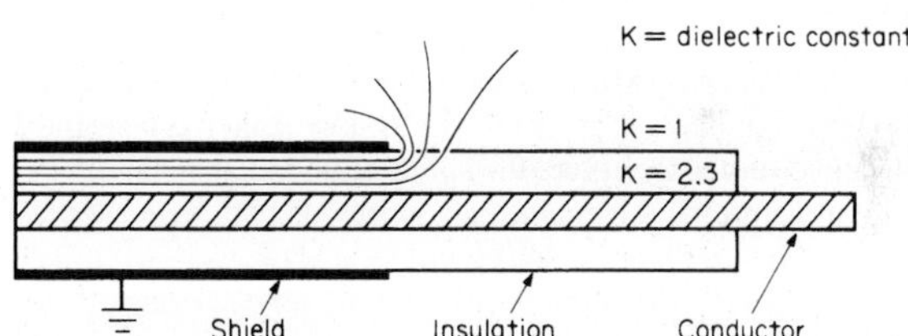

FIGURE 33.30 Electric stress distribution for high-voltage solid-dielectric insulated shielded cable with shield and insulation partially removed from cable conductor. Note high concentration of electrical stress at the edge of the cubic shield.

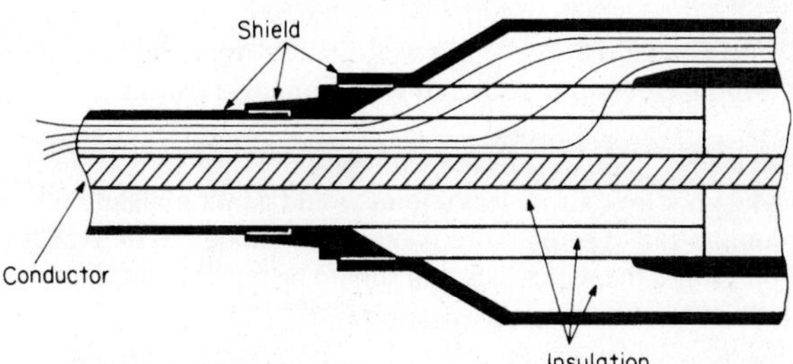

FIGURE 33.31 Schematic of electric stress distribution in a completed prefabricated splice installed on a solid-dielectric insulated shielded high-voltage cable. Note electric stress in transition area from cable to splice is evenly distributed.

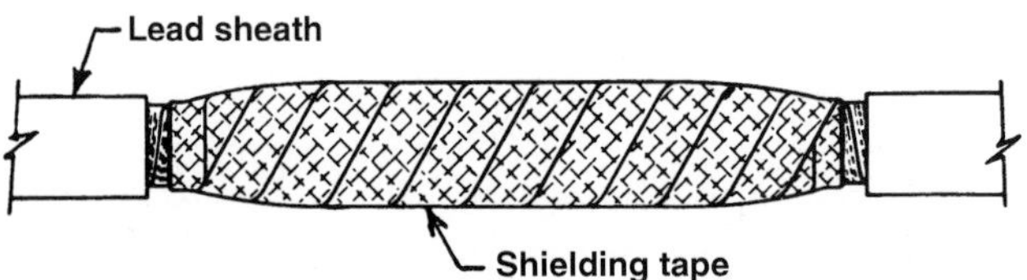

FIGURE 33.32 Cable splice with shielding braid tape covering the insulated conductor or conductors.

This shielding is made with a butt-lapp wrapping of $^3/_4$-in copper-tinsel braid applied from end to end where it is fastened to the factory shielding or to the lead sheath with a soldering iron. Adjacent turns are tacked together with solder, as necessary, to prevent the braid from sliding on the slopes of the applied insulation. Shielding braid tape may not be used in connection with splicing one-conductor 13-kV cables. These cables are shielded with a metallic paper tape, which is torn off at the end of the lead-sheath trim. In this case, the sheath end is belled out with a special belling tool. The ends of the lead-sheath trims on other shielded cables are not belled at all, other than that resulting from the removing of the sheath.

Applying Cable Spacer and Binding Tapes. In three-conductor splices about five or six layers of dry varnished-cambric tape are wrapped around each conductor over the outside of the shielding tape and near the center of the joint or at two locations, as shown on some drawings. This tape is installed to serve as a separator between conductors which permits the compound to flow freely between conductors. Then from three to six layers of dry-varnished-cambric tape are wrapped around all three conductors and directly over the spacer tape in order to hold the conductors tightly together.

Installing Cable Lead Sleeve. After all the taping of the splice is completed, the lead sleeve which was slipped on the cable previously is now centered over the splice. As mentioned before, the sleeve serves to protect the splice from mechanical injury; seals the joint, thereby keeping out moisture; and furnishes a continuous path in the cable sheath for fault currents.

Wiping the Cable Sleeve. With the sleeve centered over the splice, mark the cable at the ends of the sleeve. Coat the scraped surface with stearine flux. Beat down the ends of the sleeve with a hardwood tool called a dresser until the sleeve fits snugly around the cable sheath with the sleeve centered over the tapped area.

Wipe the sleeve ends to the cable sheath, using the stick-and-torch method. Melt small amounts of solder from the end of the stick, and apply over the scraped area (Fig. 33.33). Heat the solder with a torch, and scrub the wiping surface with a solder stick until the surface of the cable sheath and the sleeve are well tinned. Apply additional flux and solder, and work with a wiping cloth until a smooth solder joint is formed between the sleeve and sheath (Fig. 33.34).

The wiped sleeve joint should be tested with gas or oil pressure to ensure that the wipes are not porous and do not leak air. A leaky joint would admit moisture which would eventually cause the joint to fail. The pressure is built up to about 10 lb. Apply soapy water to the wipes. If bubbles form, the wipes leak and should be repaired by additional torching and rewiping until tight.

Filling Cable Sleeve with Compound. If the sleeve is not provided with fittings for filling as shown in Fig. 33.35, make two V-shaped cuts in the top of the sleeve, one near each end. Lift up the V flap formed. One opening is for pouring in the hot compound, and the

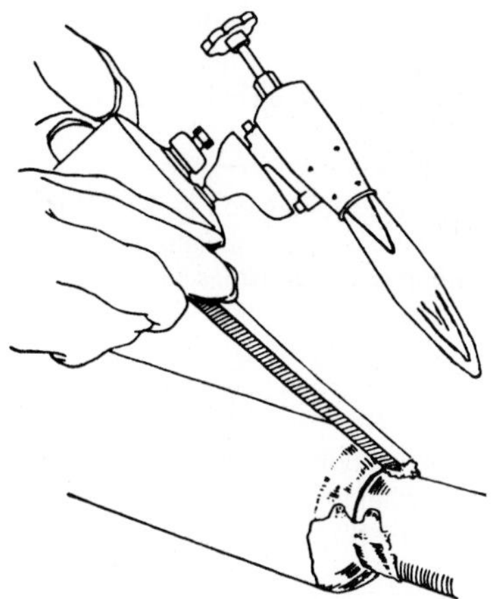

FIGURE 33.33 Wiping lead sleeve onto lead sheath, using stick and torch method.

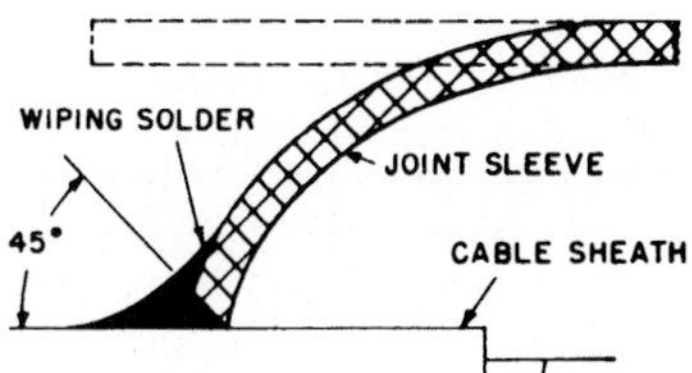

FIGURE 33.34 Completed wipe showing smoothly tapered shoulder of solder joining sleeve to cable sheath.

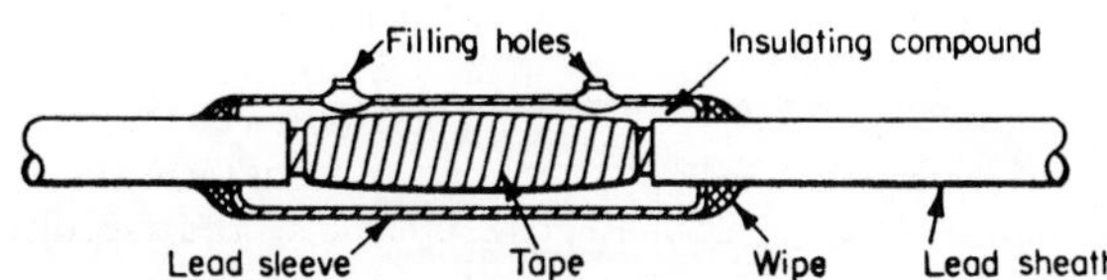

FIGURE 33.35 Cross section of sleeve joint showing sheath, sleeve, tape, wipe, filling holes, and insulating compound.

other is for venting the air in the sleeve. Before the hot compound is poured in, the joint should be tilted slightly, that is, raised about 1 in at the vent end. Place the funnel into the filling hole at the lower end of the sleeve (Fig. 33.36). When the compound is poured in, the air in the sleeve is driven upward and out through the venting hole.

The compound usually consists of 70 percent Vinsol and 30 percent castor oil. It is first heated to about 250°F to 300°F. It is then slowly and steadily poured into the funnel. Allow the compound to flow through the venting hole until it is free from bubbles. Then allow the joint to cool to about 150°F. Now level the joint and refill the sleeve every 15 min. When the joint has cooled to 100°F, replace the fittings or seal the holes with solder. Wipes and seals are then painted with Victolac to prevent corrosion.

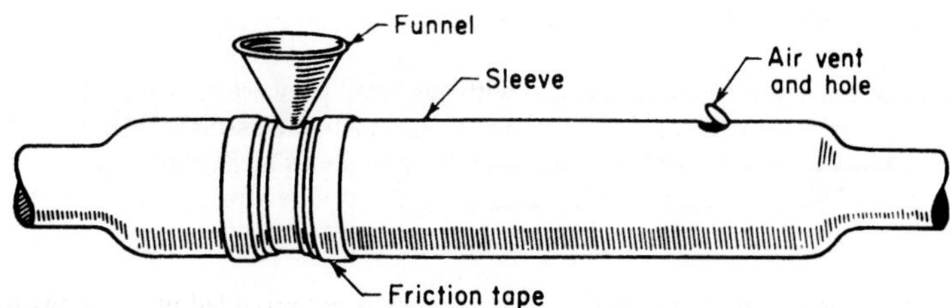

FIGURE 33.36 Funnel in place in lower hole of sleeve ready for pouring of insulating compound. Air will escape through overflow hole. Funnel can be scaled with friction tape.

When the joint cools to the temperature in the manhole, voids will form inside the sleeve. However, when the cable carries electric power, it will become heated again and the voids will become filled. If the voids were not present, the joint would burst when it became heated by the cable load.

Place Cable Splice on Rack. The cable with its completed joint (see Fig. 33.37) is now moved into its final position onto porcelain saddles on the rack. Lead sheaths are subject to electrolysis and therefore must be supported on porcelain or other insulating blocks. Some adjustment of the rack may be necessary.

Cable Bonding and Grounding. The purpose of bonding and grounding the cable sheaths is to maintain them at or near ground potential. A No. 2 AWG copper wire is generally used. It must be attached to the sheaths with a special bond clip which is soldered to the wire and sheath and connected to a low-resistance ground.

Bonding and grounding reduces the likelihood of arcing between the sheath of a faulted cable and other nearby sheaths. It thus reduces the danger to cablemen who may be in a manhole when a cable fault occurs. It also minimizes the harmful effects of corrosive action due to stray currents.

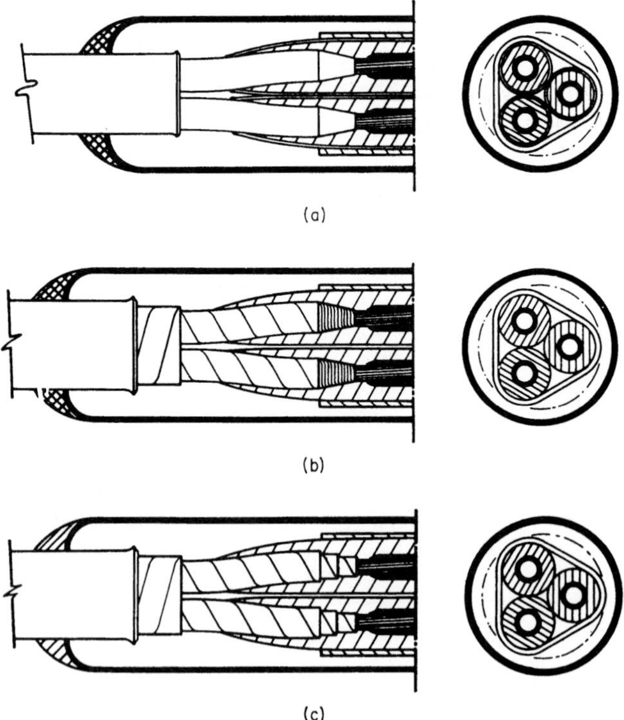

FIGURE 33.37 Sectional views of completed splices on straight two-way three-conductor cables. (*a*) Rubber-insulated, (*b*) varnished-cambric-insulated, (*c*) impregnated-paper-insulated. (*Courtesy G. & W. Electric Specialty Co.*)

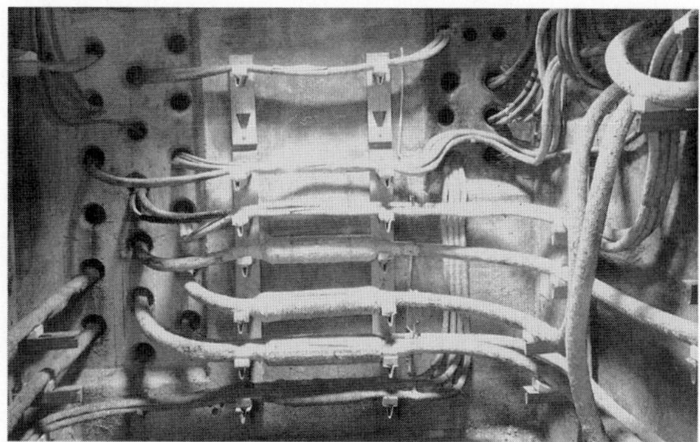

FIGURE 33.38 Cable joints protected with fireproofing tape to prevent spreading of damage from faulted cable to adjacent cables. (*Courtesy Public Service Electric and Gas Co. of New Jersey.*)

Fireproofing Cable Joint. Fireproofing tape is applied after bonding to limit the damage that could result from a failure on other cables in the manhole. Fireproofing is necessary only where there is the probability of damage to one or more cables due to failure of one of them. A group of fireproofed cables is shown in Fig. 33.38.

Tagging Cable. Every cable should bear an identifying mark or tag in each manhole through which it passes. This is usually in the form of a metal tag tied to the joint. The tag should indicate the size, voltage, origin, feeder designation, and any other pertinent information. A typical cable tag is shown in Fig. 33.39.

Cable Terminations. Cable terminations are completed after preparing the end of the cables in a manner similar to the procedure described for preparing the cable ends for splicing to other cables. Paper-insulated lead-covered cables are usually terminated with a porcelain-insulated pothead as described and illustrated in Chap. 29. Solid-dielectric insulated cables are usually terminated with a molded-rubber terminator (Figs. 33.40 through 33.42). The molded-rubber cable termination is usually delivered in a kit which contains a one-piece molded-rubber stress cone, a stainless steel grounding strap, a molded silicone one-piece skirted rubber insulator, a stem-cable connector or cable aerial lug, a cable support, silicone lubricant, and complete instructions.

Cable Shield Bonding. Cable shields must be grounded and maintained near ground potential for the safety of linemen and cablemen and protection of the cable. The cable sheaths are normally bonded together and to ground at multiple locations such as splices, terminations, and in each manhole. The bonding and grounding of the cable sheaths form

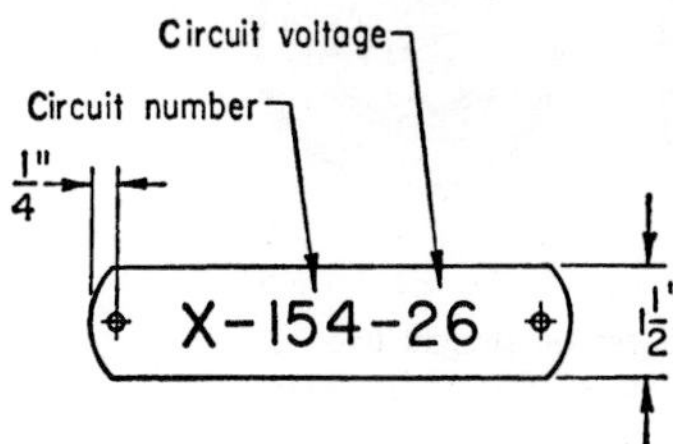

FIGURE 33.39 Typical cable tag showing circuit number on left and voltage on right. (*Courtesy Public Service Electric and Gas Co. of New Jersey.*)

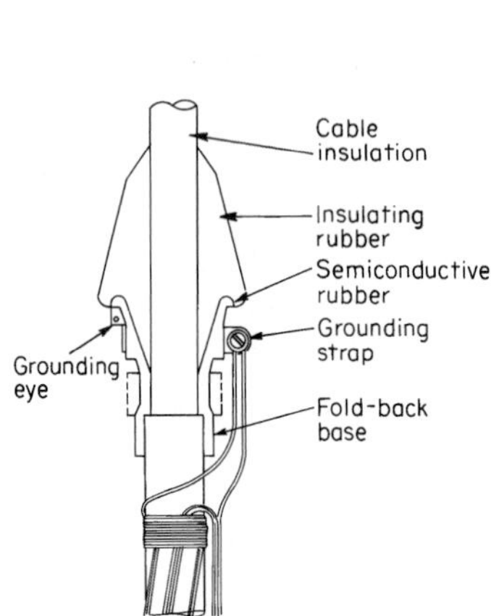

FIGURE 33.40 Schematic diagram of a portion of a molded-rubber termination for a solid-dielectric insulated cable. (*Courtesy 3M Co.*)

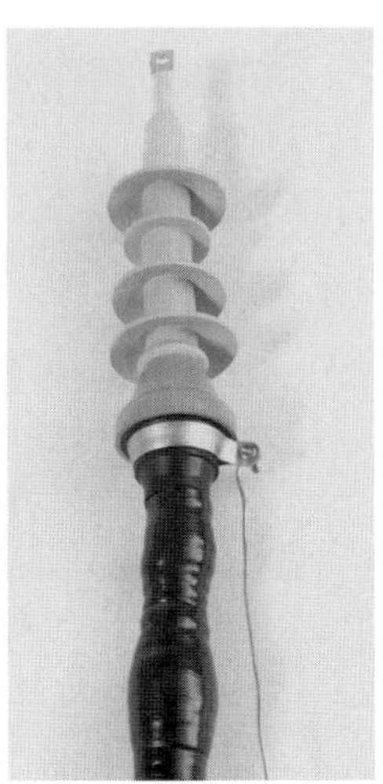

FIGURE 33.41 Solid-dielectric insulated cable terminated with a molded-rubber terminator using a cable aerial lug. (*Courtesy 3M Co.*)

multiple short-circuited loops. If the cable conductors carry large load currents, the circulating currents induced in the sheath-grounding can generate enough losses to cause the current-carrying capacity of the cable to be reduced. These conditions would normally occur on electric generator cable leads or underground transmission cable circuits.

The cable sheaths can be bonded to ground at special locations to eliminate the short-circuit cable sheath loops. If the cable sheaths are grounded at a single point, the length of cable from the ground must be kept short enough to limit the induced voltage between the

FIGURE 33.42 A solid-dielectric insulated cable molded-rubber termination with a stem-cable connector on a cable riser pole. The terminator pin is installed directly into the bottom of the fused cutout.

shield and ground to a safe value. Limiting the induced voltage during short-circuit condition or surges as a result of lightning or switching is difficult. Surge arresters of the proper voltage rating installed at the insulated end of cable sheaths can provide protection for personnel and limit induced voltages to a value that will prevent puncturing of the polymeric jacket over the metal cable shield.

Splicing Paper-Insulated Lead-Covered Cable to Solid-Dielectric Insulated Cable. Solid-dielectric insulated cables such as ethylene propylene rubber (EPR) and cross-linked polyethylene are generally being installed in place of paper-insulated lead-covered cables. The solid-dielectric insulated cables are easily installed, eliminate the problems associated with electrolysis of the lead cable sheaths, and have established a good record of reliable service. The use of solid-dielectric insulated cables has made it necessary to splice them to lead-covered cables where a circuit is being extended or if a section of cable is being replaced as a result of a cable failure. The splicing of solid-dielectric insulated cable to paper-insulated lead-covered cable requires special precautions because of the oil that saturates the paper insulation and the pressure on the oil when the cable temperature increases as a result of load current flowing in the conductor.

The splice manufacturer's instructions must be followed to complete the transition splice successfully. The ends of the cable are generally prepared for splicing in the manner previously described and illustrated in Figs. 33.43 and 33.44. The cables are trained into place and overlapped 6 in. The cables are then cut squarely at the centerline of the overlap to form a butt joint. The cable jackets, tape shield, and/or lead sheath and semiconducting shield are removed. Do not intentionally bell the lead cable sheath. Mark the sheath of the lead-covered cable at a point $1^1/_4$ in back of the edge of the lead.

A stress-relieving mastic is applied over the lead sheath, starting at the reference mark, and the paper insulation. The mastic is applied under tension which reduces the width to one-half its original dimension. It is applied evenly over the sheath and the paper insulation with a slight taper from the lead to the paper insulation (Fig. 33.45). Two layers of the half-lapped mastic are applied. A clear oil barrier tube is placed on the cable so that it extends beyond the mastic surface. The oil barrier tube is shrunk with heat to provide a smooth surface before proceeding (Fig. 33.46). Black stress control, red insulating, red/black insulating/conductive, and black outer tubes are used to complete the splice. The paper-insulated lead-covered cable jacket should be cleaned with a solvent a distance back from the end of the cable sufficient to permit the tubes to be slid back over the cable without becoming contaminated. The tubes are nested inside of each other and placed onto the cleaned lead cable.

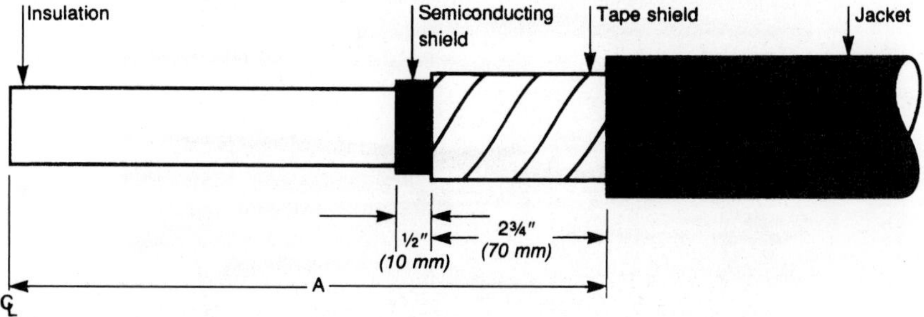

FIGURE 33.43 Sketch of a 15-kV solid-dielectric insulated cable with a polymeric jacket and metallic tape shielding properly prepared for the transition splice. See manufacturer's instructions for detailed dimensions. (*Courtesy Raychem Corporation, Energy Division.*)

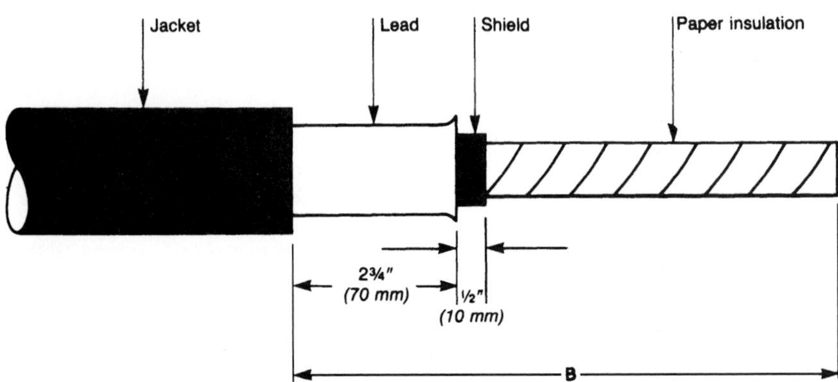

FIGURE 33.44 Sketch of a 15-kV paper-insulated lead-covered cable with metallic shielding tape and a polymeric jacket properly prepared for the transition splice. See manufacturer's instructions for detailed dimensions. (*Courtesy Raychem Corporation, Energy Division.*)

The cable insulation is removed from each cable end a sufficient distance to permit the installation of the conductor splicing connector. The paper insulation and the oil barrier tube are chamfered for a distance of 1/2 in or 10 mm. The connector is compressed onto the conductors, and any excess oxide inhibitor is removed (Fig. 33.47). The connector is deburred and cleaned with solvent. Clean the prepared paper-insulation oil barrier tube and the polymeric cable insulation with a solvent, wiping toward the conductive layer without reusing any section of the cloth or tissue.

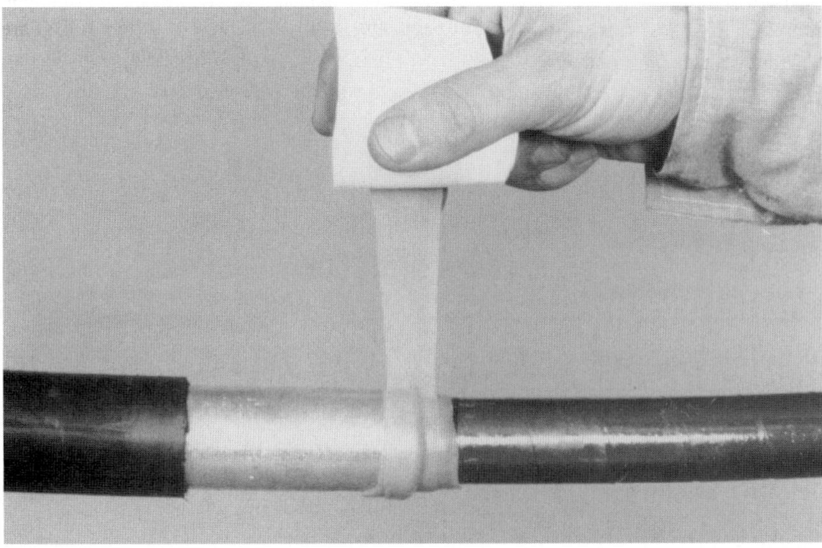

FIGURE 33.45 Mastic has been applied over a portion of the lead sheath and paper insulation of the cable. (*Courtesy Raychem Corporation, Energy Division.*)

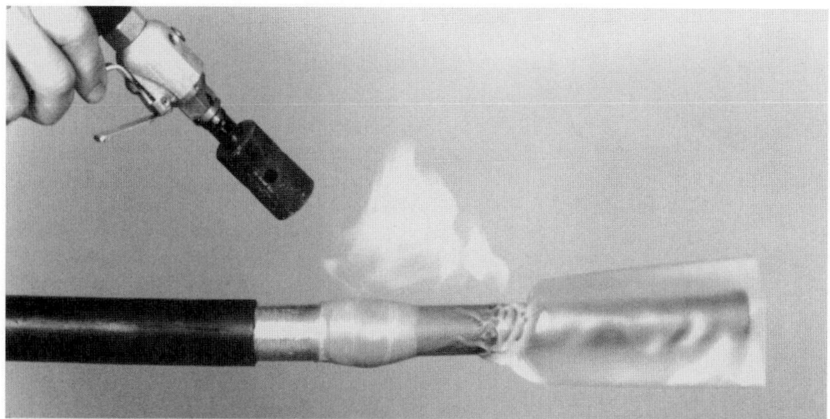

FIGURE 33.46 The oil barrier tube is shrunk over the lead sheath and paper insulation, starting at the lead sheath end and continuing until the cable surface is smooth and sealed. (*Courtesy Raychem Corporation, Energy Division.*)

Apply stress-relieving mastic over the polymeric cable insulation and the semiconducting shield with sufficient tension to reduce its width by one-half with sufficient wraps to fill the step. Apply the mastic over the connector, starting on the end near the paper insulation, and onto the polymeric insulation a distance of $^1/_4$ in (5 mm) (Fig. 33.48). The compression connector indents, and the gap between the connector edge and the insulation must be filled. Wrap the mastic back across the connector and onto the oil barrier tube a distance of $^3/_4$ in (20 mm). The chamfered area on the paper insulation must be filled. The mastic over the connector should completely cover the connector with a uniform thickness of mastic from paper to polymeric insulations (Fig. 33.49). If the connector diameter is larger than the insulation, apply two half-laps of the mastic only. If the cable insulation outer diameters (ODs) are uneven, apply the mastic in a uniform taper so that the mastic buildup is equal to the insulation diameter of each cable.

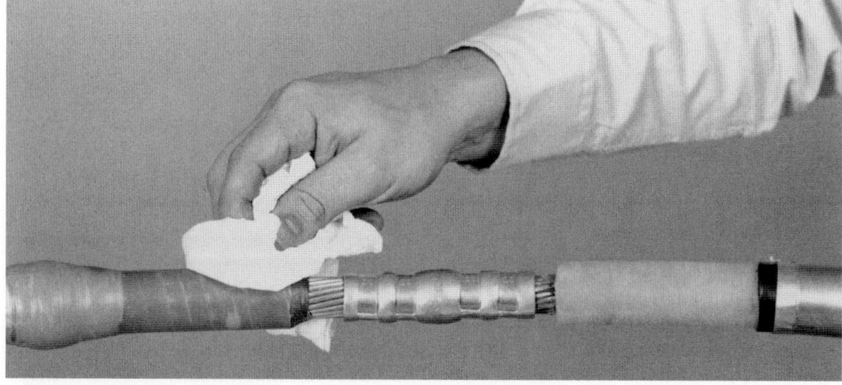

FIGURE 33.47 The compression connector has been installed, and the excess oxide inhibitor is being removed. (*Courtesy Raychem Corporation, Energy Division.*)

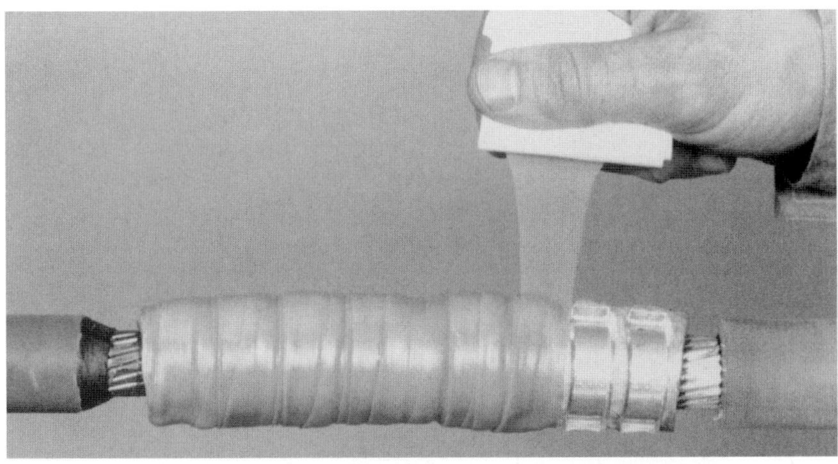

FIGURE 33.48 The mastic is stretched to one-half its width as it is applied to the connector. Start on the paper-insulated cable side, and extend the mastic onto the polymeric insulation and then back over the oil barrier and the end of the paper insulation. (*Courtesy Raychem Corporation, Energy Division.*)

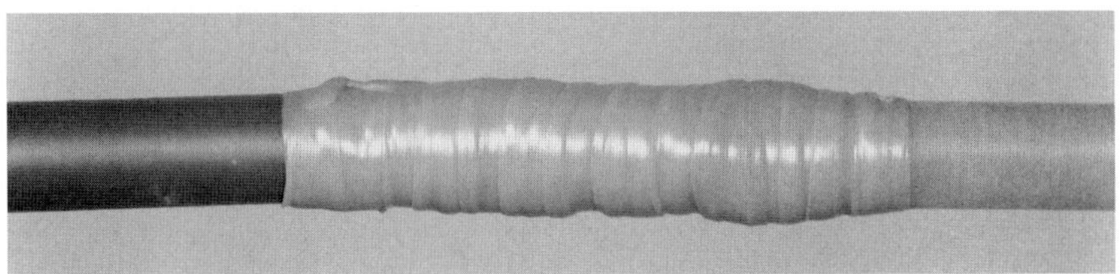

FIGURE 33.49 The cable splice connector has been completely wrapped with mastic. (*Courtesy Raychem Corporation, Energy Division.*)

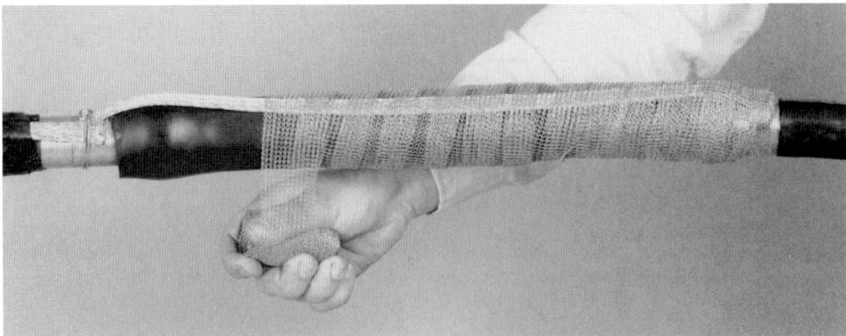

FIGURE 33.50 Shielding mesh is being applied over red/black/insulating/conductive sleeve and the ground braid connecting it to the lead sheath and the metal shield of the polymeric-insulated cable. (*Courtesy Raychem Corporation, Energy Division.*)

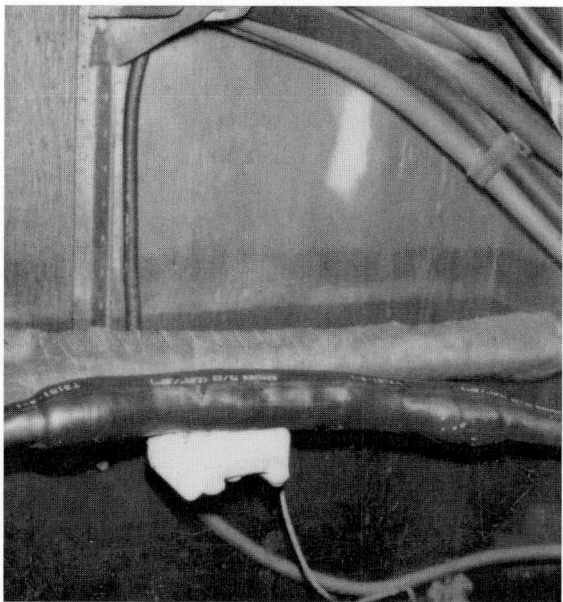

FIGURE 33.51 Completed splice of paper-insulated lead-covered cable to solid-dielectric insulated cable. (*Courtesy MidAmerican Energy Company.*)

The innermost black stress control tube temporarily parked on the lead cable should be guided over the mastic so that it completely overlaps the oil barrier tube on the paper-insulated cable. Beginning in the center of the tube with a torch, shrink the tube around the splice with a brushing motion around the tube until it is fully shrunk. Center the red insulating tube over the joint, and shrink it in the same manner as described. Center the red/black insulating/conductive tube, and shrink it into place. Solder or clamp the ground braid from the lead sheath to the polymeric cable metal shield. Half-lap a copper-shielding mesh with tension across the joint from cable shield to cable shield (Fig. 33.50). With solvent clean the cable jackets at least 3 in (75 mm) beyond the cutbacks. Center the outer black jacketing tube over the joint and shrink it, starting in the center, and complete the shrinking in the manner previously described. The finished splice is shown in Fig. 33.51.

URD PRO'S

IMMUNE TO:
- POWER POLE ACCIDENTS
- ICE
- WIND
- LIGHTNING DAMAGE
- SNOWSTORM
- CONTACT BY TREES OR OTHER FOREIGN OBJECTS

URD CONS
- ENTRANCE OF MOISTURE
- CORROSION
- CABLE DIG IN'S
- INSULATION FAILURE

CHAPTER 34
UNDERGROUND DISTRIBUTION

Electric distribution circuits have been installed underground in large cities for many years to serve the central business areas. The electric load density in the central business areas is high enough to justify the expenses associated with the conventional underground system employing the use of conduits encased in concrete, manholes, vaults with submersible transformers, network protectors, and paper-insulated, lead-covered, or polyethylene-insulated, shielded, and jacketed high-voltage cables.

The conventional underground system cannot be economically justified to serve the low electric load densities found in residential areas. Direct-buried cable systems with pad-mounted transformers, submersible transformers in fiber vaults, or direct-buried transformers can be installed in residential areas at costs that are economically feasible. The converting of overhead electric distribution circuits to underground in residential areas is generally not economically feasible. Most of the conversion programs executed have been associated with renewal projects where it was necessary to remove the overhead electric facilities as a part of the demolition work.

Underground lines are relatively immune to some of the major causes of failures in overhead circuits, such as power-pole accidents; damage from lighting, wind, ice, and snowstorms; or contacts to the wires by trees or other foreign objects. Underground circuits have their own problems, such as entrance of moisture, corrosion, cable dig-ins, insulation failures as a result of switching surges or corona, and damage during installation. Underground equipment must be designed for long life in below-ground enclosures that may be filled with water containing contaminants. Underground cables and equipment are both vulnerable to the entrance of moisture as a result of flaws in splices, terminations, gaskets, and connecting devices. Chemical corrosion or electrolysis can damage any exposed metal installed as a part of the underground distribution system.

Power Source. In most cases the underground distribution circuits will be supplied from an overhead main feeder circuit installed adjacent to the subdivision. The loads in the area may be served by several underground cable laterals connected to the open-wire overhead system at several cable riser poles. A typical riser pole with cable, cable guard, pothead, lightning arrester, and fused disconnect is shown in Fig. 34.1. Figure 34.2 illustrates the essential components required to connect the underground cable to an overhead distribution circuit.

Total underground distribution circuits have been constructed in some locations. A total underground distribution circuit is shown schematically in Fig. 34.3. The three-phase main feeder circuit extends from substation A to substation B underground. The main feeder circuit has a capacity of 10 MVA (megavoltamperes). The underground cables consist of 750-kCMIL aluminum conductor with cross-linked polyethylene insulation rated 15 kV, shielding

FIGURE 34.1 Single-phase underground cable riser installed on pole. The cable is connected to the overhead circuit through a fused cutout and protected by a lightning arrester. (*Courtesy S&C Electric Co.*)

consisting of semiconducting tapes and copper tapes and a polyethylene jacket. Taps to the main circuit to serve residential areas or commercial customers are made in switching modules installed in pad-mounted switchgear or a submersible container.

Most of the underground distribution circuits tapping the main feeder are single-phase loop feeds with fuses to protect the main feeder circuit. Extending the three-phase main feeder circuit between two substations makes it possible to isolate a faulted cable section and restore service to customers prior to completing the cable repairs. The loop-type feed serving the underground distribution area makes it possible to isolate faulted cable circuits in order to restore service to the customers in the subdivision.

Types of Systems. Direct-buried underground distribution systems can be installed in several different ways. A common method of installation places the primary and secondary cables along front or rear lot lines and utilizes pad-mounted transformers. The pad-mounted transformers would normally serve from 4 to 12 customers depending upon the terrain, the customer loads, and the size of transformers installed. This type of installation is illustrated in Fig. 34.3.

Many times it is desirable to install the cables along streets in front of the homes. This type of installation is commonly referred to as front-lot line. The front-lot line type of installation can use pad-mounted transformers, submersible transformers, direct-buried transformers, or unit residential transformers (see Fig. 34.4). The installation of pad-mounted transformers along the streets may be objectionable from an environmental point of view.

Submersible transformers are usually installed in fiber-material vaults. The submersible transformer must be protected from corrosion. Maintenance of submersible transformers and the cost of installation are usually greater than the costs associated with pad-mounted transformers. Direct-buried transformers have not been used extensively to date. It is necessary to provide corrosion protection for this type of installation.

The unit residential transformer is usually installed adjacent to a house and back from the street, where it is inconspicuous. The unit residential transformer usually consists of a dry-type core and coils installed in a metal housing for mounting on a pad. Each house is served by an individual transformer. The cost of installing this type of equipment can be

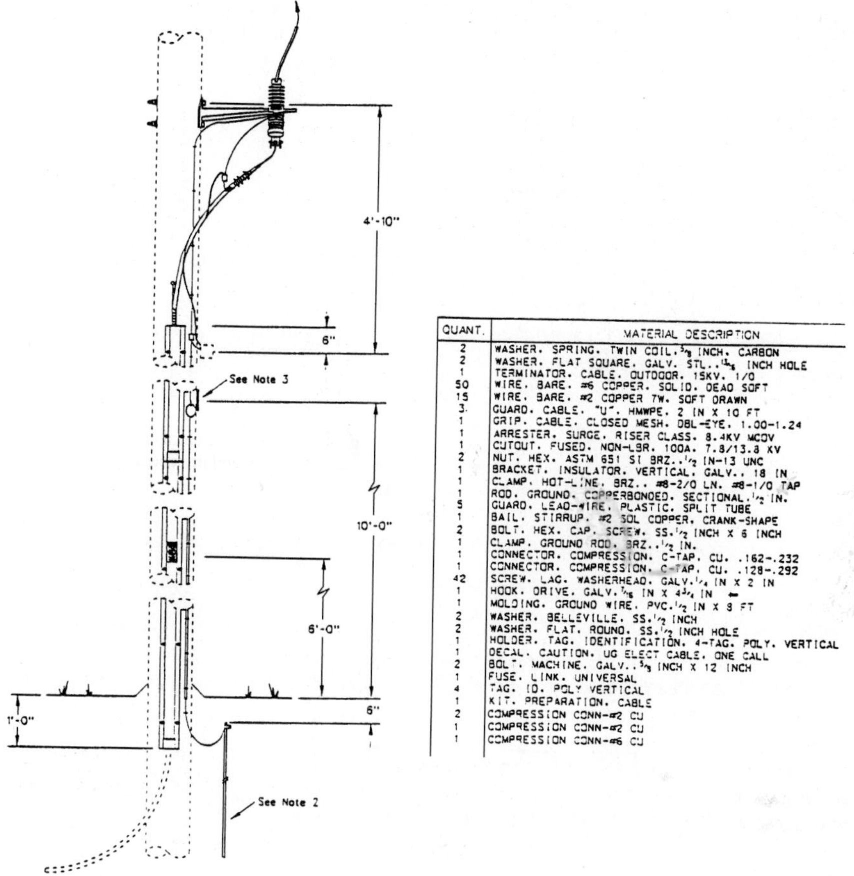

QUANT.	MATERIAL DESCRIPTION
2	WASHER. SPRING. TWIN COIL. 5/8 INCH. CARBON
2	WASHER. FLAT SQUARE. GALV. STL.. 11/16 INCH HOLE
1	TERMINATOR. CABLE. OUTDOOR. 15KV. 1/0
50	WIRE. BARE. #6 COPPER. SOLID. DEAD SOFT
15	WIRE. BARE. #2 COPPER 7W. SOFT DRAWN
3.	GUARD. CABLE. "U". HMWPE. 2 IN X 10 FT
1	GRIP. CABLE. CLOSED MESH. DBL-EYE. 1.00-1.24
1	ARRESTER. SURGE. RISER CLASS. 8.4KV MCOV
1	CUTOUT. FUSED. NON-LBR. 100A. 7.8/13.8 KV
2	NUT. HEX. ASTM 651 SI BRZ.. 1/2 IN-13 UNC
1	BRACKET. INSULATOR. VERTICAL. GALV.. 18 IN
1	CLAMP. HOT-LINE. BRZ.. #8-2/0 LN. #8-1/0 TAP
1	ROD. GROUND. COPPERBONDED. SECTIONAL. 1/2 IN.
5	GUARD. LEAD-WIRE. PLASTIC. SPLIT TUBE
1	BAIL. STIRRUP. #2 SOL COPPER. CRANK-SHAPE
2	BOLT. HEX. CAP. SCREW. SS. 1/2 INCH X 6 INCH
1	CLAMP. GROUND ROD. BRZ.. 1/2 IN.
1	CONNECTOR. COMPRESSION. C-TAP. CU. .162-.232
1	CONNECTOR. COMPRESSION. C-TAP. CU. .128-.292
42	SCREW. LAG. WASHERHEAD. GALV. 1/4 IN X 2 IN
1	HOOK. DRIVE. GALV. 7/16 IN X 4 3/4 IN
1	MOLDING. GROUND WIRE. PVC. 1/2 IN X 8 FT
2	WASHER. BELLEVILLE. SS. 1/2 INCH
2	WASHER. FLAT. ROUND. SS. 1/2 INCH HOLE
1	HOLDER. TAG. IDENTIFICATION. 4-TAG. POLY. VERTICAL
1	DECAL. CAUTION. UG ELECT CABLE. ONE CALL
2	BOLT. MACHINE. GALV.. 5/8 INCH X 12 INCH
1	FUSE. LINK. UNIVERSAL
4	TAG. ID. POLY VERTICAL
1	KIT. PREPARATION. CABLE
2	COMPRESSION CONN-#2 CU
1	COMPRESSION CONN-#2 CU
1	COMPRESSION CONN-#6 CU

FIGURE 34.2 Typical underground distribution riser pole for 7620-volt single-phase primary.

justified if the lots are large and the terrain is such that rear-lot line construction is not practical. Different types of high-voltage switching arrangements have been installed to vary the flexibility and the cost associated with each different type of underground distribution installation.

Cables. Cable should be capable of withstanding tests applied in accordance with an applicable standard issued by a recognized organization such as the Association of Edison Illuminating Companies, the Insulated Power Cables Engineers Association, the National Electrical Manufacturers Association, or the American Society for Testing and Materials. The design and construction of conductors, insulation, sheath, jacket, and shielding shall include consideration of mechanical, thermal, environmental, and electric stresses which are expected during installation and operation. Cable shall be designed and manufactured to retain specified dimensions and structural integrity during manufacture, reeling, storage, handling, and installation.

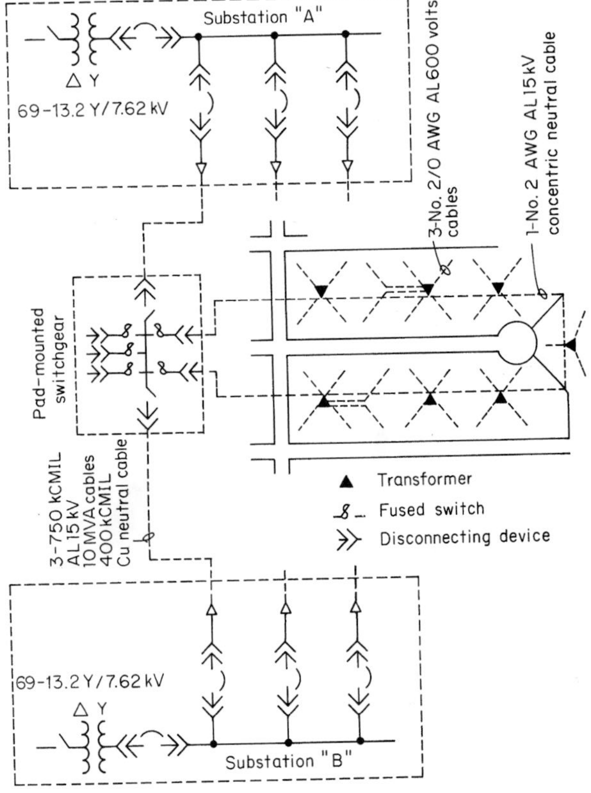

FIGURE 34.3 Total underground distribution system.

Cable shall be designed and constructed in such a manner that each component is protected from harmful effects of other components. The conductor, insulation, and shielding shall be designed to withstand the effects of the expected magnitude and duration of fault current except in the immediate vicinity of the fault. Sheaths and/or jackets shall be provided when necessary to protect the insulation or shielding from moisture or other adverse environmental conditions. Conductor shielding should be provided in accordance with an applicable standard issued by a recognized organization such as the Association of Edison Illuminating Companies, the Insulated Power Cable Engineers Association, and the National Electrical Manufacturers Association.

Insulation shielding shall be provided for cable operating at more than 5 kV to ground and is recommended for cables operating above 2 kV to ground. Shielding is not required for short jumpers which do not contact a grounded surface within enclosures or vaults provided they are guarded or isolated. Insulation shielding may be sectionalized provided that each section is effectively grounded. The shielding system may consist of semiconducting materials, nonmagnetic metal, or both. The shielding adjacent to the insulation shall be designed to remain in intimate contact with the insulation under all operating conditions. Shielding material shall be designed to resist excessive corrosion under the expected operating conditions and shall be protected.

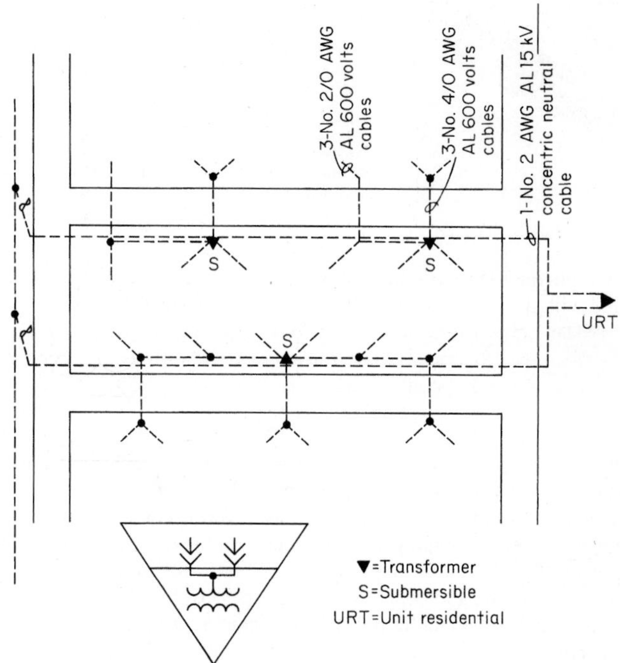

FIGURE 34.4 Underground distribution front-lot line with submersible and unit residential transformers.

Voltages. The primary voltages used for underground circuits vary from 2400 to 34,500 volts. One of the most common primary voltages is 13,200Y/7620 volts. Other voltages used extensively are 4160Y/2400, 12,500Y/7200, 13,800Y/7960, 24,900Y/14,400, and 34,500Y/19,920 volts.

Utilization voltages for residential installations are normally 120/240 volts, single-phase, three-wire, grounded neutral system. Utilization voltages for commercial installations may be 120/240-volt single-phase three-wire grounded neutral system; 208Y/120-volt three-phase four-wire grounded neutral system; 240-volt three-phase three-wire ungrounded delta system; 480Y/277-volt three-phase four-wire grounded neutral system, or 480-volt three-phase three-wire ungrounded delta system. In some instances a commercial or an industrial customer may receive service at the primary voltage which may be the same voltage used to supply power to a residential area. Large industrial customers may receive electric service at voltages commonly used for subtransmission or transmission circuits such as 69,000 to 345,000 volts three-phase.

Primary Cable. The high-voltage cables for a total underground system main feeder circuit usually consist of an aluminum or copper conductor with semiconducting tapes or an extruded conducting polyethylene conductor shielding; insulation consisting of cross-linked polyethylene, high-molecular-weight polyethylene, or ethylene propylene rubber; insulation shielding consisting of a semiconducting compound and copper shielding tapes or an extruded conducting polyethylene shield with copper shielding tapes and jacketed

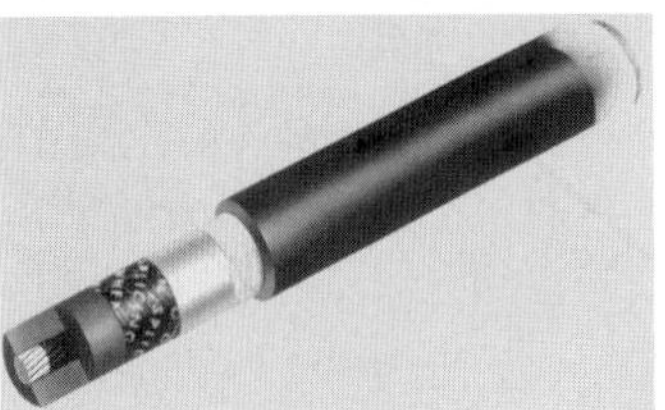

FIGURE 34.5 Single-conductor high-voltage cable used for underground main feeder circuit. (*Courtesy Rome Cable Co.*)

with an extruded polyethylene (see Fig. 34.5). A tinned, bare copper neutral conductor is usually installed with the high-voltage cables.

The main feeder cables are usually installed direct-buried at a depth greater than required by the National Electrical Safety Code to provide protection from dig-ins. The main feeder circuit cables may be installed in polyethylene plastic pipe. If ducts encased in concrete are used for the primary feeder circuits, paper-insulated lead-covered cables with an extruded polyethylene jacket are often used. The primary cables supplying residential areas or commercial customers from the main feeder circuits, either overhead or underground, are usually the two-conductor concentric neutral type.

The high-voltage concentric neutral cable usually consists of aluminum or copper conductor with an extruded conducting polyethylene conductor shielding; insulation with cross-linked polyethylene, or ethylene propylene rubber; insulation shielding consisting of a semiconducting compound and copper shielding tapes covered by conducting polyethylene jacket or an extruded conducting polyethylene shielding jacket surrounded by spiral-wound tinned copper neutral conductors. The typical primary UD (underground distribution) cable is illustrated in Fig. 34.6. Many of the high-voltage cables with a concentric neutral conductor have a semiconducting polyethylene jacket extruded over the concentric neutral conductors to prevent electrolysis of the concentric neutral conductors.

Low-Voltage Cables. The secondary and service cables for underground electric distribution are generally larger than those applied on overhead systems because of the reduced capacity for a given conductor size as a result of the temperature limitations of the cable insulation, greater load density, and the need to provide for load growth. Aluminum conductors are generally used for underground distribution because of the lower cost compared to copper. The secondary cables often have a conductor of No. 4/0 AWG or 350-kCMIL aluminum. Service cables often have a conductor of No. 2/0 AWG aluminum to provide 200-amp capacity where National Electric Code ratings are not applicable. The underground residential distribution services will vary in size from No. 1/0 AWG to No. 4/0 AWG aluminum conductor. A nominal three-wire 120/240-volt single-phase grounded neutral system is normally installed to serve homes in areas supplied by UD circuits.

FIGURE 34.6 Typical two-conductor concentric cable used for underground primary distribution. Aluminum-stranded conductor has a semiconducting extruded shielding, cross-linked polyethylene insulation, semiconducting extruded insulation shielding, surrounded by spiral-wound exposed tinned copper neutral conductor. The copper neutral conductor is jacketed with polyethylene. (*Courtesy Southwire Corporation.*)

A twin-concentric neutral cable is constructed with two 600-volt cross-linked insulated aluminum conductors in parallel configuration with an overall spirally applied concentric neutral of No. 14, 12, or 10 AWG tinned copper wires in sufficient number to provide the conductivity required. This type of cable construction is used where telephone and power conductors are installed randomly in a single trench.

Triplexed cables may be used for secondaries and services. This cable construction consists of three single-conductor aluminum cables with 600-volt cross-linked polyethylene insulation twisted together. This insulated aluminum neutral cable may be color coded, providing ease of identification and distinguishability from other cables (Fig. 34.7).

Secondary and service cables constructed in a flat parallel configuration are used in some installations. This cable assembly consists of two aluminum phase conductors and a neutral conductor, each individually insulated with cross-linked polyethylene and covered with a tight, contour-fitted polyethylene jacket. The jacket assembles the three cables in a flat parallel configuration, providing a web between the conductors for ease of separation when splicing or terminating the cable, without danger of damaging the insulation.

Transformers. Many component parts of transformers used for underground electric distribution are similar to those installed in transformers designed for pole mounting. The core and coil assemblies are interchangeable in many instances. The transformers can be equipped with no-load tap changers, high-voltage fuses, low-voltage circuit breakers, and pressure relief devices. The core and coils assembly is normally immersed in insulating oil to provide the major insulation and a cooling medium. A typical core and coils assembly is shown in Fig. 34.8.

FIGURE 34.7 Reel of low-voltage insulated triplexed cables mounted on back of truck to facilitate installation of underground secondaries and services. Cable is manufactured for direct-buried installation. Light-colored insulation identifies neutral conductor.

FIGURE 34.8 Core and coils assembly for installation in transformer. (*Courtesy ABB Power T&D Company Inc.*)

FIGURE 34.9 Single-phase pad-mounted transformer installed along rear lot line of a subdivision. (*Courtesy A. B. Chance Co.*)

Pad-mounted transformers are used more extensively than other types of UD transformers. The transformers are built in different sizes and configurations depending upon the kVA capacity of the transformer, the type of connections, and the accessory equipment installed. The pad-mounted transformers vary in height from 18 to 48 in or more. A low-profile pad-mounted transformer installed along the rear lot line in a new residential area is shown in Fig. 34.9.

Figure 34.10 illustrates a pad-mounted transformer with the hinged cover open. Many of the accessories can be seen in Figs. 34.10 and 34.11. The transformers are constructed with two high-voltage load-break bushings. In Fig. 34.11 one of the high-voltage

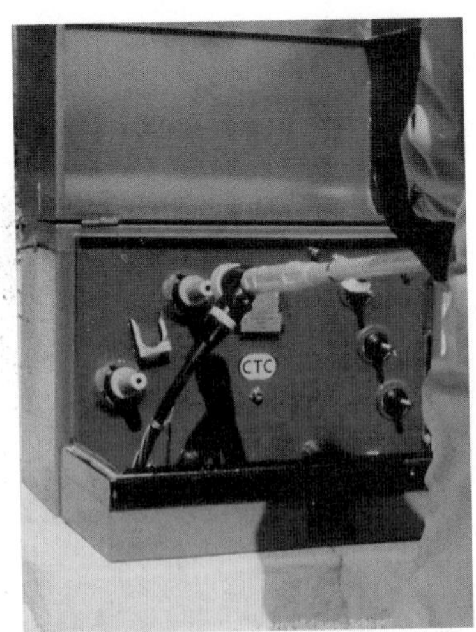

FIGURE 34.10 Pad-mounted transformer with cover open. Lineman is installing high-voltage cable load-break elbow connector with a hot-line tool on the high-voltage load-break bushing. (*Courtesy I. T. T. Blackburn Co.*)

FIGURE 34.11 Close-up view of pad-mounted transformer with cover open, showing high-voltage and low-voltage terminal equipment and accessories. Lineman is installing a grounding device on high-voltage terminator parking stand bracket with hot-line tool. (*Courtesy A. B. Chance Co.*)

load-break bushings is marked L_1. The high-voltage bushing is shown in Fig. 34.10 without the high-voltage cable elbow connector installed and in Fig. 34.11 after the connections have been made. The low-voltage bushings are marked X_1, X_2, and X_0. A parking stand bracket for the high-voltage cable connections is located between the two high-voltage load-break bushings.

In Fig. 34.11 the lineman is installing a grounded parking stand that is used to ground a high-voltage cable for maintenance work. It is very important that the cables be tested with a high-voltage meter to be sure they are de-energized prior to placing the elbow connector on the grounded device installed on the parking stand bracket. Other devices that may be accessible with the transformer cover open include draw-out load-break fuses, operating handle for low-voltage circuit breaker, and pressure relief device.

Submersible distribution transformers are similar in appearance to pole-mounted transformers and contain similar equipment (Fig. 34.12). High-voltage cable connections are completed through bushings by using elbow connectors installed on the high-voltage cables. The low-voltage connections are completely insulated. The transformers are designed to operate under water. Tanks of the transformers are made with stainless steel or mild steel and protected by special enamel finishes to prevent corrosion. Special care must be taken to protect the transformer finish from being scratched or damaged while the unit is being stored, transported, and installed. It is desirable to install cathodic protection equipment connected to the transformer tank to prevent electrolysis.

Direct-buried single-phase distribution transformers are available in sizes of 10 kVA through 37.5 kVA. The transformers are usually installed as a part of a front-lot-line system with one transformer per customer. The direct-buried transformer reduces the cost of installation by eliminating the vault and minimizes the maintenance compared to the requirements associated with submersible vault-installed transformers. The transformers are constructed with a metallic tank or a polyester tank with a metallic liner. Figure 34.13 illustrates a direct-buried transformer with a metallic tank, and Fig. 34.14 illustrates a direct-buried transformer polymer tank with a metal lining.

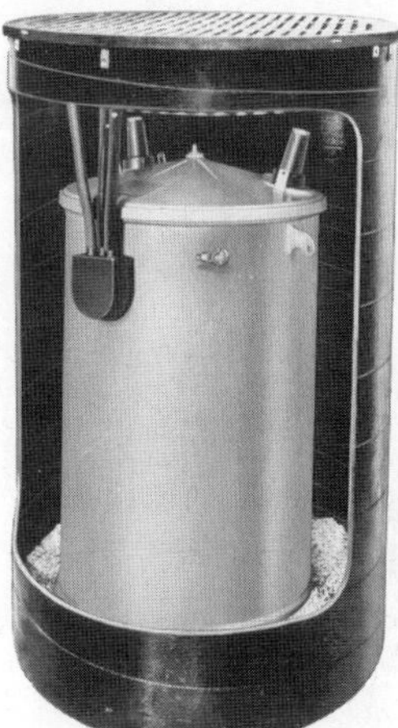

FIGURE 34.12 Cutaway view showing submersible single-phase distribution transformer installed in fiber vault. A steel grate on top of the vault provides protection and ventilation openings. (*Courtesy ABB Power T&D Company Inc.*)

FIGURE 34.13 Direct-buried transformer with a metallic tank. (*Courtesy General Electric Co.*)

FIGURE 34.14 Direct-buried single-phase distribution transformer with a nonmetallic tank. The transformer tank is molded from reinforced polyester and lined with metal. (*Courtesy General Electric Co.*)

Three-phase pad-mounted transformers are often used to serve shopping centers, schools, hospitals, churches, apartment complexes, and office buildings located in the vicinity of electric UD circuits serving several residential areas. The three-phase pad-mounted transformers are available in sizes of 75 kVA through 2500 kVA, with primary voltages of 4160Y/2400 through 34,500Y/19,920 and secondary voltages of 208Y/120, 240, 480Y/277, and 480. Standard accessories such as high-voltage bushings for use with load-break elbow connectors, fuses, no-load tap changer, secondary breaker, pressure relief device, liquid-level gauge, and lightning arresters are available. A three-phase pad-mounted transformer installed at a shopping center is illustrated in Fig. 34.15.

FIGURE 34.15 Three-phase pad-mounted transformer with underground connecting cables. (*Courtesy MidAmerican Energy Co.*)

Isolating Devices. Equipment used to sectionalize three-phase underground main feeder circuits and provide a means for isolating single-phase and three-phase taps for underground residential and underground commercial circuits may be constructed for pad-mounting or submersible installation. The pad-mounted gear may be equipped with air, vacuum, or oil switches. The submersible isolating devices are constructed with vacuum or oil switches. The use of air or vacuum switches minimizes the maintenance of the equipment. The single- and three-phase lateral taps originating at the switching cubicles are usually fused to protect the main feeder circuit from an outage as a result of failure of some of the equipment installed as part of one of the lateral circuits.

Installation. An electric UD system serving a residential subdivision normally consists of single-phase primary cable originating on riser poles (as shown in Figs. 34.1 and 34.2) connected to an overhead distribution circuit through fused cutouts and protected by lightning arresters. Mechanical protection for supply conductors or cables installed on a riser pole shall be provided. This protection should extend at least 1 ft below ground level. Supply conductors or cable should rise vertically from the cable trench with only such deviation as necessary to permit a reasonable cable bending radius. Exposed conductive pipes or guards containing supply conductors or cables shall be grounded.

The installation should be designed so that water does not stand in riser pipes above the frost line. Conductors or cables shall be supported in a manner designed to prevent damage to conductors, cables, or terminals. Where conductors or cables enter the riser pipe or connecting bend, they shall be installed in a manner that shall minimize the possibility of damage due to relative movement of the cable and pipe. Risers should be located on the pole in the safest available position with respect to climbing space and possible exposure to traffic damage. The number, size, and location of riser ducts or guards shall be limited to allow adequate access for climbing.

The single-phase primary cables may be supplied by a UD circuit (as shown schematically in Fig. 34.3) and originating at a sectionalizing switching device similar to those illustrated in Figs. 34.16 through 34.23. The UD equipment may be installed along rear lot lines as shown schematically in Fig. 34.24. Figure 34.25 is an aerial view of a rear lot line underground residential distribution system installation. A front lot line underground residential distribution system is shown schematically in Fig. 34.4.

Most electric underground residential distribution systems employ direct-buried cables. The trenches for the direct-buried cables are usually dug with a small trencher, or directional boring equipment as shown in Figs. 34.26 and 34.27. In rural areas and urban locations free of obstacles under ground, cables are frequently installed with plowing equipment. If the cables are to be installed in conduits, cable preassembled in plastic pipe is usually installed

FIGURE 34.16 Interior view of dead-front pad-mounted switchgear switch termination components. The polycarbonate windows allow visual verification of switch position. (*Courtesy Federal Pacific*)

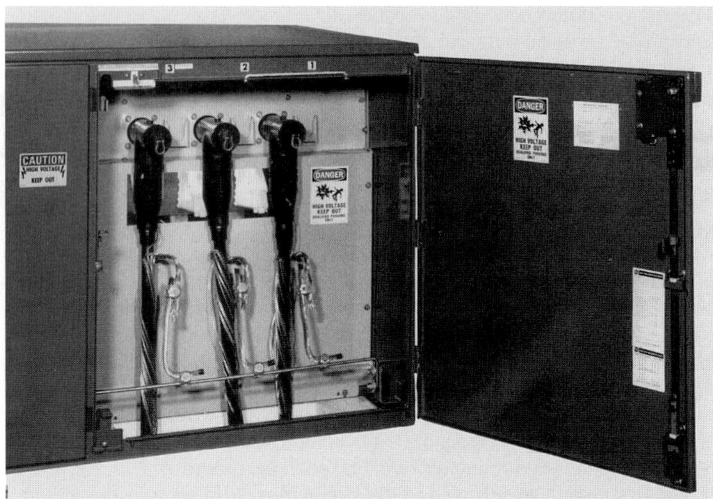

FIGURE 34.17 Open-door view of one main underground circuit compartment of pad-mounted switchgear. The main circuit cables and terminators have a capacity of 600 amps. Viewing window in the compartment permits the lineman to verify the switch position and check that the switch, when open, has visible open gaps. (*Courtesy S&C Electric Company.*)

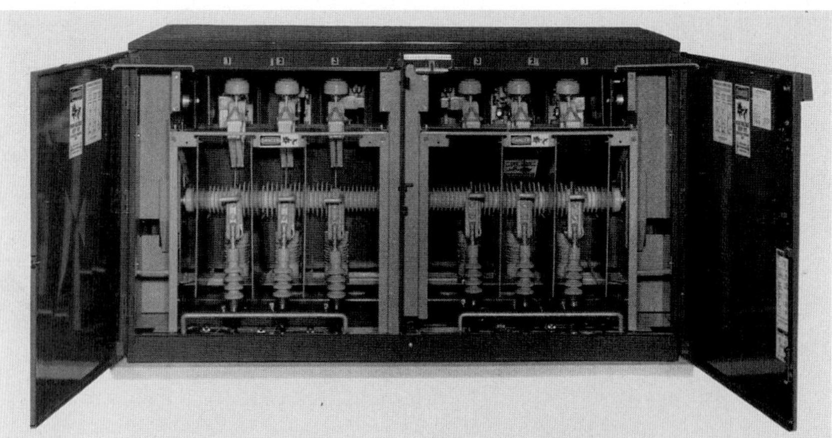

FIGURE 34.18 Pad-mounted switchgear showing interrupter switches. The interrupter switches are connected to the main underground distribution circuit to permit sectionalizing the circuit. A bus connected between the two switches in the switchgear permits the feeder circuits and their protective equipment to be tapped to the main circuit between the two switches. Switch on the right is open with dual-purpose front barriers inserted in the open gaps to isolate main feeder cables from the upper switch contacts and the bus conductors. (*Courtesy S&C Electric Company.*)

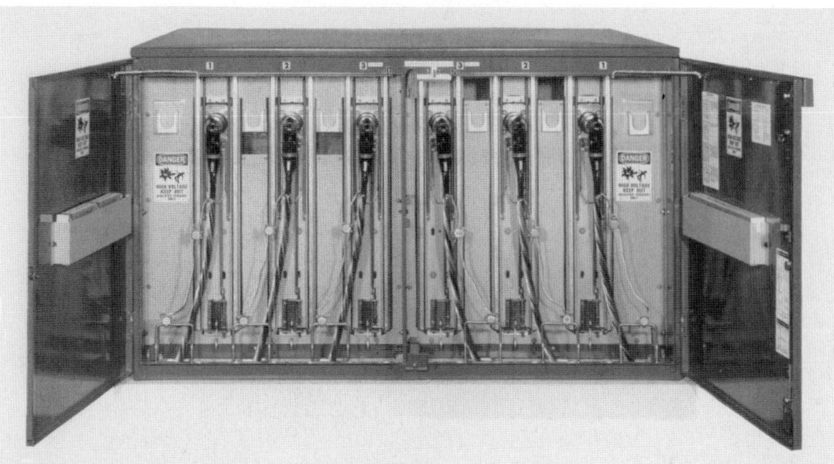

FIGURE 34.19 Open-door view of fuse compartments for underground distribution circuits tapped to main circuit in pad-mounted switchgear with totally enclosed components. The feeder cables and elbow connectors have a capacity of 200 amps. The switchgear contains power fuses and switches in series with the feeder cables. (*Courtesy S&C Electric Company.*)

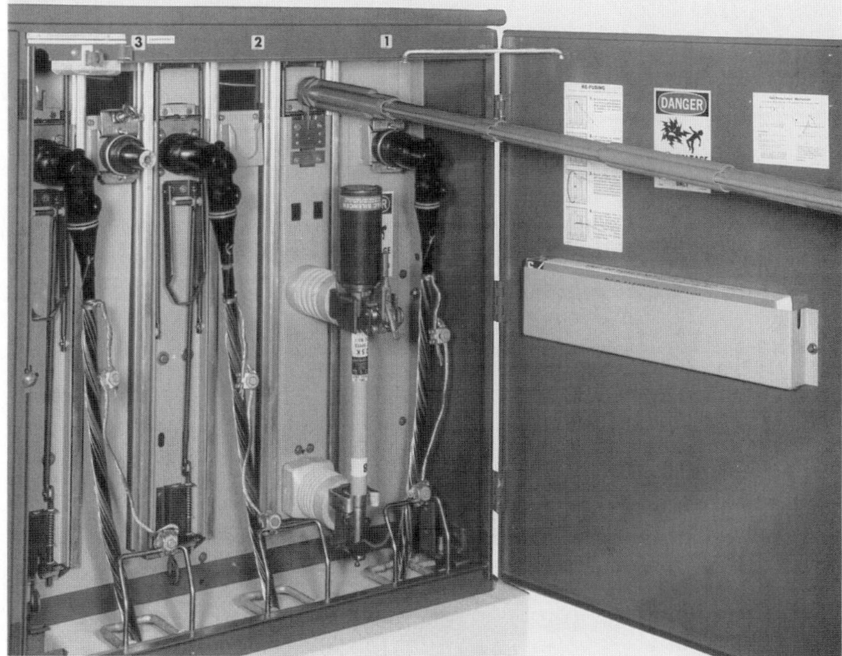

FIGURE 34.20 Open-door view of one fuse compartment of switchgear shown in Fig. 34.19. Standoff insulators have been installed on parking stands. The phase-1 feeder cable and elbow connector have been connected to a standoff insulator. A lineman has rotated the phase-1 door with a hot stick, opening the switch in series with the phase-1 fuse and placing the fuse so that it is available. The phase-1 fuse is disconnected and isolated from the switchgear bus and the phase-1 underground feeder cable. (*Courtesy S&C Electric Company.*)

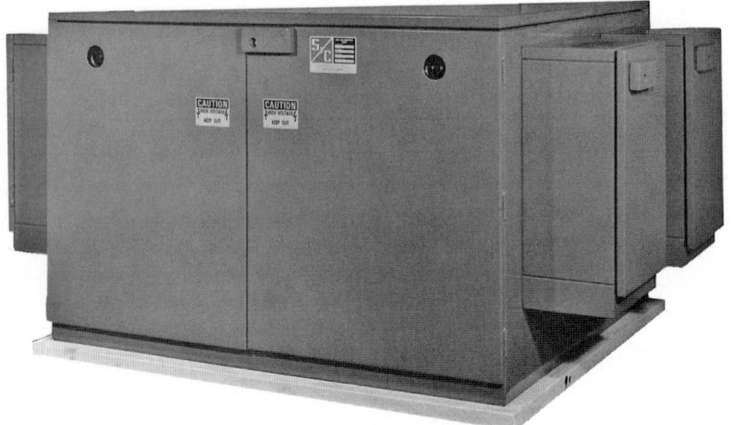

FIGURE 34.21 Pad-mounted switchgear similar to units shown in Figs. 34.16 through 34.20 with power-operated equipment installed to operate the two three-pole interrupter switches. The power-operated equipment permits the interrupter switches to be operated remotely by supervisory control equipment to sectionalize the main underground feeder circuit. (*Courtesy S&C Electric Company.*)

FIGURE 34.22 Submersible three-phase gang-operated vacuum sectionalizing switch for an underground main feeder circuit.

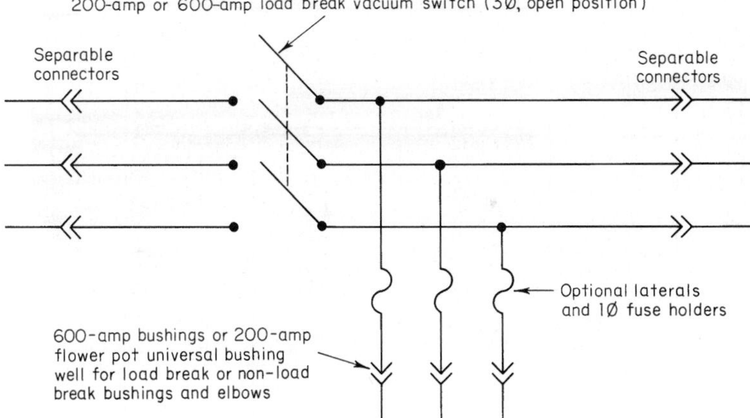

FIGURE 34.23 Schematic diagram of submersible three-phase gang-operated vacuum sectionalizing switch for the underground main feeder circuit pictured in Fig. 34.22.

by trenching or plowing. Both concentric neutral primary cables and secondary cables are available with this type of construction. The pipe or conduit may be either plain or corrugated for greater flexibility.

The distribution facilities should be installed on utility easements or along the streets in public property where the utility has a franchise authorization to make the installation.

Cables should be located so as to be subject to the least disturbance practical. Cables to be installed parallel to other subsurface structures should not be located directly over or under other subsurface structures; but if this is not practical, the rules on clearances should be followed. Cables should be installed in as straight and direct a line as practical. Where bends are required, the minimum radius shall be sufficiently large to prevent damage to the cable being installed. Cable systems should be routed so as to allow safe access for construction, inspection, and maintenance. The location of structures in the path of projected cable route shall, as far as practical, be determined prior to the trenching, plowing, or boring operation.

Routes through unstable soil such as mud, shifting soils, corrosive soils, or other natural hazards should be avoided. If burying is required through areas with natural hazards, the cable should be constructed and installed in such a manner as to protect it from damage. Such protective measures should be compatible with other installations in the area.

Supply cables should not be installed within 5 ft of a swimming pool or its auxiliary equipment. Cables should not be installed directly under building or storage tank foundations. Where a cable must be installed under such a structure, the structure shall be suitably supported to prevent the transfer of a harmful load onto the cable. The installation of cable longitudinally under the ballast section for railroad tracks should be avoided. Where cable must be installed longitudinally under the ballast section of a railroad, it should be located at a depth not less than 60 in below the top of the rail. Where a cable crosses under railroad tracks, the same clearances shall apply.

The installation of cable longitudinally under traveled surfaces of highways and streets should be avoided. When cable must be installed longitudinally under the roadway, it should be installed in the shoulder or, if this is not practical, within the limits of one lane of traffic to the extent practical. Submarine crossings should be routed and/or installed so that they are protected from erosion by tidal action or currents. They should not be located where ships normally anchor.

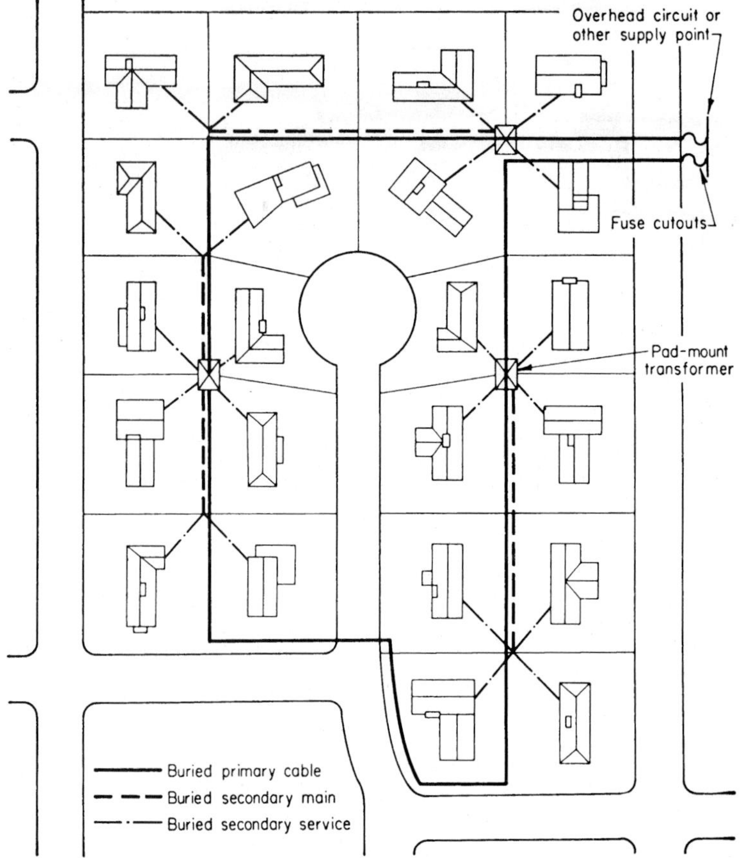

FIGURE 34.24 Typical underground residential distribution system showing transformers and primary and secondary cables located on rear property lines. This makes it possible to serve four customers directly from each transformer and from most pedestals.

The horizontal clearance between direct-buried cable and other underground structures shall be controlled at a minimum of 12 in or larger as necessary to permit access to and maintenance of either facility without damage to the other.

Where a cable crosses under another underground structure, the structure shall be suitably supported to prevent transfer of a harmful load onto the cable system. Where a cable crosses over another underground structure, the cable shall be suitably supported to prevent transfer of a harmful load onto the structure. Adequate support may be provided by installing the facilities with sufficient vertical separation. Adequate vertical clearance shall be maintained to permit access to and maintenance of either facility without damage to the other. A vertical clearance of 12 in is, in general, considered adequate, but the parties involved may agree to a smaller separation.

If conditions require a cable system to be installed with less than 12-in horizontal separation or directly over and parallel to another underground structure, or if another underground structure is to be installed directly over and parallel to a cable, it may be done providing all

FIGURE 34.25 Aerial view of residential area showing cables and transformers installed on rear lot lines, thus making service connections to homes short and convenient. ▶—transformers, O—secondary service pedestals.

parties are in agreement as to the method. Adequate vertical clearance shall be maintained to permit access to maintenance of either facility without damage to the other.

Cable should be installed with sufficient clearance from other underground structures, such as steam or cryogenic lines, to avoid thermal damage to the cable. Where it is not practical to provide adequate clearance, a suitable thermal barrier shall be placed between the two facilities.

The bottom of a trench receiving direct-buried cable should be relatively smooth, undisturbed earth, well-tamped earth, or sand. When excavation is in rock or rocky soils, the

FIGURE 34.26 Combination trencher and backhoe used for installation of direct-buried underground electric distribution cables. Linemen are reviewing plans for installation. (*Courtesy MidAmerican Energy Co.*)

FIGURE 34.27 Lineman operating directional boring equipment to establish a bored hole for placement of underground distribution cables. (*Courtesy Vermeer Manufacturing Co.*)

cable should be laid on a protective layer of well-tamped backfill. Backfill within 4 in of the cable should be free of materials that may damage the cable. Backfill should be adequately compacted. Machine compaction should not be used within 6 in of the cable.

Plowing in of cable in soil containing rock or other solid material should be done in such a manner that the solid material will not damage the cable, either during the plowing operation or afterward. The design of cable plowing equipment and the plowing-in operation should be such that the cable is not damaged by bending, sidewall pressure, or excessive cable tension.

Where a cable system is to be installed by boring and the soil and surface loading conditions are such that solid material in the region may damage the cable, the cable should be adequately protected.

The distance between the top of the cable and the surface under which it is installed should be sufficient to protect the cable from injury or damage imposed by expected surface usage. Burial depths as indicated in Table 34.1 are generally considered adequate.

TABLE 34.1 Burial Depths for Supply Conductors or Cables

Volts	Depth of burial, in
600 and below	24
601 to 22,000	30
22,001 to 40,000	36
40,001 and above	42

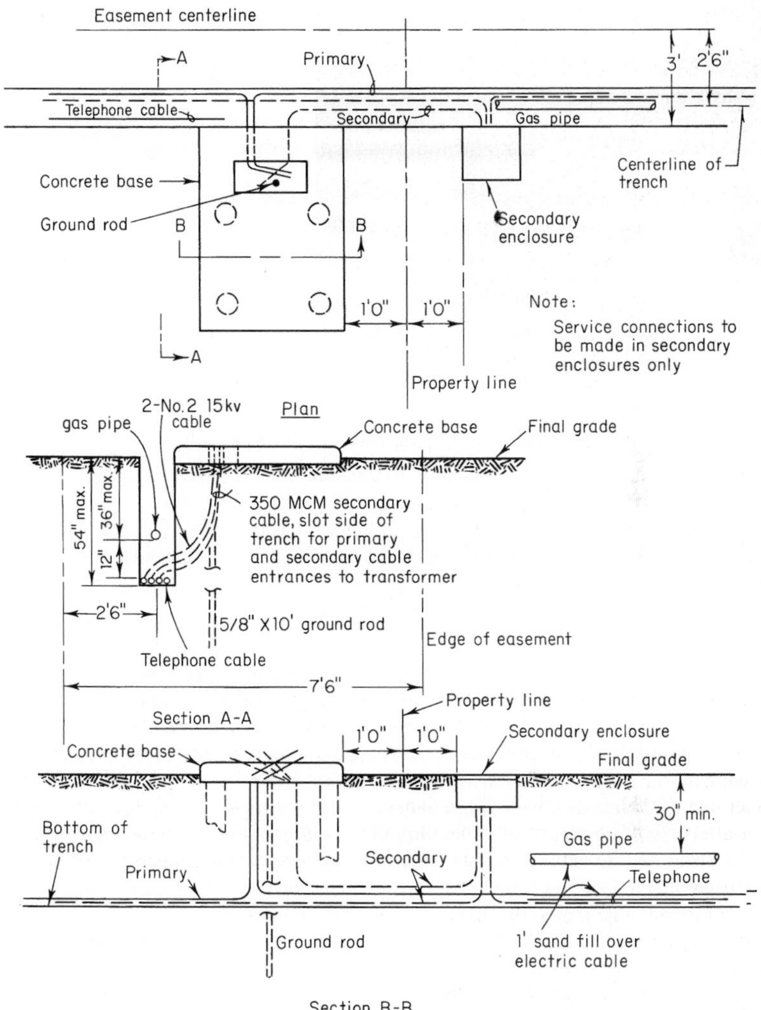

FIGURE 34.28 Underground distribution transformer, pedestal, and cable layout for joint construction of telephone, gas, and 13,200Y/7620-volt electric distribution circuit.

In areas where frost conditions could damage cables, greater burial depths than indicated in Table 34.1 may be desirable. Lesser depths than indicated may be used where supplemental protection is provided. Where the surface under which a cable is to be installed is not to final grade, the cable should be placed so as to meet or exceed the requirements indicated in Table 34.1 both at the time of installation and subsequent thereto.

The conductors or cables of a supply circuit and those of another supply circuit may be buried together at the same depth with no deliberate separation between facilities provided all parties involved are in agreement (Fig. 34.28).

The conductors or cables of a communication circuit and those of another communication circuit may be buried together and at the same depth with no deliberate separation between facilities provided all parties involved are in agreement.

Supply cables or conductors and communication cables or conductors may be buried together at the same depth with no deliberate separation between facilities provided all parties involved are in agreement and the following requirements are met:

1. Grounded supply systems shall not be operated in excess of 22,000 volts to ground.
2. Ungrounded supply systems shall not be operated in excess of 5300 volts phase to phase.

A supply facility operating above 300 volts to ground must include a bare, grounded conductor in continuous contact with earth. This conductor, adequate for the expected magnitude and duration of the fault current which may be imposed, shall be one of the following:

1. A sheath and/or shield.
2. Multiple concentric conductors closely spaced circumferentially.
3. A separate bare conductor in contact with the earth and in close proximity to the cable where such cable or cables also have a grounded sheath or shield not necessarily in contact with the earth. The sheath and/or shield as well as the bare conductor shall be adequate for the expected magnitude and duration of the fault currents which may be imposed.

Where a buried cable passes through a short section of conduit, as under a roadway, the contact with earth of the grounded conductor can be omitted provided the grounded conductor is continuous through the conduit. The bare conductors in contact with the earth shall be of suitable corrosion-resistant material. Cables of an ungrounded system operating above 300 volts shall be effectively grounded concentric shield construction in continuous contact with the earth. Such cables shall be maintained in close proximity to each other.

More than one cable system buried in random separation may be treated as one system when considering clearance from other underground structures or facilities.

The transformer pads and riser poles should be installed prior to placing the cable in the ground. Care exercised in handling the cable during installation, as illustrated in Figs. 34.29 and 34.30, will help to avoid trouble later, for damage sustained by the cable during installation has proved to be a major cause of subsequent cable failure.

The trench should be backfilled carefully. The cable can be best protected by using screened dirt or sand to provide several inches of cover over the cables. The backfill should

FIGURE 34.29 Linemen installing electric, telephone, and cable-television cables in a common trench with no deliberate separation between cables. (*Courtesy MidAmerican Energy Co.*)

FIGURE 34.30 Primary and secondary mains buried directly in ground. Primary cable is two-conductor concentric cable having neutral ground conductor spiraled around outside of cable. (*Courtesy Commonwealth Edison Co.*)

FIGURE 34.31 Helper is tamping backfill of trench with a hydraulic tamping tool. Cable must be protected by adequate backfill before tamping tool is used.

be tamped to prevent settling (see Fig. 34.31). Tamping tools should not be allowed to hit the cables.

The cables installed on riser poles are usually terminated with a terminator as shown in Fig. 34.32. The manufacturer's instructions must be followed carefully to prepare the cable for insertion in the terminator body. It is necessary to remove the insulation shielding the correct distance and insert the cable in the terminator properly.

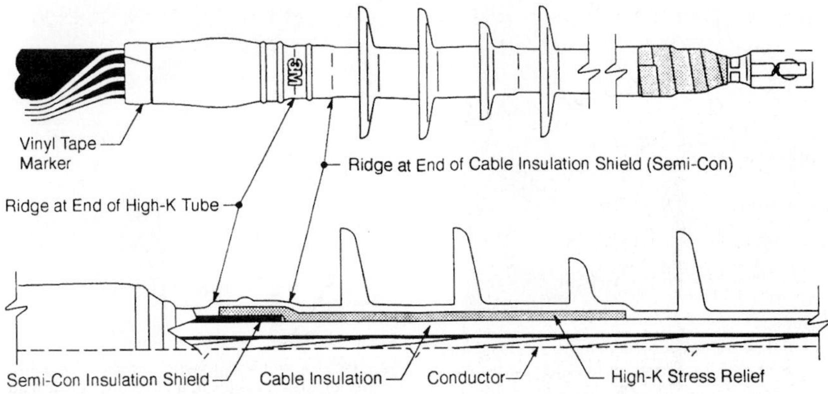

FIGURE 34.32 Typical termination makeup for use with a two-conductor No. 2 AWG concentric single-phase 7620-volt cable used for underground distribution. (*Courtesy 3M Co.*)

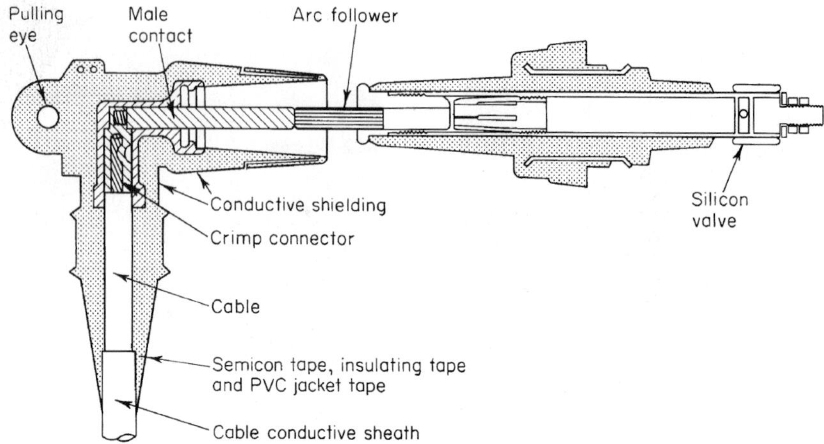

FIGURE 34.33 Cross-sectional view of load-break terminator with component parts identified.

The cables in switching cubicles and transformer enclosures are normally terminated with load-break elbow terminators as shown in Figs. 34.10 and 34.11. Figure 34.33 is a cross-sectional sketch of a typical load-break elbow terminator illustrating how cable is prepared and installed.

Figures 34.34 and 34.35 show a 7620-volt load-break elbow terminator being operated by a lineman with a hot-line tool to energize a pad-mounted transformer.

Aluminum and copper cable is available in desired lengths, so that splices are seldom required. Prefabricated high-voltage splices are available. It is necessary to follow the manufacturer's recommendations to properly prepare the cable for installing a prefabricated splice. The semiconducting insulation shielding must be removed from a section of the insulation to permit the insulation buildup at the splice location. Stress cones in the prefabricated splice must contact the insulation shielding to maintain the shielding across the splice. Aluminum connectors and terminating devices should be used with aluminum conductors

FIGURE 34.34 Load-break elbow termination in position to be connected to bushing on a pad-mounted transformer. Lineman holds energized terminator in position with hot-line tool. (*Courtesy I. T. T. Blackburn Co.*)

FIGURE 34.35 Load-break elbow terminator installed or closed in on a bushing of pad-mounted transformer by lineman using hot-line tool. (*Courtesy I. T. T. Blackburn Co.*)

FIGURE 34.36 Pad-mounted transformer installed on concrete pad located on rear property line.

FIGURE 34.37 Plastic pad kept in position with screw anchors being installed by lineman's helper. (*Courtesy A. B. Chance Co.*)

so as to avoid differential expansion and contraction that could result from the use of dissimilar metals.

Pad-mounted transformers are installed on concrete pads as illustrated in Fig. 34.36 or plastic pads as illustrated in Figs. 34.37 and 34.38.

Submersible transformers installed in vaults (see Fig. 34.12) and direct-buried transformers (see Figs. 34.13 and 34.14) must be protected for corrosion. Figure 34.39 is a suggested installation diagram for a cathodic protection system designed to prevent transformer tank corrosion. If the tank of the transformer is connected to the system ground, the copper neutral is part of the galvanic cell and the protective anode will attempt to protect it as well as the tank; thus, the anode will be depleted rapidly. If the tank of the transformer is isolated from the system ground, the anode will protect the transformer tank only, giving satisfactory anode life.

The low-voltage leads from the transformer to the customer may go directly as shown schematically in Fig. 34.3, or secondaries may be installed to service pedestals and the services

FIGURE 34.38 Plastic pad in position with pad-mounted transformer installed. Ground preparation is being completed by lineman and helper. (*Courtesy A. B. Chance Co.*)

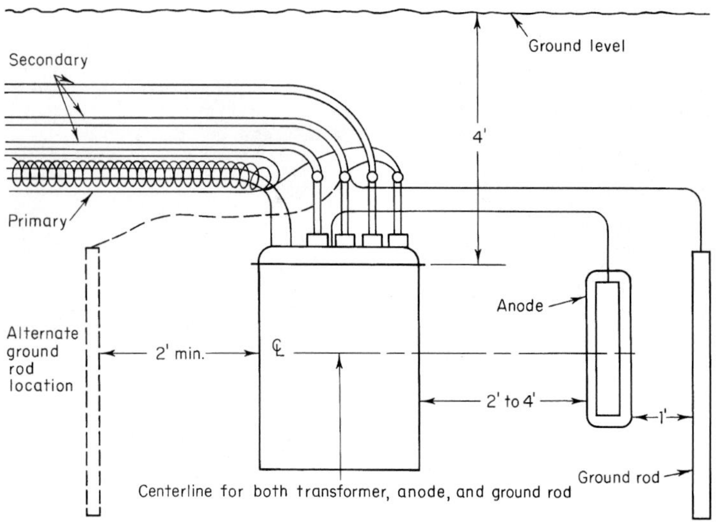

FIGURE 34.39 Sketch of suggested direct-buried transformer cathodic protection system. (*Courtesy General Electric Co.*)

to the customer extended from the service pedestals as shown in Fig. 34.25. If secondaries are installed, they will be in the same trench with the primary cable in most instances, with no definite separation. Figures 34.40 through 34.45 illustrate the installation of a below-grade service enclosure and the secondary and service wire connections completed in the enclosure.

The service wires originating at a below-grade secondary enclosure or a transformer such as the one illustrated in Fig. 34.46 extend directly to the customer's house or meter pedestal. At the customer's house, the service wires usually enter a conduit and terminate in a meter socket. The diagram (Fig. 34.47) shows a typical method of terminating the service wires at a house.

An above-ground metering pedestal for terminating the underground services for two mobile homes is in the foreground of Fig. 34.48.

Street Lights. Electric UD areas usually have street lights installed on ornamental poles with underground supply conductors. The standard is usually fabricated from aluminum or precast concrete. The luminaire is generally constructed for use with mercury vapor or high-pressure sodium lamps. The underground supply cables are connected to the terminals of a transformer or the connectors in a service enclosure. Figure 34.49 shows a typical street light installation.

Fault Indicators. When a fault occurs on an underground circuit, it is important to locate the source of the trouble quickly and restore service with minimum outage time. Fault indicators installed at strategic points in the system will aid in identifying the trouble location. The fault indicators are usually equipped with a neon light that glows, indicating fault current passed through the cable. The fault locators are waterproof and reset automatically after normal operating current is detected.

FIGURE 34.40 Typical convergence of secondary and service cables in a trench, below grade, below where the service pedestal is installed. (*Courtesy I. T. T. Blackburn Co.*)

FIGURE 34.41 The service enclosure (not visible) has been placed in the ground. The lineman has slightly penciled the cable insulation and is slipping sealing sleeves on the cable. (*Courtesy I. T. T. Blackburn Co.*)

Protection. Supply circuits operating above 300 volts to ground or 600 volts between conductors should be so constructed, operated, and maintained that, when faulted, they are promptly de-energized initially or following subsequent protective-device operation. Ungrounded supply circuits operating above 300 volts shall be equipped with a ground-fault indication system. Communication protective devices shall be adequate for the voltage and

FIGURE 34.42 The insulation is being stripped away by the lineman to expose the bare conductor. The stripped length is determined by the size of the terminal lug to be installed. (*Courtesy I. T. T. Blackburn Co.*)

FIGURE 34.43 The conductors must be wire-brushed and covered with inhibitor compound prior to installing the terminal lug. Some terminals have the inhibitor in the barrel of the conductor lug. The lineman is crimping the terminal lug on the conductor. (*Courtesy I. T. T. Blackburn Co.*)

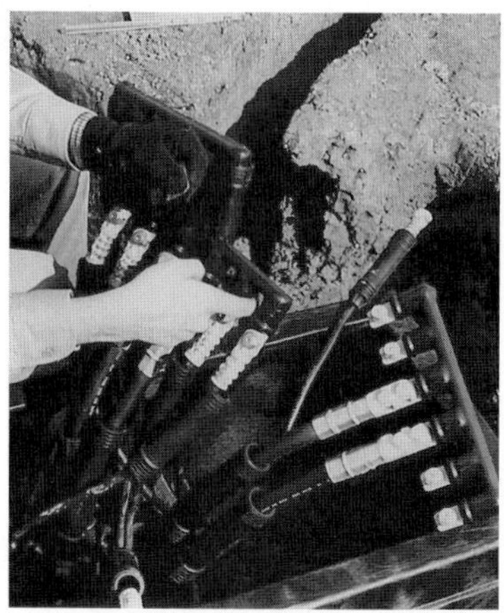

FIGURE 34.44 The lineman is applying silicon lubricant to the insulated neck of the bus bar and to the inside of the sealing sleeve to facilitate removal of the sealing sleeve at some future time. (*Courtesy I. T. T. Blackburn Co.*)

FIGURE 34.45 The lineman is preparing to bend the completed connections down into the below-grade service enclosure. A stainless steel plate will be locked on top of the pedestal. (*Courtesy I. T. T. Blackburn Co.*)

currents expected to be impressed on them in the event of contact with the supply conductors. Adequate bonding shall be provided between the effectively grounded supply conductors and the communication cable shield or sheath, preferably at intervals not to exceed 1000 ft. In the vicinity of supply stations where large ground currents flow, the effect of these currents on communication circuits should be evaluated before communication cables are placed in random separation with supply cables.

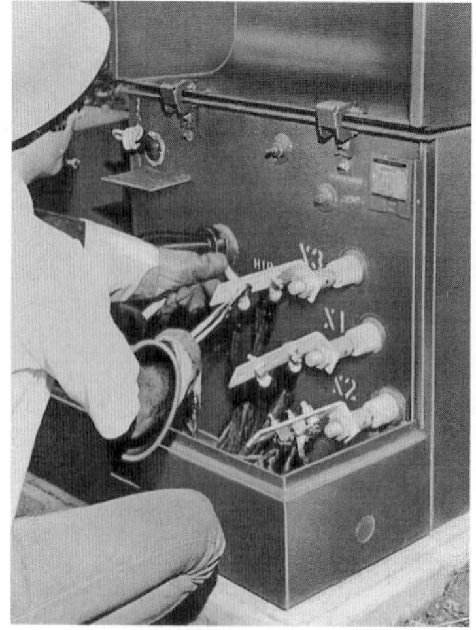

FIGURE 34.46 Lineman connecting service wires to the low-voltage terminals of a pad-mounted transformer. (*Courtesy General Electric Co.*)

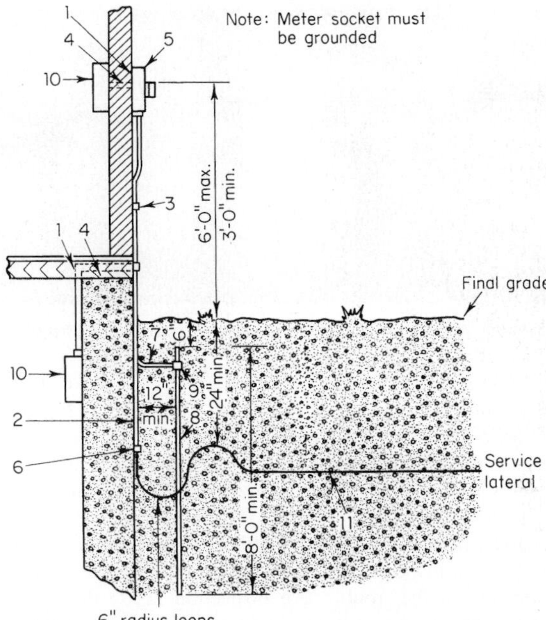

FIGURE 34.47 Typical specification diagram for installing underground service wires at customer's house for 120/240-volt three-wire single-phase supply. (1) Service entrance conductors. Minimum recommended capacity 200 amps. (2) Galvanized steel rigid conduit $2^1/_2$ in. (3) Galvanized steel conduit strap $2^1/_2$ in. (4) Galvanized steel rigid conduit $2^1/_2$ in. (5) Outdoor meter socket with $2^1/_2$ in. conduit hubs. Minimum recommended capacity 200 amps. (6) End bushing for $2^1/_2$ in conduit. (7) Ground wire. (8) Ground rod copperweld $^1/_2$ in minimum or galvanized water pipe $^3/_4$ in. minimum. (9) Connector. (10) Fused switch, circuit breaker, or fuse box. Minimum recommended size 200 amps. Two possible locations shown. (11) Underground service conductors. Minimum recommended capacity 200 amps.

Troubleshooting Underground Distribution Circuits. Underground electric distribution systems have advantages compared to overhead systems and some disadvantages which, if not recognized, could present hazards to linemen, cablemen, and the public. UD systems are innocent in appearance without exposed bare conductors and insulators to remind the lineman of the presence of primary voltages. Switchgear, transformers, and pedestals are easily accessible at ground level, rather than being located at a height which minimizes the likelihood that unqualified personnel will be in close proximity to them. All ground-level equipment must be securely locked and/or protected to prevent the public, especially children, from gaining access to the energized parts. The relatively small dimensions of equipment used in UD systems result in confined working space and close clearances between energized parts. Linemen working on UD systems are generally in contact with ground, providing one necessary ingredient for an accidental electric shock.

The first indication that trouble exists on an underground electric distribution system is generally a call from a customer stating that he has a power outage and/or the operation of a substation circuit breaker or circuit recloser. The reclosing of a substation circuit breaker or a recloser, supplying power to an underground circuit, should be kept to a minimum to limit the

FIGURE 34.48 Meter pedestal for terminating underground service for two mobile homes.

damage to the equipment. Test reclosing of the source equipment seldom restores service to the customers and may result in high fault currents that can damage cables or other facilities.

A fuse in series with the equipment damaged by a dig-in or an insulation failure will normally blow to isolate the trouble from the main circuit, thus permitting the restoration of service to most customers.

Information received from the customer's calls can normally provide the approximate location of the trouble. An inspection by the lineman of the route of the UD circuit will often reveal the location of the trouble. An inspection inside the switchgear and pad-mounted devices may reveal to the lineman or cableman damaged components or problems caused by animals entering the compartments. Cable equipped with fault indicators will help the lineman locate the circuit failure. The fault indicators indicate the flow of fault current even after the circuit has been deenergized. The fault indicators reset automatically when the circuit has been repaired and reenergized.

Most UD circuits use looped cable routes to permit the lineman to isolate the failed cable between two transformers and restore electric service to the customers. A secondary cable failure, a transformer failure, or trouble on a radial system may make it necessary to install temporary facilities such as cables across the surface of the ground to restore electric service.

The location of underground cables and other equipment should be determined from the maps. The cables can normally be precisely located with a device called a *line fault locator*. The device can be used by the lineman to trace the path of an energized underground cable (Fig. 34.50). A faulted cable can be isolated and tested to ensure the cable is de-energized

FIGURE 34.49 Street light installation in an area served by electric underground distribution.

FIGURE 34.50 Lineman is using cable locator to determine the location of the underground circuit. (*Courtesy A. B. Chance Co.*)

FIGURE 34.51 Lineman is testing an underground cable in pad-mounted transformer compartment through a feed-through device with a high-voltage tester to be sure it is de-energized after it had been isolated.

(Fig. 34.51). A signal transmitter is used to impress a signal on the isolated, de-energized faulted cable. The fault locator can be used to follow the signal on the cable to the fault location by observing the instrument or with the use of an earphone headset (Fig. 34.52).

When the cable is exposed at the fault location (Fig. 34.53), it should not be handled until it is grounded, ensuring that it is de-energized. The cable can be grounded by using a

FIGURE 34.52 Lineman using fault locator to determine cable fault location between two pad-mounted transformers. (*Courtesy A. B. Chance Co.*)

FIGURE 34.53 Faulted cable has been exposed by the linemen or cablemen at the indicated fault location.

FIGURE 34.54 Lineman is applying cable spear to concentric neutral cable to ensure it is de-energized and grounded before cutting into cable to splice in a replacement piece of cable to repair a cable fault. The jacket covering the concentric neutral conductor has been removed by the lineman, using rubber gloves to expose the grounded neutral conductor. Note lineman wears safety glasses while performing the work.

"cable spear." The cable spear is operated with a hot-line tool (shotgun). The cable spear has a C-clamp arrangement with a pointed threaded stud. The clamp is placed around the faulted cable thought to be de-energized, and the stud is twisted with the shotgun until the C-clamp is in contact with the concentric neutral conductor and the pointed stud has penetrated the insulation and made contact with the insulated conductor (Fig. 34.54).

The faulted cable may be repaired with a prefabricated splice (Fig. 34.55). If it is necessary to replace the cable, precautions should be taken to minimize the damage to the terrain (Figs. 34.56 and 34.57).

FIGURE 34.55 Lineman holds prefabricated splice and end of replacement cable that will be used to repair faulted cable.

FIGURE 34.56 Lineman uses cable plow to install replacement cable for faulted circuit, minimizing the damage to the customer's lawn.

FIGURE 34.57 Augering device is used to bore a hole for replacement cable, minimizing the damage to the customer's lawn.

CHAPTER 35
VEGETATION MANAGEMENT—TREE TRIMMING

Vegetation management is a specialized field that goes far beyond the trimming of trees. Vegetation management involves right-of-way ground clearance, as well as, line clearance issues. However, for overhead distribution facilities, the lineman normally only has to resolve tree trimming concerns. Tree trimming must be completed in a manner that is environmentally acceptable. The reader is encouraged to review ANSI Z133.1, *Pruning, Trimming, Repairing, Maintaining, and Removing and Cutting Brush-Safety Requirements* standard for tree care.

Line Clearance Objectives. Line clearance is preventive line maintenance to ensure that the utility's service to its customers is not interrupted as a result of interference with conductors or circuit equipment by growing trees. Elements include removal of dangerous trees and overhangs, trimming to clear the conductors, and clearing transmission right-of-way. Natural tree growth and storm-tossed branches can ground or break distribution and transmission lines, interrupting electric service and endangering the public. The trimming process is intended to anticipate such a possibility by removing this hazard. Lines are checked and cleared on a planned time cycle. The amount of clearance sought should provide hazard-free operation for at least 2 years. It should be accomplished while maintaining the health and beauty of the trees involved, the goodwill of property owners and public authorities, and the safety of the trimming crew and the public.

Clearing the Right-of-Way. Practically all overhead lines will traverse through some brush or timberlands. Even lines built in the Middle West, an area generally thought of as level prairie, may pass through forests or hilly country covered with shrubs and underbrush. Lines built in the mountainous country of the West or the hills of the East naturally pass through country in which the right-of-way must be cleared before construction can be started.

In clearing the right-of-way, all stumps should be cut low. All logs and brush should be removed for a distance of at least 25 ft under each conductor so that there is ample room to assemble and erect poles or towers and later to string the line conductors. Any underbrush or piles of dead wood should be removed. Otherwise, they may catch fire and burn the line or anneal the conductors and cause them to sag abnormally.

If it is possible, permission should be obtained to cut extremely tall trees immediately adjacent to the right-of-way. All dead limbs and branches in the adjoining strip also should be cut down, since a high wind may blow them into the line. In case of a high-voltage line, it is more important to remove so-called danger trees along the side of the right-of-way.

Trees which would reach within 5 ft of a point underneath the outside conductor in falling are known as danger trees.

Chemical Brush Control. Chemicals have become available by means of which the growth of brush may be controlled. The chemical is sprayed onto the basal area of the shrubs or small trees from the ground line up to a height of 12 to 15 in above ground. The rest of the shrub need not be sprayed. However, complete encirclement of the stem or trunk with the spray solution is necessary. When applied to stumps, the spray prevents resprouting. Herbicides registered by the Environmental Protection Agency have been tested by power companies and found to be effective. The herbicides can be mixed with oil carriers for summer or winter basal bark injection or cut-stump treatments. In warm weather, they can be mixed with water for injection or cut-stump treatments.

Factors in Trimming Techniques. There are fundamentals essential to safe and competent trimming operations. Primary among these are (1) knowing how to climb and use a rope, (2) knowing how to tie essential knots, (3) knowledge of tree species—their growth characteristics and wood strengths, (4) knowledge of electrical conductors, (5) knowledge of alternative trimming methods, and (6) knowing how to cut limbs and lower them under full control. Ideally, climbers should be trained to recognize structural problems such as weak crotches and chronic-disease symptoms, so that trimming and tree removal decisions result in optimum accomplishment.

Time for Pruning. Trees may be pruned at any time of the year, and utility line clearance operations are maintained throughout the year. In the past, work often was concentrated into a particular season. However, today, it is recognized that there are compensating values and factors, so that trimming operations run year round except where there are unique, limiting local considerations.

A Basic Set of Pruning Tools

1. One light power chain saw (16-in cutter bar length)
2. One handsaw and scabbard
3. One 12-ft pole saw
4. One 12-ft pole clip or pruner
5. One $^1/_2$-in polyester climbing rope (120 ft)
6. Several hand lines ($^1/_2$-in polyester)
7. One $^3/_4$-in bull rope (150 ft)
8. One wooden 24-ft extension ladder
9. One dielectric hard hat for each crew member
10. Tree-wound paint in a pot or spray can
11. Belt snaps for handsaw and paint
12. A set of road signs and flags warning of men working in the trees

Some of the pruning tools appear in Figs. 35.1 and 35.2.

Other items should be considered, especially when working in residential areas where cleanup is vital: a pair of loppers, a sledge and wedges, brush rake, street broom, and scoop shovel.

Climbing Equipment. For safety and proficiency, climbers should always tie in to a crotch before beginning to trim. The elements involved are a body sling or saddle and a

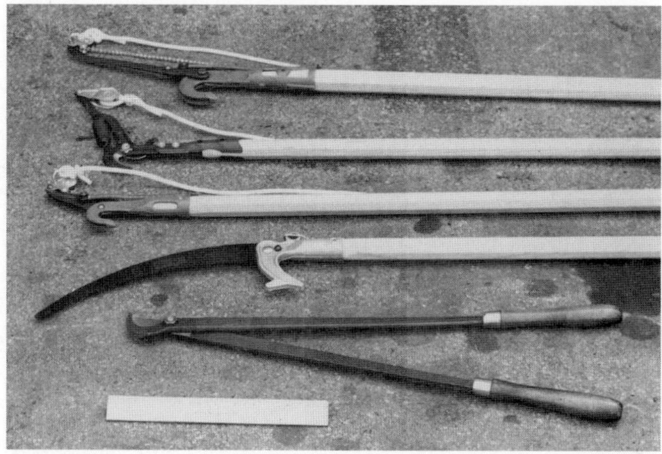

FIGURE 35.1 Trimming hand tools. Starting at the top, tools are a center-cut bull clip, a side-cut pole clip, a center-cut pole clip, a pole saw, and a pair of loppers. (*Courtesy Asplundh Tree Expert Co.*)

climbing rope. The climbing rope is a $^1/_2$-in polyester rope 120 ft long, attached to the snap of the sling which is in front at the center and then secured by a taut-line hitch. The method of tying the rope with a taut-line hitch is covered in detail in Fig. 35.3. This combination permits the climber to move around a tree freely and with safety. Proper use distributes the climber's weight, so he will tire less easily and be able to move onto branches which would not otherwise support his full weight.

Permission. No trimming should be started until permission is obtained from the property owner or the state, county, or municipal agency with jurisdiction (Fig. 35.4*a* and 35.4*b*). Verbal permission is usually satisfactory for routine trimming, but written permission may be advisable for heavy pruning or tree removals. If the property owner or the public authorities refuse to give permission to trim or remove trees endangering the power lines, a permanent record should be maintained for future reference. The trimming should

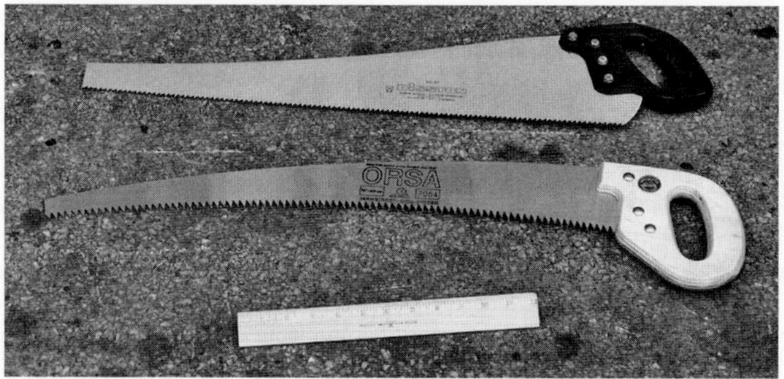

FIGURE 35.2 Two types of handsaws. Upper saw cuts on the push, lower saw cuts on the pull. Use is primarily a matter of personal preference. (*Courtesy Asplundh Tree Expert Co.*)

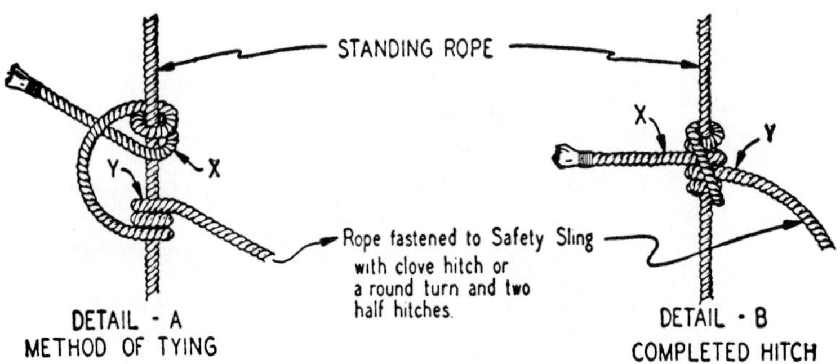

FIGURE 35.3 Tying a taut-line hitch to a standing rope. (*Courtesy Edison Electric Institute.*)

then be done under the direction of the person who secured the permission, since he can best interpret any special directions or limitations imposed. To ensure goodwill necessary for future trimming cycles, the work must be professional and the cleanup thorough.

Good and Bad Methods. Before providing the details of specific tree-trimming techniques, it will be useful to review good and bad general methods. Pollarding and shearing or rounding over are undesirable because the visual effect is ugly and contrary to normal tree form and because these methods represent uneconomical line clearance. Many small cuts take unnecessary time, create an unhealthy tree condition, and stimulate rapid regrowth back into the conductors. Natural trimming is desirable. In this method, branches are flush-cut at a suitable parent limb back toward the center of the tree. This method is also called *drop-crotching* or *lateral trimming*. It involves fewer but heavier cuts, usually made with a saw, not a pole pruner. This natural pruning generally permits more clearance, so that trimming cycles can be lengthened. In addition, better tree form and value are maintained, and regrowth direction can be influenced. Note the examples in Figs. 35.5 through 35.9.

Removing Large Branches. The procedure in the removal of a large side limb is shown in Fig. 35.10. The first cut is an undercut 10 to 12 in out from the place where the final flush cut is to be made. This cut should be one-quarter to one-half the distance through. The second cut is made about 1 to 6 in farther out on the upper side of the branch. The reason for shifting out a few inches is to prevent the bark from being stripped back beyond the point of the final cut when the limb falls. When the bulk of the weight has been removed, the finish cut is made flush with the parent limb or trunk in two steps. First an undercut to prevent stripping is made, and finally the limb is sawed through from above.

Lowering Large Limbs. Large branches that might cause damage to the line conductors or other property in falling should be carefully lowered with tackle. Figure 35.11 illustrates the manner of lowering a large limb which would otherwise fall directly into the line conductors. Before the branch is sawed off, it is supported with two ropes, one at the butt and the other well out toward the end. The ropes are run through crotches in the branches above and secured to the trunk of the tree. A third rope–known as the guide rope–is fastened to the branch at such a point that it can be used to swing the branch out from over the line conductors. After the sawing is completed, the branch is swung out to clear the line and then gradually lowered by means of the two supporting ropes.

(a)

(b)

FIGURE 35.4*a* and 35.4*b* Tree crews discussing tree removal plans with the local tree warden. (*Courtesy Eastern Utilities.*)

Removing Small Branches. In the case of small branches, it is not necessary to make three separate cuts. One cut with the saw close up against the limb while it is held in place, as shown in Fig. 35.12, is all that is necessary. Small branches can also be cut off with a pole pruner. When trees are being trimmed, all dead branches should be pruned out.

Brush Disposal. All brush and wood trimmed in line clearance operations must be disposed of in such a way as to maintain the goodwill of property owners. In residential trimming, it is usually most economical to chip the brush. Note Figs. 35.13 and 35.14. Public relations generally requires that the work area be raked or swept up. Material suitable for fireplaces should be cut to size and neatly stacked. It should first be offered to the property owner, and if he rejects it, it may be left available for the general public. However, some

FIGURE 35.5 *Topping.* Topping is cutting back large portions of the upper crown of the tree. Topping is often required when a tree is located directly beneath a line. The main leader or leaders are cut back to suitable lateral. (The lateral should be at least one-third the diameter of the limb being removed.) Most cuts should be made with a saw; the pole pruner is used only to get some of the high lateral branches. For the sake of appearance and the amount of regrowth, it is best not to remove more than one-fourth of the crown when topping. In certain species—sugar maple, for example—removal of too much of the crown will result in the death of the tree.

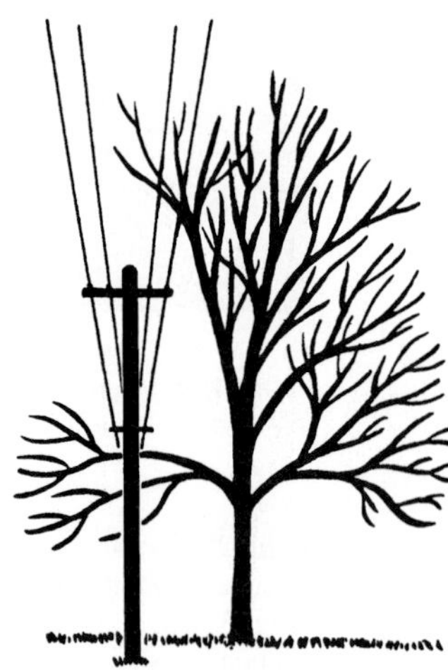

FIGURE 35.6 *Side trimming.* Side trimming consists of cutting back or removing the side branches that are threatening the conductors. Side trimming is required where trees are growing adjacent to utility lines. Limbs should be removed at a lateral branch. Unsightly notches in the tree should be avoided if possible. Shortening branches above and below the indented area or balancing the opposite side of the crown will usually improve the appearance of the tree. When trimming, remove all dead branches above the wires, since this dead wood could easily break off and cause an interruption.

Before trimming After trimming

FIGURE 35.7 *Under trimming*. Under trimming involves removing limbs beneath the tree crown to allow wires to pass below the tree. To preserve the symmetry of the tree, lower limbs on the opposite side of the tree should also be removed. All cuts should be flush to avoid leaving unsightly stubs. The natural shape of the tree is retained in this type of trimming, and the tree can continue its normal growth. Overhangs are a hazard, however, when a line passes beneath a tree. Overhangs should be removed in accordance with the species of tree, location, and the general policy of the utility that you work for. When trimming, remove all dead branches above the wires, since this dead wood could easily break off and cause an interruption. Many utilities have a removal program set up for trees that overhang important lines.

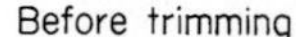

Before trimming After trimming

FIGURE 35.8 *Through trimming*. Through trimming is the removal of branches within the crown to allow lines to pass through the tree. It is best suited for secondaries, street light circuits, and cables, although it is often used on primary circuits where there is no other way of trimming the tree. Cuts should be made at crotches to encourage growth away from the lines.

FIGURE 35.9 *Natural Trimming.* Natural trimming is a method by which branches are cut flush at a suitable parent limb back toward the center of the tree. This method of trimming is sometimes called drop-crotching or lateral trimming. Large branches should be removed to laterals at least one-third the diameter of the branch being removed. Natural trimming is especially adapted to the topping of large trees, where a great deal of wood must be removed. In natural trimming, almost all cuts are made with a saw, and very little pole pruner work is required. This, when finished, results in a natural-looking tree, even if a large amount of wood has been removed. Natural trimming is also directional trimming, since it tends to guide the growth of the tree away from the wires. Stubbing or pole-clip clearance, on the other hand, tends to promote rapid sucker growth right back into the conductors. The big factor to remember is that natural clearance does work and that two or three trimming cycles done in this manner will bring about an ideal situation for both the utility and the tree owner. Most shade trees lend themselves easily to this type of trimming. Elm, Norway maple, red oak, red maple, sugar maple, silver maple, and European linden are the our most common street trees, and these species react especially well to natural clearance methods.

situations may require that the wood be hauled away and dumped. On right-of-way clearance or when trimming in remote areas, it is sometimes possible to pile brush and leave it as game cover. Generally, burning is no longer permitted because of concern about smoke and fly ash (Figs. 35.15 through 35.17).

Power Equipment. Technology responds to operational needs, and various pieces of specialized equipment are designed to solve special problems or to increase productivity.

Trim Lift. Wherever a truck can be driven, an aerial device can be effectively used to position a man for trimming, thus saving the trouble of climbing. The mechanism

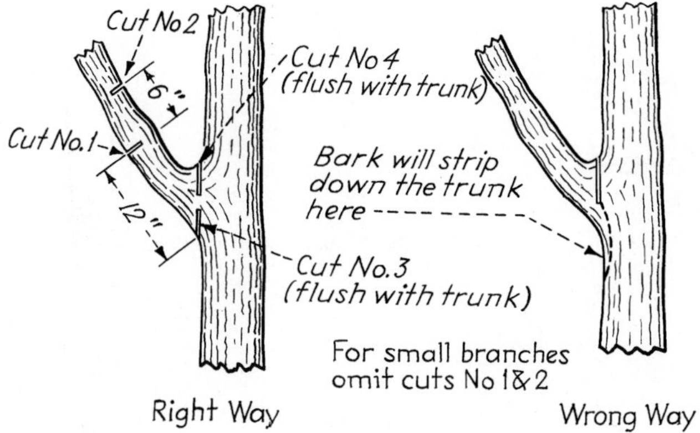

FIGURE 35.10 Steps in right and wrong removal of a large side limb. (*Courtesy REA.*)

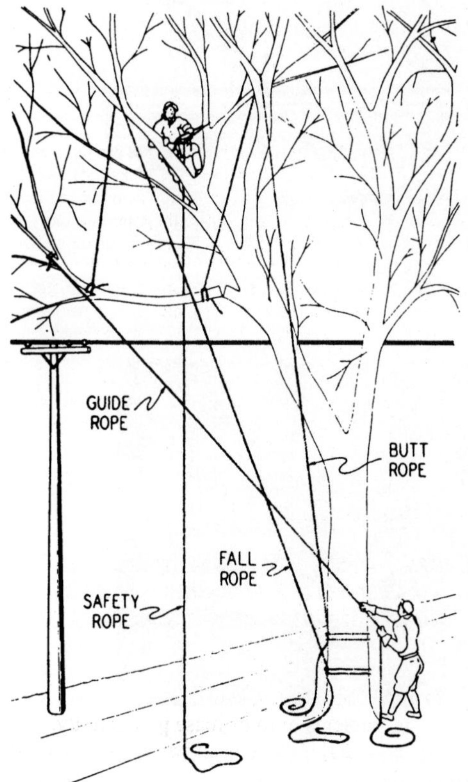

FIGURE 35.11 Lowering large or hazardous limb with rope tackle. The two supporting ropes are called butt and fall ropes; the third rope is the guide rope. (*Courtesy Edison Electric Institute.*)

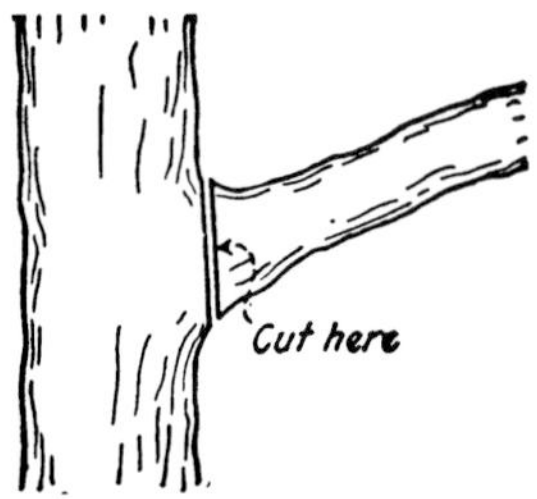

FIGURE 35.12 How a small branch can be removed with a single cut.

FIGURE 35.13 A typical chipper unit utilized by the tree crew.

FIGURE 35.14 A manual climbing crew in operation. Note the groundman chipping brush behind the climbers. The chute of the chipper directs the chips into the back of the dump van. (*Courtesy Asplundh Tree Expert Co.*)

FIGURE 35.15 A helicopter is being used to transport large limbs removed from an electric transmission line right-of-way to a central location for disposal. (*Courtesy Asplundh Tree Expert Co.*)

involves two connected booms, with a work platform or bucket at one end, mounted on a truck chassis. The upper boom is supported by the lower boom through the use of cables. A hydraulic power system runs off the truck engine and lifts the booms, operates the tools, and lifts the dump bodies to discharge chips and other materials gathered in trimming (Figs. 35.18 and 35.19). Mechanical outriggers are used to stabilize the unit during operation. Generally, the booms and the work platform are fully insulated from ground to protect the operator from shock and the truck body from becoming energized if the boom or platform should come in contact with a conductor.

Tree Crane. This type of aerial device is used primarily for taking down trees too tall to reach with a trim lift (Fig. 35.20). The equipment utilizes a boom that telescopes and a climber working from a "saddle" and a rope swung from the boom tip.

Power Tools. Chain and circular-blade saws are powered by compact gasoline engines and designed in a variety of sizes and weights to adapt to the many conditions

FIGURE 35.16 A large tree limb chipper equipped with a mechanical in-feed system is being used while clearing the right-of-way for an electric transmission line. (*Courtesy Asplundh Tree Expert Co.*)

encountered, from take-downs to tree trimming to brush cutting (Figs. 35.21 and 35.22). Noise levels generally dictate hearing protection for the operator.

A variety of cutting tools can be hydraulically powered or air-operated by a system connected with the bucket of a trim lift. These include pole chain saws, circular saws, and limb loppers illustrated in Fig. 35.23. Some situations, such as hospital areas, may require the use of electrically powered tools operated by a portable gasoline generator. Power

FIGURE 35.17 The large tree limb chipper in the picture is blowing chips onto an electric transmission line right-of-way to dispose of the brush and provide mulch for the ground in the right-of-way. The helicopter in the background of the picture delivered a large tree limb from a remote location along the right-of-way and is now returning to get another large limb. (*Courtesy Asplundh Tree Expert Co.*)

FIGURE 35.18 A trim lift and chipper is use by crew to trim trees and dispose of brush. (*Courtesy Asplundh Tree Expert Co.*)

equipment requires systematic inspection and preventive maintenance to ensure crew safety and efficiency.

Poisonous Plants. When trimming trees and removing brush there is always the possibility that a lineman may encounter plants which may cause skin poisoning. These may include the following:

Poison Ivy. This plant sometimes climbs trees and poles. It is a crawling plant. Poison ivy has broad and glossy leaves that always grow in clusters of three leaflets, two branching from the stem 1 in or less below the center one.

Poison Oak. A low-growing erect plant, poison oak never climbs. The leaves are broad but not always glossy and grow in clusters of three. The dark-green leaves are permanently and very heavily hairy on the undersurface.

Poison Sumac. This shrub or small tree may grow to a height of 20 ft. Its smooth glossy leaves are in the form of 7 to 13 oblong leaflets (always an odd number). Poison sumac grows in swamps and low areas.

The poisoning from these plants is caused by an oily substance which gets on a person's skin. The poison can be conveyed by smoke, insects, clothing, and direct contact, and it is more likely to occur when the skin is covered with perspiration. No one should ever consider himself immune to these poisons, even after having had an attack.

FIGURE 35.19 A specialized right-of-way trim lift with the boom mechanism. (*Courtesy Eastern Utilities.*)

If a worker suspects that he has been exposed to these plaints or their oil, he should wash the exposed skin with warm water and ordinary laundry soap. Another way to prevent poisoning is to wash the exposed skin in rubbing alcohol, rinse in clear water, and dry the skin. No brush should be used, because it would irritate the skin.

Tree Characteristics. Characteristics of trees are given in Table 35.1. The table lists the common tree species, growth form, average annual normal growth, annual sucker growth, mature height, modulus of rupture, and remarks. Tree growth forms are upright, spreading, and horizontal (Fig. 35.24).

The annual sucker growth and total growth of the same species of tree will vary considerably as a result of soil conditions and available moisture.

FIGURE 35.20 A tree crane in operation. (*Courtesy Eastern Utilities.*)

Modulus of rupture is the breaking point of green wood of the various species. Modulus of rupture tests are conducted by placing a piece of green wood over a beam and subjecting each end of the wood to increasing pressure until it finally breaks. These breaking points give a comparison of wood strength. However, there is a difference between a test of strength and actual strength for climbing. Deformities or weak points may cause a premature break on normally strong wood. Table 35.1 should be used as a comparison between species only to arrive at a sound working knowledge. Table 35.2 gives the suggested tree-trimming clearances for some electric distribution system voltages.

Tree Growth Retardants. The use of high-volume growth-retardant foliage spray in rural areas has been successful in reducing tree growth without significantly changing the tree characteristics.

FIGURE 35.21 Several gasoline-powered chain saws, each a size and weight to be used easily by one man. It is unlikely that the model at the top of the picture would be used off the ground. (*Courtesy Asplundh Tree Expert Co.*)

FIGURE 35.22 Large portable power-driven saw. (*Courtesy Asplundh Tree Expert Co.*)

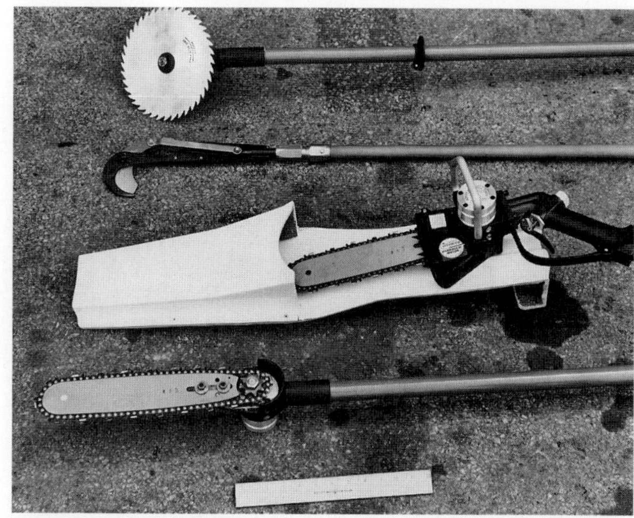

FIGURE 35.23 Hydraulic tools of a type used from a trim lift bucket. (*Courtesy Asplundh Tree Expert Co.*)

TABLE 35.1 Characteristics of Various Trees

Species (common name)	Growth form	Avg. ann. normal growth, in	Avg. ann. sucker growth, in	Mature height, ft	Modulus of rupture (grn.), lb/in^2	Remarks
Ailanthus (tree of heaven)	S	48	60	50		Fast-growing, weak-wooded. Remove if possible.
Alder, Red	U	36	84	120	6,500	Weak-wooded.
Ash	U	18	36	80	9,600	Strong wood, but tends to split. Will not stand heavy topping.
Banyan	S	24	60	70		Rather strong branching. Good crotches for lateral trimming.
Basswood (linden)	S	18	27	75	5,000	Wood not very strong. Tends to split. Watch dead wood.
Beech, American	S	12	30	60	8,600	Strong wood. Mature trees do not stand topping well.
Birch, White	S	21	52	50	6,400	Very sensitive. Intolerant of heavy trimming. Form is spoiled by topping.
Box-elder	H	26	72	50	5,000	Weak-wooded. Suckers grow rapidly. Remove if possible.
Catalpa	S	12		60	5,000	Soft, weak wood. Slow-growing.
Cedar, Eastern Red (juniper)	U	15	15	50	7,000	Host to the Cedar-Apple rust.
Cherry, Wild Black	S	14	24	60	8,000	Found mostly in rural areas. Does not stand heavy topping. Dies back easily. Watch dead wood.
Elm, American	U	26	60	85	7,200	Strong, flexible wood. Avoid tight crotches.
Elm, Chinese	U	40	72	65	5,200	Wood rather brittle. Strips easily.
Eucalyptus (blue gum)	U	24	50	125	11,200	
Ficus	H	36	60	50	5,800	
Fir, Douglas	U	18	21	125	7,600	
Gum, Sweet	U	12	20	90	6,800	A strong slow-growing tree.
Hackberry	S	18	30	60	6,500	Strong, stringy wood. Similar to elm.
Hickory	U	14	21	65	11,000	Wood very strong and tough. Will not stand heavy topping.

TABLE 35.1 Characteristics of Various Trees (*Continued*)

Species (common name)	Growth form	Avg. ann. normal growth, in	Avg. ann. sucker growth, in	Mature height, ft	Modulus of rupture (grn.), lb/in^2	Remarks
Locust, Black	S	18	80	80	13,800	Normal growth is moderate. Vigorous sucker growth when topped. Remove small trees if possible. Watch thorns.
Locust, Honey	U	22	33	80	10,200	Watch thorns.
Madrona	S	8	24	50	7,600	
Magnolia	U	20	42	90	7,400	Wood splits easily.
Mango	S	18	36	50		
Maple, Big Leaf	S	30	72	60	7,400	
Maple, Norway	H	15	35	50	9,400	Wood tends to split.
Maple, Red	S	18	42	75	7,700	Wide-spreading as a mature tree. Sucker growth rapid. Can stand heavy trimming.
Maple, Silver	U	25	65	65	5,800	Suckers grow rapidly. A rather weak and brittle tree.
Maple, Sugar	S	18	40	75	9,400	Mature trees cannot stand heavy topping.
Melaleuca (punk tree)	U	36	96	50		Soft wood. Weak branches. Suckers grow very fast. Remove if possible or top under wires.
Oak, Australian Silk	U	36	72	50		Not too strong. Rather irregular growth.
Oak, Black and Red	S	18	30	85	6,900	Very sturdy trees.
Oak, Live	S	30	45	70	11,900	Very hard strong wood.
Oak, Pin	U	24	36	100	8,300	Strong wood. Topping ruins shape of tree.
Oak, Water	S	30	45	75	8,900	Moderately strong branches.
Oak, White	S	9	18	75	8,300	A sturdy and slow-growing tree. Dies back when topped heavily. Sucker growth thin.
Oak, Willow	H	24	40	50	7,400	Strong. Tough. Can be topped safely.
Palm, Coconut	U	60		90		
Palm, Queen	U	36		65		
Palm, Royal	U	48		80		

TABLE 35.1 Characteristics of Various Trees (*Continued*)

Species (common name)	Growth form	Avg. ann. normal growth, in	Avg. ann. sucker growth, in	Mature height, ft	Modulus of rupture (grn.), lb/in^2	Remarks
Palm, Washington	U	36		70		
Pecan	U			140	9,800	
Pine, Australian	U	60	144	100		Has hard wood. Grows rapidly. Watch weak crotches. Suckers profusely.
Pine, Shortleaf	U	36	48	80	7,300	Wood rather strong, but weak branching.
Pithecellobium (ape's earring)	S	24	72	50		Hard brittle wood. Dangerous thorns.
Plane (sycamore)	U	34	72	100	6,500	Rank growth of suckers when topped.
Poinciana, Royal	H	18	36	30		
Poplar, Carolina (cottonwood)	U	52	80	85	5,300	Wood very brittle. Breaks abruptly. Suckers grow at tremendous rate. Remove if possible.
Poplar, Lombardy	U	45	72	60	5,000	Very brittle. Remove if possible. A short-lived tree.
Sassafras		24	36	50	6,000	Does not stand heavy topping. Branches break easily.
Tulip Tree	U	30	52	100	5,400	Wood splits easily.
Walnut, Black	S	20	40	80	9,500	Strong wood. Dies back under heavy topping.
Willow, Black	H	40	70	50	3,800	Very weak and brittle. Breaks easily in storms.
Willow, Weeping	S	48	72	50	3,800	A weak-wooded tree. Remove if possible.

Source: Asplundh Tree Expert Co.

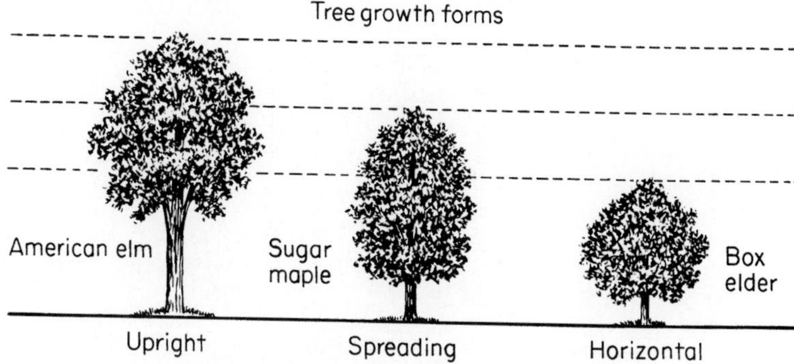

FIGURE 35.24 Tree-form types that are basic in planning and maintaining clearance for overhead lines. (*Courtesy Asplundh Tree Expert Co.*)

TABLE 35.2 Tree-Trimming Clearances, ft

Clearances	Secondary, 100–600 volts	Primary, 2400–4800 volts	Primary, 7200–13,800 volts
Topping			
Fast growers	6	8	9
Slow growers	4	6	7
Side			
Fast growers	4	6	8
Slow growers	2	4	6
Overhang			
Fast growers	4	8	12*
Slow growers	2	6	12*

*Remove if possible.
Source: Asplundh Tree Expert Co.

CHAPTER 36
DISTRIBUTION TRANSFORMER INSTALLATION

Transformers constructed to reduce the primary voltage down to the customer's utilization voltage are commonly referred to as *distribution transformers*. The distribution transformers may be single-phase, normally used for residential customers, or three-phase, normally used for commercial or industrial customers. The transformers are constructed for mounting on poles or racks, in vaults, or on pads constructed with concrete, plastic, or fiberglass and for direct burial in the earth. The construction, theory of operation, and the method of making connections to distribution transformers are described in Chap. 15, Distribution Transformers.

Distribution Transformer Size. The transformer capacity required to supply any given load may be estimated as follows:

Residential Loads. Residential loads may be estimated in accordance with Table 36.1, which gives the demand per house. As the table shows, a small house is taken as having a demand of 3500 watts, a medium-sized house 6000 watts, and a large house 15,000 watts.

A large home utilizing electricity for heating may have a demand of 20,000 to 30,000 watts. When several houses are served from the same transformer, there will be diversity between the peak use of energy, as shown in Table 36.2.

The size of distribution transformer required to serve serveral houses is determined by totaling the estimated watts for each house obtained from Table 36.1, dividing the total by 1000 to obtain kilowatts, and multiplying by the diversity factor obtained from Table 36.2.

TABLE 36.1 Wattage Demand for Residences to Be Used in Determining Size of Transformer Required

Size of house	Demand, watts
Small	3500
Average	6000
Large	15000

TABLE 36.2 Diversity Factor for Selecting Transformer Capacity Required to Serve Several Houses

Number of houses	Diversity factor
1	1
2	0.9
4	0.8
6	0.75
8	0.7
12	0.67

TABLE 36.3 Sample Calculation for Determining Proper Size of Transformer

Houses		Watts		Watts
2	×	3,500	=	7,000
2	×	6,000	=	12,000
2	×	15,000	=	30,000
6				49,000
49,000 watts/1000			=	49 kW
49 kW × 0.75 diversity factor			=	36.8 kW

If two small, two medium, and two large houses are to be served from one transformer, the size of the transformer can be determined by calculations itemized in Table 36.3.

The nearest size transformer adequate to serve the six houses with the loads shown in Table 36.3 would be 37.5 kVA. A distribution transformer can carry considerable overload without damage. A distribution transformer with a rated capacity of 37.5 kVA should be adequate to serve the six houses for several years even with normal load growth.

Commercial Lighting. In determining the size of transformer necessary for commercial lighting customers, such as stores of all kinds, it is first necessary to add the total connected load. This requires making a list of all lights, motors, air conditioners, etc., together with the wattage of each. The sum is the total connected load on the transformer. The average demand on the transformer will be 40 to 60 percent of the connected load, depending on individual cases. The demand represents the necessary transformer capacity.

Motor Loads. The transformer capacity required for motor loads is best found from the motor ratings. Add the ratings of all the motors to be supplied from one transformer or bank of transformers. Then from Table 36.4 obtain the proper percentage of the number of motors. Multiply the actual connected load by this percentage. The result is the required transformer capacity. Select the next larger standard rating.

If the transformer bank is to be open-delta-connected, multiply the rating as found above by 1.35 before selecting the transformer rating required.

The foregoing load estimates should be used only if there is no local information available for arriving at an approximate transformer rating. Load demands vary considerably for

TABLE 36.4 Percentages by Which to Multiply Total Connected Motor Load to Obtain Required Transformer Kilovoltampere Capacity

(Percentages for various numbers of motors and magnitude of connected horsepower load)

	Percentages for calculating kilovoltampere demand from the normal rated capacity of motors					
	Number of motors					
Total connected load, hp	1	2	3–5	6–10	11–20	20 or more
Less than 10	83	81	75	73		
10 to 50	78	76	70	68	67	65
50 to 100	74	72	65	62	61	60
100 to 300	72	70	61	59	58	57
Over 300	70	69	58	56	55	54

the various classes of customers throughout the country, and therefore local data should be collected to form a basis for estimating.

Load Checks. When a new subdivision in an urban area is first supplied with electric service, the smallest transformer used by the utility is generally installed. This may be a 15- or 25-kVA size. This is completed when the first house is connected to the secondary. When the load has built up to nearly full transformer capacity, load measurements are made on the transformer to see what the actual load is.

The customer's monthly kilowatthour consumption as determined by meter readings made for billing the customer can be analyzed automatically by a computer to determine the distribution transformer loading. If a transformer is over- or underloaded by 165 percent or 50 percent, respectively, or other values selected, the computer can print an exception report. The report can be used to initiate action to change the distribution transformers to the proper size. The distribution transformers must be properly loaded to efficiently utilize the investment and to provide the customer reliable service with good voltage. If a distribution transformer is overloaded, it may be desirable to install an additional transformer and divide the secondary and services to distribute the load properly on each transformer.

Distribution Transformer Location. The distribution transformer should be installed as near as possible to the center of the load area. The transformer secondary leads and services should be kept as short as possible to minimize the voltage drop and line losses. Transformers installed on a rear lot line at the point where the corners of four lots are common permit four houses to be served by services direct from the transformer whether the transformer is mounted on a pole, on a pad, or in a vault. If additional houses are to be served from one transformer, the secondaries can be extended along the rear lot line.

Commercial loads are usually served by an individual transformer or a three-phase transformer bank. The transformers should be installed as close as possible to the load to minimize the secondary and service voltage drop.

Figure 36.1 illustrates the proper location for installing a distribution transformer on a rear lot line. The transformer pictured is connected to serve six houses and will serve six additional houses when the subdivision is completed. The same plan is followed when the distribution transformers are installed along the street in front of the houses.

Distribution Transformer Grounding. The term *ground* is used in electrical work to refer to the earth as the zero potential. A ground actually consists of an artificial electrical

FIGURE 36.1 Pad-mounted distribution transformer installed on rear lot at the corner of the lots. (*Courtesy MidAmerican Energy Co.*)

connection to the earth, having a very low resistance to the flow of electric current. To ground a circuit or apparatus, therefore, means to connect it to earth. All distribution transformer tanks and one conductor of all transformer secondary circuits should be grounded to protect life and property. The points to be grounded in the various kinds of circuits are illustrated in Chap. 15, Distribution Transformers.

If the distribution transformer has a two-wire secondary, either of the two wires can be grounded. In a three-phase transformer bank with a delta-connected secondary, the center of one of the transformer windings should be grounded. If the secondaries are Y-connected, the neutral of the Y is grounded.

The ground connection starts at the ground rod, usually driven 8 ft or more into the earth, and the connecting wire extends to the proper secondary terminal of the distribution transformer. Figure 36.2 illustrates a rod used in making a ground connection. The rod must be driven deep enough to penetrate moist soil all year round.

All service neutrals are grounded at the customer's service entrance. The usual ground is the customer's steel water main and/or a driven ground rod. If the water system is not a good ground source, the customer must have a connection to a driven rod at the service entrance.

The low-voltage secondary circuit is grounded to guard against an excessive voltage being impressed on that circuit from an external source. A conductor of the primary mains may break and fall upon the secondary conductors below it. The greatest danger, however, exists in the breaking down of the insulation in the distribution transformer, thereby bringing the secondary winding in contact with a high-voltage primary winding. If a person should then touch a bare part of the secondary circuit and should also happen to be in contact with a radiator, bathtub, gas range, or water faucet, he would be likely to receive a very severe shock. Furthermore, any excessive voltage on the low-voltage fittings in lamp sockets, etc., is very apt to cause fire. If the secondary is grounded, however, a breakdown in the transformer insulation is very apt to blow the primary transformer fuses, thereby clearing the defective circuit from the source.

Distribution Transformer Mounting. The installation of distribution transformers in vaults, on pads, and buried in the earth is covered in the chapters describing underground

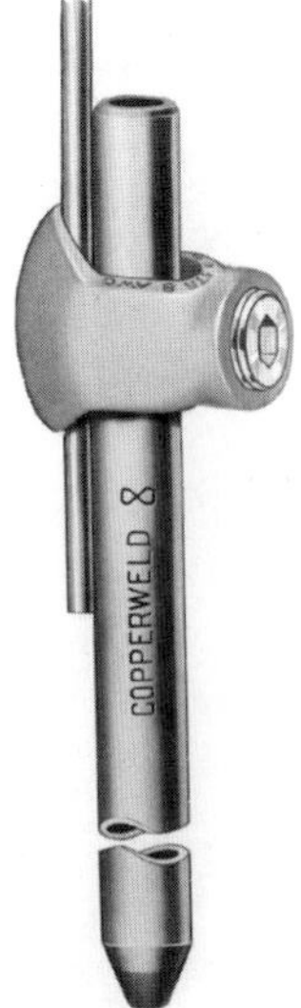

FIGURE 36.2 Ground rod and clamp used in making ground connection.

distribution installations. Small distribution transformers normally installed on overhead circuits are often referred to as pole-type transformers. These transformers may be fastened directly to the pole, hung from crossarms, mounted on racks or platforms suspended from poles, or placed on pads (usually concrete), enclosed with a fence. Figures 36.3 and 36.4 illustrate a distribution transformer bolted directly to the pole, Fig. 36.5 illustrates a bank of three transformers supported by means of cluster pole mounts, and Fig. 36.6 illustrates the installation of three transformers mounted on a platform. Platforms are built of any shape or size to suit local conditions. If the structure is to be permanent, the supporting members should be galvanized steel. Steel has a good appearance and requires no maintenance. The supporting members should be covered with wood flooring. The development of three-phase pad-mounted transformers has practically eliminated new installations of transformers on platforms supported by two poles. The costs of the two-pole platform transformer installations are excessive, and the platforms are visually unacceptable to the public. Pad-mounted transformers are safer for the linemen to install and maintain, and they reduce the hazard to the public. Very large transformers with exposed live parts, however, are set on pads on the ground (Fig. 36.7). A high fence must then be provided to keep persons from coming in contact with the live parts.

Small distribution transformers can be bolted directly to the pole by means of brackets. The transformer shown in Fig. 36.8 is direct mounted on a pole. The transformer is equipped with a partial-range current-limiting fuse and a direct-connected surge arrester. The transformer is suspended by a line truck boom-tip winch line which was used to lift the transformer from the material trailer fastened to the back of the line truck.

The pole-type transformers are normally raised to the mounting position with a boom and winch on the line truck (Fig. 36.9). Installation of distribution transformers on poles located on rear lot lines in residential areas or other locations inaccessible for line trucks requires the use of a block and tackle or a portable winch to raise the transformers (Figs. 36.10 and 36.11).

The distribution transformer should be handled carefully. Bushings and other equipment on the transformer can be damaged easily. The windings of the transformer may be

FIGURE 36.3 Distribution transformer mounted directly on pole. Transformer is rated 25 kVA, 7620-120/ 240 volts. Note service wires extending from the pole near the transformer to the house in the background of the picture. The service wires are dead-ended on the side of the house. The customer's service-entrance wires protruding from the weatherhead and conduit on the side of the house connect to the 120/240-volt, single-phase service wires.

FIGURE 36.4 Close-up view of a distribution transformer mounted directly on the pole. Note the insulated protective wildlife covering installed on the current-limiting fuse.

FIGURE 36.5 Three-phase transformer bank supported on pole by means of cluster-mount brackets. Single pole must be set 1 to 3 ft deeper than normal. Note three-phase four-wire secondary extending along the line from a bracket on the pole below the transformer bank.

FIGURE 36.6 Three single-phase transformers connected into a three-phase bank and mounted on a platform supported by two poles. The three-phase bank of transformers provides service to a commercial customer.

FIGURE 36.7 Three-phase transformer bank with exposed live parts set on ground. Note enclosing fence.

FIGURE 36.8 Distribution transformer designed to be supported by direct-to-pole mounting. Transformer is fastened to pole by bolts passing through brackets and pole.

damaged if the transformer is dropped from the truck or severely jolted. The lifting equipment, including slings used on the transformer, should be carefully inspected before the operation is started. Linemen and groundmen should stay in the clear while the transformers are raised into position. Protective equipment should be installed on the pole to make the primary area safe. Linemen must wear rubber gloves, protectors, and other appropriate personal protective equipment while working near the primary wires.

Figure 36.12 illustrates a complete conventional pole-type single-phase distribution transformer installation. Conventional distribution transformers should always be protected on the primary side with distribution fuses (cutouts) to disconnect the transformer in case of trouble. A current-limiting fuse can be mounted on each cutout to prevent the transformer lid from blowing off if a high-current fault starts to develop in the distribution transformer. A lineman could be seriously injured when working on the transformer pole if a lid blew off the distribution transformer and he was sprayed with burning oil. The cutouts are mounted near the transformer so they will be readily accessible to the lineman.

When all equipment is mounted in the proper location, the ground and neutral connections should be completed. The secondary connections should be completed last, to eliminate the danger of a back feed. The primary connections should be completed by the lineman, wearing rubber gloves and rubber sleeves or using hotline tools determined by the voltage of the primary circuit and associated working practice. The lineman should not stand on the transformer or other grounded device to work on the energized circuits.

When the transformer has been energized from the primary feeder, the secondary connections can be completed. The lineman should check the secondary with a voltmeter to be sure it is deenergized before completing the connections. If the transformer is to be connected in parallel with an existing transformer, the secondary circuits must be properly phased out by using the voltmeter to be sure they can be connected without causing a short circuit. The voltmeter should always indicate zero voltage between the transformer low-voltage terminal and the secondary wire before the connections are completed. Linemen

FIGURE 36.9 Lineman installing distribution transformer on pole. Transformer is lifted in place by boom and winch line of line truck. Transformer is mounted on bracket clamped to the pole. (*Courtesy A. B. Chance Co.*)

must be especially careful to be sure the customer's neutral, or ground, wire is not connected to an energized transformer terminal or secondary wire, to prevent a short circuit and damage to the customer's equipment.

A voltmeter or test lamp can be used to check connections if a single-phase transformer is to be connected in parallel with an existing transformer installation. Connect the voltmeter or lamp across a live secondary lead to test for proper operation. After the ground or neutral connection has been completed and the transformer is energized, test for voltage between one of the remaining secondary mains as shown in Fig. 36.13. If this is the proper connection, the voltmeter will read zero or very near zero. To avoid confusion, it is best to connect this lead immediately to the secondary main. Now proceed to the last lead and last secondary main. This, of course, should also read zero. After completing this connection, the new transformer is in parallel with the other transformers connected to the same secondary mains and will share the load connected thereto.

Figure 36.14 shows the procedure to be followed in connecting a two-wire 120-volt transformer to a three-wire 120/240-volt secondary. All connections should be completed carefully in accordance with the instructions and checked for proper tightness.

Three-phase transformer connections must be completed properly to obtain the voltages desired and to prevent an accidental short circuit. The primary connections for a wye-wye,

FIGURE 36.10 Capstan portable hoist designed to be fastened to a pole. The hoist can be powered by a motor operated electrically, hydraulically, or with gasoline. The electric motors can be constructed for 120-volt ac or 12-volt dc operation. (*Courtesy A. B. Chance Co.*)

FIGURE 36.11 Lineman is using a capstan hoist to raise a distribution transformer to the proper mounting height on a pole with the aid of a block mounted temporarily near the top of the pole. The hoist is used to lift equipment at a location not accessible by a vehicle. (*Courtesy A. B. Chance Co.*)

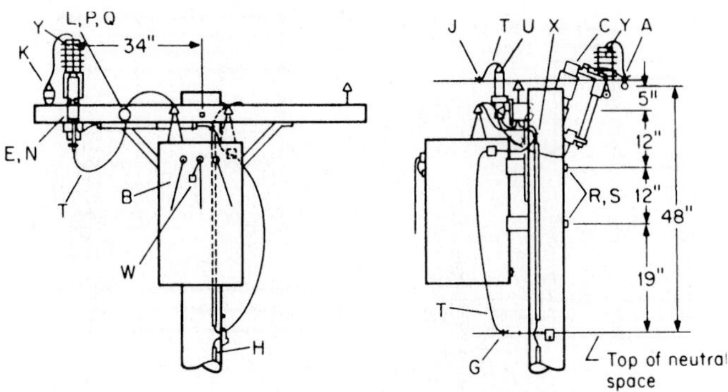

Phase to neutral 2400–7620 volts

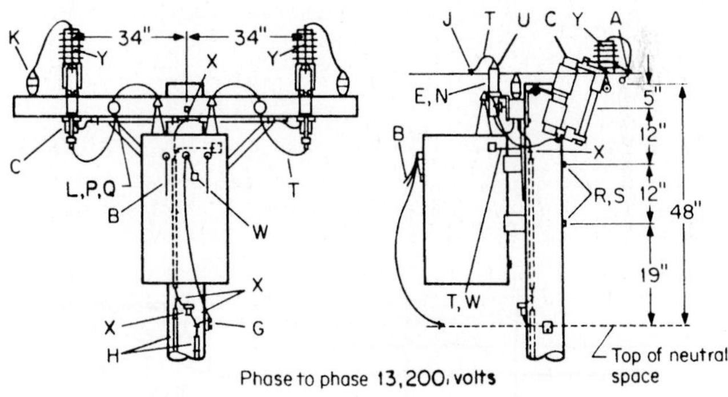

ITEM

A. Live-line clamp connections
B. Single-phase transformer secondary taps
C. Cutouts and fuses, 15 kV and below
E. Arresters
G. Conn. wedge (#4 Sol CU-neut)
J. Conn. wedge (4 Sol CU to pri.)
K. Assembly, insulator, post type, 12.5 kV
L. Conductor ties (conventional)
N. Bracket, cutout and arrester, Arm MTG
P. INS, pin, 12 kV top groove
Q. Pin, steel INS, lag screw, for XA
R. Bolt, machine, 5/8 x 12 in.
S. Washer, SQ flat, 2 1/4 in.
T. Wire, #4 WR SD Sol CU
U. Lightning arrester
V. Screw, lag, 1/4 x 2 1/2 in.
W. Ground transformer case
X. Conn. wedge, (4 Sol CU – Sol CU)
Y. Partial-range current-limiting fuse

FIGURE 36.12 Single-phase 25-kVA conventional distribution transformer installation for 13,200-volt or lower voltage distribution primary feeder circuit.

wye-delta, or delta-delta transformer bank can be completed and the transformer energized without danger of a short circuit providing one corner of the delta secondary is left open. A voltmeter must be used on the wye-wye transformer bank to measure the phase-to-phase voltages for proper magnitude and balance. The voltmeter or test lamp must have a voltage rating equal to twice the secondary line voltage to safely check the delta-delta voltage. The voltmeter or test lamp is connected across the open corner of the secondary delta windings (Fig. 36.15). If the connection of the delta is correct, the lamp will not burn and the voltmeter will read zero or almost zero. But if one of the transformer's connections is reversed,

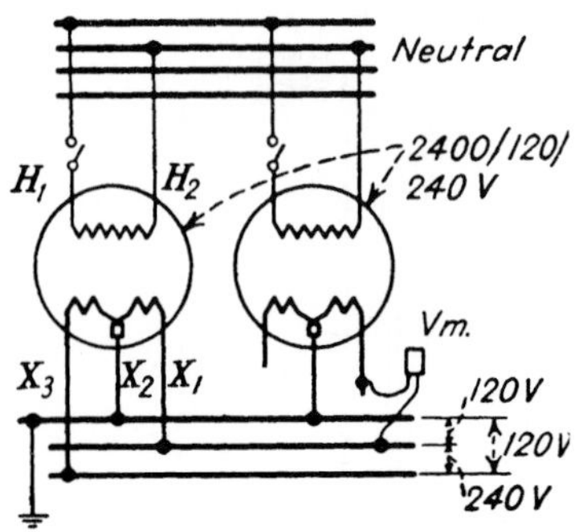

FIGURE 36.13 Paralleling two single-phase transformers each for 120/240-volt three-wire service by using voltmeter or test lamp.

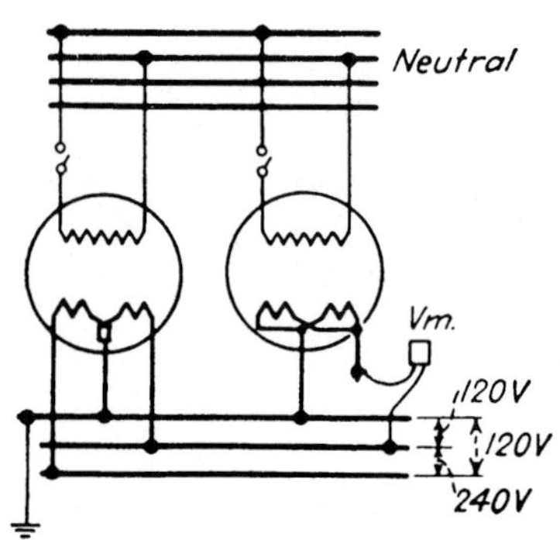

FIGURE 36.14 Paralleling a single-phase transformer two-wire, 120 volts with another connected three-wire, 120/240 volts by using voltmeter or test lamp. One of two possible connections.

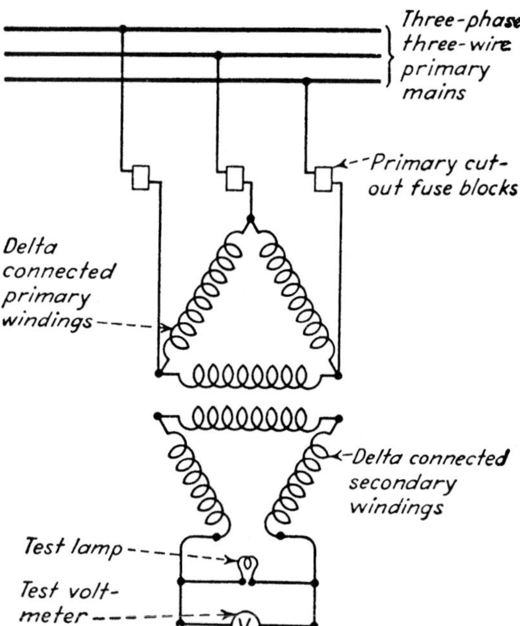

FIGURE 36.15 Use of lamp or voltmeter in testing delta connection on three-phase transformer bank. When lamp tests dark or voltmeter reads zero, the delta connection is correct.

FIGURE 36.16 Exterior view of a CSP distribution transformer showing primary bushing, lightning arrester, secondary terminals, tank ground connection, and secondary circuit breaker operating handle. (*Courtesy ABB Power T&D Company Inc.*)

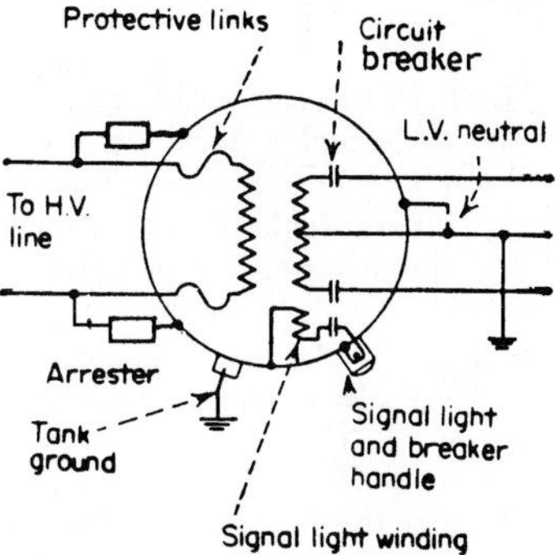

FIGURE 36.17 Schematic diagram of connections of a typical CSP distribution transformer. Note primary protective links, secondary circuit breaker, lightning arresters, signal lamp winding, and circuit breaker handle.

the lamp will burn brightly. If the voltmeter is used instead of the lamp, it will indicate a voltage equal to twice the voltage of one of the secondary windings. Under these conditions the delta must not be closed, for a severe short circuit would result. To correct the connection, reverse the leads from one of the transformer secondaries and repeat the test. If this does not correct the connection, reverse the secondary leads of another transformer and repeat the test. When the lamp fails to burn or the voltmeter reads zero, the delta connection is correct.

On higher-voltage secondaries, a small step-down potential transformer with a lamp or voltmeter connected to its secondary can be used while following the procedure described above.

The CSP Distribution Transformer. The *completely self-protected* (CSP) distribution transformer requires no auxiliary protective equipment except for current-limiting fuses. Primary fuses and lightning arresters are included with the transformer and therefore do not have to be supplied or mounted separately. Figure 36.16 shows an exterior view of a CSP transformer, and Fig. 36.17 is a diagrammatic sketch of the transformer and its component parts. Such transformers are equipped with support brackets for direct mounting to a pole; but when they are provided with hanger irons, they can also be hung from crossarms.

The following features can be incorporated in a CSP transformer:

1. Primary-voltage lightning arresters. These are supported externally on the tank.
2. Primary protective links or fuses in series with line leads serve to disconnect the transformer in case of internal transformer fault.
3. Overcurrent protection in the secondary leads provided by a low-voltage circuit breaker. This breaker is mounted above the core and coils but operates under oil.

4. Overload signal lamp, which becomes lighted when the maximum safe operating load on the transformer is exceeded or when the secondary circuit breaker has tripped automatically.
5. An external operating handle for opening or closing the secondary circuit breaker or for resetting the signal lamp and breaker-tripping mechanism.
6. No-load tap changer. This tap changer provides four 2½ percent voltage taps, which give a choice of four values of transformation ratio and compensate for voltage drop on long distribution lines. The operating handle is above the oil level, but the taps and contacts are under oil. The taps should be changed only when the transformer is de-energized.

To install a transformer of this type, it is only necessary to ground the neutral conductor, connect the high-voltage terminals to the primary line, and connect the secondary terminals to the secondary mains. Figures 36.18 through 36.20 illustrate typical installation details to obtain single-phase and three-phase secondary service using CSP single-phase distribution transformers manufactured for line-to-ground primary connections.

Phase Sequence. Phase sequence is the sequence or order in which the three voltages of a three-phase system appear. The phase sequence of a three-phase system is often desired in order to:

1. Determine the direction of rotation of polyphase motors
2. Determine the proper connections for paralleling three-phase transformer banks, generators, and power buses
3. Determine that the phase sequence is not changed when a three-phase transformer installation is replaced
4. Determine the proper connections for watthour meters, instruments, and relays

One type of phase-sequence indicator, like that illustrated in Fig. 36.21, makes use of a small capacitor connected in wye with two small neon lamps. When used, the terminals of the wye are connected to the conductors of the three-phase circuit of which the phase sequence is to be determined. A schematic diagram of the connections is shown in Fig. 36.22.

When the phase-sequence indicator is connected across a three-phase line as shown in Fig. 36.22, an unequal distribution of voltage occurs in the three arms of the wye network. This makes the voltage across one of the neon lamps considerably greater than that across the other for a given phase rotation. In fact, if the voltage across the darker lamp falls below the minimum breakdown or ignition value, the lamp will cease to glow completely. Consequently, only one of the neon lamps glow brightly at a time. The one that glows brightest indicates the phase rotation of the line voltages.

Potential transformers or potential devices must be connected in series with the phase-sequence instrument to check phase rotation of high-voltage lines.

Phase sequence or rotation is designated as 1-2-3 or 3-2-1. The first order is engraved on the case under the left-hand (*A*) lamp, and the second order is engraved under the right-hand (*B*) lamp. The leads which are used to measure the phase sequence of the three-phase line are identified with engraved markings of 1, 2, and 3.

To use the phase-sequence indicator, the voltage selector switch is first set to the desired voltage. The three leads of the instrument are then connected to the lines to be measured. One lamp should then glow much more brightly than the other. If the left-hand (*A*) lamp glows brightest, the phase sequence is 1-2-3, and if the right-hand (*B*) lamp glows brightest, the phase sequence is 3-2-1.

In order to check the sequence indicated, interchange any two of the leads. This changes the order, and therefore the first lamp should become dark and the second lamp should become bright. This also serves as a check on the condition of the lamps. If one of the lamps

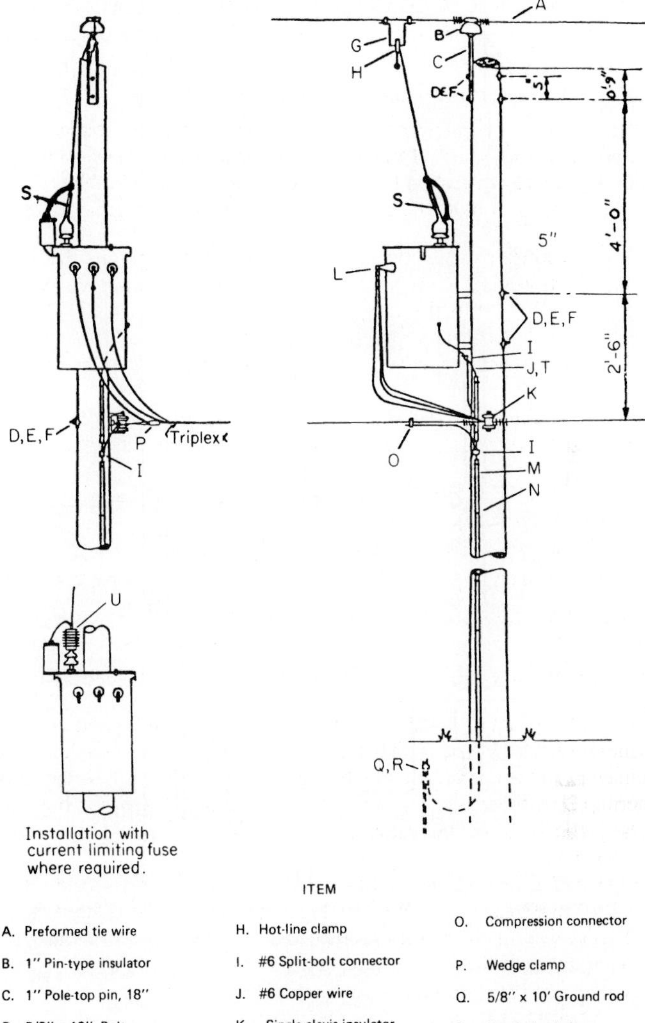

ITEM

- A. Preformed tie wire
- B. 1" Pin-type insulator
- C. 1" Pole-top pin, 18"
- D. 5/8" x 10" Bolt
- E. 2-1/2" Curved washer
- F. 5/8" Palnut
- G. Compression-type stirrup
- H. Hot-line clamp
- I. #6 Split-bolt connector
- J. #6 Copper wire
- K. Single clevis insulator
- L. Terminal pin
- M. Molding, ground wire
- N. Staples molding
- O. Compression connector
- P. Wedge clamp
- Q. 5/8" x 10' Ground rod
- R. Ground-rod clamp
- S. Squirrel & bird guard
- T. Staples, fence
- U. Partial-range current-limiting fuse

FIGURE 36.18 Specifications for installation of a single-phase CSP distribution transformer connected line to ground on a single-phase tap to a 13.2Y/7.62-kV three-phase four-wire grounded-neutral primary circuit.

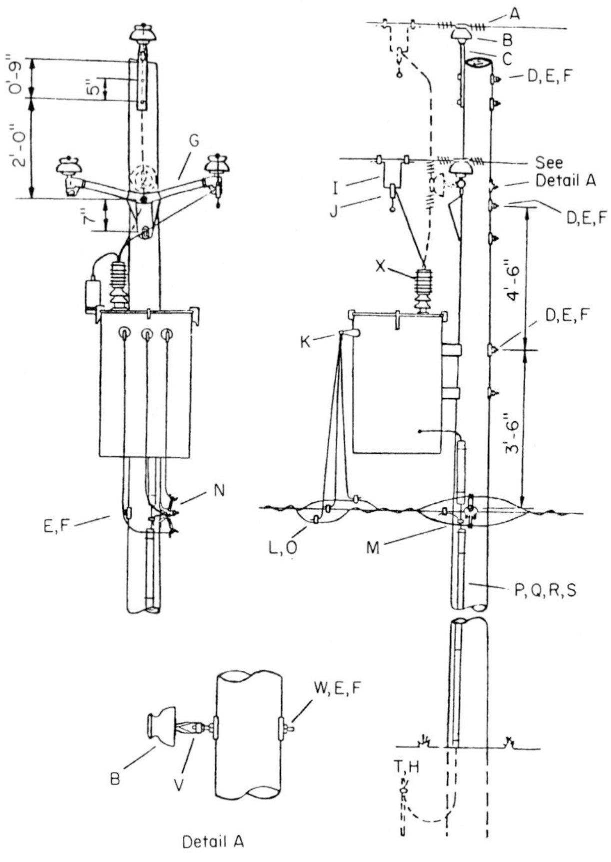

ITEM

A. Preformed tie wire
B. 1" Pin insulator
C. 1" Pole-top pin, 18"
D. 5/8" Bolt
E. 2-1/2" Curved washer
F. 5/8" Palnut
G. Epoxy-rod bi-unit assembly
H. Ground-rod clamp
I. Stirrup, compression type
J. Hot-line clamp
K. Pin, terminal
L. Compression connector
M. #6 Split-bolt connector
N. Aerial cable clamp
O. Connector cover
P. Staples, fence
Q. #6 Copper wire
R. Molding, ground wire
S. Staples, molding
T. 5/8" x 10' Ground rod
V. Transformer lead adapter pin
W. 5/8" Double-arming bolt
X. Partial-range current-limiting fuse

Details: Additional material required if transformer is connected to center-phase conductor

FIGURE 36.19 Specifications for installation of a single-phase CSP distribution transformer connected line to ground on a 13.2Y/7.62-kV three-phase four-wire grounded-neutral primary circuit.

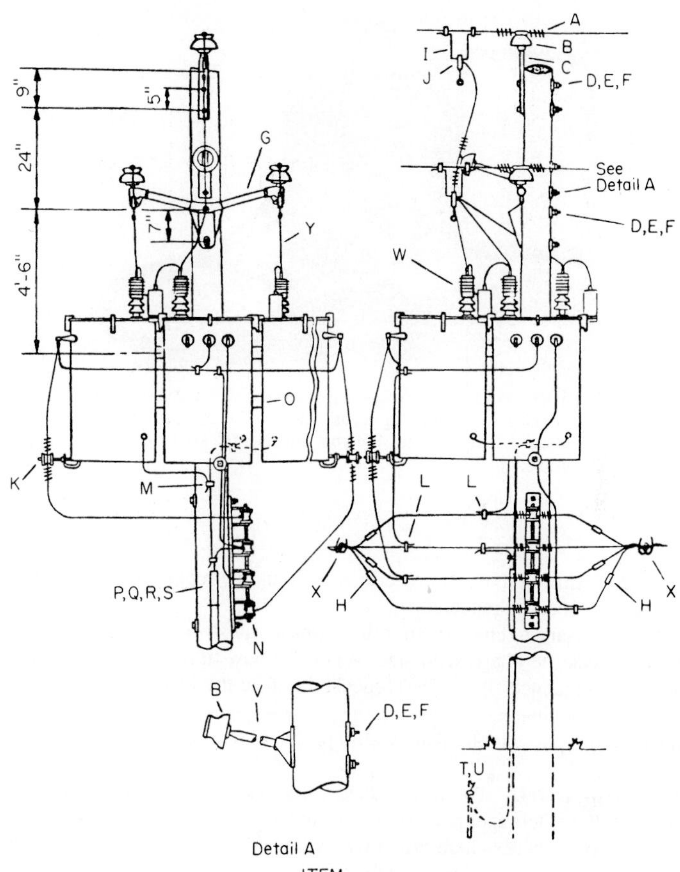

ITEM

A. Preformed tie wire

B. 1" x 5-1/2" Pin insulator

C. 1" Pole-top pin, 18"

D. 5/8" Bolt

E. 2-1/2" Curved washer

F. 5/8" Palnut

G. Epoxy-rod biunit assembly
1-1/2" D rod, 36" between conductors

H. Non-tension sleeve

I. Stirrup, compression type

J. Hot-line clamp

K. Transformer secondary lead bracket

L. Compression connector

M. #6 Split-bolt connector

N. 4 Spool rack

O. Transformer mounting bracket

P. #6 Copper wire

Q. Molding, ground wire

R. Staples, fence

S. Staples, moulding

T. 5/8" x 10' Ground rod

U. Ground-rod clamp

V. Small Epoxy-glass standoff

W. Partial-range current-limiting fuse

X. #33 Scotch electrical tape

Y. #4 Standed copper wire

FIGURE 36.20 Specifications for installation of single-phase CSP distribution transformers manufactured for line-to-ground primary connections to obtain three-phase four-wire grounded-neutral secondary voltages by using a wye-wye transformer connection from a 13.2Y/7.62-kV three-phase four-wire grounded-neutral primary circuit.

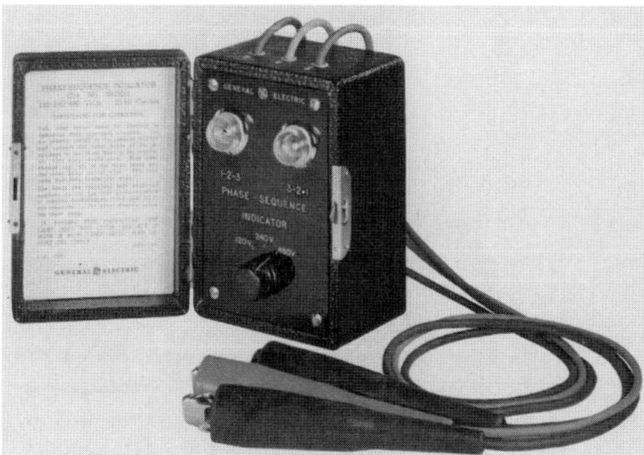

FIGURE 36.21 Phase-sequence indicator showing two neon lamps and three connecting leads. This particular model can be used on 120-, 240-, and 480-volt circuits. (*Courtesy General Electric Co.*)

should fail to glow during this test procedure, it should be replaced with a new neon lamp and the checking procedure repeated.

When it is necessary to change out a three-phase transformer installation, the lineman should always check the phase sequence. When the new transformer installation is completed, the phase sequence should be rechecked before the load is connected. The phase sequence must not be changed. Reversing the phase sequence would reverse the rotation of the customer's three-phase motors and would probably damage the customer's equipment.

Distribution Secondaries. The low-voltage wires from the distribution transformer low-voltage terminals to the terminal point of the customer's service-wire connections at a pole or an underground system pedestal are called *secondaries* or *secondary mains*. In well-designed distribution systems the voltage drop from the transformer secondary terminals to the end of the

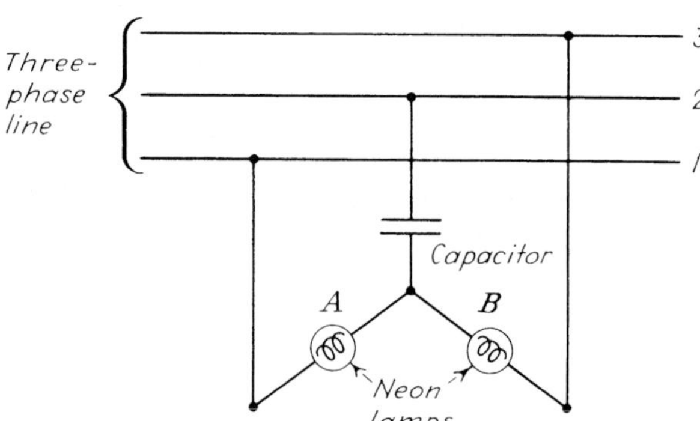

FIGURE 36.22 Phase-sequence indicator connected to three-phase line. Indicator consists of a capacitor and two neon lamps, *A* and *B*, connected in wye.

secondary in any direction should not exceed reasonable values. For secondary lighting mains, this drop is usually limited to 3 percent and for secondary power mains to 5 percent. These percentages apply to the voltage of the circuit. On a 120-volt two-wire circuit a 3 percent drop would equal 3.60 volts, and on a 240-volt circuit a 3 percent drop would equal 7.20 volts. A 5 percent drop would be equal to 12.0 volts on a 240-volt circuit.

The size of wire required to keep within the 3 percent drop in ordinary lighting circuits can be roughly determined from Table 36.5. This table is based on the number of "house spans" on either side of the transformer. By *house spans* is meant the product of the number of houses served from each pole times the number of spans between the pole and the transformer. The services, of course, will have to be connected so as to balance the three-wire secondaries, that is, the same number of houses should be connected between each outside wire and neutral. In addition, the houses connected to both sides should be distributed along the length of the secondary, so as to make the house spans connected to each outside wire equal if possible.

The estimating chart for size of three-wire secondary mains shown in Fig. 36.23 can be used to find the size of secondary mains required for residential areas.

Make a secondary layout sketch similar to that shown in Fig. 36.24, and show the kilowatt demand at each pole for the residences served from that pole. A demand of 3.5 kW is assumed for each residential customer. The kilowatt demand at each pole is calculated by multiplying the demand per customer by the number of services connected to that pole. The amount is shown on the sketch beside the pole. As an example, take the No. 2 pole. This pole serves four customers. At 3.5 kW each, the total demand at that pole is $3.5 \times 4 = 14.0$ kW.

Then the kilowatt spans are calculated for each pole by multiplying the demand at each pole by the number of pole spans between the pole and the transformer. The No. 2 pole in the sketch, for example, has a demand of 14.0 kW and is two spans from the transformer pole. The kilowatt spans therefore are $14.0 \times 2 = 28$ kW spans. These values of kilowatt spans are tabulated as shown in Table 36.6.

Now add the kilowatt spans served to the right and to the left of the transformer pole. The total is 38.5 each, for both left and right, showing that the transformer is correctly located. If these left and right values are not approximately equal, the transformer should be moved to the next pole in the direction of the larger kilowatt span. The kilowatt spans should be as nearly equal as possible. This is the same as saying that the transformer should be located as nearly as possible in the center of the load.

Now, to determine the size of wire required for the three-wire secondary mains, refer to the chart of Fig. 36.23 and find the value of kilowatt spans along the horizontal scale, which in this case is 38.5. Then move vertically upward on the chart until you reach the line marked "maximum allowable." The No. 1/0 copper or 3/0 aluminum wire is the proper size as it will give a voltage drop below 3 percent, which corresponds to 7.2 volts as shown on the right-hand scale of the chart.

The chart is based on a span length of 100 ft. If the actual average span length varies very much from this, the total kilowatt spans as calculated should be multiplied by the ratio of the actual average span to 100 ft before using the chart. Likewise, if the demand per customer is more than 3500 watts, this should also be taken into account by increasing the kilowatts per pole accordingly.

Accurate determination of the size of wire required for the secondary mains of lighting circuits can be made by use of the curves in Figs. 36.25 and 36.26. To use these curves, the kilowatt demand for each house and the distance in feet between the house and the transformer must be determined. Then the demand in kilowatts and the distance in feet are multiplied to get the kilowatt distance for that house. The kilowatt distance is likewise obtained for each house and the total for all the houses on one side of the transformer. This total may be called *KD*. Next add the total kilowatt demand on the same side of the transformer and

TABLE 36.5 Approximate Number of Allowable House Spans on One Run of Secondary Main for Any Given Wire Size and Type of Secondary

(The demand per house is assumed at about 3500 watts, and the span length is taken at 125 ft. If larger houses are to be supplied, they can be counted as two or three houses in the calculations. If longer or shorter spans are being used, the number of house spans can be changed in direct proportion. In this way the table can be used for various conditions.)

Wire size, AWG	Maximum allowable number of house spans	
	120-volt, two-wire	120/240-volt, three-wire
No. 4 copper or No. 2 aluminum	2	6
No. 2 copper or No. 1/0 aluminum	3	10
No. 1/0 copper or No. 3/0 aluminum	4	16
No. 2/0 copper or No. 4/0 aluminum	5	20
No. 4/0 copper or 350 kCMIL aluminum	8	31

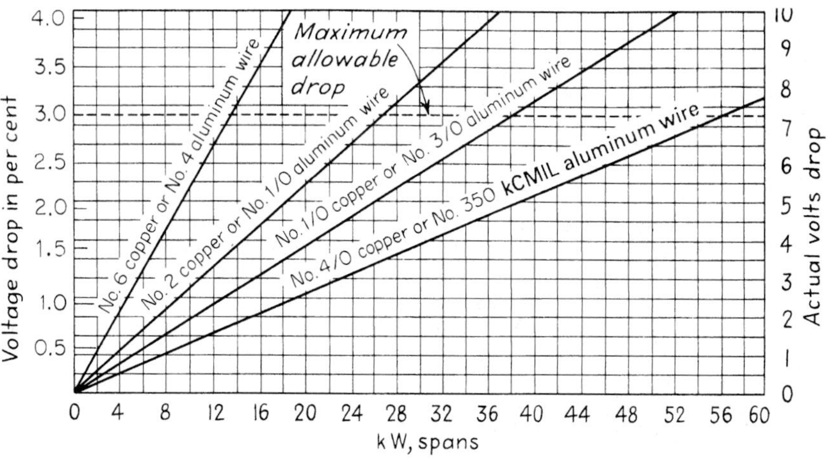

FIGURE 36.23 Chart for selecting wire size for 120/240-volt single-phase 60-Hz three-wire secondary mains. Chart is based on the use of conductors spaced 12 in apart. Assumed power factor is 80 percent. Span length is 100 ft. Unbalance factor is 1.2.

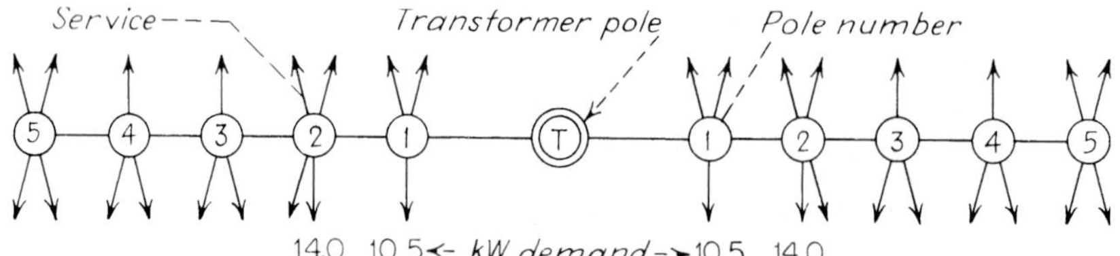

FIGURE 36.24 Secondary layout sketch showing transformer pole, services at each pole, and demand at each pole. Demand per customer is assumed to be 3.5 kW, or 3500 watts.

TABLE 36.6 Kilowatt Span Calculations for Secondary Layout

(The values shown in the kilowatt spans column are the products of the kilowatt demands at each pole multiplied by the number of spans from the transformer pole. If any load is connected to the transformer pole, it does not enter the calculation for determining the wire size. However, this load must be considered when selecting the size of the transformer required.)

Left side				Right side			
kW	Spans		kW spans	kW	Spans		kW spans
10.5 ×	1	=	10.5	10.5 ×	1	=	10.5
14.0 ×	2	=	28.0	14.0 ×	2	=	28.0
			38.5				38.5

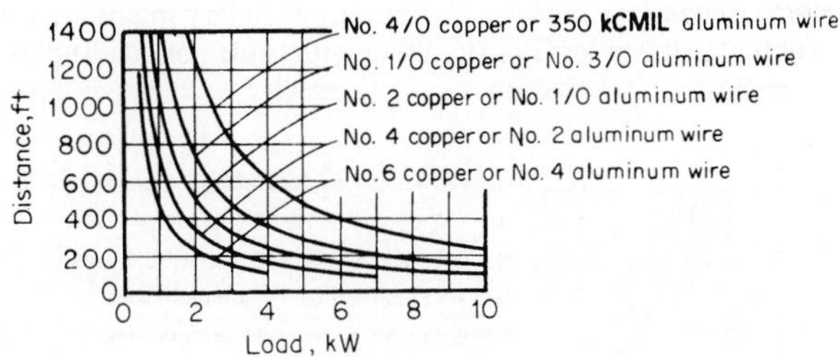

FIGURE 36.25 Load curves for 120-volt single-phase lighting secondary. Curves show distance that any load in kilowatts K can be transmitted with 3 percent drop on lines. Voltage at load, 120 volts; spacing of wires, 8 in; power factor of load, 95 percent; 60-Hz current.

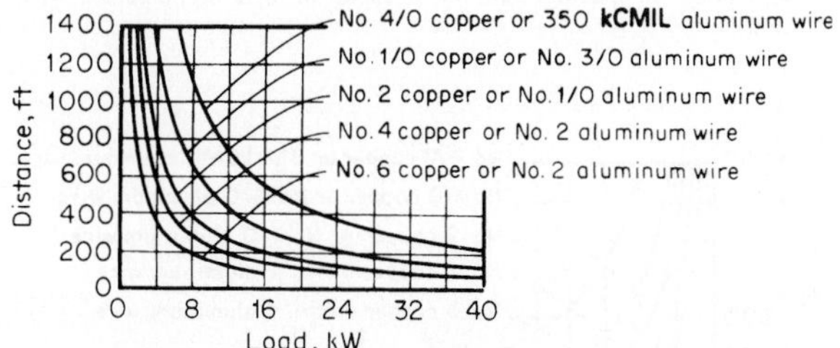

FIGURE 36.26 Load curve for 120/240-volt single-phase three-wire lighting secondary. Curve shows the distance in feet D that any load in kilowatts K can be transmitted with 3 percent voltage drop on line, allowing for unbalance of three-wire circuit. Voltage at load, 120/240 volts; spacing of wires, 4 in; power factor of load, 95 percent; 60-Hz current.

call it K. Now divide KD by K and get D, which is the equivalent distance in feet to which a concentrated load equal to K may be transmitted. With K and D known, refer to Fig. 36.25 for 120-volt two-wire mains or to Fig. 36.26 for 120/240-volt three-wire mains, and find the intersection of the lines from K and D. If the point of intersection does not fall directly on one of the curves, use the wire size of the curve next above the point. These curves are based on a 3 percent voltage drop.

In calculating the wire size for three-phase power secondaries, the procedure is the same as for lighting secondaries outlined above. In the case of power loads, however, the load can be more easily determined, as the motors to be supplied and their ratings are known. The procedure consists in getting the value of each motor load K and the distance D from the transformer. Then obtain the total of KD products, divide by K, and obtain the equivalent distance D. Now refer to Fig. 36.27 for 240-volt three-phase secondaries, and obtain the intersection of K and D. The curve on which this point of intersection falls is the proper wire size to use for a 5 percent voltage drop.

The procedure for single-phase power secondaries is the same as that already outlined for lighting and three-phase secondaries, except that the curves in Fig. 36.28 must be used. These curves are for 240 volts, single-phase, and allow for a 3 percent voltage drop.

Installing Overhead Secondary Mains. Overhead secondary mains are usually installed on vertical racks bolted to the poles (Fig. 36.29) or with triplex or quadruplex cable secured to the poles (Fig. 36.30).

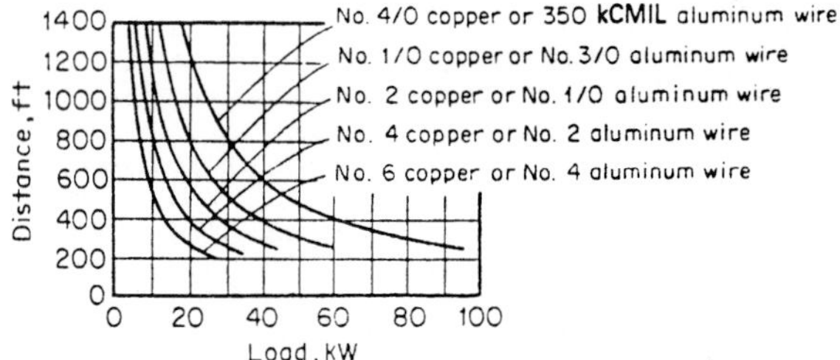

FIGURE 36.27 Load curves for 240-volt three-phase power secondary. Curves show distance in feet D that any load in kilowatts K can be transmitted with 5 percent voltage drop on lines. Voltage at load, 240 volts; equivalent spacing, 5 in; power factor of load, 80 percent; 60-Hz current.

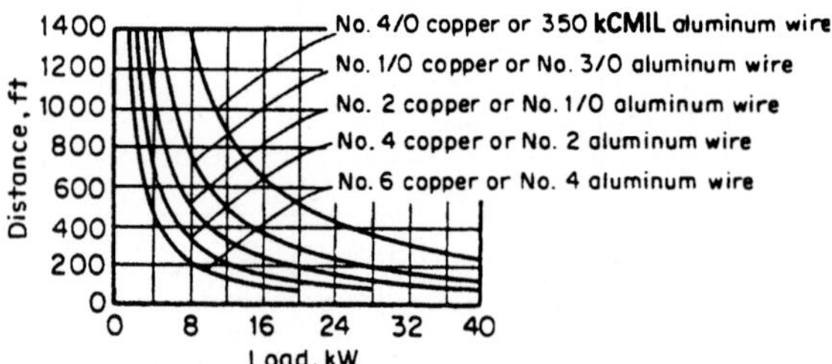

FIGURE 36.28 Load curve for 240-volt single-phase power secondary. Curves show distance in feet D that any load in kilowatts K can be transmitted with 3 percent drop on lines. Voltage at load, 240 volts; spacing of wires, 4 in; power factor of load, 94 percent.

FIGURE 36.29 Lineman working from bucket mounted on a truck making a connection to a conductor. Secondary mains above the lineman's head are secured to pole with vertical rack. The neutral or grounded conductor of the 120/240-volt three-wire single-phase secondary main is located on the white center insulator. Telephone cables are attached to pole near bottom of bucket supporting the lineman. CATV cable is located one foot below telephone cable.

FIGURE 36.30 A triplex secondary main operating at 120/240 volts single-phase extends from the transformer pole to adjacent poles along the street. The secondary conductors are attached to the pole below the transformer tank. An overhead service and an underground service connect to the secondary conductors on the transformer pole. The services supply power to the residences next to the transformer pole.

In stringing wires on bracket construction, the conductors are unreeled and passed through the rack, as shown in Fig. 36.31*a*. When the desired number of pole spans has been laid in place, the conductors are drawn up and dead-ended. As many as 10 spans can be drawn up at one time in this manner. Figure 36.32 shows a lineman pulling up a secondary main by means of block and tackle, and Fig. 36.33 shows the conductors properly dead-ended on the rack. Note the use of the terminal guy to counterbalance the pull of the line conductors.

The conductors are then lifted up to the insulator groove and tied in. The regular Western Union tie is generally used (see Fig. 36.31*b* and *c*). If the conductors are to be tied to the outside of the insulator, the Western Union tie is also used (see Fig. 36.31*d*).

In turning corners and at angles in the line, the position of the line wires on the insulators will be determined by the direction of the strain. They should always be placed so that the conductor is pulled against the insulator and not away from it. Figure 36.34 illustrates the correct positions for corner and angle construction.

Distribution Services. The low-voltage wires that connect to the secondary main conductors and extend to the customer's service entrance on the building are called *services*. The services may be overhead or underground, usually depending on the type of distribution system. However, underground services are frequently installed from an overhead secondary

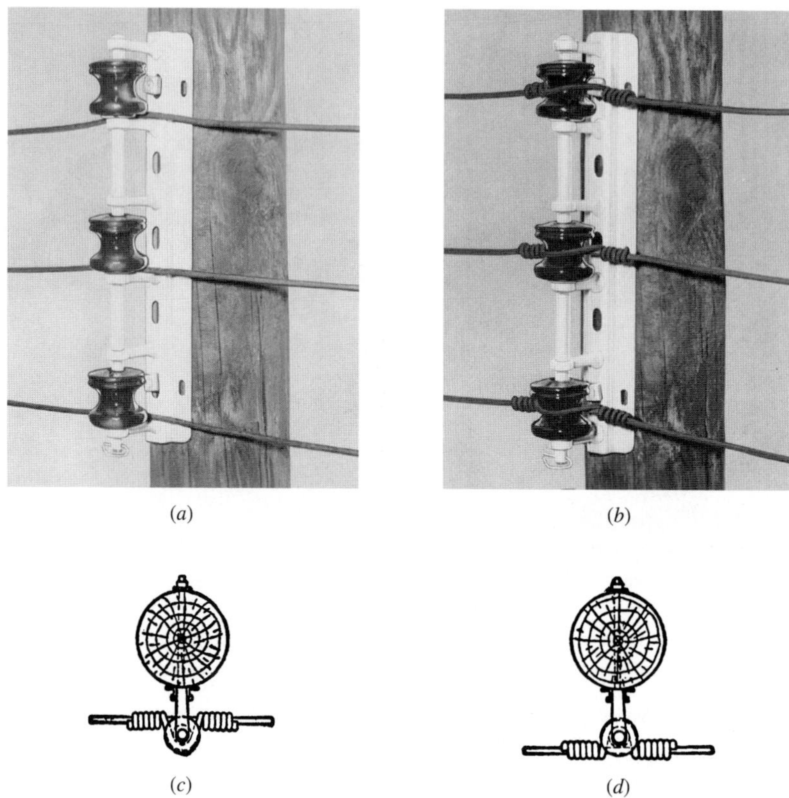

FIGURE 36.31 Stringing secondary mains. (*a*) Pulling line wires through rack, (*b*) line wires tied to inside of insulator, (*c*) top view of line wire tied to inside of insulator, (*d*) top view of line wire tied to outside of insulator.

main to the customer's building to eliminate the overhead wires across the customer's property. The size of the wire used for the customer's service is determined by the size of the electrical load. Single-phase 120/240-volt three-wire services are installed to all new homes.

Overhead services are usually No. 4 AWG aluminum conductor with a polyethylene insulation assembled in a triplex configuration (Fig. 36.35). Copper conductor No. 6 AWG is used in some instances instead of No. 4 AWG aluminum. Single-conductor wire may be used with supporting racks. The use of single-conductor wires on racks permits the use of weatherproof wire instead of insulated wire for overhead services, reducing wire costs. However, using triplexed cables with the close conductor spacing reduces the reactance and improves the voltage for the customer.

Underground services to new homes are usually installed with 2/0 aluminum conductor or larger to provide a minimum capacity of 200 amps. Underground services are installed large enough to prevent replacing them when the customer's load increases. Overhead services can be installed appropriately for the customer's load and replaced with larger conductors when the load increases, minimizing the investment.

FIGURE 36.32 Lineman pulling up a secondary-line wire by means by block and tackle. The pole is a dead-end pole from which a service connection is to be strung. Note terminal guy, which takes up the pull of the secondary mains.

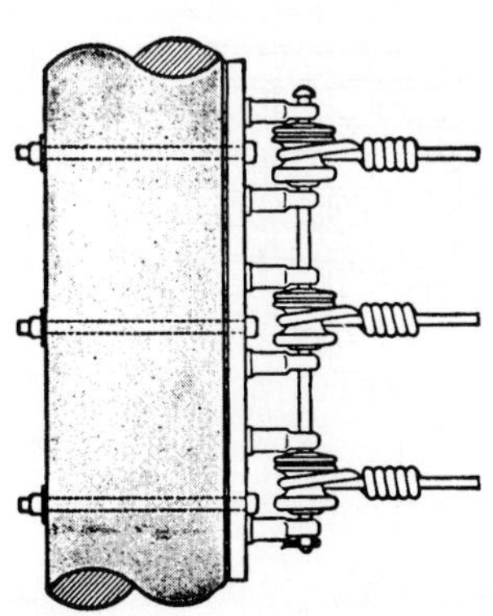

FIGURE 36.33 Completed dead-end on secondary-rack construction. Two or more through-bolts should be used on each rack.

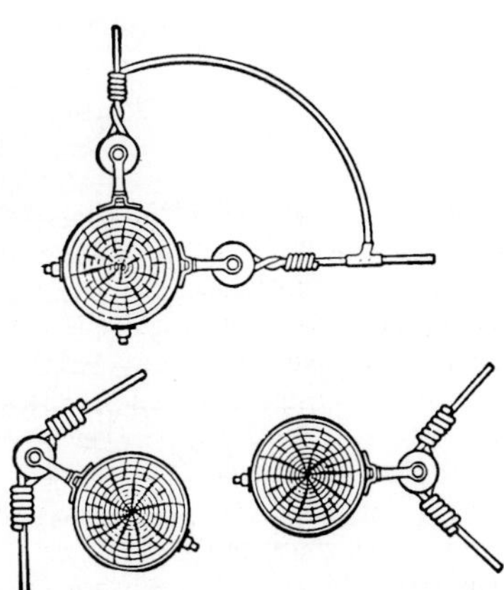

FIGURE 36.34 Suggested rack construction on corners and angles. Line conductor is always placed on insulator so that the pull is against the insulator as shown.

FIGURE 36.35 Lineman holding ends of three-conductor, No. 4 AWG aluminum, polyethylene covered triplex cable. Light-colored conductor will be used for neutral or grounded wire. Similar cable is used for overhead and underground low-voltage services.

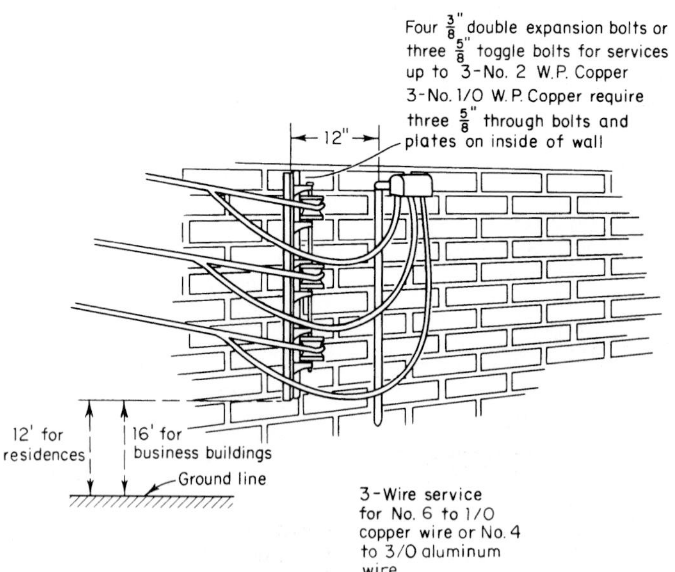

FIGURE 36.36 Details of service connection for three-wire services for residences and business buildings.

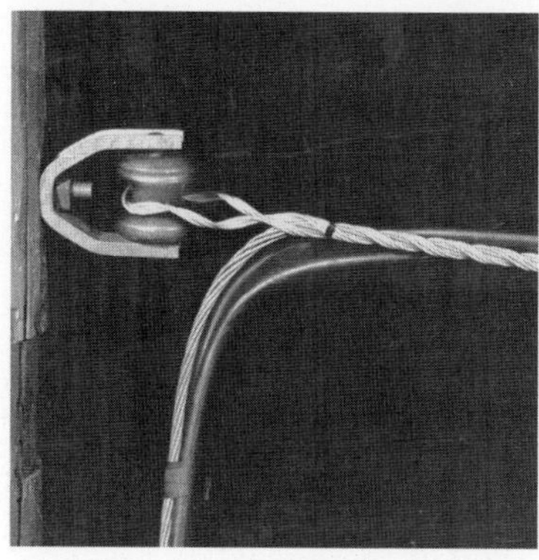

FIGURE 36.37 Typical service-drop cable terminated on the pole near residence with prefabricated grip. (*Courtesy Preformed Line Products Co.*)

Installing Overhead Services. Service brackets are installed on the customer's building, that is, the house, store, or factory, as the case may be, and on the pole to support the service (Figs. 36.36 and 36.37). Service supports should be mounted about 15 ft above the ground. The practice varies slightly with different companies. Figure 36.36 shows the practice of one company, which requires 12 ft for residences and 16 ft for business buildings, and Fig. 36.38 illustrates the requirement of another company.

The minimum required clearance to ground of the service-drop wires varies with the use made of the ground below the wires. The minimum clearance is 12 ft for voltages from 0 to 750 volts in areas restricted to pedestrian traffic. If the service wires cross an alley or street,

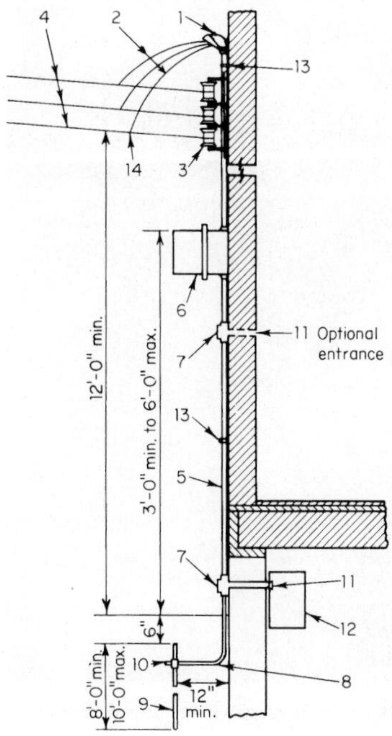

FIGURE 36.38 Installation of customer's outside meter and service entrance, three-wire 120/240-volt single-phase, from overhead distribution system. (1) Weatherhead-must extend above service wires. (2) Building service wires-size of service wires is determined by application of the National Electrical Code. Service wires must extend at least 36 in from weatherhead. The neutral wire must be identified. (3) Service Rack. (4) Service wires. (5) Galvanized rigid conduit (not water pipe). (6) Outdoor meter socket. (7) Galvanized conduit fitting with threaded hubs. (8) Galvanized conduit or water pipe of size to carry neutral wire. (9) A 1/2-in or larger copperweld ground rod or 3/4-in galvanized water pipe. Must be driven a minimum of 8 ft in ground. (10) Solderless connector (copper or bronze) of proper size to connect ground wire to rod or pipe. (11) Galvanized conduit nipple and end bushing. (12) Service-entrance fused switch, circuit breaker, or fuse box. (13) Galvanized conduit straps spaced approximately 4 ft apart. (14) Compression sleeve type connector. (*Courtesy MidAmerican Energy Co.*)

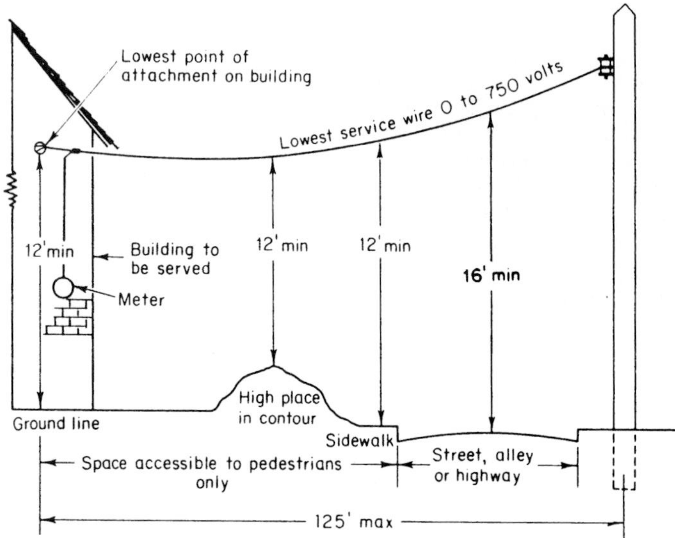

FIGURE 36.39 Sketch showing minimum ground clearances of service-drop wires, having voltages of 0 to 750 volts, above open spaces, sidewalks, and streets or alleys as required by the National Electrical Safety Code. Note that the span length is limited to 125 ft.

FIGURE 36.40 Meterman/Lineman preparing to insert watthour meter for newly connected customer's residence.

the minimum ground clearance is 16 ft. The span length is limited to 125 ft. The minimum clearances must be maintained under all loading, wind, and ice conditions. Figure 36.39 illustrates the minimum required ground clearances.

The service connection should be made at the nearest pole in the distribution line. If this distance should exceed 125 ft, an intermediate support pole should be provided.

After the service wires are unreeled, one of the linemen dead-ends the wires onto the service bracket, near the customer's service entrance wiring (Fig. 36.40). As soon as this is done, the other lineman climbs the service pole, pulls up the service wires, and dead-ends them on the distribution pole (Fig. 36.41). These wires need to be pulled up by hand only, as little strain should be placed on the service supports. If the service-drop wires are pulled otherwise than by hand, too great a strain may be put on the service brackets which may cause them to break or be pulled out of the building structure.

If the service drop consists of a self-supporting service-drop cable (Fig. 36.42), only the messenger wire need be dead-ended. Usually a special feed-through dead-end fixture is employed as shown in Figs. 36.43 and 36.44.

The type of support for the service wires on the distribution pole depends on the type of secondary main construction. If the secondary mains are supported by secondary racks, the same racks may also be used for the support of the service drops, as shown in Fig. 36.45.

The installation of the watthour meter is the next step in the running of the service. The meter is, as a rule, the property of the company and is therefore installed by the company at the time of making the service connection. The details of the installation of the meter socket (Fig. 36.46) will not be discussed here as this work is generally done by an electrician working for the customer.

FIGURE 36.41 Dead-ending service drop to new house on distribution pole. (*Courtesy American Electric Power System.*)

FIGURE 36.42 Self-supporting service-drop cable insulated with polyethylene or neoprene. Messenger serves as neutral conductor.

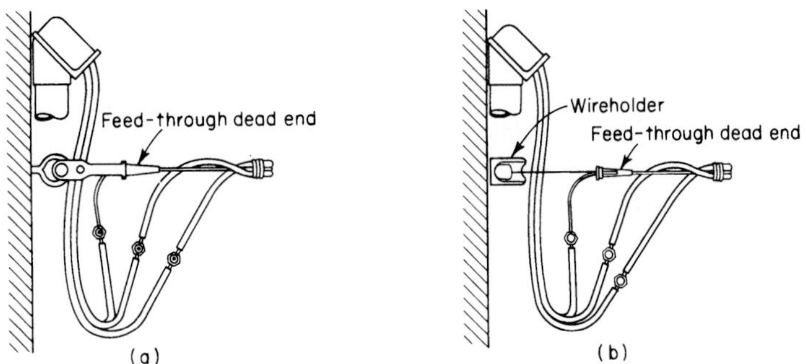

FIGURE 36.43 Dead-ending service-drop cable on consumer's building by means of (*a*) feed-through dead-end fixture attached to eye bolt, (*b*) feed-through dead-end fixture attached to wire holder.

Testing Service Wires for Polarity and Ground. Before the service wires can be connected onto the secondary mains, they must be tested with a voltmeter or *lamped out* for polarity. This must be done to avoid connecting the customer ground to either of the *hot* wires of the secondaries. To do so would produce a *short circuit* on the secondary mains. The customer ground wire, usually white or gray or bare, should always be connected to the neutral or grounded wire of the service.

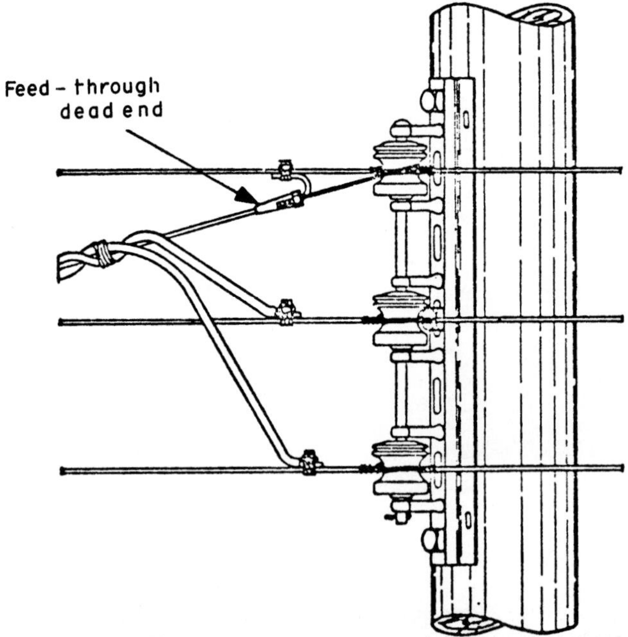

FIGURE 36.44 Dead-ending service-drop cable on distribution service pole by means of special feed-through dead-end attached to secondary-rack insulator spool.

FIGURE 36.45 Service-drop wires supported on secondary main bracket. In this instance neutral conductor of three-wire service is middle conductor. Note cable television (CATV) wires below power wires. (*Courtesy MidAmerican Energy Co.*)

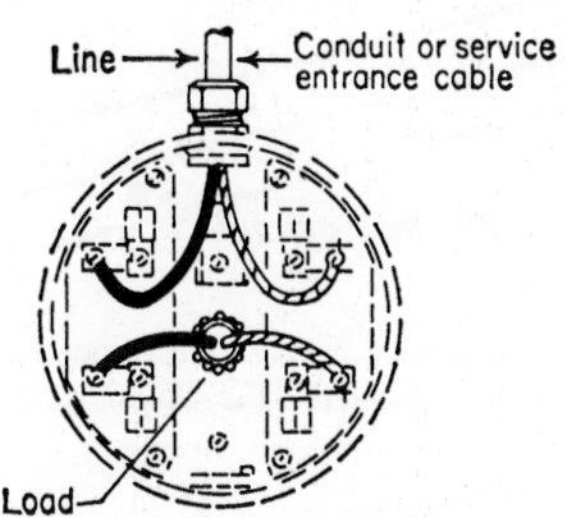

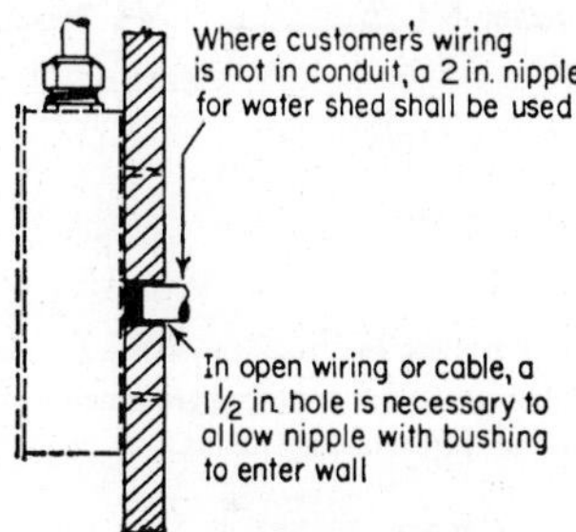

FIGURE 36.46 Details of typical meter socket installation. Leave 12 in of wire or cable for socket connections which will be made at the time meter is installed. Entrance and outlet to sockets shall be with conduit or cable and watertight connector terminating in such a manner as to be weather- and insect-proof.

To perform the polarity, or *lamping-out,* test, an incandescent lamp or voltmeter is used. This should have a voltage rating equal to the voltage across the *hot* wires of the secondary. If the service is a two-wire service, the test is made as shown in Fig. 36.47. The service wire, which is grounded in the customer's premises should be connected to the neutral of the secondary main. To determine which of the service-drop wires is grounded, the lamp is connected from one of the *hot* or *live* secondary main wires to one of the service wires. If the lamps burns, the service-drop wire to which the lamp is then connected is grounded. If the lamp does not burn, that service-drop wire is not grounded. If the service is a three-wire service, the procedure is the same except that three wires of the service must be *lamped out* (see Fig. 36.48). All one needs to remember is that the service-drop wire that makes the lamp burn is the grounded wire and should be connected to the secondary main neutral.

Connecting Service Drop to Secondary Mains. When the service-drop leads are lamped out, they are ready to be permanently connected to the secondary mains.

Connectors are used to obtain a good electrical connection. Taps to secondary mains, service entrance connections, jumper loop connections, etc., are generally made with

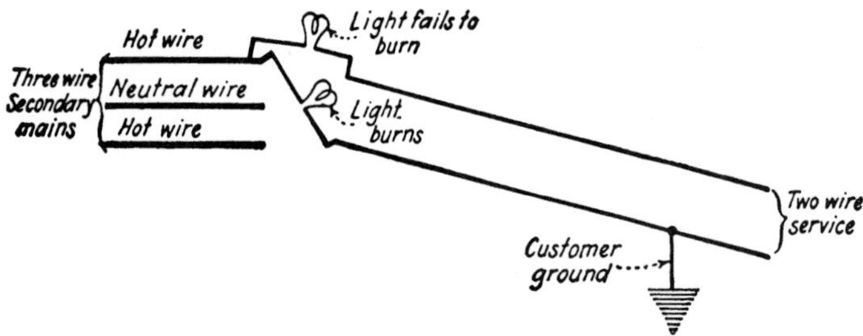

FIGURE 36.47 Lamp test for polarity on two-wire service. The service-drop wire that makes the lamp burn is the grounded wire which should be connected to the neutral or grounded wire of the secondary mains.

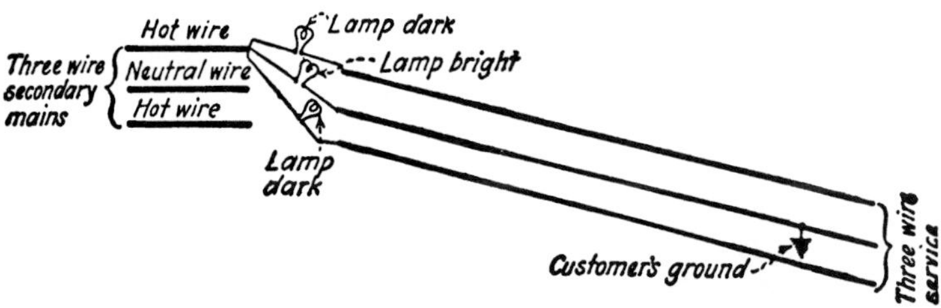

FIGURE 36.48 Lamp test for polarity on three-wire service. The service-drop wire that makes the lamp burn is the grounded wire which should be connected to the neutral or grounded wire of the secondary mains.

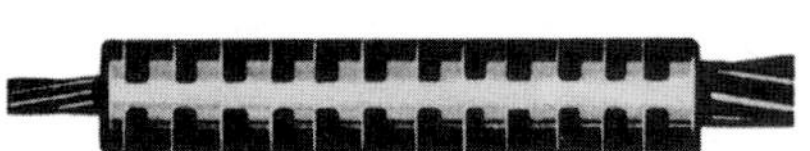

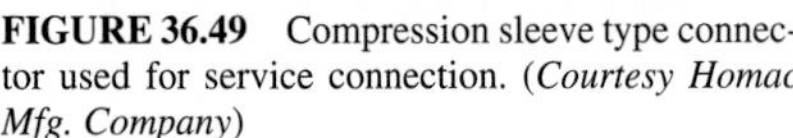

FIGURE 36.49 Compression sleeve type connector used for service connection. (*Courtesy Homac Mfg. Company*)

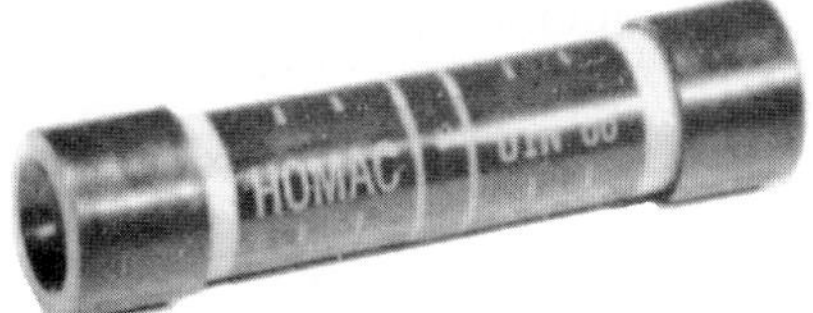

FIGURE 36.50 Insulated compression sleeve type connector used for service drop termination point. (*Courtesy Homac Mfg. Company*)

compression sleeve type connectors. The compression-type connector is a form of open-sleeve connector (Figs. 36.49 through 36.53) that is installed with a compression tool.

Connectors are easily and quickly applied. If the conductor has a weatherproof covering, this is first removed and the exposed conductor scraped clean and bright. The two ends are then inserted into the compression type connector and the proper compression tool is used to crimp the sleeve around the two conductors. Most compression connectors are covered and have inhibitor grease in the tube of the connector when received from the manufacturer. If the compression connector is not covered, it is coated with inhibitor grease and taped so that it will conform to the service-drop wires and secondary mains.

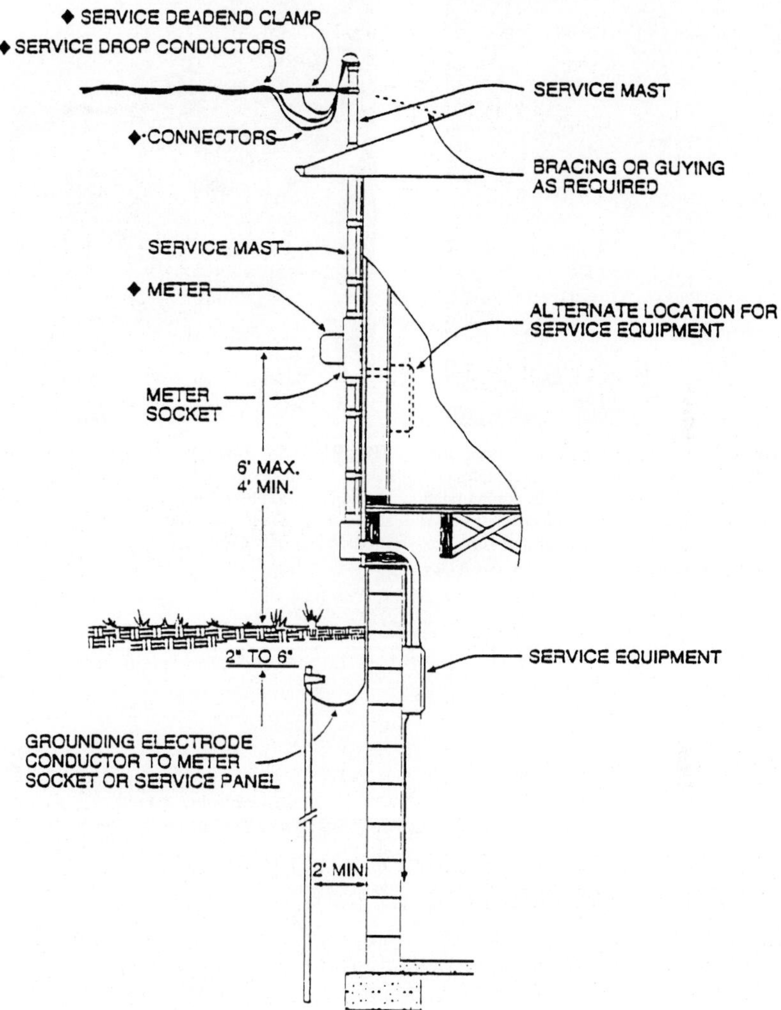

FIGURE 36.51 Sketch of service-drop cable connected to customer's service entrance loop with compression sleeve connector. (*Courtesy MidAmerican Energy Company*).

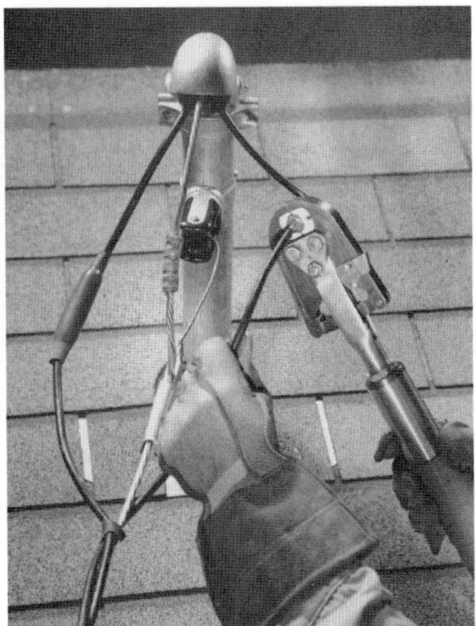

FIGURE 36.52 Lineman connecting service-drop cable onto service entrance loop with compression sleeve and power compressor. Note compression sleeve on bare neutral conductor and covered sleeve joint on left conductor. Bare neutral also serves as messenger supporting cable. (*Courtesy A. B. Chance Co.*)

FIGURE 36.53 Lineman connecting service-drop cable to secondary mains by use of crimpit tap sleeve and compression tool. Note bare neutral, which also serves as messenger wire and supports service drop. (*Courtesy Burndy Corporation.*)

CHAPTER 37

ELECTRICAL DRAWING SYMBOLS

Diagrams of electric circuits show the manner in which electrical devices are connected. Since it would be impossible to make a pictorial drawing of each device shown in a diagram, the device is represented by a symbol.

A list of some of the most common electrical symbols used is given below. It will be noted that the symbol is a sort of simplified picture of the device represented.

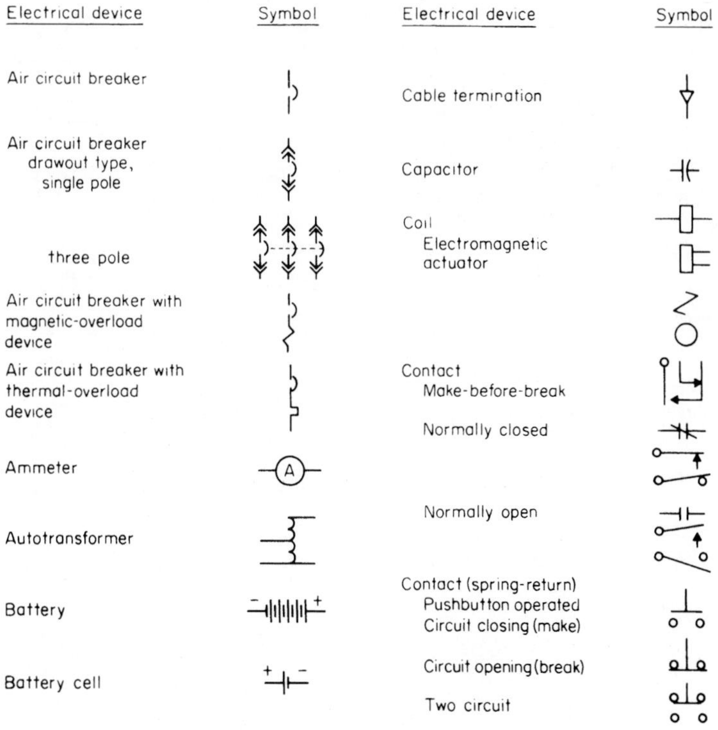

Electrical device	Symbol
Air circuit breaker	
Air circuit breaker drawout type, single pole	
three pole	
Air circuit breaker with magnetic-overload device	
Air circuit breaker with thermal-overload device	
Ammeter	
Autotransformer	
Battery	
Battery cell	

Electrical device	Symbol
Cable termination	
Capacitor	
Coil	
Electromagnetic actuator	
Contact	
Make-before-break	
Normally closed	
Normally open	
Contact (spring-return)	
Pushbutton operated	
Circuit closing (make)	
Circuit opening (break)	
Two circuit	

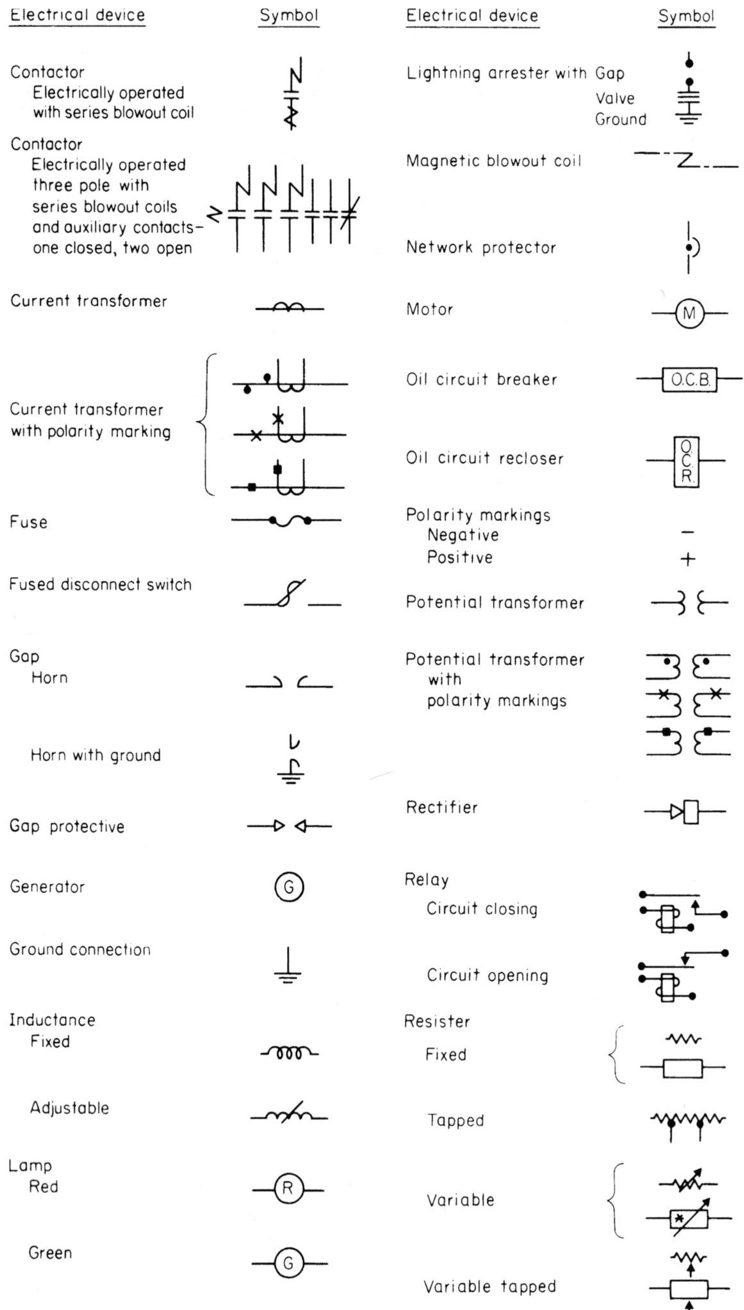
Electrical device
Symbol
Contactor
Electrically operated
with series blowout coil
Contactor
Electrically operated
three pole with
series blowout coils
and auxiliary contacts-
one closed, two open
Current transformer
Current transformer
with polarity marking
Fuse
Fused disconnect switch
Gap
Horn
Horn with ground
Gap protective
Generator
G
Ground connection
Inductance
Fixed
Adjustable
Lamp
Red
R
Green
G
Electrical device
Symbol
Lightning arrester with Gap
Valve
Ground
Magnetic blowout coil
Network protector
Motor
M
Oil circuit breaker
O.C.B.
Oil circuit recloser
O.
C.
R.
Polarity markings
Negative
Positive
Potential transformer
Potential transformer
with
polarity markings
Rectifier
Relay
Circuit closing
Circuit opening
Resister
Fixed
Tapped
Variable
Variable tapped

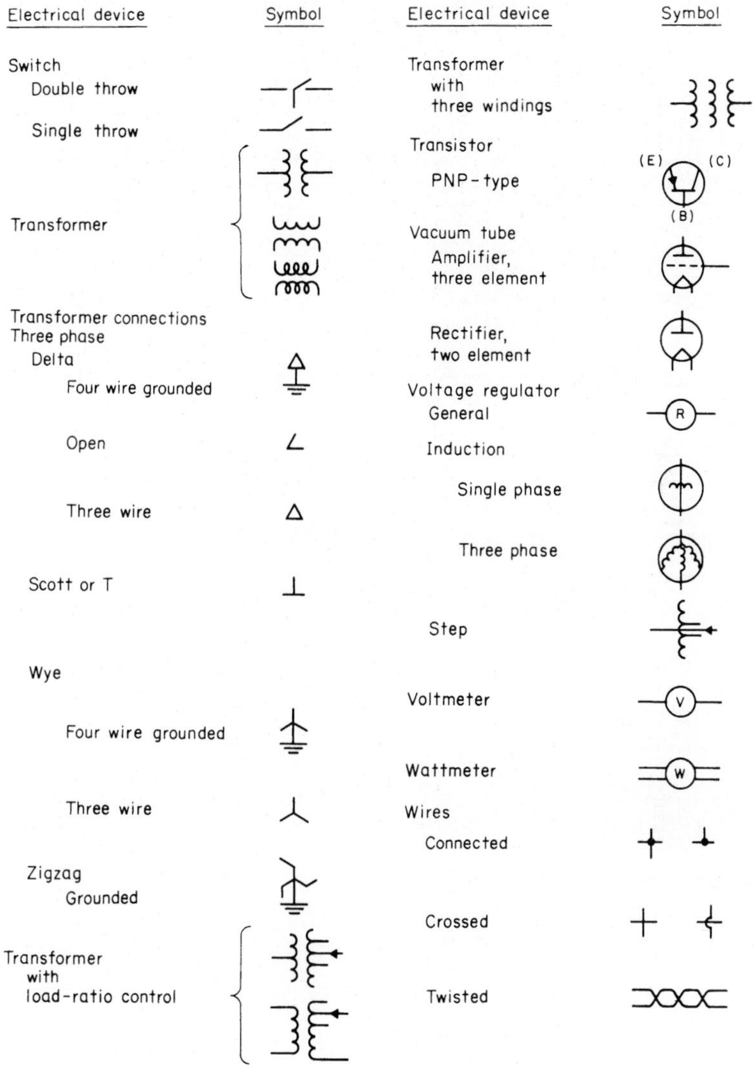

Electrical device	Symbol
Switch	
Double throw	
Single throw	
Transformer	
Transformer connections	
Three phase	
Delta	
Four wire grounded	
Open	
Three wire	
Scott or T	
Wye	
Four wire grounded	
Three wire	
Zigzag	
Grounded	
Transformer with load-ratio control	

Electrical device	Symbol
Transformer with three windings	
Transistor	
PNP-type	(E) (C) (B)
Vacuum tube	
Amplifier, three element	
Rectifier, two element	
Voltage regulator	
General	R
Induction	
Single phase	
Three phase	
Step	
Voltmeter	V
Wattmeter	W
Wires	
Connected	
Crossed	
Twisted	

The reader is encouraged to refer to American National Standards Institute Standard ANSI Y14.15-1966 (R1988), Electrical and Electronic Diagrams.

CHAPTER 38
SINGLE-LINE DIAGRAMS

Purpose. Most electric transmission and distribution circuits are three-phase and require three lines to show them schematically in detail. For many functions it is not necessary to have all the detailed information shown on the schematic diagram. Therefore single-line diagrams are utilized to represent three-phase circuits and equipment. A single-line diagram provides a quick method whereby lineman or cableman may analyze the complete facilities in schematic form.

Distribution Unit Substation Single-Line Diagram. A 13,200-4160Y/2400-volt unit-type distribution substation is illustrated by Fig. 38.1 in single-line form. Analyzing the unit substation single-line diagram, it can be determined that the substation has as its source an overhead 13,200-volt three-phase subtransmission electric circuit designated 13-61-T-1. The overhead line connects to the substation transformer through 300-kCMIL copper conductor, paper-insulated, lead-covered cables with pothead terminations on each end of the cable circuit. A dotted line representing the cable circuit indicates that it is installed underground. The power transformer has three potential transformers connected to the primary leads through fuses.

The high-voltage winding of the power transformer is connected delta, has a tap changer, and the low-voltage winding is connected wye with the neutral grounded. The transformer is rated 13,200-4160Y/2400 volts, 5000/6250 kVA. The capacity varies with the type of cooling. Oil-air cooling provides 5000 kVA of capacity, and forced-air cooling provides 6250 kVA of capacity. The tap changer operates under load. The current transformers in series with the 4160-volt bus connections to the power transformer provide a means of measuring the current flow in the power transformer secondary connections.

Cubicle No. 1 of the metal-clad switchgear contains a potential transformer that connects to the 4160-volt leads through a fuse. Cubicle No. 2 of the switchgear has three fused potential transformers that connect to the transformer secondary leads and three current transformers rated 1000/5 amps utilized for relaying and metering circuits. The power transformer's secondary leads connect to a 1200-amp air circuit breaker through draw-out contacts. The transformer's secondary circuit breaker provides a means of isolating the 4160-volt bus from the power transformer. Switchgear cubicle No. 3 contains an operating transformer to provide low-voltage power for auxiliary devices and three potential transformers complete with fuses for metering and relaying purposes. Sections 4, 5, 6, and 7 of the metal-clad switchgear each contain a feeder air circuit breaker and current transformers. Each of the feeder circuits exits from the substation through underground cables to supply distribution circuits operating at 4160Y/2400 volts.

Distribution Substation Single-Line Diagrams. Figure 38.2 is a single-line diagram of a distribution substation supplied by two 69,000-volt three-phase subtransmission circuits designated 66-58-59-1 and 66-58-A-1. The subtransmission circuits of 336,400-cir-mil

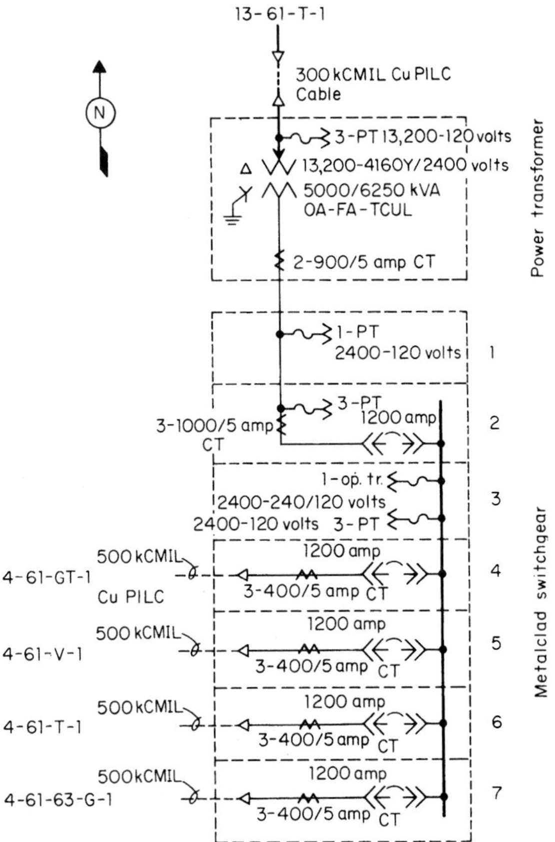

FIGURE 38.1 Distribution-unit-type substation single-line diagram.

aluminum conductor connect to the high-voltage structure of the substation overhead. Lightning arresters and potential transformers in the substation structure connect to the overhead line terminals. The potential transformers provide inputs for meters, relays, and auxiliary devices.

The subtransmission circuits are switched in and out of service utilizing the 69,000-volt 1200-amp oil circuit breakers connected in series with group-operated disconnect switches. The group-operated disconnect switches provide a means of isolating the oil circuit breakers. The disconnect switches are operated after the oil circuit breaker contacts have been opened. The substation 69,000-volt bus has a normally closed group-operated disconnect switch, making it possible to manually isolate the bus into two sections. A 69,000-volt group-operated disconnect switch provides a means for connecting a mobile substation to the 69,000-volt bus. The fused potential transformer connected to the bus provides a means of measuring bus voltage.

The two power transformers in the substation (rated 69,000-13,200Y/7620 volts, 20,000 kVA oil air-cooled, 26,700 kVA forced-oil air-cooled with tap-changing under-load equipment) are each switched in and out of service by an oil circuit breaker connected to

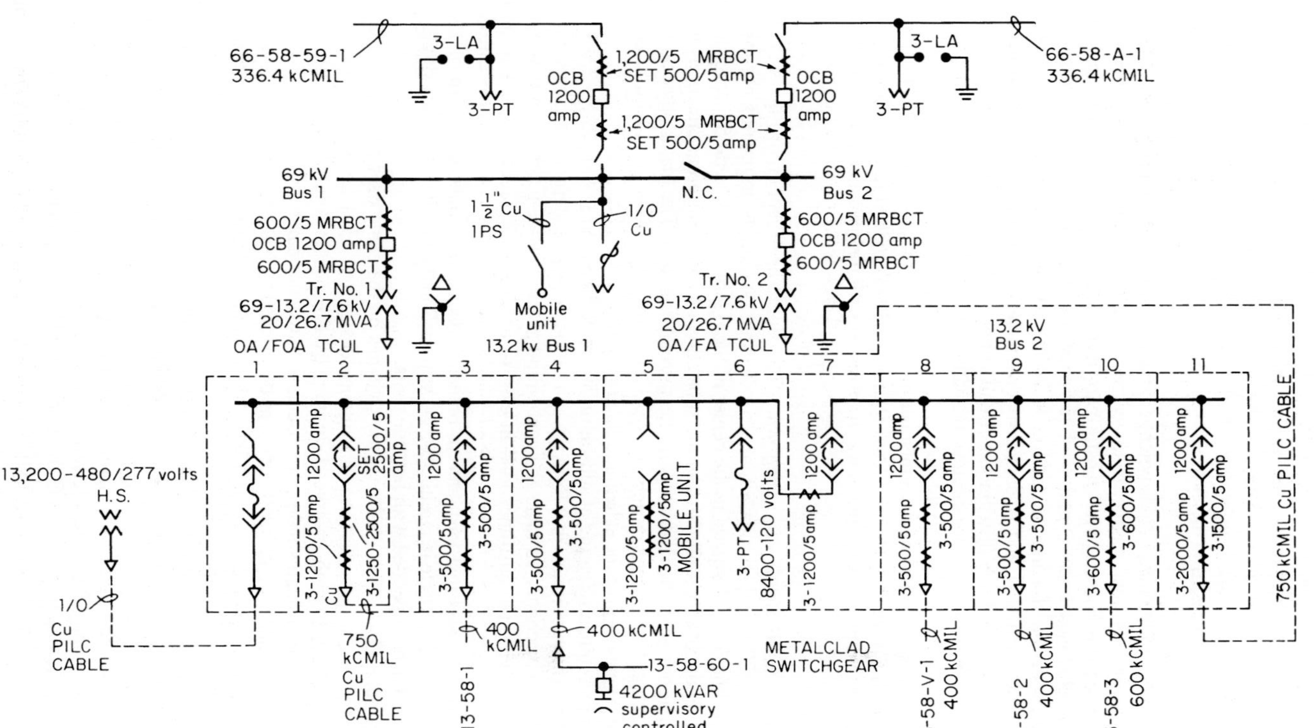

FIGURE 38.2 Distribution substation single-line diagram.

the high-voltage winding and an air circuit breaker in the metal-clad switchgear connected to the low-voltage winding. The 13,200Y/7620-volt leads from the power transformers connect to the circuit breakers in the metal-clad switchgear through 750-kCMIL copper conductor, paper-insulated, lead-covered cables installed underground and terminated with potheads. The current transformers in series with each power transformer air circuit breaker provide a means of measuring the current magnitudes and detecting short circuits with protective relays.

Cubicle No. 1 of the metal-clad switchgear has a group-operated disconnect switch and draw-out power fuses that connect to underground cables terminated with potheads providing a source to the house service transformers rated 13,200-480Y/277 volts. Cubicle No. 7 of the metal-clad switchgear contains a bus-tie air circuit breaker, providing a means for sectionalizing the bus for maintenance or to isolate a bus section in case of a bus fault. Switchgear units 3, 4, 8, 9, and 10 each provide a means of switching a distribution circuit in and out of service. Each distribution circuit connects to an air circuit breaker through underground cables. The feeder circuit originating in cubicle No. 4 has a capacitor bank connected to it in the yard of the substation.

Switchgear unit No. 5 provides a means for connecting the 13,200-volt bus to a mobile substation with a power transformer rated 69,000-13,200Y/7620 volts. The mobile substation is utilized to replace one of the power transformers for maintenance work or in the case of a transformer failure.

Distribution Circuit Single-Line Diagram. The single-line diagram of Fig. 38.3 shows the 13,200Y/7620-volt three-phase four-wire distribution circuits originating at substation 59 in single-line form. A square in the center of the figure with numeral 59 represents the substation. Four circuits originating at the substation extend in different directions to supply electrical power to the customers in the surrounding area. The major physical terrain details—such as Duck Creek, Kimberly Road, Hickory Grove Road, Lombard Street, and Wisconsin Avenue—give the schematic diagram physical significance.

The four circuits originating at substation 59 are designated 13-59-1, 13-59-2, 13-59-3, and 13-59-5. All four circuits exit from the substation through underground cables. Each of the circuits can be connected to an adjacent circuit at the substation perimeter to provide a means of supplying the customer load in case of an exit cable failure, as shown on the schematic tie insert at the left edge of the figure. Circuit 13-59-1 exits from the substation by underground cables that terminate in potheads on the riser pole, as shown on the tie insert schematic. The circle in circuit 13-59-1 with an S in the middle represents single-pole disconnect switches operated with hot-line tools.

The number 3314 is installed near the switches on the cable riser pole to identify the switches. The single-pole disconnect switches designated 3314 are used to isolate the underground cables from the overhead circuit in case of a cable failure. The circuit extends from the riser pole through single-pole disconnect switches number 3191. Single-pole disconnect switches 3315 are normally open, as designated by NO on the tie insert schematic. Circuit 13-59-1 can be tied to circuit 13-59-5 by closing the normally open single-pole disconnects numbers 3315 and 3319.

Circuit 13-59-1 extends across Duck Creek to Kimberly Road. The circle with an F followed by the number 3 indicates a fused three-phase tap to the main circuit extending north from Kimberly Road. At Pine Street a capacitor bank number C-279 with the capacity of 1050 kVAR is connected to the main circuit through switching equipment. The circuit extending north along Pine Street terminates in a group-operated disconnect switch designated by the circle with the letter G in the middle. The group-operated disconnect switch number 3076 is normally open, designated by NO.

Circuit 13-78-1 terminates on the north side of the group-operated disconnect switch. This switch provides a method of connecting circuit 13-78-1 to circuit 13-59-1. The main

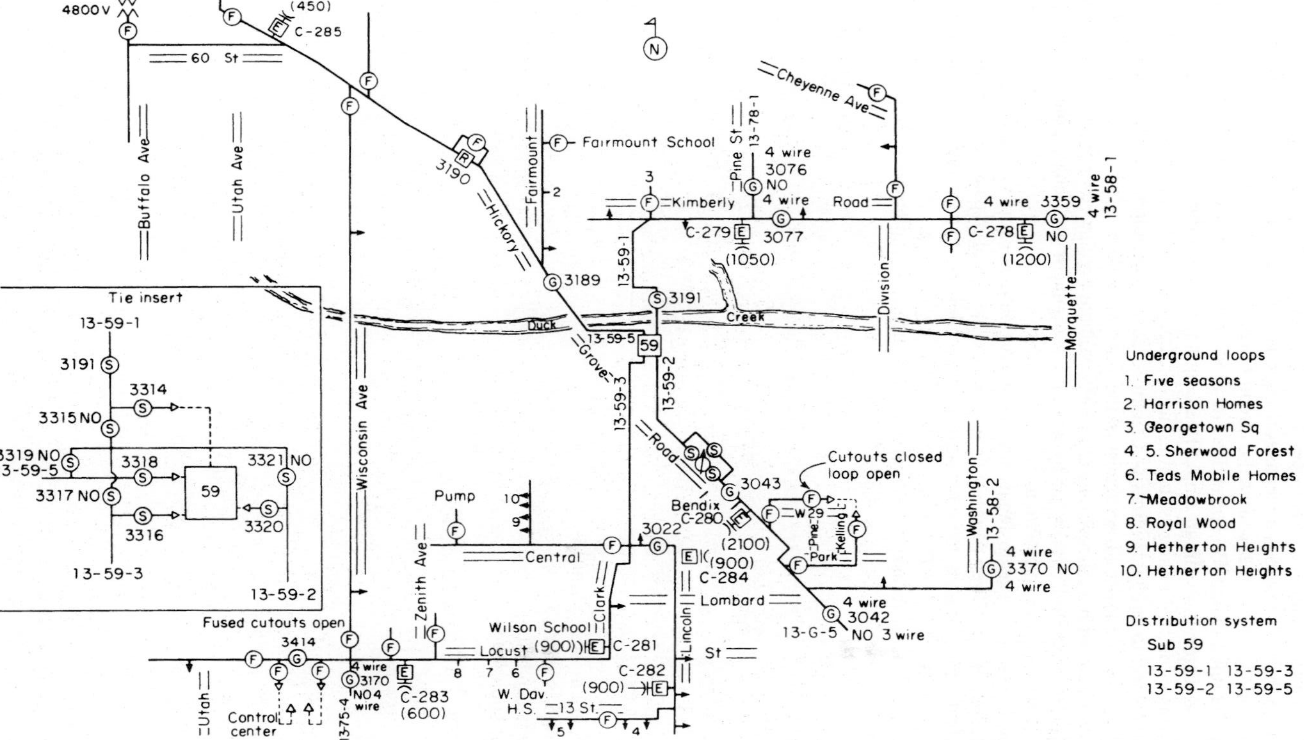

FIGURE 38.3 Distribution circuit single-line diagram.

feeder portion of circuit 13-59-1 extends east on Kimberly Road through group-operated switch 3077 and terminates at group-operated switch 3359, which is normally open. Group-operated switch 3077 provides a means for sectionalizing circuit 13-59-1 in case of a sustained fault on the circuit that causes the circuit breaker at Substation 59 to remain open.

Group-operated switch 3359 provides a method for connecting circuit 13-59-1 to circuit 13-58-1. The distribution circuit single-line diagram provides the linemen and cablemen with a means of understanding switching instructions and locating major pieces of equipment with respect to known landmarks. Circuit 13-59-2 has a voltage regulator installed in it along Hickory Grove Road, southeast of the substation. The voltage regulator is represented by a circle with an arrow through it. The voltage regulator has single-pole disconnect switches in parallel with it for bypassing the regulator and single-pole switches in series with the regulator to isolate it.

Circuit 13-59-5 has a reclosing device in the line along Hickory Grove Road northwest of the substation. The three-phase reclosing device has fused bypass switches in parallel with it. The fused bypass switches are utilized to permit removing the recloser from service for maintenance purposes. The distribution circuit single-line diagram is very valuable for analyzing the circuits to assist linemen and cablemen to complete their work safely and effectively.

CHAPTER 39
SCHEMATIC DIAGRAMS

Purpose. The purpose of a schematic diagram is to make it possible to picture the operation of an entire control circuit for a piece of equipment, so that it will be easy to think in terms of an entire circuit instead of a particular component part of it. Once the control schematic diagram is understood, it is easy to apply the information contained on it to some portion of the associated wiring diagram.

Feeder Circuit Control Schematic. The control schematic diagram for a distribution substation feeder circuit breaker installed in a metal-clad switchgear unit (Fig. 39.1) is traced by utilizing the numbers appearing by the component parts of equipment. Each piece of equipment and the purpose it serves are described.

The schematic diagram of Fig. 39.1 is drawn for the condition of all isolating devices open and the circuit breaker in the open position. This is an accepted standard method of drawing schematic control diagrams. The symbol —< <— represents a circuit breaker draw-out contact in the switchgear. All equipment shown between the draw-out contacts is located on the feeder circuit breaker.

The control source for closing the feeder circuit breaker is 240 volts alternating current. The lead designations for the control power source are X_1 and Y_1. Starting with wire X_1, a circuit extends to item (1), a knife switch which isolates and deenergizes the control circuit. From the knife switch, the circuit is completed to item (2), which is a low-voltage fuse used for isolating the control circuit for a fault in the control wiring, protecting the control power source and preventing a fault in the wiring of one circuit breaker's control wiring from causing all circuit breakers to be inoperative.

From the fuse, the circuit is completed through wires X_2 and $4X_2$ to item (3), 86, which is a contact on the lockout relay operated by the bus fault relays. The purpose of the 86 contact is to open the closing circuit for the feeder breaker to prevent closing the breaker after a bus fault until the lockout relay has been manually reset. From the lockout relay contact, the circuit is complete through wires 4W and W to item (4), CSO, the control-switch-off contact. This contact on the control switch is closed when the control switch handle is in the off or straight up-and-down position. The purpose of this contact is to keep the 152Y relay coil deenergized until the control switch is returned from the closed position to the off position if the breaker tripped immediately after a close operation.

From the CSO contact, the circuit is completed through wire 2a and the breaker draw-out contact to item (5), 152b, a contact on the circuit breaker auxiliary switch which is closed when the circuit breaker is open. This contact opens this portion of the control circuit so that current will not flow through the resistor continuously while the breaker is closed and completes the circuit to energize the 152Y relay when the breaker is in the open position. From the 152b auxiliary switch contact, the circuit is completed through the breaker draw-out contact, wire la, and a resistor to item (6), 152Y, closing cutoff relay coil. The purpose of the 152Y closing cutoff relay is to make the control scheme antipump, or electrically trip free.

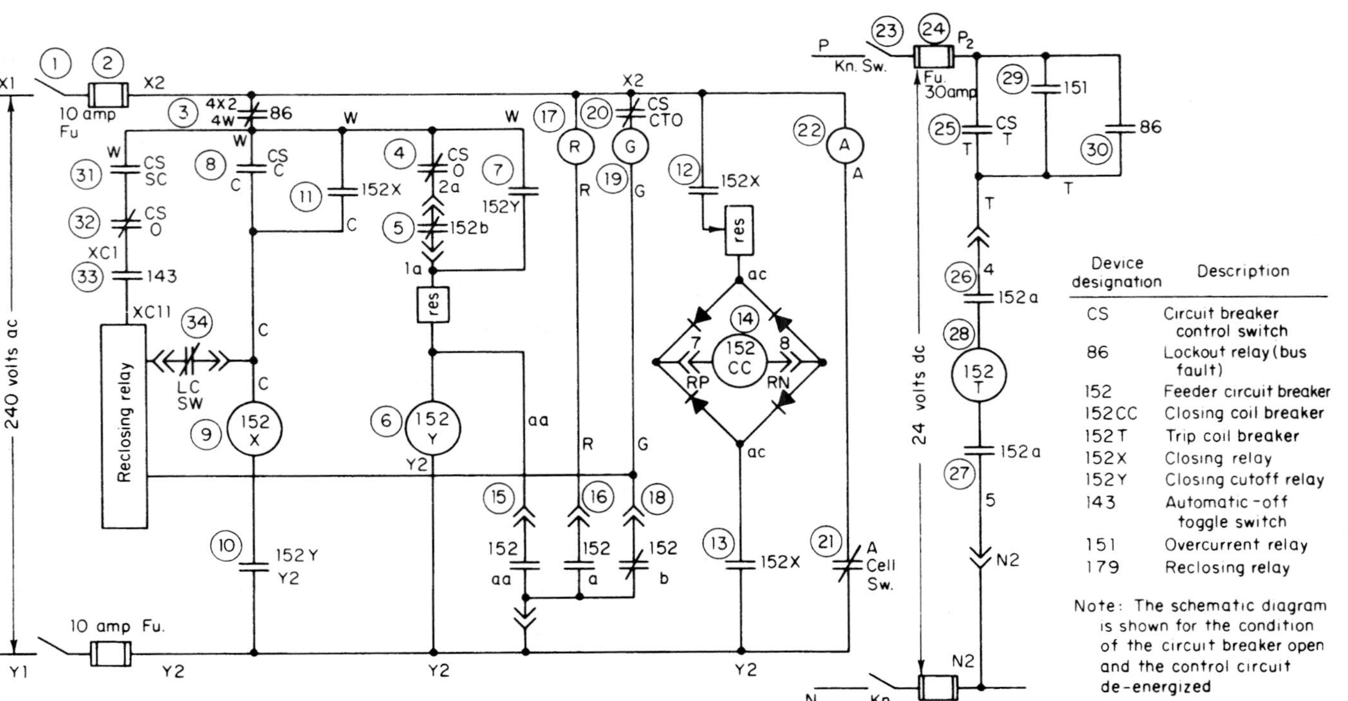

Device designation	Description
CS	Circuit breaker control switch
86	Lockout relay (bus fault)
152	Feeder circuit breaker
152CC	Closing coil breaker
152T	Trip coil breaker
152X	Closing relay
152Y	Closing cutoff relay
143	Automatic-off toggle switch
151	Overcurrent relay
179	Reclosing relay

Note: The schematic diagram is shown for the condition of the circuit breaker open and the control circuit de-energized

FIGURE 39.1 Unit-type substation, feeder circuit, control schematic diagram.

From the 152Y relay coil, the circuit is completed through wire Y2, a fuse and a knife switch to wire Y1, the other side of the alternating-current control power source.

If the knife switches are closed, a current will flow through the 152Y closing cutoff relay coil, operating the associated relay contacts, items (7) and (10). Item (7), 152Y, closing cutoff relay contact, completes the circuit through wires W and 1a to establish a second path for the current to flow through the 152Y relay coil. The purpose of this 152Y relay contact is to keep the 152Y closing cutoff relay coil energized when the control switch is turned to the closed position or when the breaker starts to close until the breaker is completely closed.

Starting with wire X_1, one side of the alternating-current source, a circuit exists through items (1), (2), (3) and wires 4W and W to item (8), CSC, which is the control-switch-close contact. This contact on the control switch is closed when the control switch handle is turned to the closed position. This contact provides a means of closing the circuit breaker by operating the control switch manually. From the CSC contact, a circuit is completed through wire C to item (9), 152X, the closing relay coil. The purpose of this relay is to energize the circuit breaker closing solenoid. From the closing relay coil, the circuit is complete to item (10), 152Y, which is a contact on the closing cutoff relay. The purpose of this contact is to make the control scheme antipump or electrically trip-free.

From the closing cutoff relay contact, the circuit is completed through wire Y2, the fuse, and the knife switch to wire Y1, the other side of the alternating-current source. If the CSC contact is closed, the closing relay coil 152X will have a current flow through it, since the closing cutoff relay is energized and the 152Y contact, item (10), is in the closed position. When a current flows through the 152X closing relay coil, the relay will operate its contacts, items (11), (12), and (13). Item (11), 152X, is a closing relay contact which short-circuits the CSC contact to keep the 152X closing relay energized once the breaker closing operation is initiated until the breaker has closed.

Starting with wire X_1, one side of the alternating-current source, the circuit is completed through the knife switch, item (1); the fuse, item (2); and wire X_2 to item (12), 152X, which is the closing relay contact. The purpose of this contact is to supply a high-current contact capable of energizing the circuit breaker closing coil. From the 152X contact, a circuit extends through the resistor, the rectifier elements, and the circuit breaker draw-out contact to item (14), 152cc, the circuit breaker closing coil. The purpose of this coil is to supply the energy to close the main contacts of the circuit breaker.

From the circuit breaker closing coil, a circuit is completed to the breaker draw-out contact, the copper oxide rectifier, and to item (13), 152X, a closing relay contact which serves the same purpose as item (12). From the closing relay contact, the circuit is established to wire Y2, the other side of the alternating-current source. With the closing relay energized and its associated contacts, items (12) and (13), closed, a current flows through the circuit breaker closing coil, causing the circuit breaker to close its main contacts.

Located on the circuit breaker mechanism is an auxiliary switch which is connected to the linkage which operates the circuit breaker main contacts. The auxiliary switch has contacts which are referred to as "a" contacts and "b" contacts. The auxiliary switch "a" contacts operate the same as the circuit breaker main contacts; that is, when the circuit breaker main contacts are closed, the "a" contacts are closed and when the circuit breaker main contacts are open, the "a" contacts are open. The auxiliary switch "b" contacts operate opposite to that of the circuit breaker main contacts; that is, when the circuit breaker main contacts are closed, the "b" contacts are open, and when the circuit breaker main contacts are open, the "b" contacts are closed.

Starting with wire Y1, one side of the alternating-current source, a circuit is established through the knife switch, fuse, wire Y2, and the circuit breaker draw-out contact to item (15), 152aa, a circuit breaker auxiliary "a" switch contact adjusted to close late in the closing cycle. The 152aa switch serves the purpose of cutting off the circuit breaker closing power. The 152aa switch is adjusted to close late in the closing operation to provide adequate power for closing the circuit breaker in case a fault should exist in the distribution line supplied through the breaker. From the 152aa contact, the circuit is completed through the breaker draw-out contact to a point between the resistor and the 152Y closing cutoff relay coil.

With the 152aa auxiliary switch contact closed, the 152Y closing cutoff relay coil is short-circuited and the 152Y relay contacts, items (7) and (10), will be opened. The 152X closing cutoff relay will be deenergized and will open its contacts, items (12) and (13), deenergizing the 152cc circuit breaker closing coil. The 152*X* closing relay, as described above, was de-energized even though CSC contact was maintained closed by the operator; therefore, the circuit breaker cannot reclose if the trips out on a closing operation due to a fault until the control switch handle is returned to the off position, closing contact CSO, item (4), and thus reenergizing the 152Y closing cutoff relay coil, item (6).

From the Y1 wire of the alternating-current source through the knife switch, the fuse, wire Y2, the breaker draw-out contact, a circuit is established to item (16), 152a, which is an "a" contact on the circuit breaker auxiliary switch. This contact serves the purpose of energizing the red light when the circuit breaker closes. From the 152a contact, a circuit continues through the breaker draw-out contact, wire R, to item (17), R, which is the red indicating light. The red indicating light serves the purpose of informing the operator that the circuit breaker is closed. From the red light, the circuit is completed through wire X_2, the fuse, and the knife switch to wire X_1, the other side of the alternating-current source.

When the circuit breaker closes and the 152a auxiliary switch contact closes, the red light will be energized, indicating that the circuit breaker is closed. From wire Y1, one side of the alternating-current source, a circuit exists through the knife switch, the fuse, wire Y2, the breaker draw-out contact to item (18), 152b, which is a circuit breaker auxiliary switch "b" contact. The 152b contact serves the purpose of energizing the green light and the reclosing relay. From the 152b contact, the circuit is complete through a circuit breaker draw-out contact, wire G, to item (19), G, which is the green indicating light.

The green indicating light serves the purpose of telling the operator that the circuit breaker is open. From the green light, a circuit extends to item (20), CS CTO, which is the control switch lamp cutout contact. This contact on the control switch is closed except when the control switch handle is turned to the trip position and pulled out, placing the control switch in the lockout position. The purpose of the CS CTO contact is to deenergize the green light so that the green indicating light will not be lighted when the circuit breaker is out of service due to a planned outage.

From the CS CTO contact, the circuit is complete to the X side of the alternating-current source. When the circuit breaker is open and the 152b contact is closed, the green light will be energized, indicating that the circuit breaker is open. From the Y side of the alternating-current source, a circuit is established to item (21), cell switch, which is a contact that is closed when the circuit breaker is in the operating position. The purpose of the cell switch is to energize the amber light.

From the cell switch, the circuit is complete through wire A to item (22), A, the amber light. The purpose of the amber light is to indicate that the circuit breaker is in the operating position. From the amber light, the circuit is complete to the X side of the alternating-current source. When the circuit breaker is in the jacked-in or operating position, the cell switch will be closed and the amber light will be lighted indicating the breaker is in the operating position.

From the wire P, the positive side of the 24-volt direct-current control power, a circuit extends to item (23), the knife switch. The knife switch can be used to isolate the circuit breaker trip circuit. From item (23), the circuit is completed to item (24), a fuse. The fuse serves the purpose of protecting the direct-current control source from a fault in the trip circuit control wiring of the circuit breaker. From the fuse, the circuit is complete through wire

P2 to item (25), CST, the control switch trip contact. The purpose of the CST contact is to initiate a trip operation when the control switch handle is turned to the trip position.

From the CST contact, the circuit is complete through wire T, the breaker draw-out contact, and wire 4, to item (26), 152a, a circuit breaker auxiliary switch "a" contact. The purpose of the circuit breaker auxiliary switch "a" contact is to deenergize the circuit breaker trip coil after the tripping operation has been initiated and the circuit breaker starts to open. From the 152a contact, the circuit is complete to item (28), 152T, which is the circuit breaker trip coil. The circuit breaker trip coil serves the purpose of unlatching the circuit breaker mechanism so that the circuit breaker main contacts will open.

From the 152T trip coil, the circuit is complete to item (27), 152a contact, which serves the same purpose as item (26). From the 152a contact, the circuit extends through wire 5, the breaker draw-out contact, wire N2, the fuse, the knife switch, and wire N to the negative side of the 24-volt direct-current control source. If the control switch handle is turned to the trip position when the circuit breaker is closed, the CST contact will close and the trip coil will be energized, initiating a tripping operation. Item (29), 151, is a trip contact on an overcurrent protective relay.

The overcurrent relay trip contact is connected to wires P2 and T so that when the contact is closed and the circuit breaker is closed, a tripping operation will be initiated. The purpose of the overcurrent relay trip contact is to trip the circuit breaker when a fault exists on the feeder connected to the circuit breaker. Item (30), 86, is the lockout relay trip contact which serves the purpose of tripping the feeder circuit breaker when a bus fault occurs in the substation. The lockout relay trip contact is wired in parallel with the overcurrent relay trip contact.

Starting with the X_1 wire of the alternating-current source, a circuit is completed through the knife switch, the fuse, wire X_2, wire $4X_2$, 86 lockout relay contact, and wire 4W to item (31), CSSC, which is the control switch sliding contact. This contact on the control switch closes when the control switch handle is turned to the closed position, and it remains closed until the control switch handle is turned to the trip position. The purpose of the CSSC control switch contact is to open the reclosing circuit to prevent an automatic reclose when the circuit breaker is tripped by the control switch.

From the CSSC contact, a circuit is established to item (32), CSO, an off contact on the control switch. This contact on the control switch is closed when the control switch handle is in the off or the straight up-and-down position. The purpose of the CSO contact is to open the reclosing circuit when the circuit breaker is closed by the operation of the circuit breaker control switch, thus preventing the circuit breaker from reclosing automatically on a fault when the circuit breaker is closed manually.

From the CSO contact, the circuit is complete through wire XC1 to item (33), 143, which is the automatic on/off toggle switch. The purpose of the 143 toggle switch is to deenergize the automatic reclosing relay. From the 143 contact, the circuit is complete through wire XC11 to the reclosing relay. Item (34), LCSW, is the latch check switch contact. The LCSW contact is open except when the circuit breaker is in the full-open position. The LCSW is connected between the 152X closing relay and the automatic reclosing relay to prevent an automatic reclosure after a breaker trip operation until the breaker has completely opened. Wire G connects from the 152b auxiliary switch contact, item (18), to the reclosing relay to energize the relay and initiate a reclosing operation when the circuit breaker opens due to a fault on the line.

The reclosing relay has a timing motor that is connected to wires XC11 and G through cam switches operated by the reclosing relay timing motor. The reclosing relay timing motor is energized after an automatic trip operation if the manual/automatic toggle switch 143 contact is closed. The timing motor operates the reclosing relay cam switches that complete a circuit from wire XC11 to the draw-out contact and the LCSW, item (34). This circuit energizes the closing relay coil, 152X and initiates an automatic reclosing of the circuit breaker after an automatic trip operation initiated by the overcurrent relay, 151.

Voltage Regulator or Substation Transformer Tap Changer Control Schematic. Distribution substation transformers are supplied by the subtransmission system. The high-voltage windings of most of the transformers have a no-load tap changer and an under-load tap changer.

The purpose of the no-load tap changer is to change the number of turns in the primary winding when the transformer is not energized, so that the number of turns of the winding is in accordance with the general range of voltage that exists on the system at the location of the transformer. For example, if the voltage at a particular site varies from 71,000 to 67,000 volts, the no-load tap changer would probably be set on a tap which is rated 69,000 volts. This means that the actual turns utilized are in proportion to 69,000 volts.

The purpose of the tap-changing under-load equipment is to further change the number of turns in the primary winding to automatically keep the voltage on the secondary of the transformer at the proper value when the primary supply voltage varies to some value other than the value for which the no-load tap changer is adjusted. The under-load tap changer consists of a switching arrangement designed to operate when the transformer is energized and carrying load. This switching arrangement is operated through an insulated mechanism by means of a motor.

The schematic control diagram (Fig. 39.2) illustrates the operation of the control equipment necessary to operate the tap changer motor which supplies the energy to operate the tap-changing under-load switching equipment. Each item of equipment on the schematic control diagram (Fig. 39.2) is numbered. The circuit is traced, each piece of equipment is described, and the purpose that each piece of equipment serves is outlined, utilizing the numbers on the diagram. The circuit elements are shown for the condition of the transformer deenergized and all operating devices open. The control schematic for a distribution-line voltage regulator would be similar.

Item (1) is a potential transformer connected to the regulated voltage supply. The purpose of the potential transformer is to reduce the voltage level so that it can be safely connected to relays and other control equipment. The output voltage of the transformer will normally be approximately 120 volts. From terminal V7 on the secondary of the potential transformer, the circuit is completed to item (2), AB1, which is an air circuit breaker. The circuit breaker protects the potential transformer from a short circuit in the control wiring. From item (2), the circuit is completed to item (3), FTS1, which is a test switch. The test switch provides a means of testing the voltage-regulating relay. From item (3), the circuit is completed to item (4), NV, which is the coil of the no-voltage relay.

The coil of the no-voltage relay operates its contacts to prevent operation of the tap changer if the regulated voltage supply is removed from the voltage-regulating relay coil. From item (4) the circuit is complete to item (5), FTS1, another test switch, and item (6), AB1, another pole of the air circuit breaker. From item (6), the circuit is complete to terminal VO on the other end of the potential transformer secondary. As long as the potential transformer is energized and the air circuit breaker is closed, the no-voltage relay coil will be energized and the relay contacts will be closed to permit automatic operation of the tap changer.

From item (3), the circuit is also complete to item (7), P the voltage-regulating relay coil. The voltage-regulating relay coil actuates the raising and lowering contacts which controls the direction of operation of the tap changer motor. From item (7) the circuit is complete to item (8), the compensator. The compensator elements connect also to the secondary of a current transformer, which has its primary winding connected in series with the load.

The compensator varies the voltage applied to the voltage-regulating relay coil, in accordance with the settings of the resistance and reactance elements, and the current flowing in the secondary of the transformer. Therefore, it is possible to have the regulated voltage at a higher level during times of heavy load conditions to compensate for line voltage drops without readjusting the voltage-regulating relay. From item (8) the circuit is completed to item (5), etc., previously described.

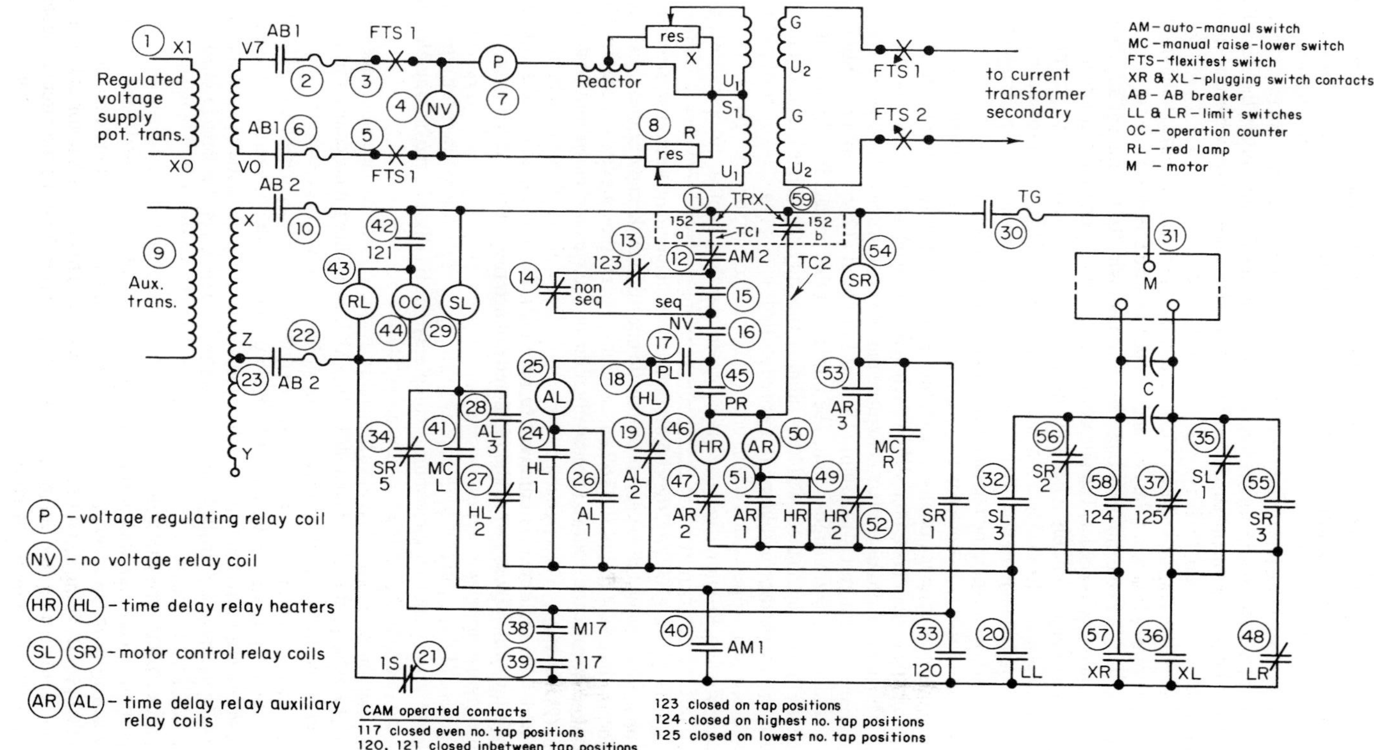

FIGURE 39.2 Unit substation transformer tap changer control, schematic diagram.

Item (9) is an auxiliary power transformer. This transformer supplies the power to operate the tap changer motor and other equipment. From terminal X on the secondary of the auxiliary power transformer, the circuit is complete to item (10), AB2, an air circuit breaker which protects the auxiliary power transformer from a fault in the tap changer control equipment. From item (10), the circuit is complete through wire TRX to item (11), 152a, a circuit breaker auxiliary switch contact located on the transformer secondary breaker. The 152a switches are closed when the circuit breaker is closed and open when the circuit breaker is open.

The purpose of the auxiliary switch 152a contact is to prevent automatic voltage regulation when the transformer secondary breaker is open and the transformer is not connected to a load. From item (11) the circuit is completed through wire TC1 to item (12), AM2, an automatic/manual switch contact. The purpose of this contact is to prevent automatic operation of the tap changer controls when the tap changer is being operated by the manual-control switch. From item (12) the circuit is completed to item (13), 123, a contact operated by a cam on the tap changer motor drive. This contact is closed when the tap changer is on a tap position and open when the tap changer is between tap positions.

The purpose of the 123 contact is to de-energize the tap changer control circuit after each tap changer operation is initiated to give nonsequential operation. From item (13) the circuit is complete to item (14), non SEQ, which is the nonsequential contact on the transfer switch. The purpose of this contact is to place the 123 cam switch in the circuit. The circuit is complete from item (14) to item (15) and from item (12) to item (15), SEQ, the sequential contact on the transfer switch. The purpose of this contact is to bypass the 123 cam switch so that the tap changer will keep running once an operation is initiated until the requirements of the voltage-regulating relay are fulfilled.

If the transformer is operated in a network, it is best to have the tap changers set on nonsequential operation to allow first one transformer and then another to operate its tap changer to fulfill the requirements of the voltage-regulating relays. From item (15) the circuit is complete to item (16), NV, the contact on the no-voltage relay previously discussed. The purpose of this contact is to prevent automatic operation of the tap changer if the voltage is removed from the coil of the voltage-regulating relay, thus preventing operation of the tap changer to raise the voltage excessively if the protective devices operate on the regulated potential supply.

From item (16) the circuit is complete to item (17), PL, the contact on the voltage-regulating relay, which closes when the regulated voltage rises above the value for which the voltage-regulating relay is adjusted.

The purpose of this contact is to initiate a tap-changing operation to lower the regulated voltage level. From item (17) the circuit is complete to item (18), HL, the heating element of the timing relay for the tap changer lower operation.

The heating element supplies heat to a bimetal strip which closes a contact after time delay. From item (18) the circuit is complete to item (19), AL2, an auxiliary relay contact, which is a component part of the timing relay. The purpose of this contact is to de-energize the heating element and permit the bimetal strip to return to its original position, opening contact HL1 and closing contact HL2 again. From item (19) the circuit is complete to item (20), LL, the tap changer lower-limit switch. The purpose of this limit switch is to prevent the tap changer motor from operating to change taps in the lower direction after the lowest tap has been reached. From item (20) the circuit is complete to item (21), IS, the interlock switch, which is opened when the crank is inserted in its socket to manually operate the tap changer.

The purpose of the interlock switch is to prevent simultaneous operation of the tap changer electrically and mechanically. From item (21) the circuit is completed to item (22), AB2, another pole of the air circuit breaker. From item (22) the circuit is completed to item (23), Z, the midpoint on the auxiliary power transformer, thus completing a circuit. Therefore, if all the conditions described previously are fulfilled for an automatic lower operation, a current will flow through item (18), HL, causing it to close its contact item (24), HL1, the time-delay relay heating element contact that closes after time delay. The purpose of the HL1 contact is to provide a circuit to energize the lower time-delay relay auxiliary relay coil item (25), AL.

The circuit is then complete from item (17) through items (25) and (24) to item (20), and a current will flow through the auxiliary relay coil, AL, operating its contacts. The purpose of the auxiliary relay is to permit the timing relay to complete its cycle. From item (25) the circuit is complete to item (26), AL1, a contact on the auxiliary relay which closed when the AL coil was energized. The purpose of this contact is to complete a circuit in parallel with the HL1 contact to keep the AL coil energized. When the AL coil was energized, item (19), AL2 contact opened, deenergizing the HL heating element, which causes contact item (24), HL1, to open, and causes contact HL2, item (27), to close after time delay.

The HL2 contact is a time-delay relay heating element contact, which is closed when the heating element is de-energized. The purpose of this contact is to help energize the relay which energizes the tap changer motor to operate in the lower direction. Item (28), AL3, is another auxiliary relay contact. The purpose of this contact is to energize the tap changer motor relay with the help of item (27), HL2 contact.

From item (10) circuit is complete to item (29), SL, the tap changer lower motor control relay coil. The purpose of this coil is to operate the relay contacts to energize the tap changer motor. From item (29) the circuit is complete through item (28), AL3 contact, which closed when the timing cycle was halfway completed; item (27), HL2 contact, which opened when the timing cycle was initiated by the voltage-regulating relay and closed when the timing cycle was completed; and item (20), etc., previously discussed.

From item (10) a circuit is established to item (30), TG, the motor thermal guard contact and heater element. The purpose of the thermal guard is to protect the motor if it is overloaded or the limit switches fail, and the motor is stopped by the mechanical stops. From item (30) the circuit is complete to item (31), the tap changer motor, which supplies the power to operate the tap changer. From item (31) the circuit is complete to item (32), SL3 contact on the motor control relay. The purpose of the SL3 contact is to energize the motor so that it will operate the tap changer to a lower tap position. From item (32) the circuit is complete to item (20), etc., as previously described. When the tap changer has moved off position, item (33), 120 cam switch contact, closes.

The purpose of the cam switch is to keep the motor control relay energized until the tap changer is back on position, at which time the 120 cam switch opens again. This prevents a partial tap change due to the voltage-regulating relay opening its contacts or due to the 123 cam switch, item (13), opening if the control is switched to nonsequential operation. This is accomplished through item (34), SR5, a contact on the raise motor control relay which is closed when the raise relay coil is de-energized. The purpose of this contact is twofold: first, to prevent operation of the SL and SR motor control relays simultaneously through the 120 cam switch and, second, to cause the tap changer to run to the next lower tap if it is stopped in a middle position between taps.

Item (35), SL1, a lower motor control relay contact, is opened when the motor is energized to operate the tap changer in the lower direction and closes as the tap change is completed. The purpose of this contact is to assist in placing power on the tap changer motor in the reverse direction to brake the motor when the operation is completed. Item (36), XL, is the lower direction plugging switch. This contact closes when the motor is operated in the direction to lower taps. Therefore, when the lower direction motor control relay coil, SL, is de-energized, power is supplied through contacts XL and SL1, which tends to rotate the motor in the reverse direction, serving the purpose of a brake.

When the motor stops, contact XL, item (36), opens, removing power from the motor. If the motor is energized in the lower direction beyond the lowest tap position, item (20), LL, the lower-limit switch contact, opens, de-energizing the motor, and item (37), 125 cam switch, closes to supply breaking power. The purpose of the 125 cam switch, which closes after the tap changer has reached the lowest position, is to apply breaking power immediately to prevent operation of the tap changer against the mechanical stop.

Item (38), M17, is a contact on a transfer switch. The purpose of the M17 transfer switch is to energize item (39), 117 cam switch. The 117 cam switch is open only on the odd-numbered tap positions. The purpose of items (39) and (38) is to short-circuit the 120 cam switch on the even-numbered tap positions so that when a tap change is initiated, it will complete two operations instead of one.

Item (40), AM1, is an automatic/manual transfer switch contact. When the transfer switch is operated to manual, contact item (12) is opened and item (40), AM1, is closed. The purpose of this is to permit manual electrical control of the tap changer. From item (29), SL, the lower motor control relay coil, a circuit is completed to item (41), MCL, the manual control switch lower contact. The purpose of this switch is to operate the tap changer electrically by means of the manually operated control switch. From item (41) the circuit is completed through items (40), (21), etc.

From item (10) a circuit is established to item (42), 121, a cam switch which is closed between each tap position. The purpose of this switch is to energize the interposition light and the operation counter. From item (42) a circuit is complete to item (43), RL, a red light. The purpose of the light is to indicate that the tap changer is between positions. From item (43), the circuit is complete to item (22) and the other side of the control power source. From item (42), a circuit is established to item (44), OC, the operation counter. The purpose of the operation counter is to record each tap changer operation to help determine the need for maintenance work and operational inspections.

The tap changer is operated in the raised direction in similar manner as previously described for the lower direction. A circuit exists from the auxiliary transformer through the circuit breaker, item (10), the circuit breaker 152a auxiliary switch contact, item (11), through wire TC1 to item (12), the automatic/manual switch contact, through item (13), 123 cam switch, and item (14), the nonsequential transfer switch contact, or item (15), the sequential switch contact; then through item (16), the no-voltage relay contact, to item (45), PR, the contact on the voltage-regulating relay which closes when the regulated voltage decreases below the value for which the voltage-regulating relay is adjusted.

The purpose of this contact is to initiate a tap-changing operation to raise the regulated voltage level. From item (45) the circuit is complete to item (46), HR, the heating element of the timing relay for the tap changer raise operation. The heating element supplies heat to a bimetal strip which closes a contact after time delay. From item (46) the circuit is complete to item (47), AR2, an auxiliary relay contact which is a component part of the timing relay. The purpose of this contact is to de-energize the heating element and permit the bimetal strip to return to its original position, opening contact HR1 and closing contact HR2 again. From item (47) the circuit is complete to item (48), LR, the tap changer raise limit switch.

The purpose of the limit switch is to prevent a tap changer motor from operating to change taps in the raise direction after the highest tap has been reached. From item (48) the circuit is complete to item (21), the interlock switch, item (22), the air circuit breaker, and item (23), the secondary of the operating transformer. Therefore, if all the conditions described previously are fulfilled for an automatic raise operation, a current will flow through item (46), HR, causing it to close its contact, item (49), HR1, the time-delay relay heating element contact that closes after a predetermined time delay.

The purpose of the HR1 contact is to provide a circuit to energize the raise time-delay relay, auxiliary relay coil item (50), AR. The circuit is complete from item (45) through items (50) and (51) to item (48), and a current will flow through the auxiliary relay coil AR, causing it to operate its contacts. The purpose of the auxiliary relay is to permit the timing relay to complete its cycle. From item (50) the circuit is complete to item (51), AR1, a contact on the auxiliary relay which closed when the AR coil was energized. The purpose of this contact is to complete a circuit in parallel with the HR1 contact to keep the AR coil energized. When the AR coil was energized, item (47), AR2 contact, opened, deenergizing the HR heating element, which opened contact, item (49), HR1, and closed contact HR2, item (52), after time delay. The HR2 contact is the time-delay relay heating element contact which is closed when the heating element is de-energized.

The purpose of this contact is to help energize the relay which energizes the tap changer motor to operate in the raise direction. Item (53), AR3, is another auxiliary relay contact. The purpose of this contact is to energize the tap changer motor relay with the help of item (52), HR2 contact. From item (10), the circuit is complete to item (54), SR, the tap changer raise motor control relay coil. The purpose of this coil is to operate the relay contacts to energize the tap changer motor. From item (54) the circuit is complete through item (53), AR3 contact, which closed when the timing cycle was halfway completed, item (52), HR2 contact, which opened when the timing cycle was initiated by the voltage-regulating relay and closed when the timing cycle was completed, and item (48), etc., previously discussed.

As previously described, a circuit is established from item (10) through item (30), TG, the motor thermal guard contact and heater element, item (31), M, the tap changer motor, to item (55), SR3 contact on the motor control relay. The purpose of the SR3 contact is to energize the motor so that it will operate the tap changer to raise the regulated voltage. From item (55) the circuit is complete to item (48), LR, etc., as previously described. When the tap changer motor has moved off position, item (33), 120 cam switch contact, closes to keep the raise motor control relay energized until the tap changer is back on position, at which time the 120 cam switch opens. This prevents a partial tap change due to the voltage-regulating relay opening its contact prior to the time the tap changer completes its operation.

Item (56), SR2 raise motor control relay contact, is opened when the motor is energized and closes as the tap change is completed. The purpose of this contact is to assist in placing power on the tap changer motor in the reverse direction to brake the motor when the operation is completed. Item (57), XR, is the raise direction plugging switch. Therefore, when the raise motor control relay coil, SR, is de-energized, power is supplied through contacts XR and SR2, which tends to rotate the motor in the reverse direction, serving the purpose of a brake. When the motor stops, contact XR, item (57), opens, removing the power from the motor.

If the motor is energized in the raise direction beyond the highest tap position, item (48), LR, the raise limit switch contact, opens, de-energizing the motor, and item (58), 124 cam switch contact, closes to supply braking power. The purpose of the 124 cam switch contact, which closes after the tap changer has reached the highest position, is to apply braking power to stop the rotation of the tap changer motor to prevent operation of the tap changer against the mechanical stop.

Item (59), 152b, is a circuit breaker auxiliary switch contact located on the power transformer secondary breaker. This contact is closed when the secondary breaker is open. The purpose of the contact is, if the control power is on, to operate the tap changer in the raise direction and thus to assist the secondary breaker network relays to close the breaker automatically after an outage of the circuit serving as a source to the power transformer if the transformer is operated in a network system.

CHAPTER 40
VOLTAGE REGULATION

Voltage regulation on any electrical system is defined as the difference in voltage between periods of no load and periods of full load on the system. Characteristics of the system and the amount of load will determine the extent of the voltage regulation. To provide good electric service to customers served from the distribution system, it is necessary to maintain voltage variations within a very limited range. The voltage variation to residences has been standardized from a high limit of 126 volts to a low limit of 114 volts for a nominal 120-volt secondary service voltage. The term *nominal voltage* designates the line-to-line voltage, as distinguished from the line-to-neutral voltage. It applies to all parts of the system or circuit.

Electrical equipment is designed to operate on a certain voltage, and best operation is obtained when that rated voltage is applied. The effect of low voltage on heating-type appliances such as heaters, toasters, irons, and ranges is an increase in heating time. Low-voltage conditions cause motors to operate at decreased efficiency. They draw more current from the line, have less starting and running torque, and may overheat to the extent that they might experience thermal failure. Incandescent lights operate less efficiently, give less light, and in the case of fluorescent lights may not operate at all.

High-voltage conditions are also detrimental to the operation of electrical equipment. Heating-type elements, light bulbs, and electron tubes have a shorter life span because of high voltage. Motors and other magnetic equipment will have excess speed or torque. Television sets are particularly susceptible to damage from high voltage.

The best service to customers and the greatest amount of revenue would result from a distribution system which provides rated voltage at every point of the system. Since this is not economically feasible, a compromise is made and the voltage variation is confined to the standard limits established.

The regulation or control of voltage on a distribution system is usually accomplished by a combination of several different options. One option utilizes transformer tap-changing equipment, step voltage regulators, or induction voltage regulators. With this equipment, distribution-system line voltage may be held constant while its supply voltage varies over a certain range. This equipment may also raise or lower the line voltage to compensate for load changes on a circuit. Still another function of this equipment is to provide, in the case of a distribution tie line, some control of kilovar flow over the tie line. To accomplish any of these functions with the above voltage-regulating equipment, an in-phase voltage or approximately in-phase voltage is either added or subtracted from the normal line voltage. Voltage control is accomplished over an approximate ± 10 percent range in easily controlled steps without interruption in service to the customer.

Another voltage control option normally used on distribution systems makes use of capacitors. The control of voltage with capacitors is accomplished by the ability of a capacitor to supply a leading current to a distribution line to counteract the lagging current which the line must supply to magnetize motors, transformers, and other magnetic equipment. The leading current and the lagging current combine to produce a smaller current more nearly in phase with the voltage. The smaller current produces less voltage drop on the line. Capacitors are usually switched on and off in banks, and therefore the control of voltage with capacitors is less flexible than with voltage-regulating equipment. However, capacitors have other advantages, probably the most important of which is their ability to reduce the kVA load on lines, transformers, and generators, making them available for load growth.

The use of voltage-regulating equipment is made necessary because of a characteristic common to all electric circuits. That characteristic is called *impedance*. Impedance is defined as the opposition to the flow of current in an electric circuit. When current flows through an electric circuit, a voltage drop occurs between the input and output terminals of the circuit. This voltage drop is a function of the impedance in the circuit as well as the amount of current flowing in the circuit.

In the circuit having a constant impedance, the voltage drop will be proportional to the current flowing in the circuit. This can be determined by the relation

$$V = IZ$$

where V = voltage drop
I = current flowing in circuit
Z = impedance of circuit

The voltage drop through a transformer can be visualized by considering a transformer energized on the primary side with 100 percent voltage and no connected load on the secondary terminals. For this particular condition, the voltage on the secondary terminals would be 100 percent rated voltage also. If a load were now applied to the secondary terminals of the transformer while the primary voltage remained constant, something other than rated voltage would be obtained on the secondary terminals. This change in voltage would be caused by current flowing through the internal impedance of the transformer.

The voltage drop on a transmission or distribution line is probably a little easier to visualize because no transformation ratio is involved as in the case of a transformer. A transmission or distribution line has an impedance distributed along its entire length. This impedance is made up of resistance and inductive and capacitive reactance. Each foot of conductor of a line has a certain amount of resistance depending on the conductor size and the material used to make up the conductor (copper, aluminum, etc.). Each 1-ft length of a line has a certain amount of inductive reactance because of the spacing between conductors which is required for electrical clearance. Also, each foot length of line has a certain amount of capacitive reactance due to the conductor size, distance between conductors, and the height of the conductor above ground. The capacitive reactance of a distribution line is small compared to the inductive reactance and is usually neglected. Current flowing through the line impedance causes a voltage drop so that the voltage at the load end of the line is less than the voltage at the source end.

Load current flowing through the impedance of a line produces a voltage drop. To help illustrate the effect of the voltage drop, an electrical representation of a line is shown in Fig. 40.1. Distributed along the line are resistance, inductance, and capacitors. To simplify

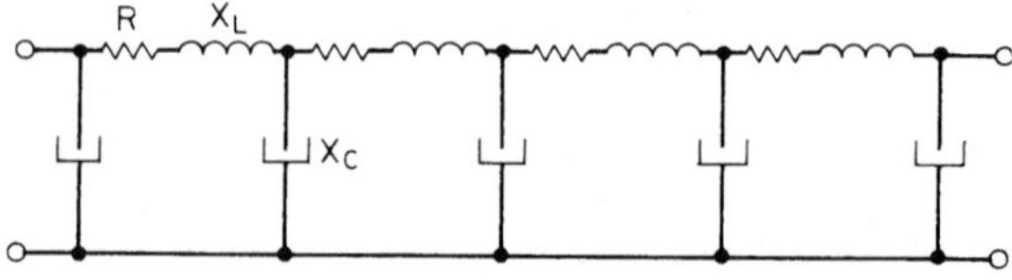

FIGURE 40.1 Representation of a transmission or distribution line.

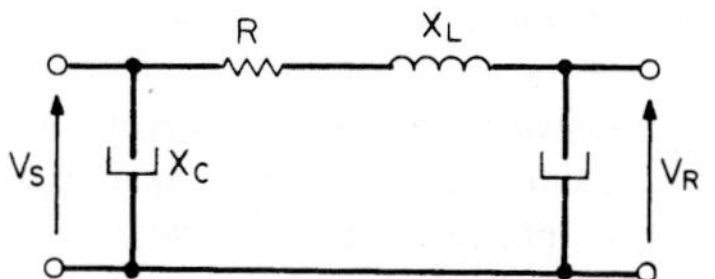

FIGURE 40.2 Representation of a transmission line.

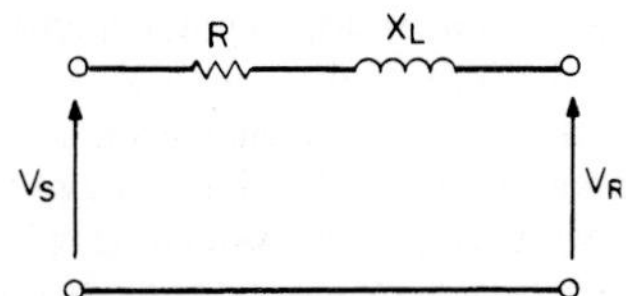

FIGURE 40.3 Short line circuit.

the representation still further, the values of resistance, inductance, and capacitance can be lumped together to form a network as shown in Fig. 40.2.

If the line is a short transmission line or a distribution feeder or tie line, the capacitive effect of the line can usually be neglected. This results in a still further simplification of the representation of a line as shown in Fig. 40.3.

The voltage drop, which would occur on this line because of the load current and impedance, is shown in the following vector diagram, which assumes a lagging power-factor load (Fig. 40.4).

From the vector diagram (Fig. 40.4) it is apparent that the sending end voltage is of greater magnitude than the receiving end voltage and the difference in the two voltages is the *IZ* or impedance voltage drop. The *IZ* component is made up of two parts—the drop due to the current flowing through the resistance and the drop due to the current flowing through the line reactance.

As has previously been noted, capacitance exists between conductors of a line and between the conductors and ground. On a long line that is lightly loaded, the capacitive charging current may exceed the load current and result in the line operating with a leading power factor. In such cases, the receiving end voltage will rise, and it may exceed the voltage at the sending end of the line.

From the vector diagram (Fig. 40.4) several important points can be seen. The first is that the receiving end voltage of a line can be made more nearly equal to the sending end voltage by reducing the current on the line and thereby reducing the impedance drop. This is often done to a distribution feeder when the load on the line becomes so great that excessive voltage drop is encountered. A portion of the load is then transferred to some other line which is not so heavily loaded.

The second point, which is apparent from the diagram, is that the impedance drop may be reduced by decreasing the impedance of the line. To reduce the resistance, wires of a larger cross-sectional area would have to be installed. To reduce the reactance of a line

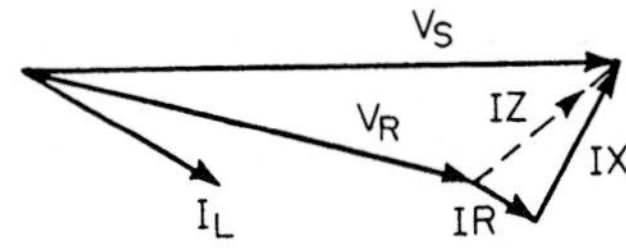

V_S = Sending end or source-end voltage
V_R = Receiving end voltage
I_L = Load current
IR = Voltage drop due to the resistance in the line
IX = Voltage drop due to the reactance in the line

FIGURE 40.4 Voltages on a short line.

would mean either decreasing the conductor spacing or using a series capacitor, neither of which is too practical.

A third point, which can be seen from the vector diagram, is that the power factor of the load could be improved. This could be done by installing shunt capacitors near the load. This method of voltage improvement is used quite extensively. For the best effect, the capacitors should be switched off or on as the load changes. This is done with time, temperature, or voltage controls. The effect of adding capacitors to a circuit is shown in Fig. 40.5.

There are times and conditions where the voltage regulation on a line, that is, the difference between the no-load voltage and the full-load voltage, becomes so great that other means must be used to maintain suitable voltages. Under theses conditions step-type voltage regulators, tap-changing transformers, or induction voltage regulators are employed. In general, voltage regulation of substation buses is accomplished through the use of under-load tap-changing equipment on transformers, while the voltage regulation on distribution circuits is controlled by shunt capacitors or step voltage regulators. The step voltage regulator is primarily an autotransformer with tap-changing under-load switching available to change the ratio between the primary and the secondary windings.

The purpose of under-load tap-changing (LTC) equipment is to change the primary-to-secondary turns ratio of a transformer winding while the transformer is carrying load. The *turns ratio* of a transformer is equal to its voltage ratio, that is, a transformer with a turns ratio of 100:1 also has voltage ratio of 100:1. This means that by controlling the turns ratio of a transformer one can regulate the voltage on its "low" side.

The heart of LTC equipment is the automatic control, which operates a motor-driven mechanical tap changer, thus maintaining a controlled voltage at the load as the input voltage to the power transformer fluctuates or as the load conditions vary.

As the distribution-line load increases, the voltage at its load end drops due to transformer and line impedances (both reactive and resistive components). This line voltage is monitored at the low-voltage side of the power transformer, and when the voltage changes by a prescribed amount, this automatic control device senses the change and causes the under-load tap-changing equipment to either increase or decrease the turns ratio, thus changing the voltage and bringing it back to its desired value.

There are compensating devices in the automatic control unit that correct for the reactive and resistive loss components of the distribution line. This *line-drop compensating device* raises the voltage at the transformer as the line current increases, thus maintaining a nearly constant voltage at the regulating point on the line.

The typical under-load tap selector (see Fig. 40.6) consists of three (one per phase) dial-type switches (A). Each switch has stationary contacts (B) which are connected to the taps in the transformer winding. These contacts are mounted on an insulating panel in a circle

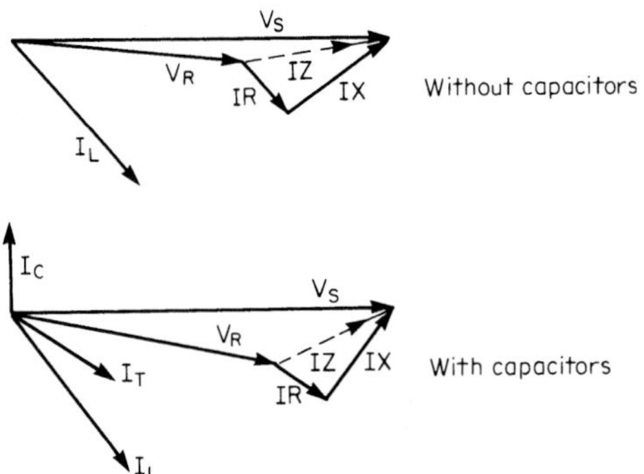

FIGURE 40.5 Effect of installing capacitors.

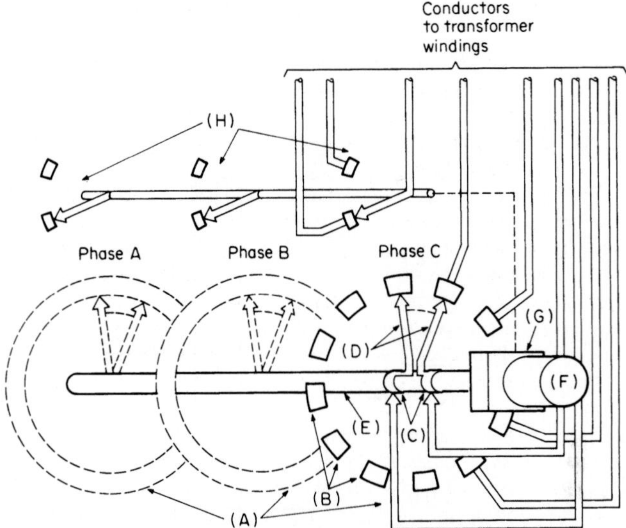

FIGURE 40.6 Transformer under-load tap selector diagram.

around two concentric slip rings (C). The two movable contact fingers (D) for each switch are mounted as an assembly on the main rotor shaft (E), each finger making contact with its own slip ring. The rotor shaft which extends through all three phases of the tap selector is turned by the tap changer motor (F) through reduction gears and switch controls (G). The two fingers move together, stopping either with both on the same contact or with one on each of two adjacent contacts.

Each of the three dial-type tap selector switches is provided with a separate single-pole double-throw switch (H) used to reverse the polarity of the under-load tap-changing windings.

Tap leads from the transformer windings are connected to stationary contacts (see Fig. 40.7) on the tap selector. When the motor operates the selector, it will change taps to vary the transformer turns ratio and thereby regulate the voltage.

Two movable contacts are used on the tap selector so that one can be moved to an adjacent tap while the other contact maintains the power circuit. These movable contacts are connected together through a reactor winding to prevent excessive circulating current through the bridged section of the winding. By designing the winding and reactor to share the circulating current, the bridged position can be used as an intermediate operating position, thus doubling the number of tap-changing steps.

A two-position switch, integral with the tap selector, is used to double the number of steps again. This switch is used to reverse the polarity taken from the tapped section of the transformer, permitting the tapped section to be used twice—one time adding to and the other time subtracting from the number of turns on the main winding.

When the automatic control unit calls for a change of tap position, the drive motor is energized and, through a series of gears, the tap selector is operated. In the first step in a sequence of operation, the movable fingers are slid toward the next stationary contact. The first finger breaks contact, and the current is carried by the second finger (see Fig. 40.8). In the second step, the movable fingers are slid so that the first finger is on the next stationary contact while the second finger is still on the original contact. In this step the currents are shared by both fingers and the center-tapped reactor to which the fingers are connected through the slip rings providing for an intermediate voltage between the two stationary contacts. In the third step,

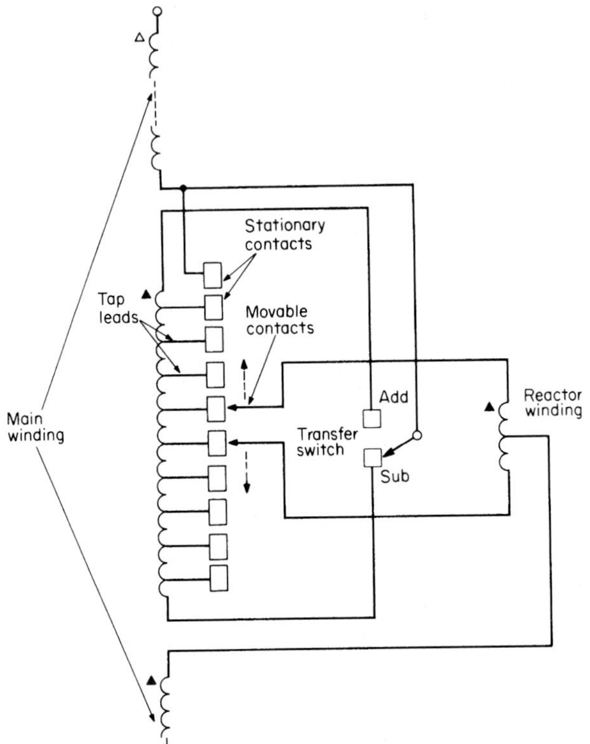

FIGURE 40.7 Under-load tap changer schematic.

the second finger is slid off the original contact and the current is carried by the first finger. In the last step in the sequence of operation, the fingers are slid so that both fingers are on the second stationary contact, thus completing the sequence.

Typical automatic controls which govern operation of the load tap changer consist of a line-drop compensating unit, a voltage-sensing unit, and a time-delay unit.

Line-Drop Compensator. Sometimes it is desirable to maintain a constant voltage at some other point on the circuit than at the source end of the line. This point, for example, may be the center of the load on the circuit. It would be possible to hold a constant voltage at this point by placing a potential transformer on the line and extending a pair of wires back to the voltage-regulating relay for the secondary voltage. This method is not too practical, and a device called a *line-drop compensator* is used instead. The line-drop compensator consists of a resistance and a reactance element wired in series with a current transformer which is in series with one of the regulator terminals, as shown in Fig. 40.9. In this circuit the values of resistance and reactance are adjusted so that they form a miniature line. Current flow through the current transformer primary winding to the load causes a proportional current to flow through the compensator. This current causes a voltage to appear across the resistance and reactance elements which is proportional to the voltage drop appearing on the line at some previously selected point. The voltages produced across the resistance and reactance elements are then introduced into the voltage-regulating relay circuit. The resistor voltage is applied so as to subtract from the voltage applied to the regulating relay by the potential

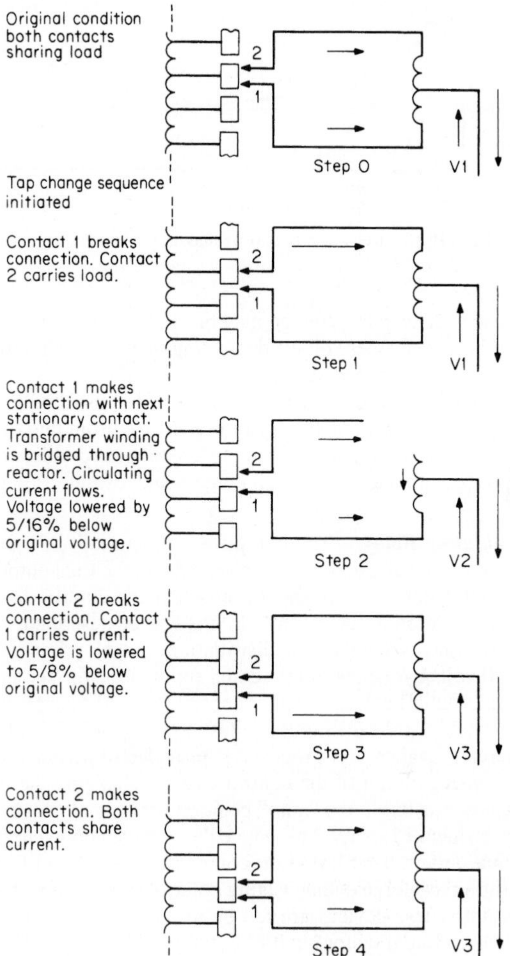

FIGURE 40.8 Under-load tap changer operation sequence.

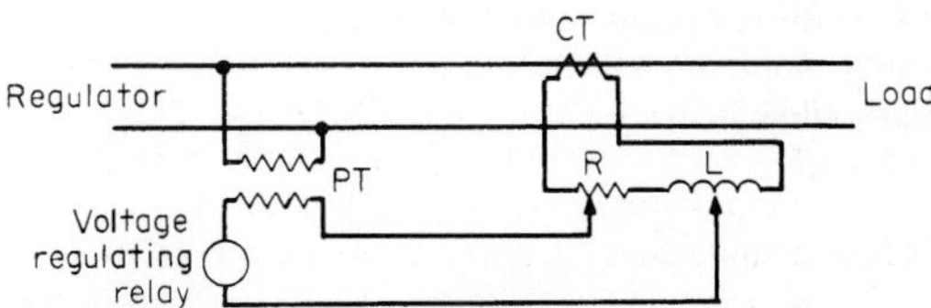

FIGURE 40.9 Simplified connections of a line-drop compensator.

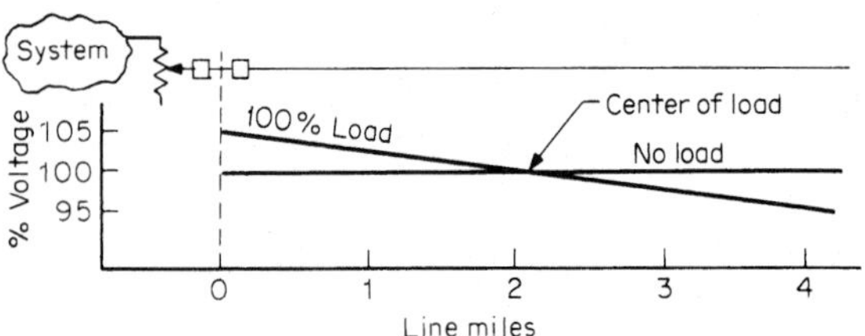

FIGURE 40.10 Regulated line using line-drop compensation.

transformer. The reactance element voltage may be applied so that it either subtracts from or adds to the potential transformer voltage depending on what condition the element is set for. Normally it would be set to subtract.

With this type of equipment, the voltage regulation of a distribution-system feeder for various load conditions is shown graphically in Fig. 40.10. From this figure it can be seen that this method of regulation is a good solution to a regulation problem because it supplies rated voltage or voltage within a very limited range to all customers on the line.

Static Voltage-Regulating Relay. The line-drop compensator unit, the voltage-sensing unit, the time-delay unit, and all auxiliary functions except the final output to the motor can be performed with solid-state devices shown on schematic control diagrams (Figs.40.11 through 40.14). The operation of the control circuits will be described.

The power transformer secondary or distribution-system primary output voltage is stepped down to 120 volts by the control potential supply transformer and is applied to the input of the automatic static-load tap changer control (Fig. 40.11). At this point, a normal test power supply switch permits the input of a test power supply for calibration and maintenance. Past the circuit breaker, a test rheostat is provided to permit variation of the voltage applied to the sensing circuit of the control without changing the transformer output voltage. The test rheostat is left in the "auto" position for normal operation.

A 5:1 ratio transformer (T1) is used to reduce the voltage applied to the line-drop compensator and sensing circuit. This line-drop compensator consists of a tapped line-drop reactance compensating reactor, a compensating polarity switch, and a line-drop resistance compensating rheostat. These devices are fed by the paralleling current transformer, the load current transformer, and transformer T1.

The purpose of the line-drop compensating unit is to develop voltages proportional to the reactive (*IX*) and resistive (*IR*) drops in the distribution line fed by the transformer. The final settings for the line-drop compensator may be made by field adjustment; however, initial settings may be determined from line data by using the following expressions:

d = miles from unit to load center or regulating point
r = resistance per conductor from unit to load center, ohms/mile
x = inductive reactance per conductor from unit to load center, ohms/mile
I = load current transformer primary current rating
E = control potential supply transformer ratio
R = resistance dial setting
X = reactance dial setting

For a three-phase balanced unit, then:

$$R = \frac{rdI}{E} \quad \text{and} \quad X = \frac{xdI}{E}$$

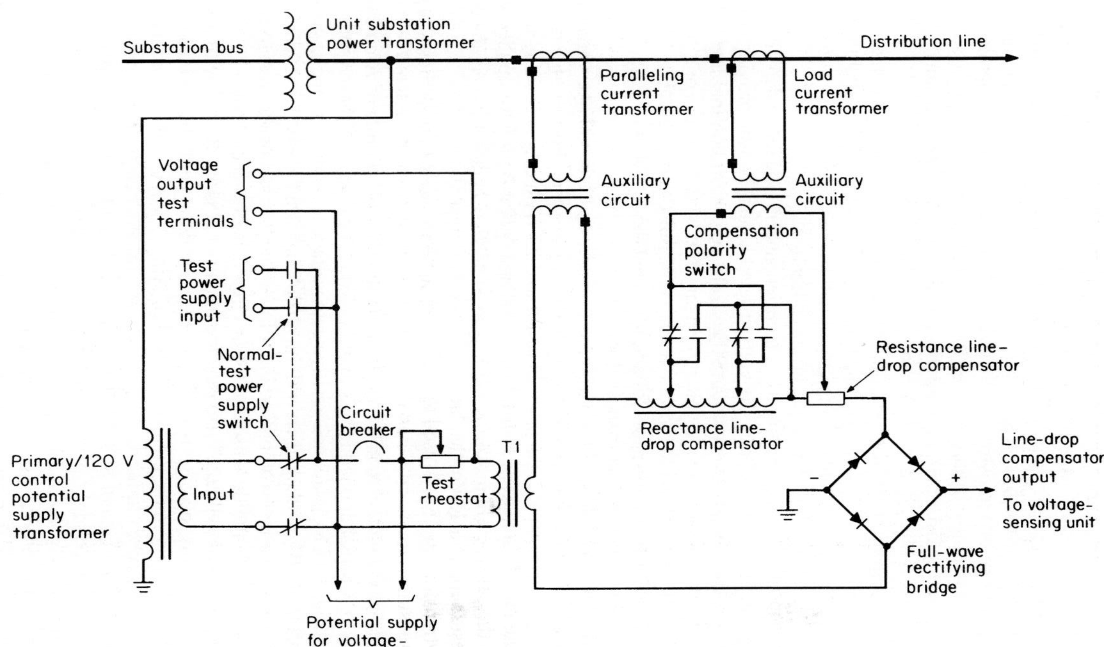

FIGURE 40.11 Line-drop compensating unit.

For normal operation, the compensation polarity switch is left in the "normal" position. Reversed reactance compensation is obtained by placing it in the "reverse" position. This reverses the current supplied by the load current transformer in the reactance element.

Reversed reactance compensation is a method used to reduce the circulating current that might flow when two or more transformers are paralleled. Instead of running toward opposite extreme positions, tap changers having reversed reactance compensation tend to move toward whatever position causes the least amount of transformer circulating current to flow. This is accomplished at some sacrifice to the normal line-drop compensation but is generally satisfactory when units paralleled are not located in close proximity to each other or where the supply is from different sources.

For the paralleling of two transformers located in close proximity and supplied from the same source, the circulating currents are used to control the line-drop compensator through the paralleling current transformer, causing the tap changers to operate in such a direction as to reduce the transformer circulating currents to a minimum.

For normal compensation applications, the voltage developed across the line-drop compensator is subtracted from the output of transformer T1. This forces the output of T1 to be higher for a given voltage into the sensing circuit and thus across the transformer to compensate for the drop created by increased current in the line.

The output of the line-drop compensator is rectified and filtered into a dc voltage by the full-wave bridge rectifier. At this point the voltage-level adjust rheostat provides for transformer voltage-level settings between about 90 and 110 percent of nominal line voltage.

The adjusted voltage is fed into the voltage-sensing unit which consists of a zener diode bridge network and the control windings of two magnetic pulse amplifiers (Fig. 40.12). The basic elements of the bridge are resistors R1 and R2, zener diodes Z1 and Z2, the minimum-bandwidth rheostat, and the ganged-bandwidth rheostat.

The zener diodes serve as reference voltage devices, and any voltage variations are reflected as differences in voltages across the resistors. The zener diode bridge is in a balanced state when the potential at the junction of R1 and Z1 (point A) is equal to the center of the active portion of the minimum-bandwidth rheostat (point B). As the applied voltage increases, the potential at point B rises with respect to that at point A. As the voltage decreases, the reverse occurs.

The output of the voltage-sensing bridge is fed through the control windings of a pair of magnetic amplifiers (Amp 1 and Amp 2). At balance, the current in the control winding of Amp 1 is from point C to point A, and the current in the control winding of Amp 2 is from point A to point D. Current applied in this direction causes each amplifier to produce negative pulses to the gate of SCR1. Since positive gate voltage is required to make the silicon controlled rectifier (SCR) conduct, there will be no output from the control. At equilibrium the ac windings of the magnetic amplifier are of low impedance; thus the transformer secondary is short-circuited, and the raise and lower circuits are de-energized.

When the applied input voltage increases, the potential at point B rises with respect to that at point A. When this increase is sufficient to cause the potential at point D to equal that at point A, the current in the control winding of Amp 2 will cut off. When the control current is cut off, the ac winding of Amp 2 changes impedance from low to high. This causes the voltage at the gate of SCR1 to become positive and causes it to conduct on the half-cycle. Since the secondary of transformer T2 is shorted out in only one direction, through diode D1, the voltage builds up across the ac winding of Amp 2 and voltage is applied to the "lower" portion of the relay circuit.

This voltage is applied through diode D4 to the time-delay unit (Fig. 40.13)—the time-delay circuit input voltage is regulated by zener diode Z3—and the time-delay control rheostat varies the amount of time delay. This time-delay unit has a range of about 10 to 100 sec and is usually set at 30 sec.

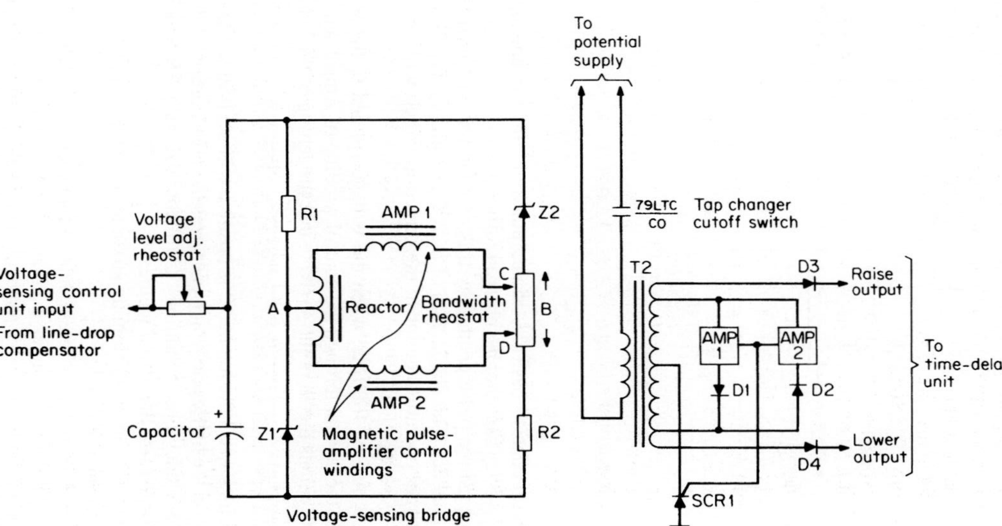

FIGURE 40.12 LTC voltage-sensing unit.

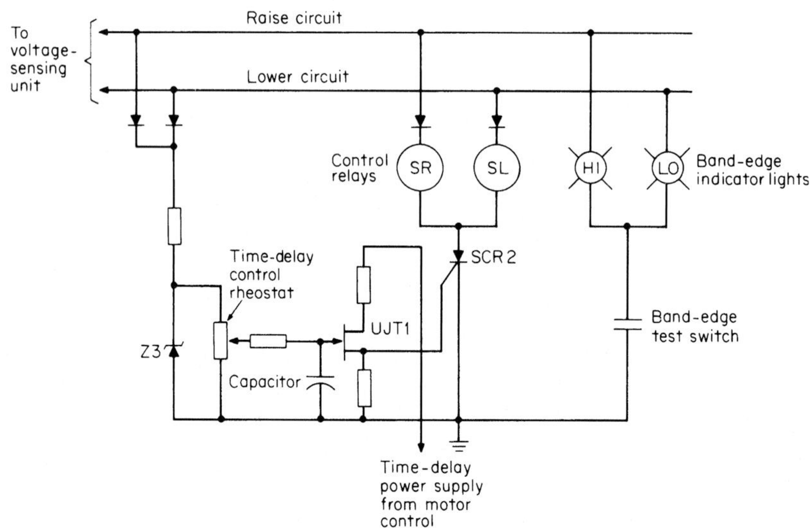

FIGURE 40.13 LTC time-delay unit.

The time-delay circuit consists basically of a unijunction transistor and an RC charging network. When the time-delay capacitor has charged sufficiently to cause the unijunction transistor to conduct, the positive pulse created by this conduction is applied to the gate of SCR 2, causing it to conduct. The conduction of SCR 2 energizes relay SL which in turn energizes the tap changer operating motor in the direction which will lower the output voltage (Fig. 40.14).

A similar sequence may be followed when the voltage applied to the sensing circuit decreases. In this case, the current through the control winding of Amp 1 will cut off when the potential at point C equals that at point A. (Note that the setting of the bandwidth rheostat determines the amount of deviation from balance required to initiate a controlling action.)

Since Amp 1 functions when the top of T2 is positive with respect to the bottom, SCR 1 is caused to conduct on the half-cycle which energizes the raise portion of the circuit, this time through diode D3. Under these conditions it will energize relay SR which will in turn operate the drive motor to raise the output voltage.

The tap changer motor is operated from an auxiliary potential supply transformer and has the following operational sequence.

Potential is supplied through the 84I crank-handle interlocking switch. When the crank handle is used to operate the tap changer manually, the 84I contact is open, preventing simultaneous operation of the tap changer manually and electrically. Potential is supplied from this point to the lower/auto/raise switch and to the remote control switch. When the remote switch is activated, the tap changer can be raised, lowered, or set to automatic control from a point outside the tap changer control compartment on the transformer.

The lower/auto/raise switch permits the tap changer to be raised, lowered, or set to automatic control at the tap changer.

In the auto position, the SL and SR contacts of the automatic static control device are employed. If the lower circuit is energized by the sensing unit and the SL relay picks up, the SL contact to the lower terminal of the motor is closed, completing a circuit through the 84/LS lower-end limit switch which cuts off motor operation if the tap changer has

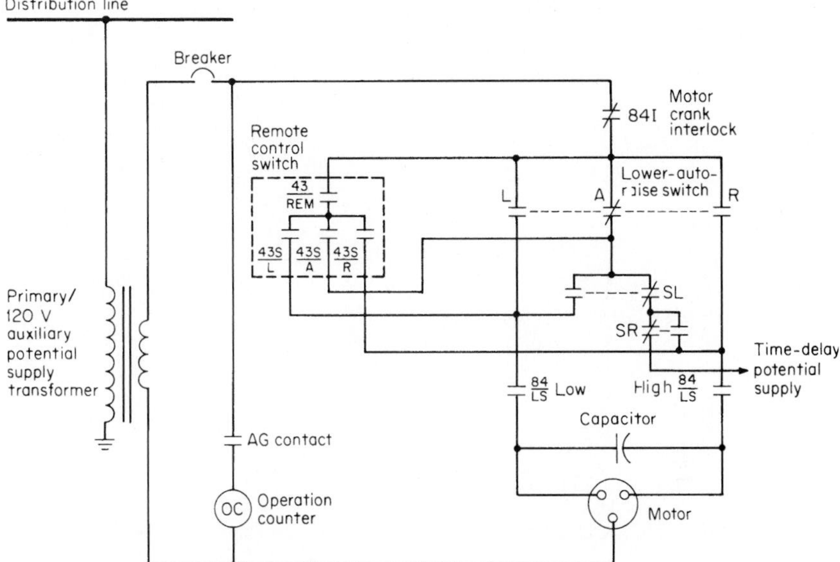

FIGURE 40.14 LTC motor control schematic.

advanced to its lowest position. The motor is now energized and will continue to run until it lowers the transformer voltage to satisfy the conditions set by the voltage-sensing unit.

When SL is energized, the contact supplying potential to the SR contacts is opened, preventing simultaneous raise and lower operations. When SL or SR is energized, the power supply to the time-delay unit is cut off, permitting it to reset.

In a similar manner, when the raise circuit is energized by the sensing unit and the SR relay picks up, the SR contact to the raise terminal of the motor is closed, completing a circuit through the 84/LS upper-end limit switch which cuts off motor operation if the tap changer has advanced to its highest position. The motor is again energized and will continue to run until it raises the transformer voltage to satisfy the conditions set by the voltage-sensing unit.

When a tap change occurs, the AG contact is closed, thus causing the operation counter OC to record that an operation has taken place.

CHAPTER 41

UNITS OF MEASUREMENT, ELECTRICAL DEFINITIONS, ELECTRICAL FORMULAS, AND CALCULATIONS

General. In this chapter, the most common units of measurement, definitions, and electrical formulas with example calculations are provided.

Units of Measurement

Time

1 minute	=	60 seconds
1 hour	=	60 minutes
1 day	=	24 hours

Length

1 foot	=	12 inches
1 yard	=	3 feet
1 mile	=	5280 feet
1 mile	=	1760 yards

Decimal units

milli- 10^{-3}
centi- 10^{-2}
deci- 10^{-1}
kilo- 10^{3}
mega- 10^{6}

Metric equivalents

U.S. to Metric

1 inch	=	0.0254 meters	=	2.54 centimeters	=	25.4 millimeters
2 inches	=	0.0508 meters	=	5.08 centimeters	=	50.8 millimeters
3 inches	=	0.0762 meters	=	7.62 centimeters	=	76.2 millimeters
4 inches	=	0.1016 meters	=	10.16 centimeters	=	101.6 millimeters
5 inches	=	0.1270 meters	=	12.70 centimeters	=	127.0 millimeters
6 inches	=	0.1524 meters	=	15.24 centimeters	=	152.4 millimeters
7 inches	=	0.1778 meters	=	17.78 centimeters	=	177.8 millimeters
8 inches	=	0.2032 meters	=	20.32 centimeters	=	203.2 millimeters
9 inches	=	0.2286 meters	=	22.86 centimeters	=	228.6 millimeters
10 inches	=	0.2540 meters	=	25.40 centimeters	=	254.0 millimeters

11 inches	=	0.2794 meters	=	27.94 centimeters	=	279.4 millimeters
1 foot	=	0.3048 meters	=	30.48 centimeters	=	304.8 millimeters
1 yard	=	0.9144 meters	=	91.44 centimeters	=	914.4 millimeters
1 mile	=	1609.35 meters	=	1.60935 kilometers		

Metric to U.S.

1 millimeter = 0.03937 inches
1 centimeter = 0.39370 inches
1 meter = 1.09361 yards = 3.28083 feet = 39.3700 inches
1 kilometer = 0.62137 miles

Temperature

Centigrade to Fahrenheit			Fahrenheit to Centigrade		
$C = 5/9(F - 32)$			$F = (9/5)C + 32$		
0°C	=	32°F	0°F	=	−17.8°C
10°C	=	50°F	10°F	=	−12.2°C
20°C	=	68°F	20°F	=	−6.7°C
30°C	=	86°F	30°F	=	−1.1°C
40°C	=	104°F	32°F	=	0°C
50°C	=	122°F	40°F	=	4.4°C
60°C	=	140°F	50°F	=	10.0°C
70°C	=	158°F	60°F	=	15.6°C
80°C	=	176°F	70°F	=	21.1°C
90°C	=	194°F	80°F	=	26.7°C
100°C	=	212°F	90°F	=	32.2°C
			100°F	=	37.8°C

DEFINITIONS

- **Voltage (E)** The electric force of work required to move current through an electric circuit, measured in units of volts.
- **Volt (V)** The unit of measurement of voltage in an electric circuit, denoted by the symbol V.
- **Current (I)** The flow of free electrons in one general direction, measured in units of amps.
- **Amp (A)** The unit of measurement of electric current, denoted by the symbol A.
- **Direct Current (dc)** Current that flows continually in one direction.
- **Alternating Current (ac)** Current that flows in a circuit in a positive direction and then reverses itself to flow in a negative direction.
- **Resistance (R)** The electrical "friction" that must be overcome through a device in order for current to flow when voltage is applied.
- **Ohm (Ω)** The unit of measurement of resistance in an electric circuit, denoted by the symbol "Ω".
- **Energy** The amount of electric work (real power) consumed or utilized in an hour. The unit of measurement is the watt-hour.
- **Power (P)** The combination of electric current and voltage causing electricity to produce work. Power is composed of two components: real power and reactive power.
- **Voltampere (VA)** The unit of both real and reactive power in an electric circuit.

- **Real Power** The resistive portion of a load found by taking the cosine (Θ) of the angle that the current and voltage are out of phase.
- **Watt (W)** The unit of real power in an electric circuit.
- **Reactive Power** The reactive portion of a load, found by taking the sine (Θ) of the angle that the current and voltage are out of phase.
- **Var (Q)** The unit of reactive power in an electric circuit.
- **Power Factor (p.f.)** The ratio of real power to reactive power.
- **Frequency (f)** The number of complete cycles made per second, measured in units of hertz.
- **Hertz (Hz)** Units of frequency (equal to 1 cycle per second).
- **Conductors** Materials that have many free electrons and are good transporters for the flow of electric current.
- **Insulators** Materials that have hardly any free electrons and inhibit or restrict the flow of electric current.
- **Series Resistive Circuit** All of the resistive devices are connected to each other so that the same current flows through all of the devices.
- **Parallel Resistive Circuit** Each resistive device is connected across a voltage source. The current in the parallel path divides and only a portion of the current flows through each of the parallel paths.

FORMULAS AND CALCULATIONS

In the material that follows, the most common electrical formulas are given and their use is illustrated with a problem. In each case, the formula is first stated by using the customary symbols; the same expression is then stated in words; a problem is given; and, finally, the substitutions are made for the symbols in the formula from which the answer is calculated. Only those formulas that the lineman is apt to use are given. The formulas are divided into three groups: direct-current circuits, alternating-current circuits, and electrical apparatus. A summary of circuit calculations is shown in Table 41.1.

Direct-current circuits

Ohm's Law. The formula for Ohm's law is the most fundamental of all electrical formulas. It expresses the relationship that exists in an electric circuit containing resistance only between the current flowing in the resistance, the voltage impressed on the resistance, and the resistance of the circuit.

$$I = \frac{E}{R}$$

where I = current, amps
E = voltage, volts
R = resistance, ohms

Expressed in words, the formula states that current equals voltage divided by resistance, or

$$\text{Amperes} = \frac{\text{volts}}{\text{ohms}}$$

TABLE 41.1 Summary of Power Formulas

To find	Direct current	Alternating current	
		Single phase	Three phase
Amperes when kilovolt-amperes and voltage are known	$\frac{\text{Kilovolt-amperes} \times 1000}{\text{Volts}}$	$\frac{\text{Kilovolt-amperes} \times 1000}{\text{Volts}}$	$\frac{\text{Kilovolt-amperes} \times 1000}{1.73 \times \text{volts}}$
Kilovolt-amperes when amperes and voltage are known	$\frac{\text{Amperes} \times \text{volts}}{1000}$	$\frac{\text{Amperes} \times \text{volts}}{1000}$	$\frac{1.73 \times \text{volts} \times \text{amperes}}{1000}$
Amperes when kilowatts, volts, and power factors are known	$\frac{\text{Kilowatts} \times 1000}{\text{Volts}}$	$\frac{\text{Kilowatts} \times 1000}{\text{Volts} \times \text{power factor}}$	$\frac{\text{Kilowatts} \times 1000}{1.73 \times \text{volts} \times \text{power factor}}$
Kilowatts when amperes, volts, and power factor are known	$\frac{\text{Amperes} \times \text{volts}}{1000}$	$\frac{\text{Amperes} \times \text{volts} \times \text{power factor}}{1000}$	$\frac{1.73 \times \text{amperes} \times \text{volts} \times \text{power factor}}{1000}$

EXAMPLE How much direct current will flow through a resistance of 10 ohms if the pressure applied is 120 volts direct current?

Solution

$$I = \frac{E}{R} = \frac{120}{10} = 12 \text{ amps}$$

Ohm's law involves three quantities: current, voltage, and resistance. Therefore, when any two are known, the third one can be found. The procedure for solving for current has already been illustrated. To find the voltage required to circulate a given amount of current through a known resistance. Ohm's law is written as

$$E = IR$$

In words, the formula says that the voltage is equal to the current multiplied by the resistance; thus

$$\text{Volts} = \text{amperes} \times \text{ohms}$$

EXAMPLE How much direct-current (dc) voltage is required to circulate 5 amps direct current through a resistance of 15 ohms?

Solution

$$E = IR = 5 \times 15 = 75 \text{ volts}$$

In like manner, if the voltage and current are known and the value of resistance is to be found, the formula is

$$R = \frac{E}{I}$$

Expressed in words, the formula says that the resistance is equal to the voltage divided by the current; thus

$$\text{Ohms} = \frac{\text{volts}}{\text{amperes}}$$

EXAMPLE What is the resistance of an electric circuit of 120 volts direct current that causes a current of 5 amps to flow through it?

Solution

$$R = \frac{E}{I} = \frac{120}{5} = 24 \text{ ohms}$$

Series resistive circuit

EXAMPLE The 12-volt battery has three lamps connected in series (Fig. 41.1). The resistance of each lamp is 8 ohms. How much current is flowing in the circuit?

Solution In a series-resistive circuit, all values of resistance are added together, as defined by the formula

$$R_{series} = R_1 + R_2 + \cdots\cdots = R_{total}$$

Therefore: $R = 8 + 8 + 8 = 24$ ohms

And because $I = E/R$
$= 12/24$
$= 0.5$ amps

FIGURE 41.1 A series circuit. The same current flows through all the lamps.

Parallel resistive circuit

EXAMPLE The 12-volt battery has three lamps connected in parallel (Fig. 41.2). The resistance of each lamp is 8 ohms. How much current is flowing in the circuit?

Solution In a parallel-resistive circuit, all values of resistance can be added together, as defined by the formula

$$1/R_{\text{parallel}} = 1/R_1 + 1/R_2 + \cdots\cdots + 1/R_n$$

In our three-resistor example

$$1/R_{\text{parallel}} = 1/R_1 + 1/R_2 + 1/R_3$$
$$1/R_{\text{parallel}} = (R_1R_2 + R_2R_3 + R_1R_3)/(R_1 \times R_2 \times R_3)$$
$$R_{\text{parallel}} = (R_1 \times R_2 \times R_3)/(R_1R_2 + R_2R_3 + R_1R_3)$$

Therefore: $R = (8 \times 8 \times 8)/[(8 \times 8) + (8 \times 8) + (8 \times 8)]$
$= 512/192$
$= 2.67$ ohms

And because $I = E/R_{\text{parallel}}$
$= 12/2.67$
$= 4.5$ amps

Power Formula. The expression for the power drawn by a dc circuit is

$$P = EI$$

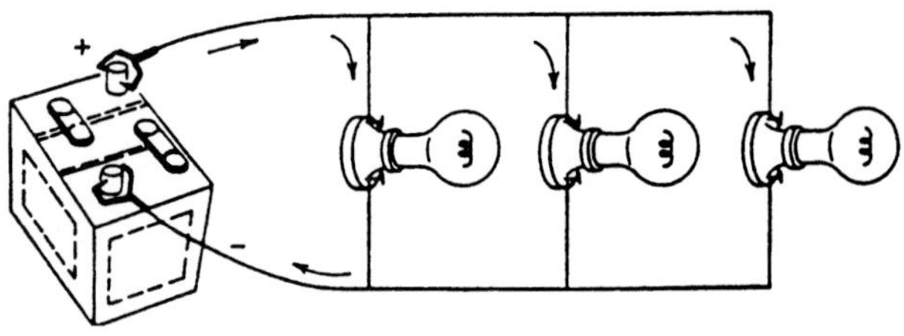

FIGURE 41.2 A parallel circuit. Each lamp is independent of the other lamps and draws its own current.

where E and I are the symbols for voltage in volts and current in amperes and P is the symbol for power in watts.

Expressed in words, the formula says that the power in watts drawn by a dc circuit is equal to the product of volts and amperes; thus

$$\text{Power} = \text{volts} \times \text{amperes}$$

EXAMPLE How much power is taken by a 120-volt dc circuit when the current flowing is 8 amps direct current?

Solution

$$P = EI = 120 \times 8 = 960 \text{ watts}$$

The power formula given above also contains three quantities or terms, watts, volts, and amperes. Therefore, when any two of the three are known, the third one can be found. The procedure for finding the power when the voltage and current are given has already been illustrated. When the power and voltage are given and the current is to be found, the formula is

$$I = \frac{P}{E}$$

In words, the formula states that the current equals the power divided by the voltage; thus

$$\text{Amperes} = \frac{\text{watts}}{\text{volts}}$$

EXAMPLE How much direct current would a 1000-watt load draw when connected to a 120-volt dc circuit?

Solution

$$I = \frac{P}{E} = \frac{1000}{120} = 8.33 \text{ amps}$$

In like manner, when the power and current are known, the voltage can be found by writing the formula thus

$$E = \frac{P}{I}$$

Expressed in words, the formula says that the voltage is equal to the power divided by the current; thus

$$\text{Volts} = \frac{\text{watts}}{\text{amperes}}$$

EXAMPLE What dc voltage would be required to deliver 660 watts with 6 amps direct current flowing in the circuit?

Solution

$$E = \frac{P}{I} = \frac{660}{6} = 110 \text{ volts}$$

Line Loss or Resistance Loss. The formula for computing the power lost in a resistance when current flows through it is

$$P = I^2R$$

where the symbols have the same meaning as in the foregoing formulas.

Expressed in words, the formula says that the power in watts lost in a resistance is equal to the square* of the current in amperes multiplied by the resistance in ohms; thus

$$\text{Watts} = \text{amperes squared} \times \text{ohms}$$

EXAMPLE Compute the watts lost in a line having a resistance of 4 ohms when 8 amps direct current is flowing in the line.

Solution $$P = I^2R = 8 \times 8 \times 4 = 256 \text{ watts}$$

Energy (electrical work). The formula for computing the amount of energy consumed is defined as

$$\text{Energy} = \text{Power} \times \text{time}$$

Power is measured in units of watts and time is typically measured in units of hours, so the units for energy will be in watt-hours.

EXAMPLE How much energy does a 60-watt load consume in a 10-hour time period?

The formula for energy is given by

$$\text{Energy} = \text{Power} \times \text{Time}$$

Therefore

$$\begin{aligned} \text{Energy} &= 60 \text{ watts} \times 10 \text{ hours} \\ &= 600 \text{ watt-hours} \\ &= 0.6 \text{ kilowatt-hours} \\ &= 0.6 \text{ kwh} \end{aligned}$$

ALTERNATING-CURRENT CIRCUITS

Ohm's Law. Ohm's law is the same for resistance circuits when alternating voltage is applied as when direct voltage is applied, namely

$$I = \frac{E}{R}$$

and

$$E = RI$$

and

$$R = \frac{E}{I}$$

In the preceding equations, E is the effective value of the alternating voltage and I the *effective* value of the alternating current. (See *Direct-Current Circuits* for examples.)

Ohm's Law for Other than Resistance Circuits. When alternating currents flow in circuits, these circuits may exhibit additional characteristics besides resistance. They may

* To *square* means to multiply by itself.

exhibit inductive reactance or capacitive reactance or both. The total opposition offered to the flow of current is then called *impedance* and is represented by the symbol Z. Ohm's law then becomes

$$I = \frac{E}{Z}$$

and

$$E = IZ$$

and

$$Z = \frac{E}{I}$$

where I = current, amps
E = voltage, volts
Z = impedance, ohms

EXAMPLE Find the impedance of an alternating-current (ac) circuit if a 120-volt alternating voltage source causes 30 amps alternating current to flow.

Solution $Z = \frac{E}{I} = \frac{120}{30} = 4\text{-ohms impedance}$

Impedance. The impedance of a series circuit is given by the expression

$$Z = \sqrt{R^2 + (X_L - X_c)^2}$$

where Z = impedance, ohms
R = resistance, ohms
X_L = inductive reactance, ohms
X_c = capacitive reactance, ohms

EXAMPLE If an ac series circuit contains a resistance of 5 ohms, an inductive reactance of 10 ohms, and a capacitive reactance of 6 ohms, what is its impedance in ohms?

$$\begin{aligned} Z &= \sqrt{R^2 + (X_L - X_c)^2} \\ &= \sqrt{5 \times 5 + (10 - 6)^2} \\ &= \sqrt{25 + (4 \times 4)} \\ &= \sqrt{41} = 6.4 \text{ ohms} \end{aligned}$$

Note: The values of inductive and capacitive reactance of a circuit depend upon the frequency of the current, the size, spacing, and length of the conductors making up the circuit, etc. For distribution circuits and transmission lines, values may be found in appropriate tables.

In the expression for Z, when neither inductive reactance nor capacitive reactance is present, Z reduces to the value R; thus

$$Z = \sqrt{R^2 + (0 - 0)^2} = \sqrt{R^2} = R$$

Likewise, when only resistance and inductive reactance are present in the circuit, the expression of Z becomes

$$Z = \sqrt{R^2 + (X_L - 0)^2} = \sqrt{R^2 + X_L^2}$$

EXAMPLE Compute the impedance of an ac circuit containing 3 ohms of resistance and 4 ohms of inductive reactance connected in series.

SOLUTION

$$Z = \sqrt{R^2 + X_L^2} = \sqrt{3^2 + 4^2}$$
$$= \sqrt{3 \times 3 + 4 \times 4} = \sqrt{9 + 16}$$
$$= \sqrt{25} = \text{5-ohms impedance}$$

Line Loss or Resistance Loss. The formula for computing the power lost in a resistance or line is

$$P = I^2R$$

where the symbols have the same meaning as above. Expressed in words, the formula says that the power in watts lost in a resistance is equal to the current in amperes squared multiplied by the resistance in ohms, thus

$$\text{Watts} = \text{amperes squared} \times \text{ohms}$$

EXAMPLE Compute the power lost in a line having a resistance of 3 ohms if 20 amps is flowing in it.

Solution $P = I^2R = 20 \times 20 \times 3 = 1200$ watts

Power Formula. The power formula for single-phase ac circuits is

$$P = E \times I \times \text{pf}$$

where P = power, watts
E = voltage, volts
I = current, amps
pf = power factor of circuit

Expressed in words, the formula says that the power in watts drawn by a single-phase ac circuit is equal to the product of volts, amperes, and power factor. If the power factor is unity, or 1, then the power equals the product of volts and amperes.

EXAMPLE What is the power drawn by a 20-amp purely resistive load, with a power factor of 1, operating at 120 volts?

Solution

$$P = E \times I \times \text{pf}$$
$$= 120 \times 20 \times 1$$
$$= 2400 \text{ watts}$$
$$= 2.4 \text{ kilowatts}$$
$$= 2.4 \text{ kW}$$

EXAMPLE How much power is delivered to a single-phase ac circuit operating at 120 volts if the circuit draws 10 amps at 80 percent power factor?

Solution

$$P = E \times I \times \text{pf}$$
$$= 120 \times 10 \times 0.80$$
$$= 960 \text{ watts}$$

The power formula given previously contains four quantities: P, E, I, and pf. Therefore, when any three are known, the fourth one can be determined. The case where the voltage, current, and power factor are known has already been illustrated. When the voltage, power, and power factor are known, the current can be computed from the expression

$$I = \frac{P}{E \times \text{pf}}$$

EXAMPLE How many amperes are required to light an ordinary 60-watt, 120-volt incadescent lamp? (Assume that the power factor of a purely resistive load, such as the lamp, is 1.00)

Solution

$$I = \frac{P}{E \times \text{pf}}$$

$$= \frac{60}{120 \times 1}$$

$$= 0.5 \text{ amps}$$

EXAMPLE How much current is drawn by a 10-kW load from a 220-volt ac circuit if the power factor is 0.80?

Solution

$$I = \frac{P}{E \times \text{pf}}$$

$$= \frac{10{,}000}{220 \times 0.80} = 56.8 \text{ amps}$$

In similar manner, the power, voltage, and current are known, the power factor can be computed from the following formula

$$\text{pf} = \frac{P}{E \times I}$$

EXAMPLE What is the power factor of a 4-kW load operating at 230 volts if the current drawn is 20 amps?

Solution

$$\text{pf} = \frac{P}{E \times I} = \frac{4000}{230 \times 20}$$

$$= 0.87 \text{ or } 87\% \text{ power factor}$$

Three-Phase. The power formula for a three-phase ac circuit is

$$P = \sqrt{3} \times E \times I \times pf$$

where E is the voltage between phase wires and the other symbols have the same meaning as for single-phase. The additional quantity in the three-phase formula is the factor $\sqrt{3}$, expressed "square root of three," the value of which is

$$\sqrt{3} = 1.73$$

The voltage between phase wires is equal to $\sqrt{3}$ or 1.73, multiplied by the phase-to-neutral voltage of the three-phase four-wire systems.

EXAMPLE How much power in watts is drawn by a three-phase load at 230 volts if the current is 10 amps and the power factor is 80 percent?

Solution

$$P = \sqrt{3} \times E \times I \times \text{ pf}$$
$$= \sqrt{3} \times 230 \times 10 \times 0.80$$
$$= 3183 \text{ watts}$$

The three-phase power formula also contains four quantities, as in the single-phase formula. Therefore, if any three are known, the fourth one can be determined. The case where the voltage, current, and power factor are known was illustrated above. When the power, voltage, and power factor are known, the current in the three-phase line can be computed from the formula

$$I = \frac{P}{\sqrt{3}E \times \text{pf}}$$

EXAMPLE If a 15-kW three-phase load operates at 2300 volts and has a power factor of 1, how much current flows in each line of the three-phase circuit?

Solution

$$I = \frac{P}{\sqrt{3}E \times \text{pf}}$$
$$= \frac{15{,}000}{\sqrt{3} \times 2300 \times 1.0} = 3.77 \text{ amps}$$

If the only quantity not known is the power factor, this can be found by using the following formula

$$\text{pf} = \frac{P}{\sqrt{3} \times E \times I}$$

EXAMPLE Find the power factor of a 100-kW three-phase load operating at 2300 volts if the line current is 40 amps.

Solution

$$\text{pf} = \frac{P}{\sqrt{3} \times E \times I}$$
$$= \frac{100{,}000}{\sqrt{3} \times 2300 \times 40}$$
$$= 0.628 \text{ or } 62.8\% \text{ power factor}$$

Voltamperes. Often loads are given in voltamperes instead of watts or in kVA instead of kW. The relationships then become

$$\text{VA} = E \times I \qquad \text{For single-phase}$$

and

$$\text{VA} = \sqrt{3}E \times I \quad \text{for three-phase}$$

In using these quantities it is not necessary to know the power factor of the load.

Power factor correction

To determine the kVars required to improve the power factor for a known load, the following steps are followed:

Steps 1: Determine the existing KVA load

$$KVA_{existing} = \frac{KW}{pf_{existing}}$$

Step 2: Determine the existing reactive power load, expressed in kVARs

$$kVAR_{existing} = \sqrt{KVA^2_{existing} - KW^2}$$

Step 3: Determine the corrected KVA load

$$KVA_{corrected} = \frac{KW}{PF_{corrected}}$$

Step 4: Determine the corrected reactive power load, expressed in kVARs

$$kVAR_{corrected} = \sqrt{KVA^2_{corrected} - KW^2}$$

Step 5: The required corrective reactive power is equal to the existing reactive power load minus the corrected reactive power load, expressed in kVARs.

$$kVAR_{required} = kVAR_{existing} - kVAR_{corrected}$$

EXAMPLE For a 40-kW load at 80% power factor, how much kVAR is needed to correct the power factor to 90 percent?

The first step is to determine the kilovolt load in the existing condition:

$$\begin{aligned} KVA_{existing} &= KW/pf_{existing} \\ &= 40/0.8 \\ &= 50\ kVA \end{aligned}$$

Step 2 is to determine the reactive power load in the existing condition:

$$\begin{aligned} kVAR_{existing} &= \sqrt{KVA^2 - KW^2} \\ &= \sqrt{50^2 - 40^2} \\ &= 30\ kVARs \end{aligned}$$

Having found the existing condition reactive power load, determine what the corrected final values for the total and reactive power will be

Step 3:

$$\begin{aligned} KVA_{corrected} &= KW/pf_{corrected} \\ &= 40/0.9 \\ &= 44.4\ kVA \end{aligned}$$

Step 4:

$$\begin{aligned} kVAR_{corrected} &= \sqrt{KVA^2 - KW^2} \\ &= \sqrt{44.4^2 - 40^2} \\ &= 19.37\ kVARs \end{aligned}$$

To determine the amount of reactive power needed on the system to correct the power factor from 80 to 90 percent, the final step of the problem is to subtract the corrected reactive power value desired from the existing value of reactive power:

Step 5:

$$\begin{aligned} \text{kVAR}_{\text{required}} &= \text{kVAR}_{\text{existing}} - \text{kVAR}_{\text{corrected}} \\ &= 30 - 19.37 \\ &= 10.63 \text{ kVARs} \end{aligned}$$

Line Loss or Resistance Loss (Three-Phase). The power loss in a three-phase line is given by the expression

$$P = 3 \times I^2R$$

where I = current in each line wire and R = resistance of each line wire. The factor 3 is present so that the loss in a line wire will be taken 3 times to account for the loss in all three wires.

EXAMPLE If a three-phase line carries a current of 12 amps and each line wire has a resistance of 3 ohms, how much power is lost in the resistance of the line?

Solution $P = 3 \times I^2R = 3 \times 12 \times 12 \times 3 = 1296$ watts

ELECTRICAL APPARATUS

Motors

Direct-Current Motor. The two quantities that are usually desired in an electric motor are its output in horsepower (hp) and its input current rating at full load. The following expression for current is used when the motor is a direct-current motor

$$I = \frac{\text{hp} \times 746}{E \times \text{eff}}$$

The formula says that the full-load current is obtained by multiplying the horse-power by 746 and dividing the result by the product of voltage and percent efficiency. (The number of watts in 1 hp is 746. A kilowatt equal $1\frac{1}{3}$ horsepower.)

EXAMPLE How much current will a 5-hp 230-volt dc motor draw at full load if its efficiency at full load is 90 percent?

Solution $$I = \frac{\text{hp} \times 746}{E \times \text{eff}} = \frac{5 \times 746}{230 \times 0.90} = 18.0 \text{ amps}$$

Alternating-Current Motor, Single-Phase. If the motor is a single-phase ac motor, the expression for current is almost the same, except that it must allow for power factor. The formula for full-load current is

$$I = \frac{\text{hp} \times 746}{E \times \text{eff} \times \text{pf}}$$

EXAMPLE How many amperes will a $\frac{1}{4}$-hp motor take at full load if the motor is rated 110 volts and has a full-load efficiency of 85 percent and a full-load power factor of 80 percent?

Solution $$I = \frac{\text{hp} \times 746}{E \times \text{eff} \times \text{pf}} = \frac{0.25 \times 746}{110 \times 0.85 \times 0.80} = 2.5 \text{ amps}$$

Alternating-Current Motor, Three-Phase. The formula for full-load current of a three-phase motor is the same as for a single-phase motor except that it has the factor $\sqrt{3}$ in it thus

$$I = \frac{\text{hp} \times 746}{\sqrt{3} \times E \times \text{eff} \times \text{pf}}$$

EXAMPLE Calculate the full-load current rating of a 5-hp three-phase motor operating at 220 volts and having an efficiency of 80 percent and a power factor of 85 percent.

Solution $$I = \frac{\text{hp} \times 746}{\sqrt{3} \times E \times \text{eff} \times \text{pf}} = \frac{5 \times 746}{\sqrt{3} \times 220 \times 0.80 \times 0.85}$$
$$= 14.4 \text{ amps}$$

Direct-Current Motor. To find the horsepower rating of a dc motor if its voltage and current rating are known, the same quantities are used as for current except the formula is rearranged thus

$$\text{hp} = \frac{E \times I \times \text{eff}}{746}$$

EXAMPLE How many horsepower can a 220-volt dc motor deliver if it draws 15 amps and has an efficiency of 90 percent?

Solution $$\text{hp} = \frac{E \times I \times \text{eff}}{746} = \frac{220 \times 15 \times 0.90}{746} = 4.0 \text{ hp}$$

Alternating-Current Motor, Single-Phase. The horsepower formula for a single-phase ac motor is

$$\text{hp} = \frac{E \times I \times \text{pf} \times \text{eff}}{746}$$

EXAMPLE What is the horsepower rating of a single-phase motor operating at 480 volts and drawing 25 amps at a power factor of 88 percent if it has a full-load efficiency of 90 percent?

Solution $$\text{hp} = \frac{E \times I \times \text{pf} \times \text{eff}}{746}$$
$$= \frac{480 \times 25 \times 0.88 \times 0.90}{746}$$
$$= 12.74 \text{ hp}$$

Alternating-Current Motor, Three-Phase. The horsepower formula for a three-phase ac motor is

$$\text{hp} = \frac{\sqrt{3} \times E \times I \times \text{pf} \times \text{eff}}{746}$$

EXAMPLE What horsepower does a three-phase 240-volt motor deliver if it draws 10 amps, has a power factor of 80 percent and has an efficiency of 85 percent?

Solution

$$\text{hp} = \frac{\sqrt{3} \times E \times I \times \text{pf} \times \text{eff}}{746}$$

$$= \frac{\sqrt{3} \times 240 \times 10 \times 0.80 \times 0.85}{746} = 3.8 \text{ hp}$$

Alternating-Current Generator

Frequency. The frequency of the voltage generated by an ac generator depends on the number of poles in its field and the speed at which it rotates:

$$f = \frac{p \times \text{rpm}}{120}$$

where f = frequency, Hz
p = number of poles in field
rpm = revolutions of field per minute

EXAMPLE Compute the frequency of the voltage generated by an alternator having two poles and rotating at 3600 rpm.

Solution

$$f = \frac{p \times \text{rpm}}{120} = \frac{2 \times 3600}{120} = 60 \text{ Hz}$$

Speed. To determine the speed at which an alternator should be driven to generate a given frequency, the following expression is used:

$$\text{rpm} = \frac{f \times 120}{p}$$

EXAMPLE At what speed must a four-pole alternator be driven to generate 60 Hz?

Solution

$$\text{rpm} = \frac{f \times 120}{p} = \frac{60 \times 120}{4} = 1800 \text{ rpm}$$

Number of Poles. If the frequency and speed of an alternator are known, the number of poles in its field can be calculated by use of the following formula:

$$p = \frac{f \times 120}{\text{rpm}}$$

EXAMPLE How many poles does an alternator have if it generates 60 Hz at 1200 rpm?

Solution

$$p = \frac{f \times 120}{\text{rpm}} = \frac{60 \times 120}{1200} = 6 \text{ poles}$$

Transformer (Single-Phase)

Primary Current. The full-load primary current can be readily calculated if the kVA rating of the transformer and the primary voltage are known:

$$I_P = \frac{\text{kVA} \times 1000}{E_P}$$

where E_P = rated primary voltage and I_P = rated primary current.

EXAMPLE Find the rated full-load primary current of a 10-kVA 2300-volt distribution transformer.

Solution

$$I_P = \frac{\text{kVA} \times 1000}{E_P} = \frac{10 \times 1000}{2300} = 4.3 \text{ amps}$$

Secondary Current. The expression for secondary current is similar to that for primary current. Thus

$$I_S = \frac{\text{kVA} \times 1000}{E_s}$$

where I_s = rated secondary current and E_s = rated secondary voltage.

EXAMPLE A 3-kVA distribution transformer is rated 2300 volts primary and 110 volts secondary. What is its full-load secondary current?

Solution

$$I_s = \frac{\text{kVA} \times 1000}{E_s}$$

$$= \frac{3 \times 1000}{110} = 27\text{-amps secondary}$$

SUMMARY OF ELECTRICAL FORMULAS

For a summary of circuit calculations, see Table 41.1.

Current

1. *Direct current.* To find the current when the voltage and resistance are given:

$$I = \frac{E}{R}$$

2. *Alternating current.* To find the current when the voltage and resistance are given:

$$I = \frac{E}{R}$$

3. *Alternating current.* To find the current when the voltage and impedance are given:

$$I = \frac{E}{Z}$$

4. *Direct current.* To find the current when the voltage and power are given:

$$I = \frac{P}{E}$$

5. *Alternating current, single-phase.* To find the current when the voltage, power, and power factor are given:

$$I = \frac{P}{E \times \text{pf}}$$

6. *Alternating current, three-phase.* To find the current when the voltage, power, and power factor are given:

$$I = \frac{P}{\sqrt{3}E \times \text{pf}}$$

7. *Alternating current, single-phase.* To find the current when the voltamperes and volts are given:

$$I = \frac{\text{VA}}{E}$$

and

$$I = \frac{\text{kVA} \times 1000}{E}$$

8. *Alternating current, three-phase.* To find the current when the voltamperes and volts are given:

$$I = \frac{\text{VA}}{\sqrt{3}E}$$

and

$$I = \frac{\text{kVA} \times 1000}{\sqrt{3}E}$$

Voltage

9. *Direct current.* To find the voltage when the current and resistance are given:

$$E = IR$$

10. *Alternating current.* To find the voltage when the current and resistance are given:

$$E = IR$$

11. *Alternating current.* To find the voltage when the current and impedance are given:

$$E = IZ$$

Resistance

12. *Direct current.* To find the resistance when the voltage and current are given:

$$R = \frac{E}{I}$$

13. *Alternating current.* To find the resistance when the voltage and current are given:

$$R = \frac{E}{I}$$

Power Loss

14. *Direct current.* To find the power loss when the current and resistance are given:

$$P = I^2R$$

15. *Alternating current, single-phase.* To find the power loss when the current and resistance are given:

$$P = I^2R$$

16. *Alternating current, three-phase.* To find the power loss when the current and resistance are given:

$$P = 3 \times I^2R$$

Impedance

17. *Alternating current, single-phase.* To find the impedance when the current and voltage are given:

$$Z = \frac{E}{I}$$

18. *Alternating current, single-phase.* To find the impedance when the resistance and inductive reactance are given:

$$Z = \sqrt{R^2 + X_L^2}$$

19. *Alternating current, single-phase.* To find the impedance when the resistance, inductive reactance, and capacitive reactance are given:

$$Z = \sqrt{R^2 + (X_L - X_c)^2}$$

Power

20. *Direct current.* To find the power when the voltage and current are given:

$$P = EI$$

21. *Alternating current, single-phase.* To find the power when the voltage, current, and power factor are given:

$$P = E \times I \times \text{pf}$$

22. *Alternating current, three-phase.* To find the power when the voltage, current, and power factor are given:

$$P = \sqrt{3}E \times I \times \text{pf}$$

Voltamperes

23. *Alternating current, single-phase.* To find the voltamperes when the voltage and current are given:

$$\text{VA} = E \times I$$

and

$$\text{kVA} = \frac{E \times I}{1000}$$

24. *Alternating current, three-phase.* To find the voltamperes when the voltage and current are given:

$$\mathrm{VA} = \sqrt{3}E \times I$$

and

$$\mathrm{kVA} = \frac{\sqrt{3}E \times I}{1000}$$

Power Factor

25. *Alternating current, single-phase.* To find the power factor when the power, volts, and amperes are given:

$$\mathrm{pf} = \frac{P}{E \times I}$$

26. *Alternating current, three-phase.* To find the power factor when the power, volts, and amperes are given:

$$\mathrm{pf} = \frac{P}{\sqrt{3} \times E \times I}$$

Motor

27. *Direct-current motor.* To find the current when the horsepower of the motor, the voltage, and the efficiency are given:

$$I = \frac{\mathrm{hp} \times 746}{E \times \mathrm{eff}}$$

28. *Alternating-current motor, single-phase.* To find the current when the horsepower of the motor, the voltage, the efficiency, and the power factor are given:

$$I = \frac{\mathrm{hp} \times 746}{E \times \mathrm{pf} \times \mathrm{eff}}$$

29. *Alternating-current motor, three-phase.* To find the current when the horse-power of the motor, the voltage, the power factor, and the efficiency are given:

$$I = \frac{\mathrm{hp} \times 746}{\sqrt{3}E \times \mathrm{pf} \times \mathrm{eff}}$$

30. *Direct-current motor.* To find the horsepower of the motor when the voltage, current, and efficiency are given:

$$\mathrm{hp} = \frac{E \times I \times \mathrm{eff}}{746}$$

31. *Alternating-current motor, single-phase.* To find the horsepower of the motor when the voltage, current, power factor, and efficiency are given:

$$\mathrm{hp} = \frac{E \times I \times \mathrm{pf} \times \mathrm{eff}}{746}$$

32. *Alternating-current motor, three-phase.* To find the horsepower of the motor when the voltage, current, power factor, and efficiency are given:

$$\text{hp} = \frac{\sqrt{3}E \times I \times \text{pf} \times \text{eff}}{746}$$

Generator

33. *Alternating-current generator.* To find the frequency of the generator when the number of poles and the rpm are given:

$$f = \frac{p \times \text{rpm}}{120}$$

34. *Alternating-current generator.* To find the rpm of the generator when the number of poles and the frequency are given:

$$\text{rpm} = \frac{f \times 120}{p}$$

35. *Alternating-current generator.* To find the number of poles of the generator when the frequency and the rpm are given:

$$p = \frac{f \times 120}{\text{rpm}}$$

Transformer

36. *Alternating-current transformer, single-phase.* To find the primary current when the kVA and the primary voltage are given:

$$I_P = \frac{\text{kVA} \times 1000}{E_P}$$

37. *Alternating-current transformer, single-phase.* To find the secondary current when the kVA and the secondary voltage are given:

$$I_s = \frac{\text{kVA} \times 1000}{E_s}$$

CHAPTER 42
MAINTENANCE OF TRANSMISSION AND DISTRIBUTION LINES

TRANSMISSION-LINE INSPECTION

Initial Inspection. All transmission lines should be inspected after construction is completed before energizing the line (Fig. 42.1). Linemen should climb each structure and check the following:

1. Conductor condition.
2. Conductor sag and clearance to ground, trees, and structures.
3. Insulator conditions.
4. Line hardware for roughness and tightness. Excess inhibitor found should be removed from conductors to prevent corona discharges.
5. Structure vibration and alignment.
6. Guys for anchors that are pulling out, guy wire conditions, and missing guy guards.

FIGURE 42.1 Extra-high-voltage ac electric transmission line completely constructed. Line is ready for climbing inspection prior to being energized. (*Courtesy Aluminum Co. of America.*)

7. Ground-wire connections and conditions.
8. Ground resistance at each structure.
9. Structure footings for washouts or damage.
10. Obstruction light operations for aircraft warning.

Aerial Inspections. Visual patrols from a helicopter or fixed-wing airplane, especially manufactured for this type of operation with a lineman observer, are normally completed on all-high-voltage electric transmission lines several times a year (Fig. 42.2). The lineman observer checks the items listed under initial inspection that can be seen from the air for problems. Aerial patrols are often initiated if a transmission line has an unscheduled outage, to locate the problem (Fig. 42.3). Damaged insulators, broken or damaged conductors, structures that are leaning or down, and tree clearance problems can easily be seen by an observer in a helicopter or airplane flying along the route of the line (Figs. 42.4 and 42.5).

Regular Inspections. Linemen normally repeat the initial climbing inspection after a line has been in service for a year and at approximately 10-year intervals thereafter. Inspections by linemen walking along the line are normally completed at 10-year intervals in between

FIGURE 42.2 A helicopter flying along an electric transmission line, with a lineman as an observer, can be used to inspect the line for deficiencies. A defective conductor spacer was located, by an aerial inspection, on the 345-kV ac transmission line in the picture. A lineman wearing a conductive suit is working from a platform on the helicopter to replace the defective spacer while the line is energized. The lineman is following procedures for bare-handed live-line work discussed in Chaps. 25 and 46.

FIGURE 42.3 Aerial inspection plane in the process of patrolling transmission line.

FIGURE 42.4 Aerial patrol observer located ice- and snow-damaged 161-kV transmission line.

FIGURE 42.5 Leaning 161-kV transmission-line angle structure located by observer during aerial patrol of line after circuit was removed from service by a snow and ice storm. (*Courtesy L. E. Myers Co.*)

the climbing inspections. These inspections provide close scrutinizing of the high-voltage transmission lines on a 5-year cycle. An instrument capable of measuring radio or TV interference voltage should be used by the linemen while inspecting the line to detect corona discharges and defective insulators.

DISTRIBUTION-LINE INSPECTION

Linemen work on distribution lines frequently to install facilities for new customers while making modifications required by load increases, and as a result of government agencies' modification of roadways. The lineman should thoroughly check the pole line and equipment each time work is performed. The distribution lines should be inspected at scheduled intervals not to exceed a period of 10 years (Fig. 42.6). Circuits with a record of poor electrical reliability should have special scheduled inspections. Inspectors should follow the route of the line, using a circuit map. Lines located remote from the streets should be inspected by the inspector walking along the circuit. Binoculars are desirable to permit the lineman to complete a thorough inspection. Operating characteristics and defects of distribution lines and equipment should be recorded in sufficient detail to permit the scheduling of repairs (Fig. 42.7). The lineman can complete the minor repairs as time permits (Fig. 42.8).

Common defects or conditions which the inspector should observe, record, and recommend to be repaired are as follows:

1. Poles

   a. Broken or damaged poles needing replacement

   b. Leaning poles in unstable soil

   c. Poles in hazardous locations

FIGURE 42.6 Lineman is checking ground connection to distribution circuit pole-top switch operating handle during scheduled switch inspection. (*Courtesy A. B. Chance Co.*)

FIGURE 42.7 A handled recording instrument utilized by the inspector used to record operating characteristics, manufacturer's recommended maintenance, and defects located by the linemen and to schedule maintenance and inspections. (*Courtesy Electric Power Research Institute*)

FIGURE 42.8 Lineman tightening loose connection causing radio interference discovered while inspecting distribution line. Notice use of coverup material for protection while performing the work. (*Courtesy W. H. Salisbury & Co.*)

2. Guys
   a. Slack, broken, or damaged guys
   b. Guy too close to primary conductors or equipment
   c. Guy insulator not installed as required
3. Pole-top assemblies
   a. Broken, burned, or damaged pins and/or cross-arm requiring replacement
   b. Broken skirt of pin-type insulators
4. Line conductors
   a. Too much slack in primary conductors
   b. Floating or loose conductors
   c. Conductors burning in trees
   d. Foreign objects on line
   e. Insufficient vertical clearance over and/or horizontal clearance from:
      (1) Other wires
      (2) Buildings, parks, playgrounds, roads, driveways, loading docks, silos, signs, windmills, and spaces and ways accessible to pedestrians
      (3) Waterways—sailboat-launching area
      (4) Railways
      (5) Airports
      (6) Swimming pools
5. Equipment installations
   a. Equipment leaking oil
   b. Arresters blown or operation of ground lead isolation device
   c. Blown fuses
   d. Switch contacts not properly closed

DISTRIBUTION-LINE EQUIPMENT MAINTENANCE

Capacitor Maintenance. All switched capacitor banks should be inspected and checked for proper operation once each year prior to the time period they are automatically switched on and off to meet system requirements (Fig. 42.9). Mechanical capacitor-bank oil switches should be maintained on a schedule related with the type of on/off controls installed at each bank. The maximum number of open and close operations between maintenance of the

Type of control	Maintenance schedule, years
Time clock	3
Voltage	3
Dual temperature	5
Temperature only	8
Time clock and temperature	8

FIGURE 42.9 Switched capacitor bank mounted on pole in distribution line. Note fuses and connections have been opened to permit removing oil switches for maintenance.

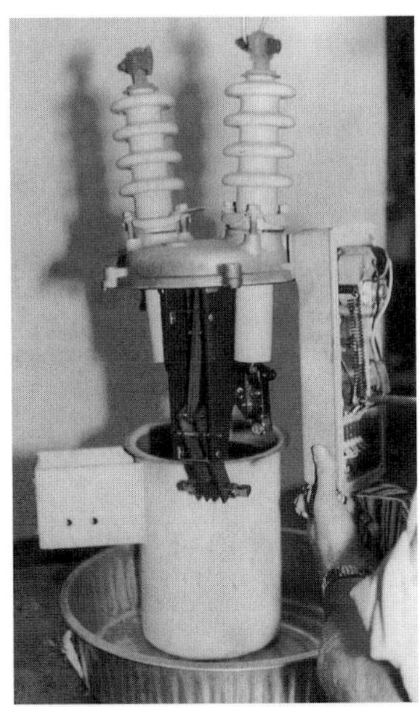

FIGURE 42.10 Capacitor oil switch mechanism removed from tank in shop for maintenance.

switches should not normally exceed 2500. Experience has shown that the schedule in the previous page will normally keep the equipment operating properly.

The capacitor switches are usually removed from the distribution line by the lineman and replaced with a spare unit during the season they are not normally operated. The capacitor switches can be maintained efficiently in a distribution shop by the lineman (Fig. 42.10).

Capacitor Precautions. Capacitors and transformers are entirely different in their operation. When a transformer is disconnected from the line, it is electrically dead. Unlike the transformer and other devices, the capacitor is not dead immediately after it is disconnected from the line. It has the peculiar property of holding its charge because it is essentially a device for storing electrical energy. It can hold this charge for a considerable length of time. There is a voltage difference across its terminals after the switch is opened.

Capacitors for use on electrical lines, however, are equipped with an internal-discharge resistor. This resistor connected across the capacitor terminals will gradually discharge the capacitor and reduce the voltage across its terminals. After 5 min, the capacitor should be discharged.

To be perfectly safe, however, proceed as follows: Before working on a disconnected capacitor, wait 5 min. Then test the capacitor with a high-voltage tester rated for the circuit voltage. If the voltage is zero, short-circuit the terminals externally using hot-line tools and ground the terminals to the case. Now you can proceed with the work.

Recloser Maintenance. Hydraulically controlled oil circuit reclosers should normally be maintained after 50 operations, or every 2 years. Electronically controlled oil reclosers should be maintained after approximately 150 operations, or every 2 years. The hydraulically or

FIGURE 42.11 Oil circuit recloser mechanism being removed from tank in shop for repairs.

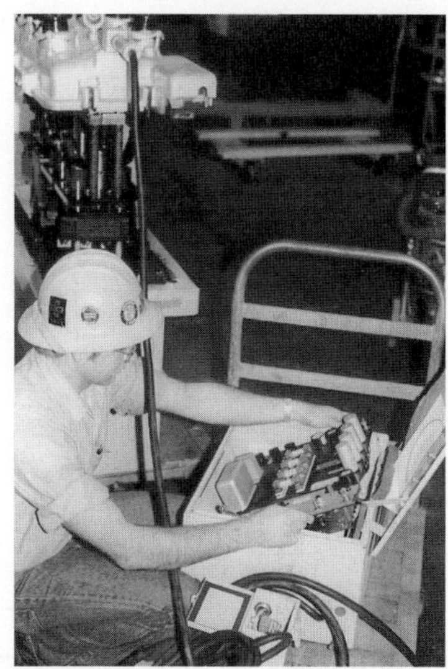

FIGURE 42.12 Oil circuit recloser electronic control being checked in shop. Note mechanism has been removed from tank for repairs and observation of operation.

electronically controlled oil reclosers can be maintained in the field or taken to the distribution shop (Figs. 42.11 and 42.12). The dielectric strength of the insulating oil should be preserved by filtering the oil until tests ensure good results. The recloser manufacturer's instructions should be followed to complete the inspection and repairs.

Vacuum reclosers require field testing on a 6-year schedule. An excessive number of recloser operations may make it necessary to reduce the scheduled maintenance time period. High-potential testing of the vacuum interrupter bottles can be completed in the field by the lineman. The electronic controls for the vacuum recloser can be tested in the field by following the manufacturer's instructions.

All line reclosers should be installed with bypass provisions and a means for easy isolation of the equipment for maintenance work (Fig. 42.13).

Pole-Mounted Switch Maintenance. Group-operated or single-pole switches installed in overhead distribution lines should be checked each time they are operated by the lineman for the following:

1. Burned contacts
2. Damaged interrupters or arcing horns
3. Proper alignment
4. Worn parts
5. Defective insulators
6. Adequate lightning arrester protection
7. Proper ground connections
8. Loose hardware

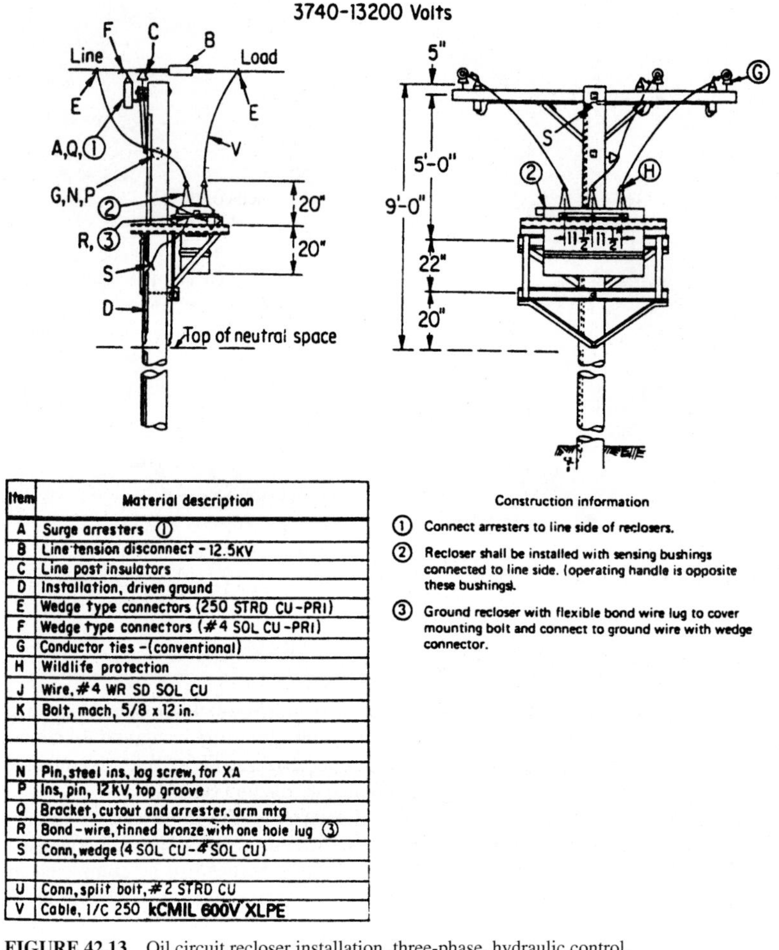

Item	Material description
A	Surge arresters ①
B	Line tension disconnect - 12.5KV
C	Line post insulators
D	Installation, driven ground
E	Wedge type connectors (250 STRD CU-PRI)
F	Wedge type connectors (#4 SOL CU-PRI)
G	Conductor ties -(conventional)
H	Wildlife protection
J	Wire, #4 WR SD SOL CU
K	Bolt, mach, 5/8 x 12 in.
N	Pin, steel ins, lag screw, for XA
P	Ins, pin, 12 KV, top groove
Q	Bracket, cutout and arrester, arm mtg
R	Bond-wire, tinned bronze with one hole lug ③
S	Conn, wedge (4 SOL CU-4 SOL CU)
U	Conn, split bolt, #2 STRD CU
V	Cable, 1/C 250 kCMIL 600V XLPE

Construction information

① Connect arresters to line side of reclosers.

② Recloser shall be installed with sensing bushings connected to line side. (operating handle is opposite these bushings).

③ Ground recloser with flexible bond wire lug to cover mounting bolt and connect to ground wire with wedge connector.

FIGURE 42.13 Oil circuit recloser installation, three-phase, hydraulic control.

Switches that have not been adequately inspected while being operated in a 5-year period should be scheduled for a maintenance inspection (Fig. 42.6). It may be necessary for the lineman to make temporary line modifications to permit operation, inspection, repairs, or adjustments of the switches.

Underground Distribution Circuit Maintenance. Circuits for underground distribution originate at a substation or a riser pole in the overhead distribution lines. Maintenance of substation equipment is discussed in *Standard Handbook for Electrical Engineers* (McGraw-Hill).

Riser Pole Maintenance. The riser pole for underground distribution circuits should be inspected when overhead lines are inspected and maintained. The inspection should include the disconnect switches and/or fused cutouts, the lightning arresters, and the operation of the arrester ground lead isolation devices, the riser cables and potheads or termination, support of the cables, conduit and/or U-guard, and identification of the circuit and pole conditions.

Switchgear Maintenance. The switchgear for underground electric distribution circuits should be inspected annually. The inspection should start with external items such as condition of the pad or foundation. Pads that are tipped or cracked or that have the dirt washed out from under them should be repaired. The paint on metal surfaces of the gear should be in good condition to prevent rusting. The cabinet doors should be locked and in most cases bolted shut. Cleanliness of the internal areas of the switchgear is important to maintain the dielectric strength of the insulation. Equipment with insulating oil must be checked for moisture or contamination of the oil. Switches, fuses, and other internal equipment should be inspected in accordance with the manufacturer's instructions. Cable terminations, grounds, and other connections to the switchgear equipment must be kept in good condition.

Thermal measuring hand-held tools provide clues as to potential hotspots. The temperature measurement tools are lightweight and simple to use (Figs. 42.14 and 42.15). Advanced devices provide additional details such as expanded temperature ranges. Even more sophisticated models allow the user to digitally record temperatures at greater distances and download the information to a personal computer for detailed analysis.

Distribution Transformer Maintenance. Transformers should be checked for proper loading. Transformer leadings are normally monitored by use of a transformer load management

FIGURE 42.14 A temperature-sensing device being utilized on connections within metal-enclosed switchgear on a periodic basis according to the maintenance schedule. Equipment hot-spots are detected and repaired. (*Courtesy Raytek Corp.*)

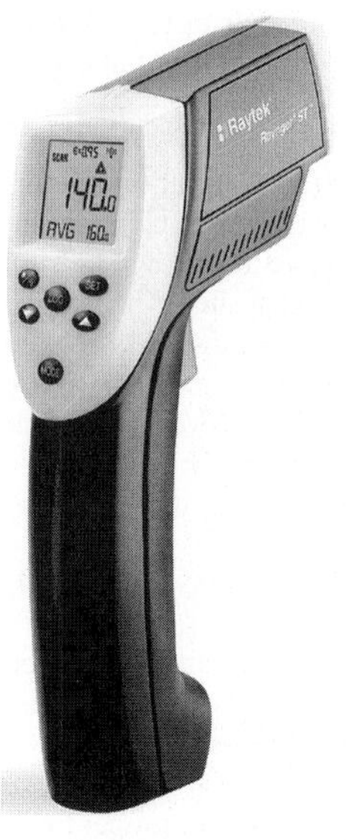

FIGURE 42.15 The lightweight easy to handle temperature sensing tool is utilized to quickly detect abnormal temperature conditions at electrical equipment connection points. (*Courtesy Raytek Corp.*)

system. The system takes monthly kilowatthour meter readings, used for billing purposes for all customers served from a transformer, and converts them to a value of kVA load that is compared to the kVA rating of the transformer. Overloaded transformers should be replaced with ones of adequate capacity to prevent the transformer from failing and to prevent fires.

Pole-type transformers should be inspected visually for defects as a part of the overhead line inspection (Fig. 42.16). Pad-mounted transformers are inspected for the items discussed previously for the underground distribution circuit switchgear. Annual inspections are desirable.

Underground Cable Maintenance. The terminations of underground cables are inspected as a part of the maintenance procedures for riser poles, switchgear, and pad-mounted transformers. The route of underground cable circuits should be walked by the lineman on a 2-year cycle. The route of the cable should be inspected for obstacles that could damage the cables, changes in grade of the ground that could cause the cables to be exposed or damaged, and washouts. Cables that fail while in operation must be located and repaired or replaced. Secondary enclosures for low-voltage cables should be inspected while checking the route of primary cables or while connecting or disconnecting services to customers (Fig. 42.17). The lid should be removed from the enclosure and checked visually for cable damage, rodent problems, and washouts. Required repairs can usually be completed by the lineman or cableman at the time of the inspection.

Underground primary distribution cables with solid-dielectric insulation have experienced a high rate of electrical failure after several years of operation as a result of electrochemical trees. The electrochemical treeing is generally caused by the presence of water in the conductor that penetrates the insulation. It has been successfully demonstrated that the life of the cables can be extended by flushing the conductor with dry nitrogen to remove the moisture and then filling the conductor with a pressurized high-dielectric fluid compatible with the insulation. Contractors are available to complete this maintenance for those companies inexperienced in the procedure.

FIGURE 42.16 Lineman has climbed pole to inspect distribution transformer and associated equipment.

FIGURE 42.17 Lineman inspecting underground secondary connections in enclosure. (*Courtesy A. B. Chance Co.*)

Street Light Maintenance. The lumen output of any electric lamp decreases with use, and accumulated dirt interferes with light reflection and transmission. Street light installations are designed such that the average footcandle levels recommended by the *ASA Standards for Roadway Lighting* represent the average illumination delivered on the roadway when the illuminating source is at its lowest output and when the luminaire is in its dirtiest condition. The luminaire refractor should be washed and the lamp replaced when the lumen output of the lamp has fallen to approximately 80 percent of its initial value, to maintain the designed footcandle level.

Lamp burnout should be a consideration in the maintenance of a street light system. The *rated lamp life* is defined as the time required for half of the lamps to fail. The mortality curves for the various types of lamps are relatively flat for the initial 60 to 70 percent of rated life. Using a 6000-hour incandescent lamp as an example, only 10 percent of the lamps will fail in the first 4000 hours of burning time, while 40 percent will fail in the next 2000 hours of burning time.

Proper maintenance of street and highway lighting is important to provide a high degree of service quality. Street lights should be inspected on a 10-year schedule (Fig. 42.18). Refractors for enclosed-type street light luminaires should be cleaned and washed at the time of inspection. The luminaires should be checked for proper level in both the transverse and horizontal directions. If the street light is installed on a steep grade, the transverse direction of the luminaire should be leveled parallel to the street grade. Defective or broken refractors or luminaires must be replaced. Many street light problems and failures are reported by the public, police, and employees of the concern owning the system. Lamps that fail, defective photocells, and ballasts should be replaced promptly to restore the lighting service. Group replacement of mercury-vapor lamps on a 4-year schedule and high-pressure sodium-vapor

FIGURE 42.18 Lineman maintaining street light. (*Courtesy Asplundh Tree Expert Co.*)

FIGURE 42.19 Three single-phase voltage regulators installed on platform in overhead distribution circuit with bypass switches to facilitate maintenance.

lamps on a 10-year schedule should minimize the number of lamp failures and maintain light output in accordance with established standards. The group replacement can be coordinated with the inspection, cleaning, and washing schedule to minimize the effort required.

Voltage Regulator Maintenance. Distribution voltage regulators should be inspected externally by the inspector while patrolling the overhead lines (Fig. 42.19). Problems such as damaged insulators, leaking oil, or unusual operating noise should be corrected by the lineman as soon as practical. Operation of the voltage regulators should be observed by checking the operation counter and reading the voltage with the aid of recording voltmeters at regular intervals set at various locations along the circuit beyond the regulator. The voltage regulator should be removed from service by the lineman and maintained in accordance with the manufacturer's recommendation as determined by the number of operations of the tap changer or time in service. Normally, the voltage regulators can operate several years between maintenance periods. All manufacturers require that the insulating oil be kept dry and free from carbon to maintain the dielectric strength of the oil.

MAINTENANCE OF EQUIPMENT COMMON TO TRANSMISSION AND DISTRIBUTION LINES

Tree trimming. See Chap. 35.

Pole Maintenance. All wood poles should be pounded with a hammer by the lineman and visibly inspected before being climbed. A good pole will sound solid when hit with a hammer. A decayed pole will sound hollow or like a drum. The visible inspection should include checking for general condition and position, cracks, shell rot, knots that might weaken the pole, setting depth, soil conditions, hollow spots or woodpecker holes, and burned spots. Poles that reveal possible defects should not be climbed by the linemen until the poles are made safe by repairs or by securing them.

All wood poles should be pressure- or thermal-treated with preservatives when they are new. New treated wood poles should resist decay for 10 to 20 years, depending upon the ambient soil conditions. Scheduled inspection should be based on experience. Distribution poles are often replaced before scheduled maintenance is required because of a change in voltage or replacement due to street-widening projects. When a lineman detects defects in wood poles of a particular type and vintage, similar poles installed at the same time should be inspected and maintained, if necessary. A pole hit with a hammer by the lineman will not always reveal decay.

The Electrical Power Research Institute (EPRI) has developed an instrument that uses nondestructive test methods to predict the life of a pole. The PoleTest instrument is shown being used by a lineman in Fig. 42.20. The lineman keys in the wood species and the diameter of the pole on the instrument and then strikes the pole a few times with a marble-size steel ball that has been attached to the pole. A microprocessor in the instrument analyzes the sonic wave patterns passing through the wood pole and indicates the current strength of the pole.

If the instrument indicates the pole strength is deficient or if a pole-testing instrument is not available, the ground around the pole can be excavated to a depth of 12 to 18 in. A probe test can be completed by hand-drilling a $^3/_8$-in hole below the ground line into the pole butt at a 30° to 45° downward angle to a depth sufficient to reach the center of the pole. The wood shavings are checked for decay. The drilled holes are plugged with a treated $^3/_8$-in dowel to keep moisture out and to prevent further decay. Poles that have sufficient decay to weaken them should be reinforced or replaced. External decay of poles should be shaved away. Poles in acceptable condition should have a preservative

FIGURE 42.20 Lineman using PoleTest instrument to predict the residual strength of the wood pole.

FIGURE 42.21 Lineman applying preservative and moisture barrier to stop external decay of wood pole. (*Courtesy Asplundh Tree Expert Co.*)

applied at the ground line and be wrapped with a moistureproof barrier (Fig. 42.21). The wood poles can be treated internally with chemicals to kill early decay before actual voids form. Poles can be strengthened to delay replacement by placing a section of galvanized steel plate adjacent to the pole. The steel plate is driven into the ground and banded to the pole above the ground line (Figs. 42.22 and 42.23).

Wood poles installed near woodpecker's feeding areas are often damaged by the birds. Woodpecker damage can be prevented by wrapping the pole with extruded sheets of high-density polyethylene with a high-gloss finish. The plastic wrap prevents the woodpeckers

FIGURE 42.22 Galvanized steel reinforcing is being driven into the ground adjacent to partially decayed wood distribution pole.

FIGURE 42.23 Lineman is completing the banding of a galvanized steel reinforcing member to a partially decayed pole to prolong its life.

from perching on the pole. Linemen can climb the wrapped pole by penetrating the wrap with climber's gaffs. If the linemen must do considerable work on the pole, the pole should be rewrapped when the work is completed to prevent sites of gaff penetration of the plastic wrap from becoming perches for the birds.

Wooden transmission-line poles are usually scheduled for inspection after 20 years of service and treated if pole conditions are adequate. Repeat inspections and treatment are normally scheduled at 8- to 10-year intervals. Experience indicates that poles properly treated when new and maintained on an adequate schedule should have an average life of 45 years.

Insulator Maintenance. Transmission- and distribution-line inspections detect damaged insulators. Most insulator damage result from gun shots, lightning flashovers, contamination flashovers, and wind damage. Broken, cracked, or damaged insulators should be replaced as soon as they are detected (Figs. 42.24 and 42.25). Defective insulators may cause visible corona and interference voltage propagation. Radio interference conditions can be detected by the lineman by using instruments designed for this purpose while completing the scheduled inspections.

Insulators in severely contaminated atmospheres may require frequent cleaning. These insulators can be washed with high-pressure water, using special hoses and hose nozzles (Figs. 42.26 and 42.27). Procedures developed for safe washing of energized insulators must be followed.

Conductor Maintenance. Conductors can be damaged by aeolian vibration, galloping, sway oscillation, unbalanced loading, lightning discharges, or short-circuit conditions. Inspections of the transmission and distribution lines by the lineman will reveal the need for conductor repairs (Fig. 42.28).

Abrasion to the outside surfaces of the conductor is visible to the inspector. Abrasion damage is a chafing, impact wear that accompanies relative movement between a loose tie or other conductor hardware and the conductor or armor rods. Abrasion is a surface damage and can be identified by black deposits on the conductor or tie wire.

FIGURE 42.24 Linemen changing out defective insulators on energized 345-kV transmission line, using bare-hand method and hot-line tools. Lineman working from insulated ladder is wearing a conductive suit.

FIGURE 42.25 Lineman changing out defective insulator on distribution line. (*Courtesy MidAmerican Energy Co.*)

FIGURE 42.26 Linemen washing insulators on energized 138-kV transmission line to remove insulator contamination.

FIGURE 42.27 Water tank, gasoline-engine-driven high-pressure water pump, and hoses mounted on trailer for washing contaminated insulator on energized lines.

Fatigue results in the failure of the conductor strands or tie wire (Fig. 42.29). If the damaged conductor is detected before serious damage or complete failure occurs, it can be repaired by the lineman by installing armor rods or a line splice (Figs. 42.30 through 42.33).

Broken conductor tie wires at an insulator discovered by inspection or as a result of severe radio interference complaints require immediate repairs (Fig. 42.34). The damaged conductor can be repaired even if some of the conductor strands are broken by removing the old tie wires and armor rods, if present, and installing armor rods or line guards and retying the assemblies with distribution ties or wraplock ties (Fig. 42.35).

Transmission and distribution lines where experience indicates the conductors have prolonged periods of vibration should have vibration dampers installed to prevent fatigue failure of the conductor (Fig. 42.36). Spiral vibration dampers will prevent conductor fatigue, inner wire fretting, or scoring of the distribution-line insulator glaze.

FIGURE 42.28 Lineman riding in carriage suspended on conductor to closely inspect conductors for damage or to complete repairs. (*Courtesy Sherman & Reilly, Inc.*)

FIGURE 42.29 Excessive conductor damage and broken tie wire at insulator on a distribution circuit. (*Courtesy Preformed Line Products Co.*)

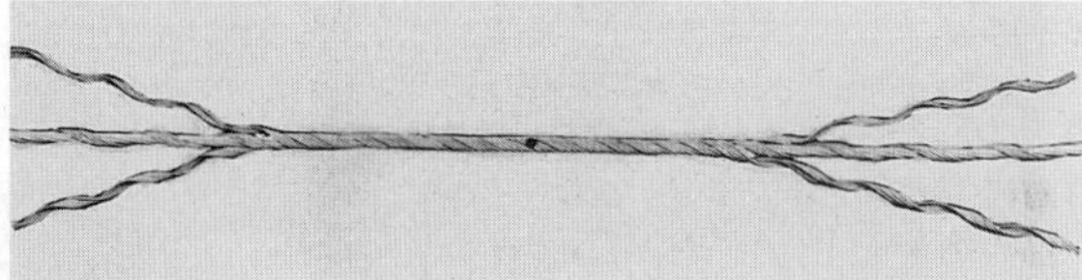

FIGURE 42.30 Damaged conductor being repaired with a splice. Splice will restore original conductivity and strength to all-aluminum, aluminum-alloy, and copper conductors of homogeneous stranding, provided core conductor of ACSR is not damaged. (*Courtesy Preformed Line Products Co.*)

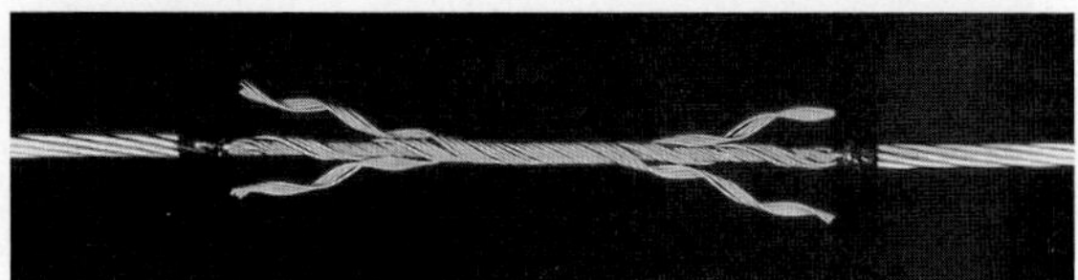

FIGURE 42.31 Damaged core and aluminum strands of ACSR conductor being repaired with ACSR full-tension splice. Splice components include core splice, filler rods, and outer splice. Conductor original strength and conductivity are restored. (*Courtesy Preformed Line Products Co.*)

FIGURE 42.32 Insulated ladder supported by cable between towers so that lineman can repair damaged conductor at a lower elevation.

FIGURE 42.33 Lineman repairing conductor damage on high-voltage energized transmission line. Lineman works from a fiber-glass-insulated ladder supported by cable above the damaged conductor.

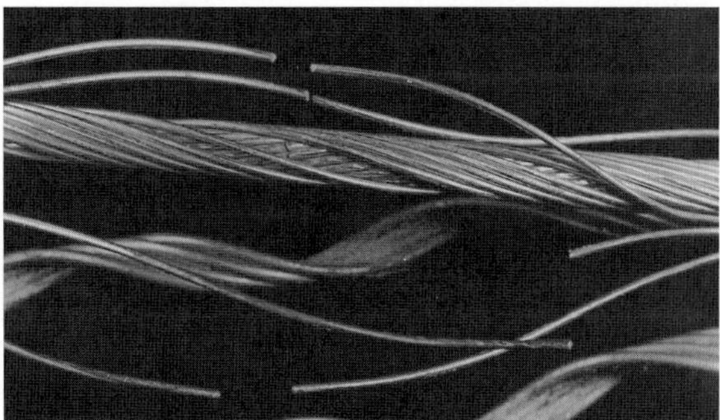

FIGURE 42.34 Picture reveals conductor damage as a result of fatigue breaks. (*Courtesy Preformed Line Products Co.*)

FIGURE 42.35 Line guard installed on damaged distribution conductor at insulator support. Conductance and strength are completely restored to aluminum or ACSR conductors. Conductor must be retied to insulator to complete repairs. (*Courtesy Preformed Line Products Co.*)

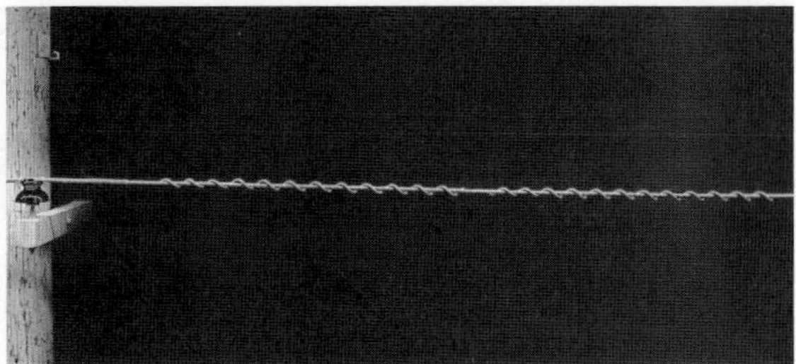

FIGURE 42.36 Spiral vibration dampers installed on distribution-line conductor after damage at insulator had been repaired. (*Courtesy Preformed Line Products Co.*)

Transmission and distribution lines can be tested with thermovision equipment to locate hot spots on conductors and equipment caused by loose connections or hardware and other defects. Thermovision equipment locates hot spots by using infrared scanning to thermally inspect the lines. Some of the equipment provides a photographic record as well as a visual display of the hot-spot defects. The thermovision equipment can be operated from a helicopter, from a van-type vehicle, or by an operator carrying the equipment (Fig. 42.37). Examples of normal black and white still photography and infrared images of detected hot spots are shown in Figs. 42.38 through 42.41.

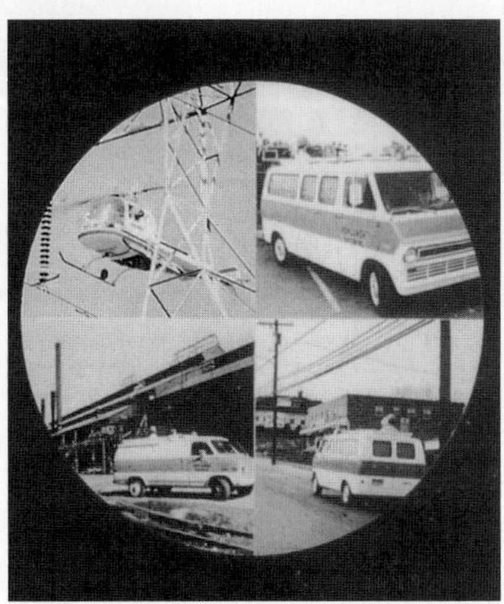

FIGURE 42.37 Helicopter and van-type vehicles equipped with thermovision equipment for infrared scanning of transmission and distribution lines. (*Courtesy Asplundh Tree Expert Co.*)

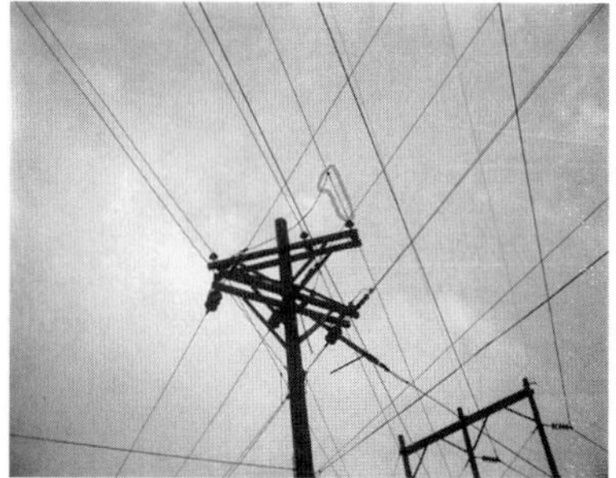

FIGURE 42.38 A black and white image of a conductor jumper tying one phase of two (2)-three phase distribution circuits joined by a hot-line clamp. The area that is circled was found to require maintenance to replace the connector that was the source of the heating problem. The problem is not visible to the naked eye. (*Courtesy MidAmerican Energy Co.*)

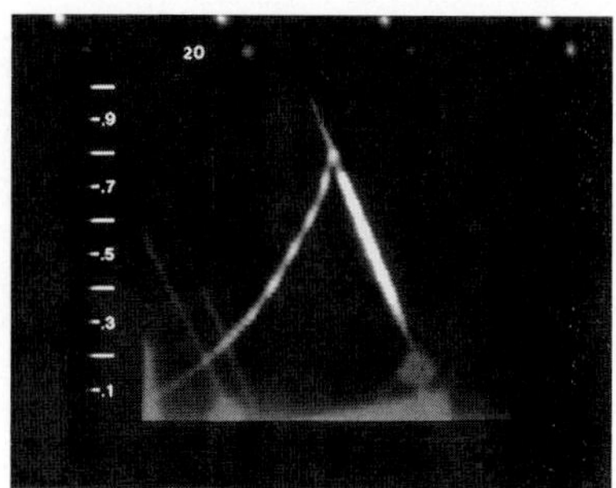

FIGURE 42.39 A close-up infrared image of the area in Fig. 42.38 needing maintenance. Heating is apparent in both directions from the source. The source was found to be a hot-line clamp. (*Courtesy MidAmerican Energy Co.*)

FIGURE 42.40 A black and white image of a capacitor bank installed on a distribution circuit. No problem is apparent, however, the arrow points to a source of heating. (*Courtesy MidAmerican Energy Co.*)

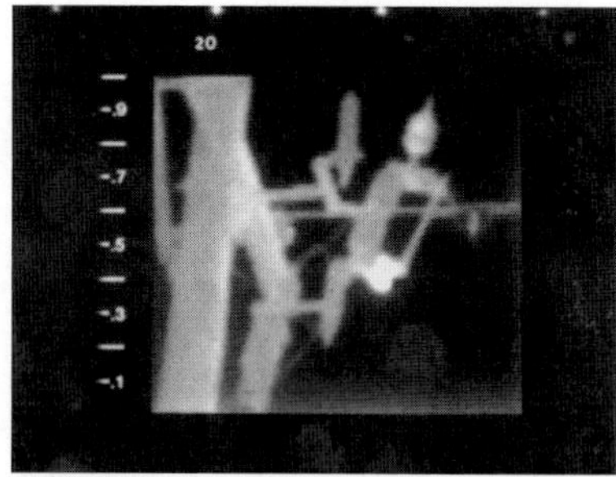

FIGURE 42.41 A close-up infrared image of a heating problem detected at the solid blade distribution cut-out door and hinge point (Fig. 42.40). If left unattended, the device will eventually fail and potentially result in a more expensive repair. (*Courtesy MidAmerican Energy Co.*)

CHAPTER 43

ROPE, KNOTS, SPLICES, AND GEAR

Ropes used by linemen and cablemen to build and maintain electric transmission and distribution lines are made from synthetic fibers consisting of nylon, polyester, polyolefin (polypropylene), aramid (kevlar), and vegetable fibers consisting of abaca (manila).

ROPE CONSTRUCTION

There are several types of rope constructions manufactured. The twisted rope is usually made with three or four strands (Fig. 43.1). A twisted rope is manufactured by first producing the strands, using a specific number of yarns per strand which are placed in a laying machine which controls the strand twist as the finished rope is formed. Braided ropes are mainly 2-in-1 or double-braided, single or hollow braid, 8-strand or 12-strand braided or plaited ropes. The double-braided rope contains a braided core over which a cover is braided, the single or hollow braid consists of one single braid containing no core, and the 8-strand and 12-strand braided or plaited rope consists of eight or twelve strands braided in pairs in a maypole fashion on a plaiting machine. Parallel ropes are produced by utilizing parallel yarns as the strength members that are bonded and/or jacketed to form the finished rope (Fig. 43.2). Parallel yarns will provide almost 100 percent of the available strength from the fibers. Parallel rope is stronger, size for size, than twisted or braided rope, but it has much lower elongation and is stiffer (Fig. 43.3). The 8-strand and 12-strand braided synthetic fiber ropes are commonly used for slings, pilot lines, and hand lines. The 2-in-1 or double-braided synthetic fiber ropes are frequently used for winchlines and slings and for stringing electrical conductors. The strengths of the fiber ropes, when new, are given by the manufacturer in Table 43.1. A rope's strength will deteriorate slowly or quickly depending on how it is used and/or abused. Rope must be stored properly to prevent deterioration. All ropes should be inspected closely each time before they are used.

CARE AND HANDLING OF ROPES

Fiber rope must not be kinked. It should be removed carefully from a reel or coil. Kinks can substantially weaken the rope. The rope should be inspected each time before it is used for excessive wear or damage, such as cuts, gouges, badly abraded or fused or glazed spots, and fiber breakage (inside as well as on the surface of the rope); light fuzzing is acceptable.

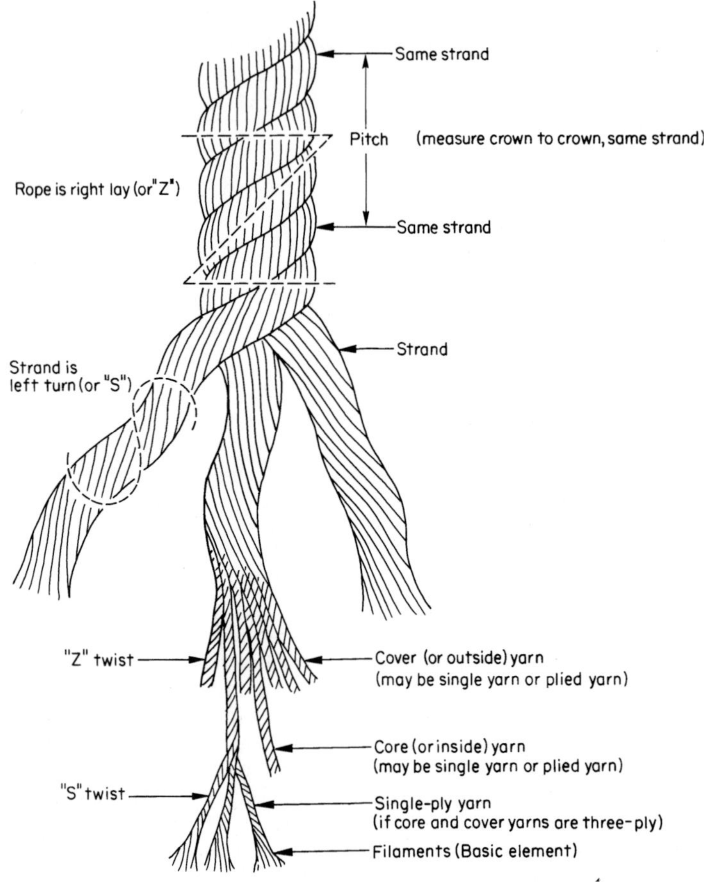

FIGURE 43.1 Three-strand rope construction. (*Courtesy Cordage Institute*)

FIGURE 43.2 Rope constructed with nylon filaments laid parallel and a nylon braided jacket. (*Courtesy Sherman & Reilly, Inc.*)

Proper terminations and attachments should be used. Rope splicing can maintain 100 percent of the rope strength. Knots can reduce rope strength 60 percent or more and can cause slipping. Avoid running a rope over sharp edges or corners, or around diameters less than 8 times the rope diameter. When tying rope to an object that has sharp corners, the rope should be padded to prevent the fibers from being cut. Rope should never be dragged over the ground or over other sharp objects. Dragging one rope over another can result in damage to both ropes. Rope that has become wet should not be permitted to freeze. If a rope has become

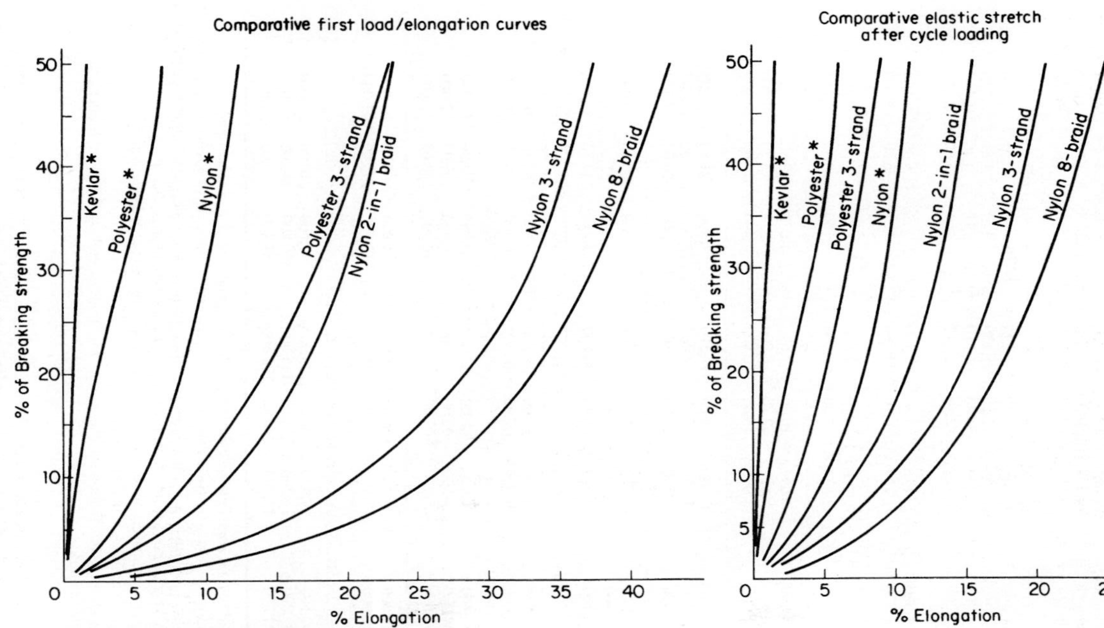

FIGURE 43.3 Stress-strain curves at various percentages of breaking strength for common types of ropes discussed. (*Courtesy Cordage Institute.*)

TABLE 43.1 Three-Strand Laid and Eight-Strand Plaited Rope Data

Nominal size, in		Manila				Polypropylene				Composite			
Dia.	Circ.	Linear density, lb/100 ft (1)	Ultimate strength, lb (2)	Safety factor	Allowable load, lb (1)	Linear density, lb/100 ft (3)	Ultimate strength, lb (2)	Safety factor	Allowable load, lb (1)	Linear density, lb/100 ft (3)	Ultimate strength, lb (2)	Safety factor	Allowable load, lb (1)
3/16	5/8	1.44	406	←10→	41	.70	720	←10→	72	.94	720	←10→	72
1/4	3/4	1.90	540		54	1.20	1,130		113	1.61	1,130		113
5/16	1	2.76	900		90	1.80	1,710		171	2.48	1,710		171
3/8	1 1/8	3.90	1,200		122	2.80	2,440		244	3.60	2,440		244
7/16	1 1/4	6.00	1,580	←9→	176	3.80	3,160	←9→	352	5.00	3,160	←9→	352
1/2	1 1/2	7.15	2,380		264	4.70	3,780		420	6.50	3,760		440
9/16	1 3/4	9.90	3,110	←8→	388	6.10	4,600	←8→	575	8.00	4,860	←8→	610
5/8	2	12.70	3,760		476	7.50	5,600		700	9.50	5,750		720
3/4	2 1/4	15.90	4,860	←7→	695	10.70	7,630	←7→	1,070	12.50	7,550	←7→	1,080
13/16	2 1/2	18.60	5,850		835	12.70	8,900		1,270	15.20	9,200		1,310
7/8	2 3/4	21.40	6,950		995	15.00	10,400		1,490	18.00	10,800		1,540
1	3	25.60	8,100		1,160	18.00	12,600		1,800	21.80	13,100		1,870
1 1/16	3 1/4	27.80	9,450		1,350	20.40	14,400		2,060	25.60	15,200		2,180
1 1/8	3 1/2	34.20	10,800		1,540	23.80	16,500		2,360	29.00	17,400		2,480
1 1/4	3 3/4	37.60	12,200		1,740	27.00	18,900		2,700	33.40	19,800		2,820
1 5/16	4	45.60	13,500		1,930	30.40	21,200		3,020	35.60	21,200		3,000
1 1/2	4 1/2	57.00	16,700		2,380	38.40	26,800		3,820	43.00	26,800		3,820
1 5/8	5	71.00	20,200		2,870	47.60	32,400		4,620	55.50	32,400		4,620
1 3/4	5 1/2	85.00	23,800		3,400	57.00	38,800		5,550	66.50	38,800		5,550
2	6	102.00	28,000		4,000	67.00	46,800		6,700	78.00	46,800		6,700
2 1/8	6 1/2	119.00	32,400		4,620	80.00	55,000		7,850	92.00	55,000		7,850
2 1/4	7	137.00	37,000		5,300	92.00	62,000		8,850	105.00	62,000		8,850
2 1/2	7 1/2	159.00	41,800		5,950	107.00	72,000		10,300	122.00	72,000		10,300
2 5/8	8	182.00	46,800		6,700	130.00	81,000		11,600	138.00	81,000		11,600
2 7/8	8 1/2	204.00	52,900		7,400	137.00	91,000		13,000	155.00	91,000		13,000
3	9	230.00	57,500		8,200	153.00	103,000		14,700	174.00	103,000		14,700
3 1/4	10	284.00	67,500	←7→	9,950	190.00	123,000	←7→	17,600	210.00	123,000	←7→	17,600
3 1/2	11	348.00	82,000		11,700	232.00	146,000		20,800	256.00	146,000		20,800
4	12	414.00	94,500		13,500	276.00	171,000		24,400	300.00	171,000		24,400

(1) Allowable loads are for rope in good condition with appropriate splices, in noncritical applications, and under normal service conditions. Allowable loads should be exceeded only with expert knowledge of conditions and professional estimates of risk. Allowable loads should be reduced where life, limb, or valuable property are involved, and/or for exceptional service conditions, such as shock loads, sustained loads, etc.

(2) Ultimate strengths are based on tests of new and unused rope in accordance with Cordage Institute methods and are exceeded at a 98 confidence percent level.

(3) Linear density (lb per 100 ft) shown is "average." Maximum weight is 5 percent higher.

Source: Cordage Institute.

full of mud or sand, it should be flushed with water and permitted to dry. Rope hand lines and blocks should be done up right and hung in the truck in the proper place when not in use. It is best to place large rope, when it is stored, on wood gratings above the floor in a ventilated location. All ropes used in line work should be kept dry. Keep ropes away from harmful chemicals. Avoid excessive exposure of the rope to sunlight and moisture. Keep the ropes away from fire and exposure to temperatures exceeding 150°F. Select the proper rope for the

Polyester				Nylon				
Linear density, lb/100 ft (3)	Ultimate strength, lb (2)	Safety factor	Allowable load, lb (1)	Linear density, lb/100 ft (3)	Ultimate strength, lb (2)	Safety factor	Allowable load, lb (1)	Nominal dia, in
1.20	900		90	1.00	900		75	3/16
2.00	1,490	←10→	149	1.50	1,490	←12→	124	1/4
3.10	2,300		230	2.56	2,300		192	5/16
4.50	3,340		334	3.50	3,340		278	3/8
6.20	4,500	←9→	500	5.00	4,500	←11→	410	7/16
8.00	5,750		640	6.50	5,750		525	1/2
10.20	7,200	←8→	900	8.15	7,200	←10→	720	9/16
15.00	9,000		1,130	10.50	9,350		935	5/8
17.50	11,300		1,610	14.50	12,800		1,430	3/4
21.00	14,000	←7→	2,000	17.00	15,300	←9→	1,700	13/16
25.00	16,200		2,320	20.00	18,000		2,000	7/8
30.40	19,800		2,820	26.40	22,600		2,520	1
34.40	23,000		3,280	29.00	26,000		2,880	1 1/16
40.00	26,600		3,800	34.00	29,800		3,320	1 1/8
46.20	29,800		4,260	40.00	33,800		3,760	1 1/4
52.50	33,800		4,820	45.00	38,800		4,320	1 5/16
67.00	42,200		6,050	55.00	47,800		5,320	1 1/2
82.00	51,500		7,350	66.50	58,500		6,500	1 5/8
98.00	61,000		8,700	83.00	70,000		7,800	1 3/4
118.00	72,000		10,300	95.00	83,000		9,200	2
135.00	83,000		11,900	109.00	95,500		10,600	2 1/8
157.00	96,500		13,800	127.00	113,000		12,600	2 1/4
181.00	110,000		15,700	147.00	126,000		14,000	2 1/2
204.00	123,000		17,600	168.00	146,000		16,200	2 5/8
230.00	139,000		19,900	189.00	162,000		18,000	2 7/8
258.00	157,000		22,400	210.00	180,000		20,000	3
318.00	189,000		27,000	264.00	226,000		25,200	3 1/4
384.00	228,000	←7→	32,600	312.00	270,000	←9→	30,000	3 1/2
454.00	270,000		38,600	380.00	324,000		36,000	4

load to be handled. Different ropes have better characteristics for shock and sustained or dynamic loading. A safety factor of 5 is recommended for manila rope, 6 for polypropylene rope, and 9 for nylon and polyester ropes. It is unsafe to substitute a small synthetic rope for a large manila rope on the basis of the two having the same ultimate strength.

ELECTRICAL CONDUCTIVITY OF ROPES

Clean, dry polypropylene rope is an insulator; all other ropes are conductors. As the rope (and this includes polypropylene rope) becomes wet, dirty, and contaminated, its conductivity increases progressively. All insulators conduct electricity if they are not kept clean and dry. All fibers used in ropes, except polypropylene, contain moisture as a part of the makeup of the fiber. The moisture regain of fibers varies from 0 percent for polypropylene to over 7.5 percent for manila. Polyester fiber has a 0.4 percent and nylon fiber a 4.5 percent moisture regain.

The fiber moisture content rises with increased relative humidity. All rope structures trap water between fibers, yarns, and strands when exposed to rain, mist, puddles, or other sources of moisture which increases the electrical conductivity of the rope. The Cordage industry has developed a rope construction that minimizes the conductivity of the finished rope. The rope is described in "A Specification for Unused Polypropylene Rope with Special Electrical Properties for Use by Electrical Utilities," available from the Cordage Institute.

ROPE SLINGS

Figure 43.4 provides a pictorial representation of basic sling configurations referred to in Tables 43.2 through 43.5, which give the rated capacity of synthetic and natural fiber rope slings. Spliced-fiber rope slings must be capable of meeting the minimum requirements of Occupational Safety and Health Administration (OSHA) Standard 1910.184.

VARIOUS USES OF ROPE

Bull Rope. Bull ropes are used for raising or lowering heavy pieces of equipment, for fastening temporary guys, for setting poles, for holding out heavy transformers, and for lowering large limbs and trunks of trees.

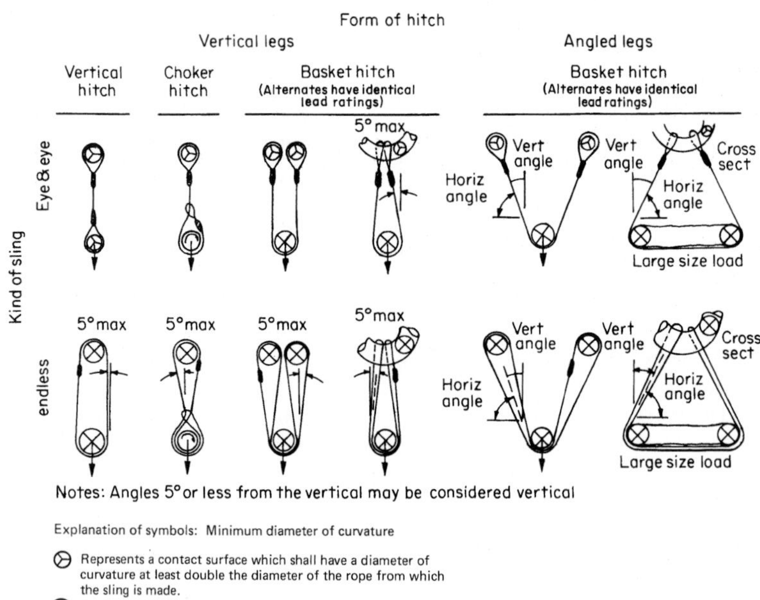

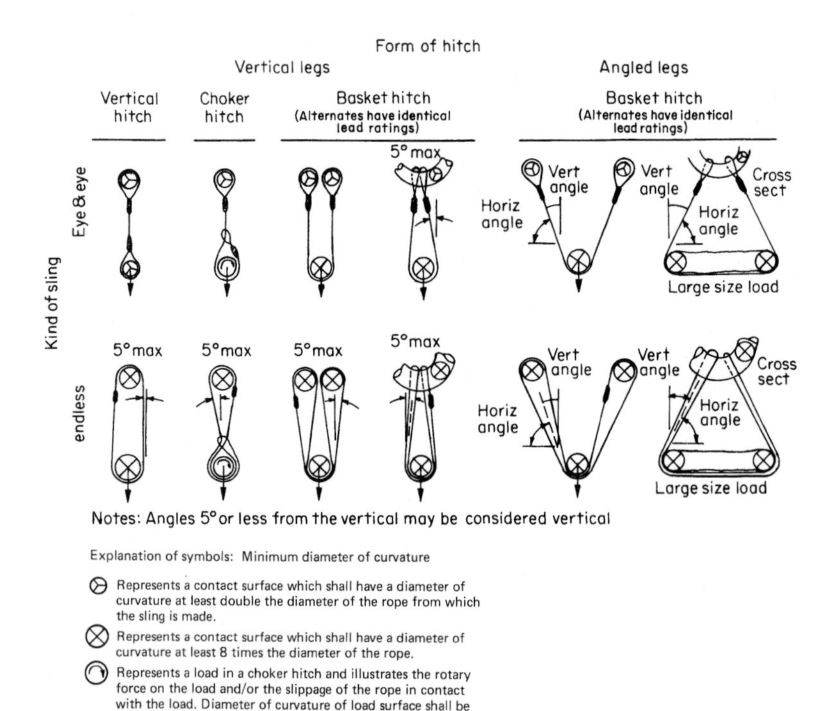

FIGURE 43.4 Sling configurations.

TABLE 43.2 Polypropylene Rope Sling, Rated Capacity in Pounds* (Safety Factor—6)

			Eye and eye sling						Endless sling					
					Basket hitch						Basket hitch			
					Angle of rope to horizontal						Angle of rope to horizontal			
					90°	60°	45°	30°			90°	60°	45°	30°
					Angle of rope to vertical						Angle of rope to vertical			
Rope diameter (nominal), in	Nominal weight per 100 ft. lb	Minimum breaking strength, lb	Vertical hitch	Choker hitch	0°	30°	45°	60°	Vertical hitch	Choker hitch	0°	30°	45°	60°
1/2	4.7	3,990	650	350	1,300	1,200	950	650	1,200	600	2,400	2,100	1,700	1,200
9/16	6.1	4,845	800	400	1,600	1,400	1,000	800	1,500	750	2,900	2,500	2,100	1,500
5/8	7.5	5,890	1,000	500	2,000	1,700	1,400	1,000	1,800	900	3,500	3,100	2,500	1,800
3/4	10.7	8,075	1,300	700	2,700	2,300	1,900	1,300	2,400	1,200	4,900	4,200	3,400	2,400
13/16	12.7	9,405	1,600	800	3,100	2,700	2,200	1,600	2,800	1,400	5,600	4,900	4,000	2,800
7/8	15.0	10,925	1,800	900	3,600	3,200	2,600	1,800	3,300	1,600	6,600	5,700	4,600	3,300
1	18.0	13,300	2,200	1,100	4,400	3,800	3,100	2,200	4,000	2,000	8,000	6,900	5,600	4,000
1 1/16	20.4	15,200	2,500	1,300	5,100	4,400	3,600	2,500	4,600	2,300	9,100	7,900	6,500	4,600
1 1/8	23.7	17,385	2,900	1,500	5,800	5,000	4,100	2,900	5,200	2,600	10,500	9,000	7,400	5,200
1 1/4	27.0	19,950	3,300	1,700	6,700	5,800	4,700	3,300	6,600	3,000	12,000	10,500	8,500	6,000
1 5/16	30.5	22,325	3,700	1,900	7,400	6,400	5,300	3,700	6,700	3,400	13,500	11,500	9,500	6,700
1 1/2	38.5	28,215	4,700	2,400	9,400	8,100	6,700	4,700	8,500	4,200	17,000	14,500	12,000	8,500
1 5/8	47.5	34,200	5,700	2,900	11,500	9,900	8,100	5,700	10,500	5,100	20,500	18,000	14,500	10,500
1 3/4	57.0	40,850	6,800	3,400	13,500	12,000	9,600	6,800	12,500	6,100	24,500	21,000	17,500	12,500
2	69.0	49,400	8,200	4,100	16,500	14,500	11,500	8,200	15,000	7,400	29,500	25,500	21,000	15,000
2 1/8	80.0	57,950	9,700	4,800	19,500	16,500	13,500	9,700	17,500	8,700	35,000	30,100	24,500	17,500
2 1/4	92.0	65,550	11,000	5,500	22,000	19,000	15,500	11,000	19,500	9,900	39,500	34,000	28,000	19,500
2 1/2	107.0	76,000	12,500	6,300	25,500	22,000	18,000	12,500	23,000	11,500	45,500	39,500	32,500	23,000
2 5/8	120.0	85,500	14,500	7,100	28,500	24,500	20,000	14,500	25,500	13,000	51,500	44,500	36,500	25,500

*See Fig. 43.4 for sling configuration descriptions.
Source: Occupational Safety and Health Administration, Rules and Regulations #1910.184 Slings, 1976.

TABLE 43.3 Polyester Rope Sling, Rated Capacity in Pounds* (Safety Factor—9)

Rope diameter (nominal), in	Nominal weight per 100 ft. lb	Minimum breaking strength, lb	Eye and eye sling						Endless sling					
					Basket hitch						Basket hitch			
					Angle of rope to horizontal						Angle of rope to horizontal			
					90°	60°	45°	30°			90°	60°	45°	30°
					Angle of rope to vertical						Angle of rope to vertical			
			Vertical hitch	Choker hitch	0°	30°	45°	60°	Vertical hitch	Choker hitch	0°	30°	45°	60°
1/2	8.0	6,080	700	350	1,400	1,200	950	700	1,200	600	2,400	2,100	1,700	1,200
9/16	10.2	7,600	850	400	1,700	1,500	1,200	850	1,500	750	3,000	2,600	2,200	1,500
5/8	13.0	9,500	1,100	550	2,100	1,800	1,500	1,100	1,900	950	3,800	3,300	2,700	1,900
3/4	17.5	11,875	1,300	650	2,600	2,300	1,900	1,300	2,400	1,200	4,800	4,100	3,400	2,400
13/16	21.0	14,725	1,600	800	3,300	2,800	2,300	1,500	2,900	1,500	5,900	5,100	4,200	2,900
7/8	25.0	17,100	1,900	950	3,800	3,300	2,700	1,900	3,400	1,700	6,800	5,900	4,800	3,400
1	30.5	20,900	2,300	1,200	4,600	4,000	3,300	2,300	4,200	2,100	8,400	7,200	5,900	4,200
1 1/16	34.5	24,225	2,700	1,300	5,400	4,700	3,800	2,700	4,800	2,400	9,700	8,400	6,900	4,800
1 1/8	40.0	28,025	3,100	1,600	6,200	5,400	4,400	3,100	5,600	2,800	11,000	9,700	7,900	5,600
1 1/4	46.3	31,540	3,500	1,800	7,000	6,100	5,000	3,500	6,300	3,200	12,500	11,000	8,900	6,300
1 5/16	52.5	35,625	4,000	2,000	7,900	6,900	5,600	4,000	7,100	3,600	14,500	12,500	10,000	7,100
1 1/2	66.8	44,460	4,900	2,500	9,900	8,600	7,000	4,900	8,900	4,400	18,000	15,500	12,500	8,900
1 5/8	82.0	54,150	6,000	3,000	12,000	10,400	8,500	6,000	11,000	5,400	21,500	19,000	15,500	11,000
1 3/4	98.0	64,410	7,200	3,600	14,500	12,500	10,000	7,200	13,000	6,400	26,000	22,500	18,000	13,000
2	118.0	76,000	8,400	4,200	17,000	14,500	12,000	8,400	15,000	7,600	30,500	26,500	21,500	15,000
2 1/8	135.0	87,400	9,700	4,900	19,500	17,000	13,500	9,700	17,500	8,700	35,000	30,500	24,500	17,500
2 1/4	157.0	101,650	11,500	5,700	22,500	19,500	16,000	11,500	20,500	10,000	40,500	35,000	29,000	20,500
2 1/2	181.0	115,900	13,000	6,400	26,000	22,500	18,000	13,000	23,000	11,500	46,500	40,000	33,000	23,000
2 5/8	205.0	130,150	14,500	7,200	29,000	25,000	20,500	14,500	26,000	13,000	52,000	45,000	37,000	26,000

*See Fig. 43.4 for sling configuration descriptions.
Source: Occupational Safety and Health Administration, Rules and Regulations #1910.184 Slings, 1976.

TABLE 43.4 Nylon Rope Sling, Rated Capacity in Pounds* (Safety Factor—9)

Rope diameter (nominal), in	Nominal weight per 100 ft. lb	Minimum breaking strength, lb	Eye and eye sling						Endless sling					
					Basket hitch						Basket hitch			
					Angle of rope to horizontal						Angle of rope to horizontal			
					90°	60°	45°	30°			90°	60°	45°	30°
					Angle of rope to vertical						Angle of rope to vertical			
			Vertical hitch	Choker hitch	0°	30°	45°	60°	Vertical hitch	Choker hitch	0°	30°	45°	60°
1/2	6.5	6,000	700	350	1,400	1,200	950	700	1,200	600	2,400	2,100	1,700	1,200
9/16	8.3	7,600	850	400	1,700	1,500	1,200	850	1,500	750	3,000	2,600	2,200	1,500
5/8	10.5	9,800	1,100	550	2,200	1,900	1,600	1,100	2,000	1,000	4,000	3,400	2,800	2,000
3/4	14.5	13,490	1,500	750	3,000	2,600	2,100	1,500	2,700	1,400	5,400	4,700	3,800	2,700
13/16	17.0	16,150	1,800	900	3,600	3,100	2,600	1,800	3,200	1,600	6,400	5,600	4,600	3,200
7/8	20.0	19,000	2,100	1,100	4,200	3,700	3,000	2,100	3,800	1,900	7,600	6,600	5,400	3,800
1	26.0	23,750	2,600	1,300	5,300	4,600	3,700	2,600	4,800	2,400	9,500	8,200	6,700	4,800
1 1/16	29.0	27,360	3,000	1,500	6,100	5,300	4,300	3,000	5,500	2,700	11,000	9,500	7,700	5,500
1 1/8	34.0	31,350	3,500	1,700	7,000	6,000	5,000	3,500	6,300	3,100	12,500	11,000	8,900	6,300
1 1/4	40.0	35.625	4,000	2,000	7,900	6,900	5,600	4,000	7,100	3,600	14,500	12,500	10,000	7,100
1 5/16	45.0	40,350	4,500	2,300	9,100	7,500	6,400	4,500	8,200	4,100	16,500	14,000	12,000	8,200
1 1/2	55.0	50,350	5,600	2,800	11,000	9,700	7,900	5,600	10,000	5,000	20,000	17,500	14,000	10,000
1 5/8	68.0	61,750	6,900	3,400	13,500	12,000	9,700	6,900	12,500	6,200	24,500	21,500	17,500	12,500
1 3/4	83.0	74,100	8,200	4,100	16,500	14,500	11,500	8,200	15,000	7,400	29,500	27,500	21,000	15,000
2	95.0	87,400	9,700	4,900	19,500	17,000	13,500	9,700	17,500	8,700	35,000	30,500	24,500	17,500
2 1/8	109.0	100,700	11,000	5,600	22,500	19,500	16,000	11,000	20,000	10,000	40,500	35,000	28,500	20,000
2 1/4	129.0	118,750	13,000	6,600	26,500	23,000	18,500	13,000	24,000	12,000	47,500	41,000	33,500	24,000
2 1/2	149.0	138,000	15,000	7,400	29,000	25,500	21,000	15,000	26,500	13,500	53,000	46,000	37,500	26,500
2 5/8	160.0	153,900	17,100	8,000	34,000	29,500	34,000	17,000	31,000	15,300	61,500	53,500	43,500	31,000

*See Fig. 43.4 for sling configuration descriptions.

Source: Occupational Safety and Health Administration, Rules and Regulations #1910.184 Slings, 1976.

TABLE 43.5 Manila Rope Sling, Rated Capacity in Pounds* (Safety Factor—5)

Rope diameter (nominal), in	Nominal weight per 100 ft. lb	Minimum breaking strength, lb	Eye and eye sling						Endless sling					
					Basket hitch						Basket hitch			
					Angle of rope to horizontal						Angle of rope to horizontal			
					90°	60°	45°	30°			90°	60°	45°	30°
					Angle of rope to vertical						Angle of rope to vertical			
			Vertical hitch	Choker hitch	0°	30°	45°	60°	Vertical hitch	Choker hitch	0°	30°	45°	60°
1/2	7.5	2,650	550	250	1,100	900	750	550	950	500	1,900	1,700	1,400	950
9/16	10.4	3,450	700	350	1,400	1,200	1,000	700	1,200	600	2,500	2,200	1,800	1,200
5/8	13.3	4,400	900	450	1,800	1,500	1,200	900	1,600	800	3,200	2,700	2,200	1,600
3/4	16.7	5,400	1,100	550	2,200	1,900	1,500	1,100	2,000	950	3,900	3,400	2,800	2,000
13/16	19.5	6,500	1,300	630	2,600	2,300	1,800	1,300	2,300	1,200	4,700	4,100	3,300	2,300
7/8	22.5	7,700	1,500	750	3,100	2,700	2,200	1,500	2,800	1,400	5,600	4,800	3,900	2,800
1	27.0	9,000	1,800	900	3,600	3,100	2,600	1,800	3,200	1,600	6,500	5,600	4,600	3,200
1 1/16	31.3	10,500	2,100	1,100	4,200	3,600	3,000	2,100	3,800	1,900	7,600	6,600	5,400	3,800
1 1/8	36.0	12,000	2,400	1,200	4,800	4,200	3,400	2,400	4,300	2,200	8,600	7,500	6,100	4,300
1 1/4	41.7	13,500	2,700	1,400	5,400	4,700	3,800	2,700	4,900	2,400	9,700	8,400	6,900	4,900
1 5/16	47.9	15,000	3,000	1,500	6,000	5,200	4,300	3,000	5,400	2,700	11,000	9,400	7,700	5,400
1 1/2	59.9	18,500	3,700	1,850	7,400	6,400	5,200	3,700	6,700	3,300	13,500	11,500	9,400	6,700
1 5/8	74.6	22,500	4,500	2,300	9,000	7,800	6,400	4,500	8,100	4,100	16,000	14,000	11,500	8,000
1 3/4	89.3	26,500	5,300	2,700	10,500	9,200	7,500	5,300	9,500	4,800	19,000	16,500	13,500	9,500
2	107.5	31,000	6,200	3,100	12,500	10,500	8,800	6,200	11,000	5,600	22,500	19,500	16,000	11,000
2 1/8	125.0	36,000	7,200	3,600	14,500	12,500	10,000	7,200	13,000	6,500	26,000	22,500	18,500	13,000
2 1/4	146.0	41,000	8,200	4,100	16,500	14,000	11,500	8,200	15,000	7,400	29,500	25,500	21,000	15,000
2 1/2	166.7	46,500	9,300	4,700	18,500	16,000	13,000	9,300	16,500	8,400	33,500	29,000	23,500	16,500
2 5/8	190.8	52,000	10,500	5,200	21,000	18,000	14,500	10,500	18,500	9,500	37,500	32,500	26,500	18,500

*See Fig. 43.4 for sling configuration descriptions.
Source: Occupational Safety and Health Administration, Rules and Regulations #1910.184 Slings, 1976.

Hand Lines. Hand lines are used for raising and lowering light material and tools or for holding small transformers away from a pole while the latter is being raised.

Throw Line. A small-diameter rope is used to throw a line over a crossarm, tree limb, or other support object. The throw line is normally used to pull a larger rope into place for performing a task.

Running Line. The running line is used for pulling in several span lengths of wire at one time.

Safety Line. The safety line is used only for lowering a man to the ground.

Slings. Slings are used for lashing tools or material in place, for attaching blocks and snatch blocks to a pole, for lashing an old pole to a new pole temporarily, and for tying line wires up temporarily.

KNOTS AND KNOT TYING

Knots are used for fastening a rope to an object or for joining two ends of a rope. The knot or hitch used must hold the strain to be applied without damaging the rope or the load. The knot used must also be one that can be tied or loosened easily and quickly.

Terms Used in Knot Tying. All knots and hitches are a combination of the three different kinds of bends: the bight, the loop, and the round turn (Fig. 43.5).

For convenience in describing the method of making various knots, the following terms will be used: *standing part*, *bight*, and *running end* (Fig. 43.5*a*). The standing part is the principal portion, or longest part of the rope; the bight is a loop formed with the rope so that the two parts lie alongside each other; and the running end of a rope is the free end that is used in forming the knot. Figures 43.6 through 43.27 illustrate various kinds of knots.

SPLICES

Eye Splice. When it is desirable to have a permanent eye at the end of a rope, such as a hand line, it can be formed by splicing the end of the rope into its side, thereby making an eye or side splice.

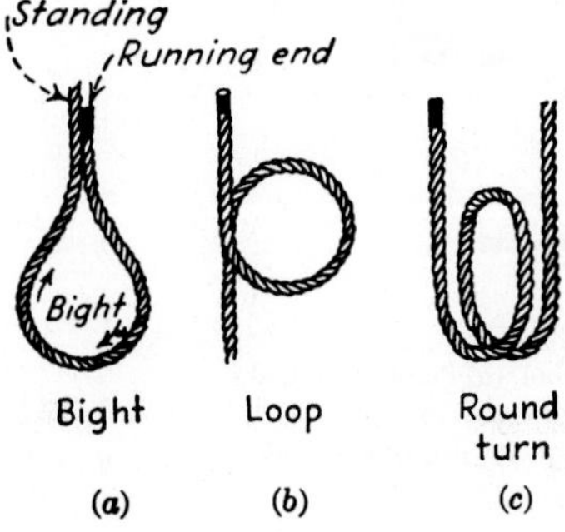

FIGURE 43.5 Terms used in knot tying.

FIGURE 43.6 Overhand knot. The overhand knot is the simplest knot made and forms a part of many other knots. This knot is often tied in the end of a rope to prevent the strands from unraveling or as a stop knot to prevent rope from slipping through a block.

FIGURE 43.7 Half hitch. A half hitch is used to throw around the end of an object to guide it or keep it erect while hoisting. A half hitch is ordinarily used with another knot or hitch. A half hitch or two thrown around the standing part of a line after tying a clove hitch makes a very secure knot.

FIGURE 43.8 Two half hitches. This knot is used in attaching a rope for anchoring or snubbing. It is easily and quickly made and easily untied.

FIGURE 43.9 Square knot. The square knot is used to tie two ropes together of approximately the same size. It will not slip and can usually be untied even after a heavy strain has been put on it. Linemen use the square knot to bind light leads, lash poles together on changeovers, on slings to raise transformers, and for attaching blocks to poles and crossarms.

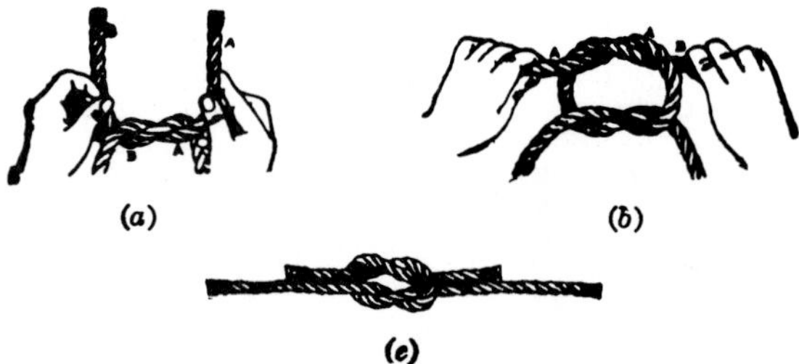

FIGURE 43.10 Method of making square knot. (*a*) Passing left end *A* over right end *B* and under. (*b*) Passing right end *B* over left end *A* and under. (*c*) The completed knot drawn up. (*Courtesy Plymouth Cordage Co.*)

FIGURE 43.11 Granny knot. Care must be taken that the standing and running parts of each rope pass through the loop of the other in the same direction, i.e., from above downward, or vice versa; otherwise a granny knot is made which will not hold.

FIGURE 43.12 Thief knot. In tying the square knot the standing part of both ropes must cross, as otherwise a useless knot known as the *thief knot* is formed.

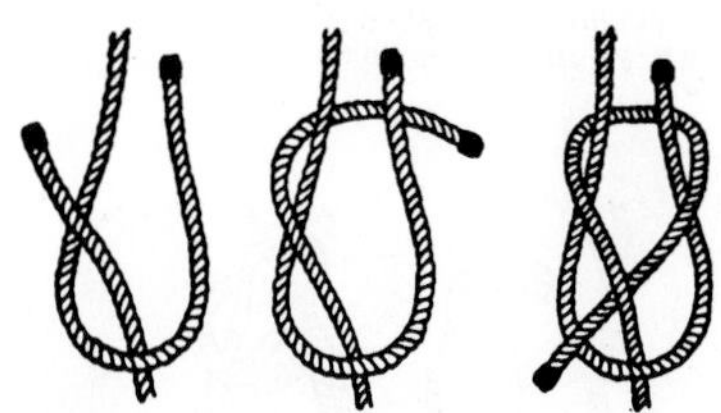

FIGURE 43.13 Single sheet bend. The single sheet bend is used in joining ropes, especially those of unequal size. It is more secure than the square knot but is more difficult to untie. It is made by forming a loop in one end of a rope; the end of the other rope is passed up through the loop and underneath the end and standing part, then down through the loop thus formed.

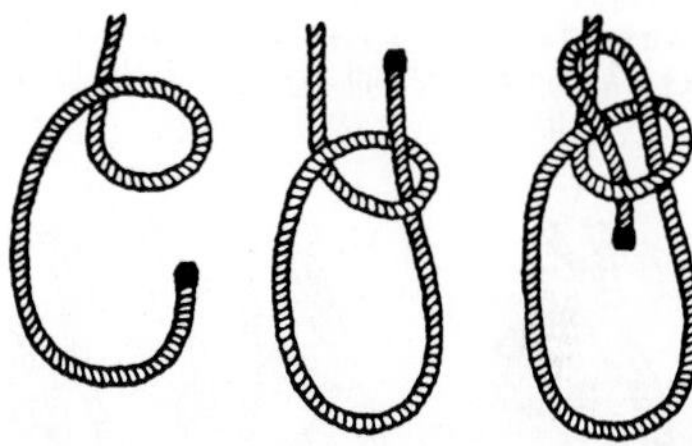

FIGURE 43.14 The bowline. The bowline is used to place a loop in the end of a line. It will not slip or pull tight. Linemen use the bowline to attach come-alongs (wire grips) to rope, to attach tail lines to hook ladders, and as a loose knot to throw on conductors to hold them in the clear while working on poles.

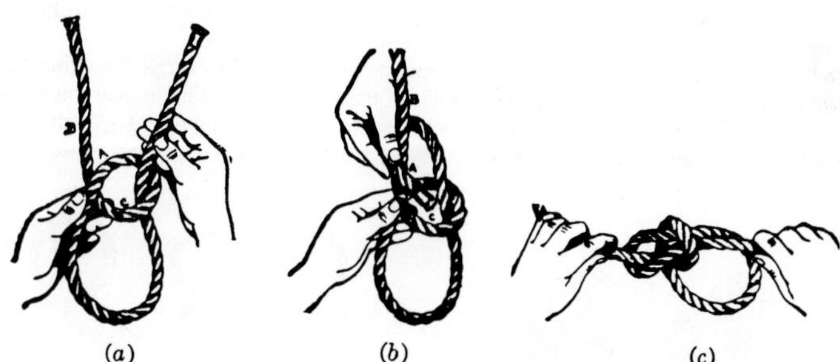

FIGURE 43.15 Method of making bowline knot. (*a*) Threading the bight from below. (*b*) Leading around standing part and back through bight C. (*c*) The completed bowline. (*Courtesy Plymouth Cordage Co.*)

The steps in making the eye splice are illustrated and described in Fig. 43.28.

Short Splice. The short splice is sometimes called a *butt splice* because the line is unlaid and the ends are butted together. This splice can be used where it is desirable to splice together two ropes which are not required to pass over a pulley. This splice can be made quickly and is nearly as strong as the rope. As the diameter of the rope is nearly doubled, this type of splice is too bulky to pass through a sheave block.

The steps in making the short splice are illustrated and described in Fig. 43.29.

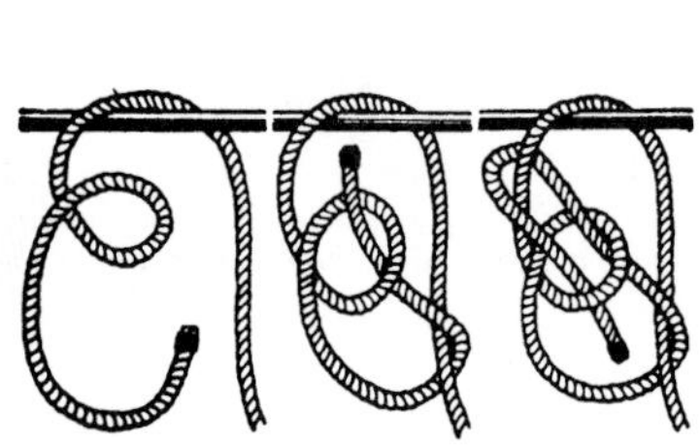

FIGURE 43.16 Running bowline. This knot is used when a hand line or bull rope is to be tied around an object at a point that cannot be safely reached, such as the end of a limb.

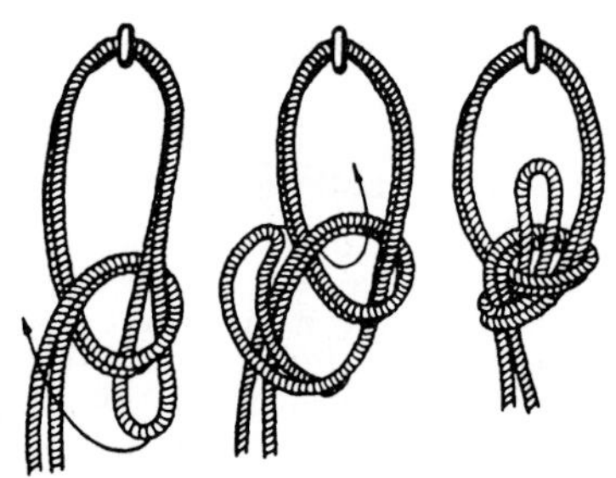

FIGURE 43.17 Double bowline. This knot is used to form a loop in the middle of a rope that will not slip when a strain is put upon it.

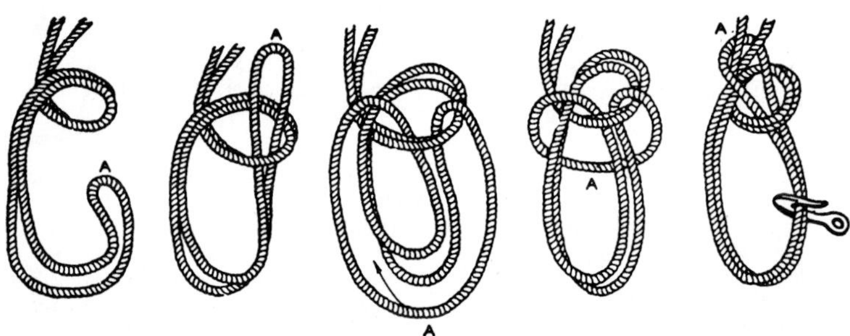

FIGURE 43.18 Bowline on a bight. The bowline on a bight is used to place a loop in a line somewhere away from the end of the rope. It can be used to gain mechanical advantage in a rope guy by doubling back through the bowline on a bight much as a set of blocks. The bowline on a bight also makes a good seat for a man when he is suspended on a rope.

To tie this bowline, take the bight of the rope and proceed as with the simple bowline; only instead of tucking the end down through the bight of the knot, carry the bight over the whole and draw up, thus leaving it double in the knot and double in the standing part. The loop under the standing part is single.

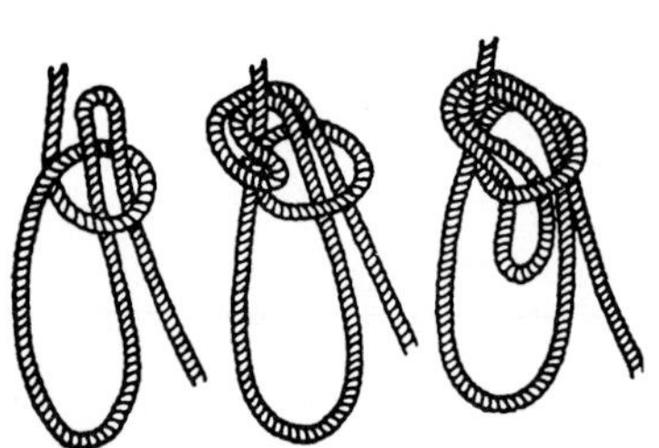

FIGURE 43.19 Single intermediate bowline. This knot is used in attaching rope to the hook of a block where the end of the rope is not readily available.

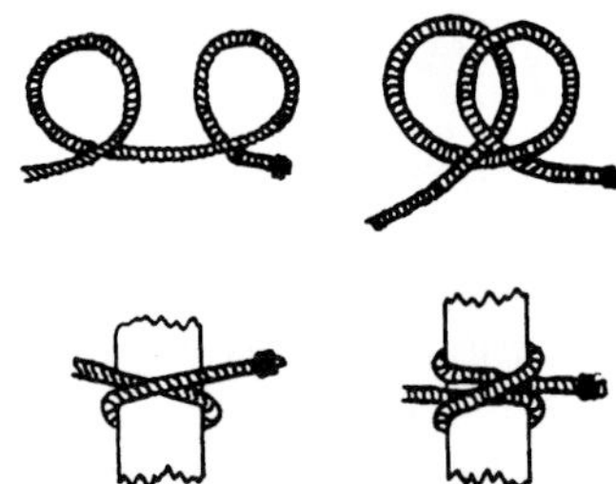

FIGURE 43.20 Clove hitch. The clove hitch is used to attach a rope to an object such as a crossarm or pole where a knot that will not slip along the object is desired. Linemen use the clove hitch for side lines, temporary guys, and hoisting steel.

To make this hitch, pass the end of the rope around the spar or timber, then over itself, over and around the spar, and pass the end under itself and between the rope and spar as shown.

FIGURE 43.21 Clove hitch used for lifting.

FIGURE 43.22 Timber hitch. The timber hitch is used to attach a rope to a pole when the pole is to be towed by hand along the ground in places where it would be impossible to use a truck or its winch line to spot it. The timber hitch is sometimes used to send crossarms aloft. This hitch forms a secure temporary fastening which may be easily undone. It is similar to the half hitch but is more secure. Instead of the end being passed under the standing part of the rope once, it is wound around the standing part three or four times, as shown in the figure.

FIGURE 43.23 Rope timber hitch and half hitch. The timber hitch will not slip under a steady pull but may slip when slack. To make the timber hitch more secure, a single half hitch may be taken a little farther along on the spar.

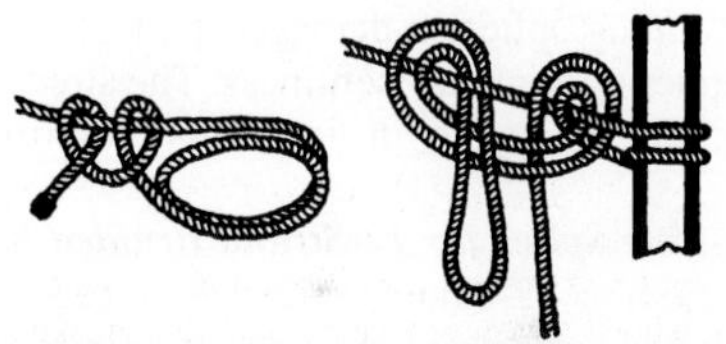

FIGURE 43.24 Snubbing hitches. Knots used for attaching a rope for anchoring or snubbing purposes.

FIGURE 43.25 Taut-rope hitch. This knot is used in attaching one rope to another for snubbing a load.

FIGURE 43.26 Rolling bend. The rolling bend is often used for attaching the rope to wires that are too large for the wire-pulling grips. It is also used for skidding poles or timber.

FIGURE 43.27 Blackwall hitch. This hitch consists of a loop with the end of the rope passed under the standing part and across the hook. Under load the hauling part jams the end against the hook. This hitch should be used only where the strain is steady and there is no hazard if the hitch slips. It is exceedingly useful when the hitch has to be made quickly or where the hitch must be changed frequently.

Long Splice. The long splice is sometimes known as a *running splice*. It is used to splice rope where it is undesirable to increase the diameter of the rope. A good long splice is hard to detect without close examination and will run freely through a sheave or blocks. Using a running splice, a damaged part of a long line can be removed from the line without detracting from its usefulness. The strength will be reduced, however. The steps in making the long splice are illustrated and described in Fig. 43.30.

In-Line Splice for 12-Strand Braided Rope.* The in-line splice is useful to repair synthetic fiber 12-strand braided damaged rope or to join two separate pieces of rope. Start the splicing by securing the ends of both ropes with tape. Lay the ropes side by side and measure one fid length, 20 times rope diameter, from the ends of the rope and place a mark on each rope (Fig. 43.31*a*). Fit rope 1 into the fid and insert through the middle of rope 2 at the mark to start making the splice (Fig. 43.31*b*). Pull the end of rope 1 through rope 2 until both marks meet (Fig. 43.31*c*). Place the end of rope 2 into the fid, and insert the fid and the end of rope 2 through the middle of rope 1 at a point three pairs of strands beyond the marks (Fig. 43.31*d*). Pull rope 2 through rope 1 until the ropes are tight together (Fig. 43.31*e*). Place rope 1 into the fid, and insert the fid and the end of rope 1 into the center of rope 2 at a point three pairs of strands beyond the marks. Run the fid and the end of rope 1 an entire fid length through the center core of rope 2, and at this point push the fid and the end of the rope out through the exterior of rope 2. Pull rope 1 through the exit point on the surface of rope 2 (Fig. 43.31*f*). Place rope 2 into the fid, and insert the fid and the end of rope 2 into the center core of rope 1 at a point three pairs of strands beyond the previous exit point from rope 1. Run the fid and the end of rope 2 an entire fid length through the center core of rope 1, and at this point push the fid out of rope 1 and pull the end of rope 2 through the exit point (Fig. 43.31*g*). Finish the splice by removing the tape from the ends of the ropes and, using a scissors or a knife, trim the rope ends at staggered intervals, near the points the ropes exit from the line (Fig. 43.31*h*). Hold the rope on each side of the splice and smooth the line away from the splice. The tapered ends of the ropes will disappear into the core of the line, leaving a clean splice (Fig. 43.31*i*).

---

* Source: Yale Cordage, Inc., instructions for splicing braided rope.

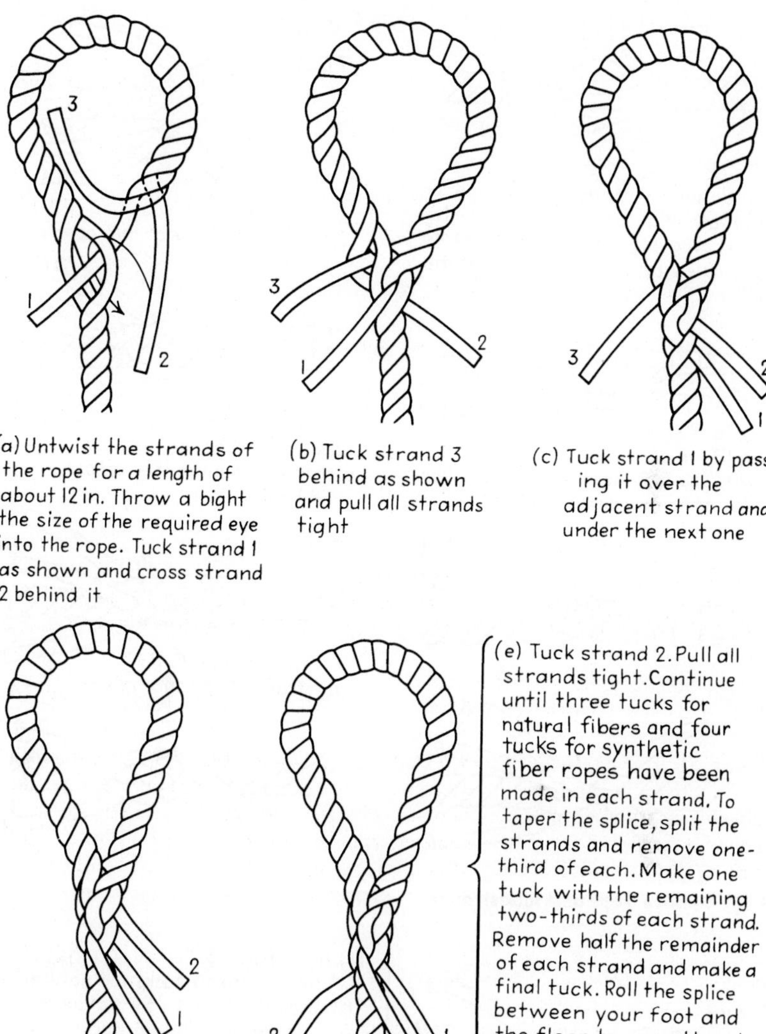

FIGURE 43.28 Eye splice, fiber rope.

Eye Splice for 12-strand Braided Rope.* Start the splicing by securing the end of the rope with tape. Lay the rope on a flat surface and measure one fid length, 20 times rope diameter, and allowing some extra rope from the end, mark the rope with an X. Starting at the X mark, form the desired size eye and mark the rope with a dot opposite the X (Fig. 43.32*a*). Fit the tapered end of the rope into the fid, and insert the fid and the rope end through the middle of

* Source: Yale Cordage, Inc., instructions for splicing braided rope.

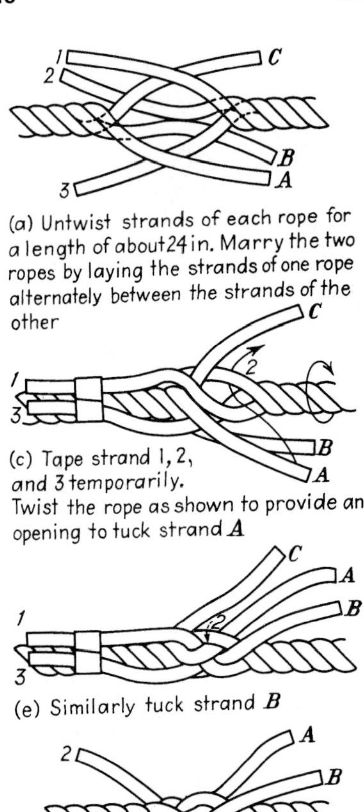

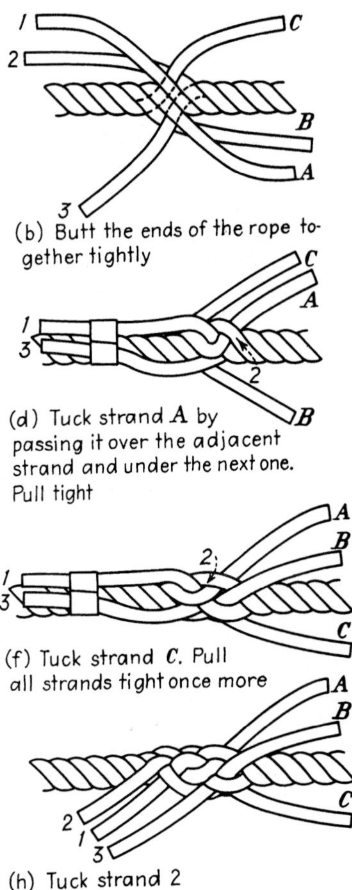

(i) Tuck strand 3. Pull all strands tight. Continue until six tucks for natural fiber and eight tucks for synthetic fiber ropes have been made in each strand. To taper the splice, split the strands and remove one-third of each. Make one tuck with the remaining two-thirds of each strand. Remove half of the remainder of each strand and make a final tuck. Roll the splice between your foot and the floor to smooth out. Cut off surplus ends flush with outside strands

FIGURE 43.29 Short splice, fiber rope.

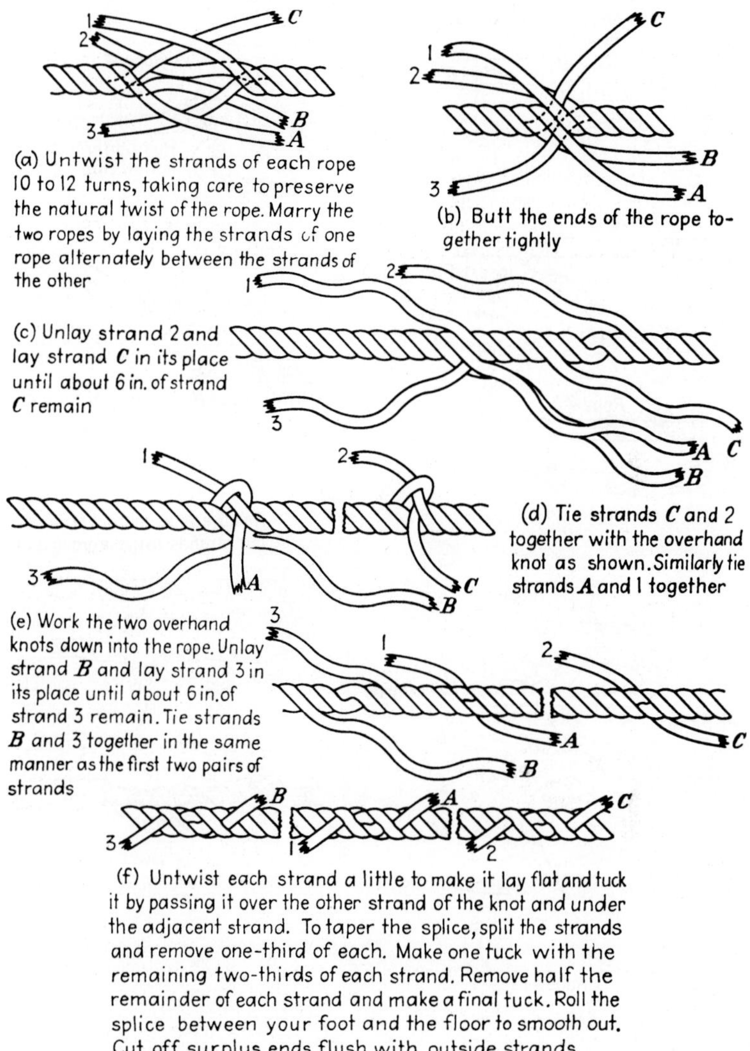

FIGURE 43.30 Long splice, fiber rope. The long splice is used for permanently joining two ropes which must pass through a close fitting pulley.

the line at the dot (Fig. 43.32*b*). Pull the end of the rope through the line until the X and the dot meet (Fig. 43.32*c*). Insert the fid and the end of the rope back through the middle of the line three pairs of strands beyond the junction of the X and the dot. Pull the end of the rope through the exit point (Fig. 43.32*d*). Insert the fid and the end of the rope back into the middle of the line three pairs of strands beyond the previous exit from the line, and run the fid and end of rope down through the center core of the line a fid length. Push the fid and end of the rope through the outside of the line. Pull the end of the rope through the exit point (Fig. 43.32*e*). Remove the tape from the end of the rope, and, using a scissors or a knife, trim the rope end at

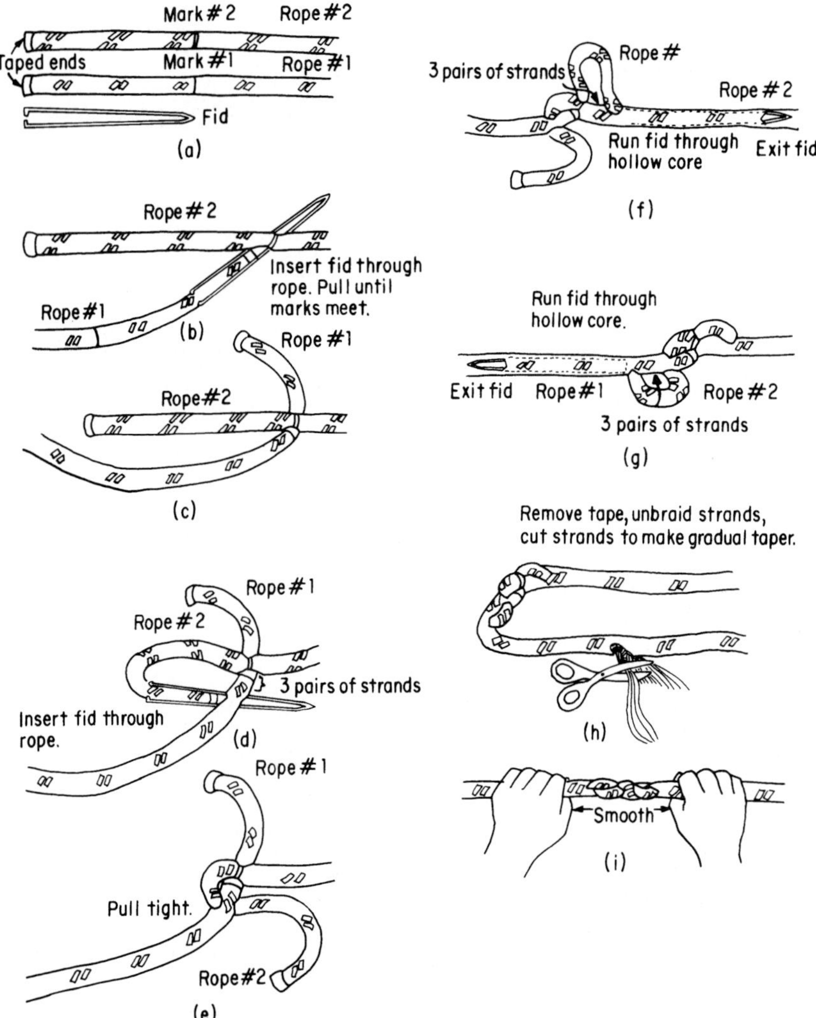

FIGURE 43.31 In-line splice for 12-strand braided synthetic fiber rope. (*Courtesy Yale Cordage, Inc.*)

staggered intervals near the point where it exits from the line (Fig. 43.32*f*). Hold the line beyond the eye splice, and smooth the line toward the eye until the tapered end of the rope disappears into the center of the line (Fig. 43.32*g*). The throat of the eye splice can be whipped if desired (Fig. 43.32*h*).

Eye Splice for Double-Braided Rope.* Start the splice by placing one thin layer of tape on the end of the rope to be spliced. Lay the rope on a flat surface and measure two fid lengths,

* Hybrid double-braided constructed ropes may require special splicing techniques. The user should consult the rope manufacturer for instructions.

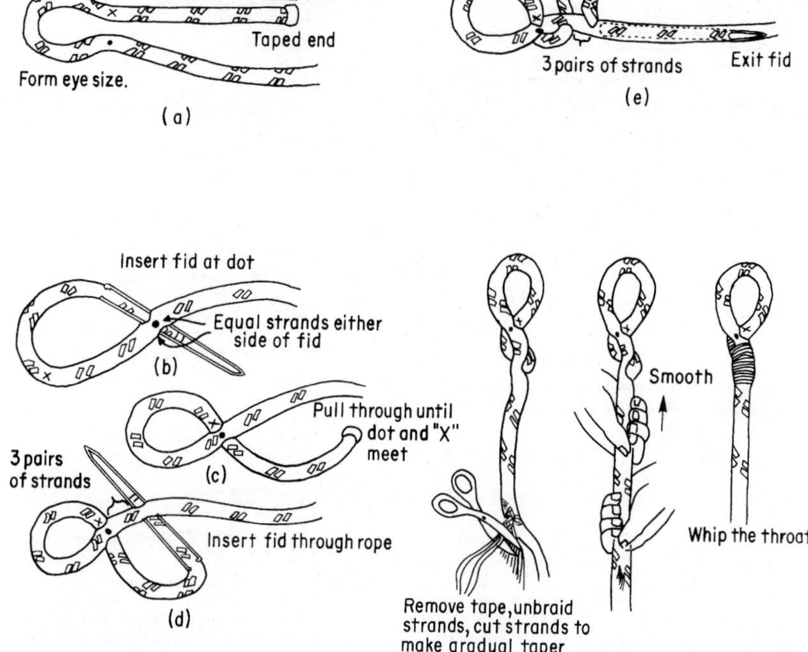

FIGURE 43.32 Eye splice for 12-strand braided synthetic fiber rope. (*Courtesy Yale Cordage, Inc.*)

20 times the rope diameter, and mark the line with the reference point R. Form a loop of the size desired from reference point R in the line and mark this point X. Tie a slipknot in the line approximately 10 fid lengths, 100 times the rope diameter, beyond point X (Fig. 43.33*a*). Bend the rope sharply at point X. Spread the braided cover of the line at point X with any sharp tool such as an ice pick, awl, or marlin spike, and pry the core of the rope out through the cover (Fig. 43.33*b*). Pull the core of the rope out through the cover at point X from the taped end of the rope. Place one layer of tape on the end of the core. Hold the exposed core and slide the cover as far back as possible toward the tightly tied slipknot. Firmly smooth the cover back from the slipknot toward the taped end until all cover slack is removed. Mark the core of the rope I at the point where it exits from the line (Fig. 43.33*c*). Slide the cover of the rope again toward the slipknot to expose more of the rope core. Measure along the core of the rope from mark I one-half fid length, and mark the rope core II. Measure along the core of the rope from mark II 2.5 fid lengths, and mark the core of the rope III (Fig. 43.33*d*). Inspection of the braided cover of the rope reveals that half of the strands revolve to the right around the rope and half revolve to the left. Measure one-half fid from reference mark R toward the end of the rope, and mark this point T and mark a line completely around the braided cover of the rope. Start at point T and mark toward the taped end of the rope every fifth right and left strand (single or paired) to the end of the taped braided cover (Fig. 43.33*e*). If the rope has 20 strands in braided cover, start at point T and mark fifth right and left strands, then fourth right and left, then fifth, etc., until reaching the end of braided cover. Press prongs of fid into the end of braided cover, and tape over the cover and the fid prongs. Hold the core of the rope lightly

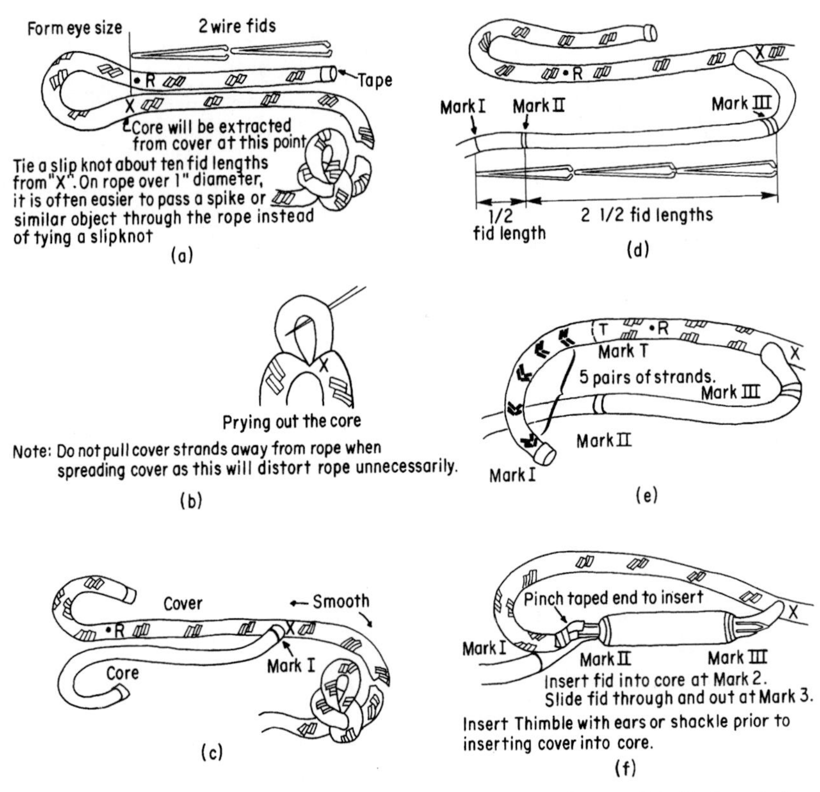

FIGURE 43.33 Eye splice for double-braided synthetic fiber rope. (*Courtesy Yale Cordage, Inc.*)

at mark III, and push the fid with the braided cover attached into the middle of the core at mark II and out at mark III (Fig. 43.33*f*). Pull the fid and the end of the braided cover from the core at mark III until reference point R appears. Remove tape and fid from the end of the braided cover of the rope (Fig. 43.33*g*). Start with the last marked pair of the braided cover strands near the end of the rope, and cut and pull the strands out. Proceed to cut and remove each right and left strand at the marks until point T is reached (Fig. 43.33*h*). The braided cover from point T will have a gradual taper to a point at the end of the rope (Fig. 43.33*i*). Measure one fid length on the braided cover from point T toward the slipknot in the rope, and mark this point Z. Place the taped end of the core into the fid and insert the fid into the middle of the braided cover at point T, and milk the braided cover over the fid and the end of the core of the rope while pulling them through the braided cover exiting at point Z, making sure the fid does not catch on any internal core strands (Fig. 43.33*j*). Hold the core tail at the exit point Z in one hand and the core with the tapered braid inserted at mark III in the other hand, and alternately work the core tail through the exit of the braided cover until the rope core at mark III is approximately one diameter of the rope from the braided cover at point X. Mark the core tail with a dot at the point it enters the braided cover at point X (Fig. 43.33*k*). Smooth out braided cover of eye completely, from mark T to the dot on the core, and mark X on the braided cover, to get all the slack out of the eye area. The tapered end of the braided cover will disappear into the core of the rope. Pull the core tail out of the braided cover at point Z until the dot on the core is exposed. Measure a distance of three-fourths of a

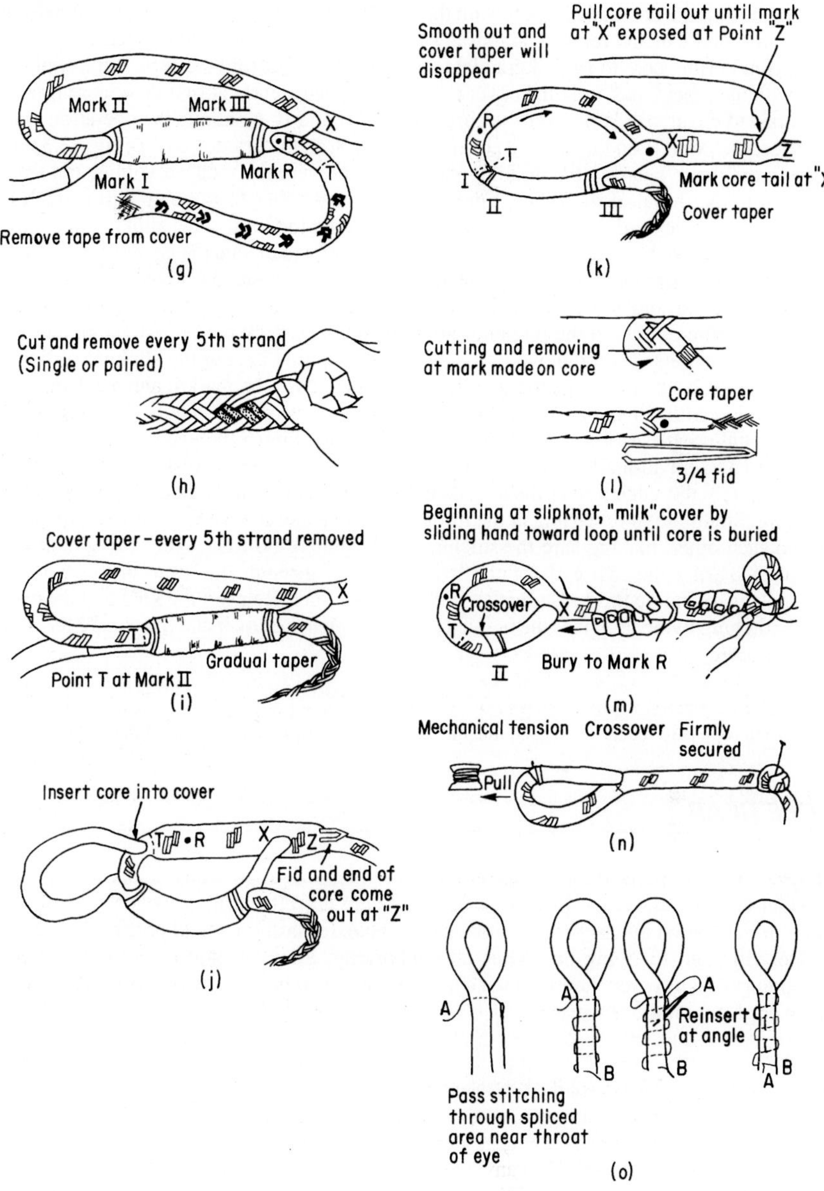

FIGURE 43.33 (*Continued*)

fid length from the dot on the core toward the end, and cut the core, removing the unneeded end of the core. Cut and remove strands at staggered intervals between the dot on the core and the end of the core to form a gradual taper (Fig. 43.33*l*). Hold the rope at the slipknot and, with the other hand, milk the cover toward the splice, gently at first then more firmly. The braided cover will slide over the core at mark III and II. Continue the operation until the reference point R on the braided cover is at point X on the braided cover (Fig. 43.33*m*). Smooth out the eye during the milking operation to prevent the reduced volume tail from catching in the throat of the splice. Smooth the braided cover from point T to point X if bunching occurs at mark I. It may be necessary to anchor the loop of the slipknot to a fixed object and then use both hands and the weight of the body to extend the cover over the core of the rope. The rope cover can be hammered at point X to loosen the strands, and the rope should be flexed and loosened at the crossover point mark I during the final burying process. Continue milking the line until all braided cover slack between the knot and the throat of the eye has been removed. It is helpful to securely anchor the slipknot of larger ropes and attach a small line to the braided core at the crossover, mark I, and mechanically apply tension with either a block and tackle or capstan. Tension will reduce the diameter of the core and crossover for easier burying (Fig. 43.33*n*). Finish the eye splice with stitch locking to prevent no-load opening due to mishandling. Nylon or polyester whipping twine approximately the same size as the strands in the rope being spliced should be used for the stitching. Stitch the line, starting at the throat of the eye and in turn two planes perpendicular to each other, making sure the stitching is not pulled too tight. After the stitching is completed, bring the two ends of the twine together through the same opening in the braided cover, and tie the ends together with a square knot. Reinsert the knot and the ends of the stitching twine back into the braided cover between the cover and the core of the rope (Fig. 43.33*o*).

Reduction in Strength due to Knots and Splices. Table 43.6 shows the extent to which the strength of rope is reduced when used in connection with some of the common knots and splices.

ROPE GEAR

Slings. The simplest sling consists of a short piece of rope whose ends are spliced together to form an endless piece of rope. Slings are used for lashing tools or material in place, for attaching blocks or snatch blocks to a pole, for lashing an old pole to a new pole temporarily, and for tying line wires up temporarily. Slings should be 6 to 10 ft long, depending, of course, on their planned use. Slings can also be made with an eye on one end and a dog knot on the other end.

TABLE 43.6 Percentage Strength of Spliced or Knotted Rope

Type or splice or knot	Percentage strength
Straight rope	100
Eye splice	90
Short splice	80
Timber hitch or half hitch	65
Bowline or clove hitch	60
Square knot or sheet bend	50
Overhand knot (half or square knot)	45

Slings are also made of rope, spliced for a hook at one end and a hook or ring at the other.

Safe Loads. Before attempting to lift a load with a rope sling, the weight of the load should be carefully estimated and a sling of the proper size and labeled strength selected. The greatest load can be lifted when all legs of the sling are in a vertical position and the least when the legs are nearly horizontal. When a rope sling has been in service for 6 months or more, even though it shows no signs of wear or damage, the loads placed on it should be reduced.

Figure 43.34 illustrates the manner in which the strain on the rope sling is increased when the angle with the horizontal decreases. Thus, at 60° the strain in the legs of the sling is only 578 lb for a 1000-lb weight, whereas it is 1930 lb for a 1000-lb load when the angle is reduced to 15°. This is almost twice as great as the weight of the load lifted. At 30° the strain is 1000 lb, which is the same as the weight of the load. Sling angles smaller than 30° are therefore not recommended.

If the sling is attached around a sharp corner, it should be carefully padded to prevent cutting.

Block and Tackle. Block and tackle are used for applying tension to line conductors when sagging in, for applying tension to guy wires, when hoisting transformers, and for other general-purpose hoisting (see Fig. 43.35).

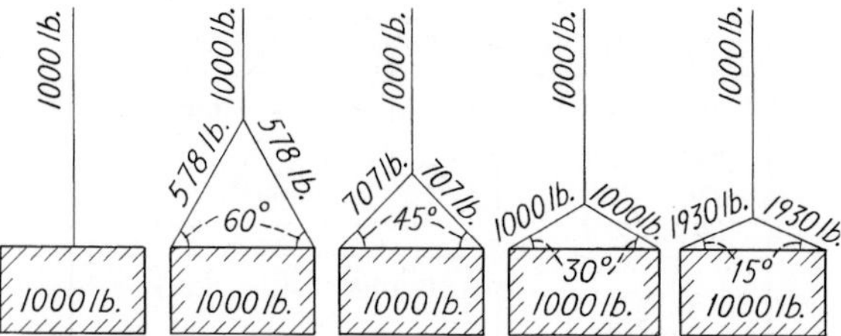

FIGURE 43.34 Sample loading of parts of lifting slings. For loads other than 1000 lb, use direct ratios. Sling angles less than 30° from horizontal are not recommended.

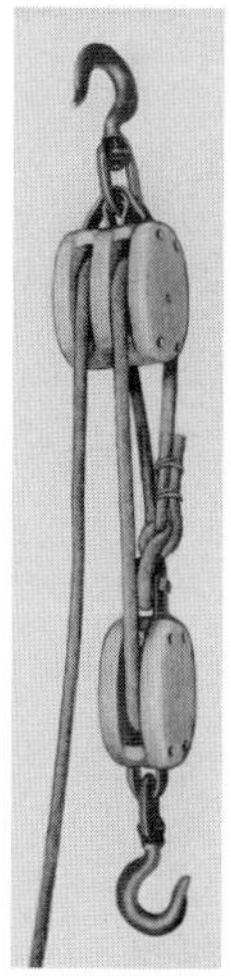

FIGURE 43.35 A typical three-part block and tackle. (*Courtesy A. B. Chance Co.*)

The use of block and tackle has two advantages: (1) the user can stand on the ground and pull downward while hoisting or lifting a load; and (2) the manual force applied need only be a fractional part of the load lifted.

Mechanical Advantage. To find the pull required to lift a given weight with a block and tackle, divide the weight by the number of ropes running from the movable block. The lead line or fall line is not to be counted unless the line comes from the moving block. There is always some friction loss around the sheaves. This can be estimated at 10 percent per sheave and added to the load to be lifted. The sketches in Fig. 43.36 show the various combinations of block and tackle employed to give mechanical advantages from 2 to 7. As mentioned above, the ratio of load to pull on the fall line is given by the number of ropes running from the movable block. The fall line is not included in the calculation of mechanical advantage because the lineman is pulling on the rope from the fixed block. The load that may be lifted is therefore the mechanical advantages times the safe load on the rope. Normal uses for 4-in blocks are sagging No. 4 and No. 6 conductors and raising distribution transformers rated up to $37\frac{1}{2}$ kVA single-phase. The 6-in blocks are used for heavier work up to the limit of the tackle. Blocks having mechanical advantages of 4, 5, and 6 are illustrated in Fig. 43.37.

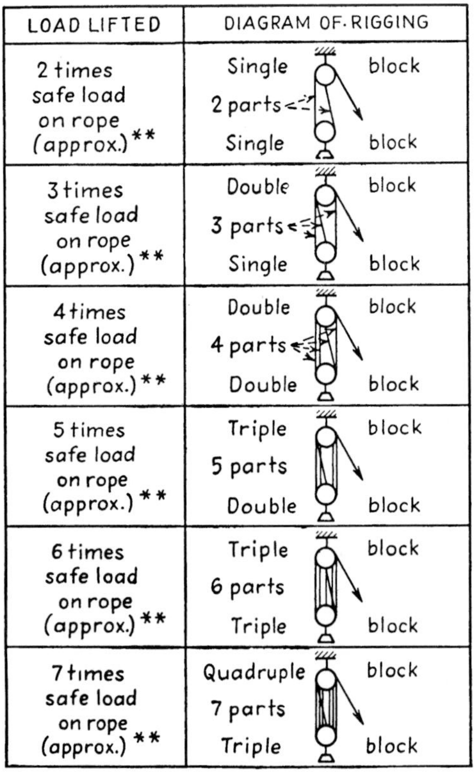

LOAD LIFTED	DIAGRAM OF RIGGING
2 times safe load on rope (approx.)**	Single block 2 parts Single block
3 times safe load on rope (approx.)**	Double block 3 parts Single block
4 times safe load on rope (approx.)**	Double block 4 parts Double block
5 times safe load on rope (approx.)**	Triple block 5 parts Double block
6 times safe load on rope (approx.)**	Triple block 6 parts Triple block
7 times safe load on rope (approx.)**	Quadruple block 7 parts Triple block

FIGURE 43.36 Lifting capacity of block and tackle.
**Less 20 percent approximately for friction.

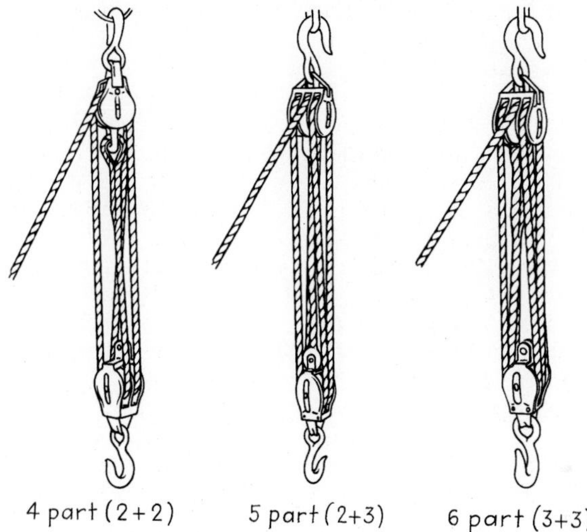

FIGURE 43.37 Four-, five-, and six-part blocks and tackle.

CHAPTER 44
CLIMBING WOOD POLES

USE AND CARE OF POLE-CLIMBING EQUIPMENT*

Pole-Climbing Equipment. Pole-climbing equipment consists of a body belt and safety strap and a pair of climbers. Figure 44.1 shows a workman wearing this equipment.

The equipment allows a workman to climb, stand, or change position on a pole when no other suitable means of support is available. It also allows the free use of both hands while in any position on the pole.

Suggested specifications for pole-climbing equipment are detailed in Edison Electric Institute (EEI) Publication No. AP-2-1973. General descriptions of components of the equipment and their use and care follow.

The Body Belt. The body belt consists of a cushion section, a belt section with tongue and buckle ends, a tool saddle, and D rings which are attached solidly to the cushion, or on shifting D-ring belts, attached solidly to a D-ring saddle. The body belt usually has provisions made for a holster which will carry one or more tools in addition to the tools carried in the tool loops.

Tool loops should be of proper size to prevent the tools from slipping through the loops and falling. There should be no tool loops for 2 in on either side of the center in the back, in accordance with EEI's Specifications for Linemen's Climbing Equipment, AP-2-1973.

The belt, as a general rule, is marked in "D" sizes. The D size is the distance between the heels of the D rings when the belt is laid flat (Fig. 44.2).

FIGURE 44.1 Worker with climbing equipment.

*Reprinted with the permission of the Edison Electric Institute.

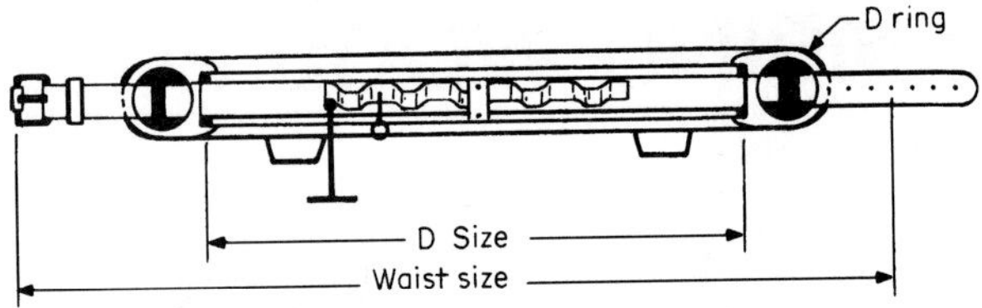

FIGURE 44.2 Body belt.

The waist measurement of the belt is found by measuring from the roller on the buckle to the center hole on the tongue end (Fig. 44.2).

Proper measurement for the D-ring size is the distance around behind the body at the point where the belt will be worn between the prominent points of the hip bones, *plus 1 in* (Fig. 44.3).

The proper waist size of body belt is determined by measuring the distance around the body at the point where the belt will be worn. The measurement should be made outside of any clothing normally worn while working (Fig. 44.4).

The body belt should be worn snugly but not too tightly. The end of the strap should always be passed through the keeper and kept clear of the D ring when the belt is being worn.

Manufacturers have standardized on a relationship between D sizes and waist sizes. In the event your measurements do not coincide with these standard sizes, the belt should be ordered by D size, as the waist size is adjustable.

The Safety Strap. The safety strap is used for support while working on poles, towers, or platforms. Snap hooks on each end are provided for attachment to the D rings in the body belt.

When climbing poles, under normal conditions, both snaps should be engaged in the *same* D ring for safety. The snap on the double end should have the keeper facing outward, and the other snap should face inward (Fig. 44.5). Right-handed men and some left-handed men usually carry the strap on the left as shown.

When in use, one snap hook should be securely engaged in each D ring, never both snaps in the same D ring. The user should look to be sure that snaps are properly engaged. He should never depend on sound or feel for security.

Safety straps are adjustable for length by means of a buckle in the strap to suit the workman and the size of the pole. When in use, the side of the strap to which the buckle is attached should be next to the pole with the buckle tongue outward.

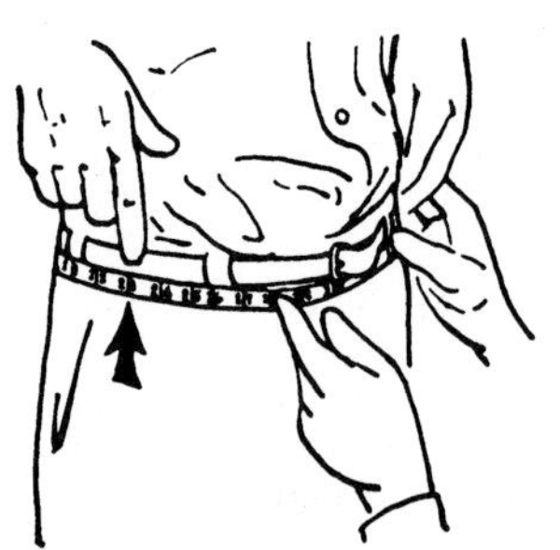

FIGURE 44.3 Measuring individual for D size.

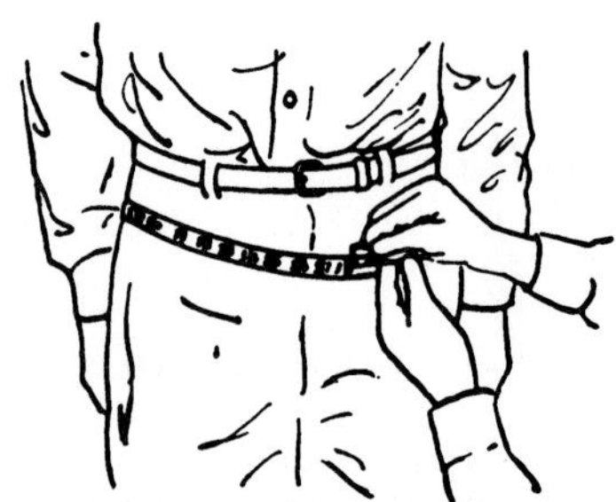

FIGURE 44.4 Measuring individual for waist size.

FIGURE 44.5 Position of belt and strap when climbing.

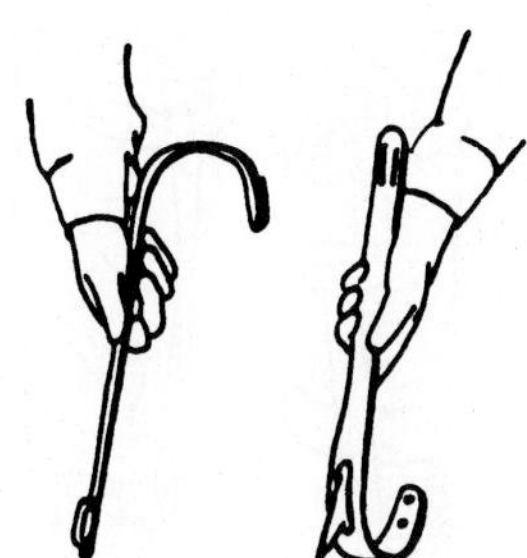

FIGURE 44.6 Climbers.

The Climbers. Climbers (Fig. 44.6) are used for ascending, descending, and maintaining the working position on poles when no other means of support is available. The condition, length, and shape of the gaffs of the climbers are of the utmost importance. The gaffs support the workman as he climbs or does his work. Defective gaffs are dangerous and can cause inadvertent electrical contacts or falls from poles.

Climbers are made in adjustable or fixed lengths from 14 to 20 in by 1/2-in steps. Gaffs are furnished in either solid or replaceable type. Proper fit requires a leg iron to reach about 1/2 in below the inside prominence of the knee joint.

Climbers are secured to the workman's legs by foot and leg straps. These straps should be drawn up to a snug fit, but *not* so tight as to be tiring. High-top shoes with heavy soles and heels should be worn for climbing.

The buckle on the foot strap should lie just outside the shoe lacing. Several types of pads are furnished for the upper end of the leg irons. One type of pad is shown in Fig. 44.7. All leg and foot strap ends should be snugged down in their keepers after buckling, and strap ends should point to the rear and outside.

Before the leg straps are fastened, pull up the pant legs so that they bag at the knees and do not bind. Fold the pant legs snugly against the calf, toward the outside, as shown in Fig. 44.8. This prevents the pant leg from tripping the workman while climbing. With climbers properly adjusted the workman will feel comfortable and confident.

Climbing. Arms and hands should not be bare, but properly protected when climbing. Before ascending a pole, inspect it carefully for unsafe conditions, such as rake, rotted places, nails, tacks, cracks, knots, foreign attachments, pole steps, or ice. Remove rocks and other objects from the ground at the base of the pole. Unauthorized attachments, such as signs, radio aerials, or clotheslines, should be reported to the workman's supervisor or removed, as company instructions may require.

Inspect the pole as you ascend and descend to avoid placing gaffs in cracks, knots, woodpecker holes, etc., which might cause a fall.

When ascending the pole, keep the arms and body relaxed with the hips, shoulders, and knees a comfortable distance away from the pole. Take it easy; favor short steps (step length should be natural for each workman); use the hands and arms for balance only. Climb with the legs (do not be tempted to pull up with the hands or arms); let the legs take the initiative over the hands (the legs "push" the hands). It is necessary that the gaffs be

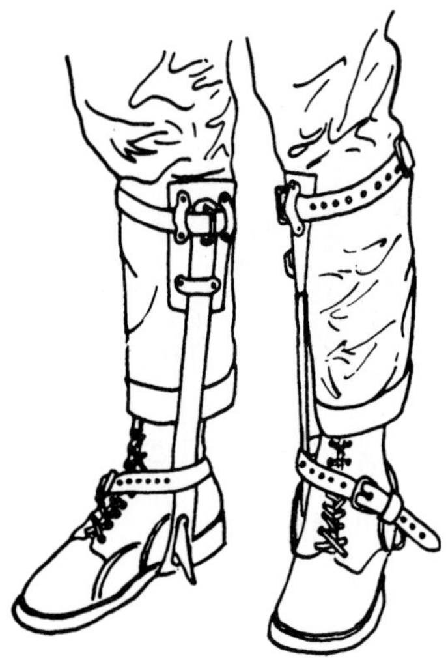

FIGURE 44.7 Climbers properly worn.

FIGURE 44.8 Adjusting pant leg.

directed toward the center (or heart) of the pole in natural manner. The size of the pole and the length of leg between hip and knee will determine automatically the amount of gaff separation on the pole (Fig. 44.9).

The effective leg stroke is that angle or stroke which will cause the gaff to cut effectively into the pole wood without side thrust of any sort. The effective stroke results when the knee is thrown comfortably away from the pole (without straining the hip), the gaff is aimed at the target (the imaginary line down the center of the pole), and the leg force and travel are made to parallel the climber shank until proper penetration is accomplished. Kicking or slapping the gaffs against the pole should be avoided. The hands and feet should work in coordination; the right hand is raised with the right foot and the left hand with the left foot. Weight should be shifted gradually and easily from one foot to the other (Fig. 44.10).

When descending, the hands are lowered first. Each leg is relaxed and straightended before lowering. When the straightened and relaxed leg is lined up with the center of the pole and the body weight has been shifted above the gaff, drop the gaff into the pole. In descent, the leg is not stroked; it is merely lowered into position with the body weight behind it. The hands and arms take the initiative over the feet. The hands "push" the feet, which is opposite from ascending the pole when the feet push the hands. Keep hips, shoulders, and knees away from the pole. Do not take long steps or try to coast or slide when descending (Fig. 44.11).

When ascending, gaff removal is facilitated by a twisting action of the ankle (outward) and slight prying action of the inside of the footwear against the pole. When descending, the climber gaffs should naturally break out with the outward and lowering movement of the knee. Removal of the climber when the last step to ground is taken is accomplished by a slight twisting and prying action as in ascending.

Wear climbers only when necessary for climbing or working on poles. Always remove them when working on the ground, riding in or driving a vehicle, etc. Gaff guards should be installed when climbers are not being worn.

No one should stand at the base of a pole while a man is ascending, descending, or working on it. There is always a possibility that the workman may fall or drop something. All persons, especially children, should be warned to keep away.

FIGURE 44.9 Position while climbing. **FIGURE 44.10** Ascending pole. **FIGURE 44.11** Descending pole. **FIGURE 44.12** Releasing safety strap.

If a second man is to ascend the pole, he should wait until the first man has placed his safety in his working position. When descending a pole, one man should remain in his working position with his safety in use until the other has reached the ground and is out of the way.

Whenever possible, a slippery pole or one partly coated with snow or ice should be ascended with the gaffs in the slippery side and the hand-holds on the less slippery side. Under very slippery conditions or when a strong wind is blowing, the safety strap may be placed around the pole and worked upward or downward in ascending or descending.

Do not hold to pins, crossarm braces, insulators, and other hardware in ascending, descending, or changing position on a pole.

Always ascend and descend on the high side of a leaning, raked, or bent pole.

Use of the Safety Strap. In placing the safety strap around a pole for support, the following steps constitute the best procedure.

Place both climber gaffs firmly in the pole at or near the same level. Keep the knees and hips away from the pole, and unsnap the single end of the safety strap with the left hand (Fig. 44.12).

Pass the single end of the safety strap to the right hand around the back of the pole, and grasp the pole with the left hand. With the right hand, carry the snap hook to the right D ring, and engage the snap with the D ring with the keeper facing outward (Figs. 44.13 and 44.14). Both snap keepers should face outward, and the strap should lie flat, without twists, against the pole, with the buckle tongue side out when in use (Fig. 44.14). Always visually check and be sure that the snaps are securely and properly engaged in the D rings before trusting your weight on the safety strap. Never depend on the sound of the snap keeper. It will make the same sound if the snap is engaged in an insecure attachment, such as a plier handle or wire hook.

Keep plier pockets and other objects well clear of D rings to avoid accidentally hooking the safety-belt snap into them. There should be a minimum of 4 in of clearance between D rings and tool pockets. No wire hooks should be used on body belts.

Always use the belt and safety strap when working aloft on a pole or structure. It is good insurance against falling and provides a safe place to carry tools normally needed. *Never* place the safety strap around the top of a pole above the top crossarm or in any other place where it can accidentally slip off. If it is necessary to place the safety strap high on a new bare pole, place a long through-bolt in the top gain hole to keep the safety strap from slipping off the pole.

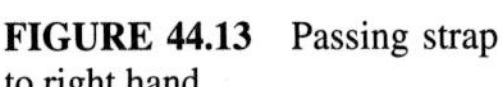

FIGURE 44.13 Passing strap to right hand.

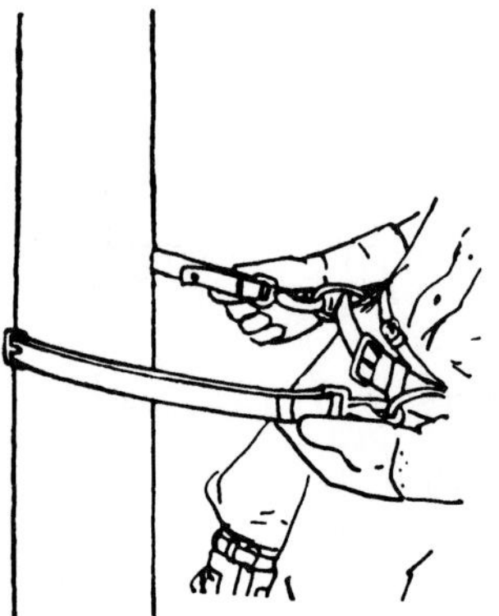

FIGURE 44.14 Position of strap in use.

FIGURE 44.15 Moving safety strap position.

To move up or down on a pole with the safety strap in use, hold the pole with one hand, as you release tension on the strap, and move the strap up or down with the other hand (Fig. 44.15).

Before a person is permitted to climb, he should be taught the proper method of handling a safety strap. This can be accomplished by standing on the ground at the base of the pole without wearing climbers and going through the operations shown in Figs. 44.12 through 44.14.

Practice at the base of a pole until the proper climbing position for the feet, arms, and body is acquired. Place the feet on the ground at about a 90° angle with each other and with the side of the arch of each foot against the sides of the pole. Extend both arms out forward in a horizontal plane from the shoulders, and hold on to the back of the pole with the hands. Thrust the hips well back from the pole with the legs and arms straight until the body is in a position as shown in Fig. 44.16.

Practice climbing on a medium-size, smooth pole which has not been badly cut by previous climbing. Start using climbers by first confining climbing to a section between the ground and 3 ft above the ground. This permits one to learn to climb without fear of falling. Before starting to climb, inspect the pole from the ground for cracks, knots, holes, tacks,

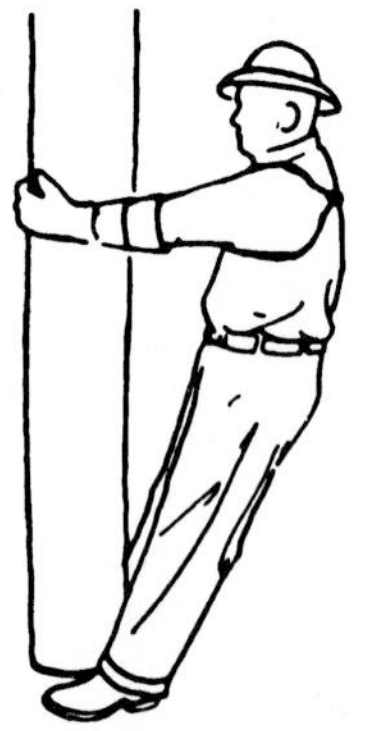

FIGURE 44.16 Practice position.

FIGURE 44.17 Practice climbing.

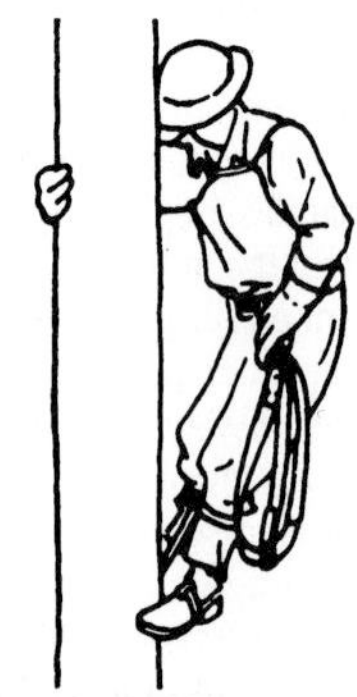

FIGURE 44.18 Ready to place strap.

and nails, as the presence of any of these may deflect a gaff and result in an insecure foothold (Fig. 44.17).

Practice placing the safety strap around the pole and fastening it to the D ring while on the pole. Stand on the pole with both legs straight and both gaffs firmly in the pole at or near the same level and at a point approximately 2 ft above the ground level for the first trials. This is the position for placing the strap when in the working position at any location on the pole. Then proceed to place it around the pole as described (Figs. 44.12 (or 44.18), 44.13, and 44.14).

Inspection and Care. The user of climbing equipment should carefully inspect it before each use.

Inspect leather or fabric parts for cuts, cracks, tears, enlarged buckle tongue holes, narrowing down due to stretch, or for hard and dry leather. Inspect stitching for broken, ragged, or rotted threads. Inspect metal parts for breaks, cracks, loose attachments, or wear that might affect strength. Care should be taken to determine that the keepers of the snaps are firmly and securely seated in the recess at the end of the hooks and that there is a reasonable resistance to depressing the keeper. Any distortion of the hook may change the tension of the spring so that the keeper is not in firm contact with, or securely seated in, the nose of the hook. If the keeper is not securely seated, it is possible that the D ring may become disengaged by slipping between the nose of the hook and the keeper and be the cause of a serious fall. Check the gaffs and determine whether the cutting edges are properly sharpened and shaped.

Defective equipment should be repaired or replaced and gaffs properly shaped as soon as possible. If defects affect safety of the equipment, it should be removed from service until repaired or replaced.

Cleaning and Dressing Leather. Leather parts of climbing equipment should be cleaned approximately every 3 months and dressed approximately every 6 months and on return from vacation periods to keep them pliable and in good condition. Cleaning and dressing may be done more often if necessary due to excessive moisture or perspiration. Paint stains should be removed before the paint dries. Never, however, use gasoline or similar solvent on leather, as this will dry out and severely damage the leather.

To Clean Leather. Wipe off surface dirt with a damp sponge.

Using a neutral soap, such as castile, and a clean, moist sponge, work up a good lather to remove embedded dirt and perspiration. Wipe off with a clean cloth.

Work up a good lather with saddle soap and a clean sponge, and rub it well into all parts of the leather. Wipe off with a clean cloth.

To Dress Leather. Never dress or oil leather before cleaning as just outlined.

While the leather is still damp after cleaning, apply a good leather dressing or neat's-foot oil, working it into the leather with the hands. Do not use an excess amount of oil, as this will saturate and weaken the leather.

Allow the leather parts to dry in a cool, shady place for 24 hours. Never dry leather near a source of heat, as heat destroys leather.

Remove any excess dressing by rubbing vigorously with a clean, soft cloth. Never use any mineral oil or grease for dressing leather.

Cleaning Nylon Fabric Safety Straps. Since there are differences in the methods of fabricating nylon safety straps, it is recommended that each manufacturer be consulted on the proper cleaning procedure to be used in cleaning the particular nylon safety straps.

Storage. Store leather equipment away from sources of heat, such as stoves, radiators, steam pipes, or open fires. Keep leather goods away from sharp-edged objects. Before storing the climbers, it is good practice to install approved climber gaff guards to protect the gaff points and edges and to protect other objects and persons from accidental injury or damage.

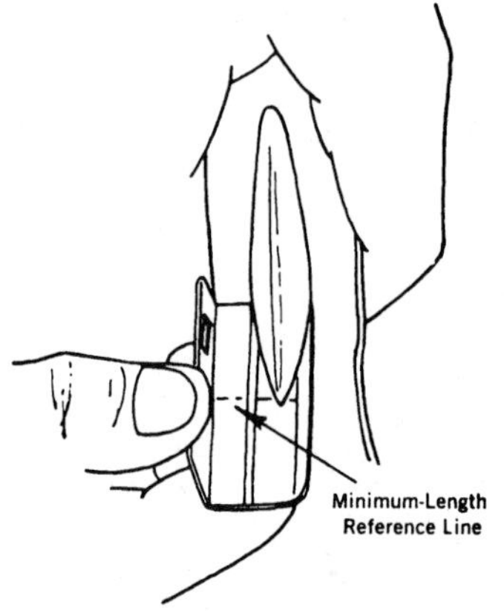

FIGURE 44.19 Checking for minimum length.

Inspection and Maintenance of Climber Gaffs. Gaffs of climbers should be inspected and checked frequently for length, width, thickness, profile of the point, and sharpness of the cutting edges. If a cutout or any other difficulty is experienced in getting the gaffs to hold in a pole, investigate immediately to determine the cause. The trouble may be that the gaffs have been accidentally damaged since the last inspection.

The gaff gauge can be used for checking climber gaffs and as an aid in shaping the gaff properly. One type of gaff gauge is shown in the following sketches (Figs. 44.19 through 44.29).

Be sure to check the instructions issued by the manufacturers of the gauge you are using since there may be differences in the exact procedures to be followed.

The reference line across the gauge indicates the minimum length of the gaff, measured on the underside from the heel of the gaff to the point (Fig. 44.19).

Company requirements as to minimum length of gaffs should be observed.

The TH slots are used to check the thickness of the gaff 1 in and 1/2 in from the point. The face of the gauge should lie flat against the back or ridge of the gaff. If within acceptable limits, the point of the gaff will lie on or between one pair of the reference lines on the gauge, as indicated in Figs. 44.20 and 44.21.

The W slots are used to check the width of the gaff 1 in and 1/2 in from the point, as shown in Figs. 44.22 and 44.23. The face of the gauge should be flat against the back or ridge of the gaff, as in the previous check. If within acceptable limits, the point of the gaff will lie on or between one pair of the reference lines on the gauge.

If necessary to shape the gaff for proper thickness, it should be filed on the flat underside which lies between the gaff and the stirrup. Care should be used not to cut or notch the

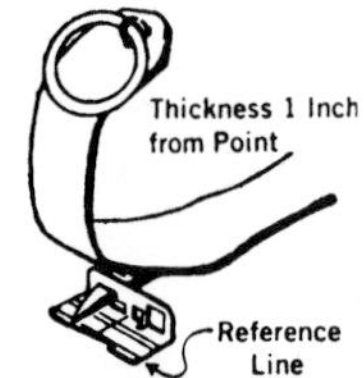

FIGURE 44.20
Checking thickness of gaff.

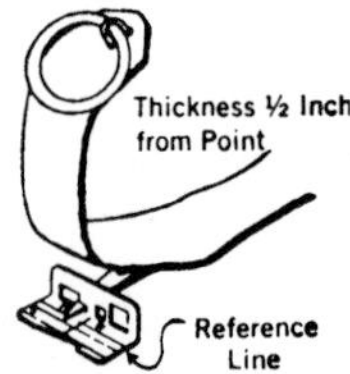

FIGURE 44.21
Checking thickness of gaff.

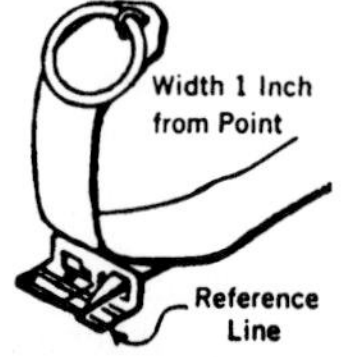

FIGURE 44.22
Checking width of gaff.

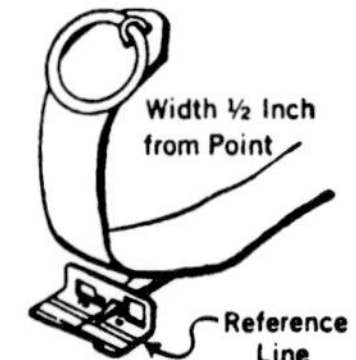

FIGURE 44.23
Checking width of gaff.

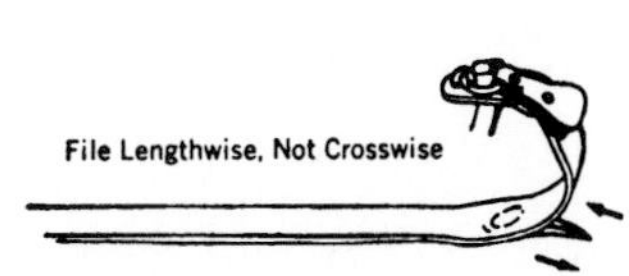

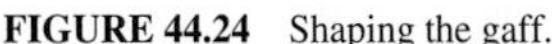

FIGURE 44.24 Shaping the gaff.

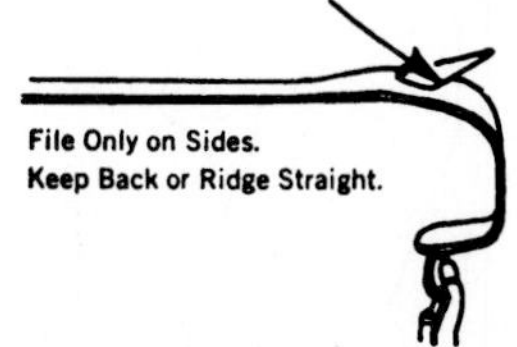

FIGURE 44.25 Shaping the gaff.

leg iron or stirrup with the file. Also, there should be no file marks left on the gaff (Fig. 44.24). A 10-in mill bastard flat file or an 8-in smooth knife file will do a nice job, if kept clean by frequent use of a file card brush. File marks on a gaff, particularly if they are crosswise, may weaken the steel and result in a broken gaff. Final filing, therefore, should be carefully done with light strokes, lengthwise of the gaffs.

If necessary to shape the gaff to proper gauge width, the filing should be done on the two outside rounded surfaces, as shown in Fig. 44.25. Care should be taken not to file the ridge or back edge of the gaff, for it is important to keep it absolutely straight, as shown in Fig. 44.26. The back of the gaff gauge can be used as a straightedge to check this.

If the back edge is rounded off, as shown in Fig. 44.27, the gaff will have a tendency to cut out of the pole.

As the filing progresses, the work should be checked with the gaff gauge until the gaff is within the indicated gauge limits.

The point of a gaff should function as a chisel (see Fig. 44.26), cutting its way into a pole, instead of as a spike or needle (see Fig. 44.28) to be driven into the pole. Spike-pointed gaffs will cut out easier than chisel points, if the knee is brought too close to the pole for any reason. Spike points also reduce the cross-sectional area near the tip and may result in gaff breakage.

The chisel point penetrates the pole easier and deeper with less effort and, if properly shaped with the proper gaff angle, will not cut out.

To obtain a chisel point, the flat undersurface should be rounded off toward the point with a slight curve, as shown in Fig. 44.26. This filing should be done with light, even strokes.

If the gauge has a profile recess, use it to check the work, as in Fig. 44.29. If not, use Fig. 44.29 as a guide to the proper profile.

After filing and shaping are completed and the gaff checks with all the proper gauge dimensions, the flat underside of the gaff and the underside of the point should be honed, using a small carborundum pocket stone. Honing removes any burrs on the edges of the gaff and sharpens the edges for easier penetration into the pole. It is this penetration of the wood which supports the workman as he ascends, descends, or does his work on the pole. The edges of the flat underside should be knife-sharp and *not* rounded.

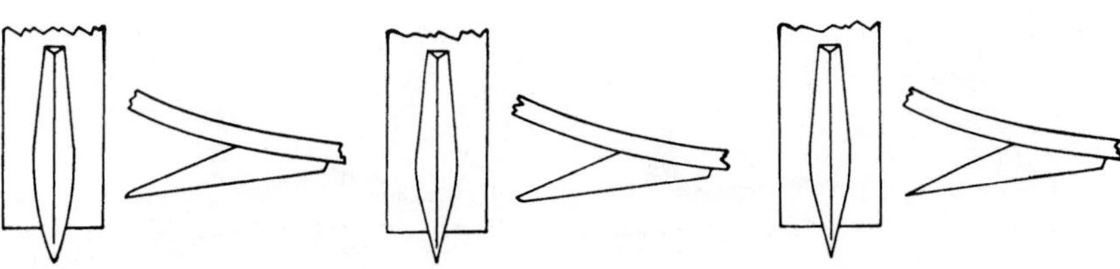

FIGURE 44.26 Correct (chisel point).

FIGURE 44.27 Incorrect (hook-nose point).

FIGURE 44.28 Incorrect (spike point).

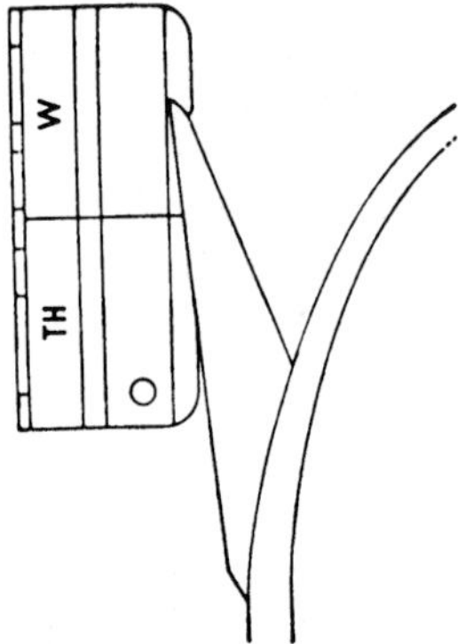

FIGURE 44.29 Checking the gaff profile.

Testing Climber Gaffs for Proper Shape and Sharpness

The Plane Test. The plane test may be used to determine that the climber gaff is properly sharpened to cut into the pole like a chisel.

Place the climber on a soft pine or cedar board as shown in Fig. 44.30. Holding it upright, but with *no* pressure on the stirrup, push the climber forward along the board. If the gaff is properly shaped and sharpened, and if the gaff angle with the wood is sufficient, the gaff point will dig into the wood and begin to hold within a distance of approximately 1 in.

If the climber gaff slides along the wood without digging in or merely leaves a mark or groove in the wood as shown in Fig. 44.31, either the gaff is not properly sharpened and shaped, or the gaff angle is too small. The gaff angle is built into the climber by the manufacturer and should be between 9° and 16° with the leg iron placed parallel to the surface of the wood, according to EEI's Specifications for Linemen's Climbing Equipment, AP-2-1973.

The Cutout Test. The cutout test may be made to determine whether the gaff will cut out of the wood when the angel between the leg iron and the surface of the wood is reduced beyond the critical point.

When the gaff is not properly sharpened or has a hook-nosed point or a convex ridge, it will cut out.

To make the cutout test, jab the climber gaff into the soft pine or cedar board at an angle of about 30° and with sufficient force to penetrate the wood to a depth of 1/4 in or more, as shown in Fig. 44.32. Hold the gaff in place in the wood, and bring the leg iron down against the wood surface, applying forward pressure on the stirrup with one hand while holding the leg iron with the other hand. If the gaff is in proper condition, it will tend to cut in deeper rather than cut out.

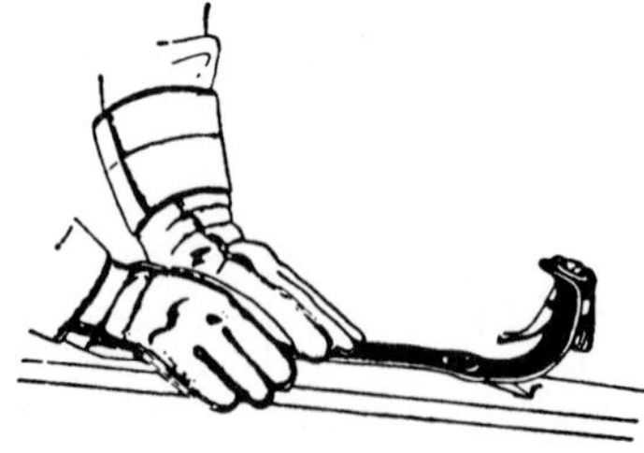

FIGURE 44.30 Plane test.

FIGURE 44.31 Plane test.

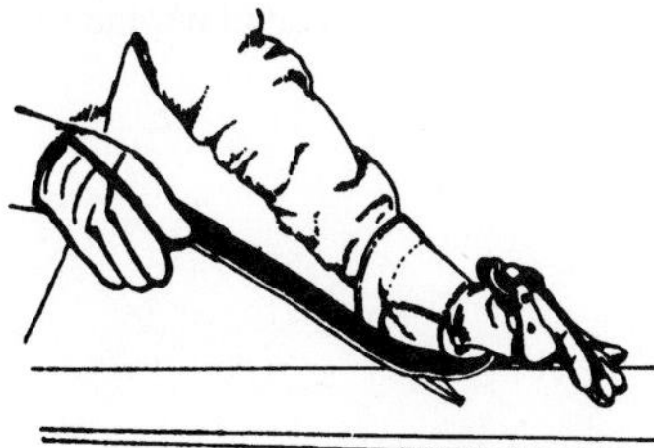

FIGURE 44.32 Cutout test.

If the gaff is not properly shaped and sharpened, it will cut out within a horizontal distance of 3 in or less, leaving a typical gaff cutout mark or groove in the wood.

The Pole Cutout Test. In addition to the above-mentioned tests, the pole cutout test may be used to check the climber gaffs in the field.

To make the test, place the climber on the leg, holding the sleeve with the hand, palm facing the pole, as shown in Fig. 44.33. With the leg at about a 30° angle to the pole and the foot about 12 in off the ground, lightly jab the gaff into the pole approximately 1/4 in. Keeping enough pressure on the stirrup to keep the gaff in the pole but not so much as to cause the gaff to penetrate any deeper, push the climber and the hand toward the pole by moving the knee until the strap loop of the sleeve is against the pole, as shown in Fig. 44.34. Making certain that the strap loop is held against the pole with pressure from the leg, gradually exert full pressure of the foot straight down on the stirrup without raising the other foot of the ground, so as to maintain balance if the gaff does not hold.

The point of the gaff should cut into the wood and hold (dig itself in) within a distance of not more than 2 in, measured from the point of gaff entry into the pole to the bottom of the cut on the pole surface.

Caution: To prevent hand injury from wood splinters, good work gloves should be worn when testing climbers.

FIGURE 44.33 Pole cutout test.

FIGURE 44.34 Pole cutout test.

Causes of Climber Cutouts. The causes of climber cutouts can be grouped under the following five general headings:

1. Condition of poles
2. Conditions on poles
3. Clothing
4. Climbing practices
5. Climbers

A more detailed listing of causes is given in the following discussion of the five general types. Also typical accidents are cited and preventive measures discussed.

1. Condition of Poles
   - **a.** Specific Causes
     - (1) Knots
     - (2) Knotholes
     - (3) Broken poles
     - (4) Weather cracks
     - (5) Wet creosoted poles
     - (6) Excessive gaff cuts
     - (7) Splinters
     - (8) Poles coated with ice
     - (9) Rotted outer surface
     - (10) Poles hardened by preservative treatments
     - (11) Crooked or leaning poles
     - (12) Bird holes
     - (13) Improperly guyed poles
     - (14) Pole splintered by lightning
     - (15) Old gains and drilled holes
     - (16) Flat surfaces on poles
     - (17) Small and limber poles
     - (18) Cemented bird holes
     - (19) Checks
   - **b.** Examples of Accidents
     - (1) As a lineman was descending a distribution pole, a climber gaff cut out and he fell to the ground, breaking one leg. The pole surface was badly checked and covered with deep gaff cuts.
     - (2) After a lineman had unsnapped his safety strap to descend a pole, a piece of the shell of the pole in which one of his gaffs was placed broke off. He fell to the ground, breaking a leg, spraining a wrist, and bruising his chest. The pole was a native cedar and was weathered and checked.
     - (3) A lineman's gaff struck a knot in a chestnut pole as he was climbing, and he cut out and fell 12 ft. He injured his right leg so severely that he may be permanently crippled.

     These are typical instances in which the condition of poles contributes to accidents.
   - **c.** Prevention

     Prevention of accidents due to condition of poles can be aided in several ways. For example, poles can be purchased to meet better specifications which will help to eliminate some of the defects. A good inspection schedule will help to eliminate other causes. Poles that are to be climbed frequently can perhaps be equipped with steps. Workers themselves should become alert to these hazards, and be continually on the lookout for the conditions on the poles they climb which might contribute to cutouts and injuries.

2. Conditions on Poles

a. Specific Causes

(1) Conduit or cable on pole
(2) Strain plates
(3) Nails and tacks
(4) Telephone wires and brackets
(5) Ground wires
(6) Ground wire molding
(7) Signs and posters
(8) Metal pole numbers
(9) Dating nails
(10) Street light controls
(11) House knob screws
(12) Insect nests
(13) Metering equipment on pole
(14) Guy hooks, pole plates, and lag screws no longer in use
(15) Tree limbs against the pole
(16) Inadequate climbing space
(17) Telephone company attachments
(18) Excessive number of service drops
(19) Telephone terminal boxes
(20) Vines on poles
(21) Fences attached to poles
(22) Clotheslines and radio aerials attached to poles
(23) Loose and bent pole steps

b. Examples of Accidents

(1) A lineman climbed a pole to a height of 14 ft, where one gaff cut out and he fell. His right heel struck the curb and was fractured. He also suffered a fracture of the right leg. The pole was excessively gaff-marked, and no steps were provided. Climbing space was restricted by a street light riser cable conduit running up the pole and a government mailbox mounted on the lower part of the pole.

(2) A new cedar pole had been set and supply attachments transferred to it. The old chestnut pole was lashed against the new pole pending transfer of the telephone attachments. As the lineman descended the new pole, one of his climber gaffs struck a strain plate and he cut out and fell 12 ft to the ground. He landed on his right side and rolled over on his back. His back was severely injured by the impact of a hammer he carried in his belt.

c. Prevention

Preventing cutouts and falls from these causes require the attention of both management and workers. Many of these causes can be removed, or at least minimized, by planning installations and setting construction standards with due regard for the men who must climb the pole, sometimes under very adverse conditions. When a pole is called a *clean pole* by men who climb, it is inferred that hardware, fixtures, and conductors are installed with an eye to adequate and safe climbing space. However, when unsafe conditions are found, they should be properly reported and extra precautions taken if the pole must be climbed.

3. Clothing

These cases illustrate how falls can occur because of unsafe conditions of the workman's clothing.

a. Specific Causes

(1) Greasy clothing or gloves
(2) Worn or loose heels on shoes

(3) Shoes with nailed soles or heels
(4) Low-cut shoes
(5) Loose or ragged clothes
(6) Wrong-size gloves
(7) Gloves with holes or torn places
(8) Soft-soled shoes
(9) Shoes with poor arch or ankle supports
(10) Buckles on overalls
(11) Loose soles on shoes
(12) High-heel boots

b. Examples of Accidents

(1) A lineman climbing a pole cut out and fell from about an 18-ft height. The pole was in good condition, but his gloves were greasy. He stated that he lost his hold while he was climbing.

(2) A lineman was attempting to belt off at his working position. As he reached around the pole to grasp his safety, his left gaff cut out and he fell off the pole. Examination revealed that the whole heel had come off the left shoe.

c. Prevention

Elimination of cutouts and other accidents due to unsafe conditions of clothing is largely a problem of the worker himself. Nevertheless, management and supervisors can be alerted to recognize such hazards and insist that their workers wear suitable and safe clothing. Many individuals will appreciate having their attention called to any unsafe condition of their work clothing. In some cases, conversely, it may be necessary for supervisors to take stronger action.

4. Climbing Practices

a. Specific Causes

(1) Climbing or descending too fast
(2) Using low side of pole
(3) Fatigue
(4) Gaffs not stuck in pole hard enough
(5) Climbing too close to pole
(6) Climbing through or past unprotected conductors
(7) Not watching placement of gaffs
(8) Safety belt too loose
(9) Aiming gaffs at pole at improper angle
(10) Using too-long steps in climbing or descending
(11) Not inspecting pole before climbing
(12) Inattention while ascending or descending
(13) Improper balancing of body weight on gaffs
(14) Belting off to pole at wrong position
(15) Physically unfit for climbing
(16) Horseplay
(17) Climbing too close to ropes and hand lines
(18) Catching material thrown from ground (while working on pole)
(19) Placing climber gaffs on ground wires
(20) Failure to get a good handhold
(21) Holding to braces, etc.
(22) Climbing in a mechanical manner
(23) Sliding climber gaffs when climbing (sneaking)
(24) Showing off (fancy dancing)

b. Examples of Accidents

(1) A lineman was climbing a 40-ft cedar pole which was stepped nearly all the way up. He said he was climbing as was customary on a stepped pole, that is, he used the steps for handholds and placed his gaffs between the pole steps. He guessed he cut out at a time when he was reaching for a new handhold with his right hand and fell 15 ft to the ground. In trying to break his fall, he extended both hands. He fractured the radius bone in both arms near the wrist and suffered minor abrasions on the face. The pole, climbers, gaffs, and straps were in safe condition. This lineman was known as a "fast climber."

(2) A lineman climbed a transformer pole which was raked, or leaning. He climbed the low side of the pole. On completing his work, he started to climb down the low side of the pole. About 15 ft from the ground he cut out and fell, receiving a broken right leg, right foot, hip, back, and left foot, and other injuries to his left arm, wrist, hand, and knee.

(3) A lineman had been working for 2 weeks getting transformer nameplate data. He kept a record of the number of poles he climbed each day. The accident occurred late in the afternoon of a very hot day. As he was descending a pole, his right gaff cut out and he fell 18 ft, fracturing his right leg and permanently disabling his back. He stated that he had climbed more poles that day than ever accomplished by an employee in a single day and that he was just about "all in" before he cut out.

c. Prevention

Experienced veteran climbers, as well as new apprentices, need to be reminded from time to time about the pitfalls involved in these unsafe climbing practices.

Foremen should watch the climbing practices of their men and aid them in correcting any unsafe practices that are observed before accidents occur.

Elimination of unsafe climbing practices is based on sound, thorough training of workers in the safe methods of conducting themselves while working on poles.

5. Climbers

The importance of the climbers themselves in the prevention of cutout accidents cannot be emphasized too strongly.

a. Specific Causes

(1) "Duckbill" gaffs
(2) Loose gaffs
(3) Gaffs too short
(4) Dulls gaffs
(5) Gaffs improperly sharpened
(6) Straps too tight
(7) Straps too loose
(8) Broken straps
(9) Bent or broken gaffs
(10) Wrong-size climbers
(11) Foreign objects under gaffs
(12) Needlepoint gaffs
(13) Faulty leg irons

b. Examples of Accidents

(1) A lineman started to climb a pole to attach a guy. About 15 ft above the ground he cut out and fell, breaking his left hip and left arm. The pole appeared to be in reasonably good condition. The man's left gaff had been sharpened in a duckbill shape, and wood splinters were found jammed under the gaff.

(2) A lineman rearranging dead-ends on a pole unfastened his safety belt to swing around to the other side of the pole. When he had almost reached his new working position, his right climber gaff apparently cut out and he fell. He struck the ground on his neck, shoulders, and back, suffering a broken neck and left shoulder blade, fractured ribs, punctured lungs, internal hemorrhages, and many other bruises. He died a few minutes after the accident. The gaff of one climber was found to be loose; otherwise his equipment was in good condition.

(3) As a lineman was preparing to descend to a lower position before removing the line hose from the conductors, he unfastened his safety strap. He then cut out and fell 30 ft to the ground, landing on his hands and arms. Both wrists, both arms, right leg, and lower leg were broken, and he suffered multiple cuts and bruises. He will be disfigured for life and partially disabled. Inspection of his climbers after the accident revealed that the gaffs were very short.

c. Prevention

Prevention of accidents due to climber defects can be accomplished with the proper effort.

First Rigid and frequent inspections of climbers, leg straps, leg iron, etc., with special attention being given to the gaffs, the shape and sharpness of the gaff points, the proper angle of the gaff with the leg iron, etc.

Second An acceptable gaff gauge should be standardized. It should be simple and easy to use. Minimum lengths of gaffs should be strictly enforced. The smallest deviation below the minimum-length requirement should not be permitted. Gaff shape and size should be similarly dealt with.

Third There should be a standard procedure for sharpening gaffs. There is universal agreement that gaffs should never be sharpened on ordinary dry grinding wheels, as this destroys the proper heat treatment of the metal. Only the proper type of file should be used. A professional job of precision sharpening should be the goal. Gaffs should always be honed after sharpening to remove file marks which may initiate cracks in the gaff steel and to improve the cutting edges of the point for better penetration of poles and ease in climbing.

Fourth Workers who use climbing equipment must be thoroughly trained in the fundamentals for proper care and use of the equipment.

Fifth Workers who use climbing equipment should accept responsibility for inspecting, sharpening, and shaping climbing equipment—and, if necessary, rejecting climbing equipment that is unsafe without waiting for supervisors or safety department personnel to make inspections.

CLIMBING POLES

Testing Wood Poles for Climbing. Precautions must be taken before linemen climb wood poles or before strains are changed in a way that could create a hazard, if the pole is defective. All poles should be visually inspected before the lineman starts to climb. An unusual angle or buckling at the ground line or along the pole may show that the pole is rotted or broken. Horizontal cracks across the grain of the wood pole may make the pole weak. Vertical cracks seldom weaken the pole. Vertical cracks should be avoided when inserting the climber gaffs into the wood. The lineman should use a hammer to check the soundness of the pole before starting to climb. Pounding the pole with a hammer may show that the pole is shell-rotted or hollow. A solid-wood sound should be heard when the pole is hit with a hammer. A hollow or drumlike sound indicates the pole has internal decay. The surface of the wood pole should not be penetrated by the hammer blows. A shell-rotted or decayed outside layer of the wood pole may cause the lineman's climber gaffs to cut out during climbing. A pole that is shell-rotted or decayed should not be climbed by the lineman using gaffs. Several knots at one location or a single large knot may signify a weak point in the pole. The pole should be set to the proper depth, and the soil must be in condition to support the pole before it is climbed. Soft, wet, or loose soil may permit the pole to shift position, especially if the strains change while the lineman is on the pole. Woodpecker holes or burn spots on the pole may indicate that the pole has been weakened. The lineman should secure the pole before climbing, if inspection reveals that the pole may be weak or if the conductors are to be removed, changing the strain on the pole. The pole can be secured with the line truck boom. A pole may be lashed to a new pole when it will be replaced. Ropes, or guy wires, secured to stationary objects in the proper direction may provide necessary support to permit the lineman to work on the pole. The lineman should use a bucket truck to do the work on a pole that is shell-rotted or decayed, or if the inspection determines that the pole is unstable. If the pole is in an area unaccessible to a bucket truck, then a new pole should be set next to the defective pole and the lineman should climb the new pole and lash the new pole to the defective pole. The lineman should do all his work while he is supported by the new pole. The defective pole should be removed form the line and disposed of.

Fall-Arrest Device. An auxiliary safety device has been developed for linemen to use while climbing poles (Fig. 44.35). The device is manufactured from a resin that provides dielectric protection and makes it easier for the lineman to place himself in the proper position to do the work safely. If the lineman's gaff should cut out while he is climbing or working on a pole, the device prevents the lineman from falling (Fig. 44.36). The climbing device can be used to lower an injured lineman to the ground safely, simplifying pole-top rescue procedures.

FIGURE 44.35 Lineman has climbed a pole, using the fall-arrest device.

FIGURE 44.36 The lineman's gaffs cut out while he was climbing the pole. The fall-arrest device prevented him from falling and suffering a serious injury.

CHAPTER 45
PROTECTIVE EQUIPMENT

Linemen and cablemen must use protective equipment to safely complete work on energized equipment. Safety rules based upon the Occupational Safety and Health Act (OSHA) rules 1926.950 and 1910.269 (Chap. 46), the National Electrical Safety Code (NESC) and work procedures developed from years of experience and corrective measures adopted from the investigation of serious electrical accidents all require the use of proper protective equipment to cover energized conductors and equipment as well as grounded surfaces in the work area. Proper procedures, the right tools, and correctly selected and applied protective equipment are all necessary to complete energized electric distribution work safely and efficiently.

Each work project and especially those projects requiring linemen and cablemen to handle energized electrical conductors should be planned in detail. The plan should be developed to ensure the safety of the workers and the reliability of electric service. Hazards to the linemen, the cablemen, and the public should be identified as a part of the planning process. As work procedures are developed, they should include safeguards for all hazards identified. Prior to starting the work, a tailgate conference should be held to communicate the plan and work procedures to all crew members. A tailgate conference is pictured in Fig. 45.1. When the work is initiated, the plan must be followed without shortcuts to ensure safe and reliable completion of the assignment.

Plan the work and work the plan for safety, efficiency, and reliability.

The safest way of completing electric distribution work is to deenergize, isolate, test, and ground the facilities to be worked on. Figure 45.2 illustrates the proper application of grounding equipment to an electric distribution circuit. The lineman must correctly identify the circuit to be grounded and obtain from the proper authority assurances that the circuit has been deenergized and isolated.

Hold-off tags must be applied to be switches isolating the circuit, and these must be identified with the name of the person in charge of the work. A high-voltage tester mounted on hot-line tools can be used to verify that the circuit is deenergized. The grounding conductors are connected first to a low-resistance ground, usually the neutral conductor, and then to the deenergized phase conductors by the lineman with a hot-line tool. Once the work is completed and all personnel are clear of the circuit, the grounds are removed in the reverse sequence.

The public's dependence on continuous electric service to maintain their health, welfare, and safety often makes it impossible to deenergize circuits to complete modifications and perform maintenance. Proper application of the protective equipment is necessary for the lineman to safeguard the work area, that is, to prevent electric shock or flash burn injuries. Correct application of the protective equipment described is essential for injury prevention and the maintenance of reliable electric service. The linemen pictured in Fig. 45.3 are properly protected for the work in progress. The linemen are wearing hard hats designed for electrical work, safety glasses, body belts with safety straps attached to the work platform, rubber gloves with protectors, and rubber sleeves. The work is being performed while the men are standing on

FIGURE 45.1 Supervisor is reviewing work plan with linemen prior to starting the job. The tailgate conference is being held in a conference room prior to driving to the work-site. (*Courtesy Northwest Lineman College.*)

an insulated platform. Low primary voltages may be worked directly from the pole by linemen. The higher primary voltages must be worked from an insulated platform attached to the pole or an insulated aerial device. The work area is guarded by conductor covers, insulated covers, crossarm covers, rubber blankets, and dead-end covers. The energized conductors have been elevated and supported clear of the crossarm with hot-line tools.

Safety Rules. The basis for an Investor-owned Utility Company's Safety rules are: OSHA 29 CFR 1926 Subpart V, Power Transmission, OSHA Federal Register 1910.269

FIGURE 45.2 Lineman completing installation of grounding conductors to electric distribution circuit with hot-line tool. The common-neutral conductor is used as a ground source. (*Courtesy A. B. Chance Co.*)

FIGURE 45.3 Lineman performing energized circuit work with proper protective equipment. (*Courtesy A. B. Chance Co.*)

and the National Electrical Safety Code (NESC). Rule 1926.950 is applicable to the construction of electric transmission and distribution lines and equipment. Subpart V was issued in 1970. Rule 1910.269, issued in 1994, is applicable to the operation and maintenance of electric power generation, control, transformation, transmission, and distribution lines and equipment. The NESC's purpose is for the practical safeguarding of persons during the installation, operation, or maintenance of electric supply and communication lines and associated equipment.

The rules are not intended to be of a conflicting nature. OSHA Rule 1910.269 fills the gap not completely covered by OSHA Rule 1926 or the NESC (last revision). Rule 1910.269 is provided in Chap. 46.

PROTECTIVE EQUIPMENT

Rubber Gloves. The most important article of protection for a lineman or a cableman is a good pair of rubber gloves with the proper dielectric strength for the voltage of the circuit to be worked on. Leather protector gloves must always be worn over the rubber gloves to prevent physical damage to the rubber while work is being performed. When the rubber gloves are not in use, they should be stored in a canvas bag to protect them from mechanical damage or deterioration from ozone generated by sun rays. Rubber gloves should always be given an air test by the lineman or cableman each day before the work is started or if the workman encounters an object that may have damaged the rubber gloves. The lineman pictured in Fig. 45.4 is properly applying an air test to his rubber gloves before putting them on.

Five classes of rubber gloves are manufactured as shown in Table 45.1.

The American National Standards Institute standard ANSI/ASTM D 120, "Rubber Insulating Gloves," covers lineman's rubber glove specifications.

The proof-test voltage of the rubber gloves should not be construed to mean the safe voltage on which the glove can be used.

The maximum voltage on which gloves can safely be used depends on many factors including the care exercised in their use; the care followed in handling, storing, and inspecting the gloves in the field; the routine established for periodic laboratory inspection

FIGURE 45.4 Lineman giving his rubber gloves an air test before starting to work to be sure they do not have a hole in them. (*Courtesy MidAmerican Energy Co.*)

and test; the quality and thickness of the rubber; the design of the gloves; and other factors such as age, usage, or weather conditions.

Inasmuch as gloves are used for personal protection and a serious personal injury may result if they fail while in use, an adequate factor of safety should be provided between the maximum voltage on which they are permitted to be used and the voltage at which they are tested.

It is common practice for the rubber protective equipment manufacturer to prepare complete instructions and regulations to govern in detail the correct and safe use of such equipment. These should include provision for proper-fitting leather protector gloves to protect rubber gloves against mechanical injury and to reduce ozone cutting. The lineman in Fig. 45.5 is wearing rubber gloves with leather protectors and rubber insulating sleeves.

Rubber insulating gloves should be thoroughly cleaned, inspected, and tested by competent personnel regularly (Fig. 45.6). A procedure for accomplishing the work is as follows:

1. Rubber labels showing the glove size, the glove number, and the test date are cemented to each rubber glove.
2. The rubber gloves are washed in a washing machine, using a detergent to remove all dirt, prior to testing.

TABLE 45.1 Classes of Rubber Gloves Manufactured, Proof-Test Voltages, and Maximum Use Voltages as Specified by ANSI/ASTM D 120-77*

Class of glove	AC proof-test voltage, rms	DC proof-test voltage, average	Maximum use voltage, ac rms
0	5,000	20,000	1,000
1	10,000	40,000	7,500
2	20,000	50,000	17,000
3	30,000	60,000	26,500
4	40,000	70,000	36,000

*Reprinted by permission of the American Society for Testing and Materials, 1916 Race Street, Philadelphia, Pa. 19103. Copyright.

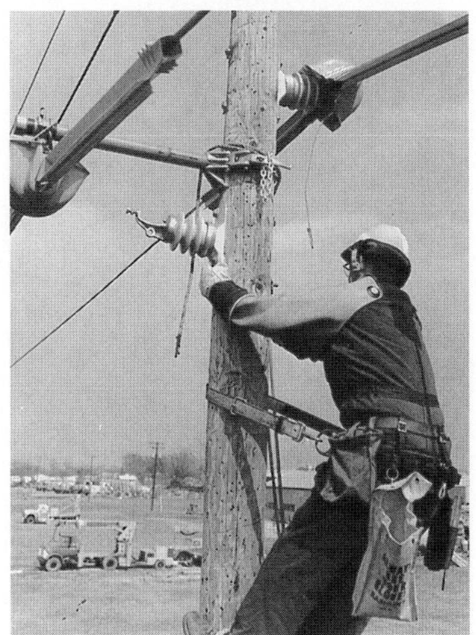

FIGURE 45.5 Lineman working on pole with proper personal protective equipment–rubber gloves with leather protectors, rubber sleeves, plastic hard hat, safety glasses, and a body belt with safety strap around pole. Canvas bag for storage of rubber gloves is snapped to the lineman's body belt. (*Courtesy A. B. Chance Co.*)

FIGURE 45.6 Rubber gloves placed in washing machine and washed with warm water and a mild detergent. (*Courtesy MidAmerican Energy Co.*)

3. The rubber gloves are inflated by utilizing glove inflator equipment, checked for air leaks, and inspected (Fig. 45.7). The inspection includes checking the surface of the rubber gloves for defects such as cuts, scratches, blisters, or embedded foreign material. The cuff area of the inside of the gloves is visually inspected after the glove is removed from the inflator. All rubber gloves having surface defects are rejected or repaired.
4. After the inflated inspection, the rubber gloves are stored for 24 hours to permit the rubber to relax, thus minimizing the possibility of corona damage while the electrical tests are completed (Fig. 45.8).
5. The rubber gloves are supported in a testing tank by means of plastic clothes-pins on the supporting rack. A ground electrode is placed inside each glove (Fig. 45.9).
6. The rubber gloves to be tested and the testing tank are filled with water. The height of the water in the glove and in the testing tank depends upon the voltage test to be applied (Fig. 45.10).

Class 0 gloves are tested at 5000-volt ac or 20,000-volt dc. The flashover clearances at the top edge of the cuff of the gloves during the test must be 1 1/2 in for both the ac and dc test.
Class 1 gloves are tested at 10,000-volt ac or 40,000-volt dc. The flashover clearances at the top edge of the cuff of the gloves during the test must be 1 1/2 in for the ac test and 2 in for the dc test.
Class 2 gloves are tested at 20,000-volt ac or 50,000-volt dc. The flashover clearances at the top edge of the cuff of the gloves during the test must be 2 1/2 in for the ac test and 3 in for the dc test.
Class 3 gloves are tested at 30,000-volt ac or 60,000-volt dc. The flashover clearances at the top edge of the cuff of the gloves during the test must be 3 1/2 in for the ac test and 4 in for the dc test.

FIGURE 45.7 Rubber gloves inflated for visual inspection. (*Courtesy MidAmerican Energy Co.*)

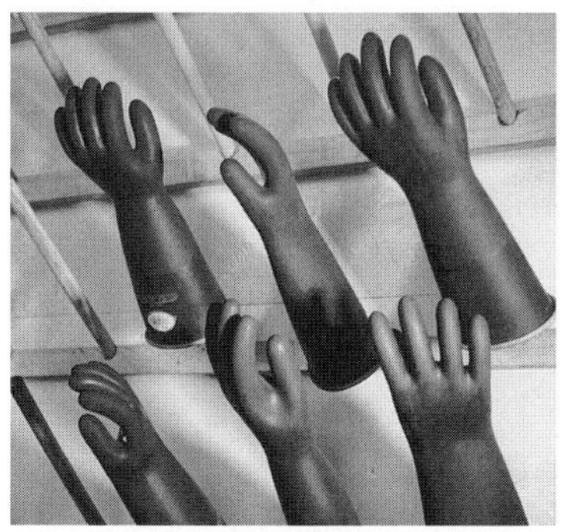

FIGURE 45.8 Air-drying and storage rack for rubber gloves. Note rubber label cemented on cuff of glove. (*Courtesy MidAmerican Energy Co.*)

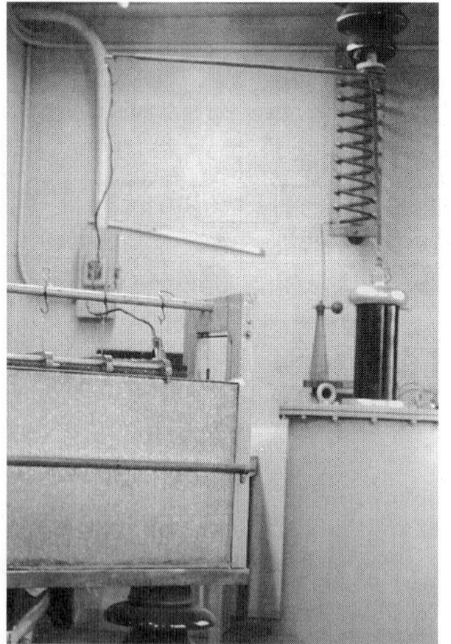

FIGURE 45.9 The rubber-glove-testing electrodes consist of a galvanized steel testing tank mounted on insulators and ground connections which are supported in the rubber gloves in the tank. The ground electrodes are connected to ground through relay contacts which can be used to switch a milliammeter in series with the ground lead. The testing equipment is designed to test six rubber gloves at a time.

FIGURE 45.10 Rubber gloves in test tank, filled with water to proper level. (*Courtesy MidAmerican Energy Co.*)

Class 4 gloves are tested at 40,000-volt ac or 70,000-volt dc. The flashover clearances at the top edge of the cuff of the gloves during the test must be 5 in for the ac test and 6 in for the dc test.

7. Sphere gap setting for the test voltage to be applied (Fig. 45.11).

Test voltage ac, volts	Sphere gap setting, mm
5,000	2.5
10,000	5.0
20,000	10.0
30,000	15.0
40,000	22.0

Sphere gap settings will change with variations of temperature and humidity.

8. Connect the high-voltage lead of the testing transformer to the metal tank portion of the rubber-glove-testing equipment.
9. Connect the ground electrode in each rubber glove to its associated ground lead.
10. Evacuate all personnel from the enclosed testing area (Fig. 45.12).
11. Close the power-supply-isolating switch to the testing equipment.
12. By means of the remote controls, gradually raise the voltage output of the testing transformer from zero to the proper level at a rate of approximately 1000 volts/sec. This voltage should be applied to the rubber gloves for a period of 3 min (Fig. 45.13).

FIGURE 45.11 Sphere gap and series resistor connected to ground are in parallel with high-voltage testing transformer to protect the gloves under test from overvoltages if gap is adjusted correctly. (*Courtesy MidAmerican Energy Co.*)

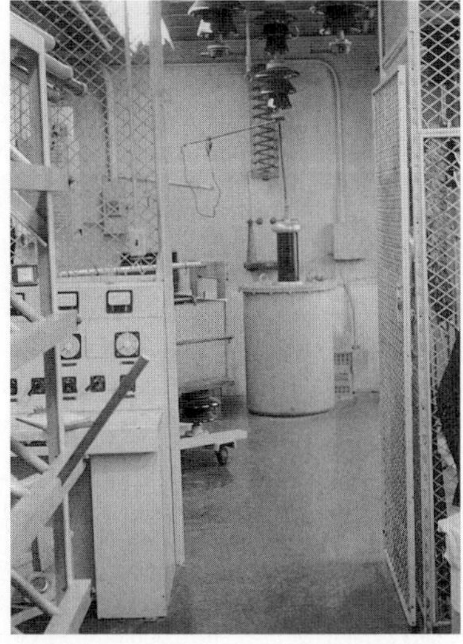

FIGURE 45.12 The testing equipment is interlocked with gate to prevent energizing the testing transformer until all personnel have left the area and the gate has been closed. (*Courtesy MidAmerican Energy Co.*)

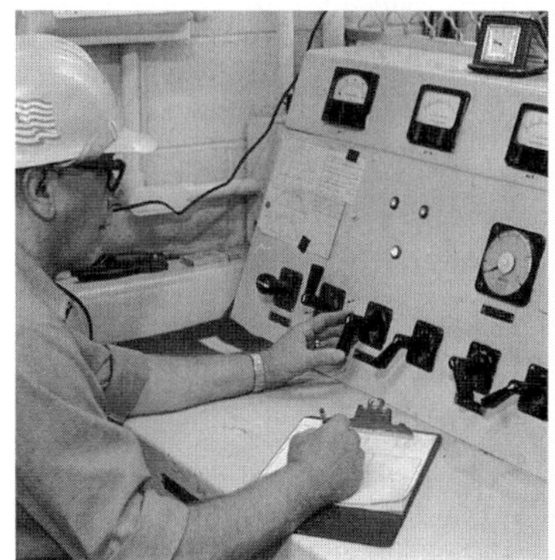

FIGURE 45.13 Control panel for test operation. (*Courtesy MidAmerican Energy Co.*)

FIGURE 45.14 Automatic test equipment being used to test linemen's rubber gloves. (*Courtesy MidAmerican Energy Co.*)

13. After the high voltage has been applied to the rubber gloves for 3 min, switch the milliammeter to read the leakage current in each ground electrode utilized for each rubber glove. Table 45.2 lists the maximum leakage current for five classes of rubber gloves.
14. The output voltage of the test transformer is gradually reduced to zero at a rate of approximately 1000 volts/sec.
15. All rubber gloves which failed during the test or had a current leakage greater than the maximum permissible value are rejected. In some instances the gloves can be repaired and used if they successfully pass a retest.
16. All rubber gloves which pass the high-voltage test are dried on a drying rack and powdered, and a tag indicating the leakage current recorded for each glove is placed in each glove. The gloves are then returned to the linemen or cablemen for use. In-service care of insulating gloves is specified in ANSI/ASTM F 496-77 Standard.

Automatic equipment available can be used to test protective equipment. The test equipment is capable of testing insulated rubber gloves, rubber sleeves, rubber boots, and hard hats (Figs. 45.14 and 45.15). The automatic test equipment is programmed by the operator for the tests to be completed. The tests can be repeated without reprogramming the equipment.

TABLE 45.2 Maximum Leakage Current for Five Classes of Rubber Gloves*

		Maximum leakage current, milliamps			
Class of glove	Test voltage, ac	10½ in†	14 in†	16 in†	18 in†
Class 0	5,000	8	12	14	16
Class 1	10,000	—	14	16	18
Class 2	20,000	—	16	18	20
Class 3	30,000	—	18	20	22
Class 4	40,000	—	—	22	24

*Reprinted by permission of the American Society for Testing and Materials, 1916 Race Street, Philadelphia, PA 19103. Copyright.

†Glove cuff length

FIGURE 45.15 Lineman is removing rubber gloves from automatic testing equipment after tests were completed. (*Courtesy MidAmerican Energy Co.*)

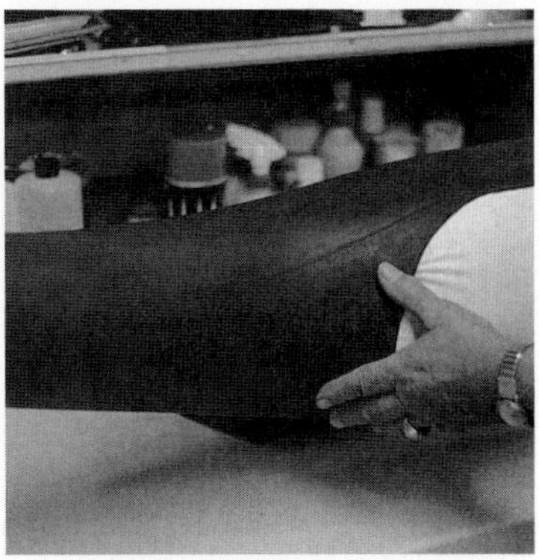

FIGURE 45.16 Rubber insulating sleeve being inspected while inflated after cleaning by washing. (*Courtesy MidAmerican Energy Co.*)

Rubber Sleeves. Rubber sleeves should be worn with rubber gloves to protect the arms and shoulders of the lineman, while he is working on high-voltage distribution circuits, from electrical contacts on the arms or shoulders. Figures 45.3 to 45.5 illustrate the proper use of rubber sleeves by linemen. Rubber insulating sleeves must be treated with care and inspected regularly by the linemen in a manner similar to that described for rubber insulating gloves.

The rubber insulating sleeves should be thoroughly cleaned, inspected, and tested by competent personnel regularly. The inspecting and testing procedures are similar to those described for rubber insulating gloves (Fig. 45.16).

In-service care of insulating sleeves is specified in ANSI/ASTM F 496 Standard.

Rubber Insulating Line Hose. Primary distribution conductors can be covered with rubber insulating line hose to protect the lineman from an accidental electrical contact. The lineman shown in Fig. 45.17 is installing a line hose on a 4160Y/2400-volt three-phase primary distribution circuit conductor. The line hoses are manufactured in various lengths, with inside-diameter measurements that vary from 1 to $1\frac{1}{2}$ in and with different test voltage withstand values. The lineman should be sure that the voltage rating of the line hose provides an ample safety factor for the voltage applied to the conductors to be covered.

All line hoses should be cleaned and inspected regularly. A hand crank wringer, as illustrated in Fig. 45.18, can be used to spread the line hose to clean and inspect it for cuts or corona damage.

The rubber insulating line hose should be tested for electric withstand in accordance with the specifications at scheduled intervals (Fig. 45.19).

In-service care of insulating line hose and covers is specified in ANSI/ASTM F 478 Standard.

Rubber Insulating Insulator Hoods. Pin-type or post-type distribution primary insulators can be covered by hoods (Fig. 45.20).

The insulator hood properly installed overlaps the line hose, providing the lineman with complete shielding from the energized conductors (Fig. 45.21).

Insulator hoods, like all other rubber insulating protective equipment, must be treated with care, kept clean, and inspected at regular intervals. Canvas bags of the proper size attached to a hand line should be used to raise and lower the protective equipment when it

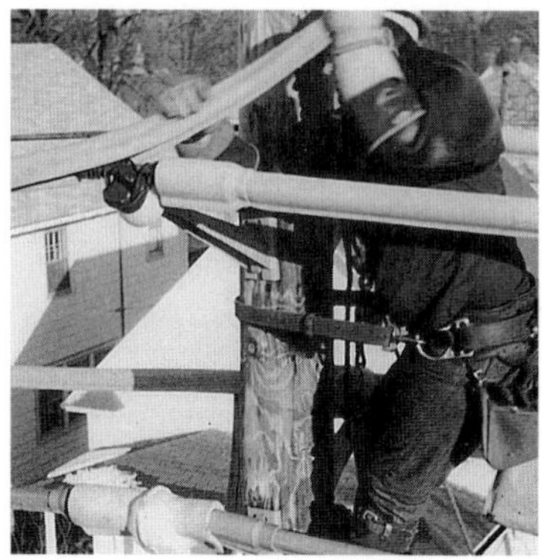

FIGURE 45.17 Rubber insulating line hose used by lineman to cover primary conductors. (*Courtesy MidAmerican Energy Co.*)

FIGURE 45.18 Rubber insulating line hose is being inspected and cleaned. (*Courtesy MidAmerican Energy Co.*)

is to be installed and removed. The lineman pictured in Fig. 45.22 has partially completed the installation of the rubber protective equipment, and the canvas bags are being lowered to the ground to permit the groundman to provide additional line hose and insulator hoods.

Conductor Covers. A conductor cover, fabricated from high-dielectric polyethylene, clips on and covers conductors up to 2 in in diameter. A positive air gap is maintained by a swinging latch that can be loosened only by a one-quarter turn with a clamp stick (Fig. 45.23).

Insulator Covers. Insulator covers are fabricated from high-dielectric polyethylene and are designed to be used in conjunction with two conductor covers. The insulator cover fits over the insulator and locks with a conductor cover on each end. A polypropylene rope swings under the crossarm and hooks with a clamp stick, thus preventing the insulator cover from being removed upward by bumping or wind gusts (Figs. 45.24).

FIGURE 45.19 Rubber insulating line hose mounted on high-voltage electrode with ground electrode clamped on it for withstand test. (*Courtesy MidAmerican Energy Co.*)

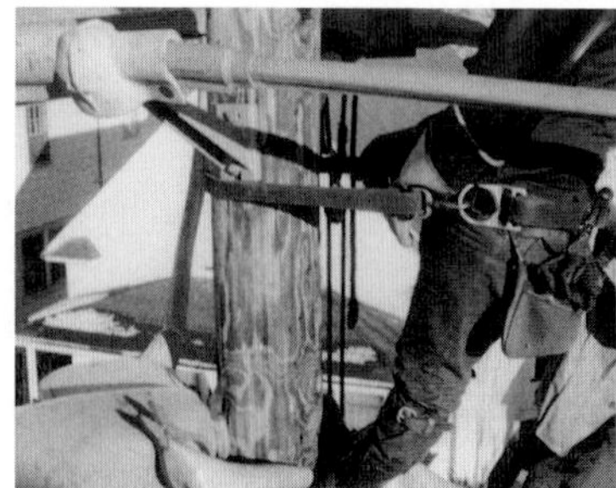

FIGURE 45.20 Rubber insulating insulator hood provides protection for lineman. (*Courtesy MidAmerican Energy Co.*)

FIGURE 45.21 Rubber insulating line hose and insulator hoods properly installed to cover primary conductors and insulators. (*Courtesy MidAmerican Energy Co.*)

FIGURE 45.22 Canvas bags used to raise rubber insulating line hose and insulator hoods to lineman on pole completing repairs to circuit after icestorm. (*Courtesy L. E. Myers Co.*)

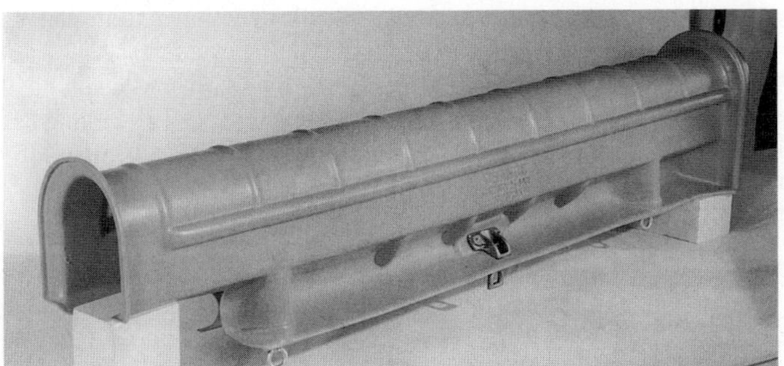

FIGURE 45.23 Conductor cover fabricated from high-dielectric polyethylene rated 26,000 volts. (*Courtesy A. B. Chance Co.*)

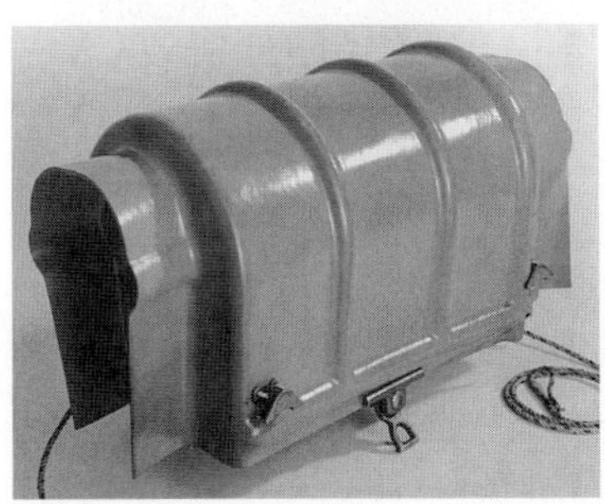

FIGURE 45.24 Insulator cover fabricated from high-dielectric polyethylene rated 26,000 volts. (*Courtesy A. B. Chance Co.*)

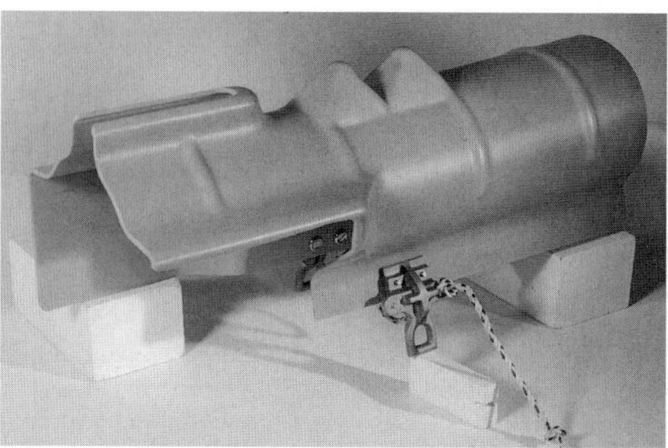

FIGURE 45.25 Crossarm cover fabricated from high-dielectric-strength polyethylene rated 26,000 volts. (*Courtesy A. B. Chance Co.*)

Crossarm Covers. High-dielectric-strength polyethylene crossarm covers are used to prevent tie wires from contacting the crossarm when conductors adjacent to insulators are being tied or untied. It is designed for single- or double-arm construction, with slots provided for the double-arm bolts. Flanges above the slots shield the ends of the double-arm bolts (Figs. 45.25).

Pole Covers. Polyethylene-constructed pole covers are designed to insulate the pole in the area adjacent to high-voltage conductors. The pole covers are available in various lengths. Positive-hold polypropylene rope handles are knotted through holes in the overlap area of the cover (Figs. 45.26 and 45.27).

FIGURE 45.26 Pole covers provide protection from ground to lineman working on insulated platforms. (*Courtesy A. B. Chance Co.*)

FIGURE 45.27 Pole, crossarms, and conductors not being worked on have been covered by the lineman. (*Courtesy A. B. Chance Co.*)

FIGURE 45.28 Rubber insulating blanket prepared for dielectric test after it has been cleaned and visually inspected. (*Courtesy MidAmerican Energy Co.*)

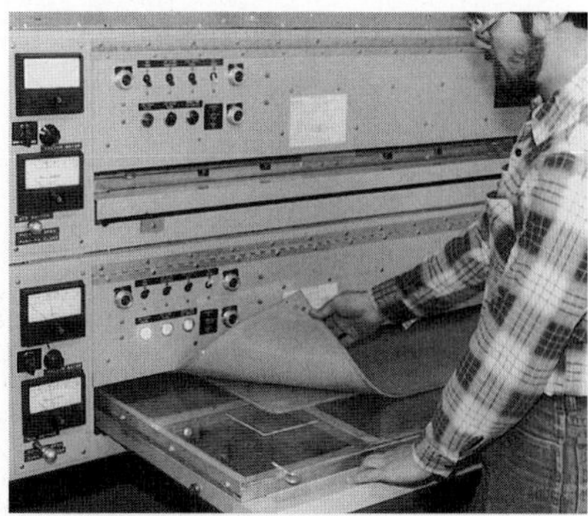

FIGURE 45.29 Lineman checking proper placement of rubber insulating blanket in automatic testing equipment before closing the equipment drawer and initiating the tests. (*Courtesy MidAmerican Energy Co.*)

Rubber Insulating Blankets. Odd-shaped conductors are equipment that can usually be covered best with rubber insulating blankets. Figure 45.28 shows a rubber blanket on a metal ground plate electrode. The blanket is covered with wet felt, and a copper disk with test lead connected is lying on the felt in preparation for testing the dielectric strength of the blanket. Figure 45.29 is a picture of automatic equipment used to test rubber insulating blankets.

The rubber insulating blankets are stored in canvas rolls or metal canisters to protect them when they are not in use. The blankets can be held in place by ropes or large wooden clamps (Fig. 45.30).

In-service care of insulating blankets is specified in ANSI/ASTM F 479 Standard.

Safety Hat. Hard hats, or safety hats, are worn by linemen, cablemen, and groundmen to protect the worker against the impact from falling or moving objects and against accidental electrical contact of the head and energized equipment, as well as to protect the worker from sun rays, cold, rain, sleet, and snow. The first combined impact-resisting and electrical insulating hat was introduced in 1952. The hat was designed "to roll with the punch" by distributing the force of a blow over the entire head. This is accomplished by a suspension

FIGURE 45.30 Rubber insulating blankets in use to cover bushing, lightning arrester, and top of distribution transformer. The blankets are held in place with large wooden clamps. Lineman is installing rubber line hose on primary conductor connecting to transformer. (*Courtesy W. H. Salisbury & Co.*)

FIGURE 45.31 Cutaway view of electrical worker's protective helmet showing suspension band which ensures proper crown clearance.

band which holds the hat about an inch away from the head and lets the hat work as a shock absorber (Fig. 45.31).

The hat is made of fiberglass, or plastic material, and has an insulating value of approximately 20,000 volts. New helmets are manufactured to withstand a test of 30,000 volts without failure. The actual voltage that the hat sustains while being worn depends upon the cleanliness of the hat, weather conditions, the type of electrode contacted, and other variables. The wearing of safety hats by linemen and cablemen has greatly reduced electrical contacts.

Physical injuries to the head have been practically eliminated as a result of workers on the ground wearing protective helmets. The Occupational Safety and Health Act of 1970 and most companies' safety rules require linemen, cablemen, and groundmen to wear safety hats while performing physical work. Specifications for safety hats are found in ANSI Standard Z89.2, "Safety Requirements for Industrial Protective Helmets for Electrical Workers, Class B."

Work Zone Safety and Traffic Control. Many projects involve positioning vehicles in or just off the traveled roadways. 25,000 serious injuries occur in work zones each year. The importance of barricading worksites properly while working in these situations is critical to the safety of the public as well as the lineman. The lineman should be knowledgeable with traffic control requirements in the jurisdiction worked. Familiarity with rules contained in the National Manual on Uniform Traffic Control Devices (MUTCD) maintained by the Federal Highway Administration (FHWA) aids the lineman in his understanding of the concepts. Permits may be required as well depending upon the agency that controls the highway. These agencies include the Department of Transportation, the County Highway Department or the City Public Works Department. Information required on the permit includes the nature and duration of the work, proposed positioning of vehicles, and a diagram of the traffic sign spacing for the work zone. In some cases, flagmen are required in addition to barricades and traffic signs or signals. The lineman may also be required to wear a highly visible shirt or illuminated vest.

Clothing. The lineman's clothing requirements are becoming more defined based upon the job's requirements and work situations. Several guidelines pertaining to outerwear, underwear, winter-wear, and rainwear have been developed. Consideration must be given as to a clothing material's flame resistance, arc resistance, and visibility. Standards written on this topic include ASTM F1506-2000 *Clothing for Electric Arc,* and ASTM F1891-99 *Specification for Arc Resistant Raingear.* Clothing manufacturers utilize fabrics that comply with standards and are comfortable, fashionable, and stylish. The clothing is subjected to rigorous testing including arc exposure and flame resistance. Care must be taken when cleaning in order to maintain the quality and protective properties of the clothing.

Full Body Harnesses. Falls are a leading cause of injury for the lineman. The industry has implemented solutions to minimize or eliminate the severity of fall-related injuries experienced. ANSI Standards A10.14 and Z359.1, ASTM Standard F887, and OSHA Standards, have been written addressing the topic in more detail. The full body harness and associated connecting straps are now required in more situations. Safety systems are in place for fall restraint, fall arrest, work positioning, suspension, and retrieval. Shock absorbing lanyards reduce the impact of a free fall from elevated work positions. The equipment is designed with worker comfort and safety in mind. The equipment is lightweight and durable. Back and total body support are included. Unnecessary chafing is minimized. As with any piece of safety equipment, the harnesses, lanyards, body belts and straps are to be thoroughly examined before use.

Summary. Protective equipment provides the lineman a margin of safety for error without injury while working on energized distribution conductors. Application of the protective equipment should be planned as with all other aspects of work. When the plan has been established, it should be followed without shortcuts. The safe way of completing the work is the most efficient and reliable method. The skilled lineman and cableman is an expert in the application of protective equipment. Figure 45.32 pictures two skilled linemen utilizing protective equipment to good advantage while working on energized distribution circuits.

FIGURE 45.32 The linemen illustrated are working on a 13,200Y/7620-volt energized distribution circuit, utilizing proper protective equipment and tools. The linemen are wearing rubber gloves, rubber sleeves, insulated hard hats, safety glasses, and body belts with safety strap secured. They are working on an insulated platform. Hot-line tools are in use to maintain proper clearances. The protective cover-up equipment installed includes conductor covers, crossarm covers, pole covers, dead-end insulator covers and rubber blankets. (*Courtesy A. B. Chance Co.*)

CHAPTER 46
OSHA § 1910.269

OSHA REGULATIONS (STANDARDS-29 CFR) ELECTRIC POWER GENERATION, TRANSMISSION, AND DISTRIBUTION—1910.269

- **Standard Number:** § 1910.269
- **Standard Title:** Electric Power Generation, Transmission, and Distribution
- **Subpart Number:** R
- **Subpart Title:** Special Industries

INTERPRETATION(S)

(a) *General*

(a)(1) *Application*

(a)(1)(i) This section covers the operation and maintenance of electric power generation, control, transformation, transmission, and distribution lines and equipment. These provisions apply to:

(a)(1)(i)(A) Power generation, transmission, and distribution installations, including related equipment for the purpose of communication or metering, which are accessible only to qualified employees;

Note: The types of installations covered by this paragraph include the generation, transmission, and distribution installations of electric utilities, as well as equivalent installations of industrial establishments. Supplementary electric generating equipment that is used to supply a workplace for emergency, standby, or similar purposes only is covered under Subpart S of this Part. (See paragraph (a)(1)(ii)(B) of this section.)

(a)(1)(i)(B) Other installations at an electric power generating station, as follows:

(a)(1)(i)(B)(1) Fuel and ash handling and processing installations, such as coal conveyors,

(a)(1)(i)(B)(2) Water and steam installations, such as penstocks, pipelines, and tanks, providing a source of energy for electric generators, and

(a)(1)(i)(B)(3) Chlorine and hydrogen systems:

(a)(1)(i)(C) Test sites where electrical testing involving temporary measurements associated with electric power generation, transmission, and distribution is performed in laboratories, in the field, in substations, and on lines, as opposed to metering, relaying, and routine line work;

(a)(1)(i)(D) Work on or directly associated with the installations covered in paragraphs (a)(1)(i)(A) through (a)(1)(i)(C) of this section; and

(a)(1)(i)(E) Line-clearance tree-trimming operations, as follows:

(a)(1)(i)(E)(1) Entire § 1910.269 of this Part, except paragraph (r)(1) of this section, applies to line-clearance tree-trimming operations performed by qualified employees (those who are knowledgeable in the construction and operation of electric power generation, transmission, or distribution equipment involved, along with the associated hazards).

(a)(1)(i)(E)(2) Paragraphs (a)(2), (b), (c), (g), (k), (p), and (r) of this section apply to line-clearance tree-trimming operations performed by line-clearance tree trimmers who are not qualified employees.

(a)(1)(ii) Notwithstanding paragraph (A)(1)(i) of this section, § 1910.269 of this Part does not apply:

(a)(1)(ii)(A) To construction work, as defined in § 1910.12 of this Part; or

(a)(1)(ii)(B) To electrical installations, electrical safety-related work practices, or electrical maintenance considerations covered by Subpart S of this Part.

Note 1: Work practices conforming to §§ 1910.332 through 1910.335 of this Part are considered as complying with the electrical safety-related work practice requirements of this section identified in Table 1 of Appendix A.2 to this section, provided the work is being performed on a generation or distribution installation meeting § 1910.303 through § 1910.308 of this Part. This table also identifies provisions in this section that apply to work by qualified persons directly on or associated with installations of electric power generation, transmission, and distribution lines or equipment, regardless of compliance with § 1910.332 through § 1910.335 of this Part.

Note 2: Work practices performed by qualified persons and conforming to § 1910.269 of this Part are considered as complying with § 1910.333(c) and § 1910.335 of this Part.

(a)(1)(iii) This section applies to addition to all other applicable standards contained in this Part § 1910. Specific references in this section to other sections of Part § 1910 are provided for emphasis only.

(a)(2) *Training*

(a)(2)(i) Employees shall be trained in and familiar with the safety-related work practices, safety procedures, and other safety requirements in this section that pertain to their respective job assignments. Employees shall also be trained in and familiar with any other safety practices, including applicable emergency procedures (such as pole top and manhole rescue), that are not specifically addressed by this section but that are related to their work and are necessary for their safety.

(a)(2)(ii) Qualified employees shall also be trained and competent in:

(a)(2)(ii)(A) The skills and techniques necessary to distinguish exposed live parts from other parts of electric equipment.

(a)(2)(ii)(B) The skills and techniques necessary to determine the nominal voltage of exposed live parts.

(a)(2)(ii)(C) The minimum approach distances specified in this section corresponding to the voltages to which the qualified employee will be exposed, and

(a)(2)(ii)(D) The proper use of the special precautionary techniques, personal protective equipment, insulating and shielding materials, and insulated tools for working on or near exposed energized parts of electric equipment.

Note: For the purposes of this section, a person must have this training in order to be considered a qualified person.

(a)(2)(iii) The employer shall determine, through regular supervision and through inspections conducted on at least an annual basis, that each employee is complying with the safety-related work practices required by this section.

(a)(2)(iv) An employee shall receive additional training (or retraining) under any of the following conditions:

(a)(2)(iv)(A) If the supervision and annual inspections required by paragraph (a)(2)(iii) of this section indicate that the employee is not complying with the safety-related work practices required by this section, or

(a)(2)(iv)(B) If new technology, new types of equipment, or changes in procedures necessitate the use of safety-related work practices that are different from those which the employee would normally use, or

(a)(2)(iv)(C) If he or she must employ safety-related work practices that are not normally used during his or her regular job duties.

Note: OSHA would consider tasks that are performed less often than once per year to necessitate retraining before the performance of the work practices involved.

(a)(2)(v) The training required by paragraph (a)(2) of this section shall be of the classroom or on-the-job type.

(a)(2)(vi) The training shall establish employee proficiency in the work practices required by this section and shall introduce the procedures necessary for compliance with this section.

(a)(2)(vii) The employer shall certify that each employee has received the training required by paragraph (a)(2) of this section. This certification shall be made when the employee demonstrates proficiency in the work practices involved and shall be maintained for the duration of the employee's employment.

Note: Employment records that indicate that an employee has received the required training are an acceptable means of meeting this requirement.

(a)(3) *Existing conditions.* Existing conditions related to the safety of the work to be performed shall be determined before work on or near electric lines or equipment is started. Such conditions include, but are not limited to, the nominal voltages of lines and equipment, the maximum switching transient voltages, the presence of hazardous induced voltages, the presence and condition of protective grounds and equipment grounding conductors, the condition of poles, environmental conditions relative to safety, and the locations of circuits and equipment, including power and communication lines and fire protective signaling circuits.

(b) *Medical services and first aid.* The employer shall provide medical services and first aid as required in § 1910.151 of this Part. In addition to the requirements of § 1910.151 of this Part, the following requirements also apply:

(b)(1) *Cardiopulmonary resuscitation and first aid training.* When employees are performing work on or associated with exposed lines or equipment energized at 50 volts or more, persons trained in first aid including cardiopulmonary resuscitation (CPR) shall be available as follows:

(b)(1)(i) For field work involving two or more employees at a work location, at least two trained persons shall be available. However, only one trained person need be available if all new employees are trained in first aid, including CPR, within 3 months of their hiring dates.

(b)(1)(ii) For fixed work locations such as generating stations, the number of trained persons available shall be sufficient to ensure that each employee exposed to electric shock can be reached within 4 minutes by a trained person. However, where the existing number of employees is insufficient to meet this requirement (at a remote substation, for example), all employees at the work location shall be trained.

(b)(2) *First aid supplies.* First aid supplies required by § 1910.151(b) of this Part shall be placed in weatherproof containers if the supplies could be exposed to the weather.

(b)(3) *First aid kits.* Each first aid kit shall be maintained, shall be readily available for use, and shall be inspected frequently enough to ensure that expended items are replaced but at least once per year.

(c) *Job briefing.* The employer shall ensure that the employee in charge conducts a job briefing with the employees involved before they start each job. The briefing shall cover at least the following subjects: hazards associated with the job, work procedures involved, special precautions, energy source controls, and personal protective equipment requirements.

(c)(1) *Number of briefings.* If the work or operations to be performed during the work day or shift are repetitive and similar, at least one job briefing shall be conducted before the start of the first job of each day or shift. Additional job briefings shall be held if significant changes, which might affect the safety of the employees, occur during the course of the work.

(c)(2) *Extent of briefing.* A brief discussion is satisfactory if the work involved is routine and if the employee, by virtue of training and experience, can reasonably be expected to recognize and avoid the hazards involved in the job. A more extensive discussion shall be conducted:

(c)(2)(i) If the work is complicated or particularly hazardous, or

(c)(2)(ii) If the employee cannot be expected to recognize and avoid the hazards involved in the job.

Note: The briefing is always required to touch on all the subjects listed in the introductory text to paragraph (c) of this section.

(c)(3) *Working alone.* An employee working alone need not conduct a job briefing. However, the employer shall ensure that the tasks to be performed are planned as if a briefing were required.

(d) *Hazardous energy control (lockout/tagout) procedures*

(d)(1) *Application.* The provisions of paragraph (d) of this section apply to the use of lockout/tagout procedures for the control of energy sources in installations for the purpose of electric power generation, including related equipment for communication or metering. Locking and tagging procedures for the deenergizing of electric energy sources which are used exclusively for purposes of transmission and distribution are addressed by paragraph (m) of this section.

Note 1: Installations in electric power generation facilities that are not an integral part of, or inextricably commingled with, power generation processes or equipment are covered under § 1910.147 and Subpart S of this Part.

Note 2: Lockout and tagging procedures that comply with paragraphs (c) through (f) of § 1910.147 of this Part will also be deemed to comply with paragraph of this section if the procedures address the hazards covered by paragraph (d) of this section.

(d)(2) *General*

(d)(2)(i) The employer shall establish a program consisting of energy control procedures, employee training, and periodic inspections to ensure that, before any employee performs any servicing or maintenance on a machine or equipment where the unexpected energizing, start up, or release of stored energy could occur and cause injury, the machine or equipment is isolated from the energy source and rendered inoperative.

(d)(2)(ii) The employer's energy control program under paragraph (d)(2) of this section shall meet the following requirements:

(d)(2)(ii)(A) If an energy isolating device is not capable of being locked out, the employer's program shall use a tagout system.

(d)(2)(ii)(B) If an energy isolating device is capable of being locked out, the employer's program shall use lockout, unless the employer can demonstrate that the use of a tagout system will provide full employee protection as follows:

(d)(2)(ii)(B)(1) When a tagout device is used on an energy isolating device which is capable of being locked out, the tagout device shall be attached at the same location that the lockout device would have been attached, and the employer shall demonstrate that the tagout program will provide a level of safety equivalent to that obtained by the use of a lockout program.

(d)(2)(ii)(B)(2) In demonstrating that a level of safety is achieved in the tagout program equivalent to the level of safety obtained by the use of a lockout program, the employer shall demonstrate full compliance with all tagout-related provisions of this standard together with such additional elements as are necessary to provide the equivalent safety available from the use of a lockout device. Additional means to be considered as part of the demonstration of full employee protection shall include the implementation of additional safety measures such as the removal of an isolating circuit element, blocking of a controlling switch, opening of an extra disconnecting device, or the removal of a valve handle to reduce the likelihood of inadvertent energizing.

(d)(2)(ii)(C) After November 1, 1994, whenever replacement or major repair, renovation, or modification of a machine or equipment is performed, and whenever new machines or equipment are installed, energy isolating devices for such machines or equipment shall be designed to accept a lockout device.

(d)(2)(iii) Procedures shall be developed, documented, and used for the control of potentially hazardous energy covered by paragraph (d) of this section.

(d)(2)(iv) The procedure shall clearly and specifically outline the scope, purpose, responsibility, authorization, rules, and techniques to be applied to the control of hazardous energy, and the measures to enforce compliance including, but not limited to, the following:

(d)(2)(iv)(A) A specific statement of the intended use of this procedure;

(d)(2)(iv)(B) Specific procedural steps for shutting down, isolating, blocking and securing machines or equipment to control hazardous energy;

(d)(2)(iv)(C) Specific procedural steps for the placement, removal, and transfer of lockout devices or tagout devices and the responsibility for them; and

(d)(2)(iv)(D) Specific requirements for testing a machine or equipment to determine and verify the effectiveness of lockout devices, tagout devices, and other energy control measures.

(d)(2)(v) The employer shall conduct a periodic inspection of the energy control procedure at least annually to ensure that the procedure and the provisions of paragraph (d) of this section are being followed.

(d)(2)(v)(A) The periodic inspection shall be performed by an authorized employee who is not using the energy control procedure being inspected.

(d)(2)(v)(B) The periodic inspection shall be designed to identify and correct any deviations or inadequacies.

(d)(2)(v)(C) If lockout is used for energy control, the periodic inspection shall include a review, between the inspector and each authorized employee, of that employee's responsibilities under the energy control procedure being inspected.

(d)(2)(v)(D) Where tagout is used for energy control, the periodic inspection shall include a review, between the inspector and each authorized and affected employee, of that employee's responsibilities under the energy control procedure being inspected, and the elements set forth in paragraph (d)(2)(vii) of this section.

(d)(2)(v)(E) The employer shall certify that the inspections required by paragraph (d)(2)(v) of this section have been accomplished. The certification shall identify the machine or equipment on which the energy control procedure was being used, the date of the inspection, the employees included in the inspection, and the person performing the inspection.

Note: If normal work schedule and operation records demonstrate adequate inspection activity and contain the required information, no additional certification is required.

(d)(2)(vi) The employer shall provide training to ensure that the purpose and function of the energy control program are understood by employees and that the knowledge and skills required for the safe application, usage, and removal of energy controls are acquired by employees. The training shall include the following:

(d)(2)(vi)(A) Each authorized employee shall receive training in the recognition of applicable hazardous energy sources, the type and magnitude of energy available in the workplace, and in the methods and means necessary for energy isolation and control.

(d)(2)(vi)(B) Each affected employee shall be instructed in the purpose and use of the energy control procedure.

(d)(2)(vi)(C) All other employees whose work operations are or may be in an area where energy control procedures may be used shall be instructed about the procedures and about the prohibition relating to attempts to restart or reenergize machines or equipment that are locked out or tagged out.

(d)(2)(vii) When tagout systems are used, employees shall also be trained in the following limitations of tags:

(d)(2)(vii)(A) Tags are essentially warning devices affixed to energy isolating devices and do not provide the physical restraint on those devices that is provided by a lock.

(d)(2)(vii)(B) When a tag is attached to an energy isolating means, it is not to be removed without authorization of the authorized person responsible for it, and it is never to be bypassed, ignored, or otherwise defeated.

(d)(2)(vii)(C) Tags must be legible and understandable by all authorized employees, affected employees, and all other employees whose work operations are or may be in the area, in order to be effective.

(d)(2)(vii)(D) Tags and their means of attachment must be made of materials which will withstand the environmental conditions encountered in the workplace.

(d)(2)(vii)(E) Tags may evoke a false sense of security, and their meaning needs to be understood as part of the overall energy control program.

(d)(2)(vii)(F) Tags must be securely attached to energy isolating devices so that they cannot be inadvertently or accidentally detached during use.

(d)(2)(viii) Retraining shall be provided by the employer as follows:

(d)(2)(viii)(A) Retraining shall be provided for all authorized and affected employees whenever there is a change in their job assignments, a change in machines, equipment, or processes that present a new hazard or whenever there is a change in the energy control procedures.

(d)(2)(viii)(B) Retraining shall also be conducted whenever a periodic inspection under paragraph (d)(2)(v) of this section reveals, or whenever the employer has reason to believe, that there are deviations from or inadequacies in an employee's knowledge or use of the energy control procedures.

(d)(2)(viii)(C) The retraining shall reestablish employee proficiency and shall introduce new or revised control methods and procedures, as necessary.

(d)(2)(ix) The employer shall certify that employee training has been accomplished and is being kept up to date. The certification shall contain each employee's name and dates of training.

(d)(3) *Protective materials and hardware*

(d)(3)(i) Locks, tags, chains, wedges, key blocks, adapter pins, self-locking fasteners, or other hardware shall be provided by the employer for isolating, securing, or blocking of machines or equipment from energy sources.

(d)(3)(ii) Lockout devices and tagout devices shall be singularly identified; shall be the only devices used for controlling energy; may not be used for other purposes; and shall meet the following requirements:

(d)(3)(ii)(A) Lockout devices and tagout devices shall be capable of withstanding the environment to which they are exposed for the maximum period of time that exposure is expected.

(d)(3)(ii)(A)(1) Tagout devices shall be constructed and printed so that exposure to weather conditions or wet and damp locations will not cause the tag to deteriorate or the message on the tag to become illegible.

(d)(3)(ii)(A)(2) Tagout devices shall be so constructed as not to deteriorate when used in corrosive environments.

(d)(3)(ii)(B) Lockout devices and tagout devices shall be standardized within the facility in at least one of the following criteria: color, shape, size. Additionally, in the case of tagout devices, print and format shall be standardized.

(d)(3)(ii)(C) Lockout devices shall be substantial enough to prevent removal without the use of excessive force or unusual techniques, such as with the use of bolt cutters or metal cutting tools.

(d)(3)(ii)(D) Tagout devices, including their means of attachment, shall be substantial enough to prevent inadvertent or accidental removal. Tagout device attachment means shall be of a non-reusable type, attachable by hand, self-locking, and non-releasable with a minimum unlocking strength of no less than 50 pounds and shall have the general design and basic characteristics of being at least equivalent to a one-piece, all-environment-tolerant nylon cable tie.

(d)(3)(ii)(E) Each lockout device or tagout device shall include provisions for the identification of the employee applying the device.

(d)(3)(ii)(F) Tagout devices shall warn against hazardous conditions if the machine or equipment is energized and shall include a legend such as the following: Do Not Start, Do Not Open, Do Not Close, Do Not Energize, Do Not Operate.

Note: For specific provisions covering accident prevention tags, see § 1910.145 of this Part.

(d)(4) *Energy isolation.* Lockout and tagout device application and removal may only be performed by the authorized employees who are performing the servicing or maintenance.

(d)(5) *Notification.* Affected employees shall be notified by the employer or authorized employee of the application and removal of lockout or tagout devices. Notification shall be given before the controls are applied and after they are removed from the machine or equipment.

Note: See also paragraph (d)(7) of this section, which requires that the second notification take place before the machine or equipment is reenergized.

(d)(6) *Lockout/tagout application.* The established procedures for the application of energy control (the lockout or tagout procedures) shall include the following elements and actions, and these procedures shall be performed in the following sequence:

(d)(6)(i) Before an authorized or affected employee turns off a machine or equipment, the authorized employee shall have knowledge of the type and magnitude of the energy, the hazards of the energy to be controlled, and the method or means to control the energy.

(d)(6)(ii) The machine or equipment shall be turned off or shut down using the procedures established for the machine or equipment. An orderly shutdown shall be used to avoid any additional or increased hazards to employees as a result of the equipment stoppage.

(d)(6)(iii) All energy isolating devices that are needed to control the energy to the machine or equipment shall be physically located and operated in such a manner as to isolate the machine or equipment from energy sources.

(d)(6)(iv) Lockout or tagout devices shall be affixed to each energy isolating device by authorized employees.

(d)(6)(iv)(A) Lockout devices shall be attached in a manner that will hold the energy isolating devices in a "safe" or "off" position.

(d)(6)(iv)(B) Tagout devices shall be affixed in such a manner as will clearly indicate that the operation or movement of energy isolating devices from the "safe" or "off" position is prohibited.

(d)(6)(iv)(B)(1) Where tagout devices are used with energy isolating devices designed with the capability of being locked out, the tag attachment shall be fastened at the same point at which the lock would have been attached.

(d)(6)(iv)(B)(2) Where a tag cannot be affixed directly to the energy isolating device, the tag shall be located as close as safely possible to the device, in a position that will be immediately obvious to anyone attempting to operate the device.

(d)(6)(v) Following the application of lockout or tagout devices to energy isolating devices, all potentially hazardous stored or residual energy shall be relieved, disconnected, restrained, or otherwise rendered safe.

(d)(6)(vi) If there is a possibility of reaccumulation of stored energy to a hazardous level, verification of isolation shall be continued until the servicing or maintenance is completed or until the possibility of such accumulation no longer exists.

(d)(6)(vii) Before starting work on machines or equipment that have been locked out or tagged out, the authorized employee shall verify that isolation and deenergizing of the machine or equipment have been accomplished. If normally energized parts will be exposed to contact by an employee while the machine or equipment is deenergized, a test shall be performed to ensure that these parts are deenergized.

(d)(7) *Release from lockout/tagout.* Before lockout or tagout devices are removed and energy is restored to the machine or equipment, procedures shall be followed and actions taken by the authorized employees to ensure the following:

(d)(7)(i) The work area shall be inspected to ensure that nonessential items have been removed and that machine or equipment components are operationally intact.

(d)(7)(ii) The work area shall be checked to ensure that all employees have been safely positioned or removed.

(d)(7)(iii) After lockout or tagout devices have been removed and before a machine or equipment is started, affected employees shall be notified that the lockout or tagout devices have been removed.

(d)(7)(iv) Each lockout or tagout device shall be removed from each energy isolating device by the authorized employee who applied the lockout or tagout device. However, if that employee is not available to remove it, the device may be removed under the direction of the employer, provided that specific procedures and training for such removal have been developed, documented, and incorporated into the employer's energy control program. The employer shall demonstrate that the specific procedure provides a degree of safety equivalent to that provided by the removal of the device by the authorized employee who applied it. The specific procedure shall include at least the following elements:

(d)(7)(iv)(A) Verification by the employer that the authorized employee who applied the device is not at the facility;

(d)(7)(iv)(B) Making all reasonable efforts to contact the authorized employee to inform him or her that his or her lockout or tagout device has been removed; and

(d)(7)(iv)(C) Ensuring that the authorized employee has this knowledge before he or she resumes work at that facility.

(d)(8) *Additional requirements*

(d)(8)(i) If the lockout or tagout devices must be temporarily removed from energy isolating devices and the machine or equipment must be energized to test or position the machine, equipment, or component thereof, the following sequence of actions shall be followed:

(d)(8)(i)(A) Clear the machine or equipment of tools and materials in accordance with paragraph (d)(7)(i) of this section;

(d)(8)(i)(B) Remove employees from the machine or equipment area in accordance with paragraphs (d)(7)(ii) and (d)(7)(iii) of this section;

(d)(8)(i)(C) Remove the lockout or tagout devices as specified in paragraph (d)(7)(iv) of this section;

(d)(8)(i)(D) Energize and proceed with the testing or positioning; and

(d)(8)(i)(E) Deenergize all systems and reapply energy control measures in accordance with paragraph (d)(6) of this section to continue the servicing or maintenance.

(d)(8)(ii) When servicing or maintenance is performed by a crew, craft, department, or other group, they shall use a procedure which affords the employees a level of protection equivalent to that provided by the implementation of a personal lockout or tagout device. Group lockout or tagout devices shall be used in accordance with the procedures required by paragraphs (d)(2)(iii) and (d)(2)(iv) of this section including, but not limited to, the following specific requirements:

(d)(8)(ii)(A) Primary responsibility shall be vested in an authorized employee for a set number of employees working under the protection of a group lockout or tagout device (such as an operations lock);

(d)(8)(ii)(B) Provision shall be made for the authorized employee to ascertain the exposure status of all individual group members with regard to the lockout or tagout of the machine or equipment;

(d)(8)(ii)(C) When more than one crew, craft, department, or other group is involved, assignment of overall job-associated lockout or tagout control responsibility shall be given to an authorized employee designated to coordinate affected work forces and ensure continuity of protection; and

(d)(8)(ii)(D) Each authorized employee shall affix a personal lockout or tagout device to the group lockout device, group lockbox, or comparable mechanism when he or she begins work and shall remove those devices when he or she stops working on the machine or equipment being serviced or maintained.

(d)(8)(iii) Procedures shall be used during shift or personnel changes to ensure the continuity of lockout or tagout protection, including provision for the orderly transfer of lockout or tagout device protection between off-going and on-coming employees, to minimize their exposure to hazards from the unexpected energizing or start-up of the machine or equipment or from the release of stored energy.

(d)(8)(iv) Whenever outside servicing personnel are to be engaged in activities covered by paragraph (d) of this section, the on-site employer and the outside employer shall inform each other of their respective lockout or tagout procedures, and each employer shall ensure that his or her personnel understand and comply with restrictions and prohibitions of the energy control procedures being used.

(d)(8)(v) If energy isolating devices are installed in a central location and are under the exclusive control of a system operator, the following requirements apply:

(d)(8)(v)(A) The employer shall use a precedure that affords employees a level of protection equivalent to that provided by the implementation of a personal lockout or tagout device.

(d)(8)(v)(B) The system operator shall place and remove lockout and tagout devices in place of the authorized employee under paragraphs (d)(4), (d)(6)(iv), and (d)(7)(iv) of this section.

(d)(8)(v)(C) Provisions shall be made to identify the authorized employee who is responsible for (that is, being protected by) the lockout or tagout device, to transfer responsibility for lockout and tagout devices, and to ensure that an authorized employee requesting removal or transfer of a lockout or tagout device is the one responsible for it before the device is removed or transferred.

(e) *Enclosed spaces.* This paragraph covers enclosed spaces that may be entered by employees. It does not apply to vented vaults if a determination is made that the ventilation system is operating to protect employees before they enter the space. This paragraph applies to routine entry into enclosed spaces in lieu of the permit-space entry requirements contained in paragraphs (d) through (k) of § 1910.146 of this Part. If, after the precautions given in paragraphs (e) and (t) of this section are taken, the hazards remaining in the enclosed space endanger the life of an entrant or could interfere with escape from the space, then entry into the enclosed space shall meet the permit-space entry requirements of paragraphs (d) through (k) of § 1910.146 of this Part.

Note: Entries into enclosed spaces conducted in accordance with permit-space entry requirements of paragraphs (d) through (k) of § 1910.146 of this Part are considered as complying with paragraph (e) of this section.

(e)(1) *Safe work practices.* The employer shall ensure the use of safe work practices for entry into and work in enclosed spaces and for rescue of employees from such spaces.

(**e**)(**2**) *Training.* Employees who enter enclosed spaces or who serve as attendants shall be trained in the hazards of enclosed space entry, in enclosed space entry procedures, and in enclosed space rescue procedures.

(**e**)(**3**) *Rescue equipment.* Employers shall provide equipment to ensure that prompt and safe rescue of employees from the enclosed space.

(**e**)(**4**) *Evaluation of potential hazards.* Before any entrance cover to an enclosed space is removed, the employer shall determine whether it is safe to do so by checking for the presence of any atmospheric pressure or temperature differences and by evaluating whether there might be a hazardous atmosphere in the space. Any conditions making it unsafe to remove the cover shall be eliminated before the cover is removed.

Note: The evaluation called for in this paragraph may take the form of a check of the conditions expected to be in the enclosed space. For example, the cover could be checked to see if it is hot and, if it is fastened in place, could be loosened gradually to release any residual pressure. A determination must also be made of whether conditions at the site could cause a hazardous atmosphere, such as an oxygen deficient or flammable atmosphere, to develop within the space.

(**e**)(**5**) *Removal of covers.* When covers are removed from enclosed spaces, the opening shall be promptly guarded by a railing, temporary cover, or other barrier intended to prevent an accidental fall through the opening and to protect employees working in the space from objects entering the space.

(**e**)(**6**) *Hazardous atmosphere.* Employees may not enter any enclosed space while it contains a hazardous atmosphere, unless the entry conforms to the generic permit-required confined spaces standard in § 1910.146 of this Part.

Note: The term "entry" is defined in § 1910.146(b) of this Part.

(**e**)(**7**) *Attendants.* While work is being performed in the enclosed space, a person with first aid training meeting paragraph (b) of this section shall be immediately available outside the enclosed space to render emergency assistance if there is reason to believe that a hazard may exist in the space or if a hazard exists because of traffic patterns in the area of the opening used for entry. That person is not precluded from performing other duties outside the enclosed space if these duties do not distract the attendant from monitoring employees within the space.

Note: See paragraph (t)(3) of this section for additional requirements on attendants for work in manholes.

(**e**)(**8**) *Calibration of test instruments.* Test instruments used to monitor atmospheres in enclosed spaces shall be kept in calibration, with a minimum accuracy of + or – 10 percent.

(**e**)(**9**) *Testing for oxygen deficiency.* Before an employee enters an enclosed space, the internal atmosphere shall be tested for oxygen deficiency with a direct-reading meter or similar instrument, capable of collection and immediate analysis of data samples without the need for off-site evaluation. If continuous forced air ventilation is provided, testing is not required provided that the procedures used ensure that employees are not exposed to the hazards posed by oxygen deficiency.

(**e**)(**10**) *Testing for flammable gases and vapors.* Before an employee enters an enclosed space, the internal atmosphere shall be tested for flammable gases and vapors with a direct-reading meter or similar instrument capable of collection and immediate analysis of data samples without the need for off-site evaluation. This test shall be performed after the oxygen testing and ventilation required by paragraph (e)(9) of this section demonstrate that there is sufficient oxygen to ensure the accuracy of the test for flammability.

(**e**)(**11**) *Ventilation and monitoring.* If flammable gases or vapors are detected or if an oxygen deficiency is found, forced air ventilation shall be used to maintain oxygen at a safe level and to prevent a hazardous concentration of flammable gases and vapors from accumulating. A continuous monitoring program to ensure that no increase in flammable gas or vapor concentration occurs may be followed in lieu of ventilation, if flammable gases or vapors are detected at safe levels.

Note: See the definition of hazardous atmosphere for guidance in determining whether or not a given concentration or a substance is considered to be hazardous.

(e)(12) *Specific ventilation requirements.* If continuous forced air ventilation is used, it shall begin before entry is made and shall be maintained long enough to ensure that a safe atmosphere exists before employees are allowed to enter the work area. The forced air ventilation shall be so directed as to ventilate the immediate area where employees are present within the enclosed space and shall continue until all employees leave the enclosed space.

(e)(13) *Air supply.* The air supply for the continuous forced air ventilation shall be from a clean source and may not increase the hazards in the enclosed space.

(e)(14) *Open flames.* If open flames are used in enclosed spaces, a test for flammable gases and vapors shall be made immediately before the open flame device is used and at least once per hour while the device is used in the space. Testing shall be conducted more frequently if conditions present in the enclosed space indicate that once per hour is insufficient to detect hazardous accumulations of flammable gases or vapors.

Note: See the definition of hazardous atmosphere for guidance in determining whether or not a given concentration of a substance is considered to be hazardous.

(f) *Excavations.* Excavation operations shall comply with Subpart P of Part § 1926 of this chapter.

(g) *Personal protective equipment*

(g)(1) *General.* Personal protective equipment shall meet the requirements of Subpart I of this Part.

(g)(2) *Fall protection.*

(g)(2)(i) Personal fall arrest equipment shall meet the requirements of Subpart M of Part § 1926 of this Chapter.

(g)(2)(ii) Body belts and safety straps for work positioning shall meet the requirements of § 1926.959 of this Chapter.

(g)(2)(iii) Body belts, safety straps, lanyards, lifelines, and body harnesses shall be inspected before use each day to determine that the equipment is in safe working condition. Defective equipment may not be used.

(g)(2)(iv) Lifelines shall be protected against being cut or abraded.

(g)(2)(v) Fall arrest equipment, work positioning equipment, or travel restricting equipment shall be used by employees working at elevated locations more than 4 feet (1.2 m) above the ground on poles, towers, or similar structures if other fall protection has not been provided. Fall protection equipment is not required to be used by a qualified employee climbing or changing location on poles, towers, or similar structures, unless conditions, such as, but not limited to, ice, high winds, the design of the structure (for example, no provision for holding on with hands), or the presence of contaminants on the structure, could cause the employee to lose his or her grip or footing.

Note 1: This paragraph applies to structures that support overhead electric power generation, transmission, and distribution lines and equipment. It does not apply to portions of buildings, such as loading docks, to electric equipment, such as transformers and capacitors, nor to aerial lifts. Requirements for fall protection associated with walking and working surfaces are contained in Subpart D of this Part; requirements for fall protection associated with aerial lifts are contained in § 1910.67 of this Part.

Note 2: Employees undergoing training are not considered "qualified employees" for the purposes of this provision. Unqualified employees (including trainees) are required to use fall protection any time they are more than 4 feet (1.2 m) above the ground.

(g)(2)(vi) The following requirements apply to personal fall arrest systems:

(g)(2)(vi)(A) When stopping or arresting a fall, personal fall arrest systems shall limit the maximim arresting force on an employee to 900 pounds (4 kN) if used with a body belt.

(g)(2)(vi)(B) When stopping or arresting a fall, personal fall arrest systems shall limit the maximum arresting force on an employee to 1800 pounds (8 kN) if used with a body harness.

(g)(2)(vi)(C) Personal fall arrest systems shall be rigged such that an employee can neither free fall more than 6 feet (1.8 m) nor contact any lower level.

(g)(2)(vii) If vertical lifelines or droplines are used, not more than one employee may be attached to any one lifeline.

(g)(2)(viii) Snaphooks may not be connected to loops made in webbing-type lanyards.

(g)(2)(ix) Snaphooks may not be connected to each other.

(h) *Ladders, platforms, step bolts, and manhole steps.*

(h)(1) *General.* Requirements for ladders contained in Subpart D of this Part apply, except as specifically noted in paragraph (h)(2) of this section.

(h)(2) Special ladders and platforms. Portable ladders and platforms used on structures or conductors in conjunction with overhead line work need not meet paragraphs (d)(2)(i) and (d)(2)(iii) of § 1910.25 of this Part or paragraph (c)(3)(iii) of § 1910.26 of this Part. However, these ladders and platforms shall meet the following requirements:

(h)(2)(i) Ladders and platforms shall be secured to prevent their becoming accidentally dislodged.

(h)(2)(ii) Ladders and platforms may not be loaded in excess of the working loads for which they are designed.

(h)(2)(iii) Ladders and platforms may be used only in applications for which they were designed.

(h)(2)(iv) In the configurations in which they are used, ladders and platforms shall be capable of supporting without failure at least 2.5 times the maximum intended load.

(h)(3) *Conductive ladders.* Portable metal ladders and other portable conductive ladders may not be used near exposed energized lines or equipment. However, in specialized high-voltage work, conductive ladders shall be used where the employer can demonstrate that nonconductive ladders would present a greater hazard than conductive ladders.

(i) *Hand and portable power tools*

(i)(1) *General.* Paragraph (i)(2) of this section applies to electric equipment connected by cord and plug. Paragraph (i)(3) of this section applies to portable and vehicle-mounted generators used to supply cord- and plug-connected equipment. Paragraph (i)(4) of this section applies to hydraulic and pneumatic tools.

(i)(2) *Cord- and plug-connected equipment*

(i)(2)(i) Cord- and plug-connected equipment supplied by premises wiring is covered by Subpart S of this Part.

(i)(2)(ii) Any cord- and plug-connected equipment supplied by other than premises wiring shall comply with one of the following in lieu of § 1910.243(a)(5) of this Part:

(i)(2)(ii)(A) It shall be equipped with a cord containing an equipment grounding conductor connected to the tool frame and to a means for grounding the other end (however, this option may not be used where the introduction of the ground into the work environment increases the hazard to an employee); or

(i)(2)(ii)(B) It shall be of the double-insulated type conforming to Subpart S of this Part; or

(i)(2)(ii)(C) It shall be connected to the power supply through an isolating transformer with an ungrounded secondary.

(i)(3) *Portable and vehicle-mounted generators.* Portable and vehicle-mounted generators used to supply cord- and plug-connected equipment shall meet the following requirements:

(i)(3)(i) The generator may only supply equipment located on the generator or the vehicle and cord- and plug-connected equipment through receptables mounted on the generator or the vehicle.

(i)(3)(ii) The non-current-carrying metal parts of equipment and the equipment grounding conductor terminals of the receptacles shall be bonded to the generator frame.

(i)(3)(iii) In the case of vehicle-mounted generators, the frame of the generator shall be bonded to the vehicle frame.

(i)(3)(iv) Any neutral conductor shall be bonded to the generator frame.

(i)(4) *Hydraulic and pneumatic tools*

(i)(4)(i) Safe operating pressures for hydraulic and pneumatic tools, hoses, valves, pipes, filters, and fittings may not be exceeded.

Note: If any hazardous defects are present, no operating pressure would be safe, and the hydraulic or pneumatic equipment involved may not be used. In the absence of defects, the maximum rated operating pressure is the maximum safe pressure.

(i)(4)(ii) A hydraulic or pneumatic tool used where it may contact exposed live parts shall be designed and maintained for such use.

(i)(4)(iii) The hydraulic system supplying a hydraulic tool used where it may contact exposed live parts shall provide protection against loss of insulating value for the voltage involved due to the formation of a partial vacuum in the hydraulic line.

Note: Hydraulic lines without check valves having a separation of more than 35 feet (10.7 m) between the oil reservoir and the upper end of the hydraulic system promote the formation of a partial vacuum.

(i)(4)(iv) A pneumatic tool used on energized electric lines or equipment or used where it may contact exposed live parts shall provide protection against the accumulation of moisture in the air supply.

(i)(4)(v) Pressure shall be released before connections are broken, unless quick acting, self-closing connectors are used. Hoses may not be kinked.

(i)(4)(vi) Employees may not use any part of their bodies to locate or attempt to stop a hydraulic leak.

(j) *Live-line tools*

(j)(1) *Design of tools.* Live-line tool rods, tubes, and poles shall be designed and constructed to withstand the following minimum tests:

(j)(1)(i) 100,000 volts per foot (3281 volts per centimeter) of length for 5 minutes if the tool is made of fiberglass-reinforced plastic (FRP), or

(j)(1)(ii) 75,000 volts per foot (2461 volts per centimeter) of length for 3 minutes if the tool is made of wood, or

(j)(1)(iii) Other tests that the employer can demonstrate are equivalent.

Note: Live-line tools using rod and tube that meet ASTM F711-89, Standard Specification for Fiberglass-Reinforced Plastic (FRP) Rod and Tube Used in Live-Line Tools, conform to paragraph (j)(1)(i) of his section.

(j)(2) *Condition of tools*

(j)(2)(i) Each live-line tool shall be wiped clean and visually inspected for defects before use each day.

(j)(2)(ii) If any defect or contamination that could adversely affect the insulating qualities or mechanical integrity of the live-line tool is present after wiping, the tool shall be removed from service and examined and tested according to paragraph (j)(2)(iii) of this section before being returned to service.

(j)(2)(iii) Live-line tools used for primary employee protection shall be removed from service every 2 years and whenever required under paragraph (j)(2)(ii) of this section for examination, cleaning, repair, and testing as follows:

(j)(2)(iii)(A) Each tool shall be thoroughly examined for defects.

(j)(2)(iii)(B) If a defect or contamination that could adversely affect the insulating qualities or mechanical integrity of the live-line tool is found, the tool shall be repaired and refinished or shall be permanently removed from service. If no such defects or contamination is found, the tool shall be cleaned and waxed.

(j)(2)(iii)(C) The tool shall be tested in accordance with paragraphs (j)(2)(iii)(D) and (j)(2)(iii)(E) of this section under the following conditions:

(j)(2)(iii)(C)(1) After the tool has been repaired or refinished; and

(j)(2)(iii)(C)(2) After the examination if repair or refinishing is not performed, unless the tool is made of FRP rod or foam-filled FRP tube and the employer can demonstrate that the tool has no defects that could cause it to fail in use.

(j)(2)(iii)(D) The test method used shall be designed to verify the tool's integrity along its entire working length and, if the tool is made of fiberglass-reinforced plastic, its integrity under wet conditions.

(j)(2)(iii)(E) The voltage applied during the tests shall be as follows:

(j)(2)(iii)(E)(1) 75,000 volts per foot (2461 volts per centimeter) of length for 1 minute if the tool is made of fiberglass, or

(j)(2)(iii)(E)(2) 50,000 volts per foot (1640 volts per centimeter) of length for 1 minute if the tool is made of wood, or

(j)(3) Other tests that the employer can demonstrate are equivalent.

Note: Guidelines for the examination, cleaning, repairing, and in-service testing of live-line tools are contained in the Institute of Electrical and Electronics Engineers Guide for In-Service Maintenance and Electrical Testing of Live-Line Tools, IEEE Std. 978-1984.

(k) *Materials handling and storage*

(k)(1) *General.* Material handling and storage shall conform to the requirements of Subpart N of this Part.

(k)(2) *Materials storage near energized lines or equipment*

(k)(2)(i) In areas not restricted to qualified persons only, materials or equipment may not be stored closer to energized lines or exposed energized parts of equipment than the following distances plus an amount providing for the maximum sag and side swing of all conductors and providing for the height and movement of material handling equipment:

(k)(2)(i)(A) For lines and equipment energized at 50 kV or less, the distance is 100 feet (305 cm).

(k)(2)(i)(B) For lines and equipment energized at more than 50 kV, the distance is 10 feet (305 cm) plus 4 inches (10 cm) for every 10 kV over 50 kV.

(k)(2)(ii) In areas restricted to qualified employees, material may not be stored within the working space about energized lines or equipment.

Note: Requirements for the size of the working space are contained in paragraphs (u)(1) and (v)(3) of this section.

(l) *Working on or near exposed energized parts.* This paragraph applies to work on exposed live parts, or near enough to them, to expose the employee to any hazard they present.

(l)(1) *General.* Only qualified employees may work on or with exposed energized lines or parts of equipment. Only qualified employees may work in areas containing unguarded, uninsulated energized lines or parts of equipment operating at 50 volts or more. Electric lines and equipment shall be considered and treated as energized unless the provisions of paragraph (d) or paragraph (m) of this section have been followed.

(l)(1)(i) Except as provided in paragraph (l)(I)(ii) of this section, at least two employees shall be present while the following types of work are being performed:

(l)(1)(i)(A) Installation, removal, or repair of lines that are energized at more than 600 volts,

(l)(1)(i)(B) Installation, removal, or repair of deenergized lines if an employee is exposed to contact with other parts energized at more than 600 volts,

(l)(1)(i)(C) Installation, removal or repair of equipment, such as transformers, capacitors, and regulators, if an employee is exposed to contact with parts energized at more than 600 volts,

(l)(1)(i)(D) Work involving the use of mechanical equipment, other than insulated aerial lifts, near parts energized at more than 600 volts, and

(l)(1)(i)(E) Other work that exposes an employee to electrical hazards greater than or equal to those posed by operations that are specifically listed in paragraphs (l)(I)(i)(A) through (l)(I)(i)(D) of this section.

(l)(1)(ii) Paragraph (l)(1)(i) of this section does not apply to the following operations:

(l)(1)(ii)(A) Routine switching of circuits, if the employer can demonstrate that conditions at the site allow this work to be performed safely,

(l)(1)(ii)(B) Work performed with live-line tools if the employee is positioned so that he or she is neither within reach of nor otherwise exposed to contact with energized parts, and

(l)(1)(ii)(C) Emergency repairs to the extent necessary to safeguard the general public.

(l)(2) *Minimum approach distances* The employer shall ensure that no employee approaches or takes any conductive object closer to exposed energized parts than set forth in Table R.6 through Table R.10, unless:

(l)(2)(i) The employee is insulated from the energized part (insulating gloves or insulating gloves and sleeves worn in accordance with paragraph (l)(3) of this section are considered insulation of the employee only with regard to the energized part upon which work is being performed), or

(l)(2)(ii) The energized part is insulated from the employee and from any other conductive object at a different potential, or

(l)(2)(iii) The employee is insulated from any other exposed conductive object, as during live-line bare-hand work.

TABLE R.6 AC Live-Line Work Minimum Approach Distance

Nominal voltage in kilovolts phase to phase	Distance			
	Phase to ground exposure		Phase to phase exposure	
	(ft-in)	(m)	(ft-in)	(m)
0.05 to 1.0	(4)	(4)	(4)	(4)
1.1 to 15.0	2–1	0.64	2–2	0.66
15.1 to 36.0	2–4	0.72	2–7	0.77
36.1 to 46.0	2–7	0.77	2–10	0.85
46.1 to 72.5	3–0	0.90	3–6	1.05
72.6 to 121	3–2	0.95	4–3	1.29
138 to 145	3–7	1.09	4–11	1.50
161 to 169	4–0	1.22	5–8	1.71
230 to 242	5–3	1.59	7–6	2.27
345 to 362	8–6	2.59	12–6	3.80
500 to 550	11–3	3.42	18–1	5.50
765 to 800	14–11	4.53	26–0	7.91

[1]These distances take into consideration the highest switching surge an employee will be exposed to on any system with air as the insulating medium and the maximum voltages shown.

[2]The clear live-line tool distance shall equal or exceed the values for the indicated voltage ranges.

[3]See Appendix B to this section for information on how the minimum approach distances listed in the tables were derived.

[4]Avoid contact.

TABLE R.7 AC Live-Line Work Minimum Approach Distance with Overvoltage Factor Phase-to-Ground Exposure

Maximum anticipated per-unit transient over voltage	Distance in feet-inches						
	Maximum phase-to-phase voltage in kilovolts						
	121	145	169	242	362	552	800
1.5	....	....	....	....	....	6–0	9–8
1.6	....	....	....	....	....	6–6	10–8
1.7	....	....	....	....	....	7–0	11–8
1.8	....	....	....	....	....	7–7	12–8
1.9	....	....	....	....	....	8–1	13–9
2.0	2–5	2–9	3–0	3–10	5–3	8–9	14–11
2.1	2–6	2–10	3–2	4–0	5–5	9–4	.....
2.2	2–7	2–11	3–3	4–1	5–9	9–11	.....
2.3	2–8	3–0	3–4	4–3	6–1	10–6	.....
2.4	2–9	3–1	3–5	4–5	6–4	11–3	.....
2.5	2–9	3–2	3–6	4–6	6–8	.....	.....
2.6	2–10	3–3	3–8	4–8	7–1	.....	.....
2.7	2–11	3–4	3–9	4–10	7–5	.....	.....
2.8	3–0	3–5	3–10	4–11	7–9	.....	.....
2.9	3–1	3–6	3–11	5–1	8–2	.....	.....
3.0	3–2	3–7	4–0	5–3	8–6	.....	.....

Note 1: The distance specified in this table may be applied only where the maximum anticipated per-unit transient overvoltage has been determined by engineering analysis and has been supplied by the employer. Table R.6 applies otherwise.

Note 2: The distances specified in the table are the air, bare-band, and live-line tool distances.

Note 3: See Appendix B to this section for information on how the minimum approach distances listed in the tables were derived and on how to calculate revised minimum approach distances based on the control of transient overvoltages.

Note: Paragraphs (u)(5)(i) and (v)(5)(i) of this section contain requirements for the guarding and isolation of live parts. Parts of electric circuits that meet these two provisions are not considered as "exposed" unless a guard is removed or an employee enters the space intended to provide isolation from the live parts.

(l)(3) *Type of insulation.* If the employee is to be insulated from energized parts by the use of insulating gloves (under paragraph (l)(2)(i) of this section), insulating sleeves shall also be used. However, insulating sleeves need not be used under the following conditions:

(l)(3)(i) If exposed energized parts on which work is not being performed are insulated from the employee and

(l)(3)(ii) If such insulation is placed from a position not exposing the employee's upper arm to contact with other energized parts.

(l)(4) *Working position.* The employer shall ensure that each employee, to the extent that other safety-related conditions at the worksite permit, works in a position from which a slip or shock will not bring the employee's body into contact with exposed, uninsulated parts energized at a potential different from the employee.

(l)(5) *Making connections.* The employer shall ensure that connections are made as follows:

(l)(5)(i) In connecting deenergized equipment or lines to an energized circuit by means of a conducting wire or device, an employee shall first attach the wire to the deenergized part;

TABLE R.8 AC Live-Line Work Minimum Approach Distance with Overvoltage Factor Phase-to-Phase Exposure

Maximum anticipated per-unit transient over voltage	Distance in feet-inches						
	Maximum phase-to-phase voltage in kilovolts						
	121	145	169	242	362	552	800
1.5	....	....	....	....	....	7–4	12–1
1.6	....	....	....	....	....	8–9	14–6
1.7	....	....	....	....	....	10–2	17–2
1.8	....	....	....	....	....	11–7	19–11
1.9	....	....	....	....	....	13–2	22–11
2.0	3–7	4–1	4–8	6–1	8–7	14–10	26–0
2.1	3–7	4–2	4–9	6–3	8–10	15–7	.....
2.2	3–8	4–3	4–10	6–4	9–2	16–4	.....
2.3	3–9	4–4	4–11	6–6	9–6	17–2	.....
2.4	3–10	4–5	5–0	6–7	9–11	18–1	.....
2.5	3–11	4–6	5–2	6–9	10–4	.....	.....
2.6	4–0	4–7	5–3	6–11	10–9	.....	.....
2.7	4–1	4–8	5–4	7–0	11–2	.....	.....
2.8	4–1	4–9	5–5	7–2	11–7	.....	.....
2.9	4–2	4–10	5–6	7–4	12–1	.....	.....
3.0	4–3	4–11	5–8	7–6	12–6	.....	.....

Note 1: The distance specified in this table may be applied only where the maximum anticipated per-unit transient overvoltage has been determined by engineering analysis and has been supplied by the employer. Table R.6 applies otherwise.

Note 2: The distances specified in this table are the air, bare-band, and live-line tool distances.

Note 3: See Appendix B to this section for information on how the minimum approach distances listed in the tables were derived and on how to calculate revised minimum approach distances based on the control of transient overvoltages.

TABLE R.9 DC Live-Line Work Minimum Approach Distance with Overvoltage Factor

Maximum anticipated per- unit transient overvoltage	Distance in feet-inches				
	Maximum line-to-ground voltage in kilovolts				
	250	400	500	600	750
1.5 or lower	3–8	5–3	6–9	8–7	11–10
1.6	3–10	5–7	7–4	9–5	13–1
1.7	4–1	6–0	7–11	10–3	14–4
1.8	4–3	6–5	8–7	11–2	15–9

Note 1: The distance specified in this table may be applied only where the maximum anticipated per-unit transient overvoltage has been determined by engineering analysis and has been supplied by the employer. However, if the transient overvoltage factor is not known, a factor of 1.8 shall be assumed.

Note 2: The distances specified in this table are the air, bare-hand, and live-line tool distances.

TABLE R.10 Altitude Correction Factor

Altitude		
ft	m	Correction Factor
3000	900	1.00
4000	1200	1.02
5000	1500	1.05
6000	1800	1.08
7000	2100	1.11
8000	2400	1.14
9000	2700	1.17
10000	3000	1.20
12000	3600	1.25
14000	4200	1.30
16000	4800	1.35
18000	5400	1.39
20000	6000	1.44

Note: If the work is performed at elevations greater than 3000 ft (900 m) above mean sea level, the minimum approach distance shall be determined by multiplying the distances in Table R.6 through Table R.9 by the correction factor corresponding to the altitude at which work is performed.

(l)(5)(ii) When disconnecting equipment or lines from an energized circuit by means of a conducting wire or device, an employee shall remove the source end first; and

(l)(5)(iii) When lines or equipment are connected to or disconnected from energized circuits, loose conductors shall be kept away from exposed energized parts.

(l)(6) Apparel

(l)(6)(i) When work is performed within reaching distance of exposed energized parts of equipment, the employer shall ensure that each employee removes or renders nonconductive all exposed conductive articles, such as key or watch chains, rings, or wrist watches or bands, unless such articles do not increase the hazards associated with contact with the energized parts.

(l)(6)(ii) The employer shall train each employee who is exposed to the hazards of flames or electric arcs in the hazards involved.

(l)(6)(iii) The employer shall ensure that each employee who is exposed to the hazards of flames or electric arcs does not wear clothing that, when exposed to flames or electric arcs, could increase the extent of injury that would be sustained by the employee.

Note: Clothing made from the following types of fabrics, either alone or in blends, is prohibited by this paragraph, unless the employer can demonstrate that the fabric has been treated to withstand the conditions that may be encountered or that the clothing is worn in such a manner as to eliminate the hazard involved: acetate, nylon, polyester, rayon.

(l)(7) *Fuse handling*. When fuses must be installed or removed with one or both terminals energized at more than 300 volts or with exposed parts energized at more than 50 volts, the employer shall ensure that tools or gloves rated for the voltage are used. When expulsion-type fuses are installed with one or both terminals energized at more than 300 volts, the employer shall ensure that each employee wears eye protection meeting the requirements of Subpart I of this Part, uses a tool rated for the voltage, and is clear of the exhaust path of the fuse barrel.

(l)(8) *Covered (noninsulated) conductors.* The requirements of this section which pertain to the hazards of exposed live parts also apply when work is performed in the proximity of covered (noninsulated) wires.

(l)(9) *Noncurrent-carrying metal parts.* Noncurrent-carrying metal parts of equipment or devices, such as transformer cases and circuit breaker housings, shall be treated as energized at the highest voltage to which they are exposed, unless the employer inspects the installation and determines that these parts are grounded before work is performed.

(l)(10) *Opening circuits under load.* Devices used to open circuits under load conditions shall be designed to interrupt the current involved.

(m) *Deenergizing lines and equipment for employee protection*

(m)(1) *Application.* Paragraph (m) of this section applies to the deenergizing of transmission and distribution lines and equipment for the purpose of protecting employees. Control of hazardous energy sources used in the generation of electric energy is covered in paragraph (d) of this section. Conductors and parts of electric equipment that have been deenergized under procedures other than those required by paragraph (d) or (m) of this section, as applicable, shall be treated as energized.

(m)(2) *General*

(m)(2)(i) If a system operator is in charge of the lines or equipment and their means of disconnection, all of the requirements of paragraph (m)(3) of this section shall be observed, in the order given.

(m)(2)(ii) If no system operator is in charge of the lines or equipment and their means of disconnection, one employee in the crew shall be designated as being in charge of the clearance. All of the requirements of paragraph (m)(3) of this section apply, in the order given, except as provided in paragraph (m)(2)(iii) of this section. The employee in charge of the clearance shall take the place of the system operator, as necessary.

(m)(2)(iii) If only one crew will be working on the lines or equipment and if the means of disconnection is accessible and visible to and under the sole control of the employee in charge of the clearance, paragraphs (m)(3)(i), (m)(3)(iii), (m)(3)(iv), (m)(3)(viii), and (m)(3)(xii) of this section do not apply. Additionally, tags required by the remaining provisions of paragraph (m)(3) of this section need not be used.

(m)(2)(iv) Any disconnecting means that are accessible to persons outside the employer's control (for example, the general public) shall be rendered inoperable while they are open for the purpose of protecting employees.

(m)(3) *Deenergizing lines and equipment*

(m)(3)(i) A designated employee shall make a request of the system operator to have the particular section of line or equipment deenergized. The designated employee becomes the employee in charge (as this term is used in paragraph (m)(3) of this section) and is responsible for the clearance.

(m)(3)(ii) All switches, disconnectors, jumpers, taps, and other means through which known sources of electric energy may be supplied to the particular lines and equipment to be deenergized shall be opened. Such means shall be rendered inoperable, unless its design does not so permit, and tagged to indicate that employees are at work.

(m)(3)(iii) Automatically and remotely controlled switches that could cause the opened disconnecting means to close shall also be tagged at the point of control. The automatic or remote control feature shall be rendered inoperable, unless its design does not so permit.

(m)(3)(iv) Tags shall prohibit operation of the disconnecting means and shall indicate that employees are at work.

(m)(3)(v) After the applicable requirements in paragraphs (m)(3)(i) through (m)(3)(iv) of this section have been followed and the employee in charge of the work has been given a clearance by the system operator, the lines and equipment to be worked shall be tested to ensure that they are deenergized.

(m)(3)(vi) Protective grounds shall be installed as required by paragraph (n) of this section.

(m)(3)(vii) After the applicable requirements of paragraphs (m)(3)(i) through (m)(3)(vi) of this section have been followed, the lines and equipment involved may be worked as deenergized.

(m)(3)(viii) If two or more independent crews will be working on the same lines or equipment, each crew shall independently comply with the requirements in paragraph (m)(3) of this section.

(m)(3)(ix) To transfer the clearance, the employee in charge (or, if the employee in charge is forced to leave the worksite due to illness or other emergency, the employee's supervisor) shall inform the system operator; employees in the crew shall be informed of the transfer; and the new employee in charge shall be responsible for the clearance.

(m)(3)(x) To release a clearance, the employee in charge shall:

(m)(3)(x)(A) Notify employees under his or her direction that the clearance is to be released;

(m)(3)(x)(B) Determine that all employees in the crew are clear of the lines and equipment;

(m)(3)(x)(C) Determine that all protective grounds installed by the crew have been removed; and

(m)(3)(x)(D) Report this information to the system operator and release the clearance.

(m)(3)(xi) The person releasing a clearance shall be the same person that requested the clearance, unless responsibility has been transferred under paragraph (m)(3)(ix) of this section.

(m)(3)(xii) Tags may not be removed unless the associated clearance has been released under paragraph (m)(3)(x) of this section.

(m)(3)(xiii) Only after all protective grounds have been removed, after all crews working on the lines or equipment have released their clearances, after all employees are clear of the lines and equipment, and after all protective tags have been removed from a given point of disconnection, may action be initiated to reenergize the lines or equipment at that point of disconnection.

(n) *Grounding for the protection of employees.*

(n)(1) *Application.* Paragraph (n) of this section applies to the grounding of transmission and distribution lines and equipment for the purpose of protecting employees. Paragraph (n)(4) of this section also applies to the protective grounding of other equipment as required elsewhere in this section.

(n)(2) *General.* For the employee to work lines or equipment as deenergized, the lines or equipment shall be deenergized under the provisions of paragraph (m) of this section and shall be grounded as specified in paragraphs (n)(3) through (n)(9) of this section. However, if the employer can demonstrate that installation of a ground is impracticable or that the conditions resulting from the installation of a ground would present greater hazards than working without grounds, the lines and equipment may be treated as deenergized provided all of the following conditions are met:

(n)(2)(i) The lines and equipment have been deenergized under the provision of paragraph (m) of this section.

(n)(2)(ii) There is no possibility of contact with another energized source.

(n)(2)(iii) The hazard of induced voltage is not present.

(n)(3) *Equipotential zone.* Temporary protective grounds shall be placed at such locations and arranged in such a manner as to prevent each employee from being exposed to hazardous differences in electrical potential.

(n)(4) *Protective grounding equipment*

(n)(4)(i) Protective grounding equipment shall be capable of conducting the maximum fault current that could flow at the point of grounding for the time necessary to clear the fault. This equipment shall have an ampacity greater than or equal to that of No. 2 AWG copper.

Note: Guidelines for protective grounding equipment are contained in American Society for Testing and Materials Standard Specifications for Temporary Grounding Systems to be Used on De-Energized Electric Power Lines and Equipment, ASTM F855–1990.

(n)(4)(ii) Protective grounds shall have an impedance low enough to cause immediate operation of protective devices in case of accidental energizing of the lines or equipment.

(n)(5) *Testing.* Before any ground is installed, lines and equipment shall be tested and found absent of nominal voltage, unless a previously installed ground is present.

(n)(6) *Order of connection.* When a ground is to be attached to a line or to equipment, the ground-end connection shall be attached first, and then the other end shall be attached by means of a live-line tool.

(n)(7) *Order of removal.* When a ground is to be removed, the grounding device shall be removed from the line or equipment using a live-line tool before the ground-end connection is removed.

(n)(8) *Additional precautions.* When work is performed on a cable at a location remote from the cable terminal, the cable may not be grounded at the cable terminal if there is a possibility of hazardous transfer of potential should a fault occur.

(n)(9) *Removal of grounds for test.* Grounds may be removed temporarily during tests. During the test procedure, the employer shall ensure that each employee uses insulating equipment and is isolated from any hazards involved, and the employer shall institute any additional measures as may be necessary to protect each exposed employee in case the previously grounded lines and equipment become energized.

(o) *Testing and test facilities*

(o)(1) *Application.* Paragraph (o) of this section provides for safe work practices for high-voltage and high-power testing performed in laboratories, shops, and substations, and in the field and on electric transmission and distribution lines and equipment. It applies only to testing involving interim measurements utilizing high voltage, high power, or combinations of both, and not to testing involving continuous measurements as in routine metering, relaying, and normal line work.

Note: Routine inspection and maintenance measurements made by qualified employees are considered to be routine line work and are not included in the scope of paragraph (o) of this section, as long as the hazards related to the use of intrinsic high-voltage or high-power sources require only the normal precautions associated with routine operation and maintenance work required in the other paragraphs of this section. Two typical examples of such excluded test work procedures are "phasing-out" testing and testing for a "no-voltage" condition.

(o)(2) *General requirements*

(o)(2)(i) The employer shall establish and enforce work practices for the protection of each worker from the hazards of high-voltage or high-power testing at all test areas, temporary and permanent. Such work practices shall include, as a minimum, test area guarding, grounding, and the safe use of measuring and control circuits. A means providing for periodic safety checks of field test areas shall also be included. (See paragraph (o)(6) of this section.)

(o)(2)(ii) Employees shall be trained in safe work practices upon their initial assignment to the test area, with periodic reviews and updates provided as required by paragraph (a)(2) of this section.

(o)(3) *Guarding of test areas*

(o)(3)(i) Permanent test areas shall be guarded by walls, fences, or barriers designed to keep employees out of the test areas.

(o)(3)(ii) In field testing, or at a temporary test site where permanent fences and gates are not provided, one of the following means shall be used to prevent unauthorized employees from entering:

(o)(3)(ii)(A) The test area shall be guarded by the use of distinctively colored safety tape that is supported approximately waist high and to which safety signs are attached.

(o)(3)(ii)(B) The test area shall be guarded by a barrier or barricade that limits access to the test area to a degree equivalent, physically and visually, to the barricade specified in paragraph (o)(3)(ii)(A) of this section, or

(o)(3)(ii)(C) The test area shall be guarded by one or more test observers stationed so that the entire area can be monitored.

(o)(3)(iii) The barriers required by paragraph (o)(3)(ii) of this section shall be removed when the protection they provide is no longer needed.

(o)(3)(iv) Guarding shall be provided within test areas to control access to test equipment or to apparatus under test that may become energized as part of the testing by either direct or inductive coupling, in order to prevent accidental employee contact with energized parts.

(o)(4) *Grounding practices*

(o)(4)(i) The employer shall establish and implement safe grounding practices for the test facility.

(o)(4)(i)(A) All conductive parts accessible to the test operator during the time the equipment is operating at high voltage shall be maintained at ground potential except for portions of the equipment that are isolated from the test operator by guarding.

(o)(4)(i)(B) Wherever ungrounded terminals of test equipment or apparatus under test may be present, they shall be treated as energized until determined by tests to be deenergized.

(o)(4)(ii) Visible grounds shall be applied, either automatically or manually with properly insulated tools, to the high-voltage circuits after they are deenergized and before work is performed on the circuit or item or apparatus under test. Common ground connections shall be solidly connected to the test equipment and the apparatus under test.

(o)(4)(iii) In high-power testing, an isolated ground-return conductor system shall be provided so that no intentional passage of current, with its attendant voltage rise, can occur in the ground grid or in the earth. However, an isolated ground-return conductor need not be provided if the employer can demonstrate that both the following conditions are met:

(o)(4)(iii)(A) An isolated ground-return conductor cannot be provided due to the distance of the test site from the electric energy source, and

(o)(4)(iii)(B) Employees are protected from any hazardous step and touch potentials that may develop during the test.

Note: See Appendix C to this section for information on measures that can be taken to protect employees from hazardous step and touch potentials.

(o)(4)(iv) In tests in which grounding of test equipment by means of the equipment grounding conductor located in the equipment power cord cannot be used due to increased hazards to test personnel or the prevention of satisfactory measurements, a ground that the employer can demonstrate affords equivalent safety shall be provided, and the safety ground shall be clearly indicated in the test set-up.

(o)(4)(v) When the test area is entered after equipment is deenergized, a ground shall be placed on the high-voltage terminal and any other exposed terminals.

(o)(4)(v)(A) High capacitance equipment or apparatus shall be discharged through a resistor rated for the available energy.

(o)(4)(v)(B) A direct ground shall be applied to the exposed terminals when the stored energy drops to a level at which it is safe to do so.

(o)(4)(vi) If a test trailer or test vehicle is used in field testing, its chassis shall be grounded. Protection against hazardous touch potentials with respect to the vehicle, instrument panels, and other conductive parts accessible to employees shall be provided by bonding, insulation, or isolation.

(o)(5) *Control and measuring circuits*

(o)(5)(i) Control wiring, meter connections, test leads and cables may not be run from a test area unless they are contained in a grounded metallic sheath and terminated in a grounded metallic enclosure or unless other precautions are taken that the employer can demonstrate as ensuring equivalent safety.

(o)(5)(ii) Meters and other instruments with accessible terminals or parts shall be isolated from test personnel to protect against hazards arising from such terminals and parts becoming energized during testing. If this isolation is provided by locating test equipment in metal compartments with viewing windows, interlocks shall be provided to interrupt the power supply if the compartment cover is opened.

(o)(5)(iii) The routing and connections of temporary wiring shall be made secure against damage, accidental interruptions and other hazards. To the maximum extent possible, signal, control, ground, and power cables shall be kept separate.

(o)(5)(iv) If employees will be present in the test area during testing, a test observer shall be present. The test observer shall be capable of implementing the immediate deenergizing of test circuits for safety purposes.

(o)(6) *Safety check*

(o)(6)(i) Safety practices governing employee work at temporary or field test areas shall provide for a routine check of such test areas for safety at the beginning of each series of tests.

(o)(6)(ii) The test operator in charge shall conduct these routine safety checks before each series of tests and shall verify at least the following conditions:

(o)(6)(ii)(A) That barriers and guards are in workable condition and are properly placed to isolate hazardous areas;

(o)(6)(ii)(B) That system test status signals, if used, are in operable condition;

(o)(6)(ii)(C) That test power disconnects are clearly marked and readily available in an emergency;

(o)(6)(ii)(D) That ground connections are clearly identifiable;

(o)(6)(ii)(E) That personal protective equipment is provided and used as required by Subpart I of this Part and by this section; and

(o)(6)(ii)(F) That signal, ground, and power cables are properly separated.

(p) *Mechanical equipment*

(p)(1) *General requirements*

(p)(1)(i) The critical safety components of mechanical elevating and rotating equipment shall receive a thorough visual inspection before use on each shift.

Note: Critical safety components of mechanical elevating and rotating equipment are components whose failure would result in a free fall or free rotation of the boom.

(p)(1)(ii) No vehicular equipment having an obstructed view to the rear may be operated on off-highway jobsites where any employee is exposed to the hazards created by the moving vehicle, unless:

(p)(1)(ii)(A) The vehicle has a reverse signal alarm audible above the surrounding noise level, or

(p)(1)(ii)(B) The vehicle is backed up only when a designated employee signals that it is safe to do so.

(p)(1)(iii) The operator of an electric line truck may not leave his or her position at the controls while a load is suspended, unless the employer can demonstrate that no employee (including the operator) might be endangered.

(p)(1)(iv) Rubber-tired, self-propelled scrapers, rubber-tired front-end loaders, rubber-tired dozers, wheel-type agricultural and industrial tractors, crawler-type tractors, crawler-type loaders, and motor graders, with or without attachments, shall have roll-over protective structures that meet the requirements of Subpart W of Part § 1926 of this chapter.

(p)(2) *Outriggers*

(p)(2)(i) Vehiclular equipment, if provided with outriggers, shall be operated with the outriggers extended and firmly set as necessary for the stability of the specific configuration of the equipment. Outriggers may not be extended or retracted outside of clear view of the operator unless all employees are outside the range of possible equipment motion.

(p)(2)(ii) If the work area or the terrain precludes the use of outriggers, the equipment may be operated only within its maximum load ratings for the particular configuration of the equipment without outriggers.

(p)(3) *Applied loads.* Mechanical equipment used to lift or move lines or other material shall be used within its maximum load rating and other design limitations for the conditions under which the work is being performed.

(p)(4) *Operations near energized lines or equipment*

(p)(4)(i) Mechanical equipment shall be operated so that the minimum approach distances of Table R.6 through Table R.10 are maintained for exposed energized lines and equipment. However, the insulated portion of an aerial lift operated by a qualified employee in the lift is exempt from this requirement.

(p)(4)(ii) A designated employee other than the equipment operator shall observe the approach distance to exposed lines and equipment and give timely warnings before the minimum approach distance required by paragraph (p)(4)(i) is reached, unless the employer can demonstrate that the operator can accurately determine that the minimum approach distance is being maintained.

(p)(4)(iii) If, during operation of the mechanical equipment, the equipment could become energized, the operation shall also comply with at least one of paragraphs (p)(4)(iii)(A) through (p)(4)(iii)(C) of this section.

(p)(4)(iii)(A) The energized lines exposed to contact shall be covered with insulating protective material that will withstand the type of contact that might be made during the operation.

(p)(4)(iii)(B) The equipment shall be insulated for the voltage involved. The equipment shall be positioned so that its uninsulated portions cannot approach the lines or equipment any closer than the minimum approach distances specified in Table R.6 through Table R.10.

(p)(4)(iii)(C) Each employee shall be protected from hazards that might arise from equipment contact with the energized lines. The measures used shall ensure that employees will not be exposed to hazardous differences in potential. Unless the employer can demonstrate that the methods in use protect each employee from the hazards that might arise if the equipment contacts the energized line, the measures used shall include all of the following techniques:

(p)(4)(iii)(C)(1) Using the best available ground to minimize the time the lines remain energized.

(p)(4)(iii)(C)(2) Bonding equipment together to minimize potential differences,

(p)(4)(iii)(C)(3) Providing ground mats to extend areas of equipotential, and

(p)(4)(iii)(C)(4) Employing insulating protective equipment or barricades to guard against any remaining hazardous potential differences.

Note: Appendix C to this section contains information on hazardous step and touch potentials and on methods of protecting employees from hazards resulting from such potentials.

(q) *Overhead lines.* This paragraph provides additional requirements for work performed on or near overhead lines and equipment.

(q)(1) *General*

(q)(1)(i) Before elevated structures, such as poles or towers, are subjected to such stresses as climbing or the installation or removal of equipment may impose, the employer shall ascertain that the structures are capable of sustaining the additional or unbalanced stresses. If the pole or other structure cannot withstand the loads which will be imposed, it shall be braced or otherwise supported so as to prevent failure.

Note: Appendix D to this section contains test methods that can be used in ascertaining whether a wood pole is capable of sustaining the forces that would be imposed by an employee climbing the pole. This paragraph also requires the employer to ascertain that the pole can sustain all other forces that will be imposed by the work to be performed.

(q)(1)(ii) When poles are set, moved, or removed near exposed energized overhead conductors, the pole may not contact the conductors.

(q)(1)(iii) When a pole is set, moved, or removed near an exposed energized overhead conductor, the employer shall ensure that each employee wears electrical protective equipment or uses insulated devices when handling the pole and that no employee contacts the pole with uninsulated parts of his or her body.

(q)(1)(iv) To protect employees from falling into holes into which poles are to be placed, the holes shall be attended by employees or physically guarded whenever anyone is working nearby.

(q)(2) *Installing and removing overhead lines.* The following provisions apply to the installation and removal of overhead conductors or cable.

(q)(2)(i) The employer shall use the tension stringing method, barriers, or other equivalent measures to minimize the possibility that conductors and cables being installed or removed will contact energized power lines or equipment.

(q)(2)(ii) The protective measures required by paragraph (p)(4)(iii) of this section for mechanical equipment shall also be provided for conductors, cables, and pulling and tensioning equipment when the conductor or cable is being installed or removed close enough to energized conductors that any of the following failures could energize the pulling or tensioning equipment or the wire or cable being installed or removed:

(q)(2)(ii)(A) *Failure of the pulling or tensioning equipment.*

(q)(2)(ii)(B) *Failure of the wire or cable being pulled*, or

(q)(2)(ii)(C) *Failure of the previously installed lines or equipment.*

(q)(2)(iii) If the conductors being installed or removed cross over energized conductors in excess of 600 volts and if the design of the circuit-interrupting devices protecting the lines so permits, the automatic-reclosing feature of these devices shall be made inoperative.

(q)(2)(iv) Before lines are installed parallel to existing energized lines, the employer shall make a determination of the approximate voltage to be induced in the new lines, or work shall proceed on the assumption that the induced voltage is hazardous. Unless the employer can demonstrate that the lines being installed are not subject to the induction of a hazardous voltage or unless the lines are treated as energized, the following requirements also apply:

(q)(2(iv)(A) Each bare conductor shall be grounded in increments so that no point along the conductor is more than 2 miles (3.22 km) from a ground.

(q)(2)(iv)(B) The grounds required in paragraph (q)(2(iv)(A) of this section shall be left in place until the conductor installation is completed between dead ends.

(q)(2)(iv)(C) The grounds required in paragraph (q)(2)(iv)(A) of this section shall be removed as the last phase of aerial cleanup.

(q)(2)(iv)(D) If employees are working on bare conductors, grounds shall also be installed at each location where these employees are working, and grounds shall be installed at all open dead-end or catch-off points or the next adjacent structure.

(q)(2)(iv)(E) If two bare conductors are to be spliced, the conductors shall be bonded and grounded before being spliced.

(q)(2)(v) Reel handling equipment, including pulling and tensioning devices, shall be in safe operating condition and shall be leveled and aligned.

(q)(2)(vi) Load ratings of stringing lines, pulling lines, conductor grips, load-bearing hardware and accessories, rigging, and hoists may not be exceeded.

(q)(2)(vii) Pulling lines and accessories shall be repaired or replaced when defective.

(q)(2)(viii) Conductor grips may not be used on wire rope, unless the grip is specifically designed for this application.

(q)(2)(ix) Reliable communications, through two-way radios or other equivalent means, shall be maintained between the reel tender and the pulling rig operator.

(q)(2)(x) The pulling rig may only the operated when it is safe to do so.

Note: Examples of unsafe conditions include employees in locations prohibited by paragraph (q)(2)(xi) of this section, conductor and pulling line hang-ups, and slipping of the conductor grip.

(q)(2)(xi) While the conductor or pulling line is being pulled (in motion) with a power-driven device, employees are not permitted directly under overhead operations or on the cross arm, except as necessary to guide the stringing sock or board over or through the stringing sheave.

(q)(3) *Live-line bare-hand work.* In addition to other applicable provisions contained in this section, the following requirements apply to live-line bare-hand work:

(q)(3)(i) Before using or supervising the use of the live-line bare-hand technique on energized circuits, employees shall be trained in the technique and in the safety requirements of paragraph (q)(3) of this section. Employees shall receive refresher training as required by paragraph (a)(2) of this section.

(q)(3)(ii) Before any employee uses the live-line bare-hand technique on energized high-voltage conductors or parts, the following information shall be ascertained:

(q)(3)(ii)(A) The nominal voltage rating of the circuit on which the work is to be performed.

(q)(3)(ii)(B) The minimum approach distances to ground of lines and other energized parts on which work is to be performed, and

(q)(3)(ii)(C) The voltage limitations of equipment to be used.

(q)(3)(iii) The insulated equipment, insulated tools, and aerial devices and platforms used shall be designed, tested, and intended for live-line bare-hand work. Tools and equipment shall be kept clean and dry while they are in use.

(q)(3)(iv) The automatic-reclosing feature of circuit-interrupting devices protecting the lines shall be made inoperative, if the design of the devices permits.

(q)(3)(v) Work may not be performed when adverse weather conditions would make the work hazardous even after the work practices required by this section are employed. Additionally, work may not be performed when winds reduce the phase-to-phase or phase-to-ground minimum approach distances at the work location below that specified in paragraph (q)(3)(xiii) of this section, unless the grounded objects and other lines and equipment are covered by insulating guards.

Note: Thunderstorms in the immediate vicinity, high winds, snow storms, and ice storms are examples of adverse weather conditions that are presumed to make live-line bare-hand work too hazardous to perform safely.

(q)(3)(vi) A conductive bucket liner or other conductive device shall be provided for bonding the insulated aerial device to the energized line or equipment.

(q)(3)(vi)(A) The employee shall be connected to the bucket liner or other conductive device by the use of conductive shoes, leg clips, or other means.

(q)(3)(vi)(B) Where differences in potentials at the worksite pose a hazard to employees, electrostatic shielding designed for the voltage being worked shall be provided.

(q)(3)(vii) Before the employee contacts the energized part, the conductive bucket liner or other conductive device shall be bonded to the energized conductor by means of a positive connection. This connection shall remain attached to the energized conductor until the work on the energized circuit is completed.

(q)(3)(viii) Aerial lifts to be used for live-line bare-hand work shall have dual controls (lower and upper) as follows:

(q)(3)(viii)(A) The upper controls shall be within easy reach of the employee in the bucket. On a two-bucket-type lift, across to the controls shall be within easy reach from either bucket.

(q)(3)(viii)(B) The lower set of controls shall be located near the base of the boom, and they shall be so designed that they can override operation of the equipment at any time.

(q)(3)(ix) Lower (ground-level) lift controls may not be operated with an employee in the lift, except in case of emergency.

(q)(3)(x) Before employees are elevated into the work position, all controls (ground level and bucket) shall be checked to determine that they are in proper working condition.

(q)(3)(xi) Before the boom of an aerial lift is elevated, the body of the truck shall be grounded, or the body of the truck shall be barricaded and treated as energized.

(q)(3)(xii) A boom-current test shall be made before work is started each day, each time during the day when higher voltage is encountered, and when changed conditions indicate a need for an additional test. This test shall consist of placing the bucket in contact with an energized source equal to the voltage to be encountered for a minimum of 3 minutes. The leakage current may not exceed 1 microampere per kilovolt of nominal phase-to-ground voltage. Work from the aerial lift shall be immediately suspended upon indication of a malfunction in the equipment.

(q)(3)(xiii) The minimum approach distances specified in Table R.6 through Table R. 10 shall be maintained from all grounded objects and from lines and equipment at a potential different from that to which the live-line bare-hand equipment is bonded, unless such grounded objects and other lines and equipment are covered by insulating guards.

(q)(3)(xiv) While an employee is approaching, leaving, or bonding to an energized circuit, the minimum approach distances in Table R.6 through Table R.10 shall be maintained between the employee and any grounded parts, including the lower boom and portions of the truck.

(q)(3)(xv) While the bucket is positioned alongside an energized bushing or insulator string, the phase-to-ground minimum approach distances to Table R.6 through Table R.10 shall be maintained between all parts of the bucket and the grounded end of the bushing or insulator string or any other grounded surface.

(q)(3)(xvi) Hand lines may not be used between the bucket and the boom or between the bucket and the ground. However, non-conductive-type hand lines may be used from conductor to ground if not supported from the bucket. Ropes used for live-line bare-hand work may not be used for other purposes.

(q)(3)(xvii) Uninsulated equipment or material may not be passed between a pole or structure and an aerial lift while an employee working from the bucket is bonded to an energized part.

(q)(3)(xviii) A minimum approach distance table reflecting the minimum approach distances listed in Table R.6 through Table R.10 shall be printed on a plate of durable nonconductive material. This table shall be mounted so as to be visible to the operator of the boom.

(q)(3)(xix) A non-conductive measuring device shall be readily accessible to assist employees in maintaining the required minimum approach distance.

(q)(4) *Towers and structures*. The following requirements apply to work performed on towers or other structures which support overhead lines.

(q)(4)(i) The employer shall ensure that no employee is under a tower or structure while work is in progress, except where the employer can demonstrate that such a working position is necessary to assist employees working above.

(q)(4)(ii) Tag lines or other similar devices shall be used to maintain control of tower sections being raised or positioned, unless the employer can demonstrate that the use of such devices would create a greater hazard.

(q)(4)(iii) The loadline may not be detached from a member or section until the load is safely secured.

(q)(4)(iv) Except during emergency restoration procedures, work shall be discontinued when adverse weather conditions would make the work hazardous in spite of the work practices required by this section.

Note: Thunderstorms in the immediate vicinity, high winds, snow storms, and ice storms are examples of adverse weather conditions that are presumed to make this work too hazardous to perform, except under emergency conditions.

(r) *Line-clearance tree trimming operations.* This paragraph provides additional requirements for line-clearance tree-trimming operations and for equipment used in these operations.

(r)(1) *Electrical hazards.* This paragraph does not apply to qualified employees.

(r)(1)(i) Before an employee climbs, enters, or works around any tree, a determination shall be made of the nominal voltage of electric power lines posing a hazard to employees. However, a determination of the maximum nominal voltage to which an employee will be exposed may be made instead, if all lines are considered as energized at this maximum voltage.

(r)(1)(ii) There shall be a second line-clearance tree trimmer within normal (that is, unassisted) voice communication under any of the following conditions:

(r)(1)(ii)(A) If a line-clearance tree trimmer is to approach more closely than 10 feet (305 cm) any conductor or electric apparatus energized at more than 750 volts or

(r)(1)(ii)(B) If branches or limbs being removed are closer to lines energized at more than 750 volts than the distances listed in Table R.6, Table R.9, and Table R.10 or

(r)(1)(ii)(C) If roping is necessary to remove branches or limbs from such conductors or apparatus.

(r)(1)(iii) Line-clearance tree trimmers shall maintain the minimum approach distances from energized conductors given in Table R.6, Table R.9, and Table R.10.

(r)(1)(iv) Branches that are contacting exposed energized conductors or equipment or that are within the distances specified in Table R.6, Table R.9, and Table R.10 may be removed only through the use of insulating equipment.

Note: A tool constructed of a material that the employer can demonstrate has insulating qualities meeting paragraph (j)(1) of this section is considered as insulated under this paragraph if the tool is clean and dry.

(r)(1)(v) Ladders, platforms, and aerial devices may not be brought closer to an energized part than the distances listed in Table R.6, Table R.9, and Table R.10.

(r)(1)(vi) Line-clearance tree-trimming work may not be performed when adverse weather conditions make the work hazardous in spite of the work practices required by this section. Each employee performing line-clearance tree trimming work in the aftermath of a storm or under similar emergency conditions shall be trained in the special hazards related to this type of work.

Note: Thunderstorms in the immediate vicinity, high winds, snow storms, and ice storms are examples of adverse weather conditions that are presumed to make line-clearance tree trimming work too hazardous to perform safely.

(r)(2) *Brush chippers*

(r)(2)(i) Brush chippers shall be equipped with a locking device in the ignition system.

(r)(2)(ii) Access panels for maintenance and adjustment of the chipper blades and associated drive train shall be in place and secure during operation of the equipment.

(r)(2)(iii) Brush chippers are equipped with a mechanical infeed system shall be equipped with an infeed hopper of length sufficient to prevent employees from contacting the blades or knives of the machine during operation.

(r)(2)(iv) Trailer chippers detached from trucks shall be chocked or otherwise secured.

(r)(2)(v) Each employee in the immediate area of an operating chipper feed table shall wear personal protective equipment as required by Subpart I of this Part.

(r)(3) *Sprayers and related equipment*

(r)(3)(i) Walking and working surfaces of sprayers and related equipment shall be covered with slip-resistant material. If slipping hazards cannot be eliminated, slip-resistant footwear or handrails and stair rails meeting the requirements of Subpart D may be used instead of slip-resistant material.

(r)(3)(ii) Equipment on which employees stand to spray while the vehicle is in motion shall be equipped with guardrails around the working area. The guardrail shall be constructed in accordance with Subpart D of this Part.

(r)(4) *Stump cutters*

(r)(4)(i) Stump cutters shall be equipped with enclosures or guards to protect employees.

(r)(4)(ii) Each employee in the immediate area of stump grinding operations (including the stump cutter operator) shall wear personal protective equipment as required by Subpart I of this Part.

(r)(5) *Gasoline-engine power saws.* Gasoline-engine power saw operations shall meet the requirements of § 1910.266(c)(5) of tl part and the following:

(r)(5)(i) Each power saw weighing more than 15 pounds (6.8 kilograms, service weight) that is used in trees shall be supported by a separate line, except when work is performed from an aerial lift and except during topping or removing operations where no supporting limb will be available.

(r)(5)(ii) Each power saw shall be equipped with a control that will return the saw to idling speed when released.

(r)(5)(iii) Each power saw shall be equipped with a clutch and shall be so adjusted that the clutch will not engage the chain drive at idling speed.

(r)(5)(iv) A power saw shall be started on the ground or where it is otherwise firmly supported. Drop starting of saws over 15 pounds (6.8 kg) is permitted outside of the bucket of an aerial lift only if the area below the lift is clear of personnel.

(r)(5)(v) A power saw engine may be started and operated only when all employees other than the operator are clear of the saw.

(r)(5)(vi) A power saw may not be running when the saw is being carried up into a tree by an employee.

(r)(5)(vii) Power saw engines shall be stopped for all cleaning, refueling, adjustments, and repairs to the saw or motor, except as the manufacturer's servicing procedures require otherwise.

(r)(6) *Backpack power units for use in pruning and clearing*

(r)(6)(i) While a backpack power unit is running, no one other than the operator may be within 10 feet (305 cm) of the cutting head of a brush saw.

(r)(6)(ii) A backpack power unit shall be equipped with a quick shutoff switch readily accessible to the operator.

(r)(6)(iii) Backpack power unit engines shall be stopped for all cleaning, refueling, adjustments, and repairs to the saw or motor, except as the manufacturer's servicing procedures require otherwise.

(r)(7) *Rope*

(r)(7)(i) Climbing ropes shall be used by employees working aloft in trees. These ropes shall have a minimum diameter of 0.5 inch (1.2 cm) with a minimum breaking strength of 2300 pounds (10.2 kN). Synthetic rope shall have elasticity of not more than 7 percent.

(r)(7)(ii) Rope shall be inspected before each use and, if unsafe (for example, because of damage or defect), may not be used.

(r)(7)(iii) Rope shall be stored away from cutting edges and sharp tools. Rope contact with corrosive chemicals, gas, and oil shall be avoided.

(r)(7)(iv) When stored, rope shall be coiled and piled, or shall be suspended, so that air can circulate through the coils.

(r)(7)(v) Rope ends shall be secured to prevent their unraveling.

(r)(7)(vi) Climbing rope may not be spliced to effect repair.

(r)(7)(vii) A rope that is wet, that is contaminated to the extent that its insulating capacity is impaired, or that is otherwise not considered to be insulated for the voltage involved may not be used near exposed energized lines.

(r)(8) *Fall protection.* Each employee shall be tied in with a climbing rope and safety saddle when the employee is working above the ground in a tree, unless he or she is ascending into the tree.

(s) *Communication facilities*

(s)(1) *Microwave transmission*

(s)(1)(i) The employer shall ensure that no employee looks into an open waveguide or antenna that is connected to an energized microwave source.

(s)(1)(ii) If the electromagnetic radiation level within an accessible area associated with microwave communications systems exceeds the radiation protection guide given in § 1910.97(a)(2) of this Part, the area shall be posted with the warning symbol described in § 1910.97(a)(3) of this Part. The lower half of the warning symbol shall include the following statements or ones that the employer can demonstrate are equivalent:

Radiation in this area may exceed hazard limitations and special precautions are required. Obtain specific instruction before entering.

(s)(1)(iii) When an employee works in an area where the electromagnetic radiation could exceed the radiation protection guide, the employer shall institute measures that ensure that the employee's exposure is not greater than that permitted by that guide. Such measures may include administrative and engineering controls and personal protective equipment.

(s)(2) *Power line carrier.* Power line carrier work, including work on equipment used for coupling carrier current to power line conductors, shall be performed in accordance with the requirements of this section pertaining to work on energized lines.

(t) *Underground electrical installations.* This paragraph provides additional requirements for work on underground electrical installations.

(t)(1) *Access.* A ladder or other climbing device shall be used to enter and exit a manhole or subsurface vault exceeding 4 feet (122 cm) in depth. No employee may climb into or out of a manhole or vault by stepping on cables or hangers.

(t)(2) *Lowering equipment into manholes.* Equipment used to lower materials and tools into manholes or vaults shall be capable of supporting the weight to be lowered and shall be checked for defects before use. Before tools or material are lowered into the opening for a manhole or vault, each employee working in the manhole or vault shall be clear of the area directly under the opening.

(t)(3) *Attendants for manholes*

(t)(3)(i) While work is being performed in a manhole containing energized electric equipment, an employee with first aid and CPR training meeting paragraph (b)(1) of this section shall be available on the surface in the immediate vicinity to render emergency assistance.

(t)(3)(ii) Occasionally, the employee on the surface may briefly enter a manhole to provide assistance, other than emergency.

Note 1: An attendant may also be required under paragraph (e)(7) of this section. One person may serve to fulfill both requirements. However, attendants required under paragraph (e)(7) of this section are not permitted to enter the manhole.

Note 2: Employees entering manholes containing unguarded, uninsulated energized lines or parts of electric equipment operating at 50 volts or more are required to be qualified under paragraph (1)(1) of this section.

(t)(3)(iii) For the purpose of inspection, housekeeping, taking readings, or similar work, an employee working alone may enter, for brief periods of time, a manhole where energized cables or equipment are in service, if the employer can demonstrate that the employee will be protected from all electrical hazards.

(t)(3)(iv) Reliable communications, through two-way radios or other equivalent means, shall be maintained among all employees involved in the job.

(t)(4) *Duct rods.* If duct rods are used, they shall be installed in the direction presenting the least hazard to employees. An employee shall be stationed at the far end of the duct line being rodded to ensure that the required minimum approach distances are maintained.

(t)(5) *Multiple cables.* When multiple cables are present in a work area, the cable to be worked shall be identified by electrical means, unless its identity is obvious by reason of distinctive appearance or location or by other readily apparent means of identification. Cables other than the one being worked shall be protected from damage.

(t)(6) *Moving cables.* Energized cables that are to be moved shall be inspected for defects.

(t)(7) *Defective cables.* Where a cable in a manhole has one or more abnormalities that could lead to or be an indication of an impending fault, the defective cable shall be deenergized before any employee may work in the manhole, except when service load conditions and a lack of feasible alternatives require that the cable remain energized. In that case, employees may enter the manhole provided they are protected from the possible effects of a failure by shields or other devices that are capable of containing the adverse effects of a fault in the joint.

Note: Abnormalities such as oil or compound leaking from cable or joints, broken cable sheaths or joint sleeves, hot localized surface temperatures of cables or joints, or joints that are swollen beyond normal tolerance are presumed to lead to or be an indication of an impending fault.

(t)(8) *Sheath continuity.* When work is performed on buried cable or on cable in manholes, metallic sheath continuity shall be maintained or the cable sheath shall be treated as energized.

(u) *Substations.* This paragraph provides additional requirements for substations and for work performed in them.

(u)(1) *Access and working space.* Sufficient access and working space shall be provided and maintained about electric equipment to permit ready and safe operation and maintenance of such equipment.

Note: Guidelines for the dimensions of access and workspace about electric equipment in substations are contained in American National Standard-National Electrical Safety Code, ANSI C2-1987. Installations meeting the ANSI provisions comply with paragraph (u)(1) of this section. An installation that does not conform to this ANSI standard will, nonetheless, be considered as complying with paragraph (u)(1) of this section if the employer can demonstrate that the installation provides ready and safe access based on the following evidence:

[1] That the installation conforms to the edition of ANSI C2 that was in effect at the time the installation was made,

[2] That the configuration of the installation enables employees to maintain the minimum approach distances required by paragraph (1)(2) of this section while they are working on exposed, energized parts, and

[3] That the precautions taken when work is performed on the installation provide protection equivalent to the protection that would be provided by access and working space meeting ANSI C2-1987.

(u)(2) *Draw-out-type circuit breakers.* When draw-out-type circuit breakers are removed or inserted, the breaker shall be in the open position. The control circuit shall also be rendered inoperative, if the design of the equipment permits.

(u)(3) *Substation fences.* Conductive fences around substations shall be grounded. When a substation fence is expanded or a section is removed, fence grounding continuity shall be maintained, and bonding shall be used to prevent electrical discontinuity.

(u)(4) *Guarding of rooms containing electric supply equipment.*

(u)(4)(i) Rooms and spaces in which electric supply lines or equipment are installed shall meet the requirements of paragraphs (u)(4)(ii) through (u)(4)(v) of this section under the following conditions:

(u)(4)(i)(A) If exposed live parts operating at 50 to 150 volts to ground are located within 8 feet of the ground or other working surface inside the room or space.

(u)(4)(i)(B) If live parts operating at 151 to 600 volts and located within 8 feet of the ground or other working surface inside the room or space are guarded only by location, as permitted under paragraph (u)(5)(i) of this section, or

(u)(4)(i)(C) If live parts operating at more than 600 volts are located within the room or space, unless:

(u)(4)(i)(C)(1) The live parts are enclosed within grounded, metal-enclosed equipment whose only openings are designed so that foreign objects inserted in these openings will be deflected from energized parts, or

(u)(4)(i)(C)(2) The live parts are installed at a height above ground and any other working surface that provides protection at the voltage to which they are energized corresponding to the protection provided by an 8-foot height at 50 volts.

(u)(4)(ii) The rooms and spaces shall be so enclosed within fences, screens, partitions, or walls as to minimize the possibility that unqualified persons will enter.

(u)(4)(iii) Signs warning unqualified persons to keep out shall be displayed at entrances to the rooms and spaces.

(u)(4)(iv) Entrances to rooms and spaces that are not under the observation of an attendant shall be kept locked.

(u)(4)(v) Unqualified persons may not enter the rooms or spaces while the electric supply lines or equipment are energized.

(u)(5) *Guarding of energized parts*

(u)(5)(i) Guards shall be provided around all live parts operating at more than 150 volts to ground without an insulating covering, unless the location of the live parts gives sufficient horizontal or vertical or a combination of these clearances to minimize the possibility of accidental employee contact.

Note: Guidelines for the dimensions of clearance distances about electric equipment in substations are contained in American National Standard-National Electrical Safety Code, ANSI C2-1987. Installations meeting the ANSI provisions comply with paragraph (u)(5)(i) of this section. An installation that does not conform to this ANSI standard will, nonetheless, be considered as complying with paragraph (u)(5)(i) of this section if the employer can demonstrate that the installation provides sufficient clearance based on the following evidence:

[1] That the installation conforms to the edition of ANSI C2 that was in effect at the time the installation was made,

[2] That each employee is isolated from energized parts at the point of closest approach, and

[3] That the precautions taken when work is performed on the installation provide protection equivalent to the protection that would be provided by horizontal and vertical clearances meeting ANSI C2-1987.

(u)(5)(ii) Except for fuse replacement and other necessary access by qualified persons, the guarding of energized parts within a compartment shall be maintained during operation and maintenance functions to prevent accidental contact with energized parts and to prevent tools or other equipment from being dropped on energized parts.

(u)(5)(iii) When guards are removed from energized equipment, barriers shall be installed around the work area to prevent employees who are not working on the equipment, but who are in the area, from contacting the exposed live parts.

(u)(6) *Substation entry*

(u)(6)(i) Upon entering an attended substation, each employee other than those regularly working in the station shall report his or her presence to the employee in charge in order to receive information on special system conditions affecting employee safety.

(u)(6)(ii) The job briefing required by paragraph (c) of this section shall cover such additional subjects as the location of energized equipment in or adjacent to the work area and the limits of any deenergized work area.

(v) *Power generation.* This paragraph provides additional requirements and related work practices for power generating plants.

(v)(1) *Interlocks and other safety devices*

(v)(1)(i) Interlocks and other safety devices shall be maintained in a safe, operable condition.

(v)(1)(ii) No interlock or other safety device may be modified to defeat its function, except for test, repair, or adjustment of the device.

(v)(2) *Changing brushes.* Before exciter or generator brushes are changed while the generator is in service, the exciter or generator field shall be checked to determine whether a ground condition exists. The brushes may not be changed while the generator is energized if a ground condition exists.

(v)(3) *Access and working space.* Sufficient access and working space shall be provided and maintained about electric equipment to permit ready and safe operation and maintenance of such equipment.

Note: Guidelines for the dimensions of access and workspace about electric equipment in generating stations are contained in American National Standard-National Electrical Safety Code, ANSI C2-1987. Installations meeting the ANSI provisions comply with paragraph (v)(3) of this section. An installation that does not conform to this ANSI standard will, nonetheless, be considered as complying with paragraph (v)(3) of this section if the employer can demonstrate that the installation provides ready and safe access based on the following evidence:

[1] That the installation conforms to the edition of ANSI C2 that was in effect at the time the installation was made,

[2] That the configuration of the installation enables employees to maintain the minimum approach distances required by paragraph (1)(2) of this section while they are working on exposed, energized parts, and

[3] That the precautions taken when work is performed on the installation provide protection equivalent to the protection that would be provided by access and working space meeting ANSI C2-1987.

(v)(4) *Guarding of rooms containing electric supply equipment.*

(v)(4)(i) Rooms and spaces in which electric supply lines or equipment are installed shall meet the requirements of paragraphs (v)(4)(ii) through (v)(4)(v) of this section under the following conditions:

(v)(4)(i)(A) If exposed live parts operating at 50 to 150 volts to ground are located within 8 feet of the ground or other working surface inside the room or space.

(v)(4)(i)(B) If live parts operating at 151 to 600 volts and located within 8 feet of the ground or other working surface inside the room or space are guarded only by location, as permitted under paragraph (v)(5)(i) of this section, or

(v)(4)(i)(C) If live parts operating at more than 600 volts are located within the room or space, unless:

(v)(4)(i)(C)(1) The live parts are enclosed within grounded, metal-enclosed equipment whose only openings are designed so that foreign objects inserted in these openings will be deflected from energized parts, or

(v)(4)(i)(C)(2) The live parts are installed at a height above ground and any other working surface that provides protection at the voltage to which they are energized corresponding to the protection provided by an 8-foot height at 50 volts.

(v)(4)(ii) The rooms and spaces shall be so enclosed within fences, screens, partitions, or walls as to minimize the possibility that unqualified persons will enter.

(v)(4)(iii) Signs warning unqualified persons to keep out shall be displayed at entrances to the rooms and spaces.

(v)(4)(iv) Entrances to rooms and spaces that are not under the observation of an attendant shall be kept locked.

(v)(4)(v) Unqualified persons may not enter the rooms or spaces while the electric supply lines or equipment are energized.

(v)(5) *Guarding of energized parts*

(v)(5)(i) Guards shall be provided around all live parts operating at more than 150 volts to ground without an insulating covering, unless the location of the live parts gives sufficient horizontal or vertical or a combination of these clearances to minimize the possibility of accidental employee contact.

Note: Guidelines for the dimensions of clearance distances about electric equipment in generating stations are contained in American National Standard-National Electrical Safety Code, ANSI C2-1987. Installations meeting the ANSI provisions comply with paragraph (v)(5)(i) of this section. An installation that does not conform to this ANSI standard will, nonetheless, be considered as complying with paragraph (v)(5)(i) of this section if the employer can demonstrate that the installation provides sufficient clearance based on the following evidence:

[1] That the installation conforms to the edition of ANSI C2 that was in effect at the time the installation was made,

[2] That each employee is isolated from energized parts at the point of closest approach, and

[3] That the precautions taken when work is performed on the installation provide protection equivalent to the protection that would be provided by horizontal and vertical clearances meeting ANSI C2-1987.

(v)(5)(ii) Except for fuse replacement and other necessary access by qualified persons, the guarding of energized parts within a compartment shall be maintained during operation and maintenance functions to prevent accidental contact with energized parts and to prevent tools or other equipment from being dropped on energized parts.

(v)(5)(iii) When guards are removed from energized equipment, barriers shall be installed around the work area to prevent employees who are not working on the equipment, but who are in the area, from contacting the exposed live parts.

(v)(6) *Water or steam spaces.* The following requirements apply to work in water and steam spaces associated with boilers:

(v)(6)(i) A designated employee shall inspect conditions before work is permitted and after its completion. Eye protection, or full face protection if necessary, shall be worn at all times when condenser, heater, or boiler tubes are being cleaned.

(v)(6)(ii) Where it is necessary for employees to work near tube ends during cleaning, shielding shall be installed at the tube ends.

(v)(7) *Chemical cleaning of boilers and pressure vessels.* The following requirements apply to chemical cleaning of boilers and pressure vessels:

(v)(7)(i) Areas where chemical cleaning is in progress shall be cordoned off to restrict access during cleaning. If flammable liquids, gases, or vapors or combustible materials will be used or might be produced during the cleaning process, the following requirements also apply:

(v)(7)(i)(A) The area shall be posted with signs restricting entry and warning of the hazards of fire and explosion; and

(v)(7)(i)(B) Smoking, welding, and other possible ignition sources are prohibited in these restricted areas.

(v)(7)(ii) The number of personnel in the restricted area shall be limited to those necessary to accomplish the task safely.

(v)(7)(iii) There shall be ready access to water or showers for emergency use.

Note: See § 1910.141 of this Part for requirements that apply to the water supply and to washing facilities.

(v)(7)(iv) Employees in restricted areas shall wear protective equipment meeting the requirements of Subpart 1 of this Part and including, but not limited to, protective clothing, boots, goggles, and gloves.

(v)(8) *Chlorine systems*

(v)(8)(i) Chlorine system enclosures shall be posted with signs restricting entry and warning of the hazard to health and the hazards of fire and explosion.

Note: See Subpart Z of this Part for requirements necessary to protect the health of employees from the effects of chlorine.

(v)(8)(ii) Only designated employees may enter the restricted area. Additionally, the number of personnel shall be limited to those necessary to accomplish the task safely.

(v)(8)(iii) Emergency repair kits shall be available near the shelter or enclosure to allow for the prompt repair of leaks in chlorine lines, equipment, or containers.

(v)(8)(iv) Before repair procedures are started, chlorine tanks, pipes, and equipment shall be purged with dry air and isolated from other sources of chlorine.

(v)(8)(v) The employer shall ensure that chlorine is not mixed with materials that would react with the chlorine in a dangerously exothermic or other hazardous manner.

(v)(9) *Boilers*

(v)(9)(i) Before internal furnace or ash hopper repair work is started, overhead areas shall be inspected for possible falling objects. If the hazard of falling objects exists, overhead protection such as planking or nets shall be provided.

(v)(9)(ii) When opening an operating boiler door, employees shall stand clear of the opening of the door to avoid the heat blast and gases which may escape from the boiler.

(v)(10) *Turbine generators*

(v)(10)(i) Smoking and other ignition sources are prohibited near hydrogen or hydrogen sealing systems, and signs warning of the danger of explosion and fire shall be posted.

(v)(10)(ii) Excessive hydrogen makeup or abnormal loss of pressure shall be considered as an emergency and shall be corrected immediately.

(v)(10)(iii) A sufficient quantity of inert gas shall be available to purge the hydrogen from the largest generator.

(v)(11) *Coal and ash handling*

(v)(11)(i) Only designated persons may operate railroad equipment.

(v)(11)(ii) Before a locomotive or locomotive crane is moved, a warning shall be given to employees in the area.

(v)(11)(iii) Employees engaged in switching or dumping cars may not use their feet to line up drawheads.

(v)(11)(iv) Drawheads and knuckles may not be shifted while locomotives or cars are in motion.

(v)(11)(v) When a railroad car is stopped for unloading, the car shall be secured from displacement that could endanger employees.

(v)(11)(vi) An emergency means of stopping dump operations shall be provided at railcar dumps.

(v)(11)(vii) The employer shall ensure that employees who work in coal- or ash-handling conveyor areas are trained and knowledgeable in conveyor operation and in the requirements of paragraphs (v)(11)(viii) through (v)(11)(xii) of this section.

(v)(11)(viii) Employees may not ride a coal- or ash-handling conveyor belt at any time. Employees may not cross over the conveyor belt, except at walkways, unless the conveyor's energy source has been deenerzied and has been locked out or tagged in accordance with paragraph (d) of this section.

(v)(11)(ix) A conveyor that could cause injury when started may not be started until personnel in the area are alerted by a signal or by a designated person that the conveyor is about to start.

(v)(11)(x) If a conveyor that could cause injury when started in automatically controlled or is controlled from a remote location, an audible device shall be provided that sounds an alarm that will be recognized by each employee as a warning that the conveyor will start and that can be clearly heard at all points along the conveyor where personnel may be present. The warning device shall be actuated by the device starting the conveyor and shall continue for a period of time before the conveyor starts that is long enough to allow employees to move clear of the conveyor system. A visual warning may be used in place of the audible device if the employer can demonstrate that it will provide an equally effective warning in the particular circumstances involved.

Exception: If the employer can demonstrate that the system's function would be seriously hindered by the required time delay, warning signs may be provided in place of the audible warning device. If the system was installed before January 31, 1995, warning signs may be provided in place of the audible warning device until such time as the conveyor or its control system is rebuilt or rewired. These warning signs shall be clear, concise, and legible and shall indicate that conveyors and allied equipment may be started at any time, that danger exists, and the personnel must keep clear. These warning signs shall be provided along the conveyor at areas not guarded by position or location.

(v)(11)(xi) Remotely and automatically controlled conveyors, and conveyors that have operating stations which are not manned or which are beyond voice and visual contact from drive areas, loading areas, transfer points, and other locations on the conveyor path not guarded by location, position, or guards shall be furnished with emergency stop buttons, pull cords, limit switches, or similar emergency stop devices. However, if the employer can demonstrate that the design, function, and operation of the conveyor do not expose an employee to hazards, an emergency stop device is not required.

(v)(11)(xi)(A) Emergency stop devices shall be easily identifiable in the immediate vicinity of such locations.

(v)(11)(xi)(B) An emergency stop device shall act directly on the control of the conveyor involved and may not depend on the stopping of any other equipment.

(v)(11)(xi)(C) Emergency stop devices shall be installed so that they cannot be overridden from other locations.

(v)(11)(xii) Where coal-handling operations may produce a combustible atmosphere from fuel sources or from flammable gases or dust, sources of ignition shall be eliminated or safely controlled to prevent ignition of the combustible atmosphere.

Note: Locations that are hazardous because of the presence of combustible dust are classified as Class II hazardous locations. See § 1910.307 of this Part.

(v)(11)(xiii) An employee may not work on or beneath overhanging coal in coal bunkers, coal silos, or coal storage areas, unless the employee is protected from all hazards posed by shifting coal.

(v)(11)(xiv) An employee entering a bunker or silo to dislodge the contents shall wear a body harness with lifeline attached. The lifeline shall be secured to a fixed support outside the bunker and shall be attended at all times by an employee located outside the bunker or facility.

(v)(12) *Hydroplants and equipment.* Employees working on or close to water gates, valves, intakes, forebays, flumes, or other locations where increased or decreased water flow or levels may pose a significant hazard shall be warned and shall vacate such dangerous areas before water flow changes are made.

(w) *Special conditions.*

(w)(1) *Capacitors.* The following additional requirements apply to work on capacitors and on lines connected to capacitors.

Note: See paragraphs (m) and (n) of this section for requirements pertaining to the deenergizing and grounding of capacitor installations.

(w)(1)(i) Before employees work on capacitors, the capacitors shall be disconnected from energized sources and, after a wait of at least 5 minutes from the time of disconnection, short-circuited.

(w)(1)(ii) Before the units are handled, each unit in series-parallel capacitor banks shall be short-circuited between all terminals and the capacitor case or its rack. If the cases of capacitors are on ungrounded substation racks, the racks shall be bonded to ground.

(w)(1)(iii) Any line to which capacitors are connected shall be short-circuited before it is considered deenergized.

(w)(2) *Current transformer secondaries.* The secondary of a current transformer may not be opened while the transformer is energized. If the primary of the current transformer cannot be deenergized before work is performed on an instrument, a relay, or other section of a current transformer secondary circuit, the circuit shall be bridged so that the current transformer secondary will not be opened.

(w)(3) *Series streetlighting*

(w)(3)(i) If the open-circuit voltage exceeds 600 volts, the series streetlighting circuit shall be worked in accordance with paragraph (q) or (t) of this section, as appropriate.

(w)(3)(ii) A series loop may only be opened after the streetlighting transformer has been deenergized and isolated from the source of supply or after the loop is bridged to avoid an open-circuit condition.

(w)(4) *Illumination.* Sufficient illumination shall be provided to enable the employee to perform the work safely.

(w)(5) *Protection against drowning*

(w)(5)(i) Whenever an employee may be pulled or pushed or may fall into water where the danger of drowning exists, the employee shall be provided with and shall use U.S. Coast Guard approved personal flotation devices.

(w)(5)(ii) Each personal flotation device shall be maintained in safe condition and shall be inspected frequently enough to ensure that it does not have rot, mildew, water saturation, or any other condition that could render the device unsuitable for use.

(w)(5)(iii) An employee may cross streams or other bodies of water only if a safe means of passage, such as a bridge, is provided.

(w)(6) *Employee protection in public work areas*

(w)(6)(i) Traffic control signs and traffic control devices used for the protection of employees shall meet the requirements of § 1926.200(g)(2) of this Chapter.

(w)(6)(ii) Before work is begun in the vicinity of vehicular or pedestrian traffic that may endanger employees, warning signs or flags and other traffic control devices shall be placed in conspicuous locations to alert and channel approaching traffic.

(w)(6)(iii) Where additional employee protection is necessary, barricades shall be used.

(w)(6)(iv) Excavated areas shall be protected with barricades.

(w)(6)((v) At night, warning lights shall be prominently displayed.

(w)(7) *Backfeed.* If there is a possibility of voltage backfeed from sources of cogeneration or from the secondary system (for example, backfeed from more than one energized phase feeding a common load), the requirements of paragraph (1) of this section apply if the lines or equipment are to be worked as energized, and the requirements of paragraphs (m) and (n) of this section apply if the lines or equipment are to be worked as deenergized.

(w)(8) *Lasers.* Laser equipment shall be installed, adjusted, and operated in accordance with § 1926.54 of this Chapter.

(w)(9) *Hydraulic fluids.* Hydraulic fluids used for the insulated sections of equipment shall provide insulation for the voltage involved.

(x) Definitions

Affected employee. An employee whose job requires him or her to operate or use a machine or equipment on which servicing or maintenance is being performed under lockout or tagout, or whose job requires him or her to work in an area in which such servicing or maintenance is being performed.

Attendant. An employee assigned to remain immediately outside the entrance to an enclosed or other space to render assistance as needed to employees inside the space.

Authorized employee. An employee who locks out or tags out machines or equipment in order to perform servicing or maintenance on that machine or equipment. An affected employee becomes an authorized employee when that employee's duties include performing servicing or maintenance covered under this section.

Automatic circuit recloser. A self-controlled device for interrupting and reclosing an alternating current circuit with a predetermined sequence of opening and reclosing followed by resetting, hold-closed, or lockout operation.

Barricade. A physical obstruction such as tapes, cones, or A-frame type wood or metal structures intended to provide a warning about and to limit access to a hazardous area.

Barrier. A physical obstruction which is intended to prevent contact with energized lines or equipment or to prevent unauthorized access to a work area.

Bond. The electrical interconnection of conductive parts designed to maintain a common electrical potential.

Bus. A conductor or a group of conductors that serve as a common connection for two or more circuits.

Bushing. An insulating structure, including a through conductor or providing a passageway for such a conductor, with provision for mounting on a barrier, conducting or otherwise, for the purposes of insulating the conductor from the barrier and conducting current from one side of the barrier to the other.

Cable. A conductor with insulation, or a stranded conductor with or without insulation and other coverings (single-conductor cable), or a combination of conductors insulated from one another (multiple-conductor cable).

Cable sheath. A conductive protective covering applied to cables.

Note: A cable sheath may consist of multiple layers of which one or more is conductive.

Circuit. A conductor or system of conductors through which an electric current is intended to flow.

Clearance (between objects). The clear distance between two objects measured surface to surface.

Clearance (for work). Authorization to perform specified work or permission to enter a restricted area.

Communication lines. (*See* Lines, communication.)

Conductor. A material, usually in the form of a wire, cable, or bus bar, used for carrying an electric current.

Covered conductor. A conductor covered with a dielectric having no rated insulating strength or having a rated insulating strength less than the voltage of the circuit in which the conductor is used.

Current-carrying part. A conducting part intended to be connected in an electric circuit to a source of voltage. Non-current-carrying parts are those not intended to be so connected.

Deenergized. Free from any electrical connection to a source of potential difference and from electric charge; not having a potential different from that of the earth.

Note: The term is used only with reference to current-carrying parts, which are sometimes energized (alive).

Designated employee (*designated person*). An employee (or person) who is designated by the employer to perform specific duties under the terms of this section and who is knowledgeable in the construction and operation of the equipment and the hazards involved.

Electric line truck. A truck used to transport personnel, tools, and material for electric supply line work.

Electric supply equipment. Equipment that produces, modifies, regulates, controls, or safeguards a supply of electric energy.

Electric supply lines. (*See* Lines, electric supply.)

Electric utility. An organization responsible for the installation, operation, or maintenance of an electric supply system.

Enclosed space. A working space, such as a manhole, vault, tunnel, or shaft, that has a limited means of egress or entry, that is designed for periodic employee entry under normal operating conditions, and that under normal conditions does not contain a hazardous atmosphere, but that may contain a hazardous atmosphere under abnormal conditions.

Note: Spaces that are enclosed but not designed for employee entry under normal operating conditions are not considered to be enclosed spaces for the purposes of this section. Similarly, spaces that are enclosed and that are expected to contain a hazardous atmosphere are not considered to be enclosed spaces for the purposes of this section. Such spaces meet the definition of permit spaces in § 1910.146 of this Part, and entry into them must be performed in accordance with that standard.

Energized (*alive, live*). Electrically connected to a source of potential difference, or electrically charged so as to have a potential significantly different from that of earth in the vicinity.

Energy isolating device. A physical device that prevents the transmission or release of energy, including, but not limited to, the following: a manually operated electric circuit breaker, a disconnect switch, a manually operated switch, a slide gate, a slip blind, a line valve, blocks, and any similar device with a visible indication of the position of the device. (Push buttons, selector switches, and other control-circuit-type devices are not energy isolating devices.)

Energy source. Any electrical, mechanical, hydraulic, pneumatic, chemical, nuclear, thermal, or other energy source that could cause injury to personnel.

Equipment (*electric*). A general term including material, fittings, devices, appliances, fixtures, apparatus, and the like used as part of or in connection with an electrical installation.

Exposed. Not isolated or guarded.

Ground. A conducting connection, whether intentional or accidental, between an electric circuit or equipment and the earth, or to some conducing body that serves in place of the earth.

Grounded. Connected to earth or to some conducting body that serves in place of the earth.

Guarded. Covered, fenced, enclosed, or otherwise protected, by means of suitable covers or casings, barrier rails or screens, mats, or platforms, designed to minimize the possibility, under normal conditions, of dangerous approach or accidental contact by persons or objects.

Note: Wires which are insulated, but not otherwise protected, are not considered as guarded.

Hazardous atmosphere means an atmosphere that may expose employees to the risk of death, incapacitation, impairment of ability to self-rescue (that is, escape unaided from an enclosed space), injury, or acute illness from one or more of the following causes:

(1) Flammable gas, vapor, or mist in excess of 10 percent of its lower flammable limit (LFL);

(2) Airborne combustible dust at a concentration that meets or exceeds its LFL

Note: This concentration may be approximated as a condition in which the dust obscures vision at a distance of 5 feet (1.52 m) or less.

(3) Atmospheric oxygen concentration below 19.5 percent or above 23.5 percent;

(4) Atmospheric concentration of any substance for which a dose or a permissible exposure limit is published in Subpart G, "Occupational Health and Environmental Control," or in Subpart Z, "Toxic and Hazardous Substances," of this Part and which could result in employee exposure in excess of its dose or permissible exposure limit;

Note: An atmospheric concentration of any substance that is not capable of causing death, incapacitation, impairment of ability to self-rescue, injury, or acute illness due to its health effects is not covered by this provision.

(5) Any other atmospheric condition that is immediately dangerous to life or health.

Note: For air contaminants for which OSHA has not determined a dose or permissible exposure limit, other sources of information, such as Material Safety Data Sheets that comply with the Hazard Communication Standard, § 1910.1200 of this Part, published information, and internal documents can provide guidance in establishing acceptable atmospheric conditions.

High-power tests. Tests in which fault currents, load currents, magnetizing currents, and line-dropping currents are used to test equipment, either at the equipment's rated voltage or at lower voltages.

High-voltage tests. Tests in which voltages of approximately 1000 volts are used as a practical minimum and in which the voltage source has sufficient energy to cause injury.

High wind. A wind of such velocity that the following hazards would be present:

[1] An employee would be exposed to being blown from elevated locations, or

[2] An employee or material handling equipment could lose control of material being handled, or

[3] An employee would be exposed to other hazards not controlled by the standard involved.

Note: Winds exceeding 40 miles per hour (64.4 kilometers per hour), or 30 miles per hour (48.3 kilometers per hour) if material handling is involved, are normally considered as meeting this criteria unless precautions are taken to protect employees from the hazardous effects of the wind.

Immediately dangerous to life or health (*IDLH*) means any condition that poses an immediate or delayed threat to life or that would cause irreversible adverse health effects or that would interfere with an individual's ability to escape unaided from a permit space.

Note: Some materials—hydrogen fluoride gas and cadmium vapor, for example—may produce immediate transient effects that, even if severe, may pass without medical attention, but are followed by sudden, possibly fatal collapse 12–72 hours after exposure. The victim "feels normal" from recovery from transient effects until collapse. Such materials in hazardous quantities are considered to be "immediately" dangerous to life or health.

Insulated. Separated from other conducting surfaces by a dielectric (including air space) offering a high resistance to the passage of current.

Note: When any object is said to be insulated, it is understood to be insulated for the conditions to which it is normally subjected. Otherwise, it is, within the purpose of this section, uninsulated.

Insulation (*cable*). That which is relied upon to insulate the conductor from other conductors or conducting parts or from ground.

Line-clearance tree trimmer. An employee who, through related training or on-the-job experience or both, is familiar with the special techniques and hazards involved in line-clearance tree trimming.

Note 1: An employee who is regularly assigned to a line-clearance tree-trimming crew and who is undergoing on-the-job training and who, in the course of such training, has demonstrated an ability to perform duties safely at his or her level of training and who is under the direct supervision of a line-clearance tree trimmer is considered to be a line-clearance tree trimmer for the performance of those duties.

Note 2: A line-clearance tree trimmer is not considered to be a "qualified employee" under this section unless he or she has the training required for a qualified employee under paragraph (a)(2)(ii) of this section. However, under the electrical safety-related work practices standard in Subpart S of this Part, a line-clearance tree trimmer is considered to be a "qualified employee." Tree trimming performed by such "qualified employees" is not subject to the electrical safety-related work practice requirements contained in § 1910.331 through § 1910.335 of this Part. (*See also* the note following § 1910.332(b)(3) of this Part for information regarding the training an employee must have to be considered a qualified employee under § 1910.331 through § 1910.335 of this part.)

Line-clearance tree trimming. The pruning, trimming, repairing, maintaining, removing, or clearing of trees or the cutting of brush that is within 10 feet (305 cm) of electric supply lines and equipment.

Lines. [1] *Communication lines.* The conductors and their supporting or containing structures which are used for public or private signal or communication service, and which operate at potentials not exceeding 400 volts to ground or 750 volts between any two points of the circuit, and the transmitted power of which does not exceed 150 watts. If the lines are operating at less than 150 volts, no limit is placed on the transmitted power of the system. Under certain conditions, communication cables may include communication circuits exceeding these limitations where such circuits are also used to supply power solely to communication equipment.

Note: Telephone, telegraph, railroad signal, data, clock, fire, police alarm, cable television, and other systems conforming to this definition are included. Lines used for signaling purposes, but not included under this definition, are considered as electric supply lines of the same voltage.

[2] *Electric supply lines.* Conductors used to transmit electric energy and their necessary supporting or containing structures. Signal lines of more than 400 volts are always supply lines within this section, and those of less than 400 volts are considered as supply lines, if so run and operated throughout.

Manhole. A subsurface enclosure which personnel may enter and which is used for the purpose of installing, operating, and maintaining submersible equipment or cable.

Manhole steps. A series of steps individually attached to or set into the walls of a manhole structure.

Minimum approach distance. The closest distance an employee is permitted to approach an energized or a grounded object.

Qualified employee (qualified person). One knowledgeable in the construction and operation of the electric power generation, transmission, and distribution equipment involved, along with the associated hazards.

Note 1: An employee must have the training required by paragraph (a)(2)(ii) of this section in order to be considered a qualified employee.

Note 2: Except under paragraph (g)(2)(v) of this section, an employee who is undergoing on-the-job training and who, in the course of such training, has demonstrated an ability to perform duties safely at his or her level of training and who is under the direct supervision of a qualified person is considered to be a qualified person for the performance of those duties.

Step bolt. A bolt or rung attached at intervals along a structural member and used for foot placement during climbing or standing.

Switch. A device for opening and closing or for changing the connection of a circuit. In this section, a switch is understood to be manually operable, unless otherwise stated.

System operator. A qualified person designated to operate the system or its parts.

Vault. An enclosure, above or below ground, which personnel may enter and which is used for the purpose of installing, operating, or maintaining equipment or cable.

Vented vault. A vault that has provision for air changes using exhaust flue stacks and low level air intakes operating on differentials of pressure and temperature providing for airflow which precludes a hazardous atmosphere from developing.

Voltage. The effective (rms) potential difference between any two conductors or between a conductor and ground. Voltages are expressed in nominal values unless otherwise indicated. The nominal voltage of a system or circuit is the value assigned to a system or circuit of a given voltage class for the purpose of convenient designation. The operating voltage of the system may vary above or below this value.

[59 FR 40672, Aug. 9, 1994; 59 FR 51672, Oct. 12, 1994]

APPENDIX A TO § 1910.269—FLOW CHARTS

This appendix presents information, in the form of flow charts, that illustrates the scope and application of § 1910.269. This appendix addresses the interface between § 1910.269 and Subpart S of this Part (Electrical), between § 1910.269 and § 1910.146 of this Part (Permit-required confined spaces), and between § 1910.269 and § 1910.147 of this Part (The control of hazardous energy (lockout//tagout)). These flow charts provide guidance for employers trying to implement the requirements of § 1910.269 in combination with other General Industry Standards contained in Part § 1910.

Appendix A.1 to § 1910.269—Application of § 1910.269 and Subpart S of this Part to Electrical Installations.

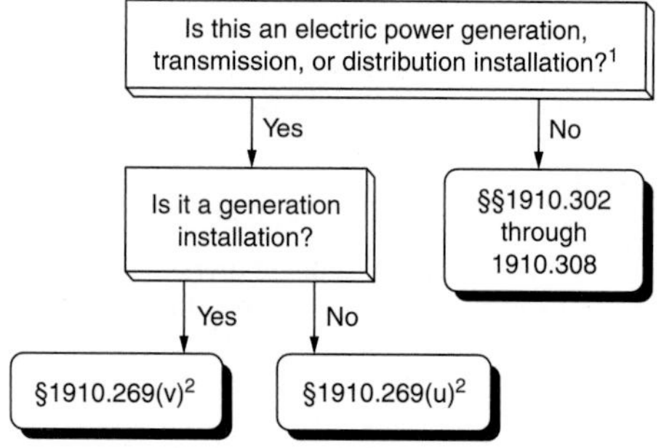

[1] Electrical installation design requirements only. See Appendix 1B for electrical safety-related work practices. Supplementary electric generating equipment that is used to supply a workplace for emergency, standby, or similar purposes only is not considered to be an electric power generation installation.

[2] See Table 1 of Appendix A-2 for requirements that can be met through compliance with subpart S.

Appendix A.2 to § 1910.269—Application of § 1910.269 and Subpart S to Electrical Safety-Related Work Practices.

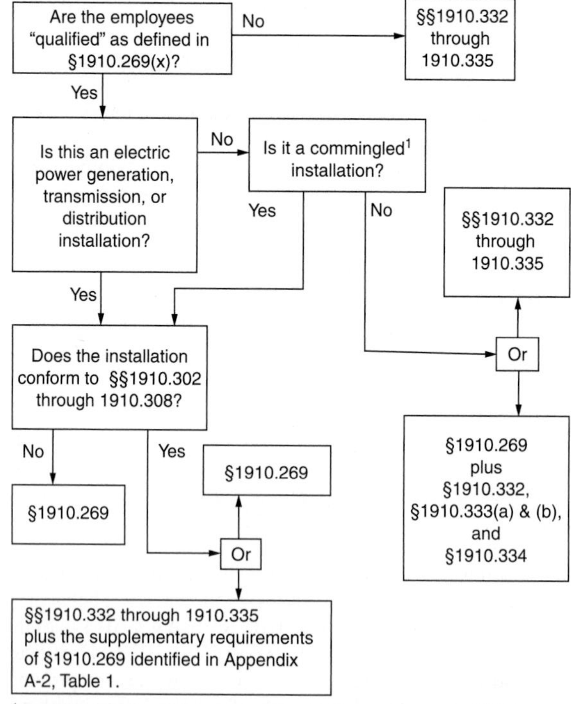

[1] Commingled to the extent that the electric power generation, transmission, or distribution installation poses the greater hazard.

APPENDIX A.2, TABLE 1 Electrical Safety-Related Work Practices in Section 1910.269

Compliance with subpart S is considered as compliance with § 1910.269[1]	Paragraphs that apply regardless of compliance with subpart S
(d), electric shock hazards only	(a)(2)[2] and (a)(3)[2].
(h)(3)	(b)[2].
(i)(2)	(c)[2].
(k)	(d), other than electric shock hazards.
(l)(1) through (l)(4), (l)(6)(i), and (l)(8) through (l)(10)	(e).
	(f).
(m)	(g).
(p)(4)	(h)(1) and (h)(2).
(s)(2)	(i)(3)[2] and (i)(4)[2].
(u)(1) and (u)(3) through (u)(5)	(j)[2].
(v)(3) through (v)(5)	(l)(5)[2], (l)(6)(iii)[2], (l)(6)(iii)[2], and (l)(7)[2].
(w)(1) and (w)(7)	(n)[2].
	(o)[2].
	(p)(1) through (p)(3).
	(q)[2].
	(r).
	(s)(1).
	(t)[2].
	(u)(2)[2] and (u)(6)[2].
	(v)(1), (v)(2)[2], and (v)(6) through (v)(12).
	(w)(2) through (w)(6)[2], (w)(8), and (w)(9)[2].

[1]If the electrical installation meets the requirements of §§ 1910.332 through 1910.308 of this Part, then the electrical installation and any associated electrical safety-related work practices conforming to §§ 1910.332 through 1910.335 of this Part are considered to comply with these provisions of § 1910.269 of this Part.

[2] These provisions include electrical safety requirements that must be met reqardless of compliance with Subpart S of this Part.

APPENDIX B TO § 1910.269—WORKING ON EXPOSED ENERGIZED PARTS

I. Introduction

Electric transmission and distribution line installations have been designed to meet National Electrical Safety Code (NESC), ANSI C2, requirements and to provide the level of line outage performance required by system reliability criteria. Transmission and distribution lines are also designed to withstand the maximum overvoltages expected to be impressed on the system. Such overvoltages can be caused by such conditions as switching surges, faults, or lightning. Insulator design and lengths and the clearances to structural parts (which, for low voltage through extra-high voltage, or EHV, facilities, are generally based on the performance of the line as a result of contamination of the insulation or during storms) have, over the years, come closer to the minimum approach distances used by workers (which are generally based on nonstorm conditions). Thus, as minimum approach (working) distances and structural distances (clearances) converge, it is increasingly important that basic considerations for establishing safe approach distances for performing work be understood by the designers and the operating and maintenance personnel involved.

Appendix A.3 to § 1910.269—Application of § 1910.269 and Subpart S of this Part to Tree-Trimming Operations.

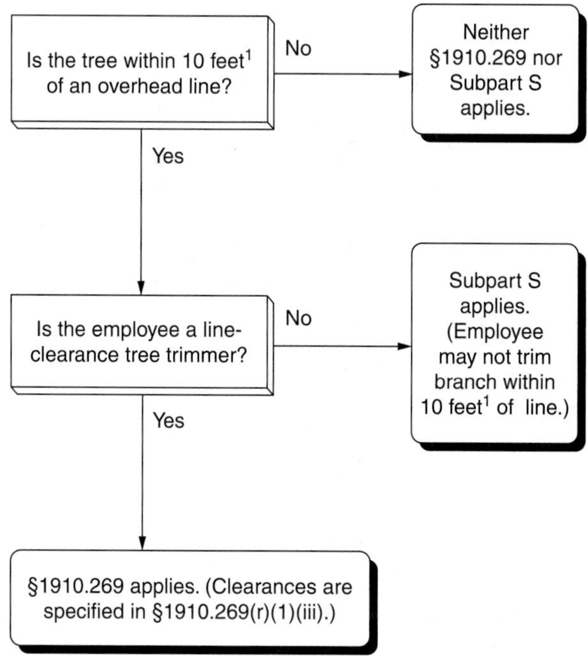

[1]10 feet plus 4 inches for every 10 kilovolts over 50 kilovolts.

The information in this appendix will assist employers in complying with the minimum approach distance requirements contained in paragraphs (1)(2) and (q)(3) of this section. The technical criteria and methodology presented herein is mandatory for employers using reduced minimum approach distances as permitted in Table R.7 and Table R.8. This appendix is intended to provide essential background information and technical criteria for the development or modification, if possible, of the safe minimum approach distances for electric transmission and distribution live-line work. The development of these safe distances must be undertaken by persons knowledgeable in the techniques discussed in this appendix and competent in the field of electric transmission and distribution system design.

II. General

A. Definitions

The following definitions from § 1910.269(x) relate to work on or near transmission and distribution lines and equipment and the electrical hazards they present.

Exposed. Not isolated or guarded.

Guarded. Covered, fenced, enclosed, or otherwise protected, by means of suitable covers or casings, barrier rails or screens, mats, or platforms, designed to minimize the possibility, under normal conditions, or dangerous approach or accidental contact by persons or objects.

Note: Wires which are insulated, but not otherwise protected, are not considered as guarded.

Insulated. Separated from other conducting surfaces by a dielectric (including air space) offering a high resistance to the passage of current.

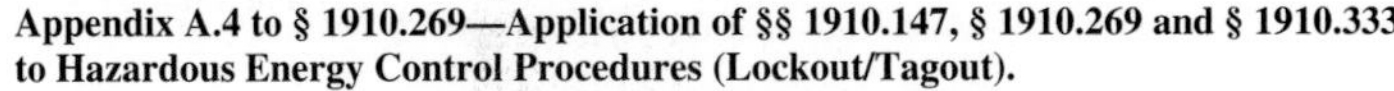

Appendix A.4 to § 1910.269—Application of §§ 1910.147, § 1910.269 and § 1910.333 to Hazardous Energy Control Procedures (Lockout/Tagout).

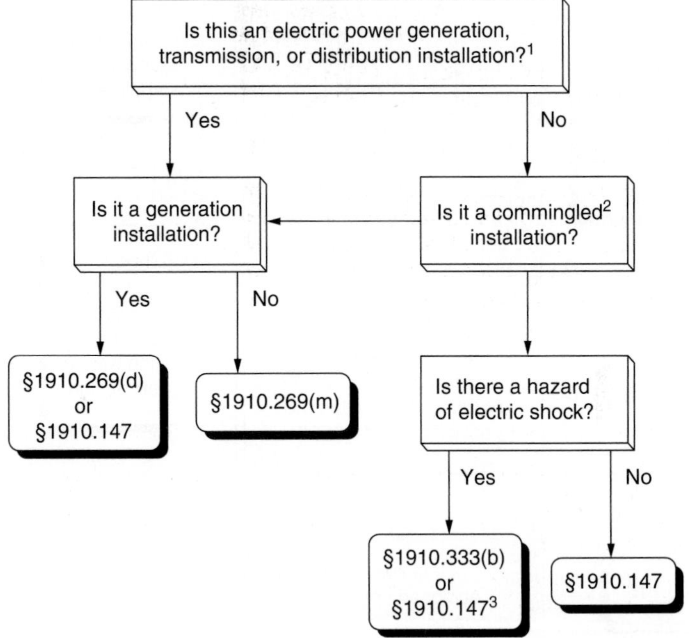

[1] If the installation conforms to §§ 1910.303 through 1910.308, the lockout and tagging procedures of 1910.333(b) may be followed for electric shock hazards.

[2] Commingled to the extent that the electric power generation, transmission, or distribution installation poses the greater hazard.

[3] §1910.333(b)(2)(iii)(D) and (b)(2)(iv)(B) still apply.

Note: When any object is said to be insulated, it is understood to be insulated for the conditions to which it is normally subjected. Otherwise, it is, within the purpose of this section, uninsulated.

B. Installations Energized at 50 to 300 Volts

The hazards posed by installations energized at 50 to 300 volts are the same as those found in many other workplaces. That is not to say that there is no hazard, but the complexity of electrical protection required does not compare to that required for high voltage systems. The employee must avoid contact with the exposed parts, and the protective equipment used (such as rubber insulating gloves) must provide insulation for the voltages involved.

C. Exposed Energized Parts Over 300 Volts AC

Table R.6, Table R.7, and Table R.8 of § 1910.269 provide safe approach and working distances in the vicinity of energized electric apparatus so that work can be done safely without risk of electrical flashover.

Appendix A.5 to § 1910.269—Application of §§ 1910.146 and § 1910.269 to Permit-Required Confined Spaces.

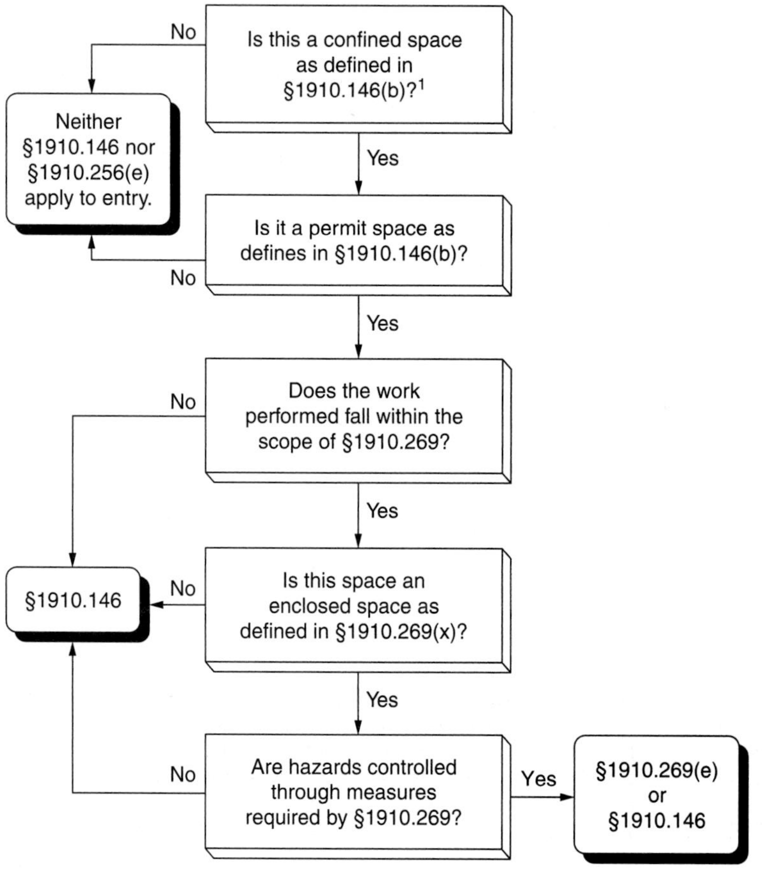

[1]See §1910.146(c) for general non-entry requirements that apply to all confined spaces.

The working distances must withstand the maximum transient overvoltage that can reach the work site under the working conditions and practices in use. Normal system design may provide or include a means to control transient overvoltages, or temporary devices may be employed to achieve the same result. The use of technically correct practices or procedures to control overvoltages (for example, portable gaps or preventing the automatic control from initiating breaker reclosing) enables line design and operation to be based on reduced transient overvoltage values. Technical information for U.S. electrical systems indicates that current design provides for the following maximum transient overvoltage values (usually produced by switching surges): 362 kV and less—3.0 per unit; 552 kV—2.4 per unit; 800 kV—2.0 per unit.

Additional discussion of maximum transient overvoltages can be found in paragraph IV.A.2, later in this Appendix.

III. Determination of the Electrical Component of Minimum Approach Distances

A. Voltages of 1.1 kV to 72.5 kV

For voltages of 1.1 kV to 72.5 kV, the electrical component of minimum approach distances is based on American National Standards Institute (ANSI)/American Institute of Electrical Engineers (AIEE) Standard No. 4, March 1943, Tables III and IV. (AIEE is the predecessor technical society to the Insitute of Electrical and Electronic Engineers (IEEE).) These distances are calculated by the following formula:

Equation (1)—For Voltages of 1.1 kV to 72.5 kV

$$D = \frac{(V_{max} \times pu)^{1.63}}{124}$$

where D = Electrical component of the minimum approach distance in air in feet

$V_{(max)}$ = Maximum rated line-to-ground rms voltage in kV

pu = Maximum transient overvoltage factor in per unit

Source: AIEE Standard No. 4, 1943.

This formula has been used to generate Table 1.

B. Voltages of 72.6 kV to 800 kV

For voltages of 72.6 to 800 kV, the electrical component of minimum approach distances in based on ANSI/IEEE Standard 516-1987, "IEEE Guide for Maintenance Methods on Energized Power Lines." This standard gives the electrical component of the minimum approach distance based on power frequency rod-gap data, supplemented with transient overvoltage information and a saturation factor for high voltages. The distances listed in ANSI/IEEE Standard 516 have been calculated according to the following formula:

Equation (2)—For Voltages of 72.6 kV to 800 kV

$$D = (C + a)\text{pu}\, V_{(max)}$$

TABLE 1 AC Energized Line-Work Phase-to-Ground Electrical Component of the Minimum Approach Distance—1.1 to 72.5 kV

Maximum anticipated per-unit transient overvoltage	Phase to phase voltage			
	15,000	36,000	46,000	72,500
3.0	0.08	0.33	0.49	1.03

Note: The distances given (in feet) are for air as the insulating medium and provide no additional clearance for inadvertent movement.

where D = Electrical component of the minimum approach distance in air in feet

C = 0.01 to take care of correction factors associated with the variation of gap sparkover with voltage

a = A factor relating to the saturation of air at voltages of 345 kV or higher

pu = Maximum anticipated transient overvoltage, in per unit (pu)

$V_{(max)}$ = Maximum rms system line-to-ground voltage in kilovolts—it should be the "actual" maximum, or the normal highest voltage for the range (for example, 10 percent above the nominal voltage)

Source: Formula developed from ANSI/IEEE Standard No. 516, 1987.

This formula is used to calculate the electrical component of the minimum approach distances in air and is used in the development of Table 2 and Table 3.

C. Provisions for Inadvertent Movement

The minimum approach distances (working distances) must include an "adder" to compensate for the inadvertent movement of the worker relative to an energized part or the movement of the part relative to the worker. A certain allowance must be made to account for this possible inadvertent movement and to provide the worker with a comfortable and safe zone in which to work. A distance for inadvertent movement (called the "ergonomic component of the minimum approach distance") must be added to the electrical component to determine the total safe minimum approach distances used in live-line work.

One approach that can be used to estimate the ergonomic component of the minimum approach distance is response time-distance analysis. When this technique is used, the total response time to a hazardous incident is estimated and converted to distance travelled. For example, the driver of a car takes a given amount of time to respond to a "stimulus" and stop the vehicle. The elapsed time involved results in a distance being travelled before the car comes to a complete stop. This distance is dependent on the speed of the car at the time the stimulus appears.

TABLE 2 AC Energized Line-Work Phase-to-Ground Electrical Component of the Minimum Approach Distance—121 to 242 kV

Maximum anticipated per-unit transient overvoltage	Phase to phase voltage			
	121,000	145,000	169,000	242,000
2.0	1.40	1.70	2.00	2.80
2.1	1.47	1.79	2.10	2.94
2.2	1.54	1.87	2.20	3.08
2.3	1.61	1.96	2.30	3.22
2.4	1.68	2.04	2.40	3.35
2.5	1.75	2.13	2.50	3.50
2.6	1.82	2.21	2.60	3.64
2.7	1.89	2.30	2.70	3.76
2.8	1.96	2.38	2.80	3.92
2.9	2.03	2.47	2.90	4.05
3.0	2.10	2.55	3.00	4.29

Note: The distances given (in feet) are for air as the insulating medium and provide no additional clearance for inadvertent movement.

TABLE 3 AC Energized Line-Work Phase-to-Ground Electrical Component of the Minimum Approach Distance—362 to 800 kV

Maximum anticipated per-unit transient overvoltage	Phase to phase voltage		
	362,000	552,000	800,000
1.5	....	4.97	8.66
1.6	....	5.46	9.60
1.7	....	5.98	10.60
1.8	....	6.51	11.64
1.9	....	7.08	12.73
2.0	4.20	7.68	13.86
2.1	4.41	8.27	....
2.2	4.70	8.87	....
2.3	5.01	9.49	....
2.4	5.34	10.21	....
2.5	5.67	....	....
2.6	6.01	....	....
2.7	6.36	....	....
2.8	6.73	....	....
2.9	7.10	....	....
3.0	7.48	....	....

Note: The distances given (in feet) are for air as the insulating medium and provide no additional clearance for inadvertent movement.

In the case of live-line work, the employee must first perceive that he or she is approaching the danger zone. Then, the worker responds to the danger and must decelerate and stop all motion toward the energized part. During the time it takes to stop, a distance will have been traversed. It is this distance that must be added to the electrical component of the minimum approach distance to obtain the total safe minimum approach distance.

At voltages below 72.5 kV, the electrical component of the minimum approach distance is smaller than the ergonomic component. At 72.5 kV the electrical component is only a little more than 1 foot. An ergonomic component of the minimum approach distance is needed that will provide for all the worker's expected movements. The usual live-line work method for these voltages is the use of rubber insulating equipment, frequently rubber gloves. The energized object needs to be far enough away to provide the worker's face with a safe approach distance, as his or her hands and arms are insulated. In this case, 2 feet has been accepted as a sufficient and practical value.

For voltages between 72.6 and 800 kV, there is a change in the work practices employed during energized line work. Generally, live-line tools (hot sticks) are employed to perform work while equipment is energized. These tools, by design, keep the energized part at a constant distance from the employee and thus maintain the appropriate minimum approach distance automatically.

The length of the ergonomic component of the minimum approach distance is also influenced by the location of the worker and by the nature of the work. In these higher voltage ranges, the employees use work methods that more tightly control their movements than when the workers perform rubber glove work. The worker is farther from energized line or equipment and needs to be more precise in his or her movements just to perform the work.

For these reasons, a smaller ergonomic component of the minimum approach distance is needed, and a distance of 1 foot has been selected for voltages between 72.6 and 800 kV.

TABLE 4 Ergonomic Component of Minimum Approach Distance

Voltage range (kV)	Distance (feet)
1.1 to 72.5	2.0
72.6 to 800	1.0

Note: This distance must be added to the electrical component of the minimum approach distance to obtain the full minimum approach distance.

Table 4 summarizes the ergonomic component of the minimum approach distance for the two voltage ranges.

D. Bare-Hand Live-Line Minimum Approach Distances

Calculating the strength of phase-to-phase transient overvoltages is complicated by the varying time displacement between overvoltages on parallel conductors (electrodes) and by the varying ratio between the positive and negative voltages on the two electrodes. The time displacement causes the maximum voltage between phases to be less than the sum of the phase-to-ground voltages. The International Electrotechnical Commission (IEC) Technical Committee 28, Working Group 2, has developed the following formula for determining the phase-to-phase maximum transient overvoltage, based on the per unit (pu) of the system nominal voltage phase-to-ground crest:

$$\text{pu(p)} = \text{pu(g)} + 1.6$$

Where pu(g) = pu phase-to-ground maximum transient overvoltage
pu(p) = pu phase-to-phase maximum transient overvoltage

This value of maximum anticipated transient overvoltage must be used in Equation (2) to calculate the phase-to-phase minimum approach distances for live-line bare-hand work.

E. Compiling the Minimum Approach Distance Tables

For each voltage involved, the distance in Table 4 in this appendix has been added to the distance in Table 1, Table 2 or Table 3 in this appendix to determine the resulting minimum approach distances in Table R.6, Table R.7, and Table R.8 in § 1910.269.

F. Miscellaneous Correction Factors

The strength of an air gap is influenced by the changes in the air medium that forms the insulation. A brief discussion of each factor follows, with a summary at the end.

1. *Dielectric strength of air.* The dielectric strength of air in a uniform electric field at standard atmospheric conditions is approximately 31 kV (crest) per cm at 60 Hz. The disruptive gradient is affected by the air pressure, temperature, and humidity, by the shape, dimensions, and separation of the electrodes, and by the characteristics of the applied voltage (wave shape).

2. *Atmospheric effect.* Flashover for a given air gap is inhibited by an increase in the density (humidity) of the air. The empirically determined electrical strength of a given gap is normally applicable at standard atmospheric conditions (20 deg. C, 101.3 kPa, 11 g/cm^3 humidity).

The combination of temperature and air pressure that gives the lowest gap flash-over voltage is high temperature and low pressure. These are conditions not likely to occur simultaneously. Low air pressure is generally associated with high humidity, and this causes increased electrical strength. An average air pressure is more likely to be associated with low humidity. Hot and dry working conditions are thus normally associated with reduced electrical strength.

The electrical component of the minimum approach distances in Table 1, Table 2, and Table 3 has been calculated using the maximum transient overvoltages to determine withstand voltages at standard atmospheric conditions.

3. *Altitude.* The electrical strength of an air gap is reduced at high altitude, due principally to the reduced air pressure. An increase of 3 percent per 300 meters in the minimum approach distance for altitudes above 900 meters is required. Table R.10 of § 1910.269 presents the information in tabular form.

Summary. After taking all these correction factors into account and after considering their interrelationships relative to the air gap insulation strength and the conditions under which live work is performed, one finds that only a correction for altitude need be made. An elevation of 900 meters is established as the base elevation, and the values of the electrical component of the minimum approach distances have been derived with this correction factor in mind. Thus, the values used for elevations below 900 meters are conservative without any change; corrections have to be made only above this base elevation.

IV. Determination of Reduced Minimum Approach Distances

A. Factors Affecting Voltage Stress at the Work Site

1. *System voltage (nominal).* The nominal system voltage range sets the absolute lower limit for the minimum approach distance. The highest value within the range, as given in the relevant table, is selected and used as a reference for per unit calculations.

2. *Transient overvoltages.* Transient overvoltages may be generated on an electrical system by the operation of switches or breakers, by the occurrence of a fault on the line or circuit being worked or on an adjacent circuit, and by similar activities. Most of the overvoltages are caused by switching, and the term "switching surge" is often used to refer generically to all types of overvoltages. However, each overvoltage has an associated transient voltage wave shape. The wave shape arriving at the site and its magnitude vary considerably.

The information used in the development of the minimum approach distances takes into consideration the most common wave shapes; thus, the required minimum approach distances are appropriate for any transient overvoltage level usually found on electric power generation, transmission, and distribution systems. The values of the per unit (p.u.) voltage relative to the nominal maximum voltage are used in the calculation of these distances.

3. *Typical magnitude of overvoltages.* The magnitude of typical transient overvoltages is given in Table 5.

4. *Standard deviation—air-gap withstand.* For each air gap length, and under the same atmospheric conditions, there is a statistical variation in the breakdown voltage. The probability of the breakdown voltage is assumed to have a normal (Gaussian) distribution. The standard deviation of this distribution varies with the wave shape, gap geometry, and atmospheric conditions. The withstand voltage of the air gap used in calculating the electrical component of the minimum approach distance has been set at three standard deviations (3 sigma[1]) below the critical flashover voltage. (The critical flashover voltage is the crest value of the impulse wave that, under specified conditions, causes flashover on 50 percent

[1] Sigma is the symbol for standard deviation.

TABLE 5 Magnitude of Typical Transient Overvoltages

Cause	Magnitude (per unit)
Energized 200 mile line without closing resistors	3.5
Energized 200 mile line with one step closing resistor	2.1
Energized 200 mile line with multi-step resistor	2.5
Reclosed with trapped charge one step resistor	2.2
Opening surge with single restrike	3.0
Fault initiation unfaulted phase	2.1
Fault initiation adjacent circuit	2.5
Fault clearing	1.7–1.9

Source: ANSI/IEEE Standard No. 516, 1987.

of the applications. An impulse wave of three standard deviations below this value, that is, the withstand voltage, has a probability of flashover of approximately 1 in 1000.)

5. *Broken insulators.* Tests have shown that the insulation strength of an insulator string with broken skirts is reduced. Broken units may have lost up to 70 percent of their withstand capacity. Because the insulating capability of a broken unit cannot be determined without testing it, damaged units in an insulator are usually considered to have no insulating value. Additionally, the overall insulating strength of a string with broken units may be further reduced in the presence of a live-line tool alongside it. The number of good units that must be present in a string is based on the maximum overvoltage possible at the worksite.

B. Minimum Approach Distances Based on Known Maximum Anticipated Per-Unit Transient Overvoltages

1. *Reduction of the minimum approach distance for AC systems.* When the transient overvoltage values are known and supplied by the employer, Table R.7 and Table R.8 of § 1910.269 allow the minimum approach distances from energized parts to be reduced. In order to determine what this maximum overvoltage is, the employer must undertake an engineering analysis of the system. As a result of this engineering study, the employer must provide new live work procedures, reflecting the new minimum approach distances, the conditions and limitations of application of the new minimum approach distances, and the specific practices to be used when these procedures are implemented.

2. *Calculation of reduced approach distance values.* The following method of calculating reduced minimum approach distances is based on ANSI/IEEE Standard 516:

Step 1. Determine the maximum voltage (with respect to a given nominal voltage range) for the energized part.

Step 2. Determine the maximum transient overvoltage (normally a switching surge) that can be present at the work site during work operation.

Step 3. Determine the technique to be used to control the maximum transient overvoltage. (See paragraphs IV.C and IV.D of this appendix.) Determine the maximum voltage that can exist at the work site with that form of control in place and with a confidence level of 3 sigma. This voltage is considered to be the withstand voltage for the purpose of calculating the appropriate minimum approach distance.

Step 4. Specify in detail the control technique to be used, and direct its implementation during the course of the work.

Step 5. Using the new value of transient overvoltage in per unit (p.u.), determine the required phase-to-ground minimum approach distance from Table R.7 or Table R.8 of § 1910.269.

C. Methods of Controlling Possible Transient Overvoltage Stress Found on a System

1. *Introduction.* There are several means of controlling overvoltages that occur on transmission systems. First, the operation of circuit breakers or other switching devices may be modified to reduce switching transient overvoltages. Second, the overvoltage itself may be forcibly held to an acceptable level by means of installation of surge arresters at the specific location to be protected. Third, the transmission system may be changed to minimize the effect of switching operations.

2. ***Operation of circuit breakers.***[2] The maximum transient overvoltage that can reach the work site is often due to switching on the line on which work is being performed. If the automatic-reclosing is removed during energized line work so that the line will not be reenergized after being opened for any reason, the maximum switching surge overvoltage is then limited to the larger of the opening surge or the greatest possible fault-generated surge, provided that the devices (for example, insertion resistors) are operable and will function to limit the transient overvoltage. It is essential that the operating ability of such devices be assured when they are employed to limit the overvoltage level. If it is prudent not to remove the reclosing feature (because of system operating conditions), other methods of controlling the switching surge level may be necessary.

Transient surges on an adjacent line, particularly for double circuit construction, may cause a significant overvoltage on the line on which work is being performed. The coupling to adjacent lines must be accounted for when minimum approach distances are calculated based on the maximum transient overvoltage.

3. *Surge arresters.* The use of modern surge arresters has permitted a reduction in the basic impulse-insulation levels of much transmission system equipment. The primary function of early arresters was to protect the system insulation from the effects of lightning. Modern arresters not only dissipate lightning-caused transients, but may also control many other system transients that may be caused by switching or faults.

It is possible to use properly designed arresters to control transient overvoltages along a transmission line and thereby reduce the requisite length of the insulator string. On the other hand, if the installation of arresters has not been used to reduce the length of the insulator string, it may be used to reduce the minimum approach distance instead.[3]

4. *Switching Restrictions.* Another form of overvoltage control is the establishment of switching restrictions, under which breakers are not permitted to be operated until certain system conditions are satisfied. Restriction of switching is achieved by the use of a tagging system, similar to that used for a "permit," except that the common term used for this activity is a "hold-off" or "restriction." These terms are used to indicate that operation is not prevented, but only modified during the live-work activity.

D. Minimum Approach Distance Based on Control of Voltage Stress (Overvoltages) at the Work Site

Reduced minimum approach distances can be calculated as follows:

1. *First Method—Determining the reduced minimum approach distance from a given withstand voltage.*[4]

---

[2] The detailed design of a circuit interrupter, such as the design of the contacts, of resistor insertion, and of breaker timing control, are beyond the scope of this appendix. These features are routinely provided as part of the design for the system. Only features that can limit the maximum switching transient overvoltage on a system are discussed in this appendix.

[3] Surge arrestor application is beyond the scope of this appendix. However, if the arrestor is installed near the work site, the application would be similar to protective gaps as discussed in paragraph IV.D. of this appendix.

[4] Since a given rod gap of a given configuration corresponds to a certain withstand voltage, this method can also be used to determine the minimum approach distance for a known gap.

Step 1. Select the appropriate withstand voltage for the protective gap based on system requirements and an acceptable probability of actual gap flashover.

Step 2. Determine a gap distance that provides a withstand voltage[5] greater than or equal to the one selected in the first step.[6]

Step 3. Using 110 percent of the gap's critical flashover voltage, determine which electrical component of the minimum approach distance from Equation (2) or Table 6, which is a tabulation of distance vs. withstand voltage based on Equation (2).

Step 4. Add the 1-foot ergonomic component to obtain the total minimum approach distance to be maintained by the employee.

2. *Second Method—Determining the necessary protective gap length from a desired (reduced) minimum approach distance.*

Step 1. Determine the desired minimum approach distance for the employee. Subtract the 1-foot ergonomic component of the minimum approach distance.

Step 2. Using this distance, calculate the air gap withstand voltage from Equation (2). Alternatively, find the voltage corresponding to the distance in Table 6.[7]

Step 3. Select a protective gap distance corresponding to a critical flashover voltage that, when multiplied by 110 percent, is less than or equal to the withstand voltage from Step 2.

Step 4. Calculate the withstand voltage of the protective gap (85 percent of the critical flashover voltage) to ensure that it provides an acceptable risk of flashover during the time the gap is installed.

3. *Sample protective gap calculations.*

Problem 1: Work is to be performed on a 500-kV transmission line that is subject to transient overvoltages of 2.4 pu. The maximum operating voltage of the line is 552 kV. Determine the length of the protective gap that will provide the minimum practical safe approach distance. Also, determine what that minimum approach distance is.

Step 1. Calculate the smallest practical maximum transient overvoltage (1.25 times the crest line-to-ground voltage):[8]

$$552 \text{ kV} \times \frac{\sqrt{2}}{\sqrt{3}} \times 1.25 = 563 \text{ kV}$$

This will be the withstand voltage of the protective gap.

Step 2. Using test data for a particular protective gap, select a gap that has a critical flashover voltage greater than or equal to:

$$563 \text{ kV}/0.85 = 662 \text{ kV}$$

For example, if a protective gap with a 4.0-foot spacing tested to a critical flashover voltage of 665 kV, crest, select this gap spacing.

[5] The withstand voltage for the gap is equal to 85 percent of its critical flashover voltage.

[6] Switch steps 1 and 2 if the length of the protective gap is known. The withstand voltage must then be checked to ensure that it provides an acceptable probability of gap flashover. In general, it should be at least 1.25 times the maximum crest operating voltage.

[7] Since the value of the saturation factor, a, in Equation (2) is dependent on the maximum voltage, several iterative computations may be necessary to determine the correct withstand voltage using the equation. A graph of withstand voltage vs. distance is given in ANSI/IEEE Std. 516, 1987. This graph could also be used to determine the appropriate withstand voltage for the minimum approach distance involved.

[8] To eliminate unwanted flashovers due to minor system disturbances, it is desirable to have the crest withstand voltage no lower than 1.25 pu.

TABLE 6 Withstand Distance for Transient Overvoltages

Crest voltage (kV)	Withstand distance (in feet) air gap
100	0.71
150	1.06
200	1.41
250	1.77
300	2.12
350	2.47
400	2.83
450	3.18
500	3.54
550	3.89
600	4.24
650	4.60
700	5.17
750	5.73
800	6.31
850	6.91
900	7.57
950	8.23
1000	8.94
1050	9.65
1100	10.42
1150	11.18
1200	12.05
1250	12.90
1300	13.79
1350	14.70
1400	15.64
1450	16.61
1500	17.61
1550	18.63

Source: Calculations are based on Equation (2).
Note: The air gap is based on the 60-Hz rod-gap withstand distance.

Step 3. This protective gap corresponds to a 110 percent of critical flashover voltage value of:

$$665 \text{ kV} \times 1.10 = 732 \text{ kV}$$

This corresponds to the withstand voltage of the electrical component of the minimum approach distance.

Step 4. Using this voltage in Equation (2) results in an electrical component of the minimum approach distance of:

$$D = (0.01 + 0.0006) \times \frac{552 \text{ kV}}{\sqrt{3}} = 5.5 \text{ ft}$$

Step 5. Add 1 foot to the distance calculated in step 4, resulting in a total minimum approach distance of 6.5 feet.

Problem 2: For a line operating at a maximum voltage of 552 kV subject to a maximum transient overvoltage of 2.4 pu, find a protective gap distance that will permit the use of a 9.0-foot minimum approach distance. (A minimum approach distance of 11 feet 3 inches is normally required.)

Step 1. The electrical component of the minimum approach distance is 8.0 feet (9.0–1.0).

Step 2. From Table 6, select the withstand voltage corresponding to a distance of 8.0 feet. By interpolation:

$$900\text{kV} + \left[50 \times \frac{(8.00 - 7.57)}{(8.23 - 7.57)}\right] = 933 \text{ kV}$$

Step 3. The voltage calculated in Step 2 corresponds to 110 percent of the critical flashover voltage of the gap that should be employed. Using test data for a particular protective gap, select a gap that has a critical flashover voltage less than or equal to:

$$933 \text{ kV}/1.10 = 848 \text{ kV}$$

For example, if a protective gap with a 5.8-foot spacing tested to a critical flashover voltage of 820 kV, crest, select this gap spacing.

Step 4. The withstand voltage of this protective gap would be:

$$820 \text{ kV} \times 0.85 = 697 \text{ kV}$$

The maximum operating crest voltage would be:

$$552 \text{ kV} \times \frac{\sqrt{2}}{\sqrt{3}} = 449 \text{ kV}$$

The crest withstand voltage of the protective gap in per unit is thus

$$697 \text{ kV} + 449 \text{ kV} = 1.55 \text{ pu}$$

If this is acceptable, the protective gap could be installed with a 5.8-foot spacing, and the minimum approach distance could then be reduced to 9.0 feet.

4. *Comments and variations.* The 1-foot ergonomic component of the minimum approach distance must be added to the electrical component of the minimum approach distance calculated under paragraph IV.D of this appendix. The calculations may be varied by starting with the protective gap distance or by starting with the minimum approach distance.

E. Location of Protective Gaps

1. Installation of the protective gap on a structure adjacent to the work site is an acceptable practice, as this does not significantly reduce the protection afforded by the gap.

2. Gaps installed at terminal stations of lines or circuits provide a given level of protection. The level may not, however, extend throughout the length of the line to the worksite. The use of gaps at terminal stations must be studied in depth. The use of substation terminal gaps raises the possibility that separate surges could enter the line at opposite ends, each with low enough magnitude to pass the terminal gaps without flashover. When voltage surges are initiated simultaneously at each end of a line and travel toward each other, the total voltage on the line at the point where they meet is the arithmetic sum of the two surges. A gap that is installed within 0.5 mile of the work site will protect against such intersecting waves. Engineering studies of a particular line or system may indicate that adequate protection can be provided by even more distant gaps.

3. If protective gaps are used at the work site, the work site impulse insulation strength is established by the gap setting. Lightning strikes as much as 6 miles away from the worksite may cause a voltage surge greater than the insulation withstand voltage, and a gap flashover may occur. The flashover will not occur between the employee and the line, but across the protective gap instead.

4. There are two reasons to disable the automatic-reclosing feature of circuit-interrupting devices while employees are performing live-line maintenance:

- To prevent the reenergizing of a circuit faulted by actions of a worker, which could possibly create a hazard or compound injuries or damage produced by the original fault:
- To prevent any transient overvoltage caused by the switching surge that would occur if the circuit were reenergized.

However, due to system stability considerations, it may not always be feasible to disable the automatic-reclosing feature.

APPENDIX C TO § 1910.269—PROTECTION FROM STEP AND TOUCH POTENTIALS

I. Introduction

When a ground fault occurs on a power line, voltage is impressed on the "grounded" object faulting the line. The voltage to which this object riases depends largely on the voltage on the line, on the impedance of the faulted conductor, and on the impedance to "true," or "absolute," ground represented by the object. If the object causing the fault represents a relatively large impedance, the voltage impressed on it is essentially the phase-to-ground system voltage. However, even faults to well grounded transmission towers or substation structures can result in hazardous voltages.[1] The degree of the hazard depends upon the magnitude of the fault current and the time of exposure.

II. Voltage-Gradient Distribution

A. Voltage-Gradient Distribution Curve

The dissipation of voltage from a grounding electrode (or from the grounded end of an energized grounded object) is called the ground potential gradient. Voltage drops associated with this dissipation of voltage are called ground potentials. Figure 1 is a typical voltage-gradient distribution curve (assuming a uniform soil texture). This graph shows that voltage decreases rapidly with increasing distance from the grounding electrode.

---

[1] This appendix provides information primarily with respect to employee protection from contact between equipment being used and an energized power line. The information presented is also relevant to ground faults to transmission towers and substation structures; however, grounding systems for these structures should be designed to minimize the step and touch potentials involved.

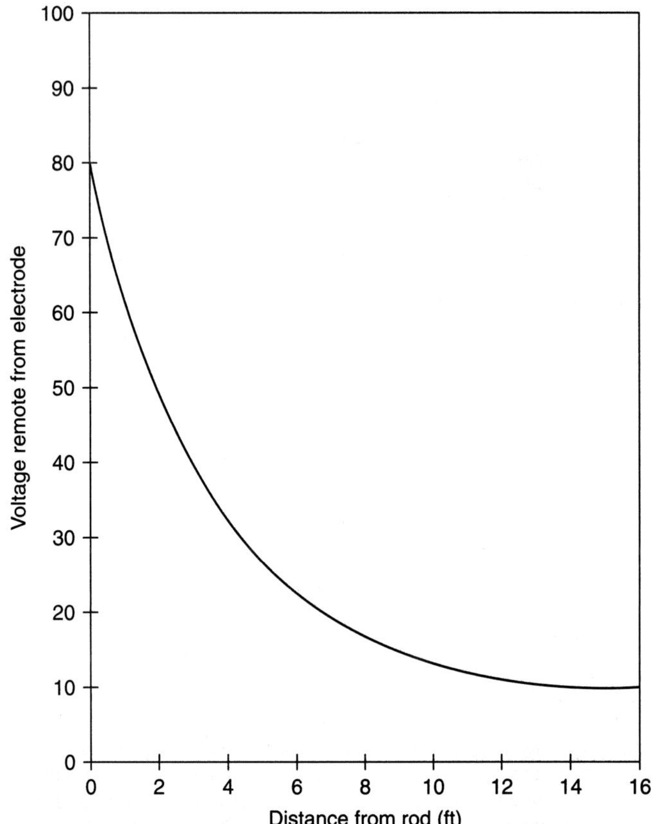

FIGURE 1 Typical voltage-gradient distribution curve.

B. Step and Touch Potentials

"Step potential" is the voltage between the feet of a person standing near an energized grounded object. It is equal to the difference in voltage, given by the voltage distribution curve, between two points at different distances from the "electrode." A person could be at risk of injury during a fault simply by standing near the grounding point.

"Touch potential" is the voltage between the energized object and the feet of a person in contact with the object. It is equal to the difference in voltage between the object (which is at a distance of 0 feet) and a point some distance away. It should be noted that the touch potential could be nearly the full voltage across the grounded object if that object is grounded at a point remote from the place where the person is in contact with it. For example, a crane that was grounded to the system neutral and that contacted an energized line would expose any person in contact with the crane or its uninsulated load line to a touch potential nearly equal to the full fault voltage.

Step and touch potentials are illustrated in Figure 2.

C. Protection from the Hazards of Ground-Potential Gradients

An engineering analysis of the power system under fault conditions can be used to determine whether or not hazardous step and touch voltages will develop. The result of this

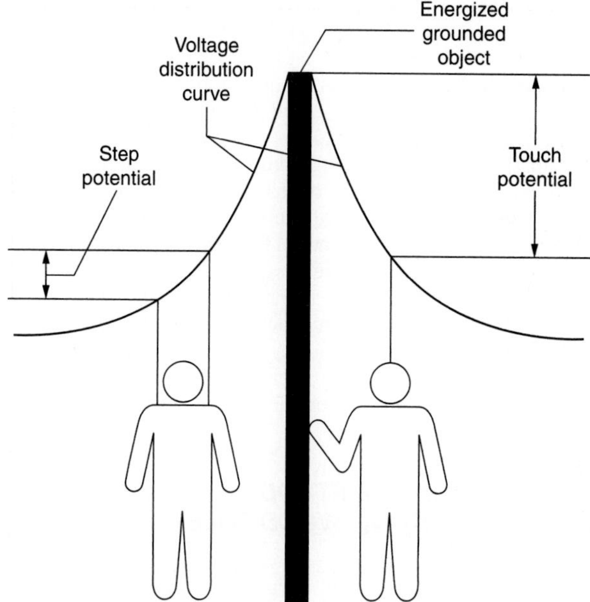

FIGURE 2 Step and touch potentials.

analysis can ascertain the need for protective measures and can guide the selection of appropriate precautions.

Several methods may be used to protect employees from hazardous ground-potential gradients, including equipotential zones, insulating equipment, and restricted work areas.

1. The creation of an equipotential zone will protect a worker standing within it from hazardous step and touch potentials. (See Figure 3.) Such a zone can be produced through

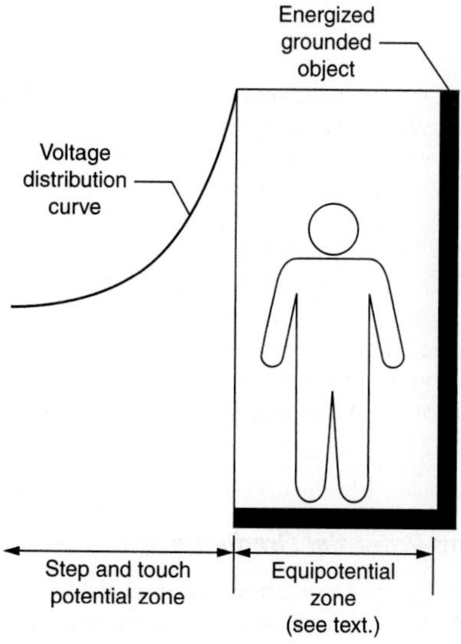

FIGURE 3 Protection from ground-potential gradients.

the use of a metal mat connected to the grounded object. In some cases, a grounding grid can be used to equalize the voltage within the grid. Equipotential zones will not, however, protect employees who are either wholly or partially outside the protected area. Bonding conductive objects in the immediate work area can also be used to minimize the potential between the objects and between each object and ground. (Bonding an object outside the work area can increase the touch potential to that object in some cases, however.)

2. The use of insulating equipment, such as rubber gloves, can protect employees handling grounded equipment and conductors from hazardous touch potentials. The insulating equipment must be rated for the highest voltage that can be impressed on the grounded objects under fault conditions (rather than for the full system voltage).

3. Restricting employees from areas where hazardous step or touch potentials could arise can protect employees not directly involved in the operation being performed. Employees on the ground in the vicinity of transmission structures should be kept at a distance where step voltages would be insufficient to cause injury. Employees should not handle grounded conductors or equipment likely to become energized to hazardous voltages unless the employees are within an equipotential zone or are protected by insulating equipment.

APPENDIX D TO § 1910.269—METHODS OF INSPECTING AND TESTING WOOD POLES

I. Introduction

When work is to be performed on a wood pole, it is important to determine the condition of the pole before it is climbed. The weight of the employee, the weight of equipment being installed, and other working stresses (such as the removal or retensioning of conductors) can lead to the failure of a defective pole or one that is not designed to handle the additional stresses.[1] For these reasons, it is essential that an inspection and test of the condition of a wood pole be performed before it is climbed.

If the pole is found to be unsafe to climb or to work from, it must be secured so that it does not fail while an employee is on it. The pole can be secured by a line truck boom, by ropes or guys, or by lashing a new pole alongside it. If a new one is lashed alongside the defective pole, work should be performed from the new one.

II. Inspection of Wood Poles

Wood poles should be inspected by a qualified employee for the following conditions:[2]

A. General Condition

The pole should be inspected for buckling at the ground line and for an unusual angle with respect to the ground. Buckling and odd angles may indicate that the pole has rotted or is broken.

B. Cracks

The pole should be inspected for cracks. Horizontal cracks perpendicular to the grain of the wood may weaken the pole. Vertical ones, although not considered to be a sign of a defective

[1] A properly guyed pole in good condition should, at a minimum, be able to handle the weight of an employee climbing it.

[2] The presence of any of these conditions is an indication that the pole may not be safe to climb or to work from. The employee performing the inspection must be qualified to make a determination as to whether or not it is safe to perform the work without taking additional precautions.

pole, can pose a hazard to the climber, and the employee should keep his or her gaffs away from them while climbing.

C. Holes

Hollow spots and woodpecker holes can reduce the strength of a wood pole.

D. Shell Rot and Decay

Rotting and decay are cutout hazards and are possible indications of the age and internal condition of the pole.

E. Knots

One large knot or several smaller ones at the same height on the pole may be evidence of a weak point on the pole.

F. Depth of Setting

Evidence of the existence of a former ground line substantially above the existing ground level may be an indication that the pole is no longer buried to a sufficient extent.

G. Soil Conditions

Soft, wet, or loose soil may not support any changes of stress on the pole.

H. Burn Marks

Burning from transformer failures or conductor faults could damage the pole so that it cannot withstand mechanical stress changes.

III. Testing of Wood Poles

The following tests, which have been taken from § 1910.268(n)(3), are recognized as acceptable methods of testing wood poles:

A. Hammer Test

Rap the pole sharply with a hammer weighing about 3 pounds, starting near the ground line and continuing upwards circumferentially around the pole to a height of approximately 6 feet. The hammer will produce a clear sound and rebound sharply when striking sound wood. Decay pockets will be indicated by a dull sound or a less pronounced hammer rebound. Also, prod the pole as near the ground line as possible using a pole prod or a screwdriver with a blade at least 5 inches long. If substantial decay is encountered, the pole is considered unsafe.

B. Rocking Test

Applying a horizontal force to the pole and attempt to rock it back and forth in a direction perpendicular to the line. Caution must be exercised to avoid causing power lines to swing together. The force may be applied either by pushing with a pike pole or pulling with a rope. If the pole cracks during the test, it shall be considered unsafe.

APPENDIX E TO § 1910.269—REFERENCE DOCUMENTS

The references contained in this appendix provide information that can be helpful in understanding and complying with the requirements contained in § 1910.269. The national consensus standards referenced in this appendix contain detailed specifications that employers may follow in complying with the more performance-oriented requirements of OSHA's final rule. Except as specifically noted in § 1910.269, however, compliance with the national consensus standards is not a substitute for compliance with the provisions of the OSHA standard.

ANSI/SIA A92.2-1990, American National Standard for Vehicle-Mounted Elevating and Rotating Aerial Devices.

ANSI C2-1993, National Electrical Safety Code.

ANSI Z133.1-1988, American National Standard Safety Requirements for Pruning, Trimming, Repairing, Maintaining, and Removing Trees, and for Cutting Brush.

ANSI/ASME B20.1-1990, Safety Standard for Conveyors and Related Equipment.

ANSI/IEEE Std. 4-1978 (Fifth Printing), IEEE Standard Techniques for High-Voltage Testing.

ANSI/IEEE Std. 100-1988, IEEE Standard Dictionary of Electrical and Electronic Terms.

ANSI/IEEE Std. 516-1987, IEEE Guide for Maintenance Methods on Energized Power-Lines.

ANSI/IEEE Std. 935-1989, IEEE Guide on Terminology for Tools and Equipment to Be Used in Live Line Working.

ANSI/IEEE Std. 957-1987, IEEE Guide for Cleaning Insulators.

ANSI/IEEE Std. 978-1984 (R1991), IEEE Guide for In-Service Maintenance and Electrical Testing of Live-Line Tools.

ASTM D 120-87, Specification for Rubber Insulating Gloves.

ASTM D 149-92, Test Method for Dielectric Breakdown Voltage and Dielectric Strength of Solid Electrical Insulating Materials at Commercial Power Frequencies.

ASTM D 178-93, Specification for Rubber Insulating Matting.

ASTM D 1048-93, Specification for Rubber Insulating Blankets.

ASTM D 1049-93, Specification of Rubber Insulating Covers.

ASTM D 1050-90, Specification for Rubber Insulating Line Hose.

ASTM D 1051-87, Specification for Rubber Insulating Sleeves.

ASTM F 478-92, Specification for In-Service Care of Insulating Line Hose and Covers.

ASTM F 479-93, Specification for In-Service Care of Insulating Blankets.

ASTM F 496-93B, Specification for In-Service Care of Insulating Gloves and Sleeves.

ASTM F 711-89, Specification for Fiberglass-Reinforced Plastic (FRP) Rod and Tube Used in Live Line Tools.

ASTM F 712-88, Test Methods for Electrically Insulating Plastic Guard Equipment for Protection of Workers.

ASTM F 819-83a (1988), Definitions of Terms Relating to Electrical Protective Equipment for Workers.

ASTM F 855-90, Specifications for Temporary Grounding Systems to Be Used on De-Energized Electric Power Lines and Equipment.

ASTM F 887-91a, Specifications for Personal Climbing Equipment.

ASTM F 914-91, Test Method for Acoustic Emission for Insulated Aerial Personnel Devices.

ASTM F 968-93, Specification for Electrically Insulating Plastic Guard Equipment for Protection of Workers.

ASTM F 1116-88, Test Method for Determining Dielectric Strength of Overshoe Footwear.

ASTM F 1117-87, Specification for Dielectric Overshoe Footwear.

ASTM F 1236-89, Guide for Visual Inspection of Electrical Protective Rubber Products.

ASTM F 1505-94, Standard Specification for Insulated and Insulating Hand Tools.

ASTM F 1506-94, Standard Performance Specification for Textile Materials for Wearing Apparel for Use by Electrical Workers Exposed to Momentary Electric Arc and Related Thermal Hazards.

IEEE Std. 62-1978, IEEE Guide for Field Testing Power Apparatus Insulation.

IEEE Std. 524-1992, IEEE Guide to the Installation of Overhead Transmission Line Conductors.

IEEE Std. 1048-1990, IEEE Guide for Protective Grounding of Power Lines.

IEEE Std. 1067-1990, IEEE Guide for the In-Service Use, Care, Maintenance, and Testing of Conductive Clothing for Use on Voltages up to 765 kV AC.

[59 FR 4437, Jan. 31, 1994; 59 FR 33658, June 30, 1994; 59 FR 40729, Aug. 9, 1994]

CHAPTER 47
RESUSCITATION

INTRODUCTION

Chapters 47 and 48 were compiled and edited by Ellen Thomas of the (Southeastern Line Constructors Apprenticeship Training) (***SELCAT***) Joint Apprenticeship and Training program, with assistance from Dennis Moody, Director of the ***SELCAT*** program, Wilma Barger of SOS Technologies, Local Unions 84 and 222 of the International Brotherhood of Electrical Workers ***IBEW***, Johnny Atwood, Grant Barker, Billy Watkins and Durwood Burks.

These chapters will present the steps you should take in an emergency by offering a review and explanation of what you have learned in professionally given first aid, cardiopulmonary resuscitation (CPR) and pole top rescue classes. It is not intended as a replacement for taking the Occupational Safety and Health Act *OSHA* required American Red Cross, the American Heart Association, the U.S. Bureau of Mines or an equivalent training that can be verified by documentary evidence.

The procedures in these chapters are based on the most current recommendations available at the time of printing. We make no guarantees as to and assume no responsibility for the correctness, or completeness of the presented information or recommendations.

To be prepared for an emergency, you should take a review course yearly and practice the skills on a mannequin in the presence of a trained instructor. Techniques are being updated, and you need to learn the current skills. The American Red Cross, the American Heart Association, the U.S. Bureau of Mines and others have courses which you may take.

BE PREPARED FOR AN EMERGENCY

The best way to care for an emergency is to PREVENT it.

THINK SAFETY

Have and **use** protective equipment.
Inspect rubber gloves daily and have them electrically tested as per industry recommendations.
Be aware of what is going on around you.
Keep your mind on what you are doing.
Do not use any substance that impairs your ability to act and to react.
Wear clothes made of 100% cotton and rated at 11-lb weave.
Avoid wearing the orange traffic vest when not directing traffic.
Wear clothing that is nonconductive and flame resistant.

FIRST AID EQUIPMENT

OSHA 1926.50 defines the medical services and first aid which should be available to the employee.

Prompt medical attention should be available within a reasonable access time (4 min). If not available, a person who has a valid first aid, CPR certificate must be at the work site go give first aid.

First-aid kits and supplies should be in a weatherproof container with individual sealed packages for each type of item. The contents shall be checked by the employer before being sent out on each job and at least weekly thereafter to see that expended items are replaced.

First-aid kits are for your use and protection on the job. When they are taken off of the truck for unauthorized use and an emergency occurs, the missing supplies could mean the difference between an incident of small proportions and a life-threatening incident.

The employer is responsible for the first-aid kits and so are you. You need to make sure they are present, and that the supplies are not out of date or soiled. Make sure the appropriate company personnel knows that supplies need to be replaced.

This is your responsibility. Your life could depend on it.

YOU HAVE AN EMERGENCY WHAT DO YOU DO?

Be prepared.

Remain Calm.

If the victim is on a pole or in a bucket, shout at your victim to see if they are conscious and can respond.

If they cannot respond, immediately send someone to call 911 or your local emergency system.

Use common sense.

We do not want two victims.

Do not approach the victim until it is safe for you to do so.

Assess your victim.

Use protective barriers.

Give rescue breathing. CPR or Heimlich maneuver as needed (these skills will be reviewed for you in the next pages.)

EFFECTS OF ELECTRIC SHOCK ON THE HUMAN BODY

The effect of electric shock on a human being is rather unpredictable and may manifest itself in a number of ways.

1. *Asphyxia.* Electric shock may cause a cessation of respiration (asphyxia). Current passing through the body may temporarily paralyze (or destroy) either the nerves or the area of the brain which controls respiration.
2. *Burns—contact and flash.* Contact burns are a common result of electric current passing through the body. The burns are generally found at the points where the current entered and left the body and vary in severity, the same as thermal burns. The seriousness of these burns may not be immediately evident because their appearance may not indicate the depth to which they have penetrated.

   In some accidents there is a flash or electric arc, the rays and heat from which may damage the eyes or result in thermal burns to exposed parts of the body.

3. *Fibrillation.* Electric shock may disturb the natural rhythm of the heartbeat. When this happens, the muscles of the heart are thrown into a twitching or trembling state, and the actions of the individual muscle fibers are no longer coordinated. The pulse disappears, and circulation ceases. This condition is known as ventricular fibrillation and is serious.
4. *Muscle spasm.* A series of erratic movements of a limb or limbs may occur owing to alternating contractions and relaxations of the muscles. This muscle spasm action on the muscles of respiration may be a factor in the stoppage of breathing.

RESCUE

Because a person may receive electric shock in many different locations—on the ground, in buildings, on poles, or on steel structures—it is neither possible nor desirable to lay down definite methods of rescue. However, certain facts should be remembered.

FREEING THE VICTIM

Because of the muscle spasm at the time of shock, most victims are thrown clear of contact. However, in some instances (usually low voltage) the victim is still touching live equipment. In either situation, the rescuer must be extremely careful not to get in contact with the live equipment or to touch the victim while he or she is still in contact. The rescuer should "free" the victim as soon as possible so that artificial respiration can be applied without hazard. This may involve opening switches or cutting wires so that equipment within reach is de-energized or using rubber gloves or other approved insulation to move the victim out of danger.

If the victim is to be lowered from a pole, tower, or other structure, a hand line of adequate strength, tied or looped around him and placed over a crossarm or tower member, is a simple and satisfactory method. Procedures for lowering a lineman from a pole or substation structure developed by personnel of the Commonwealth Edison Company are illustrated in Chap. 49—"Pole-Top and Bucket Truck Rescue."

ARTIFICIAL RESPIRATION METHODS

Early methods of artificial respiration involved applying heat by building a fire on the victim's stomach, heating the victim with a whip, inverting the victim by suspending him by his feet or rolling the victim over a barrel. Modern methods of artificial respiration include the Schafer prone method, the chest-pressure-arm-lift method, the back-pressure-arm-lift method and the mouth-to-mouth method. The mouth-to-mouth method is currently advocated by the Edison Electric Institute and the American Red Cross. The mouth-to-mouth method accepted by most utility companies is described in this chapter. Artificial respiration should be used when breathing stops as a result of electric shock, exposure to gases, drowning, or physical injury. If the heart has stopped functioning as a result of cardiac arrest or ventricular fibrillation, heart-lung resuscitation should be used with mouth-to-mouth resuscitation. Heart-lung resuscitation methods are illustrated in Chap. 48.

APPLYING ARTIFICIAL RESPIRATION

To be successful, artificial respiration must be applied within the shortest time possible to a victim who is not breathing. Without oxygen, the brain begins to die. The first four minutes are critical. Irreversible brain damage begins. The best chance of survival is for the victim to receive rescue breathing—CPR immediately and advanced medical care from an Emergency Medical System (EMS).

Normally, the type of resuscitation in which the victim is placed on the ground or floor is the best because the victim can be given additional treatment (stoppage of bleeding, wrapping in blanket to keep warm, etc.) and the rescuer can be easily relieved without an interruption of the breathing rhythm or cycle. However, pole-top resuscitation or any adaption of it should be started if there is any appreciable delay in getting the victim into the prone position.

In choosing the type of resuscitation to be used, the rescuer must consider the obvious injuries suffered by the victim. Broken ribs, for example, might make inadvisable the use of certain types of resuscitation; burns on the arms or on the face might exclude other types. However, no time should be lost in searching for injuries; artificial respiration should be started at once.

THE NEED FOR SPEED

Artificial respiration must be started at once. Time is the most important single factor in successful artificial respiration. The human brain can exist for only a few minutes, probably not more than 4 or 5, without oxygenated blood; therefore, every second counts. The greater the delay, the smaller the chance of successful recovery. There should be no delay to loosen clothing, warm the victim, get him down from the pole, or move him to a more comfortable position. However, an immediate check of the victim's mouth should be made by a quick pass of the fingers (with a protective barrier such as rubber gloves) to remove tobacco, chewing gum, etc. Do not remove dentures unless they are loose. They help to support the facial structure.

USE PROTECTIVE BARRIERS

In today's society, we run the risk of being exposed and exposing the victim to diseases such as HIV/AIDS, TB, Hepatitis, viral infections, and some forms of pneumonia. Use protective barriers when possible.

Resuscitation masks vary in size, shape, and features. The most effective are ones made of transparent, pliable material that allows you to make a tight seal on the person's face and have a one-way valve for releasing a person's exhaled air.

Use rubber gloves when available.

ASSESS YOUR VICTIM

You must act immediately. Send someone to call 911 or your local emergency system immediately. Determine and use the best method of rescuing the victim. If the injured person is on a pole or in a bucket, make sure the area is safe for you to approach and then begin pole-top or bucket-rescue procedures.

Shout at the victim on the pole to see if he can respond and is conscious. If he cannot respond, and the area is safe to reach the victim, determine if your victim is breathing. If not breathing, give two rescue breaths before you lower him to the ground (Figs. 47.1 and 47.2). Use protective barriers if possible.

LOWER VICTIM TO GROUND OPEN THE AIRWAY DETERMINE IF VICTIM IS BREATHING

To open the airway, you should tilt the head back by placing one hand on the victim's forehead. Using your other hand, place two fingers under the bony part of the lower jaw near the chin. Tilt head back and lift jaw. Avoid closing the victim's mouth. Avoid pushing on soft parts under the chin (Fig. 47.3).

The head-tilt/chin-lift will remove the tongue away from the back of the throat. Many times a victim will begin breathing on his or her own once you have done this (Fig. 47.4).

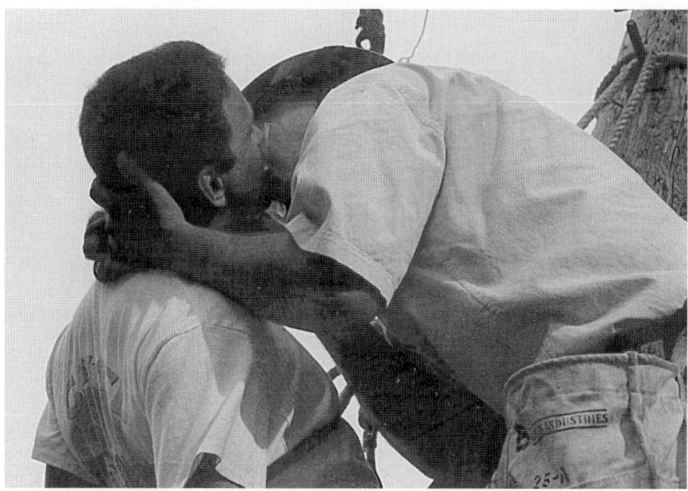

FIGURE 47.1 Check for breathing. If not breathing, give two rescue breaths before lowering to the ground.

IF NOT BREATHING, GIVE TWO RESCUE BREATHS

If the victim is not breathing, give two rescue breaths. A rescue breath is one breath given slowly for approximately $1^1/_2$ to 2 seconds. If you find that air will not enter or that the chest is not rising and falling, you should reposition the head and attempt giving a breath. Improper chin and head positioning is the most common cause of difficulty with ventilations. If after

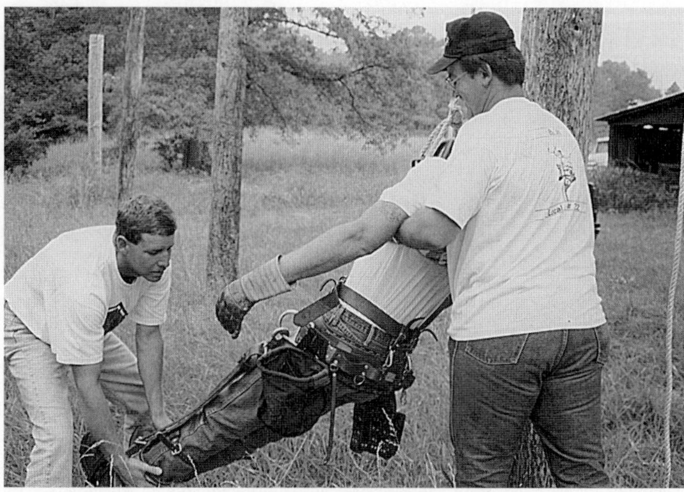

FIGURE 47.2 Lower the victim to the ground using the pole-top rescue techniques. Lay victim on back. Open the airway. Use the head-tilt/chin-lift method.

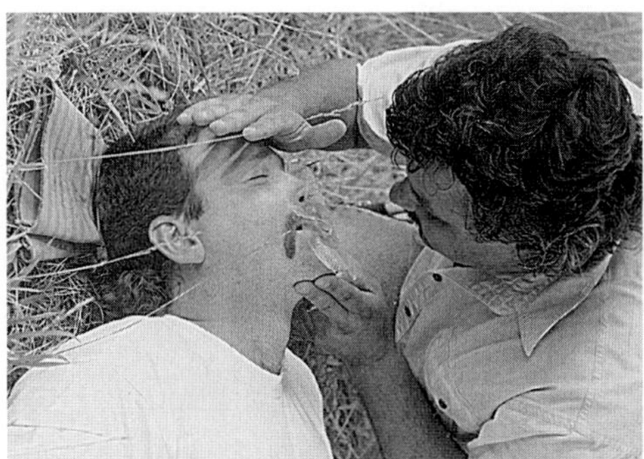

FIGURE 47.3 Proper method to be used to open the victim's airway.

you have repositioned the airway and air is not going in, you may have a foreign object trapped in the airway. You should begin airway obstruction maneuvers (Heimlich maneuver).

While maintaining the head-tilt/chin-lift, the rescuer should take a deep breath and seal his lips around the victim's mouth, creating an airtight seal. The rescuer gently pinches the nose closed with the thumb and index finger of the hand on the forehead, thereby preventing air from escaping through the victim's nose (Fig. 47.5).

Gently blow into the victim's mouth for $1^1/_2$ to 2 seconds. Look for the chest to rise. Release the nose so that the air can escape. Pause only long enough for you to take a breath, then repeat the procedure for a second rescue breath.

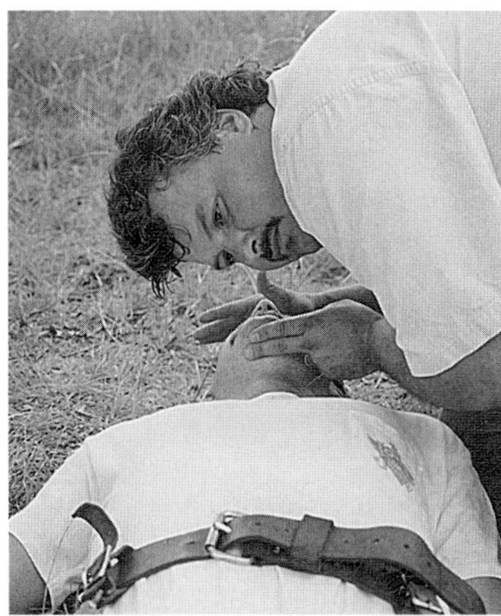

FIGURE 47.4 Look for the rise and fall of the chest, listen and feel for escaping air for 3 to 5 sec.

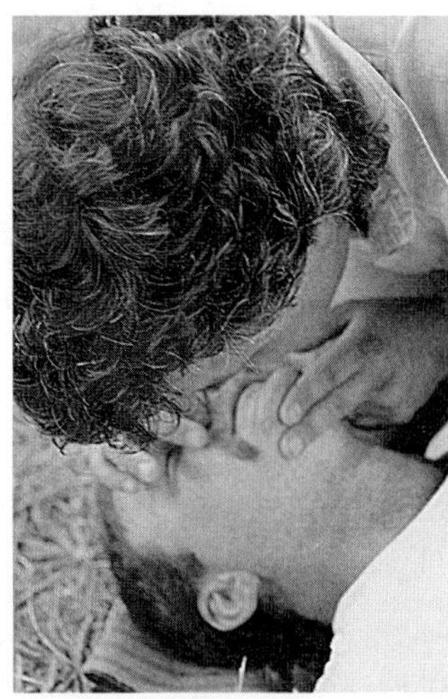

FIGURE 47.5 The rescuer gently pinches the nose closed with the thumb and index finger of the hand on the forehead.

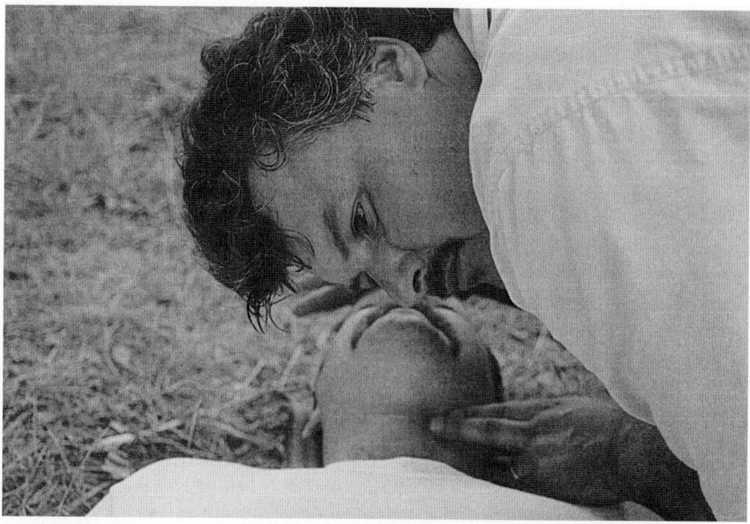

FIGURE 47.6 To find the carotid artery, find the Adam's apple and slide your fingers down into the groove closest to you. Feel for 5 to 10 seconds for a pulse.

CHECK FOR PULSE

After you have given two rescue breaths, check for pulse at the carotid artery. (Fig. 47.6). Feel the pulse for approximately 5 to 10 seconds. If pulse is present, but the victim is not breathing, begin rescue breathing. If pulse is absent, begin CPR.

MOUTH TO NOSE RESCUE BREATHING

The *Journal of the American Medical Association* recommends that mouth-to-nose breathing be used when it is impossible to ventilate through the victim's mouth, the mouth cannot be opened, the mouth is seriously injured or a tight mouth-to-mouth seal is difficult to achieve.

Keep the victim's head tilted back with one hand on the forehead and use the other hand to lift the victim's lower jaw (as in the head-tilt/chin-lift) and close the mouth. The rescuer takes a deep breath and then breathes into the victim's nose. The rescuer removes his mouth and allows the air to escape.

IF PULSE IS PRESENT BUT PERSON IS NOT BREATHING BEGIN RESCUE BREATHING

Give the victim one rescue breath about every 5 seconds.
Do this for about 1 minute (12 breaths).
Recheck the pulse and breathing about every minute.
You should continue rescue breathing as long as a pulse is present and the victim is not breathing, or until one of the following things happen:

The victim begins to breathe without your help.

The victim has no pulse (begin CPR).

Another trained rescuer takes over for you.

You are too tired to continue.

REVIEW

Recognize that you have an emergency.
Call for help immediately.
Assess your victim for unresponsiveness.
Make sure scene is safe.
Begin pole top or bucket rescue.
Check for breathing; if not breathing, give two breaths.
Lower victim to the ground.
Open airway-use head-tilt/chin-lift.
If not breathing, give two rescue breaths.
(remember use protective barrier if available)
Check for pulse.
If pulse is present and breathing is absent—
begin rescue breathing (one breath every 5 sec).
If there is no pulse, start CPR.

AIRWAY OBSTRUCTION CONSCIOUS ADULT

Ask the victim if they are choking and if you can help him or her (Fig. 47.7). If they are coughing forcefully, the rescuer should not interfere with the victim's own attempt to expel the foreign object, but stay with them and encourage them to continue coughing.

If the victim has a weak, ineffective cough or is making high-pitched noises while inhaling or is unable to speak, breathe, or cough, this should be treated as a complete airway obstruction. Stand behind the victim. Place your leg between the victim's legs (this will help you control the victim if he passes out (Fig. 47.8).

FIGURE 47.7 Universal signal for an obstructed airway is one or both hands clutching the throat.

FIGURE 47.8 Proper position for the rescuer to assume.

FIGURE 47.9 Rescuer places hand on victim's abdomen.

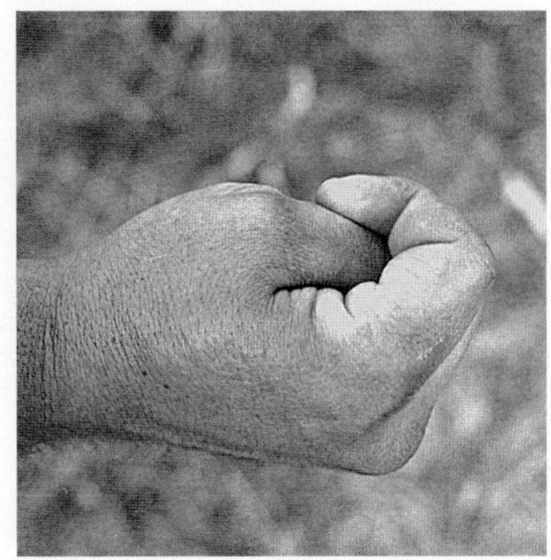

FIGURE 47.10 The thumb should be folded into the fist so that there is a flat surface placed on the abdomen.

Wrap your arms around the victim's waist, make a fist with one hand, the thumb side of the fist is placed against the victim's abdomen just slightly above the navel and well below the tip of the breastbone (Figs. 47.9 and 47.10).

The fist is grasped with the other hand and pressed into the victim's abdomen with a quick upward thrust (Fig. 47.11). Each new thrust should be a separate and distinct movement administered with the intent of relieving the obstruction. The thrusts should be repeated and continued until the object is expelled from the airway or the patient becomes unconscious.

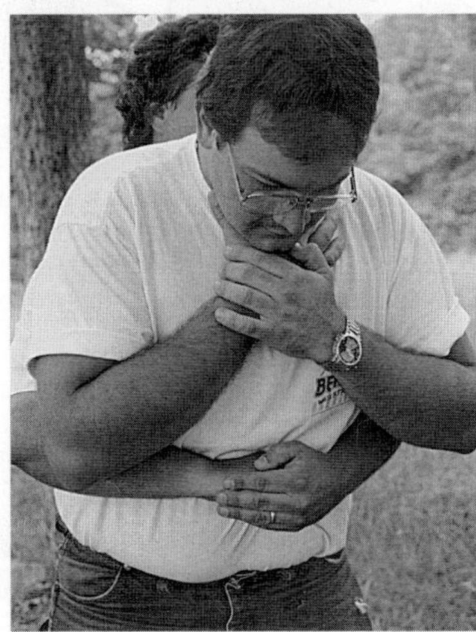

FIGURE 47.11 The fist is pressed into the victim's abdomen with a quick upward thrust.

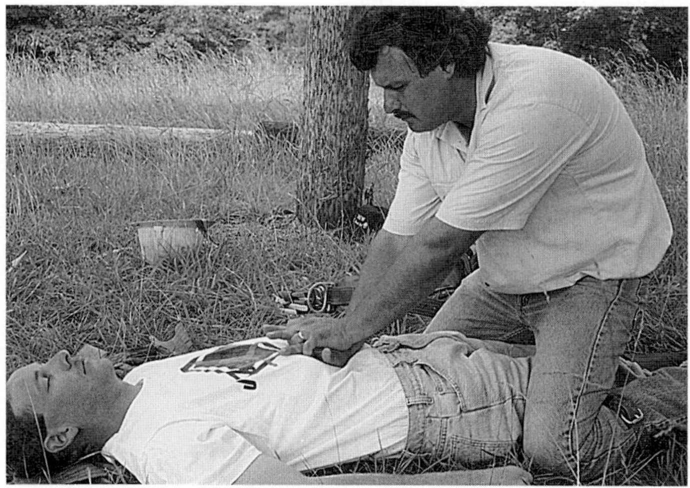

FIGURE 47.12 The rescuer straddles unconscious victim.

CONSCIOUS ADULT CHOKING VICTIM WHO BECOMES UNCONSCIOUS

If the victim becomes unconscious, lower to the floor or ground. The rescuer straddles the victim and places the heel of one hand slightly above the navel and well below the tip of the breastbone. The second hand is placed directly on top of the fist. The rescuer presses into the abdomen with quick upward motion (Fig. 47.12).

Give up to five abdominal thrusts. The victim's mouth should be opened and a finger sweep done after you have given five abdominal thrusts (Heimlich maneuvers). Give the victim a rescue breath, if air does not go in, repeat the sequence of five abdominal thrusts, finger sweep, and attempt to ventilate. Repeat as long as necessary.

HOW TO DO A FINGER SWEEP

A finger sweep is done by opening the mouth-grasp both the tongue and the lower jaw between the thumb and fingers of one hand, lift jaw. Insert index finger of other hand into the mouth, running it along the inside of the cheek and deep into throat to the base of the tongue. Use a "hooking" action to dislodge any object that might be there. Remove the object. Be careful not to force the object deeper into the airway.

WHAT TO DO WHEN THE OBJECT IS CLEARED

If the object is cleared and the victim is conscious, the victim may want to leave the area and go to the restroom, etc. Do not let them leave by themselves. The throat has been irritated and may swell and the airway may close. Stay with the victim until EMS personnel arrive.

FIGURE 47.13 Applying Heimlich maneuver to yourself.

IF YOU ARE ALONE AND CHOKING

If you are alone and choking, you can perform the Heimlich maneuver on yourself (Fig. 47.13). If this is unsuccessful, the upper abdomen should be pressed quickly over any firm surface such as the back of a chair, side of a table, or porch railing. Several thrusts may be needed to clear the airway.

WHAT TO DO IF YOUR VICTIM IS PREGNANT OR LARGE

If you are not able to get your arms around the waist of a choking victim who is conscious, you must give chest thrusts.

Check thrusts are given by getting behind the victim, the victim can be sitting or standing. Place your arms under the victim's armpits and around the chest. Place the thumb side of your fist on the middle of the breastbone. Be sure that your fist is centered right on the breastbone and not on the ribs. Make sure that your fist is not near the lower tip of the breastbone.

Grasp your fist with your other hand and give backward thrusts.

Give thrusts until the obstruction is cleared or until the person loses consciousness. You should think of each thrust as a separate attempt to dislodge the object.

REVIEW

Ask if victim is choking.
Ask for permission to help.
If victim is coughing forcefully,
encourage to continue coughing.
If victim has obstructed airway and is conscious, do abdominal thrusts until object is expelled or victim becomes unconscious.
If victim becomes unconscious,

lower to floor.
Give five abdominal thrusts.
Do a finger sweep.
Give a rescue breath, if air still does not go in,
repeat sequence of five thrusts, finger sweep, and breath.
If alone, give Heimlich maneuver (abdominal thrust) to yourself,
or lean over firm surface.

CHAPTER 48
HEART-LUNG RESUSCITATION

CARDIOPULMONARY RESUSCITATION (CPR)

Step A: Kneel facing the victim's chest. Find the correct hand position by sliding your fingers up the rib cage to the breastbone. *Step B:* Place your middle finger in the notch and the index finger next to it on the lower end of the breastbone. *Step C:* Place your other hand beside the two fingers. *Step D:* Place hand used to locate notch on top of the other hand. Lace the fingers together. Keep fingers off of the chest. Position shoulders over hands, with elbows locked and arms straight (Fig. 48.1).

There are varying sizes and shapes of hands. Arthritic or injured hands and wrists will not allow some rescuers to lace their fingers together. An acceptable alternative hand position is to grasp the wrist of the hand on the chest with the hand that has been locating the lower end of their sternum. Another acceptable alternative is to hold your fingers up off the chest. Use the method that works best for you. Just remember to keep your fingers off the chest. Use the heel of your hand to do the compressions (Fig. 48.2).

Compress the breastbone $1^1/_2$ to 2 in. Give 15 compressions. Count 1 and 2 and 3 and 4 . . . . Give two rescue breaths after the 15th compression. This cycle of 15 compressions and 2 breaths should take about 15 sec. Do four cycles of 15 compressions and two breaths, which should take about 1 min. Check the pulse at the end of the fourth cycle.

When you press down, the weight of your upper body creates the force you need to compress the chest. It is important to keep your shoulders directly over your hands and your elbows locked so that you use the weight of your upper body to compress the chest. Push with the weight of your upper body, not with your arm muscles. Push straight down; don't rock (Fig. 48.3).

If the thrust is not in a straight downward direction, the torso has a tendency to roll; part of the force is lost, and the chest compression may be less effective.

Once you have started CPR you must continue until one of the following happens:

Another trained person takes over CPR for you.

EMS personnel arrive and take over the care of the victim.

You are too exhausted and unable to continue.

The scene becomes unsafe.

REVIEW

Call for advance help—911.
Determine that victim has no pulse.
Find correct hand compression.

A

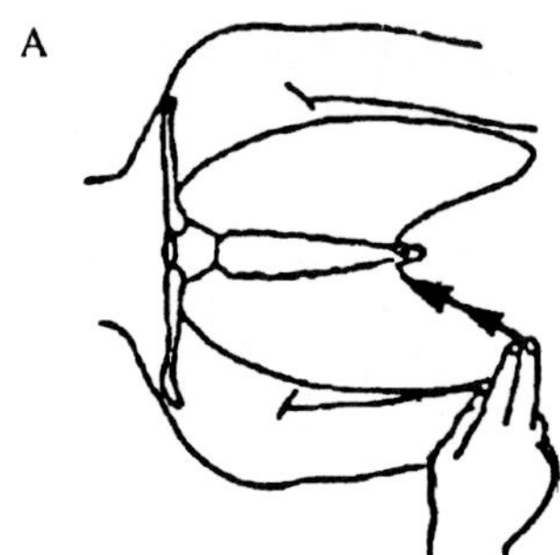

Kneel facing the victim's chest. Find the correct hand position by sliding your fingers up the rib cage to the breastbone.

B

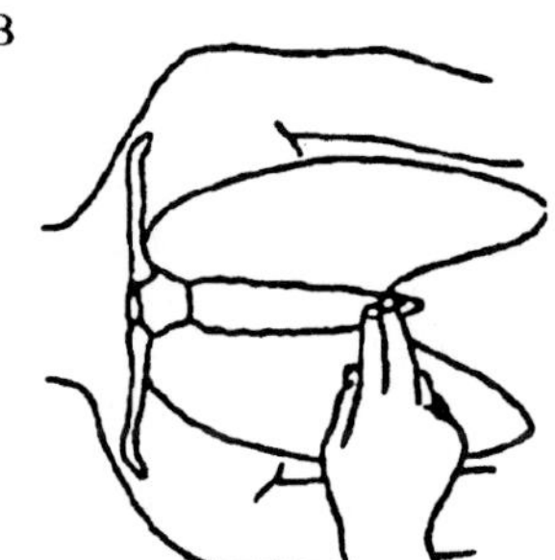

Place your middle finger in the notch and the index finger next to it on the lower end of the breastbone.

C

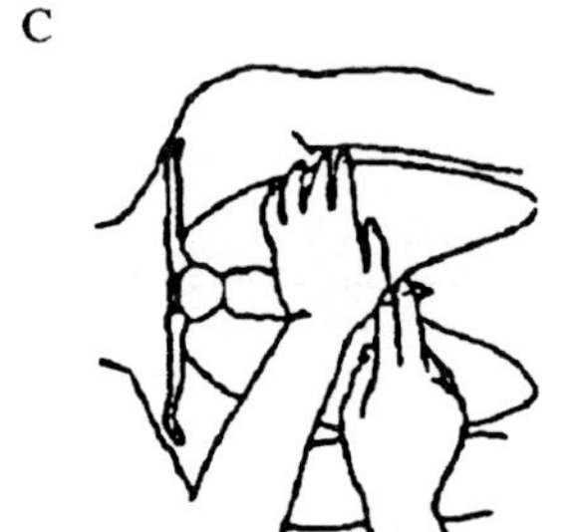

Place your other hand beside the two fingers.

D

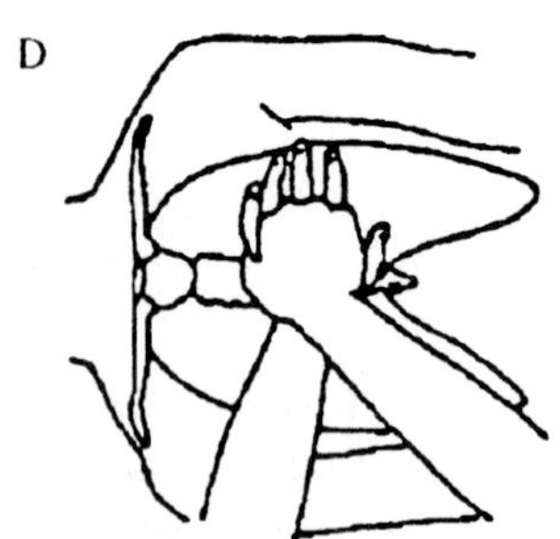

Place hand used to locate notch on top of the other hand. Lace the fingers together. Keep fingers off of the chest. Position shoulders over hands, with elbows locked and arms straight.

FIGURE 48.1 Steps A, B, C, and D illustrate procedures to properly position hands prior to applying CPR.

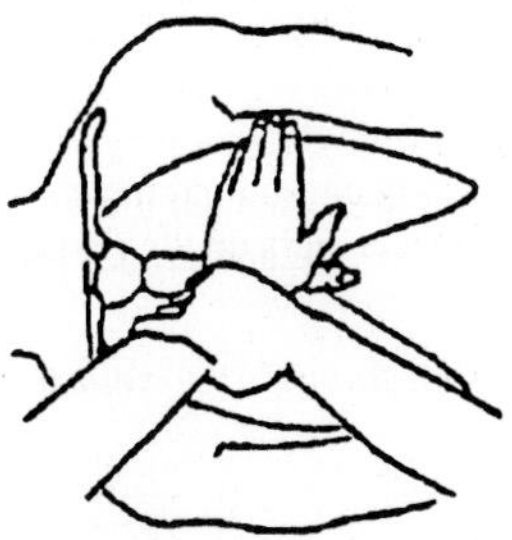

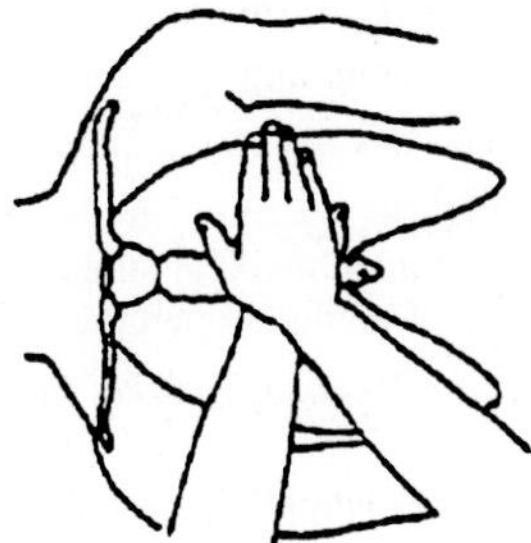

FIGURE 48.2 Alternate positions for hands while applying CPR.

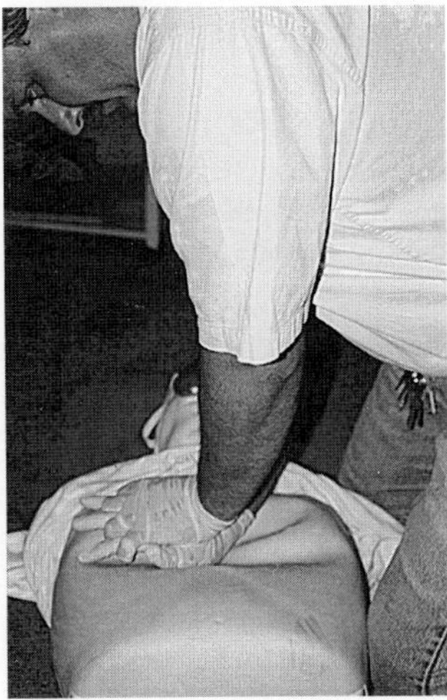

FIGURE 48.3 The weight of the rescuer's upper body is used to compress the chest of the injured.

Do 4 cycles of 15 compressions and 2 breaths.
Recheck pulse for 5 seconds.
If pulse is not present continue with CPR.

WHY DO I DO THAT, AND WHY DID THEY CHANGE?

1. *We used to give CPR for a minute and then call for help.*
   Why has it been changed to—Call for help immediately and then begin CPR?

   It was found that trained single rescuers often performed CPR much longer than 1 minute. Thereby, the call to an EMS system or an Advanced Cardiac Life Support (ACLS) care was seriously delayed. It is important for the victim to receive the advanced life saving medical support as soon as possible.

2. *Why do we use the head-tilt/chin-lift now instead of the old way of pushing up on the neck?*

Using the head-tilt/chin-lift method helps to minimize further damage to a victim. It is more efficient in opening and sealing the airway.

3. *Why do we ask if a victim is choking when it appears obvious that they are?*

   An allergic reaction can also cause the airway to close. In such a case, the victim needs advanced medical support immediately and all of the abdominal thrusts in the world is not going to open the airway.

4. *What if the jaw or mouth is injured, the mouth is shut too tight, or your mouth is too small?*

   You should do mouth-to-nose breathing.

5. *What if a victim is breathing on his own but is unconscious?*

   It is vital that professional emergency personnel be called immediately. Victims who may be breathing on their own may still go into shock if counteractive measures are not begun immediately. There is a point when shock becomes irreversible and the victim will die.

6. *Why has the time taken during ventilation for filling the lungs of an adult increased to $1^1/_2$ to 2 seconds per breath?*

   The increased time further decreases the likelihood of the stomach filling with the air and the victim then vomiting.

7. *What's the big deal if a victim vomits?*

   When an unconscious victim vomits, stomach contents may get into the lungs, obstructing breathing, which may hamper rescue breathing attempts, thus being fatal.

   Residue in the lungs can also cause complications after breathing has been restored.

8. *Why do we no longer give hard breaths but slow breaths?*

   By giving slow breaths, you can feel when the lungs are full and you meet resistance. Over inflating the lungs can cause air to go into the stomach.

9. *Do we remove dentures?*

   No. The dentures should be removed only if they cannot be kept in place.

10. *What if the victim has something in his or her mouth like chewing gum, tobacco, or vomit?*

    If foreign material or vomit is visible in the mouth, it should be removed quickly. Liquids or semiliquids should be wiped out with the index and middle fingers. The fingers should be covered by a protective barrier such as rubber gloves or by a piece of cloth. Solid material should be extracted with a hooked index finger. Remember to use *protective barriers whenever possible.*

11. *What causes respiratory arrest?*

    Many things can cause respiratory arrest. Shock, allergic reactions, drowning, and electrocution.

12. *Why is it imperative that two rescue breaths be given before you bring the victim down from the pole?*

    Extensive soft-tissue swelling may develop rapidly and close the airway. The heart may stop beating soon after breathing has stopped. The heart will continue to beat in some instances after breathing has stopped. Giving rescue breathing before bringing the victim down helps delay the heart stopping and brain damage from lack of oxygen.

13. *Why do you stop breathing after an electrocution?*

    In electrocution, depending upon the path of the current, several things may occur. Your victim may continue to have a heart rhythm, but not be breathing due to:

a. The electrical current having passed through the brain and disrupting the area of the brain that controls breathing.
b. The diaphragm and chest muscles which control breathing may have been damaged.
c. The respiratory muscles may have been paralyzed.

If artificial respiration is not started immediately, the heart will stop beating.

14. *I know CPR, and I am good at my skills. Why was I not able to save my victim? Did I do something wrong?*

There are some instances when all the skills you posses will not be able to save the victim. With your skills they had a chance, whereas if you did nothing, the victim had no chance. A hand-to-hand pathway is more likely to be fatal than a hand-to-foot or a foot to foot.

Electricity in the body causes organs (the heart, etc.) and bones to explode, rupture and break as well as tissue destruction. When the organ has ruptured, there is very little if anything you would have been able to do. This is why it is so important that you use safe work procedures and protective equipment to prevent accidents.

15. *How do I know that I am giving adequate ventilations?*

a. You should be able to observe the rise and fall of the chest with each breath you give.
b. You should be able to hear and feel the escape of air during the passive exhalation of your victim.

16. *What causes a victim to vomit when they are not even breathing?*

a. During rescue breathing, you may not have the airway opened properly. You are forcing air into the stomach and not into the lungs.
b. The breath you give is given too long, at too fast of a rate, or at too much pressure. The air is forced into the stomach.

17. *Is it a problem if the stomach is full of air?*

a. A stomach filled with air reduces lung volume by elevating the diaphragm.
b. The victim may also have the stomach contents forced up by the air in the stomach.

18. *What do I do if the victim starts to vomit?*

Turn the victim onto one side, wipe out his or her mouth with a protective barrier. Reassess your victim to see if they are breathing on their own. If not breathing, continue with the rescue procedure you are using (i.e., rescue breathing, CPR). If you are alone and must leave an unconscious victim, position the victim on his or her side in case the victim vomits while you are gone.

19. *How can someone get enough oxygen out of the air we exhale? Didn't I use all of the oxygen?*

The air we breathe has approximately 21 percent oxygen. We use about 5 percent of the oxygen. That means that we exhale air containing approximately 16 percent of oxygen. This shows you that there is enough oxygen in the air you breathe out to support life until advanced medical care arrives.

20. *Do chest compressions circulate the same amount of blood as the heart does normally?*

Under the best of conditions, chest compressions only circulate approximately $^1/_3$ of the normal blood flow. Body tissues are receiving only the barest minimum of oxygen required for short term survival. That is why it is so important to get professional help immediately.

21. *What does CPR stand for and what does it do?*

CPR stands for cardiopulmonary resuscitation. Cardio refers to the heart and pulmonary refers to the lungs. CPR is a combination of chest compressions and rescue breathing. The purpose of CPR is to keep the lungs supplied with oxygen when breathing has stopped, and to keep blood circulating with oxygen to the brain, heart, and other parts of the body.

22. *Do we ever slap an adult choking victim on the back?*

No. You could lodge the item more securely.

23. *Why does a victim who is going to be given CPR need to be on a level surface?*

When the head is elevated above the heart, blood flow to the brain is further reduced or eliminated.

24. *What do I do if my hands come off of the victim's chest or move from the correct compression point when I am giving CPR?*

You must go up the rib cage and find the correct hand position again.

25. *Does the victim's chest need to be bared to perform compressions?*

No. Clothing should remain on unless it interferes with finding the proper location for chest compressions. Do not waste time or delay the starting of the compressions by removing clothing.

26. *Can or should you do CPR on someone who has a pacemaker?*

Yes. The pacemaker is placed to the side of the heart and not directly below the breastbone, it will not get in the way of chest compressions.

27. *Do I give CPR to a victim who has a very slow or weak pulse?*

No. Never perform CPR on a victim who has a pulse. It can cause serious medical complications.

28. *What if a victim regains a pulse and resumes breathing during or following resuscitation?*

The rescuer should continue to help maintain an open airway. The rescuer should then place the patient in the recovery position. To place the victim in the recovery position, the victim is rolled onto his or her side so that the head, shoulders, and torso move simultaneously without twisting. If the victim has sustained trauma or trauma is suspected, the victim should not be moved.

29. *How hard do I press when doing the head-tilt/chin-lift maneuver?*

The fingers must not press deeply into the soft tissue under the chin, put them on the bone. Pressing into the soft tissue might obstruct the airway. The thumb should not be used for lifting the chin.

30. *If there are several people who can help the victim, should one of them take off the tool belt and hooks?*

Yes. Smoldering shoes, belts, and hooks should be removed to prevent further thermal damage and possible injury to the rescuer. NEVER remove adhered particles of charred clothing from the body. It will pull the skin off with it. If you are the only rescuer, do not take time to remove items unless it interferes with the ability to administer effective compressions. It is critical to begin CPR immediately.

CHAPTER 49

POLE-TOP AND BUCKET TRUCK RESCUE

TIME IS CRITICAL

You may have to help a man on a pole reach the ground safely when he:

- Becomes ill
- Is injured
- Loses consciousness

You must know:

- When he needs help
- When and why time is critical
- The approved method of lowering

BASIC STEPS IN POLE-TOP RESCUE

- Evaluate the situation.

- Provide for your protection.
- Climb to rescue position.
- Determine the injured's condition.

Then, if necessary

- Give first aid.
- Lower the injured.
- Give follow-up care.
- Call for help.

Evaluate the Situation

Call to the man on the pole. If he does not answer or appears stunned or dazed, do the following:

- Prepare to go to his aid.

Time is extremely important.

Provide for your protection

Your safety is vital to the rescue. Use personnel tools and rubber gloves (also rubber sleeves, if required).

LINEMEN PLIERS

- Checklist of Equipment and Site Conditions
  - Extra rubber goods
  - Live-line tools
  - Physical condition of the pole
    - Damaged conductors or equipment?
    - Fire on the pole?
    - Broken pole?
    - Hand line on pole and in good condition?

DO NOT LOSE YOU KNIFE.

Climb to the Rescue Position

Climb carefully and position yourself to

- Insure your safety.
- Clear the injured from hazard.
- Determine the injured's condition.
- Render aid as required.
- Start mouth-to-mouth, if required.
- Lower the injured, if necessary.

The best position will usually be slightly above the injured.

Determine the Injured's Condition

He may be . . .

- Conscious
- Unconscious but breathing
- Unconscious and not breathing
- Unconscious, not breathing, and heart stopped

If the injured is conscious,

- Time may no longer be critical.
- Give necessary first aid on the pole.
- Reassure the injured.
- Help him descend the pole.
- Give first aid on the ground.
- Call for help, if necessary.

If the injured is unconscious but breathing,

- Watch him closely in case the breathing stops.
- Lower him to the ground.
- Give first aid on the ground.
- Call for help.

If the injured is unconscious and not breathing,

- Provide an open airway.
- Give him two rescue breaths.
- If he responds, continue mouth-to-mouth until he is breathing without help.
- If he does not respond to the first two rescue breaths, check skin color, check for pupil dilation.
- If pupil contracts and color is good, continue mouth-to-mouth until he is breathing without help.
- Then, help victim descend pole.
- Watch him closely; his breathing may stop again.
- Give any additional first aid needed.
- Call for help.
- If pupil does not contract and skin color is bad, the victim's **HEART has STOPPED.**
- Prepare to lower him immediately.
- Give two rescue breaths just before lowering.
- Lower him to the ground.
- Start heart-lung resuscitation.
- Call for help.

The method of lowering an injured man is

- Safe
- Simple
- Available

Equipment needed:

- $^1/_2$-in hand line

Procedure:

- Position hand line
- Tie injured
- Remove slack in hand line
- Take firm grip on fall line
- Cut injured's safety strap
- Lower injured

Rescuer

Position the hand line over crossarm or other part of the structure.

Position the line for a clear path to ground (usually the best position is 2 or 3 ft from the pole).

Wrap the short end of the line around the fall line twice (two wraps around the fall line as shown in Fig. 49.1).

Tie the hand line around the victim's chest using three half-hitches.

Tie the injured as follows:

- Pass the hand line around the injured, high on the chest.
- Tie a three half-hitches knot in front, near one armpit.
- Snug the knot.

Remove the slack in the hand line.

- If only one rescuer—he removes slack while on pole
- If two rescuers—man on ground removes slack

Important Give **TWO** quick breaths . . . if necessary, then

Take firm grip on fall line

If only one rescuer—he holds fall line with one hand

If two rescuers—man on ground holds fall line

Cut injured's safety strap

Cut strap on side opposite desired swing

Caution: Do not cut your own safety strap on the hand line.

SHORT END OF LINE IS WRAPPED AROUND FALL LINE TWICE. (TWO WRAPS AROUND FALL LINE.)

FIGURE 49.1

Lower INJURED

ONE RESCUER

- **GUIDE LOAD LINE WITH ONE HAND**
- **CONTROL RATE OF DESCENT WITH THE OTHER HAND**

TWO RESCUERS

- **MAN ON THE POLE GUIDES THE LOAD LINE**
- **MAN ON THE GROUND CONTROLS RATE OF DESCENT**

IF A CONSCIOUS MAN IS BEING ASSISTED IN CLIMBING DOWN . . . THE ONLY DIFFERENCE IS THAT ENOUGH SLACK IS FED INTO THE LINE TO PERMIT HIM CLIMBING FREEDOM

FIGURE 49.2

Lower the Injured (See Fig. 49.2.)

One rescuer—guides load line with one hand
Controls rate of descent with the other hand
Two rescuers—man on the pole guides the load line
Man on the ground controls rate of descent

(If a conscious man is being assisted in climbing down, the only difference is that enough slack is fed into the line to permit him climbing freedom.)

One-Man Rescue versus Two-Man Rescue

These rescues differ only in control of the fall line.

Remember the approved method of lowering an injured man is

- Position hand line
- Tie injured
- Remove slack in hand line
- Take firm grip on fall line
- Cut injured's safety strap
- Lower the injured

If the pole does not have a crossarm, the rescuer places the hand line over the fiberglass bracket insulator support, or other substantial piece of equipment such as a secondary rack, neutral bracket, or guy wire attachment, strong enough to support the weight of the injured. The short end of the hand line is wrapped around the fall line twice and tied around the victim's chest using three half-hitches. The injured's safety strap is cut and rescuer lowers injured to the ground.

The line must be removed and the victim eased onto the ground.

Lay the victim on his back and observe if the victim is conscious. If the victim is conscious, time may no longer be critical. Give necessary first aid. Call for help.

If the injured is unconscious and not breathing, provide an open airway. Give him two rescue breaths. If he responds, continue mouth-to-mouth resuscitation until he is breathing without help. If no pulse is present, pupil does not contract, and skin color is not normal, his heart has stopped. Restore circulation. Use heart-lung resuscitation.

BUCKET TRUCK RESCUE

Time is Critical

- Equip a portion of the insulated boom of the truck with rope blocks designed for hot-line work.
- The strap is placed around the insulated boom approximately 10 ft from the bucket to support the rope blocks.
- The blocks are held taut on the boom from the strap to the top of the boom.
- The rescuer on the ground evaluates the conditions when an emergency arises.
- The bucket is lowered using the lower controls.
- Obstacles in the path of the bucket must be avoided.
- The hook on the rope blocks is engaged in a ring on the lineman's safety strap.
- The safety strap is released from the boom of the truck.
- The rope blocks are drawn taut by the rescuer on the ground.
- The unconscious victim is raised out of the bucket with rope blocks.
- The rescuer eases the victim onto the ground.
- Care should be taken to protect the victim from further injury.
- Release the rope blocks from the victim.
- The basket or bucket on the aerial device may be constructed to tilt after being released, eliminating the need for special rigging, to remove an injured person from the basket or bucket (Fig. 49.3).
- Lay the victim on his back and observe if the victim is conscious.
- If the victim is conscious, time may no longer be critical.
- Give necessary first aid.
- Call for help.
- If the injured is unconscious and not breathing, provide an open airway.
- Give the victim two rescue breaths.
- If he responds, continue mouth-to-mouth resuscitation until he is breathing without help.
- If no pulse is present, pupil does not contract, and skin color is not normal, **The victim's HEART has STOPPED.**

BASKET OR BUCKET ON AERIAL DEVICE MAY BE CONSTRUCTED TO TILT AFTER BEING RELEASED, ELIMINATING THE NEED FOR SPECIAL RIGGING, TO REMOVE AN INJURED PERSON FROM THE BASKET OR BUCKET.

FIGURE 49.3

- Restore circulation.
- Use heart-lung resuscitation.
  - Mouth-to-mouth resuscitation + external cardiac compression = heart-lung resuscitation
  - After every 15 compressions, give 2 breaths—ratio 15 to 2
- Continue heart-lung resuscitation until he recovers or he reaches a hospital or a doctor takes over.

Important Continue heart-lung resuscitation on the way to hospital.

Do not allow any pressure-cycling mechanical resuscitator to be used with cardiac compression.

CHAPTER 50
SELF-TESTING QUESTIONS AND EXERCISES

The chapter number and page number for the answer to the question or a description of the exercise requested is indicated for each item.

1. Describe the electron theory. 1.1
2. Define an electric current. 1.1, 1.4
3. Define a conductor. 1.2
4. Define an insulator. 1.2
5. What are three samples of good insulators? 1.2
6. What is the unit of electric current? 1.5
7. What instrument is used to measure electric current? 1.5
8. What is the unit of electric pressure? 1.6
9. Draw a schematic diagram to illustrate the proper methods of connecting instruments to a single-phase motor to measure electric current and the electric pressure supplied to the motor. 1.7
10. What is the unit of electric power? 1.8
11. Draw a schematic diagram illustrating the method of connecting a wattmeter in a single-phase circuit to measure the power consumed. 1.9
12. What is the unit of electric energy? 1.9
13. What instrument is used to measure electric energy? 1.9
14. What is the formula for Ohm's law? 1.14
15. Define an alternating current. 1.16
16. What is the unit of an alternating-current frequency? 1.16
17. Define the power factor of an alternating-current circuit. 1.22
18. Describe the theory of operation of an electric transformer. 1.29
19. Describe a typical electric system, itemizing the major parts of the system. 2.1
20. What is a transmission line? 2.13
21. How do the voltage levels of transmission, subtransmission, and distribution lines compare? 2.2 and 2.14
22. When are direct-current transmission lines used? 2.15
23. List the functions that can be accomplished in an electric substation. 2.16

24. List the component parts of the electric distribution system. 2.19
25. Describe a typical distribution feeder circuit. 2.19
26. What is a primary circuit? 2.27
27. What is the purpose of the distribution transformer installed on an electric distribution circuit? 2.28
28. Describe an electric service. 2.31
29. What is the purpose of an electric substation? 3.1
30. What types of equipment can be found in a typical substation? 3.5
31. What are power transformers used for in an electric substation? 3.5
32. What function is accomplished with circuit breakers installed in electric substations? 3.10
33. Why are disconnect switches installed in series with circuit breakers? 3.12
34. What function is accomplished with protective relays installed in a substation? 3.15
35. What normally limits the capacity of a transmission line? 4.9
36. What are the advantages of using dc transmission for underground circuits rather than ac transmission? 4.16
37. What is the relationship between the phase-to-neutral voltage and the phase-to-phase voltage of a three-phase four-wire system? 5.4
38. How would the secondary windings of the substation transformer be connected to provide a source for a three-phase four-wire distribution circuit? 5.4
39. Which conductor of a three-phase four-wire distribution system is grounded? 5.10
40. Draw a schematic connection diagram for connecting single-phase, pole-type distribution transformers to a three-phase four-wire primary circuit to obtain three-phase four-wire secondary voltages. 5.8
41. What are harmonic voltages? 5.11
42. How are wood poles classified? 7.2
43. What types of lines are commonly constructed with wood pole structures? 7.5
44. What are the advantages of using crossarms on a subtransmission structure? 7.7
45. What types of steel structures are used for transmission lines? 8.9
46. List the advantages of monitoring devices. 9.1
47. List some of the disaster response assessment data utilized to access a storm situation. 10.5
48. Describe a safe method of unloading poles from a railroad flar car. 11.1
49. What is the rule for determining the setting depth for a pole to be installed in soil? 11.12
50. Where should guys be installed? 12.1
51. When is it necessary to insulate a guy? 12.12
52. Describe an insulator. 13.1
53. Where are double crossarms used? 13.4
54. What types of circuits permit the installation of post-type insulators? 13.10
55. How many porcelain 10-in suspension insulator units are required in a string on a circuit operating at 13,200-volt ac? 13.27
56. Where are strain insulators used? 13.18

57. What is the difference between a strain insulator and a suspension insulator? 13.18

58. What factors must be taken into account when selecting line conductors? 14.16

59. What is the purpose of a distribution transformer? 15.1

60. What equipment is used with a self-protected type of distribution transformer? 15.1

61. Draw a schematic diagram of a single-phase distribution transformer connected to a primary circuit and a load through a switch, and describe its operation without load and with the switch closed and the load connected. 15.2

62. How is the capacity of a distribution transformer determined? 15.8

63. Draw a schematic diagram of a distribution transformer secondary winding connected for 120-volt two-wire operation and 120/240-volt three-wire operation. 15.9

64. How can the polarity of a transformer be determined? 15.14

65. Why are some distribution transformers manufactured with taps on the primary windings? 15.16

66. Describe the theory of operation of an elementary lightning arrester. 16.5

67. Describe the operation of a metal-oxide valve-type lightning arrester. 16.6

68. How can the distribution system be isolated from the failure of a distribution-class lightning arrester? 16.3

69. What is the purpose of installing a fuse in a circuit? 17.1

70. Describe the operation of an expulsion fuse cutout. 17.2

71. How do you know when the fuse in an indicating enclosed cutout is blown? 17.3

72. Refer to the time-current curve (Fig. 17.17) and determine the minimum clearing time for a 15K fuse link if 100 amps of current flowed through the fuse link. 17.12

73. Describe the operation of a liquid fuse. 17.11

74. Describe the operation of a current-limiting fuse. 17.17

75. What is likely to happen if a disconnect switch is opened while current is flowing through the switch? 18.5

76. Describe the operation of an oil circuit recloser. 18.7

77. What is the purpose of using a voltage regulator in a distribution circuit? 19.2

78. Describe the operation of a distribution step voltage regulator. 19.3

79. Why must the voltage regulator be in the neutral position to operate a voltage regulator bypass switch with the circuit energized? 19.10

80. What is the order of operations for transmission-line tower erection? 20.1

81. Why must the surface of high-voltage transmission-line conductors be protected from scratches while they are being installed? 21.1

82. Why is it necessary to effectively ground conductors while they are being installed? 21.2

83. Describe the procedures to be followed to reconductor distribution lines. 21.11

84. When should conductors be sagged? 22.1

85. Describe how a lineman can use a transit to complete sagging operations. 22.7

86. Why are armor rods installed on some line conductors at the point of support? 22.23

87. Why are spacers installed on bundled-conductor transmission lines? 22.26

88. What types of splices can be used on line conductors under tension? 23.1

89. When can bolted connectors be used to join line conductors? 23.23

90. What types of work can be completed to maintain lines while energized with hot-line tools? 24.1

91. Describe the hot-line tool called a wire tong. 24.4

92. Describe the method used to phase out two circuits while they are energized before connecting them. 24.27

93. Describe the procedure for testing insulators in a string with a high-voltage tester to determine if one of the insulators is defective. 24.29

94. What methods can be used to perform live-line maintenance while using an insulated aerial platform? 25.1

95. How deep should a ground rod be driven? 26.2

96. If the normal ground rod installation does not establish an acceptable value of ground resistance, what methods can be used to reduce the ground resistance? 26.2

97. Protective grounds can guard the lineman from what types of hazards? 27.1

98. How can a high-voltage circuit be tested to be sure it is deenergized? 27.1

99. What is the proper sequence for installing protective grounds? 27.2

100. What are the requirements necessary for a lineman to establish an adequate protective ground? 27.3

101. Where should protective grounds be installed to establish the "man-shorted-out" concept that provides the best protection for the lineman? 27.5

102. Describe the best procedures for establishing a temporary protective ground on a cable in a dead-front-type pad-mounted distribution transformer. 27.7

103. Define the following street lighting terms: lamp, luminaire, lumen, footcandle, ballast, mast arm. 28.1

104. What types of lamps are in common use for street lighting? 28.5

105. What types of streetlight lamps require a ballast? 28.2

106. How is the light emitted from a streetlight controlled? 28.12

107. Name two common streetlight distribution patterns. 28.12

108. Name the parts of an underground system. 29.1

109. At what circuit voltages are cables that have an effective insulation shield normally installed? 29.7

110. What is the advantage of sector cable construction? 29.13

111. What is the purpose of having shielding over the insulation of a high-voltage cable? 29.16

112. Why are potheads installed on cables? 29.22

113. What purpose do vaults serve in an underground system? 31.11

114. What precautions must be taken before cables are installed in a duct? 32.2

115. Describe precautions that must be taken to remove the insulation shielding from a cable to prepare the cable for splicing. 33.5

116. Describe the types of underground residential distribution systems normally installed. 34.2

117. What are the voltages normally used for underground residential distribution circuits? 34.5

118. Describe the triplexed-type cable assemblies used for underground residential distribution secondaries and services. 34.7

119. Identify the component parts of transformers used for underground electric distribution systems. 34.7

120. What is the minimum depth allowable for direct-buried cables operating in the range of 601 to 22,000 volts? 34.18

121. What requirements are necessary to install electric supply cables and communication cables at the same depth with no deliberate separation? 34.20

122. What is the maximum distance permitted between the points where effectively grounded electric cables are bonded to communication cable shields if they are installed with random separation? 34.26

123. What are the objectives of line clearance tree trimming? 35.1

124. What are the fundamentals essential for safe and competent tree-trimming operations? 35.2

125. Where should distribution transformers be installed? 36.3

126. Describe the proper location of grounds for various types of distribution transformer secondary wiring. 36.3

127. Describe the procedure to be followed if two single-phase distribution transformers are to be connected in parallel. 36.9

128. When is it desirable to determine the phase sequence of an ac circuit? 36.14

129. What is the minimum height allowable for attaching a low-voltage service to a building? 36.27

130. Describe a method for testing the service wires to avoid connecting the customer's ground wire to either of the hot wires of a secondary. 36.30

131. Draw the symbol of a single-pole draw-out of air circuit breaker. 37.1

132. Draw the symbol for a cable termination. 37.1

133. Draw the symbol for a fused disconnect switch. 37.2

134. Draw the symbol for a lightning arrester with gap, valve, and ground. 37.2

135. Draw the symbol for a transformer. 37.3

136. Draw the symbol for the three-phase wye-connected four-wire grounded transformer connection. 37.3

137. Describe the components and significant features shown on the distribution circuit single-line diagram in Fig. 38.3. 38.3

138. Describe the operation and identify the component parts shown on the unit substation transformer tap changer control schematic diagram in Fig. 39.2. 39.7

139. Define voltage regulation on an electrical system. 40.1

140. What are the high limit and the low limit of the voltages of an electric service to a residence? 40.1

141. What is the effect of low voltage on residential electrical equipment? 40.1

142. What is the effect of high voltage on residential electrical equipment? 40.1

143. How can the voltage on an electric distribution system be controlled? 40.1

144. What is the purpose of a line-drop compensator on a voltage regulator? 40.4

145. What is the formula for Ohm's law? 41.3

146. What is the formula for calculating the impedance of an alternating-current series circuit? 41.9

147. What is the formula for calculating power for a single-phase alternating-current circuit? 41.4

148. What is the formula for calculating power for a three-phase alternating-current circuit? 41.4

149. What is the horsepower rating of a single-phase ac motor operating at 480 volts and drawing 25 amps at a power factor of 88 percent if it has a full-load efficiency of 90 percent? 41.15

150. What items are important to check as part of the initial inspection of transmission lines before they are energized? 42.1

151. How can most transmission lines be inspected quickly after the line experiences a sustained unscheduled outage? 42.2

152. What conditions normally occur when insulators are defective? 42.3

153. What common problems or defective conditions should a lineman check for when inspecting distribution circuits? 42.3

154. What procedures are normally followed to maintain capacitor bank switches? 42.5

155. What items are important to check when inspecting switches? 42.7

156. What items should be checked when inspecting and maintaining switchgear for underground distribution circuits? 42.9

157. How can distribution transformer loading be checked? 42.9

158. What work should be performed by the lineman to keep street lights properly maintained? 42.11

159. Using a piece of rope, tie examples of the following knots:

two half-hitches	43.12
a square knot	43.12
a bowline	43.13
a clove hitch	43.14

160. Describe the proper inspection procedures that should be executed prior to climbing a pole. 44.3

161. Describe the proper methods of inspecting and caring for climbing equipment. 44.7

162. Describe procedures for maintaining climber gaffs. 44.8

163. What is the safest way of completing electric distribution work? 45.1

164. Itemize the protective equipment that should be used by the lineman or cableman when working on energized electric circuits. 45.1

165. How often should rubber gloves be given an air test by the lineman or cableman? 45.3

166. How soon must resuscitation be started after a person stops breathing to prevent brain damage? 47.3

167. Describe the proper procedure of performing mouth-to-mouth resuscitation. 47.5

168. Describe the procedures necessary to perform heart-lung resuscitation. 48.1

169. Describe the method one lineman can use to rescue an accident victim from the top of a pole. 49.1 through 49.7

INDEX